MATHEMATICAL SYMBOLS

$=$	is equal to
$\neq$	is not equal to
$\approx$	is approximately equal to
$\propto$	is proportional to
$>$	is greater than
$\geq$	is greater than or equal to
$\gg$	is much greater than
$<$	is less than
$\leq$	is less than or equal to
$\ll$	is much less than
$\pm$	plus or minus
$\mp$	minus or plus
x_{av} or $\bar{x}$	average value of x
Δx	change in x $(x_f - x_i)$
$\lvert x \rvert$	absolute value of x
Σ	sum of
$\to 0$	approaches 0
∞	infinity

USEFUL MATHEMATICAL FORMULAS

Area of a triangle	$\frac{1}{2}bh$
Area of a circle	πr^2
Circumference of a circle	$2\pi r$
Surface area of a sphere	$4\pi r^2$
Volume of a sphere	$\frac{4}{3}\pi r^3$
Pythagorean theorem	$r^2 = x^2 + y^2$
Quadratic formula	if $ax^2 + bx + c = 0$, then

$$x = \frac{-b \pm \sqrt{b^2 - 4ac}}{2a}$$

EXPONENTS AND LOGARITHMS

$x^n x^m = x^{n+m}$

$x^{-n} = \dfrac{1}{x^n}$ 　　　　$\ln(xy) = \ln x + \ln y$

$\dfrac{x^n}{x^m} = x^{n-m}$ 　　　$\ln\left(\dfrac{x}{y}\right) = \ln x - \ln y$

$(xy)^n = x^n y^n$ 　　　　　$\ln x^n = n \ln x$

$(x^n)^m = x^{nm}$

VALUES OF SOME USEFUL NUMBERS

$\pi = 3.14159\ldots$　　$\log 2 = 0.30103\ldots$　　$\sqrt{2} = 1.41421\ldots$

$e = 2.71828\ldots$　　$\ln 2 = 0.69315\ldots$　　$\sqrt{3} = 1.73205\ldots$

TRIGONOMETRIC RELATIONSHIPS

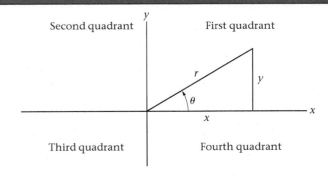

Definitions of Trigonometric Functions

$$\sin\theta = \frac{y}{r} \qquad \cos\theta = \frac{x}{r} \qquad \tan\theta = \frac{\sin\theta}{\cos\theta} = \frac{y}{x}$$

Trigonometric Functions of Important Angles

$\theta°$(rad)	$\sin\theta$	$\cos\theta$	$\tan\theta$
$0°(0)$	0	1	0
$30°(\pi/6)$	0.500	$\sqrt{3}/2 \approx 0.866$	$\sqrt{3}/3 \approx 0.577$
$45°(\pi/4)$	$\sqrt{2}/2 \approx 0.707$	$\sqrt{2}/2 \approx 0.707$	1.00
$60°(\pi/3)$	$\sqrt{3}/2 \approx 0.866$	0.500	$\sqrt{3} \approx 1.73$
$90°(\pi/2)$	1	0	∞

Trigonometric Identities

$\sin(\theta + \phi) = \sin\theta\cos\phi + \sin\phi\cos\theta$

$\cos(\theta + \phi) = \cos\theta\cos\phi - \sin\theta\sin\phi$

$\sin 2\theta = 2\sin\theta\cos\theta$

VECTOR MANIPULATIONS

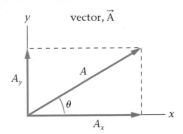

Magnitude and direction	$\longleftrightarrow$	x and y components
A, θ	$\longrightarrow$	$A_x = A\cos\theta$
		$A_y = A\sin\theta$
$A = \sqrt{A_x{}^2 + A_y{}^2}$	$\longleftarrow$	A_x, A_y
$\theta = \tan^{-1}\left(\dfrac{A_y}{A_x}\right)$		

CONVERSION FACTORS

Mass

$1 \text{ kg} = 10^3 \text{ g}$
$1 \text{ g} = 10^{-3} \text{ kg}$
$1 \text{ u} = 1.66 \times 10^{-24} \text{ g} = 1.66 \times 10^{-27} \text{ kg}$
$1 \text{ slug} = 14.6 \text{ kg}$
$1 \text{ metric ton} = 1000 \text{ kg}$

Length

$1 \text{ Å} = 10^{-10} \text{ m}$
$1 \text{ nm} = 10^{-9} \text{ m}$
$1 \text{ cm} = 10^{-2} \text{ m} = 0.394 \text{ in.}$
$1 \text{ yd} = 3 \text{ ft}$
$1 \text{ m} = 10^{-3} \text{ km} = 3.281 \text{ ft} = 39.4 \text{ in.}$
$1 \text{ km} = 10^3 \text{ m} = 0.621 \text{ mi}$
$1 \text{ in.} = 2.54 \text{ cm} = 2.54 \times 10^{-2} \text{ m}$
$1 \text{ ft} = 0.305 \text{ m} = 30.5 \text{ cm}$
$1 \text{ mi} = 5280 \text{ ft} = 1609 \text{ m} = 1.609 \text{ km}$
$1 \text{ ly (light year)} = 9.46 \times 10^{12} \text{ km}$
$1 \text{ pc (parsec)} = 3.09 \times 10^{13} \text{ km}$

Area

$1 \text{ cm}^2 = 10^{-4} \text{ m}^2 = 0.1550 \text{ in.}^2$
$\qquad = 1.08 \times 10^{-3} \text{ ft}^2$
$1 \text{ m}^2 = 10^4 \text{ cm}^2 = 10.76 \text{ ft}^2 = 1550 \text{ in.}^2$
$1 \text{ in.}^2 = 6.94 \times 10^{-3} \text{ ft}^2 = 6.45 \text{ cm}^2$
$\qquad = 6.45 \times 10^{-4} \text{ m}^2$
$1 \text{ ft}^2 = 144 \text{ in.}^2 = 9.29 \times 10^{-2} \text{ m}^2 = 929 \text{ cm}^2$

Volume

$1 \text{ cm}^3 = 10^{-6} \text{ m}^3 = 3.35 \times 10^{-5} \text{ ft}^3$
$\qquad = 6.10 \times 10^{-2} \text{ in.}^3$
$1 \text{ m}^3 = 10^6 \text{ cm}^3 = 10^3 \text{ L} = 35.3 \text{ ft}^3$
$\qquad = 6.10 \times 10^4 \text{ in.}^3 = 264 \text{ gal}$
$1 \text{ liter} = 10^3 \text{ cm}^3 = 10^{-3} \text{ m}^3 = 1.056 \text{ qt}$
$\qquad = 0.264 \text{ gal}$
$1 \text{ in.}^3 = 5.79 \times 10^{-4} \text{ ft}^3 = 16.4 \text{ cm}^3$
$\qquad = 1.64 \times 10^{-5} \text{ m}^3$
$1 \text{ ft}^3 = 1728 \text{ in.}^3 = 7.48 \text{ gal} = 0.0283 \text{ m}^3$
$\qquad = 28.3 \text{ L}$
$1 \text{ qt} = 2 \text{ pt} = 946 \text{ cm}^3 = 0.946 \text{ L}$
$1 \text{ gal} = 4 \text{ qt} = 231 \text{ in.}^3 = 0.134 \text{ ft}^3 = 3.785 \text{ L}$

Time

$1 \text{ h} = 60 \text{ min} = 3600 \text{ s}$
$1 \text{ day} = 24 \text{ h} = 1440 \text{ min} = 8.64 \times 10^4 \text{ s}$
$1 \text{ y} = 365 \text{ days} = 8.76 \times 10^3 \text{ h}$
$\qquad = 5.26 \times 10^5 \text{ min} = 3.16 \times 10^7 \text{ s}$

Speed

$1 \text{ m/s} = 3.60 \text{ km/h} = 3.28 \text{ ft/s}$
$\qquad = 2.24 \text{ mi/h}$
$1 \text{ km/h} = 0.278 \text{ m/s} = 0.621 \text{ mi/h}$
$\qquad = 0.911 \text{ ft/s}$
$1 \text{ ft/s} = 0.682 \text{ mi/h} = 0.305 \text{ m/s}$
$\qquad = 1.10 \text{ km/h}$
$1 \text{ mi/}h = 1.467 \text{ ft/s} = 1.609 \text{ km/h}$
$\qquad = 0.447 \text{ m/s}$
$60 \text{ mi/h} = 88 \text{ ft/s}$

Force

$1 \text{ N} = 0.225 \text{ lb}$
$1 \text{ lb} = 4.45 \text{ N}$
Equivalent weight of a mass of 1 kg
$\quad$ on Earth's surface $= 2.2 \text{ lb} = 9.8 \text{ N}$
$1 \text{ dyne} = 10^{-5} \text{ N} = 2.25 \times 10^{-6} \text{ lb}$

Pressure

$1 \text{ Pa} = 1 \text{ N/m}^2 = 1.45 \times 10^{-4} \text{ lb/in.}^2$
$\qquad = 7.5 \times 10^{-3} \text{ mm Hg}$
$1 \text{ mm Hg} = 133 \text{ Pa}$
$\qquad = 0.02 \text{ lb/in.}^2 = 1 \text{ torr}$
$1 \text{ atm} = 14.7 \text{ lb/in.}^2 = 101.3 \text{ kPa}$
$\qquad = 30 \text{ in. Hg} = 760 \text{ mm Hg}$
$1 \text{ lb/in.}^2 = 6.89 \text{ kPa}$
$1 \text{ bar} = 10^5 \text{ Pa} = 100 \text{ kPa}$
$1 \text{ millibar} = 10^2 \text{ Pa}$

Energy

$1 \text{ J} = 0.738 \text{ ft} \cdot \text{lb} = 0.239 \text{ cal}$
$\qquad = 9.48 \times 10^{-4} \text{ Btu} = 6.24 \times 10^{18} \text{ eV}$
$1 \text{ kcal} = 4186 \text{ J} = 3.968 \text{ Btu}$
$1 \text{Btu} = 1055 \text{ J} = 778 \text{ ft} \cdot \text{lb} = 0.252 \text{ kcal}$
$1 \text{ cal} = 4.186 \text{ J} = 3.97 \times 10^{-3} \text{ Btu}$
$\qquad = 3.09 \text{ ft} \cdot \text{lb}$
$1 \text{ ft} \cdot \text{lb} = 1.36 \text{ J} = 1.29 \times 10^{-3} \text{ Btu}$
$1 \text{ eV} = 1.60 \times 10^{-19} \text{ J}$
$1 \text{ kWh} = 3.6 \times 10^6 \text{ J}$
$1 \text{ erg} = 10^{-7} \text{ J} = 7.38 \times 10^{-6} \text{ ft} \cdot \text{lb}$

Power

$1 \text{ W} = 1 \text{ J/s} = 0.738 \text{ ft} \cdot \text{lb/s}$
$\qquad = 1.34 \times 10^{-3} \text{ hp} = 3.41 \text{ Btu/h}$
$1 \text{ ft} \cdot \text{lb/s} = 1.36 \text{ W} = 1.82 \times 10^{-3} \text{ hp}$
$1 \text{ hp} = 550 \text{ ft} \cdot \text{lb/s} = 745.7 \text{ W}$
$\qquad = 2545 \text{ Btu/h}$

Mass–Energy Equivalents

$1 \text{ u} = 1.66 \times 10^{-27} \text{ kg} \leftrightarrow 931.5 \text{ MeV}$
$1 \text{ electron mass} = 9.11 \times 10^{-31} \text{ kg}$
$\qquad = 5.49 \times 10^{-4} \text{ u} \leftrightarrow 0.511 \text{ MeV}$
$1 \text{ proton mass} = 1.673 \times 10^{-27} \text{ kg}$
$\qquad = 1.007\,267 \text{ u} \leftrightarrow 938.28 \text{ MeV}$
$1 \text{ neutron mass} = 1.675 \times 10^{-27} \text{ kg}$
$\qquad = 1.008\,665 \text{ u} \leftrightarrow 939.57 \text{ MeV}$

Temperature

$T_{\text{F}} = \frac{9}{5} T_{\text{C}} + 32$
$T_{\text{C}} = \frac{5}{9} (T_{\text{F}} - 32)$
$T_{\text{K}} = T_{\text{C}} + 273.15$

Angle

$1 \text{ rad} = 57.3°$

$1° = 0.0175 \text{ rad}$	$60° = \pi/3 \text{ rad}$
$15° = \pi/12 \text{ rad}$	$90° = \pi/2 \text{ rad}$
$30° = \pi/6 \text{ rad}$	$180° = \pi \text{ rad}$
$45° = \pi/4 \text{ rad}$	$360° = 2\pi \text{ rad}$

$1 \text{ rev/min} = (\pi/30) \text{ rad/s} = 0.1047 \text{ rad/s}$

APPENDICES IN THE TEXT

PHYSICS

VOLUME 1

PHYSICS

FIFTH EDITION

JAMES S. WALKER

Western Washington University

PEARSON

Editor-in-Chief: Jeanne Zalesky
Acquisitions Editor: Darien Estes
Director of Marketing: Christy Lesko
Marketing Manager: Elizabeth Ellsworth
Project Manager: Chandrika Madhavan
Program Manager: Katie Conley
Development Manager: Cathy Murphy
Program and Project Management Team Lead: Kristen Flathman
Copyeditor: Carol Reitz
Compositor: Cenveo Publisher Services
Vendor Production Manager: Rose Kernan
Illustrations: Rolin Graphics
Design Manager: Derek Bacchus
Interior and Cover Designer: Gary Hespenheide
Rights & Permissions Project Manager: Tim Nicholls
Photo Researcher: Eric Schrader
Media Producer: Sarah Kaubisch
Manufacturing Buyer: Maura Zaldivar-Garcia
Printer and Binder: LSC Communications
Cover Printer: LSC Communications

THIS BOOK IS DEDICATED TO MY PARENTS, IVAN AND JANET WALKER, AND TO MY WIFE, BETSY.

Cover Photo Credits: The Corona on grassland: David Hu/Getty Images; spotted deer: Nishant Shah/Getty Images; thermal image of young male athlete training: Joseph Giacomin/Getty Images; Sydney summer lightning strike: Australian Land, City, People Scape Photographer/Getty Images; F/A-18 Hornet: US Navy; Hurricane Arthur's cloud-covered eye: NASA Goddard MODIS Rapid Response Team

Library of Congress Cataloging-in-Publication Data
Names: Walker, James S., 1950- author.
Title: Physics/James S. Walker, Western Washington University.
Description: Fifth edition. | San Francisco : Pearson Addison-Wesley, [2016]
 | Includes index.
Identifiers: LCCN 2015044999| ISBN 9780321976444 | ISBN 0321976444
Subjects: LCSH: Physics--Textbooks.
Classification: LCC QC23.2 .W35 2016 | DDC 530--dc23
LC record available at http://lccn.loc.gov/2015044999

ISBN 10: 0-321-97644-4; ISBN 13: 978-0-321-97644-4 (Student edition)
ISBN 10: 0-134-03124-5; ISBN 13: 978-0-134-03124-8 (Volume One)
ISBN 10: 0-134-03125-3; ISBN 13: 978-0-134-03125-5 (Volume Two)

Acknowledgements of third party content appear on pages C-1-C-3, which constitutes an extension of this copyright page.

www.pearsonhighered.com

ABOUT THE AUTHOR

JAMES S. WALKER

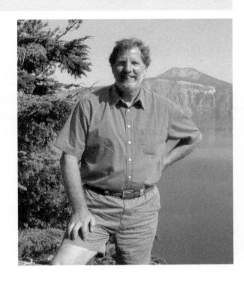

James Walker obtained his Ph.D. in theoretical physics from the University of Washington in 1978. He subsequently served as a post-doc at the University of Pennsylvania, the Massachusetts Institute of Technology, and the University of California at San Diego before joining the physics faculty at Washington State University in 1983. Professor Walker's research interests include statistical mechanics, critical phenomena, and chaos. His many publications on the application of renormalization-group theory to systems ranging from absorbed monolayers to binary-fluid mixtures have appeared in *Physical Review, Physical Review Letters, Physica,* and a host of other publications. He has also participated in observations on the summit of Mauna Kea, looking for evidence of extra-solar planets.

Jim Walker likes to work with students at all levels, from judging elementary school science fairs to writing research papers with graduate students, and has taught introductory physics for many years. Through his enjoyment of this course and his empathy for students, Jim has earned a reputation as an innovative, enthusiastic, and effective teacher. Jim's educational publications include "Reappearing Phases" (*Scientific American*, May 1987) as well as articles in the *American Journal of Physics* and *The Physics Teacher*. In recognition of his contributions to the teaching of physics at Washington State University, Jim was named the Boeing Distinguished Professor of Science and Mathematics Education for 2001–2003.

When he is not writing, conducting research, teaching, or developing new classroom demonstrations and pedagogical materials, Jim enjoys amateur astronomy, eclipse chasing, bird and dragonfly watching, photography, juggling, unicycling, boogie boarding, and kayaking. Jim is also an avid jazz pianist and organist. He has served as ballpark organist for a number of Class A minor league baseball teams, including the Bellingham Mariners, an affiliate of the Seattle Mariners, and the Salem-Keizer Volcanoes, an affiliate of the San Francisco Giants. He can play "Take Me Out to the Ball Game" in his sleep.

BRIEF CONTENTS

An accessible, problem-solving approach to physics, grounded in real-world applications

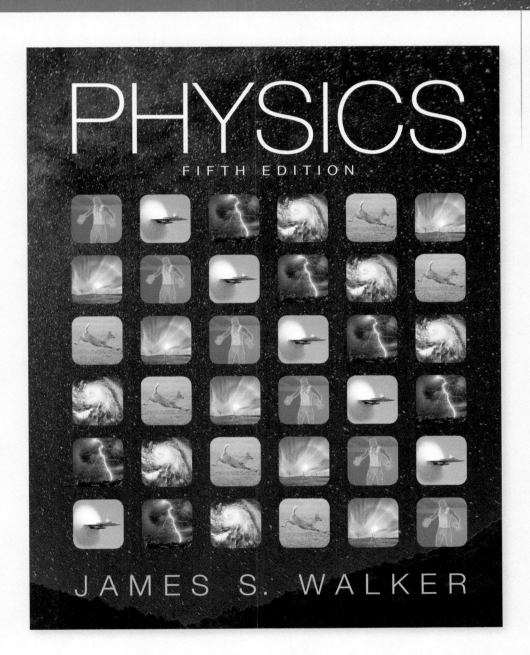

PHYSICS

FIFTH EDITION

JAMES S. WALKER

What's the Big Idea?

James Walker's **Physics**, Fifth Edition engages students by connecting conceptual and quantitative physics to the world around them, making complex concepts understandable, and helping students build problem-solving skills. New "just in time" learning aids, such as the "Big Ideas," quickly orient students to the overarching principles of each chapter, while new Real-World Physics and Biological Applications expose students to physics they can observe in their own lives. A revised problem-solving pedagogy allows students to build a deep understanding of the relationship between the conceptual and the quantitative.

Ample "just in time" learning aids help students where they need it, when they need it

NEW! **BIG IDEAS** appear on each chapter-opening page to quickly orient students to the overarching principles of each chapter. Details highlighting key takeaways for students appear in the chapter margins next to where these ideas are introduced in the narrative.

Big Ideas

1 Work is force times distance.

2 Kinetic energy is one-half mass times velocity squared.

3 Power is the rate at which wo...

Big Idea **1** Work is force times distance when force and displacement are in the same direction. More generally, work is the component of force in the direction of displacement times the distance.

NEW! **PHYSICS IN CONTEXT** calls attention to a related or supporting concept covered in a previous chapter ("Physics in Context: Looking Back"); or alerts students to a concept to be covered in a future chapter that relates to what they're reading ("Physics in Context: Looking Ahead").

PHYSICS IN CONTEXT
Looking Back

Conservation of energy, first introduced in Chapter 8, is just as important in rotational motion as it is in linear motion.

PHYSICS IN CONTEXT
Looking Ahead

In Chapter 5 we will introduce one of the most important concepts in all of physics—force. It is a vector quantity. Other important vector quantities to be introduced in later chapters include linear momentum (Chapter 9), angular momentum (Chapter 11), electric field (Chapter 19), and magnetic field (Chapter 22).

NEW! **REAL-WORLD PHYSICS AND BIOLOGICAL APPLICATIONS** woven into the text have been updated with fresh topics relevant to today's student. These applications bring abstract physics principles to life with real-life examples from the world around us.

Enhance Your Understanding   (Answers given at the end of the chapter)

5. An object moves along the brown path in **FIGURE 3-30** in the direction indicated. Which physical quantity (position, acceleration, velocity) is represented by the following vectors: (a) $\vec{A}$, (b) $\vec{B}$, (c) $\vec{C}$?

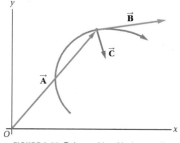

▲ FIGURE 3-30 Enhance Your Understanding 5

NEW! **ENHANCE YOUR UNDERSTANDING** qualitative multiple-choice and ranking questions appear before each Section Review to give students an opportunity to practice what they've just learned. Answers are listed at the end of the chapter.

Section Review

• Many physical quantities are represented by vectors. Among these are position, displacement, velocity, and acceleration.

NEW! **SECTION REVIEWS** briefly synthesize the key ideas covered in the preceding section for a quick at-a-glance summary.

Thorough problem-solving instruction and multiple opportunities for practice

The Fifth Edition continues the Walker tradition of providing ample opportunity for students to develop problem-solving skills with a greater variety of example types and more thoroughly stepped out explanations and guidance.

NEW! KNOWNS AND UNKNOWNS have been added to worked examples to model how scientists think about setting up a problem before they solve it.

NEW! 50% UPDATED AND REVISED FULLY WORKED EXAMPLES provide a systematic process for solving problems:

- **Picture the Problem** encourages students to visualize the situation, identify and label important quantities, and set up a coordinate system. This step is accompanied by a figure and free-body diagram when appropriate.

- **UPDATED! Reasoning and Strategy** helps students learn to analyze the problem, identify the key physical concepts, and map a plan for the solution, including the Known and Unknown quantities.

- **Solution** is presented in a *two-column format* to help students translate the words of the problem on the left to the equations they will use to solve it on the right.

- **Insight** points out interesting or significant features of the problem, solution process, or the result.

- **Practice Problems** give students the opportunity to test their understanding and skills on similar problems to the one just worked.

NEW! END-OF-CHAPTER HOMEWORK set includes 10 new problems per chapter and more than 15% revised problems. Hundreds of additional problems are available in Mastering as an alternate problem set as well as the Instructor's Solutions Manual.

NEW! BIO PASSAGE PROBLEMS have been thoroughly rewritten to better reflect the new MCAT exam, released in 2015. With a focus on skills and core competencies, rather than rote knowledge, each Bio Passage Problem offers opportunities to practice the types of questions pre-meds will encounter on the exam.

EXAMPLE 7-2 HEADING FOR THE ER

An intern pushes an 87-kg patient on an 18-kg gurney, producing an acceleration of 0.55 m/s². **(a)** How much work does the intern do in pushing the patient and gurney through a distance of 1.9 m? Assume the gurney moves without friction. **(b)** How far must the intern push the gurney to do 140 J of work?

PICTURE THE PROBLEM
Our sketch shows the physical situation for this problem. Notice that the force exerted by the intern is in the same direction as the displacement of the gurney; therefore, we know that the work is $W = Fd$.

$a = 0.55\ \text{m/s}^2$

$d = 1.9\ \text{m}$

REASONING AND STRATEGY
We are not given the magnitude of the force, F, so we cannot apply Equation 7-1 ($W = Fd$) directly. However, we are given the mass and acceleration of the patient and gurney, and from them we can calculate the force with $F = ma$. The work done by the intern is then $W = Fd$, where $d = 1.9$ m.

Known Mass of patient, 87 kg; mass of gurney, 18 kg; acceleration, $a = 0.55$ m/s²; **(a)** pushing distance, $d = 1.9$ m; **(b)** work, $W = 140$ J.

Unknown **(a)** Work done, $W = ?$ **(b)** Pushing distance, $d = ?$

SOLUTION

Part (a)
1. First, find the force F exerted by the intern: $F = ma = (87\ \text{kg} + 18\ \text{kg})(0.55\ \text{m/s}^2) = 58\ \text{N}$
2. The work done by the intern, W, is the force times the distance: $W = Fd = (58\ \text{N})(1.9\ \text{m}) = 110\ \text{J}$

Part (b)
3. Use $W = Fd$ to solve for the distance d: $d = \dfrac{W}{F} = \dfrac{140\ \text{J}}{58\ \text{N}} = 2.4\ \text{m}$

INSIGHT
You might wonder whether the work done by the intern depends on the speed of the gurney. The answer is no. The work done on an object, $W = Fd$, doesn't depend on whether the object moves through the distance d quickly or slowly. What does depend on the speed of the gurney is the *rate* at which work is done, which we discuss in detail in Section 7-4.

PRACTICE PROBLEM — PREDICT/CALCULATE
(a) If the total mass of the gurney plus patient is halved and the acceleration is doubled, does the work done by the intern increase, decrease, or remain the same? Explain. **(b)** Determine the work in this case. [**Answer: (a)** The work remains the same because the two changes offset one another; that is, $F = ma = (m/2)(2a)$. **(b)** The work is 110 J, as before.]

Some related homework problems: Problem 3, Problem 4

PASSAGE PROBLEMS

Bam!—Apollo 15 Lands on the Moon

The first word spoken on the surface of the Moon after *Apollo 15* landed on July 30, 1971, was "Bam!" This was James Irwin's involuntary reaction to their rather bone-jarring touchdown. "We did hit harder than any of the other flights!" says Irwin. "And I was startled, obviously, when I said, 'Bam!'"

The reason for the "firm touchdown" of *Apollo 15*, as pilot David Scott later characterized it, was that the rocket engine was shut off a bit earlier than planned, when the lander was still 4.30 ft above the lunar surface and moving downward with a speed of 0.500 ft/s. From that point on the lander descended in lunar free fall, with an acceleration of 1.62 m/s². As a result, the landing speed of *Apollo 15* was by far the largest of any of the *Apollo* missions. In comparison, Neil Armstrong's landing speed on *Apollo 11* was the lowest at 1.7 ft/s—he didn't shut off the engine until the footpads were actually on the surface. *Apollos 12, 14,* and *17* all landed with speeds between 3.0 and 3.5 ft/s.

To better understand the descent of *Apollo 15*, we show its trajectory during the final stages of landing in FIGURE 2-47 (a). In FIGURE 2-47 (b) we show a variety of speed-versus-time plots.

Personalize learning with MasteringPhysics®

BEFORE CLASS

NEW! INTERACTIVE PRELECTURE VIDEOS address the
rapidly growing movement toward prelecture teaching and flipped classrooms. These whiteboard-style animations provide an introduction to key topics with embedded assessment to help students prepare before lecture and to help professors identify student misconceptions.

NEW! DYNAMIC STUDY MODULES (DSMs) leverage
research from the fields of cognitive psychology, neurobiology, and game studies to help students study on their own by continuously assessing their activity and performance, then using data and analytics to provide personalized content in real time to reinforce concepts that target each student's particular strengths and weaknesses. Assignable for Pre-Class Prep or Self-Study, physics DSMs include the mathematics of physics, conceptual definitions, relationships for topics across all of mechanics and electricity and magnetism, and more.

DURING CLASS

NEW! LEARNING CATALYTICS™ is an interac-
tive classroom tool that uses students' devices to engage them in class and provide "just in time" analytics to inform your lecture/active classroom. A new math palette for the mathematical expression question type allows for Greek symbols, PI, Euler's number, Logarithm, Exponent, Trigonometric functions, Absolute value, Square root, Nth square root, and Fractions with the exception of vector or unit vector.

AFTER CLASS

NEW! ENHANCED END-OF-CHAPTER QUESTIONS offer students

instructional support when and where they need it including links to the eText, math remediation, and wrong-answer specific feedback for an improved student experience and greater learning gains.

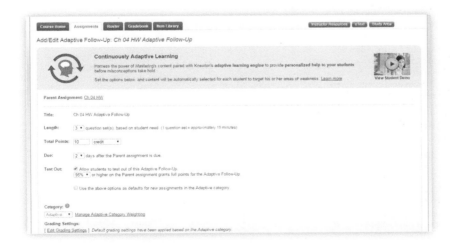

ADAPTIVE FOLLOW-UPS are personalized assignments that pair Mastering's powerful content with Knewton's adaptive learning engine to provide individualized help to students before misconceptions take hold. These adaptive follow-ups address topics students struggled with on assigned homework, including core prerequisite topics.

Note: This list includes applied topics that receive significant discussion in the chapter text or a worked Example, as well as topics that are touched on in - end-of-chapter Conceptual Questions, Conceptual Exercises, and Problems. Topics of particular relevance to the life sciences or medicine are marked BIO. Topics related to Passage Problems are marked PP.

Teaching introductory algebra-based physics can be a most challenging—and rewarding—experience. Students enter the course with a wide range of backgrounds, interests, and skills and we, the instructors, strive not only to convey the basic concepts and fundamental laws of physics, but also to give students an appreciation for its relevance and appeal.

I wrote this book to help with that task. It incorporates a number of unique and innovative pedagogical features that evolved from years of teaching experience. The materials have been tested extensively in the classroom and in focus groups, and refined based on comments from students and teachers who used the earlier editions of the text. The enthusiastic response I've received from users of the first four editions is both flattering and motivating. The Fifth Edition has been improved in direct response to this feedback with new and revised examples; modern biological and real-world physics applications woven into the text; a refreshed homework problem set, including new biological passage problems that align with the new MCAT exam; and "just in time" learning aids throughout the text.

Learning Tools in the Text

A key goal of this text is to help students make the connection between a conceptual understanding of physics and the various skills necessary to solve quantitative problems. One of the chief means to that end is an integrated system of learning tools—including fully worked Examples with solutions in two-column format, two-column Quick Examples, Conceptual Examples, and Exercises. Each of these tools is specialized to meet the needs of students at a particular point in the development of a chapter.

These needs are not always the same. Sometimes students benefit from a detailed explanation of how to tackle a particular problem; at other times it is helpful for them to explore a key idea in a conceptual context. And sometimes, all that is required is practice using a new equation or definition.

This text emulates the teaching style of successful instructors by providing the right tool at the right time and place. This "just in time" approach helps students master the new ideas and concepts as they are encountered.

In a similar spirit, two new features appear at the end of every section in each chapter to give students immediate, timely feedback on the material just covered. These features are the **Enhance Your Understanding** and the **Section Review**. The Enhance Your Understanding feature is a conceptual question designed to solidify the concepts presented in the section. Answers to the Enhance Your Understanding questions are given at the end of each chapter. The Section Review gives a brief review of the key concepts covered in that section.

WORKED EXAMPLES WITH SOLUTIONS IN TWO-COLUMN FORMAT

Examples provide students with a complete and detailed method of solving a particular type of problem. The Examples in this text are presented in a format that focuses on the basic strategies and thought processes involved in problem solving. This focus on the intimate connection between conceptual insights and problem-solving techniques encourages students to view the ability to solve problems as a logical outgrowth of conceptual understanding. In addition, the Examples encourage students to think of solving physics problems as an opportunity to exercise their innate creativity.

Each **Example** has the same basic structure:

- **Picture the Problem** This first step discusses how the physical situation can be represented visually and what such a representation can tell us about how to analyze and solve the problem. At this step, we set up a coordinate system where appropriate, and label important quantities.

- **Reasoning and Strategy**   The Reasoning and Strategy section addresses the commonly asked question, "How do I get started?" It does this by providing a clear overview of the problem and helping students to identify the relevant physical principles. It then guides the student in using known relationships to map out a step-by-step path to the solution. In the Fifth Edition I've fleshed out this step to give students more thorough guidance.   UPDATED

- **Knowns and Unknowns**   Before the process of solving the problem begins, we list the quantities that are known (given), and those that are to be found. This new feature in the Reasoning and Strategy step helps students organize their thoughts, and sets a clear goal for their calculations.   NEW

- **Solution in a Two-Column Format**   In the step-by-step Solution of the problem, each step is presented with a prose statement in the left-hand column and the corresponding mathematical implementation in the right-hand column. Each step clearly translates the idea described in words into the appropriate equations.

- **Insight**   Each Example wraps up with an Insight—a comment regarding the solution just obtained. Some Insights deal with possible alternative solution techniques, others with new ideas suggested by the results.

- **Practice Problem**   Following the Insight is a Practice Problem, which gives the student a chance to practice the type of calculation just presented. The Practice Problems are always accompanied by answers, and provide students with a valuable check on their understanding of the material. **Some of the Practice Problems are of the new Predict/Calculate type** in the end-of-chapter homework section. These problems ask for a prediction based on physical concepts, and then present a numerical problem to verify the prediction. Finally, each Example ends with a reference to some related end-of-chapter Problems to allow students to test their skills further.   UPDATED

QUICK EXAMPLES

Quick Examples are streamlined versions of the full Examples. By streamlining the process, this new feature allows for more sample problems to be covered in the text without taking up too much space or becoming redundant in the details.   NEW

CONCEPTUAL EXAMPLES

Conceptual Examples help students sharpen their insight into key physical principles. A typical Conceptual Example presents a thought-provoking question that can be answered by logical reasoning based on physical concepts rather than by numerical calculations. The statement of the question is followed by a detailed discussion and analysis in the section titled Reasoning and Discussion, and the Answer is given at the end of the checkpoint for quick and easy reference.

NEW to this edition are **Conceptual Examples that prepare students to solve Predict/Explain** problems in the end-of-chapter homework section. These problems ask for a prediction, and then ask the student to pick the best explanation from those provided.

EXERCISES

Exercises present brief one-step calculations designed to illustrate the application of important new relationships, without the expenditure of time and space required by a fully worked Example or Quick Example. Exercises generally give students an opportunity to practice the use of a new equation, become familiar with the units of a new physical quantity, and get a feeling for typical magnitudes.

PROBLEM-SOLVING NOTES

Each chapter includes a number of Problem-Solving Notes presented in a "just in time" fashion in the margin. These practical hints are designed to highlight useful problem-solving methods while helping students avoid common pitfalls and misconceptions.

End-of-Chapter Learning Tools

The end-of-chapter material in this text also includes a number of innovations, along with refinements of more familiar elements.

- Each chapter concludes with a **Chapter Summary** presented in an easy-to-use outline style. Key concepts, equations, and important figures are organized by topic for convenient reference.

- The homework for each chapter begins with a section of **Conceptual Questions**. Answers to the odd-numbered Questions can be found in the back of the book. Answers to even-numbered Conceptual Questions are available in the Instructor's Solutions Manual.

- Following the Conceptual Questions is a complete set of **Problems and Conceptual Exercises. Conceptual Exercises (CE)** consist of multiple-choice and ranking questions. Answers to the odd-numbered Exercises can be found in the back of the book. Answers to even-numbered Conceptual Exercises are available in the online Instructor's Solutions Manual.

UPDATED
- The Fifth Edition boasts a refreshed set of homework problems: at least 10 new problems per chapter and more than 15% of the previous edition problems have been revised and improved. **Problems** are divided into sections, with increasing difficulty levels indicated with one, two, or three colored bullets. Problems of particular real-world interest are indicated with titles. In addition, a section titled **General Problems** presents a variety of problems that use material from two or more sections within the chapter, or refer to material covered in earlier chapters.

- **Bio/Med Problems** are homework problems of special biological or medical relevance. These problems are indicated with the symbol **BIO**.

NEW
- **Predict/Calculate** is a new type of problem that combines a conceptual question (prediction) with a numerical problem (calculation). Problems of this type, which stress the importance of reasoning from basic principles, show how conceptual insight and numerical calculation go hand in hand in physics.

- **Predict/Explain** problems ask students to predict what will happen in a given physical situation, and then to choose the best explanation for their prediction.

UPDATED
- **Passage Problems** are based on an extended multi-paragraph description of a physical situation. These problems are similar to those found on MCAT exams, and have associated multiple-choice questions. All of the Passage Problems in the book are either new or revised and support the new MCAT released in 2015.

Perspective Across Chapters

It's easy for students to miss the forest for the trees—to overlook the unifying concepts that are central to physics and make the details easier to learn, understand, and retain. To address this difficulty, the Fifth Edition adds a NEW feature called **Physics In Context**. This feature, which appears in the margin at appropriate locations in the chapters, comes in two varieties—**Looking Back** and **Looking Ahead**. The Looking Back variety connects material just developed to related material from earlier chapters. This helps students apply their understanding of earlier material to a new situation, and provides a greater perspective on physics as a whole. The Looking Ahead variety gives students a "heads up" that the material presented in this chapter is important at a specific point later in the text. With these two varieties working together, Physics In Context helps students develop connections between different topics in physics that share a common central theme.

Scope and Organization

The presentation of physics in this text follows the standard practice for introductory courses, with only a few well-motivated refinements.

**PHYSICS
IN CONTEXT
Looking Back**

Conservation of energy, first introduced in Chapter 8, is just as important in rotational motion as it is in linear motion.

**PHYSICS
IN CONTEXT
Looking Ahead**

In Chapter 5 we will introduce one of the most important concepts in all of physics—force. It is a vector quantity. Other important vector quantities to be introduced in later chapters include linear momentum (Chapter 9), angular momentum (Chapter 11), electric field (Chapter 19), and magnetic field (Chapter 22).

First, note that Chapter 3 is devoted to **vectors and their application to physics**. My experience has been that students benefit greatly from a full discussion of vectors early in the course. Most students have seen vectors and trigonometric functions before, but rarely from the point of view of physics. Thus, including a chapter on vectors sends a message that this is important material, and gives students an opportunity to develop and improve their understanding of vectors in a physics context.

Note also that **additional time is given to some of the more fundamental aspects of physics**, such as Newton's laws and energy. Presenting such material in two chapters gives the student a better opportunity to assimilate and master these crucial topics.

REAL-WORLD AND BIOLOGICAL PHYSICS

UPDATED

Since physics applies to everything in nature, it's only reasonable to point out applications of physics that students may encounter in the real world. Each chapter presents a number of discussions focusing on "Real-World Physics." Those of general interest are designated by RWP at the start of the paragraph. Applications that pertain more specifically to biology or medicine are indicated by BIO.

These applications have been thoughtfully updated with topics that are current and relevant to today's student. For example, new RWP features include timely discussions of accelerometers in cell phones and game controllers (Chapter 2), traffic collision avoidance systems (Chapter 4), the tension in a tether connecting astronauts in the movie *Gravity* (Chapter 6), and the use of magnetic forces to improve recycling (Chapter 22). Updated BIO topics include the deadly blow delivered by a mantis shrimp (Chapter 8), the 1000-rpm mid-air spinning behavior of dragonflies (Chapter 10), how starfish use torque to open clamshells (Chapter 11), and how radioactivity is used to treat hyperactive thyroids (Chapter 32).

These new applications are in addition to the many classic, and often surprising, examples of physics that have always been an important part of Walker *Physics*. Examples like these engage a student's interest, and help motivate them to think more deeply about physics concepts. Students are often intrigued, for example, to discover that they are shorter at the end of the day than when they get up in the morning (Chapter 5), that humming next to a spider web can cause a resonance effect that sends the spider into a tizzy (Chapter 13), and that scorpions in the nighttime desert are brightly fluorescent when illuminated by an ultraviolet flashlight (Chapter 31). With real-world applications like these, it's easy to show students that physics is relevant to their everyday lives.

VISUALS

One of the most fundamental ways in which we learn is by comparing and contrasting. A new feature called **Visualizing Concepts** helps with this process by presenting a selection of photos that illustrate a physical concept in a variety of different contexts. Grouping carefully chosen photographs in this way helps students to see the universality of physics. We have also included a number of **demonstration photos** that use high-speed time-lapse photography to dramatically illustrate topics, such as standing waves, static versus kinetic friction, and the motion of center of mass, in a way that reveals physical principles in the world around us.

Resources

The Fifth Edition is supplemented by an ancillary package developed to address the needs of both students and instructors.

FOR THE INSTRUCTOR

Instructor's Solutions Manual by Kenneth L. Menningen (University of Wisconsin–Stevens Point) is available online at the Instructor's Resource Center: www.pearsonhighered.com/educator, on the Instructor's Resource DVD, and in the Instructor's Resource Area on Mastering. You will find detailed, worked solutions to every Problem and Conceptual Exercise in the text, all solved using the step-by-step problem-solving strategy of the in-chapter Examples (Picture the Problem, Reasoning and Strategy, two-column Solutions, and Insight). The solutions also contain answers to the even-numbered Conceptual Questions as well as problem statements and solutions for the hundreds of problems in the alternate problem set in Mastering.

The cross-platform **Instructor's Resource DVD** (ISBN 978-0-321-76570-3) provides a comprehensive library of applets from ActivPhysics OnLine as well as PhET simulations. All line figures, photos, and examples from the textbook are provided in JPEG format and the key equations are available in editable Word files. A revised set of Lecture Outlines and Clicker questions, both in PowerPoint, are included for use in lecture. Assets available in Mastering are provided here, too: Pause and Predict Video Tutor Demonstrations and Author Demonstration Videos. And it includes the Test Bank in Word and Test-Gen formats and the Instructor's Solutions Manual in Word and pdf.

MasteringPhysics® (www.mastering-physics.com) is the most advanced, educationally effective, and widely used physics homework and tutorial system in the world. Ten years in development, it provides instructors with a library of extensively pre-tested end-of-chapter problems and rich multipart, multistep tutorials that incorporate a wide variety of answer types, wrong answer feedback, individualized help (comprising hints or simpler sub-problems upon request), all driven by the largest metadatabase of student problem-solving in the world.

New to MasteringPhysics:

NEW **Interactive Prelecture Videos** address the rapidly growing movement toward prelecture teaching and flipped classrooms. These whiteboard-style animations provide an introduction to key topics with embedded assessment to help students prepare before lecture and to help professors identify student misconceptions.

NEW **Dynamic Study Modules (DSMs)** leverage research from the fields of cognitive psychology, neurobiology, and game studies to help students study on their own by continuously assessing their activity and performance, then using data and analytics to provide personalized content in real-time to reinforce concepts that target each student's particular strengths and weaknesses. Assignable for Pre-Class Prep or Self-Study, physics DSMs include the mathematics of physics, conceptual definitions, relationships for topics across all of mechanics and electricity and magnetism, and more.

Learning Catalytics™ is an inter- NEW active classroom tool that uses students' devices to engage them in class and provide "just in time" analytics to inform your lecture/active classroom. A new math palette for the mathematical expression question type allows for Greek symbols, PI, Euler's number, Logarithm, Exponent, Trigonometric functions, Absolute value, Square root, Nth square root, and Fractions with the exception of vector or unit vector.

Enhanced End-of-Chapter Ques- NEW **tions** offer students instructional support when and where they need it including links to the eText, math remediation, and wrong-answer specific feedback for an improved student experience and greater learning gains.

Adaptive Follow-Ups are personalized assignments that pair Mastering's powerful content with Knewton's adaptive learning engine to provide individualized help to students before misconceptions take hold. These adaptive follow-ups address topics students struggled with on assigned homework, including core prerequisite topics.

An **alternate set of hundreds of** NEW **homework problems** not included in the textbook gives instructors more assignable homework options than before. Solutions and problem statements are available in the Instructor's Solutions Manual.

MCAT questions are a set of multi- NEW part passage problems and standalone problems in a biological context that cover the key topics of the new 2015 MCAT format.

Video Tutor Demomonstration NEW **coaching activities** give brief demonstrations that include "Pause and Predict" questions with Mastering assessment containing wrong answer feedback and hints.

The **Test Bank** contains more than 2000 high-quality problems assignable as auto-graded items in MasteringPhysics. Test files are provided both in TestGen (an easy-to-use, fully networkable program for creating and editing quizzes and exams) and Word formats. Available in the

MasteringPhysics Instructor's Area, on the Instructor's Resource Center (www.pearsonhighered.com/irc), and on the Instructor's Resource DVD.

FOR THE STUDENT

MasteringPhysics® (www.masteringphysics.com) is the most advanced physics homework and tutorial system available. This online homework and tutoring system guides students through the most important topics in physics with self-paced tutorials that provide individualized coaching. These assignable, in-depth tutorials are designed to coach students with hints and feedback specific to their individual errors. Instructors can also assign end-of-chapter problems from every chapter including multiple-choice questions, section-specific exercises, and general problems. Quantitative problems can be assigned with numerical answers and randomized values (with sig fig feedback) or solutions. Mastering is available either bundled with the textbook, as an online download—with or without the eText—and as a standalone item available for purchase at the campus bookstore.

Pearson eText is available through MasteringPhysics®, either automatically when MasteringPhysics® is packaged with new books or as a purchased upgrade online. Allowing students access to the text wherever they have access to the Internet, Pearson eText comprises the full text with additional interactive features. Users can search for words or phrases, create notes, highlight text, bookmark sections, and click on definitions to key terms as they read.

Acknowledgments

I would like to express my sincere gratitude to colleagues at Washington State University and Western Washington University, as well as to many others in the physics community, for their contributions to this project.

In particular, I would like to thank Professor Ken Menningen of the University of Wisconsin–Stevens Point for his painstaking attention to detail in producing the Instructor's Solutions Manual. Ken has also contributed a number of new excellent Real-World Applications, Biological Applications, and end-of-chapter homework problems to the Fifth Edition.

My thanks are also due to the many wonderful and talented people at Pearson who have been such a pleasure to work with during the development of the Fifth Edition, and especially to Katie Conley, Jeanne Zalesky, Darien Estes, Cathy Murphy and Chandrika Madhavan.

In addition, I am grateful for the dedicated efforts of Rose Kernan, who choreographed a delightfully smooth production process.

Finally, I owe a great debt to all my students over the years. My interactions with them provided the motivation and inspiration that led to this book.

We are grateful to the following instructors for their thoughtful comments on the revision of this text.

REVIEWERS OF THE FIFTH EDITION

Ugur Akgun
University of Iowa
Kapila Castoldi
Oakland University
Joshua Colwell
University of Central Florida
Diana Driscoll
Case Western Reserve University
Daniel Giese
University of Wisconsin, Milwaukee
Klaus Honscheid
Ohio State University
Darrin Johnson
University of Minnesota, Duluth
Weining Man
San Francisco State University
Kin Mon
University of Georgia

Kriton Papavasiliou
Virginia Polytechnic Institute
Arnd Pralle
SUNY, University at Buffalo
Andrew Richter
Valparaiso University
Andreas Shalchi
University of Manitoba
Douglas Sherman
San Jose State University
Maxim Sukharev
Arizona State University
Michael Tammaro
University of Rhode Island
Orhan Yenen
University of Nebraska, Lincoln
Ataollah Zamani
Houston Community College

REVIEWERS OF THE FOURTH EDITION

Raymond Benge
Tarrant County College–NE Campus
Matthew Bigelow
Saint Cloud University
Edward J. Brash
Christopher Newport University
Michaela Burkardt
New Mexico State University
Jennifer Chen
University of Illinois at Chicago
Eugenia Ciocan
Clemson University
Shahida Dar
University of Delaware
Joseph Dodoo
University of Maryland, Eastern Shore
Thomas Dooling
University of Northern Carolina at Pembroke
Hui Fang
Sam Houston State University
Carlos E. Figueroa
Cabrillo College
Lyle Ford
University of Wisconsin, Eau Claire

Darrin Johnson
University of Minnesota, Duluth
Paul Lee
California State University, Northridge
Sheng-Chiang (John) Lee
Mercer University
Nilanga Liyanage
University of Virginia
Michael Ottinger
Missouri Western State University
Melodi Rodrigue
University of Nevada
Claudiu Rusu
Richland College of DCCCD
Mark Sprague
East Carolina University
Richard Szwerc
Montgomery College
Lisa Will
San Diego City College
Guanghua Xu
University of Houston
Bill Yen
University of Georgia

REVIEWERS OF PREVIOUS EDITIONS

Daniel Akerib, *Case Western Reserve University*
Richard Akerib, *Queens College*
Alice M. Hawthorne Allen, *Virginia Tech*
Barbara S. Andereck, *Ohio Wesleyan University*
Eva Andrei, *Rutgers University*
Bradley C. Antanaitis, *Lafayette College*
Michael Arnett, *Kirkwood Community College*
Robert W. Arts, *Pikeville College*

David Balogh, *Fresno City College*
David T. Bannon, *Oregon State University*
Rama Bansil, *Boston University*
Anand Batra, *Howard University*
Paul Beale, *University of Colorado–Boulder*
Mike Berger, *Indiana University*
David Berman, *University of Iowa*
S. M. Bhagat, *University of Maryland*

James D. Borgardt, *Juniata College*

James P. Boyle, *Western Connecticut State University*

David Branning, *Trinity College*

Jeff Braun, *University of Evansville*

Matthew E. Briggs, *University of Wisconsin-Madison*

Jack Brockway, *State University of New York-Oswego*

Neal Cason, *University of Notre Dame*

Thomas B. Cobb, *Bowling Green State University*

Lattie Collins, *Eastern Tennessee State University*

James Cook, *Middle Tennessee State University*

Stephen Cotanch, *North Carolina State University*

David Craig, *LeMoyne College*

David Curott, *University of North Alabama*

William Dabby, *Edison Community College*

Robert Davie, *St. Petersburg Junior College*

Steven Davis, *University of Arkansas–Little Rock*

N. E. Davison, *University of Manitoba*

Duane Deardorff, *University of North Carolina at Chapel Hill*

Edward Derringh, *Wentworth Institute of Technology*

Martha Dickinson, *Maine Maritime Academy*

Anthony DiStefano, *University of Scranton*

David C. Doughty, Jr., *Christopher Newport University*

F. Eugene Dunnam, *University of Florida*

John J. Dykla, *Loyola University-Chicago*

Eldon Eckard, *Bainbridge College*

Donald Elliott, *Carroll College*

David Elmore, *Purdue University*

Robert Endorf, *University of Cincinnati*

Raymond Enzweiler, *Northern Kentucky University*

John Erdei, *University of Dayton*

David Faust, *Mt. Hood Community College*

Frank Ferrone, *Drexel University*

John Flaherty, *Yuba College*

Curt W. Foltz, *Clarion University*

Lewis Ford, *Texas A&M University*

Armin Fuchs, *Florida Atlantic University*

Joseph Gallant, *Kent State University, Trumbull*

Asim Gangopadhyaya, *Loyola University–Chicago*

Thor Garber, *Pensacola Junior College*

David Gerdes, *University of Michigan*

John D. Gieringer, *Alvernia College*

Karen Gipson, *Grand Valley State University*

Barry Gilbert, *Rhode Island College*

Fred Gittes, *Washington State University*

Michael Graf, *Boston College*

William Gregg, *Louisiana State University*

Rainer Grobe, *Illinois State University*

Steven Hagen, *University of Florida*

Mitchell Haeri, *Saddleback College*

Parameswar Hari, *California State University–Fresno*

Xiaochun He, *Georgia State University*

Timothy G. Heil, *University of Georgia*

J. Erik Hendrickson, *University of Wisconsin-Eau Claire*

Scott Holmstrom, *University of Tulsa*

John Hopkins, *The Pennsylvania State University*

Manuel A. Huerta, *University of Miami*

Zafar Ismail, *Daemen College*

Adam Johnston, *Weber State University*

Gordon O. Johnson, *Washington State University*

Nadejda Kaltcheva, *University of Wisconsin-Oshkosh*

William Karstens, *Saint Michael's College*

Sanford Kern, *Colorado State University*

Dana Klinck, *Hillsborough Community College*

Ilkka Koskelo, *San Francisco State University*

Laird Kramer, *Florida International University*

R. Gary Layton, *Northern Arizona University*

Kevin M. Lee, *University of Nebraska-Lincoln*

Michael Lieber, *University of Arkansas*

Ian M. Lindevald, *Truman State University*

Mark Lindsay, *University of Louisville*

Jeff Loats, *Fort Lewis College*

Daniel Ludwigsen, *Kettering University*

Lorin Matthews, *Baylor University*

Hilliard Macomber, *University of Northern Iowa*

Trecia Markes, *University of Nebraska–Kearny*

William McNairy, *Duke University*

Kenneth L. Menningen, *University of Wisconsin-Stevens Point*

Joseph Mills, *University of Houston*

Anatoly Miroshnichenko, *University of Toledo*

Wouter Montfrooij, *University of Missouri*

Gary Morris, *Valparaiso University*

Paul Morris, *Abilene Christian University*

David Moyle, *Clemson University*

Ashok Muthukrishnan, *Texas A&M University*

K. W. Nicholson, *Central Alabama Community College*

Robert Oman, *University of South Florida*

Michael Ottinger, *Missouri Western State College*

Larry Owens, *College of the Sequoias*

A. Ray Penner, *Malaspina University*

Francis Pichanick, *University of Massachusetts, Amherst*

Robert Piserchio, *San Diego State University*

Anthony Pitucco, *Pima Community College*

William Pollard, *Valdosta State University*

Jerry Polson, *Southeastern Oklahoma State University*

Robert Pompi, *Binghamton University*

David Procopio, *Mohawk Valley Community College*

Earl Prohofsky, *Purdue University*

Jia Quan, *Pasadena City College*

David Raffaelle, *Glendale Community College*

Michele Rallis, *Ohio State University*

Michael Ram, *State University of New York-Buffalo*

Prabha Ramakrishnan, *North Carolina State University*

Rex Ramsier, *University of Akron*

John F. Reading, *Texas A&M University*

Lawrence B. Rees, *Brigham Young University*

M. Anthony Reynolds, *Embry Riddle University*

Dennis Rioux, *University of Wisconsin-Oshkosh*

John A. Rochowicz, Jr., *Alvernia College*

STUDENT REVIEWERS

As a student preparing to take an algebra-based physics course, you are probably aware that physics applies to absolutely everything in the natural world, from raindrops and people, to galaxies and atoms. Because physics is so wide-ranging and comprehensive, it can sometimes seem a bit overwhelming. This text, which reflects nearly two decades of classroom experience, is designed to help you deal with a large body of information and develop a working understanding of the basic concepts in physics. Now in its fifth edition, it incorporates many refinements that have come directly from interacting with students using the first four editions. As a result of these interactions, I am confident that as you develop a deeper understanding of physics, you will also enrich your experience of the world in which you live.

Now, I must admit that I like physics, and so I may be a bit biased in this respect. Still, the reason I teach and continue to study physics is that I enjoy the insight it gives into the physical world. I can't help but notice—and enjoy—aspects of physics all around me each and every day. I would like to share some of this enjoyment and delight in the natural world with you. It is for this reason that I undertook the task of writing this book.

To assist you in the process of studying physics, this text incorporates a number of learning aids, including two-column Examples, Quick Examples, Conceptual Examples, and Exercises. These and other elements work together in a unified way to enhance your understanding of physics on both a conceptual and a quantitative level—they have been developed to give you the benefit of what we know about how students learn physics, and to incorporate strategies that have proven successful to students over the years. The pages that follow will introduce these elements to you, describe the purpose of each, and explain how they can help you.

As you progress through the text, you will encounter many interesting and intriguing applications of physics drawn from the world around you. Some of these, such as magnetically levitated trains or the Global Positioning System (GPS), are primarily technological in nature. Others focus on explaining familiar or not-so-familiar phenomena, such as why the Moon has no atmosphere, how sweating cools the body, or why flying saucer shaped clouds often hover over mountain peaks even on a clear day. Still others, such as countercurrent heat exchange in animals and humans, or the use of sound waves to destroy kidney stones, are of particular relevance to students of biology and those interested in pursuing a career in medicine.

In many cases, you may find the applications to be a bit surprising. Did you know, for example, that you are shorter at the end of the day than when you first get up in the morning? (This is discussed in Chapter 5.) That an instrument called the ballistocardiograph can detect the presence of a person hiding in a truck, just by registering the minute recoil from the beating of the stowaway's heart? (This is discussed in Chapter 9.) That if you hum next to a spider's web at just the right pitch you can cause a resonance effect that sends the spider into a tizzy? (This is discussed in Chapter 13.) That powerful magnets can exploit the phenomenon of diamagnetism to levitate living creatures? (This is discussed in Chapter 22.) That scorpions in the nighttime desert are brightly fluorescent when illuminated by ultraviolet light? (This is discussed in Chapter 31.) The natural world truly is filled with marvelous things, and physics applies to all of them.

Writing this textbook was a rewarding and enjoyable experience for me. I hope using it will prove equally rewarding to you, and that it will inspire an appreciation for physics that will give you a lifetime of enjoyment.

James S. Walker

DETAILED CONTENTS

PART I MECHANICS

PART II THERMAL PHYSICS

PHYSICS

Big Ideas

1 Physics is the study of the laws of nature.

2 The basic units in physics are length, mass, and time.

3 Valid physics equations must be dimensionally consistent.

4 Some physical quantities are scalars; others are vectors.

Introduction to Physics

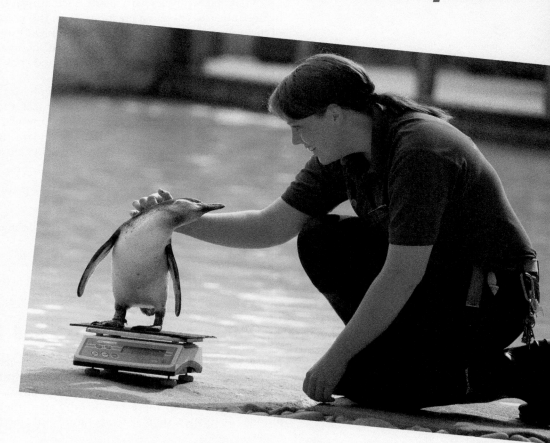

▲ Physics is a quantitative science, based on careful measurements of quantities such as mass, length, and time. In the measurement shown here, a naturalist determines the mass of a cooperative penguin. The length of the penguin, and its age, are also important data in monitoring the penguin's health.

The goal of physics is to gain a deeper understanding of the world in which we live. For example, the laws of physics allow us to predict the behavior of everything from rockets sent to the Moon, to integrated chips in computers, to lasers used to perform eye surgery. In short, everything in nature—from atoms and subatomic particles to solar systems and galaxies—obeys the laws of physics. In this chapter, we develop a common "language" of physics that will be used throughout this book.

1-1 Physics and the Laws of Nature

Physics is the study of the fundamental laws of nature, which, simply put, are the laws that underlie all physical phenomena in the universe. Remarkably, we have found that these laws can be expressed in terms of mathematical equations. As a result, it is possible to make precise, quantitative comparisons between the predictions of theory—derived from the mathematical form of the laws—and the observations of experiments. Physics, then, is a science rooted equally firmly in both theory and experiment. As physicists make new observations, they constantly test and—if necessary—refine the present theories.

What makes physics particularly fascinating is the fact that it relates to everything in the universe. There is a great beauty in the vision that physics brings to our view of the universe—namely, that all the complexity and variety that we see in the world around us, and in the universe as a whole, are manifestations of a few fundamental laws and principles. That we can discover and apply these basic laws of nature is both astounding and exhilarating.

For those not familiar with the subject, physics may seem to be little more than a confusing mass of formulas. Sometimes, in fact, these formulas can be the trees that block the view of the forest. For a physicist, however, the many formulas of physics are simply different ways of expressing a few fundamental ideas. It is the forest—the basic laws and principles of physical phenomena in nature—that is the focus of this text.

Enhance Your Understanding (Answers given at the end of the chapter)

1. The laws of physics apply to which of the following: **(a)** gravity, **(b)** electricity, **(c)** magnetism, **(d)** light, **(e)** atoms, or **(f)** all of these?

Section Review

- Physics combines both theory and experiment in the study of the laws of nature.

1-2 Units of Length, Mass, and Time

To make quantitative comparisons between the laws of physics and our experience of the natural world, certain basic physical quantities must be measured. The most common of these quantities are **length** (L), **mass** (M), and **time** (T). In fact, in the next several chapters these are the only quantities that arise. Later in the text, additional quantities, such as temperature and electric current, will be introduced as needed.

We begin by defining the units in which each of these quantities is measured. Once the units are defined, the values obtained in specific measurements can be expressed as multiples of them. For example, our unit of length is the **meter** (m). It follows, then, that a person who is 1.5 m tall has a height 1.5 times this unit of length. Similar comments apply to the unit of mass, the **kilogram,** and the unit of time, the **second.**

The detailed system of units used in this book was established in 1960 at the Eleventh General Conference on Weights and Measures in Paris, France, and goes by the name Système International d'Unités, or SI for short. Thus, when we refer to **SI units,** we mean units of meters (m), kilograms (kg), and seconds (s). Taking the first letter from each of these units leads to an alternate name that is often used—the **mks system.**

In the remainder of this section we define each of the SI units.

Length

Early units of length were often associated with the human body. For example, the Egyptians defined the cubit to be the distance from the elbow to the tip of the middle finger. Similarly, the foot was at one time defined to be the length of the royal

TABLE 1-1 Typical Distances

Distance from Earth to the nearest large galaxy (the Andromeda galaxy, M31)	2×10^{22} m
Diameter of our galaxy (the Milky Way)	1×10^{21} m
Distance from Earth to the nearest star (other than the Sun)	4×10^{16} m
One light-year	9.46×10^{15} m
Average radius of Pluto's orbit	6×10^{12} m
Distance from Earth to the Sun	1.5×10^{11} m
Radius of Earth	6.37×10^{6} m
Length of a football field	10^{2} m
Height of a person	2 m
Diameter of a CD	0.12 m
Diameter of the aorta	0.018 m
Diameter of a period in a sentence	5×10^{-4} m
Diameter of a red blood cell	8×10^{-6} m
Diameter of the hydrogen atom	10^{-10} m
Diameter of a proton	2×10^{-15} m

(a)

(b)

▲ **FIGURE 1-1 (a)** The size of these viruses, seen here attacking a bacterial cell, is about 10^{-7} m. **(b)** The diameter of this typical galaxy is about 10^{21} m. (How many viruses would it take to span the galaxy?)

foot of King Louis XIV. As colorful as these units may be, they are not particularly reproducible—at least not to great precision.

In 1793 the French Academy of Sciences, seeking a more objective and reproducible standard, decided to define a unit of length equal to one ten-millionth the distance from the North Pole to the equator. This new unit was named the metre (from the Greek *metron* for "measure"). The preferred spelling in the United States is *meter*. This definition was widely accepted, and in 1799 a "standard" meter was produced. It consisted of a platinum-iridium alloy rod with two marks on it one meter apart.

Since 1983 we have used an even more precise definition of the meter, based on the speed of light in a vacuum:

> One meter is defined to be the distance traveled by light in a vacuum in 1/299,792,458 of a second.

No matter how its definition is refined, however, a meter is still about 3.28 feet, which is roughly 10 percent longer than a yard. A list of typical lengths is given in Table 1-1, and an illustration of the range of lengths is given in **FIGURE 1-1**.

Mass

In SI units, mass is measured in kilograms. Unlike the meter, the kilogram is not based on any natural physical quantity. By convention, the kilogram is defined as follows:

> The kilogram is the mass of a particular platinum-iridium alloy cylinder at the International Bureau of Weights and Measures in Sèvres, France. The cylinder, referred to as the standard kilogram, is shown in **FIGURE 1-2**.

To put the kilogram in everyday terms, a quart of milk has a mass slightly less than 1 kilogram. Additional masses, in kilograms, are given in Table 1-2.

It's important to note that we do not define the kilogram to be the *weight* of the platinum-iridium cylinder. In fact, weight and mass are quite different quantities, even though they are often confused in everyday language. Mass is an intrinsic, unchanging property of an object. Weight, in contrast, is a measure of the gravitational force acting on an object, which can vary depending on the object's location. For example, if you are fortunate enough to travel to Mars someday, you will find that your weight is less than on Earth, though your mass is unchanged. The force of gravity will be discussed in detail in Chapter 12.

TABLE 1-2 Typical Masses

Galaxy (Milky Way)	4×10^{41} kg
Sun	2×10^{30} kg
Earth	5.97×10^{24} kg
Space shuttle	2×10^{6} kg
Elephant	5400 kg
Automobile	1200 kg
Human	70 kg
Baseball	0.15 kg
Honeybee	1.5×10^{-4} kg
Red blood cell	10^{-13} kg
Bacterium	10^{-15} kg
Hydrogen atom	1.67×10^{-27} kg
Electron	9.11×10^{-31} kg

▲ **FIGURE 1-2** The standard kilogram, a cylinder of platinum and iridium 0.039 m in height and diameter, is kept under carefully controlled conditions in Sèvres, France. Exact replicas are maintained in other laboratories around the world.

PHYSICS IN CONTEXT
Looking Ahead

The three physical dimensions introduced in this chapter—mass, length, and time—are the only ones we'll use until Chapter 16, when temperature is introduced. Other quantities found in the next several chapters, like force, momentum, and energy, are combinations of these three basic dimensions.

▲ **FIGURE 1-3** This atomic clock, which keeps time on the basis of radiation from cesium atoms, is accurate to about two ten-millionths of a second per year. (How much time would it take for it to gain or lose an hour?)

Time

Nature has provided us with a fairly accurate timepiece in the revolving Earth. In fact, prior to 1956 the mean solar day was defined to consist of 24 hours, with 60 minutes per hour, and 60 seconds per minute, for a total of $(24)(60)(60) = 86,400$ seconds. Even the rotation of the Earth is not completely regular, however.

Today, the most accurate timekeepers are "atomic clocks," like the one in FIGURE 1-3, which are based on characteristic frequencies of radiation emitted by certain atoms. These clocks have typical accuracies of about 1 second in 6 million years. The atomic clock used for defining the second operates with cesium-133 atoms. In particular, the second is defined as follows:

> One second is defined to be the time it takes for radiation from a cesium-133 atom to complete 9,192,631,770 cycles of oscillation.

A range of characteristic time intervals is given in Table 1-3.

Atomic Clocks and Official Time   The nation's time and frequency standard is determined by a *cesium fountain atomic clock* developed at the National Institute of Standards and Technology (NIST) in Boulder, Colorado. The fountain atomic clock, designated NIST-F2, produces a "fountain" of cesium atoms that are projected upward in a vacuum to a height of about a meter. It takes roughly a second for the atoms to rise and fall through this height (as we shall see in the next chapter), and during this relatively long period of time, the frequency of their oscillation can be measured with great precision. In fact, the NIST-F2 will gain or lose no more than one second in every 300 million years of operation.

Atomic clocks are almost commonplace these days. For example, the satellites that participate in the Global Positioning System (GPS) actually carry atomic clocks with them in orbit. This allows them to make the precise time measurements that are needed for an equally precise determination of a GPS user's position and speed. Similarly, the "atomic clocks" that are advertised for use in the home, while not atomic in their operation, nonetheless get their time from radio signals sent out from the atomic clocks at NIST in Boulder. You can access the official U.S. time on your computer by going to time.gov on the Web.

Other Systems of Units and Standard Prefixes

Although SI units are used throughout most of this book, and are used almost exclusively in scientific research and in industry, we will occasionally refer to other systems that you may encounter from time to time.

For example, a system of units similar to the mks system, though comprised of smaller units, is the **cgs system**, which stands for centimeter (cm), gram (g), and second (s). In addition, the British engineering system is often encountered in everyday usage in the United States. Its basic units are the slug for mass, the foot (ft) for length, and the second (s) for time.

Finally, multiples of the basic units are common no matter which system is used. Standard prefixes are used to designate common multiples in powers of ten. For example, the prefix *kilo* means one thousand, or, equivalently, 10^3. Thus, 1 kilogram is 10^3 grams, and 1 kilometer is 10^3 meters. Similarly, *milli* is the prefix for one thousandth, or 10^{-3}. Thus, a millimeter is 10^{-3} meter, and so on. The most common prefixes are listed in Table 1-4.

EXERCISE 1-1   CONVERTING UNITS

   a. A baseball has a diameter of 75 millimeters. Give the diameter in meters and kilometers.

   b. A house sells for 450,000 dollars. Express the price of the house in kilodollars and megadollars.

REASONING AND SOLUTION
Refer to Table 1-4 for a list of prefixes and their corresponding powers of ten.

   a. 0.075 m, 0.000075 km

   b. 450 kilodollars, 0.450 megadollars

2. Rank the following lengths (A, B, C, and D) in order of increasing value. Indicate ties where appropriate. A = 2 millimeters; B = 10 micrometers; C = 0.1 kilometers; D = 10 centimeters.

Section Review

- The Sl units of length, mass, and time are the meter, kilogram, and second, respectively.

TABLE 1-3 Typical Times

Age of the universe	4×10^{17} s
Age of the Earth	1.4×10^{17} s
Existence of human species	6×10^{12} s
Human lifetime	2×10^9 s
One year	3×10^7 s
One day	8.6×10^4 s
Time between heart-beats	0.8 s
Human reaction time	0.1 s
One cycle of a high-pitched sound wave	5×10^{-5} s
One cycle of an AM radio wave	10^{-6} s
One cycle of a visible light wave	2×10^{-15} s

1-3 Dimensional Analysis

In physics, when we speak of the **dimension** of a physical quantity, we refer to the *type* of quantity in question, regardless of the units used in the measurement. For example, a distance measured in cubits and another distance measured in light-years both have the same dimension—length. The same is true of compound units such as velocity, which has the dimensions of length per unit time (length/time). A velocity measured in miles per hour has the same dimensions—length/time—as a velocity measured in inches per century.

Now, any valid formula in physics must be **dimensionally consistent**; that is, each term in the equation must have the same dimensions. It simply doesn't make sense to add a distance to a time, for example, any more than it makes sense to add apples and oranges. They are different things.

Checking Dimensional Consistency and Dimensional Analysis To check the dimensional consistency of an equation, it's convenient to introduce a special notation for the dimension of a quantity. We will use square brackets, [ ], for this purpose. Thus, if x represents a distance, which has dimensions of length $[L]$, we write this as $x = [L]$. Similarly, a velocity, v, has dimensions of length $[L]$ per time $[T]$; thus we write $v = [L]/[T]$ to indicate its dimensions. Acceleration, a, which is the change in velocity per time, has the dimensions $a = ([L]/[T])/[T] = [L]/[T^2]$. The dimensions of some common physical quantities are listed in Table 1-5.

Let's use this notation to check the dimensional consistency of a simple equation. Consider the following formula:

$$x = x_0 + vt$$

In this equation, x and x_0 represent distances, v is a velocity, and t is time. Writing out the dimensions of each term, we have

$$[L] = [L] + \frac{[L]}{[T]}[T]$$

It might seem that the last term has different dimensions than the other two. However, dimensions obey the same rules of algebra as other quantities. Thus the dimensions of time cancel in the last term:

$$[L] = [L] + \frac{[L]}{[T]}[T] = [L] + [L]$$

As a result, we see that each term in this formula has the same dimension, $[L]$. This type of calculation with dimensions is referred to as **dimensional analysis**.

TABLE 1-4 Common Prefixes

Power	Prefix	Abbreviation
10^{15}	peta	P
10^{12}	tera	T
10^9	giga	G
10^6	mega	M
10^3	kilo	k
10^2	hecto	h
10^1	deka	da
10^{-1}	deci	d
10^{-2}	centi	c
10^{-3}	milli	m
10^{-6}	micro	μ
10^{-9}	nano	n
10^{-12}	pico	p
10^{-15}	femto	f

Big Idea ③ Valid equations in physics must be dimensionally consistent. This means that each term in the equation must have the same dimensions.

EXERCISE 1-2 ANALYZING DIMENSIONS

Show that $x = x_0 + v_0 t + \frac{1}{2}at^2$ is dimensionally consistent. The quantities x and x_0 are distances, v_0 is a velocity, and a is an acceleration.

CONTINUED

TABLE 1-5 Dimensions of Some Common Physical Quantities

Quantity	Dimension
Distance	$[L]$
Area	$[L^2]$
Volume	$[L^3]$
Velocity	$[L]/[T]$
Acceleration	$[L]/[T^2]$
Energy	$[M][L^2]/[T^2]$

PHYSICS IN CONTEXT
Looking Ahead

Dimensional analysis is used frequently in the coming chapters to verify that each term in an equation has the correct dimensions. We will also use dimensional analysis to help derive some results, such as the speed of waves on a string in Chapter 14.

REASONING AND SOLUTION

Use the dimensions given in Table 1-5 for each term in the equation. Cancel dimensions where appropriate.

Original equation: $\quad x = x_0 + v_0 t + \frac{1}{2}at^2$

Dimensions: $\quad [L] = [L] + \dfrac{[L]}{[T]}[T] + \dfrac{[L]}{[T^2]}[T^2] = [L] + [L] + [L]$

Notice that $\frac{1}{2}$ is ignored in this analysis because it is simply a numerical factor, and has no dimensions.

Later in this text you will derive your own formulas from time to time. As you do so, it is helpful to check dimensional consistency at each step of the derivation. If at any time the dimensions don't agree, you will know that a mistake has been made, and you can go back and look for it. If the dimensions check, however, it's not a guarantee the formula is correct—after all, dimensionless numerical factors, like $\frac{1}{2}$ or 2, don't show up in a dimensional check.

Enhance Your Understanding   (Answers given at the end of the chapter)

3. Give the dimensions of each of the following quantities (A, B, C, D), given that x is a distance, v is a velocity, a is an acceleration, and t is a time. $A = at$; $B = v/t$; $C = 0.5x/t^2$; $D = 2ax$.

Section Review

- The dimension of a quantity is the type of quantity it is—such as length, mass, or time.

- Each term in a valid physics equation must have the same dimensions.

▲ **FIGURE 1-4** Every measurement has some degree of uncertainty associated with it. How precise would you expect this measurement to be?

1-4 Significant Figures, Scientific Notation, and Round-Off Error

In this section we discuss a few issues regarding the numerical values that arise in scientific measurements and calculations.

Significant Figures

When a length, a mass, or a time is measured in a scientific experiment, the result is known only to within a certain accuracy. The inaccuracy or uncertainty can be caused by a number of factors, as illustrated in **FIGURE 1-4**, ranging from limitations of the measuring device itself to limitations associated with the senses and the skill of the person performing the experiment. In any case, the fact that observed values of experimental quantities have inherent uncertainties should always be kept in mind when performing calculations with those values.

The level of uncertainty in a numerical value is indicated by the number of **significant figures** it contains. We define significant figures as follows:

Number of Significant Figures

The number of significant figures ("sig figs" for short) in a physical quantity is equal to the number of digits in it that are known with certainty.

As an example, suppose you want to determine the walking speed of your pet tortoise. To do so, you measure the time, t, it takes for the tortoise to walk a distance, d, and then you calculate the quotient, d/t. When you measure the distance with a ruler, which has one tick mark per millimeter, you find that $d = 21.4$ cm. Each of these three digits is known with certainty, though the digit that follows the 4 is uncertain. Hence, we say that d has *three* significant figures. Similarly, you measure the time with an old pocket watch, and determine that $t = 8.5$ s. It follows that t is known to only *two* significant figures, because you are not certain of the digit that follows the 5.

We would now like to calculate the speed of the tortoise. Using the above values for d and t, and a calculator with eight digits in its display, we find $(21.4 \text{ cm})/(8.5 \text{ s}) = 2.5176470$ cm/s. Clearly, such an accurate value for the speed is unjustified, considering the limitations of our measurements. In general, the number of significant figures that result when we multiply or divide physical quantities is given by the following rule of thumb:

Significant Figures When Multiplying or Dividing
The number of significant figures after multiplication or division is equal to the number of significant figures in the *least* accurately known quantity.

In our speed calculation, for example, we know the distance to three significant figures, but we know the time to only two significant figures. As a result, the speed should be given with just two significant figures: $d/t = (21.4 \text{ cm})/(8.5 \text{ s}) = 2.5$ cm/s. Thus, 2.5 cm/s is our best estimate for the tortoise's speed, given the uncertainty in our measurements.

EXAMPLE 1-3   IT'S THE TORTOISE BY A HARE

A tortoise races a rabbit by walking with a constant speed of 2.51 cm/s for 12.27 s. How much distance does the tortoise cover?

PICTURE THE PROBLEM
The race between the rabbit and the tortoise is shown in our sketch. The rabbit pauses to eat a carrot while the tortoise walks with a constant speed.

2.51 cm/s

REASONING AND STRATEGY
The distance covered by the tortoise is the speed of the tortoise multiplied by the time during which it walks.

Known   Constant speed $= 2.51$ cm/s; walking time $= 12.27$ s.
Unknown Distance, $d = ?$

SOLUTION

1. Multiply the speed by the time
   to find the distance d:

$d = (\text{speed})(\text{time})$
$= (2.51 \text{ cm/s})(12.27 \text{ s}) = 30.8$ cm

INSIGHT
If we simply multiply 2.51 cm/s by 12.27 s, we obtain 30.7977 cm. We don't give all of these digits in our answer, however. In particular, because the quantity that is known with the least accuracy (the speed) has only three significant figures, we give a result with three significant figures. In addition, notice that the third digit in our answer has been rounded up from 7 to 8 because the number that follows 7 is greater than or equal to 5.

PRACTICE PROBLEM
How much time does it take for the tortoise to walk 17 cm?   [**Answer:** $t = (17 \text{ cm})/(2.51 \text{ cm/s}) = 6.8$ s]

Some related homework problems: Problem 14, Problem 18

The distance of 17 cm in the above Practice Problem has only two significant figures, because we don't know the digits to the right of the decimal place. If the distance were given as 17.0 cm, on the other hand, it would have three significant figures. The role of significant figures in a real-world situation is illustrated in **FIGURE 1-5**.

When physical quantities are added or subtracted, we use a slightly different rule of thumb. In this case, the rule involves the number of decimal places in each of the terms:

Significant Figures When Adding or Subtracting
The number of decimal places after addition or subtraction is equal to the smallest number of decimal places in any of the individual terms.

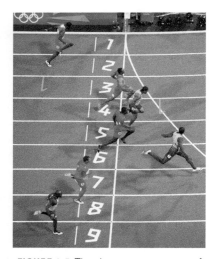

▲ **FIGURE 1-5** The time measurements for a race like this must be accurate to several significant figures. This level of accuracy is needed to determine both the winning time, and the order in which the other runners finished.

Thus, if you make a time measurement of 16.74 s and then a subsequent time measurement of 5.1 s, the total time of the two measurements should be given as 16.74 s + 5.1 s = 21.8 s, rather than 21.84 s.

EXERCISE 1-4 FINDING THE TOTAL WEIGHT

You and a friend go fishing. You catch a fish that weighs 10.7 lb, and your friend catches a fish that weighs 8.35 lb. What is the combined weight of the two fish?

REASONING AND SOLUTION
Simply adding the two numbers gives 19.05 lb. According to our rule of thumb, however, the final result must have only a single decimal place (corresponding to the term with the smallest number of decimal places). Rounding off to one decimal place, then, gives 19.1 lb as the accepted result.

Scientific Notation

The number of significant figures in a given quantity may be ambiguous due to the presence of zeros at the beginning or end of the number. For example, if a distance is stated to be 2500 m, the two zeros could be significant figures, or they could be zeros that simply show where the decimal point is located. If the two zeros are significant figures, the uncertainty in the distance is roughly a meter; if they are not significant figures, however, the uncertainty is about 100 m.

To remove this type of ambiguity, we can write the distance in **scientific notation**—that is, as a number of order unity times an appropriate power of ten. Thus, in this example, we would express the distance as 2.5×10^3 m if there are only two significant figures, or as 2.500×10^3 m to indicate four significant figures. Likewise, a time given as 0.000036 s has only two significant figures—the preceding zeros only serve to fix the decimal point. If this quantity were known to three significant figures, we would write it as 0.0000360 s, or equivalently as 3.60×10^{-5} s, to remove any ambiguity. See Appendix A for a more detailed discussion of scientific notation.

EXERCISE 1-5 SIGNIFICANT FIGURES

How many significant figures are there in **(a)** 0.0210, **(b)** 5.060, **(c)** 7.10×10^{-4}, **(d)** 1.0×10^5?

REASONING AND SOLUTION
Preceding zeros don't count when determining the number of significant figures, but trailing zeros and zeros within a number do count. As a result, the number of significant figures is as follows: **(a)** 3, **(b)** 4, **(c)** 3, **(d)** 2.

Round-Off Error

Finally, even if you perform all your calculations to the same number of significant figures as in the text, you may occasionally obtain an answer that differs in its last digit from that given in the book. In most cases this is not an issue as far as understanding the physics is concerned—usually it is due to **round-off error**.

Round-off error occurs when numerical results are rounded off at different times during a calculation. To see how this works, let's consider a simple example. Suppose you are shopping for knickknacks, and you buy one item for $2.21, plus 8 percent sales tax. The total price is $2.3868 or, rounded off to the nearest penny, $2.39. Later, you buy another item for $1.35. With tax this becomes $1.458 or, again to the nearest penny, $1.46. The total expenditure for these two items is $2.39 + $1.46 = $3.85.

Now, let's do the rounding off in a different way. Suppose you buy both items at the same time for a total before-tax price of $2.21 + $1.35 = $3.56. Adding in the 8 percent tax gives $3.8448, which rounds off to $3.84, one penny different from the previous amount. This type of discrepancy can also occur in physics problems. In general, it's a good idea to keep one extra digit throughout your calculations whenever possible, rounding off only the final result. But, while this practice can help to reduce the likelihood of round-off error, there is no way to avoid it in every situation.

Enhance Your Understanding   (Answers given at the end of the chapter)

4. Rank the following numbers in order of increasing number of significant figures. Indicate ties where appropriate. A = 1; B = 1001; C = 10.01; D = 123.

Section Review

- The number of significant figures in a physical quantity is the number of digits known with certainty. The greater the number of significant figures, the more accurately the quantity is known.

- Round-off error occurs when numbers are rounded off at different points in a calculation. The small discrepancy that results is not significant.

1-5 Converting Units

It's often convenient to convert from one set of units to another. For example, suppose you want to convert 316 ft to its equivalent in meters. Looking at the conversion factors on the inside front cover of the text, we see that

$$1\ m = 3.281\ ft$$

Equivalently,

$$\frac{1\ m}{3.281\ ft} = 1$$

Now, to make the conversion, we simply multiply 316 ft by this expression, which is equivalent to multiplying by 1:

$$(316\ ft)\left(\frac{1\ m}{3.281\ ft}\right) = 96.3\ m$$

We write the conversion factor in this particular way, as 1 m divided by 3.281 ft, so that the units of feet cancel out, leaving the final result in the desired units of meters.

Of course, we can just as easily convert from meters to feet if we use the reciprocal of this conversion factor—which is also equal to 1:

$$1 = \frac{3.281\ ft}{1\ m}$$

For example, a distance of 26.4 m is converted to feet by canceling out the units of meters, as follows:

$$(26.4\ m)\left(\frac{3.281\ ft}{1\ m}\right) = 86.6\ ft$$

Thus, we see that converting units is as easy as multiplying by 1—because that's really what you're doing. A real-world example of unit conversion is shown in **FIGURE 1-6**.

▲ **FIGURE 1-6** From this sign, you can calculate factors for converting miles to kilometers and vice versa. (Why do you think the conversion factors seem to vary for different destinations?)

EXAMPLE 1-6   A HIGH-VOLUME ITEM

The interior of a popular microwave oven has a width $W = 15.5$ in., a depth $D = 14.5$ in., and a height $H = 9.25$ in. What is the interior volume of the oven in SI units?

PICTURE THE PROBLEM
In our sketch we picture the oven, and indicate the dimensions given in the problem statement.

REASONING AND STRATEGY
We begin by converting the width, depth, and height of the oven to meters. Once this is done, the volume in SI units is simply the product of the three dimensions.

CONTINUED

Known   Width, $W = 15.5$ in.; depth, $D = 14.5$ in.; height, $H = 9.25$ in.
Unknown   Volume, $V = ?$

SOLUTION

1. Convert the width of the oven to meters:

$$W = (15.5 \text{ in.})\left(\frac{1 \text{ ft}}{12 \text{ in.}}\right)\left(\frac{1 \text{ m}}{3.281 \text{ ft}}\right) = 0.394 \text{ m}$$

2. Convert the depth to meters:

$$D = (14.5 \text{ in.})\left(\frac{1 \text{ ft}}{12 \text{ in.}}\right)\left(\frac{1 \text{ m}}{3.281 \text{ ft}}\right) = 0.368 \text{ m}$$

3. Convert the height to meters:

$$H = (9.25 \text{ in.})\left(\frac{1 \text{ m}}{39.4 \text{ in.}}\right) = 0.235 \text{ m}$$

4. Calculate the volume of the oven in cubic meters:

$$V = W \times D \times H = (0.394 \text{ m})(0.368 \text{ m})(0.235 \text{ m}) = 0.0341 \text{ m}^3$$

INSIGHT
This volume is just 3.41 percent of a cubic meter. In terms of cubic feet, which is the unit used by manufacturers, the volume is 1.20 cu ft. (Note: We used a different conversion factor in Step 3, just to illustrate that conversions can be carried out in different ways.)

PRACTICE PROBLEM
What is the volume of the oven if its width is one-half of a yard, and the other dimensions are unchanged?
[**Answer:** $V = 0.0395 \text{ m}^3$]

Some related homework problems: Problem 21, Problem 22

Converting More Than One Unit at a Time   The procedure outlined above can be applied to conversions involving any number of units. For instance, if you walk at 3.00 mi/h, how fast is that in m/s? In this case we need the following additional conversion factors:

$$1 \text{ mi} = 5280 \text{ ft} \qquad 1 \text{ h} = 3600 \text{ s}$$

With these factors at hand, we carry out the conversion as follows:

$$(3.00 \text{ mi/h})\left(\frac{5280 \text{ ft}}{1 \text{ mi}}\right)\left(\frac{1 \text{ m}}{3.281 \text{ ft}}\right)\left(\frac{1 \text{ h}}{3600 \text{ s}}\right) = 1.34 \text{ m/s}$$

In each conversion factor, the numerator is equal to the denominator. In addition, each conversion factor is written in such a way that the unwanted units cancel, leaving just meters per second in our final result.

QUICK EXAMPLE 1-7   FIND THE SPEED OF BLOOD

Blood in the human aorta (**FIGURE 1-7**) can have speeds of 39.7 cm/s. How fast is this in **(a)** ft/min and **(b)** mi/h?

REASONING AND SOLUTION
To convert a speed to different units it may be necessary to convert both the units of length and those of time.

Part (a) Convert centimeters to feet, and seconds to minutes:

$$39.7 \text{ cm/s} = (39.7 \text{ cm/s})\left(\frac{1 \text{ m}}{100 \text{ cm}}\right)\left(\frac{3.281 \text{ ft}}{1 \text{ m}}\right)\left(\frac{60 \text{ s}}{1 \text{ min}}\right)$$

$$= 78.2 \text{ ft/min}$$

Part (b) Convert centimeters to miles, and seconds to hours:

$$39.7 \text{ cm/s} = (39.7 \text{ cm/s})\left(\frac{1 \text{ m}}{100 \text{ cm}}\right)\left(\frac{3.281 \text{ ft}}{1 \text{ m}}\right)\left(\frac{1 \text{ mi}}{5280 \text{ ft}}\right)\left(\frac{3600 \text{ s}}{1 \text{ h}}\right)$$

$$= 0.888 \text{ mi/h}$$

▲ **FIGURE 1-7** Major blood vessels branch from the aorta (bottom), the artery that receives blood directly from the heart.

Enhance Your Understanding   (Answers given at the end of the chapter)

5. To convert a length of 6 ft to its equivalent in meters, which of the following conversion factors (A, B, C, D) should we use? $A = \left(\dfrac{1 \text{ mi}}{5280 \text{ ft}}\right)$; $B = \left(\dfrac{3.281 \text{ ft}}{1 \text{ m}}\right)$; $C = \left(\dfrac{3.281 \text{ m}}{1 \text{ ft}}\right)$; $D = \left(\dfrac{1 \text{ m}}{3.281 \text{ ft}}\right)$.

Section Review

* To convert units, multiply by a conversion factor that cancels one unit and replaces it with another.

1-6 Order-of-Magnitude Calculations

An **order-of-magnitude** calculation is a rough "ballpark" estimate designed to be accurate to within a factor of about 10. One purpose of such a calculation is to give a quick idea of what order of magnitude should be expected from a more complete calculation. If an order-of-magnitude calculation indicates that a distance should be on the order of 10^4 m, for example, and your calculator gives an answer on the order of 10^7 m, then there is an error somewhere that needs to be resolved.

For example, suppose you would like to estimate the speed of a cliff diver on entering the water. First, the cliff may be 20 or 30 feet high; thus in SI units we would say that the order of magnitude of the cliff's height is 10 m—certainly not 1 m or 10^2 m. Next, the diver hits the water something like a second later—certainly not 0.1 s later or 10 s later. Thus, a reasonable order-of-magnitude estimate of the diver's speed is 10 m/1 s = 10 m/s, or roughly 20 mi/h. If you do a detailed calculation and your answer is on the order of 10^4 m/s, you probably entered one of your numbers incorrectly. Order-of-magnitude calculations like this are often referred to as "Fermi problems" in honor of Italian physicist Enrico Fermi (FIGURE 1-8), who was a master of such calculations.

Another reason for doing an order-of-magnitude calculation is to get a feeling for what size numbers we are talking about in situations where a precise count is not possible. This is illustrated in the following Example.

▲ **FIGURE 1-8** Enrico Fermi (1901–1954) was renowned for his ability to pose and solve interesting order-of-magnitude problems. A winner of the 1938 Nobel Prize in physics, Fermi would ask his classes to obtain order-of-magnitude estimates for questions such as "How many piano tuners are there in Chicago?"

EXAMPLE 1-8   ESTIMATION: HOW MANY RAINDROPS IN A STORM?

A thunderstorm drops half an inch (~0.01 m) of rain on Washington, DC, which covers an area of about 70 square miles (~10^8 m²). Estimate the number of raindrops that fell during the storm.

PICTURE THE PROBLEM
Our sketch shows an area $A = 10^8$ m² covered to a depth $d = 0.01$ m by rainwater from the storm. Each drop of rain is approximated by a small sphere with a radius of $r = (4 \text{ mm})/2 = 2 \text{ mm}$.

REASONING AND STRATEGY
To find the number of raindrops, we first calculate the volume of water required to cover 10^8 m² to a depth of 0.01 m. Next, we calculate the volume of an individual drop of rain, recalling that the volume of a sphere of radius r is $4\pi r^3/3$. We estimate the radius of a raindrop to be about 2 mm. Finally, dividing the volume of a drop into the volume of water that fell during the storm gives the number of drops.

Known   Area of rainfall, $A = 10^8$ m²; depth of rain, $d = 0.01$ m; radius of raindrop, $r = 2$ mm.

Unknown   Number of raindrops = ?

CONTINUED

SOLUTION

1. Calculate the order of magnitude of the volume of water, V_{water}, that fell during the storm:

$$V_{water} = Ad = (10^8 \text{ m}^2)(0.01 \text{ m}) \approx 10^6 \text{ m}^3$$

2. Calculate the order of magnitude of the volume of a drop of rain, V_{drop}. Recall that if the diameter of a drop is 4 mm, its radius is $r = 2$ mm $= 0.002$ m:

$$V_{drop} = \frac{4}{3}\pi r^3 \approx \frac{4}{3}\pi(0.002 \text{ m})^3 \approx 10^{-8} \text{ m}^3$$

3. Divide V_{drop} into V_{water} to find the order of magnitude of the number of drops that fell during the storm:

$$number\ of\ raindrops \approx \frac{V_{water}}{V_{drop}} \approx \frac{10^6 \text{ m}^3}{10^{-8} \text{ m}^3} = 10^{14}$$

INSIGHT

Thus, the number of raindrops in this one small storm is roughly 10,000 times greater than the current population of the Earth.

PRACTICE PROBLEM

If a storm pelts Washington, DC, with 10^{15} raindrops, how many inches of rain fall on the city?   [**Answer:** About 5 inches]

Some related homework problems: Problem 35, Problem 38

Appendix B provides a number of interesting "typical values" for length, mass, speed, acceleration, and many other quantities. You may find these to be of use in making your own order-of-magnitude estimates.

Enhance Your Understanding   (Answers given at the end of the chapter)

6. Give an order-of-magnitude estimate for the speed at which fingernails grow. Express your answer in meters per second.

Section Review

- Order-of-magnitude estimates are rough calculations that give answers accurate to within a factor of 10.

1-7 Scalars and Vectors

Physical quantities are sometimes defined solely in terms of a number and the corresponding unit, like the volume of a room or the temperature of the air it contains. Other quantities require both a numerical value *and* a direction. For example, suppose a car is traveling at a rate of 25 m/s in a direction that is due north. Both pieces of information—the rate of travel (25 m/s) and the direction (north)—are required to fully specify the motion of the car. The rate of travel is given the name **speed**; the rate of travel combined with the direction is referred to as the **velocity**.

In general, quantities that are specified by a numerical value only are referred to as **scalars**; quantities that require both a numerical value and a direction are called **vectors**:

- A scalar is a numerical value, expressed in terms of appropriate units. An example is the temperature of a room or the speed of a car.

- A vector is a physical quantity with both a numerical value and a direction. An example is the velocity of a car.

All the physical quantities discussed in this text are either vectors or scalars. The properties of numbers (scalars) are well known, but the properties of vectors are sometimes less well known—though no less important. For this reason, you will find that Chapter 3 is devoted entirely to a discussion of vectors in two and three dimensions and, more specifically, to how they are used in physics.

Big Idea **4** Some physical quantities are described by scalars, which are simply numbers. Other physical quantities are described by vectors, which combine both a numerical value and a direction.

Vectors in One Dimension   The rather straightforward special case of vectors in one dimension is discussed in Chapter 2. There, we shall see that the direction of a velocity vector, for example, can be only to the left or to the right, up or down, and so on. That is, only two choices are available for the direction of a vector in one dimension.

This is illustrated in FIGURE 1-9, where we see two cars, each traveling with a speed of 25 m/s. Notice that the cars are traveling in opposite directions, with car 1 moving to the right and car 2 moving to the left. We indicate the direction of travel with a plus sign for motion to the right, and a negative sign for motion to the left. Thus, the velocity of car 1 is written $v_1 = +25$ m/s, and the velocity of car 2 is $v_2 = -25$ m/s. The speed of each car is the absolute value, or **magnitude**, of the velocity; that is, speed $= |v_1| = |v_2| = 25$ m/s.

Whenever we deal with one-dimensional vectors, we shall indicate their direction with the appropriate sign. Many examples are found in Chapter 2 and, again, in later chapters where the simplicity of one dimension can again be applied.

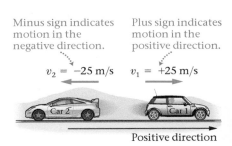

Minus sign indicates motion in the negative direction.

Plus sign indicates motion in the positive direction.

$v_2 = -25$ m/s $v_1 = +25$ m/s

Positive direction

▲ **FIGURE 1-9 Velocity vectors in one dimension** The two cars shown in this figure have equal speeds of 25 m/s but are traveling in opposite directions. To indicate the direction of travel, we first choose a positive direction (to the right in this case), and then give appropriate signs to the velocity of each car. For example, car 1 moves to the right, and hence its velocity is positive, $v_1 = +25$ m/s; the velocity of car 2 is negative, $v_2 = -25$ m/s, because it moves to the left.

Enhance Your Understanding

(Answers given at the end of the chapter)

7. **(a)** Can two cars have the same speed but different velocities? Explain. **(b)** Can two cars have the same velocity but different speeds? Explain.

Section Review

- Scalars are numbers. Vectors have both a numerical value and a direction.
- In one dimension, the direction of a vector is designated by its sign.

1-8 Problem Solving in Physics

Physics is a lot like swimming—you have to learn by doing. You could read a book on swimming and memorize every word in it, but when you jump into a pool the first time you are going to have problems. Similarly, you could read this book carefully, memorizing every formula in it, but when you finish, you still haven't learned physics. To learn physics, you have to go beyond passive reading; you have to interact with physics and experience it by doing problems.

Problem-Solving Tips In what follows, we present a general overview of problem solving in physics. The suggestions given below, which apply to problems in all areas of physics, should help to develop a systematic approach.

We should emphasize at the outset that there is no recipe for solving problems in physics—it is a creative activity. In fact, the opportunity to be creative is one of the attractions of physics. The following suggestions, then, are not intended as a rigid set of steps that must be followed like the steps in a computer program. Rather, they provide a general guideline that experienced problem solvers find to be effective.

Problem-Solving Suggestions

- **Read the problem carefully** Before you can solve a problem, you need to know exactly what information it gives and what it asks you to determine. Some information is given explicitly, as when a problem states that a person has a mass of 70 kg. Other information is implicit; for example, saying that a ball is dropped from rest means that its initial speed is zero. Clearly, a *careful* reading is the essential first step in problem solving.

- **Sketch the system** This may seem like a step you can skip—but don't. A sketch helps you to acquire a physical feeling for the system. It also provides an opportunity to label those quantities that are known and those that are to be determined.

- **Visualize the physical process** Try to visualize what is happening in the system as if you were watching it in a movie. Your sketch should help. This step ties in closely with the next step.

- **Strategize** This may be the most difficult, but at the same time the most creative, part of the problem-solving process. From your sketch and visualization, try to identify the physical processes at work in the system. Ask yourself what concepts or principles are involved in this situation. Then, develop a strategy—a game plan—for solving the problem.

PHYSICS IN CONTEXT
Looking Ahead

In this chapter we introduced the idea of a vector in one dimension and showed how the direction of the vector can be indicated by its sign. These concepts are developed in more detail in Chapter 2.

PHYSICS IN CONTEXT
Looking Ahead

Vectors are extended to two and three dimensions in Chapter 3. Vectors are a standard tool throughout the remaining chapters on mechanics, and they appear again in electricity and magnetism.

- **Identify appropriate equations** Once a strategy has been developed, find the specific equations that are needed to carry it out.

- **Solve the equations** Use basic algebra to solve the equations identified in the previous step. Work with symbols such as x or y for the most part, substituting numerical values near the end of the calculations. Working with symbols will make it easier to go back over a problem to locate and identify mistakes, if there are any, and to explore limits and special cases.

- **Check your answer** Once you have an answer, check to see if it makes sense: (i) Does it have the correct dimensions? (ii) Is the numerical value reasonable?

- **Explore limits/special cases** Getting the correct answer is nice, but it's not all there is to physics. You can learn a great deal about physics and about the connection between physics and mathematics by checking various limits of your answer. For example, if you have two masses in your system, m_1 and m_2, what happens in the special case where $m_1 = 0$, or in the case where $m_1 = m_2$? Check to see whether your answer and your physical intuition agree.

Following these suggestions will help to strengthen your ability to solve problems—in physics and in other disciplines as well.

Pedagogical Features The pedagogical features in this text are designed to deepen your understanding of physics concepts, and at the same time to develop your problem-solving skills. Perhaps the most important of these are the **Examples,** which present a complete solution to a substantial problem and include these helpful sections:

- A sketch of the system, along with a **Picture the Problem** discussion to help you visualize the problem.

- A **Reasoning and Strategy** section to develop a "game plan" before carrying out the calculations.

- A **Known** and **Unknown** section to specify what is given and what is to be calculated.

- A two-column, step-by-step **Solution** with the reasoning given in words in the left-hand column and the corresponding math given in the right-hand column.

- An **Insight** section to point out interesting aspects of the calculation and its results.

- A **Practice Problem** to give you a chance to try out the calculations just given on a similar problem.

These sections work together to give a thorough understanding of a problem and its solution. You will also find **Quick Examples,** which are streamlined versions of the full Examples, and **Exercises,** which present simple one-line calculations.

In addition, **Conceptual Examples** are designed to strengthen your understanding of the basic concepts of physics, and to develop your skill in using these concepts to "make sense" of the numerical aspects of physics. Some of the homework problems help with making these connections as well. **Predict/Calculate** problems ask for a prediction based on physical concepts, followed by a calculation to verify the prediction. **Predict/Explain** problems ask for a prediction and the corresponding explanation. You will also find problems of these types in some of the Examples and Conceptual Examples in each chapter.

Finally, it's often tempting to look for shortcuts when doing a problem—to look for a formula that seems to fit and some numbers to plug into it. It may seem harder to think ahead, to be systematic as you solve the problem, and then to think back over what you've done at the end of the problem. The extra effort is worth it, however. It will help you develop powerful problem-solving skills that can be applied to unexpected problems you may encounter on exams—and in life in general.

Section Review

- Problem solving is an important part of learning physics. Practice is required, as in learning to ride a bicycle.

CHAPTER 1 REVIEW

CHAPTER SUMMARY

1–1 PHYSICS AND THE LAWS OF NATURE

Physics is based on a small number of fundamental laws and principles.

1–2 UNITS OF LENGTH, MASS, AND TIME

Length

One meter is defined as the distance traveled by light in a vacuum in 1/299,792,458 second.

Mass

One kilogram is the mass of a metal cylinder kept at the International Bureau of Weights and Measures.

Time

One second is the time required for a particular type of radiation from cesium-133 to undergo 9,192,631,770 oscillations.

1–3 DIMENSIONAL ANALYSIS

Dimension

The dimension of a quantity is the type of quantity it is—for example, length [L], mass [M], or time [T].

Dimensional Consistency

An equation is dimensionally consistent if each term in it has the same dimensions. All valid physical equations are dimensionally consistent.

Dimensional Analysis

A calculation based on the dimensional consistency of an equation.

1–4 SIGNIFICANT FIGURES, SCIENTIFIC NOTATION, AND ROUND-OFF ERROR

Significant Figures

The number of digits reliably known, excluding digits that simply indicate the decimal place.

Round-off Error

Discrepancies caused by rounding off numbers in intermediate results.

1–5 CONVERTING UNITS

Multiply by the ratio of two units to convert from one to another.

1–6 ORDER-OF-MAGNITUDE CALCULATIONS

A ballpark estimate designed to be accurate to within the nearest power of ten.

1–7 SCALARS AND VECTORS

A physical quantity that can be represented by a numerical value only is called a scalar. Quantities that require a direction in addition to the numerical value are called vectors.

$v_2 = -25 \text{ m/s}$ $v_1 = +25 \text{ m/s}$

Positive direction

1–8 PROBLEM SOLVING IN PHYSICS

A good general approach to problem solving is as follows: read, sketch, visualize, strategize, identify equations, solve, check, explore limits.

ANSWERS TO ENHANCE YOUR UNDERSTANDING QUESTIONS

1. **(f)** all of these.
2. B < A < D < C.
3. A = [L]/[T]; B = [L]/[T^2];
   C = [L]/[T^2]; D = [L^2]/[T^2].
4. A < D < B = C.

5. D.
6. 10^{-9} m/s.
7. **(a)** Yes. They can travel equally fast in different directions.
   (b) No. Both the speed and direction must be the same.

CONCEPTUAL QUESTIONS

For instructor-assigned homework, go to www.masteringphysics.com. (MP)

Conceptual Questions are meant to stimulate discussion of significant physics concepts. Answers to odd-numbered Conceptual Questions are given in the back of the book.

1. Can dimensional analysis determine whether the area of a circle is πr^2 or $2\pi r^2$? Explain.
2. If a distance d has units of meters, and a time T has units of seconds, does the quantity $T + d$ make sense physically? What about the quantity d/T? Explain in both cases.
3. Is it possible for two quantities to **(a)** have the same units but different dimensions or **(b)** have the same dimensions but different units? Explain.
4. The frequency of a pendulum is the number of oscillations it makes per second. Is frequency a scalar or a vector quantity? Explain.

5. Albert runs 34 m to the right in 10 s. Isaac runs 17 m to the left in 5 s. **(a)** Is Albert's speed greater than, less than, or equal to Isaac's speed? **(b)** Is Albert's velocity equal to Isaac's velocity? Explain.
6. Give an order-of-magnitude estimate for the time in seconds of the following: **(a)** a year; **(b)** a baseball game; **(c)** a heartbeat; **(d)** the age of the Earth; **(e)** the age of a person.
7. Give an order-of-magnitude estimate for the length in meters of the following: **(a)** a person; **(b)** a fly; **(c)** a car; **(d)** a 747 airplane; **(e)** an interstate freeway stretching coast-to-coast.

PROBLEMS AND CONCEPTUAL EXERCISES

Answers to odd-numbered Problems and Conceptual Exercises can be found in the back of the book. **BIO** *identifies problems of biological or medical interest;* **CE** *indicates a conceptual exercise.* Predict/Explain *problems ask for two responses: (a) your prediction of a physical outcome, and (b) the best explanation among three provided; and Predict/Calculate problems ask for a prediction of a physical outcome, based on fundamental physics concepts, and follow that with a numerical calculation to verify the prediction. On all problems, bullets (•, ••, •••) indicate the level of difficulty.*

SECTION 1-2 UNITS OF LENGTH, MASS, AND TIME

1. • **The Hunger Games** The movie *The Hunger Games* brought in about $152,000,000 in its opening weekend. Express this amount in **(a)** gigadollars and **(b)** teradollars.

2. • **BIO The Thickness of Hair** A human hair has a thickness of about 85 μm. What is this in **(a)** meters and **(b)** millimeters?

3. • The speed of light in a vacuum is approximately 0.3 Gm/s. Express the speed of light in meters per second.

4. • **A Fast Computer** IBM has a computer it calls the Blue Gene/L that can do 136.8 teracalculations per second. How many calculations can it do in a nanosecond?

SECTION 1-3 DIMENSIONAL ANALYSIS

5. • **CE** Which of the following equations are dimensionally consistent? **(a)** $x = \frac{1}{2}at^2$, **(b)** $t = v/x$, **(c)** $t = (2x/a)^{1/2}$.

6. • **CE** Which of the following quantities have the dimensions of a time? **(a)** x/v, **(b)** a/v, **(c)** $(2x/a)^{1/2}$, **(d)** v^2/a.

7. • **CE** Which of the following quantities have the dimensions of a distance? **(a)** vt, **(b)** $\frac{1}{2}at^2$, **(c)** $2at$, **(d)** v^2/a.

8. • **CE** Which of the following quantities have the dimensions of a speed? **(a)** $\frac{1}{2}at^2$, **(b)** at, **(c)** $(2x/a)^{1/2}$, **(d)** $(2ax)^{1/2}$.

9. •• Velocity is related to acceleration and distance by the following expression: $v^2 = 2ax^p$. Find the power p that makes this equation dimensionally consistent.

10. •• Acceleration is related to distance and time by the following expression: $a = 2xt^p$. Find the power p that makes this equation dimensionally consistent.

11. •• The time t required for an object to fall from rest through a height h is given by

$$t = h^p\sqrt{\frac{2}{g}}$$

In this expression, g is the acceleration due to gravity. Find the power p that makes this equation dimensionally consistent.

12. •• Newton's second law (to be discussed in Chapter 5) states that the acceleration a of an object is proportional to the force F acting on it, and inversely proportional to its mass m. That is,

$$a = \frac{F}{m}$$

What are the dimensions of force?

13. •• The time T required for one complete oscillation of a mass m on a spring of force constant k is

$$T = 2\pi\sqrt{\frac{m}{k}}$$

Find the dimensions k must have for this equation to be dimensionally correct.

SECTION 1-4 SIGNIFICANT FIGURES, SCIENTIFIC NOTATION, AND ROUND-OFF ERROR

14. • The speed of light to five significant figures is 2.9979×10^8 m/s. What is the speed of light to three significant figures?

15. • A parking lot is 124.3 m long and 41.06 m wide. What is the perimeter of the lot?

16. • On a fishing trip you catch a 2.77-lb bass, a 14.3-lb rock cod, and a 13.43-lb salmon. What is the total weight of your catch?

17. •• How many significant figures are there in **(a)** 0.0000303 and **(b)** 6.201×10^5?

18. •• What is the area of a circle of radius **(a)** 11.37 m and **(b)** 6.8 m?

19. •• The first several digits of π, the ratio of the circumference to the diameter of a circle, are $\pi = 3.14159265358979\ldots$. What is π to **(a)** three significant figures, **(b)** five significant figures, and **(c)** seven significant figures?

SECTION 1-5 CONVERTING UNITS

20. • Rank the following speeds in order of increasing magnitude. Indicate ties where appropriate. **(a)** 0.25 m/s, **(b)** 0.75 km/h, **(c)** 12 ft/s, **(d)** 16 cm/s.

21. • The Ark of the Covenant is described as a chest of acacia wood 2.5 cubits in length and 1.5 cubits in width and height. Given that a cubit is equivalent to 17.7 in., find the volume of the ark in cubic feet.

22. • A car drives on a highway with a speed of 68 mi/h. What is this speed in km/h?

23. • **Angel Falls** Water going over Angel Falls, in Venezuela, the world's highest waterfall, drops through a distance of 3212 ft. What is this distance in km?

24. • An electronic advertising sign repeats a message every 9 seconds, day and night, for a week. How many times did the message appear on the sign?

25. • **BIO Blue Whales** The blue whale (*Balaenoptera musculus*) is thought to be the largest animal ever to inhabit the Earth. The longest recorded blue whale had a length of 108 ft. What is this length in meters?

Angel Falls (Problem 23)

26. • **The Star of Africa** The Star of Africa, a diamond in the royal scepter of the British crown jewels, has a mass of 530.2 carats, where 1 carat = 0.20 g. Given that 1 kg has an approximate weight of 2.21 lb, what is the weight of this diamond in pounds?

27. • **BIO Woodpecker Impact** When red-headed woodpeckers (*Melanerpes erythrocephalus*) strike the trunk of a tree, they can experience an acceleration ten times greater than the acceleration of gravity, or about 98.1 m/s^2. What is this acceleration in ft/s^2?

28. •• Predict/Calculate Many highways have a speed limit of 55 mi/h. **(a)** Is this speed greater than, less than, or equal to 55 km/h? Explain. **(b)** Find the speed limit in km/h that corresponds to 55 mi/h.

29. •• **BIO Mantis Shrimp** Peacock mantis shrimps (*Odontodactylus scyllarus*) feed largely on snails. They shatter the shells of their prey by delivering a sharp blow with their front legs, which have been observed to reach peak speeds of 23 m/s. What is this speed in **(a)** feet per second and **(b)** miles per hour?

30. •• **A Jiffy** The American physical chemist Gilbert Newton Lewis (1875–1946) proposed a unit of time called the "jiffy." According to Lewis, 1 jiffy = the time it takes light to travel one centimeter. **(a)** If you perform a task in a jiffy, how much time has it taken in seconds? **(b)** How many jiffys are in one minute? (Use the fact that the speed of light is approximately $2.9979 \times 10^8 \text{ m/s.}$)

31. •• **The Mutchkin and the Noggin (a)** A mutchkin is a Scottish unit of liquid measure equal to 0.42 L. How many mutchkins are required to fill a container that measures one foot on a side? **(b)** A noggin is a volume equal to 0.28 mutchkin. What is the conversion factor between noggins and gallons?

32. •• Suppose 1.0 cubic meter of oil is spilled into the ocean. Find the area of the resulting slick, assuming that it is one molecule thick, and that each molecule occupies a cube 0.50 μm on a side.

33. •• The acceleration of gravity is approximately 9.81 m/s^2 (depending on your location). What is the acceleration of gravity in feet per second squared?

34. •• **BIO Squid Nerve Impulses** Nerve impulses in giant axons of the squid can travel with a speed of 25.0 m/s. How fast is this in **(a)** ft/s and **(b)** mi/h?

SECTION 1-6 ORDER-OF-MAGNITUDE CALCULATIONS

35. • Give a ballpark estimate of the number of seats in a typical major-league ballpark.

36. • Estimate the speed at which your hair grows. Give your answer in meters per second.

37. •• Milk is often sold by the gallon in plastic containers. **(a)** Estimate the number of gallons of milk that are purchased in the United States each year. **(b)** What approximate weight of plastic does this represent?

38. •• New York is roughly 3000 miles from Seattle. When it is 10:00 A.M. in Seattle, it is 1:00 P.M. in New York. Using this information, estimate **(a)** the rotational speed of the surface of Earth, **(b)** the circumference of Earth, and **(c)** the radius of Earth.

GENERAL PROBLEMS

39. • CE Which of the following equations are dimensionally consistent? **(a)** $v = at$, **(b)** $v = \frac{1}{2}at^2$, **(c)** $t = a/v$, **(d)** $v^2 = 2ax$.

40. • CE Which of the following quantities have the dimensions of an acceleration? **(a)** xt^2, **(b)** v^2/x, **(c)** x/t^2, **(d)** v/t.

41. • **BIO Photosynthesis** The light that plants absorb to perform photosynthesis has a wavelength that peaks near 675 nm. Express this distance in **(a)** millimeters and **(b)** inches.

42. • **Glacial Speed** On June 9, 1983, the lower part of the Variegated Glacier in Alaska was observed to be moving at a rate of 210 feet per day. What is this speed in meters per second?

Alaska's Variegated Glacier (Problem 42)

43. • One liter of pure gold weighs 42.5 lb. If you discovered 1.0 ft^3 of pure gold, how much would it weigh in pounds?

44. • What is the speed in miles per second of a beam of light traveling at 3.00×10^8 m/s?

45. • **BIO Rattlesnake Frequency** A timber rattlesnake (*Crotalus horridus*) shakes its rattle at a characteristic frequency of about 3300 shakes per minute. What is this frequency in shakes per second?

46. •• **BIO** A single human red blood cell has a mass of about 27 pg. What is this mass in **(a)** kilograms and **(b)** nanograms?

47. •• **BIO Corn Growth** Sweet corn plants can grow 4.1 cm in height per day. What is this speed in **(a)** millimeters per second and **(b)** feet per week?

48. •• **BIO Mosquito Courtship** Male mosquitoes in the mood for mating find female mosquitoes of their own species by listening for the characteristic "buzzing" frequency of the female's wing beats. This frequency is about 605 wing beats per second. **(a)** How many wing beats occur in one minute? **(b)** How many cycles of oscillation does the radiation from a cesium-133 atom complete during one mosquito wing beat?

49. •• Predict/Calculate A Porsche sports car can accelerate at 14 m/s². **(a)** Is this acceleration greater than, less than, or equal to 14 ft/s²? Explain. **(b)** Determine the acceleration of a Porsche in ft/s².

50. •• **A Speeding Bullet** The fastest commercially available rifle bullet claims a muzzle velocity of 4225 ft/s. **(a)** What is this speed in miles per hour? **(b)** How far (in meters) can this bullet travel in 5.0 ms?

51. •• **BIO Human Nerve Fibers** Type A nerve fibers in humans can conduct nerve impulses at speeds up to 140 m/s. **(a)** How fast are the nerve impulses in miles per hour? **(b)** How far (in meters) can the impulses travel in 5.0 ms?

52. •• **BIO Brain Growth** The mass of a newborn baby's brain has been found to increase by about 1.6 mg per minute. **(a)** How much does the brain's mass increase in one day? **(b)** How much time does it take for the brain's mass to increase by 0.0075 kg?

53. •• **The Huygens Probe** NASA's *Cassini* mission to Saturn released a probe on December 25, 2004, that landed on the Saturnian moon Titan on January 14, 2005. The probe, which was named *Huygens*, was released with a gentle relative speed of 31 cm/s. As *Huygens* moved away from the main spacecraft, it rotated at a rate of seven revolutions per minute. **(a)** How many revolutions had *Huygens* completed when it was 150 yards from the mother ship? **(b)** How far did *Huygens* move away from the mother ship during each revolution? Give your answer in feet.

54. •• **BIO Spin-Dry Dragonflies** Some dragonflies are observed to splash into the water to bathe, and then spin rapidly to shed the water. Slow-motion video shot at 240 frames per second shows that dragonflies complete one spin revolution every 14 frames. What is the spin rate of the dragonfly in revolutions per minute (rpm)?

55. ••• Acceleration is related to velocity and time by the following expression: $a = v^p t^q$. Find the powers p and q that make this equation dimensionally consistent.

56. ••• The period T of a simple pendulum is the amount of time required for it to undergo one complete oscillation. If the length of the pendulum is L and the acceleration of gravity is g, then T is given by $T = 2\pi L^p g^q$. Find the powers p and q required for dimensional consistency.

57. ••• Driving along a crowded freeway, you notice that it takes a time t to go from one mile marker to the next. When you increase your speed by 7.9 mi/h, the time to go one mile decreases by 13 s. What was your original speed?

PASSAGE PROBLEMS

BIO Using a Cricket as a Thermometer

All chemical reactions, whether organic or inorganic, proceed at a rate that depends on temperature—the higher the temperature, the higher the rate of reaction. This can be understood in terms of molecules moving with increased energy as the temperature is increased, and colliding with other molecules more frequently. In the case of organic reactions, the result is that metabolic processes speed up with increasing temperature.

An increased or decreased metabolic rate can manifest itself in a number of ways. For example, a cricket trying to attract a mate chirps at a rate that depends on its overall rate of metabolism. As a result, the chirping rate of crickets depends directly on temperature. In fact, some people even use a pet cricket as a thermometer.

The cricket that is most accurate as a thermometer is the snowy tree cricket (*Oecanthus fultoni* Walker). Its rate of chirping is described by the following formula:

$$N = \text{number of chirps per 13.0 seconds} = T - 40.0$$

In this expression, T is the temperature in degrees Fahrenheit.

58. • Which plot in **FIGURE 1-10**—A, B, C, D, or E—represents the chirping rate of the snowy tree cricket?

Chirping rate versus temperature

FIGURE1-10 Problem 58

59. •• If the temperature is 43 °F, how much time does it take for the cricket to produce 12 chirps?

    A. 12 s   B. 24 s   C. 43 s   D. 52 s

60. •• Your pet cricket chirps 112 times in one minute (60.0 s). What is the temperature in degrees Fahrenheit?

    A. 41.9   B. 47.0   C. 64.3   D. 74.7

61. •• Suppose a snowy tree cricket is chirping when the temperature is 65.0 °F. How many oscillations does the radiation from a cesium-133 atom complete between successive chirps?

    A. 7.98×10^7 B. 3.68×10^8
    C. 4.78×10^9 D. 9.58×10^9

Big Ideas

1 Position is measured relative to a coordinate system.

2 Velocity is the rate of change of position with time.

3 Acceleration is the rate of change of velocity with time.

4 Speed increases (decreases) when velocity and acceleration are in the same (opposite) direction.

5 Equations of motion relate position, velocity, acceleration, and time.

6 Free fall is motion with a constant downward acceleration of magnitude g.

One-Dimensional Kinematics

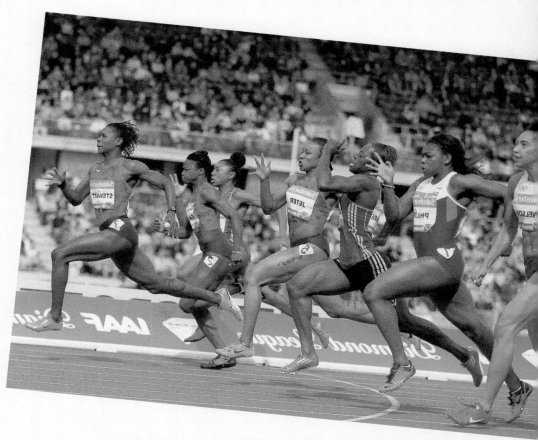

▲ These sprinters provide a good illustration of one-dimensional, straight-line motion. They accelerated as they left the starting blocks, and now they maintain a constant velocity as they approach the finish line.

We begin our study of physics with **mechanics**, the area of physics most apparent to us in our everyday lives. Every time you raise an arm, stand up or sit down, throw a ball, or open a door, your actions are governed by the laws of mechanics. In this chapter we focus on kinematics—the basic properties of motion—in one dimension.

2-1 Position, Distance, and Displacement

In physics, the terms *position, distance,* and *displacement* have specific meanings. This section gives the physics definitions of these terms and shows how they are used to describe the motion of a particle.

Position The first step in describing the motion of a particle is to set up a **coordinate system** that defines its location—that is, its **position.** An example of a coordinate system in one dimension is shown in FIGURE 2-1. This is simply an x axis, with an origin (where $x = 0$) and an arrow pointing in the positive direction—the direction in which x increases. In setting up a coordinate system, we are free to choose the origin and the positive direction as we like, but once we make a choice, we must be consistent with it throughout any calculations that follow.

▶ FIGURE 2-1 **A one-dimensional coordinate system** You are free to choose the origin and positive direction as you like, but once your choice is made, stick with it.

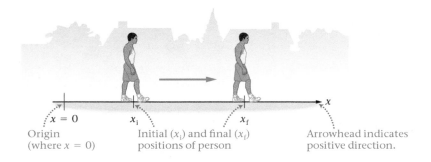

Origin (where $x = 0$)   Initial (x_i) and final (x_f) positions of person   Arrowhead indicates positive direction.

The "particle" in Figure 2-1 is a person who has moved to the right from an initial position, x_i, to a final position, x_f. Because the positive direction is to the right, it follows that x_f is greater than x_i; that is, $x_f > x_i$.

Distance Now that we've seen how to set up a coordinate system, let's use one to investigate the situation shown in FIGURE 2-2. Suppose you leave your house, drive to the grocery store, and then return home. The **distance** you've covered in your trip is 4.3 mi + 4.3 mi = 8.6 mi. In general, distance is defined as the length of a trip:

Definition of Distance

distance = total length of travel

SI unit: meter, m

Using SI units, we find that the distance in this case is

$$8.6 \text{ mi} = (8.6 \text{ mi})\left(\frac{1609 \text{ m}}{1 \text{ mi}}\right) = 1.4 \times 10^4 \text{ m}$$

▶ FIGURE 2-2 **One-dimensional coordinates** The locations of your house, your friend's house, and the grocery store in terms of a one-dimensional coordinate system.

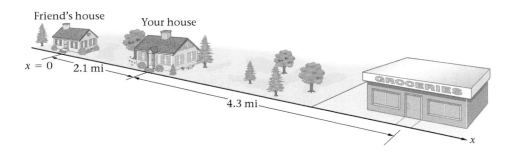

In a car, the distance traveled is indicated by the odometer. Distance is always positive, and is a scalar quantity (see Chapter 1) with no associated direction.

Displacement Another useful way to characterize a particle's motion is in terms of its **displacement,** Δx, which is simply the change in position:

Definition: Displacement, Δx

displacement = change in position = final position − initial position

$$\text{displacement} = \Delta x = x_f - x_i \qquad \text{2-1}$$

SI unit: meter, m

The SI units of displacement are meters—the same as for distance—but displacement and distance are really quite different. For example, in the round trip from your house to the grocery store and back, the distance traveled is 8.6 mi, whereas the displacement is zero because $x_f = x_i = 2.1$ mi, and hence $\Delta x = x_f - x_i = 0$.

Notice that we use the delta notation, Δx, as a convenient shorthand for the quantity $x_f - x_i$. (See Appendix A for a complete discussion of delta notation.) Also, notice that Δx can be positive (if the final position is to the right of the initial position, $x_f > x_i$), negative (if the final position is to the left of the initial position, $x_f < x_i$), or zero (if the final and initial positions are the same, $x_f = x_i$). In fact, the displacement is a one-dimensional vector, as defined in Chapter 1, and its direction (right or left) is given by its sign (positive or negative, respectively).

For example, suppose you go from your house to the grocery store and then to your friend's house in Figure 2-2. On this trip the distance is 10.7 mi, but the displacement is

$$\Delta x = x_f - x_i = (0) - (2.1 \text{ mi}) = -2.1 \text{ mi}$$

The minus sign means your displacement is in the negative direction; that is, to the left.

QUICK EXAMPLE 2-1 DISPLACEMENT AND DISTANCE

Predict/Calculate

Referring to Figure 2-2, suppose you take a trip from your friend's house to the grocery store and then to your house. **(a)** Is the displacement for this trip positive, negative, or zero? Explain. **(b)** Find the displacement for this trip. **(c)** What is the distance covered in this trip?

REASONING AND SOLUTION
Displacement is the final position minus the initial position; it can be positive or negative. Distance is the length of travel, which is always positive.

1. **Part (a)** The displacement is positive because the final position is to the right (positive direction) of the initial position.

2. **Part (b)** Determine the initial position for the trip, using Figure 2-2:
   $$x_i = 0$$

3. Determine the final position for the trip, using Figure 2-2:
   $$x_f = 2.1 \text{ mi}$$

4. Subtract x_i from x_f to find the displacement. Notice that the result is positive, as expected:
   $$\Delta x = x_f - x_i = 2.1 \text{ mi} - 0 = 2.1 \text{ mi}$$

5. **Part (c)** Add the distances for the various parts of the trip:
   $$2.1 \text{ mi} + 4.3 \text{ mi} + 4.3 \text{ mi} = 10.7 \text{ mi}$$

Enhance Your Understanding   (Answers given at the end of the chapter)

1. For each of the following questions, give an example if your answer is yes. Explain why not if your answer is no. **(a)** Is it possible to take a trip in which the distance covered is less than the magnitude of the displacement? **(b)** Is it possible to take a trip in which the distance covered is greater than the magnitude of the displacement?

Section Review

- Distance is the total length of a trip.
- Displacement is the change in position; displacement $= \Delta x = x_f - x_i$.

PHYSICS IN CONTEXT
Looking Back

In this chapter we make extensive use of the sign conventions for one-dimensional vectors introduced in Chapter 1—positive for one direction, negative for the opposite direction.

Big Idea 1 Position and displacement are measured relative to a coordinate system. The coordinate system must include an origin and a positive direction. Displacement has a direction and a magnitude, and hence it is a vector. Distance has only a numerical value, and hence it is a scalar.

2-2 Average Speed and Velocity

The next step in describing motion is to consider how rapidly an object moves. For example, how much time does it take for a major-league fastball to reach home plate? How far does an orbiting satellite travel in one hour? These are examples of some of the most basic questions regarding motion, and in this section we learn how to answer them.

Average Speed The simplest way to characterize the rate of motion is with the **average speed**:

$$\text{average speed} = \frac{\text{distance}}{\text{elapsed time}} \qquad 2\text{-}2$$

The dimensions of average speed are distance per time or, in SI units, meters per second, m/s. Both distance and elapsed time are positive; thus average speed is always positive.

EXAMPLE 2-2   DOG RUN

A dog trots back to its owner with an average speed of 1.40 m/s from a distance of 2.3 m. How much time does it take for the dog to reach its owner?

PICTURE THE PROBLEM
The dog moves in a straight line through a distance of $d = 2.3$ m. The average speed of the dog is $v = 1.40$ m/s.

REASONING AND STRATEGY
Equation 2-2 $\left(\text{average speed} = \dfrac{\text{distance}}{\text{elapsed time}}\right)$ relates average speed, distance, and elapsed time. We can solve for the elapsed time by rearranging this equation.

Known   Average speed, $v = 1.40$ m/s; distance, $d = 2.3$ m.
Unknown   Elapsed time = ?

SOLUTION

1. Rearrange Equation 2-2 to solve for the elapsed time: $\text{elapsed time} = \dfrac{\text{distance}}{\text{average speed}} = \dfrac{d}{v}$

2. Substitute numerical values to find the time: $\text{elapsed time} = \dfrac{2.3\ \cancel{m}}{1.40\ \cancel{m}/s} = \dfrac{2.3}{1.40}\ s = 1.6\ s$

INSIGHT
As this example shows, Equation 2-2 is not just a formula for calculating the average speed. It relates speed, time, and distance. Any one of these quantities can be determined if the other two are known.

PRACTICE PROBLEM
A dog trots with an average speed of 1.44 m/s for 1.9 s. What is the distance it covers?
[**Answer:** distance = (average speed) (elapsed time) = (1.44 m/s) (1.9 s) = 2.7 m]

Some related homework problems: Problem 11, Problem 13

Next, we calculate the average speed for a trip consisting of two parts of equal length, each traveled with a different speed.

CONCEPTUAL EXAMPLE 2-3   AVERAGE SPEED   PREDICT/EXPLAIN

You drive 4.00 mi at 30.0 mi/h and then another 4.00 mi at 50.0 mi/h. **(a)** Is your average speed for the 8.00-mi trip greater than, less than, or equal to 40.0 mi/h? **(b)** Which of the following is the *best explanation* for your prediction?

  I. The average of 30.0 mi/h and 50.0 mi/h is 40.0 mi/h, and hence this is the average speed for the trip.

 II. You go farther during the 50.0 mi/h part of the trip, and hence your average speed is greater than 40.0 mi/h.

III. You spend more time during your trip traveling at 30.0 mi/h, and hence your average speed is less than 40.0 mi/h.

CONTINUED

REASONING AND DISCUSSION

At first glance it might seem that the average speed is definitely 40.0 mi/h. On further reflection, however, it is clear that it takes more time to travel 4.00 mi at 30.0 mi/h than it does to travel 4.00 mi at 50.0 mi/h. Therefore, you will be traveling at the lower speed for a greater period of time, and hence your average speed will be *less* than 40.0 mi/h—that is, closer to 30.0 mi/h than to 50.0 mi/h.

ANSWER

(a) The average speed is less than 40.0 mi/h. (b) The best explanation is III.

To confirm the conclusion of Conceptual Example 2-3, we simply apply the definition of average speed to find its value for the entire trip. We already know that the distance traveled is 8.00 mi; what we need now is the total elapsed time. The elapsed time on the first 4.00 mi is

$$t_1 = \frac{4.00 \text{ mi}}{30.0 \text{ mi/h}} = \left(\frac{4.00}{30.0}\right) \text{h} = 0.133 \text{ h}$$

The time required to cover the second 4.00 mi is

$$t_2 = \frac{4.00 \text{ mi}}{50.0 \text{ mi/h}} = \left(\frac{4.00}{50.0}\right) \text{h} = 0.0800 \text{ h}$$

Therefore, the elapsed time for the entire trip is

$$t_1 + t_2 = 0.133 \text{ h} + 0.0800 \text{ h} = 0.213 \text{ h}$$

This gives the following average speed:

$$\text{average speed} = \frac{8.00 \text{ mi}}{0.213 \text{ h}} = 37.6 \text{ mi/h} < 40.0 \text{ mi/h}$$

It's important to note that a "guess" will never give a detailed result like 37.6 mi/h; a systematic, step-by-step calculation is required.

Average Velocity   There is another physical quantity that is often more useful than the average speed. It is the **average velocity**, v_{av}, and it is defined as displacement per time:

Definition: Average Velocity, v_{av}

$$\text{average velocity} = \frac{\text{displacement}}{\text{elapsed time}}$$

$$v_{av} = \frac{\Delta x}{\Delta t} = \frac{x_f - x_i}{t_f - t_i} \qquad 2\text{-}3$$

SI unit: meter per second, m/s

Not only does the average velocity tell us, on average, how fast something is moving, it also tells us the *direction* the object is moving. For example, if an object moves in the positive direction, then $x_f > x_i$ and the average velocity is positive, $v_{av} > 0$. On the other hand, if an object moves in the negative direction, then $x_f < x_i$ and $v_{av} < 0$. As with displacement, the average velocity is a one-dimensional vector, and its direction

Balls Take High and Low Tracks

PROBLEM-SOLVING NOTE

"Coordinate" the Problem

The first step in solving a physics problem is to produce a simple sketch of the system. Your sketch should include a coordinate system, along with an origin and a positive direction. Next, you should identify quantities that are given in the problem, such as initial position, initial velocity, acceleration, and so on. These preliminaries will help you produce a mathematical representation of the problem.

is given by its sign. Average velocity gives more information than average speed; hence it is used more frequently in physics.

In the next Example, pay close attention to the sign of each quantity.

EXAMPLE 2-4 SPRINT TRAINING

An athlete sprints in a straight line for 50.0 m in 8.00 s, and then walks slowly back to the starting line in 40.0 s. If the "sprint direction" is taken to be positive, what are (a) the average sprint velocity, (b) the average walking velocity, and (c) the average velocity for the complete round trip?

PICTURE THE PROBLEM
In our sketch we set up a coordinate system with the sprint going in the positive x direction, as described in the problem. For convenience, we choose the origin to be at the starting line. The finish line, then, is at $x = 50.0$ m.

REASONING AND STRATEGY
In each part of the problem we are asked for the average velocity, and we are given information for times and distances. All that is needed, then, is to determine $\Delta x = x_f - x_i$ and $\Delta t = t_f - t_i$ in each case, and to apply Equation 2-3 $\left(v_{av} = \dfrac{\Delta x}{\Delta t}\right)$

Known   Sprint distance = 50.0 m; sprint time = 8.00 s; walking distance = 50.0 m; walking time = 40.0 s.

Unknown   (a) Average sprint velocity, $v_{av} = ?$ (b) Average walking velocity, $v_{av} = ?$ (c) Average velocity for round trip, $v_{av} = ?$

SOLUTION

Part (a)

1. Apply $v_{av} = \dfrac{\Delta x}{\Delta t}$ to the sprint, with $x_f = 50.0$ m, $x_i = 0$, $t_f = 8.00$ s, and $t_i = 0$:

$$v_{av} = \frac{\Delta x}{\Delta t} = \frac{x_f - x_i}{t_f - t_i} = \frac{50.0\ \text{m} - 0}{8.00\ \text{s} - 0} = \frac{50.0}{8.00}\ \text{m/s} = 6.25\ \text{m/s}$$

Part (b)

2. Apply $v_{av} = \dfrac{\Delta x}{\Delta t}$ to the walk. In this case, $x_f = 0$, $x_i = 50.0$ m, $t_f = 48.0$ s, and $t_i = 8.00$ s:

$$v_{av} = \frac{x_f - x_i}{t_f - t_i} = \frac{0 - 50.0\ \text{m}}{48.0\ \text{s} - 8.00\ \text{s}} = -\frac{50.0}{40.0}\ \text{m/s} = -1.25\ \text{m/s}$$

Part (c)

3. For the round trip, $x_f = x_i = 0$; thus $\Delta x = 0$:

$$v_{av} = \frac{\Delta x}{\Delta t} = \frac{0}{48.0\ \text{s}} = 0$$

INSIGHT
The sign of the velocities in parts (a) and (b) indicates the direction of motion: positive for motion to the right, negative for motion to the left. In addition, notice that the average *speed* for the entire 100.0-m trip (100.0 m/48.0 s = 2.08 m/s) is nonzero, even though the average velocity vanishes.

PRACTICE PROBLEM
If the average velocity during the walk is −1.50 m/s, how much time does it take the athlete to walk back to the starting line?
[**Answer:** $\Delta t = \Delta x/v_{av} = (-50.0\ \text{m})/(-1.50\ \text{m/s}) = 33.3$ s]

Some related homework problems: Problem 9, Problem 14, Problem 16

TABLE 2-1 Time and Position Values for Figure 2-3

t (s)	x (m)
0	1
1	3
2	4
3	2
4	−1

Graphical Interpretation of Average Velocity

It's often useful to "visualize" a particle's motion by sketching its position as a function of time. For example, suppose a particle moves back and forth along the x axis, with the positions and times listed in Table 2-1. This data is plotted in FIGURE 2-3 (a), which is certainly a better way to "see" the motion than a table of numbers.

Even so, this way of showing a particle's position and time is a bit messy, so let's replot the same information with a different type of graph. In FIGURE 2-3 (b) we again plot the motion shown in Figure 2-3 (a), but this time with the vertical axis representing the position, x, and the horizontal axis representing time, t. This is referred to as an **x-versus-t graph**, and it makes it much easier to visualize a particle's motion, as we shall see.

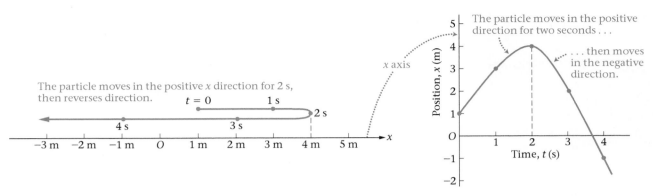

(a) The particle's path shown on a coordinate axis

(b) The same path as a graph of position x versus time t

▲ **FIGURE 2-3 Two ways to visualize one-dimensional motion (a)** Position plotted for different times. Although the path is shown with a "U" shape for clarity, the particle actually moves straight back and forth along the x axis. **(b)** Plot of position (vertical) versus time (horizontal).

Slope Is Equal to Average Velocity   An x-versus-t plot leads to a particularly useful interpretation of average velocity. To see how, suppose you would like to know the average velocity of the particle in Figure 2-3 from $t = 0$ to $t = 3$ s. From our definition of average velocity in Equation 2-3, we know that $v_{av} = \Delta x/\Delta t$, or $v_{av} = (2\,\text{m} - 1\,\text{m})/(3\,\text{s} - 0) = +0.3$ m/s for this particle. To relate this to the x-versus-t plot, draw a straight line connecting the position at $t = 0$ (call this point A) and the position at $t = 3$ s (point B). The result is shown in **FIGURE 2-4 (a)**.

The slope of the straight line from A to B is equal to the rise over the run, which in this case is $\Delta x/\Delta t$. But $\Delta x/\Delta t$ is the average velocity. Thus, we conclude the following:

> The *slope* of a line connecting two points on an x-versus-t plot is equal to the *average velocity* during that time interval.

As an additional example, let's calculate the average velocity of this particle between times $t = 2$ s and $t = 3$ s. A line connecting the corresponding points is shown in **FIGURE 2-4 (b)**. The first thing we notice about this line is that it has a negative slope; thus $v_{av} < 0$ and the particle is moving to the left. We can also see that this line is inclined more steeply than the line in Figure 2-4 (a); hence the magnitude of its slope (its speed) is greater. In fact, if we calculate the slope of this line we find that $v_{av} = -2$ m/s for this time interval.

Thus, connecting points on an x-versus-t plot gives an immediate "feeling" for the average velocity over a given time interval. This type of graphical analysis will be particularly useful in the next section.

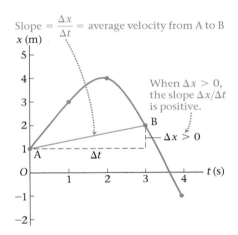

(a) Average velocity between $t = 0$ and $t = 3$ s

Enhance Your Understanding   (Answers given at the end of the chapter)

2. The position of an object as a function of time is shown in **FIGURE 2-5**. For the time intervals between **(a)** points 1 and 2, **(b)** points 2 and 3, and **(c)** points 3 and 4, state whether the average velocity is positive, negative, or zero.

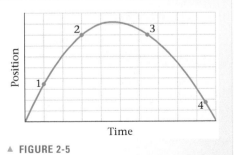

▲ **FIGURE 2-5**

Section Review

- Average speed is distance divided by time.
- Average velocity is displacement divided by time; $v_{av} = \dfrac{\Delta x}{\Delta t}$.
- The slope of a line connecting two points on an x-versus-t plot is the average velocity between those two points.

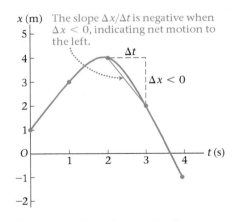

(b) Average velocity between $t = 2$ s and $t = 3$ s

▲ **FIGURE 2-4 Average velocity on an x-versus-t graph** The slope of a straight line between any two points on an x-versus-t graph equals the average velocity between those points. Positive slopes indicate net motion to the right; negative slopes indicate net motion to the left.

▲ **FIGURE 2-6** A speedometer indicates a car's instantaneous speed but gives no information about its *direction*. Thus, the speedometer is truly a "speed meter," not a velocity meter.

Big Idea 2 Velocity is the rate of change of position with time, and speed is the magnitude of velocity. Velocity has a magnitude and a direction, and hence it is a vector; speed has only a magnitude, and hence it is a scalar.

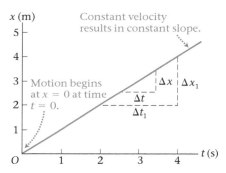

▲ **FIGURE 2-7 Constant velocity corresponds to constant slope on an *x*-versus-*t* graph** The slope $\Delta x_1/\Delta t_1$ is equal to $(4\text{ m} - 2\text{ m})/(4\text{ s} - 2\text{ s}) = (2\text{ m})/(2\text{ s}) = 1$ m/s. Because the *x*-versus-*t* plot is a straight line, the slope $\Delta x/\Delta t$ is also equal to 1 m/s for any value of Δt.

TABLE 2-2 *x*-versus-*t* Values for Figure 2-8

t (s)	x (m)
0	0
0.25	9.85
0.50	17.2
0.75	22.3
1.00	25.6
1.25	27.4
1.50	28.1
1.75	28.0
2.00	27.4

2-3 Instantaneous Velocity

Though average velocity is a useful way to characterize motion, it can miss a lot. For example, suppose you travel by car on a long, straight highway, covering 92 mi in 2.0 hours. Your average velocity is 46 mi/h. Even so, there may have been only a few times during the trip when you were actually driving at 46 mi/h.

To have a more accurate representation of your trip, you should average your velocity over shorter periods of time. If you calculate your average velocity every 15 minutes, you have a better picture of what the trip was like. An even better, more realistic picture of the trip is obtained if you calculate the average velocity every minute or every second. Ideally, when you deal with the motion of any particle, it's desirable to know the velocity of the particle at each *instant* of time.

This idea of a velocity corresponding to an instant of time is what is meant by the **instantaneous velocity**. Mathematically, we define the instantaneous velocity as follows:

Definition: Instantaneous Velocity, *v*

$$v = \lim_{\Delta t \to 0} \frac{\Delta x}{\Delta t}$$

2-4

SI unit: meter per second, m/s

In this expression the notation $\lim_{\Delta t \to 0}$ means "evaluate the average velocity, $\Delta x/\Delta t$, over shorter and shorter time intervals, approaching zero in the limit." Notice that the instantaneous velocity can be positive, negative, or zero, just like the average velocity— and just like the average velocity, the instantaneous velocity is a one-dimensional vector. The magnitude of the instantaneous velocity is called the **instantaneous speed**. In a car, the speedometer gives a reading of the vehicle's instantaneous speed. **FIGURE 2-6** shows a speedometer reading 60 mi/h.

Calculating the Instantaneous Velocity As Δt becomes smaller, Δx becomes smaller as well, but the ratio $\Delta x/\Delta t$ approaches a constant value. To see how this works, consider first the simple case of a particle moving with a constant velocity of $+1$ m/s. If the particle starts at $x = 0$ at $t = 0$, then its position at $t = 1$ s is $x = 1$ m, its position at $t = 2$ s is $x = 2$ m, and so on. Plotting this motion in an *x*-versus-*t* plot gives a straight line, as shown in **FIGURE 2-7**.

Now, suppose we want to find the instantaneous velocity at $t = 3$ s. To do so, we calculate the average velocity over small intervals of time centered at 3 s, and let the time intervals become arbitrarily small, as shown in the Figure. Because the *x*-versus-*t* plot is a straight line, it's clear that $\Delta x/\Delta t = \Delta x_1/\Delta t_1$, no matter how small the time interval Δt. As Δt becomes smaller, so does Δx, but the ratio $\Delta x/\Delta t$ is simply the slope of the line, 1 m/s. Thus, the instantaneous velocity at $t = 3$ s is 1 m/s.

Of course, in this example the instantaneous velocity is 1 m/s for any instant of time, not just $t = 3$ s. Therefore:

> When the velocity is constant, the average velocity over any time interval is equal to the instantaneous velocity at any time.

In general, a particle's velocity varies with time, and the *x*-versus-*t* plot is not a straight line. An example is shown in **FIGURE 2-8**, with the corresponding numerical values of *x* and *t* given in Table 2-2.

In this case, what is the instantaneous velocity at, say, $t = 1.00$ s? As a first approximation, let's calculate the average velocity for the time interval from $t_i = 0$ to $t_f = 2.00$ s. Notice that this time interval is centered at $t = 1.00$ s. From Table 2-2 we see that $x_i = 0$ and $x_f = 27.4$ m; thus $v_{av} = 13.7$ m/s. The corresponding straight line connecting these two points is the lowest straight line in Figure 2-8.

The next three lines, in upward progression, refer to time intervals from 0.250 s to 1.75 s, 0.500 s to 1.50 s, and 0.750 s to 1.25 s, respectively. The corresponding average velocities, given in Table 2-3, are 12.1 m/s, 10.9 m/s, and 10.2 m/s. Table 2-3 also gives results for even smaller time intervals. In particular, for the interval from 0.900 s to 1.10 s

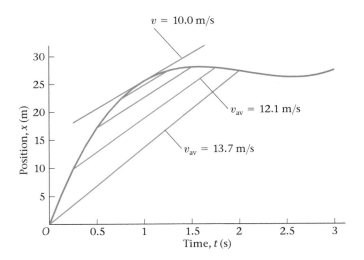

TABLE 2-3 Calculating the Instantaneous Velocity at $t = 1$ s for Figure 2-8

t_i (s)	t_f (s)	Δt (s)	x_i (m)	x_f (m)	Δx (m)	$v_{av} = \Delta x/\Delta t$ (m/s)
0	2.00	2.00	0	27.4	27.4	13.7
0.250	1.75	1.50	9.85	28.0	18.2	12.1
0.500	1.50	1.00	17.2	28.1	10.9	10.9
0.750	1.25	0.50	22.3	27.4	5.10	10.2
0.900	1.10	0.20	24.5	26.5	2.00	10.0
0.950	1.05	0.10	25.1	26.1	1.00	10.0

the average velocity is 10.0 m/s. Smaller intervals also give 10.0 m/s. Thus, we can conclude that the instantaneous velocity at $t = 1.00$ s is $v = 10.0$ m/s.

Tangent Lines and the Instantaneous Velocity The uppermost straight line in Figure 2-8 is the tangent line to the x-versus-t curve at the time $t = 1.00$ s; that is, it is the line that touches the curve at just a single point. Its slope is 10.0 m/s. Clearly, the average-velocity lines have slopes that approach the slope of the tangent line as the time intervals become smaller. This is an example of the following general result:

> The instantaneous velocity at a given time is equal to the slope of the tangent line at that point on an x-versus-t graph.

Thus, a visual inspection of an x-versus-t graph gives information not only about the location of a particle, but also about its velocity.

CONCEPTUAL EXAMPLE 2-5   INSTANTANEOUS VELOCITY

Referring to Figure 2-8, is the instantaneous velocity at $t = 0.500$ s **(a)** greater than, **(b)** less than, or **(c)** the same as the instantaneous velocity at $t = 1.00$ s?

REASONING AND DISCUSSION
From the x-versus-t graph in Figure 2-8 it is clear that the slope of a tangent line drawn at $t = 0.500$ s is greater than the slope of the tangent line at $t = 1.00$ s. It follows that the particle's velocity at 0.500 s is greater than its velocity at 1.00 s.

ANSWER
(a) The instantaneous velocity is greater at $t = 0.500$ s.

In the remainder of the book, when we say *velocity* it is to be understood that we mean *instantaneous* velocity. If we want to refer to the average velocity, we will specifically say average velocity.

▲ FIGURE 2-9 Graphical interpretation of average and instantaneous velocity
Average velocities correspond to the slope of straight-line segments connecting different points on an *x*-versus-*t* graph. Instantaneous velocities are given by the slope of the tangent line at a given time.

*In the figure margin:*
x
Slope = average velocity between times t_1 and t_2
Slope = average velocity between times t_2 and t_3
Slope of tangent line = instantaneous velocity at time t_3
O t_1 t_2 t_3 *t*

Graphical Interpretation of Average and Instantaneous Velocity

Let's summarize the graphical interpretations of average and instantaneous velocity on an *x*-versus-*t* graph:

- Average velocity is the slope of the straight line connecting two points corresponding to a given time interval.

- Instantaneous velocity is the slope of the tangent line at a given instant of time.

These relationships are illustrated in **FIGURE 2-9**.

Enhance Your Understanding   (Answers given at the end of the chapter)

3. **FIGURE 2-10** shows the position-versus-time graph for an object. Rank the instantaneous velocity of the object at points A, B, C, and D in order of increasing velocity, from most negative to most positive. Indicate ties where appropriate.

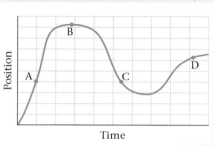

▲ FIGURE 2-10

Section Review

- Instantaneous velocity is the limit of average velocity over shorter and shorter time intervals; $v = \lim_{\Delta t \to 0} \dfrac{\Delta x}{\Delta t}$. The magnitude of the instantaneous velocity is the instantaneous speed.

- The instantaneous velocity on an *x*-versus-*t* plot is the slope of a tangent line at a given time.

2-4 Acceleration

Just as velocity is the rate of change of *displacement* with time, **acceleration** is the rate of change of *velocity* with time. Thus, an object accelerates whenever its velocity *changes,* no matter what the change—it accelerates when its velocity increases, it accelerates when its velocity decreases. Of all the concepts discussed in this chapter, perhaps none is more central to physics than acceleration. Galileo, for example, showed that falling bodies move with constant acceleration. Newton showed that acceleration and force are directly related, as we shall see in Chapter 5. Thus, it is particularly important to have a clear, complete understanding of acceleration before leaving this chapter.

Average Acceleration   We begin with the definition of **average acceleration**, which is the change in velocity divided by the change in time:

Definition: Average Acceleration, a_{av}

$$a_{av} = \frac{\Delta v}{\Delta t} = \frac{v_f - v_i}{t_f - t_i}$$ 2-5

SI unit: meter per second per second, m/s^2

Notice that the dimensions of average acceleration are the dimensions of velocity per time, or (meters per second) per second:

$$\frac{\text{meters per second}}{\text{second}} = \frac{m/s}{s} = \frac{m}{s^2}$$

This is generally spoken as "meters per second squared." For example, the acceleration of gravity on the Earth's surface is approximately 9.81 m/s², which means that the velocity of a falling object changes by 9.81 meters per second (m/s) every second (s). In addition, we see that the average acceleration can be positive, negative, or zero. In fact, it is a one-dimensional vector, just like displacement, average velocity, and instantaneous velocity. Typical magnitudes of acceleration are given in Table 2-4. A real-world example of linear acceleration is shown in FIGURE 2-11.

EXERCISE 2-6 AVERAGE ACCELERATION

a. A certain car can go from 0 to 60.0 mi/h in 7.40 s. What is the average acceleration of this car in meters per second squared?

b. An airplane has an average acceleration of 2.19 m/s² during takeoff. If the airplane starts at rest, how much time does it take for it to reach a speed of 174 mi/h?

REASONING AND SOLUTION

(a) Average acceleration is the change in velocity divided by the elapsed time, $a_{av} = \Delta v/\Delta t$.

(b) The equation for average acceleration can be rearranged to solve for the elapsed time, $\Delta t = \Delta v/a_{av}$.

   a. average acceleration $= a_{av} = (60.0 \text{ mi/h})/(7.40 \text{ s})$

$$= (60.0 \text{ mi/h})\left(\frac{1609 \text{ m}}{1 \text{ mi}}\right)\left(\frac{1 \text{ h}}{3600 \text{ s}}\right)/(7.40 \text{ s})$$

$$= (26.8 \text{ m/s})/(7.40 \text{ s}) = 3.62 \text{ m/s}^2$$

   b. $\Delta t = \Delta v/a_{av} = (174 \text{ mi/h})\left(\frac{1609 \text{ m}}{1 \text{ mi}}\right)\left(\frac{1 \text{ h}}{3600 \text{ s}}\right)/(2.19 \text{ m/s}^2)$

$$= (77.8 \text{ m/s})/(2.19 \text{ m/s}^2) = 35.5 \text{ s}$$

Instantaneous Acceleration We considered the limit of smaller and smaller time intervals to find an instantaneous velocity, and we can do the same to define the **instantaneous acceleration:**

Definition: Instantaneous Acceleration, *a*

$$a = \lim_{\Delta t \to 0} \frac{\Delta v}{\Delta t} \qquad\qquad 2\text{-}6$$

SI unit: meter per second per second, m/s²

As you might expect, the instantaneous acceleration is a vector, just like the average acceleration, and its direction in one dimension is given by its sign. For simplicity, when we say acceleration in this text, we are referring to the instantaneous acceleration.

   One final note before we go on to some examples. If the acceleration is constant, it has the same value at all times. Therefore:

> When acceleration is constant, the instantaneous and average accelerations are the same.

We shall make use of this fact when we return to the special case of constant acceleration in the next section.

Graphical Interpretation of Acceleration

To see how acceleration can be interpreted graphically, suppose a particle has a constant acceleration of −0.50 m/s². This means that the velocity of the particle *decreases* by 0.50 m/s each second. Thus, if its velocity is 1.0 m/s at $t = 0$, then at $t = 1$ s its velocity is 0.50 m/s, at $t = 2$ s its velocity is 0, at $t = 3$ s its velocity is −0.50 m/s, and so on. This is illustrated by curve I in FIGURE 2-12, where we see that a plot of v versus

▶ FIGURE 2-12 ***v*-versus-*t* plots for motion with constant acceleration** Curve I represents the movement of a particle with constant acceleration $a = -0.50$ m/s². Curve II represents the motion of a particle with constant acceleration $a = +0.25$ m/s².

▲ FIGURE 2-11 The space shuttle *Columbia* accelerates upward on the initial phase of its journey into orbit. The astronauts on board experienced an approximately linear acceleration that may have been as great as 20 m/s².

Big Idea **3** Acceleration is the rate of change of velocity with time. Acceleration has both a magnitude and a direction, and hence it is a vector.

TABLE 2-4 Typical Accelerations (m/s²)

Ultracentrifuge	3×10^6
Bullet fired from a rifle	4.4×10^5
Batted baseball	3×10^4
Click beetle righting itself	400
Acceleration required to deploy air bags	60
Bungee jump	30
High jump	15
Acceleration of gravity on Earth	9.81
Emergency stop in a car	8
Airplane during takeoff	5
An elevator	3
Acceleration of gravity on the Moon	1.62

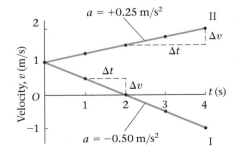

t results in a straight line with a negative slope. Curve II in Figure 2-12 has a positive slope, corresponding to a constant acceleration of +0.25 m/s². Thus, in terms of a *v*-versus-*t* plot, a constant acceleration results in a straight line with a slope equal to the acceleration.

CONCEPTUAL EXAMPLE 2-7   SPEED AS A FUNCTION OF TIME

The speed of a particle with the *v*-versus-*t* graph shown by curve II in Figure 2-12 increases steadily with time. Consider, instead, a particle whose *v*-versus-*t* graph is given by curve I in Figure 2-12. As a function of time, does the speed of this particle **(a)** increase, **(b)** decrease, or **(c)** decrease and then increase?

REASONING AND DISCUSSION
Recall that speed is the *magnitude* of velocity. In curve I of Figure 2-12 the speed starts out at 1.0 m/s, then *decreases* to 0 at *t* = 2 s. After *t* = 2 s the speed *increases* again. For example, at *t* = 3 s the speed is 0.50 m/s, and at *t* = 4 s the speed is 1 m/s.

Did you realize that the particle represented by curve I in Figure 2-12 changes direction at *t* = 2 s? It certainly does. Before *t* = 2 s the particle moves in the positive direction; after *t* = 2 s it moves in the negative direction. At precisely *t* = 2 s the particle is momentarily at rest. However, regardless of whether the particle is moving in the positive direction, moving in the negative direction, or instantaneously at rest, it still has the same constant acceleration. Acceleration has to do only with the way the velocity is *changing* at a given moment.

ANSWER
(c) The speed decreases and then increases.

The graphical interpretations for velocity presented in Figure 2-9 apply equally well to acceleration, with just one small change: Instead of an *x*-versus-*t* graph, we use a *v*-versus-*t* graph, as in **FIGURE 2-13**. Thus, the average acceleration in a *v*-versus-*t* plot is the slope of a straight line connecting points corresponding to two different times. Similarly, the instantaneous acceleration is the slope of the tangent line at a particular time.

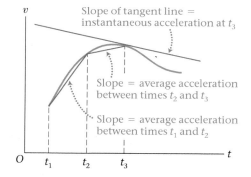

▶ **FIGURE 2-13 Graphical interpretation of average and instantaneous acceleration** Average accelerations correspond to the slopes of straight-line segments connecting different points on a *v*-versus-*t* graph. Instantaneous accelerations are given by the slope of the tangent line at a given time.

EXAMPLE 2-8   ACCELERATION OF A BICYCLE

A cyclist riding in a straight line has an initial velocity of 3.5 m/s, and accelerates at −1.0 m/s² for 2.0 s. The cyclist then coasts with zero acceleration for 3.0 s, and finally accelerates at 1.5 m/s² for an additional 2.0 s. **(a)** What is the final velocity of the cyclist? **(b)** What is the average acceleration of the cyclist for these seven seconds?

PICTURE THE PROBLEM
We begin by sketching a *v*-versus-*t* plot for the cyclist. The basic idea is that each interval of constant acceleration is represented by a straight line of the appropriate slope. Therefore, we draw a straight line with the slope −1.0 m/s² from *t* = 0 to *t* = 2.0 s, a line with zero slope from *t* = 2.0 s to *t* = 5.0 s, and a line with the slope 1.5 m/s² from *t* = 5.0 s to *t* = 7.0 s. The line connecting the initial and final points determines the average acceleration.

CONTINUED

REASONING AND STRATEGY

During each period of constant acceleration the change in velocity is $\Delta v = a_{av}\Delta t = a\Delta t$.

a. Adding the individual changes in velocity gives the total change, $\Delta v = v_f - v_i$. Because v_i is known, this expression can be solved for the final velocity, $v_f = \Delta v + v_i$.

b. The average acceleration can be calculated using Equation 2-5, $a_{av} = \Delta v/\Delta t$. Notice that Δv has been obtained in part (a), and the total time interval is $\Delta t = 7.0$ s.

Known   Initial velocity, $v_i = 3.5$ m/s; accelerations, $a_1 = -1.0$ m/s², $a_2 = 0$, $a_3 = 1.5$ m/s².

Unknown   (a) Final velocity, $v_f = $? (b) Average acceleration, $a_{av} = $?

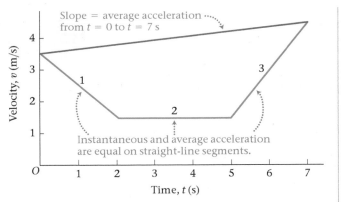

SOLUTION

Part (a)

1. Find the change in velocity during each of the three periods of constant acceleration:

$$\Delta v_1 = a_1\Delta t_1 = (-1.0 \text{ m/s}^2)(2.0 \text{ s}) = -2.0 \text{ m/s}$$
$$\Delta v_2 = a_2\Delta t_2 = (0)(3.0 \text{ s}) = 0$$
$$\Delta v_3 = a_3\Delta t_3 = (1.5 \text{ m/s}^2)(2.0 \text{ s}) = 3.0 \text{ m/s}$$

2. Sum the change in velocity for each period to obtain the total Δv:

$$\Delta v = \Delta v_1 + \Delta v_2 + \Delta v_3$$
$$= -2.0 \text{ m/s} + 0 + 3.0 \text{ m/s} = 1.0 \text{ m/s}$$

3. Use Δv to find v_f, recalling that $v_i = 3.5$ m/s:

$$\Delta v = v_f - v_i$$
$$v_f = \Delta v + v_i = 1.0 \text{ m/s} + 3.5 \text{ m/s} = 4.5 \text{ m/s}$$

Part (b)

4. The average acceleration is $\Delta v/\Delta t$:

$$a_{av} = \frac{\Delta v}{\Delta t} = \frac{1.0 \text{ m/s}}{7.0 \text{ s}} = 0.14 \text{ m/s}^2$$

INSIGHT

The average acceleration for these seven seconds is *not* the average of the individual accelerations, -1.0 m/s², 0, and 1.5 m/s². The reason is that different amounts of time are spent with each acceleration.

PRACTICE PROBLEM

What is the average acceleration of the cyclist between $t = 3.0$ s and $t = 6.0$ s?

[**Answer:** $a_{av} = \Delta v/\Delta t = (3.0 \text{ m/s} - 1.5 \text{ m/s})/(6.0 \text{ s} - 3.0 \text{ s}) = 0.50 \text{ m/s}^2$]

Some related homework problems: Problem 32, Problem 34

Relating the Signs of Velocity and Acceleration to the Change in Speed   In one dimension, nonzero velocities and accelerations are either positive or negative, depending on whether they point in the positive or negative direction of the coordinate system chosen. Thus, the velocity and acceleration of an object may have the same or opposite signs. (Of course, in two or three dimensions the relationship between velocity and acceleration can be much more varied, as we shall see in the next several chapters.) This leads to the following two possibilities in one dimension:

- When the velocity and acceleration of an object have the same sign (point in the same direction), the speed of the object increases.

- When the velocity and acceleration of an object have opposite signs (point in opposite directions), the speed of the object decreases.

These two possibilities are illustrated in **FIGURE 2-14**. Notice that when a particle's speed increases, it means either that its velocity becomes more positive, as in **FIGURE 2-14 (a)**, or more negative, as in **FIGURE 2-14 (d)**. In either case, it is the magnitude of the velocity—the speed—that increases. **FIGURE 2-15** shows an example where the velocity and the acceleration are definitely in the same direction.

When a particle's speed decreases, it is often said to be *decelerating*. A common misconception is that deceleration implies a negative acceleration. This is not true. Deceleration can be caused by a positive or a negative acceleration, depending on the direction of the initial velocity. For example, the car in **FIGURE 2-14 (b)** has a positive velocity and

PHYSICS IN CONTEXT
Looking Ahead

The distinctions developed in this chapter between velocity and acceleration play a key role in our understanding of Newton's laws of motion in Chapters 5 and 6.

Big Idea **4** Speed increases when velocity and acceleration are in the same direction; speed decreases when velocity and acceleration are in opposite directions.

▶ FIGURE 2-14 **Cars accelerating or decelerating** A car's speed increases when its velocity and acceleration point in the same direction, as in cases **(a)** and **(d)**. When the velocity and acceleration point in opposite directions, as in cases **(b)** and **(c)**, the car's speed decreases.

(a) Speed increases

(b) Speed decreases

(c) Speed decreases

(d) Speed increases

▶ FIGURE 2-15 The winner of this race was traveling at a speed of 313.91 mi/h at the end of the quarter-mile course. The winning time was just 4.607 s, and hence the *average* acceleration during this race was approximately three times the acceleration of gravity (which is covered in Section 2-7).

a negative acceleration, while the car in FIGURE 2-14 (c) has a negative velocity and a positive acceleration. In both cases, the speed of the car decreases. Again, all that is required for deceleration in one dimension is that the velocity and acceleration have *opposite signs*; that is, they must point in *opposite directions*, as in parts (b) and (c) of Figure 2-14.

Velocity-versus-time plots for the four situations shown in Figure 2-14 are presented in FIGURE 2-16. In each of the four plots in Figure 2-16 we assume constant acceleration. Be sure to understand clearly the connection between the *v*-versus-*t* plots in Figure 2-16 and the corresponding physical motions indicated in Figure 2-14.

EXAMPLE 2-9 THE FERRY DOCKS

A ferry makes a short run between two docks: one in Anacortes, Washington, the other on Guemes Island. As the ferry approaches the Guemes dock (traveling in the positive *x* direction), its speed is 6.8 m/s. **(a)** If the ferry slows to a stop in 13.3 s, what is its average acceleration? **(b)** As the ferry returns to the Anacortes dock, its speed is 6.1 m/s. If it comes to rest in 12.9 s, what is its average acceleration?

PICTURE THE PROBLEM
Our sketch shows the locations of the two docks and the positive direction indicated in the problem. The distance between docks is not given, nor is it needed.

REASONING AND STRATEGY
We are given the initial and final velocities (the ferry comes to a stop in each case, so its final speed is zero) and the relevant times. Therefore, we can find the average acceleration using $a_{av} = \Delta v/\Delta t$, being careful to get the signs right.

Known (a) Initial velocity, $v_i = 6.8$ m/s; time to stop, $\Delta t = 13.3$ s. (b) Initial velocity, $v_i = -6.1$ m/s; time to stop, $\Delta t = 12.9$ s.

Unknown (a) Average acceleration, $a_{av} = ?$ (b) Average acceleration, $a_{av} = ?$

SOLUTION

Part (a)

1. Calculate the average acceleration, noting that $v_i = 6.8$ m/s and $v_f = 0$:

$$a_{av} = \frac{\Delta v}{\Delta t} = \frac{v_f - v_i}{\Delta t} = \frac{0 - 6.8 \text{ m/s}}{13.3 \text{ s}} = -0.51 \text{ m/s}^2$$

CONTINUED

Part (b)

2. In this case, $v_i = -6.1$ m/s and $v_f = 0$:

$$a_{av} = \frac{\Delta v}{\Delta t} = \frac{v_f - v_i}{\Delta t} = \frac{0 - (-6.1 \text{ m/s})}{12.9 \text{ s}} = 0.47 \text{ m/s}^2$$

INSIGHT

In each case, the acceleration of the ferry is opposite in sign to its velocity; therefore the ferry decelerates.

PRACTICE PROBLEM

When the ferry leaves Guemes Island, its speed increases from 0 to 4.8 m/s in 9.05 s. What is its average acceleration?
[**Answer:** $a_{av} = -0.53 \text{ m/s}^2$]

Some related homework problems: Problem 30, Problem 36

RWP* The ability to detect and measure acceleration has become an important capability of many new technologies. A device that measures acceleration is called an *accelerometer*. In recent years a number of accelerometers have been developed that are contained in tiny integrated circuit chips, such as the one shown in FIGURE 2-17. These accelerometers allow devices such as smartphones, video game consoles, automotive air bag sensors, aircraft flight stabilization systems, and numerous others to detect acceleration and respond accordingly.

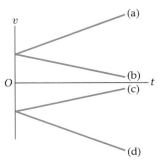

▲ FIGURE 2-16 *v*-versus-*t* plots with constant acceleration Four plots of *v* versus *t* corresponding to the four situations shown in Figure 2-14. Notice that the speed increases in cases **(a)** and **(d)**, but decreases in cases **(b)** and **(c)**.

Enhance Your Understanding   (Answers given at the end of the chapter)

4. At a certain time, object 1 has an initial velocity of -2 m/s and an acceleration of 3 m/s^2. At the same time, object 2 has an initial velocity of 5 m/s and an acceleration of -1 m/s^2. **(a)** Is the speed of object 1 increasing or decreasing? Explain. **(b)** Is the velocity of object 1 increasing or decreasing? Explain. **(c)** Is the speed of object 2 increasing or decreasing? Explain. **(d)** Is the velocity of object 2 increasing or decreasing? Explain.

Section Review

- The average acceleration is the change in velocity divided by the time; $a_{av} = \frac{\Delta v}{\Delta t}$.

- The instantaneous acceleration is the average acceleration over shorter and shorter time intervals; $a = \lim_{\Delta t \to 0} \frac{\Delta v}{\Delta t}$.

- Average acceleration is the slope between two points on a *v*-versus-*t* plot; instantaneous acceleration is the slope of a tangent line on a *v*-versus-*t* plot.

▲ FIGURE 2-17 **Integrated circuit accelerometer** Microscopic image of the ADXL330, the accelerometer incorporated into the Nintendo *Wii* remote controller. When the chip is accelerated in any direction, it generates an electrical signal that is proportional to the magnitude of the acceleration. It also indicates the direction of acceleration.

2-5 Motion with Constant Acceleration

In this section, we derive equations that describe the motion of particles moving with **constant acceleration**. These "equations of motion" can be used to describe a wide range of everyday phenomena. For example, in an idealized world with no air resistance, falling bodies have constant acceleration.

Velocity as a Function of Time   As mentioned in the previous section, if a particle has constant acceleration—that is, the same acceleration at every instant of time—then its instantaneous acceleration, a, is equal to its average acceleration, a_{av}. Recalling the definition of average acceleration (Equation 2-5), we have

$$a_{av} = \frac{v_f - v_i}{t_f - t_i} = a$$

*Real World Physics applications are denoted by the acronym RWP.

The initial and final times may be chosen arbitrarily in this equation. For example, let $t_i = 0$ for the initial time, and let $v_i = v_0$ denote the velocity at time zero.* For the final time and velocity we drop the subscripts to simplify the notation; thus we let $t_f = t$ and $v_f = v$. With these identifications we have

$$a_{av} = \frac{v - v_0}{t - 0} = a$$

With a slight rearrangement we find

$$v - v_0 = a(t - 0) = at$$

This yields our first equation of motion:

Constant-Acceleration Equation of Motion: Velocity as a Function of Time

$$v = v_0 + at \qquad\qquad 2\text{-}7$$

Equation 2-7 describes a straight line on a v-versus-t plot. The line crosses the velocity axis at the value v_0 and has a slope a, in agreement with the graphical interpretations discussed in the previous section. For example, in curve I of Figure 2-12, the equation of motion is $v = v_0 + at = (1\,\text{m/s}) + (-0.5\,\text{m/s}^2)t$. Also, notice that $(-0.5\,\text{m/s}^2)t$ has the units $(\text{m/s})^2\,(\text{s}) = \text{m/s}$; thus each term in Equation 2-7 has the same dimensions (as it must to be a valid physical equation).

PHYSICS IN CONTEXT
Looking Back

We are careful to check the dimensional consistency of our equations in this chapter. This concept was introduced in Chapter 1.

EXERCISE 2-10 VELOCITY WITH CONSTANT ACCELERATION

A ball is thrown straight upward with an initial velocity of $+7.3\,\text{m/s}$. If the acceleration of the ball is downward, with the value $-9.81\,\text{m/s}^2$, find the velocity of the ball after (a) 0.45 s and (b) 0.90 s.

REASONING AND SOLUTION

For constant acceleration, velocity is related to acceleration and time by Equation 2-7 ($v = v_0 + at$). Given the initial velocity ($v_0 = 7.3\,\text{m/s}$) and the acceleration ($a = -9.81\,\text{m/s}^2$), the final velocity can be found by substituting the desired time.

a. Substituting $t = 0.45\,\text{s}$ in $v = v_0 + at$ yields

$$v = 7.3\,\text{m/s} + (-9.81\,\text{m/s}^2)(0.45\,\text{s}) = 2.9\,\text{m/s}$$

The ball is still moving upward at this time.

b. Similarly, using $t = 0.90\,\text{s}$ in $v = v_0 + at$ gives

$$v = 7.3\,\text{m/s} + (-9.81\,\text{m/s}^2)(0.90\,\text{s}) = -1.5\,\text{m/s}$$

The ball is moving downward (negative direction) at this time.

Position as a Function of Time and Velocity How far does a particle move in a given time if its acceleration is constant? To answer this question, recall the definition of average velocity:

$$v_{av} = \frac{\Delta x}{\Delta t} = \frac{x_f - x_i}{t_f - t_i}$$

Using the same identifications given previously for initial and final times, and letting $x_i = x_0$ and $x_f = x$, we have

$$v_{av} = \frac{x - x_0}{t - 0}$$

Multiplying through by $(t - 0)$ gives

$$x - x_0 = v_{av}(t - 0) = v_{av}t$$

Finally, a simple rearrangement results in

$$x = x_0 + v_{av}t \qquad\qquad 2\text{-}8$$

---

*We often use the subscript i to denote the initial value of a quantity, as in v_i. In the special case where the initial value corresponds to zero time, $t = 0$, we use the subscript 0 to be more specific, as in v_0.

▶ **FIGURE 2-18 Average velocity (a)** When acceleration is constant, the velocity varies linearly with time. As a result, the average velocity, v_{av}, is simply the average of the initial velocity, v_0, and the final velocity, v. **(b)** The velocity curve for nonconstant acceleration is nonlinear. In this case, the average velocity is no longer midway between the initial and final velocities.

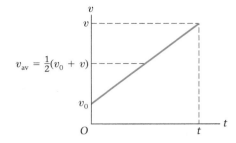

(a)

Now, Equation 2-8 is fine as it is. In fact, it applies whether the acceleration is constant or not. A more useful expression for the case of constant acceleration is obtained by writing v_{av} in terms of the initial and final velocities. This can be done by referring to **FIGURE 2-18 (a)**. Here the velocity changes linearly (since a is constant) from v_0 at $t = 0$ to v at some later time t. The average velocity during this period of time is simply the average of the initial and final velocities—that is, the sum of the two velocities divided by two:

Constant-Acceleration Equation of Motion: Average Velocity

$$v_{av} = \tfrac{1}{2}(v_0 + v) \qquad\qquad 2\text{-}9$$

The average velocity is indicated in the figure. Notice that if the acceleration is not constant, as in **FIGURE 2-18 (b)**, this simple averaging of initial and final velocities is no longer valid.

Substituting the expression for v_{av} from Equation 2-9 into Equation 2-8 yields

Constant-Acceleration Equation of Motion: Position as a Function of Time

$$x = x_0 + \tfrac{1}{2}(v_0 + v)t \qquad\qquad 2\text{-}10$$

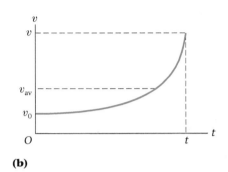

(b)

This equation, like Equation 2-7, is valid *only* for constant acceleration. The utility of Equations 2-7 and 2-10 is illustrated in the next Example.

EXAMPLE 2-11   **FULL SPEED AHEAD**

A boat moves slowly inside a marina (so as not to leave a wake) with a constant speed of 1.50 m/s. As soon as it passes the breakwater, leaving the marina, it throttles up and accelerates at 2.40 m/s². **(a)** How fast is the boat moving after accelerating for 5.00 s? **(b)** How far has the boat traveled in these 5.00 s?

PICTURE THE PROBLEM
In our sketch we choose the origin to be at the breakwater, and the positive x direction to be the direction of motion. With this choice the initial position is $x_0 = 0$, and the initial velocity is $v_0 = 1.50$ m/s.

REASONING AND STRATEGY
The acceleration is constant, so we can use Equations 2-7 to 2-10. In part (a) we want to relate velocity to time, so we use Equation 2-7, $v = v_0 + at$. In part (b) our knowledge of the initial and final velocities allows us to relate position to time using Equation 2-10, $x = x_0 + \tfrac{1}{2}(v_0 + v)t$.

Known   Velocity at time $t = 0$, $v_0 = 1.50$ m/s; acceleration, $a = 2.40$ m/s².
Unknown   **(a)** Velocity at $t = 5.00$ s, $v = ?$ **(b)** Distance traveled at $t = 5.00$ s, $x = ?$

SOLUTION

Part (a)

1. Use Equation 2-7 ($v = v_0 + at$) to find the final velocity, with $v_0 = 1.50$ m/s and $a = 2.40$ m/s²:

$$v = v_0 + at = 1.50 \text{ m/s} + (2.40 \text{ m/s}^2)(5.00 \text{ s})$$
$$= 1.50 \text{ m/s} + 12.0 \text{ m/s} = 13.5 \text{ m/s}$$

Part (b)

2. Apply Equation 2-10 $\left(x = x_0 + \tfrac{1}{2}(v_0 + v)t\right)$ to find the distance covered, using the result for v obtained in part (a):

$$x = x_0 + \tfrac{1}{2}(v_0 + v)t$$
$$= 0 + \tfrac{1}{2}(1.50 \text{ m/s} + 13.5 \text{ m/s})(5.00 \text{ s})$$
$$= (7.50 \text{ m/s})(5.00 \text{ s}) = 37.5 \text{ m}$$

CONTINUED

Position as a Function of Time and Acceleration   The velocity of the boat in Example 2-11 is plotted as a function of time in FIGURE 2-19, with the acceleration starting at time $t = 0$ and ending at $t = 5.00$ s. We will now show that the *distance* traveled by the boat from $t = 0$ to $t = 5.00$ s *is equal to the corresponding area under the velocity-versus-time curve*. This is a general result, valid for any velocity curve and any time interval:

> The distance traveled by an object from a time t_1 to a time t_2 is equal to the area under the velocity curve between those two times.

▲ FIGURE 2-19 **Velocity versus time for the boat in Example 2-11** The distance traveled by the boat between $t = 0$ and $t = 5.00$ s is equal to the corresponding area under the velocity curve.

In this case, the area is the sum of the areas of a rectangle and a triangle. The rectangle has a base of 5.00 s and a height of 1.50 m/s, which gives an area of $(5.00 \text{ s})(1.50 \text{ m/s}) = 7.50$ m. Similarly, the triangle has a base of 5.00 s and a height of $(13.5 \text{ m/s} - 1.50 \text{ m/s}) = 12.0$ m/s, for an area of $\frac{1}{2}(5.00 \text{ s})(12.0 \text{ m/s}) = 30.0$ m. Clearly, the total area is 37.5 m, which is the same distance found in Example 2-11.

Staying with Example 2-11 for a moment, let's repeat the calculation of part (b), only this time for the general case. First, we use the final velocity from part (a), calculated with $v = v_0 + at$, in the expression for the average velocity, $v_{av} = \frac{1}{2}(v_0 + v)$. Symbolically, this gives the following result:

$$\frac{1}{2}(v_0 + v) = \frac{1}{2}[v_0 + (v_0 + at)] = v_0 + \frac{1}{2}at \qquad \textit{(constant acceleration)}$$

Next, we substitute this result into Equation 2-10 $\left(x = x_0 + \frac{1}{2}(v_0 + v)t\right)$, which yields

$$x = x_0 + \frac{1}{2}(v_0 + v)t = x_0 + \left(v_0 + \frac{1}{2}at\right)t$$

Simplifying this expression yields the following result:

Constant-Acceleration Equation of Motion: Position as a Function of Time

$$x = x_0 + v_0 t + \frac{1}{2}at^2 \qquad\qquad\qquad 2\text{-}11$$

Here we have an expression for position versus time that is explicitly in terms of the acceleration, a.

Notice that each term in Equation 2-11 has the same dimensions, as they must. For example, the velocity term, $v_0 t$, has the units $(m/s)(s) = m$. Similarly, the acceleration term, $\frac{1}{2} at^2$, has the units $(m/s^2)(s^2) = m$.

EXERCISE 2-12 POSITION WITH CONSTANT ACCELERATION

Repeat part (b) of Example 2-11 using Equation 2-11, $x = x_0 + v_0 t + \frac{1}{2} at^2$.

REASONING AND SOLUTION

Given the initial position, initial velocity, and acceleration, the position at any given time can be found with $x = x_0 + v_0 t + \frac{1}{2} at^2$. In this case, we have

$$x = x_0 + v_0 t + \tfrac{1}{2} at^2 = 0 + (1.50 \text{ m/s})(5.00 \text{ s}) + \tfrac{1}{2}(2.40 \text{ m/s}^2)(5.00 \text{ s})^2 = 37.5 \text{ m}$$

The next Example gives further insight into the physical meaning of Equation 2-11.

EXAMPLE 2-13 PUT THE PEDAL TO THE METAL

A drag racer starts from rest and accelerates at 7.40 m/s^2. How far has it traveled in **(a)** 1.00 s, **(b)** 2.00 s, **(c)** 3.00 s?

PICTURE THE PROBLEM

We set up a coordinate system in which the drag racer starts at the origin and accelerates in the positive x direction. With this choice, it follows that $x_0 = 0$ and $a = +7.40 \text{ m/s}^2$. Also, the racer starts from rest, and hence its initial velocity is zero, $v_0 = 0$. Incidentally, the positions of the racer in the sketch have been drawn to scale.

REASONING AND STRATEGY

This problem gives the acceleration, which is constant, and asks for a relationship between position and time. Therefore, we use Equation 2-11, $x = x_0 + v_0 t + \frac{1}{2} at^2$.

Known   Velocity at time $t = 0$, $v_0 = 0$; acceleration, $a = 7.40 \text{ m/s}^2$.
Unknown   Distance traveled, $x = ?$, at **(a)** $t = 1.00$ s, **(b)** $t = 2.00$ s, **(c)** $t = 3.00$ s

SOLUTION

Part (a)

1. Evaluate $x = x_0 + v_0 t + \frac{1}{2} at^2$ with $a = 7.40 \text{ m/s}^2$ and $t = 1.00$ s:

$$x = x_0 + v_0 t + \tfrac{1}{2} at^2 = 0 + 0 + \tfrac{1}{2} at^2 = \tfrac{1}{2} at^2$$
$$x = \tfrac{1}{2}(7.40 \text{ m/s}^2)(1.00 \text{ s})^2 = 3.70 \text{ m}$$

Part (b)

2. From the calculation in part (a), we see that $x = x_0 + v_0 t + \frac{1}{2} at^2$ reduces to $x = \frac{1}{2} at^2$ in this situation. Evaluate $x = \frac{1}{2} at^2$ at $t = 2.00$ s:

$$x = \tfrac{1}{2} at^2 = \tfrac{1}{2}(7.40 \text{ m/s}^2)(2.00 \text{ s})^2 = 14.8 \text{ m} = 4(3.70 \text{ m})$$

Part (c)

3. Evaluate $x = \frac{1}{2} at^2$ at $t = 3.00$ s:

$$x = \tfrac{1}{2} at^2$$
$$= \tfrac{1}{2}(7.40 \text{ m/s}^2)(3.00 \text{ s})^2 = 33.3 \text{ m} = 9(3.70 \text{ m})$$

INSIGHT

This Example illustrates one of the key features of accelerated motion—position does not change uniformly with time when an object accelerates. In this case, the distance traveled in the first two seconds is 4 times the distance traveled in the first second, and the distance traveled in the first three seconds is 9 times the distance traveled in the first second. This kind of behavior is a direct result of the fact that x depends on t^2 when the acceleration is nonzero.

PRACTICE PROBLEM

We've seen that in one second the racer travels 3.70 m. How much time does it take for the racer to travel twice this distance, $2(3.70 \text{ m}) = 7.40 \text{ m}$?   [**Answer:** $t = \sqrt{2}$ s $= 1.41$ s]

Some related homework problems: Problem 43, Problem 46

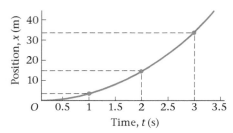

▲ FIGURE 2-20 **Position versus time for Example 2-13** The upward-curving, parabolic shape of this x-versus-t plot indicates a positive, constant acceleration. The dots on the curve show the position of the drag racer in Example 2-13 at the times 1.00 s, 2.00 s, and 3.00 s.

PHYSICS
IN CONTEXT
Looking Ahead

The equations developed for motion with constant acceleration in this chapter are used again with slightly different symbols when we study motion in two dimensions in Chapter 4 and rotational motion in Chapter 10.

FIGURE 2-20 shows a graph of x versus t for Example 2-13. Notice the parabolic shape of the x-versus-t curve, which is due to the $\frac{1}{2}at^2$ term and is characteristic of constant acceleration. In particular, if acceleration is positive ($a > 0$), then a plot of x versus t curves upward; if acceleration is negative ($a < 0$), a plot of x versus t curves downward. The greater the magnitude of a, the greater the curvature. In contrast, if a particle moves with constant velocity ($a = 0$), the t^2 dependence vanishes, and the x-versus-t plot is a straight line.

Velocity as a Function of Position Our final equation of motion with constant acceleration relates velocity to position. We start by solving for the time, t, in Equation 2-7 ($v = v_0 + at$):

$$v = v_0 + at \quad \text{or} \quad t = \frac{v - v_0}{a}$$

Next, we substitute this result into Equation 2-10, $x = x_0 + \frac{1}{2}(v_0 + v)t$, thus eliminating t:

$$x = x_0 + \frac{1}{2}(v_0 + v)t = x_0 + \frac{1}{2}(v_0 + v)\left(\frac{v - v_0}{a}\right)$$

Noting that $(v_0 + v)(v - v_0) = v_0v - v_0^2 + v^2 - vv_0 = v^2 - v_0^2$, we have

$$x = x_0 + \frac{v^2 - v_0^2}{2a}$$

Finally, a straightforward rearrangement of terms yields the following:

Constant-Acceleration Equation of Motion: Velocity in Terms of Displacement

$$v^2 = v_0^2 + 2a(x - x_0) = v_0^2 + 2a\Delta x \qquad \text{2-12}$$

This equation allows us to relate the velocity at one position to the velocity at another position, without knowing how much time is involved. The next Example shows how Equation 2-12 is used.

EXAMPLE 2-14   TAKEOFF DISTANCE FOR AN AIRLINER

RWP Jets at John F. Kennedy International Airport accelerate from rest at one end of a runway, and must attain takeoff speed before reaching the other end of the runway. **(a)** Plane A has acceleration a and takeoff speed v_{to}. What is the minimum length of runway, Δx_A, required for this plane? Give a symbolic answer. **(b)** Plane B has the same acceleration as plane A, but requires twice the takeoff speed. Find Δx_B and compare with Δx_A. **(c)** Find the minimum runway length for plane A if $a = 2.20 \text{ m/s}^2$ and $v_{to} = 95.0 \text{ m/s}$. (These values are typical for a 747 jetliner. For purposes of comparison, the shortest runway at JFK International Airport is 04R/22L, which has a length of 2560 m.)

PICTURE THE PROBLEM
In our sketch, we choose the positive x direction to be the direction of motion. With this choice, it follows that the acceleration of the plane is positive, $a = +2.20 \text{ m/s}^2$. Similarly, the takeoff velocity is positive as well, $v_{to} = +95.0 \text{ m/s}$.

REASONING AND STRATEGY
From the sketch it's clear that we want to express Δx, the distance the plane travels in attaining takeoff speed, in terms of the acceleration, a, and the takeoff speed, v_{to}. Equation 2-12 ($v^2 = v_0^2 + 2a\Delta x$), which relates distance to velocity, allows us to do this.

Known   Acceleration, $a = 2.20 \text{ m/s}^2$; takeoff velocity, $v_{to} = 95.0 \text{ m/s}$.
Unknown   Takeoff distance, $\Delta x = ?$

SOLUTION

Part (a)

1. Solve $v^2 = v_0^2 + 2a(x - x_0) = v_0^2 + 2a\Delta x$ for Δx.
   To find Δx_A, set $v_0 = 0$ and $v = v_{to}$:

$$\Delta x = \frac{v^2 - v_0^2}{2a}$$

$$\Delta x_A = \frac{v_{to}^2}{2a}$$

CONTINUED

Part (b)

2. To find Δx_B, simply change v_{to} to $2v_{to}$ in part (a):

$$\Delta x_B = \frac{(2v_{to})^2}{2a} = \frac{4v_{to}^2}{2a} = 4\Delta x_A$$

Part (c)

3. Substitute numerical values into the result found in part (a):

$$\Delta x_A = \frac{v_{to}^2}{2a} = \frac{(95.0 \text{ m/s})^2}{2(2.20 \text{ m/s}^2)} = 2050 \text{ m}$$

INSIGHT

This Example illustrates the fact that there are many advantages to obtaining symbolic results before substituting numerical values. In this case, we found that the takeoff distance is proportional to v^2; hence, we conclude immediately that doubling v results in a fourfold increase of Δx.

PRACTICE PROBLEM

Find the minimum acceleration needed for a takeoff speed of $v_{to} = (95.0 \text{ m/s})/2 = 47.5 \text{ m/s}$ on a runway of length $\Delta x = (2050 \text{ m})/4 = 513 \text{ m}$.   [**Answer:** $a = v_{to}^2/2\Delta x = 2.20 \text{ m/s}^2$]

Some related homework problems: Problem 40, Problem 90

Enhance Your Understanding   (Answers given at the end of the chapter)

5. The equation of motion for an object moving with constant acceleration is
$x = 6 \text{ m} - (5 \text{ m/s})t + (4 \text{ m/s}^2)t^2$. (a) What is the position of this object at $t = 0$?
(b) What is the velocity of this object at $t = 0$? (c) What is the acceleration of this object?

Section Review

• Motion with constant acceleration can be described by equations of motion relating quantities like position, velocity, time, and acceleration.

2-6 Applications of the Equations of Motion

We devote this section to a variety of examples that further illustrate the use of the constant-acceleration equations of motion. For convenience, all of our constant-acceleration equations of motion are collected in Table 2-5.

Big Idea 5 Equations of motion for constant acceleration relate position, velocity, acceleration, and time.

TABLE 2-5 **Constant-Acceleration Equations of Motion**

Variables Related	Equation	Number
velocity, time, acceleration	$v = v_0 + at$	2-7
initial, final, and average velocity	$v_{av} = \frac{1}{2}(v_0 + v)$	2-9
position, time, velocity	$x = x_0 + \frac{1}{2}(v_0 + v)t$	2-10
position, time, acceleration	$x = x_0 + v_0 t + \frac{1}{2}at^2$	2-11
velocity, position, acceleration	$v^2 = v_0^2 + 2a(x - x_0) = v_0^2 + 2a\Delta x$	2-12

In our first Example, we consider the distance and time needed to brake a vehicle to a complete stop.

EXAMPLE 2-15   HIT THE BRAKES!

A park ranger driving on a back country road suddenly sees a deer "frozen" in the headlights. The ranger, who is driving at 11.4 m/s, immediately applies the brakes and slows with an acceleration of 3.80 m/s^2. **(a)** If the deer is 20.0 m from the ranger's vehicle when the brakes are applied, how close does the ranger come to hitting the deer? **(b)** How much time is needed for the ranger's vehicle to stop?

CONTINUED

PICTURE THE PROBLEM

We choose the positive x direction to be the direction of motion. With this choice it follows that $v_0 = +11.4$ m/s. In addition, the fact that the ranger's vehicle is slowing down means its acceleration points in the *opposite* direction to that of the velocity [see Figure 2-14 (b) and (c)]. Therefore, the vehicle's acceleration is $a = -3.80$ m/s^2. Finally, when the vehicle comes to rest its velocity is zero, $v = 0$.

REASONING AND STRATEGY

The acceleration is constant, so we can use the equations listed in Table 2-5. In part (a) we want to find a distance when we know the velocity and acceleration, so we use a rearranged version of Equation 2-12 ($v^2 = v_0^2 + 2a\Delta x$). In part (b) we want to find a time when we know the velocity and acceleration, so we use a rearranged version of Equation 2-7 ($v = v_0 + at$).

Known   Velocity at time $t = 0$, $v_0 = 11.4$ m/s; acceleration, $a = -3.80$ m/s^2.
Unknown **(a)** Distance required to stop, $\Delta x = ?$ **(b)** Time required to stop, $t = ?$

SOLUTION

Part (a)

1. Solve $v^2 = v_0^2 + 2a\Delta x$ for Δx:

$$\Delta x = \frac{v^2 - v_0^2}{2a}$$

2. Set $v = 0$, and substitute numerical values:

$$\Delta x = -\frac{v_0^2}{2a} = -\frac{(11.4 \text{ m/s})^2}{2(-3.80 \text{ m/s}^2)} = 17.1 \text{ m}$$

3. Subtract Δx from 20.0 m to find the distance between the stopped vehicle and the deer:

$$20.0 \text{ m} - 17.1 \text{ m} = 2.9 \text{ m}$$

Part (b)

4. Set $v = 0$ in $v = v_0 + at$ and solve for t:

$$v = v_0 + at = 0$$
$$t = -\frac{v_0}{a} = -\frac{11.4 \text{ m/s}}{(-3.80 \text{ m/s}^2)} = 3.00 \text{ s}$$

INSIGHT

Notice the different ways that t and Δx depend on the initial speed. If the initial speed is doubled, for example, the time needed to stop also doubles. On the other hand, the distance needed to stop increases by a factor of four. This is one reason speed on the highway has such a great influence on safety.

PRACTICE PROBLEM

Show that using $t = 3.00$ s in Equation 2-11 $\left(x = x_0 + v_0 t + \frac{1}{2} at^2\right)$ results in the same distance needed to stop.

[**Answer:** $x = x_0 + v_0 t + \frac{1}{2} at^2 = 0 + (11.4 \text{ m/s})(3.00 \text{ s}) + \frac{1}{2}(-3.80 \text{ m/s}^2)(3.00 \text{ s})^2 = 17.1$ m, as expected]

Some related homework problems: Problem 55, Problem 56

In Example 2-15, we calculated the distance necessary for a vehicle to come to a complete stop. But how does the speed v vary with distance as the vehicle slows down? The next Conceptual Example deals with this topic.

CONCEPTUAL EXAMPLE 2-16   STOPPING DISTANCE

The ranger in Example 2-15 brakes for 17.1 m. After braking for only half that distance, $\frac{1}{2}(17.1 \text{ m}) = 8.55$ m, is the ranger's speed **(a)** equal to $\frac{1}{2}v_0$, **(b)** greater than $\frac{1}{2}v_0$, or **(c)** less than $\frac{1}{2}v_0$?

REASONING AND DISCUSSION

As pointed out in the Insight for Example 2-15, the fact that the stopping distance, Δx, depends on v_0^2 means that this distance increases by a factor of four when the speed is doubled. For example, the stopping distance with an initial speed of v_0 is four times the stopping distance when the initial speed is $v_0/2$.

CONTINUED

To apply this observation to the ranger, suppose that the stopping distance with an initial speed of v_0 is Δx. It follows that the stopping distance for an initial speed of $v_0/2$ is $\Delta x/4$. This means that as the ranger slows from v_0 to 0, it takes a distance $\Delta x/4$ to slow from $v_0/2$ to 0, and the remaining distance, $3\Delta x/4$, to slow from v_0 to $v_0/2$. Thus, at the halfway point the ranger has not yet slowed to half of the initial velocity—the speed at this point is greater than $v_0/2$.

ANSWER
(b) The ranger's speed is greater than $\frac{1}{2}v_0$.

Clearly, v does not decrease uniformly with distance. A plot showing v as a function of x for Example 2-15 is shown in **FIGURE 2-21**. As we can see from the graph, v changes more in the second half of the braking distance than in the first half.

We close this section with a familiar everyday example: a police car accelerating to overtake a speeder. This is the first time we use two equations of motion for two different objects to solve a problem—but it won't be the last. Problems of this type are often more interesting than problems involving only a single object, and they relate to many types of situations in everyday life.

PROBLEM-SOLVING NOTE

Strategize

Before you attempt to solve a problem, it is a good idea to have some sort of plan, or "strategy," for how to proceed. It may be as simple as saying, "The problem asks me to relate velocity and time; therefore I will use Equation 2-7." In other cases the strategy is a bit more involved. Producing effective strategies is one of the most challenging—and creative—aspects of problem solving.

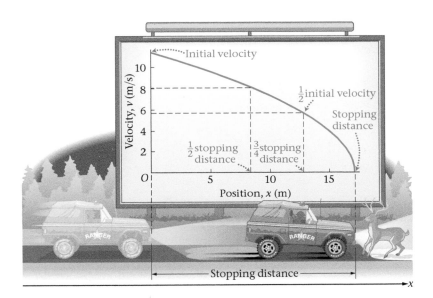

◄ **FIGURE 2-21 Velocity as a function of position for the ranger in Example 2-15** The ranger's vehicle comes to rest with constant acceleration, which means that its velocity decreases uniformly with time. The velocity *does not* decrease uniformly with distance, however. In particular, note how rapidly the velocity decreases in the final one-quarter of the stopping distance.

EXAMPLE 2-17   CATCHING A SPEEDER

A speeder doing 40.0 mi/h (about 17.9 m/s) in a 25 mi/h zone approaches a parked police car. The instant the speeder passes the police car, the police begin their pursuit. If the speeder maintains a constant velocity, and the police car accelerates with a constant acceleration of 4.51 m/s², **(a)** how much time does it take for the police car to catch the speeder, **(b)** how far have the two cars traveled in this time, and **(c)** what is the velocity of the police car when it catches the speeder?

PICTURE THE PROBLEM
Our sketch shows the two cars at the moment the speeder passes the resting police car. At this instant, which we take to be $t = 0$, both the speeder and the police car are at the origin, $x = 0$. In addition, we choose the positive x direction to be the direction of motion; therefore, the speeder's initial velocity is given by $v_s = +17.9$ m/s, and the police car's initial velocity is zero. The speeder's acceleration is zero, but the police car has an acceleration given by $a = +4.51$ m/s². Finally, our figure shows the linear x-versus-t plot for the speeder, and the parabolic x-versus-t plot for the police car.

REASONING AND STRATEGY
To solve this problem, we first write a position-versus-time equation $\left(x = x_0 + v_0t + \frac{1}{2}at^2\right)$ for the police car, x_p, and a separate equation for the speeder, x_s. Next, we find the time it takes the police car to catch the speeder by setting $x_p = x_s$ and solving the resulting equation for t. Once the catch time is determined, it's straightforward to calculate the distance traveled and the velocity of the police car.

CONTINUED

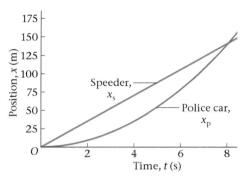

Known Speeder velocity, v_s = +17.9 m/s; initial police velocity = 0; police acceleration, a = +4.51 m/s^2.

Unknown (a) Time to catch speeder, t = ? (b) Distance to catch speeder? (c) Velocity of police when speeder is caught, v_p = ?

SOLUTION

Part (a)

1. Use $x = x_0 + v_0 t + \frac{1}{2}at^2$ to write equations of motion for the two vehicles. For the police car, v_0 = 0 and a = 4.51 m/s^2. For the speeder, v_0 = 17.9 m/s = v_s and a = 0:

$$x_p = \frac{1}{2}at^2$$
$$x_s = v_s t$$

2. Set $x_p = x_s$ and solve for the time:

$$\frac{1}{2}at^2 = v_s t \quad \text{or} \quad \left(\frac{1}{2}at - v_s\right)t = 0$$

two solutions: $t = 0$ and $t = \dfrac{2v_s}{a}$

3. Clearly, $t = 0$ corresponds to the initial conditions, because both vehicles started at $x = 0$ at that time. The time of interest is obtained by substituting numerical values into the other solution:

$$t = \frac{2v_s}{a} = \frac{2(17.9 \text{ m/s})}{4.51 \text{ m/s}^2} = 7.94 \text{ s}$$

Part (b)

4. Substitute t = 7.94 s into the equations of motion for x_p and x_s. Notice that $x_p = x_s$, as expected:

$$x_p = \frac{1}{2}at^2 = \frac{1}{2}(4.51 \text{ m/s}^2)(7.94 \text{ s})^2 = 142$$
$$x_s = v_s t = (17.9 \text{ m/s})(7.94 \text{ s}) = 142 \text{ m}$$

Part (c)

5. To find the velocity of the police car use Equation 2-7 ($v = v_0 + at$), which relates velocity to time:

$$v_p = v_0 + at = 0 + (4.51 \text{ m/s}^2)(7.94 \text{ s}) = 35.8 \text{ m/s}$$

INSIGHT

When the police car catches up with the speeder, its velocity is 35.8 m/s, which is exactly twice the velocity of the speeder. A coincidence? Not at all. When the police car catches the speeder, both have traveled the same distance (142 m) in the same time (7.94 s); therefore, they have the same average velocity. Of course, the average velocity of the speeder is simply v_s. The average velocity of the police car is $\frac{1}{2}(v_0 + v)$, since its acceleration is constant, and thus $\frac{1}{2}(v_0 + v) = v_s$. Noting that $v_0 = 0$ for the police car, we see that $v = 2v_s$. This result is independent of the acceleration of the police car, as we show in the following Practice Problem.

PRACTICE PROBLEM

Repeat this Example for the case where the acceleration of the police car is a = 3.25 m/s^2. [**Answer:** (a) t = 11.0 s, (b) $x_p = x_s$ = 197 m, (c) v_p = 35.8 m/s]

Some related homework problems: Problem 62, Problem 64

Enhance Your Understanding   (Answers given at the end of the chapter)

6. A submerged alligator swims directly toward two unsuspecting ducks. The equation of motion of the alligator is $x_a = (0.25 \text{ m/s}^2)t^2$, the equation of motion of duck 1 is x_{d1} = 3 m − (1.5 m/s)t, and the equation of motion of duck 2 is x_{d2} = 3 m + (1 m/s)t. Which duck does the alligator encounter first?

CONTINUED

Section Review

- Equations of motion for constant acceleration are listed in Table 2-5. As an example, the equation relating velocity and time is $v = v_0 + at$.

2-7 Freely Falling Objects

The most famous example of motion with constant acceleration is **free fall**—the motion of an object falling freely under the influence of gravity. It was Galileo (1564–1642) who first showed, with his own experiments, that falling bodies move with constant acceleration. His conclusions were based on experiments done by rolling balls down inclines of various steepness. By using an incline, Galileo was able to reduce the acceleration of the balls, thus producing motion slow enough to be timed with the instruments available at the time.

Galileo also pointed out that objects of different weight fall with the *same* constant acceleration—provided air resistance is small enough to be ignored. Whether he dropped objects from the Leaning Tower of Pisa to demonstrate this fact, as legend has it, will probably never be known for certain, but we do know that he performed extensive experiments to support his claim.

Today it is easy to verify Galileo's assertion by dropping objects in a vacuum chamber, where the effects of air resistance are essentially removed. In a standard classroom demonstration, a feather and a coin are dropped in a vacuum, and both fall at the same rate. In 1971, a novel version of this experiment was carried out on the Moon by astronaut David Scott. In the near-perfect vacuum on the Moon's surface he dropped a feather and a hammer and showed a worldwide television audience that they fell to the ground in the same time. Examples of free fall in different contexts are shown in FIGURE 2-22.

To illustrate the effect of air resistance in everyday terms, drop a sheet of paper and a rubber ball from the same height (FIGURE 2-23). The paper drifts slowly to the ground, taking much more time to fall than the ball. Now, wad the sheet of paper into a tight ball and repeat the experiment. This time the ball of paper and the rubber ball reach the ground in nearly the same time. What was different in the two experiments? Clearly, when the sheet of paper was wadded into a ball, the effect of air resistance on it was greatly reduced, so that both objects fell almost as they would in a vacuum.

Characteristics of Free Fall Before considering a few examples, let's discuss exactly what is meant by "free fall." To begin, the word *free* in free fall means free from any effects other than gravity. For example, in free fall we assume that an object's motion is not influenced by any form of friction or air resistance.

Free fall is the motion of an object subject *only* to the influence of gravity.

(a) Dropping a sheet of paper and a rubber ball

(b) Dropping a wadded-up sheet of paper and a rubber ball

▲ FIGURE 2-23 **Free fall and air resistance**

(a)

(b)

(c)

▲ FIGURE 2-22 **Visualizing Concepts Free fall (a)** In the absence of air resistance, all objects fall with the same acceleration, regardless of their mass. **(b)** Whether she is on the way up, at the peak of her flight, or on the way down, this girl is in free fall, accelerating downward with the acceleration of gravity. Only when she is in contact with the blanket does her acceleration change. **(c)** In the absence of air resistance, these lava bombs from the Kilauea volcano on the Big Island of Hawaii would strike the water with the same speed they had when they were blasted into the air. It's an example of the symmetry of free fall.

Big Idea 6 Free fall is motion with a constant downward acceleration of magnitude g, where g is the acceleration due to gravity.

Though free fall is an idealization—which does not apply to many real-world situations—it is still a useful approximation in many other cases. In the following examples we assume that the motion may be considered as free fall.

Next, it should be realized that the word *fall* in free fall does not mean the object is necessarily moving downward. By free fall, we mean *any* motion under the influence of gravity alone. If you drop a ball, it is in free fall. If you throw a ball upward or downward, it is in free fall as soon as it leaves your hand.

> An object is in free fall as soon as it is released, whether it is dropped from rest, thrown downward, or thrown upward.

Finally, the acceleration produced by gravity on the Earth's surface (sometimes called the gravitational strength) is denoted with the symbol g. As a shorthand name, we will frequently refer to g simply as "the acceleration due to gravity." In fact, as we shall see in Chapter 12, the value of g varies according to one's location on the surface of the Earth, as well as one's altitude above it. Table 2-6 shows how g varies with latitude.

In all the calculations that follow in this book, we shall use $g = 9.81 \text{ m/s}^2$ for the acceleration due to gravity. Note, in particular, that g always stands for $+9.81 \text{ m/s}^2$, never -9.81 m/s^2. For example, if we choose a coordinate system with the positive direction upward, the acceleration in free fall is $a = -g$. If the positive direction is downward, then free-fall acceleration is $a = g$.

With these comments in mind, we're ready to explore a variety of free-fall examples.

TABLE 2-6 Values of g at Different Locations on Earth (m/s^2)

Location	Latitude	g
North Pole	90° N	9.832
Oslo, Norway	60° N	9.819
Hong Kong	30° N	9.793
Quito, Ecuador	0°	9.780

EXAMPLE 2-18 LEMON DROP

A lemon drops from a tree and falls to the ground 3.15 m below. **(a)** How much time does it take for the lemon to reach the ground? **(b)** What is the lemon's speed just before it hits the ground?

PICTURE THE PROBLEM
In our sketch we choose the origin to be at the drop height of the lemon, and we let the positive x direction be downward. With these choices, $x_0 = 0$, $a = g$, and the ground is at $x = h = 3.15$ m. Of course, the initial velocity is zero, $v_0 = 0$, because the lemon drops from rest out of the tree.

REASONING AND STRATEGY
We can ignore air resistance in this case and model the motion as free fall. This means we can assume a constant acceleration equal to g and use the equations of motion in Table 2-5. For part (a) we want to find the time of fall when we know the distance and acceleration, so we use Equation 2-11 $\left(x = x_0 + v_0 t + \frac{1}{2} a t^2\right)$. For part (b) we can relate velocity to time by using Equation 2-7 $(v = v_0 + at)$, or we can relate velocity to position by using Equation 2-12 $(v^2 = v_0^2 + 2a\Delta x)$. We will implement both approaches and show that the results are the same.

$x_0 = 0$

$x = h$

Known Initial position of lemon, $x_0 = 0$; final position of lemon, $x = h = 3.15$ m; initial velocity of lemon, $v_0 = 0$; acceleration of lemon, $a = g = 9.81 \text{ m/s}^2$.

Unknown (a) Drop time, $t = ?$ (b) Landing speed, $v = ?$

SOLUTION

Part (a)

1. Write $x = x_0 + v_0 t + \frac{1}{2} a t^2$, with $x_0 = 0$, $v_0 = 0$, and $a = g$:

$$x = x_0 + v_0 t + \frac{1}{2} a t^2 = 0 + 0 + \frac{1}{2} g t^2 = \frac{1}{2} g t^2$$

2. Solve for the time, t, and set $x = h = 3.15$ m:

$$t = \sqrt{\frac{2x}{g}} = \sqrt{\frac{2h}{g}} = \sqrt{\frac{2(3.15 \text{ m})}{9.81 \text{ m/s}^2}} = 0.801 \text{ s}$$

Part (b)

3. To find the velocity, use the time found in part (a), $t = 0.801$ s, in $v = v_0 + at$:

$$v = v_0 + gt = 0 + (9.81 \text{ m/s}^2)(0.801 \text{ s}) = 7.86 \text{ m/s}$$

4. We can also find the velocity without knowing the time by using $v^2 = v_0^2 + 2a\Delta x$ with $\Delta x = 3.15$ m:

$$v^2 = v_0^2 + 2a\Delta x = 0 + 2g\Delta x$$
$$v = \sqrt{2g\Delta x} = \sqrt{2(9.81 \text{ m/s}^2)(3.15 \text{ m})} = 7.86 \text{ m/s}$$

CONTINUED

INSIGHT
We could just as well solve this problem with the origin at ground level, the drop height at $x = h$, and the positive x direction upward, which means that $a = -g$. The results are the same, of course.

PRACTICE PROBLEM — PREDICT/CALCULATE
Consider the distance the lemon drops in half the time required to reach the ground—that is, in the time $t = (0.801\,\text{s})/2$. **(a)** Is this distance greater than, less than, or equal to half the distance to the ground? Explain. **(b)** Find the distance the lemon drops in this time.   [**Answer: (a)** The distance is less than half the distance to the ground, because the average speed of the lemon in the first half of its drop is less than its average speed in the second half of its drop. **(b)** The distance dropped is

$$x = \tfrac{1}{2}gt^2 = \tfrac{1}{2}(9.81\,\text{m/s}^2)\left(\frac{0.801\,\text{s}}{2}\right)^2 = 0.787\,\text{m, which is only one-quarter the distance to the ground.]}$$

Some related homework problems: Problem 70, Problem 73

Free Fall from Rest   The special case of free fall from rest occurs so frequently, and in so many different contexts, that it deserves special attention. If we take x_0 to be zero, and positive to be downward, then position as a function of time is $x = x_0 + v_0t + \tfrac{1}{2}gt^2 = 0 + 0 + \tfrac{1}{2}gt^2$, or

$$x = \tfrac{1}{2}gt^2 \qquad\qquad \textbf{2-13}$$

Similarly, velocity as a function of time is

$$v = gt \qquad\qquad \textbf{2-14}$$

In addition, velocity as a function of position is

$$v = \sqrt{2gx} \qquad\qquad \textbf{2-15}$$

The behavior of these functions is illustrated in FIGURE 2-24. Notice that position increases with time squared, whereas velocity increases linearly with time.

Next we consider two objects that drop from rest, one after the other, and discuss how their separation varies with time.

$t = 0$	$v = 0$		$x = 0$
$t = 1\,\text{s}$	$v = 9.81\,\text{m/s}$		$x = 4.91\,\text{m}$
$t = 2\,\text{s}$	$v = 19.6\,\text{m/s}$		$x = 19.6\,\text{m}$
$t = 3\,\text{s}$	$v = 29.4\,\text{m/s}$		$x = 44.1\,\text{m}$
$t = 4\,\text{s}$	$v = 39.2\,\text{m/s}$		$x = 78.5\,\text{m}$

▶ FIGURE 2-24 **Free fall from rest** Position and velocity are shown as functions of time. It is apparent that velocity depends linearly on t, whereas position depends on t^2.

CONCEPTUAL EXAMPLE 2-19   FREE-FALL SEPARATION

Drops of water detach from the tip of an icicle and fall from rest. When one drop separates, the drop ahead of it has already fallen through a distance d, as shown below. As these two drops continue to fall, does their separation **(a)** increase, **(b)** decrease, or **(c)** stay the same?

REASONING AND DISCUSSION
It might seem that the separation between the drops will remain the same, because both are in free fall. This is not so. The drop that has a head start always has a greater velocity than the one that comes next. Therefore, the first drop covers a greater distance in any interval of time, and as a result, the separation between the drops increases.

ANSWER
(a) The separation between the drops increases.

An erupting volcano shooting out fountains of lava is an impressive sight. In the next Example we show how a simple timing experiment can determine the initial velocity of the erupting lava.

EXAMPLE 2-20   BOMBS AWAY: CALCULATING THE SPEED OF A LAVA BOMB

RWP A volcano shoots out blobs of molten lava, called lava bombs, from its summit. A geologist observing the eruption uses a stopwatch to time the flight of a particular lava bomb that is projected straight upward. If the time for the bomb to rise and fall back to its launch height is 4.75 s, and its acceleration is 9.81 m/s² downward, what is its initial speed?

PICTURE THE PROBLEM

Our sketch shows a coordinate system with upward as the positive x direction. For clarity, we offset the upward and downward trajectories slightly in our sketch. In addition, we choose $t = 0$ to be the time at which the lava bomb is launched. With these choices it follows that $x_0 = 0$ and the acceleration is $a = -g = -9.81$ m/s². The initial speed to be determined is v_0.

REASONING AND STRATEGY

Once again, we can ignore air resistance and model the motion of the lava bomb as free fall—this time with an initial upward velocity. We know that the lava bomb starts at $x = 0$ at the time $t = 0$ and returns to $x = 0$ at the time $t = 4.75$ s. This means that we know the bomb's position, time, and acceleration ($a = -g$), from which we would like to determine the initial velocity. A reasonable approach is to use Equation 2-11 $\left(x = x_0 + v_0 t + \frac{1}{2} a t^2\right)$ and solve it for the one unknown it contains, v_0.

Known   Flight time, $t = 4.75$ s; acceleration, $a = -g = -9.81$ m/s².
Unknown   Initial velocity of the lava bomb, $v_0 = ?$

SOLUTION

1. Write out $x = x_0 + v_0 t + \frac{1}{2} a t^2$ with $x_0 = 0$ and $a = -g$. Factor out a time, t, from the two remaining terms:

$$x = x_0 + v_0 t + \tfrac{1}{2} a t^2 = v_0 t - \tfrac{1}{2} g t^2 = \left(v_0 - \tfrac{1}{2} g t\right)t$$

2. Set x equal to zero, because this is the position of the lava bomb at $t = 0$ and $t = 4.75$ s:

$$x = \left(v_0 - \tfrac{1}{2} g t\right)t = 0$$

Two solutions: (i) $t = 0$ (ii) $v_0 - \tfrac{1}{2} g t = 0$

3. The first solution is simply the initial condition; that is, $x = 0$ at $t = 0$. Solve the second solution for the initial speed:

$$v_0 - \tfrac{1}{2} g t = 0 \quad \text{or} \quad v_0 = \tfrac{1}{2} g t$$

4. Substitute numerical values for g and the time the lava bomb lands:

$$v_0 = \tfrac{1}{2} g t = \tfrac{1}{2}(9.81 \text{ m/s}^2)(4.75 \text{ s}) = 23.3 \text{ m/s}$$

INSIGHT

A geologist can determine a lava bomb's initial speed by simply observing its flight time. Knowing the lava bomb's initial speed can help geologists determine how severe a volcanic eruption will be, and how dangerous it might be to people in the surrounding area.

PRACTICE PROBLEM

A second lava bomb is projected straight upward with an initial speed of 25 m/s. What is its flight time?
[**Answer:** $t = 5.1$ s]

Some related homework problems: Problem 71, Problem 78

What is the speed of a lava bomb when it returns to Earth; that is, when it returns to the same level from which it was launched? Physical intuition might suggest that, in the absence of air resistance, it should be the same as its initial speed. To show that this hypothesis is indeed correct, write out Equation 2-7 ($v = v_0 + at$) for this case:

$$v = v_0 - gt$$

Substituting numerical values, we find

$$v = v_0 - gt = 23.3 \text{ m/s} - (9.81 \text{ m/s}^2)(4.75 \text{ s}) = -23.3 \text{ m/s}$$

Thus, the velocity of the lava bomb when it lands is just the negative of the velocity it had when launched upward. Or, put another way, when the lava bomb lands, it has the same speed as when it was launched; it's just traveling in the opposite direction.

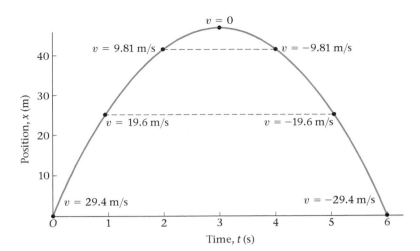

It's instructive to verify this result symbolically. Recall from Example 2-20 that $v_0 = \frac{1}{2}gt$, where t is the time the bomb lands. Substituting this result into Equation 2-7, we find

$$v = \tfrac{1}{2}gt - gt = -\tfrac{1}{2}gt = -v_0$$

The advantage of the symbolic solution lies in showing that the result is not a fluke—no matter what the initial velocity, no matter what the acceleration, the bomb lands with the velocity $-v_0$.

The Symmetry of Free Fall   The preceding results suggest a symmetry relating the motion on the way up to the motion on the way down. To make this symmetry more apparent, we first solve for the time when the lava bomb lands. Using the result $v_0 = \frac{1}{2}gt$ from Example 2-20, we find

$$t = \frac{2v_0}{g} \qquad \text{(time of landing)}$$

Next, we find the time when the velocity of the lava bomb is zero, which is at its highest point. Setting $v = 0$ in Equation 2-7 ($v = v_0 + at$), we have $v = v_0 - gt = 0$, or

$$t = \frac{v_0}{g} \qquad \text{(time when } v = 0\text{)}$$

This is exactly half the time required for the lava to make the round trip. Thus, the velocity of the lava bomb is zero and the height of the bomb is greatest exactly halfway between launch and landing.

This symmetry is illustrated in **FIGURE 2-25**. In this case we consider a lava bomb that is in the air for 6.00 s, moving without air resistance. Note that at $t = 3.00$ s the lava bomb is at its highest point and its velocity is zero. At times equally spaced before and after $t = 3.00$ s, the lava bomb is at the same height and has the same speed, but is moving in opposite directions. As a result of this symmetry, a movie of the lava bomb's flight would look the same whether run forward or in reverse.

FIGURE 2-26 shows the time dependence of position, velocity, and acceleration for an object in free fall without air resistance after being thrown upward. As soon as the object is released, it begins to accelerate downward—as indicated by the negative slope of the velocity-versus-time plot—though it isn't necessarily moving downward. For example, if you throw a ball *upward*, it begins to accelerate *downward* the moment it leaves your hand. It continues moving upward, however, until its speed diminishes to zero. Because gravity is causing the downward acceleration, and gravity doesn't turn off just because the ball's velocity goes through zero, the ball continues to accelerate downward even when it is momentarily at rest.

Similarly, in the next Example we consider a sandbag that falls from an ascending hot-air balloon. This means that before the bag is in free fall it was moving upward— just like a ball thrown upward. And just like the ball, the sandbag continues moving upward for a brief time before momentarily stopping and then moving downward.

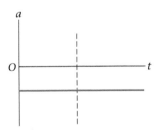

▲ **FIGURE 2-26 Position, velocity, and acceleration of a lava bomb as functions of time** The fact that the x-versus-t plot is curved indicates an acceleration; the downward curvature shows that the acceleration is negative. This is also clear from the v-versus-t plot, which has a negative slope. The constant slope of the straight line in the v-versus-t plot indicates a constant acceleration, as shown in the a-versus-t plot.

EXAMPLE 2-21   LOOK OUT BELOW! A SANDBAG IN FREE FALL

A hot-air balloon is rising straight upward with a constant speed of 6.5 m/s. When the basket of the balloon is 20.0 m above the ground, a bag of sand tied to the basket comes loose. **(a)** How much time elapses before the bag of sand hits the ground? **(b)** What is the greatest height of the bag of sand during its fall to the ground?

PICTURE THE PROBLEM

We choose the origin to be at ground level and positive to be upward. This means that, for the bag, we have $x_0 = 20.0$ m, $v_0 = 6.5$ m/s, and $a = -g$. Our sketch also shows snapshots of the balloon and bag of sand at three different times, starting at $t = 0$ when the bag comes loose. Notice that the bag is moving upward with the balloon at the time it comes loose. It therefore continues to move upward for a short time after it separates from the basket, exactly as if it had been thrown upward.

REASONING AND STRATEGY

The effects of air resistance on the sandbag can be ignored. As a result, we can use the equations in Table 2-5 with a constant acceleration $a = -g$.

    In part (a) we want to relate position and time—knowing the initial position and initial velocity—so we use Equation 2-11 $\left(x = x_0 + v_0 t + \frac{1}{2} a t^2\right)$. To find the time the bag hits the ground, we set $x = 0$ and solve for t.

    For part (b) we have no expression that gives the maximum height of a particle—so we will have to come up with something on our own. We can start with the fact that $v = 0$ at the greatest height, since it is there the bag momentarily stops as it changes direction. Therefore, we can find the time t when $v = 0$ by using Equation 2-7 $(v = v_0 + at)$, and then substitute t into Equation 2-11 to find x_{max}.

Known   Drop height, $x_0 = 20.0$ m; upward initial velocity of basket, $v_0 = 6.5$ m/s; acceleration due to gravity, $a = -g = -9.81$ m/s^2.

Unknown   **(a)** Time to reach the ground, $t = ?$ **(b)** Maximum height of sandbag, $x_{max} = ?$

SOLUTION

Part (a)

1. Apply $x = x_0 + v_0 t + \frac{1}{2} a t^2$ to the bag of sand. Do this by letting $a = -g$ and setting $x = 0$, which corresponds to ground level:

$$x = x_0 + v_0 t - \tfrac{1}{2} g t^2 = 0$$

2. Notice that we have a quadratic equation for t, with $A = -\frac{1}{2}g = -\frac{1}{2}(9.81 \text{ m/s}^2)$, $B = v_0 = 6.5$ m/s, and $C = x_0 = 20.0$ m. Solve this equation for t. The positive solution, 2.78 s, applies to this problem: (Quadratic equations and their solutions are discussed in Appendix A. In general, one can expect two solutions to a quadratic equation.)

$$t = \frac{-v_0 \pm \sqrt{v_0{}^2 - 4\left(-\frac{1}{2}g\right)(x_0)}}{2\left(-\frac{1}{2}g\right)}$$

$$= \frac{-(6.5 \text{ m/s}) \pm \sqrt{(6.5 \text{ m/s})^2 + 2(9.81 \text{ m/s}^2)(20.0 \text{ m})}}{(-9.81 \text{ m/s}^2)}$$

$$= \frac{-(6.5 \text{ m/s}) \pm 20.8 \text{ m/s}}{(-9.81 \text{ m/s}^2)} = 2.78 \text{ s}, -1.46 \text{ s}$$

Part (b)

3. Apply $v = v_0 + at$ to the bag of sand, and solve for the time when the velocity equals zero:

$$v = v_0 + at = v_0 - gt$$

$$v_0 - gt = 0 \quad \text{or} \quad t = \frac{v_0}{g} = \frac{6.5 \text{ m/s}}{9.81 \text{ m/s}^2} = 0.66 \text{ s}$$

4. Use $t = 0.66$ s in $x = x_0 + v_0 t + \frac{1}{2} a t^2$ to find the maximum height:

$$x_{max} = 20.0 \text{ m} + (6.5 \text{ m/s})(0.66 \text{ s}) - \tfrac{1}{2}(9.81 \text{ m/s}^2)(0.66 \text{ s})^2$$
$$= 22 \text{ m}$$

INSIGHT

Thus, the bag of sand continues to move upward for 0.66 s after it separates from the basket, reaching a maximum height of 22 m above the ground. It then begins to move downward, and hits the ground 2.78 s after detaching from the basket.

PRACTICE PROBLEM

What is the velocity of the sandbag just before it hits the ground?   **[Answer:** $v = v_0 - gt = (6.5 \text{ m/s}) - (9.81 \text{ m/s}^2) \times (2.78 \text{ s}) = -20.8$ m/s; the minus sign indicates the bag is moving downward, as expected.**]**

Some related homework problems: Problem 83, Problem 84

Enhance Your Understanding   (Answers given at the end of the chapter)

7. On a distant, airless planet, an astronaut drops a rock to test the planet's gravitational pull. The astronaut finds that in the first second of falling (from $t = 0$ to $t = 1$ s) the rock drops a distance of 1 m. How far does the rock drop from $t = 1$ s to $t = 2$ s? **(a)** 1 m, **(b)** 2 m, **(c)** 3 m, **(d)** 4 m.

Section Review

- Free fall is motion with a constant downward acceleration of magnitude $g = 9.81$ m/s^2.

CHAPTER 2 REVIEW

CHAPTER SUMMARY

2-1   POSITION, DISTANCE, AND DISPLACEMENT

Position
Position is the location of an object as measured on a coordinate system.

Distance
Distance is the total length of travel, from beginning to end. Distance is always positive.

Displacement
Displacement, Δx, is the change in position; that is, $\Delta x = x_f - x_i$.

Positive and Negative Displacement
The *sign* of the displacement indicates the *direction* of motion.

2-2   AVERAGE SPEED AND VELOCITY

Average Speed
Average speed is *distance* divided by elapsed time.

Average Velocity
Average velocity is *displacement* divided by elapsed time. Average velocity is positive for motion in the positive direction, negative for motion in the negative direction.

Graphical Interpretation
In an *x*-versus-*t* plot, the average velocity is the slope of a line connecting two points.

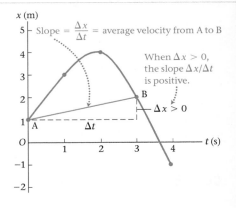

2-3   INSTANTANEOUS VELOCITY

The velocity at an instant of time, v, is the average velocity over shorter and shorter time intervals: $v = \lim\limits_{\Delta t \to 0} \dfrac{\Delta x}{\Delta t}$.

Graphical Interpretation
In an *x*-versus-*t* plot, the instantaneous velocity is the slope of a tangent line at a given instant of time.

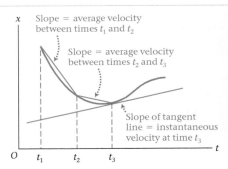

2-4 ACCELERATION

Average Acceleration
Average acceleration is the change in velocity divided by the elapsed time.

Instantaneous Acceleration
The acceleration at an instant of time, a, is the limit of the average acceleration over shorter and shorter time intervals: $a = \lim_{\Delta t \to 0} \frac{\Delta v}{\Delta t}$.

Graphical Interpretation
In a v-versus-t plot, the instantaneous acceleration is the slope of a tangent line at a given instant of time.

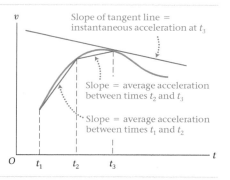

2-5 MOTION WITH CONSTANT ACCELERATION

Several different equations are used to describe the motion of objects moving with constant acceleration. These "equations of motion" are listed in Table 2-5. The ones that are used most often are (1) velocity as a function of time ($v = v_0 + at$), (2) position as a function of time $\left(x = x_0 + v_0 t + \frac{1}{2} at^2\right)$, and (3) velocity as a function of position ($v^2 = v_0^2 + 2a\Delta x$).

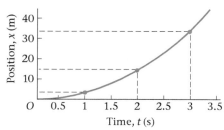

2-7 FREELY FALLING OBJECTS

Objects in free fall move under the influence of gravity alone. An object is in free fall as soon as it is released, whether it is thrown upward, thrown downward, or released from rest. As a result, objects in free fall move with a constant downward acceleration of magnitude $g = 9.81 \text{ m/s}^2$.

ANSWERS TO ENHANCE YOUR UNDERSTANDING QUESTIONS

1. **(a)** No. Distance is always increasing on a trip, whereas the magnitude of displacement can increase or decrease. **(b)** Yes. For example, in a 5-mi round trip the distance is 5 mi but the magnitude of the displacement is zero.

2. **(a)** Positive. **(b)** Zero. **(c)** Negative.

3. $C < B < D < A$.

4. **(a)** The speed is decreasing because velocity and acceleration have opposite signs. **(b)** The velocity is increasing (becoming less negative) because the acceleration is positive. **(c)** The speed is decreasing because velocity and acceleration have opposite signs. **(d)** The velocity is decreasing (becoming less positive) because the acceleration is negative.

5. **(a)** $x = 6$ m. **(b)** $v = -5$ m/s. **(c)** $a = 8 \text{ m/s}^2$.

6. The alligator encounters duck 1 first.

7. **(c)** 3 m.

CONCEPTUAL QUESTIONS

For instructor-assigned homework, go to www.masteringphysics.com. (MP)

(Answers to odd-numbered Conceptual Questions can be found in the back of the book.)
(The effects of air resistance are to be ignored in this chapter.)

1. You take your dog on a walk to a nearby park. On the way, your dog takes many short side trips to chase squirrels, examine fire hydrants, and so on. When you arrive at the park, do you and your dog have the same displacement from home? Have you and your dog traveled the same distance? Explain.

2. Does an odometer in a car measure distance or displacement? Explain.

3. An astronaut orbits Earth in the space shuttle. In one complete orbit, is the magnitude of the displacement the same as the distance traveled? Explain.

4. After a tennis match the players dash to the net to congratulate one another. If they both run with a speed of 3 m/s, are their velocities equal? Explain.

5. Does a speedometer measure speed or velocity? Explain.

6. Is it possible for a car to circle a racetrack with constant velocity? Can it do so with constant speed? Explain.

7. For what kinds of motion are the instantaneous and average velocities equal?

8. Assume that the brakes in your car create a constant deceleration, regardless of how fast you are going. If you double your driving speed, how does this affect **(a)** the time required to come to a stop, and **(b)** the distance needed to stop?

9. The velocity of an object is zero at a given instant of time. **(a)** Is it possible for the object's acceleration to be zero at this time? Explain. **(b)** Is it possible for the object's acceleration to be nonzero at this time? Explain.

10. If the velocity of an object is nonzero, can its acceleration be zero? Give an example if your answer is yes; explain why not if your answer is no.

11. Is it possible for an object to have zero average velocity over a given interval of time, yet still be accelerating during the interval? Give an example if your answer is yes; explain why not if your answer is no.

12. A batter hits a pop fly straight up. **(a)** Is the acceleration of the ball on the way up different from its acceleration on the way down? **(b)** Is the acceleration of the ball at the top of its flight different from its acceleration just before it lands?

13. A person on a trampoline bounces straight upward with an initial speed of 4.5 m/s. What is the person's speed when she returns to her initial height?

14. A volcano shoots a lava bomb straight upward. Does the displacement of the lava bomb depend on **(a)** your choice of origin for your coordinate system, or **(b)** your choice of a positive direction? Explain in each case.

PROBLEMS AND CONCEPTUAL EXERCISES

Answers to odd-numbered Problems and Conceptual Exercises can be found in the back of the book. **BIO** *identifies problems of biological or medical interest;* **CE** *indicates a conceptual exercise.* Predict/Explain *problems ask for two responses: (a) your prediction of a physical outcome, and (b) the best explanation among three provided; and* Predict/Calculate *problems ask for a prediction of a physical outcome, based on fundamental physics concepts, and follow that with a numerical calculation to verify the prediction. On all problems, bullets (•, ••, •••) indicate the level of difficulty.*

SECTION 2-1 POSITION, DISTANCE, AND DISPLACEMENT

1. • Referring to **FIGURE 2-27**, you walk from your home to the library, then to the park. **(a)** What is the distance traveled? **(b)** What is your displacement?

FIGURE 2-27 Problems 1 and 2

2. • In Figure 2-27, you walk from the park to your friend's house, then back to your house. What are your **(a)** distance traveled, and **(b)** displacement?

3. • The two tennis players shown in **FIGURE 2-28** walk to the net to congratulate one another. **(a)** Find the distance traveled and the displacement of player A. **(b)** Repeat for player B.

FIGURE 2-28 Problem 3

4. • The golfer in **FIGURE 2-29** sinks the ball in two putts, as shown. What are **(a)** the distance traveled by the ball, and **(b)** the displacement of the ball?

First putt

Second putt

FIGURE 2-29 Problem 4

5. • A jogger runs on the track shown in **FIGURE 2-30**. Neglecting the curvature of the corners, **(a)** what are the distance traveled and the displacement in running from point A to point B? **(b)** Find the distance and displacement for a complete circuit of the track.

FIGURE 2-30 Problem 5

6. •• Predict/Calculate A child rides a pony on a circular track whose radius is 5.25 m. **(a)** Find the distance traveled and the displacement after the child has gone halfway around the track. **(b)** Does the distance traveled increase, decrease, or stay the same when the child completes one circuit of the track? Explain. **(c)** Does the displacement increase, decrease, or stay the same when the child completes one circuit of the track? Explain. **(d)** Find the distance and displacement after a complete circuit of the track.

SECTION 2-2 AVERAGE SPEED AND VELOCITY

7. • **CE** Predict/Explain You drive your car in a straight line at 15 m/s for 10 kilometers, then at 25 m/s for another 10 kilometers. **(a)** Is your average speed for the entire trip greater than, less than, or equal to 20 m/s? **(b)** Choose the *best* explanation from among the following:
   I. More time is spent at 15 m/s than at 25 m/s.
   II. The average of 15 m/s and 25 m/s is 20 m/s.
   III. Less time is spent at 15 m/s than at 25 m/s.

8. • **CE** Predict/Explain You drive your car in a straight line at 15 m/s for 10 minutes, then at 25 m/s for another 10 minutes. **(a)** Is your average speed for the entire trip greater than, less than, or equal to 20 m/s? **(b)** Choose the *best* explanation from among the following:
   I. More time is required to drive at 15 m/s than at 25 m/s.
   II. Less distance is covered at 25 m/s than at 15 m/s.
   III. Equal time is spent at 15 m/s and 25 m/s.

9. • Usain Bolt of Jamaica set a world record in 2009 for the 200-meter dash with a time of 19.19 seconds. What was his average speed? Give your answer in meters per second and miles per hour.

10. • **BIO** Kangaroos have been clocked at speeds of 65 km/h. **(a)** How far can a kangaroo hop in 3.2 minutes at this speed? **(b)** How much time will it take a kangaroo to hop 0.25 km at this speed?

11. • **Rubber Ducks** A severe storm on January 10, 1992, caused a cargo ship near the Aleutian Islands to spill 29,000 rubber ducks

and other bath toys into the ocean. Ten months later hundreds of rubber ducks began to appear along the shoreline near Sitka, Alaska, roughly 1600 miles away. What was the approximate average speed of the ocean current that carried the ducks to shore in **(a)** m/s and **(b)** mi/h? (Rubber ducks from the same spill began to appear on the coast of Maine in July 2003.)

12. • Radio waves travel at the speed of light, approximately 186,000 miles per second. How much time does it take for a radio message to travel from Earth to the Moon and back? (See the inside back cover for the necessary data.)

13. • It was a dark and stormy night, when suddenly you saw a flash of lightning. Six and a half seconds later you heard the thunder. Given that the speed of sound in air is about 340 m/s, how far away was the lightning bolt? (Ignore the travel time for the flash of light.)

14. • **BIO Nerve Impulses** The human nervous system can propagate nerve impulses at about 10^2 m/s. Estimate the time it takes for a nerve impulse to travel 2 m from your toes to your brain.

15. •• A finch rides on the back of a Galapagos tortoise, which walks at the stately pace of 0.060 m/s. After 1.5 minutes the finch tires of the tortoise's slow pace, and takes flight in the same direction for another 1.5 minutes at 11 m/s. What was the average speed of the finch for this 3.0-minute interval?

16. •• You jog at 9.1 km/h for 5.0 km, then you jump into a car and drive an additional 13 km. With what average speed must you drive your car if your average speed for the entire 18 km is to be 25 km/h?

17. •• A dog runs back and forth between its two owners, who are walking toward one another (**FIGURE 2-31**). The dog starts running when the owners are 8.2 m apart. If the dog runs with a speed of 2.7 m/s, and the owners each walk with a speed of 1.3 m/s, how far has the dog traveled when the owners meet?

1.3 m/s   1.3 m/s
,2.7 m/s
———————— 8.2 m ————————

FIGURE 2-31 Problem 17

18. •• **BIO** Predict/Calculate Blood flows through a major artery at 1.0 m/s for 0.50 s, then at 0.60 m/s for another 0.50 s through a smaller artery. **(a)** Is the average speed of the blood 0.80 m/s, greater than 0.80 m/s, or less than 0.80 m/s? Explain. **(b)** Verify your answer to part (a) by calculating the average speed.

19. •• **BIO** Predict/Calculate Blood flows through a major artery at 1.0 m/s over a distance of 0.50 m, then at 0.60 m/s for another 0.50 m through a smaller artery. **(a)** Is the average speed of the blood 0.80 m/s, greater than 0.80 m/s, or less than 0.80 m/s? Explain. **(b)** Verify your answer to part (a) by calculating the average speed.

20. •• In heavy rush-hour traffic you drive in a straight line at 12 m/s for 1.5 minutes, then you have to stop for 3.5 minutes, and finally you drive at 15 m/s for another 2.5 minutes. **(a)** Plot a position-versus-time graph for this motion. Your plot should extend from $t = 0$ to $t = 7.5$ minutes. **(b)** Use your plot from part (a) to calculate the average velocity between $t = 0$ and $t = 7.5$ minutes.

21. •• Predict/Calculate An expectant father paces back and forth, producing the position-versus-time graph shown in **FIGURE 2-32**. Without performing a calculation, indicate whether the father's velocity is positive, negative, or zero on each of the following segments of the graph: **(a)** A, **(b)** B, **(c)** C, and **(d)** D. Calculate the numerical value of the father's velocity for the segments **(e)** A, **(f)** B, **(g)** C, and **(h)** D, and show that your results verify your answers to parts (a)–(d).

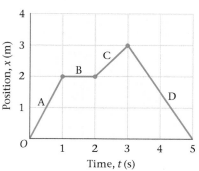

FIGURE 2-32 Problem 21

22. •• The position of a particle as a function of time is given by $x = (-5 \text{ m/s})t + (3 \text{ m/s}^2)t^2$. **(a)** Plot x versus t for $t = 0$ to $t = 2$ s. **(b)** Find the average velocity of the particle from $t = 0$ to $t = 1$ s. **(c)** Find the average speed from $t = 0$ to $t = 1$ s.

23. •• The position of a particle as a function of time is given by $x = (6 \text{ m/s})t + (-2 \text{ m/s}^2)t^2$. **(a)** Plot x versus t for $t = 0$ to $t = 2$ s. **(b)** Find the average velocity of the particle from $t = 0$ to $t = 1$ s. **(c)** Find the average speed from $t = 0$ to $t = 1$ s.

24. •• Predict/Calculate A tennis player moves back and forth along the baseline while waiting for her opponent to serve, producing the position-versus-time graph shown in **FIGURE 2-33**. **(a)** Without performing a calculation, indicate on which of the segments of the graph, A, B, or C, the player has the greatest speed. Calculate the player's speed for **(b)** segment A, **(c)** segment B, and **(d)** segment C, and show that your results verify your answers to part (a).

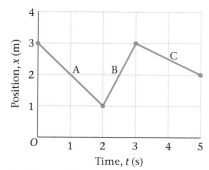

FIGURE 2-33 Problem 24

25. ••• On your wedding day you leave for the church 30.0 minutes before the ceremony is to begin, which should be plenty of time since the church is only 10.0 miles away. On the way, however, you have to make an unanticipated stop for construction work on the road. As a result, your average speed for the first 15 minutes is only 5.0 mi/h. What average speed do you need for the rest of the trip to get you to the church on time?

SECTION 2-3 INSTANTANEOUS VELOCITY

26. • CE The position-versus-time plot of a boat positioning itself next to a dock is shown in FIGURE 2-34. Rank the six points indicated in the plot in order of increasing value of the velocity v, starting with the most negative. Indicate a tie with an equal sign.

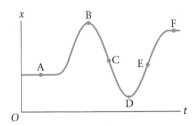

FIGURE 2-34 Problem 26

27. •• The position of a particle as a function of time is given by $x = (2.0 \text{ m/s})t + (-3.0 \text{ m/s}^3)t^3$. (a) Plot x versus t for time from $t = 0$ to $t = 1.0$ s. (b) Find the average velocity of the particle from $t = 0.35$ s to $t = 0.45$ s. (c) Find the average velocity from $t = 0.39$ s to $t = 0.41$ s. (d) Do you expect the instantaneous velocity at $t = 0.40$ s to be closer to 0.54 m/s, 0.56 m/s, or 0.58 m/s? Explain.

28. •• The position of a particle as a function of time is given by $x = (-2.00 \text{ m/s})t + (3.00 \text{ m/s}^3)t^3$. (a) Plot x versus t for time from $t = 0$ to $t = 1.00$ s. (b) Find the average velocity of the particle from $t = 0.150$ s to $t = 0.250$ s. (c) Find the average velocity from $t = 0.190$ s to $t = 0.210$ s. (d) Do you expect the instantaneous velocity at $t = 0.200$ s to be closer to -1.62 m/s, -1.64 m/s, or -1.66 m/s? Explain.

SECTION 2-4 ACCELERATION

29. • CE Predict/Explain On two occasions you accelerate uniformly from rest along an on-ramp in order to merge into interstate traffic at 65 mi/h. On-ramp A is 200 m long and on-ramp B is 400 m long. (a) Is your acceleration along on-ramp A greater than, less than, or equal to your acceleration along ramp B? (b) Choose the *best explanation* from among the following:
    I. The shorter acceleration distance on ramp A requires a greater acceleration.
    II. In each case, you start from rest and have the same final speed.
    III. The greater distance traveled on ramp B requires a greater acceleration.

30. • A 747 airliner reaches its takeoff speed of 156 mi/h in 35.2 s. What is the magnitude of its average acceleration?

31. • At the starting gun, a runner accelerates at 1.9 m/s² for 5.2 s. The runner's acceleration is zero for the rest of the race. What is the speed of the runner (a) at $t = 2.0$ s, and (b) at the end of the race?

32. • A jet makes a landing traveling due east with a speed of 70.6 m/s. If the jet comes to rest in 13.0 s, what are the magnitude and direction of its average acceleration?

33. • A car is traveling due north at 23.6 m/s. Find the velocity of the car after 7.10 s if its acceleration is (a) 1.30 m/s² due north, or (b) 1.15 m/s² due south.

34. •• A motorcycle moves according to the velocity-versus-time graph shown in FIGURE 2-35. Find the average acceleration of the motorcycle during each of the following segments of the motion: (a) A, (b) B, and (c) C.

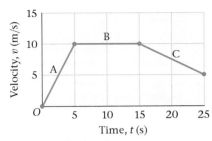

FIGURE 2-35 Problem 34

35. •• A person on horseback moves according to the velocity-versus-time graph shown in FIGURE 2-36. Find the displacement of the person for each of the following segments of the motion: (a) A, (b) B, and (c) C.

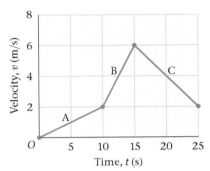

FIGURE 2-36 Problem 35

36. •• Running with an initial velocity of +9.2 m/s, a horse has an average acceleration of -1.81 m/s². How much time does it take for the horse to decrease its velocity to +5.5 m/s?

37. •• Predict/Calculate Assume that the brakes in your car create a constant deceleration of 4.2 m/s² regardless of how fast you are driving. If you double your driving speed from 18 m/s to 36 m/s, (a) does the time required to come to a stop increase by a factor of two or a factor of four? Explain. Verify your answer to part (a) by calculating the stopping times for initial speeds of (b) 18 m/s and (c) 36 m/s.

38. •• As a train accelerates away from a station, it reaches a speed of 4.7 m/s in 5.0 s. If the train's acceleration remains constant, what is its speed after an additional 5.0 s has elapsed?

39. •• A particle has an acceleration of +6.24 m/s² for 0.450 s. At the end of this time the particle's velocity is +9.31 m/s. What was the particle's initial velocity?

SECTION 2-5 MOTION WITH CONSTANT ACCELERATION

40. • Landing with a speed of 71.4 m/s, and traveling due south, a jet comes to rest in 949 m. Assuming the jet slows with constant acceleration, find the magnitude and direction of its acceleration.

41. • When you see a traffic light turn red, you apply the brakes until you come to a stop. If your initial speed was 18 m/s, and you were heading due west, what was your average velocity during braking? Assume constant deceleration.

42. CE •• A ball is released at the point $x = 2$ m on an inclined plane with a nonzero initial velocity. After being released, the ball moves with constant acceleration. The acceleration and initial velocity of the ball are described by one of the following four cases: case 1, $a > 0$, $v_0 > 0$; case 2, $a > 0$, $v_0 < 0$; case 3, $a < 0$, $v_0 > 0$; case

4, $a < 0$, $v_0 < 0$. **(a)** In which of these cases will the ball definitely pass $x = 0$ at some later time? **(b)** In which of these cases is more information needed to determine whether the ball will cross $x = 0$? **(c)** In which of these cases will the ball come to rest momentarily at some time during its motion?

43. •• Starting from rest, a boat increases its speed to 4.82 m/s with constant acceleration. **(a)** What is the boat's average speed? **(b)** If it takes the boat 4.77 s to reach this speed, how far has it traveled?

44. •• The position of a car as a function of time is given by $x = (50 \text{ m}) + (-5.0 \text{ m/s})t + (-10 \text{ m/s}^2)t^2$. **(a)** What are the initial position, initial velocity, and acceleration of the car? **(b)** Plot x versus t for $t = 0$ to $t = 2.0$ s. **(c)** What distance does the car travel during the first 1.0 s? **(d)** What is the average velocity of the car between $t = 1.0$ s and $t = 2.0$ s?

45. •• The position of a ball as a function of time is given by $x = (5.0 \text{ m/s})t + (-10 \text{ m/s}^2)t^2$. **(a)** What are the initial position, initial velocity, and acceleration of the ball? **(b)** Plot x versus t for $t = 0$ to $t = 2.0$ s. **(c)** Find the average velocity of the ball from $t = 0$ to $t = 1.0$ s. **(d)** Find the average speed of the ball between $t = 1.0$ s and $t = 2.0$ s.

46. •• **BIO** A cheetah can accelerate from rest to 25.0 m/s in 6.22 s. Assuming constant acceleration, **(a)** how far has the cheetah run in this time? **(b)** After sprinting for just 3.11 s, is the cheetah's speed 12.5 m/s, greater than 12.5 m/s, or less than 12.5 m/s? Explain. **(c)** What is the cheetah's average speed for the first 3.11 s of its sprint? For the second 3.11 s of its sprint? **(d)** Calculate the distance covered by the cheetah in the first 3.11 s and the second 3.11 s.

47. •• A sled slides from rest down an icy slope. Measurements taken from a video show that the distance from the starting point to the sled at various times is as follows:

Time (s)	Distance from starting point (m)
0	0
0.50	0.36
1.0	1.3
1.5	3.2
2.0	5.7
2.5	8.8

(a) Plot the position of the sled versus time. **(b)** From your plot, determine the approximate average speed of the sled from $t = 0.25$ s to $t = 1.3$ s. **(c)** Determine the approximate acceleration of the sled, assuming it to be constant.

SECTION 2-6 APPLICATIONS OF THE EQUATIONS OF MOTION

48. • A child slides down a hill on a toboggan with an acceleration of 1.6 m/s². If she starts at rest, how far has she traveled in **(a)** 1.0 s, **(b)** 2.0 s, and **(c)** 3.0 s?

49. • **The Detonator** On a ride called the Detonator at Worlds of Fun in Kansas City, passengers accelerate straight downward from rest to 45 mi/h in 2.2 seconds. What is the average acceleration of the passengers on this ride?

The Detonator (Problem 49)

50. • **Jules Verne** In his novel *From the Earth to the Moon* (1865), Jules Verne describes a spaceship that is blasted out of a cannon, called the *Columbiad*, with a speed of 12,000 yards/s. The *Columbiad* is 900 ft long, but part of it is packed with explosives, so the spaceship accelerates over a distance of only 700 ft. Estimate the acceleration experienced by the occupants of the spaceship during launch. Give your answer in m/s². (Verne realized that the "travelers would . . . encounter a violent recoil," but he probably didn't know that people generally lose consciousness if they experience accelerations greater than about $7g \sim 70$ m/s².)

51. •• **BIO Bacterial Motion** Approximately 0.1% of the bacteria in an adult human's intestines are *Escherichia coli*. These bacteria have been observed to move with speeds up to 15 μm/s and maximum accelerations of 166 μm/s². Suppose an *E. coli* bacterium in your intestines starts at rest and accelerates at 156 μm/s². How much **(a)** time and **(b)** distance are required for the bacterium to reach a speed of 12 μm/s?

52. •• Two cars drive on a straight highway. At time $t = 0$, car 1 passes mile marker 0 traveling due east with a speed of 20.0 m/s. At the same time, car 2 is 1.0 km east of mile marker 0 traveling at 30.0 m/s due west. Car 1 is speeding up with an acceleration of magnitude 2.5 m/s², and car 2 is slowing down with an acceleration of magnitude 3.2 m/s². **(a)** Write x-versus-t equations of motion for both cars, taking east as the positive direction. **(b)** At what time do the cars pass next to one another?

53. •• **A Meteorite Strikes** On October 9, 1992, a 27-pound meteorite struck a car in Peekskill, NY, leaving a dent 22 cm deep in the trunk. If the meteorite struck the car with a speed of 130 m/s, what was the magnitude of its deceleration, assuming it to be constant?

54. •• A rocket blasts off and moves straight upward from the launch pad with constant acceleration. After 2.8 s the rocket is at a height of 91 m. **(a)** What are the magnitude and direction of the rocket's acceleration? **(b)** What is its speed at this time?

55. •• **Predict/Calculate** You are driving through town at 12.0 m/s when suddenly a ball rolls out in front of you. You apply the brakes and begin decelerating at 3.5 m/s². **(a)** How far do you travel before stopping? **(b)** When you have traveled only half the distance in part (a), is your speed 6.0 m/s, greater than 6.0 m/s, or less than 6.0 m/s? **(c)** Find the speed of your car when you have traveled half the distance in part (a).

56. •• Predict/Calculate You are driving through town at 16 m/s when suddenly a car backs out of a driveway in front of you. You apply the brakes and begin decelerating at 3.2 m/s². **(a)** How much time does it take to stop? **(b)** After braking half the time found in part (a), is your speed 8.0 m/s, greater than 8.0 m/s, or less than 8.0 m/s? **(c)** Find your speed after braking half the time found in part (a).

57. •• **BIO** Predict/Calculate **A Tongue's Acceleration** When a chameleon captures an insect, its tongue can extend 16 cm in 0.10 s. **(a)** Find the magnitude of the tongue's acceleration, assuming it to be constant. **(b)** In the first 0.050 s, does the tongue extend 8.0 cm, more than 8.0 cm, or less than 8.0 cm? **(c)** Find the extension of the tongue in the first 0.050 s.

It's not polite to reach! (Problem 57)

58. •• **BIO** **Surviving a Large Deceleration** On July 13, 1977, while on a test drive at Britain's Silverstone racetrack, the throttle on David Purley's car stuck wide open. The resulting crash subjected Purley to the greatest "g-force" ever survived by a human—he decelerated from 173 km/h to zero in a distance of only about 0.66 m. Calculate the magnitude of the acceleration experienced by Purley (assuming it to be constant), and express your answer in units of the acceleration of gravity, $g = 9.81$ m/s².

59. •• A boat is cruising in a straight line at a constant speed of 2.6 m/s when it is shifted into neutral. After coasting 12 m the engine is engaged again, and the boat resumes cruising at the reduced constant speed of 1.6 m/s. Assuming constant acceleration while coasting, **(a)** how much time did it take for the boat to coast 12 m? **(b)** What was the boat's acceleration while it was coasting? **(c)** What was the speed of the boat when it had coasted for 6.0 m? Explain.

60. •• A model rocket rises with constant acceleration to a height of 4.2 m, at which point its speed is 26.0 m/s. **(a)** How much time does it take for the rocket to reach this height? **(b)** What was the magnitude of the rocket's acceleration? **(c)** Find the height and speed of the rocket 0.10 s after launch.

61. •• The infamous chicken is dashing toward home plate with a speed of 5.7 m/s when he decides to hit the dirt. The chicken slides for 1.2 s, just reaching the plate as he stops (safe, of course). **(a)** What are the magnitude and direction of the chicken's acceleration? **(b)** How far did the chicken slide?

62. •• A bicyclist is finishing his repair of a flat tire when a friend rides by with a constant speed of 3.5 m/s. Two seconds later the bicyclist hops on his bike and accelerates at 2.4 m/s² until he catches his friend. **(a)** How much time does it take until he catches his friend? **(b)** How far has he traveled in this time? **(c)** What is his speed when he catches up?

63. •• A car in stop-and-go traffic starts at rest, moves forward 22 m in 8.0 s, then comes to rest again. The velocity-versus-time plot for this car is given in **FIGURE 2-37**. What distance does the car cover in **(a)** the first 4.0 seconds of its motion and **(b)** the last 2.0 seconds of its motion? **(c)** What is the constant speed V that characterizes the middle portion of its motion?

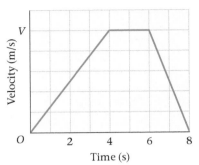

FIGURE 2-37 Problem 63

64. ••• A car and a truck are heading directly toward one another on a straight and narrow street, but they avoid a head-on collision by simultaneously applying their brakes at $t = 0$. The resulting velocity-versus-time graphs are shown in **FIGURE 2-38**. What is the separation between the car and the truck when they have come to rest, given that at $t = 0$ the car is at $x = 15$ m and the truck is at $x = -35$ m? (Note that this information determines which line in the graph corresponds to which vehicle.)

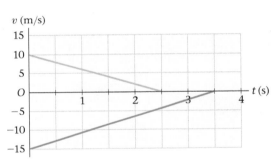

FIGURE 2-38 Problem 64

65. ••• Suppose you use videos to analyze the motion of penguins as they slide from rest down three different frictionless ramps (A, B, C) and into the water. The distances they travel and the times elapsed are recorded in the following table:

Ramp	A	B	C
distance, m	4.09	1.96	1.08
time, s	2.19	1.08	0.663

(a) Find the magnitude of the penguin's acceleration for each ramp. Assume the acceleration is uniform. **(b)** Determine the inclination angle θ above horizontal for each ramp, assuming that the accelerations for part (a) are in close agreement with the formula, $a = g \sin \theta$. (We will derive this formula in Chapter 5.)

SECTION 2-7 FREELY FALLING OBJECTS

66. • **CE** At the edge of a roof you throw ball 1 upward with an initial speed v_0; a moment later you throw ball 2 downward with the same initial speed. The balls land at the same time. Which of the following statements is true for the instant just before the balls hit the ground? **A.** The speed of ball 1 is greater than the speed of ball

2; **B.** The speed of ball 1 is equal to the speed of ball 2; **C.** The speed of ball 1 is less than the speed of ball 2.

67. • A cliff diver drops from rest to the water below. How many seconds does it take for the diver to go from 0 to 60; that is, to go from 0 mph to 60 mph? (For comparison, it takes about 3.5 s to 4.0 s for a powerful car to go from 0 to 60.)

68. • For a flourish at the end of her act, a juggler tosses a single ball high in the air. She catches the ball 3.2 s later at the same height from which it was thrown. What was the initial upward speed of the ball?

69. • **Soaring Shaun** During the 2014 Olympic games, snowboarder Shaun White rose 6.4 m vertically above the rim of the half-pipe. With what speed did he launch from the rim?

70. • **BIO** Gulls are often observed dropping clams and other shellfish from a height to the rocks below, as a means of opening the shells. If a seagull drops a shell from rest at a height of 17 m, how fast is the shell moving when it hits the rocks?

71. • A volcano launches a lava bomb straight upward with an initial speed of 28 m/s. Taking upward to be the positive direction, find the speed and direction of motion of the lava bomb **(a)** 2.0 seconds and **(b)** 3.0 seconds after it is launched.

72. • **An Extraterrestrial Volcano** The first active volcano observed outside the Earth was discovered in 1979 on Io, one of the moons of Jupiter. The volcano was observed to be ejecting material to a height of about 3.00×10^5 m. Given that the acceleration of gravity on Io is 1.80 m/s^2, find the initial velocity of the ejected material.

73. • **BIO** **Measure Your Reaction Time** Here's something you *can* try at home—an experiment to measure your reaction time. Have a friend hold a ruler by one end, letting the other end hang down vertically. At the lower end, hold your thumb and index finger on either side of the ruler, ready to grip it. Have your friend release the ruler without warning. Catch it as quickly as you can. If you catch the ruler 5.2 cm from the lower end, what is your reaction time?

How fast are your reactions? (Problem 73)

74. •• **CE** Predict/Explain A carpenter on the roof of a building accidentally drops her hammer. As the hammer falls it passes two windows of equal height, as shown in **FIGURE 2-39**. **(a)** Is the *increase* in speed of the hammer as it drops past window 1 greater than, less than, or equal to the *increase* in speed as it drops past window 2? **(b)** Choose the *best explanation* from among the following:
   I. The greater speed at window 2 results in a greater increase in speed.
   II. Constant acceleration means the hammer speeds up the same amount for each window.
   III. The hammer spends more time dropping past window 1.

FIGURE 2-39 Problem 74

75. •• **CE** Predict/Explain **FIGURE 2-40** shows a *v*-versus-*t* plot for the hammer dropped by the carpenter in Problem 74. Notice that the times when the hammer passes the two windows are indicated by shaded areas. **(a)** Is the area of the shaded region corresponding to window 1 greater than, less than, or equal to the area of the shaded region corresponding to window 2? **(b)** Choose the *best explanation* from among the following:

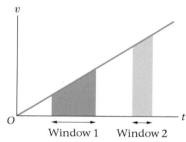

FIGURE 2-40 Problem 75

   I. The shaded area for window 2 is higher than the shaded area for window 1.
   II. The windows are equally tall.
   III. The shaded area for window 1 is wider than the shaded area for window 2.

76. •• **CE** A ball is thrown straight upward with an initial speed v_0. When it reaches the top of its flight at height h, a second ball is thrown straight upward with the same initial speed. Do the balls cross paths at height $\frac{1}{2}h$, above $\frac{1}{2}h$, or below $\frac{1}{2}h$?

77. •• On a hot summer day in the state of Washington while kayaking, I saw several swimmers jump from a railroad bridge into the Snohomish River below. The swimmers stepped off the bridge, and I estimated that they hit the water 1.5 s later. **(a)** How high was the bridge? **(b)** How fast were the swimmers moving when they hit the water? **(c)** What would the swimmers' drop time be if the bridge were twice as high?

78. •• **Highest Water Fountain** The USA's highest fountain of water is located, appropriately enough, in Fountain Hills, Arizona. The fountain rises to a height of 560 ft (5 feet higher than the Washington Monument). **(a)** What is the initial speed of the water? **(b)** How much time does it take for water to reach the top of the fountain?

79. •• Wrongly called for a foul, an angry basketball player throws the ball straight down to the floor. If the ball bounces straight up and returns to the floor 3.2 s after first striking it, what was the ball's greatest height above the floor?

80. •• To celebrate a victory, a pitcher throws her glove straight upward with an initial speed of 6.5 m/s. **(a)** How much time does it take for the glove to return to the pitcher? **(b)** How much time does it take for the glove to reach its maximum height?

81. •• Predict/Calculate Standing at the edge of a cliff 30.5 m high, you drop a ball. Later, you throw a second ball downward with an initial speed of 11.2 m/s. **(a)** Which ball has the greater *increase* in speed when it reaches the base of the cliff, or do both balls speed up by the same amount? **(b)** Verify your answer to part (a) with a calculation.

82. •• You shoot an arrow into the air. Two seconds later (2.00 s) the arrow has gone straight upward to a height of 30.0 m above its launch point. **(a)** What was the arrow's initial speed? **(b)** How much time did it take for the arrow to first reach a height of 15.0 m above its launch point?

83. •• While riding on an elevator descending with a constant speed of 3.0 m/s, you accidentally drop a book from under your arm. **(a)** How much time does it take for the book to reach the elevator floor, 1.2 m below your arm? **(b)** What is the book's speed relative to you when it hits the elevator floor?

84. •• A hot-air balloon is descending at a rate of 2.3 m/s when a passenger drops a camera. If the camera is 41 m above the ground when it is dropped, **(a)** how much time does it take for the camera to reach the ground, and **(b)** what is its velocity just before it lands? Let upward be the positive direction for this problem.

85. •• A model rocket blasts off and moves upward with an acceleration of 12 m/s² until it reaches a height of 29 m, at which point its engine shuts off and it continues its flight in free fall. **(a)** What is the maximum height attained by the rocket? **(b)** What is the speed of the rocket just before it hits the ground? **(c)** What is the total duration of the rocket's flight?

86. •• **BIO** The southern flying squirrel (*Glaucomys volans*) can reduce its free-fall acceleration by extending a furry membrane called a *patagium* and gliding through the air. **FIGURE 2-41** shows the vertical position-versus-time plot of one squirrel's flight. Use the information given in the plot to determine the acceleration of the squirrel. Assume that its initial height is $x_0 = 4.0$ m, its initial vertical speed is $v_0 = 0$, and the elapsed time is $t = 1.10$ s.

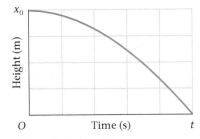

FIGURE 2-41 Problem 86

87. ••• **Hitting the "High Striker"** A young woman at a carnival steps up to the "high striker," a popular test of strength where the contestant hits one end of a lever with a mallet, propelling a small metal plug upward toward a bell. She gives the mallet a mighty swing and sends the plug to the top of the striker, where it rings the bell. **FIGURE 2-42** shows the corresponding position-versus-time plot for the plug. Using the information given in the plot, answer the following questions: **(a)** What is the average speed of the plug during its upward journey? **(b)** By how much does the speed of the plug decrease during its upward journey? **(c)** What is the initial speed of

the plug? (Assume the plug is in free fall during its upward motion, with no effects of air resistance or friction.)

FIGURE 2-42 Problem 87

88. ••• While sitting on a tree branch 10.0 m above the ground, you drop a chestnut. When the chestnut has fallen 2.5 m, you throw a second chestnut straight down. What initial speed must you give the second chestnut if they are both to reach the ground at the same time?

GENERAL PROBLEMS

89. • An astronaut on the Moon drops a rock straight downward from a height of 1.25 m. If the acceleration of gravity on the Moon is 1.62 m/s², what is the speed of the rock just before it lands?

90. • **Taipei 101** An elevator in the Taipei 101 skyscraper can start from rest and accelerate to a maximum speed of 16 m/s with a constant acceleration of 1.1 m/s². Through what distance does the elevator move as it accelerates to its maximum speed?

91. • **A Supersonic Waterfall** Geologists have learned of periods in the past when the Strait of Gibraltar closed off, and the Mediterranean Sea dried out and become a desert. Later, when the strait reopened, a massive saltwater waterfall was created. According to geologists, the water in this waterfall was supersonic; that is, it fell with speeds in excess of the speed of sound. Ignoring air resistance, what is the minimum height necessary to create a supersonic waterfall? (The speed of sound may be taken to be 340 m/s.)

92. • A juggler throws a ball straight up into the air. If the ball returns to its original position in the juggler's hand in 0.98 s, what was its maximum height above the juggler's hand?

93. •• **CE** At the edge of a roof you drop ball A from rest, and at the same instant your friend throws ball B upward from the ground with an initial velocity of v_0 sufficient to reach your position on the roof. What are the signs (positive or negative) of the velocity and the acceleration of each ball at a point halfway between the ground and the roof? Assume upward is the positive direction.

94. •• **CE** Two balls start their motion at the same time, with ball A dropped from rest and ball B thrown upward with an initial speed v_0. Which of the five plots shown in **FIGURE 2-43** corresponds to **(a)** ball A and **(b)** ball B?

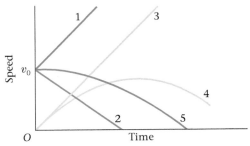

FIGURE 2-43 Problem 94

95. •• **CE** Refer to the position-versus-time plot in **FIGURE 2-44** for the following questions: **(a)** Is the average speed for the time interval between points 1 and 2 greater than, less than, or equal to the average speed for the time interval between points 1 and 3? Explain. **(b)** Is the average velocity for the time interval between points 2 and 4 greater than, less than, or equal to the average velocity for the time interval between points 3 and 4? Explain.

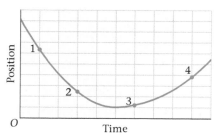

FIGURE 2-44 Problem 95

96. •• **Drop Tower** NASA operates a 2.2-second drop tower at the Glenn Research Center in Cleveland, Ohio. At this facility, experimental packages are dropped from the top of the tower, on the 8th floor of the building. During their 2.2 seconds of free fall, the packages experience a microgravity environment similar to that of a spacecraft in orbit. **(a)** What is the drop distance of a 2.2-s tower? **(b)** How fast are the packages traveling when they hit the air bags at the bottom of the tower? **(c)** If an experimental package comes to rest over a distance of 0.75 m upon hitting the air bags, what is the average stopping acceleration?

97. •• The velocity-versus-time graph for an object moving in a straight line is given in **FIGURE 2-45**. **(a)** What is the acceleration of this object? **(b)** If this object is at $x = 12.0$ m at time $t = 0$, what is its position at $t = 1.00$ s? **(c)** Assuming the acceleration of the object remains constant, what is its position at $t = 5.00$ s?

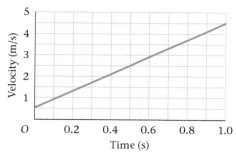

FIGURE 2-45 Problem 97

98. •• At the 18th green of the U.S. Open you need to make a 23.5-ft putt to win the tournament. When you hit the ball, giving it an initial speed of 1.54 m/s, it stops 6.00 ft short of the hole. **(a)** Assuming the deceleration caused by the grass is constant, what should the initial speed have been to just make the putt? **(b)** What initial speed do you need to make the remaining 6.00-ft putt?

99. •• A glaucous-winged gull, ascending straight upward at 5.20 m/s, drops a shell when it is 12.5 m above the ground. **(a)** What are the magnitude and direction of the shell's acceleration just after it is released? **(b)** Find the maximum height above the ground reached by the shell. **(c)** How much time does it take for the shell to reach the ground? **(d)** What is the speed of the shell just before it hits the ground?

100. •• A doctor, preparing to give a patient an injection, squirts a small amount of liquid straight upward from a syringe. If the liquid emerges with a speed of 1.5 m/s, **(a)** how much time does it take for it to return to the level of the syringe? **(b)** What is the maximum height of the liquid above the syringe?

101. •• A hot-air balloon has just lifted off and is rising at the constant rate of 2.0 m/s. Suddenly one of the passengers realizes she has left her camera on the ground. A friend picks it up and tosses it straight upward with an initial speed of 13 m/s. If the passenger is 2.5 m above her friend when the camera is tossed, how high is she when the camera reaches her?

102. •• Astronauts on a distant planet throw a rock straight upward and record its motion with a video camera. After digitizing their video, they are able to produce the graph of height, y, versus time, t, shown in **FIGURE 2-46**. **(a)** What is the acceleration of gravity on this planet? **(b)** What was the initial speed of the rock?

FIGURE 2-46 Problem 102

103. •• **BIO** **A Jet-Propelled Squid** Squids can move through the water using a form of jet propulsion. Suppose a squid jets forward from rest with constant acceleration for 0.170 s, moving through a distance of 0.179 m. The squid then turns off its jets and coasts to rest with constant acceleration. The total time for this motion (from rest to rest) is 0.400 s, and the total distance covered is 0.421 m. What is the magnitude of the squid's acceleration during **(a)** the time it is jetting and **(b)** the time it coasts to a stop?

104. ••• A ball, dropped from rest, covers three-quarters of the distance to the ground in the last second of its fall. **(a)** From what height was the ball dropped? **(b)** What was the total time of fall?

105. ••• You drop a ski glove from a height h onto fresh snow, and it sinks to a depth d before coming to rest. **(a)** In terms of g and h, what is the speed of the glove when it reaches the snow? **(b)** What are the magnitude and direction of the glove's acceleration as it moves through the snow, assuming it to be constant? Give your answer in terms of g, h, and d.

106. ••• To find the height of an overhead power line, you throw a ball straight upward. The ball passes the line on the way up after 0.75 s, and passes it again on the way down 1.5 s after it was tossed. What are the height of the power line and the initial speed of the ball?

107. ••• Sitting in a second-story apartment, a physicist notices a ball moving straight upward just outside her window. The ball is visible for 0.25 s as it moves a distance of 1.05 m from the bottom to the top of the window. **(a)** How much time does it take before the ball reappears? **(b)** What is the greatest height of the ball above the top of the window?

PASSAGE PROBLEMS

Bam!—*Apollo 15* Lands on the Moon

The first word spoken on the surface of the Moon after *Apollo 15* landed on July 30, 1971, was "Bam!" This was James Irwin's involuntary reaction to their rather bone-jarring touchdown. "We did hit harder than any of the other flights!" says Irwin. "And I was startled, obviously, when I said, 'Bam!'"

The reason for the "firm touchdown" of *Apollo 15*, as pilot David Scott later characterized it, was that the rocket engine was shut off a bit earlier than planned, when the lander was still 4.30 ft above the lunar surface and moving downward with a speed of 0.500 ft/s. From that point on the lander descended in lunar free fall, with an acceleration of 1.62 m/s². As a result, the landing speed of *Apollo 15* was by far the largest of any of the *Apollo* missions. In comparison, Neil Armstrong's landing speed on *Apollo 11* was the lowest at 1.7 ft/s—he didn't shut off the engine until the footpads were actually on the surface. *Apollos 12, 14,* and *17* all landed with speeds between 3.0 and 3.5 ft/s.

To better understand the descent of *Apollo 15*, we show its trajectory during the final stages of landing in **FIGURE 2-47 (a)**. In **FIGURE 2-47 (b)** we show a variety of speed-versus-time plots.

(a)

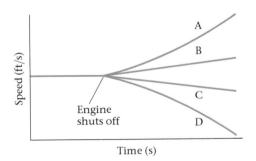

(b)

FIGURE 2-47 Problems 108, 109, 110, and 111

---

108. • How much time did it take for the lander to drop the final 4.30 ft to the Moon's surface?

A. 1.18 s   B. 1.37 s
C. 1.78 s   D. 2.36 s

109. •• What was the impact speed of the lander when it touched down? Give your answer in feet per second (ft/s), the same units used by the astronauts.

A. 2.41 ft/s   B. 6.78 ft/s
C. 9.95 ft/s   D. 10.6 ft/s

110. • Which of the speed-versus-time plots in Figure 2-48 (b) correctly represents the speed of the *Apollo 15* lander?

A   B   C   D

111. • Suppose, instead of shutting off the engine, the astronauts had increased its thrust, giving the lander a small, but constant, upward acceleration. Which speed-versus-time plot in Figure 2-48 (b) would describe this situation?

A   B   C   D

112. •• **Referring to Example 2-17** Suppose the speeder (red car) is traveling with a constant speed of 25 m/s, and the maximum acceleration of the police car (blue car) is 3.8 m/s². If the police car is to start from rest and catch the speeder in 15 s or less, what is the maximum head-start distance the speeder can have? Measure time from the moment the police car starts.

113. •• **Referring to Example 2-17** The speeder passes the position of the police car with a constant speed of 15 m/s. The police car immediately starts from rest and pursues the speeder with constant acceleration. What acceleration must the police car have if it is to catch the speeder in 7.0 s? Measure time from the moment the police car starts.

114. •• Predict/Calculate **Referring to Example 2-21** (a) In Example 2-21, the bag of sand is released at 20.0 m and reaches a maximum height of 22 m. If the bag had been released at 30.0 m instead, with everything else remaining the same, would its maximum height be 32 m, greater than 32 m, or less than 32 m? (b) Find the speed of the bag just before it lands when it is released from 30.0 m.

115. •• **Referring to Example 2-21** Suppose the balloon is descending with a constant speed of 4.2 m/s when the bag of sand comes loose at a height of 35 m. (a) What is the amount of time the bag is in the air? (b) What is the speed of the bag when it is 15 m above the ground?

Vectors in Physics

1 A vector has both a magnitude and a direction.

2 The component of a vector is a measure of how much it extends in a certain direction.

3 The most precise way to add or subtract vectors is to add or subtract their components.

4 Unit vectors are dimensionless, have a magnitude of 1, and point in a specific direction.

5 Position, displacement, velocity, and acceleration are all vectors.

6 Relative-motion problems are an application of vector addition and subtraction.

▲ The points of the compass have long been used as a framework for indicating directions. In physics, we more frequently indicate directions in terms of *x* and *y* rather than N, S, E, and W. Either way, specifying a direction as well as a magnitude is essential to defining one of the physicist's most useful tools, the vector.

O f all the mathematical tools used in this book, perhaps none is more important than the vector. In this chapter we discuss what a vector is, how it differs from a scalar, and how it can describe a physical quantity.

3-1 Scalars Versus Vectors

Numbers can represent many quantities in physics. For example, a numerical value, together with the appropriate units, can specify the volume of a container, the temperature of the air, or the time of an event. In physics, a number with its units is referred to as a **scalar**:

> A scalar is a number with units. It can be positive, negative, or zero.

Sometimes, however, a scalar isn't enough to adequately describe a physical quantity—in many cases, a direction is needed as well. For example, suppose you're walking in an unfamiliar city and you want directions to the library. You ask a passerby, "Do you know where the library is?" If the person replies "Yes," and walks on, he hasn't been too helpful. If he says, "Yes, it is half a mile from here," that is more helpful, but you still don't know where it is. The library could be anywhere on a circle of radius one-half mile, as shown in FIGURE 3-1. To pin down the location, you need a reply such as, "Yes, the library is half a mile northwest of here." With both a distance *and* a direction, you know the location of the library.

Thus, if you walk northwest for half a mile you arrive at the library, as indicated by the upper left arrow in Figure 3-1. The arrow points in the direction traveled, and its **magnitude,** 0.5 mi in this case, represents the distance covered. In general, a quantity that is specified by both a *magnitude* and a *direction* is represented by a **vector**:

> A vector is a mathematical quantity with both a magnitude and a direction.

In the example of walking to the library, the vector corresponding to the trip is the displacement vector. Additional examples of displacement vectors are given on the directional signpost shown in FIGURE 3-2.

Symbols Used to Represent Vectors   When we indicate a vector on a diagram or a sketch, we draw an arrow, as in Figure 3-1. To indicate a vector with a written symbol, we use **boldface** for the vector itself, with a small arrow above it to remind us of its vector nature, and we use *italics* to indicate its magnitude. For example, the upper-left vector in Figure 3-1 is designated by the symbol $\vec{\mathbf{r}}$, and its magnitude is $r = 0.5$ mi. (When we represent a vector in a graph, we sometimes label it with the corresponding boldface symbol, and sometimes with the appropriate magnitude.) It is common in handwritten material to draw a small arrow over the vector's symbol, which is very similar to the way vectors are represented in this text.

Enhance Your Understanding   (Answers given at the end of the chapter)

1. Is it possible for two vectors to be different from one another, but still have the same magnitude? State why if your answer is no; give an example if your answer is yes.

Section Review

- A scalar is a number with a magnitude and units, but no direction.
- A vector has both a magnitude and a direction.

3-2 The Components of a Vector

When we discussed the directions for finding a library in the previous section, we pointed out that knowing the magnitude and direction angle—0.5 mi northwest—gives its precise location. We left out one key ingredient in actually *getting* to the library, however. In most cities it would not be possible to simply walk in a straight line for 0.5 mi directly to the library, because to do so would take you through buildings where there are no doors, through people's backyards, and through all kinds of other obstructions. In fact, if the city streets are laid out along north-south and east-west directions, you might instead walk west for a certain distance, then turn and proceed north an equal distance until you reach the library, as illustrated in FIGURE 3-3. What you have

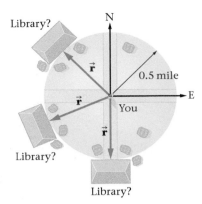

▲ FIGURE 3-1 **Distance and direction** If you know only that the library is 0.5 mi from you, it could lie anywhere on a circle of radius 0.5 mi. If, instead, you are told that the library is 0.5 mi northwest, you know its precise location.

Big Idea **1** A vector has both a magnitude and a direction; a scalar is simply a number with appropriate units.

▲ FIGURE 3-2 This directional signpost gives both a distance and a direction to each city. In effect, the sign defines a displacement vector for each of these destinations.

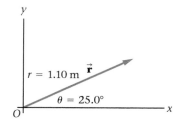

▲ FIGURE 3-3 **A walk along city streets to the library** By taking the indicated path, we have "resolved" the vector $\vec{r}$ into east-west and north-south components.

▲ FIGURE 3-4 **A two-dimensional coordinate system** The positive x and y directions are indicated in this shorthand form.

Big Idea **2** The component of a vector is a measure of how much it extends in a certain direction.

▶ FIGURE 3-5 **A vector and its scalar components** **(a)** The vector $\vec{r}$ is defined by its length ($r = 1.10$ m) and its direction angle ($\theta = 25.0°$) measured counterclockwise from the positive x axis. **(b)** Alternatively, the vector $\vec{r}$ can be defined by its x component, $r_x = 0.997$ m, and its y component, $r_y = 0.465$ m.

just done is "resolved" the displacement vector $\vec{r}$, between you and the library into east-west and north-south "components."

Defining the Components of a Vector In general, to find the components of a vector we need to set up a coordinate system. In two dimensions we choose an origin, O, and a positive direction for both the x and the y axes, as in FIGURE 3-4. If the system were three-dimensional, we would also indicate a z axis.

Now, a vector is defined by its magnitude (indicated by the length of the arrow representing the vector) and its direction. For example, suppose an ant leaves its nest at the origin and, after foraging for some time, is at the location given by the vector $\vec{r}$, in FIGURE 3-5 (a). This vector has a magnitude $r = 1.10$ m and points in a direction $\theta = 25.0°$ above the x axis. Equivalently, $\vec{r}$, can be defined by saying that it extends a distance r_x in the x direction and a distance r_y in the y direction, as shown in FIGURE 3-5 (b). The quantities r_x and r_y are referred to as the x and y **scalar components** of the vector $\vec{r}$.

Converting from Magnitude and Angle to Components We can find r_x and r_y by using standard trigonometric relationships, as summarized in the Problem-Solving Note on this page. Referring to Figure 3-5 (b), we see that r_x and r_y have the following values:

$$r_x = r\cos 25.0° = (1.10\text{ m})(0.906) = 0.997\text{ m}$$

$$r_y = r\sin 25.0° = (1.10\text{ m})(0.423) = 0.465\text{ m}$$

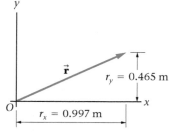

(a) A vector defined in terms of its length and direction angle

(b) The same vector defined in terms of its x and y components

Thus, we can say that the ant's final displacement is equivalent to what it would be if the ant had simply walked 0.997 m in the x direction and then 0.465 m in the y direction.

Converting from Components to Magnitude and Direction Sometimes you are given the components of a vector, and from them you would like to find the corresponding magnitude and direction. As an example, let's start with the components $r_x = 0.997$ m and $r_y = 0.465$ m, just determined above, and verify that $r = 1.10$ m and $\theta = 25.0°$. First, notice that r_x, r_y, and r form a right triangle with r as the hypotenuse. Thus, we can use the Pythagorean theorem (see Appendix A) to find r in terms of r_x and r_y. This gives the expected result:

$$r = \sqrt{r_x^2 + r_y^2} = \sqrt{(0.997\text{ m})^2 + (0.465\text{ m})^2} = \sqrt{1.21\text{ m}^2} = 1.10\text{ m}$$

Second, we can use any two sides of the triangle to obtain the angle θ, as shown in the next three calculations:

$$\theta = \sin^{-1}\left(\frac{0.465\text{ m}}{1.10\text{ m}}\right) = \sin^{-1}(0.423) = 25.0°$$

$$\theta = \cos^{-1}\left(\frac{0.997\text{ m}}{1.10\text{ m}}\right) = \cos^{-1}(0.906) = 25.0°$$

$$\theta = \tan^{-1}\left(\frac{0.465\text{ m}}{0.997\text{ m}}\right) = \tan^{-1}(0.466) = 25.0°$$

The final expression, the one involving the inverse tangent ($\tan^{-1}$), is the one that is most commonly used.

In some situations we know a vector's magnitude and direction; in other cases we are given the vector's components. You will find it useful to be able to convert quickly and easily from one description of a vector to the other using trigonometric functions and the Pythagorean theorem.

PROBLEM-SOLVING NOTE

A Vector and Its Components

Given the magnitude and direction of a vector, find its components:

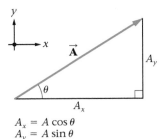

$A_x = A \cos\theta$
$A_y = A \sin\theta$

Given the components of a vector, find its magnitude and direction:

$$A = \sqrt{A_x^2 + A_y^2}$$
$$\theta = \tan^{-1}\frac{A_y}{A_x}$$

PHYSICS IN CONTEXT
Looking Back

In Chapter 2 we indicated the direction of a vector with a plus sign ($+$) or a minus sign ($-$), because only two directions were possible. With the results from this chapter we can now deal with vectors that point in any direction.

EXAMPLE 3-1 ZIP-A-DEE ZIPLINE

RWP* A neighborhood zipline is set up from a large tree to the ground. The zipline makes an angle of $\theta = 18°$ with the horizontal, and is anchored to the ground 12 m from the base of the tree. **(a)** What is the height of the top of the zipline above the ground? **(b)** What length of wire is needed to make the zipline?

PICTURE THE PROBLEM
Our sketch shows an adventurer heading toward the ground on the zipline. The angle of the zipline relative to the horizontal is $\theta = 18°$, and the horizontal distance from the base of the tree to the anchor point is $r_x = 12$ m. The opposite side of the zipline triangle is the height to the top of the zipline, r_y, and the hypotenuse is the length of the zipline, r.

REASONING AND STRATEGY
The tangent of θ is the height of the zipline triangle divided by the base: $\tan\theta = r_y/r_x$. We know both θ and the base of the triangle, r_x, and hence we can rearrange this relationship to find the height r_y. Similarly, the length of the zipline can be obtained by solving $\cos\theta = r_x/r$ for r.

Known Angle of zipline, $\theta = 18°$; horizontal distance to the tree, $r_x = 12$ m.
Unknown (a) Height of zipline, $r_y = ?$ **(b)** Length of zipline, $r = ?$

CONTINUED

---

*Real World Physics applications are denoted by the acronym RWP.

SOLUTION

Part (a)

1. Use $\tan \theta = r_y/r_x$ to solve for the height of the zipline, r_y:

$$r_y = r_x \tan \theta = (12 \text{ m}) \tan 18° = 3.9 \text{ m}$$

Part (b)

2. Use $\cos \theta = r_x/r$ to solve for the length of the zipline:

$$r = \frac{r_x}{\cos \theta} = \frac{12 \text{ m}}{\cos 18°} = 13 \text{ m}$$

INSIGHT

Another way to solve part (b) is to use the Pythagorean theorem: $r = \sqrt{r_x^2 + r_y^2} = \sqrt{(12 \text{ m})^2 + (3.9 \text{ m})^2} = 13 \text{ m}$. Thus, if we let $\vec{\mathbf{r}}$ denote the vector from the bottom to the top of the zipline, its magnitude is 13 m and its direction above the horizontal is 18°. Equivalently, the x component of $\vec{\mathbf{r}}$ is 12 m and its y component is 3.9 m.

PRACTICE PROBLEM — PREDICT/CALCULATE

(a) If the zipline is anchored to the ground 24 m from the base of the tree, but the top is at the same location, is the angle to the top of the zipline greater than, less than, or equal to 18°? Explain. **(b)** Find the angle to the top of the zipline in this case.   [**Answer: (a)** The angle is less than 18° because the base of the triangle is greater, while the height remains the same. **(b)** The angle is $\theta = \tan^{-1}\left(\dfrac{3.9 \text{ m}}{24 \text{ m}}\right) = 9.2°$, which is less than 18°, as expected.]

Some related homework problems: Problem 5, Problem 19

EXERCISE 3-2 COMPONENTS, MAGNITUDE, AND DIRECTION

a. Find the components A_x and A_y for the vector $\vec{\mathbf{A}}$, whose magnitude and direction are given by $A = 3.5$ m and $\theta = 66°$, respectively.

b. Find B and θ for the vector $\vec{\mathbf{B}}$ with components $B_x = 75.5$ m and $B_y = 6.20$ m.

REASONING AND SOLUTION

(a) Multiply the magnitude A by $\cos \theta$ to find the x component; multiply the magnitude by $\sin \theta$ to find the y component. **(b)** Use the Pythagorean theorem to find the magnitude B, and the inverse tangent to find the angle θ.

a. $A_x = (3.5 \text{ m}) \cos 66° = 1.4 \text{ m}$, $A_y = (3.5 \text{ m}) \sin 66° = 3.2 \text{ m}$

b. $B = \sqrt{(75.5 \text{ m})^2 + (6.20 \text{ m})^2} = 75.8 \text{ m}$, $\theta = \tan^{-1}\left(\dfrac{6.20 \text{ m}}{75.5 \text{ m}}\right) = 4.69°$

Determining the Sign of Vector Components How do we determine the correct signs for the x and y components of a vector? This is done by considering the right triangle formed by A_x, A_y, and $\vec{\mathbf{A}}$, as shown in **FIGURE 3-6**. To determine the sign of A_x, start at the tail of the vector and move along the x axis toward the right angle. If you are moving in the positive x direction, then A_x is positive ($A_x > 0$); if you are moving in the negative x direction, then A_x is negative ($A_x < 0$). For the y component, start at the right angle and move toward the tip of the arrow. A_y is positive or negative depending on whether you are moving in the positive or negative y direction.

For example, consider the vector shown in **FIGURE 3-7 (a)**. In this case, $A_x > 0$ and $A_y < 0$, as indicated in the figure. Similarly, the signs of A_x and A_y are given in **FIGURE 3-7 (b, c, d)** for the vectors shown there. Be sure to verify each of these cases by applying the rules just given. As we continue our study of physics, it's important to be able to find the components of a vector *and* to assign to them the correct signs.

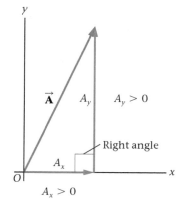

▲ **FIGURE 3-6 A vector whose *x* and *y* components are positive**

EXERCISE 3-3 FIND THE COMPONENTS

The vector $\vec{\mathbf{A}}$ has a magnitude of 9.75 m. Find its components for direction angles of

a. $\theta = 7.25°$ **c.** $\theta = 235°$

b. $\theta = 115°$ **d.** $\theta = 355°$

CONTINUED

REASONING AND SOLUTION

Multiplying the magnitude A by $\cos\theta$ gives the x component; multiplying the magnitude by $\sin\theta$ gives the y component.

a. $A_x = (9.75\text{ m})\cos 7.25° = 9.67\text{ m}$, $A_y = (9.75\text{ m})\sin 7.25° = 1.23\text{ m}$

b. $A_x = (9.75\text{ m})\cos 115° = -4.12\text{ m}$, $A_y = (9.75\text{ m})\sin 115° = 8.84\text{ m}$

c. $A_x = (9.75\text{ m})\cos 235° = -5.59\text{ m}$, $A_y = (9.75\text{ m})\sin 235° = -7.99\text{ m}$

d. $A_x = (9.75\text{ m})\cos 355° = 9.71\text{ m}$, $A_y = (9.75\text{ m})\sin 355° = -0.850\text{ m}$

Using Care with a Calculator Be careful when using your calculator to determine the direction angle, θ, because you may need to add 180° to get the correct angle. For example, if $A_x = -0.50\text{ m}$ and $A_y = 1.0\text{ m}$, your calculator will give the following result:

$$\theta = \tan^{-1}\left(\frac{1.0\text{ m}}{-0.50\text{ m}}\right) = \tan^{-1}(-2.0) = -63°$$

Does this angle correspond to the specified vector? The way to check is to sketch $\vec{A}$. When you do, your drawing is similar to Figure 3-7 (c), and thus the direction angle of $\vec{A}$ should be between 90° and 180°. To obtain the correct angle, add 180° to the calculator's result:

$$\theta = -63° + 180° = 117°$$

This, in fact, is the direction angle for the vector $\vec{A}$. Why the need to add 180°? Well, the tangents of θ and $(\theta + 180°)$ are the same for any θ. A calculator doesn't know which of these answers is the one that's appropriate for your situation—you have to make that decision.

EXERCISE 3-4 DIRECTION ANGLE

The vector $\vec{B}$ has components $B_x = -2.10\text{ m}$ and $B_y = -1.70\text{ m}$. Find the direction angle, θ, for this vector.

REASONING AND SOLUTION

The direction angle is found using the inverse tangent. If the resulting angle is not appropriate for the vector, it may be necessary to add 180°. This is illustrated in the following calculations:

$$\tan^{-1}[(-1.70\text{ m})/(-2.10\text{ m})] = \tan^{-1}(1.70/2.10) = 39.0°, \theta = 39.0° + 180° = 219°$$

Enhance Your Understanding (Answers given at the end of the chapter)

2. A vector has the components $A_x = 5\text{ m}$ and $A_y = 3\text{ m}$. (a) Is the magnitude of $\vec{A}$ greater than, less than, or equal to 8 m? Explain. (b) Is the direction angle of $\vec{A}$ greater than, less than, or equal to 45°? Explain.

Section Review

- The x and y components of a vector tell how far the vector extends in those directions. The components are found using the sine and cosine functions.

- The magnitude of a vector is found from its components by using the Pythagorean theorem. The angle is found from the components by using the inverse tangent.

3-3 Adding and Subtracting Vectors

One important reason for determining the components of a vector is that they are useful in adding and subtracting vectors. In this section we begin by defining vector addition graphically, and then show how the same addition can be performed more concisely and accurately with components.

(a)

(b)

(c)

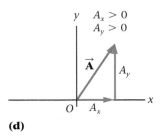

(d)

▲ FIGURE 3-7 **Examples of vectors with components of different signs** To determine the signs of a vector's components, it is only necessary to observe the direction in which they point. For example, in part (a) the x component points in the positive direction; hence $A_x > 0$. Similarly, the y component in part (a) points in the negative y direction; therefore $A_y < 0$.

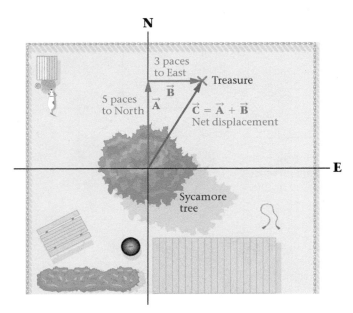

▲ **FIGURE 3-8 The sum of two vectors** To go from the sycamore tree to the treasure, one must first go 5 paces north ($\vec{A}$) and then 3 paces east ($\vec{B}$). The net displacement from the tree to the treasure is $\vec{C} = \vec{A} + \vec{B}$.

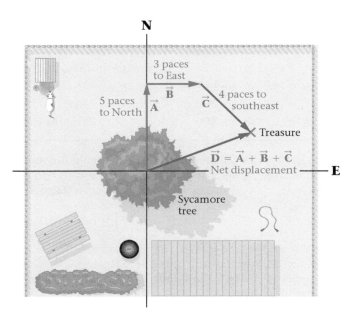

▲ **FIGURE 3-9 Adding several vectors** Searching for a treasure that is 5 paces north ($\vec{A}$), 3 paces east ($\vec{B}$), and 4 paces southeast ($\vec{C}$) of the sycamore tree. The net displacement from the tree to the treasure is $\vec{D} = \vec{A} + \vec{B} + \vec{C}$.

Adding Vectors Graphically

One day you open an old chest in the attic and find a treasure map inside. To locate the buried treasure, the map says that you must "Go to the sycamore tree in the backyard, march 5 paces north, then 3 paces east, and dig there." If these two displacements are represented by the vectors $\vec{A}$ and $\vec{B}$ in **FIGURE 3-8**, the total displacement from the tree to the treasure is given by the vector $\vec{C}$. We say that $\vec{C}$ is the *vector sum* of $\vec{A}$ and $\vec{B}$; that is, $\vec{C} = \vec{A} + \vec{B}$. In general, vectors are added graphically according to the following rule:

> To add the vectors $\vec{A}$ and $\vec{B}$, place the tail of $\vec{B}$ at the head of $\vec{A}$. The sum, $\vec{C} = \vec{A} + \vec{B}$, is the vector extending from the tail of $\vec{A}$ to the head of $\vec{B}$.

If the instructions to find the treasure were a bit more complicated—5 paces north, 3 paces east, then 4 paces southeast, for example—the path from the sycamore tree to the treasure would be like that shown in **FIGURE 3-9**. In this case, the total displacement, $\vec{D}$, is the sum of the three vectors $\vec{A}$, $\vec{B}$, and $\vec{C}$; that is, $\vec{D} = \vec{A} + \vec{B} + \vec{C}$. It follows that to add more than two vectors, we just keep placing the vectors head-to-tail, head-to-tail, and then draw a vector from the tail of the first vector to the head of the last vector, as in Figure 3-9.

In order to place a given pair of vectors head-to-tail, it may be necessary to move the corresponding arrows. This is fine, as long as you don't change their length or their direction. After all, a vector is defined by its length and direction; if these are unchanged, so is the vector.

> A vector is defined by its magnitude and direction, regardless of its location.

For example, in **FIGURE 3-10** all of the vectors are the same, even though they are at different locations on the graph. Similarly, the geese in **FIGURE 3-11** all have the same velocity vectors, even though they are in different locations.

As an example of moving vectors to different locations, consider two vectors, $\vec{A}$ and $\vec{B}$, and their vector sum, $\vec{C}$, as illustrated in **FIGURE 3-12 (a)**:

$$\vec{C} = \vec{A} + \vec{B}$$

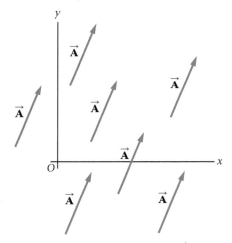

▲ **FIGURE 3-10 Identical vectors $\vec{A}$ at different locations** A vector is defined by its direction and length; its location is immaterial.

◄ **FIGURE 3-11** To a good approximation, these snow geese are all moving in the same direction with the same speed. As a result, their velocity vectors are equal, even though their positions are different.

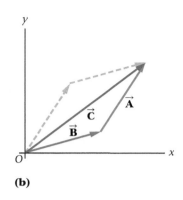

By moving the arrow representing $\vec{B}$ so that its tail is at the origin, and moving the arrow for $\vec{A}$ so that its tail is at the head of $\vec{B}$, we obtain the construction shown in **FIGURE 3-12 (b)**. From this graph we see that $\vec{C}$, which is $\vec{A} + \vec{B}$, is also equal to $\vec{B} + \vec{A}$:

$$\vec{C} = \vec{A} + \vec{B} = \vec{B} + \vec{A}$$

That is, the sum of vectors is independent of the order in which the vectors are added.

Now, suppose that $\vec{A}$ has a magnitude of 5.00 m and a direction angle of 60.0° above the x axis, and that $\vec{B}$ has a magnitude of 4.00 m and a direction angle of 20.0° above the x axis. These two vectors and their sum, $\vec{C}$, are shown in **FIGURE 3-13**. The question is: What are the length and direction angle of $\vec{C}$?

A graphical way to answer this question is to simply measure the length and direction of $\vec{C}$ in Figure 3-13. With a ruler, we find the length of $\vec{C}$ to be approximately 1.75 times the length of $\vec{A}$, which means that $\vec{C}$ is roughly 1.75 (5.00 m) = 8.75 m. Similarly, with a protractor we measure the angle θ to be about 45.0° above the x axis.

Adding Vectors Using Components

The graphical method of adding vectors yields approximate results, limited by the accuracy with which the vectors can be drawn and measured. In contrast, exact results can be obtained by adding $\vec{A}$ and $\vec{B}$ in terms of their components. To see how this is done, consider **FIGURE 3-14 (a)**, which shows the components of $\vec{A}$ and $\vec{B}$, and **FIGURE 3-14 (b)**, which shows the components of $\vec{C}$. Clearly,

$$C_x = A_x + B_x$$

and

$$C_y = A_y + B_y$$

Thus, to add vectors, you simply add the components.

▲ **FIGURE 3-13 Graphical addition of vectors** The vector $\vec{A}$ has a magnitude of 5.00 m and a direction angle of 60.0°; the vector $\vec{B}$ has a magnitude of 4.00 m and a direction angle of 20.0°. The magnitude and direction of $\vec{C} = \vec{A} + \vec{B}$ can be measured on the graph with a ruler and a protractor.

(a)

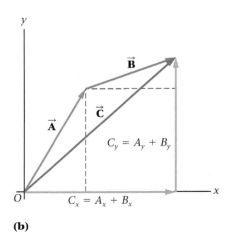

(b)

◀ **FIGURE 3-14 Component addition of vectors (a)** The x and y components of $\vec{A}$ and $\vec{B}$. **(b)** The x and y components of $\vec{C}$. Notice that $C_x = A_x + B_x$ and $C_y = A_y + B_y$.

Returning to our example in Figure 3-13, we find that the components of $\vec{A}$ are

$$A_x = (5.00\,\text{m})\cos 60.0° = 2.50\,\text{m}, \quad A_y = (5.00\,\text{m})\sin 60.0° = 4.33\,\text{m}$$

Similarly, the components of $\vec{B}$ are

$$B_x = (4.00\,\text{m})\cos 20.0° = 3.76\,\text{m}, \quad B_y = (4.00\,\text{m})\sin 20.0° = 1.37\,\text{m}$$

Adding component by component yields the components of $\vec{C} = \vec{A} + \vec{B}$:

$$C_x = A_x + B_x = 2.50\,\text{m} + 3.76\,\text{m} = 6.26\,\text{m}$$

and

$$C_y = A_y + B_y = 4.33\,\text{m} + 1.37\,\text{m} = 5.70\,\text{m}$$

With these results, we can now find *precise* values for C, the magnitude of vector $\vec{C}$, and its direction angle θ. In particular,

$$C = \sqrt{C_x{}^2 + C_y{}^2} = \sqrt{(6.26\,\text{m})^2 + (5.70\,\text{m})^2} = \sqrt{71.7\,\text{m}^2} = 8.47\,\text{m}$$

and

$$\theta = \tan^{-1}\!\left(\frac{C_y}{C_x}\right) = \tan^{-1}\!\left(\frac{5.70\,\text{m}}{6.26\,\text{m}}\right) = \tan^{-1}(0.911) = 42.3°$$

Big Idea **3** The most precise way to add or subtract vectors is to add or subtract their components.

Note that the approximate results found by graphical addition are in rough agreement with these exact results.

In the future, we will always add vectors using components—graphical addition is useful primarily as a rough check on the results obtained with components.

EXAMPLE 3-5   TREASURE HUNT: FIND THE DIRECTION AND MAGNITUDE

What are the magnitude D and direction angle θ of the net displacement for the treasure hunt illustrated in the figure?

PICTURE THE PROBLEM
Our drawing shows the direction and magnitude for each of the three vectors, $\vec{A}$, $\vec{B}$, and $\vec{C}$, that lead to the location of the buried treasure. The net displacement is the vector sum, $\vec{D} = \vec{A} + \vec{B} + \vec{C}$.

REASONING AND STRATEGY
We'll start by writing out the components for $\vec{A}$, $\vec{B}$, and $\vec{C}$. Notice that $\vec{A}$ has only a positive y component (in the N direction), $\vec{B}$ has only a positive x component (in the E direction), and $\vec{C}$ has both a positive x component and a negative y component. Adding the vectors component by component yields the components of $\vec{D}$. We then convert the components of $\vec{D}$ to the magnitude D and angle θ.

Known   The lengths of the segments of the path are $A = 3.75\,\text{m}$, $B = 2.25\,\text{m}$, and $C = 3.00\,\text{m}$. The directions are given in the figure.

Unknown   Length of final displacement, $D = ?$ Direction angle of final displacement, $\theta = ?$

SOLUTION

1. Find the components of $\vec{A}$: $A_x = 0, A_y = 3.75\,\text{m}$

2. Find the components of $\vec{B}$: $B_x = 2.25\,\text{m}, B_y = 0$

3. Find the components of $\vec{C}$: $C_x = (3.00\,\text{m})\cos(-45°) = 2.12\,\text{m}$
   $C_y = (3.00\,\text{m})\sin(-45°) = -2.12\,\text{m}$

4. Sum the components of $\vec{A}$, $\vec{B}$, and $\vec{C}$ to find the components of $\vec{D}$: $D_x = 0 + 2.25\,\text{m} + 2.12\,\text{m} = 4.37\,\text{m}$
   $D_y = 3.75\,\text{m} + 0 - 2.12\,\text{m} = 1.63\,\text{m}$

CONTINUED

5. Use the Pythagorean theorem to find D and the inverse tangent to find θ:

$$D = \sqrt{(4.37\text{ m})^2 + (1.63\text{ m})^2} = 4.66\text{ m}$$

$$\theta = \tan^{-1}\left(\frac{1.63\text{ m}}{4.37\text{ m}}\right) = 20.5°$$

INSIGHT

A direct route to the treasure would be to walk 4.66 m from the tree in a direction 20.5° above the x axis.

PRACTICE PROBLEM — PREDICT/CALCULATE

(a) If the magnitudes of $\vec{\mathbf{A}}$, $\vec{\mathbf{B}}$, and $\vec{\mathbf{C}}$ are all decreased by a factor of two, by what factors do you expect D and θ to change? **(b)** Verify your predictions with a numerical calculation. [**Answer: (a)** The distance D will be halved; the angle θ will stay the same. **(b)** $D = (4.66\text{ m})/2 = 2.33\text{ m}$, $\theta = 20.5°$.]

Some related homework problems: Problem 22, Problem 28, Problem 74

Subtracting Vectors

Next, how do we subtract vectors? Suppose, for example, that we would like to determine the vector $\vec{\mathbf{D}}$, where

$$\vec{\mathbf{D}} = \vec{\mathbf{A}} - \vec{\mathbf{B}}$$

As an example, we'll consider the vectors $\vec{\mathbf{A}}$ and $\vec{\mathbf{B}}$ that are shown in Figure 3-13. To find $\vec{\mathbf{D}}$, we start by rewriting the equation for it as follows:

$$\vec{\mathbf{D}} = \vec{\mathbf{A}} + (-\vec{\mathbf{B}})$$

That is, $\vec{\mathbf{D}}$ is the sum of $\vec{\mathbf{A}}$ and $-\vec{\mathbf{B}}$. Now the negative of a vector has a very simple graphical interpretation:

> The *negative* of a vector is represented by an arrow of the same length as the original vector, but pointing in the *opposite* direction. That is, multiplying a vector by minus one *reverses its direction.*

For example, the vectors $\vec{\mathbf{B}}$ and $-\vec{\mathbf{B}}$ are indicated in **FIGURE 3-15 (a)**. Thus, to subtract $\vec{\mathbf{B}}$ from $\vec{\mathbf{A}}$, simply reverse the direction of $\vec{\mathbf{B}}$ and add it to $\vec{\mathbf{A}}$, as indicated in **FIGURE 3-15 (b)**.

In terms of components, you subtract vectors by simply subtracting the components. For example, consider the following subtraction:

$$\vec{\mathbf{D}} = \vec{\mathbf{A}} - \vec{\mathbf{B}}$$

The components of $\vec{\mathbf{D}}$ are given by subtracting the components of $\vec{\mathbf{B}}$ from the components of $\vec{\mathbf{A}}$:

$$D_x = A_x - B_x$$

$$D_y = A_y - B_y$$

Once the components of $\vec{\mathbf{D}}$ are found, its magnitude and direction angle can be calculated as usual.

(a)

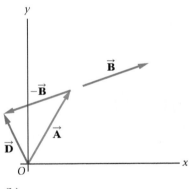

(b)

▲ **FIGURE 3-15 Vector subtraction (a)** The vector $\vec{\mathbf{B}}$ and its negative $-\vec{\mathbf{B}}$. **(b)** A vector construction for $\vec{\mathbf{D}} = \vec{\mathbf{A}} - \vec{\mathbf{B}}$.

EXERCISE 3-6 SUBTRACTING VECTORS

a. For the vectors $\vec{\mathbf{A}}$ and $\vec{\mathbf{B}}$ shown in Figure 3-13, find the components of $\vec{\mathbf{D}} = \vec{\mathbf{A}} - \vec{\mathbf{B}}$.

b. Find D and θ for the vector in part (a), and compare your answer with the vector $\vec{\mathbf{D}}$ shown in Figure 3-15 (b).

REASONING AND SOLUTION

(a) Subtract the components of $\vec{\mathbf{B}}$ from the components of $\vec{\mathbf{A}}$ to find the components of $\vec{\mathbf{D}}$:

$$D_x = A_x - B_x = (2.50\text{ m}) - (3.76\text{ m}) = -1.26\text{ m},$$

$$D_y = A_y - B_y = (4.33\text{ m}) - (1.37\text{ m}) = 2.96\text{ m}$$

CONTINUED

(b) Converting components to magnitude and direction, using the Pythagorean theorem and the inverse tangent, we find $D = 3.22$ m and $\theta = -66.9° + 180° = 113°$. In Figure 3-15 (b) we see that $\vec{D}$ is shorter than $\vec{B}$, which has a magnitude of 4.00 m, and its direction angle is somewhat greater than 90°, in agreement with our numerical results.

Enhance Your Understanding   (Answers given at the end of the chapter)

3. Several vectors are shown in **FIGURE 3-16**. The red vectors have a magnitude equal to a diagonal across the square formed by the blue vectors. Which of the given vectors represents the following quantities: **(a)** $\vec{A} + \vec{B}$, **(b)** $\vec{D} - \vec{A}$, **(c)** $\vec{G} - \vec{C}$, and **(d)** $\vec{B} + \vec{H} - \vec{D}$?

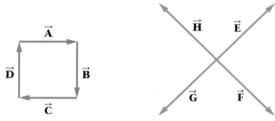

▲ **FIGURE 3-16** Enhance Your Understanding 3

Section Review

- Vectors are added graphically by placing them head-to-tail.
- Vectors are added with components by adding the components.

3-4 Unit Vectors

▲ **FIGURE 3-17 Unit vectors** The unit vectors $\hat{\mathbf{x}}$ and $\hat{\mathbf{y}}$ point in the positive x and y directions, respectively.

Big Idea 4 Unit vectors are dimensionless, have a magnitude of 1, and point in a specific direction. The most commonly used unit vectors point in the positive x or positive y direction.

Unit vectors provide a convenient way of expressing an arbitrary vector in terms of its components, as we shall see. But first, let's define what we mean by a unit vector. In particular, the unit vectors $\hat{\mathbf{x}}$ and $\hat{\mathbf{y}}$ are defined to be dimensionless vectors of unit magnitude pointing in the positive x and y directions, respectively:

- The x unit vector, $\hat{\mathbf{x}}$, is a dimensionless vector of unit length pointing in the positive x direction.

- The y unit vector, $\hat{\mathbf{y}}$, is a dimensionless vector of unit length pointing in the positive y direction.

FIGURE 3-17 shows $\hat{\mathbf{x}}$ and $\hat{\mathbf{y}}$ on a two-dimensional coordinate system. Unit vectors have no physical dimensions—like mass, length, or time—and hence they are used to specify direction only.

Multiplying Unit Vectors by Scalars

To see the utility of unit vectors, consider the effect of multiplying a vector by a scalar. For example, multiplying a vector by 3 increases its magnitude by a factor of 3 but does not change its direction, as shown in **FIGURE 3-18**. Multiplying by -3 increases the magnitude by a factor of 3 *and* reverses the direction of the vector. This is also shown in Figure 3-18. In the case of unit vectors—which have a magnitude of 1 and are dimensionless—multiplication by a scalar results in a vector with the same magnitude and dimensions as the scalar.

For example, if a vector $\vec{A}$ has the scalar components $A_x = 5$ m and $A_y = 3$ m, we can write it as

$$\vec{A} = (5\text{ m})\hat{\mathbf{x}} + (3\text{ m})\hat{\mathbf{y}}$$

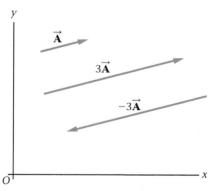

▲ **FIGURE 3-18 Multiplying a vector by a scalar** Multiplying a vector by a positive scalar different from 1 will change the length of the vector but leave its direction the same. If the vector is multiplied by a negative scalar its direction is reversed.

We refer to the quantities $(5\text{ m})\hat{\mathbf{x}}$ and $(3\text{ m})\hat{\mathbf{y}}$ as the x and y **vector components** of the vector $\vec{A}$. In general, an arbitrary two-dimensional vector $\vec{A}$ can always be written as the sum of a vector component in the x direction and a vector component in the y direction:

$$\vec{A} = A_x\hat{\mathbf{x}} + A_y\hat{\mathbf{y}}$$

This is illustrated in **FIGURE 3-19 (a)**. An equivalent way of representing the vector components of a vector is illustrated in **FIGURE 3-19 (b)**. In this case we see that the vector components are the *projection* of a vector onto the x and y axes. The sign of the vector components is positive if they point in the positive x or y direction, and negative if they point in the negative x or y direction. This is how vector components will generally be shown in later chapters.

Finally, notice that vector addition and subtraction are straightforward with unit vector notation. For example,

$$\vec{\mathbf{C}} = \vec{\mathbf{A}} + \vec{\mathbf{B}} = (A_x + B_x)\hat{\mathbf{x}} + (A_y + B_y)\hat{\mathbf{y}}$$

and

$$\vec{\mathbf{D}} = \vec{\mathbf{A}} - \vec{\mathbf{B}} = (A_x - B_x)\hat{\mathbf{x}} + (A_y - B_y)\hat{\mathbf{y}}$$

Clearly, unit vectors provide a useful way to keep track of the x and y components of a vector.

Enhance Your Understanding (Answers given at the end of the chapter)

4. Which of the vectors shown in **FIGURE 3-20** has the same direction as
   (a) $(1\text{ m})\hat{\mathbf{x}} + (1\text{ m})\hat{\mathbf{y}}$,
   (b) $(1\text{ m})\hat{\mathbf{x}} - (1\text{ m})\hat{\mathbf{y}}$,
   (c) $-(1\text{ m})\hat{\mathbf{x}} + (1\text{ m})\hat{\mathbf{y}}$,
   (d) $-(1\text{ m})\hat{\mathbf{x}} - (1\text{ m})\hat{\mathbf{y}}$?

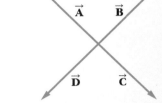

▲ **FIGURE 3-20** Enhance Your Understanding 4

Section Review

- Unit vectors provide a convenient way to represent the vector components of a vector.

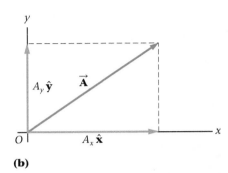

▲ **FIGURE 3-19 Vector components (a)** A vector $\vec{\mathbf{A}}$ can be written in terms of unit vectors as $\vec{\mathbf{A}} = A_x\hat{\mathbf{x}} + A_y\hat{\mathbf{y}}$. **(b)** Vector components can be thought of as the projection of the vector onto the x and y axes. This method of representing vector components will be used frequently in subsequent chapters.

3-5 Position, Displacement, Velocity, and Acceleration Vectors

In Chapter 2 we discussed four different one-dimensional vectors: position, displacement, velocity, and acceleration. Each of these quantities had a direction associated with it, indicated by its sign; positive meant in the positive direction, negative meant in the negative direction. Now we consider these vectors again, this time in two dimensions, where the possibilities for direction are not so limited.

Position Vectors

To begin, imagine a two-dimensional coordinate system, as in **FIGURE 3-21**. Position is indicated by a vector from the origin to the location in question. We refer to the position vector as $\vec{\mathbf{r}}$; its units are meters, m.

> **Definition: Position Vector, $\vec{\mathbf{r}}$**
>
> position vector $= \vec{\mathbf{r}}$ 3-1
>
> SI unit: meter, m

In terms of unit vectors, the position vector is simply $\vec{\mathbf{r}} = x\,\hat{\mathbf{x}} + y\,\hat{\mathbf{y}}$.

Displacement Vectors

Now, suppose that initially you are at the location indicated by the position vector $\vec{\mathbf{r}}_i$ (subscript i stands for initial), and that later you are at the final position represented by

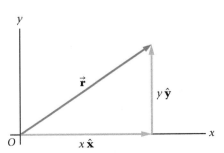

▲ **FIGURE 3-21 Position vector** The position vector $\vec{\mathbf{r}}$ points from the origin to the current location of an object. The x and y vector components of $\vec{\mathbf{r}}$ are $x\hat{\mathbf{x}}$ and $y\hat{\mathbf{y}}$, respectively.

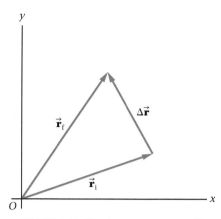

▲ **FIGURE 3-22 Displacement vector** The displacement vector $\Delta\vec{\mathbf{r}}$ is the change in position. It points from the head of the initial position vector $\vec{\mathbf{r}}_i$ to the head of the final position vector $\vec{\mathbf{r}}_f$. Thus $\vec{\mathbf{r}}_f = \vec{\mathbf{r}}_i + \Delta\vec{\mathbf{r}}$ or $\Delta\vec{\mathbf{r}} = \vec{\mathbf{r}}_f - \vec{\mathbf{r}}_i$.

▲ **FIGURE 3-23** A map can be used to determine the direction and magnitude of the displacement vector from your initial position to your desired final position.

the position vector $\vec{\mathbf{r}}_f$ (subscript f stands for final). Your displacement vector, $\Delta\vec{\mathbf{r}}$, is the change in position:

> **Definition: Displacement Vector, $\Delta\vec{\mathbf{r}}$**
>
> $$\Delta\vec{\mathbf{r}} = \vec{\mathbf{r}}_f - \vec{\mathbf{r}}_i$$ 3-2
>
> SI unit: meter, m

Rearranging this definition slightly, we see that

$$\vec{\mathbf{r}}_f = \vec{\mathbf{r}}_i + \Delta\vec{\mathbf{r}}$$

That is, the final position is equal to the initial position plus the change in position. This is illustrated in **FIGURE 3-22**, where we see that $\Delta\vec{\mathbf{r}}$ extends from the head of $\vec{\mathbf{r}}_i$ to the head of $\vec{\mathbf{r}}_f$. This is similar to how a map and compass can be used to determine the displacement vector from your present position to your destination, as in **FIGURE 3-23**.

Velocity Vectors

Next, the average velocity vector is defined as the displacement vector $\Delta\vec{\mathbf{r}}$ divided by the elapsed time Δt:

> **Definition: Average Velocity Vector, $\vec{\mathbf{v}}_{av}$**
>
> $$\vec{\mathbf{v}}_{av} = \frac{\Delta\vec{\mathbf{r}}}{\Delta t}$$ 3-3
>
> SI unit: meter per second, m/s

Since $\Delta\vec{\mathbf{r}}$ is a vector, it follows that $\vec{\mathbf{v}}_{av}$ is also a vector; it is the vector $\Delta\vec{\mathbf{r}}$ multiplied by the scalar $(1/\Delta t)$. Thus $\vec{\mathbf{v}}_{av}$ is parallel to $\Delta\vec{\mathbf{r}}$ and has the units m/s.

QUICK EXAMPLE 3-7   A DRAGONFLY'S AVERAGE VELOCITY

A dragonfly is initially at the position $\vec{\mathbf{r}}_i = (4.60\,\text{m})\hat{\mathbf{x}} + (0.960\,\text{m})\hat{\mathbf{y}}$, as shown in the sketch. After 3.00 s, the dragonfly is at the position $\vec{\mathbf{r}}_f = (0.520\,\text{m})\hat{\mathbf{x}} + (2.40\,\text{m})\hat{\mathbf{y}}$. What was the dragonfly's average velocity during this time?

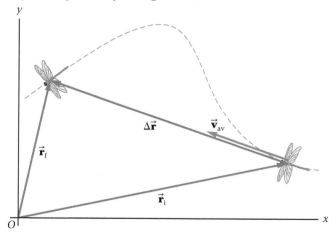

REASONING AND SOLUTION
We show the initial and final positions in our sketch, labeled $\vec{\mathbf{r}}_i$ and $\vec{\mathbf{r}}_f$, respectively. The corresponding change in position is $\Delta\vec{\mathbf{r}} = \vec{\mathbf{r}}_f - \vec{\mathbf{r}}_i$. Dividing $\Delta\vec{\mathbf{r}}$ by the time $\Delta t = 3.00\,\text{s}$ yields the average velocity.

1. Calculate the change in position: $\Delta\vec{\mathbf{r}} = \vec{\mathbf{r}}_f - \vec{\mathbf{r}}_i$
$$= (0.520\,\text{m} - 4.60\,\text{m})\hat{\mathbf{x}} + (2.40\,\text{m} - 0.960\,\text{m})\hat{\mathbf{y}}$$
$$= (-4.08\,\text{m})\hat{\mathbf{x}} + (1.44\,\text{m})\hat{\mathbf{y}}$$

CONTINUED

2. Divide by the elapsed time to find the average velocity:

$$\vec{\mathbf{v}}_{av} = \frac{\Delta\vec{\mathbf{r}}}{\Delta t} = \frac{(-4.08\text{ m})\hat{\mathbf{x}} + (1.44\text{ m})\hat{\mathbf{y}}}{3.00\text{ s}}$$
$$= (-1.36\text{ m/s})\hat{\mathbf{x}} + (0.480\text{ m/s})\hat{\mathbf{y}}$$

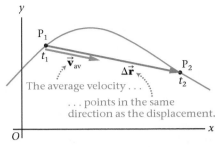

▲ **FIGURE 3-24 Average velocity vector** The average velocity, $\vec{\mathbf{v}}_{av}$, points in the same direction as the displacement, $\Delta\vec{\mathbf{r}}$, for any given interval of time.

To help visualize $\vec{\mathbf{v}}_{av}$, imagine a particle moving in two dimensions along the blue path shown in **FIGURE 3-24**. If the particle is at point P_1 at time t_1, and at P_2 at time t_2, its displacement is indicated by the vector $\Delta\vec{\mathbf{r}}$. The average velocity is parallel to $\Delta\vec{\mathbf{r}}$, as indicated in Figure 3-24. It makes sense physically that $\vec{\mathbf{v}}_{av}$ is parallel to $\Delta\vec{\mathbf{r}}$; after all, on *average* you have moved in the direction of $\Delta\vec{\mathbf{r}}$ during the time from t_1 to t_2. To put it another way, a particle that starts at P_1 at the time t_1 and moves with the velocity $\vec{\mathbf{v}}_{av}$ until the time t_2 will arrive in precisely the same location as the particle that follows the blue path.

If we consider smaller and smaller time intervals, as in **FIGURE 3-25**, it is possible to calculate the instantaneous velocity vector:

Definition: Instantaneous Velocity Vector, $\vec{\mathbf{v}}$

$$\vec{\mathbf{v}} = \lim_{\Delta t \to 0} \frac{\Delta\vec{\mathbf{r}}}{\Delta t} \qquad\qquad 3\text{-}4$$

SI unit: meter per second, m/s

As can be seen in Figure 3-25, the instantaneous velocity at a given time is tangential to the path of the particle at that time. In addition, the magnitude of the velocity vector is the instantaneous speed of the particle. Thus, the instantaneous velocity vector tells you both how fast a particle is moving and in what direction.

EXERCISE 3-8 A TRAVELING TROUT

Find the speed and direction of motion for a rainbow trout whose velocity is $\vec{\mathbf{v}} = (2.6\text{ m/s})\hat{\mathbf{x}} + (-1.9\text{ m/s})\hat{\mathbf{y}}$.

REASONING AND SOLUTION
The velocity components are given, and hence we can find the magnitude of the velocity (the speed) using the Pythagorean theorem. Therefore, the speed of the trout is

$$v = \sqrt{(2.6\text{ m/s})^2 + (-1.9\text{ m/s})^2} = 3.2\text{ m/s}$$

We can find the direction of motion using the inverse tangent, as follows:

$$\theta = \tan^{-1}\!\left(\frac{-1.9\text{ m/s}}{2.6\text{ m/s}}\right) = -36°$$

We conclude that the trout moves at the rate of 3.2 m/s in a direction 36° *below* the x axis.

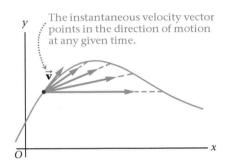

▲ **FIGURE 3-25 Instantaneous velocity vector** The instantaneous velocity vector $\vec{\mathbf{v}}$ is obtained by calculating the average velocity vector over smaller and smaller time intervals. In the limit of vanishingly small time intervals, the average velocity approaches the instantaneous velocity, which points in the direction of motion.

Acceleration Vectors

Finally, the average acceleration vector over an interval of time, Δt, is defined as the change in the velocity vector, $\Delta\vec{\mathbf{v}}$, divided by the elapsed time Δt:

Definition: Average Acceleration Vector, $\vec{\mathbf{a}}_{av}$

$$\vec{\mathbf{a}}_{av} = \frac{\Delta\vec{\mathbf{v}}}{\Delta t} \qquad\qquad 3\text{-}5$$

SI unit: meter per second per second, m/s^2

An example is given in **FIGURE 3-26 (a)**, where we show the initial and final velocity vectors corresponding to two different times. The change in velocity is defined as

$$\Delta\vec{\mathbf{v}} = \vec{\mathbf{v}}_f - \vec{\mathbf{v}}_i$$

It follows that

$$\vec{\mathbf{v}}_f = \vec{\mathbf{v}}_i + \Delta\vec{\mathbf{v}}$$

▶ **FIGURE 3-26 Average acceleration vector (a)** As a particle moves along the blue path its velocity changes in magnitude and direction. At the time t_i the velocity is $\vec{\mathbf{v}}_i$; at the time t_f the velocity is $\vec{\mathbf{v}}_f$. **(b)** The average acceleration vector $\vec{\mathbf{a}}_{av} = \Delta\vec{\mathbf{v}}/\Delta t$ points in the direction of the change in velocity vector $\Delta\vec{\mathbf{v}}$. We obtain $\Delta\vec{\mathbf{v}}$ by moving $\vec{\mathbf{v}}_f$ so that its tail coincides with the tail of $\vec{\mathbf{v}}_i$, and then drawing the arrow that connects the head of $\vec{\mathbf{v}}_i$ to the head of $\vec{\mathbf{v}}_f$. Note that $\vec{\mathbf{a}}_{av}$ need not point in the direction of motion, and in general it doesn't.

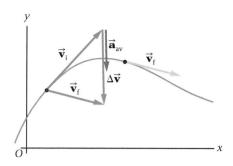

(a) The instantaneous velocity at two different times

(b) The average acceleration vector points in the same direction as the change in velocity.

This is indicated in **FIGURE 3-26 (b)**. Notice that $\Delta\vec{\mathbf{v}}$ is the vector extending from the head of $\vec{\mathbf{v}}_i$ to the head of $\vec{\mathbf{v}}_f$, just as $\Delta\vec{\mathbf{r}}$ extends from the head of $\vec{\mathbf{r}}_i$ to the head of $\vec{\mathbf{r}}_f$ in Figure 3-22. The direction of $\vec{\mathbf{a}}_{av}$ is the direction of $\Delta\vec{\mathbf{v}}$, as shown in Figure 3-26 (b).

Can an object accelerate if its speed is constant? Absolutely—if its direction changes. Consider a car traveling with a constant speed on a circular track, as shown in **FIGURE 3-27**. Suppose that the initial velocity of the car is $\vec{\mathbf{v}}_i = (12\,\text{m/s})\hat{\mathbf{x}}$, and that 10.0 s later its final velocity is $\vec{\mathbf{v}}_f = (-12\,\text{m/s})\hat{\mathbf{y}}$. The speed is 12 m/s in each case, but the velocity is different because the *direction* has changed. Calculating the average acceleration, we find a nonzero acceleration:

$$\vec{\mathbf{a}}_{av} = \frac{\Delta\vec{\mathbf{v}}}{\Delta t} = \frac{\vec{\mathbf{v}}_f - \vec{\mathbf{v}}_i}{10.0\,\text{s}}$$

$$= \frac{(-12\,\text{m/s})\hat{\mathbf{y}} - (12\,\text{m/s})\hat{\mathbf{x}}}{10.0\,\text{s}} = (-1.2\,\text{m/s}^2)\hat{\mathbf{x}} + (-1.2\,\text{m/s}^2)\hat{\mathbf{y}}$$

Thus, a change in direction is just as important as a change in speed in producing an acceleration. We shall study circular motion in detail in Chapter 6.

Finally, by going to infinitesimally small time intervals, $\Delta t \rightarrow 0$, we can define the instantaneous acceleration vector:

Definition: Instantaneous Acceleration Vector, $\vec{\mathbf{a}}$

$$\vec{\mathbf{a}} = \lim_{\Delta t \to 0} \frac{\Delta\vec{\mathbf{v}}}{\Delta t} \qquad\qquad 3\text{-}6$$

SI unit: meter per second per second, m/s^2

PHYSICS IN CONTEXT
Looking Ahead

In Chapter 5 we will introduce one of the most important concepts in all of physics—force. It is a vector quantity. Other important vector quantities to be introduced in later chapters include linear momentum (Chapter 9), angular momentum (Chapter 11), electric field (Chapter 19), and magnetic field (Chapter 22).

▶ **FIGURE 3-27 Average acceleration for a car traveling in a circle with constant speed** Although the speed of this car never changes, it is still accelerating—due to the change in its direction of motion. For the time interval depicted, the car's average acceleration is in the direction of $\Delta\vec{\mathbf{v}}$, which is toward the center of the circle. (As we shall see in Chapter 6, the car's acceleration is always toward the center of the circle.)

The car's speed is constant . . .

. . . but the car accelerates because its direction of motion changes.

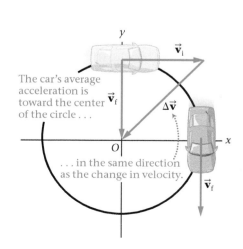

The car's average acceleration is toward the center of the circle . . .

. . . in the same direction as the change in velocity.

QUICK EXAMPLE 3-9   FIND THE AVERAGE ACCELERATION

A car is traveling with an initial velocity of $\vec{\mathbf{v}}_i = (-6.36\,\text{m/s})\hat{\mathbf{x}} + (6.36\,\text{m/s})\hat{\mathbf{y}}$. Eight seconds later, after following a bend in the road, its final velocity is $\vec{\mathbf{v}}_f = (15.00\,\text{m/s})\hat{\mathbf{y}}$. What are the magnitude and direction of its average acceleration during those 8.00 seconds?

REASONING AND SOLUTION

Given the initial and final velocities, we can find the corresponding change in velocity, $\Delta\vec{\mathbf{v}} = \vec{\mathbf{v}}_f - \vec{\mathbf{v}}_i$. Dividing $\Delta\vec{\mathbf{v}}$ by the time $\Delta t = 8.00\,\text{s}$ yields the average acceleration vector. We can find its magnitude and direction using the Pythagorean theorem and the inverse tangent, respectively.

1. Calculate the change in velocity, $\Delta\vec{\mathbf{v}}$:

$$\Delta\vec{\mathbf{v}} = \vec{\mathbf{v}}_f - \vec{\mathbf{v}}_i$$
$$= (15.00\,\text{m/s})\hat{\mathbf{y}} - [(-6.36\,\text{m/s})\hat{\mathbf{x}} + (6.36\,\text{m/s})\hat{\mathbf{y}}]$$
$$= (6.36\,\text{m/s})\hat{\mathbf{x}} + (8.64\,\text{m/s})\hat{\mathbf{y}}$$

2. Divide the change in velocity by the time to determine the average acceleration, $\vec{\mathbf{a}}_{av}$:

$$\vec{\mathbf{a}}_{av} = \frac{\Delta\vec{\mathbf{v}}}{\Delta t} = \frac{(6.36\,\text{m/s})\hat{\mathbf{x}} + (8.64\,\text{m/s})\hat{\mathbf{y}}}{(8.00\,\text{s})}$$
$$= (0.795\,\text{m/s}^2)\hat{\mathbf{x}} + (1.08\,\text{m/s}^2)\hat{\mathbf{y}}$$

3. Find the magnitude, a_{av}, and direction angle, θ, of the average acceleration:

$$a_{av} = \sqrt{(0.795\,\text{m/s}^2)^2 + (1.08\,\text{m/s}^2)^2}$$
$$= 1.34\,\text{m/s}^2$$
$$\theta = \tan^{-1}\left(\frac{1.08\,\text{m/s}^2}{0.795\,\text{m/s}^2}\right)$$
$$= 53.6° \quad \text{(above the positive } x \text{ axis)}$$

It's important to recognize the following distinctions between the velocity vector and the acceleration vector:

- The velocity vector, $\vec{\mathbf{v}}$, always points in the direction of a particle's motion.

- The acceleration vector, $\vec{\mathbf{a}}$, can point in directions other than the direction of motion, and in general it does.

In one dimension, the acceleration is either in the same direction as the velocity or in the opposite direction. In two dimensions, the acceleration can be in any direction relative to the velocity.

An example of a particle's motion, showing the velocity and acceleration vectors at various times, is presented in **FIGURE 3-28**. Notice that in all cases the velocity is tangential to the motion, though the acceleration points in various directions. When the acceleration is perpendicular to the velocity of an object, as at points (2) and (3) in Figure 3-28, the speed of the object remains constant while its direction of motion changes. At points (1) and (4) in Figure 3-28 the acceleration is antiparallel (opposite) or parallel to the velocity of the object, respectively. In such cases, the direction

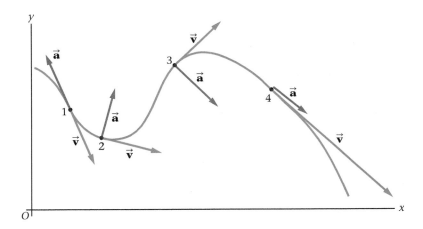

◀ **FIGURE 3-28 Velocity and acceleration vectors for a particle moving along a winding path** The acceleration of a particle need not point in the direction of motion. At point (1) the particle is slowing down, at (2) it is turning to the left, at (3) it is turning to the right, and, finally, at point (4) it is speeding up.

► **FIGURE 3-29** The velocities of these cyclists change in both magnitude and direction as they slow to negotiate a series of sharp curves and then speed up again. Both kinds of velocity change involve an acceleration.

Big Idea ⑤ Position, displacement, velocity, and acceleration are all vectors; each has a specific direction and magnitude.

PHYSICS IN CONTEXT
Looking Ahead

The vectors developed in this section play a key role in Chapter 4, where we study two-dimensional kinematics. In particular, vectors and their components will allow us to analyze two-dimensional motion as a combination of two completely independent one-dimensional motions.

of motion remains the same while the speed changes. A similar real-world example is provided in **FIGURE 3-29**. Throughout the next chapter we shall see further examples of motion in which the velocity and acceleration are in different directions.

Enhance Your Understanding   (Answers given at the end of the chapter)

5. An object moves along the brown path in **FIGURE 3-30** in the direction indicated. Which physical quantity (position, acceleration, velocity) is represented by the following vectors: **(a)** $\vec{\mathbf{A}}$, **(b)** $\vec{\mathbf{B}}$, **(c)** $\vec{\mathbf{C}}$?

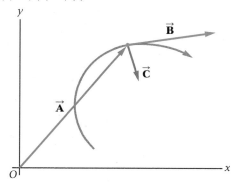

▲ **FIGURE 3-30** Enhance Your Understanding 5

Section Review

• Many physical quantities are represented by vectors. Among these are position, displacement, velocity, and acceleration.

3-6 Relative Motion

A good example of the use of vectors is in the description of relative motion. Suppose, for example, that you are standing on the ground as a train goes by at 15.0 m/s, as shown in **FIGURE 3-31**. Inside the train, a person is walking in the forward direction at 1.2 m/s relative to the train. How fast is the person moving relative to you? Clearly, the answer is 1.2 m/s + 15.0 m/s = 16.2 m/s. What if the person had been walking with the same speed, but toward the back of the train? In this case, you would see the person going by with a speed of −1.2 m/s + 15.0 m/s = 13.8 m/s.

Let's generalize these results. Call the velocity of the *train* relative to the *ground* $\vec{\mathbf{v}}_{tg}$, the velocity of the *passenger* relative to the *train* $\vec{\mathbf{v}}_{pt}$, and the velocity of the *passenger*

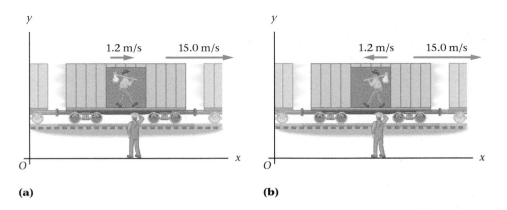

(a)　　　　　　　　　　　**(b)**

relative to the ground $\vec{\mathbf{v}}_{pg}$. As we saw in the previous paragraph, the velocity of the *passenger relative to the ground is*

$$\vec{\mathbf{v}}_{pg} = \vec{\mathbf{v}}_{pt} + \vec{\mathbf{v}}_{tg} \qquad\qquad 3\text{-}7$$

This vector addition is illustrated in **FIGURE 3-32** for the two cases we discussed.

Though this example dealt with one-dimensional motion, Equation 3-7 is valid for velocity vectors pointing in any directions. For example, instead of walking on the car's floor, the person might be climbing a ladder to the roof of the car, as in **FIGURE 3-33**. In this case $\vec{\mathbf{v}}_{pt}$ is vertical, $\vec{\mathbf{v}}_{tg}$ is horizontal, and $\vec{\mathbf{v}}_{pg}$ is simply the vector sum $\vec{\mathbf{v}}_{pt} + \vec{\mathbf{v}}_{tg}$.

QUICK EXAMPLE 3-10　RELATIVE MOTION

Suppose the person in Figure 3-33 is climbing a vertical ladder with a speed of 0.24 m/s, and the train is slowly coasting forward at 0.68 m/s. Find the speed and direction of motion of the person relative to the ground.

REASONING AND SOLUTION
The velocity of the person relative to the ground (pg) is equal to the velocity of the person relative to the train (pt) plus the velocity of the train relative to the ground (tg). The velocity of the person relative to the train is in the positive y direction, $\vec{\mathbf{v}}_{pt} = (0.24\text{ m/s})\hat{\mathbf{y}}$, and the velocity of the train relative to the ground is in the positive x direction, $\vec{\mathbf{v}}_{tg} = (0.68\text{ m/s})\hat{\mathbf{x}}$. Thus, the desired velocity is

$$\vec{\mathbf{v}}_{pg} = \vec{\mathbf{v}}_{pt} + \vec{\mathbf{v}}_{tg}$$
$$= (0.24\text{ m/s})\hat{\mathbf{y}} + (0.68\text{ m/s})\hat{\mathbf{x}}$$

In the more usual order of writing things, with the x component first, we have

$$\vec{\mathbf{v}}_{pg} = (0.68\text{ m/s})\hat{\mathbf{x}} + (0.24\text{ m/s})\hat{\mathbf{y}}$$

It follows that $v_{pg} = \sqrt{(0.68\text{ m/s})^2 + (0.24\text{ m/s})^2} = 0.72\text{ m/s}$ and $\theta = \tan^{-1}(0.24/0.68) = 19°$.

◄ **FIGURE 3-31 Relative velocity of a passenger on a train with respect to a person on the ground (a)** The passenger walks toward the front of the train. **(b)** The passenger walks toward the rear of the train.

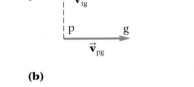

(a)

(b)

▲ **FIGURE 3-32 Adding velocity vectors** Vector addition to find the velocity of the passenger with respect to the ground for **(a)** Figure 3-31 (a) and **(b)** Figure 3-31 (b).

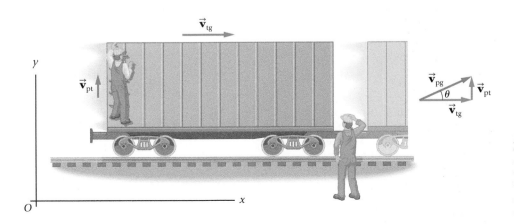

◄ **FIGURE 3-33 Relative velocity in two dimensions** A person climbs up a ladder on a moving train with velocity $\vec{\mathbf{v}}_{pt}$ relative to the train. If the train moves relative to the ground with a velocity $\vec{\mathbf{v}}_{tg}$, the velocity of the person on the train relative to the ground is $\vec{\mathbf{v}}_{pg} = \vec{\mathbf{v}}_{pt} + \vec{\mathbf{v}}_{tg}$.

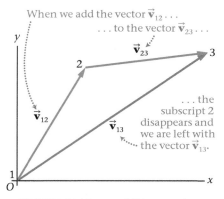

When we add the vector $\vec{\mathbf{v}}_{12}$ ...

... to the vector $\vec{\mathbf{v}}_{23}$ ...

...the subscript 2 disappears and we are left with the vector $\vec{\mathbf{v}}_{13}$.

▲ FIGURE 3-34 **Vector addition used to determine relative velocity**

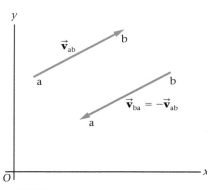

▲ FIGURE 3-35 **Reversing the subscripts of a velocity reverses the corresponding velocity vector**

Subscripts Are the Key to Relative Motion Notice that the subscripts in Equation 3-7 and Exercise 3-10 follow a definite pattern. On the left-hand side of the equation we have the subscripts pg. On the right-hand side we have two sets of subscripts, pt and tg; notice that a pair of t's has been inserted between the p and the g. This pattern always holds for any relative-motion problem, though the subscripts will be different when referring to different objects. Thus, we can say quite generally that

$$\vec{\mathbf{v}}_{13} = \vec{\mathbf{v}}_{12} + \vec{\mathbf{v}}_{23} \qquad \text{3-8}$$

In the train example, we can identify 1 as the *passenger*, 2 as the *train*, and 3 as the *ground*.

The vector addition in Equation 3-8 is shown in **FIGURE 3-34**. For convenience in seeing how the subscripts are ordered in the equation, we have labeled the tail of each vector with its first subscript and the head of each vector with its second subscript.

One final note about velocities and their subscripts: Reversing the subscripts reverses the velocity. This is indicated in **FIGURE 3-35**, where we see that

$$\vec{\mathbf{v}}_{ba} = -\vec{\mathbf{v}}_{ab}$$

Physically, what we are saying is that if you are riding in a car due *north* at 20 m/s relative to the ground, then the ground, relative to you, is moving due *south* at 20 m/s.

Let's apply these results to a real-world, two-dimensional example.

EXAMPLE 3-11   CROSSING A RIVER

RWP You ride in a boat whose speed relative to the water is 6.1 m/s. The boat points at an angle of 25° upstream on a river flowing at 1.4 m/s. **(a)** What is the velocity of the boat relative to the ground? **(b)** Suppose the speed of the boat relative to the water remains the same, but the direction in which it points is changed. What angle is required for the boat to go straight across the river—that is, for the boat to move perpendicular to the shores of the river?

PICTURE THE PROBLEM
We choose the x axis to be perpendicular to the river, and the y axis to point upstream. With these choices the velocity of the water relative to the ground has a magnitude of 1.4 m/s and points in the negative y direction. In part (a), the velocity of the boat relative to the water has a direction angle of $\theta = 25°$ above the x axis. In part (b), the angle θ is to be determined.

REASONING AND STRATEGY
If the water were still, the boat would move in the direction in which it is pointed. With the water flowing downstream, as shown, the boat will move in a direction closer to the x axis. **(a)** To find the velocity of the boat we use $\vec{\mathbf{v}}_{13} = \vec{\mathbf{v}}_{12} + \vec{\mathbf{v}}_{23}$ with 1 referring to the boat (b), 2 referring to the water (w), and 3 referring to the ground (g). **(b)** To go "straight across the river" means that the velocity of the boat relative to the ground should be in the x direction. Thus, we choose the angle θ that cancels the y component of velocity.

CONTINUED

Known	The speed of the boat relative to the water, $v_{bw} = 6.1$ m/s; the angle of the boat above the x axis, $\theta = 25°$; the speed of the water relative to the ground, $v_{wg} = 1.4$ m/s.
Unknown	**(a)** Velocity of the boat relative to the ground, $\vec{\mathbf{v}}_{bg} = ?$
	(b) Angle of motion to go straight across the river, $\theta = ?$

SOLUTION

Part (a)

1. Rewrite $\vec{\mathbf{v}}_{13} = \vec{\mathbf{v}}_{12} + \vec{\mathbf{v}}_{23}$ with $1 \rightarrow b$, $2 \rightarrow w$, and $3 \rightarrow g$: $\vec{\mathbf{v}}_{bg} = \vec{\mathbf{v}}_{bw} + \vec{\mathbf{v}}_{wg}$

2. From our sketch we see that the water flows at 1.4 m/s in the negative $\vec{\mathbf{v}}_{wg} = (-1.4 \text{ m/s})\hat{\mathbf{y}}$
y direction relative to the ground:

3. The velocity of the boat relative to the water is given in $\vec{\mathbf{v}}_{bw} = (6.1 \text{ m/s}) \cos 25° \hat{\mathbf{x}} + (6.1 \text{ m/s}) \sin 25° \hat{\mathbf{y}}$
the problem statement: $= (5.5 \text{ m/s})\hat{\mathbf{x}} + (2.6 \text{ m/s})\hat{\mathbf{y}}$

4. Carry out the vector sum in Step 1 to find $\vec{\mathbf{v}}_{bg}$: $\vec{\mathbf{v}}_{bg} = (5.5 \text{ m/s})\hat{\mathbf{x}} + (2.6 \text{ m/s} - 1.4 \text{ m/s})\hat{\mathbf{y}}$
   $= (5.5 \text{ m/s})\hat{\mathbf{x}} + (1.2 \text{ m/s})\hat{\mathbf{y}}$

Part (b)

5. To cancel the y component of $\vec{\mathbf{v}}_{bg}$, we choose the angle that gives $(6.1 \text{ m/s}) \sin \theta = 1.4 \text{ m/s}$
1.4 m/s for the y component of $\vec{\mathbf{v}}_{bw}$:

6. Solve for θ. With this angle, we see that the y component of $\vec{\mathbf{v}}_{bg}$ in $\theta = \sin^{-1}(1.4/6.1) = 13°$
Step 4 will be zero:

INSIGHT
(a) The speed of the boat relative to the ground is $\sqrt{(5.5 \text{ m/s})^2 + (1.2 \text{ m/s})^2} = 5.6$ m/s, and the direction angle of its motion is $\theta = \tan^{-1}(1.2/5.5) = 12°$ upstream. **(b)** The speed of the boat relative to the ground in this case is equal to the x component of its velocity, because the y component is zero. Therefore, its speed is $(6.1 \text{ m/s}) \cos 13° = 5.9$ m/s.

PRACTICE PROBLEM
Find the speed and direction of the boat relative to the ground in part (a) of this Example if the river flows at 4.5 m/s. Everything else remains the same.   [**Answer:** $v_{bg} = 5.8$ m/s, $\theta = -19°$. In this case, a person on the ground sees the boat going slowly downstream, even though the boat itself points upstream.]

Some related homework problems: Problem 50, Problem 55, Problem 57

Enhance Your Understanding   (Answers given at the end of the chapter)

6. Suppose the speed of the boat is increased in part (a) of Example 3-11, while everything else remains the same. In this case, is the angle of the boat's path across the river greater than, less than, or equal to 12°? Explain.

Section Review

- Relative motion is analyzed with vector addition. For example, the velocity of object 1 relative to object 3 is $\vec{\mathbf{v}}_{13} = \vec{\mathbf{v}}_{12} + \vec{\mathbf{v}}_{23}$, where object 2 can be any other object.

CHAPTER 3 REVIEW

CHAPTER SUMMARY

3-1   SCALARS VERSUS VECTORS

Scalar
A number with appropriate units. Examples of scalar quantities include time and length.

Vector
A quantity with both a magnitude and a direction. Examples include displacement, velocity, and acceleration.

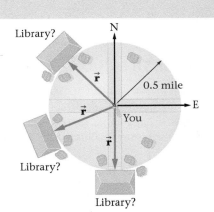

3-2   THE COMPONENTS OF A VECTOR

x Component of Vector $\vec{A}$
$A_x = A \cos \theta$, where θ is measured relative to the x axis.

y Component of Vector $\vec{A}$
$A_y = A \sin \theta$, where θ is measured relative to the x axis.

Magnitude of Vector $\vec{A}$
The magnitude of $\vec{A}$ is $A = \sqrt{A_x^2 + A_y^2}$.

Direction Angle of Vector $\vec{A}$
The direction angle of $\vec{A}$ is $\theta = \tan^{-1}(A_y/A_x)$, where θ is measured relative to the x axis.

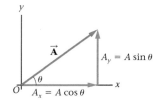

3-3   ADDING AND SUBTRACTING VECTORS

Graphical Method
To add $\vec{A}$ and $\vec{B}$, place them so that the tail of $\vec{B}$ is at the head of $\vec{A}$. The sum $\vec{C} = \vec{A} + \vec{B}$ is the arrow from the tail of $\vec{A}$ to the head of $\vec{B}$.

Component Method
If $\vec{C} = \vec{A} + \vec{B}$, then $C_x = A_x + B_x$ and $C_y = A_y + B_y$. If $\vec{C} = \vec{A} - \vec{B}$, then $C_x = A_x - B_x$ and $C_y = A_y - B_y$.

3-4   UNIT VECTORS

x Unit Vector
Written $\hat{\mathbf{x}}$, the x unit vector is a dimensionless vector of unit length in the positive x direction.

y Unit Vector
Written $\hat{\mathbf{y}}$, the y unit vector is a dimensionless vector of unit length in the positive y direction.

3-5   POSITION, DISPLACEMENT, VELOCITY, AND ACCELERATION VECTORS

Position Vector
The position vector $\vec{\mathbf{r}}$ points from the origin to a particle's location.

Displacement Vector
The displacement vector $\Delta\vec{\mathbf{r}}$ is the change in position; $\Delta\vec{\mathbf{r}} = \vec{\mathbf{r}}_f - \vec{\mathbf{r}}_i$.

Velocity Vector
The velocity vector $\vec{\mathbf{v}}$ points in the direction of motion and has a magnitude equal to the speed.

Acceleration Vector
The acceleration vector $\vec{\mathbf{a}}$ indicates how quickly and in what direction the velocity is changing. It need not point in the direction of motion.

3-6   RELATIVE MOTION

Velocity of Object 1 Relative to Object 3
$\vec{\mathbf{v}}_{13} = \vec{\mathbf{v}}_{12} + \vec{\mathbf{v}}_{23}$, where object 2 can be anything.

Reversing the Subscripts on a Velocity
$\vec{\mathbf{v}}_{12} = -\vec{\mathbf{v}}_{21}$.

ANSWERS TO ENHANCE YOUR UNDERSTANDING QUESTIONS

1. Yes, if the vectors point in different directions.

2. **(a)** Less than 8 m because the x and y components are at right angles to one another. **(b)** Less than 45° because the vertical side of the corresponding triangle is shorter than the horizontal side.

3. **(a)** $\vec{F}$, **(b)** $\vec{H}$, **(c)** $\vec{B}$, **(d)** $\vec{G}$.

4. **(a)** $\vec{B}$, **(b)** $\vec{C}$, **(c)** $\vec{A}$, **(d)** $\vec{D}$.

5. **(a)** Position. **(b)** Velocity. **(c)** Acceleration.

6. Greater than 12°. The angle will be closer to the boat's heading of 25° because as the boat speeds up the river has less effect.

CONCEPTUAL QUESTIONS

For instructor-assigned homework, go to www.masteringphysics.com.

(Answers to odd-numbered Conceptual Questions can be found in the back of the book.)

1. For the following quantities, indicate which is a scalar and which is a vector: **(a)** the time it takes for you to run the 100-yard dash; **(b)** your displacement after running the 100-yard dash; **(c)** your average velocity while running; **(d)** your average speed while running.

2. Which, if any, of the vectors shown in FIGURE 3-36 are equal?

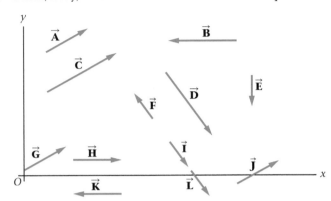

FIGURE 3-36 Conceptual Question 2

3. Given that $\vec{A} + \vec{B} = 0$, **(a)** how does the magnitude of $\vec{B}$ compare with the magnitude of $\vec{A}$? **(b)** How does the direction of $\vec{B}$ compare with the direction of $\vec{A}$?

4. Can a component of a vector be greater than the vector's magnitude?

5. Suppose that $\vec{A}$ and $\vec{B}$ have nonzero magnitude. Is it possible for $\vec{A} + \vec{B}$ to be zero?

6. Can a vector with zero magnitude have one or more components that are nonzero? Explain.

7. Given that $\vec{A} + \vec{B} = \vec{C}$, and that $A^2 + B^2 = C^2$, how are $\vec{A}$ and $\vec{B}$ oriented relative to one another?

8. Given that $\vec{A} + \vec{B} = \vec{C}$, and that $A + B = C$, how are $\vec{A}$ and $\vec{B}$ oriented relative to one another?

9. Given that $\vec{A} + \vec{B} = \vec{C}$, and that $A - B = C$, how are $\vec{A}$ and $\vec{B}$ oriented relative to one another?

10. Vector $\vec{A}$ has x and y components of equal magnitude. What can you say about the possible directions of $\vec{A}$?

11. The components of a vector $\vec{A}$ satisfy the relation $A_x = -A_y \neq 0$. What are the possible directions of $\vec{A}$?

12. Use a sketch to show that two vectors of unequal magnitude cannot add to zero, but that three vectors of unequal magnitude can.

13. Rain is falling vertically downward and you are running for shelter. To keep driest, should you hold your umbrella vertically, tilted forward, or tilted backward? Explain.

14. When sailing, the wind feels stronger when you sail upwind ("beating") than when you are sailing downwind ("running"). Explain.

PROBLEMS AND CONCEPTUAL EXERCISES

Answers to odd-numbered Problems and Conceptual Exercises can be found in the back of the book. **BIO** *identifies problems of biological or medical interest;* **CE** *indicates a conceptual exercise.* Predict/Explain *problems ask for two responses: **(a)** your prediction of a physical outcome, and **(b)** the best explanation among three provided; and* Predict/Calculate *problems ask for a prediction of a physical outcome, based on fundamental physics concepts, and follow that with a numerical calculation to verify the prediction. On all problems, bullets (•, ••, •••) indicate the level of difficulty.*

SECTION 3-2 THE COMPONENTS OF A VECTOR

1. • **CE** Suppose that each component of a certain vector is doubled. **(a)** By what multiplicative factor does the magnitude of the vector change? **(b)** By what multiplicative factor does the direction angle of the vector change?

2. • **CE** Rank the vectors in FIGURE 3-37 in order of increasing magnitude.

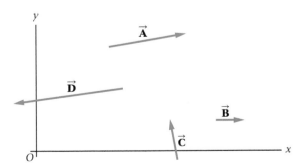

FIGURE 3-37 Problems 2, 3, and 4

3. • **CE** Rank the vectors in Figure 3-37 in order of increasing value of their x component.

4. • **CE** Rank the vectors in Figure 3-37 in order of increasing value of their y component.

5. • The press box at a baseball park is 44.5 ft above the ground. A reporter in the press box looks at an angle of 13.4° below the horizontal to see second base. What is the horizontal distance from the press box to second base?

6. • You are driving up a long, inclined road. After 1.2 miles you notice that signs along the roadside indicate that your elevation has increased by 530 ft. **(a)** What is the angle of the road above the horizontal? **(b)** How far do you have to drive to gain an additional 150 ft of elevation?

7. • **A One-Percent Grade** A road that rises 1 ft for every 100 ft traveled horizontally is said to have a 1% grade. Portions of the Lewiston grade, near Lewiston, Idaho, have a 6% grade. At what angle is this road inclined above the horizontal?

8. • You walk in a straight line for 95 m at an angle of 162° above the positive x axis. What are the x and y components of your displacement?

9. •• Find the x and y components of a position vector $\vec{r}$ of magnitude $r = 88$ m, if its angle relative to the x axis is **(a)** 32.0° and **(b)** 64.0°.

10. •• A vector has the components $A_x = 22$ m and $A_y = 13$ m. **(a)** What is the magnitude of this vector? **(b)** What angle does this vector make with the positive x axis?

11. •• A vector has the components $A_x = -36$ m and $A_y = 43$ m. **(a)** What is the magnitude of this vector? **(b)** What angle does this vector make with the positive x axis?

12. •• A baseball "diamond" (**FIGURE 3-38**) is a square with sides 90 ft in length. If the positive x axis points from home plate to first base, and the positive y axis points from home plate to third base, find the displacement vector of a base runner who has just hit **(a)** a double, **(b)** a triple, or **(c)** a home run.

FIGURE 3-38 Problem 12

13. •• A lighthouse that rises 49 ft above the surface of the water sits on a rocky cliff that extends 19 ft from its base, as shown in **FIGURE 3-39**. A sailor on the deck of a ship sights the top of the lighthouse at an angle of 30.0° above the horizontal. If the sailor's eye level is 14 ft above the water, how far is the ship from the rocks?

FIGURE 3-39 Problem 13

14. •• **H₂O** A water molecule is shown schematically in **FIGURE 3-40**. The distance from the center of the oxygen atom to the center of a hydrogen atom is 0.96 Å, and the angle between the hydrogen atoms is 104.5°. Find the center-to-center distance between the hydrogen atoms. (1 Å = 10^{-10} m.)

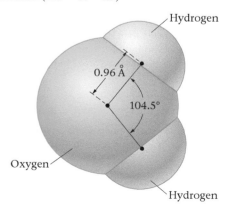

FIGURE 3-40 Problem 14

15. •• **Predict/Calculate** The x and y components of a vector $\vec{r}$ are $r_x = 14$ m and $r_y = -9.5$ m, respectively. Find **(a)** the direction and **(b)** the magnitude of the vector $\vec{r}$. **(c)** If both r_x and r_y are doubled,

how do you predict your answers to parts (a) and (b) will change? **(d)** Verify your prediction in part (c) by calculating the magnitude and direction for this new case.

16. •• You drive a car 680 ft to the east, then 340 ft to the north. **(a)** What is the magnitude of your displacement? **(b)** Using a sketch, estimate the direction of your displacement. **(c)** Verify your estimate in part (b) with a numerical calculation of the direction.

17. •• Vector $\vec{A}$ has a magnitude of 50 units and points in the positive x direction. A second vector, $\vec{B}$, has a magnitude of 120 units and points at an angle of 70° below the x axis. Which vector has **(a)** the greater x component, and **(b)** the greater y component?

18. •• A treasure map directs you to start at a palm tree and walk due north for 15.0 m. You are then to turn 90° and walk 22.0 m; then turn 90° again and walk 5.00 m. Give the distance from the palm tree, and the direction relative to north, for each of the four possible locations of the treasure.

19. •• A whale comes to the surface to breathe and then dives at an angle of 20.0° below the horizontal (**FIGURE 3-41**). If the whale continues in a straight line for 215 m, **(a)** how deep is it, and **(b)** how far has it traveled horizontally?

FIGURE 3-41 Problem 19

SECTION 3-3 ADDING AND SUBTRACTING VECTORS

20. • **CE** Consider the vectors $\vec{A}$ and $\vec{B}$ shown in **FIGURE 3-42**. Which of the other four vectors in the figure ($\vec{C}$, $\vec{D}$, $\vec{E}$, and $\vec{F}$) best represents the *direction* of **(a)** $\vec{A} + \vec{B}$, **(b)** $\vec{A} - \vec{B}$, and **(c)** $\vec{B} - \vec{A}$?

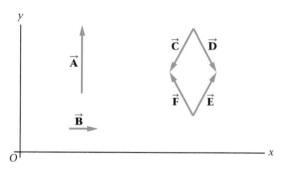

FIGURE 3-42 Problems 20 and 21

21. • **CE** Refer to Figure 3-42 for the following questions: **(a)** Is the magnitude of $\vec{A} + \vec{D}$ greater than, less than, or equal to the magnitude of $\vec{A} + \vec{E}$? **(b)** Is the magnitude of $\vec{A} + \vec{E}$ greater than, less than, or equal to the magnitude of $\vec{A} + \vec{F}$?

22. •• A vector $\vec{A}$ has a magnitude of 40.0 m and points in a direction 20.0° below the positive x axis. A second vector, $\vec{B}$, has a magnitude of 75.0 m and points in a direction 50.0° above the positive x axis. **(a)** Sketch the vectors $\vec{A}$, $\vec{B}$, and $\vec{C} = \vec{A} + \vec{B}$. **(b)** Using the component method of vector addition, find the magnitude and direction of the vector $\vec{C}$.

23. •• An air traffic controller observes two airplanes approaching the airport. The displacement from the control tower to plane 1 is given by the vector $\vec{\mathbf{A}}$, which has a magnitude of 220 km and points in a direction 32° north of west. The displacement from the control tower to plane 2 is given by the vector $\vec{\mathbf{B}}$, which has a magnitude of 140 km and points 65° east of north. **(a)** Sketch the vectors $\vec{\mathbf{A}}$, $-\vec{\mathbf{B}}$, and $\vec{\mathbf{D}} = \vec{\mathbf{A}} - \vec{\mathbf{B}}$. Notice that $\vec{\mathbf{D}}$ is the displacement from plane 2 to plane 1. **(b)** Find the magnitude and direction of the vector $\vec{\mathbf{D}}$.

24. •• The initial velocity of a car, $\vec{\mathbf{v}}_i$, is 45 km/h in the positive x direction. The final velocity of the car, $\vec{\mathbf{v}}_f$, is 66 km/h in a direction that points 75° above the positive x axis. **(a)** Sketch the vectors $-\vec{\mathbf{v}}_i$, $\vec{\mathbf{v}}_f$, and $\Delta\vec{\mathbf{v}} = \vec{\mathbf{v}}_f - \vec{\mathbf{v}}_i$. **(b)** Find the magnitude and direction of the change in velocity, $\Delta\vec{\mathbf{v}}$.

25. •• Vector $\vec{\mathbf{A}}$ points in the positive x direction and has a magnitude of 75 m. The vector $\vec{\mathbf{C}} = \vec{\mathbf{A}} + \vec{\mathbf{B}}$ points in the positive y direction and has a magnitude of 95 m. **(a)** Sketch $\vec{\mathbf{A}}$, $\vec{\mathbf{B}}$, and $\vec{\mathbf{C}}$. **(b)** Estimate the magnitude and direction of the vector $\vec{\mathbf{B}}$. **(c)** Verify your estimate in part (b) with a numerical calculation.

26. •• Vector $\vec{\mathbf{A}}$ points in the negative x direction and has a magnitude of 22 units. The vector $\vec{\mathbf{B}}$ points in the positive y direction. **(a)** Find the magnitude of $\vec{\mathbf{B}}$ if $\vec{\mathbf{A}} + \vec{\mathbf{B}}$ has a magnitude of 37 units. **(b)** Sketch $\vec{\mathbf{A}}$ and $\vec{\mathbf{B}}$.

27. •• Vector $\vec{\mathbf{A}}$ points in the negative y direction and has a magnitude of 5 units. Vector $\vec{\mathbf{B}}$ has twice the magnitude and points in the positive x direction. Find the direction and magnitude of **(a)** $\vec{\mathbf{A}} + \vec{\mathbf{B}}$, **(b)** $\vec{\mathbf{A}} - \vec{\mathbf{B}}$, and **(c)** $\vec{\mathbf{B}} - \vec{\mathbf{A}}$.

28. •• A basketball player runs down the court, following the path indicated by the vectors $\vec{\mathbf{A}}$, $\vec{\mathbf{B}}$, and $\vec{\mathbf{C}}$ in **FIGURE 3-43**. The magnitudes of these three vectors are $A = 10.0$ m, $B = 20.0$ m, and $C = 7.0$ m. Find the magnitude and direction of the net displacement of the player using **(a)** the graphical method and **(b)** the component method of vector addition. Compare your results.

FIGURE 3-43 Problem 28

SECTION 3-4 UNIT VECTORS

29. • A particle undergoes a displacement $\Delta\vec{\mathbf{r}}$ of magnitude 54 m in a direction 42° below the x axis. Express $\Delta\vec{\mathbf{r}}$ in terms of the unit vectors $\hat{\mathbf{x}}$ and $\hat{\mathbf{y}}$.

30. • A vector has a magnitude of 3.50 m and points in a direction that is 145° counterclockwise from the x axis. Find the x and y components of this vector.

31. • A vector $\vec{\mathbf{A}}$ has a length of 6.1 m and points in the negative x direction. Find **(a)** the x component and **(b)** the magnitude of the vector $-3.7\,\vec{\mathbf{A}}$.

32. • The vector $-5.2\,\vec{\mathbf{A}}$ has a magnitude of 34 m and points in the positive x direction. Find **(a)** the x component and **(b)** the magnitude of the vector $\vec{\mathbf{A}}$.

33. • Find the direction and magnitude of the vectors.
 (a) $\vec{\mathbf{A}} = (5.0\,\text{m})\hat{\mathbf{x}} + (-2.0\,\text{m})\hat{\mathbf{y}}$,
 (b) $\vec{\mathbf{B}} = (-2.0\,\text{m})\hat{\mathbf{x}} + (5.0\,\text{m})\hat{\mathbf{y}}$, and **(c)** $\vec{\mathbf{A}} + \vec{\mathbf{B}}$.

34. • Find the direction and magnitude of the vectors.
 (a) $\vec{\mathbf{A}} = (22\,\text{m})\hat{\mathbf{x}} + (-14\,\text{m})\hat{\mathbf{y}}$,
 (b) $\vec{\mathbf{B}} = (2.5\,\text{m})\hat{\mathbf{x}} + (13\,\text{m})\hat{\mathbf{y}}$, and **(c)** $\vec{\mathbf{A}} + \vec{\mathbf{B}}$.

35. • For the vectors given in Problem 34, express **(a)** $\vec{\mathbf{A}} - \vec{\mathbf{B}}$ and **(b)** $\vec{\mathbf{B}} - \vec{\mathbf{A}}$ in unit vector notation.

36. • Express each of the vectors in **FIGURE 3-44** in unit vector notation.

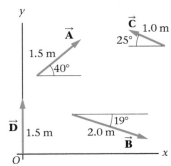

FIGURE 3-44 Problems 36 and 37

37. •• Referring to the vectors in Figure 3-44, express the sum $\vec{\mathbf{A}} + \vec{\mathbf{B}} + \vec{\mathbf{C}}$ in unit vector notation.

SECTION 3-5 POSITION, DISPLACEMENT, VELOCITY, AND ACCELERATION VECTORS

38. • **CE** The blue curves shown in **FIGURE 3-45** display the constant-speed motion of two different particles in the x-y plane. For each of the eight vectors in Figure 3-45, state whether it is **(a)** a position vector, **(b)** a velocity vector, or **(c)** an acceleration vector for the particles.

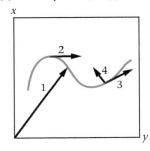

FIGURE 3-45 Problem 38

39. • What are the direction and magnitude of your total displacement if you have traveled due west with a speed of 23 m/s for 175 s, then due south at 12 m/s for 285 s?

40. •• Predict/Calculate **Moving the Knight** Two of the allowed chess moves for a knight are shown in **FIGURE 3-46**. **(a)** Is the magnitude of displacement 1 greater than, less than, or equal to the magnitude of displacement 2? Explain. **(b)** Find the magnitude and direction of the knight's displacement for each of the two moves. Assume that the checkerboard squares are 3.5 cm on a side.

FIGURE 3-46 Problem 40

41. •• To visit your favorite ice cream shop, you must travel 490 m west on Main Street and then 950 m south on Division Street. **(a)** Find the total distance you traveled. **(b)** Write your displacement vector in unit vector notation, taking the $\hat{\mathbf{x}}$ direction to be east and the $\hat{\mathbf{y}}$ direction to be north. **(c)** Write the displacement vector required to return to your starting point in unit vector notation.

42. •• Referring to Problem 41, suppose you take 44 s to complete the 490-m displacement and 73 s to complete the 950-m displacement. **(a)** What are the magnitude and direction of your average velocity during this 117-second period of time? **(b)** What is your average speed during the trip?

43. •• You drive a car 1500 ft to the east, then 2500 ft to the north. If the trip took 3.0 minutes, what were the direction and magnitude of your average velocity?

44. •• Predict/Calculate A jogger runs with a speed of 3.25 m/s in a direction 30.0° above the x axis. **(a)** Find the x and y components of the jogger's velocity. **(b)** How will the velocity components found in part (a) change if the jogger's speed is halved? **(c)** Calculate the components of the velocity for the case where the jogger's speed is halved.

45. •• You throw a ball upward with an initial speed of 4.5 m/s. When it returns to your hand 0.92 s later, it has the same speed in the downward direction (assuming air resistance can be ignored). What was the average acceleration vector of the ball?

46. •• Consider a skateboarder who starts from rest at the top of a ramp that is inclined at an angle of 17.5° to the horizontal. Assuming that the skateboarder's acceleration is $g \sin 17.5°$, find his speed when he reaches the bottom of the ramp in 3.25 s.

47. ••• In a soccer game a midfielder kicks the ball from one touchline directly toward the other with a velocity of $(14.0 \text{ m/s}) \, \hat{\mathbf{x}}$. After the ball travels for 2.50 s, a striker makes an acrobatic kick that gives the ball a new velocity in the $\hat{\mathbf{y}}$ direction, and after 0.900 s the ball passes the diving goalkeeper and goes into the net. If the total displacement of the ball is 40.2 m at 29.5° above the x axis, with what speed did the striker kick the ball?

SECTION 3-6 RELATIVE MOTION

48. • CE The accompanying photo shows a KC-10A Extender using a boom to refuel an aircraft in flight. If the velocity of the KC-10A is 125 m/s due east relative to the ground, what is the velocity of the aircraft being refueled relative to **(a)** the ground, and **(b)** the KC-10A?

Air-to-air refueling. (Problem 48)

49. • As an airplane taxies on the runway with a speed of 16.5 m/s, a flight attendant walks toward the tail of the plane with a speed

of 1.22 m/s. What is the flight attendant's speed relative to the ground?

50. • Referring to part (a) of Example 3-11, find the time it takes for the boat to reach the opposite shore if the river is 35 m wide.

51. •• A police car travels at 38.0 m/s due east while in pursuit of a vehicle that is traveling at 33.5 m/s due east. **(a)** What is the velocity of the vehicle relative to the police car? **(b)** What is the velocity of the police car relative to the vehicle?

52. •• Consider the river crossing problem in Example 3-11. Suppose we would like the boat to move directly across the river (in the positive x direction) with a speed of 5.0 m/s. What is the corresponding speed and direction of the boat's velocity relative to the water?

53. •• As you hurry to catch your flight at the local airport, you encounter a moving walkway that is 61 m long and has a speed of 2.2 m/s relative to the ground. If it takes you 49 s to cover 61 m when walking on the ground, how much time will it take you to cover the same distance on the walkway? Assume that you walk with the same speed on the walkway as you do on the ground.

54. •• In Problem 53, how much time would it take you to cover the 61-m length of the walkway if, once you get on the walkway, you immediately turn around and start walking in the opposite direction with a speed of 1.3 m/s relative to the walkway?

55. •• Predict/Calculate The pilot of an airplane wishes to fly due north, but there is a 65-km/h wind blowing toward the east. **(a)** In what direction relative to north should the pilot head her plane if its speed relative to the air is 340 km/h? **(b)** Draw a vector diagram that illustrates your result in part (a). **(c)** If the pilot decreases the air speed of the plane, but still wants to head due north, should the angle found in part (a) be increased or decreased?

56. •• A passenger walks from one side of a ferry to the other as it approaches a dock. If the passenger's velocity is 1.50 m/s due north relative to the ferry, and 4.50 m/s at an angle of 30.0° west of north relative to the water, what are the direction and magnitude of the ferry's velocity relative to the water?

57. •• You are riding on a Jet Ski at an angle of 35° upstream on a river flowing with a speed of 2.8 m/s. If your velocity relative to the ground is 9.5 m/s at an angle of 20.0° upstream, what is the speed of the Jet Ski relative to the water? (*Note*: Angles are measured relative to the x axis shown in Example 3-11.)

58. ••• Predict/Calculate In Problem 57, suppose the Jet Ski is moving at a speed of 11 m/s relative to the water. **(a)** At what angle must you point the Jet Ski if your velocity relative to the ground is to be perpendicular to the shore of the river? **(b)** If you increase the speed of the Jet Ski relative to the water, does the angle in part (a) increase, decrease, or stay the same? Explain. **(c)** Calculate the new angle if you increase the Jet Ski speed to 15 m/s. (*Note*: Angles are measured relative to the x axis shown in Example 3-11.)

59. ••• Predict/Calculate Two people take identical Jet Skis across a river, traveling at the same speed relative to the water. Jet Ski A heads directly across the river and is carried downstream by the current before reaching the opposite shore. Jet Ski B travels in a direction that is 35° upstream and arrives at the opposite shore directly across from the starting point. **(a)** Which Jet Ski reaches the opposite shore in the least amount of time? **(b)** Confirm your answer to part (a) by finding the ratio of the time it takes for the two Jet Skis to cross the river. (*Note*: Angles are measured relative to the x axis shown in Example 3-11.)

GENERAL PROBLEMS

60. • **CE** Predict/Explain Consider the vectors $\vec{A} = (1.2\,\text{m})\hat{x}$ and $\vec{B} = (-3.4\,\text{m})\hat{x}$. **(a)** Is the magnitude of vector $\vec{A}$ greater than, less than, or equal to the magnitude of vector $\vec{B}$? **(b)** Choose the *best explanation* from among the following:
I. The number 3.4 is greater than the number 1.2.
II. The component of $\vec{B}$ is negative.
III. The vector $\vec{A}$ points in the positive x direction.

61. • **CE** Predict/Explain Two vectors are defined as follows: $\vec{A} = (-3.8\,\text{m})\hat{x}$ and $\vec{B} = (2.1\,\text{m})\hat{y}$. **(a)** Is the magnitude of $\vec{A} + \vec{B}$ greater than, less than, or equal to the magnitude of $\vec{A} - \vec{B}$? **(b)** Choose the *best explanation* from among the following:
I. A vector sum is always greater than a vector difference.
II. $\vec{A} + \vec{B}$ and $\vec{A} - \vec{B}$ produce similar triangles because $\vec{A}$ is perpendicular to $\vec{B}$.
III. The vector $-\vec{B}$ is in the same direction as $\vec{A}$.

62. • To be compliant with regulations the inclination angle of a wheelchair ramp must not exceed 4.76°. If the wheelchair ramp must have a vertical height of 1.22 m, what is its minimum horizontal length?

63. • Find the direction and magnitude of the vector $2\vec{A} + \vec{B}$, where $\vec{A} = (12.1\,\text{m})\hat{x}$ and $\vec{B} = (-32.2\,\text{m})\hat{y}$.

64. •• **CE** The components of a vector $\vec{A}$ satisfy $A_x < 0$ and $A_y < 0$. Is the direction angle of $\vec{A}$ between 0° and 90°, between 90° and 180°, between 180° and 270°, or between 270° and 360°?

65. •• **CE** The components of a vector $\vec{B}$ satisfy $B_x > 0$ and $B_y < 0$. Is the direction angle of $\vec{B}$ between 0° and 90°, between 90° and 180°, between 180° and 270°, or between 270° and 360°?

66. •• It is given that $\vec{A} - \vec{B} = (-51.4\,\text{m})\hat{x}$, $\vec{C} = (62.2\,\text{m})\hat{x}$, and $\vec{A} + \vec{B} + \vec{C} = (13.8\,\text{m})\hat{x}$. Find the vectors $\vec{A}$ and $\vec{B}$.

67. •• You pilot an airplane with the intent to fly 392 km in a direction 55.0° south of east from the starting airport. Air traffic control requires you to first fly due south for 85.0 km. **(a)** After completing the 85.0-km flight due south, how far are you from your destination? **(b)** In what direction should you point your airplane to complete the trip? Give your answer as an angle relative to due east. **(c)** If your airplane has an average ground speed of 485 km/h, what estimated time of travel should you give to your passengers?

68. •• Find the x, y, and z components of the vector $\vec{A}$ shown in FIGURE 3-47, given that $A = 65$ m.

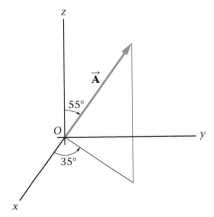

FIGURE 3-47 Problem 68

69. •• Observer 1 rides in a car and drops a ball from rest straight downward, relative to the interior of the car. The car moves horizontally with a constant speed of 3.50 m/s relative to observer 2 standing on the sidewalk. **(a)** What is the speed of the ball 1.00 s after it is released, as measured by observer 2? **(b)** What is the direction of travel of the ball 1.00 s after it is released, as measured relative to the horizontal by observer 2?

70. •• A person riding in a subway train drops a ball from rest straight downward, relative to the interior of the train. The train is moving horizontally with a constant speed of 6.7 m/s. A second person standing at rest on the subway platform observes the ball drop. From the point of view of the person on the platform, the ball is released at the position $x = 0$ and $y = 1.2$ m. Make a plot of the position of the ball for the times $t = 0$, 0.1 s, 0.2 s, 0.3 s, and 0.4 s. (Your plot is parabolic in shape, as we shall see in Chapter 4.)

71. •• A football is thrown horizontally with an initial velocity of $(16.6\,\text{m/s})\hat{x}$. Ignoring air resistance, the average acceleration of the football over any period of time is $(-9.81\,\text{m/s}^2)\hat{y}$. **(a)** Find the velocity vector of the ball 1.75 s after it is thrown. **(b)** Find the magnitude and direction of the velocity at this time.

72. •• As a function of time, the velocity of the football described in Problem 71 can be written as $\vec{v} = (16.6\,\text{m/s})\hat{x} - [(9.81\,\text{m/s}^2)t]\hat{y}$. Calculate the average acceleration vector of the football for the time periods **(a)** $t = 0$ to $t = 1.00$ s, **(b)** $t = 0$ to $t = 2.50$ s, and **(c)** $t = 0$ to $t = 5.00$ s.

73. •• Two airplanes taxi as they approach the terminal. Plane 1 taxies with a speed of 12 m/s due north. Plane 2 taxies with a speed of 7.5 m/s in a direction 20° north of west. **(a)** What are the direction and magnitude of the velocity of plane 1 relative to plane 2? **(b)** What are the direction and magnitude of the velocity of plane 2 relative to plane 1?

74. •• A shopper at the supermarket follows the path indicated by vectors $\vec{A}$, $\vec{B}$, $\vec{C}$, and $\vec{D}$ in FIGURE 3-48. Given that the vectors have the magnitudes $A = 51$ ft, $B = 45$ ft, $C = 35$ ft, and $D = 13$ ft, find the total displacement of the shopper using **(a)** the graphical method and **(b)** the component method of vector addition. Give the direction of the displacement relative to the direction of vector $\vec{A}$.

FIGURE 3-48 Problem 74

75. •• **BIO** A food particle from your breakfast takes a circuitous path through your digestive system. Suppose its motion over a period of time can be represented by the four displacement vectors depicted in FIGURE 3-49. Let the vectors have magnitudes $A = 8.0$ cm, $B = 16$ cm, $C = 23$ cm, and $D = 5.6$ cm. **(a)** What is the total displacement of the food particle? Give its direction relative to the direction of vector $\vec{A}$. **(b)** If the average speed of the particle is 0.010 mm/s, what is its average velocity over the time interval of the four displacements?

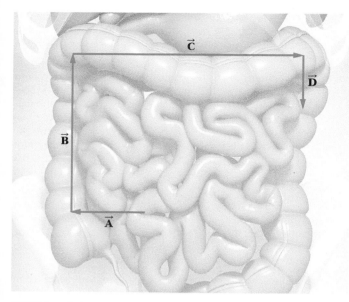

FIGURE 3-49 Problem 75

76. •• Initially, a particle is moving at 4.10 m/s at an angle of 33.5° above the horizontal. Two seconds later, its velocity is 6.05 m/s at an angle of 59.0° below the horizontal. What was the particle's average acceleration during these 2.00 seconds?

77. •• A passenger on a stopped bus notices that rain is falling vertically just outside the window. When the bus moves with constant velocity, the passenger observes that the falling raindrops are now making an angle of 15° with respect to the vertical. (a) What is the ratio of the speed of the raindrops to the speed of the bus? (b) Find the speed of the raindrops, given that the bus is moving with a speed of 18 m/s.

78. •• Predict/Calculate Suppose we orient the x axis of a two-dimensional coordinate system along the beach at Waikiki, as shown in FIGURE 3-50. Waves approaching the beach have a velocity relative to the shore given by $\vec{v}_{ws} = (-1.3\ \text{m/s})\hat{y}$. Surfers move more rapidly than the waves, but at an angle θ to the beach. The angle is chosen so that the surfers approach the shore with the same speed as the waves along the y direction. (a) If a surfer has a speed of 7.2 m/s relative to the water, what is his direction of motion θ relative to the beach? (b) What his the surfer's velocity

relative to the wave? (c) If the surfer's speed is increased, will the angle in part (a) increase or decrease? Explain.

79. •• Predict/Calculate **The Longitude Problem** In 1759, John Harrison (1693–1776) completed his fourth precision chronometer, the H4, which eventually won the celebrated Longitude Prize. (For the human drama behind the Longitude Prize, see *Longitude*, by Dava Sobel.) When the minute hand of the H4 indicates 10 minutes past the hour, it extends 3.0 cm in the horizontal direction. (a) How long is the H4's minute hand? (b) At 10 minutes past the hour, is the extension of the minute hand in the vertical direction more than, less than, or equal to 3.0 cm? Explain. (c) Calculate the vertical extension of the minute hand at 10 minutes past the hour.

Not just a watch! The Harrison H4.
(Problem 79)

80. ••• Referring to Example 3-11, (a) what heading must the boat have if it is to land directly across the river from its starting point? (b) How much time is required for this trip if the river is 25.0 m wide? (c) Suppose the speed of the boat is increased, but it is still desired to land directly across from the starting point. Should the boat's heading be more upstream, more downstream, or the same as in part (a)? Explain.

81. ••• Vector $\vec{A}$ points in the negative x direction. Vector $\vec{B}$ points at an angle of 30.0° above the positive x axis. Vector $\vec{C}$ has a magnitude of 15 m and points in a direction 40.0° below the positive x axis. Given that $\vec{A} + \vec{B} + \vec{C} = 0$, find the magnitudes of $\vec{A}$ and $\vec{B}$.

FIGURE 3-50 Problem 78

82. ••• As two boats approach the marina, the velocity of boat 1 relative to boat 2 is 2.15 m/s in a direction 47.0° east of north. If boat 1 has a velocity that is 0.775 m/s due north, what is the velocity (magnitude and direction) of boat 2?

PASSAGE PROBLEMS

BIO **Motion Camouflage in Dragonflies**

Dragonflies, whose ancestors were once the size of hawks, have prowled the skies in search of small flying insects for over 250 million years. Faster and more maneuverable than any other insect, they even fold their front two legs in flight and tuck them behind their head to be as streamlined as possible. They also employ an intriguing stalking strategy known as "motion camouflage" to approach their prey almost undetected.

The basic idea of motion camouflage is for the dragonfly to move in such a way that the line of sight from the prey to the dragonfly is always in the same direction. Moving in this way, the dragonfly appears almost motionless to its prey, as if it were an object at infinity. Eventually the prey notices the dragonfly has grown in size and is therefore closer, but by that time it's too late for the prey to evade capture.

A typical capture scenario is shown in FIGURE 3-51, where the prey moves in the positive y direction with the constant speed $v_p = 0.750$ m/s, and the dragonfly moves at an angle $\theta = 48.5°$ to the x axis with the constant speed v_d. If the dragonfly chooses its speed correctly, the line of sight from the prey to the dragonfly will always be in the same direction—parallel to the x axis in this case.

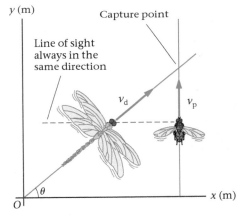

FIGURE 3-51 Problems 83, 84, 85, and 86

83. • What speed must the dragonfly have if the line of sight, which is parallel to the x axis initially, is to remain parallel to the x axis?

    A. 0.562 m/s   B. 0.664 m/s
    C. 1.00 m/s   D. 1.13 m/s

84. • Suppose the dragonfly now approaches its prey along a path with $\theta > 48.5°$, but it still keeps the line of sight parallel to the x axis. Is the speed of the dragonfly in this new case greater than, less than, or equal to its speed in Problem 83?

85. •• If the dragonfly approaches its prey with a speed of 0.950 m/s, what angle θ is required to maintain a constant line of sight parallel to the x axis?

    A. 37.9°   B. 38.3°
    C. 51.7°   D. 52.1°

86. •• In a similar situation, the dragonfly is observed to fly at a constant speed, but its angle to the x axis gradually increases from 20° to 40°. Which of the following is the best explanation for this behavior?

    A. The prey is moving in the positive y direction and maintaining a constant speed.
    B. The prey is moving in the positive y direction and increasing its speed.
    C. The prey is moving in the positive y direction and decreasing its speed.
    D. The prey is moving in the positive x direction and maintaining a constant speed.

87. •• **Referring to Example 3-11** Suppose the speed of the boat relative to the water is 7.0 m/s. **(a)** At what angle to the x axis must the boat be headed if it is to land directly across the river from its starting position? **(b)** If the speed of the boat relative to the water is increased, will the angle needed to go directly across the river increase, decrease, or stay the same? Explain.

88. ••• **Referring to Example 3-11** Suppose that the boat has a speed of 6.7 m/s relative to the water, and that the dock on the opposite shore of the river is at the location $x = 55$ m and $y = 28$ m relative to the starting point of the boat. **(a)** At what angle relative to the x axis must the boat be pointed in order to reach the other dock? **(b)** With the angle found in part (a), what is the speed of the boat relative to the ground?

4

Two-Dimensional Kinematics

Big Ideas

1 Two-dimensional motion consists of independent horizontal and vertical motions.

2 Objects in free fall move under the influence of gravity alone.

3 Objects in free fall follow parabolic paths.

▲ When you hear the word projectile, you probably think of an artillery shell or a home run into the upper deck. But the term projectile applies to any object moving under the influence of gravity alone. For example, a juggling ball undergoes projectile motion—and follows a parabolic path—as it moves from one hand to the other. In this chapter we explore the physical laws that govern projectile motion.

The main idea in this chapter is quite simple: Horizontal and vertical motions are independent. That's it. This chapter develops and applies the idea of independence of motion to many common physical systems.

4-1 Motion in Two Dimensions

In this section we develop equations of motion to describe objects moving in two dimensions. First, we consider motion with constant velocity, determining x and y as functions of time. Later, we investigate motion with constant acceleration. In both cases, we show that the one-dimensional kinematic equations of Chapter 2 can be extended in a straightforward way to two dimensions.

Constant Velocity

To begin, consider the simple situation shown in FIGURE 4-1. A turtle starts at the origin at $t = 0$ and moves with a constant speed $v_0 = 0.26$ m/s in a direction 25° above the x axis. How far has the turtle moved in the x and y directions after 5.0 seconds?

First, notice that the turtle moves in a straight line a distance given by speed multiplied by time:

$$d = v_0 t = (0.26 \text{ m/s})(5.0 \text{ s}) = 1.3 \text{ m}$$

This is indicated in Figure 4-1 (a). From the definitions of sine and cosine given in the previous chapter, we see that the horizontal (x) and vertical (y) distances are given by

$$x = d \cos 25° = 1.2 \text{ m}$$
$$y = d \sin 25° = 0.55 \text{ m}$$

An alternative way to approach this problem is to treat the x and y motions separately. First, we determine the speed of the turtle in each direction. Referring to Figure 4-1 (b), we see that the x component of velocity is

$$v_{0x} = v_0 \cos 25° = 0.24 \text{ m/s}$$

Similarly, the y component of velocity is

$$v_{0y} = v_0 \sin 25° = 0.11 \text{ m/s}$$

Next, we find the distance traveled by the turtle in the x and y directions by multiplying the speed in each direction by the time:

$$x = v_{0x}t = (0.24 \text{ m/s})(5.0 \text{ s}) = 1.2 \text{ m}$$
$$y = v_{0y}t = (0.11 \text{ m/s})(5.0 \text{ s}) = 0.55 \text{ m}$$

This is in agreement with our previous results. To summarize, we can think of the turtle's actual motion as a combination of separate x and y motions.

Basic Equations for Constant Velocity   In general, the turtle (or any object) might start at a position $x = x_0$ and $y = y_0$ at time $t = 0$. In this case, the equations of motion are

$$x = x_0 + v_{0x}t \qquad\qquad \text{4-1}$$

and

$$y = y_0 + v_{0y}t \qquad\qquad \text{4-2}$$

These equations give x and y as functions of time.

(a)

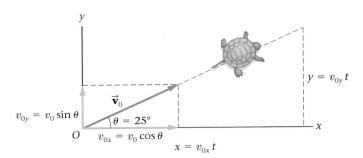

(b)

▲ FIGURE 4-1 **Constant velocity** A turtle walks from the origin with a speed of $v_0 = 0.26$ m/s. **(a)** In a time t the turtle moves through a straight-line distance of $d = v_0 t$; thus the x and y displacements are $x = d \cos \theta$ and $y = d \sin \theta$. **(b)** Equivalently, the turtle's x and y components of velocity are $v_{0x} = v_0 \cos \theta$ and $v_{0y} = v_0 \sin \theta$; hence $x = v_{0x}t$ and $y = v_{0y}t$.

Compare these equations with Equation 2-11, $x = x_0 + v_0 t + \frac{1}{2}at^2$, which gives position as a function of time in one dimension. When the acceleration is zero, as it is for the turtle, Equation 2-11 reduces to $x = x_0 + v_0 t$. Replacing v_0 with the x component of the velocity, v_{0x}, yields Equation 4-1. Similarly, replacing each x in Equation 4-1 with y converts it to Equation 4-2, the y equation of motion.

A situation illustrating the use of $x = x_0 + v_{0x}t$ and $y = y_0 + v_{0y}t$ is given in the following Example.

EXAMPLE 4-1 THE EAGLE DESCENDS

An eagle perched on a tree limb 19.5 m above the water spots a fish swimming near the surface. The eagle pushes off from the branch and descends toward the water. By adjusting its body in flight, the eagle maintains a constant speed of 3.10 m/s at an angle of 20.0° below the horizontal. **(a)** How much time does it take for the eagle to reach the water? **(b)** How far has the eagle traveled in the horizontal direction when it reaches the water?

PICTURE THE PROBLEM

We set up our coordinate system so that the eagle starts at $x_0 = 0$ and $y_0 = h = 19.5$ m. The water level is $y = 0$. As indicated in our sketch, $v_{0x} = v_0 \cos\theta$ and $v_{0y} = -v_0 \sin\theta$, where $v_0 = 3.10$ m/s and $\theta = 20.0°$. The eagle's velocity is constant, and therefore its equations of motion are given by $x = x_0 + v_{0x}t$ and $y = y_0 + v_{0y}t$.

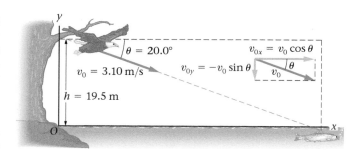

REASONING AND STRATEGY

As usual in such problems, it's best to treat the eagle's flight as a combination of separate x and y motions. Since we are given the constant speed of the eagle, and the angle at which it descends, we can find the x and y components of its velocity. We then use the y equation of motion, $y = y_0 + v_{0y}t$, to find the time t when the eagle reaches the water. Next, we use this value of t in the x equation of motion, $x = x_0 + v_{0x}t$, to find the horizontal distance the bird travels.

Known Height, $h = 19.5$ m; speed, $v_0 = 3.10$ m/s; angle, $\theta = 20.0°$.
Unknown **(a)** Time, $t = ?$ **(b)** Horizontal distance, $x = ?$

SOLUTION

Part (a)

1. Begin by determining the x and y components of the eagle's velocity, v_{0x} and v_{0y}:

$$v_{0x} = v_0 \cos\theta = (3.10 \text{ m/s}) \cos 20.0° = 2.91 \text{ m/s}$$
$$v_{0y} = -v_0 \sin\theta = -(3.10 \text{ m/s}) \sin 20.0° = -1.06 \text{ m/s}$$

2. Now, set $y = 0$ in $y = y_0 + v_{0y}t$ and solve for t:

$$y = y_0 + v_{0y}t = h + v_{0y}t = 0$$
$$t = -\frac{h}{v_{0y}} = -\frac{19.5 \text{ m}}{(-1.06 \text{ m/s})} = 18.4 \text{ s}$$

Part (b)

3. Substitute $t = 18.4$ s into $x = x_0 + v_{0x}t$ to find x:

$$x = x_0 + v_{0x}t = 0 + (2.91 \text{ m/s})(18.4 \text{ s}) = 53.5 \text{ m}$$

INSIGHT

Notice how the two minus signs in Step 2 combine to give a positive time. One minus sign comes from setting $y = 0$, the other from the fact that v_{0y} is negative. No matter where we choose the origin, or what direction we choose to be positive, the time will always have the same value.

As mentioned in the problem statement, the eagle cannot travel in a straight line by simply dropping from the tree limb—it has to adjust its wings and tail to produce enough lift and drag to balance the force of gravity. Airplanes do the same thing when they adjust their flight surfaces to make a smooth landing.

PRACTICE PROBLEM

What is the location of the eagle 2.00 s after it takes flight? **[Answer: $x = 5.82$ m, $y = 17.4$ m]**

Some related homework problems: Problem 2, Problem 3

RWP* One interesting application of the mathematical description of motion in *three* dimensions is the Traffic Collision Avoidance System (TCAS), which uses information from radio signals exchanged between equipped aircraft to warn pilots of a

---

*Real World Physics applications are denoted by the acronym RWP.

possible mid-air collision. When a possible collision is detected, the system automatically negotiates a mutual avoidance maneuver and gives the pilots instructions on a cockpit display as well as by a synthesized voice. The TCAS has prevented numerous mid-air collisions and has saved many lives.

Constant Acceleration

To study motion with constant acceleration in two dimensions we repeat what was done in one dimension in Chapter 2, but with separate equations for both x and y. For example, to obtain x as a function of time we start with $x = x_0 + v_0t + \frac{1}{2}at^2$ (Equation 2-11) and replace both v_0 and a with the corresponding x components, v_{0x} and a_x. This gives

$$x = x_0 + v_{0x}t + \frac{1}{2}a_xt^2 \qquad \text{4-3(a)}$$

To obtain y as a function of time, we write y in place of x in Equation 4-3(a):

$$y = y_0 + v_{0y}t + \frac{1}{2}a_yt^2 \qquad \text{4-3(b)}$$

These are the position-versus-time equations of motion for two dimensions. (In three dimensions we introduce a third coordinate direction and label it z. We would then simply replace x with z in Equation 4-3(a) to obtain z as a function of time.)

The same approach gives velocity as a function of time. Start with Equation 2-7, $v = v_0 + at$, and write it in terms of x and y components. This yields

$$v_x = v_{0x} + a_xt \qquad \text{4-4(a)}$$
$$v_y = v_{0y} + a_yt \qquad \text{4-4(b)}$$

Notice that we simply repeat everything we did for one dimension, only now with separate equations for the x and y components.

Finally, we can write $v^2 = v_0^2 + 2a\Delta x$ in terms of components as well:

$$v_x^2 = v_{0x}^2 + 2a_x\Delta x \qquad \text{4-5(a)}$$
$$v_y^2 = v_{0y}^2 + 2a_y\Delta y \qquad \text{4-5(b)}$$

Our results are listed in Table 4-1.

TABLE 4-1 **Constant-Acceleration Equations of Motion**

Position as a function of time	Velocity as a function of time	Velocity as a function of position
$x = x_0 + v_{0x}t + \frac{1}{2}a_xt^2$	$v_x = v_{0x} + a_xt$	$v_x^2 = v_{0x}^2 + 2a_x\Delta x$
$y = y_0 + v_{0y}t + \frac{1}{2}a_yt^2$	$v_y = v_{0y} + a_yt$	$v_y^2 = v_{0y}^2 + 2a_y\Delta y$

The Same Equations Are Used Throughout This Chapter The fundamental equations in Table 4-1 are used to obtain *all* of the results presented throughout the rest of this chapter. Though it may *appear* sometimes that we are writing new sets of equations for different problems, the equations aren't new at all—what we're actually doing is simply writing these same equations again, but with different specific values substituted for the constants that appear in them.

PHYSICS IN CONTEXT
Looking Back

The equations of one-dimensional kinematics derived in Chapter 2 are used again in this chapter, even though we are now studying kinematics in two dimensions. For example, the equations in Table 4-1 are the same as those used in Chapter 2, only now they are applied individually to the x and y directions.

Big Idea **1** Two-dimensional motion is a combination of horizontal and vertical motions. The key concept behind two-dimensional motion is that the horizontal and vertical motions are completely independent of one another; each can be considered separately as one-dimensional motion.

EXAMPLE 4-2 A CHOPPER ACCELERATES VERTICALLY

A helicopter is in level flight with a constant horizontal speed of 11 m/s. At $t = 0$ it begins to accelerate vertically at 0.96 m/s^2. Assuming the helicopter's acceleration remains constant for the time interval of interest, find **(a)** the horizontal and vertical distances through which it moves in 5.3 s and **(b)** its x and y velocity components at $t = 5.3$ s.

PICTURE THE PROBLEM
In our sketch we have placed the origin of a two-dimensional coordinate system at the location of the helicopter at the initial time, $t = 0$. In addition, we have chosen the initial direction of motion to be in the positive x direction, and the direction of acceleration to be in the positive y direction. As a result, it follows that $x_0 = y_0 = 0$, $v_{0x} = 11$ m/s, $v_{0y} = 0$, $a_x = 0$, and $a_y = 0.96$ m/s^2. As the helicopter moves upward, its y component of velocity increases, resulting in a curved path, as shown.

CONTINUED

REASONING AND STRATEGY

a. We want to relate position and time, and hence we find the horizontal position of the helicopter with $x = x_0 + v_{0x}t + \frac{1}{2}a_xt^2$, and the vertical position with $y = y_0 + v_{0y}t + \frac{1}{2}a_yt^2$. In each case we substitute $t = 5.3$ s.

b. The velocity components at $t = 5.3$ s can be found using $v_x = v_{0x} + a_xt$ and $v_y = v_{0y} + a_yt$.

Known Horizontal speed, $v_{0x} = 11$ m/s; vertical acceleration, $a_y = 0.96$ m/s^2.

Unknown (a) Horizontal distance, $x = ?$; vertical distance, $y = ?$ (b) Horizontal component of velocity, $v_x = ?$; vertical component of velocity, $v_y = ?$

SOLUTION

Part (a)

1. Use $x = x_0 + v_{0x}t + \frac{1}{2}a_xt^2$ to find x at $t = 5.3$ s: $x = x_0 + v_{0x}t + \frac{1}{2}a_xt^2 = 0 + (11 \text{ m/s})(5.3 \text{ s}) + 0 = 58 \text{ m}$

2. Use $y = y_0 + v_{0y}t + \frac{1}{2}a_yt^2$ to find y at $t = 5.3$ s: $y = y_0 + v_{0y}t + \frac{1}{2}a_yt^2 = 0 + 0 + \frac{1}{2}(0.96 \text{ m/s}^2)(5.3 \text{ s})^2 = 13 \text{ m}$

Part (b)

3. Use $v_x = v_{0x} + a_xt$ to find v_x at $t = 5.3$ s: $v_x = v_{0x} + a_xt = 11 \text{ m/s} + 0 = 11 \text{ m/s}$

4. Use $v_y = v_{0y} + a_yt$ to find v_y at $t = 5.3$ s: $v_y = v_{0y} + a_yt = 0 + (0.96 \text{ m/s}^2)(5.3 \text{ s}) = 5.1 \text{ m/s}$

INSIGHT

If we assume constant acceleration, the vertical position of the helicopter will eventually increase more rapidly with time than the horizontal position, due to the t^2 dependence of y as compared with the t dependence of x. This results in a curved, parabolic path for the helicopter, as shown in our sketch. The helicopter's velocity at 5.3 s is $v = \sqrt{v_x{}^2 + v_y{}^2} = \sqrt{(11 \text{ m/s})^2 + (5.1 \text{ m/s})^2} = 12 \text{ m/s}$ at an angle of $\theta = \tan^{-1}(v_y/v_x) = \tan^{-1}[(5.1 \text{ m/s})/(11 \text{ m/s})] = 25°$ above the x axis. It's clear that the angle of flight must be less than 45° at this time because the x component of velocity is greater than the y component.

PRACTICE PROBLEM

How much time is required for the helicopter to move 18 m vertically from its initial position?   [**Answer:** $t = 6.1$ s]

Some related homework problems: Problem 4, Problem 5, Problem 6.

Enhance Your Understanding   (Answers given at the end of the chapter)

1. The equations of motion of an object are $x = (1 \text{ m/s}^2)t^2$ and $y = (2 \text{ m}) - (3 \text{ m/s})t$. What are the x and y components of the object's (a) initial position, (b) initial velocity, and (c) acceleration?

Section Review

- Motion in two dimensions is a combination of one-dimensional motion in the x and y directions. The basic equations used in this chapter are a straightforward extension of the equations of motion in Chapter 2. They are listed in Table 4-1.

4-2 Projectile Motion: Basic Equations

We now apply the independence of horizontal and vertical motions to projectiles. Just what do we mean by a projectile? Well, a **projectile** is an object that is thrown, kicked, batted, or otherwise launched into motion and then allowed to follow a path determined solely by the influence of gravity. As you might expect, this covers a wide variety of physical systems.

In studying projectile motion we make the following assumptions:

- Air resistance is ignored.
- The acceleration due to gravity is constant, downward, and has a magnitude equal to $g = 9.81$ m/s^2.
- The Earth's rotation is ignored.

Air resistance can be significant if a projectile moves with relatively high speed or if it encounters a strong wind. In many everyday situations, however, like tossing a ball to a friend or dropping a book, air resistance is relatively insignificant. As for the acceleration due to gravity, $g = 9.81 \text{ m/s}^2$, this value varies slightly from place to place on the Earth's surface and decreases with increasing altitude. In addition, the rotation of the Earth can be significant when we consider projectiles that cover great distances. Little error is made in ignoring the variation of g or the rotation of the Earth, however, in the examples of projectile motion considered in this chapter.

Equations of Motion for Projectiles Let's incorporate the preceding assumptions into the equations of motion given in the previous section. Suppose, as in **FIGURE 4-2**, that the x axis is horizontal and the y axis is vertical, with the positive direction upward. Noting that downward is the negative direction, it follows that

$$a_y = -9.81 \text{ m/s}^2 = -g$$

Gravity causes no acceleration in the x direction. Thus, the x component of acceleration is zero:

$$a_x = 0$$

With these acceleration components substituted into the fundamental constant-acceleration equations of motion (Table 4-1) we find:

Projectile Motion ($a_x = 0$, $a_y = -g$)

$$x = x_0 + v_{0x}t \qquad v_x = v_{0x} \qquad v_x{}^2 = v_{0x}{}^2$$
$$y = y_0 + v_{0y}t - \tfrac{1}{2}gt^2 \qquad v_y = v_{0y} - gt \qquad v_y{}^2 = v_{0y}{}^2 - 2g\Delta y$$

4-6

In these expressions, the positive y direction is upward and the quantity g is positive. *All* of our studies of projectile motion use Equations 4-6 as our fundamental equations—again, special cases simply correspond to substituting different specific values for the constants.

Demonstrating Independence of Motion A simple demonstration illustrates the independence of horizontal and vertical motions in projectile motion. First, while standing still, drop a rubber ball to the floor and catch it on the rebound. Notice that the ball goes straight down, lands near your feet, and returns almost to the level of your hand in about a second.

Next, walk—or roller skate—with constant speed before dropping the ball, then observe its motion carefully. To you, its motion looks the same as before: It goes straight down, lands near your feet, bounces straight back up, and returns in about one second. This is illustrated in **FIGURE 4-3**. The fact that you were moving in the horizontal direction the whole time had no effect on the ball's vertical motion—the motions are independent.

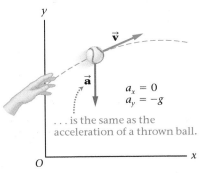

▲ **FIGURE 4-2 Acceleration in free fall** All objects in free fall have acceleration components $a_x = 0$ and $a_y = -g$ when the coordinate system is chosen as shown here. This is true regardless of whether the object is dropped, thrown, kicked, or otherwise set into motion.

Big Idea 2 Projectiles are objects that move under the influence of gravity alone. Projectiles can be dropped from rest or thrown at some angle to the horizontal. Once they are launched, they have all the characteristics of projectile motion, irrespective of how their motion started.

PHYSICS IN CONTEXT
Looking Ahead

The basic idea behind projectile motion is used again in Chapter 12, when we consider orbital motion.

PROBLEM-SOLVING NOTE

Acceleration of a Projectile

When the x axis is chosen to be horizontal and the y axis points vertically upward, it follows that the acceleration of an ideal projectile is $a_x = 0$ and $a_y = -g$.

▲ **FIGURE 4-3 Independence of vertical and horizontal motions** When you drop a ball while walking, running, or skating with constant velocity, it appears to you to drop straight down from the point where you released it. To a person at rest, the ball follows a curved path that combines horizontal and vertical motions.

(a)

(b)

(c)

▲ **FIGURE 4-4 Visualizing Concepts Independence of Motion (a)** An athlete jumps upward from a moving skateboard. The athlete retains his initial horizontal velocity, and hence remains directly above the skateboard at all times. **(b)** The pilot ejection seat of a jet fighter is being ground-tested. The horizontal and vertical motions are independent, and hence the test dummy is still almost directly above the cockpit from which it was ejected. (Notice that air resistance is beginning to reduce the dummy's horizontal velocity.) **(c)** This rollerblader may not be thinking about independence of motion, but the ball she released illustrates the concept perfectly as it falls directly below her hand.

To an observer who sees you walking by, the ball follows a curved path, as shown. The precise shape of this curved path—a parabola—is verified in the next section. Additional examples of this principle are shown in **FIGURE 4-4**.

Ball Fired Upward from Moving Cart

Ball Fired Upward from Accelerating Cart

Enhance Your Understanding   (Answers given at the end of the chapter)

2. A sailor drops a pair of binoculars from the "crow's nest" at the top of a sailing ship's mast. When the ship is at rest, the binoculars land at the base of the mast. If the ship is moving forward with constant velocity, and air resistance can be ignored, do the binoculars land in front of the mast (toward the front of the ship), at the base of the mast, or behind the mast (toward the rear of the ship)? Explain.

Section Review

- Projectiles move under the influence of gravity alone, with a constant downward acceleration. The basic equations describing projectile motion are summarized in Equations 4-6.

(a)

(b)

4-3 Zero Launch Angle

A special case of some interest is a projectile launched horizontally, so that the angle between the initial velocity and the horizontal is $\theta = 0$. We devote this section to a brief look at this type of motion.

Equations of Motion

Suppose you are walking with a speed v_0 when you release a ball from a height h. If we choose ground level to be $y = 0$, and the release point to be directly above the origin, the initial position of the ball is given by

$$x_0 = 0$$

◀ **FIGURE 4-5 Launch angle of a projectile (a)** A projectile launched at an angle above the horizontal, $\theta > 0$. A launch below the horizontal would correspond to $\theta < 0$. **(b)** A projectile launched horizontally, $\theta = 0$. In this section we consider $\theta = 0$. The next section deals with $\theta \neq 0$.

▶ **FIGURE 4-6 Trajectory of a projectile launched horizontally** In this plot, the projectile was launched from a height of 9.5 m with an initial speed of 5.0 m/s. The positions shown in the plot correspond to the times $t = 0.20$ s, 0.40 s, 0.60 s, ... Notice the uniform motion in the x direction, and the accelerated motion in the y direction.

and

$$y_0 = h$$

This is illustrated in Figure 4-3.

The initial velocity is horizontal in this case, corresponding to $\theta = 0$ in **FIGURE 4-5**. As a result, the x component of the initial velocity is simply the initial speed:

$$v_{0x} = v_0 \cos 0° = v_0$$

The y component of the initial velocity is zero:

$$v_{0y} = v_0 \sin 0° = 0$$

Substituting these specific values into our fundamental equations for projectile motion (Equations 4-6) gives the following simplified results for zero launch angle ($\theta = 0$):

$$
\begin{array}{lll}
x = v_0 t & v_x = v_0 = \text{constant} & v_x{}^2 = v_0{}^2 = \text{constant} \\
y = h - \frac{1}{2}gt^2 & v_y = -gt & v_y{}^2 = -2g\Delta y
\end{array}
$$

4-7

Snapshots of this motion at equal time intervals are shown in **FIGURE 4-6**.

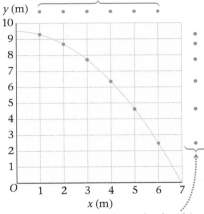

Horizontal motion is uniform—equal distance in equal time.

Vertical motion is accelerated—the object goes farther in each successive interval.

 Dropped and Thrown Balls

PROBLEM-SOLVING NOTE

Identify Initial Conditions

The launch point of a projectile determines x_0 and y_0. The initial velocity of a projectile determines v_{0x} and v_{0y}.

EXAMPLE 4-3 | DROPPING A BALL

A person skateboarding with a constant speed of 1.30 m/s releases a ball from a height of 1.25 m above the ground. Given that $x_0 = 0$ and $y_0 = h = 1.25$ m, find x and y for **(a)** $t = 0.250$ s and **(b)** $t = 0.500$ s. **(c)** Find the velocity, speed, and direction of motion of the ball at $t = 0.500$ s.

PICTURE THE PROBLEM

The ball starts at $x_0 = 0$ and $y_0 = h = 1.25$ m. Its initial velocity is horizontal; therefore $v_{0x} = v_0 = 1.30$ m/s and $v_{0y} = 0$. In addition, it accelerates with the acceleration due to gravity in the negative y direction, $a_y = -g$, and moves with constant speed in the x direction, $a_x = 0$.

REASONING AND STRATEGY

The x and y positions are given by $x = v_0 t$ and $y = h - \frac{1}{2}gt^2$, respectively. We simply substitute time into these expressions. Similarly, the velocity components are $v_x = v_0$ and $v_y = -gt$.

Known Speed, $v_0 = 1.30$ m/s; height, $h = 1.25$ m.

Unknown **(a)** $x = ?$ and $y = ?$ at $t = 0.250$ s; **(b)** $x = ?$ and $y = ?$ at $t = 0.500$ s; **(c)** velocity, $\vec{\mathbf{v}} = ?$, speed, $v = ?$, direction angle, $\theta = ?$, at $t = 0.500$ s

SOLUTION

Part (a)

1. Substitute $t = 0.250$ s into the x and y equations of motion:

$$x = v_0 t = (1.30 \text{ m/s})(0.250 \text{ s}) = 0.325 \text{ m}$$
$$y = h - \frac{1}{2}gt^2$$
$$= 1.25 \text{ m} - \frac{1}{2}(9.81 \text{ m/s}^2)(0.250 \text{ s})^2 = 0.943 \text{ m}$$

Part (b)

2. Substitute $t = 0.500$ s into the x and y equations of motion. Notice that the ball is only about an inch above the ground at this time:

$$x = v_0 t = (1.30 \text{ m/s})(0.500 \text{ s}) = 0.650 \text{ m}$$
$$y = h - \frac{1}{2}gt^2$$
$$= 1.25 \text{ m} - \frac{1}{2}(9.81 \text{ m/s}^2)(0.500 \text{ s})^2 = 0.0238 \text{ m}$$

CONTINUED

Part (c)

3. First, calculate the x and y components of the velocity at $t = 0.500$ s using $v_x = v_0$ and $v_y = -gt$:

$$v_x = v_0 = 1.30 \text{ m/s}$$
$$v_y = -gt = -(9.81 \text{ m/s}^2)(0.500 \text{ s}) = -4.91 \text{ m/s}$$

4. Use these components to determine $\vec{\mathbf{v}}$, v, and θ:

$$\vec{\mathbf{v}} = (1.30 \text{ m/s})\hat{\mathbf{x}} + (-4.91 \text{ m/s})\hat{\mathbf{y}}$$
$$v = \sqrt{v_x^2 + v_y^2}$$
$$= \sqrt{(1.30 \text{ m/s})^2 + (-4.91 \text{ m/s})^2} = 5.08 \text{ m/s}$$
$$\theta = \tan^{-1}\frac{v_y}{v_x} = \tan^{-1}\frac{(-4.91 \text{ m/s})}{1.30 \text{ m/s}} = -75.2°$$

INSIGHT

It's interesting to note that the x position of the ball doesn't depend on the acceleration of gravity, g, and its y position doesn't depend on the initial horizontal speed of the ball, v_0. For example, if the person is running when he drops the ball, the ball is moving faster in the horizontal direction, and it keeps up with the person as it falls. Its vertical motion doesn't change at all, however.

PRACTICE PROBLEM

How much time does it take for the ball to land?   [**Answer:** From the results of part (b), it's clear that the time of landing is slightly greater than 0.500 s. Setting $y = 0$ gives a precise answer: $t = \sqrt{2h/g} = 0.505$ s.]

Some related homework problems: Problem 17, Problem 18, Problem 21

CONCEPTUAL EXAMPLE 4-4   COMPARE SPLASHDOWN SPEEDS

Two youngsters dive off an overhang into a lake. Diver 1 drops straight down, and diver 2 runs off the cliff with an initial horizontal speed v_0. Is the splashdown speed of diver 2 **(a)** greater than, **(b)** less than, or **(c)** equal to the splashdown speed of diver 1?

REASONING AND DISCUSSION

Neither diver has an initial y component of velocity, and both fall with the same vertical acceleration—the acceleration due to gravity. Therefore, the two divers fall for the same amount of time, and their y components of velocity are the same at splashdown. However, because diver 2 also has a nonzero x component of velocity, unlike diver 1, the speed of diver 2 is greater.

ANSWER

(a) The speed of diver 2 is greater than that of diver 1.

Parabolic Path

RWP Just what is the shape of the curved path followed by a projectile launched horizontally? This can be found by combining $x = v_0 t$ and $y = h - \frac{1}{2}gt^2$, which allows us to express y in terms of x. First, solve for time using the x equation. This gives

$$t = \frac{x}{v_0}$$

Next, substitute this result into the y equation to eliminate t:

$$y = h - \frac{1}{2}g\left(\frac{x}{v_0}\right)^2 = h - \left(\frac{g}{2v_0^2}\right)x^2 \qquad 4\text{-}8$$

It follows that y has the form

$$y = a + bx^2$$

◄ **FIGURE 4-7** Visualizing Concepts
Parabolic Trajectories Lava bombs (left) and fountain jets (right) trace out parabolic paths, as is typical in projectile motion. The trajectories are only slightly altered by air resistance.

In this expression, $a = h$ = constant and $b = -g/2v_0^2$ = constant. This is the equation of a *parabola* that curves downward, a characteristic shape in projectile motion. Examples of parabolas produced by real-world projectiles are shown in **FIGURE 4-7**.

Landing Site

Where does a projectile land if it is launched horizontally with a speed v_0 from a height h?

The most direct way to answer this question is to set $y = 0$ in Equation 4-8, since $y = 0$ corresponds to ground level. This gives

$$0 = h - \left(\frac{g}{2v_0^2}\right)x^2$$

Solving for x yields the landing site:

$$x = v_0\sqrt{\frac{2h}{g}} \qquad \text{4-9}$$

We've chosen the positive sign for the square root because the projectile was launched in the positive x direction, and hence it lands at a positive value of x.

PROBLEM-SOLVING NOTE

Use Independence of Motion

Projectile problems can be solved by breaking the problem into its x and y components, and then solving for the motion of each component separately.

EXAMPLE 4-5   JUMPING A CREVASSE

A mountain climber encounters a crevasse in an ice field. The opposite side of the crevasse is 2.74 m lower, and is separated horizontally by a distance of 4.10 m. To cross the crevasse, the climber gets a running start and jumps in the horizontal direction. **(a)** What is the minimum speed needed by the climber to safely cross the crevasse? **(b)** If, instead, the climber's speed is 6.00 m/s, where does the climber land?

PICTURE THE PROBLEM

The mountain climber jumps from $x_0 = 0$ and $y_0 = h = 2.74$ m. The landing site for part (a) is $x = w = 4.10$ m and $y = 0$. As for the initial velocity, we are given that $v_{0x} = v_0$ and $v_{0y} = 0$. Finally, with our choice of coordinates it follows that $a_x = 0$ and $a_y = -g$.

REASONING AND STRATEGY

We can model the climber as a projectile, and apply our equations for projectile motion with a horizontal launch.

a. From Equations 4-7 we have that $x = v_0 t$ and $y = h - \frac{1}{2}gt^2$. Setting $y = 0$ determines the time of landing. Using this time in the x equation gives the horizontal landing position in terms of the initial speed.

b. We can now use the relationship from part (a) to find x in terms of $v_0 = 6.00$ m/s.

Known   Height, $h = 2.74$ m; width, $w = 4.10$ m.

Unknown   **(a)** Minimum initial speed, $v_0 = ?$ **(b)** Landing distance, $x = ?$, for $v_0 = 6.00$ m/s

SOLUTION

Part (a)

1. Set $y = h - \frac{1}{2}gt^2$ equal to zero (landing condition) and solve for the corresponding time t:

$$y = h - \frac{1}{2}gt^2 = 0$$

$$t = \sqrt{\frac{2h}{g}}$$

CONTINUED

2. Substitute this expression for t into the x equation of motion, $x = v_0 t$, and solve for the speed, v_0:

$$x = v_0 t = v_0 \sqrt{\frac{2h}{g}} \quad \text{or} \quad v_0 = x \sqrt{\frac{g}{2h}}$$

3. Substitute numerical values in this expression:

$$v_0 = x \sqrt{\frac{g}{2h}} = (4.10 \text{ m}) \sqrt{\frac{9.81 \text{ m/s}^2}{2(2.74 \text{ m})}} = 5.49 \text{ m/s}$$

Part (b)

4. Substitute $v_0 = 6.00 \text{ m/s}$ into the expression for x obtained in Step 2, $x = v_0 \sqrt{2h/g}$:

$$x = v_0 \sqrt{\frac{2h}{g}} = (6.00 \text{ m/s}) \sqrt{\frac{2(2.74 \text{ m})}{9.81 \text{ m/s}^2}} = 4.48 \text{ m}$$

INSIGHT

The minimum speed to safely cross the crevasse is 5.49 m/s. If the initial horizontal speed is 6.00 m/s, the climber will land 4.48 m − 4.10 m = 0.38 m beyond the edge of the crevasse.

PRACTICE PROBLEM

(a) When the climber's speed is the minimum needed to cross the crevasse, $v_0 = 5.49 \text{ m/s}$, for what amount of time is the climber in the air? (b) For what amount of time is the climber in the air when $v_0 = 6.00 \text{ m/s}$? [**Answer:** (a) $t = x/v_0 = (4.10 \text{ m})/(5.49 \text{ m/s}) = 0.747 \text{ s}$. (b) $t = x/v_0 = (4.48 \text{ m})/(6.00 \text{ m/s}) = 0.747 \text{ s}$. The times are exactly the same! The answer to both parts is simply the time needed to fall through a height h: $t = \sqrt{2h/g} = 0.747 \text{ s}$.]

Some related homework problems: Problem 13, Problem 14, Problem 19

CONCEPTUAL EXAMPLE 4-6   MINIMUM SPEED   PREDICT/EXPLAIN

(a) If the height h is increased in the Example 4-5, but the width w remains the same, does the minimum speed needed to cross the crevasse increase, decrease, or stay the same? (b) Which of the following is the *best explanation* for your prediction?

  I. The time of fall is greater, so the minimum speed decreases.

 II. The width remains the same, so the minimum speed stays the same as well.

III. Increasing the height increases the minimum speed.

REASONING AND DISCUSSION

If the height is greater, the time of fall is also greater. This means that the climber is in the air for a greater time, and hence the horizontal distance covered for a given initial speed is also greater. It follows that a lower initial speed allows for a safe crossing.

ANSWER

(a) The minimum speed decreases. (b) The best explanation is I.

**PHYSICS
IN CONTEXT
Looking Ahead**

Two-dimensional kinematics comes up again when we study the motion of charged particles (like electrons) in electric fields. For example, compare Figure 19-41 and Figure 22-10 with the person jumping a crevasse in Example 4-5. The same basic principles apply in each case.

Enhance Your Understanding   (Answers given at the end of the chapter)

3. Two objects, A and B, are launched horizontally, as indicated in FIGURE 4-8. The horizontal distance covered before landing is smaller for object A than for object B. (a) Is the initial speed of object A greater than, less than, or equal to the initial speed of object B? Explain. (b) Is the amount of time object A is in the air greater than, less than, or equal to the time object B is in the air? Explain.

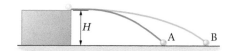

▲ **FIGURE 4-8** Enhance Your Understanding 3

Section Review

- Projectiles follow a parabolic path.
- A projectile launched horizontally falls to the ground in the same time as one that is dropped from rest from the same height.

4-4 General Launch Angle

We now consider the case of a projectile launched at an arbitrary angle with respect to the horizontal. As always, we return to our basic equations for projectile motion (Equations 4-6), and this time we simply let θ be nonzero.

FIGURE 4-9 (a) shows a projectile launched with an initial speed v_0 at an angle θ above the horizontal. Since the projectile starts at the origin, the initial x and y positions are zero:

$$x_0 = y_0 = 0$$

The components of the initial velocity are determined as indicated in FIGURE 4-9 (b):

$$v_{0x} = v_0 \cos \theta$$
$$v_{0y} = v_0 \sin \theta$$

As a quick check, notice that if $\theta = 0$, then $v_{0x} = v_0$ and $v_{0y} = 0$, as expected. Similarly, if $\theta = 90°$, we find $v_{0x} = 0$ and $v_{0y} = v_0$. These checks are depicted in FIGURE 4-9 (c).

General Equations of Motion   Substituting the preceding results into the basic equations for projectile motion yields the following results for a general launch angle:

$$x = (v_0 \cos \theta)t \qquad v_x = v_0 \cos \theta \qquad v_x{}^2 = v_0{}^2 \cos^2 \theta$$
$$y = (v_0 \sin \theta)t - \tfrac{1}{2}gt^2 \quad v_y = v_0 \sin \theta - gt \quad v_y{}^2 = v_0{}^2 \sin^2 \theta - 2g\Delta y$$

4-10

These equations, which are valid for any launch angle, reduce to the simpler Equations 4-7 when we set $\theta = 0$ and $y_0 = h$. In the next two Examples, we use Equations 4-10 to calculate a projectile's position and velocity for three equally spaced times.

QUICK EXAMPLE 4-7   POSITION OF A PROJECTILE

A projectile is launched from the origin with an initial speed of 20.0 m/s at an angle of 35.0° above the horizontal. Find the x and y positions of the projectile at the times **(a)** $t = 0.500$ s, **(b)** $t = 1.00$ s, and **(c)** $t = 1.50$ s.

REASONING AND SOLUTION
Use $x = (v_0 \cos \theta)t$ with $v_0 = 20.0$ m/s and $\theta = 35.0°$ to find the x position of the projectile. Similarly, use $y = (v_0 \sin \theta)t - \tfrac{1}{2}gt^2$ with $g = 9.81$ m/s^2 to find the y position.

  a. $x = (20.0 \text{ m/s})(\cos 35.0°)(0.500 \text{ s}) = 8.19$ m
     $y = (20.0 \text{ m/s})(\sin 35.0°)(0.500 \text{ s}) - \tfrac{1}{2}(9.81 \text{ m/s}^2)(0.500 \text{ s})^2 = 4.51$ m

  b. Repeating the calculation in part (a), but with 1.00 s for the time, we find $x = 16.4$ m and $y = 6.57$ m.

  c. Using 1.50 s for the time in part (a), we find $x = 24.6$ m and $y = 6.17$ m.

Notice that x increases steadily with time. In contrast, y first increases and then decreases.

QUICK EXAMPLE 4-8   VELOCITY OF A PROJECTILE

Find the velocity of the projectile in Quick Example 4-7 at the times **(a)** $t = 0.500$ s, **(b)** $t = 1.00$ s, and **(c)** $t = 1.50$ s.

REASONING AND SOLUTION
The x component of the velocity is constant, with the value $v_x = v_0 \cos \theta = (20.0 \text{ m/s}) \cos 35.0° = 16.4$ m/s. The y component of the velocity changes with time according to the relationship $v_y = v_0 \sin \theta - gt$.

  a. $v_x = v_0 \cos \theta = (20.0 \text{ m/s}) \cos 35.0° = 16.4$ m/s
     $v_y = (20.0 \text{ m/s}) \sin 35.0° - (9.81 \text{ m/s}^2)(0.500 \text{ s}) = 6.57$ m/s
     $\vec{v} = (16.4 \text{ m/s})\hat{x} + (6.57 \text{ m/s})\hat{y}$

CONTINUED

(a)

(b)

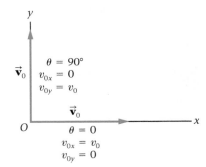

(c)

▲ FIGURE 4-9 **Projectile with an arbitrary launch angle (a)** A projectile launched from the origin at an angle θ above the horizontal. **(b)** The x and y components of the initial velocity. **(c)** Velocity components in the limits $\theta = 0$ and $\theta = 90°$.

PHYSICS IN CONTEXT
Looking Back

The vector methods developed in Chapter 3 are used in this section to resolve velocity vectors into their x and y components.

b. Repeating part (a) with 1.00 s for the time, we find $\vec{v} = (16.4 \text{ m/s})\hat{x} + (1.66 \text{ m/s})\hat{y}$.

c. Repeating part (a) with 1.50 s for the time, we find $\vec{v} = (16.4 \text{ m/s})\hat{x} + (-3.24 \text{ m/s})\hat{y}$.

Notice that the projectile is moving upward in parts (a) and (b), but is moving downward (negative v_y) in part (c).

FIGURE 4-10 shows the projectile referred to in the previous Quick Examples for a series of times spaced by 0.10 s. Notice that the points in Figure 4-10 are not evenly spaced in terms of position, even though they are evenly spaced in time. In fact, the points bunch closer together at the top of the trajectory, showing that a comparatively large fraction of the flight time is spent near the highest point. This is why it seems that a basketball player soaring toward a slam dunk, or a ballerina performing a grand jeté, is "hanging in air." This concept is explored further in **FIGURE 4-11**.

◀ **FIGURE 4-10 Snapshots of a trajectory** This plot shows a projectile launched from the origin with an initial speed of 20.0 m/s at an angle of 35.0° above the horizontal. The positions shown in the plot correspond to the times $t = 0.1$ s, 0.2 s, 0.3 s, ... Red dots mark the positions considered in Quick Examples 4-7 and 4-8.

▲ **FIGURE 4-11 Visualizing Concepts Hang Time** "Hanging in air" near the peak of a jump requires no special knack—in fact, it's an unavoidable consequence of the laws of physics. This phenomenon, which makes big leapers (such as deer and dancers) look particularly graceful, can also make life dangerous for salmon fighting their way upstream to spawn.

EXAMPLE 4-9 A ROUGH SHOT

Chipping from the rough, a golfer sends the ball over a 3.00-m-high tree that is 14.0 m away. The ball lands at the same level from which it was struck after traveling a horizontal distance of 17.8 m—on the green, of course. The ball leaves the club at 54.0° above the horizontal and lands on the green 2.24 s later. **(a)** What was the initial speed of the ball? **(b)** How high was the ball when it passed directly over the tree?

PICTURE THE PROBLEM
Our sketch shows the ball taking flight from the origin, $x_0 = y_0 = 0$, with a launch angle of 54.0°, and arcing over the tree. The individual points along the parabolic trajectory correspond to equal time intervals.

REASONING AND STRATEGY

a. The projectile moves with constant speed in the x direction, and hence the x component of velocity is simply the horizontal distance divided by time. Knowing v_x and θ, we can find v_0 from $v_x = v_0 \cos \theta$.

b. We can use $x = (v_0 \cos \theta)t$ to find the time when the ball is at $x = 14.0$ m. Substituting this time into $y = (v_0 \sin \theta)t - \frac{1}{2}gt^2$ gives the height.

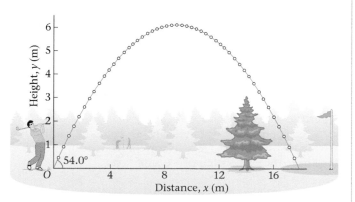

Known Horizontal distance, $d = 17.8$ m; launch angle, $\theta = 54.0°$; time of flight, $t = 2.24$ s.
Unknown **(a)** Initial speed, $v_0 = $? **(b)** Height of ball at tree, $y = $?

CONTINUED

SOLUTION

Part (a)

1. Divide the horizontal distance, d, by the time of flight, t, to obtain v_x:

$$v_x = \frac{d}{t} = \frac{17.8 \text{ m}}{2.24 \text{ s}} = 7.95 \text{ m/s}$$

2. Use $v_x = v_0 \cos \theta$ to find v_0, the launch speed:

$$v_x = v_0 \cos \theta \quad \text{or} \quad v_0 = \frac{v_x}{\cos \theta} = \frac{7.95 \text{ m/s}}{\cos 54.0°} = 13.5 \text{ m/s}$$

Part (b)

3. Use $x = (v_0 \cos \theta)t$ to find the time when $x = 14.0$ m. Recall that $x_0 = 0$:

$$x = (v_0 \cos \theta)t \quad \text{or} \quad t = \frac{x}{v_0 \cos \theta} = \frac{14.0 \text{ m}}{7.95 \text{ m/s}} = 1.76 \text{ s}$$

4. Evaluate $y = (v_0 \sin \theta)t - \frac{1}{2}gt^2$ at the time found in Step 3. Recall that $y_0 = 0$:

$$y = (v_0 \sin \theta)t - \frac{1}{2}gt^2$$
$$= [(13.5 \text{ m/s}) \sin 54.0°](1.76 \text{ s}) - \frac{1}{2}(9.81 \text{ m/s}^2)(1.76 \text{ s})^2$$
$$= 4.03 \text{ m}$$

INSIGHT

The ball clears the top of the tree by 4.03 m − 3.00 m = 1.03 m, and lands on the green 2.24 s − 1.76 s = 0.48 s later. Just before landing, its speed (in the absence of air resistance) is again 13.5 m/s—the same as when it was launched. This result will be verified in the next section.

PRACTICE PROBLEM

What are the speed and direction of the ball when it passes over the tree? **[Answer:** To find the ball's speed and direction, notice that $v_x = 7.95$ m/s and $v_y = v_0 \sin \theta - gt = -6.34$ m/s. It follows that $v = \sqrt{v_x^2 + v_y^2} = 10.2$ m/s and $\theta = \tan^{-1}(v_y/v_x) = -38.6°.$]

Some related homework problems: Problem 32, Problem 38

QUICK EXAMPLE 4-10 AN ELEVATED GREEN

Ball Fired from Cart on Incline

A golfer hits a ball from the origin with an initial speed of 15.0 m/s at an angle of 62.0° above the horizontal. The ball is in the air 2.58 s, and lands on a green that is above the level where the ball was struck.

a. What is the level of the green?

b. How far has the ball traveled in the horizontal direction when it lands?

c. What is the ball's y component of velocity just before it lands?

REASONING AND SOLUTION

The equations of projectile motion (Equations 4-10) can be applied to the golf ball, with $\theta = 62.0°$, $v_0 = 15.0$ m/s, and the landing time equal to $t = 2.58$ s. **(a)** The level of the green is given by y as a function of time, $y = (v_0 \sin \theta)t - \frac{1}{2}gt^2$. **(b)** The horizontal distance is given by x as a function of time, $x = (v_0 \cos \theta)t$. **(c)** The y component of velocity is given by $v_y = v_0 \sin \theta - gt$.

a. Evaluate $y = (v_0 \sin \theta)t - \frac{1}{2}gt^2$ at $t = 2.58$ s:

$$y = (v_0 \sin \theta)t - \frac{1}{2}gt^2$$
$$= (15.0 \text{ m/s})(\sin 62.0°)(2.58 \text{ s}) - \frac{1}{2}(9.81 \text{ m/s}^2)(2.58 \text{ s})^2$$
$$= 1.52 \text{ m (above the level where the ball was struck)}$$

b. Substitute $t = 2.58$ s into $x = (v_0 \cos \theta)t$:

$$x = (v_0 \cos \theta)t$$
$$= (15.0 \text{ m/s})(\cos 62.0°)(2.58 \text{ s}) = 18.2 \text{ m}$$

c. Substitute $t = 2.58$ s into $v_y = v_0 \sin \theta - gt$ to find v_y:

$$v_y = v_0 \sin \theta - gt$$
$$= (15.0 \text{ m/s}) \sin 62.0° - (9.81 \text{ m/s}^2)(2.58 \text{ s}) = -12.1 \text{ m/s}$$

The negative value of v_y indicates that the ball is moving downward at $t = 2.58$ s, as expected.

Colliding Projectiles The next Example presents a classic situation in which two projectiles collide. One projectile is launched from the origin, and thus its equations of motion are given by Equations 4-10. The second projectile is simply dropped from a height, which is a special case of the equations of motion in Equations 4-7 with $v_0 = 0$.

EXAMPLE 4-11   A LEAP OF FAITH

A trained dolphin leaps from the water with an initial speed of 12.0 m/s. It jumps directly toward a ball held by the trainer a horizontal distance of 5.50 m away and a vertical distance of 4.10 m above the water. In the absence of gravity, the dolphin would move in a straight line to the ball and catch it, but because of gravity the dolphin follows a parabolic path well below the ball's initial position, as shown in the sketch. If the trainer releases the ball the instant the dolphin leaves the water, show that the dolphin and the falling ball meet.

PICTURE THE PROBLEM

In our sketch we have the dolphin leaping from the water at the origin $x_0 = y_0 = 0$ with an angle above the horizontal given by $\theta = \tan^{-1}(h/d)$. The initial position of the ball is $x_0 = d = 5.50$ m and $y_0 = h = 4.10$ m, and its initial velocity is zero. The ball drops straight down with the acceleration due to gravity, $a_y = -g$.

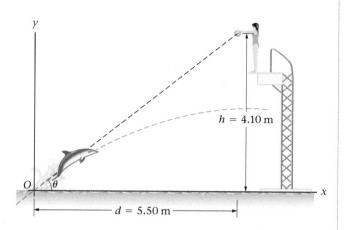

REASONING AND STRATEGY

We want to show that when the dolphin is at $x = d$, its height above the water is the same as the height of the ball above the water. To do this we first find the time when the dolphin is at $x = d$, then calculate y for the dolphin at this time. Next, we calculate y of the ball at the same time and then check to see if they are equal. Because the ball drops from rest from a height h, its y equation of motion is $y = h - \frac{1}{2}gt^2$, as in Equations 4-7 in Section 4-3.

Known   Initial speed, $v_0 = 12.0$ m/s; initial horizontal distance to ball, $d = 5.50$ m; initial vertical distance to ball, $h = 4.10$ m; acceleration of ball and dolphin, $a_y = -g$.

Unknown   Height at which dolphin and ball meet, $y_d = y_b = ?$

SOLUTION

1. Calculate the angle at which the dolphin leaves the water:

$$\theta = \tan^{-1}\left(\frac{h}{d}\right) = \tan^{-1}\left(\frac{4.10\text{ m}}{5.50\text{ m}}\right) = 36.7°$$

2. Use this angle and the initial speed to find the time t when the x position of the dolphin, x_d, is equal to 5.50 m. The x equation of motion is $x_d = (v_0 \cos\theta)t$:

$$x_d = (v_0\cos\theta)t = [(12.0\text{ m/s})\cos 36.7°]t = (9.62\text{ m/s})t$$
$$= 5.50\text{ m}$$
$$t = \frac{5.50\text{ m}}{9.62\text{ m/s}} = 0.572\text{ s}$$

3. Evaluate the y position of the dolphin, y_d, at $t = 0.572$ s. The y equation of motion is $y_d = (v_0\sin\theta)t - \frac{1}{2}gt^2$:

$$y_d = (v_0\sin\theta)t - \frac{1}{2}gt^2$$
$$= [(12.0\text{ m/s})\sin 36.7°](0.572\text{ s}) - \frac{1}{2}(9.81\text{ m/s}^2)(0.572\text{ s})^2$$
$$= 4.10\text{ m} - 1.60\text{ m} = 2.50\text{ m}$$

4. Evaluate the y position of the ball, y_b, at $t = 0.572$ s. The ball's equation of motion is $y_b = h - \frac{1}{2}gt^2$:

$$y_b = h - \frac{1}{2}gt^2 = 4.10\text{ m} - \frac{1}{2}(9.81\text{ m/s}^2)(0.572\text{ s})^2$$
$$= 4.10\text{ m} - 1.60\text{ m} = 2.50\text{ m}$$

INSIGHT

In the absence of gravity, both the dolphin and the ball are at $x = 5.50$ m and $y = 4.10$ m at $t = 0.572$ s. Because of gravity, however, the dolphin and the ball fall below their zero-gravity positions—and by the *same amount*, 1.60 m. In fact, from the point of view of the dolphin, the ball is always at the same angle of 36.7° above the horizontal until it is caught.

This is shown in the accompanying plot, where the red dots show the position of the ball at ten equally spaced times, and the blue dots show the position of the dolphin at the corresponding times. In addition, the dashed lines from the dolphin to the ball all make the same angle with the horizontal, 36.7°.

PRACTICE PROBLEM

At what height does the dolphin catch the ball if it leaves the water with an initial speed of 8.22 m/s?   [**Answer:** $y_d = y_b = 0.680$ m. If the dolphin's initial speed is less than 7.50 m/s, it reenters the water before it catches the ball.]

Some related homework problems: Problem 69, Problem 79

Enhance Your Understanding   (Answers given at the end of the chapter)

4. A projectile is launched and lands at the same level, as shown in **FIGURE 4-12**. Three points during the flight of the projectile are indicated in the figure. **(a)** Which of the four velocity vectors (shown in green) corresponds to points 1, 2, and 3? **(b)** Which of the four acceleration vectors (shown in purple) corresponds to points 1, 2, and 3?

A   B   C   D

Velocity vectors

E   F   G   H

Acceleration vectors

▲ **FIGURE 4-12** Enhance Your Understanding 4

Section Review

- The basic equations for projectile motion are given in Equations 4-10. They combine constant-acceleration motion in the vertical direction with constant-speed motion in the horizontal direction.

4-5 Projectile Motion: Key Characteristics

We conclude this chapter with a brief look at some additional characteristics of projectile motion that are both interesting and useful. In all cases, our results follow as a direct consequence of the fundamental kinematic equations (Equations 4-10) describing projectile motion.

Range

The **range**, R, of a projectile is the horizontal distance it travels before landing. We consider the case shown in **FIGURE 4-13**, where the initial and final elevations are the same ($y = 0$). The initial elevation occurs at $t = 0$, when the projectile is launched, and the projectile returns to the same elevation at a later time when it lands. One way to obtain the range, then, is as follows:

(i) Find the later time ($t > 0$) when the projectile lands by setting $y = 0$ in the expression

$$y = (v_0 \sin \theta)t - \tfrac{1}{2}gt^2.$$

(ii) Substitute the time found in (i) into the x equation of motion. Carrying out the first part of the calculation yields the following:

$$(v_0 \sin \theta)t - \tfrac{1}{2}gt^2 = 0 \quad \text{or} \quad (v_0 \sin \theta)t = \tfrac{1}{2}gt^2$$

Clearly, $t = 0$ is a solution to this equation—it corresponds to the initial condition—but the solution we seek is a time that is greater than zero. We can find the desired time by dividing both sides of the equation by t. This gives

$$(v_0 \sin \theta) = \tfrac{1}{2}gt \quad \text{or} \quad t = \left(\frac{2v_0}{g}\right)\sin \theta \qquad \text{4-11}$$

This is the time when the projectile lands—also known as the *time of flight*.

Now, substitute this time into $x = (v_0 \cos \theta)t$ to find the value of x when the projectile lands:

$$x = (v_0 \cos \theta)t = (v_0 \cos \theta)\left(\frac{2v_0}{g}\right)\sin \theta = \left(\frac{2v_0{}^2}{g}\right)\sin \theta \cos \theta$$

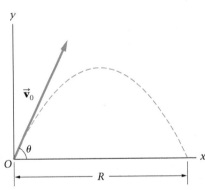

▲ **FIGURE 4-13 Range of a projectile** The range R of a projectile is the horizontal distance it travels between its takeoff and landing positions.

Use the Same Math Regardless of the Initial Conditions

Once an object is launched, its trajectory follows the kinematic equations of motion, regardless of specific differences in the initial conditions. Thus, our equations of motion can be used to derive any desired characteristic of projectile motion, including range, symmetry, and maximum height.

This value of x is the range, R; thus

$$R = \left(\frac{2v_0^2}{g}\right)\sin\theta\cos\theta$$

Using the trigonometric identity $\sin 2\theta = 2\sin\theta\cos\theta$, as given in Appendix A, we can write this more compactly as follows:

$$R = \left(\frac{v_0^2}{g}\right)\sin 2\theta \quad \text{(same initial and final elevation)} \qquad \text{4-12}$$

EXERCISE 4-12 FIND THE INITIAL SPEED

A football game begins with a kickoff in which the ball travels a horizontal distance of 37 m (approximately 40 yd) and lands at the same level from which it was kicked. If the ball was kicked at an angle of 44 ° above the horizontal, what was its initial speed?

REASONING AND SOLUTION

We know the range ($R = 37$ m) and the initial angle ($\theta = 44°.$) With this information we can rearrange the equation $R = (v_0^2/g)\sin 2\theta$ to solve for the initial speed, v_0:

$$v_0 = \sqrt{gR/\sin 2\theta}$$
$$= \sqrt{(9.81 \text{ m/s}^2)(37 \text{ m})/\sin 88°} = 19 \text{ m/s}$$

We choose the positive square root because we are interested in the *speed* of the ball, which is always positive.

RWP Notice that R depends inversely on the acceleration due to gravity, g—thus the smaller g, the larger the range. For example, a projectile launched on the Moon, where the acceleration due to gravity is only about one-sixth that on Earth, travels about six times as far as it would on Earth. It was for this reason that astronaut Alan Shepard simply couldn't resist the temptation of bringing a golf club and ball with him on the third lunar landing mission in 1971. He ambled out onto the Fra Mauro Highlands and became the first person to hit a golf shot on the Moon. His distance was undoubtedly respectable—unfortunately, his ball landed in a sand trap.

Maximum Range   Now, what launch angle gives the greatest range? From Equation 4-12 we see that R varies with angle as $\sin 2\theta$; thus R is maximum when $\sin 2\theta$ is greatest—that is, when $\sin 2\theta = 1$. Since $\sin 90° = 1$, it follows that $\theta = 45°$ gives the maximum range. Thus

$$R_{\text{max}} = \frac{v_0^2}{g} \qquad \text{4-13}$$

As expected, the range (Equation 4-12) and maximum range (Equation 4-13) depend strongly on the initial speed of the projectile—they are both proportional to v_0^2.

These results are specifically for the case where a projectile lands at the same level from which it was launched. If a projectile lands at a higher level, for example, the launch angle that gives the maximum range is greater than 45°, and if it lands at a lower level, the angle for the maximum range is less than 45°.

Finally, the range given here applies only to the ideal case of no air resistance. In cases where air resistance is significant, as in the flight of a rapidly moving golf ball, for example, the overall range of the ball is reduced. In addition, the maximum range occurs for a launch angle less than 45° (FIGURE 4-14). The reason is that with a smaller launch angle the golf ball is in the air for less time, giving air resistance less time to affect its flight.

Symmetry in Projectile Motion

There are many striking symmetries in projectile motion, beginning with the graceful symmetry of the parabola itself. As an example, recall that the time when a projectile lands at the same level from which it was launched is

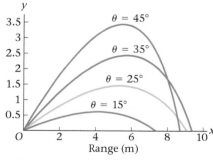

▲ FIGURE 4-14 **Projectiles with air resistance** Projectiles with the same initial speed but different launch angles showing the effects of air resistance. Notice that the maximum range occurs for a launch angle less than 45°, and that the projectiles return to the ground at a steeper angle than the launch angle.

$$t = \left(\frac{2v_0}{g}\right)\sin\theta$$

This result was obtained in Equation 4-11. Now, recall that when a projectile is launched, its y component of velocity is $v_y = v_0 \sin \theta$. When the projectile lands, at time $t = (2v_0/g) \sin \theta$, its y component of velocity is

$$v_y = v_0 \sin \theta - gt = v_0 \sin \theta - g\left(\frac{2v_0}{g}\right) \sin \theta = -v_0 \sin \theta$$

This is exactly the opposite of the y component of the velocity when it was launched. Since the x component of velocity is always the same, it follows that the projectile lands with the same *speed* it had when it was launched. The *velocities* are different, however, because the direction of motion is different at launch and landing. In fact, there's a nice symmetry with the direction of motion: If the initial velocity is *above* the horizontal by the angle θ, the landing velocity is *below* the horizontal by the same angle θ.

So far, these results have referred to launching and landing, which both occur at $y = 0$. The same symmetry occurs at any level, though. That is, at a given height the speed of a projectile is the same on the way up as on the way down. In addition, the angle of the velocity above the horizontal on the way up is the same as the angle below the horizontal on the way down. This is illustrated in FIGURE 4-15 and in the next Conceptual Example.

Big Idea **3** Projectiles follow parabolic paths that curve downward toward the ground. The motion of a projectile on the way up is mirrored by its motion on the way down.

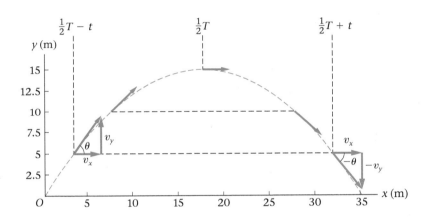

◀ **FIGURE 4-15 Velocity vectors for a projectile launched at the origin** At a given height the speed (length of velocity vector) is the same on the way up as on the way down. The direction of motion on the way up is above the horizontal by the same amount that it is below the horizontal on the way down. In this case, the total time of flight is T, and the greatest height is reached at the time $T/2$. Notice that the speed is the same at the time $(T/2) - t$ as it is at the time $(T/2) + t$.

CONCEPTUAL EXAMPLE 4-13   **COMPARE LANDING SPEEDS**

Two cats are sitting side-by-side on a fence when they're startled by a dog. The cats jump off the fence with equal speeds, but different initial angles relative to the horizontal. Cat 1 jumps at an angle of 45° above the horizontal; cat 2 jumps at an angle of 45° below the horizontal. Is the landing speed of cat 1 greater than, less than, or equal to the landing speed of cat 2?

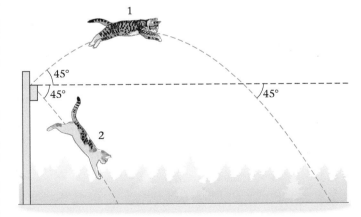

REASONING AND DISCUSSION

One consequence of symmetry in projectile motion is that when cat 1 returns to the level of the jump, its speed is the same as its initial speed. In addition, it is now moving downward at 45° below the horizontal. From that point on its motion is the same as that of cat 2. Therefore, cat 1 lands with the same speed as cat 2.

What if cat 1 jumps horizontally? Or suppose it jumps at an angle of 80° above the horizontal? In either case, the final speed is unchanged! In fact, for a given initial speed, the speed on landing simply doesn't depend on the direction at which the cat jumps. This is shown in Homework Problems 34 and 77. We return to this point in Chapter 8 when we discuss energy conservation.

ANSWER

The cats have equal speeds on landing.

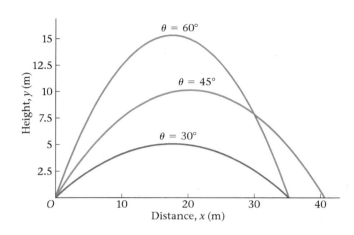

(a) Launch angles that are greater or less than 45° by the same amount give the same range.

(b) Projectiles with $\theta = 30°$ and $\theta = 60°$ follow different paths but have the same range.

Range of a Gun at Two Firing Angles

▲ **FIGURE 4-16 Range and launch angle in the absence of air resistance** **(a)** A plot of range versus launch angle for a projectile launched with an initial speed of 20 m/s. Note that the maximum range occurs at $\theta = 45°$. Launch angles equally greater than or less than 45°, such as 30° and 60°, give the same range. **(b)** Trajectories of projectiles with initial speeds of 20 m/s and launch angles of 60°, 45°, and 30°. The projectiles with launch angles of 30° and 60° land at the same location.

Range Versus Launch Angle As our final example of projectile symmetry, consider the range R. A plot of R versus launch angle θ is shown in **FIGURE 4-16 (a)** for $v_0 = 20$ m/s. Note that in the absence of air resistance, R is greatest at $\theta = 45°$, as pointed out previously. In addition, we can see from the figure that the range for angles equally above or below 45° is the same. For example, if air resistance is negligible, the range for $\theta = 30°$ is the same as the range for $\theta = 60°$, as we can see in both parts (a) and (b) of Figure 4-16.

Symmetries such as these are just some of the many reasons why physicists find physics to be "beautiful" and "aesthetically pleasing." Discovering such patterns and symmetries in nature is really what physics is all about. A physicist does not consider the beauty of projectile motion to be diminished by analyzing it in detail. Just the opposite—detailed analysis reveals deeper, more subtle, and sometimes unexpected levels of beauty.

EXAMPLE 4-14 WHAT A SHOT!

The archerfish hunts by dislodging an unsuspecting insect from its resting place with a stream of water expelled from the fish's mouth. Suppose the archerfish squirts water with an initial speed of 2.30 m/s at an angle of 19.5° above the horizontal. When the stream of water reaches a beetle on a leaf at height h above the water's surface, it is moving horizontally.

a. How much time does the beetle have to react?
b. What is the height h of the beetle?
c. What is the horizontal distance d between the fish and the beetle when the water is launched?

PICTURE THE PROBLEM

Our sketch shows the fish squirting water from the origin, $x_0 = y_0 = 0$, and the beetle at $x = d$, $y = h$. The stream of water starts off with a speed $v_0 = 2.30$ m/s at an angle $\theta = 19.5°$ above the horizontal. The water is moving horizontally when it reaches the beetle.

REASONING AND STRATEGY

a. Because the stream of water is moving horizontally when it reaches the beetle, it is at the top of its parabolic trajectory, as can be seen in Figure 4-15. This means that its y component of velocity is zero. Therefore, we can set $v_y = 0$ in $v_y = v_0 \sin \theta - gt$ and solve for the time t.

CONTINUED

b. To find the maximum height of the stream of water, and of the beetle, we substitute the time found in part (a) into $y = (v_0 \sin \theta)t - \frac{1}{2}gt^2$.

c. Similarly, we can find the horizontal distance d by substituting the time from part (a) into $x = (v_0 \cos \theta)t$.

Known Initial speed, $v_0 = 2.30$ m/s; angle above the horizontal, $\theta = 19.5°$.

Unknown **(a)** Reaction time, $t = ?$ **(b)** Height of beetle, $h = ?$
  (c) Horizontal distance to beetle, $d = ?$

SOLUTION

Part (a)

1. Set $v_y = v_0 \sin \theta - gt$ equal to zero and solve for the corresponding time t:

$$v_y = v_{0y} - gt = v_0 \sin \theta - gt = 0$$
$$t = \frac{v_0 \sin \theta}{g}$$

2. Substitute numerical values to determine the reaction time:

$$t = \frac{v_0 \sin \theta}{g} = \frac{(2.30 \text{ m/s}) \sin 19.5°}{9.81 \text{ m/s}^2} = 0.0783 \text{ s}$$

Part (b)

3. To calculate the height, we can substitute $t = (v_0 \sin \theta)/g$ or $t = 0.0783$ s into $y = (v_0 \sin \theta)t - \frac{1}{2}gt^2$. For practice, let's work symbolically:

$$y = (v_0 \sin\theta)\left(\frac{v_0 \sin\theta}{g}\right) - \frac{1}{2}g\left(\frac{v_0 \sin\theta}{g}\right)^2 = \frac{(v_0 \sin\theta)^2}{2g}$$

4. Now, substitute numerical values into our symbolic result to find the height, $y = h$:

$$h = \frac{(v_0 \sin\theta)^2}{2g} = \frac{[(2.30 \text{ m/s}) \sin 19.5°]^2}{2(9.81 \text{ m/s}^2)} = 0.0300 \text{ m}$$

Part (c)

5. We can find the horizontal distance d using x as a function of time, $x = (v_0 \cos \theta)t$:

$$x = (v_0 \cos \theta)t$$
$$d = [(2.30 \text{ m/s}) \cos 19.5°](0.0783 \text{ s}) = 0.170 \text{ m}$$

INSIGHT

To hit the beetle, the fish aims 19.5° above the horizontal. For comparison, the straight-line angle to the beetle is $\tan^{-1}(0.0300/0.170) = 10.0°$. Therefore, the fish cannot aim directly at its prey if it wants a meal. An example of an archerfish scoring a direct hit is shown in **FIGURE 4-17**.

Finally, notice that by working symbolically in Step 3 we derived a general result for the maximum height of a projectile. In particular, we found that $y_{max} = (v_0 \sin \theta)^2/2g$, a result that is valid for any launch speed and angle.

PRACTICE PROBLEM

How far does the stream of water go if it happens to miss the beetle? [**Answer:** By symmetry, the distance d is half the range. Thus the stream of water travels a distance $R = 2d = 0.340$ m.]

Some related homework problems: Problem 43, Problem 82

▶ **FIGURE 4-17** An archerfish would have trouble procuring its lunch without an instinctive grasp of projectile motion.

Enhance Your Understanding   (Answers given at the end of the chapter)

5. A baseball player throws a ball to another player. The initial speed of the ball is v_0, and it is thrown at an angle of 45° above the horizontal. Is there ever a time during its flight to the other player when its speed is $\frac{1}{2}v_0$? Explain.

Section Review

• Many symmetries characterize projectile motion. For example, the speed of a projectile at a given height is the same whether the projectile is on its way up or on its way down.

CHAPTER 4 REVIEW

CHAPTER SUMMARY

4-1 MOTION IN TWO DIMENSIONS

Independence of Motion

Motion in the x and y directions are independent of one another.

4-2 PROJECTILE MOTION: BASIC EQUATIONS

Projectile motion refers to the path of an object after it is thrown, kicked, batted, or otherwise launched into the air.

x and y as Functions of Time

The x and y equations of motion are

$$x = x_0 + v_{0x}t$$
$$y = y_0 + v_{0y}t - \frac{1}{2}gt^2$$

4-6

v_x and v_y as Functions of Time

The velocity components vary with time as follows:

$$v_x = v_{0x}$$
$$v_y = v_{0y} - gt$$

4-6

v_x and v_y as Functions of Displacement

v_x and v_y vary with displacement as

$$v_x^2 = v_{0x}^2$$
$$v_y^2 = v_{0y}^2 - 2g\Delta y$$

4-6

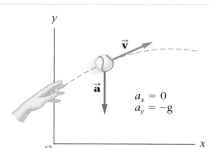

4-3 ZERO LAUNCH ANGLE

Equations of Motion

A projectile launched horizontally from $x_0 = 0$, $y_0 = h$ with an initial speed v_0 has the following equations of motion:

$$x = v_0t \qquad v_x = v_0 \qquad v_x^2 = v_0^2$$
$$y = h - \frac{1}{2}gt^2 \qquad v_y = -gt \qquad v_y^2 = -2g\Delta y$$

4-7

Parabolic Path

The parabolic path followed by a projectile launched horizontally with an initial speed v_0 is described by

$$y = h - \left(\frac{g}{2v_0^2}\right)x^2$$

4-8

4-4 GENERAL LAUNCH ANGLE

Launch from the Origin

The equations of motion for a launch from the origin with an initial speed v_0 at an angle of θ with respect to the horizontal are

$$x = (v_0\cos\theta)t \qquad v_x = v_0\cos\theta \qquad v_x^2 = v_0^2\cos^2\theta$$
$$y = (v_0\sin\theta)t - \frac{1}{2}gt^2 \qquad v_y = v_0\sin\theta - gt \qquad v_y^2 = v_0^2\sin^2\theta - 2g\Delta y$$

4-10

4-5 PROJECTILE MOTION: KEY CHARACTERISTICS

Range

The range of a projectile launched with an initial speed v_0 and a launch angle θ is

$$R = \left(\frac{v_0^2}{g}\right)\sin 2\theta \text{ (same initial and final elevation)}$$

4-12

Symmetry

The motion of a projectile on the way down mirrors its motion on the way up.

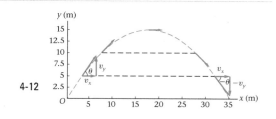

ANSWERS TO ENHANCE YOUR UNDERSTANDING QUESTIONS

1. (a) $x_0 = 0$, $y_0 = 2$ m. (b) $v_{0x} = 0$, $v_{0y} = 3$ m/s. (c) $a_x = 2$ m/s^2, $a_y = 0$.

2. At the base of the mast. The binoculars continue to move forward with constant velocity as they fall.

3. (a) Less, because object A covers a smaller horizontal distance. (b) The times are equal because the objects fall through the same vertical distance.

4. (a) Point 1, vector A; point 2, vector C; point 3, vector B. (b) Points 1 through 3, vector G.

5. No. The minimum speed is $v_x = v_0 \cos 45° = 0.7v_0$, which is greater than $0.5v_0$.

CONCEPTUAL QUESTIONS

For instructor-assigned homework, go to www.masteringphysics.com.

(Answers to odd-numbered Conceptual Questions can be found in the back of the book.)
(The effects of air resistance are to be ignored in this chapter.)

1. What is the acceleration of a projectile when it reaches its highest point? What is its acceleration just before and just after reaching this point?

2. A projectile is launched with an initial speed of v_0 at an angle θ above the horizontal. It lands at the same level from which it was launched. What was its average velocity between launch and landing? Explain.

3. A projectile is launched from level ground. When it lands, its direction of motion has rotated clockwise through 60°. What was the launch angle? Explain.

4. In a game of baseball, a player hits a high fly ball to the outfield. (a) Is there a point during the flight of the ball where its velocity is parallel to its acceleration? (b) Is there a point where the ball's velocity is perpendicular to its acceleration? Explain in each case.

5. A projectile is launched with an initial velocity of $\vec{v} = (4\,\text{m/s})\hat{x} + (3\,\text{m/s})\hat{y}$. What is the velocity of the projectile when it reaches its highest point? Explain.

6. A projectile is launched from a level surface with an initial velocity of $\vec{v} = (2\,\text{m/s})\hat{x} + (4\,\text{m/s})\hat{y}$. What is the velocity of the projectile just before it lands? Explain.

7. Do projectiles for which air resistance is nonnegligible, such as a bullet fired from a rifle, have maximum range when the launch angle is greater than, less than, or equal to 45°? Explain.

8. Two projectiles are launched from the same point at the same angle above the horizontal. Projectile 1 reaches a maximum height twice that of projectile 2. What is the ratio of the initial speed of projectile 1 to the initial speed of projectile 2? Explain.

9. A child rides on a pony walking with constant velocity. The boy leans over to one side and a scoop of ice cream falls from his ice cream cone. Describe the path of the scoop of ice cream as seen by (a) the child and (b) his parents standing on the ground nearby.

10. Driving down the highway, you find yourself behind a heavily loaded tomato truck. You follow close behind the truck, keeping the same speed. Suddenly a tomato falls from the back of the truck. Will the tomato hit your car or land on the road, assuming you continue moving with the same speed and direction? Explain.

11. A projectile is launched from the origin of a coordinate system where the positive x axis points horizontally to the right and the positive y axis points vertically upward. What was the projectile's launch angle with respect to the x axis if, at its highest point, its direction of motion has rotated (a) clockwise through 50° or (b) counterclockwise through 30°? Explain.

PROBLEMS AND CONCEPTUAL EXERCISES

Answers to odd-numbered Problems and Conceptual Exercises can be found in the back of the book. **BIO** *identifies problems of biological or medical interest;* **CE** *indicates a conceptual exercise.* Predict/Explain *problems ask for two responses: (a) your prediction of a physical outcome, and (b) the best explanation among three provided; and* Predict/Calculate *problems ask for a prediction of a physical outcome, based on fundamental physics concepts, and follow that with a numerical calculation to verify the prediction. On all problems, bullets (•, ••, •••) indicate the level of difficulty.* (The effects of air resistance can be ignored in this chapter)

SECTION 4-1 MOTION IN TWO DIMENSIONS

1. • **CE** Predict/Explain As you walk briskly down the street, you toss a small ball into the air. (a) If you want the ball to land in your hand when it comes back down, should you toss the ball straight upward, in a forward direction, or in a backward direction, relative to your body? (b) Choose the *best explanation* from among the following:
   I. If the ball is thrown straight up you will leave it behind.
   II. You have to throw the ball in the direction you are walking.
   III. The ball moves in the forward direction with your walking speed at all times.

2. • A sailboat runs before the wind with a constant speed of 4.2 m/s in a direction 32° north of west. How far (a) west and (b) north has the sailboat traveled in 25 min?

3. • As you walk to class with a constant speed of 1.75 m/s, you are moving in a direction that is 18.0° north of east. How much time does it take to change your position by (a) 20.0 m east or (b) 30.0 m north?

4. • Starting from rest, a car accelerates at 2.0 m/s^2 up a hill that is inclined 5.5° above the horizontal. How far (a) horizontally and (b) vertically has the car traveled in 12 s?

5. •• Predict/Calculate A particle passes through the origin with a velocity of $(6.2\,\text{m/s})\hat{y}$. If the particle's acceleration is $(-4.4\,\text{m/s}^2)\hat{x}$, (a) what are its x and y positions after 5.0 s? (b) What are v_x and v_y at this time? (c) Does the speed of this particle increase with time, decrease with time, or increase and then decrease? Explain.

6. •• A skateboarder travels on a horizontal surface with an initial velocity of 3.8 m/s toward the south and a constant acceleration of 2.2 m/s^2 toward the east. Let the x direction be eastward and the y direction be northward, and let the skateboarder be at the origin at $t = 0$. (a) What are her x and y positions at $t = 0.80$ s? (b) What are her x and y velocity components at $t = 0.80$ s?

7. •• A hot-air balloon is drifting in level flight due east at 2.5 m/s due to a light wind. The pilot suddenly notices that the balloon must gain 24 m of altitude in order to clear the top of a hill 120 m

to the east. **(a)** How much time does the pilot have to make the altitude change without crashing into the hill? **(b)** What minimum, constant, upward acceleration is needed in order to clear the hill? **(c)** What are the horizontal and vertical components of the balloon's velocity at the instant that it clears the top of the hill?

8. •• An electron in a cathode-ray tube is traveling horizontally at 2.10×10^9 cm/s when deflection plates give it an upward acceleration of 5.30×10^{17} cm/s^2. **(a)** How much time does it take for the electron to cover a horizontal distance of 6.20 cm? **(b)** What is its vertical displacement during this time?

9. •• Two canoeists start paddling at the same time and head toward a small island in a lake, as shown in **FIGURE 4-18**. Canoeist 1 paddles with a speed of 1.10 m/s at an angle of 45° north of east. Canoeist 2 starts on the opposite shore of the lake, a distance of 1.5 km due east of canoeist 1. **(a)** In what direction relative to north must canoeist 2 paddle to reach the island? **(b)** What speed must canoeist 2 have if the two canoes are to arrive at the island at the same time?

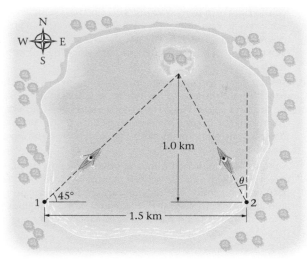

FIGURE 4-18 Problem 9

SECTION 4-3 ZERO LAUNCH ANGLE

10. • **CE** Predict/Explain Two divers run horizontally off the edge of a low cliff. Diver 2 runs with twice the speed of diver 1. **(a)** When the divers hit the water, is the horizontal distance covered by diver 2 twice as much as, four times as much as, or equal to the horizontal distance covered by diver 1? **(b)** Choose the *best explanation* from among the following:
  I. The drop time is the same for both divers.
  II. Drop distance depends on t^2.
  III. All divers in free fall cover the same distance.

11. • **CE** Predict/Explain Two youngsters dive off an overhang into a lake. Diver 1 drops straight down, and diver 2 runs off the cliff with an initial horizontal speed v_0. **(a)** Is the splashdown speed of diver 2 greater than, less than, or equal to the splashdown speed of diver 1? **(b)** Choose the *best explanation* from among the following:
  I. Both divers are in free fall, and hence they will have the same splashdown speed.
  II. The divers have the same vertical speed at splashdown, but diver 2 has the greater horizontal speed.
  III. The diver who drops straight down gains more speed than the one who moves horizontally.

12. • An archer shoots an arrow horizontally at a target 15 m away. The arrow is aimed directly at the center of the target, but it hits 52 cm lower. What was the initial speed of the arrow?

13. • **Victoria Falls** The great, gray-green, greasy Zambezi River flows over Victoria Falls in south central Africa. The falls are approximately 108 m high. If the river is flowing horizontally at 3.60 m/s just before going over the falls, what is the speed of the water when it hits the bottom? Assume the water is in free fall as it drops.

14. • A diver runs horizontally off the end of a diving board with an initial speed of 1.85 m/s. If the diving board is 3.00 m above the water, what is the diver's speed just before she enters the water?

15. • An astronaut on the planet Zircon tosses a rock horizontally with a speed of 6.95 m/s. The rock falls through a vertical distance of 1.40 m and lands a horizontal distance of 8.75 m from the astronaut. What is the acceleration due to gravity on Zircon?

16. •• Predict/Calculate **Pitcher's Mounds** Pitcher's mounds are raised to compensate for the vertical drop of the ball as it travels a horizontal distance of 18 m to the catcher. **(a)** If a pitch is thrown horizontally with an initial speed of 32 m/s, how far does it drop by the time it reaches the catcher? **(b)** If the speed of the pitch is increased, does the drop distance increase, decrease, or stay the same? Explain. **(c)** If this baseball game were to be played on the Moon, would the drop distance increase, decrease, or stay the same? Explain.

17. •• Playing shortstop, you pick up a ground ball and throw it to second base. The ball is thrown horizontally, with a speed of 18 m/s, directly toward point A (**FIGURE 4-19**). When the ball reaches the second baseman 0.54 s later, it is caught at point B. **(a)** How far were you from the second baseman? **(b)** What is the distance of vertical drop, AB?

FIGURE 4-19 Problem 17

18. •• Predict/Calculate A crow is flying horizontally with a constant speed of 2.70 m/s when it releases a clam from its beak (**FIGURE 4-20**). The clam lands on the rocky beach 2.10 s later. Just before the clam lands, what are **(a)** its horizontal component of velocity, and **(b)** its vertical component of velocity? **(c)** How would your answers to parts (a) and (b) change if the speed of the crow were increased? Explain.

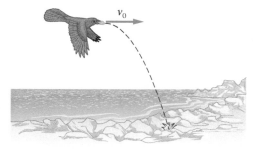

FIGURE 4-20 Problem 18

19. •• A mountain climber jumps a 2.8-m-wide crevasse by leaping horizontally with a speed of 7.8 m/s. **(a)** If the climber's direction of motion on landing is −45°, what is the height difference between the two sides of the crevasse? **(b)** Where does the climber land?

20. •• Predict/Calculate A white-crowned sparrow flying horizontally with a speed of 1.80 m/s folds its wings and begins to drop in free fall. **(a)** How far does the sparrow fall after traveling a horizontal distance of 0.500 m? **(b)** If the sparrow's initial speed is increased, does the distance of fall increase, decrease, or stay the same?

21. •• **Pumpkin Toss** In Denver, children bring their old jack-o-lanterns to the top of a tower and compete for accuracy in hitting a target on the ground (**FIGURE 4-21**). Suppose that the tower is 9.0 m high and that the bull's-eye is a horizontal distance of 3.5 m from the launch point. If the pumpkin is thrown horizontally, what is the launch speed needed to hit the bull's-eye?

9.0 m

←3.5 m→

FIGURE 4–21 Problem 21

22. •• Fairgoers ride a Ferris wheel with a radius of 5.00 m (**FIGURE 4-22**). The wheel completes one revolution every 32.0 s. **(a)** What is the average speed of a rider on this Ferris wheel? **(b)** If a rider accidentally drops a stuffed animal at the top of the wheel, where does it land relative to the base of the ride? (*Note:* The bottom of the wheel is 1.75 m above the ground.)

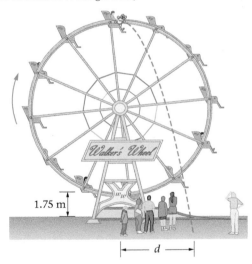

1.75 m

Walker's Wheel

w_w

|← d →|

FIGURE 4-22 Problem 22

23. •• Predict/Calculate A swimmer runs horizontally off a diving board with a speed of 3.32 m/s and hits the water a horizontal distance of 1.78 m from the end of the board. **(a)** How high above the water was the diving board? **(b)** If the swimmer runs off the board with a reduced speed, does it take more, less, or the same time to reach the water?

24. •• **Baseball and the Washington Monument** On August 25, 1894, Chicago catcher William Schriver caught a baseball thrown from the top of the Washington Monument (555 ft, 898 steps). **(a)** If the ball was thrown horizontally with a speed of 5.00 m/s, where did it land? **(b)** What were the ball's speed and direction of motion when caught?

25. ••• A basketball is thrown horizontally with an initial speed of 4.20 m/s (**FIGURE 4-23**). A straight line drawn from the release point to the landing point makes an angle of 30.0° with the horizontal. What was the release height?

30.0°

FIGURE 4-23 Problem 25

26. ••• Predict/Calculate A ball rolls off a table and falls 0.75 m to the floor, landing with a speed of 4.0 m/s. **(a)** What is the acceleration of the ball just before it strikes the ground? **(b)** What was the initial speed of the ball? **(c)** What initial speed must the ball have if it is to land with a speed of 5.0 m/s?

SECTION 4-4   GENERAL LAUNCH ANGLE

27. • **CE** A certain projectile is launched with an initial speed v_0. At its highest point its speed is $\frac{1}{2}v_0$. What was the launch angle of the projectile?

   A. 30°   **B.** 45°   **C.** 60°   **D.** 75°

28. • **CE** Three projectiles (A, B, and C) are launched with the same initial speed but with different launch angles, as shown in **FIGURE 4-24**. Rank the projectiles in order of increasing **(a)** horizontal component of initial velocity and **(b)** time of flight. Indicate ties where appropriate.

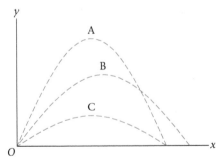

FIGURE 4-24 Problewm 28

29. • **CE** Three projectiles (A, B, and C) are launched with different initial speeds so that they reach the same maximum height, as shown in **FIGURE 4-25**. Rank the projectiles in order of increasing **(a)** initial speed and **(b)** time of flight. Indicate ties where appropriate.

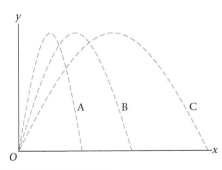

FIGURE 4-25 Problem 29

30. • A cannonball is launched at an angle above level ground, giving it an initial vertical and horizontal velocity that are each positive. The cannonball lands at a time T after it is launched. **(a)** Which plot (A, B, C, D, or E) in FIGURE 4-26 best represents the horizontal component of the cannonball's velocity? **(b)** Which plot best represents the vertical component of the cannonball's velocity?

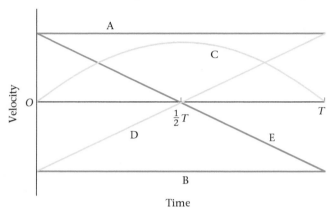

FIGURE 4-26 Problem 30

31. • A second baseman tosses the ball to the first baseman, who catches it at the same level from which it was thrown. The throw is made with an initial speed of 18.0 m/s at an angle of 37.5° above the horizontal. **(a)** What is the horizontal component of the ball's velocity just before it is caught? **(b)** For what amount of time is the ball in the air?

32. • A soccer ball is kicked with a speed of 15.6 m/s at an angle of 32.5° above the horizontal. If the ball lands at the same level from which it was kicked, for what amount of time was it in the air?

33. •• In a game of basketball, a forward makes a bounce pass to the center. The ball is thrown with an initial speed of 4.3 m/s at an angle of 15° below the horizontal. It is released 0.80 m above the floor. What horizontal distance does the ball cover before bouncing?

34. •• Predict/Calculate Snowballs are thrown with a speed of 13 m/s from a roof 7.0 m above the ground. Snowball A is thrown straight downward; snowball B is thrown in a direction 25° above the horizontal. **(a)** Is the landing speed of snowball A greater than, less than, or the same as the landing speed of snowball B? Explain. **(b)** Verify your answer to part (a) by calculating the landing speed of both snowballs.

35. •• In Problem 34, find the direction of motion of the two snowballs just before they land.

36. •• A golfer gives a ball a maximum initial speed of 51.4 m/s. **(a)** What is the longest possible hole-in-one for this golfer? Neglect any distance the ball might roll on the green and assume that the tee and the green are at the same level. **(b)** What is the minimum speed of the ball during this hole-in-one shot?

37. •• What is the highest tree the ball in the previous problem could clear on its way to the longest possible hole-in-one?

38. •• The "hang time" of a punt is measured to be 4.50 s. If the ball was kicked at an angle of 63.0° above the horizontal and was caught at the same level from which it was kicked, what was its initial speed?

39. •• In a friendly game of handball, you hit the ball essentially at ground level and send it toward the wall with a speed of 16 m/s at an angle of 37° above the horizontal. **(a)** What amount of time is required for the ball to reach the wall if it is 3.3 m away? **(b)** How high is the ball when it hits the wall?

40. •• On a hot summer day, a young girl swings on a rope above the local swimming hole (FIGURE 4-27). When she lets go of the rope her initial velocity is 2.25 m/s at an angle of 35.0° above the horizontal. If she is in flight for 0.616 s, how high above the water was she when she let go of the rope?

FIGURE 4-27 Problem 40

41. •• A certain projectile is launched with an initial speed v_0. At its highest point its speed is $v_0/4$. What was the launch angle?

SECTION 4-5 PROJECTILE MOTION: KEY CHARACTERISTICS

42. • **Punkin Chunkin** In Dover, Delaware, a post-Halloween tradition is "Punkin Chunkin," in which contestants build cannons, catapults, trebuchets, and other devices to launch pumpkins and compete for the greatest distance. Though hard to believe, pumpkins have been projected a distance of 4694 feet in this contest. What is the minimum initial speed needed for such a shot?

43. • A dolphin jumps with an initial velocity of 12.0 m/s at an angle of 40.0° above the horizontal. The dolphin passes through the center of a hoop before returning to the water. If the dolphin is moving horizontally when it goes through the hoop, how high above the water is the center of the hoop?

44. • A player passes a basketball to another player who catches it at the same level from which it was thrown. The initial speed of the ball is 7.1 m/s, and it travels a distance of 4.6 m. What were **(a)** the initial direction of the ball and **(b)** its time of flight?

45. • A golf ball is struck with a five iron on level ground. It lands 92.2 m away 4.30 s later. What were **(a)** the direction and **(b)** the magnitude of the initial velocity?

46. •• CE Predict/Explain You throw a ball into the air with an initial speed of 10 m/s at an angle of 60° above the horizontal. The ball returns to the level from which it was thrown in the time T. **(a)** Referring to FIGURE 4-28, which of the plots (A, B, or C) best represents the speed of the ball as a function of time? **(b)** Choose the *best explanation* from among the following:
I. Gravity causes the ball's speed to increase during its flight.
II. The ball has zero speed at its highest point.
III. The ball's speed decreases during its flight, but it doesn't go to zero.

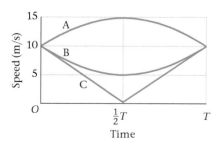

FIGURE 4-28 Problem 46

47. •• A football quarterback shows off his skill by throwing a pass 45.70 m downfield and into a bucket. The quarterback consistently launches the ball at 38.00° above horizontal, and the bucket is placed at the same level from which the ball is thrown. **(a)** What initial speed is needed so that the ball lands in the bucket? **(b)** By how much would the launch speed have to be increased if the bucket is moved to 46.50 m downfield?

48. •• A clever inventor has created a device that can launch water balloons with an initial speed of 85.0 m/s. Her goal is to pass a balloon through a small hoop mounted on the observation platform at the top of the Eiffel Tower, 276 m above the ground. If the balloon is to pass through the hoop at the peak of its flight, at what angle above horizontal should she launch the balloon?

49. •• Predict/Calculate **Volcanoes on Io** Astronomers have discovered several volcanoes on Io, a moon of Jupiter. One of them, named Loki, ejects lava to a maximum height of 2.00×10^5 m. **(a)** What is the initial speed of the lava? (The acceleration of gravity on Io is $1.80 \, \text{m/s}^2$.) **(b)** If this volcano were on Earth, would the maximum height of the ejected lava be greater than, less than, or the same as on Io? Explain.

50. •• Predict/Calculate A soccer ball is kicked with an initial speed of 10.2 m/s in a direction 25.0° above the horizontal. Find the magnitude and direction of its velocity **(a)** 0.250 s and **(b)** 0.500 s after being kicked. **(c)** Is the ball at its greatest height before or after 0.500 s? Explain.

51. •• A soccer ball is kicked with an initial speed of 8.25 m/s. After 0.750 s it is at its highest point. What was its initial direction of motion?

52. •• An archer shoots an arrow over a castle wall by launching it with an initial speed of 28 m/s at 65° above horizontal. Assume the arrow lands on the other side of the castle wall at the same elevation as the launch point. **(a)** What maximum height does the arrow attain? **(b)** What is the range of the arrow's flight? **(c)** What is the vertical component of the arrow's velocity just before it lands on the other side of the castle wall?

GENERAL PROBLEMS

53. • CE Child 1 throws a snowball horizontally from the top of a roof; child 2 throws a snowball straight down. Once in flight, is the acceleration of snowball 2 greater than, less than, or equal to the acceleration of snowball 1?

54. • CE The penguin to the left in the accompanying photo is about to land on an ice floe. Just before it lands, is its speed greater than, less than, or equal to its speed when it left the water?

This penguin behaves much like a projectile from the time it leaves the water until it touches down on the ice. (Problem 54)

55. • CE Dolphins may leap from the water just for the fun of it. At the instant a leaping dolphin lands, is its speed greater than, less than, or equal to its speed when it left the water?

56. • CE Predict/Explain A person flips a coin into the air and it lands on the ground a few feet away. **(a)** If the person were to perform an identical coin flip on an elevator rising with constant speed, would the coin's time of flight be greater than, less than, or equal to its time of flight when the person was at rest? **(b)** Choose the *best explanation* from among the following:
  I. The floor of the elevator is moving upward, and hence it catches up with the coin in mid flight.
  II. The coin has the same upward speed as the elevator when it is tossed, and the elevator's speed doesn't change during the coin's flight.
  III. The coin starts off with a greater upward speed because of the elevator, and hence it reaches a greater height.

57. • CE Predict/Explain Suppose the elevator in the previous problem is rising with a constant upward acceleration, rather than constant velocity. **(a)** In this case, would the coin's time of flight be greater than, less than, or equal to its time of flight when the person was at rest? **(b)** Choose the *best explanation* from among the following:
  I. The coin has the same acceleration once it is tossed, whether the elevator accelerates or not.
  II. The elevator's upward speed increases during the coin's flight, and hence it catches up with the coin at a greater height than before.
  III. The coin's downward acceleration is less than before because the elevator's upward acceleration partially cancels it.

58. • A train moving with constant velocity travels 170 m north in 12 s and an undetermined distance to the west. The speed of the train is 32 m/s. **(a)** Find the direction of the train's motion relative to north. **(b)** How far west has the train traveled in this time?

59. • A tennis ball is struck in such a way that it leaves the racket with a speed of 4.87 m/s in the horizontal direction. When the ball hits the court, it is a horizontal distance of 1.95 m from the racket. Find the height of the tennis ball when it left the racket.

60. • A person tosses a ball for her puppy to retrieve. The ball leaves her hand horizontally with a speed of 4.6 m/s. If the initial height of the ball is 0.95 m above the ground, how far does it travel in the horizontal direction before landing?

61. •• An osprey flies horizontally with a constant speed of 6.8 m/s when it drops the fish it was carrying. How much time elapses after the fish is dropped before the speed of the fish doubles?

62. •• Predict/Calculate A hot-air balloon rises from the ground with a velocity of $(2.00 \, \text{m/s})\hat{y}$. A champagne bottle is opened to celebrate takeoff, expelling the cork horizontally with a velocity of $(5.00 \, \text{m/s})\hat{x}$ relative to the balloon. When opened, the bottle is 6.00 m above the ground. **(a)** What is the initial velocity of the cork, as seen by an observer on the ground? Give your answer in terms of the *x* and *y* unit vectors. **(b)** What are the speed of the cork and its initial direction of motion as seen by the same observer? **(c)** Determine the maximum height above the ground attained by the cork. **(d)** For what amount of time does the cork remain in the air?

63. •• In a friendly neighborhood squirt gun contest a participant runs at 7.8 m/s horizontally off the back deck and fires her squirt gun in the plane of her motion but 45° above horizontal. The gun can shoot water at 11 m/s relative to the barrel, and she fires the gun 0.42 s after leaving the deck. **(a)** What is the initial velocity of the water particles as seen by an observer on the ground? Give your answer in terms of the horizontal and vertical components. **(b)** At the instant she fires, the gun is 1.9 m above the level ground. How far will the water travel horizontally before landing?

64. •• BIO **Spitting Llamas** An agitated llama may spit to assert dominance, or to ward off threats. Llamas can spit a considerable distance, and people handling them need to keep this in mind. If the spittle from a llama is launched from an initial height of 1.8 m with a speed of 6.1 m/s, and at an angle of 12° above horizontal, how far will it travel horizontally?

65. •• A particle leaves the origin with an initial velocity $\vec{v} = (2.40\,\text{m/s})\hat{x}$, and moves with constant acceleration $\vec{a} = (-1.90\,\text{m/s}^2)\hat{x} + (3.20\,\text{m/s}^2)\hat{y}$. **(a)** How far does the particle move in the x direction before turning around? **(b)** What is the particle's velocity at this time? **(c)** Plot the particle's position at $t = 0.500\,\text{s}$, 1.00 s, 1.50 s, and 2.00 s. Use these results to sketch position versus time for the particle.

66. •• **BIO** When the dried-up seed pod of a scotch broom plant bursts open, it shoots out a seed with an initial velocity of 2.62 m/s at an angle of 60.5° above the horizontal. If the seed pod is 0.455 m above the ground, **(a)** what amount of time does it take for the seed to land? **(b)** What horizontal distance does it cover during its flight?

67. •• **Trick Shot** In an Internet video an athlete launches a basketball from a stadium platform that is 16.2 m higher than the hoop. He makes the basket by launching the ball at an angle of 14.5° above horizontal with a speed of 11.6 m/s. What horizontal distance does the ball travel before passing through the hoop?

68. •• A shot-putter throws the shot with an initial speed of 12.2 m/s from a height of 5.15 ft above the ground. What is the range of the shot if the launch angle is **(a)** 20.0°, **(b)** 30.0°, or **(c)** 40.0°?

69. •• Two marbles are launched at $t = 0$ in the experiment illustrated in **FIGURE 4-29**. Marble 1 is launched horizontally with a speed of 4.20 m/s from a height $h = 0.950$ m. Marble 2 is launched from ground level with a speed of 5.94 m/s at an angle $\theta = 45.0°$ above the horizontal. **(a)** Where would the marbles collide in the absence of gravity? Give the x and y coordinates of the collision point. **(b)** Where do the marbles collide given that gravity produces a downward acceleration of $g = 9.81\,\text{m/s}^2$? Give the x and y coordinates.

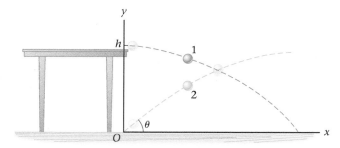

FIGURE 4-29 Problem 69

70. •• **Rescue Swimmers** Coast Guard rescue swimmers are trained to leap from helicopters into the sea to save boaters in distress. The rescuers like to step off their helicopter when it is "ten and ten," which means that it is *ten* feet above the water and moving forward horizontally at *ten* knots. What are **(a)** the speed and **(b)** the direction of motion as a rescue swimmer enters the water following a ten and ten jump?

71. •• A football player kicks a field goal, launching the ball at an angle of 48° above the horizontal. During the kick, the ball is in contact with the player's foot for 0.045 s, and the ball's acceleration is 290 m/s². What is the range of the football?

72. •• A ball thrown straight upward returns to its original level in 2.75 s. A second ball is thrown at an angle of 40.0° above the

horizontal. What is the initial speed of the second ball if it also returns to its original level in 2.75 s?

73. •• **Predict/Calculate** To decide who pays for lunch, a passenger on a moving train tosses a coin straight upward with an initial speed of 5.25 m/s and catches it again when it returns to its initial level. From the point of view of the passenger, then, the coin's initial velocity is $(5.25\,\text{m/s})\hat{y}$. The train's velocity relative to the ground is $(12.1\,\text{m/s})\hat{x}$. **(a)** What is the minimum speed of the coin relative to the ground during its flight? At what point in the coin's flight does this minimum speed occur? Explain. **(b)** Find the initial speed and direction of the coin as seen by an observer on the ground. **(c)** Use the expression for y_{max} derived in Example 4-14 to calculate the maximum height of the coin, as seen by an observer on the ground. **(d)** What is the maximum height of the coin from the point of view of the passenger, who sees only one-dimensional motion?

74. •• **Predict/Calculate** A cannon is placed at the bottom of a cliff 61.5 m high. If the cannon is fired straight upward, the cannonball just reaches the top of the cliff. **(a)** What is the initial speed of the cannonball? **(b)** Suppose a second cannon is placed at the top of the cliff. This cannon is fired horizontally, giving its cannonballs the same initial speed found in part (a). Show that the range of this cannon is the same as the maximum range of the cannon at the base of the cliff. (Assume the ground at the base of the cliff is level, though the result is valid even if the ground is not level.)

75. •• A golfer hits a shot to an elevated green. The ball leaves the club with an initial speed of 16 m/s at an angle of 58° above the horizontal. If the speed of the ball just before it lands is 12 m/s, what is the elevation of the green above the point where the ball is struck?

76. •• **Shot Put Record** A men's world record for the shot put, 23.12 m, was set by Randy Barnes of the United States on May 20, 1990. If the shot was launched from 6.00 ft above the ground at an initial angle of 42.0°, what was its initial speed?

77. •• Referring to Conceptual Example 4-13, suppose the two cats jump from an elevation of 2.5 m with an initial speed of 3.0 m/s. What is the speed of each cat when it is 1.0 m above the ground?

78. •• **A "Lob" Pass Versus a "Bullet"** A quarterback can throw a receiver a high, lazy "lob" pass or a low, quick "bullet" pass. These passes are indicated by curves 1 and 2, respectively, in **FIGURE 4-30**. **(a)** The lob pass is thrown with an initial speed of 21.5 m/s and its time of flight is 3.97 s. What is its launch angle? **(b)** The bullet pass is thrown with a launch angle of 25.0°. What is the initial speed of this pass? **(c)** What is the time of flight of the bullet pass?

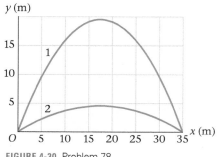

FIGURE 4-30 Problem 78

79. •• For summertime fun, you decide to combine diving from a board with shooting a basketball through a hoop. **(a)** During several practice runs you stand at the end of a diving board and launch the basketball horizontally from a position 4.50 m above

the water. If the average landing spot is 6.25 m horizontally from your initial position, what is the average launch speed? **(b)** Now you step off the diving board and launch the ball in order to make a basket in a hoop that is 3.75 m horizontally from the ball and 1.00 m above the water as shown in FIGURE 4-31. At what time after stepping off the board should you launch the ball? *Hint:* Consider the time required for the ball to travel the required horizontal distance. **(c)** What are the horizontal and vertical components of the ball's velocity at the instant of launch?

FIGURE 4-31 Problem 79

80. •• **Landing on Mars** When the twin exploration rovers, *Spirit* and *Opportunity*, landed on Mars, their method of landing was unique and elaborate. After initial braking with rockets and parachutes to a virtual standstill several meters above the ground, the rovers inflated four air bags with six lobes each. The rovers were then detached from the parachutes and allowed to drop in free fall (3.72 m/s^2) to the surface, where they bounced about 12 times before coming to rest. They then deflated their air bags, righted themselves, and began to explore the surface. FIGURE 4-32 shows a rover with its surrounding cushion of air bags making its first contact with the Martian surface. Assume that the first bounce of the rover is with an initial speed of 9.92 m/s at an angle of 75.0° above the horizontal. **(a)** What is the maximum height of a rover between its first and second bounces? **(b)** How much time elapses between the first and second bounces? **(c)** How far does a rover travel in the horizontal direction between its first and second bounces?

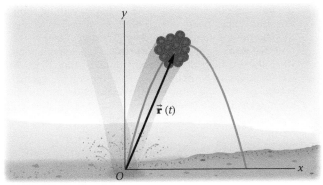

FIGURE 4-32 Problem 80

81. ••• **Collision Course** A useful rule of thumb in piloting is that if the heading from your airplane to a second airplane remains constant, the two airplanes are on a collision course. Consider the two airplanes shown in FIGURE 4-33. At time $t = 0$, airplane 1 is at the location $(X, 0)$ and moving in the positive y direction; airplane 2 is at $(0, Y)$ and moving in the positive x direction. The speed of airplane 1 is v_1. **(a)** What speed must airplane 2 have if the airplanes are to collide at the point (X, Y)? **(b)** Assuming airplane 2 has the speed found in part (a), calculate the displacement from airplane 1 to airplane 2, $\Delta \vec{r} = \vec{r}_2 - \vec{r}_1$. **(c)** Use your results from part (b) to show that $(\Delta r)_y/(\Delta r)_x = -Y/X$, independent of time. This shows that $\Delta \vec{r} = \vec{r}_2 - \vec{r}_1$ maintains a constant direction until the collision, as specified in the rule of thumb.

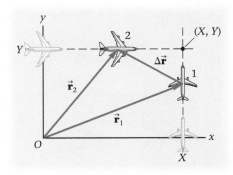

FIGURE 4-33 Problem 81

82. ••• As discussed in Example 4-14, the archerfish hunts by dislodging an unsuspecting insect from its resting place with a stream of water expelled from the fish's mouth. Suppose the archerfish squirts water with a speed of 2.15 m/s at an angle of 52.0° above the horizontal, and aims for a beetle on a leaf 3.00 cm above the water's surface. **(a)** At what horizontal distance from the beetle should the archerfish fire if it is to hit its target in the least time? **(b)** How much time will the beetle have to react?

83. ••• Find the launch angle for which the range and maximum height of a projectile are the same.

84. ••• A mountain climber jumps a crevasse of width W by leaping horizontally with speed v_0. **(a)** If the height difference between the two sides of the crevasse is h, what is the minimum value of v_0 for the climber to land safely on the other side? **(b)** In this case, what is the climber's direction of motion on landing?

85. ••• **Landing on a Different Level** A projectile fired from $y = 0$ with initial speed v_0 and initial angle θ lands on a different level, $y = h$. Show that the time of flight of the projectile is

$$T = \tfrac{1}{2}T_0\left(1 + \sqrt{1 - \frac{h}{H}}\right)$$

In this expression, T_0 is the time of flight for $h = 0$, and H is the maximum height of the projectile.

86. ••• A mountain climber jumps a crevasse by leaping horizontally with speed v_0. If the climber's direction of motion on landing is θ below the horizontal, what is the height difference h between the two sides of the crevasse?

87. ••• **Projectiles: Coming or Going?** Most projectiles continually move farther from the origin during their flight, but this is not the case if the launch angle is greater than $\cos^{-1}(\tfrac{1}{3}) = 70.5°$. For example, the projectile shown in FIGURE 4-34 has a launch angle of 75.0° and an initial speed of 10.1 m/s. During the portion of its motion shown in red, it is moving closer to the origin—it is moving away on the blue portions. Calculate the distance from the origin to the projectile **(a)** at the start of the red portion, **(b)** at the end of the

red portion, and **(c)** just before the projectile lands. Notice that the distance for part **(b)** is the smallest of the three.

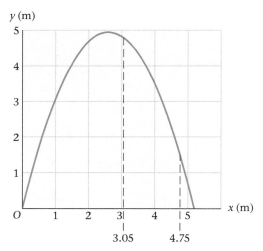

FIGURE 4-34 Problem 87

PASSAGE PROBLEMS

Caterpillar Pellets

The larvae (caterpillars) of certain species of butterflies and moths construct shelters on a host plant out of folded or rolled leaves secured with silk threads. Many of these types of caterpillars eliminate waste by ejecting fecal pellets (frass) at high speeds so that the pellets are projected far away from the caterpillar and its shelter. Various explanations for this behavior have been proposed, but some studies have shown that it may help keep predatory insects such as wasps from locating the caterpillars by homing in on the odor of caterpillar frass.

Video microscopy has been used to study the pellet ejection process. In one study, the video images reveal that a group of Brazilian skipper caterpillars that are about 50 mm long eject pellets at angles between 10° and 40° above the horizontal. There is no correlation between the size of the pellet and the angle at which it is ejected, but larger pellets are shot with lower speeds, as shown in **FIGURE 4-35**. The pellets are small and dense enough that at the speeds at which they travel, air resistance is negligible.

FIGURE 4-35 Problems 88, 89, 90, and 91

88. • If pellets shot at the same angle are compared, which of the following is true?

    A. A lighter pellet travels farther because of its higher ejection speed.
    B. A heavier pellet travels farther because it is launched more horizontally.
    C. Pellets of all masses travel equally far because the launch angle is the same.

89. •• What is the maximum height above its launch site of a 20-mg pellet that's launched at an angle of 20° above the horizontal?

    A. 5.1 cm   **B.** 4.5 cm   **C.** 1.2 cm   **D.** 6.0 mm

90. •• A pellet launched at 30° above the horizontal that spends 0.15 s in the air before returning to its original level has approximately what mass?

    A. 2 mg   **B.** 9 mg   **C.** 14 mg   **D.** 20 mg

91. •• What is the maximum range of a 5-mg pellet that lands at the same height from which it is ejected?

    A. 14 cm   **B.** 20 cm   **C.** 40 cm   **D.** 78 cm

92. •• **REFERRING TO EXAMPLE 4-9 (a)** At what launch angle *greater* than 54.0° does the golf ball just barely miss the top of the tree in front of the green? Assume that the ball has an initial speed of 13.5 m/s, and that the tree is 3.00 m high and is a horizontal distance of 14.0 m from the launch point. **(b)** Where does the ball land in the case described in part (a)? **(c)** At what launch angle *less* than 54.0° does the golf ball just barely miss the top of the tree in front of the green? **(d)** Where does the ball land in the case described in part (c)?

93. •• **REFERRING TO EXAMPLE 4-9** Suppose that the golf ball is launched with a speed of 15.0 m/s at an angle of 57.5° above the horizontal, and that it lands on a green 3.50 m above the level where it was struck. **(a)** What horizontal distance does the ball cover during its flight? **(b)** What increase in initial speed would be needed to increase the horizontal distance in part (a) by 7.50 m? Assume everything else remains the same.

94. •• **REFERRING TO EXAMPLE 4-11** Suppose the ball is dropped at the horizontal distance of 5.50 m, but from a new height of 5.00 m. The dolphin jumps with the same speed of 12.0 m/s. **(a)** What launch angle must the dolphin have if it is to catch the ball? **(b)** At what height does the dolphin catch the ball in this case? **(c)** What is the minimum initial speed the dolphin must have to catch the ball before it hits the water?

95. •• Predict/Calculate **REFERRING TO EXAMPLE 4-11** Suppose we change the dolphin's launch angle to 45.0°, but everything else remains the same. Thus, the horizontal distance to the ball is 5.50 m, the drop height is 4.10 m, and the dolphin's launch speed is 12.0 m/s. **(a)** What is the vertical distance between the dolphin and the ball when the dolphin reaches the horizontal position of the ball? We refer to this as the "miss distance." **(b)** If the dolphin's launch speed is reduced, will the miss distance increase, decrease, or stay the same? **(c)** Find the miss distance for a launch speed of 10.0 m/s.

Big Ideas

1 A force is a push or a pull.

2 An object at rest stays at rest if the net force acting on it is zero.

3 A net force causes an object to accelerate.

4 All forces come in equal but opposite pairs.

Newton's Laws of Motion

▲ This motorist is learning firsthand that it takes a force to accelerate an object—the greater the force, the greater the acceleration. What she may not realize, however, is that the car is also pushing back on her with a force of equal magnitude. These observations follow directly from Newton's laws of motion, the subject of this chapter.

We are all subject to Newton's laws of motion, whether we know it or not. In this chapter we present Newton's three laws and show how they can be applied to everyday situations.

5-1 Force and Mass

Two important elements in Newton's laws, and all of physics, are force and mass. In this section we discuss the physical meaning of these concepts.

Force Simply put, a **force** is a push or a pull. When you push on a box to slide it across the floor, for example, or pull on the handle of a wagon to give a child a ride, you are exerting a force. Similarly, when you hold this book in your hand, you exert an upward force to oppose the downward pull of gravity. If you set the book on a table, the table exerts the same upward force you exerted a moment before. Forces are truly all around us.

Now, when you push or pull on something, there are two quantities that characterize the force you are exerting. The first is the strength or **magnitude** of your force; the second is the **direction** in which you are pushing or pulling. Because a force is determined by both a magnitude and a direction, it is a vector. We consider the vector properties of forces in more detail in Section 5-5.

In general, an object has several forces acting on it at any given time. In the previous example, a book at rest on a table experiences a downward force due to gravity and an upward force due to the table. If you push the book across the table, it also experiences a horizontal force due to your push. The total, or net, force exerted on the book is the vector sum of the individual forces acting on it.

Mass A second key ingredient in Newton's laws is the **mass** of an object, which is a measure of how difficult it is to change its velocity. For example, mass determines how hard it is to start an object moving if it is at rest, to bring it to rest if it is moving, or to change its direction of motion. For example, if you throw a baseball or catch one thrown to you, the force required is not too great. But if you want to start a car moving or stop a car that is coming at you, the force involved is much greater. It follows that the mass of a car is greater than the mass of a baseball.

In agreement with everyday usage, mass can also be thought of as a measure of the quantity of matter in an object. Thus, it is clear that the mass of an elephant, for example, is much greater than the mass of a mouse, but much less than the mass of Earth. We measure mass in units of kilograms (kg), where one kilogram is defined as the mass of a standard cylinder of platinum-iridium, as discussed in Chapter 1. A list of typical masses is given in Table 5-1.

The properties of force and mass presented here are developed in detail in the next three sections.

Big Idea **1** A force is a push or a pull in a given direction. Most objects are acted on by a number of forces at any given time.

TABLE 5-1 Typical Masses in Kilograms (kg)

Earth	5.97×10^{24}
Space shuttle	2,000,000
Blue whale (largest animal on Earth)	105,000
Whale shark (largest fish)	18,000
Elephant (largest land animal)	5400
Automobile	1200
Human (adult)	70
Gallon of milk	3.9
Quart of milk	1.0
Baseball	0.145
Honeybee	0.00015
Bacterium	10^{-15}

Enhance Your Understanding   (Answers given at the end of the chapter)

1. Two forces have magnitudes F_1 and F_2. If these forces are added together, which of the following statements about the net force is correct? **A:** The net force must have a magnitude equal to $F_1 + F_2$. **B:** The net force can have a magnitude less than $F_1 + F_2$. **C:** The net force can have a magnitude greater than $F_1 + F_2$. **D:** The net force has a magnitude greater than F_1.

Section Review

- A force is a push or a pull. Forces are vectors, with specific directions and magnitudes.
- Mass is the quantity of matter contained in an object.

5-2 Newton's First Law of Motion

If you've ever stood in line at an airport, pushing your bags forward a few feet at a time, you know that as soon as you stop pushing the bags, they stop moving. Observations such as this often lead to the erroneous conclusion that a force is required for an object to move. In fact, according to Newton's first law of motion, a force is required only to *change* an object's motion.

Side view   **End view**

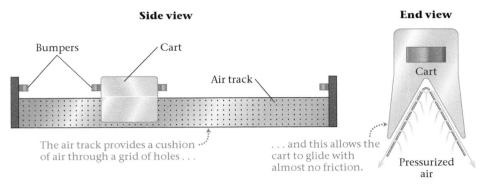

The air track provides a cushion ⋯
of air through a grid of holes . . .

. . . and this allows the
cart to glide with
almost no friction.

▲ **FIGURE 5-1 The air track** An air track provides a cushion of air on which a cart can ride with virtually no friction.

What's missing in this analysis is the force of friction between the bags and the floor. When you stop pushing the bags, it is not true that they stop moving because they no longer have a force acting on them. On the contrary, there is a rather large *frictional force* between the bags and the floor. It is this force that causes the bags to come to rest.

Reducing Friction   To see how motion is affected by reducing friction, imagine that you slide on dirt into second base during a baseball game. You won't slide very far before stopping. On the other hand, if you slide with the same initial speed on a sheet of ice—where the friction is much less than on a ball field—you slide considerably farther. If you could reduce the friction more, you would slide even farther.

In the classroom, air tracks allow us to observe motion with practically no friction. An example of such a device is shown schematically in FIGURE 5-1 and in real life in FIGURE 5-2. Notice that air is blown through small holes in the track, creating a cushion of air for a small "cart" to ride on. A cart placed at rest on a level track remains at rest—unless you push on it to get it started.

Once set in motion, the cart glides along with constant velocity—constant speed in a straight line—until it hits a bumper at the end of the track. The bumper exerts a force on the cart, causing it to change its direction of motion. After bouncing off the bumper, the cart again moves with constant velocity. If the track could be extended to infinite length, and could be made perfectly frictionless, the cart would simply keep moving with constant velocity forever.

Newton's first law of motion summarizes these observations in the following statements:

Newton's First Law

An object at rest remains at rest as long as no net force acts on it.

An object moving with constant velocity continues to move with the same speed and in the same direction as long as no net force acts on it.

Notice the recurring phrase, "no net force," in these statements. It's important to realize that this can mean one of two things: (i) no force acts on the object; or (ii) forces act on the object, but they sum to zero. We shall see examples of the second possibility later in this chapter and again in the next chapter.

Newton's first law, which was first enunciated by Galileo, is also known as the **law of inertia**. This is an appropriate name because the literal meaning of the word *inertia* is "laziness." Speaking loosely, we can say that matter is "lazy," in that it won't change its motion unless forced to do so. For example, if an object is at rest, it won't start moving on its own. If an object is already moving with constant velocity, it won't alter its speed *or* direction, unless a force causes the change. We call this property of matter its inertia.

Frames of Reference   According to Newton's first law, being at rest and moving with constant velocity are actually equivalent. To see this, imagine two observers: one is in a train moving with constant velocity; the second is standing next to the tracks, at rest

▲ **FIGURE 5-2** An air track provides a nearly frictionless environment for experiments involving linear motion.

Big Idea 2 An object at rest stays at rest if the net force acting on it is zero. An object moving with constant velocity continues to move with constant velocity if the net force acting on it is zero.

Suspended Balls:
Which String Breaks?

on the ground. The observer in the train places an ice cube on a dinner tray. From that person's point of view—in that person's **frame of reference**—the ice cube has no net force acting on it and it is at rest on the tray. It obeys the first law. In the frame of reference of the observer on the ground, the ice cube has no net force on it and it moves with constant velocity. This also agrees with the first law. Thus Newton's first law holds for both observers: They both see an ice cube with zero net force moving with constant velocity—it's just that for the first observer the constant velocity happens to be zero.

In this example, we say that each observer is in an **inertial frame of reference**—that is, a frame of reference in which the law of inertia holds. In general, if one frame is an inertial frame of reference, then any frame of reference that moves with constant velocity relative to that frame is also an inertial frame of reference. Thus, if an object moves with constant velocity in one inertial frame, it is always possible to find another inertial frame in which the object is at rest. It is in this sense that there really isn't any difference between being at rest and moving with constant velocity. It's all relative—relative to the frame of reference the object is viewed from.

This gives us a more compact statement of Newton's first law:

> If the net force on an object is zero, its velocity is constant.

As an example of a frame of reference that is not inertial, imagine that the train carrying the first observer suddenly comes to a halt. From the point of view of the observer in the train, there is still no net force on the ice cube. However, because of the rapid braking, the ice cube flies off the tray. In fact, the ice cube simply continues to move forward with the same constant velocity it had before, while the *train* comes to rest. To the observer on the train, it appears that the ice cube has accelerated forward, even though no force acts on it, which is in violation of Newton's first law. Thus the *accelerating* train is *not* an inertial frame of reference; we refer to it as a **noninertial frame of reference**.

In general, any frame of reference that accelerates relative to an inertial frame is a noninertial frame. The surface of the Earth accelerates slightly, due to its rotational and orbital motions, but because the acceleration is so small, it may be considered an excellent approximation to an inertial frame of reference. Unless specifically stated otherwise, we will always consider the surface of the Earth to be an inertial frame.

Enhance Your Understanding   (Answers given at the end of the chapter)

2. Which of the following statements is correct? **A:** A net force is required to keep an object moving with constant velocity. **B:** A net force is required to make an object accelerate. **C:** When you hit the brakes in a car, a force pushes you toward the front of the car. **D:** When you accelerate in a car, a force pushes you back into your seat. **E:** An object that is moving will continue to move with the same velocity unless a net force acts on it.

Section Review

- **Newton's First Law of Motion**   An object at rest remains at rest, as long as no net force acts on it. An object moving with constant velocity maintains its constant velocity, as long as no net force acts on it.

- An inertial frame of reference is one in which Newton's first law is valid. The surface of the Earth is a good approximation to an inertial frame of reference.

5-3 Newton's Second Law of Motion

To hold an object in your hand, you have to exert an upward force to oppose, or "balance," the force of gravity. If you suddenly remove your hand so that the only force acting on the object is gravity, the object accelerates downward, as discussed in Chapter 2. This is one example of Newton's second law, which states, basically, that unbalanced forces cause accelerations.

To explore this in more detail, consider a spring scale of the type used to weigh fish. The scale gives a reading of the force, F, exerted by the spring contained within it. If we hang one weight from the scale, it gives a reading that we will call F_1. If two identical weights are attached, the scale reads $F_2 = 2F_1$, as indicated in FIGURE 5-3. With these two forces marked on the scale, we are ready to perform some force experiments.

First, attach the scale to an air-track cart, as in FIGURE 5-4. If we pull with a force F_1, we observe that the cart accelerates at the rate a_1. If we now pull with a force $F_2 = 2F_1$, the acceleration we observe is $a_2 = 2a_1$. Thus, the acceleration is proportional to the force—the greater the force, the greater the acceleration.

▲ FIGURE 5-3 **Calibrating a "force meter"** With two weights, the force exerted by the scale is twice the force exerted when only a single weight is attached.

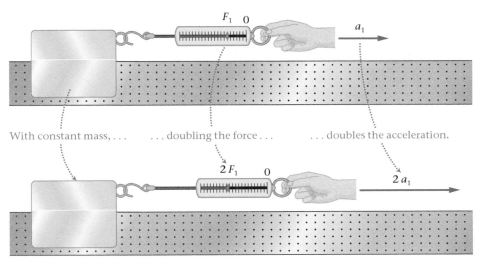

With constant mass, ... doubling the force . . .   . . . doubles the acceleration.

▲ FIGURE 5-4 **Acceleration is proportional to force** The spring calibrated in Figure 5-3 is used to accelerate a mass on a "frictionless" air track. If the force is doubled, the acceleration is also doubled.

Second, instead of doubling the force, let's double the mass m of the cart by connecting two together, as in FIGURE 5-5. In this case, if we pull a mass $2m$ with a force F_1 we find an acceleration equal to $\frac{1}{2}a_1$. Thus, the acceleration is inversely proportional to mass—the greater the mass, the less the acceleration.

◀ FIGURE 5-5 **Acceleration is inversely proportional to mass** If the mass of an object is doubled but the force remains the same, the acceleration is halved.

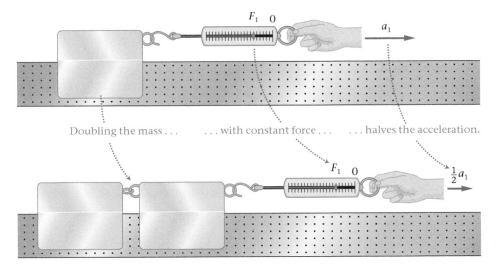

Doubling the mass . . .   . . . with constant force . . .   . . . halves the acceleration.

Combining these results, we find that in this simple case—with just one force in just one direction—the acceleration is given by

$$a = \frac{F}{m}$$

Big Idea 3 A net force acting on an object causes it to accelerate.

PHYSICS
IN CONTEXT
Looking Back

The fact that a constant force produces a constant acceleration gives special significance to the discussion of constant acceleration in Chapters 2 and 4.

Rearranging the equation yields the form of Newton's second law that is perhaps best known, $F = ma$. The dependence of acceleration on force and mass is illustrated in FIGURE 5-6.

Net Force and Newton's Second Law In general, there may be several forces acting on a given mass, and these forces may be in different directions. Thus, we replace F with the sum of the force vectors acting on a mass. We call this the net force:

$$\text{sum of force vectors} = \vec{F}_{net} = \sum \vec{F}$$

The notation $\sum \vec{F}$, which uses the Greek letter sigma ($\sum$), is read "sum of $\vec{F}$." Recalling that acceleration is also a vector, we arrive at the formal statement of Newton's second law of motion:

Newton's Second Law

$$\vec{a} = \frac{\sum \vec{F}}{m} \quad \text{or} \quad \sum \vec{F} = m\vec{a} \qquad \text{5-1}$$

In words:

If an object of mass m is acted on by a net force $\sum \vec{F}$, it will experience an acceleration $\vec{a}$ that is equal to the net force divided by the mass. Because the net force is a vector, the acceleration is also a vector. In fact, the direction of an object's acceleration is the *same* as the direction of the net force acting on it.

One should note that Newton's laws cannot be derived from anything more basic. In fact, this is what we mean by a law of nature. The validity of Newton's laws, and all other laws of nature, comes directly from comparisons with experiment.

Newton's Second Law in Component Form In terms of vector components, an equivalent statement of the second law is

$$\sum F_x = ma_x \qquad \sum F_y = ma_y \qquad \sum F_z = ma_z \qquad \text{5-2}$$

Notice that Newton's second law holds independently for each coordinate direction. This component form of the second law is particularly useful when we solve problems involving more than one direction.

Let's pause for a moment to consider an important special case of the second law. Suppose an object has zero net force acting on it. This may be because no forces act on it at all, or because it is acted on by forces whose vector sum is zero. In either case, we can state this mathematically as

$$\sum \vec{F} = 0$$

Now, according to Newton's second law, we conclude that the acceleration of this object must be zero:

$$\vec{a} = \frac{\sum \vec{F}}{m} = \frac{0}{m} = 0$$

But if an object's acceleration is zero, its velocity must be constant. In other words, if the net force on an object is zero, the object moves with constant velocity. This is Newton's first law. Thus we see that Newton's first and second laws are consistent with one another.

Units for Measuring Forces Forces are measured in a unit called, appropriately enough, the **newton (N)**. In particular, one newton is defined as the force required to give one kilogram of mass an acceleration of $1 \, \text{m/s}^2$. Thus,

$$1 \, \text{N} = (1 \, \text{kg})(1 \, \text{m/s}^2) = 1 \, \text{kg} \cdot \text{m/s}^2 \qquad \text{5-3}$$

◄ **FIGURE 5-6 Visualizing Concepts Force, Mass, and Acceleration** Even though the tugboats exert a large force on this ship, the ship's acceleration is small. This is because the acceleration of an object is inversely proportional to its mass, and the mass of the ship is enormous. The force exerted on the unfortunate hockey player is much smaller than the force exerted on the ship, but his resulting acceleration is much larger, due to his relatively small mass. The acceleration caused by an ion-thrust rocket is even smaller than that of the hockey player, but when it acts for a long period of time the final velocity of the rocket can be quite large.

In everyday terms, a newton is roughly a quarter of a pound. A force in newtons divided by a mass in kilograms has the units of acceleration:

$$\frac{1\,\text{N}}{1\,\text{kg}} = \frac{1\,\text{kg} \cdot \text{m/s}^2}{1\,\text{kg}} = 1\,\text{m/s}^2 \qquad\qquad \textbf{5-4}$$

Other common units for force are presented in Table 5-2. Typical forces and their magnitudes in newtons are listed in Table 5-3.

TABLE 5-2 Units of Mass, Acceleration, and Force

System of units	Mass	Acceleration	Force
SI	kilogram (kg)	m/s²	newton (N)
cgs	gram (g)	cm/s²	dyne (dyn)
British	slug	ft/s²	pound (lb)

(*Note:* $1\,\text{N} = 10^5\,\text{dyne} = 0.225\,\text{lb}$)

TABLE 5-3 Typical Forces in Newtons (N)

Main engines of space shuttle	31,000,000
Pulling force of locomotive	250,000
Thrust of jet engine	75,000
Force to accelerate a car	7000
Weight of adult human	700
Weight of an apple	1
Weight of a rose	0.1
Weight of an ant	0.001

EXERCISE 5-1   FIND THE ACCELERATION

The net force acting on a Cessna 172 airplane has a magnitude of 1900 N and points in the positive *x* direction. If the plane has a mass of 860 kg, what is its acceleration?

REASONING AND SOLUTION

Only *x* components are involved in this system, so we use the *x* component of Newton's second law, $\sum F_x = ma_x$. Solving this equation for the acceleration yields

$$a_x = \frac{\sum F_x}{m} = \frac{1900\,\text{N}}{860\,\text{kg}} = 2.2\,\text{m/s}^2$$

The plane accelerates in the positive *x* direction, as expected. The magnitude of the acceleration is about one-fourth the acceleration due to gravity.

PHYSICS IN CONTEXT
Looking Back

All forces are vectors, and therefore the ability to use and manipulate vectors is essential to a full and complete understanding of forces. Again, we see the importance of the vector material presented in Chapter 3.

The following Conceptual Example presents a situation in which both Newton's first and second laws play an important role.

CONCEPTUAL EXAMPLE 5-2   TIGHTENING A HAMMER

The metal head of a hammer is loose. To tighten it, you drop the hammer down onto a table. Should you **(a)** drop the hammer with the handle end down, **(b)** drop the hammer with the head end down, or **(c)** do you get the same result either way?

REASONING AND DISCUSSION

It might seem that since the same hammer hits against the same table in either case, there shouldn't be a difference. Actually, there is.

In case (a) the handle of the hammer comes to rest when it hits the table, but the head—with its large inertia—continues downward until a force acts on it to bring it to rest. The force that acts on it is supplied by the handle, which results in the head being wedged more tightly onto the handle. Because the metal head is heavy, the force wedging it onto the handle is great. In case (b) the head of the hammer comes to rest, but the handle continues to move until a force brings it to rest. The handle is lighter than the head, however; thus the force acting on it is less, resulting in less tightening.

ANSWER

(a) Drop the hammer with the handle end down.

RWP* A similar effect occurs when you walk—with each step you tamp your head down onto your spine, as when you drop a hammer handle end down. This causes you to grow shorter during the day! Try it. Measure your height first thing in the morning, then again before going to bed. If you're like many people, you'll find that you have shrunk by an inch or so during the day.

*Real World Physics applications are denoted by the acronym RWP.

External Forces

External forces acting on an object fall into two main classes: (i) forces at the point of contact with another object, and (ii) forces exerted by an external agent, such as gravity.

Free-Body Diagrams

When we solve problems involving forces and Newton's laws, it's essential to begin by making a sketch that indicates *each and every external force* acting on a given object. This type of sketch is referred to as a **free-body diagram**. If we are concerned only with nonrotational motion, as is the case in this and the next chapter, we treat the object of interest as a point particle and apply each of the forces acting on the object to that point, as **FIGURE 5-7** shows. Once the forces are drawn, we choose a coordinate system and resolve each force into components. At this point, Newton's second law can be applied to each coordinate direction separately.

For example, in Figure 5-7 there are three external forces acting on the chair. One is the force $\vec{F}$ exerted by the person. In addition, gravity exerts a downward force, $\vec{W}$, which is simply the weight of the chair. Finally, the floor exerts an upward force on the chair that prevents it from falling toward the center of the Earth. This force is referred to as the *normal force*, $\vec{N}$, because it is perpendicular (that is, normal) to the surface of the floor. We will consider the weight and the normal force in greater detail in Sections 5-6 and 5-7, respectively.

(a) Sketch the forces

Physical picture

(b) Isolate the object of interest

(c) Choose a convenient coordinate system

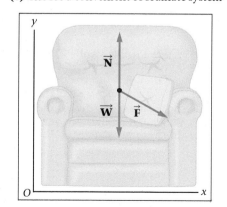

(d) Resolve forces into their components

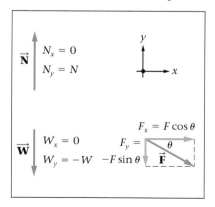

Free-body diagram

▲ **FIGURE 5-7 Constructing and using a free-body diagram** The four basic steps in constructing and using a free-body diagram. **(a)** Sketch all of the external forces acting on an object of interest. Note that only forces acting *on* the object are shown; none of the forces exerted *by* the object are included. **(b)** Isolate the object and treat it as a point particle. **(c)** Choose a convenient coordinate system. This will often mean aligning a coordinate axis to coincide with the direction of one or more forces in the system. **(d)** Resolve each of the forces into components using the coordinate system of part (c).

We can summarize the steps involved in constructing a free-body diagram as follows:

Sketch the Forces

Identify and sketch all of the external forces acting on an object. Sketching the forces roughly to scale will help in estimating the direction and magnitude of the net force.

Isolate the Object of Interest

Replace the object with a point particle of the same mass. Apply each of the forces acting on the object to that point.

Choose a Convenient Coordinate System

Any coordinate system will work; however, if the object moves in a known direction, it is often convenient to pick that direction for one of the coordinate axes. Otherwise, it is reasonable to choose a coordinate system that aligns with one or more of the forces acting on the object.

Resolve the Forces into Components

Determine the components of each force in the free-body diagram.

Apply Newton's Second Law to Each Coordinate Direction

Analyze motion in each coordinate direction using the component form of Newton's second law, as given in Equation 5-2.

These basic steps are illustrated in Figure 5-7. Notice that the figures in this chapter use the labels "Physical picture" to indicate a sketch of the physical situation and "Free-body diagram" to indicate a free-body sketch.

One-Dimensional Systems In this section, we apply free-body diagrams to one-dimensional examples; we'll save two-dimensional systems for Section 5-5. Suppose, for instance, that you hold a book at rest in your hand. What is the magnitude of the upward force that your hand must exert to keep the book at rest? From everyday experience, we expect that the upward force must be equal in magnitude to the weight of the book, but let's see how this result can be obtained directly from Newton's second law.

We begin with a sketch of the physical situation, as shown in FIGURE 5-8 (a). The corresponding free-body diagram, in FIGURE 5-8 (b), shows just the book, represented by a point, and the forces acting on it. Notice that two forces act on the book: (i) the downward force of gravity, $\vec{\mathbf{W}}$, and (ii) the upward force, $\vec{\mathbf{F}}$, exerted by your hand. Only the forces acting *on* the book are included in the free-body diagram.

Now that the free-body diagram is drawn, we indicate a coordinate system so that the forces can be resolved into components. In this case all the forces are vertical, thus we draw a y axis in the vertical direction. Note that we have chosen upward to be the positive direction. With this choice, the y components of the forces are $F_y = F$ and $W_y = -W$. It follows that

$$\sum F_y = F - W$$

Using the y component of the second law ($\sum F_y = ma_y$), we find

$$F - W = ma_y$$

The book remains at rest, and hence its acceleration is zero. Thus, $a_y = 0$, which gives

$$F - W = ma_y = 0 \quad \text{or} \quad F = W$$

This is the expected result, but we had to be careful with the signs of components to obtain it.

Next, we consider a situation where the net force acting on an object is nonzero, meaning that its acceleration is also nonzero.

PROBLEM-SOLVING NOTE

Picture the Problem

In problems involving Newton's laws, it is important to begin with a free-body diagram and to identify all the external forces that act on an object. Once these forces are identified and resolved into their components, Newton's laws can be applied in a straightforward way. It is crucial, however, that only external forces acting on the object be included, and that none of the external forces be omitted.

(a) Physical picture **(b) Free-body diagram**

▲ FIGURE 5-8 **A book supported in a person's hand (a)** The physical situation. **(b)** The free-body diagram for the book, showing the two external forces acting on it. We also indicate our choice for a coordinate system.

EXAMPLE 5-3 | THREE FORCES

Moe, Larry, and Curly push on a 752-kg boat that floats next to a dock. They each exert an 80.5-N force parallel to the dock.
(a) What is the acceleration of the boat if they all push in the same direction? Give both direction and magnitude. **(b)** What are the magnitude and direction of the boat's acceleration if Larry and Curly push in the opposite direction to Moe's push?

PICTURE THE PROBLEM
In our sketch we indicate the three relevant forces acting on the boat: $\vec{\mathbf{F}}_M$, $\vec{\mathbf{F}}_L$, and $\vec{\mathbf{F}}_C$. Notice that we have chosen the positive x direction to the right, in the direction that all three push for part (a). Therefore, all three forces have a positive x component in part (a). In part (b), however, the forces exerted by Larry and Curly have negative x components.

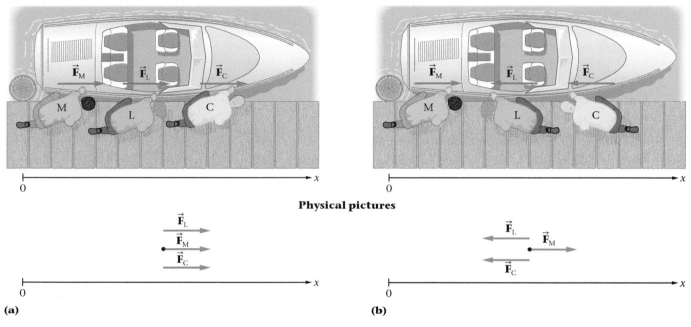

Physical pictures

(a)   **(b)**

Free-body diagrams

REASONING AND STRATEGY
We know the mass of the boat and the forces acting on it, and hence we can find the acceleration using $\sum F_x = ma_x$. Even though this problem is one-dimensional, it's important to think of it in terms of vector components. For example, when we sum the x components of the forces, we are careful to use the appropriate signs—just as we always do when dealing with vectors.

Known   Mass of the boat, $m = 752$ kg; force of each person $= 80.5$ N.
Unknown   **(a)** Acceleration of the boat, $a_x = ?$, when all forces are in the same direction
               (b) Acceleration of the boat, $a_x = ?$, when forces are in different directions

SOLUTION

Part (a)

1. Write out the x component for each of the three forces:

$$F_{M,x} = F_{L,x} = F_{C,x} = 80.5 \text{ N}$$

2. Sum the x components of force and set equal to ma_x:

$$\sum F_x = F_{M,x} + F_{L,x} + F_{C,x} = 241.5 \text{ N} = ma_x$$

3. Divide by the mass to find a_x. Because a_x is positive, the acceleration is to the right, as expected:

$$a_x = \frac{\sum F_x}{m} = \frac{241.5 \text{ N}}{752 \text{ kg}} = 0.321 \text{ m/s}^2$$

Part (b)

4. Again, start by writing the x component for each force:

$$F_{M,x} = 80.5 \text{ N}$$
$$F_{L,x} = F_{C,x} = -80.5 \text{ N}$$

5. Sum the x components of force and set equal to ma_x:

$$\sum F_x = F_{M,x} + F_{L,x} + F_{C,x}$$
$$= 80.5 \text{ N} - 80.5 \text{ N} - 80.5 \text{ N} = -80.5 \text{ N} = ma_x$$

6. Solve for a_x. In this case a_x is negative, indicating an acceleration to the left:

$$a_x = \frac{\sum F_x}{m} = \frac{-80.5 \text{ N}}{752 \text{ kg}} = -0.107 \text{ m/s}^2$$

CONTINUED

INSIGHT

The results of this Example are in agreement with everyday experience: Three forces in the same direction cause more acceleration than three forces in opposing directions. The method of using vector components and being careful about their signs gives the expected results in a simple situation like this, and also works in more complicated situations where everyday experience may be of little help.

PRACTICE PROBLEM

If Moe, Larry, and Curly all push to the right with 85.0-N forces, and the boat accelerates at 0.530 m/s², what is its mass?   [**Answer:** 481 kg]

Some related homework problems: Problem 2, Problem 4

Combining Newton's Laws with Kinematics   In some problems, we are given information that allows us to calculate an object's acceleration using the kinematic equations of Chapters 2 and 4 . Once the acceleration is known, the second law can be used to find the net force that caused the acceleration.

RWP For example, suppose an astronaut uses a jet pack to push a satellite toward the space shuttle. These jet packs, which are known to NASA as Manned Maneuvering Units, or MMUs, are basically small "one-person rockets" strapped to the back of an astronaut's spacesuit. An MMU contains pressurized nitrogen gas that can be released through varying combinations of 24 nozzles spaced around the unit, producing a force of about 10 pounds. The MMUs contain enough propellant for a six-hour EVA (extra-vehicular activity).

In **FIGURE 5-9 (a)** we show a physical situation where an astronaut pushes on a 655-kg satellite. The corresponding free-body diagram for the satellite is shown in **FIGURE 5-9 (b)**. Notice that we've chosen the x axis to point in the direction of the push. Now, if the satellite starts at rest and moves 0.675 m after 5.00 seconds of pushing, what is the force, F, exerted on it by the astronaut?

Clearly, we would like to use Newton's second law (basically, $\vec{\mathbf{F}} = m\vec{\mathbf{a}}$) to find the force, but we know only the mass of the satellite, not its acceleration. We can find the acceleration, however, by assuming constant acceleration (after all, the force is constant) and using the kinematic equation relating position to time: $x = x_0 + v_{0x}t + \frac{1}{2}a_x t^2$. We can choose the initial position of the satellite to be $x_0 = 0$, and we are given that it starts at rest; thus $v_{0x} = 0$. Hence,

$$x = \frac{1}{2}a_x t^2$$

Since we know the distance covered in a given time, we can solve for the acceleration:

$$a_x = \frac{2x}{t^2} = \frac{2(0.675\,\text{m})}{(5.00\,\text{s})^2} = 0.0540\,\text{m/s}^2$$

Now that kinematics has provided the acceleration, we use the x component of the second law to find the force. Only one force acts on the satellite, and its x component is F; thus,

$$\sum F_x = F = ma_x$$
$$F = ma_x = (655\,\text{kg})(0.0540\,\text{m/s}^2) = 35.4\,\text{N}$$

This force corresponds to a push of about 8 pounds.

Another system in which we use kinematics to find the acceleration is presented in the next Example.

(a) Physical picture

(b) Free-body diagram

▲ **FIGURE 5-9 An astronaut using a jet pack to push a satellite (a)** The physical situation. **(b)** The free-body diagram for the satellite. Only one force acts on the satellite, and it is in the positive x direction.

PHYSICS IN CONTEXT
Looking Ahead

Forces are a central theme throughout physics. In particular, we shall see in Chapters 7 and 8 that a force acting on an object over a distance changes its energy.

EXAMPLE 5-4   THE FORCE EXERTED BY FOAMCRETE

RWP Foamcrete is a substance designed to stop an airplane that has run off the end of a runway, without causing injury to passengers. It is solid enough to support a car, but it crumbles under the weight of a large airplane, as shown in **FIGURE 5-10**. By crumbling, Foamcrete slows the plane to a safe stop. For example, suppose a 747 jetliner with a mass of 1.75×10^5 kg and an initial speed of 26.8 m/s is slowed to a stop in 122 m. What is the magnitude of the retarding force exerted by the Foamcrete on the plane?

CONTINUED

PICTURE THE PROBLEM

The airplane and its stopping distance of $\Delta x = 122$ m are shown in our sketch. The initial velocity is $v_0 = 26.8$ m/s in the positive x direction, and the final velocity of the plane is $v = 0$. The retarding force is in the negative x direction, and hence we can write it as $\vec{F} = (-f)\hat{x}$, where $-f$ is the x component of the force, and f is the magnitude.

$v_0 = 26.8$ m/s

$\Delta x = 122$ m

$\vec{F}$

$v = 0$

Runway

Foamcrete

x

REASONING AND STRATEGY

First, use the kinematics equation that relates displacement to velocity to solve for the acceleration of the plane. Knowing the acceleration and mass, we can determine the retarding force acting on the plane.

Known Mass of the airplane, $m = 1.75 \times 10^5$ kg; initial speed of airplane, $v_0 = 26.8$ m/s; stopping distance, $\Delta x = 122$ m.
Unknown Magnitude of the retarding force, $f = ?$

SOLUTION

1. Solve $v^2 = v_0^2 + 2a_x \Delta x$ to find the plane's acceleration. Recall that $v = 0$, $v_0 = 26.8$ m/s, and $\Delta x = 122$ m:

$$a_x = \frac{v^2 - v_0^2}{2\Delta x} = \frac{-(26.8 \text{ m/s})^2}{2(122 \text{ m})}$$
$$= -2.94 \text{ m/s}^2$$

2. Sum the forces in the x direction (there's only one force in the x direction in this system):

$$\sum F_x = -f$$

3. Set the sum of the forces equal to mass times acceleration:

$$-f = ma_x \quad \text{or, equivalently,} \quad f = -ma_x$$

4. Solve for the magnitude of the retarding force, f:

$$f = -ma_x = -(1.75 \times 10^5 \text{ kg})(-2.94 \text{ m/s}^2)$$
$$= 5.15 \times 10^5 \text{ N}$$

INSIGHT

The plane moves in the positive direction, but the retarding force exerted on it, and its acceleration, are in the negative direction. As a result, the plane's speed decreases with time.

PRACTICE PROBLEM

What is the stopping distance if the magnitude of the retarding force is $f = 2.25 \times 10^5$ N?   [**Answer: 278 m**]

Some related homework problems: Problem 9, Problem 12

▶ **FIGURE 5-10** A technician inspects the landing gear of an airliner in a test of Foamcrete.

EXAMPLE 5-5   PITCH MAN: ESTIMATE THE FORCE ON THE BALL

A pitcher throws a 0.15-kg baseball, accelerating it from rest to a speed of about 90 mi/h. Estimate the force exerted by the pitcher on the ball.

PICTURE THE PROBLEM

We choose the x axis to point in the direction of the pitch. Also indicated in the sketch is the distance over which the pitcher accelerates the ball, Δx. Because we are interested only in the pitch, and not in the subsequent motion of the ball, we ignore the effects of gravity.

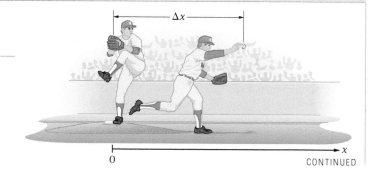

Δx

x

0

CONTINUED

REASONING AND STRATEGY

We know the mass, so we can find the force with $F_x = ma_x$ if we can estimate the acceleration. To find the acceleration, we start with the fact that $v_0 = 0$ and $v \approx 90$ mi/h. In addition, we can see from the sketch that a reasonable estimate for Δx is about 2.0 m. Combining these results with the kinematic equation $v^2 = v_0^2 + 2a_x \Delta x$ yields the acceleration, which we then use to find the force.

Known   Mass of baseball, $m = 0.15$ kg; initial speed, $v_0 = 0$; final speed, $v = 90$ mi/h.
Unknown   Force exerted by pitcher, $F_x = ?$

SOLUTION

1. Starting with the fact that 60 mi/h = 1 mi/min, perform a rough back-of-the-envelope conversion of 90 mi/h to meters per second:

$$v \approx 90 \text{ mi/h} = \frac{1.5 \text{ mi}}{\text{min}} \approx \frac{2400 \text{ m}}{60 \text{ s}} = 40 \text{ m/s}$$

2. Solve $v^2 = v_0^2 + 2a_x \Delta x$ for the acceleration, a_x. Use the estimates $\Delta x \approx 2.0$ m and $v \approx 40$ m/s:

$$a_x = \frac{v^2 - v_0^2}{2 \Delta x} \approx \frac{(40 \text{ m/s})^2 - 0}{2(2.0 \text{ m})} = 400 \text{ m/s}^2$$

3. Find the corresponding force with $F_x = ma_x$:

$$F_x = ma_x \approx (0.15 \text{ kg})(400 \text{ m/s}^2) = 60 \text{ N} \approx 10 \text{ lb}$$

INSIGHT

On the one hand, this is a sizable force, especially when you consider that the ball itself weighs only about 1/3 lb. Thus, the pitcher exerts a force on the ball that is about 30 times greater than the force exerted by Earth's gravity. It follows that ignoring gravity during the pitch is a reasonable approximation.

On the other hand, you might say that 10 lb isn't that much force for a person to exert. That's true, but this force is being exerted with an average speed of about 20 m/s, which means that the pitcher is actually generating about 1.5 horsepower—a sizable power output for a person. We will cover power in detail in Chapter 7, and relate it to human capabilities.

PRACTICE PROBLEM

What is the approximate speed of the ball if the force exerted by the pitcher is $\frac{1}{2}(60 \text{ N}) = 30 \text{ N}$?   [**Answer:** 30 m/s or 60 mi/h]

Some related homework problems: Problem 5, Problem 9

Enhance Your Understanding   (Answers given at the end of the chapter)

3. The acceleration of an object has a magnitude a. What is the magnitude of the acceleration in the following cases? **(a)** All the forces acting on the object are doubled. **(b)** The mass and the net force acting on the object are doubled. **(c)** The net force acting on the object is doubled, and its mass is halved. **(d)** The mass of the object is doubled, and the net force acting on it is halved.

Section Review

- **Newton's Second Law of Motion** The acceleration of an object is proportional to the force acting on it, and inversely proportional to its mass.

- A free-body diagram shows all the external forces acting on a given object.

5-4 Newton's Third Law of Motion

Nature never produces just one force at a time—*forces always come in pairs*. In addition, the forces in a pair always act on *different objects*, are *equal in magnitude*, and point in *opposite directions*. This is Newton's third law of motion.

Newton's Third Law
For every force that acts on an object, there is a reaction force acting on a different object that is equal in magnitude and opposite in direction.

In a somewhat more specific form:

If object 1 exerts a force $\vec{\mathbf{F}}$ on object 2, then object 2 exerts a force $-\vec{\mathbf{F}}$ on object 1.

This law, more commonly known by its abbreviated form, "For every action there is an equal and opposite reaction," completes Newton's laws of motion.

Big Idea **4** If object 1 exerts a force on object 2, then object 2 exerts an equal but opposite force on object 1. In short, all forces come in equal but opposite pairs.

Weighing a Hovering Magnet

▶ **FIGURE 5-11 Examples of action-reaction force pairs**

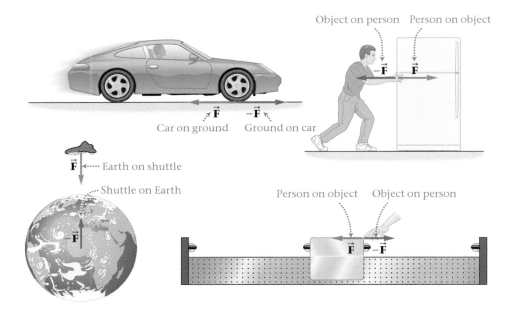

FIGURE 5-11 illustrates some action-reaction pairs. Notice that there is always a reaction force, whether the action force pushes on something hard to move, like a refrigerator, or on something that moves with no friction, like an air-track cart. In some cases, the reaction force tends to be overlooked, as when the Earth exerts a *downward* gravitational force on the space shuttle, and the shuttle exerts an equal and opposite *upward* gravitational force on the Earth. Still, the reaction force always exists.

Third-Law Forces Act on Different Objects Another important aspect of the third law is that the action-reaction forces always act on *different* objects. This, again, is illustrated in Figure 5-11. Thus, in drawing a free-body diagram, we draw only one of the action-reaction pair of forces for a given object. The other force in the pair would appear in the free-body diagram of a different object. As a result:

> The two forces in an action-reaction pair do not cancel in the free-body diagram for an object.

For example, consider a car accelerating from rest, as in Figure 5-11. As the car's engine turns the wheels, the tires exert a force on the road. By the third law, the road exerts an equal and opposite force on the car's tires. It is this second force—which acts on the car through its tires—that propels the car forward. The force exerted by the tires on the road does not accelerate the car.

Because the action-reaction forces act on different objects, they generally produce different accelerations. This is the case in the next Example.

EXAMPLE 5-6 TIPPY CANOE

Two groups of canoeists meet in the middle of a lake. After a brief visit, a person in canoe 1 pushes on canoe 2 with a force of 46 N to separate the canoes. The mass of canoe 1 and its occupants is $m_1 = 150$ kg, and the mass of canoe 2 and its occupants is $m_2 = 250$ kg. **(a)** Find the acceleration the push gives to each canoe. **(b)** What is the separation of the canoes after 1.2 s of pushing?

PICTURE THE PROBLEM
We have chosen the positive x direction to point from canoe 1 to canoe 2. With this choice, the force exerted on canoe 2 is $\vec{F}_2 = (+46\text{ N})\hat{x}$. By Newton's third law, the force exerted on the person in canoe 1, and thus on canoe 1 itself if the person is firmly seated, is $\vec{F}_1 = (-46\text{ N})\hat{x}$. For convenience, we have placed the origin at the point where the canoes touch.

REASONING AND STRATEGY
From Newton's third law, the force on canoe 1 is equal in magnitude to the force on canoe 2. The masses of the canoes are different, however, and therefore their accelerations are different as well. **(a)** We can find the acceleration of each

CONTINUED

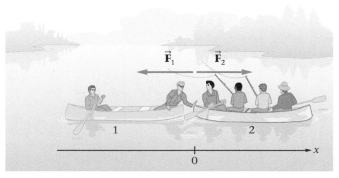

Physical picture

Free-body diagrams

canoe by solving $\Sigma F_x = ma_x$ for a_x. **(b)** The kinematic equation relating position to time, $x = x_0 + v_{0x}t + \frac{1}{2}a_xt^2$, can then be used to find the displacement of each canoe.

Known Force of separation, 46 N; mass of canoe 1, $m_1 = 150$ kg; mass of canoe 2, $m_2 = 250$ kg; time of pushing, $t = 1.2$ s.

Unknown **(a)** Acceleration of canoe 1, $a_{1,x} = ?$ Acceleration of canoe 2, $a_{2,x} = ?$ **(b)** Separation of canoes at $t = 1.2$ s, $x_2 - x_1 = ?$

SOLUTION

Part (a)

1. Use Newton's second law to find the acceleration of canoe 2:

$$a_{2,x} = \frac{\Sigma F_{2,x}}{m_2} = \frac{46\,\text{N}}{250\,\text{kg}} = 0.18\,\text{m/s}^2$$

2. Do the same calculation for canoe 1. Notice that the acceleration of canoe 1 is in the negative direction:

$$a_{1,x} = \frac{\Sigma F_{1,x}}{m_1} = \frac{-46\,\text{N}}{150\,\text{kg}} = -0.31\,\text{m/s}^2$$

Part (b)

3. Use $x = x_0 + v_{0x}t + \frac{1}{2}a_xt^2$ to find the position of canoe 2 at $t = 1.2$ s. From the problem statement, we know the canoes start at the origin ($x_0 = 0$) and at rest ($v_{0x} = 0$):

$$x_2 = \frac{1}{2}a_{2,x}t^2 = \frac{1}{2}(0.18\,\text{m/s}^2)(1.2\,\text{s})^2 = 0.13\,\text{m}$$

4. Repeat the calculation for canoe 1:

$$x_1 = \frac{1}{2}a_{1,x}t^2 = \frac{1}{2}(-0.31\,\text{m/s}^2)(1.2\,\text{s})^2 = -0.22\,\text{m}$$

5. Subtract the two positions to find the separation of the canoes:

$$x_2 - x_1 = 0.13\,\text{m} - (-0.22\,\text{m}) = 0.35\,\text{m}$$

INSIGHT

The same magnitude of force acts on each canoe; hence the lighter one has the greater acceleration and the greater displacement. If the heavier canoe were replaced by a large ship of great mass, both vessels would still accelerate as a result of the push. However, the acceleration of the large ship would be so small as to be practically imperceptible. In this case, it would appear as if only the canoe moved, whereas, in fact, both vessels move.

PRACTICE PROBLEM — PREDICT/CALCULATE

(a) If the mass of canoe 2 is increased, does its acceleration increase, decrease, or stay the same? Explain. **(b)** Check your prediction by calculating the acceleration for the case where canoe 2 is replaced by a 25,000-kg ship. [**Answer:** **(a)** Acceleration is inversely proportional to mass, so the acceleration of canoe 2 decreases. **(b)** In this case, $a = 0.0018$ m/s^2. This is one hundred times smaller than the acceleration found in this Example, because the mass is one hundred times larger.]

Some related homework problems: Problem 20, Problem 21

Contact Forces When objects are touching one another, the action-reaction forces are often referred to as **contact forces**. The behavior of contact forces is explored in the following Conceptual Example.

CONCEPTUAL EXAMPLE 5-7   CONTACT FORCES

Two boxes—one large and heavy, the other small and light—rest on a smooth, level floor. You push with a force $\vec{\mathbf{F}}$ on either the small box or the large box. Is the contact force between the two boxes **(a)** the same in either case, **(b)** larger when you push on the large box, or **(c)** larger when you push on the small box?

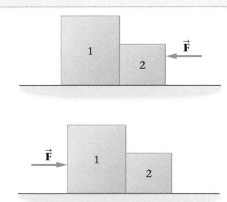

REASONING AND DISCUSSION

The same force pushes on the boxes, and so you might think the force of contact is the same in both cases. It is not. What we can conclude, however, is that the boxes have the same acceleration in either case—the same net force acts on the same total mass, so the same acceleration, a, results.

To find the contact force between the boxes, we focus our attention on each box individually, and note that *Newton's second law must be satisfied for each of the boxes, just as it is for the entire two-box system.* For example, when the external force is applied to the small box, the only force acting on the *large* box (mass m_1) is the contact force; hence, the contact force must have a magnitude equal to m_1a. In the second case, the only force acting on the *small* box (mass m_2) is the contact force, and so the magnitude of the contact force is m_2a. Noting that m_1 is greater than m_2, we conclude that the force of contact is larger when you push on the small box, m_1a, than when you push on the large box, m_2a.

To put it another way, the contact force is larger when *it* must push the larger box.

ANSWER

(c) The contact force is larger when you push on the small box.

In the next Example, we calculate a numerical value for the contact force in a system similar to that described in Conceptual Example 5-7. We also show explicitly that Newton's third law is required for a full analysis of this system.

EXAMPLE 5-8   WHEN PUSH COMES TO SHOVE

A box of mass $m_1 = 10.0$ kg rests on a smooth, horizontal floor in contact with a box of mass $m_2 = 5.00$ kg. You now push on box 1 with a horizontal force of magnitude $F = 30.0$ N. **(a)** What is the acceleration of the boxes? **(b)** What is the force of contact between the boxes?

PICTURE THE PROBLEM

We choose the x axis to be horizontal and pointing to the right. Thus, the force pushing the boxes is $\vec{\mathbf{F}} = (30.0\,\text{N})\hat{\mathbf{x}}$. The contact forces are labeled as follows: $\vec{\mathbf{F}}_1$ is the contact force exerted on box 1; $\vec{\mathbf{F}}_2$ is the contact force exerted on box 2. By Newton's third law, the contact forces have the same magnitude, f, but point in opposite directions. With our coordinate system, we have $\vec{\mathbf{F}}_1 = -f\hat{\mathbf{x}}$ and $\vec{\mathbf{F}}_2 = f\hat{\mathbf{x}}$.

REASONING AND STRATEGY

a. The two boxes are in contact, and hence they have the same acceleration. We find this acceleration with Newton's second law; that is, we divide the net horizontal force by the total mass of the two boxes.

b. In this case, we consider the system to consist solely of box 2. The mass in this case is 5.00 kg, and the only horizontal force acting on the system is $\vec{\mathbf{F}}_2$. Thus, we can find f, the magnitude of $\vec{\mathbf{F}}_2$, by requiring that box 2 has the acceleration found in part (a).

Known   Mass of box 1, $m_1 = 10.0$ kg; mass of box 2, $m_2 = 5.00$ kg; pushing force, $F = 30.0$ N.
Unknown   **(a)** Acceleration of the boxes, $a_x = ?$ **(b)** Force of contact, $f = ?$

SOLUTION

Part (a)

1. Find the net horizontal force acting on the two boxes. Notice that $\vec{\mathbf{F}}_1$ and $\vec{\mathbf{F}}_2$ are equal in magnitude but opposite in direction. Hence, they sum to zero; $\vec{\mathbf{F}}_1 + \vec{\mathbf{F}}_2 = 0$:

$$\sum_{\substack{\text{both}\\\text{boxes}}} F_x = F = 30.0\,\text{N}$$

CONTINUED

2. Divide the net force by the total mass, $m_1 + m_2$, to find the acceleration of the boxes:

$$a_x = \frac{\sum F_x}{m_1 + m_2}$$

$$= \frac{30.0\,\text{N}}{(10.0\,\text{kg} + 5.00\,\text{kg})} = \frac{30.0\,\text{N}}{15.0\,\text{kg}} = 2.00\,\text{m/s}^2$$

Part (b)

3. Find the net horizontal force acting on box 2, and set it equal to the mass of box 2 times its acceleration:

$$\sum_{\text{box2}} F_x = F_{2,x} = f = m_2 a_x$$

4. Determine the magnitude of the contact force, f, by substituting numerical values for m_2 and a_x:

$$f = m_2 a_x = (5.00\,\text{kg})(2.00\,\text{m/s}^2) = 10.0\,\text{N}$$

INSIGHT

The net horizontal force acting on box 1 is $F - f = 30.0\,\text{N} - 10.0\,\text{N} = 20.0\,\text{N}$, and hence its acceleration is $(20.0\,\text{N})/(10.0\,\text{kg}) = 2.00\,\text{m/s}^2$. Thus, as expected, box 1 and box 2 have precisely the same acceleration.

If box 2 were not present, the 30.0-N force acting on box 1 would give it an acceleration of $3.00\,\text{m/s}^2$. As it is, the contact force between the boxes slows box 1, so that its acceleration is less than $3.00\,\text{m/s}^2$. At the same time, the contact force gives box 2 an acceleration that is greater than zero. The precise value of the contact force is simply the value that gives *both boxes* the same acceleration.

PRACTICE PROBLEM

Suppose the relative positions of the boxes are reversed, so that F pushes on the small box, as shown here. Calculate the contact force for this case, and show that the force is greater than 10.0 N, as expected from Conceptual Example 5-7. [**Answer:** The contact force in this case is 20.0 N, double its previous value. This follows because the box being pushed by the contact force has twice the mass of the box that was pushed by the contact force originally.]

Some related homework problems: Problem 22, Problem 23

BIO Living animals utilize Newton's third law of motion to propel themselves. Perhaps the champion of propulsion is the flea (*Ctenocephalides felis*), which can jump 33 cm horizontally, a distance that is more than 150 times its 2-mm body length. To propel itself, the flea's legs exert a force on the surface on which it sits. The surface, in turn, exerts an equal and opposite force on the flea, in accordance with Newton's third law. This force then accelerates the flea according to Newton's second law of motion. All biological propulsion can be explained in similar terms. In Chapter 8 we will use energy concepts to reexamine the flea's amazing jumping ability.

Enhance Your Understanding   (Answers given at the end of the chapter)

4. A force F pushes on three boxes that slide without friction on a horizontal surface, as indicated in **FIGURE 5-12**. Box 1 has a mass of 1 kg, box 2 has a mass of 2 kg, and box 3 has a mass of 3 kg. The boxes are in contact, and slide as a group with an acceleration a. Is the magnitude of the force of contact between boxes 1 and 2 greater than, less than, or equal to the magnitude of the force of contact between boxes 2 and 3? Explain.

▲ **FIGURE 5-12** Enhance Your Understanding 4

Section Review

- **Newton's Third Law of Motion** Forces always come in pairs that are equal in magnitude but opposite in direction. The two forces in a pair always act on different objects.

5-5 The Vector Nature of Forces: Forces in Two Dimensions

When we presented Newton's second law in Section 5-3, we said that an object's acceleration is equal to the net force acting on it divided by its mass. If more than one force acts on an object, experiments show that its acceleration is in the direction of the vector sum of the forces.

As an example of the vector nature of forces, suppose two astronauts are using jet packs to push a 940-kg satellite toward a space station, as shown in **FIGURE 5-13**. With the coordinate system indicated in the figure, astronaut 1 pushes in the positive x direction and astronaut 2 pushes in a direction 52° above the x axis. If astronaut 1 pushes with a force of magnitude $F_1 = 26$ N and astronaut 2 pushes with a force of magnitude $F_2 = 41$ N, what are the magnitude and direction of the satellite's acceleration?

▶ **FIGURE 5-13 Two astronauts pushing a satellite with forces that differ in magnitude and direction** The acceleration of the satellite can be found by calculating a_x and a_y separately, then combining these components to find a and θ.

(a) Physical picture   **(b) Free-body diagram**   **Total force**

Different Coordinate Directions Can Be Treated Independently The easiest way to solve a problem like this is to treat each coordinate direction independently of the other, just as we did many times when studying two-dimensional kinematics in Chapter 4. Thus, we first resolve each force into its x and y components. Referring to Figure 5-13, we see that for the x direction

$$F_{1,x} = F_1$$
$$F_{2,x} = F_2 \cos 52°$$

For the y direction

$$F_{1,y} = 0$$
$$F_{2,y} = F_2 \sin 52°$$

Next, we find the acceleration in the x direction by using the x component of Newton's second law:

$$\sum F_x = ma_x$$

Applied to this system, we have

$$\sum F_x = F_{1,x} + F_{2,x} = F_1 + F_2 \cos 52° = 26 \text{ N} + (41 \text{ N}) \cos 52° = 51 \text{ N}$$
$$= ma_x$$

Solving for the acceleration yields

$$a_x = \frac{\sum F_x}{m} = \frac{51 \text{ N}}{940 \text{ kg}} = 0.054 \text{ m/s}^2$$

Similarly, in the y direction we start with

$$\sum F_y = ma_y$$

This gives

$$\sum F_y = F_{1,y} + F_{2,y} = 0 + F_2 \sin 52° = (41 \text{ N}) \sin 52° = 32 \text{ N}$$
$$= ma_y$$

As a result, the y component of acceleration is

$$a_y = \frac{\sum F_y}{m} = \frac{32\,\text{N}}{940\,\text{kg}} = 0.034\,\text{m/s}^2$$

Thus, the satellite accelerates in both the x and the y directions. Its total acceleration has a magnitude of

$$a = \sqrt{a_x{}^2 + a_y{}^2} = \sqrt{(0.054\,\text{m/s}^2)^2 + (0.034\,\text{m/s}^2)^2} = 0.064\,\text{m/s}^2$$

From Figure 5-13 we expect the total acceleration to be in a direction above the x axis but at an angle less than 52°. Straightforward calculation yields

$$\theta = \tan^{-1}\!\left(\frac{a_y}{a_x}\right) = \tan^{-1}\!\left(\frac{0.034\,\text{m/s}^2}{0.054\,\text{m/s}^2}\right) = \tan^{-1}(0.63) = 32°$$

This is the same direction as the total force in Figure 5-13, as expected.

The next two Examples give further practice with resolving force vectors and using Newton's second law in component form.

**PHYSICS
IN CONTEXT
Looking Back**

As with two-dimensional kinematics in Chapter 4, where motions in the x and y directions were seen to be independent, the x and y components of force are independent. In particular, acceleration in the x direction depends on only the x component of force, and acceleration in the y direction depends on only the y component of force.

EXAMPLE 5-9 BARELY OUT OF REACH

Campers lift a food cooler with a weight of $W = 192.0\,\text{N}$ out of reach of bears using two ropes hung over tree branches. Rope 1 exerts a force $\vec{F}_1$ with a magnitude of 166 N at an angle of 47.4° above the horizontal, as shown in the sketch. Rope 2 exerts a force $\vec{F}_2$ at an angle of 36.1° above the horizontal. **(a)** Find the magnitude of the force $\vec{F}_2$ that produces a net force on the cooler that is straight upward. **(b)** Determine the net force acting on the cooler, given that $\vec{F}_2$ has the magnitude determined in part (a).

PICTURE THE PROBLEM
Our physical picture and free-body diagram show the cooler and the three forces acting on it, as well as the angles of $\vec{F}_1$ and $\vec{F}_2$ relative to the horizontal. We also show the x and y components of $\vec{F}_1$ and $\vec{F}_2$. The weight has only a negative y component; that is, $\vec{W} = -(192.0\,\text{N})\hat{y}$.

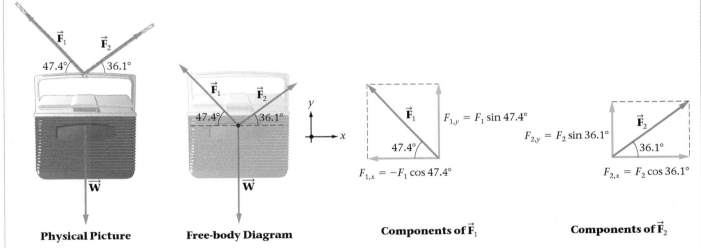

| **Physical Picture** | **Free-body Diagram** | **Components of $\vec{F}_1$** | **Components of $\vec{F}_2$** |

REASONING AND STRATEGY

a. We want the net force to be purely vertical. This means that the x component of the net force must be zero; that is, $\sum F_x = 0$. This condition can be solved to yield F_2, the magnitude of $\vec{F}_2$.

b. The x component of the net force is zero if we use F_2 from part (a). All that remains to find the net force on the cooler is to sum the y components of all three forces.

Known   Weight of cooler, $W = 192.0\,\text{N}$; force exerted by rope 1, 166 N at 47.4° above horizontal; force exerted by rope 2, 36.1° above horizontal.

Unknown   (a) Magnitude of force exerted by rope 2 to produce a straight upward net force, $F_2 = ?$
(b) Net force acting on cooler, $\sum F_y = ?$

SOLUTION

Part (a)

1. Begin by writing the x component of each force. Notice that $\vec{W}$ has no x component and that the x component of $\vec{F}_1$ points in the negative x direction:

$$F_{1,x} = -F_1 \cos 47.4°$$
$$F_{2,x} = F_2 \cos 36.1°$$
$$W_x = 0$$

CONTINUED

2. Sum the x components of force and set equal to zero. The magnitude F_2 is the only unknown in this equation:

$$\sum F_x = -F_1 \cos 47.4° + F_2 \cos 36.1° + 0 = 0 \quad \text{or}$$
$$F_1 \cos 47.4° = F_2 \cos 36.1°$$

3. Solve for F_2:

$$F_2 = \frac{F_1 \cos 47.4°}{\cos 36.1°} = \frac{(166\,\text{N}) \cos 47.4°}{\cos 36.1°} = 139\,\text{N}$$

Part (b)

4. Determine the y component of each force:

$$F_{1,y} = F_1 \sin 47.4° = (166\,\text{N}) \sin 47.4° = 122\,\text{N}$$
$$F_{2,y} = F_2 \sin 36.1° = (139\,\text{N}) \sin 36.1° = 81.9\,\text{N}$$
$$W_y = -W = -192.0\,\text{N}$$

5. Sum the y components of force:

$$\sum F_y = F_1 \sin 47.4° + F_2 \sin 36.1° - W$$
$$= 122\,\text{N} + 81.9\,\text{N} - 192.0\,\text{N} = 11.9\,\text{N}$$

INSIGHT

Thus, the net force acting on the cooler is 11.9 N in the positive y direction; that is, $\vec{F}_{net} = (11.9\,\text{N})\hat{y}$. As a double check that the x component of force is zero, we have $\sum F_x = -F_1 \cos 47.4° + F_2 \cos 36.1° + 0 = -(166\,\text{N})\cos 47.4° + (139\,\text{N})\cos 36.1° = -112\,\text{N} + 112\,\text{N} = 0$.

PRACTICE PROBLEM — PREDICT/CALCULATE

Suppose we change the angle of $\vec{F}_1$ to be 37.4° above the horizontal. **(a)** Is the magnitude F_2 that gives a zero x component of the net force greater than, less than, or equal to 139 N? Explain. **(b)** Find the new value of F_2. **[Answer: (a)** Changing the angle to 37.4° increases the magnitude of the x component of $\vec{F}_1$. This means that F_2 is greater than 139 N. **(b)** $F_2 = 163\,\text{N}$]

Some related homework problems: Problem 29, Problem 31, Problem 33

QUICK EXAMPLE 5-10 FIND THE SPEED OF THE SLED

A 4.60-kg sled is pulled across a smooth ice surface. The force acting on the sled has a magnitude of $F = 6.20\,\text{N}$ and points in a direction $\theta = 35.0°$ above the horizontal. If the sled starts at rest, how fast is it moving after being pulled for 1.15 s?

REASONING AND SOLUTION

The physical system is shown in our sketch, along with a coordinate system. In terms of this coordinate system, the force is 35.0° above the positive x axis. Newton's second law can be used to find the x component of acceleration. Once we know the acceleration and the time of pulling, we can use kinematics to find the final speed of the sled.

1. Find the x component of $\vec{F}$:

$$F_x = F \cos\theta = (6.20\,\text{N}) \cos(35.0°) = 5.08\,\text{N}$$

2. Apply Newton's second law to the x direction:

$$\sum F_x = F_x = ma_x$$

3. Solve for the x component of acceleration:

$$a_x = \frac{F_x}{m} = \frac{5.08\,\text{N}}{4.60\,\text{kg}} = 1.10\,\text{m/s}^2$$

4. Use $v_x = v_{0x} + a_x t$ to find the speed of the sled:

$$v_x = v_{0x} + a_x t$$
$$= 0 + (1.10\,\text{m/s}^2)(1.15\,\text{s}) = 1.27\,\text{m/s}$$

The y component of $\vec{F}$ has no effect on the acceleration of the sled.

FIGURE 5-14 shows a dramatic example of a case where the net force acting on a system is zero. As a result, neither climber is accelerating. Further examples of this type will be discussed in Chapter 6.

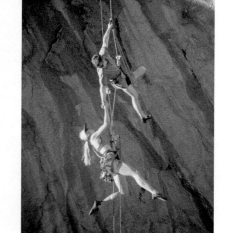

▲ **FIGURE 5-14** At the moment this picture was taken, the acceleration of both climbers was zero because the net force acting on them was zero. In particular, the upward forces exerted on the lower climber by the other climber and the ropes exactly cancel the downward force that gravity exerts on her.

Enhance Your Understanding (Answers given at the end of the chapter)

5. An object is acted on by a single force that is at an angle of 35° above the positive x axis. The object's x component of acceleration is 1 m/s². Is the y component of the object's acceleration greater than, less than, or equal to 1 m/s²? Explain.

CONTINUED

Section Review

• Newton's second law can be applied to each coordinate direction independently.

5-6 Weight

When you step onto a scale to weigh yourself, the scale gives a measurement of the pull of Earth's gravity. This is your weight, W. Similarly, the weight of any object on the Earth's surface is simply the gravitational force exerted on it by the Earth.

> The weight, W, of an object on the Earth's surface is the gravitational force exerted on it by the Earth.

As we know from everyday experience, the greater the mass of an object, the greater its weight. For example, if you put a brick on a scale and weigh it, you might get a reading of 9.0 N. If you put a second, identical brick on the scale—which doubles the mass—you will find a weight of 2(9.0 N) = 18 N. Clearly, there must be a simple connection between weight, W, and mass, m.

Relating Weight and Mass To see exactly what this connection is, consider taking one of the bricks just mentioned and letting it drop in free fall. As indicated in FIGURE 5-15, the only force acting on the brick is its weight, W, which is downward. If we choose upward to be the positive direction, we have

$$\sum F_y = -W$$

In addition, we know from Chapter 2 that the brick moves downward with an acceleration of $g = 9.81 \text{ m/s}^2$ regardless of its mass. Thus,

$$a_y = -g$$

Using these results in Newton's second law, we find

$$\sum F_y = ma_y$$
$$-W = -mg$$

Therefore, the weight of an object of mass m is $W = mg$:

> **Definition: Weight, W**
>
> $W = mg$ 5-5
>
> SI unit: newton, N

Distinguishing Weight and Mass There is a clear distinction between weight and mass. Weight is a gravitational force, measured in newtons; mass is a measure of the inertia of an object, and it is given in kilograms. For example, if you were to travel to the Moon, your mass would not change—you would have the same amount of matter in you, regardless of your location. On the other hand, the gravitational force on the Moon's surface is less than the gravitational force on the Earth's surface. As a result, you would weigh less on the Moon than on the Earth, even though your mass is the same.

To be specific, on Earth an 81.0-kg person has a weight given by

$$W_{\text{Earth}} = mg_{\text{Earth}} = (81.0 \text{ kg})(9.81 \text{ m/s}^2) = 795 \text{ N}$$

In contrast, the same person on the Moon, where the acceleration of gravity is 1.62 m/s^2, weighs only

$$W_{\text{Moon}} = mg_{\text{Moon}} = (81.0 \text{ kg})(1.62 \text{ m/s}^2) = 131 \text{ N}$$

This is roughly one-sixth the weight on Earth. If, sometime in the future, there is a Lunar Olympics, the Moon's low gravity would be a boon for pole-vaulters, gymnasts, and others.

Finally, because weight is a force—which is a vector quantity—it has both a magnitude and a direction. Its magnitude, of course, is mg, and its direction is simply the

Physical picture	Free-body diagram

▲ FIGURE 5-15 **Weight and mass** A brick of mass m has only one force acting on it in free fall—its weight, $\vec{W}$. The resulting acceleration has a magnitude $a = g$; hence $W = mg$.

PHYSICS IN CONTEXT
Looking Ahead

Another important application of forces is in the study of collisions. Central to this topic is the concept of momentum, a physical quantity that is changed when a force acts on an object over a period of time. We shall study momentum in Chapter 9.

PHYSICS IN CONTEXT
Looking Ahead

In this chapter we introduced the force law for gravity near the Earth's surface, $F = mg$. The more general law of gravity, valid at any location, is introduced in Chapter 12.

direction of gravitational acceleration. Thus, if $\vec{g}$ denotes a vector of magnitude g, pointing in the direction of free-fall acceleration, the weight of an object can be written in vector form as follows:

$$\vec{W} = m\vec{g}$$

We use the weight vector and its magnitude, mg, in the next Example.

EXAMPLE 5-11   WHERE'S THE FIRE?

The fire alarm goes off, and a 97-kg fireman slides 3.0 m down a pole to the ground floor. Suppose the fireman starts from rest, slides with constant acceleration, and reaches the ground floor in 1.2 s. What was the upward force $\vec{F}$ exerted by the pole on the fireman?

Physical picture   **Free-body diagram**

PICTURE THE PROBLEM
Our sketch shows the fireman sliding down the pole and the two forces acting on him: the upward force exerted by the pole, $\vec{F}$, and the downward force of gravity, $\vec{W}$. We choose the positive y direction to be upward; therefore $\vec{F} = F\hat{y}$ and $\vec{W} = (-mg)\hat{y}$. In addition, we choose $y = 0$ to be at ground level.

REASONING AND STRATEGY
The basic idea in approaching this problem is to apply Newton's second law to the y direction: $\Sigma F_y = ma_y$. The acceleration is not given directly, but we can find it using the kinematic equation $y = y_0 + v_{0y}t + \frac{1}{2}a_yt^2$. Substituting the result for a_y into Newton's second law, along with $W_y = -W = -mg$, allows us to solve for the unknown force, $\vec{F}$.

Known   Mass of fireman, $m = 97$ kg; sliding distance, $y_0 = 3.0$ m; sliding time starting from rest, $t = 1.2$ s.
Unknown   Upward force exerted on the fireman, $\vec{F} = ?$

SOLUTION

1. Solve $y = y_0 + v_{0y}t + \frac{1}{2}a_yt^2$ for a_y, using the fact that $v_{0y} = 0$:

$$y = y_0 + v_{0y}t + \frac{1}{2}a_yt^2 = y_0 + \frac{1}{2}a_yt^2$$
$$a_y = \frac{2(y - y_0)}{t^2}$$

2. Substitute $y = 0$, $y_0 = 3.0$ m, and $t = 1.2$ s to find the acceleration:

$$a_y = \frac{2(0 - 3.0\,\text{m})}{(1.2\,\text{s})^2} = -4.2\,\text{m/s}^2$$

3. Sum the forces in the y direction:

$$\sum F_y = F - mg$$

4. Set the sum of the forces equal to mass times acceleration:

$$F - mg = ma_y$$

5. Solve for F, the y component of the force exerted by the pole. Use the result for F to write the force vector $\vec{F}$:

$$F = mg + ma_y = m(g + a_y)$$
$$= (97\,\text{kg})(9.81\,\text{m/s}^2 - 4.2\,\text{m/s}^2) = 540\,\text{N}$$
$$\vec{F} = (540\,\text{N})\hat{y}$$

INSIGHT
How is it that the pole exerts a force on the fireman? Well, by wrapping his arms and legs around the pole as he slides, the fireman exerts a downward force on the pole. By Newton's third law, the pole exerts an upward force of equal magnitude on the fireman. These forces are due to friction, which we shall study in detail in Chapter 6.

PRACTICE PROBLEM
What is the fireman's acceleration if the force exerted on him by the pole is 650 N?   [**Answer:** $a_y = -3.1\,\text{m/s}^2$]

Some related homework problems: Problem 36, Problem 68

Apparent Weight

We've all had the experience of riding in an elevator and feeling either heavy or light, depending on its motion. For example, when an elevator accelerates upward from rest, we feel heavier. On the other hand, we feel lighter when an elevator that is moving upward comes to rest by accelerating downward. In short, the motion of an elevator can give rise to an **apparent weight** that differs from our true weight. Why?

▶ **FIGURE 5-16 Apparent weight** A person rides in an elevator that is accelerating upward. Because the acceleration is upward, the net force must also be upward. As a result, the force exerted on the person by the floor of the elevator, $\vec{W}_a$, must be greater than the person's weight, $\vec{W}$. This means that the person feels heavier than normal.

Physical picture

Free-body diagram

The reason is that our sensation of weight in this case is due to the force exerted on our feet by the floor of the elevator. If this force is greater than our weight, mg, we feel heavy; if it is less than mg, we feel light.

Riding in an Accelerating Elevator As an example, imagine you are in an elevator that is moving with an upward acceleration a, as indicated in **FIGURE 5-16**. Two forces act on you: (i) your weight, W, acting downward; and (ii) the upward normal force exerted on your feet by the floor of the elevator. Let's call the second force W_a, since it represents your apparent weight—that is, W_a is the force that pushes upward on your feet and gives you the sensation of your "weight" pushing down on the floor. We can find W_a by applying Newton's second law to the vertical direction.

To be specific, the sum of the forces acting on you is

$$\sum F_y = W_a - W$$

By Newton's second law, this sum must equal ma_y. Setting $a_y = a$, we find

$$W_a - W = ma$$

Solving for the apparent weight, W_a, yields

$$W_a = W + ma$$
$$= mg + ma = m(g + a) \qquad \text{5-6}$$

Notice that W_a is greater than your weight, mg, and hence you feel heavier. In fact, your apparent weight is precisely what it would be if you were suddenly "transported" to a planet where the acceleration due to gravity is $g + a$ instead of g.

On the other hand, if the elevator accelerates downward, so that $a_y = -a$, your apparent weight is found by simply replacing a with $-a$ in Equation 5-6:

$$W_a = W - ma$$
$$= mg - ma = m(g - a) \qquad \text{5-7}$$

In this case you feel lighter than usual.

We explore these results in the next Example, in which we consider weighing a fish on a scale. The reading on the scale is equal to the upward force it exerts on an object. Thus, the upward force exerted by the scale is the apparent weight, W_a.

EXAMPLE 5-12   SOMETHING FISHY

As part of an attempt to combine physics and biology in the same class, an instructor asks students to weigh a 5.2-kg salmon by hanging it from a fish scale attached to the ceiling of an elevator. What is the apparent weight of the salmon, $\vec{W}_a$, if the elevator **(a)** is at rest, **(b)** moves with an upward acceleration of 3.5 m/s², or **(c)** moves with a downward acceleration of 2.3 m/s²?

Physical picture

Free-body diagram

PICTURE THE PROBLEM
The free-body diagram for the salmon shows the weight of the salmon, $\vec{W}$, and the force exerted by the scale, $\vec{W}_a$, which is the apparent weight. Notice that upward is the positive direction. Therefore, the y component of the weight is $-W = -mg$ and the y component of the apparent weight is W_a.

REASONING AND STRATEGY
We know the magnitude of the weight, $W = mg$, and the acceleration, a_y. To find the apparent weight, W_a, we use Newton's second law applied to the y component:
$\sum F_y = W_a - W = ma_y$. **(a)** Set $a_y = 0$. **(b)** Set $a_y = 3.5$ m/s². **(c)** Set $a_y = -2.3$ m/s².

CONTINUED

Known Mass of salmon, $m = 5.2$ kg; acceleration of elevator, (a) $a_y = 0$, (b) $a_y = 3.5$ m/s², (c) $a_y = -2.3$ m/s².
Unknown (a), (b), and (c) Apparent weight, $\vec{W}_a = ?$

SOLUTION

Part (a)

1. Sum the y components of the forces and set equal to mass, m, times the y component of acceleration, a_y:

$$\sum F_y = W_a - W = ma_y$$

2. Solve for W_a and set $a_y = 0$. Use the numerical result to write the vector $\vec{W}$:

$$W_a = W + ma_y$$
$$= mg + 0 = (5.2 \text{ kg})(9.81 \text{ m/s}^2) = 51 \text{ N}$$
$$\vec{W}_a = (51 \text{ N})\hat{\mathbf{y}}$$

Part (b)

3. Use $W_a = W + ma_y$ again, this time with $a_y = 3.5$ m/s². Use the numerical result to write the vector $\vec{W}_a$:

$$W_a = W + ma_y$$
$$= mg + ma_y = 51 \text{ N} + (5.2 \text{ kg})(3.5 \text{ m/s}^2) = 69 \text{ N}$$
$$\vec{W}_a = (69 \text{ N})\hat{\mathbf{y}}$$

Part (c)

4. Use $W_a = W + ma_y$ again, this time with $a_y = -2.3$ m/s². Use the numerical result to write the vector $\vec{W}_a$:

$$W_a = W + ma_y$$
$$= mg + ma_y = 51 \text{ N} + (5.2 \text{ kg})(-2.3 \text{ m/s}^2) = 39 \text{ N}$$
$$\vec{W}_a = (39 \text{ N})\hat{\mathbf{y}}$$

INSIGHT

When the salmon is at rest, or moving with constant velocity, its acceleration is zero and the apparent weight is equal to the actual weight, mg. In part (b) the apparent weight is greater than the actual weight because the scale must exert an upward force capable not only of supporting the salmon, but of accelerating it upward as well. In part (c) the apparent weight is less than the actual weight. In this case the net force acting on the salmon is downward, and hence its acceleration is downward.

PRACTICE PROBLEM

What is the elevator's acceleration if the scale gives a reading of (a) 52 N, (b) 42 N? [**Answer:** (a) $a_y = 0.19$ m/s², (b) $a_y = -1.7$ m/s²]

Some related homework problems: Problem 38, Problem 40, Problem 41

▲ **FIGURE 5-17** Astronaut candidates pose for a floating class picture during weightlessness training aboard the "vomit comet."

Apparent Weight and Weightlessness Let's return for a moment to Equation 5-7:

$$W_a = m(g - a)$$

This result indicates that a person feels lighter than normal when riding in an elevator with a downward acceleration a. In particular, if the elevator's downward acceleration is g—that is, if the elevator is in free fall—it follows that $W_a = m(g - g) = 0$. Thus, a person feels "weightless" (zero apparent weight) in a freely falling elevator!

RWP NASA uses this effect when training astronauts. Trainees are sent aloft in a KC-135 airplane, as shown in **FIGURE 5-17**. This plane is affectionately known as the "vomit comet" because many trainees experience nausea along with the weightlessness. To generate an experience of weightlessness, the plane flies on a parabolic path—the same path followed by a projectile in free fall. Each round of weightlessness lasts about half a minute, after which the plane pulls up to regain altitude and start the cycle again. On a typical flight, trainees experience about 40 cycles of weightlessness. Many scenes in the movie *Apollo 13* were shot in 30-second takes aboard the vomit comet.

RWP Many amusement parks have popular rides, like drop towers and roller coasters, that give visitors a thrill by providing brief periods of weightlessness. For example, the Zumanjaro: Drop of Doom ride at Six Flags Great Adventure in Jackson, New Jersey, is a drop tower that is 126 m (415 ft) high, providing passengers with a few seconds of downward acceleration at 9.81 m/s², so that they are weightless relative to the gondola in which they are riding. Similarly, the Outlaw Run roller coaster at Silver Dollar City in Branson, Missouri, features a 49.4-m (162 ft) drop at a 81° inclination during its 87-second ride in which passengers briefly feel weightless relative to their seats.

This idea of free-fall weightlessness applies to more than just the vomit comet. In fact, astronauts in orbit experience weightlessness for the same reason—they and their craft are actually in free fall. As we shall see in detail in Chapter 12, orbital motion is just a special case of free fall.

CONCEPTUAL EXAMPLE 5-13   ELEVATOR RIDE   PREDICT/EXPLAIN

(a) If you ride in an elevator moving upward with constant speed, is your apparent weight greater than, less than, or equal to your actual weight, *mg*? (b) Which of the following is the *best explanation* for your prediction?

   I. The elevator is moving upward. This will cause a person to press down harder on the floor, giving an apparent weight greater than *mg*.

  II. The elevator moves with zero acceleration, and therefore the apparent weight is equal to *mg*.

 III. Because the elevator is moving upward it is partially lifting the person, resulting in an apparent weight less than *mg*.

REASONING AND DISCUSSION

If the elevator is moving in a straight line with constant speed, its acceleration is zero. Now, if the acceleration is zero, the net force must also be zero. Hence, the upward force exerted by the floor of the elevator, W_a, must equal the downward force of gravity, *mg*. As a result, your apparent weight is equal to *mg*.

Notice that this conclusion agrees with Equations 5-6 and 5-7, with $a = 0$.

ANSWER

(a) Your apparent weight is equal to *mg*. (b) The best explanation is II.

Enhance Your Understanding   (Answers given at the end of the chapter)

6. When a certain person steps onto a scale on solid ground, the reading is 750 N. When the same person rides in an elevator that accelerates, the reading on the scale is different. Rank the three cases (A, B, and C) shown in **FIGURE 5-18** in order of increasing acceleration (from most negative to most positive) in the *y* direction. Indicate ties where appropriate.

▲ **FIGURE 5-18** Enhance Your Understanding 6

Section Review

- Weight is the force of gravity exerted on an object.
- The apparent weight of an object depends on the acceleration of its frame of reference.

5-7 Normal Forces

As you get ready for lunch, you take a can of soup from the cupboard and place it on the kitchen counter. The can is now at rest, which means that its acceleration is zero, so the net force acting on it is also zero. Thus, you know that the downward force of gravity is being opposed by an upward force exerted by the counter, as shown in **FIGURE 5-19**. As we've mentioned before, this force is referred to as the **normal force, $\vec{N}$**. The reason the force is called normal is that it is *perpendicular to the surface,* and in mathematical terms, *normal* simply means perpendicular.

    The origin of the normal force is the interaction between atoms in a solid that acts to maintain its shape. When the can of soup is placed on the countertop, for example, it causes an imperceptibly small compression of the surface of the counter. This is similar to compressing a spring, and just like a spring, the countertop exerts a force to oppose the compression. Therefore, the greater the weight placed on the countertop, the greater the normal force it exerts to oppose being compressed.

Physical picture

Free-body diagram

▲ **FIGURE 5-19 The normal force is perpendicular to a surface** A can of soup rests on a kitchen counter, which exerts a normal (perpendicular) force, $\vec{N}$, to support it. In the special case shown here, the normal force is equal in magnitude to the weight, $W = mg$, and opposite in direction.

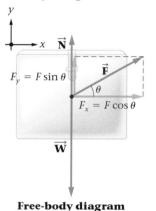

Physical picture

Free-body diagram

▲ FIGURE 5-20 **The normal force may differ from the weight** A suitcase is pulled across the floor by an applied force of magnitude F, directed at an angle θ above the horizontal. As a result of the upward component of $\vec{F}$, the normal force $\vec{N}$ will have a magnitude less than the weight of the suitcase.

The Normal Force Can Be Greater Than, Less Than, or Equal to the Weight In the example of the soup can and the countertop, the magnitude of the normal force is equal to the weight of the can. This is a special case, however. In general, the normal force can also be greater than or less than the weight of an object.

To see how this can come about, consider pulling a 12.0-kg suitcase across a smooth floor by exerting a force, $\vec{F}$, at an angle θ above the horizontal. The weight of the suitcase is $mg = (12.0\text{ kg})(9.81\text{ m/s}^2) = 118\text{ N}$. The normal force will have a magnitude less than this, however, because the force $\vec{F}$ has an upward component that supports part of the suitcase's weight. To be specific, suppose that $\vec{F}$ has a magnitude of 45.0 N and that $\theta = 20.0°$. What is the normal force exerted by the floor on the suitcase?

The situation is illustrated in **FIGURE 5-20**, where we show the three forces acting on the suitcase: (i) the weight of the suitcase, $\vec{W}$, (ii) the force $\vec{F}$, and (iii) the normal force, $\vec{N}$. We also indicate a typical coordinate system in the figure, with the x axis horizontal and the y axis vertical. Now, the key to solving a problem like this is to realize that since the suitcase does not move in the y direction, its y component of acceleration is zero; that is, $a_y = 0$. It follows, from Newton's second law, that the sum of the y components of force must also equal zero; that is, $\Sigma F_y = ma_y = 0$. Using this condition, we can solve for the one force that is unknown, $\vec{N}$.

To find $\vec{N}$, then, we start by writing out the y component of each force. For the weight we have $W_y = -mg = -118\text{ N}$; for the applied force, $\vec{F}$, the y component is $F_y = F\sin 20.0° = (45.0\text{ N})\sin 20.0° = 15.4\text{ N}$; finally, the y component of the normal force is $N_y = N$. Setting the sum of the y components of force equal to zero yields

$$\Sigma F_y = W_y + F_y + N_y = -mg + F\sin 20.0° + N = 0$$

Solving for N gives

$$N = mg - F\sin 20.0° = 118\text{ N} - 15.4\text{ N} = 103\text{ N}$$

In vector form,

$$\vec{N} = N_y\hat{y} = (103\text{ N})\hat{y}$$

Thus, as mentioned, the normal force has a magnitude less than $mg = 118\text{ N}$ because the y component of $\vec{F}$, $F_y = F\sin 20.0°$, supports part of the weight. In the following Example, however, the applied forces cause the normal force to be greater than the weight.

EXAMPLE 5-14 A MOVEABLE FEAST

A 4.53-kg holiday turkey is moved toward the center of a smooth table by two people at the feast. Person 1 exerts a force $\vec{F}_1$ with a magnitude of 35.5 N at an angle of 34.6° below the horizontal, as indicated in the sketch. Person 2 exerts a force $\vec{F}_2$ with a magnitude of 39.7 N at an angle of 21.2° above the horizontal. What is the magnitude N of the normal force exerted by the table on the turkey?

PICTURE THE PROBLEM
The sketch shows our choice of coordinate system, as well as all the forces acting on the turkey. Notice that $\vec{F}_1$ has a positive x component and a negative y component; $\vec{F}_2$ has positive x and y components. The weight $\vec{W}$ and the normal force $\vec{N}$ have only y components. For example, noting that the mass of the turkey is $m = 4.53\text{ kg}$, we have for the components of the weight, $W_x = 0$ and $W_y = -mg = -(4.53\text{ kg})(9.81\text{ m/s}^2) = -44.4\text{ N}$. The components of the normal force are $N_x = 0$ and $N_y = N$. For the moment we ignore friction with the table, which we will cover in the next chapter.

Components of $\vec{F}_1$

$F_{1,x} = F_1 \cos 34.6°$

$F_{1,y} = -F_1 \sin 34.6°$

Components of $\vec{F}_2$

$F_{2,y} = F_2 \sin 21.2°$

$F_{2,x} = F_2 \cos 21.2°$

CONTINUED

REASONING AND STRATEGY

The basic idea in this problem is to apply Newton's second law to the y direction. Specifically, the fact that there is no motion in the y direction means that the acceleration in the y direction is zero. Hence, we can find the magnitude N of the normal force by setting $\sum F_y = ma_y = 0$.

Known   Mass of turkey, $m = 4.53$ kg; force 1, 35.5 N at 34.6° below the horizontal; force 2, 39.7 N at 21.2° above the horizontal.

Unknown   Magnitude of the normal force exerted by the table, $N = ?$

SOLUTION

1. Write out the y component of each force. The only force we don't know is the normal. We represent its magnitude by N:

$F_{1,y} = -F_1 \sin 34.6° = -(35.5\,\text{N}) \sin 34.6° = -20.2\,\text{N}$
$F_{2,y} = F_2 \sin 21.2° = (39.7\,\text{N}) \sin 21.2° = 14.4\,\text{N}$
$N_y = N \quad W_y = -mg = -44.4\,\text{N}$

2. Sum the y components of force:

$\sum F_y = F_{1,y} + F_{2,y} + N_y + W_y$

$= -20.2\,\text{N} + 14.4\,\text{N} + N - 44.4\,\text{N}$

3. Set this sum equal to 0, since the acceleration in the y direction is zero, and solve for N:

$-20.2\,\text{N} + 14.4\,\text{N} + N - 44.4\,\text{N} = 0$
$N = 20.2\,\text{N} - 14.4\,\text{N} + 44.4\,\text{N} = 50.2\,\text{N}$

INSIGHT

The normal force is greater in magnitude than the weight, $mg = 44.4$ N. This is because the force $\vec{F}_1$ pushes down with a y component that is greater than the upward y component of $\vec{F}_2$, even though $\vec{F}_2$ has the greater magnitude.

In general, the normal force exerted by a surface is only as large as is necessary to prevent the motion of an object into the surface. If the required force is larger than the material can provide, the surface will break.

PRACTICE PROBLEM

What is the magnitude of the normal force if $\vec{F}_2$ is at an angle of 42.5° above the horizontal?   [**Answer:** In this case, $N = 37.8$ N, which is less than the weight of the turkey.]

Some related homework problems: Problem 45, Problem 52

Normal Forces on Tilted Surfaces   To this point, we have considered surfaces that are horizontal, in which case the normal force is vertical. When a surface is inclined, the normal force is still at right angles to the surface, even though it is no longer vertical. This is illustrated in **FIGURE 5-21**. (If friction is present, a surface may also exert a force that is parallel to its surface. This will be considered in detail in Chapter 6.)

When we choose a coordinate system for an inclined surface, it is generally best to have the x and y axes of the system parallel and perpendicular to the surface, respectively, as in **FIGURE 5-22**. One can imagine the coordinate system to be "bolted down" to the surface, so that when the surface is tilted, the coordinate system tilts along with it.

With this choice of coordinate system, there is no motion in the y direction, even on the inclined surface, and the normal force points in the positive y direction. Thus, the condition that determines the normal force is still $\sum F_y = ma_y = 0$ as before. In addition, if the object slides on the surface, its motion is purely in the x direction.

Finally, if the surface is inclined by an angle θ, notice that the weight—which is still vertically downward—is at the same angle θ with respect to the negative y axis, as shown in Figure 5-22. As a result, the x and y components of the weight are

$$W_x = W \sin \theta = mg \sin \theta \qquad \text{5-8}$$

and

$$W_y = -W \cos \theta = -mg \cos \theta \qquad \text{5-9}$$

Let's quickly check some special cases of these results. First, if $\theta = 0$ the surface is horizontal, and we find $W_x = 0$, $W_y = -mg$, as expected. Second, if $\theta = 90°$ the surface is vertical; therefore, the weight is parallel to the surface, pointing in the positive x direction. In this case, $W_x = mg$ and $W_y = 0$.

The next Example shows how to use the weight components to find the acceleration of an object on an inclined surface.

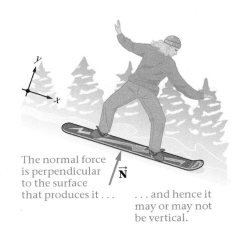

The normal force is perpendicular to the surface that produces it . . .   . . . and hence it may or may not be vertical.

▲ **FIGURE 5-21 An object on an inclined surface** The normal force $\vec{N}$ is always at right angles to the surface; hence, it is not always in the vertical direction.

▶ **FIGURE 5-22 Components of the weight on an inclined surface** Whenever a surface is tilted by an angle θ, the weight $\vec{\mathbf{W}}$ makes the same angle θ with respect to the negative y axis. This is proven in part (b), where we show that $\theta + \phi = 90°$ and that $\theta' + \phi = 90°$. From these results it follows that $\theta' = \theta$. The component of the weight perpendicular to the surface is $W_y = -W \cos \theta$; the component parallel to the surface is $W_x = W \sin \theta$.

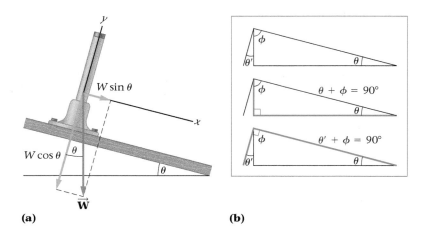

(a)   **(b)**

EXAMPLE 5-15   TOBOGGAN TO THE BOTTOM

A child of mass m rides on a toboggan down a slick, ice-covered hill inclined at an angle θ with respect to the horizontal. **(a)** What is the acceleration of the child? **(b)** What is the normal force exerted on the child by the toboggan?

PICTURE THE PROBLEM
We choose the x axis to be parallel to the slope, with the positive direction pointing downhill. Similarly, we choose the y axis to be perpendicular to the slope, pointing up and to the right. With these choices, the x component of $\vec{\mathbf{W}}$ is positive, $W_x = W \sin \theta$, and its y component is negative, $W_y = -W \cos \theta$. Finally, the x component of the normal force is zero, $N_x = 0$, and its y component is positive, $N_y = N$.

REASONING AND STRATEGY
Only two forces act on the child: (i) the weight, $\vec{\mathbf{W}}$, and (ii) the normal force, $\vec{\mathbf{N}}$. **(a)** We find the child's acceleration by solving $\Sigma F_x = ma_x$ for a_x. **(b)** Because there is no motion in the y direction, the y component of acceleration is zero. Therefore, we can find the normal force by setting $\Sigma F_y = ma_y = 0$.

Known Mass of child, m; angle of incline, θ.
Unknown **(a)** Acceleration of the child, $a_x = ?$ **(b)** Normal force exerted on the child, $\vec{\mathbf{N}} = ?$

SOLUTION

Part (a)

1. Write out the x components of the forces acting on the child:

$$N_x = 0$$
$$W_x = W \sin \theta = mg \sin \theta$$

2. Sum the x components of the forces and set equal to ma_x:

$$\Sigma F_x = N_x + W_x = mg \sin \theta = ma_x$$

3. Divide by the mass m to find the acceleration in the x direction:

$$a_x = \frac{\Sigma F_x}{m} = \frac{mg \sin \theta}{m} = g \sin \theta$$

Part (b)

4. Write out the y components of the forces acting on the child:

$$N_y = N$$
$$W_y = -W \cos \theta = -mg \cos \theta$$

5. Sum the y components of the forces and set the sum equal to zero, since $a_y = 0$:

$$\Sigma F_y = N_y + W_y = N - mg \cos \theta$$
$$= ma_y = 0$$

6. Solve for the magnitude of the normal force, N:

$$N - mg \cos \theta = 0 \quad \text{or} \quad N = mg \cos \theta$$

7. Write the normal force in vector form:

$$\vec{\mathbf{N}} = (mg \cos \theta)\hat{\mathbf{y}}$$

INSIGHT
Notice that for θ between 0 and 90° the acceleration of the child is *less* than the acceleration of gravity. This is because only a *component* of the weight is causing the acceleration.

CONTINUED

Let's check some special cases of our general result, $a_x = g \sin \theta$. First, let $\theta = 0$. In this case, we find zero acceleration: $a_x = g \sin 0 = 0$. This makes sense because with $\theta = 0$ the hill is actually level, and we don't expect an acceleration. Second, let $\theta = 90°$. In this case, the hill is vertical, and the toboggan should drop straight down in free fall. This also agrees with our general result: $a_x = g \sin 90° = g$.

PRACTICE PROBLEM

What is the child's acceleration if his mass is doubled to $2m$?   [**Answer:** The acceleration is still $a_x = g \sin \theta$. As in free fall, the acceleration produced by gravity is independent of mass.]

Some related homework problems: Problem 46, Problem 51

Enhance Your Understanding   (Answers given at the end of the chapter)

7. **FIGURE 5-23** shows four identical bricks that are at rest on surfaces tilted at different angles. Rank the bricks in order of increasing normal force exerted on them by the surface. Indicate ties where appropriate.

| A | B | C | D |

▲ **FIGURE 5-23** Enhance Your Understanding 7

Section Review

• A normal force is perpendicular (normal) to a surface.

CHAPTER 5 REVIEW

CHAPTER SUMMARY

5-1   FORCE AND MASS

Force
A push or a pull.

Mass
A measure of the difficulty in accelerating an object. Equivalently, a measure of the quantity of matter in an object.

5-2   NEWTON'S FIRST LAW OF MOTION

First Law (Law of Inertia)
If the net force on an object is zero, its velocity is constant.

Inertial Frame of Reference
A frame of reference in which the first law holds. All inertial frames of reference move with constant velocity relative to one another.

5-3   NEWTON'S SECOND LAW OF MOTION

Second Law
An object of mass m has an acceleration $\vec{a}$ given by the net force $\sum \vec{F}$ divided by m. That is,

$$\vec{a} = \sum \vec{F}/m \qquad \text{5-1}$$

Component Form

$$\sum F_x = ma_x \qquad \sum F_y = ma_y \qquad \sum F_z = ma_z \qquad \text{5-2}$$

SI Unit: Newton (N)

$$1\,\text{N} = 1\,\text{kg} \cdot \text{m/s}^2 \qquad \text{5-3}$$

Free-Body Diagram
A sketch showing all external forces acting on an object.

5-4 NEWTON'S THIRD LAW OF MOTION

Third Law
For every force that acts on an object, there is a reaction force acting on a different object that is equal in magnitude and opposite in direction.

Contact Forces
Action-reaction pair of forces produced by the physical contact of two objects.

5-5 THE VECTOR NATURE OF FORCES: FORCES IN TWO DIMENSIONS

Forces are vectors.

Newton's second law can be applied to each component of force separately and independently.

5-6 WEIGHT

Gravitational force exerted *by the Earth on an object*.

On the surface of the Earth the weight, W, of an object of mass m has the magnitude

$$W = mg \qquad\qquad 5\text{-}5$$

Apparent Weight
The force felt from contact with the floor or a scale in an accelerating system—for example, the sensation of feeling heavier or lighter in an accelerating elevator.

5-7 NORMAL FORCES

A force exerted by a surface that is *perpendicular* to the surface.

ANSWERS TO ENHANCE YOUR UNDERSTANDING QUESTIONS

1. Statement **B** is correct. The net force can have a magnitude with any value from $|F_1 - F_2|$ to $(F_1 + F_2)$ depending on the direction of the two forces.

2. Statements **B** and **E** are correct—a net force is required to make an object accelerate (change its velocity).

3. **(a)** $2a$.   **(b)** a.   **(c)** $4a$.   **(d)** $a/4$.

4. The force of contact between boxes 1 and 2 is greater than the force of contact between boxes 2 and 3. The reason is that the force between 1 and 2 must give an acceleration a to a mass of 2 kg + 3 kg = 5 kg, but the force between 2 and 3 must give an acceleration a to a mass of only 3 kg.

5. Less than 1 m/s² because the y component of a vector is less than the x component when the vector is only 35° above the x axis.

6. B < A < C.

7. B = C < D < A.

CONCEPTUAL QUESTIONS

For instructor-assigned homework, go to www.masteringphysics.com. (MP)

(Answers to odd-numbered Conceptual Questions can be found in the back of the book.)

1. Driving down the road, you hit the brakes suddenly. As a result, your body moves toward the front of the car. Explain, using Newton's laws.

2. You've probably seen pictures of someone pulling a tablecloth out from under glasses, plates, and silverware set out for a formal dinner. Perhaps you've even tried it yourself. Using Newton's laws of motion, explain how this stunt works.

3. As you read this, you are most likely sitting quietly in a chair. Can you conclude, therefore, that you are at rest? Explain.

4. When a dog gets wet, it shakes its body from head to tail to shed the water. Explain, in terms of Newton's first law, why this works.

5. A young girl slides down a rope. As she slides faster and faster she tightens her grip, increasing the force exerted on her by the rope. What happens when this force is equal in magnitude to her weight? Explain.

6. A block of mass m hangs from a string attached to a ceiling, as shown in **FIGURE 5-24**. An identical string hangs down from the bottom of the block. Which string breaks if **(a)** the lower string is pulled with a slowly increasing force or **(b)** the lower string is jerked rapidly downward? Explain.

7. An astronaut on a space walk discovers that his jet pack no longer works, leaving him stranded 50 m from the spacecraft. If the jet

pack is removable, explain how the astronaut can still use it to return to the ship.

8. Two untethered astronauts on a space walk decide to take a break and play catch with a baseball. Describe what happens as the game of catch progresses.

9. In **FIGURE 5-25** Wilbur asks Mr. Ed, the talking horse, to pull a cart. Mr. Ed replies that he would like to, but the laws of nature just won't allow it. According to Newton's third law, he says, if he pulls on the wagon it pulls back on him with an equal force. Clearly, then, the net force is zero and the wagon will stay put. How should Wilbur answer the clever horse?

FIGURE 5-24 Conceptual Question 6

FIGURE 5-25 Conceptual Question 9

10. A whole brick has more mass than half a brick, thus the whole brick is harder to accelerate. Why doesn't a whole brick fall more slowly than half a brick? Explain.

11. The force exerted by gravity on a whole brick is greater than the force exerted by gravity on half a brick. Why, then, doesn't a whole brick fall faster than half a brick? Explain.

12. Is it possible for an object at rest to have only a single force acting on it? If your answer is yes, provide an example. If your answer is no, explain why not.

13. Is it possible for an object to be in motion and yet have zero net force acting on it? Explain.

14. A bird cage, with a parrot inside, hangs from a scale. The parrot decides to hop to a higher perch. What can you say about the reading on the scale **(a)** when the parrot jumps, **(b)** when the parrot is in the air, and **(c)** when the parrot lands on the second perch? Assume that the scale responds rapidly so that it gives an accurate reading at all times.

15. Suppose you jump from the cliffs of Acapulco and perform a perfect swan dive. As you fall, you exert an upward force on the Earth equal in magnitude to the downward force the Earth exerts on you. Why, then, does it seem that you are the one doing all the accelerating? Since the forces are the same, why aren't the accelerations?

16. A friend tells you that since his car is at rest, there are no forces acting on it. How would you reply?

17. Since all objects are "weightless" in orbit, how is it possible for an orbiting astronaut to tell if one object has more mass than another object? Explain.

18. To clean a rug, you can hang it from a clothesline and beat it with a tennis racket. Use Newton's laws to explain why beating the rug should have a cleansing effect.

19. If you step off a high board and drop to the water below, you plunge into the water without injury. On the other hand, if you were to drop the same distance onto solid ground, you might break a leg. Use Newton's laws to explain the difference.

20. Is it possible for an object to be moving in one direction while the net force acting on it is in another direction? If your answer is yes, provide an example. If your answer is no, explain why not.

21. Since a bucket of water is "weightless" in space, would it hurt to kick the bucket? Explain.

22. In the movie *The Rocketeer,* a teenager discovers a jet-powered backpack in an old barn. The backpack allows him to fly at incredible speeds. In one scene, however, he uses the backpack to rapidly accelerate an old pickup truck that is being chased by "bad guys." He does this by bracing his arms against the cab of the pickup and firing the backpack, giving the truck the acceleration of a drag racer. Is the physics of this scene "Good," "Bad," or "Ugly"? Explain.

23. List three common objects that have a weight of approximately 1 N.

PROBLEMS AND CONCEPTUAL EXERCISES

Answers to odd-numbered Problems and Conceptual Exercises can be found in the back of the book. **BIO** *identifies problems of biological or medical interest;* **CE** *indicates a conceptual exercise.* Predict/Explain *problems ask for two responses: (a) your prediction of a physical outcome, and (b) the best explanation among three provided; and* Predict/Calculate *problems ask for a prediction of a physical outcome, based on fundamental physics concepts, and follow that with a numerical calculation to verify the prediction. On all problems, bullets (•, ••, •••) indicate the level of difficulty.*

SECTION 5-3 NEWTON'S SECOND LAW OF MOTION

1. • **CE** An object of mass m is initially at rest. After a force of magnitude F acts on it for a time T, the object has a speed v. Suppose the mass of the object is doubled, and the magnitude of the force acting on it is quadrupled. In terms of T, how much time does it take for the object to accelerate from rest to a speed v now?

2. • On a planet far, far away, an astronaut picks up a rock. The rock has a mass of 5.00 kg, and on this particular planet its weight is 40.0 N. If the astronaut exerts an upward force of 46.2 N on the rock, what is its acceleration?

3. • In a grocery store, you push a 15.4-kg shopping cart horizontally with a force of 13.4 N. If the cart starts at rest, how far does it move in 2.50 s?

4. • You are pulling your little sister on her sled across an icy (frictionless) surface. When you exert a constant horizontal force of 120 N, the sled has an acceleration of 2.5 m/s². If the sled has a mass of 7.4 kg, what is the mass of your little sister?

5. • A 0.53-kg billiard ball initially at rest is given a speed of 12 m/s during a time interval of 4.0 ms. What average force acted on the ball during this time?

6. • A 92-kg water skier floating in a lake is pulled from rest to a speed of 12 m/s in a distance of 25 m. What is the net force exerted on the skier, assuming his acceleration is constant?

7. • A 0.5-kg object is acted on by a force whose x component varies with time as shown in **FIGURE 5-26**. Draw the corresponding a_x-versus-time graph for this object.

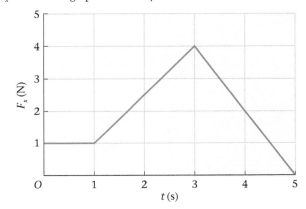

FIGURE 5-26 Problem 7

8. •• **CE** Predict/Explain You drop two balls of equal diameter from the same height at the same time. Ball 1 is made of metal and has a greater mass than ball 2, which is made of wood. The upward force due to air resistance is the same for both balls. **(a)** Is the drop time of ball 1 greater than, less than, or equal to the drop time of ball 2? **(b)** Choose the *best explanation* from among the following:
   I. The acceleration of gravity is the same for all objects, regardless of mass.
   II. The more massive ball is harder to accelerate.
   III. Air resistance has less effect on the more massive ball.

9. •• Predict/Calculate A 42.0-kg parachutist is moving straight downward with a speed of 3.85 m/s. **(a)** If the parachutist comes to rest with constant acceleration over a distance of 0.750 m, what force does the ground exert on her? **(b)** If the parachutist comes to rest over a shorter distance, is the force exerted by the ground greater than, less than, or the same as in part (a)? Explain.

10. •• Predict/Calculate In baseball, a pitcher can accelerate a 0.15-kg ball from rest to 98 mi/h in a distance of 1.7 m. **(a)** What is the average force exerted on the ball during the pitch? **(b)** If the mass of the ball is increased, is the force required of the pitcher increased, decreased, or unchanged? Explain.

11. •• A major-league catcher gloves a 92 mi/h pitch and brings it to rest in 0.15 m. If the force exerted by the catcher is 803 N, what is the mass of the ball?

12. •• Driving home from school one day, you spot a ball rolling out into the street (FIGURE 5-27). You brake for 1.20 s, slowing your 950-kg car from 16.0 m/s to 9.50 m/s. **(a)** What was the average force exerted on your car during braking? **(b)** How far did you travel while braking?

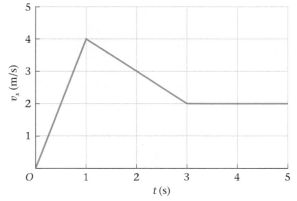

FIGURE 5-27 Problem 12

13. •• **Stopping a 747** A 747 jetliner lands and begins to slow to a stop as it moves along the runway. If its mass is 3.50×10^5 kg, its speed is 75.0 m/s, and the net braking force is 7.25×10^5 N, **(a)** what is its speed 10.0 s later? **(b)** How far has it traveled in this time?

14. •• The v_x-versus-time graph for a 1.8-kg object is shown in FIGURE 5-28. A single force acts on the object, and the force has only an x component. What is F_x at **(a)** $t = 0.50$ s, **(b)** $t = 2.0$ s, and **(c)** $t = 4.0$ s?

FIGURE 5-28 Problem 14

15. •• Predict/Calculate A drag racer crosses the finish line doing 202 mi/h and promptly deploys her drag chute (the small parachute used for braking). **(a)** What force must the drag chute exert on the 891-kg car to slow it to 45.0 mi/h in a distance of 185 m? **(b)** Describe the strategy you used to solve part (a).

SECTION 5-4 NEWTON'S THIRD LAW OF MOTION

16. • **CE** Predict/Explain A small car collides with a large truck. **(a)** Is the magnitude of the force experienced by the car greater than, less than, or equal to the magnitude of the force experienced by the truck? **(b)** Choose the *best explanation* from among the following:
   I. Action-reaction forces always have equal magnitude.
   II. The truck has more mass, and hence the force exerted on it is greater.
   III. The massive truck exerts a greater force on the lightweight car.

17. • **CE** Predict/Explain A small car collides with a large truck. **(a)** Is the acceleration experienced by the car greater than, less than, or equal to the acceleration experienced by the truck? **(b)** Choose the *best explanation* from among the following:
   I. The truck exerts a larger force on the car, giving it the greater acceleration.
   II. Both vehicles experience the same magnitude of force; therefore the lightweight car experiences the greater acceleration.
   III. The greater force exerted on the truck gives it the greater acceleration.

18. • As you catch a 0.14-kg ball it accelerates at -320 m/s^2 and comes to rest in your hand. What force does the ball exert on your hand?

19. • **BIO** **Woodpecker Concussion Prevention** A woodpecker exerts a 230-N force on a tree trunk during its effort to find hidden insects to eat. If its head has an effective mass of 0.025 kg, what is the acceleration of its head away from the tree trunk? The woodpecker's skull has a special spongy bone plate to protect itself against brain concussions.

20. •• On vacation, your 1400-kg car pulls a 560-kg trailer away from a stoplight with an acceleration of 1.85 m/s^2. **(a)** What is the net force exerted on the trailer? **(b)** What force does the trailer exert on the car? **(c)** What is the net force acting on the car?

21. •• Predict/Calculate An 85-kg parent and a 24-kg child meet at the center of an ice rink. They place their hands together and push. **(a)** Is the force experienced by the child greater than, less than, or the same as the force experienced by the parent? **(b)** Is the acceleration of the child greater than, less than, or the same as the acceleration of the parent? Explain. **(c)** If the acceleration of the child is 3.3 m/s^2 in magnitude, what is the magnitude of the parent's acceleration?

22. •• A force of magnitude 7.50 N pushes three boxes with masses $m_1 = 1.30$ kg, $m_2 = 3.20$ kg, and $m_3 = 4.90$ kg, as shown in FIGURE 5-29. Find the magnitude of the contact force **(a)** between boxes 1 and 2, and **(b)** between boxes 2 and 3.

FIGURE 5-29 Problem 22

23. •• A force of magnitude 7.50 N pushes three boxes with masses $m_1 = 1.30\,\text{kg}$, $m_2 = 3.20\,\text{kg}$, and $m_3 = 4.90\,\text{kg}$, as shown in **FIGURE 5-30**. Find the magnitude of the contact force **(a)** between boxes 1 and 2, and **(b)** between boxes 2 and 3.

FIGURE 5-30 Problem 23

24. •• **Predict/Calculate** Two boxes sit side-by-side on a smooth horizontal surface. The lighter box has a mass of 5.2 kg; the heavier box has a mass of 7.4 kg. **(a)** Find the contact force between these boxes when a horizontal force of 5.0 N is applied to the light box. **(b)** If the 5.0-N force is applied to the heavy box instead, is the contact force between the boxes the same as, greater than, or less than the contact force in part (a)? Explain. **(c)** Verify your answer to part (b) by calculating the contact force in this case.

SECTION 5-5 THE VECTOR NATURE OF FORCES

25. • **CE** A skateboarder on a ramp is accelerated by a nonzero net force. For each of the following statements, state whether it is always true, never true, or sometimes true. **(a)** The skateboarder is moving in the direction of the net force. **(b)** The acceleration of the skateboarder is at right angles to the net force. **(c)** The acceleration of the skateboarder is in the same direction as the net force. **(d)** The skateboarder is instantaneously at rest.

26. • **CE** Three objects, A, B, and C, have x and y components of velocity that vary with time as shown in **FIGURE 5-31**. What is the direction of the net force acting on **(a)** object A, **(b)** object B, and **(c)** object C, as measured from the positive x axis? (All of the nonzero slopes have the same magnitude.)

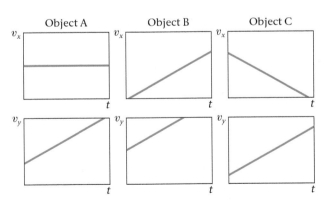

FIGURE 5-31 Problem 26

27. • A farm tractor tows a 3300-kg trailer up a 14° incline with a steady speed of 2.8 m/s. What force does the tractor exert on the trailer? (Ignore friction.)

28. • A shopper pushes a 7.5-kg shopping cart up a 13° incline, as shown in **FIGURE 5-32**. Find the magnitude of the horizontal force, $\vec{\mathbf{F}}$, needed to give the cart an acceleration of 1.41 m/s².

FIGURE 5-32 Problem 28

29. • Two crewmen pull a raft through a lock, as shown in **FIGURE 5-33**. One crewman pulls with a force of 130 N at an angle of 34° relative to the forward direction of the raft. The second crewman, on the opposite side of the lock, pulls at an angle of 45°. With what force should the second crewman pull so that the net force of the two crewmen is in the forward direction?

FIGURE 5-33 Problem 29

30. •• **CE** A hockey puck is acted on by one or more forces, as shown in **FIGURE 5-34**. Rank the four cases, A, B, C, and D, in order of the magnitude of the puck's acceleration, starting with the smallest. Indicate ties where appropriate.

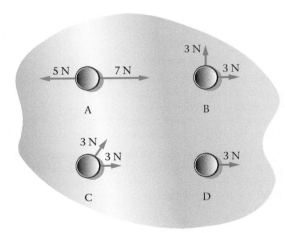

FIGURE 5-34 Problem 30

31. •• To give a 19-kg child a ride, two teenagers pull on a 3.7-kg sled with ropes, as indicated in **FIGURE 5-35**. Both teenagers pull with a force of 55 N at an angle of 35° relative to the forward direction, which is the direction of motion. In addition, the snow exerts a retarding force on the sled that points opposite to the direction of motion, and has a magnitude of 57 N. Find the acceleration of the sled and child.

FIGURE 5-35 Problem 31

32. •• Predict/Calculate A 65-kg skier speeds down a trail, as shown in **FIGURE 5-36**. The surface is smooth and inclined at an angle of 22° with the horizontal. **(a)** Find the direction and magnitude of the net force acting on the skier. **(b)** Does the net force exerted on the skier increase, decrease, or stay the same as the slope becomes steeper? Explain.

FIGURE 5-36 Problems 32 and 46

33. •• An object acted on by three forces moves with constant velocity. One force acting on the object is in the positive *x* direction and has a magnitude of 6.5 N; a second force has a magnitude of 4.4 N and points in the negative *y* direction. Find the direction and magnitude of the third force acting on the object.

34. •• A train is traveling up a 2.88° incline at a speed of 4.31 m/s when the last car breaks free and begins to coast without friction. **(a)** How much time does it take for the last car to come to rest momentarily? **(b)** How far did the last car travel before momentarily coming to rest?

35. •• **The Force Exerted on the Moon** In **FIGURE 5-37** we show the Earth, Moon, and Sun (not to scale) in their relative positions at the time when the Moon is in its third-quarter phase. Though few people realize it, the force exerted on the Moon by the Sun is actually greater than the force exerted on the Moon by the Earth. In fact, the force exerted on the Moon by the Sun has a magnitude of $F_{SM} = 4.34 \times 10^{20}$ N, whereas the force exerted by the Earth has a magnitude of only $F_{EM} = 1.98 \times 10^{20}$ N. These forces are indicated to scale in Figure 5-37. Find **(a)** the direction

and **(b)** the magnitude of the net force acting on the Moon. **(c)** Given that the mass of the Moon is $M_M = 7.35 \times 10^{22}$ kg, find the magnitude of its acceleration at the time of the third-quarter phase.

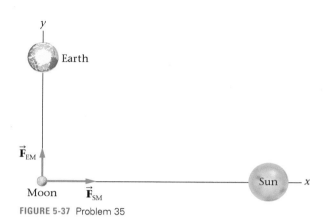

FIGURE 5-37 Problem 35

SECTION 5-6 WEIGHT

36. • You pull upward on a stuffed suitcase with a force of 105 N, and it accelerates upward at 0.725 m/s². What are **(a)** the mass and **(b)** the weight of the suitcase?

37. • **BIO Brain Growth** A newborn baby's brain grows rapidly. In fact, it has been found to increase in mass by about 1.6 mg per minute. **(a)** How much does the brain's weight increase in one day? **(b)** How much time does it take for the brain's weight to increase by 0.15 N?

38. • Suppose a rocket launches with an acceleration of 30.5 m/s². What is the apparent weight of a 92-kg astronaut aboard this rocket?

39. • During an episode of turbulence in an airplane you feel 190 N heavier than usual. If your mass is 78 kg, what are the magnitude and direction of the airplane's acceleration?

40. • At the bow of a ship on a stormy sea, a crewman conducts an experiment by standing on a bathroom scale. In calm waters, the scale reads 182 lb. During the storm, the crewman finds a maximum reading of 225 lb and a minimum reading of 138 lb. Find **(a)** the maximum upward acceleration and **(b)** the maximum downward acceleration experienced by the crewman.

41. •• Predict/Calculate As part of a physics experiment, you stand on a bathroom scale in an elevator. Though your normal weight is 610 N, the scale at the moment reads 730 N. **(a)** Is the acceleration of the elevator upward, downward, or zero? Explain. **(b)** Calculate the magnitude of the elevator's acceleration. **(c)** What, if anything, can you say about the velocity of the elevator? Explain.

42. •• When you weigh yourself on good old *terra firma* (solid ground), your weight is 142 lb. In an elevator your apparent weight is 121 lb. What are the direction and magnitude of the elevator's acceleration?

43. •• Predict/Calculate **BIO Flight of the Samara** A 1.21-g samara— the winged fruit of a maple tree—falls toward the ground with a constant speed of 1.1 m/s (**FIGURE 5-38**). **(a)** What is the force of air resistance exerted on the samara? **(b)** If the constant speed of descent is greater than 1.1 m/s, is the force of air resistance greater than, less than, or the same as in part **(a)**? Explain.

$v = 1.1 \text{ m/s}$

FIGURE 5-38 Problem 43

44. •••When you lift a bowling ball with a force of 82 N, the ball accelerates upward with an acceleration a. If you lift with a force of 92 N, the ball's acceleration is $2a$. Find **(a)** the weight of the bowling ball, and **(b)** the acceleration a.

SECTION 5-7 NORMAL FORCES

45. • A 23-kg suitcase is pulled with constant speed by a handle at an angle of 25° above the horizontal. If the normal force exerted on the suitcase is 180 N, what is the force F applied to the handle?

46. • **(a)** Draw a free-body diagram for the skier in Problem 32. **(b)** Determine the normal force acting on the skier.

47. • A 9.3-kg child sits in a 3.7-kg high chair. **(a)** Draw a free-body diagram for the child, and find the normal force exerted by the chair on the child. **(b)** Draw a free-body diagram for the chair, and find the normal force exerted by the floor on the chair.

48. •• **FIGURE 5-39** shows the normal force N experienced by a rider of weight $W = mg$ on a Ferris wheel as a function of time. **(a)** At what time, A, B, C, or D, is the rider accelerating upward? **(b)** What is the magnitude of the maximum acceleration experienced by the rider?

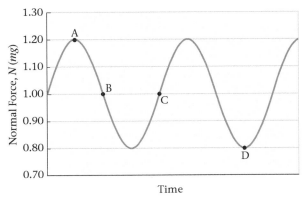

FIGURE 5-39 Problem 48

49. •• **FIGURE 5-40** shows the normal force N as a function of the angle θ for the suitcase shown in Figure 5-20. Determine the magnitude of the force $\vec{F}$ for each of the three curves shown in Figure 5-40. Give your answer in terms of the weight of the suitcase, mg.

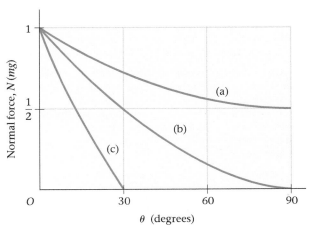

FIGURE 5-40 Problem 49

50. •• A 5.0-kg bag of potatoes sits on the bottom of a stationary shopping cart. **(a)** Sketch a free-body diagram for the bag of potatoes. **(b)** Now suppose the cart moves with a constant velocity. How does this affect your free-body diagram? Explain.

51. •• Predict/Calculate **(a)** Find the normal force exerted on a 2.9-kg book resting on a surface inclined at 36° above the horizontal. **(b)** If the angle of the incline is reduced, do you expect the normal force to increase, decrease, or stay the same? Explain.

52. •• Predict/Calculate A gardener mows a lawn with an old-fashioned push mower. The handle of the mower makes an angle of 32° with the surface of the lawn. **(a)** If a 68-N force is applied along the handle of the 17-kg mower, what is the normal force exerted by the lawn on the mower? **(b)** If the angle between the surface of the lawn and the handle of the mower is increased, does the normal force exerted by the lawn increase, decrease, or stay the same? Explain.

53. •••An ant walks slowly away from the top of a bowling ball, as shown in **FIGURE 5-41**. If the ant starts to slip when the normal force on its feet drops below one-half its weight, at what angle θ does slipping begin?

FIGURE 5-41 Problem 53

GENERAL PROBLEMS

54. • **CE** Predict/Explain Riding in an elevator moving upward with constant speed, you begin a game of darts. **(a)** Do you have to aim your darts higher than, lower than, or the same as when you play darts on solid ground? **(b)** Choose the *best explanation* from among the following:
   I. The elevator rises during the time it takes for the dart to travel to the dartboard.
   II. The elevator moves with constant velocity. Therefore, Newton's laws apply within the elevator in the same way as on the ground.
   III. You have to aim lower to compensate for the upward speed of the elevator.

55. • **CE** Predict/Explain Riding in an elevator moving with a constant upward acceleration, you begin a game of darts. **(a)** Do you have to aim your darts higher than, lower than, or the same as when you play darts on solid ground? **(b)** Choose the *best explanation* from among the following:
   I. The elevator accelerates upward, giving its passengers a greater "effective" acceleration due to gravity.

II. You have to aim lower to compensate for the upward acceleration of the elevator.

III. Since the elevator moves with a constant acceleration, Newton's laws apply within the elevator the same as on the ground.

56. • **CE** Give the direction of the net force acting on each of the following objects. If the net force is zero, state "zero." **(a)** A car accelerating northward from a stoplight. **(b)** A car traveling southward and slowing down. **(c)** A car traveling westward with constant speed. **(d)** A skydiver parachuting downward with constant speed. **(e)** A baseball during its flight from pitcher to catcher (ignoring air resistance).

57. • **CE** Predict/Explain You jump out of an airplane and open your parachute after an extended period of free fall. **(a)** To decelerate your fall, must the force exerted on you by the parachute be greater than, less than, or equal to your weight? **(b)** Choose the *best explanation* from among the following:

I. Parachutes can only exert forces that are less than the weight of the skydiver.

II. The parachute exerts a force exactly equal to the skydiver's weight.

III. To decelerate after free fall, the net force acting on a skydiver must be upward.

58. • In a tennis serve, a 0.070-kg ball can be accelerated from rest to 36 m/s over a distance of 0.75 m. Find the magnitude of the average force exerted by the racket on the ball during the serve.

59. • **BIO Human Heart Force** The left ventricle of the human heart expels about 0.070 kg of oxygenated blood straight upward into the ascending aorta with a force of 5.5 N. What is the acceleration of the blood as it enters the aorta?

60. • A 51.5-kg swimmer with an initial speed of 1.25 m/s decides to coast until she comes to rest. If she slows with constant acceleration and stops after coasting 2.20 m, what was the force exerted on her by the water?

61. • The a_x-versus-time graph for a 2.0-kg object is shown in FIGURE 5-42. Draw the corresponding F_x-versus-time graph, assuming only a single force acts on the object.

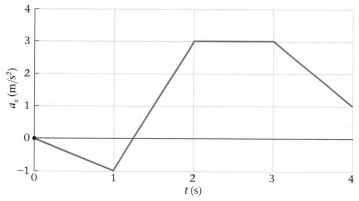

FIGURE 5-42 Problem 61

62. • A skateboarder starts from rest and rolls down a ramp that is inclined at an angle of 15° below the horizontal. Friction can be ignored. **(a)** What is the acceleration of the skateboarder? **(b)** What is the skateboarder's speed 1.5 s after starting?

63. •• The rotors of a 15,200-kg heavy-lift helicopter exert a downward force of 322,000 N in order to accelerate itself and its external cargo upward at 0.50 m/s². By itself, what upward force does the external cargo exert on the Earth?

64. •• As it pulls itself up to a branch, a chimpanzee accelerates upward at 2.2 m/s² at the instant it exerts a 260-N force downward on the branch. The situation is shown in FIGURE 5-43. **(a)** In what direction is (i) the force the branch exerts on the chimpanzee; (ii) the force the tree exerts on the branch; and (iii) the force the chimpanzee exerts on the Earth? **(b)** Find the magnitude of the force the chimpanzee exerts on the Earth.

FIGURE 5-43 Problem 64

65. •• **CE** Each of the three identical hockey pucks shown in FIGURE 5-44 is acted on by a 3-N force. Puck A moves with a speed of 7 m/s in a direction opposite to the force; puck B is instantaneously at rest; puck C moves with a speed of 7 m/s at right angles to the force. Rank the three pucks in order of the magnitude of their acceleration, starting with the smallest. Indicate ties with an equal sign.

FIGURE 5-44 Problem 65

66. •• Predict/Calculate **The VASIMR Rocket** NASA plans to use a new type of rocket, a Variable Specific Impulse Magnetoplasma Rocket (VASIMR), on future missions. If a VASIMR can produce 6.0 N of thrust (force), and has a mass of 620 kg, **(a)** what acceleration will it experience? Assume that the only force acting on the rocket is its own thrust, and that the mass of the rocket is constant. **(b)** Over what distance must the rocket accelerate from rest to achieve a speed of 9500 m/s? **(c)** When the rocket has covered one-quarter the acceleration distance found in part (b), is its average speed 1/2, 1/3, or 1/4 its average speed during the final three-quarters of the acceleration distance? Explain.

67. •• An object of mass $m = 5.95$ kg has an acceleration $\vec{a} = (1.17 \text{ m/s}^2)\hat{x} + (-0.664 \text{ m/s}^2)\hat{y}$. Three forces act on this object: $\vec{F}_1, \vec{F}_2$, and $\vec{F}_3$. Given that $\vec{F}_1 = (3.22 \text{ N})\hat{x}$ and $\vec{F}_2 = (-1.55 \text{ N})\hat{x} + (2.05 \text{ N})\hat{y}$, find $\vec{F}_3$.

68. •• At the local grocery store, you push a 14.5-kg shopping cart. You stop for a moment to add a bag of dog food to your cart. With a force of 12.0 N, you now accelerate the cart from rest through a distance of 2.29 m in 3.00 s. What was the mass of the dog food?

69. •• **BIO** Predict/Calculate **The Force of Running** Biomechanical research has shown that when a 67-kg person is running, the force exerted on each foot as it strikes the ground can be as great as 2300 N. **(a)** What is the ratio of the force exerted on the foot by the ground to the person's body weight? **(b)** If the only forces acting on the person are (i) the force exerted by the ground and (ii) the person's weight, what are the magnitude and direction of the person's acceleration?

(c) If the acceleration found in part (b) acts for 10.0 ms, what is the resulting change in the vertical component of the person's velocity?

70. •• **BIO** Predict/Calculate **Grasshopper Liftoff** To become airborne, a 2.0-g grasshopper requires a takeoff speed of 2.7 m/s. It acquires this speed by extending its hind legs through a distance of 3.7 cm. **(a)** What is the average acceleration of the grasshopper during take-off? **(b)** Find the magnitude of the average net force exerted on the grasshopper by its hind legs during takeoff. **(c)** If the mass of the grass-hopper increases, does the takeoff acceleration increase, decrease, or stay the same? **(d)** If the mass of the grasshopper increases, does the required takeoff force increase, decrease, or stay the same? Explain.

71. •• **Takeoff from an Aircraft Carrier** On an aircraft carrier, a jet can be catapulted from 0 to 155 mi/h in 2.00 s. If the average force exerted by the catapult is 9.35×10^5 N, what is the mass of the jet?

72. •• A 79-kg person stands on a bathroom scale and rides in an elevator whose v_y-versus-time graph is shown in FIGURE 5-45. What is the reading on the scale at **(a)** $t = 1.0$ s, **(b)** $t = 4.0$ s, and **(c)** $t = 8.0$ s?

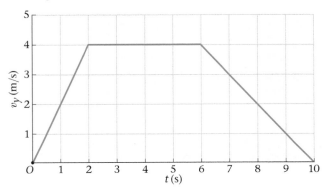

FIGURE 5-45 Problem 72

73. •• Predict/Calculate An archer shoots a 0.024-kg arrow at a target with a speed of 54 m/s. When it hits the target, it penetrates to a depth of 0.083 m. **(a)** What was the average force exerted by the target on the arrow? **(b)** If the mass of the arrow is doubled, and the force exerted by the target on the arrow remains the same, by what multiplicative factor does the penetration depth change? Explain.

74. •• An apple of mass $m = 0.13$ kg falls out of a tree from a height $h = 3.2$ m. **(a)** What is the magnitude of the force of gravity, mg, acting on the apple? **(b)** What is the apple's speed, v, just before it lands? **(c)** Show that the force of gravity times the height, mgh, is equal to $\frac{1}{2}mv^2$. (We shall investigate the significance of this result in Chapter 8.) Be sure to show that the dimensions are in agreement as well as the numerical values.

75. •• **BIO** **The Fall of *T. rex*** Paleontologists estimate that if a *Tyrannosaurus rex* had tripped and fallen, the ground would have exerted a normal force of approximately 260,000 N on its torso as it landed. Assuming the torso has a mass of 3800 kg, **(a)** find the magnitude of the torso's upward acceleration as it comes to rest. (For comparison, humans lose consciousness with an acceleration of about 7g.) **(b)** Assuming the torso of a *T. rex* is in free fall for a distance of 1.46 m as it falls to the ground, how much time is required for the torso to come to rest once it contacts the ground?

76. •• **Deep Space 1** The NASA spacecraft *Deep Space 1* was shut down on December 18, 2001, following a three-year journey to the asteroid Braille and the comet Borrelly. This spacecraft used a solar-powered ion engine to produce 0.021 pounds of thrust (force) by stripping electrons from xenon atoms and accelerating the resulting ions to 70,000 mi/h. The thrust was only as much as the weight of a couple sheets of paper, but the engine operated continuously for 16,000 hours. As a result, the speed of the spacecraft increased by

7900 mi/h. What was the mass of *Deep Space 1*? (Assume that the mass of the xenon gas is negligible.)

77. •• Your groceries are in a bag with paper handles. The two handles together will tear off if a force greater than 51.5 N is applied to them. What is the greatest mass of groceries that can be lifted safely with this bag, given that the bag is raised **(a)** with constant speed, or **(b)** with an acceleration of 1.25 m/s²?

78. •• **BIO** **A Leafhopper's Leap** The motion of jumping insects is of interest not only to biologists, but also to those who design robots. FIGURE 5-46 shows the speed-versus-time graph for two different takeoff jumps of a 19-mg green leafhopper, as determined from high-speed videos. **(a)** What is the magnitude of the average net force acting on the leafhopper's body during jump 1? **(b)** What is the ratio of the force found in part (a) to the leafhopper's weight? **(c)** Is the average force during jump 2 greater than, less than, or equal to the average force during jump 1? Explain.

FIGURE 5-46 Problem 78

79. •• Predict/Calculate At the airport, you observe some of the jets as they take off. With your watch you find that it takes about 35 seconds for a plane to go from rest to takeoff speed. In addition, you estimate that the distance required is about 1.5 km. **(a)** If the mass of a jet is 1.70×10^5 kg, what force is needed for takeoff? **(b)** Describe the strategy you used to solve part (a).

80. •• **BIO** **Gecko Feet** Researchers have found that a gecko's foot is covered with hundreds of thousands of small hairs (*setae*) that allow it to walk up walls and even across ceilings. A single foot pad, which has an area of 1.0 cm², can attach to a wall or ceiling with a force of 11 N. **(a)** How many 250-g geckos could be suspended from the ceiling by a single foot pad? **(b)** Estimate the force per square centimeter that your body exerts on the soles of your shoes, and compare with the 11 N/cm² of the sticky gecko foot.

81. •• Two boxes are at rest on a smooth, horizontal surface. The boxes are in contact with one another. If box 1 is pushed with a force of magnitude $F = 12.00$ N, the contact force between the boxes is 8.50 N; if, instead, box 2 is pushed with the force F, the contact force is 12.00 N − 8.50 N = 3.50 N. In either case, the boxes move together with an acceleration of 1.70 m/s². What is the mass of **(a)** box 1 and **(b)** box 2?

82. •••You have been hired to help improve the material movement system at a manufacturing plant. Boxes containing 22 kg of tomato sauce in glass jars must slide from rest down a frictionless roller ramp to the loading dock, but they must not accelerate at a rate that exceeds 2.5 m/s² because of safety concerns. **(a)** What is the maximum angle of inclination of the ramp? **(b)** If the vertical distance the ramp must span is 1.2 m, with what speed will the boxes exit the bottom of the ramp? **(c)** What is the normal force on a box as it moves down the ramp?

83. ••• You and some friends take a hot-air balloon ride. One friend is late, so the balloon floats a couple of feet off the ground as you wait. Before this person arrives, the combined weight of the basket and people is 1220 kg, and the balloon, which contains 5980 kg of hot air, is neutrally buoyant. When the late arrival climbs up into the basket, the balloon begins to accelerate downward at $0.10 \, \text{m/s}^2$. What was the mass of the last person to climb aboard?

84. •••A baseball of mass m and initial speed v strikes a catcher's mitt. If the mitt moves a distance Δx as it brings the ball to rest, what is the average force it exerts on the ball?

85. •••When two people push in the same direction on an object of mass m they cause an acceleration of magnitude a_1. When the same people push in opposite directions, the acceleration of the object has a magnitude a_2. Determine the magnitude of the force exerted by each of the two people in terms of m, a_1, and a_2.

86. •••An air-track cart of mass $m_1 = 0.14 \, \text{kg}$ is moving with a speed $v_0 = 1.3 \, \text{m/s}$ to the right when it collides with a cart of mass $m_2 = 0.25 \, \text{kg}$ that is at rest. Each cart has a wad of putty on its bumper, and hence they stick together as a result of their collision. Suppose the average contact force between the carts is $F = 1.5 \, \text{N}$ during the collision. (a) What is the acceleration of cart 1? Give direction and magnitude. (b) What is the acceleration of cart 2? Give direction and magnitude. (c) How much time does it take for both carts to have the same speed? (Once the carts have the same speed the collision is over and the contact force vanishes.) (d) What is the final speed of the carts, v_f? (e) Show that $m_1 v_0$ is equal to $(m_1 + m_2)v_f$. (We shall investigate the significance of this result in Chapter 9.)

PASSAGE PROBLEMS

BIO **Increasing Safety in a Collision**

Safety experts say that an automobile accident is really a succession of three separate collisions: (1) the automobile collides with an obstacle and comes to rest; (2) people within the car continue to move forward until they collide with the interior of the car, or are brought to rest by a restraint system like a seatbelt or an air bag; and (3) organs within the occupants' bodies continue to move forward until they collide with the body wall and are brought to rest. Not much can be done about the third collision, but the effects of the first two can be mitigated by increasing the distance over which the car and its occupants are brought to rest.

For example, the severity of the first collision is reduced by building collapsible "crumple zones" into the body of a car, and by placing compressible collision barriers near dangerous obstacles like bridge supports. The second collision is addressed primarily through the use of seatbelts and air bags. These devices reduce the force that acts on an occupant to survivable levels by increasing the distance over which he or she comes to rest. This is illustrated in FIGURE 5-47, where we see the force exerted on a 65.0-kg driver who slows from an initial speed of 18.0 m/s (lower curve) or 36.0 m/s (upper curve) to rest in a distance ranging from 5.00 cm to 1.00 m.

FIGURE 5-47 Problems 87, 88, 89, and 90

87. •• The combination of "crumple zones" and air bags/seatbelts might increase the distance over which a person stops in a collision to as great as 1.00 m. What is the magnitude of the force exerted on a 65.0-kg driver who decelerates from 18.0 m/s to 0.00 m/s over a distance of 1.00 m?

A. 162 N   B. 585 N
C. 1.05×10^4 N   D. 2.11×10^4 N

88. •• A driver who does not wear a seatbelt continues to move forward with a speed of 18.0 m/s (due to inertia) until something solid like the steering wheel is encountered. The driver now comes to rest in a much shorter distance—perhaps only a few centimeters. Find the magnitude of the net force acting on a 65.0-kg driver who is decelerated from 18.0 m/s to rest in 5.00 cm.

A. 3240 N   B. 1.17×10^4 N
C. 2.11×10^5 N   D. 4.21×10^5 N

89. • Suppose the initial speed of the driver is doubled to 36.0 m/s. If the driver still has a mass of 65.0 kg, and comes to rest in 1.00 m, what is the magnitude of the force exerted on the driver during this collision?

A. 648 N   B. 1170 N
C. 2.11×10^4 N   D. 4.21×10^4 N

90. • If both the speed and stopping distance of a driver are doubled, by what factor does the force exerted on the driver change?

A. 0.5   B. 1   C. 2   D. 4

91. •• Predict/Calculate **Referring to Example 5-8** Suppose that we would like the contact force between the boxes to have a magnitude of 5.00 N, and that the only thing in the system we are allowed to change is the mass of box 2; the mass of box 1 is 10.0 kg and the applied force is 30.0 N. (a) Should the mass of box 2 be increased or decreased? Explain. (b) Find the mass of box 2 that results in a contact force of magnitude 5.00 N. (c) What is the acceleration of the boxes in this case?

92. •• **Referring to Example 5-8** Suppose the force of 30.0 N pushes on two boxes of unknown mass. We know, however, that the acceleration of the boxes is $1.20 \, \text{m/s}^2$ and the contact force has a magnitude of 4.45 N. Find the mass of (a) box 1 and (b) box 2.

93. •• Predict/Calculate **Referring to Figure 5-13** Suppose the magnitude of $\vec{F}_2$ is increased from 41 N to 55 N, and that everything else in the system remains the same. (a) Do you expect the direction of the satellite's acceleration to be greater than, less than, or equal to 32°? Explain. Find (b) the direction and (c) the magnitude of the satellite's acceleration in this case.

94. •• Predict/Calculate **Referring to Figure 5-13** Suppose we would like the acceleration of the satellite to be at an angle of 25°, and the only quantity we can change in the system is the magnitude of $\vec{F}_1$. (a) Should the magnitude of $\vec{F}_1$ be increased or decreased? Explain. (b) What is the magnitude of the satellite's acceleration in this case?

Big Ideas

1 Friction is a force exerted between surfaces in contact.

2 Springs exert a force proportional to the amount of stretch or compression.

3 A system is in translational equilibrium when the total force acting on it is zero.

4 An object in circular motion accelerates toward the center of the circle.

6

Applications of Newton's Laws

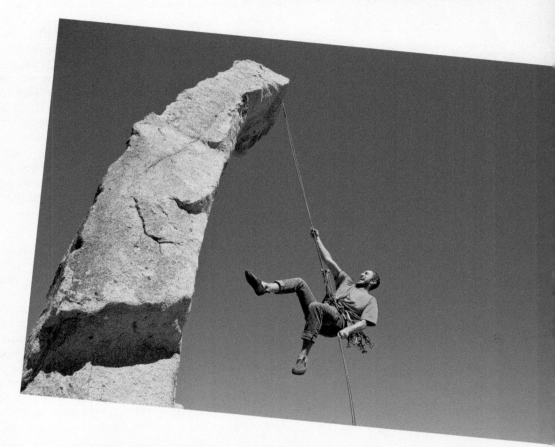

▲ The climber in this photograph may not be thinking about Newton's laws, but they are involved in every aspect of his endeavor. He relies on the forces transmitted through the ropes to support his weight, and on the pulleys through which the ropes are threaded to give those forces the desired directions. His forward progress is made possible by the substantial frictional force between his hands and the rope. These are but a few of the many real-world applications of Newton's laws that are explored in this chapter.

Newton's laws of motion can be applied to an immense variety of systems. This chapter shows how Newton's laws are applied to friction, springs, connected objects, and circular motion.

Viewed at a
microscopic scale . . .

. . . even a "smooth" surface is rough.

▲ **FIGURE 6-1** **The origin of friction** Even "smooth" surfaces have irregularities at the microscopic level that lead to friction.

Big Idea **1** Friction is a force exerted between surfaces in contact. The force of friction is different depending on whether the surfaces are in static contact, or slide relative to one another.

6-1 Frictional Forces

In Chapter 5 we assumed that surfaces are smooth and that objects slide without resistance to their motion. No surface is perfectly smooth, however. When viewed on the atomic level, even the "smoothest" surface is actually rough and jagged, as indicated in FIGURE 6-1. To slide one such surface across another requires a force large enough to overcome the resistance of microscopic hills and valleys bumping together. This is the origin of the force we call **friction.**

We often think of friction as something that should be reduced, or even eliminated if possible. For example, roughly 20% of the gasoline you buy does nothing but overcome friction within your car's engine. Clearly, reducing that friction would be desirable.

On the other hand, friction can be helpful—even indispensable—in other situations. Suppose, for example, that you are standing still and then decide to begin walking forward. The force that accelerates you is the force of friction between your shoes and the ground. We simply couldn't walk or run without friction—it's hard enough when friction is merely reduced, as on an icy sidewalk. Similarly, starting and stopping a car, and even turning a corner, all require friction. Examples of systems with small and large amounts of friction are shown in FIGURE 6-2.

Since friction is caused by the random, microscopic irregularities of a surface, and since it is greatly affected by other factors such as the presence of lubricants, there is no simple "law of nature" for friction. There are, however, some very useful rules of thumb that give accurate results for calculating frictional forces. In what follows, we describe these rules of thumb for the two most common types of friction—kinetic friction and static friction.

Kinetic Friction

As its name implies, kinetic friction is the friction produced when surfaces slide against one another. The force generated by this friction, which we designate with the symbol f_k, acts to oppose the sliding.

A series of simple experiments illustrates the main characteristics of kinetic friction. First, imagine attaching a spring scale to a rough object, like a brick, and pulling it across a table, as shown in FIGURE 6-3. If the brick moves with constant velocity, Newton's second law tells us that the net force on the brick must be zero. Hence, the force indicated on the scale, F, has the same magnitude as the force of kinetic friction, f_k. Now, if we repeat the experiment, but this time put a second brick on top of the first, we find that the force needed to pull the brick with constant velocity is doubled, to $2F$.

From this experiment we see that when we double the normal force—by stacking up two bricks, for example—the force of kinetic friction is also doubled. In general,

▶ **FIGURE 6-2** **Visualizing Concepts**
Minimizing and Maximizing Friction
Friction plays an important role in almost everything we do. Sometimes it is desirable to reduce friction; in other cases we want as much friction as possible. For example, it is more fun to ride on a water slide (left) if the friction is low. When running, however, we need friction to help us speed up, slow down, and make turns. The sole of this running shoe (right), like a car tire, is designed to maximize friction.

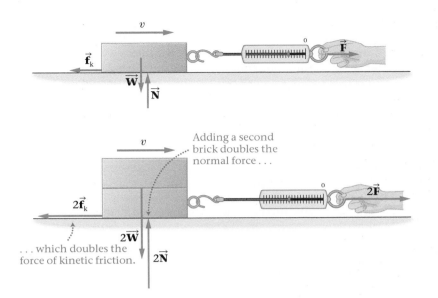

◄ **FIGURE 6-3 The force of kinetic friction depends on the normal force** In the top part of the figure, a force F is required to pull the brick with constant speed v. Thus the force of kinetic friction is $f_k = F$. In the bottom part, the normal force has been doubled, and so has the force of kinetic friction, to $f_k = 2F$.

the magnitude of the force of kinetic friction is proportional to the magnitude of the normal force, N. This is stated mathematically as follows:

$$f_k = \mu_k N \qquad \text{6-1}$$

The constant of proportionality, μ_k (pronounced "mew sub k"), is referred to as the **coefficient of kinetic friction.** In Figure 6-3 the normal force is equal to the weight of the bricks, but this is a special case. The normal force is greater than the weight if someone pushes down on the bricks, and this would cause more friction, or less than the weight of the bricks if they are placed on an incline or someone pulls upward on them.

Because f_k and N are both forces, and hence have the same units, it follows that μ_k is a dimensionless number. The coefficient of kinetic friction is always positive, and typical values range between 0 and 1, as indicated in Table 6-1. The interpretation of μ_k is simple: If $\mu_k = 0.1$, for example, the force of kinetic friction is one-tenth of the normal force. Simply put, the greater μ_k, the greater the friction.

As we know from everyday experience, the force of kinetic friction tends to oppose relative motion, as shown in Figure 6-3. Thus, $f_k = \mu_k N$ is not a vector equation, because N is perpendicular to the direction of motion. When doing calculations with the force of kinetic friction, we use $f_k = \mu_k N$ to find its magnitude, and we draw its direction so that it is opposite to the direction of motion.

There are two more friction experiments of particular interest. First, suppose that when we pull a brick, we initially pull it at the speed v, then later at the speed $2v$. What forces do we measure? It turns out that the force of kinetic friction is approximately the same in each case. Second, let's try standing the brick on end, so that it has a smaller area in contact with the table. If this smaller area is half the previous area, is the force halved? No! It turns out that the force of kinetic friction remains essentially the same, regardless of the area of contact.

We summarize our observations with the following three rules of thumb for kinetic friction:

Rules of Thumb for Kinetic Friction

The force of kinetic friction between two surfaces is:

1. Proportional to the magnitude of the normal force, N, between the surfaces:

$$f_k = \mu_k N$$

2. Independent of the relative speed of the surfaces.
3. Independent of the area of contact between the surfaces.

Before we show how to use f_k in calculations, we should make a comment regarding rule 3. This rule often seems rather surprising and counterintuitive. How is it that

TABLE 6-1 Typical Coefficients of Friction

Materials	Kinetic, μ_k	Static, μ_s
Rubber on concrete (dry)	0.80	1–4
Steel on steel	0.57	0.74
Glass on glass	0.40	0.94
Wood on leather	0.40	0.50
Copper on steel	0.36	0.53
Rubber on concrete (wet)	0.25	0.30
Steel on ice	0.06	0.10
Waxed ski on snow	0.05	0.10
Teflon on Teflon	0.04	0.04
Synovial joints in humans	0.003	0.01

PHYSICS IN CONTEXT
Looking Back

The equations of kinematics from Chapters 2 and 4 are useful in this chapter, especially in Examples 6-1 and 6-2.

Choice of Coordinate System: Incline

On an incline, align one axis (x) parallel to the surface, and the other axis (y) perpendicular to the surface. That way the motion is in the x direction. Since no motion occurs in the y direction, we know that $a_y = 0$.

a larger area of contact doesn't produce a larger force? One way to think about this is to consider that when the area of contact is large, the normal force is spread out over a large area, giving a small force per area, F/A. As a result, the microscopic hills and valleys are not pressed too deeply against one another. On the other hand, if the area is small, the normal force is concentrated in a small region, which presses the surfaces together more firmly, due to the large force per area. The net effect is roughly the same in either case.

The next Example considers a commonly encountered situation in which kinetic friction plays a decisive role.

EXAMPLE 6-1   PASS THE SALT—PLEASE

Someone at the other end of the table asks you to pass the salt. Feeling quite dashing, you slide the 50.0-g salt shaker in that direction, giving it an initial speed of 1.15 m/s. **(a)** If the shaker comes to rest with constant acceleration in 0.840 m, what is the coefficient of kinetic friction between the shaker and the table? **(b)** How much time is required for the shaker to come to rest if you slide it with an initial speed of 1.32 m/s?

PICTURE THE PROBLEM

We choose the positive x direction to be the direction of motion, and the positive y direction to be upward. Two forces act in the y direction: the shaker's weight, $\vec{W} = -W\hat{y} = -mg\hat{y}$, and the normal force, $\vec{N} = N\hat{y}$. Only one force acts in the x direction: the force of kinetic friction, $\vec{f}_k = -\mu_k N\hat{x}$. Notice that the shaker moves through a distance of 0.840 m with an initial speed $v_{0x} = 1.15$ m/s.

Physical picture

Free-body diagram

REASONING AND STRATEGY

a. The frictional force has a magnitude of $f_k = \mu_k N$, and hence it follows that $\mu_k = f_k/N$. Therefore, we need to find the magnitudes of the frictional force, f_k, and the normal force, N, to determine μ_k. To find f_k we set $\Sigma F_x = ma_x$, and find a_x with the kinematic equation $v_x^2 = v_{0x}^2 + 2a_x\Delta x$. To find N we set $a_y = 0$ (since there is no motion in the y direction) and solve for N using $\Sigma F_y = ma_y = 0$.

b. The coefficient of kinetic friction is independent of the sliding speed, and hence the acceleration of the shaker is also independent of the speed. As a result, we can use the acceleration from part (a) in the equation $v_x = v_{0x} + a_x t$ to find the sliding time.

Known   Mass of salt shaker, $m = 50.0$ g; initial speed, $v_0 = 1.15$ m/s or 1.32 m/s; sliding distance, $\Delta x = 0.840$ m.

Unknown   **(a)** Coefficient of kinetic friction, $\mu_k = ?$ **(b)** Time to come to rest, $t = ?$

SOLUTION

Part (a)

1. Set $\Sigma F_x = ma_x$ to find f_k in terms of a_x:

$$\Sigma F_x = -f_k = ma_x \quad \text{or} \quad f_k = -ma_x$$

2. Determine a_x by using the kinematic equation relating velocity to position, $v_x^2 = v_{0x}^2 + 2a_x\Delta x$:

$$v_x^2 = v_{0x}^2 + 2a_x\Delta x$$
$$a_x = \frac{v_x^2 - v_{0x}^2}{2\Delta x} = \frac{0 - (1.15 \text{ m/s})^2}{2(0.840 \text{ m})} = -0.787 \text{ m/s}^2$$

3. Set $\Sigma F_y = ma_y = 0$ to find the normal force, N:

$$\Sigma F_y = N + (-W) = ma_y = 0 \quad \text{or} \quad N = W = mg$$

4. Substitute $N = mg$ and $f_k = -ma_x$ (with $a_x = -0.787 \text{ m/s}^2$) into $\mu_k = f_k/N$ to find μ_k:

$$\mu_k = \frac{f_k}{N} = \frac{-ma_x}{mg} = \frac{-a_x}{g} = \frac{-(-0.787 \text{ m/s}^2)}{9.81 \text{ m/s}^2} = 0.0802$$

Part (b)

5. Use $a_x = -0.787 \text{ m/s}^2$, $v_{0x} = 1.32$ m/s, and $v_x = 0$ in $v_x = v_{0x} + a_x t$ to solve for the time, t:

$$v_x = v_{0x} + a_x t \quad \text{or}$$
$$t = \frac{v_x - v_{0x}}{a_x} = \frac{0 - (1.32 \text{ m/s})}{-0.787 \text{ m/s}^2} = 1.68 \text{ s}$$

INSIGHT

Notice that m canceled in Step 4, so our result for the coefficient of friction is independent of the shaker's mass. For example, if we were to slide a shaker with twice the mass, but with the same initial speed, it would slide the same distance. It's unlikely this independence would have been apparent if we had worked the problem numerically rather than symbolically. Part (b) shows that the same comments apply to the sliding time—it too is independent of the shaker's mass.

CONTINUED

In the next Example, the system is inclined at an angle θ relative to the horizontal. As a result, the normal force responsible for the kinetic friction is *less* than the weight of the object.

EXAMPLE 6-2   MAKING A BIG SPLASH

A trained sea lion slides from rest with constant acceleration down a 3.0-m-long ramp into a pool of water. If the ramp is inclined at an angle of $\theta = 23°$ above the horizontal and the coefficient of kinetic friction between the sea lion and the ramp is 0.26, how much time does it take for the sea lion to make a splash in the pool?

PICTURE THE PROBLEM
As is usual with inclined surfaces, we choose one axis to be parallel to the surface and the other to be perpendicular to it. In our sketch, the sea lion accelerates in the positive x direction ($a_x > 0$), having started from rest, $v_{0x} = 0$. We are free to choose the initial position of the sea lion to be $x_0 = 0$. There is no motion in the y direction, and therefore $a_y = 0$. Finally, we note from the free-body diagram that $\vec{\mathbf{N}} = N\hat{\mathbf{y}}$, $\vec{\mathbf{f}}_k = -\mu_k N\hat{\mathbf{x}}$, and $\vec{\mathbf{W}} = (mg\sin\theta)\hat{\mathbf{x}} + (-mg\cos\theta)\hat{\mathbf{y}}$.

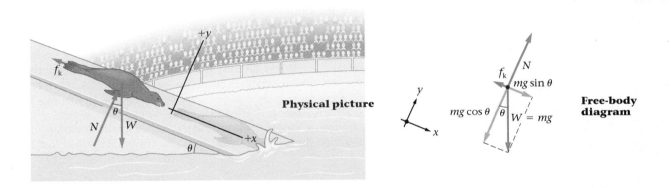

Physical picture

Free-body diagram

REASONING AND STRATEGY
We can use the kinematic equation relating position to time, $x = x_0 + v_{0x}t + \frac{1}{2}a_x t^2$, to find the time of the sea lion's slide. It will be necessary, however, to first determine the acceleration of the sea lion in the x direction, a_x.

    To find a_x we apply Newton's second law to the sea lion. First, we can find N by setting $\Sigma F_y = ma_y$ equal to zero (since $a_y = 0$). It is important to start by finding N because we need it to find the force of kinetic friction, $f_k = \mu_k N$. Using f_k in the sum of forces in the x direction, $\Sigma F_x = ma_x$, allows us to solve for a_x and, finally, for the time.

Known   Length of ramp, $x = 3.0$ m; angle of incline, $\theta = 23°$; coefficient of kinetic friction, $\mu_k = 0.26$.
Unknown Sliding time, $t = ?$

SOLUTION

1. We begin by resolving each of the three force vectors into x and y components:

$$N_x = 0 \qquad\qquad N_y = N$$
$$f_{k,x} = -f_k = -\mu_k N \qquad f_{k,y} = 0$$
$$W_x = mg\sin\theta \qquad W_y = -mg\cos\theta$$

2. Set $\Sigma F_y = ma_y = 0$ to find N.
   We see that N is less than the weight, mg:

$$\Sigma F_y = N - mg\cos\theta = ma_y = 0$$
$$N = mg\cos\theta$$

3. Next, set $\Sigma F_x = ma_x$.
   Notice that the mass cancels in this equation:

$$\Sigma F_x = mg\sin\theta - f_k$$
$$= mg\sin\theta - \mu_k mg\cos\theta = ma_x$$

4. Solve for the acceleration in the x direction, a_x:

$$a_x = g(\sin\theta - \mu_k\cos\theta)$$
$$= (9.81 \text{ m/s}^2)[\sin 23° - (0.26)\cos 23°]$$
$$= 1.5 \text{ m/s}^2$$

CONTINUED

5. Use $x = x_0 + v_{0x}t + \frac{1}{2}a_xt^2$ to find the time when the sea lion reaches the bottom. We choose $x_0 = 0$, and we are given that $v_{0x} = 0$; hence we set $x = \frac{1}{2}a_xt^2 = 3.0\,\text{m}$ and solve for t:

$$x = \frac{1}{2}a_xt^2 = 3.0\,\text{m}$$

$$t = \sqrt{\frac{2x}{a_x}} = \sqrt{\frac{2(3.0\,\text{m})}{1.5\,\text{m/s}^2}} = 2.0\,\text{s}$$

INSIGHT

We don't need the sea lion's mass to find the time. On the other hand, if we wanted the magnitude of the force of kinetic friction, $f_k = \mu_k N = \mu_k mg \cos\theta$, the mass would be needed.

It is useful to compare the sliding salt shaker in Example 6-1 with the sliding sea lion in this Example. In the case of the salt shaker, friction is the only force acting along the direction of motion (opposite to the direction of motion, in fact), and it brings the object to rest. Because of the slope on which the sea lion slides, however, it experiences both a component of its weight in the forward direction and the frictional force opposite to the motion. Because the component of the weight is the larger of the two forces, the sea lion accelerates down the slope—friction only acts to slow its progress.

PRACTICE PROBLEM

How much time would it take the sea lion to reach the water if there were no friction in this system?   **[Answer: 1.3 s]**

Some related homework problems: Problem 4, Problem 58

▲ **FIGURE 6-4** The coefficient of static friction between two surfaces depends on many factors, including whether the surfaces are dry or wet. On the desert floor of Death Valley, California, occasional rains can reduce the friction between rocks and the sandy ground to such an extent that strong winds can move the rocks over considerable distances. This results in linear "rock trails," which record the direction of the winds at different times.

Static Friction

Static friction tends to keep two surfaces from moving relative to one another. Like kinetic friction, it is due to the microscopic irregularities of surfaces that are in contact. In fact, static friction is typically stronger than kinetic friction because when surfaces are in static contact, their microscopic hills and valleys nestle down deeply into one another, forming a strong connection between the surfaces. In kinetic friction, the surfaces bounce along relative to one another and don't become as firmly enmeshed. Static friction also depends on lubrication, as illustrated in **FIGURE 6-4**.

As we did with kinetic friction, let's use the results of some simple experiments to determine the rules of thumb for static friction. We start with a brick at rest on a table, with no horizontal force pulling on it, as in **FIGURE 6-5**. Of course, in this case the force of static friction is zero; no force is needed to keep the brick from sliding.

Next, we attach a spring scale to the brick and pull with a small force of magnitude F_1, a force small enough that the brick doesn't move. Since the brick is still at rest, it follows

Static friction can have a magnitude of zero . . .

$v = 0$

$\vec{f}_s = 0$

. . . or greater than zero . . .

$v = 0$

$\vec{f}_s = -\vec{F}_1$ $\vec{F}_1$

. . . up to a maximum value.

$v = 0$

$\vec{f}_s = -\vec{F}_2$ $\vec{F}_2$

Once sliding begins, however, the friction is kinetic and has a magnitude less than the maximum value for static friction.

v

$\vec{f}_k$ $\vec{F}$

▲ **FIGURE 6-5 The maximum limit of static friction** As the force applied to an object increases, so does the force of static friction—up to a certain point. Beyond this maximum value, static friction can no longer hold the object, and it begins to slide. Now kinetic friction takes over.

that the force of static friction, f_s, is equal in magnitude to the applied force; that is, $f_s = F_1$. Now, we increase the applied force to a new value, F_2, which is still small enough that the brick stays at rest. In this case, the force of static friction has also increased so that $f_s = F_2$. If we continue increasing the applied force, we eventually reach a value beyond which the brick starts to move and kinetic friction takes over, as shown in the figure. It follows that there is a maximum force, $f_{s,max}$, that can be exerted by static friction. Thus, the force of static friction, f_s, can have any value between 0 and $f_{s,max}$:

$$0 \le f_s \le f_{s,max} \qquad \qquad \text{6-2}$$

Imagine repeating the experiment, only now with a second brick on top of the first. This doubles the normal force and it also doubles the maximum force of static friction. Thus, the maximum force is proportional to the magnitude of the normal force; that is,

$$f_{s,max} = \mu_s N \qquad \qquad \text{6-3}$$

The constant of proportionality is called μ_s (pronounced "mew sub s"), the **coefficient of static friction.** Notice that μ_s is dimensionless, just like μ_k. Typical values are given in Table 6-1. In most cases, μ_s is greater than μ_k, indicating that the force of static friction is greater than the force of kinetic friction. In some cases, like rubber in contact with dry concrete, μ_s is greater than 1.

Two additional experimental results regarding static friction are important: (i) Static friction, like kinetic friction, is independent of the area of contact. (ii) The force of static friction is parallel to the surface of contact, and opposite to the direction the object would move if there were no friction. All of these observations are summarized in the following rules of thumb:

Rules of Thumb for Static Friction

The force of static friction between two surfaces has the following properties:

1. It takes on any value between zero and the maximum possible force of static friction, $f_{s,max} = \mu_s N$:

$$0 \le f_s \le \mu_s N$$

2. It is independent of the area of contact between the surfaces.
3. It is parallel to the surface of contact, and in the direction that opposes relative motion.

The next Example presents a practical method of determining the coefficient of static friction in a real-world system.

EXAMPLE 6-3   SLIGHTLY TILTED

A flatbed truck slowly tilts its bed upward to dispose of a 95.0-kg crate. For small angles of tilt the crate stays put, but when the tilt angle exceeds $\theta = 23.3°$, the crate begins to slide. What is the coefficient of static friction between the bed of the truck and the crate?

PICTURE THE PROBLEM
We align our coordinate system with the incline, and choose the positive x direction to point down the slope. Three forces act on the crate: the normal force, $\vec{N} = N\hat{y}$, the force of static friction, $\vec{f}_s = -\mu_s N\hat{x}$, and the weight, $\vec{W} = (mg \sin \theta)\hat{x} + (-mg \cos \theta)\hat{y}$.

Physical picture　　　　　　　　　　**Free-body diagram**　　　　CONTINUED

REASONING AND STRATEGY

When the crate is on the verge of slipping, but has not yet slipped, its acceleration is zero in both the x and y directions. In addition, "on the verge of slipping" means that the magnitude of the static friction is at its maximum value, $f_s = f_{s,max} = \mu_s N$. Thus, we set $\Sigma F_y = ma_y = 0$ to find N, and then use $\Sigma F_x = ma_x = 0$ to find μ_s.

Known   Mass of crate, $m = 95.0$ kg; critical tilt angle, $\theta = 23.2°$.
Unknown   Coefficient of static friction, $\mu_s = ?$

SOLUTION

1. Resolve the three force vectors acting on the crate into x and y components:

$$N_x = 0 \qquad\qquad N_y = N$$
$$f_{s,x} = -f_{s,max} = -\mu_s N \qquad f_{s,y} = 0$$
$$W_x = mg\sin\theta \qquad\qquad W_y = -mg\cos\theta$$

2. Set $\Sigma F_y = ma_y = 0$, since $a_y = 0$.
Solve for the normal force, N:

$$\Sigma F_y = N_y + f_{s,y} + W_y = N + 0 - mg\cos\theta = ma_y = 0$$
$$N = mg\cos\theta$$

3. Set $\Sigma F_x = ma_x = 0$, since the crate is at rest, and use the result for N obtained in Step 2:

$$\Sigma F_x = N_x + f_{s,x} + W_x = ma_x = 0$$
$$= 0 - \mu_s N + mg\sin\theta$$
$$= 0 - \mu_s mg\cos\theta + mg\sin\theta$$

4. Solve the expression for the coefficient of static friction, μ_s:

$$\mu_s mg\cos\theta = mg\sin\theta$$
$$\mu_s = \frac{mg\sin\theta}{mg\cos\theta} = \tan\theta = \tan 23.2° = 0.429$$

INSIGHT

In general, if an object is on the verge of slipping when a surface is tilted at a critical angle θ_c, the coefficient of static friction between the object and the surface is $\mu_s = \tan\theta_c$. This result is independent of the mass of the object. For example, the critical angle for this crate is precisely the same whether it is filled with feathers or lead bricks. Real-world examples of the critical angle are found in hourglasses and talus slopes, as shown in **FIGURE 6-6**.

PRACTICE PROBLEM

Find the magnitude of the force of static friction acting on the crate just before it starts to slide.
[**Answer:** $f_{s,max} = \mu_s N = 367$ N]

Some related homework problems: Problem 9, Problem 11

▶ **FIGURE 6-6 Visualizing Concepts Talus Slope** The angle that the sloping sides of a sand pile (left) make with the horizontal is determined by the coefficient of static friction between the grains of sand, in much the same way that static friction determines the angle at which the crate in Example 6-3 begins to slide. The same basic mechanism determines the angle of the cone-shaped mass of rock debris at the base of a cliff—known as a talus slope (right).

Recall that static friction can have a magnitude that is less than its maximum possible value. This point is emphasized in the next Example.

QUICK EXAMPLE 6-4   THE FORCE OF STATIC FRICTION

In Example 6-3, what is the magnitude of the force of static friction acting on the crate when the truck bed is tilted at an angle of 20.0°?

REASONING AND SOLUTION

The crate is at rest, and hence its acceleration is zero in the x direction: $a_x = 0$. This condition determines the magnitude of the static friction force necessary to hold the crate in place.

1. Sum the x components of force acting on the crate:

$$\Sigma F_x = 0 - f_s + mg\sin\theta$$

CONTINUED

2. Set this sum equal to zero (since $a_x = 0$) $f_s = mg \sin \theta$
and solve for the magnitude of the
static friction force, f_s:

3. Substitute numerical values, $f_s = (95.0 \text{ kg})(9.81 \text{ m/s}^2) \sin (20.0°) = 319 \text{ N}$
including $\theta = 20.0°$:

The force of static friction in this case has a magnitude of 319 N, which is less than the value of 367 N found in the Practice Problem in Example 6-3, even though the coefficient of static friction is precisely the same.

Finally, friction often enters into problems that involve vehicles with rolling wheels. In Conceptual Example 6-5, we consider which type of friction is appropriate in such cases.

CONCEPTUAL EXAMPLE 6-5   FRICTION FOR ROLLING TIRES

A car drives with its tires rolling freely. Is the friction between the tires and the road **(a)** kinetic or **(b)** static?

REASONING AND DISCUSSION

A reasonable-sounding answer is that because the car is moving, the friction between its tires and the road must be kinetic friction—but this is not the case.

Actually, the friction is static because the bottom of the tire is in static contact with the road. To understand this, watch your feet as you walk. Even though you are moving, each foot is in static contact with the ground once you step down on it. Your foot doesn't move again until you lift it up and move it forward for the next step. A tire can be thought of as a succession of feet arranged in a circle, each momentarily in static contact with the ground.

ANSWER

(b) The friction between rolling tires and the road is static friction.

RWP* If a car skids, the friction acting on it is kinetic; if its wheels are rolling, the friction is static. Static friction is generally greater than kinetic friction, however, and hence it follows that a car can be stopped in a shorter distance if its wheels are rolling (static friction) than if its wheels are locked up (kinetic friction). This is the idea behind the antilock braking systems (ABS) that are available on many cars. When the brakes are applied in a car with ABS, an electronic rotation sensor at each wheel detects whether the wheel is about to start skidding. To prevent skidding, a small computer automatically begins to pulse the hydraulic pressure in the brake lines in short bursts, causing the brakes to release and then reapply in rapid succession. This allows the wheels to continue rotating, even in an emergency stop, and for static friction to determine the stopping distance. **FIGURE 6-7** shows a comparison of braking distances for cars with and without ABS. An added benefit of ABS is that a driver is better able to steer and control a braking car if its wheels are rotating. An illustration of the danger posed by locked wheels is presented in **FIGURE 6-8**.

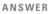

◄ **FIGURE 6-7 Stopping distance with and without ABS** Antilock braking systems (ABS) allow a car to stop with static friction rather than kinetic friction—even in a case where a person slams on the brakes. As a result, the braking distance is reduced. Professional drivers can beat the performance of ABS by carefully adjusting the force they apply to the brake pedal during a stop, but ABS provides essentially the same performance—within a few percent—for a person who simply pushes the brake pedal to the floor and holds it there.

*Real World Physics applications are denoted by the acronym RWP.

(a)

(b)

▲ **FIGURE 6-8 Visualizing Concepts Static Versus Kinetic Friction** The two time-lapse images show a toy car as it slides down an inclined surface. In **(a)** the front wheels are locked and skid on the surface, but the rear wheels roll without slipping. The front wheels experience kinetic friction and the rear wheels experience static friction. Because the force of kinetic friction is usually less than the force of static friction, the front wheels go down the incline first, pulling the rear wheels behind. The situation is reversed in **(b)**, where the rear wheels are the ones that skid and experience a smaller frictional force. As a result, the rear wheels slide down the incline more quickly than the front wheels, causing the car to spin around. This change in behavior could be dangerous in a real-life situation.

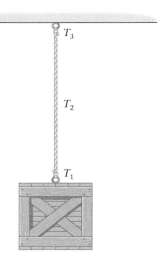

Enhance Your Understanding (Answers given at the end of the chapter)

1. A block rests on a rough, horizontal surface, as shown in **FIGURE 6-9**. A force will be applied to the block along one of the four indicated directions (A, B, C, D). Rank the directions in order of increasing force required to make the block move. Indicate ties where appropriate.

▲ **FIGURE 6-9** Enhance Your Understanding 1

Section Review

- Frictional forces are due to the microscopic roughness of surfaces in contact.
- The force of kinetic friction is proportional to the magnitude of the normal force. That is, $f_k = \mu_k N$, where μ_k is the coefficient of kinetic friction.
- The maximum force of static friction is proportional to the magnitude of the normal force. That is, $f_{s,max} = \mu_s N$, where μ_s is the coefficient of static friction. The force of static friction can always be less than the maximum amount.

▲ **FIGURE 6-10 Tension in a string** A string, pulled from either end, has a tension, T. If the string were to be cut at any point, the force required to hold the ends together is T.

6-2 Strings and Springs

A common way to exert a force on an object is to pull on it with a string, a rope, or a wire. Similarly, you can push or pull on an object if you attach it to a spring. In this section we discuss the basic features of strings and springs and how they transmit forces.

Strings and Tension

Imagine picking up a light string and holding it with one end in each hand. If you pull to the right with your right hand with a force T and to the left with your left hand with a force T, the string becomes taut. In such a case, we say that there is a **tension** T in the string. To be more specific, if your friend were to cut the string at some point, the tension T is the force pulling the ends apart, as illustrated in **FIGURE 6-10**—that is, T is the force your friend would have to exert with each hand to hold the cut ends together. At any given point in the string, the tension pulls equally to the right and to the left.

As an example, consider a rope that is attached to the ceiling at one end, and to a box with a weight of 105 N at the other end, as shown in **FIGURE 6-11**. In addition,

◀ **FIGURE 6-11 Tension in a heavy rope** Because of the weight of the rope, the tension is noticeably different at points 1, 2, and 3. In the limit of a rope of zero mass, the tension is the same throughout the rope.

suppose that the rope is uniform and has a total weight of 2.00 N. What is the tension in the rope (i) where it attaches to the box, (ii) at its midpoint, and (iii) where it attaches to the ceiling?

First, the rope holds the box at rest; thus, the tension where the rope attaches to the box is simply the weight of the box, $T_1 = 105$ N. At the midpoint of the rope, the tension supports the weight of the box, plus half the weight of the rope. Thus, $T_2 = 105$ N $+ \frac{1}{2}(2.00$ N$) = 106$ N. Similarly, at the ceiling the tension supports the box plus all of the rope, giving a tension of $T_3 = 107$ N.

From this discussion, we can see that the tension in the rope changes slightly from top to bottom because of the mass of the rope. If the rope had less mass, the difference in tension between its two ends would also be less. In particular, if the rope's mass were to be vanishingly small, the difference in tension would vanish as well. In this text we will assume that all ropes, strings, wires, and so on are essentially massless—unless specifically stated otherwise—and hence the tension is the same throughout their length.

Pulleys Change the Direction of the Tension Pulleys are often used to redirect a force exerted by a string, as indicated in **FIGURE 6-12**. In the ideal case, a pulley has no mass and no friction in its bearings. Thus, the following statement applies:

> An ideal pulley changes the direction of the tension in a string, without changing its magnitude.

If a system contains more than one pulley, however, it's possible to arrange them in such a way as to "magnify a force," even if each pulley itself merely redirects the tension in a string. The traction device considered in the next Example shows one way this can be accomplished in a system that uses three ideal pulleys.

The tension in an ideal rope has the same magnitude everywhere . . .

. . . but a pulley changes the direction in which it acts.

▲ **FIGURE 6-12 A pulley changes the direction of a tension**

EXAMPLE 6-6 SETTING A BROKEN LEG WITH TRACTION

A traction device employing three pulleys is applied to a broken leg, as shown in the sketch. The middle pulley is attached to the sole of the foot, and a mass m supplies the tension in the ropes. Find the value of the mass m if the force exerted on the sole of the foot by the middle pulley is to be 165 N.

PICTURE THE PROBLEM
Our sketch shows the physical picture as well as the tension forces acting on the middle pulley. Notice that on the upper portion of the rope the tension is $\vec{T}_1 = (T\cos 40.0°)\hat{x} + (T\sin 40.0°)\hat{y}$; on the lower portion it is $\vec{T}_2 = (T\cos 40.0°)\hat{x} + (-T\sin 40.0°)\hat{y}$.

REASONING AND STRATEGY
We begin by noting that the rope supports the hanging mass m. As a result, the tension in the rope, T, must be equal in magnitude to the weight of the mass: $T = mg$.

Next, the pulleys simply change the direction of the tension without changing its magnitude. Therefore, the net force exerted on the sole of the foot is the sum of the tension T at 40.0° above the horizontal plus the tension T at 40.0° below the horizontal. We will calculate the net force component by component.

Once we calculate the net force acting on the foot, we set it equal to 165 N and solve for the tension T. Finally, we find the mass using the relationship $T = mg$.

Known Force exerted on foot, 165 N.
Unknown Suspended mass, $m = ?$

Physical picture

Tension forces acting on middle pulley

SOLUTION

1. First, consider the tension that acts upward and to the right on the middle pulley. Resolve this tension into x and y components:

$$T_{1,x} = T\cos 40.0° \qquad T_{1,y} = T\sin 40.0°$$

CONTINUED

2. Next, consider the tension that acts downward and to the right on the middle pulley. Resolve this tension into x and y components. Notice the minus sign in the y component:

$$T_{2,x} = T\cos 40.0° \qquad T_{2,y} = -T\sin 40.0°$$

3. Sum the x and y components of the force acting on the middle pulley. We see that the net force acts only in the x direction, as one might expect from symmetry:

$$\sum F_x = T\cos 40.0° + T\cos 40.0° = 2T\cos 40.0°$$
$$\sum F_y = T\sin 40.0° - T\sin 40.0° = 0$$

4. Step 3 shows that the net force acting on the middle pulley is $2T\cos 40.0°$. Set this force equal to 165 N and solve for T:

$$2T\cos 40.0° = 165\,\text{N}$$
$$T = \frac{165\,\text{N}}{2\cos 40.0°} = 108\,\text{N}$$

5. Solve for the mass, m, using $T = mg$:

$$T = mg$$
$$m = \frac{T}{g} = \frac{108\,\text{N}}{9.81\,\text{m/s}^2} = 11.0\,\text{kg}$$

INSIGHT

As pointed out earlier, this pulley arrangement "magnifies the force" in the sense that a 108-N weight attached to the rope produces a 165-N force exerted on the foot by the middle pulley. Notice that the tension in the rope always has the same value—$T = 108$ N—as expected with ideal pulleys, but because of the arrangement of the pulleys the force applied to the foot by the rope is $2T\cos 40.0° > T$.

The force exerted on the foot by the middle pulley produces an opposing force in the leg that acts in the direction of the head (a cephalad force), as desired to set a broken leg and keep it straight as it heals.

PRACTICE PROBLEM — PREDICT/CALCULATE

(a) Would the required mass m increase or decrease if the angles in this device were changed from 40.0° to 30.0°?
(b) Find the mass m for an angle of 30.0°.   [**Answer: (a)** m would decrease. **(b)** 9.71 kg]

Some related homework problems: Problem 26, Problem 34

CONCEPTUAL EXAMPLE 6-7   COMPARE THE READINGS ON THE SCALES

The vertical scale shown in the sketch reads 9.81 N. Is the reading of the horizontal scale **(a)** greater than, **(b)** less than, or **(c)** equal to 9.81 N?

REASONING AND DISCUSSION

Because a pulley simply changes the direction of the tension in a string without changing its magnitude, it's clear that the vertical scale attached to the ceiling reads the same as the horizontal scale shown in the sketch to the right.

There is no difference, however, between attaching the left end of the horizontal scale to something rigid or attaching it to another 1.00-kg hanging mass, as in the problem statement. In either case the scale is at rest, and hence a force of 9.81 N must be exerted to the left to balance the 9.81-N force exerted by the mass on the right end of the scale. As a result, the two horizontal scales read the same.

ANSWER

(c) The reading of the horizontal scale is equal to 9.81 N.

Tension in String between
Hanging Weights

Springs and Hooke's Law

Suppose you take a spring of length L, as shown in **FIGURE 6-13 (a)**, and attach it to a block. If you pull on the spring, causing it to stretch to a length $L + x$, the spring *pulls*

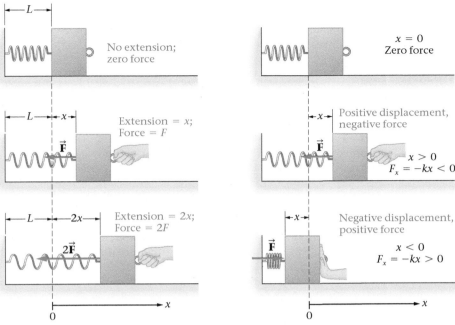

◄ **FIGURE 6-13 The force exerted by a spring** When we are dealing with a spring, it is convenient to choose the origin at the equilibrium (zero force) position.

(a) Doubling the extension doubles the force.

(b) The spring force is opposite to the displacement from equilibrium.

on the block with a force of magnitude F. If you increase the length of the spring to $L + 2x$, the force exerted by the spring increases to $2F$. Similarly, if you compress the spring to a length $L - x$, the spring *pushes* on the block with a force of magnitude F, where F is the same force given previously. As you might expect, compression to a length $L - 2x$ results in a push of magnitude $2F$.

As a result of these experiments, we can say that a spring exerts a force that is proportional to the amount, x, by which it is stretched or compressed. Thus, if F is the magnitude of the spring force, it follows that

$$F = kx$$

In this expression, k is a constant of proportionality, referred to as the **force constant.** Because F has units of newtons and x has units of meters, it follows that k has units of newtons per meter, or N/m. The larger the value of k, the stiffer the spring.

To be more precise, consider the spring shown in **FIGURE 6-13 (b).** Notice that we have placed the origin of the x axis at the equilibrium length of the spring—that is, at the position of the spring when no force acts on it. Now, if we stretch the spring so that the end of the spring is at a positive value of x ($x > 0$), we find that the spring exerts a force of magnitude kx in the negative x direction. Thus, the spring force (which has only an x component) can be written as

$$F_x = -kx$$

Similarly, consider compressing the spring so that its end is at a negative value of x ($x < 0$). In this case, the force exerted by the spring is of magnitude kx, and points in the positive x direction, as is shown in Figure 6-13 (b). Again, we can write the spring force as

$$F_x = -kx$$

To see that this is correct—that is, that F_x is positive in this case—recall that x is negative, which means that $(-x)$ is positive.

This result for the force of a spring is known as Hooke's law, after Robert Hooke (1635–1703). It's really just a good rule of thumb rather than a law of nature. Clearly, it can't work for any amount of stretching. For example, we know that if we stretch a spring far enough, it will be permanently deformed and will never return to its original length. Still, for small stretches or compressions, Hooke's law is quite accurate.

Big Idea 2 Springs exert a force proportional to the amount of stretch or compression.

PHYSICS IN CONTEXT
Looking Ahead

Our discussion of springs, and Hooke's law in particular, is important when we consider oscillations in Chapter 13.

▲ **FIGURE 6-14** Visualizing Concepts
Springs Springs come in a variety of
sizes and shapes. The large springs on
a railroad car (top) are so stiff and heavy
that you can't compress or stretch them
by hand. Still, three of them are needed
to smooth the ride of this car. In contrast,
the delicate spiral spring inside a watch
(bottom) flexes with even the slightest
touch. It exerts enough force, however, to
power the equally delicate mechanism of
the watch.

Rules of Thumb for Springs (Hooke's Law)

A spring stretched or compressed by the amount x from its equilibrium length exerts a
force whose x component is given by

$$F_x = -kx \text{ (gives magnitude and direction)} \qquad \text{6-4}$$

If we are interested only in the magnitude of the force associated with a given stretch or
compression, we use the somewhat simpler form of Hooke's law:

$$F = kx \text{ (gives magnitude only)} \qquad \text{6-5}$$

In this text, we consider only **ideal springs**—that is, springs that are massless,
and that are assumed to obey Hooke's law exactly. A variety of real-world springs is
shown in Visualizing Physics in **FIGURE 6-14**.

Hooke's Law Finds Many Applications   Recalling that the stretch of a spring and the
force it exerts are proportional, we can see how a spring scale operates. In particular,
pulling on the two ends of a scale stretches the spring inside it by an amount proportional to the applied force. Once the scale is calibrated—by stretching the spring with a
known, or reference, force—we can use it to measure other unknown forces.

It's also useful to note that Hooke's law, which we've introduced in the context of
ideal springs, is particularly important in physics because it applies to so much more
than just springs. For example, the forces that hold atoms together are often modeled
by Hooke's law—that is, as "interatomic springs"—and these are the forces that are
ultimately responsible for the normal force (Chapter 5), vibrations and oscillations
(Chapter 13), wave motion (Chapter 14), and even the thermal expansion of solids
(Chapter 16). And this just scratches the surface—Hooke's law comes up in one form or
another in virtually every field of physics, as well as chemistry, geology, and biology.

BIO Elastic Lungs For instance, your lungs obey a version of Hooke's law. The
spring-like property of your lungs arises from surface tension and tissue elasticity,
which assists with exhalation and opposes inhalation. The greater the volume of air
that you inhale (corresponding to the spring stretch distance x), the greater the force
required to do so, and the greater the force your lungs exert when you exhale. Various
diseases can change the effective force constant of lungs, and hamper their function.
For example, *pulmonary fibrosis* causes scarring of tissues and an increase in the effective force constant of the lungs—this makes it more difficult for a person to inhale.
Emphysema destroys elastic tissue and results in lungs with a lower force constant—this
makes it difficult for a person to exhale fully. In the following Quick Example, we present
another biomedical application of Hooke's law associated with breathing.

QUICK EXAMPLE 6-8   NASAL STRIPS

BIO An increasingly popular device for improving air flow through nasal passages is the nasal
strip, which consists of two flat, polyester springs enclosed by an adhesive tape covering. Measurements show that a nasal strip can exert an outward force of 0.22 N on the nose, causing it
to expand by 3.6 mm. **(a)** Treating the nose as an ideal spring, find its force constant in newtons per meter. **(b)** How much force would be required to expand the nose by 4.1 mm?

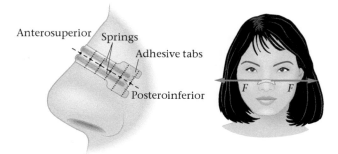

REASONING AND SOLUTION

Using Hooke's law, along with the applied force and resulting stretch, we can solve for the
force constant. Once the force constant is determined, we can use it to solve for the force
corresponding to any given stretch.

CONTINUED

Part (a)

1. Solve the magnitude form of Hooke's law, $F = kx$, for the force constant, k. Then substitute numerical values for F and x:

$$k = \frac{F}{x}$$
$$= \frac{0.22\ \text{N}}{0.0036\ \text{m}} = 61\ \text{N/m}$$

Part (b)

2. Use $F = kx$ to find the required force: $\quad F = kx = (61\ \text{N/m})(0.0041\ \text{m}) = 0.25\ \text{N}$

The human nose is certainly not an ideal spring, but Hooke's law is still a useful way to model its behavior when dealing with forces and the stretches they cause. In addition, notice the linear nature of Hooke's law—a 14% increase in the stretch distance (from 3.6 mm to 4.1 mm) requires a 14% increase in force (from 0.22 N to 0.25 N).

Enhance Your Understanding (Answers given at the end of the chapter)

2. When a mass is attached to a certain spring, the spring stretches by the amount x. Suppose we now connect two of these springs together in series, as shown in **FIGURE 6-15**. If we attach the same mass to this system, is the amount of stretch greater than, less than, or equal to x? Explain.

Section Review

- The force transmitted through a string is referred to as the tension.
- The force exerted by a spring stretched by the amount x has a magnitude $F = kx$. This is referred to as Hooke's law.

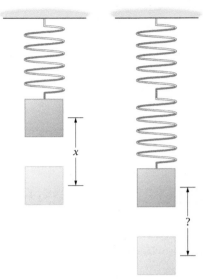

▲ **FIGURE 6-15** Enhance Your Understanding 2

Physical picture

6-3 Translational Equilibrium

When we say that an object is in **translational equilibrium**, we mean that the net force acting on it is zero. We state this mathematically as follows:

$$\sum \vec{\mathbf{F}} = 0 \qquad \text{6-6}$$

From Newton's second law, this is equivalent to saying that the object's acceleration is zero. In two-dimensional systems, translational equilibrium implies two independent conditions, one for each coordinate direction: $\sum F_x = 0$ and $\sum F_y = 0$. In one dimension, only one of these conditions applies.

As a first example, consider the one-dimensional situation illustrated in **FIGURE 6-16**. Here we see a person lifting a bucket of water from a well by pulling down on a rope that passes over a pulley. If the bucket's mass is m, and it is rising with constant speed v, what is the tension T_1 in the rope attached to the bucket? In addition, what is the tension T_2 in the chain that supports the pulley?

To answer these questions, we first note that both the bucket and the pulley are in equilibrium; that is, they both have zero acceleration. As a result, the net force on each of them must be zero.

Let's start with the bucket. In Figure 6-16, we see that just two forces act on the bucket, both along the y direction: (i) its weight, $W = mg$ downward, and (ii) the tension in the rope, T_1 upward. If we take upward to be the positive direction, we can write $\sum F_y = 0$ for the bucket as follows:

$$T_1 - mg = 0$$

Therefore, the tension in the rope is $T_1 = mg$. Notice that this is also the force the person must exert downward on the rope, as expected.

Next, we consider the pulley. In Figure 6-16, we see that three forces act on it: (i) the tension in the chain, T_2 upward, (ii) the tension in the part of the rope

Forces acting on the pulley | **Forces acting on the bucket**

▶ **FIGURE 6-16 Raising a bucket** A person lifts a bucket of water from the bottom of a well with a constant speed, v. Because the speed is constant, the net force acting on the bucket is zero.

Big Idea **3** A system is in translational equilibrium when the total force acting on it is zero.

leading to the bucket, T_1 downward, and (iii) the tension in the part of the rope leading to the person, T_1 downward. We don't include the weight of the pulley, since we consider it to be ideal—that is, massless and frictionless. If we again take upward to be positive, the statement that the net force acting on the pulley is zero ($\Sigma F_y = 0$) can be written as

$$T_2 - T_1 - T_1 = 0$$

It follows that the tension in the chain is $T_2 = 2T_1 = 2mg$, twice the weight of the bucket of water!

In the next Conceptual Example we consider a slight variation of this situation.

CONCEPTUAL EXAMPLE 6-9 **COMPARING TENSIONS**

A person hoists a bucket of water from a well and holds the rope, keeping the bucket at rest, as in (1). Next, the person ties the rope to the bucket handle so the rope holds the bucket in place, as in (2). In this case, is the tension in the rope **(a)** greater than, **(b)** less than, or **(c)** equal to the tension in the first case?

(1)　　　　　　　(2)　　　　　　　(3)　　　　　　　(4)

REASONING AND DISCUSSION

In the first case (1), the only upward force exerted on the bucket is the tension in the rope. The bucket is at rest, and hence the tension must be equal in magnitude to the weight of the bucket (3). In the second case (2), the two ends of the rope exert equal upward forces on the bucket; therefore, the tension in the rope is only half the weight of the bucket. To see this more clearly, imagine cutting the bucket in half so that each end of the rope supports half the weight, as indicated in (4).

ANSWER

(b) The tension in the second case is less than in the first.

In the next two Examples, we consider two-dimensional systems in which forces act at various angles with respect to one another. Hence, our first step is to resolve the relevant vectors into their x and y components. Following that, we apply the conditions for translational equilibrium: $\Sigma F_x = 0$ and $\Sigma F_y = 0$.

EXAMPLE 6-10 **SUSPENDED VEGETATION**

To hang a 6.20-kg pot of flowers, a gardener uses two wires—one attached horizontally to a wall, the other sloping upward at an angle of $\theta = 40.0°$ and attached to the ceiling. Find the tension in each wire.

PICTURE THE PROBLEM

We choose a typical coordinate system, with the positive x direction to the right and the positive y direction upward. With this choice, tension 1 is in the positive x direction, $\vec{\mathbf{T}}_1 = T_1\hat{\mathbf{x}}$; the weight is in the negative y direction, $\vec{\mathbf{W}} = -mg\,\hat{\mathbf{y}}$; and tension 2 has a negative x component and a positive y component, $\vec{\mathbf{T}}_2 = (-T_2\cos\theta)\hat{\mathbf{x}} + (T_2\sin\theta)\hat{\mathbf{y}}$.

CONTINUED

REASONING AND STRATEGY

The pot is at rest, and therefore the net force acting on it is zero. As a result, we can say that (i) $\sum F_x = 0$ and (ii) $\sum F_y = 0$. These two conditions allow us to determine the magnitude of the two tensions, T_1 and T_2.

Known   Mass of flower pot, $m = 6.20$ kg; angle of upper wire, $\theta = 40.0°$.

Unknown   Tension in each wire: $T_1 = ?$, $T_2 = ?$

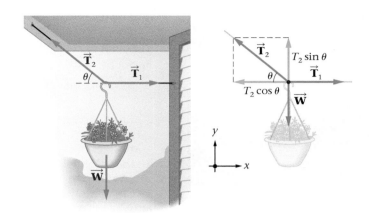

Physical picture   **Free-body diagram**

SOLUTION

1. First, resolve each of the forces acting on the pot into x and y components:

$$T_{1,x} = T_1 \qquad\qquad T_{1,y} = 0$$
$$T_{2,x} = -T_2 \cos\theta \qquad T_{2,y} = T_2 \sin\theta$$
$$W_x = 0 \qquad\qquad W_y = -mg$$

2. Now, set $\sum F_x = 0$. Notice that this condition gives a relationship between T_1 and T_2:

$$\sum F_x = T_{1,x} + T_{2,x} + W_x = T_1 + (-T_2\cos\theta) + 0 = 0$$
$$T_1 = T_2\cos\theta$$

3. Next, set $\sum F_y = 0$. This time, the resulting condition determines T_2 in terms of the weight, mg:

$$\sum F_y = T_{1,y} + T_{2,y} + W_y = 0 + T_2\sin\theta + (-mg) = 0$$
$$T_2\sin\theta = mg$$

4. Use the relationship obtained in Step 3 to find T_2:

$$T_2 = \frac{mg}{\sin\theta} = \frac{(6.20\text{ kg})(9.81\text{ m/s}^2)}{\sin 40.0°} = 94.6\text{ N}$$

5. Finally, use the connection between the two tensions (obtained from $\sum F_x = 0$ in Step 2) to find T_1:

$$T_1 = T_2\cos\theta = (94.6\text{ N})\cos 40.0° = 72.5\text{ N}$$

INSIGHT

Notice that even though two wires suspend the pot, they both have tensions *greater* than the pot's weight, $mg = 60.8$ N. This is an important point for architects and engineers to consider when designing structures.

PRACTICE PROBLEM

Find T_1 and T_2 if the second wire slopes upward at the angle **(a)** $\theta = 20°$, **(b)** $\theta = 60.0°$, or **(c)** $\theta = 90.0°$.
[**Answer: (a)** $T_1 = 167$ N, $T_2 = 178$ N; **(b)** $T_1 = 35.1$ N, $T_2 = 70.2$ N; **(c)** $T_1 = 0$, $T_2 = mg = 60.8$ N]

Some related homework problems: Problem 30, Problem 32 , Problem 72

QUICK EXAMPLE 6-11   THE FORCES IN A LOW-TECH LAUNDRY

A 1.84-kg bag of clothespins hangs in the middle of a clothesline, causing it to sag at an angle $\theta = 3.50°$. Find the tension, T, in the clothesline.

REASONING AND SOLUTION

The tension in the clothesline must result in a net vertical component of force equal in magnitude to the weight of the clothespins. The horizontal components of the tension cancel because there is no horizontal acceleration.

CONTINUED

▲ **FIGURE 6-17** Like the bag of clothespins in Quick Example 6-11, this mountain climber is in static equilibrium. Since the ropes suspending the climber are nearly horizontal, the tension in them is significantly greater than the climber's weight.

1. Find the y component for the tension on each side of the bag:

$$T_y = T \sin \theta$$

2. Find the y component of the bag's weight:

$$W_y = -mg$$

3. Set $\Sigma F_y = 0$:

$$T \sin \theta + T \sin \theta - mg = 0$$

4. Solve for T and substitute numerical values:

$$T = \frac{mg}{2 \sin \theta} = \frac{(1.84 \text{ kg})(9.81 \text{ m/s}^2)}{2 \sin 3.50°}$$
$$= 148 \text{ N}$$

At 148 N the tension in the clothesline is quite large, especially when compared with the weight of the clothespin bag itself, which is only 18.1 N. The reason for such a large value is that the vertical component of the two tensions is $2T \sin \theta$, which, for $\theta = 3.50°$, is only $(0.122)T$. If $(0.122)T$ is to equal the weight of the bag, it is clear that T must be roughly eight times the bag's weight. On the other hand, if $\theta = 90°$ the tension is just $mg/2$, as expected. A similar situation involving large tensions and a rock climber is shown in **FIGURE 6-17**.

If you and a friend were to pull on the two ends of the clothesline, in an attempt to straighten it out, you would find that no matter how hard you pulled, the line would still sag. You may be able to reduce θ to quite a small value, but as you do so the corresponding tension increases rapidly. In principle, it would take an infinite force to completely straighten the line and reduce θ to zero.

Enhance Your Understanding   (Answers given at the end of the chapter)

3. Suppose the tension in the clothesline in Quick Example 6-11 is 120 N. Is the sag angle greater than, less than, or equal to 3.50°? Explain.

Section Review

• An object in translational equilibrium has zero net force acting on it. This means it won't have a translational (left/right, up/down) acceleration.

6-4 Connected Objects

Interesting applications of Newton's laws arise when we consider accelerating objects that are tied together. Suppose, for example, that a force of magnitude F pulls two boxes—connected by a string—along a frictionless surface, as in **FIGURE 6-18**. In such a case, the string has a certain tension, T, and the two boxes have the same acceleration, a. Given the masses of the boxes and the applied force F, we would like to determine both the tension in the string and the acceleration of the boxes.

First, we sketch the free-body diagram for each box. Box 1 has two horizontal forces acting on it: (i) the tension T to the left, and (ii) the force F to the right. Box 2 has only a single horizontal force, the tension T to the right. If we take the positive direction to be to the right, Newton's second law for the two boxes can be written as follows:

$$F - T = m_1 a_1 = m_1 a \quad \text{box 1}$$
$$T = m_2 a_2 = m_2 a \quad \text{box 2}$$

6-7

Because the boxes have the same acceleration, a, we have set $a_1 = a_2 = a$.

▼ **FIGURE 6-18 Two boxes connected by a string** The string ensures that the two boxes have the same acceleration. This physical connection results in a mathematical connection, as shown in Equation 6-7. In this case we treat each box as a separate system.

Physical picture

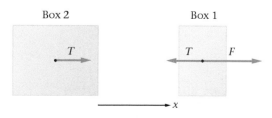

Free-body diagrams
(horizontal components only)

Physical picture

Free-body diagram
(horizontal components only)

▲ **FIGURE 6-19 Two boxes, one system** In this case we consider the two boxes together as a single system of mass $m_1 + m_2$. The only external horizontal force acting on this system is $\vec{\mathbf{F}}$; hence the horizontal acceleration of the system is $a = F/(m_1 + m_2)$, in agreement with Equation 6-8.

Next, we can eliminate the tension T by adding the two equations:

$$F - T = m_1 a$$
$$\underline{T = m_2 a}$$
$$F = (m_1 + m_2)a$$

With this result, it is straightforward to solve for the acceleration in terms of the applied force F:

$$a = \frac{F}{m_1 + m_2} \qquad \text{6-8}$$

Finally, we substitute this expression for a into either of the original equations to find the tension. The algebra is simpler if we use the equation for box 2. We find

$$T = m_2 a = \left(\frac{m_2}{m_1 + m_2}\right)F \qquad \text{6-9}$$

It is left as an exercise to show that the equation for box 1 gives the same expression for T.

A second way to approach this problem is to treat both boxes together as a single system with a mass $m_1 + m_2$, as shown in **FIGURE 6-19**. The only *external* horizontal force acting on this system is the applied force F—the two tension forces are now *internal* to the system, and internal forces are not included when applying Newton's second law. As a result, the horizontal acceleration is simply $F/(m_1 + m_2)$, as given in Equation 6-8. This is certainly a quick way to find the acceleration a, but to find the tension T we must still use one of the relationships given in Equations 6-7.

In general, we are always free to choose the "system" any way we like. We can choose any individual object, as when we considered box 1 and box 2 separately, or we can choose all the objects together. The important point is that Newton's second law is equally valid no matter what choice we make for the system, as long as we remember to include only forces *external* to *that system* in the corresponding free-body diagram.

RWP A dramatic example of tension in a rope is depicted in the 2013 movie *Gravity*, in which two astronauts orbit the Earth while tethered by a long rope (**FIGURE 6-20**). Whenever astronaut 1 fires a jetpack to go forward, the tether became taut. The tension T in the tether pulls backward on astronaut 1 and forward on astronaut 2. The fact that the astronauts are joined by a tether demands that the acceleration of astronaut 1, $a_1 = (F_{\text{jetpack}} - T)/m_1$, be exactly the same as the acceleration of astronaut 2, $a_2 = T/m_2$. When the jetpack is turned off, the tension in the tether becomes zero and both astronauts drift at constant velocity. The movie depicts the laws of physics realistically amid the drama of running out of oxygen while attempting to travel a great distance!

▲ **FIGURE 6-20 Tethered astronauts** A dramatic scene from the movie *Gravity* illustrates the physics of objects connected by a rope.

CONCEPTUAL EXAMPLE 6-12 TENSION IN THE STRING

PREDICT/EXPLAIN

Two masses, m_1 and m_2, are connected by a string that passes over a pulley. Mass m_1 slides without friction on a horizontal tabletop, and mass m_2 falls vertically downward. Both masses move with a constant acceleration of magnitude a. **(a)** Is the tension in the string greater than, less than, or equal to the weight of block 2, $m_2 g$? **(b)** Which of the following is the *best explanation* for your prediction?

CONTINUED

I. The mass m_2 has to pull the mass m_1 and this adds to the tension, making it greater than m_2g.

II. The mass m_2 accelerates downward, and hence the tension is less than m_2g so that the net force on m_2 is downward.

III. The mass m_2 is suspended by the rope, and hence the tension is equal to m_2g.

REASONING AND DISCUSSION
First, note that m_2 accelerates downward, which means that the net force acting on it is downward. Only two forces act on m_2, however: the tension in the string (upward) and its weight (downward). Because the net force is downward, the tension in the string must be less than the weight, m_2g. This is explanation II.

A common misconception is that since m_2 has to pull m_1 behind it, the tension in the string must be greater than m_2g. Certainly, attaching the string to m_1 has an effect on the tension. If the string were not attached, its tension would be zero. Hence, m_2 pulling on m_1 increases the tension to a value greater than zero, but still less than m_2g.

ANSWER
(a) The tension in the string is less than m_2g. (b) The best explanation is II.

Choose a Coordinate System That Follows the Motion In the next Example, we verify the qualitative conclusions given in Conceptual Example 6-12 with a detailed calculation. But first, a note about choosing a coordinate system for a problem such as this. Rather than apply the same coordinate system to both masses, it is useful to take into consideration the fact that a pulley simply changes the direction of the tension in a string. With this in mind, we choose a set of axes that "follow the motion" of the string, so that both masses accelerate in the positive x direction with accelerations of equal magnitude. Example 6-13 illustrates the use of this type of coordinate system.

PROBLEM-SOLVING NOTE

Choice of Coordinate System: Connected Objects

If two objects are connected by a string passing over a pulley, let the coordinate system follow the direction of the string. With this choice, both objects have accelerations of the same magnitude and in the same coordinate direction.

EXAMPLE 6-13 CONNECTED BLOCKS

A block of mass m_1 slides on a frictionless tabletop. It is connected to a string that passes over a pulley and suspends a mass m_2. Find (a) the acceleration of the masses and (b) the tension in the string.

PICTURE THE PROBLEM
Our coordinate system follows the motion of the string so that both masses move in the positive x direction. Since the masses are connected, their accelerations have the same magnitude. Thus, $a_{1,x} = a_{2,x} = a$. In addition, notice that the tension, $\vec{T}$, is in the positive x direction for mass 1, but in the negative x direction for mass 2. Its magnitude, T, is the same for each mass, however. Finally, the weight of mass 2, W_2, acts in the positive x direction, whereas the weight of mass 1 is offset by the normal force, N.

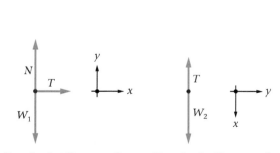

Physical picture   **Free-body diagram for** m_1 **Free-body diagram for** m_2

REASONING AND STRATEGY
Applying Newton's second law to the two masses yields the following relationships: for mass 1, $\Sigma F_{1,x} = T = m_1a_{1,x} = m_1a$, and for mass 2, $\Sigma F_{2,x} = m_2g - T = m_2a_{2,x} = m_2a$. These two equations can be solved for the two unknowns, a and T.

CONTINUED

Known Symbolic masses of the blocks, m_1, m_2.
Unknown (a) Acceleration of the blocks, $a = ?$ (b) Tension in the string, $T = ?$

SOLUTION

Part (a)

1. First, write $\sum F_{1,x} = m_1 a$. Note that the only force acting on m_1 in the x direction is T:

$$\sum F_{1,x} = T = m_1 a$$
$$T = m_1 a$$

2. Next, write $\sum F_{2,x} = m_2 a$. In this case, two forces act in the x direction: $W_2 = m_2 g$ (positive direction) and T (negative direction):

$$\sum F_{2,x} = m_2 g - T = m_2 a$$
$$m_2 g - T = m_2 a$$

3. Sum the two relationships obtained to eliminate T:

$$T = m_1 a$$
$$\underline{m_2 g - T = m_2 a}$$
$$m_2 g = (m_1 + m_2)a$$

4. Solve for a:

$$a = \left(\frac{m_2}{m_1 + m_2}\right)g$$

Part (b)

5. Substitute a into the first equation ($T = m_1 a$) to find T:

$$T = m_1 a = \left(\frac{m_1 m_2}{m_1 + m_2}\right)g$$

INSIGHT
We could just as well have determined T using $m_2 g - T = m_2 a$, though the algebra is a bit messier. Also, notice that $a = 0$ if $m_2 = 0$, and that $a = g$ if $m_1 = 0$, as expected. Similarly, $T = 0$ if either m_1 or m_2 is zero. This type of check, where you connect equations with physical situations, is one of the best ways to increase your understanding of physics.

PRACTICE PROBLEM
Find the acceleration and tension for the case $m_1 = 1.50$ kg and $m_2 = 0.750$ kg, and compare the tension to $m_2 g$. [**Answer:** $a = 3.27$ m/s^2, $T = 4.91$ N $< m_2 g = 7.36$ N]

Some related homework problems: Problem 38, Problem 41

Conceptual Example 6-12 shows that the tension in the string is less than $m_2 g$. Let's rewrite our solution for T from Example 6-13 to show that this is indeed the case. We have

$$T = \left(\frac{m_1 m_2}{m_1 + m_2}\right)g = \left(\frac{m_1}{m_1 + m_2}\right)m_2 g$$

Noting that the ratio $m_1/(m_1 + m_2)$ is always less than 1 (as long as m_2 is nonzero), it follows that $T < m_2 g$, as expected.

A Method for Measuring the Acceleration due to Gravity We conclude this section with a classic system that can be used to measure the acceleration due to gravity. It is referred to as Atwood's machine, and it is basically two blocks of different mass connected by a string that passes over a pulley. The resulting acceleration of the blocks is related to the acceleration due to gravity by a relatively simple expression, which we derive in the following Example.

EXAMPLE 6-14 ATWOOD'S MACHINE

Atwood's machine consists of two masses connected by a string that passes over a pulley, as shown in the sketch. Find the acceleration of the masses for general m_1 and m_2, and evaluate for the specific case $m_1 = 3.1$ kg, $m_2 = 4.4$ kg.

PICTURE THE PROBLEM
Our sketch shows Atwood's machine, along with our choice of coordinate directions for the two blocks. Note that both blocks accelerate in the positive x direction with accelerations of equal magnitude, a. From the free-body diagrams we can see that for mass 1 the weight is in the negative x direction and the tension is in the positive x direction. For mass 2, the tension is in the negative x direction and the weight is in the positive x direction. The tension has the same magnitude T for both masses, but their weights are different.

CONTINUED

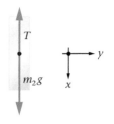

Physical picture

Free-body diagram for m_1

Free-body diagram for m_2

REASONING AND STRATEGY

To find the acceleration of the blocks, we follow the same strategy given in the previous Example. In particular, we start by applying Newton's second law to each block individually, using the fact that $a_{1,x} = a_{2,x} = a$. This gives two equations, both involving the tension T and the acceleration a. Eliminating T allows us to solve for the acceleration.

Known   Masses, $m_1 = 3.1$ kg, $m_2 = 4.4$ kg.
Unknown   Acceleration of the masses, a = ?

SOLUTION

1. Begin by writing out the expression $\Sigma F_{1,x} = m_1 a$. Note that two forces act in the x direction: T (positive direction) and $m_1 g$ (negative direction):

$$\Sigma F_{1,x} = T - m_1 g = m_1 a$$

2. Next, write out $\Sigma F_{2,x} = m_2 a$. The two forces acting in the x direction in this case are $m_2 g$ (positive direction) and T (negative direction):

$$\Sigma F_{2,x} = m_2 g - T = m_2 a$$

3. Sum the two relationships obtained above to eliminate T:

$$
\begin{aligned}
T - m_1 g &= m_1 a \\
m_2 g - T &= m_2 a \\
\hline
(m_2 - m_1)g &= (m_1 + m_2)a
\end{aligned}
$$

4. Solve for a:

$$a = \left(\frac{m_2 - m_1}{m_1 + m_2}\right)g$$

5. To evaluate the acceleration, substitute numerical values for the masses and for g:

$$
\begin{aligned}
a &= \left(\frac{m_2 - m_1}{m_1 + m_2}\right)g \\
&= \left(\frac{4.4\text{ kg} - 3.1\text{ kg}}{3.1\text{ kg} + 4.4\text{ kg}}\right)(9.81\text{ m/s}^2) = 1.7\text{ m/s}^2
\end{aligned}
$$

INSIGHT

Because m_2 is greater than m_1, we find that the acceleration is positive, meaning that the masses accelerate in the positive x direction. On the other hand, if m_1 were greater than m_2, we would find that a is negative, indicating that the masses accelerate in the negative x direction. Finally, if $m_1 = m_2$, we have $a = 0$, as expected.

PRACTICE PROBLEM — PREDICT/CALCULATE

(a) If m_1 is increased by a small amount, does the acceleration of the blocks increase, decrease, or stay the same? Explain. (b) Check your answer to part (a) by evaluating the acceleration for $m_1 = 3.3$ kg.   [**Answer: (a)** The acceleration decreases because the masses are more nearly balanced. **(b)** $a = 1.4$ m/s^2]

Some related homework problems: Problem 40, Problem 41, Problem 43

Enhance Your Understanding

(Answers given at the end of the chapter)

4. Three boxes are connected by ropes and pulled across a smooth, horizontal surface with an acceleration a, as shown in FIGURE 6-21. The force applied to the first box on the right is F, the tension in the rope connecting the first

▲ FIGURE 6-21 Enhance Your Understanding 4

CONTINUED

and second boxes is T_1, and the tension in the rope connecting the second and third boxes is T_2. Rank the three forces (F, T_1, T_2) in order of increasing magnitude. Indicate ties where appropriate.

Section Review

- Objects connected by strings have the same acceleration. This implies certain values for the tension in the strings.

6-5 Circular Motion

According to Newton's second law, if no force acts on an object, it will move with constant speed in a constant direction. A force is required to change the speed, the direction, or both. For example, if you drive a car with constant speed on a circular track, the direction of the car's motion changes continuously. A force must act on the car to cause this change in direction. We would like to know two things about a force that causes circular motion: (i) What is its direction, and (ii) What is its magnitude?

First, let's consider the direction of the force. Imagine swinging a ball tied to a string in a circle above your head, as shown in FIGURE 6-22. As you swing the ball, you feel a tension in the string pulling outward. Of course, on the other end of the string, where it attaches to the ball, the tension pulls inward, toward the center of the circle. Thus, the force the ball experiences is a force that is always directed toward the center of the circle. In summary:

> To make an object move in a circle with constant speed, a force must act on it that is directed toward the center of the circle.

Because the ball is acted on by a *force* toward the center of the circle, it follows that it must be *accelerating* toward the center of the circle. This might seem odd at first: How can a ball that moves with constant speed have an acceleration? The answer is that acceleration is produced whenever the speed or direction of the velocity changes—and in circular motion, the direction changes continuously. The resulting center-directed acceleration is called **centripetal acceleration** (centripetal is from the Latin for "center seeking").

Calculating the Centripetal Acceleration Let's calculate the magnitude of the centripetal acceleration, a_{cp}, for an object moving with a constant speed v in a circle of radius r. FIGURE 6-23 shows the circular path of an object, with the center of the circle at the origin. To calculate the acceleration at the top of the circle, at point P, we first calculate the average acceleration from point 1 to point 2:

$$\vec{\mathbf{a}}_{av} = \frac{\Delta \vec{\mathbf{v}}}{\Delta t} = \frac{\vec{\mathbf{v}}_2 - \vec{\mathbf{v}}_1}{\Delta t} \qquad 6\text{-}10$$

The instantaneous acceleration at P is the limit of $\vec{\mathbf{a}}_{av}$ as points 1 and 2 move closer to P.

Referring to Figure 6-23, we see that $\vec{\mathbf{v}}_1$ is at an angle θ above the horizontal, and $\vec{\mathbf{v}}_2$ is at an angle θ below the horizontal. Both $\vec{\mathbf{v}}_1$ and $\vec{\mathbf{v}}_2$ have a magnitude v. Therefore, we can write the two velocities in vector form as follows:

$$\vec{\mathbf{v}}_1 = (v\cos\theta)\hat{\mathbf{x}} + (v\sin\theta)\hat{\mathbf{y}}$$
$$\vec{\mathbf{v}}_2 = (v\cos\theta)\hat{\mathbf{x}} + (-v\sin\theta)\hat{\mathbf{y}}$$

Substituting these results into $\vec{\mathbf{a}}_{av}$ gives

$$\vec{\mathbf{a}}_{av} = \frac{\vec{\mathbf{v}}_2 - \vec{\mathbf{v}}_1}{\Delta t} = \frac{-2v\sin\theta}{\Delta t}\hat{\mathbf{y}} \qquad 6\text{-}11$$

Notice that $\vec{\mathbf{a}}_{av}$ points in the negative y direction—which, at point P, is toward the center of the circle.

To complete the calculation, we need to know the time, Δt, it takes the object to go from point 1 to point 2. Since the object's speed is v, and the distance from point

▲ FIGURE 6-22 **Swinging a ball in a circle** The tension in the string pulls outward on the person's hand and pulls inward on the ball.

Ball Leaves Circular Track

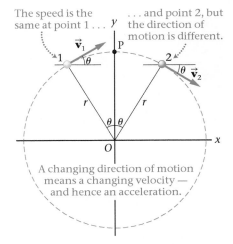

▲ FIGURE 6-23 **A particle moving with constant speed in a circular path centered on the origin** The speed of the particle is constant, but its velocity is constantly changing direction. Because the velocity changes, the particle is accelerating.

Big Idea 4 When an object moves in a circular path, it accelerates toward the center of the path. As a result, circular motion requires a force directed toward the center.

TABLE 6-2 $\dfrac{\sin \theta}{\theta}$ for Values of θ Approaching Zero

θ, radians	$\dfrac{\sin \theta}{\theta}$
1.00	0.841
0.500	0.959
0.250	0.990
0.125	0.997
0.0625	0.999

PHYSICS IN CONTEXT
Looking Back

The derivation of the direction and magnitude of centripetal acceleration makes extensive use of our knowledge of vectors from Chapter 3, and especially how to resolve vectors into components.

PHYSICS IN CONTEXT
Looking Ahead

Circular motion comes up again in a number of physical systems, especially when we consider gravitational orbital motion in Chapter 12 and the Bohr model of the hydrogen atom in Chapter 31.

PROBLEM-SOLVING NOTE
Choice of Coordinate System: Circular Motion

In circular motion, it is convenient to choose the coordinate system so that one axis points toward the center of the circle. Then, we know that the acceleration in that direction must be $a_{cp} = v^2/r$.

1 to point 2 is $d = r(2\theta)$, where θ is measured in radians (see Appendix, page A-2 for a discussion of radians and degrees), we find

$$\Delta t = \frac{d}{v} = \frac{2r\theta}{v} \qquad 6\text{-}12$$

Combining this result for Δt with the previous result for $\vec{\mathbf{a}}_{av}$ gives

$$\vec{\mathbf{a}}_{av} = \frac{-2v\sin\theta}{(2r\theta/v)}\hat{\mathbf{y}} = -\frac{v^2}{r}\left(\frac{\sin\theta}{\theta}\right)\hat{\mathbf{y}} \qquad 6\text{-}13$$

To find $\vec{\mathbf{a}}$ at point P, we let points 1 and 2 approach P, which means letting θ go to zero. Table 6-2 shows that as θ goes to zero ($\theta \to 0$), the ratio $(\sin\theta)/\theta$ goes to 1:

$$\lim_{\theta \to 0}\frac{\sin\theta}{\theta} = 1$$

It follows, then, that the instantaneous acceleration at point P is

$$\vec{\mathbf{a}} = -\frac{v^2}{r}\hat{\mathbf{y}} = -a_{cp}\hat{\mathbf{y}} \qquad 6\text{-}14$$

As mentioned, the direction of the acceleration is toward the center of the circle, and now we see that its magnitude is

$$a_{cp} = \frac{v^2}{r} \qquad 6\text{-}15$$

We can summarize these results as follows:

- When an object moves in a circular path of radius r with constant speed v, its centripetal acceleration has a magnitude given by $a_{cp} = v^2/r$.

- A centripetal force must be applied to an object to give it circular motion. For an object of mass m, the net force acting on it must have a magnitude given by

$$f_{cp} = ma_{cp} = m\frac{v^2}{r} \qquad 6\text{-}16$$

This force must be directed toward the center of the circle.

The **centripetal force**, f_{cp}, can be produced in any number of ways. For example, f_{cp} might be the tension in a string, as in the example with the ball, or it might be due to friction between tires and the road, as when a car turns a corner. In addition, f_{cp} could be the force of gravity that causes a satellite, or the Moon, to orbit the Earth. Thus, f_{cp} is a force that must be present to cause circular motion, but the specific cause of f_{cp} varies from system to system. The centripetal force in a carnival ride is illustrated in FIGURE 6-24.

We now show how these results for centripetal force and centripetal acceleration can be applied in practice.

◀ **FIGURE 6-24** The people enjoying this carnival ride are experiencing a centripetal acceleration of roughly 10 m/s² directed inward, toward the axis of rotation. The force needed to produce this acceleration, which keeps the riders moving in a circular path, is provided by the horizontal component of the tension in the chains.

EXAMPLE 6-15 ROUNDING A CORNER

A 1200-kg car rounds a corner of radius $r = 45$ m. If the coefficient of static friction between the tires and the road is $\mu_s = 0.82$, what is the greatest speed the car can have in the corner without skidding?

PICTURE THE PROBLEM

In the first sketch we show a bird's-eye view of the car as it moves along its circular path. The next sketch shows the car moving directly toward the observer. Notice that we have chosen the positive x direction to point toward the center of the circular path, and the positive y axis to point vertically upward. We also indicate the three forces acting on the car: gravity, $\vec{\mathbf{W}} = -W\hat{\mathbf{y}} = -mg\hat{\mathbf{y}}$; the normal force, $\vec{\mathbf{N}} = N\hat{\mathbf{y}}$; and the force of static friction, $\vec{\mathbf{f}}_s = \mu_s N\hat{\mathbf{x}}$.

CONTINUED

REASONING AND STRATEGY

In this system, the force of static friction provides the centripetal force required for the car to move in a circular path. That is why the force of friction is at right angles to the car's direction of motion; it is directed toward the center of the circle. In addition, the friction in this case is static because the car's tires are rolling without slipping—always making static contact with the ground. Finally, if the car moves faster, more centripetal force (i.e., more friction) is required. Thus, the greatest speed for the car corresponds to the maximum static friction, $f_s = \mu_s N$. Hence, by setting $\mu_s N$ equal to the centripetal force, $ma_{cp} = mv^2/r$, we can solve for v.

Known Mass of car, $m = 1200$ kg; corner radius, $r = 45$ m; coefficient of static friction, $\mu_s = 0.82$.
Unknown Maximum speed without skidding, $v = ?$

SOLUTION

1. Sum the x components of force to relate the force of static friction to the centripetal acceleration of the car:

$$\sum F_x = f_s = ma_x$$

2. The car moves in a circular path, with the center of the circle in the x direction, and hence it follows that $a_x = a_{cp} = v^2/r$. Make this substitution, along with $f_s = \mu_s N$ for the force of static friction:

$$\mu_s N = ma_{cp} = m\frac{v^2}{r}$$

3. Next, set the sum of the y components of force equal to zero, since $a_y = 0$:

$$\sum F_y = N - W = ma_y = 0$$

4. Solve for the normal force:

$$N = W = mg$$

5. Substitute the result $N = mg$ in Step 2 and solve for v. Notice that the mass of the car cancels:

$$\mu_s mg = m\frac{v^2}{r}$$
$$v = \sqrt{\mu_s rg}$$

6. Substitute numerical values to determine v:

$$v = \sqrt{(0.82)(45 \text{ m})(9.81 \text{ m/s}^2)} = 19 \text{ m/s}$$

INSIGHT

The maximum speed is less if the radius is smaller (tighter corner) or if μ_s is smaller (slick road). The mass of the vehicle, however, is irrelevant. For example, the maximum speed is precisely the same for a motorcycle rounding this corner as it is for a large, heavily loaded truck.

PRACTICE PROBLEM — PREDICT/CALCULATE

Suppose the situation described in this Example takes place on the Moon, where the acceleration due to gravity is less than it is on Earth. **(a)** If a lunar rover goes around this same corner, is its maximum speed greater than, less than, or the same as the result found in Step 4? Explain. **(b)** Find the maximum speed for a lunar rover when it rounds a corner with $r = 45$ m and $\mu_s = 0.82$. (On the Moon, $g = 1.62$ m/s².) [**Answer: (a)** The maximum speed is less, because gravity holds the rover to the road with a smaller force. **(b)** On the Moon we find $v = 7.7$ m/s, considerably less than the maximum speed of 19 m/s on Earth.]

Some related homework problems: Problem 47, Problem 74

RWP If you try to round a corner too rapidly, you may experience a skid; that is, your car may begin to slide sideways across the road. A common bit of road wisdom is that you should turn in the direction of the skid to regain control—which, to most people, seems counterintuitive. The advice is sound, however. Suppose you are turning to the left and begin to skid to the right. If you turn more sharply to the left to try to correct for the skid, you simply reduce the turning radius of your car, r. The result is that the centripetal acceleration, v^2/r, becomes larger, and an even larger force is required from the road to make the turn. The tendency to skid is therefore increased. On the other hand, if you turn slightly to the right when you start to skid, you *increase* your turning radius and the centripetal acceleration *decreases*. In this case your car may stop skidding, and you can regain control of your vehicle.

You may also have noticed that many roads are tilted, or banked, when they round a corner. The same type of banking is observed on many automobile racetracks as well. Next time you drive around a banked curve, notice that the banking tilts you in toward the center of the circular path you are following. This is by design. On a banked curve, the normal force exerted by the road contributes to the required centripetal force. If the tilt angle is just right, the normal force provides all of the centripetal force, so that the car can negotiate the curve even if there is no friction between its tires and the road. The next Example determines the optimum banking angle for a given speed and radius of turn.

EXAMPLE 6-16   BANK ON IT

If a roadway is banked at the proper angle, a car can round a corner without any assistance from friction between the tires and the road. Find the appropriate banking angle for a 900-kg car traveling at 20.5 m/s in a turn of radius 85.0 m.

PICTURE THE PROBLEM

We choose the positive y axis to point vertically upward and the positive x direction to point toward the center of the circular path. Since $\vec{\mathbf{N}}$ is perpendicular to the banked roadway, it is at an angle θ to the y axis. Therefore, $\vec{\mathbf{N}} = (N\sin\theta)\hat{\mathbf{x}} + (N\cos\theta)\hat{\mathbf{y}}$ and $\vec{\mathbf{W}} = -W\hat{\mathbf{y}} = -mg\hat{\mathbf{y}}$.

REASONING AND STRATEGY

In order for the car to move in a circular path, there must be a force acting on it in the positive x direction. Because the weight $\vec{\mathbf{W}}$ has no x component, it follows that the normal force $\vec{\mathbf{N}}$ must supply the needed centripetal force. Thus, we find N by setting $\Sigma F_y = ma_y = 0$, since there is no motion in the y direction. Then we use N in $\Sigma F_x = ma_x = mv^2/r$ to find the angle θ.

Known   Mass of car, $m = 900$ kg; speed, $v = 20.5$ m/s; turn radius, $r = 85.0$ m.
Unknown   Optimum banking angle, $\theta = ?$

SOLUTION

1. Start by determining N from the condition $\Sigma F_y = 0$:

$$\Sigma F_y = N\cos\theta - W = 0$$
$$N = \frac{W}{\cos\theta} = \frac{mg}{\cos\theta}$$

2. Next, set $\Sigma F_x = mv^2/r$:

$$\Sigma F_x = N\sin\theta$$
$$= ma_x = ma_{cp} = m\frac{v^2}{r}$$

3. Substitute $N = mg/\cos\theta$ (from $\Sigma F_y = 0$, Step 1) and solve for θ, using the fact that $\sin\theta/\cos\theta = \tan\theta$. Notice that, once again, the mass of the car cancels:

$$N\sin\theta = \frac{mg}{\cos\theta}\sin\theta = m\frac{v^2}{r}$$
$$\tan\theta = \frac{v^2}{gr} \quad \text{or} \quad \theta = \tan^{-1}\left(\frac{v^2}{gr}\right)$$

4. Substitute numerical values to determine θ:

$$\theta = \tan^{-1}\left[\frac{(20.5 \text{ m/s})^2}{(9.81 \text{ m/s}^2)(85.0 \text{ m})}\right] = 26.7°$$

INSIGHT

The symbolic result in Step 3 shows that the banking angle increases with increasing speed and decreasing radius of turn, as one would expect.

From the point of view of a passenger, the experience of rounding a properly banked corner is basically the same as riding on a level road—there are no "sideways forces" to make the turn uncomfortable. There is one small difference, however: The passenger feels heavier due to the increased normal force. The Visualizing Concepts feature in **FIGURE 6-25** shows a variety of situations where banking in a turn is useful.

PRACTICE PROBLEM

A turn of radius 65 m is banked at 30.0°. What speed should a car have in order to make the turn with no assistance from friction?   [**Answer:** $v = 19$ m/s]

Some related homework problems: Problem 49, Problem 78

If you've ever driven through a dip in the road, you know that you feel momentarily heavier near the bottom of the dip. This change in apparent weight is due to the approximately circular motion of the car, as we show next.

QUICK EXAMPLE 6-17 NORMAL FORCE IN A DIP

Predict/Calculate

While driving along a country lane with a constant speed of 17.0 m/s you encounter a dip in the road. The dip can be approximated as a circular arc, with a radius of 65.1 m. **(a)** At the bottom of the dip, is the magnitude of the normal force exerted by a car seat on a passenger in the car greater than, less than, or equal to the passenger's weight? Explain. **(b)** What is the normal force exerted by a car seat on an 80.0-kg passenger when the car is at the bottom of the dip?

REASONING AND SOLUTION

Part (a) At the bottom of the dip, the normal force is greater than the weight of the passenger. This is because the normal force must not only counter the weight of the passenger, but also provide the centripetal force toward the center of the circular path needed for circular motion.

Part (b) At the bottom of the dip the passenger is moving along a circular path and accelerating vertically upward—with an acceleration of v^2/r toward the center of the circle. Set the sum of the vertical forces equal to the mass times the vertical acceleration of the passenger.

1. Write $\Sigma F_y = ma_y$ for $N - mg = ma_y$
   the passenger:

2. Replace a_y with the $a_y = v^2/r$
   centripetal acceleration:

3. Solve for N: $N = mg + mv^2/r$

4. Substitute numerical $N = (80.0 \text{ kg})(9.81 \text{ m/s}^2) + (80.0 \text{ kg})(17.0 \text{ m/s})^2/(65.1 \text{ m})$
   values: $= 785 \text{ N} + 355 \text{ N}$
   $= 1140 \text{ N}$

The same physics applies to a jet pilot who pulls a plane out of a high-speed dive. In that case, the magnitude of the effect can be much larger, resulting in a decrease of blood flow to the brain and eventually to loss of consciousness. Here's a case where basic physics really can be a matter of life and death!

A similar calculation can be applied to a car going over the top of a bump. In that case, circular motion results in a reduced apparent weight. With just the right speed you can even experience momentary "weightlessness."

RWP Next, we determine the acceleration produced in a **centrifuge,** a common device in biological and medical laboratories that uses large centripetal accelerations to perform such tasks as separating red and white blood cells from serum. A simplified top view of a centrifuge is shown in **FIGURE 6-26**, and a laboratory centrifuge is shown in **FIGURE 6-27**.

▲ **FIGURE 6-25 Visualizing Concepts**
Banked Turns The steeply banked track at the Talladega Superspeedway in Alabama (top) helps to keep the rapidly moving cars from skidding off along a tangential path. Even when there is no solid roadway, however, banking can still help. Airplanes bank when making turns (center) to keep from "skidding" sideways. Banking is beneficial in another way as well. Occupants of cars on a banked roadway or of a banking airplane feel no sideways force when the banking angle is just right, so turns become a safer and more comfortable experience. For this reason, some trains use hydraulic suspension systems to bank when rounding corners (bottom), even though the tracks themselves are level.

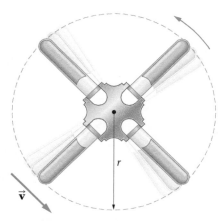

▲ FIGURE 6-26 **Simplified top view of a centrifuge in operation**

▲ FIGURE 6-27 A laboratory centrifuge of the kind commonly used to separate blood components.

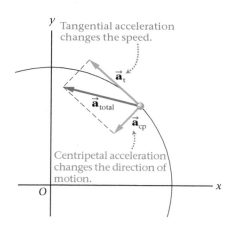

◀ FIGURE 6-28 **A particle moving in a circular path with tangential acceleration** In this case, the particle's speed is increasing at the rate given by a_t.

EXERCISE 6-18 CENTRIFUGE ACCELERATION

The centrifuge in Figure 6-26 rotates at a rate that gives the bottom of the test tube a linear speed of 90.6 m/s. If the bottom of the test tube is 9.25 cm from the axis of rotation, what is the centripetal acceleration experienced there?

REASONING AND SOLUTION
Applying the relationship $a_{cp} = v^2/r$ yields

$$a_{cp} = \frac{v^2}{r} = \frac{(90.6 \text{ m/s})^2}{0.0925 \text{ m}} = 88{,}700 \text{ m/s}^2 = 9040\,g$$

In this expression, g is the acceleration due to gravity, 9.81 m/s^2.

Thus, a centrifuge can produce centripetal accelerations that are many thousand times greater than the acceleration due to gravity. In fact, devices referred to as **ultra-centrifuges** can produce accelerations as great as 1 million g. Even in the relatively modest case considered in Exercise 6-18, the forces involved in a centrifuge can be quite significant. For example, if the contents of the test tube have a relatively modest mass of 12.0 g, the centripetal force that must be exerted by the bottom of the tube is $(0.0120 \text{ kg})(9040\,g) = 1064$ N, or about 239 lb!

Circular Motion with Variable Speed Supppose the speed of an object moving in a circular path increases or decreases. In such a case, the object has both an acceleration tangential to its path that changes its speed, $\vec{a}_t$, and a centripetal acceleration perpendicular to its path that changes its direction of motion, $\vec{a}_{cp}$. Such a situation is illustrated in FIGURE 6-28. The total acceleration of the object is the vector sum of $\vec{a}_t$ and $\vec{a}_{cp}$. We will explore this case more fully in Chapter 10.

Enhance Your Understanding   (Answers given at the end of the chapter)

5. A system consists of an object with mass m and speed v that moves on a circular path with a radius of curvature r. Rank the following systems (A, B, C, D) in order of increasing centripetal force required to cause the motion. Indicate ties where appropriate. A: $m = 1$ kg, $v = 5$ m/s, $r = 5$ m; B: $m = 6$ kg, $v = 1$ m/s, $r = 2$ m; C: $m = 6$ kg, $v = 2$ m/s, $r = 2$ m; D: $m = 10$ kg, $v = 2$ m/s, $r = 5$ m.

Section Review

- An object moving along a circular path of radius r with a speed v experiences a centripetal (cp) acceleration toward the center of the circle of magnitude $a_{cp} = v^2/r$.
- The centripetal (cp) force that must be exerted on an object of mass m to make it move in a circular path of radius r with a speed v is $f_{cp} = ma_{cp} = mv^2/r$.

CHAPTER 6 REVIEW

CHAPTER SUMMARY

6-1 FRICTIONAL FORCES

Frictional rforces are due to the microscopic roughness of surfaces in contact. As a rule of thumb, friction is independent of the area of contact and the relative speed of the surfaces.

Kinetic Friction

Kinetic friction is experienced by surfaces that are in contact and moving relative to one another. The force of kinetic friction is given by

$$f_k = \mu_k N \qquad\qquad 6\text{-}1$$

In this expression, μ_k is the coefficient of kinetic friction and N is the magnitude of the normal force.

Static Friction

Static friction is experienced by surfaces that are in static contact. The maximum force of static friction is given by

$$f_{s,max} = \mu_s N \qquad\qquad 6\text{-}3$$

In this expression, μ_s is the coefficient of static friction and N is the magnitude of the normal force. The force of static friction can have any magnitude between zero and its maximum value.

6-2 STRINGS AND SPRINGS

Strings and springs provide a common way of exerting forces on objects. Ideal strings and springs are massless.

Tension

Tension is the force transmitted through a string. The tension is the same throughout the length of an ideal string.

Hooke's Law

Hooke's law states that the force exerted by an ideal spring stretched by the amount x is proportional to the stretch. Specifically,

$$F_x = -kx \qquad\qquad 6\text{-}4$$

6-3 TRANSLATIONAL EQUILIBRIUM

An object is in translational equilibrium if the net force acting on it is zero.

6-4 CONNECTED OBJECTS

Connected objects are linked physically and mathematically. For example, objects connected by strings have the same magnitude of acceleration.

6-5 CIRCULAR MOTION

An object moving with speed v in a circle of radius r has a centripetal acceleration of magnitude $a_{cp} = v^2/r$ directed toward the center of the circle. If the object has a mass m, the force required for the circular motion is

$$f_{cp} = ma_{cp} = mv^2/r \qquad\qquad 6\text{-}16$$

ANSWERS TO ENHANCE YOUR UNDERSTANDING QUESTIONS

1. A < B = D < C.
2. Greater than x. Each of the springs stretches by the amount x, and hence the total stretch is $2x$.
3. The sag angle is greater, because with a smaller tension (120 N versus 148 N) a larger vertical component is

required to support the weight. This in turn requires a larger angle.
4. $T_2 < T_1 < F$.
5. B < A < D < C.

CONCEPTUAL QUESTIONS

For instructor-assigned homework, go to www.masteringphysics.com.

(Answers to odd-numbered Conceptual Questions can be found in the back of the book.)

1. A clothesline always sags a little, even if nothing hangs from it. Explain.
2. In the *Jurassic Park* sequel, *The Lost World*, a man tries to keep a large vehicle from going over a cliff by connecting a cable from his Jeep to the vehicle. The man then puts the Jeep in gear and spins the rear wheels. Do you expect that spinning the tires will increase the force exerted by the Jeep on the vehicle? Why or why not?
3. When a traffic accident is investigated, it is common for the length of the skid marks to be measured. How could this information be used to estimate the initial speed of the vehicle that left the skid marks?
4. In a car with rear-wheel drive, the maximum acceleration is often less than the maximum deceleration. Why?
5. A train typically requires a much greater distance to come to rest, for a given initial speed, than does a car. Why?
6. Give some everyday examples of situations in which friction is beneficial.
7. At the local farm, you buy a flat of strawberries and place them on the backseat of the car. On the way home, you begin to brake as you approach a stop sign. At first the strawberries stay put, but as you brake a bit harder, they begin to slide off the seat. Explain.
8. It is possible to spin a bucket of water in a vertical circle and have none of the water spill when the bucket is upside down. How would you explain this to members of your family?
9. Water sprays off a rapidly turning bicycle wheel. Why?
10. Can an object be in translational equilibrium if it is moving? Explain.
11. In a dramatic circus act, a motorcyclist drives his bike around the inside of a vertical circle. How is this possible, considering that the motorcycle is upside down at the top of the circle?
12. The gravitational attraction of the Earth is only slightly less at the altitude of an orbiting spacecraft than it is on the Earth's surface. Why is it, then, that astronauts feel weightless?

13. A popular carnival ride has passengers stand with their backs against the inside wall of a cylinder. As the cylinder begins to spin, the passengers feel as if they are being pushed against the wall. Explain.
14. Referring to Question 13, after the cylinder reaches operating speed, the floor is lowered away, leaving the passengers "stuck" to the wall. Explain.
15. Your car is stuck on an icy side street. Some students on their way to class see your predicament and help out by sitting on the trunk of your car to increase its traction. Why does this help?
16. The parking brake on a car causes the rear wheels to lock up. What would be the likely consequence of applying the parking brake in a car that is in rapid motion? (*Note:* Do *not* try this at home.)
17. **BIO** The foot of your average gecko is covered with billions of tiny hair tips—called spatulae—that are made of keratin, the protein found in human hair. A subtle shift of the electron distribution in both the spatulae and the wall to which a gecko clings produces an adhesive force by means of the van der Waals interaction between molecules. Suppose a gecko uses its spatulae to cling to a vertical windowpane. If you were to describe this situation in terms of a coefficient of static friction, μ_s, what value would you assign to μ_s? Is this a sensible way to model the gecko's feat? Explain.
18. Discuss the physics involved in the spin cycle of a washing machine. In particular, how is circular motion related to the removal of water from the clothes?
19. The gas pedal and the brake pedal are capable of causing a car to accelerate. Can the steering wheel also produce an acceleration? Explain.
20. In the movie *2001: A Space Odyssey*, a rotating space station provides "artificial gravity" for its inhabitants. How does this work?
21. When rounding a corner on a bicycle or a motorcycle, the driver leans inward, toward the center of the circle. Why?

PROBLEMS AND CONCEPTUAL EXERCISES

Answers to odd-numbered Problems and Conceptual Exercises can be found in the back of the book. **BIO** *identifies problems of biological or medical interest;* **CE** *indicates a conceptual exercise.* Predict/Explain *problems ask for two responses: (a) your prediction of a physical outcome, and (b) the best explanation among three provided; and* Predict/Calculate *problems ask for a prediction of a physical outcome, based on fundamental physics concepts, and follow that with a numerical calculation to verify the prediction. On all problems, bullets (•, ••, •••) indicate the level of difficulty.*

SECTION 6-1 FRICTIONAL FORCES

1. • **CE** Predict/Explain You push two identical bricks across a tabletop with constant speed, v, as shown in **FIGURE 6-29**. In case 1, you place the bricks end to end; in case 2, you stack the bricks one on top of the other. **(a)** Is the force of kinetic friction in case 1 greater than, less than, or equal to the force of kinetic friction in case 2? **(b)** Choose the *best explanation* from among the following:
   I. The normal force in case 2 is larger, and hence the bricks press down more firmly against the tabletop.
   II. The normal force is the same in the two cases, and friction is independent of surface area.

   III. Case 1 has more surface area in contact with the tabletop, and this leads to more friction.

FIGURE 6-29 Problem 1

2. • **CE** Predict/Explain Two drivers traveling side-by-side at the same speed suddenly see a deer in the road ahead of them and begin braking. Driver 1 stops by locking up his brakes and screeching to a halt; driver 2 stops by applying her brakes just to the verge of locking, so that the wheels continue to turn until her car comes to a complete stop. **(a)** All other factors being equal, is the stopping distance of driver 1 greater than, less than, or equal to the stopping distance of driver 2? **(b)** Choose the *best explanation* from among the following:

I. Locking up the brakes gives the greatest possible braking force.

II. The same tires on the same road result in the same force of friction.

III. Locked-up brakes lead to sliding (kinetic) friction, which is less than rolling (static) friction.

3. • A 1.8-kg block slides on a horizontal surface with a speed of $v = 0.80$ m/s and an acceleration of magnitude $a = 2.8$ m/s², as shown in FIGURE 6-30. **(a)** What is the coefficient of kinetic friction between the block and the surface? **(b)** When the speed of the block slows to 0.40 m/s, is the magnitude of the acceleration greater than, less than, or equal to 2.8 m/s²? Explain.

FIGURE 6-30 Problem 3

4. • A child goes down a playground slide with an acceleration of 1.26 m/s². Find the coefficient of kinetic friction between the child and the slide if the slide is inclined at an angle of 33.0° below the horizontal.

5. • What is the minimum horizontal force F needed to make the box start moving in FIGURE 6-31? The coefficients of kinetic and static friction between the box and the floor are 0.27 and 0.38, respectively.

FIGURE 6-31 Problem 5

6. •• What is the minimum horizontal force F needed to make the box start moving in FIGURE 6-32? The coefficients of kinetic and static friction between the box and the floor are 0.24 and 0.41, respectively.

FIGURE 6-32 Problem 6

7. •• **CE** The three identical boxes shown in FIGURE 6-33 remain at rest on a rough, horizontal surface, even though they are acted on by two different forces, $\vec{F}_1$ and $\vec{F}_2$. All of the forces labeled $\vec{F}_1$ have the same magnitude; all of the forces labeled $\vec{F}_2$ are identical to one another. Rank the boxes in order of increasing magnitude of the force of static friction between them and the surface. Indicate ties where appropriate.

FIGURE 6-33 Problem 7

8. •• To move a large crate across a rough floor, you push on it with a force F at an angle of 21° below the horizontal, as shown in FIGURE 6-34. Find the force necessary to start the crate moving, given that the mass of the crate is 32 kg and the coefficient of static friction between the crate and the floor is 0.57.

FIGURE 6-34 Problem 8

9. •• Predict/Calculate A 37-kg crate is placed on an inclined ramp. When the angle the ramp makes with the horizontal is increased to 23°, the crate begins to slide downward. **(a)** What is the coefficient of static friction between the crate and the ramp? **(b)** At what angle does the crate begin to slide if its mass is doubled?

10. •• **Coffee To Go** A person places a cup of coffee on the roof of his car while he dashes back into the house for a forgotten item. When he returns to the car, he hops in and takes off with the coffee cup still on the roof. **(a)** If the coefficient of static friction between the coffee cup and the roof of the car is 0.24, what is the maximum acceleration the car can have without causing the cup to slide? Ignore the effects of air resistance. **(b)** What is the smallest amount of time in which the person can accelerate the car from rest to 15 m/s and still keep the coffee cup on the roof?

11. •• A mug rests on an inclined surface, as shown in FIGURE 6-35. **(a)** What is the magnitude of the frictional force exerted on the mug? **(b)** What is the minimum coefficient of static friction required to keep the mug from sliding?

FIGURE 6-35 Problem 11

12. •• Predict/Calculate **Force Times Distance** At the local hockey rink, a puck with a mass of 0.12 kg is given an initial speed of $v = 5.3$ m/s. **(a)** If the coefficient of kinetic friction between the ice and the puck is 0.11, what distance d does the puck slide before coming to rest? **(b)** If the mass of the puck is doubled, does the frictional force F exerted on the puck increase, decrease, or stay the same? Explain. **(c)** Does the stopping distance of the puck increase, decrease, or stay the same when its mass is doubled? Explain. **(d)** For the situation considered in part (a), show that $Fd = \frac{1}{2}mv^2$. (In Chapter 7 we will see $\frac{1}{2}mv^2$ is the kinetic energy of an object.)

13. ••• Predict/Calculate The coefficient of kinetic friction between the tires of your car and the roadway is μ. **(a)** If your initial speed is v and you lock your tires during braking, how far do you skid? Give your answer in terms of v, μ, and m, the mass of your car. **(b)** If you double your speed, what happens to the stopping distance? **(c)** What is the stopping distance for a truck with twice the mass of your car, assuming the same initial speed and coefficient of kinetic friction?

SECTION 6-2 STRINGS AND SPRINGS

14. • **CE** A certain spring has a force constant k. **(a)** If this spring is cut in half, does the resulting half spring have a force constant that is greater than, less than, or equal to k? **(b)** If two of the original full-length springs are connected end to end, does the resulting double spring have a force constant that is greater than, less than, or equal to k?

15. • **CE** A certain spring has a force constant k. **(a)** If two such identical springs are connected in parallel to each other, as shown in part (a) of **FIGURE 6-36**, does the resulting double spring have an effective force constant that is greater than, less than, or equal to k? Explain. **(b)** If two of the original springs are connected as shown in part (b) of Figure 6-36, does the resulting double spring have an effective force constant that is greater than, less than, or equal to k? Explain.

(a)

(b)

FIGURE 6-36 Problem 15

16. • Pulling up on a rope, you lift a 7.27-kg bucket of water from a well with an acceleration of 1.24 m/s². What is the tension in the rope?

17. • When a 9.09-kg mass is placed on top of a vertical spring, the spring compresses 4.18 cm. Find the force constant of the spring.

18. • **Predict/Calculate** A backpack full of books weighing 52.0 N rests on a table in a physics laboratory classroom. A spring with a force constant of 150 N/m is attached to the backpack and pulled horizontally, as indicated in **FIGURE 6-37**. **(a)** If the spring is pulled until it stretches 2.00 cm and the pack remains at rest, what is the force of friction exerted on the backpack by the table? **(b)** Does your answer to part (a) change if the mass of the backpack is doubled? Explain.

FIGURE 6-37 Problem 18

19. •• Two springs, with force constants $k_1 = 150$ N/m and $k_2 = 250$ N/m, are connected in series, as shown in **FIGURE 6-38**. When a mass $m = 0.90$ kg is attached to the springs, what is the amount of stretch, x?

FIGURE 6-38 Problem 19

20. •• **Predict/Calculate** Illinois Jones is being pulled from a snake pit with a rope that breaks if the tension in it exceeds 755 N. **(a)** If Illinois Jones has a mass of 70.0 kg and the snake pit is 3.40 m deep, what is the minimum time that is required to pull our intrepid explorer from the pit? **(b)** Explain why the rope breaks if Jones is pulled from the pit in less time than that calculated in part (a).

21. •• **Predict/Calculate** A spring with a force constant of 120 N/m is used to push a 0.27-kg block of wood against a wall, as shown in **FIGURE 6-39**. **(a)** Find the minimum compression of the spring needed to keep the block from falling, given that the coefficient of static friction between the block and the wall is 0.46. **(b)** Does your answer to part (a) change if the mass of the block of wood is doubled? Explain.

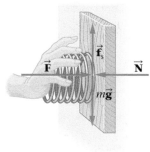

FIGURE 6-39 Problem 21

22. •• A spring is suspended vertically from the ceiling of a room. When a mass is attached to the bottom of the spring, the amount of stretch is recorded. A plot of stretch versus mass is given in **FIGURE 6-40**. Use the data given in the plot to determine the force constant of the spring.

FIGURE 6-40 Problem 22

23. •• **Mechanical Advantage** The pulley system shown in **FIGURE 6-41** is used to lift a $m = 67$-kg crate. Note that one chain connects the upper pulley to the ceiling and a second chain connects the lower pulley to the crate. Assuming the masses of the chains, pulleys, and ropes are negligible, determine **(a)** the force $\vec{F}$ required to lift the crate with constant speed, **(b)** the tension in the upper chain, and **(c)** the tension in the lower chain.

FIGURE 6-41 Problem 23

SECTION 6-3 TRANSLATIONAL EQUILIBRIUM

24. • Pulling the string on a bow back with a force of 28.7 lb, an archer prepares to shoot an arrow. If the archer pulls in the center of the string, and the angle between the two halves is 138°, what is the tension in the string?

25. • In **FIGURE 6-42** we see two blocks connected by a string and tied to a wall. The mass of the lower block is 1.0 kg; the mass of the upper block is 2.0 kg. Given that the angle of the incline is 31°, find the tensions in **(a)** the string connecting the two blocks and **(b)** the string that is tied to the wall.

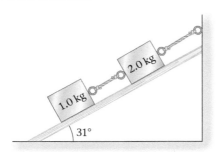

FIGURE 6-42 Problem 25

26. • **BIO** **Traction** After a skiing accident, your leg is in a cast and supported in a traction device, as shown in **FIGURE 6-43**. Find the magnitude of the force $\vec{F}$ exerted by the leg on the small pulley. (By Newton's third law, the small pulley exerts an equal and opposite force on the leg.) Let the mass m be 2.27 kg.

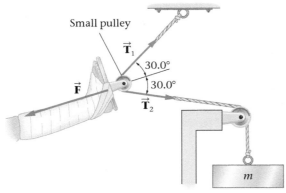

FIGURE 6-43 Problem 26

27. • Two blocks are connected by a string, as shown in **FIGURE 6-44**. The smooth inclined surface makes an angle of 42° with the horizontal, and the block on the incline has a mass of 6.7 kg. Find the mass of the hanging block that will cause the system to be in equilibrium. (The pulley is assumed to be ideal.)

FIGURE 6-44 Problem 27

28. • Predict/Calculate The system shown in **FIGURE 6-45** consists of two pulleys and two blocks. The pulleys each have a mass of 2.0 kg, the block on the left has a mass of 16 kg, and the block on the right has a mass m. The system is in equilibrium, and the string is massless. **(a)** Is the mass m greater than, less than, or equal to 16 kg? Explain. **(b)** Determine the mass m.

FIGURE 6-45 Problem 28

29. • **CE** Predict/Explain **(a)** Referring to the hanging planter in Example 6-10, which of the three graphs (A, B, or C) in **FIGURE 6-46** shows an accurate plot of the tensions T_1 and T_2 as a function of the angle θ? **(b)** Choose the *best explanation* from among the following:

   I. The two tensions must be equal at some angle between $\theta = 0$ and $\theta = 90°$.

   II. T_2 is greater than T_1 at all angles, and is equal to mg at $\theta = 90°$.

   III. T_2 is less than T_1 at all angles, and is equal to 0 at $\theta = 90°$.

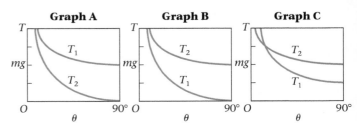

FIGURE 6-46 Problem 29

30. •• **BIO** **Spiderweb Forces** An orb-weaver spider sits in her web that is supported by three radial strands, as shown in **FIGURE 6-47**. Assume that only the radial strands contribute to supporting the weight of the spider. If the mass of the spider is 5.2×10^{-4} kg, and the radial strands are all under the same tension, find the magnitude of the tension, T.

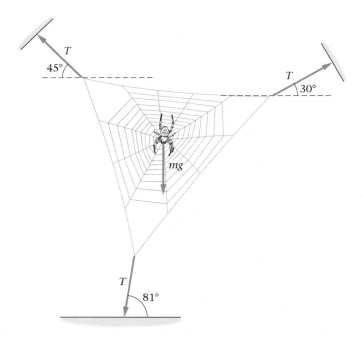

FIGURE 6-47 Problem 30

31. •• A 0.15-kg ball is placed in a shallow wedge with an opening angle of 120°, as shown in **FIGURE 6-48**. For each contact point between the wedge and the ball, determine the force exerted on the ball. Assume the system is frictionless.

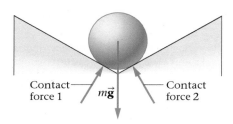

FIGURE 6-48 Problem 31

32. •• Predict/Calculate A picture hangs on the wall suspended by two strings, as shown in **FIGURE 6-49**. The tension in string 1 is 1.7 N. **(a)** Is the tension in string 2 greater than, less than, or equal to 1.7 N? Explain. **(b)** Verify your answer to part (a) by calculating the tension in string 2. **(c)** What is the weight of the picture?

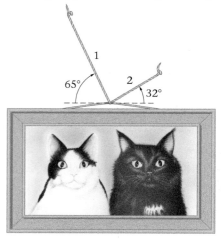

FIGURE 6-49 Problem 32

33. •• Predict/Calculate You want to nail a 1.6-kg board onto the wall of a barn. To position the board before nailing, you push it against the wall with a horizontal force $\vec{F}$ to keep it from sliding to the ground (**FIGURE 6-50**). **(a)** If the coefficient of static friction between the board and the wall is 0.79, what is the least force you can apply and still hold the board in place? **(b)** What happens to the force of static friction if you push against the wall with a force greater than that found in part (a)?

FIGURE 6-50 Problem 33

34. ••• **BIO The Russell Traction System** To immobilize a fractured femur (the thigh bone), doctors often utilize the Russell traction system illustrated in **FIGURE 6-51**. Notice that one force is applied directly to the knee, $\vec{F}_1$, while two other forces, $\vec{F}_2$ and $\vec{F}_3$, are applied to the foot. The latter two forces combine to give a force $\vec{F}_2 + \vec{F}_3$ that is transmitted through the lower leg to the knee. The result is that the knee experiences the total force $\vec{F}_{total} = \vec{F}_1 + \vec{F}_2 + \vec{F}_3$. The goal of this traction system is to have $\vec{F}_{total}$ directly in line with the fractured femur, at an angle of 20.0° above the horizontal. Find **(a)** the angle θ required to produce this alignment of $\vec{F}_{total}$ and **(b)** the magnitude of the force, $\vec{F}_{total}$, that is applied to the femur in this case. (Assume the pulleys are ideal.)

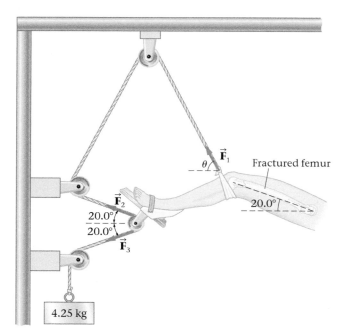

FIGURE 6-51 Problem 34

SECTION 6-4 CONNECTED OBJECTS

35. • **CE** In Example 6-13 (Connected Blocks), suppose m_1 and m_2 are both increased by a factor of 2. **(a)** Does the acceleration of the blocks increase, decrease, or stay the same? **(b)** Does the tension in the string increase, decrease, or stay the same?

36. • **CE** Predict/Explain Suppose m_1 and m_2 in Example 6-14 (Atwood's Machine) are both increased by 1 kg. Does the acceleration of the blocks increase, decrease, or stay the same? **(b)** Choose the *best explanation* from among the following:
I. The net force acting on the blocks is the same, but the total mass that must be accelerated is greater.
II. The difference in the masses is the same, and this is what determines the net force on the system.
III. The force exerted on each block is greater, leading to an increased acceleration.

37. • **CE** Three boxes of masses m, $2m$, and $3m$ are connected as shown in **FIGURE 6-52**. A force is applied to accelerate the boxes toward the right. **(a)** Rank the tensions in the ropes, T_1, T_2, and T_3, from smallest to largest, when the boxes are in motion and there is no friction between the boxes and the horizontal surface. **(b)** Rank the tensions in the ropes, from smallest to largest, when the boxes are in motion and there is a coefficient of kinetic friction μ_k between the boxes and the horizontal surface.

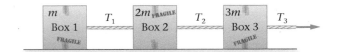

FIGURE 6-52 Problem 37

38. •• Find the acceleration of the masses shown in **FIGURE 6-53**, given that $m_1 = 1.0$ kg, $m_2 = 2.0$ kg, and $m_3 = 3.0$ kg. Assume the table is frictionless and the masses move freely.

FIGURE 6-53 Problem 38

39. •• Predict/Calculate **(a)** If the hanging mass m_3 in Figure 6-53 is decreased, will the acceleration of the system increase, decrease, or stay the same? Explain. **(b)** Find the acceleration of the system given that $m_1 = 1.0$ kg, $m_2 = 2.0$ kg, and $m_3 = 0.50$ kg.

40. •• Two blocks are connected by a string, as shown in **FIGURE 6-54**. The smooth inclined surface makes an angle of 35° with the horizontal, and the block on the incline has a mass of 5.7 kg. The mass of the hanging block is $m = 3.2$ kg. Find **(a)** the direction and **(b)** the magnitude of the hanging block's acceleration.

FIGURE 6-54 Problem 40

41. •• Predict/Calculate A 3.50-kg block on a smooth tabletop is attached by a string to a hanging block of mass 2.80 kg, as shown in **FIGURE 6-55**. The blocks are released from rest and allowed to move freely. **(a)** Is the tension in the string greater than, less than, or equal to the weight of the hanging mass? Find **(b)** the acceleration of the blocks and **(c)** the tension in the string.

FIGURE 6-55 Problem 41

42. •• Predict/Calculate A 7.7-N force pulls horizontally on a 1.6-kg block that slides on a smooth horizontal surface. This block is connected by a horizontal string to a second block of mass $m_2 = 0.83$ kg on the same surface. **(a)** What is the acceleration of the blocks? **(b)** What is the tension in the string? **(c)** If the mass of block 1 is increased, does the tension in the string increase, decrease, or stay the same?

43. •• Predict/Calculate **(a)** Find the magnitude of the acceleration of the two blocks in **FIGURE 6-56 (a)**. Assume the top block slides without friction on the horizontal surface. **(b)** If the 12-N block in Figure 6-56 (a) is replaced with a 12-N force pulling downward, as in **FIGURE 6-56 (b)**, is the acceleration of the sliding block greater than, less than, or equal to the acceleration calculated in part (a)? Explain. **(c)** Calculate the acceleration of the block in Figure 6-56 (b), assuming it slides without friction.

FIGURE 6-56 Problem 43

SECTION 6-5 CIRCULAR MOTION

44. • **CE** A car drives with constant speed on an elliptical track, as shown in **FIGURE 6-57**. Rank the points A, B, and C in order of increasing likelihood that the car might skid. Indicate ties where appropriate.

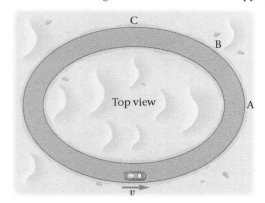

FIGURE 6-57 Problem 44

45. • **CE** A puck attached to a string undergoes circular motion on an air table. If the string breaks at the point indicated in **FIGURE 6-58**, is the subsequent motion of the puck best described by path A, B, C, or D?

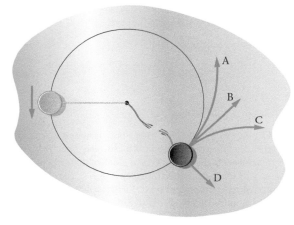

FIGURE 6-58 Problem 45

46. • **BIO Bubble Net Fishing** Humpback whales sometimes catch fish by swimming rapidly in a circle, blowing a curtain of bubbles that confuses a school of fish and traps it in a small area, where the whales can easily catch and eat them. Suppose a 28,000-kg humpback whale swims at 2.2 m/s in a circle of radius 8.5 m. What centripetal force must the whale generate?

47. • When you take your 1900-kg car out for a spin, you go around a corner of radius 53 m with a speed of 13 m/s. The coefficient of static friction between the car and the road is 0.88. Assuming your car doesn't skid, what is the force exerted on it by static friction?

48. • **BIO A Human Centrifuge** To test the effects of high acceleration on the human body, the National Aeronautics and Space Administration (NASA) constructed a large centrifuge at the Manned Spacecraft Center in Houston. In this device, astronauts were placed in a capsule that moved in a circular path with a radius of 15 m. If the astronauts in this centrifuge experienced a centripetal acceleration 9.0 times that of gravity, what was the linear speed of the capsule?

49. • A car goes around a curve on a road that is banked at an angle of 21.5°. Even though the road is slick, the car will stay on the road without any friction between its tires and the road when its speed is 19.8 m/s. What is the radius of the curve?

50. • **Clearview Screen** Large ships often have circular structures in their windshields, as shown in **FIGURE 6-59**. What is their purpose? Called clearview screens, they consist of a glass disk that is rotated at high speed by a motor (in the center of the circle) to disperse rain and spray. If the screen has a diameter of 0.39 m and rotates at 1700 rpm, the speed at the rim of the screen is 35 m/s. What is the centripetal acceleration at the rim of the screen? (A large centripetal acceleration keeps liquid from remaining on the screen. For comparison, recall that the acceleration due to gravity is 9.81 m/s².)

FIGURE 6-59 Problem 50

51. •• Predict/Calculate **(a)** As you ride on a Ferris wheel, your apparent weight is different at the top than at the bottom. Explain. **(b)** Calculate your apparent weight at the top and bottom of a Ferris wheel, given that the radius of the wheel is 7.2 m, it completes one revolution every 28 s, and your mass is 55 kg.

52. •• Driving in your car with a constant speed of $v = 22$ m/s, you encounter a bump in the road that has a circular cross section, as indicated in **FIGURE 6-60**. If the radius of curvature of the bump is $r = 52$ m, find the apparent weight of a 67-kg person in your car as you pass over the top of the bump.

FIGURE 6-60 Problem 52

GENERAL PROBLEMS

53. • **CE** If you weigh yourself on a bathroom scale at the equator, is the reading you get greater than, less than, or equal to the reading you get if you weigh yourself at the North Pole?

54. • **CE BIO Maneuvering a Jet** Humans lose consciousness if exposed to prolonged accelerations of more than about 7g. This is of concern to jet fighter pilots, who may experience centripetal accelerations of this magnitude when making high-speed turns. Suppose we would like to decrease the centripetal acceleration of a jet. Rank the following changes in flight path in order of how effective they are in decreasing the centripetal acceleration, starting with the least effective: **A**, decrease the turning radius by a factor of two; **B**, decrease the speed by a factor of three; or **C**, increase the turning radius by a factor of four.

55. • **CE BIO Gravitropism** As plants grow, they tend to align their stems and roots along the direction of the gravitational field. This tendency, which is related to differential concentrations of plant hormones known as auxins, is referred to as *gravitropism*. As an illustration of gravitropism, experiments show that seedlings placed in pots on the rim of a rotating turntable do not grow in the vertical direction. Do you expect their stems to tilt inward—toward the axis of rotation—or outward—away from the axis of rotation?

56. • **BIO Human-Powered Centrifuge** One of the hazards of prolonged weightlessness on long space missions is bone and muscle loss. To combat this problem, NASA is studying a human-powered centrifuge, like the one shown in FIGURE 6-61. In this device, one astronaut pedals a suspended, bicycle-like contraption, while a second astronaut rides standing up in a similarly suspended platform. If the standing astronaut moves with a speed of 2.7 m/s at a distance of 2.9 m from the center of the device, what is the astronaut's centripetal acceleration? Give your answer as a multiple of g, the acceleration due to gravity on the surface of the Earth.

FIGURE 6-61 Problem 56 A human-powered centrifuge

57. •• Predict/Calculate A 9.3-kg box slides across the floor of an elevator with a coefficient of kinetic friction equal to 0.38. (a) What is the force of kinetic friction acting on the box when the elevator is at rest? (b) If the elevator moves upward with constant acceleration, is the force of kinetic friction acting on the box greater than, less than, or equal to the value found in part (a)? Explain. (c) Determine the force of kinetic friction acting on the sliding box when the elevator moves upward with a constant acceleration of 1.4 m/s².

58. •• A child goes down a playground slide that is inclined at an angle of 26.5° below the horizontal. Find the acceleration of the child given that the coefficient of kinetic friction between the child and the slide is 0.315.

59. •• **Spin-Dry Dragonflies** Some dragonflies splash down onto the surface of a lake to clean themselves. After this dunking, the dragonflies gain altitude, and then spin rapidly at about 1000 rpm to spray the water off their bodies. When the dragonflies do this "spin-dry," they tuck themselves into a "ball." If they spin with a linear speed of 2.3 m/s, and curl into a ball with a radius of 1.1 cm, what is the centripetal acceleration they produce?

FIGURE 6-62 Problem 59

60. •• **The da Vinci Code** Leonardo da Vinci (1452–1519) is credited with being the first to perform quantitative experiments on friction, though his results weren't known until centuries later, due in part to the secret code (mirror writing) he used in his notebooks. Leonardo would place a block of wood on an inclined plane and measure the angle at which the block begins to slide. He reports that the coefficient of static friction was 0.25 in his experiments. At what angle did Leonardo's blocks begin to slide?

61. •• A 4.5-kg sled is pulled with constant speed across a rough, horizontal surface with various amounts of weight added to the sled. Data showing the required force for several different amounts of added weight are given in the table.

Added weight (kg)	Force (N)
0	11
1	15
2	17
3	21
4	22

Plot the data and determine the coefficient of kinetic friction between the sled and the horizontal surface.

62. •• A 0.045-kg golf ball hangs by a string from the rearview mirror of a car, as shown in FIGURE 6-63. The car accelerates in the forward direction with a magnitude of 2.1 m/s². (a) What angle θ does the string make with the vertical? (b) What is the tension in the string?

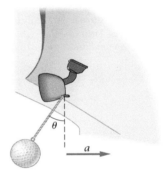

FIGURE 6-63 Problem 62

63. •• A physics textbook weighing 22 N rests on a desk. The coefficient of static friction between the book and the desk is $\mu_s = 0.60$; the coefficient of kinetic friction is $\mu_k = 0.40$. You push horizontally on the book with a force that gradually increases from 0 to 15 N, and then slowly decreases to 5.0 N, as indicated in the table below. For each value of the applied force given in the table, give the magnitude of the force of friction and state whether the book is accelerating, decelerating, at rest, or moving with constant speed.

Applied force	Friction force	Motion
0		
5.0 N		
11 N		
15 N		
11 N		
8.0 N		
5.0 N		

64. •• Predict/Calculate The blocks shown in FIGURE 6-64 are at rest. **(a)** Find the frictional force exerted on block A given that the mass of block A is 8.82 kg, the mass of block B is 2.33 kg, and the coefficient of static friction between block A and the surface on which it rests is 0.320. **(b)** If the mass of block A is doubled, does the frictional force exerted on it increase, decrease, or stay the same? Explain.

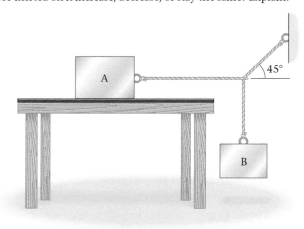

FIGURE 6-64 Problem 64

65. •• **A Conical Pendulum** A 0.075-kg toy airplane is tied to the ceiling with a string. When the airplane's motor is started, it moves with a constant speed of 4.47 m/s in a horizontal circle of radius 1.28 m, as illustrated in FIGURE 6-65. Find **(a)** the angle the string makes with the vertical and **(b)** the tension in the string.

FIGURE 6-65 Problem 65

66. •• A tugboat tows a barge at constant speed with a 3500-kg cable, as shown in FIGURE 6-66. If the angle the cable makes with the horizontal where it attaches to the barge and the tugboat is 22°, find the force the cable exerts on the barge in the forward direction.

FIGURE 6-66 Problem 66

67. •• Predict/Calculate Two blocks, stacked one on top of the other, can move without friction on the horizontal surface shown in FIGURE 6-67. The surface between the two blocks is rough, however, with a coefficient of static friction equal to 0.47. **(a)** If a horizontal force F is applied to the 5.0-kg bottom block, what is the maximum value F can have before the

FIGURE 6-67 Problem 67

2.0-kg top block begins to slip? **(b)** If the mass of the top block is increased, does the maximum value of F increase, decrease, or stay the same? Explain.

68. •• Predict/Calculate In a daring rescue by helicopter, two men with a combined mass of 172 kg are lifted to safety. **(a)** If the helicopter lifts the men straight up with constant acceleration, is the tension in the rescue cable greater than, less than, or equal to the combined weight of the men? Explain. **(b)** Determine the tension in the cable if the men are lifted with a constant acceleration of 1.10 m/s².

69. •• Predict/Calculate A light spring with a force constant of 13 N/m is connected to a wall and to a 1.2-kg toy bulldozer, as shown in FIGURE 6-68. When the electric motor in the bulldozer is turned on, it stretches the spring for a distance of 0.45 m before its tread begins to slip on the floor. **(a)** Which coefficient of friction (static or kinetic) can be determined from this information? Explain. **(b)** What is the numerical value of this coefficient of friction?

FIGURE 6-68 Problem 69

70. •• Predict/Calculate The blocks in FIGURE 6-69 have an acceleration of magnitude $a = 1.1$ m/s². **(a)** Is the tension in the string greater than, less than, or equal to $(1.5 \text{ kg}) \times (9.81 \text{ m/s}^2)$? Explain. **(b)** What is the tension in the string connecting the blocks? **(c)** What is the magnitude of the force of kinetic friction acting on the upper block? **(d)** What is the coefficient of kinetic friction between the upper block and the horizontal surface?

FIGURE 6-69 Problem 70

71. •• Predict/Calculate **Playing a Violin** The tension in a violin string is 2.7 N. When pushed down against the neck of the violin, the string makes an angle of 4.1° with the horizontal. **(a)** With what force must you push down on the string to bring it into contact with the neck? **(b)** If the angle were less than 4.1°, would the required force be greater than, less than, or the same as in part (a)? Explain.

72. •• Predict/Calculate A 9.8-kg monkey hangs from a rope, as shown in FIGURE 6-70. The tension on the left side of the rope is T_1 and the tension on the right side is T_2. **(a)** Which tension is largest in magnitude, or are they the same? Explain. **(b)** Determine the tensions T_1 and T_2.

FIGURE 6-70 Problem 72

73. •• As your plane circles an airport, it moves in a horizontal circle of radius 2300 m with a speed of 390 km/h. If the lift of the airplane's wings is perpendicular to the wings, at what angle should the plane be banked so that it doesn't tend to slip sideways?

74. •• At a playground, a 22-kg child sits on a spinning merry-go-round, as shown from above in **FIGURE 6-71**. The merry-go-round completes one revolution every 6.2 s, and the child sits at a radius of $r = 1.4$ m. **(a)** What is the force of static friction acting on the child? **(b)** What is the minimum coefficient of static friction between the child and the merry-go-round to keep the child from slipping?

FIGURE 6-71 Problem 74

75. ••• A 2.0-kg box rests on a plank that is inclined at an angle of 65° above the horizontal. The upper end of the box is attached to a spring with a force constant of 360 N/m, as shown in **FIGURE 6-72**. If the coefficient of static friction between the box and the plank is 0.22, what is the maximum amount the spring can be stretched and the box remain at rest?

FIGURE 6-72 Problem 75

76. ••• A wood block of mass m rests on a larger wood block of mass M that rests on a wooden table. The coefficients of static and kinetic friction between all surfaces are μ_s and μ_k, respectively. What is the minimum horizontal force, F, applied to the lower block that will cause it to slide out from under the upper block?

77. ••• A hockey puck of mass m is attached to a string that passes through a hole in the center of a table, as shown in **FIGURE 6-73**. The hockey puck moves in a circle of radius r. Tied to the other end of the string, and hanging vertically beneath the table, is a mass M. Assuming the tabletop is perfectly smooth, what speed must the hockey puck have if the mass M is to remain at rest?

FIGURE 6-73 Problem 77

78. ••• Predict/Calculate A popular ride at amusement parks is illustrated in **FIGURE 6-74**. In this ride, people sit in a swing that is suspended from a rotating arm. Riders are at a distance of 12 m from the axis of rotation and move with a speed of 25 mi/h. **(a)** Find the centripetal acceleration of the riders. **(b)** Find the angle θ the supporting wires make with the vertical. **(c)** If you observe a ride like that in Figure 6-74, or as shown in the photo on page 178, you will notice that all the swings are at the same angle θ to the vertical, regardless of the weight of the rider. Explain.

FIGURE 6-74 Problem 78

79. ••• **A Conveyor Belt** A box is placed on a conveyor belt that moves with a constant speed of 1.25 m/s. The coefficient of kinetic friction between the box and the belt is 0.780. **(a)** How much time does it take for the box to stop sliding relative to the belt? **(b)** How far does the box move in this time?

80. ••• As part of a circus act, a person drives a motorcycle with constant speed v around the inside of a vertical track of radius r, as indicated in **FIGURE 6-75**. If the combined mass of the motorcycle and rider is m, find the normal force exerted on the motorcycle by the track at the points **(a)** A, **(b)** B, and **(c)** C.

FIGURE 6-75 Problem 80

PASSAGE PROBLEMS

BIO Nasal Strips

People in all walks of life use nasal strips, or external nasal dilator strips (ENDS), to alleviate a number of respiratory problems. First introduced to eliminate snoring, they are now finding use in a number of other areas. For example, dentists have found that nasal strips help patients breathe better during dental procedures, making the experience considerably more pleasant for both doctor and patient.

One of the great advantages of ENDS is that no drugs are involved; the strips are a purely mechanical device, consisting of two flat, polyester springs enclosed by an adhesive tape covering. When applied to the nose, they exert an outward force that enlarges the nasal passages and reduces the resistance to air flow (see the illustration in Quick Example 6-8). The mechanism shown in **FIGURE 6-76 (a)** is used to measure the behavior of these strips. For example, if a 30-g weight is placed on the movable platform (of negligible mass), the strip is found to compress from an initial length of 50 mm to a reduced length of 19 mm, as can be seen in **FIGURE 6-76 (b)**.

81. • On the straight-line segment II in Figure 6-76 (b) we see that increasing the applied mass from 20 g to 25 g results in a reduction of the end-to-end distance from 42 mm to 24 mm. What is the force constant in N/m on segment II?

A. 0.28 N/m   B. 0.37 N/m
C. 2.7 N/m   D. 3.6 N/m

82. • Rank the straight segments I, II, and III in order of increasing "stiffness" of the nasal strips.

83. • In use on a typical human nose, the end-to-end distance is measured to be 30 mm. What force does the strip exert on the nasal passages in this situation?

A. 0.23 N   B. 0.69 N
C. 9.0 N   D. 2 N

84. •• Predict/Calculate REFERRING TO EXAMPLE 6-3 Suppose the coefficients of static and kinetic friction between the crate and the truck bed are 0.415 and 0.382, respectively. **(a)** Does the crate begin to slide at a tilt angle that is greater than, less than, or equal to 23.2°? **(b)** Verify your answer to part (a) by determining the angle at which the crate begins to slide. **(c)** Find the length of time it takes for the crate to slide a distance of 2.75 m when the tilt angle has the value found in part (b).

85. •• Predict/Calculate REFERRING TO EXAMPLE 6-3 The crate begins to slide when the tilt angle is 17.5°. When the crate reaches the bottom of the flatbed, after sliding a distance of 2.75 m, its speed is 3.11 m/s. Find **(a)** the coefficient of static friction and **(b)** the coefficient of kinetic friction between the crate and the flatbed.

86. •• REFERRING TO EXAMPLE 6-13 Suppose that the mass on the frictionless tabletop has the value $m_1 = 2.45$ kg. **(a)** Find the value of m_2 that gives an acceleration of 2.85 m/s². **(b)** What is the corresponding tension, T, in the string? **(c)** Calculate the ratio $T/m_2 g$ and show that it is less than 1, as expected.

87. •• REFERRING TO EXAMPLE 6-15 **(a)** At what speed will the force of static friction exerted on the car by the road be equal to half the weight of the car? The mass of the car is $m = 1200$ kg, the radius of the corner is $r = 45$ m, and the coefficient of static friction between the tires and the road is $\mu_s = 0.82$. **(b)** Suppose that the mass of the car is now doubled, and that it moves with a speed that again makes the force of static friction equal to half the car's weight. Is this new speed greater than, less than, or equal to the speed in part (a)?

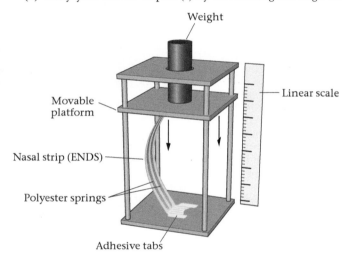

Weight

Movable platform

Linear scale

Nasal strip (ENDS)

Polyester springs

Adhesive tabs

(a)

(b)

FIGURE 6-76 Problems 81, 82, and 83

Big Ideas

1 Work is force times distance.

2 Kinetic energy is one-half mass times velocity squared.

3 Power is the rate at which work is done.

Work and Kinetic Energy

▲ The work done by this cyclist can increase his energy of motion. The relationship between work (force times distance) and energy of motion (kinetic energy) is developed in detail in this chapter.

In this chapter we introduce the idea that force times distance is an important new physical quantity, which we refer to as *work*. Closely related to work is the energy of motion, or *kinetic energy*. Together, work and kinetic energy set the stage for a deeper understanding of the physical world.

7-1 Work Done by a Constant Force

In this section we define work—in the physics sense of the word—and apply our definition to a variety of physical situations. We start with the simplest case—namely, the work done when force and displacement are in the same direction.

Force in the Direction of Displacement

When we push a shopping cart in a store or pull a suitcase through an airport, we do work. The greater the force, the greater the work; the greater the distance, the greater the work. These simple ideas form the basis for our definition of work.

To be specific, suppose we push a box with a constant force $\vec{F}$, as shown in FIGURE 7-1. If we move the box *in the direction of* $\vec{F}$ through a displacement $\vec{d}$, the **work** W we have done is Fd:

> **Definition of Work, W, When a Constant Force Is in the Direction of Displacement**
>
> $$W = Fd \qquad\qquad 7\text{-}1$$
>
> SI unit: newton-meter (N·m) = joule, J

Work is the product of two magnitudes, and hence it is a scalar. In addition, notice that a small force acting over a large distance gives the same work as a large force acting over a small distance. For example, $W = (1\,\text{N})(400\,\text{m}) = (400\,\text{N})(1\,\text{m})$.

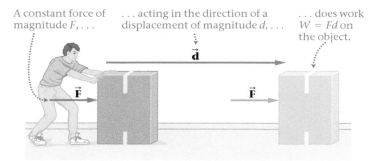

A constant force of magnitude F, ... acting in the direction of a displacement of magnitude d, ... does work $W = Fd$ on the object.

$\vec{d}$

$\vec{F}$ $\vec{F}$

▲ **FIGURE 7-1 Work: constant force in the direction of motion** A constant force $\vec{F}$ pushes a box through a displacement $\vec{d}$. In this special case, where the force and displacement are in the *same* direction, the work done on the box by the force is $W = Fd$.

The dimensions of work are newtons (force) times meters (distance), or N·m. This combination of dimensions is called the **joule** (rhymes with "school," as commonly pronounced) in honor of James Prescott Joule (1818–1889), a dedicated physicist who is said to have conducted physics experiments even while on his honeymoon. We define a joule as follows:

> **Definition of the joule, J**
>
> $$1\ \text{joule} = 1\,\text{J} = 1\,\text{N}\cdot\text{m} = 1(\text{kg}\cdot\text{m/s}^2)\cdot\text{m} = 1\ \text{kg}\cdot\text{m}^2/\text{s}^2 \qquad\qquad 7\text{-}2$$

To get a better feeling for work and the associated units, suppose you exert a force of 82.0 N on the box in Figure 7-1 and move it in the direction of the force through a distance of 3.00 m. The work you have done is

$$W = Fd = (82.0\,\text{N})(3.00\,\text{m}) = 246\,\text{N}\cdot\text{m} = 246\,\text{J}$$

Similarly, if you do 5.00 J of work to lift a book through a vertical distance of 0.750 m, the force you have exerted on the book is

$$F = \frac{W}{d} = \frac{5.00\,\text{J}}{0.750\,\text{m}} = \frac{5.00\,\text{N}\cdot\text{m}}{0.750\,\text{m}} = 6.67\,\text{N}$$

EXERCISE 7-1 WORK DONE

One species of Darwin's finch, *Geospiza magnirostris*, can exert a force of 217 N with its beak as it cracks open a *Tribulus* seed case. If its beak moves through a distance of 0.42 cm during this operation, how much work does the finch do to get the seed?

REASONING AND SOLUTION

The work the finch does is force times distance. In this case, the force is 217 N and the distance is 0.42 cm = 0.0042 m. Thus,

$$W = Fd = (217 \text{ N})(0.0042 \text{ m}) = 0.91 \text{ J}$$

This is a relatively small amount of work for a human, but significant for a small bird.

TABLE 7-1 Typical Values of Work

Activity	Equivalent work (J)
Annual U.S. energy use	8×10^{19}
Mt. St. Helens eruption	10^{18}
Burning 1 gallon of gas	10^8
Human food intake/day	10^7
Melting an ice cube	10^4
Lighting a 100-W bulb for 1 minute	6000
Heartbeat	0.5
Turning a page of a book	10^{-3}
Hop of a flea	10^{-7}
Breaking a bond in DNA	10^{-20}

Just how much work is a joule, anyway? Well, you do one joule of work when you lift a gallon of milk through a height of about an inch, or lift an apple a meter. One joule of work lights a 100-watt lightbulb for 0.01 second or heats a glass of water 0.00125 degree Celsius. Clearly, a joule is a modest amount of work in everyday terms. Additional examples of work are listed in Table 7-1.

EXAMPLE 7-2 HEADING FOR THE ER

An intern pushes an 87-kg patient on an 18-kg gurney, producing an acceleration of 0.55 m/s². **(a)** How much work does the intern do in pushing the patient and gurney through a distance of 1.9 m? Assume the gurney moves without friction. **(b)** How far must the intern push the gurney to do 140 J of work?

PICTURE THE PROBLEM

Our sketch shows the physical situation for this problem. Notice that the force exerted by the intern is in the same direction as the displacement of the gurney; therefore, we know that the work is $W = Fd$.

REASONING AND STRATEGY

We are not given the magnitude of the force, F, so we cannot apply Equation 7-1 ($W = Fd$) directly. However, we are given the mass and acceleration of the patient and gurney, and from them we can calculate the force with $F = ma$. The work done by the intern is then $W = Fd$, where $d = 1.9$ m.

Known Mass of patient, 87 kg; mass of gurney, 18 kg; acceleration, $a = 0.55$ m/s²; **(a)** pushing distance, $d = 1.9$ m; **(b)** work, $W = 140$ J.

Unknown **(a)** Work done, $W = ?$ **(b)** Pushing distance, $d = ?$

$a = 0.55 \text{ m/s}^2$

$d = 1.9 \text{ m}$

SOLUTION

Part (a)

1. First, find the force F exerted by the intern:

$$F = ma = (87 \text{ kg} + 18 \text{ kg})(0.55 \text{ m/s}^2) = 58 \text{ N}$$

2. The work done by the intern, W, is the force times the distance:

$$W = Fd = (58 \text{ N})(1.9 \text{ m}) = 110 \text{ J}$$

Part (b)

3. Use $W = Fd$ to solve for the distance d:

$$d = \frac{W}{F} = \frac{140 \text{ J}}{58 \text{ N}} = 2.4 \text{ m}$$

INSIGHT

You might wonder whether the work done by the intern depends on the speed of the gurney. The answer is no. The work done on an object, $W = Fd$, doesn't depend on whether the object moves through the distance d quickly or slowly. What does depend on the speed of the gurney is the *rate* at which work is done, which we discuss in detail in Section 7-4.

PRACTICE PROBLEM — PREDICT/CALCULATE

(a) If the total mass of the gurney plus patient is halved and the acceleration is doubled, does the work done by the intern increase, decrease, or remain the same? Explain. **(b)** Determine the work in this case. [**Answer: (a)** The work remains the same because the two changes offset one another; that is, $F = ma = (m/2)(2a)$. **(b)** The work is 110 J, as before.]

Some related homework problems: Problem 3, Problem 4

▲ **FIGURE 7-2 Visualizing Concepts Force, Work, and Displacement** The person doing push ups does positive work as he moves upward, but does no work when he holds himself in place. The weightlifter does positive work as she raises the weights, but no work is done on the weights as she holds them motionless above her head.

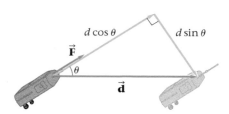

▲ **FIGURE 7-4 Force at an angle to direction of motion: another look** The displacement of the suitcase in Figure 7-3 is equivalent to a displacement of magnitude $d \cos \theta$ in the direction of the force $\vec{F}$, plus a displacement of magnitude $d \sin \theta$ perpendicular to the force. Only the displacement parallel to the force results in nonzero work; hence the total work done is $F(d \cos \theta)$ as expected.

Zero Distance Implies Zero Work Before moving on, let's note an interesting point about our definition of work. It's clear from Equation 7-1 that *the work W is zero if the distance d is zero*—and this is true regardless of how great the force might be, as illustrated in **FIGURE 7-2**. For example, if you push against a solid wall, you do no work on it, even though you may become tired from your efforts. Similarly, if you stand in one place holding a 50-pound suitcase in your hand, you do no work on the suitcase, even though you soon feel worn out. The fact that we become tired when we push against a wall or hold a heavy object is due to the repeated contraction and expansion of individual cells within our muscles. Thus, even when we are "at rest," our muscles are doing mechanical work on the microscopic level.

Force at an Angle to the Displacement

In **FIGURE 7-3** we see a person pulling a suitcase on a level surface with a strap that makes an angle θ with the horizontal—in this case the force is at an angle to the direction of motion. How do we calculate the work now? Well, instead of force times distance, we say that work is the *component* of force in the *direction* of displacement times the magnitude of the displacement. In Figure 7-3, the component of force in the direction of the displacement is $F \cos \theta$ and the magnitude of the displacement is d. Therefore, the work is $F \cos \theta$ times d:

> **Definition of Work When the Angle Between a Constant Force and the Displacement Is θ**
>
> $$W = (F \cos \theta)d = Fd \cos \theta$$
>
> SI unit: joule, J

7-3

Of course, in the case where the force is in the direction of motion, the angle θ is zero; then $W = Fd \cos 0 = Fd \cdot 1 = Fd$, in agreement with Equation 7-1.

The component of force in the direction of displacement is $F \cos \theta$. This is the only component of the force that does work.

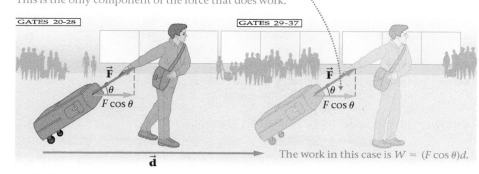

The work in this case is $W = (F \cos \theta)d$.

▲ **FIGURE 7-3 Work: force at an angle to the direction of motion** A person pulls a suitcase with a strap at an angle θ to the direction of motion. The component of force in the direction of motion is $F \cos \theta$, and the work done by the person is $W = (F \cos \theta)d$.

Equally interesting is a situation in which the force and the displacement are at right angles to one another. In this case $\theta = 90°$ and the work done by the force F is zero; $W = Fd \cos 90° = 0$.

This result leads naturally to an alternative way to think about the expression $W = Fd \cos \theta$. In **FIGURE 7-4** we show the displacement and the force for the suitcase in Figure 7-3. Notice that the displacement is equivalent to a displacement in the direction of the force of magnitude $(d \cos \theta)$ *plus* a displacement at right angles to the force of magnitude $(d \sin \theta)$. Since the displacement at right angles to the force corresponds to zero work and the displacement in the direction of the force corresponds to a work $W = F(d \cos \theta)$, it follows that the work done in this case is $Fd \cos \theta$, as given in Equation 7-3. Thus, the work done by a force can be thought of in the following two *equivalent* ways:

(i) *Work is the component of force in the direction of the displacement times the magnitude of the displacement.*

(ii) *Work is the component of displacement in the direction of the force times the magnitude of the force.*

In both interpretations, the mathematical expression for work is exactly the same, $W = Fd \cos\theta$, where θ is the angle between the force vector and the displacement vector when they are placed tail-to-tail. This definition of θ is illustrated in Figure 7-4.

Finally, we can also express work as the **dot product** of the vectors $\vec{F}$ and $\vec{d}$; that is, $W = \vec{F} \cdot \vec{d} = Fd \cos\theta$. Notice that the dot product, which is always a scalar, is simply the magnitude of one vector times the magnitude of the second vector times the cosine of the angle between them. We discuss the dot product in greater detail in Appendix A.

Big Idea 1 Work is force times distance when force and displacement are in the same direction. More generally, work is the component of force in the direction of displacement times the distance.

PHYSICS IN CONTEXT
Looking Back

Even though work is a scalar quantity, notice that vector components (Chapter 3) are used in the definition of work.

EXAMPLE 7-3   GRAVITY ESCAPE SYSTEM

RWP* In a gravity escape system (GES), an enclosed lifeboat on a large ship is deployed by letting it slide down a ramp and then continue in free fall to the water below. Suppose a 4970-kg lifeboat slides a distance of 5.00 m on a ramp, dropping through a vertical height of 2.50 m. How much work does gravity do on the boat?

PICTURE THE PROBLEM
From our sketch, we see that the force of gravity $m\vec{g}$ and the displacement $\vec{d}$ are at an angle θ relative to one another when placed tail-to-tail, and that θ is also the angle the ramp makes with the vertical. In addition, we note that the vertical height of the ramp is $h = 2.50$ m and the length of the ramp is $d = 5.00$ m.

REASONING AND STRATEGY
By definition, the work done on the lifeboat by gravity is $W = Fd \cos\theta$, where $F = mg$, $d = 5.00$ m, and θ is the angle between $m\vec{g}$ and $\vec{d}$. We are not given θ in the problem statement, but from the right triangle that forms the ramp we see that $\cos\theta = h/d$. Once θ is determined from the geometry of our sketch, it is straightforward to calculate W.

Known   Mass of lifeboat, $m = 4970$ kg; sliding distance, $d = 5.00$ m; vertical height, $h = 2.50$ m.
Unknown   Work done by gravity, $W = ?$

SOLUTION

1. First, find the component of $\vec{F} = m\vec{g}$ in the direction of motion:

$$F \cos\theta = (mg)\left(\frac{h}{d}\right)$$
$$= (4970 \text{ kg})(9.81 \text{ m/s}^2)\left(\frac{2.50 \text{ m}}{5.00 \text{ m}}\right) = 24{,}400 \text{ N}$$

2. Multiply by distance to find the work:

$$W = (F \cos\theta)d = (24{,}400 \text{ N})(5.00 \text{ m}) = 122{,}000 \text{ J}$$

3. Alternatively, cancel d algebraically before substituting numerical values:

$$W = Fd \cos\theta = (mg)(d)\left(\frac{h}{d}\right)$$
$$= mgh = (4970 \text{ kg})(9.81 \text{ m/s}^2)(2.50 \text{ m}) = 122{,}000 \text{ J}$$

INSIGHT
The work is simply $W = mgh$, exactly the same as if the lifeboat had fallen straight down through the height h.

Working the problem symbolically, as in Step 3, results in two distinct advantages. First, it makes for a simpler expression for the work. Second, and more important, it shows that the distance d cancels; hence the work depends on the height h but not on the distance. Such a result is not apparent when we work solely with numbers, as in Steps 1 and 2.

PRACTICE PROBLEM
Suppose the lifeboat slides halfway to the water, gets stuck for a moment, and then starts up again and continues to the end of the ramp. What is the work done by gravity in this case?   [**Answer:** The work done by gravity is exactly the same, $W = mgh$, independent of how the boat moves down the ramp.]

Some related homework problems: Problem 10, Problem 11

Next, we present a Conceptual Example that compares the work required to move an object along two different paths.

---

*Real World Physics applications are denoted by the acronym RWP.

CONCEPTUAL EXAMPLE 7-4 PATH DEPENDENCE OF WORK

You want to load a box into the back of a truck, as shown in the sketch. One way is to lift it straight up with constant speed through a height h, as shown, doing a work W_1. Alternatively, you can slide the box up a loading ramp with constant speed a distance L, doing a work W_2. Assuming the box slides on the ramp without friction, which of the following statements is correct: (a) $W_1 < W_2$, (b) $W_1 = W_2$, (c) $W_1 > W_2$?

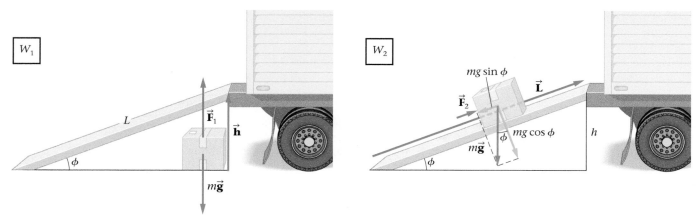

REASONING AND DISCUSSION

You might think that W_2 is less than W_1, because the force needed to slide the box up the ramp, F_2, is less than the force needed to lift it straight up. On the other hand, the distance up the ramp, L, is greater than the vertical distance, h, so perhaps W_2 should be greater than W_1. In fact, these two effects cancel exactly, giving $W_1 = W_2$.

To see this, we first calculate W_1. The force needed to lift the box with constant speed is $F_1 = mg$, and the height is h; therefore $W_1 = mgh$.

Next, the work to slide the box up the ramp with constant speed is $W_2 = F_2L$, where F_2 is the force required to push against the tangential component of gravity. In the sketch we see that $F_2 = mg \sin \phi$. The sketch also shows that $\sin \phi = h/L$; thus $W_2 = (mg \sin \phi)L = (mg)(h/L)L = mgh = W_1$: The two works are identical!

Clearly, the ramp is a useful device—it reduces the *force* required to move the box upward from $F_1 = mg$ to $F_2 = mg(h/L)$. Even so, it doesn't decrease the amount of *work* we need to do. As we have seen, the reduced force on the ramp is offset by the increased distance.

ANSWER

(b) $W_1 = W_2$

Negative Work and Total Work

Work depends on the angle between the force, $\vec{F}$, and the displacement (or direction of motion), $\vec{d}$. This dependence gives rise to three distinct possibilities, as shown in FIGURE 7-5:

   (i) *Work is positive if the force has a component in the direction of motion* ($-90° < \theta < 90°$).

  (ii) *Work is zero if the force has no component in the direction of motion* ($\theta = \pm 90°$).

 (iii) *Work is negative if the force has a component opposite to the direction of motion* ($90° < \theta < 270°$).

Thus, whenever we calculate work, we must be careful about its sign and not just assume it is positive.

▲ **FIGURE 7-5 Positive, negative, and zero work** Work is positive when the force is in the same general direction as the displacement and is negative if the force is generally opposite to the displacement. Zero work is done if the force is at right angles to the displacement.

When more than one force acts on an object, the total work is the sum of the work done by each force separately. Thus, if force $\vec{\mathbf{F}}_1$ does work W_1, force $\vec{\mathbf{F}}_2$ does work W_2, and so on, the total work is

$$W_{\text{total}} = W_1 + W_2 + W_3 + \cdots = \sum W \qquad \text{7-4}$$

Equivalently, the total work can be calculated by first performing a vector sum of all the forces acting on an object to obtain $\vec{\mathbf{F}}_{\text{total}}$ and then using our basic definition of work:

$$W_{\text{total}} = (F_{\text{total}} \cos \theta)d = F_{\text{total}}\, d \cos \theta \qquad \text{7-5}$$

In this expression, θ is the angle between $\vec{\mathbf{F}}_{\text{total}}$ and the displacement $\vec{\mathbf{d}}$. In the next two Examples we calculate the total work in each of these ways.

PROBLEM-SOLVING NOTE

Be Careful About the Angle θ

In calculating $W = Fd\cos\theta$, be sure that the angle you use in the cosine is the angle between the force and the displacement vectors when they are placed tail-to-tail. Sometimes θ may be used to label a different angle in a given problem. For example, θ is often used to label the angle of a slope, in which case it may have nothing to do with the angle between the force and the displacement. To summarize: Just because an angle is labeled θ doesn't mean it's automatically the correct angle to use in the work formula.

EXAMPLE 7-5   A COASTING CAR I

A car of mass m coasts down a hill inclined at an angle ϕ below the horizontal. The car is acted on by three forces: (i) the normal force $\vec{\mathbf{N}}$ exerted by the road, (ii) a force due to air resistance, $\vec{\mathbf{F}}_{\text{air}}$, and (iii) the force of gravity, $m\vec{\mathbf{g}}$. Find the total work done on the car as it travels a distance d along the road.

PICTURE THE PROBLEM
Because ϕ is the angle the slope makes with the horizontal, it is also the angle between $m\vec{\mathbf{g}}$ and the downward normal direction, as was shown in Figure 5-22. It follows that the angle between $m\vec{\mathbf{g}}$ and the displacement $\vec{\mathbf{d}}$ is $\theta = 90° - \phi$. Our sketch also shows that the angle between $\vec{\mathbf{N}}$ and $\vec{\mathbf{d}}$ is $\theta = 90°$, and the angle between $\vec{\mathbf{F}}_{\text{air}}$ and $\vec{\mathbf{d}}$ is $\theta = 180°$.

REASONING AND STRATEGY
For each force we calculate the work using $W = Fd\cos\theta$, where θ is the angle between that particular force and the displacement $\vec{\mathbf{d}}$. The total work is the sum of the work done by each of the three forces.

Known   Mass of car, m; angle of incline, ϕ; distance, d; forces acting on car, $\vec{\mathbf{N}}$, $\vec{\mathbf{F}}_{\text{air}}$, and $m\vec{\mathbf{g}}$.
Unknown Work done on car, $W = ?$

SOLUTION

1. We start with the work done by the normal force, $\vec{\mathbf{N}}$.
   From the figure we see that $\theta = 90°$ for this force:
   $$W_N = Nd\cos\theta = Nd\cos 90° = Nd(0) = 0$$

2. For the force of air resistance, $\theta = 180°$:
   $$W_{\text{air}} = F_{\text{air}}d\cos 180° = F_{\text{air}}d(-1) = -F_{\text{air}}d$$

3. For gravity the angle θ is $\theta = 90° - \phi$, as indicated in the sketch. Recall that $\cos(90° - \phi) = \sin\phi$ (see Appendix A):
   $$W_{mg} = mgd\cos(90° - \phi) = mgd\sin\phi$$

4. The total work is the sum of the individual works:
   $$W_{\text{total}} = W_N + W_{\text{air}} + W_{mg} = 0 - F_{\text{air}}d + mgd\sin\phi$$

INSIGHT
The normal force is perpendicular to the motion of the car, and thus does no work. Air resistance points in a direction that opposes the motion, so it does negative work. On the other hand, gravity has a component in the direction of motion; therefore, its work is positive. The physical significance of positive, negative, and zero work will be discussed in detail in the next section.

PRACTICE PROBLEM
Calculate the total work done on a 1550-kg car as it coasts 20.4 m down a hill with $\phi = 5.00°$. Let the force due to air resistance be 15.0 N.
[**Answer:** $W_{\text{total}} = W_N + W_{\text{air}} + W_{mg} = 0 - F_{\text{air}}d + mgd\sin\phi = 0 - 306\,\text{J} + 2.70 \times 10^4\,\text{J} = 2.67 \times 10^4\,\text{J}$]

Some related homework problems: Problem 15, Problem 63

In the next Example, we first sum the forces acting on the car to find F_{total}. Once the total force is determined, we calculate the total work using $W_{\text{total}} = F_{\text{total}}d \cos \theta$.

EXAMPLE 7-6 A COASTING CAR II

Consider the car described in Example 7-5. Calculate the total work done on the car using $W_{\text{total}} = F_{\text{total}}d \cos \theta$.

PICTURE THE PROBLEM

First, we choose the x axis to point down the slope, and the y axis to be at right angles to the slope. With this choice, there is no acceleration in the y direction, which means that the total force in that direction must be zero. As a result, the total force acting on the car is in the x direction. The magnitude of the total force is $mg \sin \phi - F_{\text{air}}$, as can be seen in our sketch.

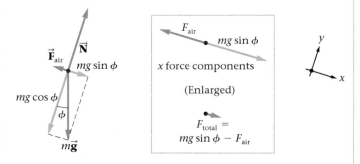

REASONING AND STRATEGY

We begin by finding the x component of each force vector, and then we sum them to find the total force acting on the car. As can be seen from the sketch, the total force points in the positive x direction—that is, in the same direction as the displacement. Therefore, the angle θ in $W = F_{\text{total}}d \cos \theta$ is zero.

Known Mass of car, m; angle of incline, ϕ; distance, d; forces acting on car, $\vec{\mathbf{N}}$, $\vec{\mathbf{F}}_{\text{air}}$, and $m\vec{\mathbf{g}}$.
Unknown Work done on car, $W = ?$

SOLUTION

1. Referring to the sketch, we see that the *magnitude* of the total force is $mg \sin \phi$ minus F_{air}:

$$F_{\text{total}} = mg \sin \phi - F_{\text{air}}$$

2. The *direction* of $\vec{\mathbf{F}}_{\text{total}}$ is the same as the direction of $\vec{\mathbf{d}}$; thus $\theta = 0°$. We can now calculate W_{total}:

$$W_{\text{total}} = F_{\text{total}}d \cos \theta = (mg \sin \phi - F_{\text{air}})d \cos 0°$$
$$= mgd \sin \phi - F_{\text{air}}d$$

INSIGHT

Notice that we were careful to calculate both the magnitude and the direction of the total force. The magnitude (which is always positive) gives F_{total} and the direction gives $\theta = 0°$, allowing us to use $W_{\text{total}} = F_{\text{total}}d \cos \theta$.

PRACTICE PROBLEM

Suppose the total work done on a 1620-kg car as it coasts 25.0 m down a hill with $\phi = 6.00°$ is $W_{\text{total}} = 3.75 \times 10^4 \, \text{J}$. Find the magnitude of the force due to air resistance. [**Answer:** $F_{\text{air}}d = -W_{\text{total}} + mgd \sin \phi = 4030 \, \text{J}$; thus $F_{\text{air}} = (4030 \, \text{J})/d = 161 \, \text{N}$]

Some related homework problems: Problem 15, Problem 63

Enhance Your Understanding
(Answers given at the end of the chapter)

1. A block slides a distance d to the right on a horizontal surface, as indicated in **FIGURE 7-6**. Rank the four forces that act on the block (F, N, mg, f_k) in order of the amount of work they do, from most negative to most positive. Indicate ties where appropriate.

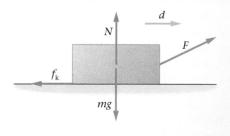

▲ **FIGURE 7-6** Enhance Your Understanding 1.

PROBLEM-SOLVING NOTE

Work Can Be Positive, Negative, or Zero

When you calculate work, be sure to keep track of whether it is positive or negative. The distinction is important, since positive work increases speed, whereas negative work decreases speed. Zero work, of course, has no effect on speed.

Section Review

• The work done by a force is $W = (F \cos \theta)d = Fd \cos \theta$, where F is the force, d is the distance, and θ is the angle between the force and displacement.

- Work is positive when the force and displacement are in the same direction, zero if the force is perpendicular to the displacement, and negative if the force and displacement are in opposite directions.

7-2 Kinetic Energy and the Work–Energy Theorem

Suppose you drop an apple. As it falls, gravity does positive work on it, as indicated in FIGURE 7-7, and its speed increases. If you toss the apple upward, gravity does negative work on it, and the apple slows down. In general, whenever the total work done on an object is positive, its speed increases; whenever the total work is negative, its speed decreases. In this section we derive an important result, the **work–energy theorem,** which makes this connection between work and change in speed precise.

Work and Kinetic Energy To begin, consider an apple of mass m falling through the air, and suppose that two forces act on the apple—gravity, $m\vec{\mathbf{g}}$, and the average force of air resistance, $\vec{\mathbf{F}}_{air}$. The total force acting on the apple, $\vec{\mathbf{F}}_{total}$, gives the apple a constant downward acceleration of magnitude

$$a = \frac{F_{total}}{m}$$

Since the total force is downward and the motion is downward, the work done on the apple is positive.

Now, suppose that the initial speed of the apple is v_i, and that after falling a distance d its speed increases to v_f. The apple falls with constant acceleration a; hence constant-acceleration kinematics (Equation 2–12) gives

$$v_f{}^2 = v_i{}^2 + 2ad$$

With a slight rearrangement we find

$$2ad = v_f{}^2 - v_i{}^2$$

Next, we substitute $a = F_{total}/m$ into this equation:

$$2\left(\frac{F_{total}}{m}\right)d = v_f{}^2 - v_i{}^2$$

Multiplying both sides by m and dividing by 2 yields

$$F_{total}d = \tfrac{1}{2}mv_f{}^2 - \tfrac{1}{2}mv_i{}^2$$

In this expression, $F_{total}d$ is simply the total work done on the apple. Thus we find

$$W_{total} = \tfrac{1}{2}mv_f{}^2 - \tfrac{1}{2}mv_i{}^2$$

This shows that the total work is directly related to the change in speed. Notice that $W_{total} > 0$ means $v_f > v_i$, $W_{total} < 0$ means $v_f < v_i$, and $W_{total} = 0$ means $v_f = v_i$.

The quantity $\tfrac{1}{2}mv^2$ in the equation for W_{total} has a special significance in physics. We call it the **kinetic energy,** K:

Definition of Kinetic Energy, K

$K = \tfrac{1}{2}mv^2$ 7-6

SI unit: $kg \cdot m^2/s^2 = joule, J$

In general, the kinetic energy of an object is the energy due to its motion. We measure kinetic energy in joules, the same units as work, and both kinetic energy and work are scalars. Unlike work, however, kinetic energy is never negative. Instead, K is always greater than or equal to zero, independent of the direction of motion or the direction of any forces.

Force is in the direction of displacement . . .

. . . so positive work is done on the apple. $\vec{\mathbf{d}}$ $m\vec{\mathbf{g}}$

This causes the apple to speed up.

Apple falling: $W > 0$, speed increases

(a)

Negative work done on the apple . . .

$\vec{\mathbf{d}}$ $m\vec{\mathbf{g}}$

. . . causes it to slow down.

Apple tossed upward: $W < 0$, speed decreases

(b)

▲ FIGURE 7-7 **Gravitational work (a)** The work done by gravity on an apple that moves downward is positive. If the apple is in free fall, this positive work will result in an increase in speed. **(b)** In contrast, the work done by gravity on an apple that moves upward is negative. If the apple is in free fall, the negative work done by gravity will result in a decrease of speed.

**PHYSICS
IN CONTEXT
Looking Back**

The kinematic equations of motion for constant acceleration (Chapters 2 and 4) are used in the derivation of kinetic energy.

**PHYSICS
IN CONTEXT
Looking Ahead**

In Chapter 8 we introduce the concept of potential energy. The combination of kinetic and potential energy is referred to as mechanical energy, which will play a central role in our discussion of the conservation of energy.

TABLE 7-2 Typical Kinetic Energies

Source	Approximate kinetic energy (J)
Jet aircraft at 500 mi/h	10^9
Car at 60 mi/h	10^6
Home-run baseball	10^2
Person at walking speed	50
Housefly in flight	10^{-4}

Big Idea **2** Kinetic energy is one-half mass times velocity squared. The total work done on an object is equal to the change in its kinetic energy.

**PHYSICS
IN CONTEXT
Looking Ahead**

The kinetic energy before and after a collision is an important characterizing feature, as we shall see in Chapter 9. In addition, look for kinetic energy to reappear when we discuss rotational motion in Chapter 10, and again when we study ideal gases in Chapter 17.

To get a feeling for typical values of kinetic energy, consider your kinetic energy when jogging. If we assume a mass of about 62 kg and a speed of 2.5 m/s, your kinetic energy is $K = \frac{1}{2}(62 \text{ kg})(2.5 \text{ m/s})^2 = 190 \text{ J}$. Additional examples of kinetic energy are given in Table 7-2.

EXERCISE 7-7 KINETIC ENERGY TAKES OFF

A small airplane moving along the runway during takeoff has a mass of 690 kg and a kinetic energy of 25,000 J. **(a)** What is the speed of the plane? **(b)** By what multiplicative factor does the kinetic energy of the plane change if its speed is tripled?

REASONING AND SOLUTION

(a) The kinetic energy of an object is given by $K = \frac{1}{2}mv^2$. We know that $K = 25,000 \text{ J}$ and $m = 690 \text{ kg}$; therefore the speed is

$$v = \sqrt{\frac{2K}{m}} = \sqrt{\frac{2(25,000 \text{ J})}{690 \text{ kg}}} = 8.5 \text{ m/s}$$

(b) Kinetic energy depends on the speed squared, and hence tripling the speed increases the kinetic energy by a factor of nine.

The Work–Energy Theorem In the preceding discussion we used a calculation of work to derive an expression for the kinetic energy of an object. The precise connection we derived between these quantities is known as the **work–energy theorem.** This theorem can be stated as follows:

Work–Energy Theorem

The *total work* done on an object is equal to the *change* in its kinetic energy:

$$W_{\text{total}} = \Delta K = \frac{1}{2}mv_{\text{f}}^2 - \frac{1}{2}mv_{\text{i}}^2 \qquad \text{7-7}$$

Thus, the work–energy theorem says that when a force acts on an object over a distance—doing work on it—the result is a change in the speed of the object, and hence a change in its energy of motion. Equation 7-7 is the quantitative expression of this connection.

We have derived the work–energy theorem for a force that is constant in direction and magnitude, but it is valid for any force. In fact, the work–energy theorem is completely general, making it one of the most important and fundamental results in physics. It is also a very handy tool for problem solving, as we shall see many times throughout this text.

EXERCISE 7-8 WORK TO ACCELERATE

A 220-kg motorcycle is cruising at 14 m/s. What is the total work that must be done on the motorcycle to increase its speed to 19 m/s?

REASONING AND SOLUTION

From the work–energy theorem, we know that the total work required to change an object's kinetic energy is $W_{\text{total}} = \Delta K = \frac{1}{2}mv_{\text{f}}^2 - \frac{1}{2}mv_{\text{i}}^2$. Given that $m = 220 \text{ kg}$, $v_{\text{i}} = 14 \text{ m/s}$, and $v_{\text{f}} = 19 \text{ m/s}$, we find

$$W_{\text{total}} = \frac{1}{2}mv_{\text{f}}^2 - \frac{1}{2}mv_{\text{i}}^2$$
$$= \frac{1}{2}(220 \text{ kg})(19 \text{ m/s})^2 - \frac{1}{2}(220 \text{ kg})(14 \text{ m/s})^2 = 18,000 \text{ J}$$

This much energy could lift a 90-kg person through a vertical distance of about 20 m.

We now present Examples showing how the work–energy theorem is used in practical situations.

EXAMPLE 7-9   HIT THE BOOKS

A 4.10-kg box of books is lifted vertically from rest a distance of 1.60 m with a constant, upward applied force of 52.7 N. Find **(a)** the work done by the applied force, **(b)** the work done by gravity, and **(c)** the final speed of the box.

PICTURE THE PROBLEM
Our sketch shows that the direction of motion of the box is upward. In addition, we see that the applied force, $\vec{\mathbf{F}}_{app}$, is upward and the force of gravity, $m\vec{\mathbf{g}}$, is downward. Finally, the box is lifted from rest ($v_i = 0$) through a distance $\Delta y = 1.60$ m.

REASONING AND STRATEGY
The applied force is in the direction of motion, so the work it does, W_{app}, is positive. Gravity is opposite in direction to the motion; thus its work, W_{mg}, is negative. The total work is the sum of W_{app} and W_{mg} and the final speed of the box is found by applying the work–energy theorem, $W_{total} = \Delta K$.

Known   Mass of box, $m = 4.10$ kg; vertical distance, $\Delta y = 1.60$ m; applied force, $F_{app} = 52.7$ N.
Unknown **(a)** Work done by applied force, $W_{app} = $? **(b)** work done by gravity, $W_{mg} = $? **(c)** Final speed of box, $v_f = $?

SOLUTION

Part (a)

1. First we find the work done by the applied force. In this case, $\theta = 0°$ and the distance is $\Delta y = 1.60$ m:

$$W_{app} = F_{app} \cos 0° \, \Delta y = (52.7 \text{ N})(1)(1.60 \text{ m}) = 84.3 \text{ J}$$

Part (b)

2. Next, we calculate the work done by gravity. The distance is $\Delta y = 1.60$ m, as before, but now $\theta = 180°$:

$$W_{mg} = mg \cos 180° \, \Delta y$$
$$= (4.10 \text{ kg})(9.81 \text{ m/s}^2)(-1)(1.60 \text{ m}) = -64.4 \text{ J}$$

Part (c)

3. The total work done on the box, W_{total}, is the sum of W_{app} and W_{mg}:

$$W_{total} = W_{app} + W_{mg} = 84.3 \text{ J} - 64.4 \text{ J} = 19.9 \text{ J}$$

4. To find the final speed, v_f, we apply the work–energy theorem. Recall that the box started at rest; thus $v_i = 0$:

$$W_{total} = \tfrac{1}{2}mv_f^2 - \tfrac{1}{2}mv_i^2 = \tfrac{1}{2}mv_f^2$$
$$v_f = \sqrt{\frac{2W_{total}}{m}} = \sqrt{\frac{2(19.9 \text{ J})}{4.10 \text{ kg}}} = 3.12 \text{ m/s}$$

INSIGHT
As a check on our result, we can find v_f in a completely different way. First, calculate the acceleration of the box with the result $a = (F_{app} - mg)/m = 3.04$ m/s^2. Next, use this result in the kinematic equation $v^2 = v_0^2 + 2a\Delta y$. With $v_0 = 0$ and $\Delta y = 1.60$ m, we find $v = 3.12$ m/s, in agreement with the results using the work–energy theorem.

PRACTICE PROBLEM — PREDICT/CALCULATE
(a) If the box is lifted only a quarter of the distance, $\Delta y = (1.60 \text{ m})/4 = 0.400$ m, is the final speed 1/8, 1/4, or 1/2 of the value found in this Example? Explain. **(b)** Determine v_f in this case.   [**Answer: (a)** Work depends linearly on Δy, and v_f depends on the square root of the work. As a result, it follows that the final speed is $\sqrt{1/4} = \tfrac{1}{2}$ the value found in Step 4. **(b)** We find $v_f = \tfrac{1}{2}(3.12 \text{ m/s}) = 1.56$ m/s.]

Some related homework problems: Problem 28, Problem 29, Problem 65

In the previous Example the initial speed was zero. This is not always the case, of course. The next Example considers a case with a nonzero initial speed.

EXAMPLE 7-10   GOING FOR A STROLL

When a father takes his daughter out for a spin in a stroller, he exerts a force $\vec{\mathbf{F}}$ with a magnitude of 44.0 N at an angle of 32.0° below the horizontal, as shown in the sketch. The stroller and child together have a mass $m = 22.7$ kg. **(a)** How much work does the father do on the stroller when he pushes it through a horizontal distance $d = 1.13$ m? **(b)** If the initial speed of the stroller is $v_i = 1.37$ m/s, what is its final speed? (Assume the stroller rolls with negligible friction.)

CONTINUED

PICTURE THE PROBLEM

Our sketch shows the direction of motion and the directions of each of the forces. The normal forces and the force due to gravity are vertical, whereas the displacement is horizontal. The force exerted by the father has a vertical component, $-F \sin 32.0°$, and a horizontal component, $F \cos 32.0°$, where $F = 44.0$ N.

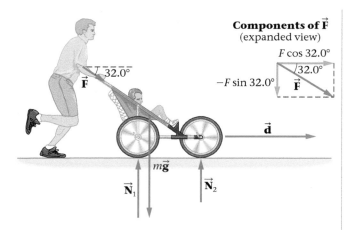

Components of $\vec{\mathbf{F}}$
(expanded view)

REASONING AND STRATEGY

a. The normal forces ($\vec{\mathbf{N}}_1$ and $\vec{\mathbf{N}}_2$) and the weight ($m\vec{\mathbf{g}}$) do no work because they are at right angles to the horizontal displacement. The force exerted by the father, however, has a horizontal component that does positive work on the stroller. Therefore, the total work is simply the work done by the father.

b. After calculating the work in part (a), we can find the final speed v_f by applying the work–energy theorem with $v_i = 1.37$ m/s.

Known   Force, $F = 44.0$ N; angle below horizontal, $\theta = 32.0°$; mass, $m = 22.7$ kg; distance, $d = 1.13$ m; initial speed, $v_i = 1.37$ m/s.

Unknown   (a) Work done by the father, $W_{father} = ?$ (b) Final speed, $v_f = ?$

SOLUTION

Part (a)

1. The work done by the father is the horizontal component of his force times the distance, $(F \cos \theta)d$. This is also the total work done on the stroller:

$$\begin{aligned} W_{father} &= (F \cos \theta)d \\ &= (44.0 \text{ N})(\cos 32.0°)(1.13 \text{ m}) \\ &= 42.2 \text{ J} = W_{total} \end{aligned}$$

Part (b)

2. Use the work–energy theorem to solve for the final speed:

$$W_{total} = \Delta K = \tfrac{1}{2}mv_f^2 - \tfrac{1}{2}mv_i^2$$

$$\tfrac{1}{2}mv_f^2 = W_{total} + \tfrac{1}{2}mv_i^2$$

$$v_f = \sqrt{\frac{2W_{total}}{m} + v_i^2}$$

3. Substitute numerical values to get the desired result:

$$v_f = \sqrt{\frac{2(42.2 \text{ J})}{22.7 \text{ kg}} + (1.37 \text{ m/s})^2}$$

$$= 2.37 \text{ m/s}$$

INSIGHT

Notice that the speed of the stroller increased by 1.00 m/s, from 1.37 m/s to 2.37 m/s. If the stroller had started from rest, would its speed increase from zero to 1.00 m/s? No. The work–energy theorem depends on the *square* of the speeds, and hence speeds don't simply add or subtract in an intuitive "linear" way. The final speed in this case is greater than 1.00 m/s, as we show in the following Practice Problem.

PRACTICE PROBLEM

Suppose the stroller starts at rest. What is its final speed in this case?   [**Answer:** $v_f = \sqrt{2W_{total}/m} = 1.93$ m/s, almost twice what simple "linear reasoning" would suggest]

Some related homework problems: Problem 29, Problem 62

PROBLEM-SOLVING NOTE

Be Careful About Linear Reasoning

Though some relationships are linear—if you *double* the mass, you *double* the kinetic energy—others are not. For example, if you *double* the speed, you *quadruple* the kinetic energy. Be careful not to jump to conclusions based on linear reasoning.

The final speeds in the previous Examples could have been found using Newton's laws and the constant-acceleration kinematics of Chapter 2, as indicated in the Insight in Example 7-9. The work–energy theorem provides an alternative method of calculation that is often much easier to apply than Newton's laws. We return to this point in Chapter 8.

CONCEPTUAL EXAMPLE 7-11   **COMPARE THE WORK**

(a) To accelerate a certain car from rest to the speed v requires the work W_1, as shown in the sketch. The work needed to accelerate the car from v to $2v$ is W_2. Which of the following statements is correct: $W_2 = W_1$, $W_2 = 2W_1$, $W_2 = 3W_1$, $W_2 = 4W_1$? **(b)** Which of the following is the *best explanation* for your prediction?

I.   The increase in speed is the same, so the work is also the same.

II. To double the speed requires double the work.

III. Kinetic energy depends on v^2, and hence it takes four times as much work to increase the speed to $2v$.

IV. Four times as much work is required to go from 0 to $2v$ as to go from 0 to v. Therefore, the work required to increase the speed from v to $2v$ is three times the original work.

REASONING AND DISCUSSION

A common misconception is to reason that because we increase the speed by the same amount in each case, the work required is the same. It is not, and the reason is that work depends on the speed squared rather than on the speed itself.

To see how this works, first calculate W_1, the work needed to go from rest to a speed v. From the work–energy theorem, with $v_i = 0$ and $v_f = v$, we find $W_1 = \frac{1}{2}mv_f^2 - \frac{1}{2}mv_i^2 = \frac{1}{2}mv^2$. Similarly, the work W_2 needed to go from $v_i = v$ to $v_f = 2v$ is $W_2 = \frac{1}{2}m(2v)^2 - \frac{1}{2}mv^2 = 3\left(\frac{1}{2}mv^2\right) = 3W_1$.

ANSWER

(a) The required work is $W_2 = 3W_1$. **(b)** The best explanation is IV.

Enhance Your Understanding   (Answers given at the end of the chapter)

2. An object has an initial kinetic energy of 100 J. A few minutes later, after a single external force has acted on the object, its kinetic energy is 200 J. Is the work done by the force positive, negative, or zero? Explain.

Section Review

• The kinetic energy of an object is one-half its mass times its velocity squared.

• The total work done on an object is equal to the change in its kinetic energy.

7-3 Work Done by a Variable Force

Thus far we have calculated work only for constant forces, yet most forces in nature vary with position. For example, the force exerted by a spring depends on how far the spring is stretched, and the force of gravity between planets depends on their separation. In this section we show how to calculate the work for a force that varies with position.

A Graphical Interpretation of Work   First, let's review briefly the case of a constant force, and develop a graphical interpretation of work. **FIGURE 7-8** shows a constant force plotted versus position, x. If the force acts in the positive x direction and moves an object a distance d, from x_1 to x_2, the work it does is $W = Fd = F(x_2 - x_1)$. Referring to the figure, we see that the work is equal to the shaded area[1] between the force line and the x axis.

Next, consider a force that has the value F_1 from $x = 0$ to $x = x_1$ and a different value F_2 from $x = x_1$ to $x = x_2$, as in **FIGURE 7-9 (a)**. The work in this case is the sum of the works done by F_1 and F_2. Therefore, $W = F_1 x_1 + F_2(x_2 - x_1)$. This, again, is the area between the force lines and the x axis. Clearly, this type of calculation can be extended to a force with any number of different values, as indicated in **FIGURE 7-9 (b)**.

If a force varies continuously with position, we can approximate it with a series of constant values that follow the shape of the curve, as shown in **FIGURE 7-10 (a)**. It follows that the work done by the continuous force is approximately equal to the area of the corresponding

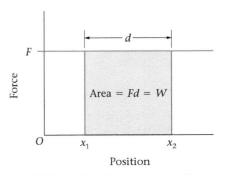

▲ **FIGURE 7-8 Graphical representation of the work done by a constant force**
A constant force F acting through a distance d does the work $W = Fd$. Note that Fd is also equal to the shaded area between the force line and the x axis.

---

[1] Usually, area has the dimensions of (length) $\times$ (length), or length^2. In this case, however, the vertical axis is force and the horizontal axis is distance. As a result, the dimensions of area are (force) $\times$ (distance), which in SI units is $N \cdot m = J$.

▼ **FIGURE 7-10 Work done by a continuously varying force (a)** A continuously varying force can be approximated by a series of constant values that follow the shape of the curve. **(b)** The work done by the continuous force is approximately equal to the area of the small rectangles corresponding to the constant values of force shown in part (a). **(c)** In the limit of an infinite number of vanishingly small rectangles, we see that the work done by the force is equal to the area between the force curve and the x axis.

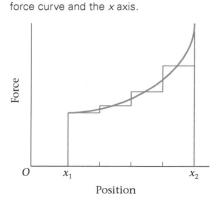

(a) Approximating a continuous force

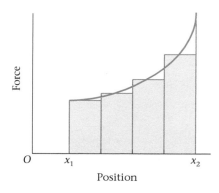

(b) Approximating the work done by a continuous force

(c) A better approximation

(a)

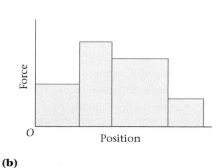

(b)

▲ **FIGURE 7-9 Work done by a nonconstant force (a)** A force with a value F_1 from 0 to x_1 and a value F_2 from x_1 to x_2 does the work $W = F_1 x_1 + F_2(x_2 - x_1)$. This is simply the combined area of the two shaded rectangles. **(b)** If a force takes on a number of different values, the work it does is still the total area between the force lines and the x axis, just as in part (a).

rectangles, as **FIGURE 7-10 (b)** shows. The approximation can be made better by using more rectangles, as illustrated in **FIGURE 7-10 (c)**. In the limit of an infinite number of vanishingly small rectangles, the area of the rectangles becomes identical to the area under the force curve. Hence this area is the work done by the continuous force. To summarize:

> The work done by a force in moving an object from one position to another is equal to the corresponding area between the force curve and the x axis.

Work for the Spring Force A case of particular interest is a spring. If we recall that the force exerted by a spring is given by $F_x = -kx$ (Section 6-2), it follows that the force we must exert to hold the spring at the position x is $+kx$. This is illustrated in **FIGURE 7-11**, where we also show that the corresponding force curve is a straight line extending from the origin. Therefore, the work we do in stretching a spring from $x = 0$ (equilibrium) to the general position x is the shaded triangular area shown in **FIGURE 7-12**. This area is equal to $\frac{1}{2}$(base)(height), where in this case the base is x and the height is kx. As a result, the work is $\frac{1}{2}(x)(kx) = \frac{1}{2}kx^2$. Similar reasoning shows that the work needed to compress a spring a distance x is also $\frac{1}{2}kx^2$. Therefore:

Work to Stretch or Compress a Spring a Distance x from Equilibrium

$$W = \frac{1}{2}kx^2$$ 7-8

SI unit: joule, J

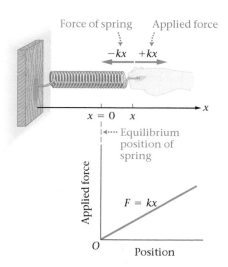

▲ **FIGURE 7-11 Stretching a spring** The force we must exert on a spring to stretch it a distance x is $+kx$. Thus, applied force versus position for a spring is a straight line of slope k.

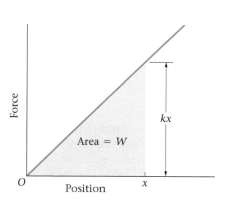

▲ **FIGURE 7-12 Work needed to stretch a spring a distance x** The work done is equal to the shaded area, which is a right triangle. The area of the triangle is $\frac{1}{2}(x)(kx) = \frac{1}{2}kx^2$.

We can get a feeling for the amount of work required to compress a typical spring in the following Exercise.

EXERCISE 7-12   SPRING WORK

(a) A toy snake has a spring with a force constant of 230 N/m. How much work is required to stretch this spring 2.5 cm? **(b)** The coiled suspension spring in a car has a force constant of 21,000 N/m. If 3.5 J of work is done to compress this spring, what is the compression distance?

REASONING AND SOLUTION
Work is related to the force constant and compression (or stretch) by $W = \frac{1}{2}kx^2$. We can apply this relationship to both questions. **(a)** Substitute $k = 230$ N/m and $x = 0.025$ m into $W = \frac{1}{2}kx^2$:

$$W = \frac{1}{2}kx^2 = \frac{1}{2}(230\,\text{N/m})(0.025\,\text{m})^2 = 0.072\,\text{J}$$

(b) Solve $W = \frac{1}{2}kx^2$ for x, then substitute $W = 3.5$ J and $k = 21,000$ N/m:

$$x = \sqrt{\frac{2W}{k}} = \sqrt{\frac{2(3.5\,\text{J})}{21,000\,\text{N/m}}} = 0.018\,\text{m}$$

The work done in compressing or expanding a spring varies with the second power of x, the displacement from equilibrium. The consequences of this dependence are explored throughout the rest of this section.

"Springs" in Other Contexts   Our results for a spring apply to more than just the classic case of a helical coil of wire. In fact, any flexible structure satisfies the relationships $F_x = -kx$ and $W = \frac{1}{2}kx^2$, given the appropriate value of the force constant, k, and small enough displacements, x. Several examples were mentioned in Section 6-2.

Here we consider an example from the field of nanotechnology—namely, the cantilevers used in **atomic-force microscopy** (AFM). As we show in Example 7-13, a typical atomic-force cantilever is basically a thin silicon bar about 250 μm in length, supported at one end like a diving board, with a sharp, hanging point at the other end. When the point is pulled across the surface of a material—like an old-fashioned phonograph needle in the groove of a record—individual atoms on the surface cause the point to move up and down, deflecting the cantilever. These deflections, which can be measured by reflecting a laser beam from the top of the cantilever, are then converted into an atomic-level picture of the surface, as shown in FIGURE 7-13.

A typical force constant for an AFM cantilever is on the order of 1 N/m, much smaller than the 100–500-N/m force constant of a common lab spring. The implications of this are discussed in the following Example.

▲ **FIGURE 7-13** Human chromosomes, as imaged by an atomic-force microscope.

EXAMPLE 7-13   FLEXING AN AFM CANTILEVER

The work required to deflect a typical AFM cantilever by 0.10 nm is 1.2×10^{-20} J. **(a)** What is the force constant of the cantilever, treating it as an ideal spring? **(b)** How much work is required to increase the deflection of the cantilever from 0.10 nm to 0.20 nm?

PICTURE THE PROBLEM
The upper sketch shows the cantilever and its sharp point being dragged across the surface of a material. In the lower sketch, we show an exaggerated view of the cantilever's deflection, and indicate that it is equivalent to the stretch of an "effective" ideal spring with a force constant k.

REASONING AND STRATEGY

a. Given that $W = 1.2 \times 10^{-20}$ J for a deflection of $x = 0.10$ nm, we can find the effective force constant k using $W = \frac{1}{2}kx^2$.

b. To find the work required to deflect from $x = 0.10$ nm to $x = 0.20$ nm, $W_{1\rightarrow2}$, we calculate the work to deflect from $x = 0$ to $x = 0.20$ nm, $W_{0\rightarrow2}$, and then subtract the work needed to deflect from $x = 0$ to $x = 0.10$ nm, $W_{0\rightarrow1}$. (Notice that we *cannot* simply assume that the work to go from $x = 0.10$ nm to $x = 0.20$ nm is the same as the work to go from $x = 0$ to $x = 0.10$ nm.)

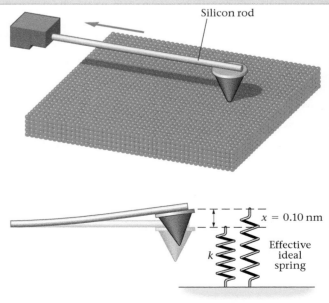

Silicon rod

$x = 0.10$ nm

k Effective ideal spring

CONTINUED

Known   Deflection distance, $x = 0.10 \text{ nm} = 0.10 \times 10^{-9} \text{ m}$; work to deflect, $W = 1.2 \times 10^{-20} \text{ J}$.
Unknown   (a) Force constant, $k = ?$ (b) Work to increase deflection, $W_{1 \to 2} = ?$

SOLUTION

Part (a)

1. Solve $W = \frac{1}{2}kx^2$ for the force constant k:

$$k = \frac{2W}{x^2} = \frac{2(1.2 \times 10^{-20} \text{ J})}{(0.10 \times 10^{-9} \text{ m})^2} = 2.4 \text{ N/m}$$

Part (b)

2. First, calculate the work needed to deflect the cantilever from $x = 0$ to $x = 0.20 \text{ nm}$:

$$W_{0 \to 2} = \frac{1}{2}kx^2$$
$$= \frac{1}{2}(2.4 \text{ N/m})(0.2 \times 10^{-9} \text{ m})^2 = 4.8 \times 10^{-20} \text{ J}$$

3. Subtract from the above result the work to deflect from $x = 0$ to $x = 0.10 \text{ nm}$, which the problem statement gives as $1.2 \times 10^{-20} \text{ J}$:

$$W_{1 \to 2} = W_{0 \to 2} - W_{0 \to 1}$$
$$= 4.8 \times 10^{-20} \text{ J} - 1.2 \times 10^{-20} \text{ J} = 3.6 \times 10^{-20} \text{ J}$$

INSIGHT

Our results show that more energy is needed to deflect the cantilever the second 0.10 nm than to deflect it the first 0.10 nm. Why? The reason is that the force of the cantilever increases with distance; thus, the average force over the second 0.10 nm is greater than the average force over the first 0.10 nm. In fact, we can see from the graph that the average force between 0.10 nm and 0.20 nm (0.36 nN) is three times the average force between 0 and 0.10 nm (0.12 nN). It follows that the work required for the second 0.10 nm is three times the work required for the first 0.10 nm.

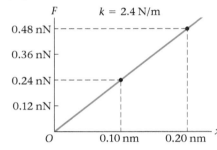

PRACTICE PROBLEM — PREDICT/CALCULATE

A second cantilever has half the force constant of the cantilever in this Example. (a) Is the work required to deflect the second cantilever by 0.20 nm greater than, less than, or equal to the work required to deflect the cantilever in this Example by 0.10 nm? Explain. (b) Determine the work required to deflect the second cantilever 0.20 nm. [**Answer:** (a) Halving the force constant halves the work, but doubling the deflection quadruples the work. The net effect is that the work increases by a factor of two. (b) The work required is $2.4 \times 10^{-20} \text{ J}$.]

Some related homework problems: Problem 31, Problem 36

BIO Another example of the work required to stretch a flexible structure is the work done inside your eye. The human eye can *accommodate*, or focus on objects at different distances, by using *ciliary muscles* to alter the shape of the flexible lens behind the pupil of the eye. (The physiology and optical properties of the eye are explored in more detail in Chapter 27.) The shorter the distance you attempt to focus your eye, the harder your ciliary muscles must work to change the shape of the lens. This is why long periods of viewing close-up objects, such as a book or a computer screen, can lead to *asthenopia*, or eye strain.

Using Average Force to Calculate Work   An equivalent way to calculate the work for a variable force is to multiply the average force, F_{av}, by the distance, d:

$$W = F_{av}d \qquad \text{7-9}$$

For a spring that is stretched a distance x from equilibrium, the force varies linearly from 0 to kx. Thus, the average force is $F_{av} = \frac{1}{2}kx$, as indicated in **FIGURE 7-14**. Therefore, the work is

$$W = \frac{1}{2}kx(x) = \frac{1}{2}kx^2$$

As expected, our result agrees with Equation 7-8.

Finally, when you stretch or compress a spring from its equilibrium position, the work you do is always positive. The work done *by* a spring, however, may be positive or negative, depending on the situation. For example, consider a block sliding to the right with an initial speed v_0 on a smooth, horizontal surface, as shown in **FIGURE 7-15 (a)**. When the block begins to compress the spring, as in **FIGURE 7-15 (b)**, the spring exerts a force

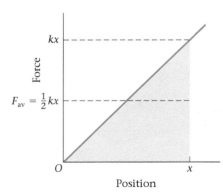

▲ **FIGURE 7-14 Work done in stretching a spring: average force** The average force to stretch a spring from $x = 0$ to x is $F_{av} = \frac{1}{2}kx$, and the work done is $W = F_{av}d = \frac{1}{2}kx(x) = \frac{1}{2}kx^2$.

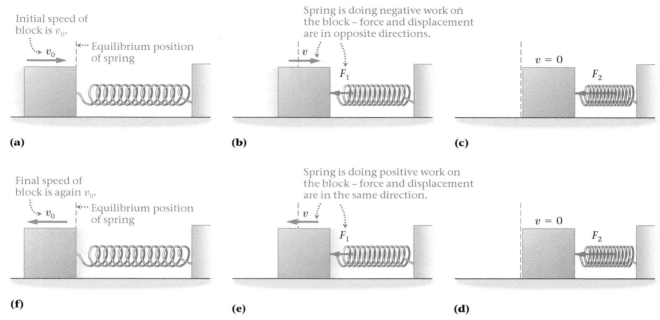

▲ **FIGURE 7-15 The work done by a spring can be positive or negative** **(a)** A block slides to the right on a frictionless surface with a speed v_0 until it encounters a spring. **(b)** The spring now exerts a force to the left—opposite to the block's motion—and hence it does negative work on the block. This causes the block's speed to decrease. **(c)** The negative work done by the spring eventually is equal in magnitude to the block's initial kinetic energy, at which point the block comes to rest momentarily. As the spring expands, **(d)** and **(e)**, it does positive work on the block and increases its speed. **(f)** When the block leaves the spring, its speed is again equal to v_0.

on the block to the left—that is, opposite to the block's direction of motion. As a result, the spring does *negative* work on the block, which causes the block's speed to decrease. Eventually the negative work done by the spring, $W = -\frac{1}{2}kx^2$, is equal in magnitude to the initial kinetic energy of the block. At this point, **FIGURE 7-15 (c)**, the block comes to rest momentarily, and $W = \Delta K = K_f - K_i = 0 - K_i = -K_i = -\frac{1}{2}mv_0^2 = -\frac{1}{2}kx^2$.

After the block comes to rest, the spring expands back to its equilibrium position, as we see in **FIGURE 7-15 (d)–(f)**. During this expansion the force exerted by the spring is in the same direction as the block's motion, and hence it does *positive* work in the amount $W = \frac{1}{2}kx^2$. As a result, the block leaves the spring with the same speed it had initially.

QUICK EXAMPLE 7-14 **A BLOCK COMPRESSES A SPRING**

A block with a mass of 1.5 kg and an initial speed of $v_i = 2.2$ m/s slides on a frictionless, horizontal surface. The block comes into contact with a spring that is in its equilibrium position, and compresses it until the block comes to rest momentarily. Find the maximum compression of the spring, assuming its force constant is 475 N/m.

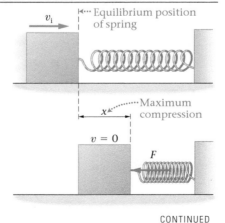

REASONING AND SOLUTION

Our sketch shows the block just as it comes into contact with the spring with an initial speed of $v_i = 2.2$ m/s. We also show the spring at maximum compression, x, when the block is momentarily at rest. As the block comes to rest, the force of the spring is opposite to the

CONTINUED

block's displacement, and hence the work done by the spring is negative. The (negative) work done by the spring, $W = -\frac{1}{2}kx^2$, is equal to the (negative) change in kinetic energy, ΔK. Thus, we can set $-\frac{1}{2}kx^2 = \Delta K$ and solve for x to find the maximum compression.

1. Calculate the initial and final kinetic energies of the block:

$$K_i = \tfrac{1}{2}mv_i^2$$
$$= \tfrac{1}{2}(1.5 \text{ kg})(2.2 \text{ m/s})^2 = 3.6 \text{ J}$$
$$K_f = 0$$

2. Calculate the change in kinetic energy of the block:

$$\Delta K = K_f - K_i = -3.6 \text{ J}$$

3. Set the negative work done by the spring equal to the negative change in kinetic energy of the block:

$$-\tfrac{1}{2}kx^2 = \Delta K = -3.6 \text{ J}$$

4. Solve for the compression, x, and substitute numerical values:

$$x = \sqrt{\frac{-2\Delta K}{k}}$$
$$= \sqrt{\frac{-2(-3.6 \text{ J})}{475 \text{ N/m}}} = 0.12 \text{ m}$$

When the spring brings the block to rest, the kinetic energy it had initially is not lost—it is stored in the spring itself and can be released later. We discuss situations like this, and their connection with energy conservation, in Chapter 8.

Enhance Your Understanding   (Answers given at the end of the chapter)

3. As an object moves along the positive x axis, the force shown in **FIGURE 7-16** acts on it. Is the work done by the force from $x = 0$ to $x = 4$ m greater than, less than, or equal to the work done by the force from $x = 5$ m to $x = 9$ m? Explain.

▲ **FIGURE 7-16** Enhance Your Understanding 3.

Section Review

- Work is equal to the area between a force curve and the x axis.

- Work is also equal to the average force times displacement.

7-4 Power

Power is a measure of how *quickly* work is done. To be precise, suppose the work W is performed in the time t. The average power delivered during this time is defined as follows:

Definition of Average Power, P

$$P = \frac{W}{t}$$

7-10

SI unit: J/s = watt, W

For simplicity of notation we drop the usual subscript av for an average quantity and simply understand that the power P refers to an average power unless stated otherwise.

The Dimensions of Power The dimensions of power are joules (work) per second (time). We define one joule per second to be a watt (W), after James Watt (1736–1819), the Scottish engineer and inventor who played a key role in the development of practical steam engines:

$$1 \text{ watt} = 1 \text{ W} = 1 \text{ J/s} \qquad 7\text{-}11$$

Of course, the watt is the unit of power used to rate the output of lightbulbs. Another common unit of power is the horsepower (hp), which is used to rate the output of car engines. It is defined as follows:

$$1 \text{ horsepower} = 1 \text{ hp} = 746 \text{ W} \qquad 7\text{-}12$$

Though it sounds like a horse should be able to produce one horsepower, in fact, a horse can generate only about 2/3 hp for sustained periods. The reason for the discrepancy is that when James Watt defined the horsepower—as a way to characterize the output of his steam engines—he purposely chose a unit that was overly generous to the horse, so that potential investors couldn't complain he was overstating the capability of his engines.

Human Power To get a feel for the magnitude of the watt and the horsepower, consider the power you might generate when walking up a flight of stairs. Suppose, for example, that an 80.0-kg person walks up a flight of stairs in 20.0 s, and that the altitude gain is 12.0 ft (3.66 m). Referring to Example 7-3 and Conceptual Example 7-4, we find that the work done by the person is $W = mgh = (80.0 \text{ kg})(9.81 \text{ m/s}^2)(3.66 \text{ m}) = 2870 \text{ J}$. To find the power, we simply divide the work by the time: $P = W/t = (2870 \text{ J})/(20.0 \text{ s}) = 144 \text{ W} = 0.193 \text{ hp}$. Thus, a leisurely stroll up the stairs requires about 1/5 hp or 150 W. Similarly, the power produced by a sprinter bolting out of the starting blocks is about 1 hp, and the greatest power most people can produce for sustained periods of time is roughly 1/3 to 1/2 hp. Further examples of power are given in Table 7-3.

Human-powered flight is a feat just barely within our capabilities, since the most efficient human-powered airplanes require a steady power output of about 1/3 hp. In 1979 the *Gossamer Albatross* became the first (and so far the only) human-powered aircraft to fly across the English Channel. This 22.25-mile flight—from Folkestone, England, to Cap Gris-Nez, France—took 2 hours 49 minutes and required a total energy output roughly equivalent to climbing to the top of the Empire State Building 10 times. The *Gossamer Albatross* is shown in midflight in FIGURE 7-17.

RWP Power output is also an important factor in the performance of a car. For example, suppose it takes a certain amount of work, W, to accelerate a car from 0 to 60 mi/h. If the average power provided by the engine is P, then according to Equation 7-10, the amount of time required to reach 60 mi/h is $t = W/P$. Clearly, the greater the power P, the less the time required to accelerate. Thus, in a loose way of speaking, we can say that of "how fast it can go fast."

Big Idea 3 Power is the rate at which work is done. The more work that is done in a shorter time, the greater the power.

TABLE 7-3 Typical Values of Power

Source	Approximate power (W)
Hoover Dam	1.34×10^9
Car moving at 40 mi/h	7×10^4
Home stove	1.2×10^4
Sunlight falling on one square meter	1380
Refrigerator	615
Television	200
Person walking up stairs	150
Human brain	20

▲ FIGURE 7-17 The *Gossamer Albatross* Twice the aircraft touched the surface of the water, but the pilot was able to maintain control.

EXAMPLE 7-15 PASSING FANCY

To pass a slow-moving truck, you want your fancy 1.30×10^3-kg car to accelerate from 13.4 m/s (30.0 mi/h) to 17.9 m/s (40.0 mi/h) in 3.00 s. What is the minimum power required for this pass?

PICTURE THE PROBLEM
Our sketch shows the car accelerating from an initial speed of $v_i = 13.4$ m/s to a final speed of $v_f = 17.9$ m/s. We assume the road is level, so that no work is done against gravity, and that friction and air resistance may be ignored.

CONTINUED

REASONING AND STRATEGY

Power is work divided by time, and work is equal to the change in kinetic energy as the car accelerates. We can determine the change in kinetic energy from the given mass of the car and its initial and final speeds. With this information at hand, we can determine the power with the relationship $P = W/t = \Delta K/t$.

Known Mass of car, $m = 1.30 \times 10^3$ kg; initial speed, $v_i = 13.4$ m/s; final speed, $v_f = 17.9$ m/s; time, $t = 3.00$ s.

Unknown Minimum power, $P = ?$

SOLUTION

1. First, calculate the change in kinetic energy:

$$\Delta K = \tfrac{1}{2}mv_f^2 - \tfrac{1}{2}mv_i^2 = \tfrac{1}{2}(1.30 \times 10^3 \text{ kg})(17.9 \text{ m/s})^2$$
$$-\tfrac{1}{2}(1.30 \times 10^3 \text{ kg})(13.4 \text{ m/s})^2$$
$$= 9.16 \times 10^4 \text{ J}$$

2. Divide by time to find the minimum power (the actual power would have to be greater to overcome frictional losses):

$$P = \frac{W}{t} = \frac{\Delta K}{t} = \frac{9.16 \times 10^4 \text{ J}}{3.00 \text{ s}} = 3.05 \times 10^4 \text{ W} = 40.9 \text{ hp}$$

INSIGHT

Suppose that your fancy car continues to produce the same 3.05×10^4 W of power as it accelerates from $v = 17.9$ m/s (40.0 mi/h) to $v = 22.4$ ms (50.0 mi/h). Is the time required more than, less than, or equal to 3.00 s? It will take more than 3.00 s. The reason is that ΔK is greater for a change in speed from 40.0 mi/h to 50.0 mi/h than for a change in speed from 30.0 mi/h to 40.0 mi/h, because K depends on speed squared. Since ΔK is greater, the time $t = \Delta K/P$ is also greater.

PRACTICE PROBLEM

Find the time required to accelerate from 40.0 mi/h to 50.0 mi/h with 3.05×10^4 W of power.
[**Answer:** First, $\Delta K = 1.18 \times 10^5$ J. Next, $P = \Delta K/t$ can be solved for time to give $t = \Delta K/P$. Thus, $t = 3.87$ s.]

Some related homework problems: Problem 43, Problem 56, Problem 72

Power and Speed Finally, consider a system in which a car, or some other object, is moving with a constant speed v. For example, a car might be traveling uphill on a road inclined at an angle θ above the horizontal. To maintain a constant speed, the engine must exert a constant force F equal to the combined effects of friction, gravity, and air resistance, as indicated in **FIGURE 7-18**. Now, as the car travels a distance d, the work done by the engine is $W = Fd$, and the power it delivers is

$$P = \frac{W}{t} = \frac{Fd}{t}$$

▼ **FIGURE 7-18 Driving up a hill** A car traveling uphill at constant speed requires a constant force, F, of magnitude $mg \sin \theta + F_{\text{air res}} + F_{\text{friction}}$ applied in the direction of motion.

Since the car has a constant speed, $v = d/t$, it follows that

$$P = \frac{Fd}{t} = F\left(\frac{d}{t}\right) = Fv \qquad\qquad \text{7-13}$$

Notice that power is directly proportional to both the force and the speed. For example, suppose you push a heavy shopping cart with a force F. You produce twice as much power when you push at 2 m/s than when you push at 1 m/s, even though you are pushing no harder. It's just that the amount of work you do in a given time period is doubled.

EXERCISE 7-16 UPHILL SPEED

It takes a force of 1450 N to keep a 1300-kg car moving with constant speed up a slope of 5.25°. If the engine delivers 4.10×10^4 W (55 hp) to the drive wheels, what is the maximum speed of the car?

REASONING AND SOLUTION
Solve the power relationship, $P = Fv$, for the speed v, and substitute $P = 4.10 \times 10^4$ W and $F = 1450$ N:

$$v = \frac{P}{F} = \frac{4.10 \times 10^4 \text{ W}}{1450 \text{ N}} = 28.3 \text{ m/s}$$

This is approximately 63 mi/h. The power delivered by the engine overcomes friction and air resistance, and also provides the force needed to lift the car up the hill.

Enhance Your Understanding   (Answers given at the end of the chapter)

4. Engine A generates the power P and operates for 10 s; engine B generates the power $2P$ and operates for 5 s; engine C generates the power $2P$ and operates for 20 s; and engine D generates the power $P/2$ and operates for 40 s. Rank the engines in order of increasing work done. Indicate ties where appropriate.

Section Review

•   Power is the rate at which work is done.

•   A force F pushing or pulling an object at a speed v produces the power $P = Fv$.

CHAPTER 7 REVIEW

CHAPTER SUMMARY

7-1   WORK DONE BY A CONSTANT FORCE

A force exerted through a distance performs mechanical work.

Force in Direction of Motion
In this, the simplest case, work is force times distance:

$$W = Fd \qquad\qquad \text{7-1}$$

Force at an Angle θ to Motion
Work is the component of force in the direction of motion, $F \cos \theta$, times distance, d:

$$W = (F \cos \theta)d = Fd \cos \theta \qquad\qquad \text{7-3}$$

Negative and Total Work
Work is negative if the force opposes the motion. If more than one force does work, the total work is the sum of the works done by each force separately:

$$W_{\text{total}} = W_1 + W_2 + W_3 + \cdots \qquad\qquad \text{7-4}$$

Equivalently, sum the forces first to find F_{total}, then

$$W_{\text{total}} = (F_{\text{total}} \cos \theta)d = F_{\text{total}}\, d \cos \theta \qquad\qquad \text{7-5}$$

Units

The SI unit of work and energy is the joule, J:

$$1\,\text{J} = 1\,\text{N}\cdot\text{m} \qquad \text{7-2}$$

7-2 KINETIC ENERGY AND THE WORK–ENERGY THEOREM

Kinetic energy is one-half mass times speed squared:

$$K = \tfrac{1}{2}mv^2 \qquad \text{7-7}$$

Total work is equal to the change in kinetic energy:

$$W_{\text{total}} = \Delta K = \tfrac{1}{2}mv_{\text{f}}^2 - \tfrac{1}{2}mv_{\text{i}}^2 \qquad \text{7-6}$$

7-3 WORK DONE BY A VARIABLE FORCE

Work is equal to the area between the force curve and the displacement on the *x* axis. For the case of a spring force, the work to stretch or compress a distance *x* from equilibrium is

$$W = \tfrac{1}{2}kx^2 \qquad \text{7-8}$$

7-4 POWER

Average power is the work divided by the time required to do the work:

$$P = \frac{W}{t} \qquad \text{7-10}$$

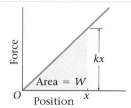

Equivalently, power is force times speed:

$$P = Fv \qquad \text{7-13}$$

Units

The SI unit of power is the watt, W:

$$1\,\text{W} = 1\,\text{J/s} \qquad \text{7-11}$$

$$746\,\text{W} = 1\,\text{hp} \qquad \text{7-12}$$

ANSWERS TO ENHANCE YOUR UNDERSTANDING QUESTIONS

1. Work done by $f_{\text{k}} <$ work done by $N =$ work done by $mg <$ work done by F.

2. The kinetic energy of the object has increased, and hence the total work done on it is positive.

3. The work done from $x = 0$ to $x = 4$ m is greater than the work done from $x = 5$ m to $x = 9$ m because the blue area is larger than the green area.

4. A = B < D < C.

CONCEPTUAL QUESTIONS

For instructor-assigned homework, go to www.masteringphysics.com.

(Answers to odd-numbered Conceptual Questions can be found in the back of the book.)

1. Is it possible to do work on an object that remains at rest?

2. A friend makes the statement, "Only the total force acting on an object can do work." Is this statement true or false? If it is true, state why; if it is false, give a counterexample.

3. A friend makes the statement, "A force that is always perpendicular to the velocity of a particle does no work on the particle." Is this statement true or false? If it is true, state why; if it is false, give a counterexample.

4. The net work done on a certain object is zero. What can you say about its speed?

5. Give an example of a frictional force doing positive work.

6. A ski boat moves with constant velocity. Is the net force acting on the boat doing work? Explain.

7. A package rests on the floor of an elevator that is rising with constant speed. The elevator exerts an upward normal force on the package, and hence does positive work on it. Why doesn't the kinetic energy of the package increase?

8. An object moves with constant velocity. Is it safe to conclude that no force acts on the object? Why, or why not?

9. Engine 1 does twice the work of engine 2. Is it correct to conclude that engine 1 produces twice as much power as engine 2? Explain.

10. Engine 1 produces twice the power of engine 2. Is it correct to conclude that engine 1 does twice as much work as engine 2? Explain.

PROBLEMS AND CONCEPTUAL EXERCISES

Answers to odd-numbered Problems and Conceptual Exercises can be found in the back of the book. **BIO** *identifies problems of biological or medical interest;* **CE** *indicates a conceptual exercise.* Predict/Explain *problems ask for two responses: (a) your prediction of a physical outcome, and (b) the best explanation among three provided; and* Predict/Calculate *problems ask for a prediction of a physical outcome, based on fundamental physics concepts, and follow that with a numerical calculation to verify the prediction. On all problems, bullets (•, ••, •••) indicate the level of difficulty.*

SECTION 7-1 WORK DONE BY A CONSTANT FORCE

1. • **CE** A pendulum bob swings from point I to point II along the circular arc indicated in **FIGURE 7-19**. **(a)** Is the work done on the bob by gravity positive, negative, or zero? Explain. **(b)** Is the work done on the bob by the string positive, negative, or zero? Explain.

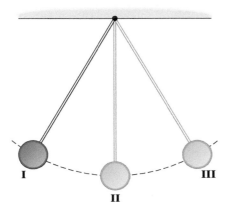

FIGURE 7-19 Problems 1 and 2

2. • **CE** A pendulum bob swings from point II to point III along the circular arc indicated in Figure 7-19. **(a)** Is the work done on the bob by gravity positive, negative, or zero? Explain. **(b)** Is the work done on the bob by the string positive, negative, or zero? Explain.

3. • A farmhand pushes a 26-kg bale of hay 3.9 m across the floor of a barn. If she exerts a horizontal force of 88 N on the hay, how much work has she done?

4. • Children in a tree house lift a small dog in a basket 3.85 m up to their house. If it takes 201 J of work to do this, what is the combined mass of the dog and basket?

5. • Early one October, you go to a pumpkin patch to select your Halloween pumpkin. You lift the 3.2-kg pumpkin to a height of 1.2 m, then carry it 50.0 m (on level ground) to the check-out stand. **(a)** Calculate the work you do on the pumpkin as you lift it from the ground. **(b)** How much work do you do on the pumpkin as you carry it from the field?

6. • The coefficient of kinetic friction between a suitcase and the floor is 0.272. If the suitcase has a mass of 71.5 kg, how far can it be pushed across the level floor with 642 J of work?

7. • **BIO Peristaltic Work** The human small intestine moves food along its 6.9-m length by means of *peristalsis*, a contraction and relaxation of muscles that propagate in a wave along the digestive tract. If the average force exerted by peristalsis is 0.18 N, how much work does the small intestine do to move food along its entire length?

8. •• Predict/Calculate A tow rope, parallel to the water, pulls a water skier directly behind the boat with constant velocity for a distance of 72 m before the skier falls. The tension in the rope is 140 N. **(a)** Is the work done on the boat by the rope positive, negative, or zero? Explain. **(b)** Calculate the work done by the rope on the boat.

9. •• A child pulls a friend in a little red wagon with constant speed. If the child pulls with a force of 16 N for 10.0 m, and the handle of the wagon is inclined at an angle of 25° above the horizontal, how much work does the child do on the wagon?

10. •• A 57-kg packing crate is pulled with constant speed across a rough floor with a rope that is at an angle of 37° above the horizontal. If the tension in the rope is 142 N, how much work is done on the crate to move it 6.1 m?

11. •• Predict/Calculate To clean a floor, a janitor pushes on a mop handle with a force of 50.0 N. **(a)** If the mop handle is at an angle of 55° above the horizontal, how much work is required to push the mop 0.50 m? **(b)** If the angle the mop handle makes with the horizontal is increased to 65°, does the work done by the janitor increase, decrease, or stay the same? Explain.

12. •• A small plane tows a glider at constant speed and altitude. If the plane does 2.00×10^5 J of work to tow the glider 145 m and the tension in the tow rope is 2560 N, what is the angle between the tow rope and the horizontal?

13. •• As a snowboarder descends a mountain slope, gravity, the normal force, and friction combine to exert a net force of 320 N in the downhill direction. The net force does 4700 J of work on her during one segment of her descent. If her path cuts across the face of the mountain at a 65° angle from the direction of the net force, over what distance does she travel?

14. •• A young woman on a skateboard is pulled by a rope attached to a bicycle. The velocity of the skateboarder is $\vec{v} = (4.1 \text{ m/s})\hat{x}$ and the force exerted on her by the rope is $\vec{F} = (17 \text{ N})\hat{x} + (12 \text{ N})\hat{y}$. **(a)** Find the work done on the skateboarder by the rope in 25 seconds. **(b)** Assuming the velocity of the bike is the same as that of the skateboarder, find the work the rope does on the bicycle in 25 seconds.

15. •• To keep her dog from running away while she talks to a friend, Susan pulls gently on the dog's leash with a constant force given by $\vec{F} = (2.2 \text{ N})\hat{x} + (1.1 \text{ N})\hat{y}$. How much work does she do on the dog if its displacement is **(a)** $\vec{d} = (0.25 \text{ m})\hat{x}$, **(b)** $\vec{d} = (0.25 \text{ m})\hat{y}$, or **(c)** $\vec{d} = (-0.50 \text{ m})\hat{x} + (-0.25 \text{ m})\hat{y}$?

16. •• Water skiers often ride to one side of the center line of a boat, as shown in **FIGURE 7-20**. In this case, the ski boat is traveling at 12 m/s and the tension in the rope is 91 N. If the boat does 3800 J of work on the skier in 50.0 m, what is the angle θ between the tow rope and the center line of the boat?

FIGURE 7-20 Problems 16 and 62

SECTION 7-2 KINETIC ENERGY AND THE WORK–ENERGY THEOREM

17. • **CE** A pitcher throws a ball at 90 mi/h and the catcher stops it in her glove. **(a)** Is the work done on the ball by the pitcher positive, negative, or zero? Explain. **(b)** Is the work done on the ball by the catcher positive, negative, or zero? Explain.

18. • How much work is needed for a 73-kg runner to accelerate from rest to 7.7 m/s?

19. • **Skylab's Reentry** When Skylab reentered the Earth's atmosphere on July 11, 1979, it broke into a myriad of pieces. One of the largest fragments was a 1770-kg lead-lined film vault, and it landed with an estimated speed of 120 m/s. What was the kinetic energy of the film vault when it landed?

20. • Predict/Calculate A 9.50-g bullet has a speed of 1.30 km/s. **(a)** What is its kinetic energy in joules? **(b)** What is the bullet's kinetic energy if its speed is halved? **(c)** If its speed is doubled?

21. •• The energy required to increase the speed of a certain car from 18 m/s to 24 m/s is 190 kJ. **(a)** What is the mass of the car? **(b)** At what speed is the car's kinetic energy equal to 190 kJ?

22. •• **CE** Predict/Explain The work W_0 accelerates a car from 0 to 50 km/h. **(a)** Is the work required to accelerate the car from 50 km/h to 150 km/h equal to $2W_0$, $3W_0$, $8W_0$, or $9W_0$? **(b)** Choose the *best explanation* from among the following:
    I. The work to accelerate the car depends on the speed squared.
    II. The final speed is three times the speed that was produced by the work W_0.
    III. The increase in speed from 50 km/h to 150 km/h is twice the increase in speed from 0 to 50 km/h.

23. •• **CE** Car A has a mass m and a speed v, car B has a mass $m/2$ and a speed $3v$, car C has a mass $3m$ and a speed $v/2$, and car D has a mass $4m$ and a speed $v/2$. Rank the cars in order of increasing kinetic energy. Indicate ties where appropriate.

24. •• Predict/Calculate A 0.14-kg pinecone falls 16 m to the ground, where it lands with a speed of 13 m/s. **(a)** With what speed would the pinecone have landed if there had been no air resistance? **(b)** Did air resistance do positive work, negative work, or zero work on the pinecone? Explain.

25. •• In the previous problem, **(a)** how much work was done on the pinecone by air resistance? **(b)** What was the average force of air resistance exerted on the pinecone?

26. •• At $t = 1.0$ s, a 0.55-kg object is falling with a speed of 6.5 m/s. At $t = 2.0$ s, it has a kinetic energy of 38 J. **(a)** What is the kinetic energy of the object at $t = 1.0$ s? **(b)** What is the speed of the object at $t = 2.0$ s? **(c)** How much work was done on the object between $t = 1.0$ s and $t = 2.0$ s?

27. •• After hitting a long fly ball that goes over the right fielder's head and lands in the outfield, the batter decides to keep going past second base and try for third base. The 62.0-kg player begins sliding 3.40 m from the base with a speed of 4.35 m/s. If the player comes to rest at third base, **(a)** how much work was done on the player by friction? **(b)** What was the coefficient of kinetic friction between the player and the ground?

28. •• Predict/Calculate A 1100-kg car coasts on a horizontal road with a speed of 19 m/s. After crossing an unpaved, sandy stretch of road 32 m long, its speed decreases to 12 m/s. **(a)** Was the net work done on the car positive, negative, or zero? Explain. **(b)** Find the magnitude of the average net force on the car in the sandy section.

29. •• A 65-kg bicyclist rides his 8.8-kg bicycle with a speed of 14 m/s. **(a)** How much work must be done by the brakes to bring the bike and rider to a stop? **(b)** How far does the bicycle travel if it takes 4.0 s to come to rest? **(c)** What is the magnitude of the braking force?

SECTION 7-3 WORK DONE BY A VARIABLE FORCE

30. • **CE** A block of mass m and speed v collides with a spring, compressing it a distance Δx. What is the compression of the spring if the force constant of the spring is increased by a factor of four?

31. • A spring with a force constant of 3.5×10^4 N/m is initially at its equilibrium length. **(a)** How much work must you do to stretch the spring 0.050 m? **(b)** How much work must you do to compress it 0.050 m?

32. • Initially sliding with a speed of 4.1 m/s, a 1.5-kg block collides with a spring and compresses it 0.27 m before coming to rest. What is the force constant of the spring?

33. • The force shown in FIGURE 7-21 moves an object from $x = 0$ to $x = 0.75$ m. **(a)** How much work is done by the force? **(b)** How much work is done by the force if the object moves from $x = 0.15$ m to $x = 0.60$ m?

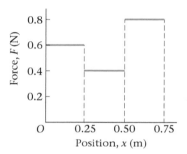

FIGURE 7-21 Problem 33

34. • An object is acted on by the force shown in FIGURE 7-22. What is the final position of the object if its initial position is $x = 0.40$ m and the work done on it is equal to **(a)** 0.21 J, or **(b)** −0.19 J?

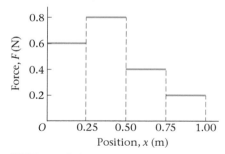

FIGURE 7-22 Problems 34 and 37

35. •• To compress spring 1 by 0.20 m takes 150 J of work. Stretching spring 2 by 0.30 m requires 210 J of work. Which spring is stiffer?

36. •• Predict/Calculate It takes 180 J of work to compress a certain spring 0.15 m. **(a)** What is the force constant of this spring? **(b)** To compress the spring an additional 0.15 m, does it take 180 J, more than 180 J, or less than 180 J? Verify your answer with a calculation.

37. •• The force shown in Figure 7-22 acts on a 1.3-kg object whose initial speed is 0.35 m/s and initial position is $x = 0.27$ m. **(a)** Find

the speed of the object when it is at the location $x = 0.99$ m. **(b)** At what location would the object's speed be 0.25 m/s?

38. ••• A block is acted on by a force that varies as $(2.0 \times 10^4 \text{ N/m})x$ for $0 \le x \le 0.21$ m, and then remains constant at 4200 N for larger x. How much work does the force do on the block in moving it **(a)** from $x = 0$ to $x = 0.30$ m, or **(b)** from $x = 0.10$ m to $x = 0.40$ m?

SECTION 7-4 POWER

39. • **CE** Force F_1 does 5 J of work in 10 seconds, force F_2 does 3 J of work in 5 seconds, force F_3 does 6 J of work in 18 seconds, and force F_4 does 25 J of work in 125 seconds. Rank these forces in order of increasing power they produce. Indicate ties where appropriate.

40. • **BIO Climbing the Empire State Building** A new record for running the stairs of the Empire State Building was set on February 4, 2003. The 86 flights, with a total of 1576 steps, was run in 9 minutes and 33 seconds. If the height gain of each step was 0.20 m, and the mass of the runner was 70.0 kg, what was his average power output during the climb? Give your answer in both watts and horsepower.

41. • Calculate the power output of a 14-mg fly as it walks straight up a windowpane at 2.3 cm/s.

42. • An ice cube is placed in a microwave oven. Suppose the oven delivers 105 W of power to the ice cube and that it takes 32,200 J to melt it. How much time does it take for the ice cube to melt?

43. • Your car produces about 34 kW of power to maintain a constant speed of 31 m/s on the highway. What average force does the engine exert?

44. • You raise a bucket of water from the bottom of a deep well. If your power output is 108 W, and the mass of the bucket and the water in it is 5.00 kg, with what speed can you raise the bucket? Ignore the weight of the rope.

45. •• **BIO Salmon Migration** As Chinook salmon swim upstream in the Yukon River to spawn, they must travel about 3000 km in about 90 days, and in the process each fish must do about 1.7×10^6 J of work. **(a)** What is the average power output of a Chinook salmon as it swims upstream? **(b)** What average force does the salmon exert as it swims?

46. •• In order to keep a leaking ship from sinking, it is necessary to pump 12.0 lb of water each second from below deck up a height of 2.00 m and over the side. What is the minimum horsepower motor that can be used to save the ship?

47. •• Predict/Calculate A kayaker paddles with a power output of 35.0 W to maintain a steady speed of 2.24 m/s. **(a)** Calculate the resistive force exerted by the water on the kayak. **(b)** If the kayaker doubles her power output, and the resistive force due to the water remains the same, by what factor does the kayaker's speed change?

48. •• **BIO Human-Powered Flight** Human-powered aircraft require a pilot to pedal, as in a bicycle, and produce a sustained power output of about 0.30 hp. The *Gossamer Albatross* flew across the English Channel on June 12, 1979, in 2 h 49 min. **(a)** How much energy did the pilot expend during the flight? **(b)** What is the minimum number of Snickers bars (280 Cal per bar) the pilot would have to consume to be "fueled up" for the flight? [*Note:* The nutritional calorie, 1 Cal, is equivalent to 1000 calories (1000 cal) as defined in physics. In addition, the conversion factor between calories and joules is as follows: 1 Cal = 1000 cal = 1 kcal = 4184 J.]

49. •• Predict/Calculate **Beating to Windward** A sailboat can be propelled into the wind by a maneuver called *beating to windward*. Beating requires the sailboat to travel in a zigzag pattern at an angle to the wind that is greater than the *no-go zone*, which is shaded red in **FIGURE 7-23**. When a sailboat is just outside the no-go zone (boats B in the figure) the wind exerts a force $\vec{\mathbf{F}}$ on the sail that has a component in the direction of motion $\vec{\mathbf{v}}$. Similar comments apply to boats C. The work done by the wind on the sail is $W = Fd\cos\theta$, and because $v = d/t$, the propulsion power $P = W/t$ delivered to the sailboat is $Fv\cos\theta$, where θ is the angle between the sail force and the direction of motion. **(a)** Assuming that F and v have the same magnitudes for each sailboat, will the propulsion power delivered to sailboats B be greater than, less than, or the same as the propulsion power delivered to sailboats C? Explain. **(b)** If $F = 870$ N and $v = 11$ m/s, what propulsion power is delivered to sailboats B, for which $\theta = 79°$? **(c)** What propulsion power is delivered to sailboats C, for which $\theta = 56°$?

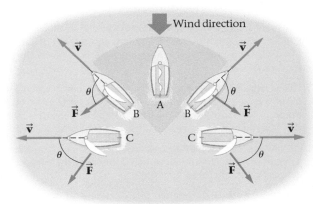

FIGURE 7-23 Problem 49

50. •• Predict/Calculate A grandfather clock is powered by the descent of a 4.35-kg weight. **(a)** If the weight descends through a distance of 0.760 m in 3.25 days, how much power does it deliver to the clock? **(b)** To increase the power delivered to the clock, should the time it takes for the mass to descend be increased or decreased? Explain.

51. ••• Predict/Calculate A certain car can accelerate from rest to the speed v in T seconds. If the power output of the car remains constant, **(a)** how much time does it take for the car to accelerate from v to $2v$? **(b)** How fast is the car moving at $2T$ seconds after starting?

GENERAL PROBLEMS

52. • **CE** As the three small sailboats shown in **FIGURE 7-24** drift next to a dock, because of wind and water currents, students pull on a line attached to the bow and exert forces of equal magnitude F. Each boat drifts through the same distance d. Rank the three boats (A, B, and C) in order of increasing work done on the boat by the force F. Indicate ties where appropriate.

FIGURE 7-24 Problem 52

53. • **CE** Predict/Explain A car is accelerated by a constant force, F. The distance required to accelerate the car from rest to the speed v is Δx. **(a)** Is the distance required to accelerate the car from the speed v to the speed $2v$ equal to Δx, $2\Delta x$, $3\Delta x$, or $4\Delta x$? **(b)** Choose the *best explanation* from among the following:
    I. The final speed is twice the initial speed.
    II. The increase in speed is the same in each case.
    III. Work is force times distance, and work depends on the speed squared.

54. • **CE** Car 1 has four times the mass of car 2, but they both have the same kinetic energy. If the speed of car 2 is v, is the speed of car 1 equal to $v/4$, $v/2$, $2v$, or $4v$? Explain.

55. • **BIO Muscle Cells** Biological muscle cells can be thought of as nanomotors that use the chemical energy of ATP to produce mechanical work. Measurements show that the active proteins within a muscle cell (such as myosin and actin) can produce a force of about 1.5 pN and displacements of 8.0 nm. How much work is done by such proteins?

56. • A small motor runs a lift that raises a load of bricks weighing 836 N to a height of 10.7 m in 23.2 s. Assuming that the bricks are lifted with constant speed, what is the minimum power the motor must produce?

57. • You push a 67-kg box across a floor where the coefficient of kinetic friction is $\mu_k = 0.55$. The force you exert is horizontal. **(a)** How much power is needed to push the box at a speed of 0.50 m/s? **(b)** How much work do you do if you push the box for 35 s?

58. • A 1300-kg elevator is lifted at a constant speed of 1.3 m/s through a height of 22 m. How much work is done **(a)** by the tension in the elevator cable and **(b)** by gravity?

59. •• **CE** The work W_0 is required to accelerate a car from rest to the speed v_0. How much work is required to accelerate the car **(a)** from rest to the speed $v_0/2$ and **(b)** from $v_0/2$ to v_0?

60. •• After a tornado, a 0.55-g straw was found embedded 2.3 cm into the trunk of a tree. If the average force exerted on the straw by the tree was 65 N, what was the speed of the straw when it hit the tree?

61. •• You throw a glove straight upward to celebrate a victory. Its initial kinetic energy is K and it reaches a maximum height h. What is the kinetic energy of the glove when it is at the height $h/2$?

62. •• The water skier in Figure 7-20 is at an angle of 32.5° with respect to the center line of the boat, and is being pulled at a constant speed of 15.6 m/s. If the tension in the tow rope is 111 N, **(a)** how much work does the rope do on the skier in 10.0 s? **(b)** How much work does the resistive force of water do on the skier in the same time?

63. •• Predict/Calculate A sled with a mass of 5.80 kg is pulled along the ground through a displacement given by $\vec{d} = (4.55 \text{ m})\hat{x}$. (Let the x axis be horizontal and the y axis be vertical.) **(a)** How much work is done on the sled when the force acting on it is $\vec{F} = (2.89 \text{ N})\hat{x} + (0.131 \text{ N})\hat{y}$? **(b)** How much work is done on the sled when the force acting on it is $\vec{F} = (2.89 \text{ N})\hat{x} + (0.231 \text{ N})\hat{y}$? **(c)** If the mass of the sled is increased, does the work done by the forces in parts **(a)** and **(b)** increase, decrease, or stay the same? Explain.

64. •• Predict/Calculate A 0.19-kg apple falls from a branch 3.5 m above the ground. **(a)** Does the power delivered to the apple by gravity increase, decrease, or stay the same during the time the apple falls to the ground? Explain. Find the power delivered by gravity to the apple when the apple is **(b)** 2.5 m and **(c)** 1.5 m above the ground.

65. •• A boy pulls a bag of baseball bats across a ball field toward the parking lot. The bag of bats has a mass of 6.80 kg, and the boy exerts a horizontal force of 24.0 N on the bag. As a result, the bag accelerates from rest to a speed of 1.12 m/s in a distance of 5.25 m. What is the coefficient of kinetic friction between the bag and the ground?

66. •• At the instant it leaves the player's hand after a jump shot, a 0.62-kg basketball is traveling 8.4 m/s. When it passes through the hoop its speed is 6.8 m/s. **(a)** How much work has gravity done on the ball? **(b)** What is the vertical component of the ball's displacement? **(c)** If the magnitude of the total displacement of the ball is 4.6 m, what is the angle between the displacement vector and the gravitational force?

67. •• The force shown in FIGURE 7-25 acts on an object that moves along the x axis. How much work is done by the force as the object moves from **(a)** $x = 0$ to $x = 2.0$ m, **(b)** $x = 1.0$ m to $x = 4.0$ m, and **(c)** $x = 3.5$ m to $x = 1.2$ m?

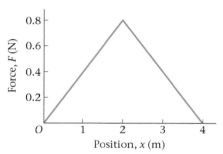

FIGURE 7-25 Problem 67

68. •• **A Compound Bow** A compound bow in archery allows the user to hold the bowstring at full draw with considerably less force than the maximum force exerted by the string. The draw force as a function of the string position x for a particular compound bow is shown in FIGURE 7-26. **(a)** How much work does the archer do on the bow in order to draw the string from $x = 0$ to $x = 0.60$ m? **(b)** If all of this work becomes the kinetic energy of a 0.065-kg arrow, what is the speed of the arrow?

FIGURE 7-26 Problem 68

69. •• **A Compound Versus a Simple Bow** The compound bow in the previous problem requires 152 J of work to draw the string back to $x = 0.540$ m, at which point the archer need only exert 250 N of force to hold the string in place. **(a)** Suppose a simple bow is used instead of a compound bow. What should its force constant k (in N/m) be so that it requires the same amount of work to draw the

string a distance of $x = 0.540$ m? Assume the simple bow behaves the same as an ideal spring. **(b)** How much force must the archer exert to draw the simple bow to $x = 0.540$ m? Compare this to the 250-N force required by the compound bow.

70. •• Calculate the power output of a 0.42-g spider as it walks up a windowpane at 2.3 cm/s. The spider walks on a path that is at 25° to the vertical, as illustrated in **FIGURE 7-27**.

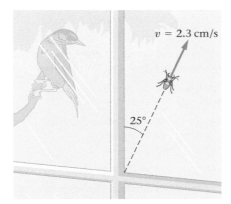

FIGURE 7-27 Problem 70

71. •• **Cookie Power** To make a batch of cookies, you mix half a bag of chocolate chips into a bowl of cookie dough, exerting a 28-N force on the stirring spoon. Assume that your force is always in the direction of motion of the spoon. **(a)** What power is needed to move the spoon at a speed of 0.44 m/s? **(b)** How much work do you do if you stir the mixture for 1.5 min?

72. •• Predict/Calculate A pitcher accelerates a 0.14-kg hardball from rest to 42.5 m/s in 0.060 s. **(a)** How much work does the pitcher do on the ball? **(b)** What is the pitcher's power output during the pitch? **(c)** Suppose the ball reaches 42.5 m/s in less than 0.060 s. Is the power produced by the pitcher in this case more than, less than, or the same as the power found in part **(b)**? Explain.

73. •• **BIO Brain Power** The human brain consumes about 22 W of power under normal conditions, though more power may be required during exams. **(a)** For what amount of time can one Snickers bar (see the note following Problem 48) power the normally functioning brain? **(b)** At what rate must you lift a 3.6-kg container of milk (one gallon) if the power output of your arm is to be 22 W? **(c)** How much time does it take to lift the milk container through a distance of 1.0 m at this rate?

74. •• **Meteorite** On October 9, 1992, a 27-pound meteorite struck a car in Peekskill, NY, creating a dent about 22 cm deep. If the initial speed of the meteorite was 550 m/s, what was the average force exerted on the meteorite by the car?

75. ••• **BIO Powering a Pigeon** A pigeon in flight experiences a force of air resistance given approximately by $F = bv^2$, where v is the flight speed and b is a constant. **(a)** What are the units of the constant b? **(b)** What is the greatest possible speed of the pigeon if its maximum power output is P? **(c)** By what factor does the greatest possible speed increase if the maximum power is doubled?

76. ••• **Springs in Series** Two springs, with force constants k_1 and k_2, are connected in series, as shown in **FIGURE 7-28**. How much work is required to stretch this system a distance x from the equilibrium position?

FIGURE 7-28 Problem 76

77. ••• **Springs in Parallel** Two springs, with force constants k_1 and k_2, are connected in parallel, as shown in **FIGURE 7-29**. How much work is required to stretch this system a distance x from the equilibrium position?

FIGURE 7-29 Problem 77

78. ••• A block rests on a horizontal frictionless surface. A string is attached to the block, and is pulled with a force of 45.0 N at an angle θ above the horizontal, as shown in **FIGURE 7-30**. After the block is pulled through a distance of 1.50 m, its speed is 2.60 m/s, and 50.0 J of work has been done on it. **(a)** What is the angle θ? **(b)** What is the mass of the block?

FIGURE 7-30 Problem 78

PASSAGE PROBLEMS

BIO *Microraptor gui*: **The Biplane Dinosaur**

The evolution of flight is a subject of intense interest in paleontology. Some subscribe to the "cursorial" (or ground-up) hypothesis, in which flight began with ground-dwelling animals running and jumping after prey. Others favor the "arboreal" (or trees-down) hypothesis, in which tree-dwelling animals, like modern-day flying squirrels, developed flight as an extension of gliding from tree to tree.

A recently discovered fossil from the Cretaceous period in China supports the arboreal hypothesis and adds a new element—it suggests that feathers on both the wings and the lower legs and feet allowed this dinosaur, *Microraptor gui*, to glide much like a biplane, as shown in **FIGURE 7-31 (a)**. Researchers have produced a detailed computer simulation of *Microraptor*, and with its help have obtained the power-versus-speed plot presented in **FIGURE 7-31 (b)**. This curve shows how much power is required for flight at speeds between 0 and 30 m/s. Notice that the power increases at high speeds, as expected, but it is also high for low speeds, where the dinosaur is almost hovering. A minimum of 8.1 W is needed for flight at 10 m/s. The lower horizontal line shows the estimated 9.8-W power output of *Microraptor*, indicating the small range of speeds for which flight would be possible. The upper horizontal line shows the wider range of flight speeds that would be available if *Microraptor* were able to produce 20 W of power.

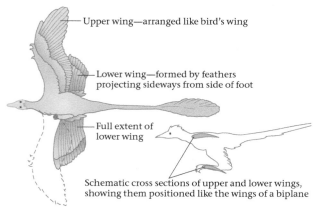

Upper wing—arranged like bird's wing

Lower wing—formed by feathers projecting sideways from side of foot

Full extent of lower wing

Schematic cross sections of upper and lower wings, showing them positioned like the wings of a biplane

(a) Possible reconstruction of *Microraptor gui* **in flight**

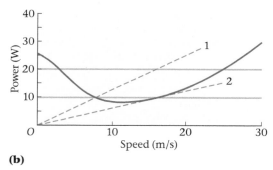

(b)

FIGURE 7-31 Problems 79, 80, 81, and 82

Also of interest are the two dashed, straight lines labeled 1 and 2. These lines represent constant ratios of power to speed—that is, a constant value for P/v. Referring to Equation 7-13, we see that $P/v = Fv/v = F$, so lines 1 and 2 correspond to lines of constant force. Line 2 is interesting in that it has the smallest slope that still touches the power-versus-speed curve.

79. • Estimate the range of flight speeds for *Microraptor gui* if its power output is 9.8 W.

    A. 0–7.7 m/s   **B.** 7.7–15 m/s
    C. 15–30 m/s   **D.** 0–15 m/s

80. • What approximate range of flight speeds would be possible if *Microraptor gui* could produce 20 W of power?

    A. 0–25 m/s   **B.** 25–30 m/s
    C. 2.5–25 m/s   **D.** 0–2.5 m/s

81. •• How much energy would *Microraptor* have to expend to fly with a speed of 10 m/s for 1.0 minute?

    A. 8.1 J   **B.** 81 J
    C. 490 J   **D.** 600 J

82. • Estimate the minimum force that *Microraptor* must exert to fly.

    A. 0.65 N   **B.** 1.3 N
    C. 1.0 N   **D.** 10 N

83. •• **REFERRING TO FIGURE 7-15** Suppose the block has a mass of 1.4 kg and an initial speed of 0.62 m/s. **(a)** What force constant must the spring have if the maximum compression is to be 2.4 cm? **(b)** If the spring has the force constant found in part (a), find the maximum compression if the mass of the block is doubled *and* its initial speed is halved.

84. •• Predict/Calculate **REFERRING TO FIGURE 7-15** In the situation shown in Figure 7-15 (d), a spring with a force constant of 750 N/m is compressed by 4.1 cm. **(a)** If the speed of the block in Figure 7-15 (f) is 0.88 m/s, what is its mass? **(b)** If the mass of the block is doubled, is the final speed greater than, less than, or equal to 0.44 m/s? **(c)** Find the final speed for the case described in part (b).

85. •• Predict/Calculate **REFERRING TO EXAMPLE 7-15** Suppose the car has a mass of 1400 kg and delivers 48 hp to the wheels. **(a)** How much time does it take for the car to increase its speed from 15 m/s to 25 m/s? **(b)** Would the time required to increase the speed from 5.0 m/s to 15 m/s be greater than, less than, or equal to the time found in part (a)? **(c)** Determine the time required to accelerate from 5.0 m/s to 15 m/s.

Potential Energy and Conservation of Energy

Big Ideas

1 Forces are either conservative (conserve energy) or nonconservative (dissipate it to other forms).

2 Potential energy is a storage mechanism for energy.

3 Mechanical energy is the sum of the potential and kinetic energies of a system.

▲ As this athlete hangs motionless in midair for a moment, he might wonder what happened to the kinetic energy he had just a few moments before? Where did that energy go? And how does it reappear as he begins to fall back toward the ground? The answers to these questions involve new types of energy that are the subject of this chapter.

One of the greatest accomplishments of physics is the concept of energy and its conservation. As we shall see, the universe has a certain amount of energy, and this energy can ebb and flow from one form to another, but the total amount remains fixed. Conservation of energy is of central importance in physics, and is also an important practical tool in problem solving.

8-1 Conservative and Nonconservative Forces

Work done by person = *mgh*

Work done by gravity = *mgh*

▲ **FIGURE 8-1 Work against gravity** Lifting a box against gravity with constant speed takes work *mgh*. When the box is released, gravity does the same work on the box as it falls. Gravity is a conservative force.

PHYSICS IN CONTEXT
Looking Back

The concept of work, first introduced in Chapter 7, is important in this chapter as well. For example, we use work (force times distance) in this section to illustrate the difference between conservative and nonconservative forces.

In physics, we classify forces according to whether they are *conservative* or *nonconservative*. The key distinction is that when a conservative force acts, the work it does is stored in a form that can be released as kinetic energy at a later time. In contrast, nonconservative forces dissipate energy to other forms, such as thermal energy, that cannot be recovered. In this section, we sharpen this distinction and explore some examples of conservative and nonconservative forces.

Gravity Is a Conservative Force Perhaps the simplest case of a conservative force is gravity. Imagine lifting a box of mass *m* from the floor to a height *h*, as in **FIGURE 8-1**. To lift the box with constant speed, the force you must exert against gravity is *mg*. Since the upward distance is *h*, the work you do on the box is $W = mgh$. If you now release the box and allow it to drop back to the floor, gravity does the same work, $W = mgh$, and in the process gives the box an equivalent amount of kinetic energy. Thus, the work done by a **conservative force** can be recovered later as kinetic energy.

Friction Is a Nonconservative Force Contrast gravity with the force of kinetic friction, which is nonconservative. To slide a box of mass *m* across the floor with constant speed, as shown in **FIGURE 8-2**, you must exert a force of magnitude $\mu_k N = \mu_k mg$. After the box has moved a distance *d*, the work you have done is $W = \mu_k mgd$. In this case, when you release the box it simply stays put—friction does no work on it after you let go. Thus, the work done by a **nonconservative force** cannot be recovered later as kinetic energy; instead, it is converted to other forms of energy, such as a slight warming of the floor and box in our example. A visual comparison of conservative and nonconservative forces is presented in **FIGURE 8-3**.

Work = $\mu_k mgd$

▲ **FIGURE 8-2 Work against friction** Pushing a box with constant speed against friction takes the work $\mu_k mgd$. When the box is released, it quickly comes to rest and friction does no further work. Friction is a nonconservative force.

Forces on a Closed Path The differences between conservative and nonconservative forces are even more apparent for an object moving around a closed path. Consider, for example, the path shown in **FIGURE 8-4**. If we move a box of mass *m* along this path, the total work done by gravity is the sum of the work done on each segment of the path; that is, $W_{total} = W_{AB} + W_{BC} + W_{CD} + W_{DA}$. The work done by gravity from A to B and from C to D is zero, because the force is at right angles to the displacement on these segments. Thus $W_{AB} = W_{CD} = 0$. On the segment from B to C, gravity does negative work (displacement and force are in opposite directions), but it does positive work from D to A (displacement and force are in the same direction). Hence, $W_{BC} = -mgh$ and $W_{DA} = mgh$. As a result, the total work done by gravity is zero:

$$W_{total} = 0 + (-mgh) + 0 + mgh = 0$$

With friction, the results are quite different. If we push the box around the closed horizontal path shown in **FIGURE 8-5**, the total work done by friction does not vanish. In fact, friction does the negative work $W = -f_k d = -\mu_k mgd$ on each segment. Therefore, the total work done by kinetic friction is

$$W_{total} = (-\mu_k mgd) + (-\mu_k mgd) + (-\mu_k mgd) + (-\mu_k mgd) = -4\mu_k mgd$$

▲ **FIGURE 8-3 Visualizing Concepts Conservative Versus Nonconservative Forces** Because gravity is a conservative force, the work done against gravity in lifting these logs (left) can, in principle, all be recovered. If the logs are released, for example, they will acquire an amount of kinetic energy exactly equal to the work done to lift them. Friction, by contrast, is a nonconservative force. Some of the work done by this spinning grindstone (right) goes into removing material from the object being ground, while the rest is transformed into sound energy and (especially) heat. Most of this work can never be recovered as kinetic energy.

These results lead to the following definition of a conservative force:

Conservative Force: Definition 1

A conservative force does zero total work on a closed path.

A roller coaster provides a good illustration of this definition. A car on a roller coaster has speed v at point A in **FIGURE 8-6**, it speeds up as it drops to point B, it slows down as it approaches point C, and so on. When the car returns to its original height, at point D, it will again have the speed v, as long as friction and other nonconservative forces can be neglected. Similarly, if the car completes a circuit of the track and returns to point A, it will again have the speed v. Hence, a roller-coaster car's kinetic energy is unchanged ($\Delta K = 0$) after *any* complete circuit of the track. From the work–energy theorem, $W_{\text{total}} = \Delta K$, it follows that the work done by gravity is zero for the closed path of the car, as expected for a conservative force.

▲ **FIGURE 8-4 Work done by gravity on a closed path is zero** Gravity does no work on the two horizontal segments of the path. On the two vertical segments, the amounts of work done are equal in magnitude but opposite in sign. Therefore, the total work done by gravity on this—or any—closed path is zero.

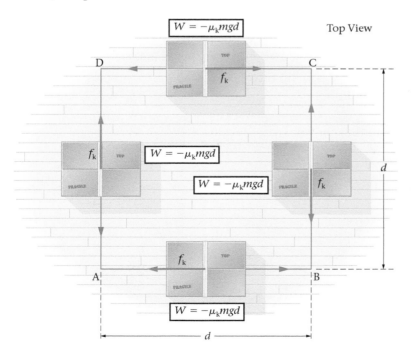

◀ **FIGURE 8-5 Work done by friction on a closed path is nonzero** The work done by friction when an object moves through a distance d is $-\mu_{\text{k}}mgd$. Thus, the total work done by friction on a closed path is nonzero. In this case, it is equal to $-4\,\mu_{\text{k}}mgd$.

▶ FIGURE 8-6 **Gravity is a conservative force** If frictional forces can be ignored, a roller-coaster car will have the same speed at points A and D, since they are at the same height. Hence, after any complete circuit of the track, the speed of the car returns to its initial value. It follows that the change in kinetic energy is zero for a complete circuit, and therefore the work done by gravity is also zero.

When the roller coaster returns to its original level . . .

. . . the net work done by gravity is zero, and the speed of the roller coaster returns to its original value.

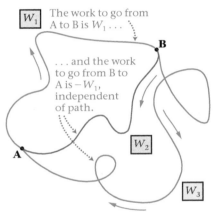

W_1 The work to go from A to B is W_1 . . .

. . . and the work to go from B to A is $-W_1$, independent of path.

W_2

W_3

▲ FIGURE 8-7 **The work done by a conservative force is independent of path** Considering paths 1 and 2, we see that $W_1 + W_2 = 0$, or $W_2 = -W_1$. From paths 1 and 3, however, we see that $W_1 + W_3 = 0$, or $W_3 = -W_1$. It follows, then, that $W_3 = W_2$, since they are both equal to $-W_1$; hence the work done in going from B to A is independent of the path.

The Work Done by a Conservative Force Is Path Independent The property of zero total work on a closed path has interesting consequences for conservative forces. For instance, consider the closed paths shown in FIGURE 8-7. On each of these paths, we know that the work done by a conservative force is zero. Thus, it follows from paths 1 and 2 that $W_{total} = W_1 + W_2 = 0$, or

$$W_2 = -W_1$$

Similarly, using paths 1 and 3 we have $W_{total} = W_1 + W_3 = 0$, or

$$W_3 = -W_1$$

As a result, we see that the work done on path 3 is the same as the work done on path 2:

$$W_3 = W_2$$

But paths 2 and 3 are arbitrary, as long as they start at point B and end at point A. This leads to an equivalent definition of a conservative force:

Conservative Force: Definition 2
The work done by a conservative force in going from point A to point B is *independent of the path* from A to B.

This definition is given an explicit check in Example 8-1.

EXAMPLE 8-1 DIFFERENT PATHS, DIFFERENT FORCES

(a) A 4.57-kg box is moved with constant speed from A to B along the two paths shown in part (a) of our sketch. Calculate the work done by gravity on each of these paths. **(b)** The same box is pushed across a floor from A to B along path 1 and path 2 in part (b) of our sketch. If the coefficient of kinetic friction between the box and the surface is $\mu_k = 0.63$, how much work is done by friction along each path?

(a)

(b)

CONTINUED

PICTURE THE PROBLEM

Part (a) of our sketch shows two different paths a box might be taken through in going from point A to point B. Path 1 is indicated by two red lines, indicating a vertical displacement of 1.0 m and a horizontal displacement of 3.0 m. Path 2, indicated in green, consists of two horizontal and two vertical displacements. In this case, we are interested in the work done by gravity. Part (b) shows the same basic paths—path 1 in orange and path 2 in purple—only this time on a rough floor. Here it is the force of kinetic friction that is of interest.

REASONING AND STRATEGY

To calculate the work done for each path, we break it down into segments. Path 1 is made up of two segments; path 2 has four segments.

a. For gravity, the work is zero on horizontal segments. On vertical segments, the work done by gravity is positive when the motion is downward and negative when the motion is upward.

b. The work done by kinetic friction is negative on all segments of both paths.

Known   Mass of box, $m = 4.57$ kg; coefficient of kinetic friction, $\mu_k = 0.63$; the paths are shown in the sketch.
Unknown (a) Work done by gravity on two paths: $W_1 = ?$; $W_2 = ?$
             (b) Work done by friction on two paths: $W_1 = ?$; $W_2 = ?$

SOLUTION

Part (a)

1. Using $W = Fd = mgy$, calculate the work done by gravity along the two segments of path 1:

$$W_1 = -(4.57 \text{ kg})(9.81 \text{ m/s}^2)(1.0 \text{ m}) + 0 = -45 \text{ J}$$

2. In the same way, calculate the work done by gravity along the four segments of path 2:

$$W_2 = 0 - (4.57 \text{ kg})(9.81 \text{ m/s}^2)(2.0 \text{ m})$$
$$+ 0 + (4.57 \text{ kg})(9.81 \text{ m/s}^2)(1.0 \text{ m}) = -45 \text{ J}$$

Part (b)

3. Using $F = \mu_k N$, calculate the work done by kinetic friction along the two segments of path 1:

$$W_1 = -(0.63)(4.57 \text{ kg})(9.81 \text{ m/s}^2)(1.0 \text{ m})$$
$$- (0.63)(4.57 \text{ kg})(9.81 \text{ m/s}^2)(3.0 \text{ m}) = -110 \text{ J}$$

4. Similarly, calculate the work done by kinetic friction along the four segments of path 2:

$$W_2 = -(0.63)(4.57 \text{ kg})(9.81 \text{ m/s}^2)(6.0 \text{ m})$$
$$- (0.63)(4.57 \text{ kg})(9.81 \text{ m/s}^2)(2.0 \text{ m})$$
$$- (0.63)(4.57 \text{ kg})(9.81 \text{ m/s}^2)(3.0 \text{ m})$$
$$- (0.63)(4.57 \text{ kg})(9.81 \text{ m/s}^2)(1.0 \text{ m}) = -340 \text{ J}$$

INSIGHT

As expected, the conservative force of gravity gives the same work in going from A to B, regardless of the path. The work done by kinetic friction, however, has a larger magnitude on the path of greater length.

PRACTICE PROBLEM — PREDICT/CALCULATE

Suppose the work done by gravity when the box is moved from point B to a new point C is 140 J. **(a)** Is point C above, below, or at the same height as point B? Explain. **(b)** Find the vertical distance between points B and C. [**Answer: (a)** The work done by gravity is positive. This means that the displacement of the box is downward—in the direction of the force of gravity. Thus point C is *below* point B. **(b)** Point C is 3.1 m below point B.]

Some related homework problems: Problem 2, Problem 3

Table 8-1 summarizes the different kinds of conservative and nonconservative forces we have encountered thus far in this text.

Enhance Your Understanding

(Answers given at the end of the chapter)

1. In **FIGURE 8-8**, the work done by a conservative force along path AB is 15 J, and the work the force does along path BC is −5 J. What is the work done by this force along the paths **(a)** CDA and **(b)** AEC?

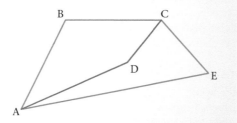

▲ **FIGURE 8-8** Enhance Your Understanding 1.

CONTINUED

TABLE 8-1 Conservative and Nonconservative Forces

Force	Section
Conservative forces	
Gravity	5-6
Spring force	6-2
Nonconservative forces	
Friction	6-1
Tension in a rope, cable, etc.	6-2
Forces exerted by a motor	7-4
Forces exerted by muscles	5-3

Big Idea **2** Potential energy is a storage mechanism for energy. Each type of conservative force has its own form of potential energy. There is no potential energy associated with nonconservative forces.

PHYSICS IN CONTEXT
Looking Back

Work (Chapter 7) is used in this section to introduce the concept of potential energy, U, and to define its change.

PROBLEM-SOLVING NOTE

Zero of Potential Energy

When working potential energy problems it is important to make a definite choice for the location where the potential energy is to be set equal to zero. Any location can be chosen, but once the choice is made, it must be used consistently.

▲ **FIGURE 8-9 Gravitational Potential Energy** A person drops from a diving board into a swimming pool. The diving board is at the height y, and the surface of the water is at $y = 0$. We choose the gravitational potential energy to be zero at $y = 0$; hence, the potential energy is mgy at the diving board.

8-2 Potential Energy and the Work Done by Conservative Forces

Work must be done to lift a bowling ball from the floor to a shelf. Once on the shelf, the bowling ball has zero kinetic energy, just as it did on the floor. Even so, the work done in lifting the ball has not been lost. If the ball is allowed to fall from the shelf, gravity does the same amount of work on it as you did to lift it in the first place. As a result, the work you did is "recovered" in the form of kinetic energy. Thus, we say that when the ball is lifted to a new position, there is an increase in **potential energy,** U, and that this potential energy can be converted to kinetic energy when the ball falls.

In a sense, potential energy is a storage system for energy. When we increase the separation between the ball and the ground, the work we do is stored in the form of an increased potential energy. Not only that, but the storage system is perfect, in the sense that the energy is never lost as long as the separation remains the same. The ball can rest on the shelf for a million years, and still, when it falls, it gains the same amount of kinetic energy.

Work done against friction, however, is not "stored" as potential energy. Instead, it is dissipated into other forms of energy such as heat or sound. The same is true of other non-conservative forces. Only conservative forces have the potential-energy storage system.

Before proceeding, we should point out an interesting difference between kinetic and potential energy. Kinetic energy is given by the expression $K = \frac{1}{2}mv^2$, no matter what force might be involved. On the other hand, each different conservative force has a different expression for its potential energy. To see how this comes about, we turn now to a precise definition of potential energy.

Potential Energy, U

When a conservative force does an amount of work W_c (the subscript c stands for conservative), the corresponding potential energy U is changed according to the following definition:

Definition of the Change in Potential Energy, U

$$W_c = U_i - U_f = -(U_f - U_i) = -\Delta U \qquad 8\text{-}1$$

SI unit: joule, J

In words, the work done by a conservative force is equal to the negative of the change in the corresponding potential energy. For example, when an object falls, gravity does *positive* work on it and the gravitational potential energy *decreases*. Similarly, when an object is lifted, gravity does *negative* work and the gravitational potential energy *increases*.

Work is a scalar with the units of joules, and hence the same is true of potential energy. In addition, our definition determines only the *change* in potential energy between two points, not the actual value of the potential energy. As a result, we are free to choose the place where the potential energy is zero ($U = 0$) in much the same way that we are free to choose the location of the origin of a coordinate system.

Gravity

Let's apply our definition of potential energy to the force of gravity near the Earth's surface. Suppose a person of mass m drops a distance y from a diving board into a pool, as shown in **FIGURE 8-9**. As the person drops, gravity does the work

$$W_c = mgy$$

When we apply the definition given in Equation 8-1, the corresponding change in potential energy is

$$-\Delta U = U_i - U_f = W_c = mgy$$

In this expression, U_i is the potential energy when the diver is on the board, and U_f is the potential energy when the diver enters the water. Rearranging slightly, we have

$$U_i = mgy + U_f \qquad\qquad \text{8-2}$$

Notice that the initial potential energy, U_i, is greater than the final potential energy, U_f.

As mentioned above, we are free to choose $U = 0$ anywhere we like; *only the change in U is important.* For example, if you slip and fall to the ground, you hit with the same thud whether you fall in Denver (altitude 1 mile) or in Honolulu (at sea level). It's the difference in height that matters, not the height itself. This is illustrated in FIGURE 8-10. The only issue to be careful about when choosing a location for $U = 0$ is to be consistent with the choice once it is made.

In general, we choose $U = 0$ in a convenient location. In Figure 8-9, a reasonable place for $U = 0$ is the surface of the water, where $y = 0$; that is, $U_f = 0$. Then, Equation 8-2 becomes $U_i = mgy$. If we omit the subscript on U_i, letting U stand for the potential energy at the arbitrary height y, we have:

Gravitational Potential Energy (Near Earth's Surface)

$$U = mgy \qquad\qquad \text{8-3}$$

The gravitational potential energy depends only on the height, y, and is independent of the horizontal position.

EXERCISE 8-2   GRAVITATIONAL POTENTIAL ENERGY

Find the gravitational potential energy associated with a 7.25-kg bowling ball that is held 1.45 m above the ground. Let $U = 0$ be at ground level.

REASONING AND SOLUTION
The gravitational potential energy is given by $U = mgh$. Substituting $m = 7.25$ kg and $h = 1.45$ m yields

$$U = mgy = (7.25 \text{ kg})(9.81 \text{ m/s}^2)(1.45 \text{ m}) = 103 \text{ J}$$

The next Example considers the change in gravitational potential energy associated with a mountain climber, given different choices for the location of $U = 0$.

▲ **FIGURE 8-10** The $U = 0$ level for the gravitational potential energy of this system can be assigned to the point where the diver starts his dive, to the water level, or to any other level. Regardless of the choice, however, his kinetic energy when he strikes the water will be exactly equal to the difference in gravitational potential energy between his launch and splashdown points.

EXAMPLE 8-3   PIKES PEAK OR BUST

An 82.0-kg mountain climber is in the final stage of the ascent of 4301-m-high Pikes Peak. What is the change in gravitational potential energy as the climber gains the last 100.0 m of altitude? Let $U = 0$ be **(a)** at sea level or **(b)** at the top of the peak.

PICTURE THE PROBLEM
Our sketch shows the mountain climber and the last 100.0 m of altitude to be climbed. We choose a typical coordinate system, with the positive y axis upward and the positive x axis to the right.

REASONING AND STRATEGY
The gravitational potential energy of the Earth–climber system depends only on the height y; the path followed in gaining the last 100.0 m of altitude is unimportant. The change in potential energy is $\Delta U = U_f - U_i = mgy_f - mgy_i$, where y_f is the altitude of the peak and y_i is 100.0 m less than y_f.

Known   Mass of climber, $m = 82.0$ kg; difference in altitude, $y_f - y_i = 100.0$ m; **(a)** $U = 0$ at sea level; **(b)** $U = 0$ at top of peak.

Unknown   **(a)** Change in potential energy, $\Delta U = $? **(b)** Change in potential energy, $\Delta U = $?

100.0 m

CONTINUED

SOLUTION

Part (a)

1. Calculate ΔU with $y_f = 4301$ m and $y_i = 4201$ m:

$$\Delta U = mgy_f - mgy_i$$
$$= (82.0 \text{ kg})(9.81 \text{ m/s}^2)(4301 \text{ m})$$
$$- (82.0 \text{ kg})(9.81 \text{ m/s}^2)(4201 \text{ m}) = 80,400 \text{ J}$$

Part (b)

2. Calculate ΔU with $y_f = 0$ and $y_i = -100.0$ m:

$$\Delta U = mgy_f - mgy_i$$
$$= (82.0 \text{ kg})(9.81 \text{ m/s}^2)(0)$$
$$- (82.0 \text{ kg})(9.81 \text{ m/s}^2)(-100.0 \text{ m}) = 80,400 \text{ J}$$

INSIGHT

As expected, the *change* in gravitational potential energy does not depend on where we choose $U = 0$. Nor does it depend on the path taken between the initial and final points.

PRACTICE PROBLEM

Find the altitude of the climber for which the gravitational potential energy of the Earth–climber system is 1.00×10^5 J less than it is when the climber is at the summit.   [**Answer: 4180 m**]

Some related homework problems: Problem 10, Problem 17

A single item of food can be converted into a surprisingly large amount of potential energy. This is shown for the case of a candy bar in the next Example.

EXAMPLE 8-4   CONVERTING FOOD ENERGY TO MECHANICAL ENERGY

BIO A bicyclist eats a 195 Cal = 195 kcal candy bar. The energy equivalent of the candy is 8.16×10^5 J, and the mass of the bicyclist and bicycle together is 70.3 kg. If the bicyclist could convert all of the candy bar's energy to gravitational potential energy, what gain in altitude would she achieve?

PICTURE THE PROBLEM

The bicyclist pedals up a slope, gaining altitude. She starts at the origin, $y = 0$, and rises in altitude to the final height $y = h$.

REASONING AND STRATEGY

The initial gravitational potential energy of the system is $U = 0$; the final potential energy is $U = mgh$. To find the altitude gain, we set $U = mgh$ equal to the energy provided by the candy bar, 8.16×10^5 J, and solve for the height h.

Known   Energy of candy bar, $U = 8.16 \times 10^5$ J; mass of bicyclist and bicycle, $m = 70.3$ kg.
Unknown   Gain in altitude, $h = ?$

SOLUTION

1. Solve $U = mgh$ for h:

$$U = mgh$$
$$h = \frac{U}{mg}$$

2. Substitute numerical values, with $U = 8.16 \times 10^5$ J:

$$h = \frac{U}{mg} = \frac{8.16 \times 10^5 \text{ J}}{(70.3 \text{ kg})(9.81 \text{ m/s}^2)} = 1180 \text{ m}$$

INSIGHT

This is almost 3/4 of a mile in elevation. Even if we take into account the fact that metabolic efficiency is only about 20%, the height is still 236 m, or nearly 0.15 mi. It's remarkable just how much our bodies can do with so little.

PRACTICE PROBLEM — PREDICT/CALCULATE

(a) If the mass of the bicyclist is increased—by adding a backpack, for example—does the possible elevation gain increase, decrease, or stay the same? Explain. (b) Find the elevation gain if the bicyclist and bicycle have a combined mass of 85.5 kg.   [**Answer: (a)** The elevation gain is inversely proportional to the mass. Hence, an increase in mass results in a smaller gain in elevation. **(b)** $h = 973$ m]

Some related homework problems: Problem 9, Problem 11

We have been careful *not* to say that the potential energy of a mountain climber—or any object—increases when its height increases. The reason is that the potential energy is a property of an entire system, not of its individual parts. The correct statement is that if an object is lifted, the potential energy of the Earth–object system is increased.

Springs

Consider a spring that is stretched from its equilibrium position a distance x. According to Equation 7-8, the work required to cause this stretch is given by $W = \frac{1}{2}kx^2$. Therefore, if the spring is released—and allowed to return to the equilibrium position—it will do the same work, $\frac{1}{2}kx^2$. From our definition of potential energy, then, we see that

$$W_c = \tfrac{1}{2}kx^2 = U_i - U_f \qquad\qquad 8\text{-}4$$

In this case, U_f is the potential energy when the spring is at $x = 0$ (equilibrium position), and U_i is the potential energy when the spring is stretched by the amount x.

A convenient choice for $U = 0$ is the equilibrium position of the spring. With this choice we have $U_f = 0$, and Equation 8-4 becomes $U_i = \frac{1}{2}kx^2$. Omitting the subscript i, so that U represents the potential energy of the spring for an arbitrary amount of stretch x, we have:

Potential Energy of a Spring

$$U = \tfrac{1}{2}kx^2 \qquad\qquad 8\text{-}5$$

Since U depends on x^2, which is positive even if x is negative, the potential energy of a spring is always greater than or equal to zero. Thus, a spring's potential energy increases whenever it is displaced from equilibrium. A dramatic example of a stretched spring is shown in FIGURE 8-11.

▲ **FIGURE 8-11** Because springs, and bungee cords, exert conservative forces, they can serve as energy storage devices. In this case, the stretched bungee cord is beginning to give up the energy it has stored, and to convert that potential energy into kinetic energy as the jumper is pulled rapidly skyward.

EXERCISE 8-5 SPRING POTENTIAL ENERGY

(a) Find the potential energy of a spring with force constant $k = 740$ N/m if it is stretched by 2.1 cm. (b) Find the force constant of a spring whose potential energy is 1.7 J when it is compressed by 6.4 cm.

REASONING AND SOLUTION

(a) The spring potential energy is $U = \frac{1}{2}kx^2$. We substitute $k = 740$ N/m and $x = 0.021$ m:

$$U = \tfrac{1}{2}(740\,\text{N/m})(0.021\,\text{m})^2 = 0.16\,\text{J}$$

(b) We can rearrange $U = \frac{1}{2}kx^2$ to find the force constant, $k = \dfrac{2U}{x^2}$. Then we substitute $U = 1.7$ J and $x = -0.064$ m:

$$k = \frac{2U}{x^2} = \frac{2(1.7\,\text{J})}{(-0.064\,\text{m})^2} = 830\,\text{N/m}$$

Notice that the potential energies for gravity $(U = mgy)$ and a spring $\left(U = \frac{1}{2}kx^2\right)$ are given by different expressions. As mentioned earlier, each conservative force has its own form of potential energy.

EXAMPLE 8-6   COMPRESSED ENERGY AND THE JUMP OF A FLEA

When a force of 120.0 N is applied to a certain spring, it causes a stretch of 2.25 cm. What is the potential energy of this spring when it is **(a)** compressed by 3.50 cm or **(b)** expanded by 7.00 cm?

PICTURE THE PROBLEM

The top sketch shows the spring stretched 2.25 cm by the force $F_1 = 120.0$ N. The lower sketch shows the same spring compressed by a second force, F_2, which causes a compression of 3.50 cm. An expansion of the spring by 7.00 cm would look similar to the top sketch.

CONTINUED

REASONING AND STRATEGY

From the first piece of information—a certain force causes a certain stretch—we can calculate the force constant using $F = kx$. Once we know k, we can find the potential energy for either a compression or an expansion with $U = \frac{1}{2}kx^2$.

Known Applied force, $F = 120.0$ N; resulting stretch of spring, $x = 0.0225$ m; **(a)** compression, $x = -0.0350$ m; **(b)** expansion, $x = 0.0700$ m.

Unknown **(a)** Potential energy of spring, $U = ?$ **(b)** Potential energy of spring, $U = ?$

SOLUTION

1. Solve $F = kx$ for the spring constant, k:

$$F = kx$$
$$k = \frac{F}{x} = \frac{120.0 \text{ N}}{0.0225 \text{ m}} = 5330 \text{ N/m}$$

Part (a)

2. Substitute $k = 5330$ N/m and $x = -0.0350$ m into the potential energy expression, $U = \frac{1}{2}kx^2$:

$$U = \tfrac{1}{2}kx^2 = \tfrac{1}{2}(5330 \text{ N/m})(-0.0350 \text{ m})^2 = 3.26 \text{ J}$$

Part (b)

3. Substitute $k = 5330$ N/m and $x = 0.0700$ m into the potential energy expression, $U = \frac{1}{2}kx^2$:

$$U = \tfrac{1}{2}kx^2 = \tfrac{1}{2}(5330 \text{ N/m})(0.0700 \text{ m})^2 = 13.1 \text{ J}$$

INSIGHT

Though this Example deals with ideal springs, the same basic physics applies to many other real-world situations. A case in point is the jump of a flea, in which a flea can propel itself up to 100 times its body length. The physics behind this feat is the slow accumulation of energy in a "springy" strip of resilin in the thigh and coxa of the flea, as shown in the accompanying sketches, and the sudden release of this energy at a later time. Specifically, as the flea's muscles flex the leg, the resilin strip is compressed, storing the work done by the muscles in the form of potential energy, $U = \frac{1}{2}kx^2$. Later, when a trigger mechanism unlocks the flexed leg, the energy stored in the resilin is released explosively—rapidly extending the leg and propelling the flea upward. See Problem 69 for a calculation using the force constant of resilin.

PRACTICE PROBLEM

What stretch is necessary for the spring in this Example to have a potential energy of 5.00 J? **[Answer: 4.33 cm]**

Some related homework problems: Problem 12, Problem 14, and Problem 16

The jump of a flea is similar in many respects to the operation of a bow and arrow. In the latter case, the work done in slowly pulling the string back is stored in the flex of the bow. The string is held in place while aim is taken, and then released to allow the bow to return to its original shape. This propels the arrow forward with great speed—a speed many times faster than could be obtained by simply throwing the arrow with the same arm muscles that pulled back on the string. An analogous process occurs in the flash unit of a camera, as we shall see in Chapter 20.

BIO Another example of a creature that has the ability to store energy in a spring is the peacock mantis shrimp (*Odontodactylus scyllarus*), shown in FIGURE 8-12. This small crustacean can kill its prey by smashing it with a club-like appendage at amazing speeds. By storing energy in a spring-like portion of exoskeleton on its appendage, it can move the club from rest to 23 m/s with an acceleration equivalent to $10,600g$, similar to the acceleration of a 0.22-caliber bullet when fired. The club strikes with a force as large as 1500 N—two thousand times the weight of the shrimp—and is capable of shattering snails, crabs, mollusks, and rock oysters, cracking their shells into pieces. Mantis shrimp have even been known to break the glass walls of the aquarium in which they are kept! The material from which its club is made is exceptionally tough, and has inspired the design of new carbon fiber–epoxy composite materials.

Enhance Your Understanding
(Answers given at the end of the chapter)

2. The work done by a conservative force on a system is 10 J, and the initial potential energy of the system is 30 J. **(a)** Is the system's final potential energy greater than, less than, or equal to 30 J? Explain. **(b)** What is the final potential energy of the system?

Section Review

• Potential energy is an energy storage mechanism.

• The work done by a conservative force is equal to the negative of the change in potential energy.

▲ **FIGURE 8-12** A peacock mantis shrimp is a formidable predator with its spring-loaded, club-like appendages.

8-3 Conservation of Mechanical Energy

In this section, we show how potential energy can be used as a powerful tool in solving a variety of problems and in gaining greater insight into the workings of physical systems. To do so, we begin by defining the **mechanical energy,** E, as the sum of the potential and kinetic energies of an object:

$$E = U + K \qquad \text{8-6}$$

Big Idea **3** Mechanical energy is the sum of the potential and kinetic energies of a system. The total amount of mechanical energy is conserved—remains constant—for systems that have only conservative forces.

The significance of mechanical energy is that it is **conserved** in systems that involve only conservative forces. By conserved, we mean that its value never changes; that is, $E = \text{constant}$. (In situations where nonconservative forces are involved, the mechanical energy can change, as when friction converts mechanical energy to thermal energy. When *all* possible forms of energy are considered, like mechanical and thermal energy, energy is always found to be conserved.)

Mechanical Energy and Conservative Forces To show that the mechanical energy E is conserved for conservative forces, we start with the work–energy theorem from Chapter 7:

$$W_{\text{total}} = \Delta K = K_{\text{f}} - K_{\text{i}}$$

Suppose for a moment that the system has only a single force and that the force is conservative. If this is the case, then the total work, W_{total}, is the work done by the conservative force, W_{c}:

$$W_{\text{total}} = W_{\text{c}}$$

▲ **FIGURE 8-13 Visualizing Concepts**
Conservation of Energy A roller coaster (top) illustrates the conservation of energy. With every descent, gravitational potential energy is converted into kinetic energy; with every rise, kinetic energy is converted back into gravitational potential energy. If friction is neglected, the total mechanical energy of the car remains constant. The same principle is exploited at a pumped-storage facility, such as this one at the Mormon Flat Dam near Phoenix, Arizona (bottom). When surplus electrical power is available, it is used to pump water uphill into the reservoir. This process, in effect, stores electrical energy as gravitational potential energy. When power demand is high, the stored water is allowed to flow back downhill through the electrical generators in the dam, converting the gravitational energy to kinetic energy and the kinetic energy to electrical energy.

PHYSICS IN CONTEXT
Looking Ahead

Conservation of energy is one of the key elements in the study of elastic collisions in Section 9-6.

PROBLEM-SOLVING NOTE

Conservative Systems

A convenient approach to problems involving energy conservation is to first sketch the system, and then label the initial and final points with i and f, respectively. To apply energy conservation, write out the energy at these two points and set $E_i = E_f$.

From the definition of potential energy, we know that $W_c = -\Delta U = U_i - U_f$. Combining these results, we have

$$W_{\text{total}} = W_c$$
$$K_f - K_i = U_i - U_f$$

With a slight rearrangement we find

$$U_f + K_f = U_i + K_i$$

Recalling that $E = U + K$, we have

$$E_f = E_i$$

The initial and final points can be chosen arbitrarily, and hence it follows that E is conserved:

$$E = \text{constant}$$

If the system has more than one conservative force, the only change to these results is to replace U with the sum of the potential energies of all the forces.

To summarize:

Conservation of Mechanical Energy

In systems with conservative forces only, the mechanical energy E is conserved; that is, $E = U + K = \text{constant}$.

In terms of physical systems, conservation of mechanical energy means that energy can be converted between potential and kinetic forms, but that the sum remains the same. As an example, in the roller coaster shown in Figure 8-6, the gravitational potential energy decreases as the car approaches point B; as it does, the car's kinetic energy increases by the same amount. Similar examples of energy conservation are presented in **FIGURE 8-13**. From a practical point of view, conservation of mechanical energy means that many physics problems can be solved by what amounts to simple bookkeeping.

Applying Energy Conservation Consider a key chain of mass m that is dropped to the floor from a height h, as illustrated in **FIGURE 8-14**. The question is, how fast are the keys moving just before they land? We know how to solve this problem using Newton's laws and kinematics, but let's see how energy conservation can be used instead.

First, we note that the only force acting on the keys is gravity—ignoring air resistance, of course—and that gravity is a conservative force. As a result, we can say that $E = U + K$ is constant during the entire time the keys are falling. To solve the problem, then, we pick two points on the motion of the keys, say i and f in Figure 8-14, and we set the mechanical energy equal at these points:

$$E_i = E_f \qquad \text{8-7}$$

Writing this out in terms of potential and kinetic energies, we have

$$U_i + K_i = U_f + K_f \qquad \text{8-8}$$

This one equation—which is nothing but bookkeeping—can be used to solve for the one unknown, the final speed.

To be specific, in Figure 8-14 (a) we choose $y = 0$ at ground level, which means that $U_i = mgh$. In addition, the fact that the keys are released from rest means that $K_i = 0$. Similarly, at point f—just before hitting the ground—the energy is all kinetic, and the potential energy is zero; that is, $U_f = 0, K_f = \frac{1}{2}mv^2$. Substituting these values into $U_i + K_i = U_f + K_f$, we find

$$mgh + 0 = 0 + \frac{1}{2}mv^2$$

Canceling m and solving for v yields the same result we get with kinematics:

$$v = \sqrt{2gh}$$

Suppose, instead, that we had chosen $y = 0$ to be at the release point of the keys, as in Figure 8-14 (b), so that the keys land at $y = -h$. Now, when the keys are released,

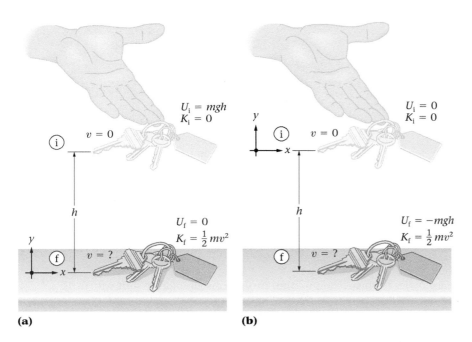

$U_i = mgh$
$K_i = 0$

$v = 0$

h

$U_f = 0$
$K_f = \frac{1}{2}mv^2$

$v = ?$

(a)

$U_i = 0$
$K_i = 0$

$v = 0$

h

$U_f = -mgh$
$K_f = \frac{1}{2}mv^2$

$v = ?$

(b)

◀ **FIGURE 8-14 Solving a kinematics problem using conservation of energy (a)** A set of keys falls to the floor. Ignoring frictional forces, we know that the mechanical energy at points i and f must be equal: $E_i = E_f$. Using this condition, we can find the speed of the keys just before they land. **(b)** The same physical situation as in part (a), except this time we have chosen $y = 0$ to be at the point where the keys are dropped. As before, we set $E_i = E_f$ to find the speed of the keys just before they land. The result is the same.

we have $U_i = 0$ and $K_i = 0$, and when they land, $U_f = -mgh$ and $K_f = \frac{1}{2}mv^2$. Substituting these results in $U_i + K_i = U_f + K_f$ yields

$$0 + 0 = -mgh + \frac{1}{2}mv^2$$

Solving for v gives the same result:

$$v = \sqrt{2gh}$$

Thus, as expected, changing the zero level has no effect on the physical results.

EXAMPLE 8-7 GRADUATION FLING

At the end of a graduation ceremony, graduates fling their caps into the air. Suppose a 0.120-kg cap is thrown straight upward with an initial speed of 7.85 m/s, and that frictional forces can be ignored. **(a)** Use kinematics to find the speed of the cap when it is 1.18 m above the release point. **(b)** Show that the mechanical energy at the release point is the same as the mechanical energy 1.18 m above the release point.

PICTURE THE PROBLEM
In our sketch, we choose $y = 0$ to be at the level where the cap is released with an initial speed of 7.85 m/s, and we choose upward to be the positive direction. In addition, we designate the release point as i (initial) and the point at which $y = 1.18$ m as f (final). It is the speed at point f that we wish to find.

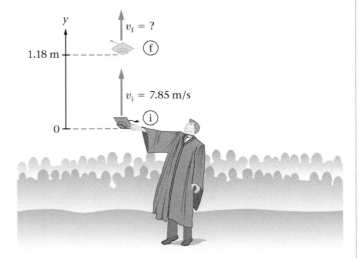

y

$v_f = ?$

f

1.18 m

$v_i = 7.85$ m/s

i

0

REASONING AND STRATEGY

a. The cap is in free fall, which justifies the use of constant-acceleration kinematics. In addition, the cap is thrown straight upward, and hence the motion is one dimensional along the y direction. We want to relate velocity to position; thus we use $v_f^2 = v_i^2 + 2a_y\Delta y$ (Section 2-5). In this case, $v_i = 7.85$ m/s, $\Delta y = 1.18$ m, and $a_y = -g$. Substituting these values gives v_f.

b. At the points i and f we calculate the mechanical energy, $E = U + K$, with $U = mgy$ for the gravitational potential energy. The kinetic energy, as always, is $K = \frac{1}{2}mv^2$.

Known   Mass of cap, $m = 0.120$ kg; initial speed, $v_i = 7.85$ m/s; height above release point, $\Delta y = 1.18$ m.
Unknown   **(a)** Speed of cap, $v_f = ?$ **(b)** Mechanical energy at initial and final points, $E_i = ?$, $E_f = ?$

CONTINUED

SOLUTION

Part (a)

1. Use kinematics to solve for v_f:

$$v_f^2 = v_i^2 + 2a_y\,\Delta y$$
$$v_f = \pm\sqrt{v_i^2 + 2a_y\,\Delta y}$$

2. Substitute $v_i = 7.85$ m/s, $\Delta y = 1.18$ m, and $a_y = -g$ to find v_f. Choose the plus sign, since we are interested only in the speed:

$$v_f = \sqrt{v_i^2 + 2a_y\,\Delta y}$$
$$= \sqrt{(7.85\ \text{m/s})^2 + 2(-9.81\ \text{m/s}^2)(1.18\ \text{m})} = 6.20\ \text{m/s}$$

Part (b)

3. Calculate E_i. At this point $y_i = 0$ and $v_i = 7.85$ m/s:

$$E_i = U_i + K_i = mgy_i + \tfrac{1}{2}mv_i^2$$
$$= 0 + \tfrac{1}{2}(0.120\ \text{kg})(7.85\ \text{m/s})^2 = 3.70\ \text{J}$$

4. Calculate E_f. At this point $y_f = 1.18$ m and $v_f = 6.20$ m/s:

$$E_f = U_f + K_f = mgy_f + \tfrac{1}{2}mv_f^2$$
$$= (0.120\ \text{kg})(9.81\ \text{m/s}^2)(1.18\ \text{m}) + \tfrac{1}{2}(0.120\ \text{kg})(6.20\ \text{m/s})^2$$
$$= 1.39\ \text{J} + 2.31\ \text{J} = 3.70\ \text{J}$$

INSIGHT

As expected, E_f is equal to E_i. In the remaining Examples in this section we turn this process around; we start with $E_f = E_i$, and use this relationship to find a final speed or a final height. As we shall see, this procedure of using energy conservation is a more powerful approach—it actually makes the calculations much simpler.

PRACTICE PROBLEM

Use energy conservation to find the height at which the speed of the cap is 5.00 m/s. [**Answer:** 1.87 m]

Some related homework problems: Problem 27, Problem 29

Finding the Speed at a Given Height An interesting extension of Example 8-7 is shown in **FIGURE 8-15**. In this case, we are given that the speed of the cap is v_i at the height y_i, and we would like to know its speed v_f when it is at the height y_f.

To find v_f, we apply energy conservation to the points i and f:

$$U_i + K_i = U_f + K_f$$

Writing out U and K specifically for these two points yields the following:

$$mgy_i + \tfrac{1}{2}mv_i^2 = mgy_f + \tfrac{1}{2}mv_f^2$$

As before, we cancel m and solve for the unknown speed, v_f:

$$v_f = \sqrt{v_i^2 + 2g(y_i - y_f)}$$

This result is in agreement with the kinematic equation, $v_y^2 = v_{0y}^2 + 2a_y\,\Delta y$.

► **FIGURE 8-15 Speed is independent of path** If the speed of the cap is v_i at the height y_i, its speed is v_f at the height y_f, independent of the path between the two heights. This assumes, of course, that frictional forces can be neglected.

Notice that v_f depends only on y_i and y_f, not on the path connecting them. This is because conservative forces such as gravity do work that is path independent. What this means physically is that the cap has the same speed v_f at the height y_f, whether it goes straight upward or follows some other trajectory, as in Figure 8-15. All that matters is the height difference.

EXAMPLE 8-8 · CATCHING A HOME RUN

In the bottom of the ninth inning, a player hits a 0.15-kg baseball over the outfield fence. The ball leaves the bat with a speed of 36 m/s, and a fan in the bleachers catches it 7.2 m above the point where it was hit. Assuming frictional forces can be ignored, find **(a)** the kinetic energy of the ball and **(b)** its speed just before it's caught.

PICTURE THE PROBLEM
Our sketch shows the ball's trajectory. We label the hit point i and the catch point f. At point i we choose $y_i = 0$; at point f, then, $y_f = h = 7.2$ m. In addition, we are given that $v_i = 36$ m/s. The final speed v_f is to be determined.

REASONING AND STRATEGY

a. Because frictional forces can be ignored, it follows that the initial mechanical energy is equal to the final mechanical energy; that is, $U_i + K_i = U_f + K_f$. Use this relationship to find K_f.

b. Once K_f is determined, use $K_f = \frac{1}{2}mv_f^2$ to find v_f.

Known Mass of baseball, $m = 0.15$ kg; initial speed of ball, $v_i = 36$ m/s; catch height, $h = 7.2$ m.
Unknown **(a)** Kinetic energy of ball just before being caught, $K_f = ?$ **(b)** Speed of ball just before being caught, $v_f = ?$

SOLUTION

Part (a)

1. Begin by writing U and K for point i:
$$U_i = 0$$
$$K_i = \tfrac{1}{2}mv_i^2 = \tfrac{1}{2}(0.15 \text{ kg})(36 \text{ m/s})^2 = 97 \text{ J}$$

2. Next, write U and K for point f:
$$U_f = mgh = (0.15 \text{ kg})(9.81 \text{ m/s}^2)(7.2 \text{ m}) = 11 \text{ J}$$
$$K_f = \tfrac{1}{2}mv_f^2$$

3. Set the total mechanical energy at point i, $E_i = U_i + K_i$, equal to the total mechanical energy at point f, $E_f = U_f + K_f$, and solve for K_f:
$$U_i + K_i = U_f + K_f$$
$$0 + 97 \text{ J} = 11 \text{ J} + K_f$$
$$K_f = 97 \text{ J} - 11 \text{ J} = 86 \text{ J}$$

Part (b)

4. Use $K_f = \frac{1}{2}mv_f^2$ to find v_f:
$$K_f = \tfrac{1}{2}mv_f^2$$
$$v_f = \sqrt{\frac{2K_f}{m}} = \sqrt{\frac{2(86 \text{ J})}{0.15 \text{ kg}}} = 34 \text{ m/s}$$

INSIGHT
To find the ball's speed just before being caught, we need to know the height of point f, but we don't need to know any details about the ball's trajectory. For example, it is not necessary to know the angle at which the ball leaves the bat or its maximum height.

The histograms show the values of U and K at the points i and f. Notice that the energy of the system is still mostly kinetic just before the ball is caught.

PRACTICE PROBLEM
If the mass of the ball were increased, would the catch speed be greater than, less than, or the same as the value we just found? [**Answer:** The same. U and K depend on mass in the same way; hence the mass cancels.]

Some related homework problems: Problem 25, Problem 27

The connection between height difference and speeds is explored further in the following two Examples. The first Example deals with the speed of swimmers on water slides, like those pictured in **FIGURE 8-16**.

CONCEPTUAL EXAMPLE 8-9   **COMPARE THE FINAL SPEEDS**

Swimmers at a water park can enter a pool using one of two frictionless slides of equal height. Slide 1 approaches the water with a uniform slope; slide 2 dips rapidly at first, then levels out. **(a)** Is the speed at the bottom of slide 1 greater than, less than, or the same as the speed at the bottom of slide 2? **(b)** Which of the following is the *best explanation* for your prediction?

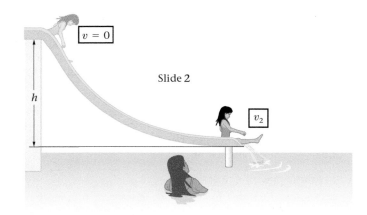

  I. The same amount of potential energy is converted to kinetic energy in each case, so the speeds are the same.

 II. Slide 2 dips rapidly, resulting in a greater speed for that slide.

III. The uniform drop of slide 1 results in a steady increase in speed, and a greater final speed for that slide.

REASONING AND DISCUSSION
In both cases, the same amount of potential energy, mgh, is converted to kinetic energy. Because the conversion of gravitational potential energy to kinetic energy is the *only* energy transaction taking place, it follows that the speed is the same for each slide.

    Interestingly, although the final speeds are the same, the time required to reach the water is less for slide 2. The reason is that swimmer 2 reaches a high speed early and maintains it, whereas the speed of swimmer 1 increases slowly and steadily.

ANSWER
(a) The speeds are the same for the two slides. **(b)** The best explanation is I.

EXAMPLE 8-10   **SKATEBOARD EXIT RAMP**

A 55-kg skateboarder enters a ramp moving horizontally with a speed of 6.5 m/s and leaves the ramp moving vertically with a speed of 4.1 m/s. Find the height of the ramp, assuming no energy loss to frictional forces.

PICTURE THE PROBLEM
We choose $y = 0$ to be the level of the bottom of the ramp; thus the gravitational potential energy is zero there. Point i indicates the skateboarder entering the ramp with a speed of 6.5 m/s; point f is the top of the ramp, where the speed is 4.1 m/s.

REASONING AND STRATEGY
To find h, we simply set the initial energy, $E_i = U_i + K_i$, equal to the final energy, $E_f = U_f + K_f$.

Known   Mass of skateboarder, $m = 55$ kg; initial speed, $v_i = 6.5$ m/s; final speed, $v_f = 4.1$ m/s.

Unknown Height of ramp, $h = ?$

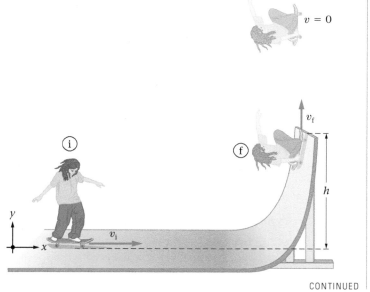

CONTINUED

SOLUTION

1. Write expressions for U_i and K_i:

$$U_i = mg \cdot 0 = 0 \qquad K_i = \frac{1}{2}mv_i^2$$

2. Write expressions for U_f and K_f:

$$U_f = mgh \qquad K_f = \frac{1}{2}mv_f^2$$

3. Set the total mechanical energy at point i, $E_i = U_i + K_i$, equal to the total mechanical energy at point f, $E_f = U_f + K_f$:

$$U_i + K_i = U_f + K_f$$
$$0 + \frac{1}{2}mv_i^2 = mgh + \frac{1}{2}mv_f^2$$

4. Solve for h. Notice that m cancels:

$$mgh = \frac{1}{2}mv_i^2 - \frac{1}{2}mv_f^2$$

$$h = \frac{v_i^2 - v_f^2}{2g}$$

5. Substitute numerical values:

$$h = \frac{(6.5 \text{ m/s})^2 - (4.1 \text{ m/s})^2}{2(9.81 \text{ m/s}^2)} = 1.3 \text{ m}$$

INSIGHT

Our result for h is independent of the shape of the ramp—it is equally valid for one with the shape shown here, or one that simply inclines upward at a constant angle. In addition, the height does not depend on the person's mass, as we see in Step 4 where the mass cancels.

The histograms show U and K to scale at the points i and f, as well as at the maximum height, where $K = 0$.

PRACTICE PROBLEM

What is the skateboarder's maximum height above the bottom of the ramp?   [**Answer:** 2.2 m]

Some related homework problems: Problem 23, Problem 29, Problem 60

It's useful to express the equation in Step 3 from Example 8-10 $\left(0 + \frac{1}{2}mv_i^2 = mgh + \frac{1}{2}mv_f^2\right)$ in words. First, the left side of the equation is the initial kinetic energy of the skateboarder, $\frac{1}{2}mv_i^2$. This is the initial energy content of the system. At point f the system still has the same amount of energy, only now part of it, mgh, is in the form of gravitational potential energy. The remainder is the final kinetic energy, $\frac{1}{2}mv_f^2$.

Conceptual Example 8-11 considers the effect of a slight change in the initial speed of an object.

▶ **FIGURE 8-16** Does the shape of the slide matter? (See Conceptual Example 8-9.)

CONCEPTUAL EXAMPLE 8-11   WHAT'S THE FINAL SPEED

A snowboarder coasts on a smooth track that rises from one level to another. If the snowboarder's initial speed is 4 m/s, the snowboarder just makes it to the upper level and comes to rest. With a slightly greater initial speed of 5 m/s, the snowboarder is still moving to the right on the upper level. Is the snowboarder's final speed in this case **(a)** 1 m/s, **(b)** 2 m/s, or **(c)** 3 m/s?

REASONING AND DISCUSSION

A plausible-sounding answer is that since the initial speed is greater by 1 m/s in the second case, the final speed should be greater by 1 m/s as well. Therefore, the answer should be 0 + 1 m/s = 1 m/s. This is incorrect, however.

As surprising as it may seem, an increase in the initial speed from 4 m/s to 5 m/s results in an increase in the final speed from 0 to 4 m/s. This is due to the fact that kinetic energy depends on v^2 rather than v; thus, it is the difference in v^2 that counts. In this case, the initial value of v^2 increases from 16 m²/s² to 25 m²/s², for a total increase of 25 m²/s² − 16 m²/s² = 9 m²/s². The final value of v^2 must increase by the same amount, 9 m²/s² = (3 m/s)². As a result, the final speed is 3 m/s.

ANSWER

(c) The final speed of the snowboarder in the second case is 3 m/s.

**PHYSICS
IN CONTEXT**
Looking Ahead

The wide-ranging importance of energy conservation is illustrated by its use in the following disparate topics: rotational motion (Section 10-6), gravitation (Section 12-5), oscillatory motion (Section 13-5), fluid dynamics (Section 15-7), and phase changes (Section 17-6).

Let's check the results of the previous Conceptual Example with a specific numerical example. Suppose the snowboarder has a mass of 74.0 kg. It follows that in the first case the initial kinetic energy is $K_i = \frac{1}{2}(74.0 \text{ kg})(4.00 \text{ m/s})^2 = 592 \text{ J}$. At the top of the hill all of this kinetic energy is converted to gravitational potential energy, mgh.

In the second case, the initial speed of the snowboarder is 5.00 m/s; thus, the initial kinetic energy is $K_i = \frac{1}{2}(74.0 \text{ kg})(5.00 \text{ m/s})^2 = 925 \text{ J}$. When the snowboarder reaches the top of the hill, 592 J of this kinetic energy is converted to gravitational potential energy, leaving the snowboarder with a final kinetic energy of $925 \text{ J} - 592 \text{ J} = 333 \text{ J}$. The corresponding speed is given by

$$\frac{1}{2}mv^2 = 333 \text{ J}$$

$$v = \sqrt{\frac{2(333 \text{ J})}{m}} = \sqrt{\frac{2(333 \text{ J})}{74.0 \text{ kg}}} = \sqrt{9.00 \text{ m}^2/\text{s}^2} = 3.00 \text{ m/s}$$

Thus, as expected, the snowboarder in the second case has a final speed of 3.00 m/s. We conclude this section with two Examples involving springs.

EXAMPLE 8-12 SPRING TIME

A 1.70-kg block slides on a horizontal, frictionless surface until it encounters a spring with a force constant of 955 N/m. The block comes to rest after compressing the spring a distance of 4.60 cm. Find the initial speed of the block. (Ignore air resistance and any energy lost when the block initially contacts the spring.)

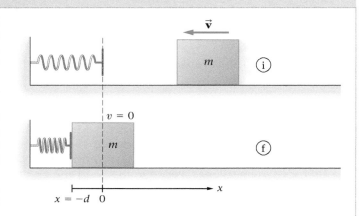

PICTURE THE PROBLEM
Point i refers to times before the block makes contact with the spring, which means the block has a speed v_i and the end of the spring is at $x = 0$. Point f refers to the time when the block has come to rest, which means $v_f = 0$, and the spring is compressed to $x = -d = -4.60 \text{ cm}$.

We can choose the center of the block to be the $y = 0$ level. With this choice, the gravitational potential energy of the sytem is zero at all times.

REASONING AND STRATEGY
Set E_i equal to E_f to find the one unknown, v_i. Notice that the initial energy, E_i, is the kinetic energy of the block before it reaches the spring. The final energy, E_f, is the potential energy of the compressed spring.

Known   Mass of block, $m = 1.70$ kg; force constant, $k = 955$ N/m; compression distance, $d = 0.0460$ m; final speed, $v_f = 0$.
Unknown   Initial speed of block, $v_i = ?$

SOLUTION

1. Write expressions for U_i and K_i. For U, we consider only the potential energy of the spring, $U = \frac{1}{2}kx^2$:

$$U_i = \frac{1}{2}k \cdot 0^2 = 0 \qquad K_i = \frac{1}{2}mv_i^2$$

2. Write expressions for U_f and K_f:

$$U_f = \frac{1}{2}k(-d)^2 = \frac{1}{2}kd^2 \qquad K_f = \frac{1}{2}m \cdot 0^2 = 0$$

3. Set the initial mechanical energy, $E_i = U_i + K_i$, equal to the final mechanical energy, $E_f = U_f + K_f$, and solve for v:

$$U_i + K_i = U_f + K_f$$
$$0 + \frac{1}{2}mv_i^2 = \frac{1}{2}kd^2 + 0$$
$$v_i = d\sqrt{\frac{k}{m}}$$

4. Substitute numerical values:

$$v_i = d\sqrt{\frac{k}{m}} = (0.0460 \text{ m})\sqrt{\frac{955 \text{ N/m}}{1.70 \text{ kg}}} = 1.09 \text{ m/s}$$

INSIGHT
After the block comes to rest, the spring expands again, converting its potential energy back into the kinetic energy of the block. When the block leaves the spring, moving to the right, its speed is once again 1.09 m/s.

PRACTICE PROBLEM
What is the compression distance, d, if the block's initial speed is 0.500 m/s?   [**Answer:** 2.11 cm]

Some related homework problems: Problem 26, Problem 28

QUICK EXAMPLE 8-13 FIND THE SPEED OF THE BLOCK

Suppose the spring and block in Example 8-12 are oriented vertically. Initially, the spring is compressed 4.60 cm and the block is at rest. When the block is released, it accelerates upward. Find the speed of the block when the spring has returned to its equilibrium position.

REASONING AND SOLUTION

The physical system is shown in our sketch, with the positive y direction upward, and $y = 0$ at the equilibrium position of the spring. The initial condition corresponds to the block at rest at a negative value of y and the spring compressed. Thus, the initial mechanical energy is the positive spring potential energy $(U = \frac{1}{2}kd^2)$, plus the negative gravitational potential energy $(U = -mgd)$, plus zero kinetic energy $(K = 0)$. The final condition corresponds to the block leaving the uncompressed spring, where the both the spring potential energy and gravitational potential energy are zero. Therefore, the final mechanical energy is simply the kinetic energy of the block. We set E_i equal to E_f to find the one unknown, which in this case is v_f.

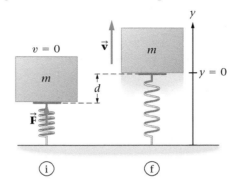

1. Write an expression for the initial mechanical energy E_i:
$$E_i = U_i + K_i = \tfrac{1}{2}kd^2 - mgd + 0$$

2. Write an expression for the final mechanical energy E_f:
$$E_f = U_f + K_f = 0 + 0 + \tfrac{1}{2}mv_f^2$$

3. Set E_i equal to E_f and solve for v_f:
$$\tfrac{1}{2}kd^2 - mgd = \tfrac{1}{2}mv_f^2$$
$$v_f = \sqrt{\frac{kd^2}{m} - 2gd}$$

4. Substitute numerical values:
$$v_f = \sqrt{\frac{(955\ \text{N/m})(0.0460\ \text{m})^2}{1.70\ \text{kg}} - 2(9.81\ \text{m/s}^2)(0.0460\ \text{m})}$$
$$= 0.535\ \text{m/s}$$

In this system, part of the initial potential energy of the spring $\left(\frac{1}{2}kd^2\right)$ goes into increasing the gravitational potential energy of the block (mgd). The remainder of the initial energy, $\frac{1}{2}kd^2 - mgd$, is converted into the block's kinetic energy.

Enhance Your Understanding   (Answers given at the end of the chapter)

3. A system with only conservative forces acting on it has a mechanical energy of 50 J and an initial kinetic energy of 25 J. If the final potential energy of the system is 15 J, what is its final kinetic energy?

Section Review

- If a system is acted on by conservative forces only, its mechanical energy, $E = U + K$, is conserved.

8-4 Work Done by Nonconservative Forces

Nonconservative forces change the amount of mechanical energy in a system. They might decrease the mechanical energy by converting it to thermal energy, or increase it by converting muscular work to kinetic or potential energy. In some systems, both types of processes occur at the same time.

Connecting Nonconservative Work and Energy Change   To see the connection between the work done by a nonconservative force, W_{nc}, and the mechanical energy, E, we return once more to the work–energy theorem, which says that the *total* work is equal to the change in kinetic energy:

$$W_{total} = \Delta K$$

Suppose, for instance, that a system has one conservative and one nonconservative force. In this case, the total work is the sum of the conservative work W_c and the nonconservative work W_{nc}:

$$W_{total} = W_c + W_{nc}$$

Recalling that conservative work is related to the change in potential energy by $W_c = -\Delta U$, we have

$$W_{total} = -\Delta U + W_{nc} = \Delta K$$

Solving this relationship for the nonconservative work yields

$$W_{nc} = \Delta U + \Delta K$$

Finally, since the total mechanical energy is $E = U + K$, it follows that the *change* in mechanical energy is $\Delta E = \Delta U + \Delta K$. As a result, *the nonconservative work is simply the change in mechanical energy:*

$$W_{nc} = \Delta E = E_f - E_i \qquad\qquad \text{8-9}$$

If more than one nonconservative force acts, we simply add the nonconservative work done by each such force to obtain W_{nc}.

At this point it may be useful to collect the three "working relationships" that have been introduced in the last two chapters:

$$\begin{aligned} W_{total} &= \Delta K \\ W_c &= -\Delta U \\ W_{nc} &= \Delta E \end{aligned} \qquad\qquad \text{8-10}$$

PROBLEM-SOLVING NOTE

Nonconservative Systems

Start by sketching the system and labeling the initial and final points with i and f, respectively. The initial and final mechanical energies are related to the nonconservative work by $W_{nc} = E_f - E_i$.

Notice that positive nonconservative work increases the total mechanical energy of a system, while negative nonconservative work decreases the mechanical energy—and converts it to other forms. In the next Example, for instance, part of the initial mechanical energy of a leaf is converted to heat and other forms of energy by air resistance as the leaf falls to the ground.

EXAMPLE 8-14   A LEAF FALLS IN THE FOREST: FIND THE NONCONSERVATIVE WORK

Deep in the forest, a 17.0-g leaf falls from a tree and drops straight to the ground. If its initial height was 5.30 m and its speed just before landing was 1.3 m/s, how much nonconservative work was done on the leaf?

PICTURE THE PROBLEM
In our sketch, the initial and final points are labeled i and f, respectively. At the initial point, the leaf drops from rest, $v_i = 0$, from a height $y_i = h = 5.30$ m. At the final point, just before landing at $y_f = 0$, the leaf's speed is $v_f = 1.3$ m/s.

REASONING AND STRATEGY
To begin, calculate the initial mechanical energy, E_i, and the final mechanical energy, E_f. Once these energies have been determined, the nonconservative work is the change in energy, $W_{nc} = \Delta E = E_f - E_i$.

Known   Mass of leaf, $m = 0.0170$ kg; initial height, $y_i = h = 5.30$ m; final height, $y_f = 0$; initial speed, $v_i = 0$; final speed, $v_f = 1.3$ m/s.

Unknown Nonconservative work done on leaf, $W_{nc} = ?$

CONTINUED

SOLUTION

1. Evaluate U_i, K_i, and E_i:

$$U_i = mgh = (0.0170 \text{ kg})(9.81 \text{ m/s}^2)(5.30 \text{ m}) = 0.884 \text{ J}$$
$$K_i = \tfrac{1}{2} m \cdot 0^2 = 0$$
$$E_i = U_i + K_i = 0.884 \text{ J}$$

2. Next, evaluate U_f, K_f, and E_f.

$$U_f = mg \cdot 0 = 0$$
$$K_f = \tfrac{1}{2} m v_f^2 = \tfrac{1}{2}(0.0170 \text{ kg})(1.3 \text{ m/s})^2 = 0.014 \text{ J}$$
$$E_f = U_f + K_f = 0.014 \text{ J}$$

3. Use $W_{nc} = \Delta E$ to find the nonconservative work:

$$W_{nc} = \Delta E = E_f - E_i = 0.014 \text{ J} - 0.884 \text{ J} = -0.870 \text{ J}$$

INSIGHT
Most of the leaf's initial mechanical energy is dissipated as it falls. This is indicated in the histograms. The small amount that remains (only about 1.6%) appears as the kinetic energy of the leaf just before it lands.

PRACTICE PROBLEM
What was the average nonconservative force exerted on the leaf as it fell?
[**Answer:** $W_{nc} = -Fh$, $F = -W_{nc}/h = 0.164 \text{ N}$, upward]

Some related homework problems: Problem 34, Problem 36, Problem 38

In the following Example, we use nonconservative work to find the drop height of a ski glove.

EXAMPLE 8-15   SNOW JOB

A skier drops her 0.312-kg glove from rest at a height h into fresh powder snow. The glove falls to a depth $d = 0.147 \text{ m}$ below the surface of the snow before coming to rest. If the nonconservative work done on the glove is $W_{nc} = -1.44 \text{ J}$, what is the drop height of the glove?

PICTURE THE PROBLEM
We choose the positive y direction to be upward. In addition, we set $y = 0$ at the level of the snow. The glove starts with zero speed at $y = h$, and ends with zero speed at $y = -d$.

REASONING AND STRATEGY
We start by calculating the initial and final mechanical energies, E_i and E_f. Given these energies, and the fact that $W_{nc} = -1.44 \text{ J}$, we can use $W_{nc} = \Delta E = E_f - E_i$ to solve for the one unknown, h.

Known   Mass of glove, $m = 0.312 \text{ kg}$; depth in snow, $d = 0.147 \text{ m}$; nonconservative work, $W_{nc} = -1.44 \text{ J}$.
Unknown   Drop height, $h = ?$

SOLUTION

1. Write the initial mechanical energy, E_i: $E_i = mgh + 0 = mgh$

2. Write the final mechanical energy, E_f: $E_f = mg(-d) + 0 = -mgd$

3. Set W_{nc} equal to ΔE: $W_{nc} = \Delta E = E_f - E_i = -mgd - mgh$

4. Solve for h: $h = -\dfrac{(W_{nc} + mgd)}{mg}$

5. Substitute numerical values: $h = -\dfrac{[-1.44 \text{ J} + (0.312 \text{ kg})(9.81 \text{ m/s}^2)(0.147 \text{ m})]}{(0.312 \text{ kg})(9.81 \text{ m/s}^2)} = 0.323 \text{ m}$

INSIGHT
Another way to write Step 3 is $E_f = E_i + W_{nc}$. In words, this equation says that the final mechanical energy is the initial mechanical energy plus the nonconservative work done on the system. In this case, $W_{nc} < 0$; hence the final mechanical energy is less than the initial mechanical energy.

CONTINUED

We now present a Conceptual Example that further examines the relationship between nonconservative work and distance.

CONCEPTUAL EXAMPLE 8-16   JUDGING A PUTT

A golfer badly misjudges a putt, sending the ball only one-quarter of the distance to the hole. The original putt gave the ball an initial speed of v_i. If the force of resistance due to the grass is constant, would an initial speed of **(a)** $2v_i$, **(b)** $3v_i$, or **(c)** $4v_i$ be needed to get the ball to the hole from its original position?

REASONING AND DISCUSSION
In the original putt, the ball started with a kinetic energy of $\frac{1}{2}mv_i^2$ and came to rest in the distance d. The kinetic energy was dissipated by the nonconservative force due to grass resistance, F, which does the work $W_{nc} = -Fd$. Since the change in mechanical energy is $\Delta E = 0 - \frac{1}{2}mv_i^2 = -\frac{1}{2}mv_i^2$, it follows from $W_{nc} = \Delta E$ that $Fd = \frac{1}{2}mv_i^2$. Therefore, to go four times the distance, $4d$, we need to give the ball four times as much kinetic energy. Noting that kinetic energy is proportional to v^2, we see that the initial speed need only be doubled.

ANSWER
(a) The initial speed should be doubled to $2v_i$.

A common example of a nonconservative force is kinetic friction. In the next Example, we show how to include the effects of friction in a system that also includes kinetic energy and gravitational potential energy.

EXAMPLE 8-17   LANDING WITH A THUD

A block of mass $m_1 = 2.40$ kg is connected to a second block of mass $m_2 = 1.80$ kg, as shown in the sketch. When the blocks are released from rest, they move through a distance $d = 0.500$ m, at which point m_2 hits the floor. Given that the coefficient of kinetic friction between m_1 and the horizontal surface is $\mu_k = 0.450$, find the speed of the blocks just before m_2 lands.

PICTURE THE PROBLEM
We choose $y = 0$ to be at floor level; therefore, the gravitational potential energy of m_2 is zero when it lands. The potential energy of m_1 doesn't change during this process; it is always m_1gh. Thus, it isn't necessary to know the value of h. We've labeled the initial and final points with i and f, respectively.

REASONING AND STRATEGY
A nonconservative force (friction) is doing work in this system, and hence we use $W_{nc} = \Delta E = E_f - E_i$. We can solve this relationship for the final mechanical energy, $E_f = E_i + W_{nc}$, and from this determine the final speed, v_f.

Known   Mass of block 1, $m_1 = 2.40$ kg; mass of block 2, $m_2 = 1.80$ kg; distance, $d = 0.500$ m; coefficient of kinetic friction, $\mu_k = 0.450$.
Unknown   Final speed (just before landing), $v_f = ?$

CONTINUED

SOLUTION

1. Evaluate U_i, K_i, and E_i. Be sure to include contributions from both masses:

$$U_i = m_1gh + m_2gd$$
$$K_i = \tfrac{1}{2}m_1 \cdot 0^2 + \tfrac{1}{2}m_2 \cdot 0^2 = 0$$
$$E_i = U_i + K_i = m_1gh + m_2gd$$

2. Next, evaluate U_f, K_f, and E_f. Note that E_f depends on the final speed, v_f:

$$U_f = m_1gh + 0$$
$$K_f = \tfrac{1}{2}m_1v_f^2 + \tfrac{1}{2}m_2v_f^2$$
$$E_f = U_f + K_f = m_1gh + \tfrac{1}{2}m_1v_f^2 + \tfrac{1}{2}m_2v_f^2$$

3. Calculate the nonconservative work, W_{nc}. Recall that the force of friction is $f_k = \mu_kN = \mu_km_1g$, and that it points opposite to the displacement of distance d:

$$W_{nc} = -f_kd = -\mu_km_1gd$$

4. Set W_{nc} equal to $\Delta E = E_f - E_i$ and solve for E_f. Notice that m_1gh cancels because it occurs in both E_i and E_f:

$$E_f = E_i + W_{nc}$$
$$\cancel{m_1gh} + \tfrac{1}{2}m_1v_f^2 + \tfrac{1}{2}m_2v_f^2 = \cancel{m_1gh} + m_2gd - \mu_km_1gd$$

5. Solve for v_f:

$$v_f = \sqrt{\frac{2(m_2 - \mu_km_1)gd}{m_1 + m_2}}$$

6. Substitute numerical values:

$$v_f = \sqrt{\frac{2[1.80\ \text{kg} - (0.450)(2.40\ \text{kg})](9.81\ \text{m/s}^2)(0.500\ \text{m})}{2.40\ \text{kg} + 1.80\ \text{kg}}}$$
$$= 1.30\ \text{m/s}$$

INSIGHT

Step 4 can be simplified to the following: $\tfrac{1}{2}m_1v_f^2 + \tfrac{1}{2}m_2v_f^2 = m_2gd - \mu_km_1gd$. Translating this into words, we can say that the final kinetic energy of the blocks is equal to the initial gravitational potential energy of m_2, minus the energy dissipated by friction on m_1.

PRACTICE PROBLEM

Find the coefficient of kinetic friction if the final speed of the blocks is 0.950 m/s. [**Answer:** $\mu_k = 0.589$]

Some related homework problems: Problem 40, Problem 42, Problem 64

Next, we present an Example for the common situation of a system in which two different nonconservative forces do work.

QUICK EXAMPLE 8-18   MARATHON MAN: FIND THE HEIGHT OF THE HILL

An 80.0-kg jogger starts from rest and runs uphill into a stiff breeze. At the top of the hill the jogger has done the work $W_{nc1} = +1.80 \times 10^4$ J, air resistance has done the work $W_{nc2} = -4420$ J, and the jogger's final speed is 3.50 m/s. Find the height of the hill.

REASONING AND SOLUTION

We choose $y = 0$ to be the level at the base of the hill. We also let the positive y direction be upward. The jogger starts with the speed $v_i = 0$ and ends with the speed $v_f = 3.50$ m/s. Nonconservative forces are present in this system, and hence we can use $W_{nc} = \Delta E = E_f - E_i$, where in this case $W_{nc} = W_{nc1} + W_{nc2}$. We can solve this relationship for the final energy, $E_f = E_i + W_{nc}$, and from this determine the height, h.

1. Calculate the initial mechanical energy, E_i:

$$E_i = U_i + K_i = 0 + 0 = 0$$

2. Calculate the final mechanical energy, E_f:

$$E_f = U_f + K_f = mgh + \tfrac{1}{2}mv_f^2$$

CONTINUED

3. Rearrange $W_{nc} = E_f - E_i$ and set E_f equal to $E_i + W_{nc}$:

$$E_f = E_i + W_{nc}$$
$$mgh + \tfrac{1}{2}mv_f^2 = 0 + W_{nc}$$

4. Rearrange to solve for h:

$$h = \frac{\left(W_{nc} - \tfrac{1}{2}mv_f^2\right)}{mg}$$

5. Calculate the total nonconservative work:

$$W_{nc} = W_{nc1} + W_{nc2} = 13{,}600\,\text{J}$$

6. Substitute numerical values to determine h:

$$h = \frac{\left[13{,}600\,\text{J} - \tfrac{1}{2}(80.0\,\text{kg})(3.50\,\text{m/s})^2\right]}{(80.0\,\text{kg})(9.81\,\text{m/s}^2)} = 16.7\,\text{m}$$

As usual when we deal with energy calculations, our final result is independent of the shape of the hill.

A dramatic example of the conservation of energy in the presence of nonconservative forces is shown in FIGURE 8-17.

Enhance Your Understanding   (Answers given at the end of the chapter)

4. A system is acted on by more than one force, resulting in the following amounts of work: $W_{total} = 100\,\text{J}$, $W_c = 35\,\text{J}$, and $W_{nc} = 65\,\text{J}$. Find the final values of the mechanical energy, kinetic energy, and potential energy of this system, given that their initial values are $E_i = 15\,\text{J}$, $K_i = 5\,\text{J}$, and $U_i = 10\,\text{J}$.

Section Review

- The work done by nonconservative forces on a system is equal to the change in the system's mechanical energy.

8-5 Potential Energy Curves and Equipotentials

FIGURE 8-18 shows a metal ball rolling on a roller coaster–like track. Initially the ball is at rest at point A. The height at A is $y = h$, and hence the ball's initial mechanical energy is $E_0 = mgh$. If friction and other nonconservative forces can be ignored, the ball's mechanical energy remains fixed at E_0 throughout its motion. Thus,

$$E = U + K = E_0$$

As the ball moves, its potential energy falls and rises in the same way as the track. After all, the gravitational potential energy, $U = mgy$, is directly proportional to the height of the track, y. In a sense, then, the track itself represents a graph of the corresponding potential energy.

This is shown explicitly in FIGURE 8-19, where we plot energy on the vertical axis and x on the horizontal axis. The potential energy U looks just like the track in Figure 8-18. In addition, we plot a horizontal line at the value E_0, indicating the constant energy of the ball. Since the potential energy plus the kinetic energy must always add up to E_0, it follows that K is the amount of energy from the potential energy curve up to the horizontal line at E_0. This is also shown in Figure 8-19.

Examining an energy plot like Figure 8-19 gives a great deal of information about the motion of an object. For example, at point B the potential energy has its lowest value, and thus the kinetic energy is greatest there. At point C the potential energy has increased, indicating a corresponding decrease in kinetic energy. As the ball continues to the right, the potential energy increases until, at point D, it is again equal to the total energy, E_0. At this point the kinetic energy is zero, and the ball comes to rest momentarily. It then "turns around" and begins to move to the left, eventually returning to point A where it again stops, changes direction, and begins a new cycle. Points A and D, then, are referred to as **turning points** of the motion.

▲ FIGURE 8-17 Steep highways often have "runaway truck" ramps that enable truck drivers whose brakes fail to bring their rigs to a safe stop. These ramps provide a perfect illustration of the conservation of energy. From a physics point of view, the driver's problem is to get rid of an enormous amount of kinetic energy in the safest possible way. The ramps run uphill, so some of the kinetic energy is simply converted back into gravitational potential energy. In addition, the ramps are typically surfaced with sand or gravel, allowing much of the initial kinetic energy to be dissipated by friction into other forms of energy, such as sound and heat.

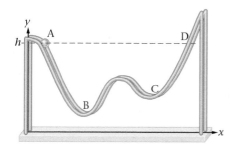

▲ FIGURE 8-18 **A ball rolling on a frictionless track** The ball starts at A, where $y = h$, with zero speed. Its greatest speed occurs at B. At D, where $y = h$ again, its speed returns to zero.

Chin Basher?

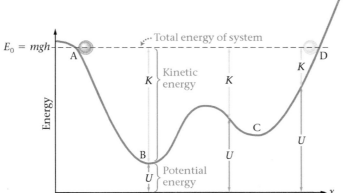

◄ **FIGURE 8-19 Gravitational potential energy versus position for the track shown in Figure 8-18** The shape of the potential energy curve is exactly the same as the shape of the track. In this case, the total mechanical energy is fixed at its initial value, $E_0 = U + K = mgh$. Because the height of the curve is U, by definition, it follows that K is the distance from the curve up to the dashed line at $E_0 = mgh$. Note that K is largest at B. In addition, K vanishes at A and D, which are turning points of the motion.

Turning points are also seen in the motion of a mass on a spring, as indicated in **FIGURE 8-20**. Figure 8-20 (a) shows a mass pulled to the position $x = A$ and then released from rest; Figure 8-20 (b) shows the potential energy of the system, $U = \frac{1}{2}kx^2$. Starting the system this way gives it an initial energy $E_0 = \frac{1}{2}kA^2$, shown by the horizontal line in Figure 8-20 (b). As the mass moves to the left, its speed increases, reaching a maximum where the potential energy is lowest, at $x = 0$. If no nonconservative forces act, the mass continues to $x = -A$, where it stops momentarily before returning to $x = A$. This type of **oscillatory motion** will be studied in detail in Chapter 13.

Side view of mountain

(a)   **(b)**

▲ **FIGURE 8-20 A mass on a spring (a)** A spring is stretched by an amount A, giving it a potential energy of $U = \frac{1}{2}kA^2$. **(b)** The potential energy curve, $U = \frac{1}{2}kx^2$, for the spring in (a). Because the mass starts at rest, its initial mechanical energy is $E_0 = \frac{1}{2}kA^2$. The mass oscillates between $x = A$ and $x = -A$.

In some cases, a two-dimensional plot of potential energy contours is useful. For instance, **FIGURE 8-21** shows a contour map of a hill. Each contour corresponds to a given altitude and, hence, to a given value of the gravitational potential energy. In general, lines corresponding to constant values of potential energy are called **equipotentials.** Since the altitude changes by equal amounts from one contour to the next, it follows that when gravitational equipotentials are packed close together, the corresponding terrain is steep. In contrast, when the equipotentials are widely spaced, the ground is nearly flat, since a large horizontal distance is required for a given change in altitude. We shall see similar plots with similar interpretations when we study electric potential energy in Chapter 20.

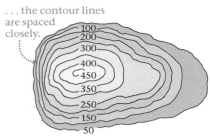

Contour map of mountain (from above)

▲ **FIGURE 8-21 A contour map** A small mountain (top, in side view) is very steep on the left, more gently sloping on the right. A contour map of this mountain (bottom) shows a series of equal-altitude contour lines from 50 ft to 450 ft. Notice that the contour lines are packed close together where the terrain is steep, but are widely spaced where it is more level.

Enhance Your Understanding   (Answers given at the end of the chapter)

5. A system consists of an object moving along the x axis subject to the potential energy curve shown in **FIGURE 8-22**. The mechanical energy of the system is E. Rank the points A, B, C, and D in order of increasing kinetic energy. Indicate ties where appropriate.

Section Review

- Potential energy curves show the dependence of potential energy on position.
- The potential energy of a system is constant along an equipotential curve.

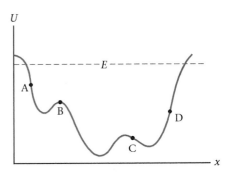

▲ **FIGURE 8-22** Enhance Your Understanding 5.

CHAPTER 8 REVIEW

CHAPTER SUMMARY

8-1 CONSERVATIVE AND NONCONSERVATIVE FORCES

Conservative forces conserve the mechanical energy of a system. Thus, in a conservative system the total mechanical energy remains constant.

Nonconservative forces convert mechanical energy into other forms of energy, or convert other forms of energy into mechanical energy.

Conservative Force, Definition

A conservative force does zero total work on any closed path. In addition, the work done by a conservative force in going from point A to point B is *independent of the path* from A to B. Examples of conservative forces are given in Table 8-1.

Nonconservative Force, Definition

The work done by a nonconservative force on a closed path is nonzero. The work is also path dependent. Examples of nonconservative forces are given in Table 8-1.

8-2 POTENTIAL ENERGY AND THE WORK DONE BY CONSERVATIVE FORCES

Potential energy, U, can "store" energy in a system. Energy in the form of potential energy can be converted to kinetic or other forms of energy.

Potential Energy, Definition

The work done by a conservative force is the negative of the change in potential energy:

$$W_c = -\Delta U = U_i - U_f \qquad \text{8-1}$$

Zero Level

Any location can be chosen for $U = 0$. Once the choice is made, however, it must be used consistently.

Gravity

Choosing $y = 0$ to be the zero level near Earth's surface,

$$U = mgy \qquad \text{8-3}$$

Spring

Choosing $x = 0$ (the equilibrium position) to be the zero level,

$$U = \frac{1}{2}kx^2 \qquad \text{8-5}$$

8-3 CONSERVATION OF MECHANICAL ENERGY

Mechanical energy, E, is conserved in systems with conservative forces only.

Mechanical Energy, Definition

Mechanical energy is the sum of the potential and kinetic energies of a system:

$$E = U + K \qquad \text{8-6}$$

8-4 WORK DONE BY NONCONSERVATIVE FORCES

Nonconservative forces can change the mechanical energy of a system.

Change in Mechanical Energy

The work done by a nonconservative force is equal to the change in the mechanical energy of a system:

$$W_{nc} = \Delta E = E_f - E_i \qquad \text{8-9}$$

8-5 POTENTIAL ENERGY CURVES AND EQUIPOTENTIALS

A potential energy curve plots U as a function of position.

An equipotential plot shows contours corresponding to constant values of U.

Turning Points

Turning points occur where an object stops momentarily before reversing direction. At turning points the kinetic energy is zero.

Oscillatory Motion

An object moving back and forth between two turning points is said to have oscillatory motion.

ANSWERS TO ENHANCE YOUR UNDERSTANDING QUESTIONS

1. **(a)** −10 J. **(b)** 10 J.
2. **(a)** Less than 30 J because positive work by a conservative force decreases the potential energy. **(b)** 20 J.
3. 35 J.
4. $E_f = 80$ J, $K_f = 105$ J, $U_f = -25$ J.
5. A < B < D < C.

CONCEPTUAL QUESTIONS

For instructor-assigned homework, go to www.masteringphysics.com.

(Answers to odd-numbered Conceptual Questions can be found in the back of the book.)

1. Is it possible for the kinetic energy of an object to be negative? Is it possible for the gravitational potential energy of an object to be negative? Explain.

2. If the stretch of a spring is doubled, the force it exerts is also doubled. By what factor does the spring's potential energy increase?

3. When a mass is placed on top of a vertical spring, the spring compresses and the mass moves downward. Analyze this system in terms of its mechanical energy.

4. If a spring is stretched so far that it is permanently deformed, its force is no longer conservative. Why?

5. An object is thrown upward to a person on a roof. At what point is the object's kinetic energy at maximum? At what point is the potential energy of the system at maximum? At what locations do these energies have their minimum values?

6. It is a law of nature that the total energy of the universe is conserved. What do politicians mean, then, when they urge "energy conservation"?

7. Discuss the various energy conversions that occur when a person performs a pole vault. Include as many conversions as you can, and consider times before, during, and after the actual vault itself.

How many energy conversions can you identify? (Conceptual Question 7)

8. Discuss the nature of the work done by the equipment shown in this photo. What types of forces are involved?

Conservative or nonconservative? (Conceptual Question 8)

9. If the force on an object is zero, does that mean the potential energy of the system is zero? If the potential energy of a system is zero, is the force zero?

10. When a ball is thrown upward, its mechanical energy, $E = mgy + \frac{1}{2}mv^2$, is constant with time if air resistance can be ignored. How does E vary with time if air resistance cannot be ignored?

11. When a ball is thrown upward, it spends the same amount of time on the way up as on the way down—as long as air resistance can be ignored. If air resistance is taken into account, is the time on the way down the same as, greater than, or less than the time on the way up? Explain.

PROBLEMS AND CONCEPTUAL EXERCISES

Answers to odd-numbered Problems and Conceptual Exercises can be found in the back of the book. **BIO** *identifies problems of biological or medical interest;* **CE** *indicates a conceptual exercise.* Predict/Explain *problems ask for two responses: (a) your prediction of a physical outcome, and (b) the best explanation among three provided; and* Predict/Calculate *problems ask for a prediction of a physical outcome, based on fundamental physics concepts, and follow that with a numerical calculation to verify the prediction. On all problems, bullets (•, ••, •••) indicate the level of difficulty.*

SECTION 8-1 CONSERVATIVE AND NONCONSERVATIVE FORCES

1. • **CE** The work done by a conservative force is indicated in **FIGURE 8-23** for a variety of different paths connecting the points A and B. What is the work done by this force **(a)** on path 1 and **(b)** on path 2?

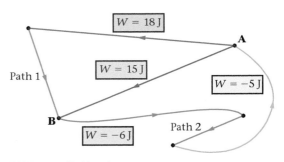

FIGURE 8-23 Problem 1

2. • Calculate the work done by gravity as a 3.2-kg object is moved from point A to point B in **FIGURE 8-24** along paths 1, 2, and 3.

FIGURE 8-24 Problem 2

3. • Calculate the work done by friction as a 3.7-kg box is slid along a floor from point A to point B in **FIGURE 8-25** along paths 1, 2, and 3. Assume that the coefficient of kinetic friction between the box and the floor is 0.26.

FIGURE 8-25 Problem 3

4. • Predict/Calculate A 2.8-kg block is attached to a spring with a force constant of 375 N/m, as shown in **FIGURE 8-26**. (a) Find the work done by the spring on the block as the block moves from A to B along paths 1 and 2. (b) How do your results depend on the mass of the block? Specifically, if you increase the mass, does the work done by the spring increase, decrease, or stay the same? (Assume the system is frictionless.)

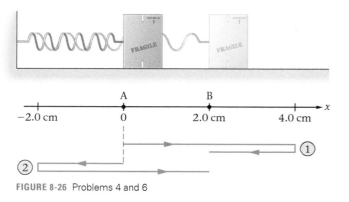

FIGURE 8-26 Problems 4 and 6

5. • Predict/Calculate (a) Calculate the work done by gravity as a 5.2-kg object is moved from A to B in **FIGURE 8-27** along paths 1 and 2. (b) How do your results depend on the mass of the block? Specifically, if you increase the mass, does the work done by gravity increase, decrease, or stay the same?

FIGURE 8-27 Problem 5

6. •• In the system shown in Figure 8-26, suppose the block has a mass of 2.7 kg, the spring has a force constant of 480 N/m, and the coefficient of kinetic friction between the block and the floor is 0.16. (a) Find the work done on the block by the spring and by friction as the block is moved from point A to point B along path 2. (b) Find the work done on the block by the spring and by friction if the block is moved directly from point A to point B.

SECTION 8-2 POTENTIAL ENERGY AND THE WORK DONE BY CONSERVATIVE FORCES

7. • **CE** Predict/Explain Ball 1 is thrown to the ground with an initial downward speed; ball 2 is dropped to the ground from rest. Assuming the balls have the same mass and are released from the same height, is the change in gravitational potential energy of ball 1 greater than, less than, or equal to the change in gravitational potential energy of ball 2? (b) Choose the *best explanation* from among the following:
  I. Ball 1 has the greater total energy, and therefore more energy can go into gravitational potential energy.
  II. The gravitational potential energy depends only on the mass of the ball and the drop height.
  III. All of the initial energy of ball 2 is gravitational potential energy.

8. • **CE** A mass is attached to the bottom of a vertical spring. This causes the spring to stretch and the mass to move downward. (a) Does the potential energy of the spring increase, decrease, or stay the same during this process? Explain. (b) Does the gravitational potential energy of the Earth–mass system increase, decrease, or stay the same during this process? Explain.

9. • Find the gravitational potential energy of an 88-kg person standing atop Mt. Everest at an altitude of 8848 m. Use sea level as the location for $y = 0$.

10. • A student lifts a 1.42-kg book from her desk to a bookshelf. If the gravitational potential energy of the book–Earth system increases by 9.08 J, how high is the bookshelf above the desk?

11. • At the local ski slope, an 82.0-kg skier rides a gondola to the top of the mountain. If the lift has a length of 2950 m and makes an angle of 13.1° with the horizontal, what is the change in the gravitational potential energy of the skier–Earth system?

12. •• **BIO The Wing of the Hawkmoth** Experiments performed on the wing of a hawkmoth (*Manduca sexta*) show that it deflects by a distance of $x = 4.8$ mm when a force of magnitude $F = 3.0$ mN is applied at the tip, as indicated in **FIGURE 8-28**. Treating the wing as an ideal spring, find (a) the force constant of the wing and (b) the energy stored in the wing when it is deflected. (c) What force must be applied to the tip of the wing to store twice the energy found in part (b)?

FIGURE 8-28 Problem 12

13. •• Predict/Calculate A vertical spring stores 0.962 J in spring potential energy when a 3.5-kg mass is suspended from it. **(a)** By what multiplicative factor does the spring potential energy change if the mass attached to the spring is doubled? **(b)** Verify your answer to part (a) by calculating the spring potential energy when a 7.0-kg mass is attached to the spring.

14. •• Pushing on the pump of a soap dispenser compresses a small spring. When the spring is compressed 0.50 cm, its potential energy is 0.0025 J. **(a)** What is the force constant of the spring? **(b)** What compression is required for the spring potential energy to equal 0.0084 J?

15. •• BIO **Mantis Shrimp Smasher** A peacock mantis shrimp smashes its prey with a hammer-like appendage by storing energy in a spring-like section of exoskeleton on its appendage. The spring has a force constant of 5.9×10^4 N/m. **(a)** If the spring stores 0.0103 J of energy, over what distance must the shrimp compress it? **(b)** What power does the shrimp generate if it unleashes the energy in 0.0018 s? (The power per mass of the shrimp appendage is more than 20 times greater than that generated by muscle alone, showcasing the shrimp's ability to harness the potential energy of its spring.)

16. •• Predict/Calculate The work required to stretch a certain spring from an elongation of 4.00 cm to an elongation of 5.00 cm is 30.5 J. **(a)** Is the work required to increase the elongation of the spring from 5.00 cm to 6.00 cm greater than, less than, or equal to 30.5 J? Explain. **(b)** Verify your answer to part (a) by calculating the required work.

17. •• A 0.33-kg pendulum bob is attached to a string 1.2 m long. What is the change in the gravitational potential energy of the system as the bob swings from point A to point B in FIGURE 8-29?

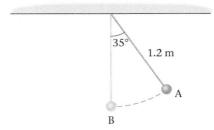

FIGURE 8-29 Problems 17 and 30

SECTION 8-3 CONSERVATION OF MECHANICAL ENERGY

18. • CE Predict/Explain You throw a ball upward and let it fall to the ground. Your friend drops an identical ball straight down to the ground from the same height. Is the change in kinetic energy of your ball greater than, less than, or equal to the change in kinetic energy of your friend's ball? **(b)** Choose the *best explanation* from among the following:
   I. Your friend's ball converts all its initial energy into kinetic energy.

   II. Your ball is in the air longer, which results in a greater change in kinetic energy.
   III. The change in gravitational potential energy is the same for each ball, which means the change in kinetic energy must be the same also.

19. • CE Predict/Explain When a ball of mass m is dropped from rest from a height h, its kinetic energy just before landing is K. Now, suppose a second ball of mass $4m$ is dropped from rest from a height $h/4$. **(a)** Just before ball 2 lands, is its kinetic energy $4K$, $2K$, K, $K/2$, or $K/4$? **(b)** Choose the *best explanation* from among the following:
   I. The two balls have the same initial energy.
   II. The more massive ball will have the greater kinetic energy.
   III. The reduced drop height results in a reduced kinetic energy.

20. • CE For an object moving along the x axis, the potential energy of the frictionless system is shown in FIGURE 8-30. Suppose the object is released from rest at the point A. Rank the other points in the figure in increasing order of the object's speed. Indicate ties where appropriate.

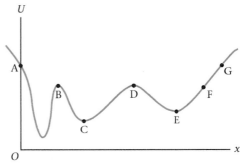

FIGURE 8-30 Problem 20

21. • At an amusement park, a swimmer uses a water slide to enter the main pool. If the swimmer starts at rest, slides without friction, and descends through a vertical height of 2.31 m, what is her speed at the bottom of the slide?

22. • Predict/Calculate A player passes a 0.600-kg basketball down-court for a fast break. The ball leaves the player's hands with a speed of 8.30 m/s and slows down to 7.10 m/s at its highest point. **(a)** Ignoring air resistance, how high above the release point is the ball when it is at its maximum height? **(b)** How would doubling the ball's mass affect the result in part (a)? Explain.

23. • A skateboarder at a skate park rides along the path shown in FIG-URE 8-31. If the speed of the skateboarder at point A is $v = 1.4$ m/s, what is her speed at point B? Assume that friction is negligible.

FIGURE 8-31 Problem 23

24. •• **CE** Three balls are thrown upward with the same initial speed v_0, but at different angles relative to the horizontal, as shown in FIGURE 8-32. Ignoring air resistance, indicate which of the following statements is correct: At the dashed level, (A) ball 3 has the lowest speed; (B) ball 1 has the lowest speed; (C) all three balls have the same speed; (D) the speed of the balls depends on their mass.

FIGURE 8-32 Problem 24

25. •• A 0.21-kg apple falls from a tree to the ground, 4.0 m below. Ignoring air resistance, determine the apple's kinetic energy, K, the gravitational potential energy of the system, U, and the total mechanical energy of the system, E, when the apple's height above the ground is (a) 4.0 m, (b) 3.0 m, (c) 2.0 m, (d) 1.0 m, and (e) 0 m. Take ground level to be $y = 0$.

26. •• Predict/Calculate A 2.9-kg block slides with a speed of 2.1 m/s on a frictionless horizontal surface until it encounters a spring. (a) If the block compresses the spring 5.6 cm before coming to rest, what is the force constant of the spring? (b) What initial speed should the block have to compress the spring by 1.4 cm?

27. •• A 0.26-kg rock is thrown vertically upward from the top of a cliff that is 32 m high. When it hits the ground at the base of the cliff, the rock has a speed of 29 m/s. Assuming that air resistance can be ignored, find (a) the initial speed of the rock and (b) the greatest height of the rock as measured from the base of the cliff.

28. •• A 1.40-kg block slides with a speed of 0.950 m/s on a frictionless horizontal surface until it encounters a spring with a force constant of 734 N/m. The block comes to rest after compressing the spring 4.15 cm. Find the spring potential energy, U, the kinetic energy of the block, K, and the total mechanical energy of the system, E, for compressions of (a) 0 cm, (b) 1.00 cm, (c) 2.00 cm, (d) 3.00 cm, and (e) 4.00 cm.

29. •• A 5.76-kg rock is dropped and allowed to fall freely. Find the initial kinetic energy, the final kinetic energy, and the change in kinetic energy for (a) the first 2.00 m of fall and (b) the second 2.00 m of fall.

30. •• Predict/Calculate Suppose the pendulum bob in Figure 8-29 has a mass of 0.33 kg and is moving to the right at point B with a speed of 2.4 m/s. Air resistance is negligible. (a) What is the change in the system's gravitational potential energy when the bob reaches point A? (b) What is the speed of the bob at point A? (c) If the mass of the bob is increased, does your answer to part (a) increase, decrease, or stay the same? Explain. (d) If the mass of the bob is increased, does your answer to part (b) increase, decrease, or stay the same? Explain.

31. ••• The two masses in the Atwood's machine shown in FIGURE 8-33 are initially at rest at the same height. After they are released, the large mass, m_2, falls through a height h and hits the floor, and the small mass, m_1, rises through a height h. (a) Find the speed of the masses just before m_2 lands, giving your answer in terms of m_1, m_2, g, and h. Assume the ropes and pulley have negligible mass and that friction can be ignored. (b) Evaluate your answer to part (a) for the case $h = 1.2$ m, $m_1 = 3.7$ kg, and $m_2 = 4.1$ kg.

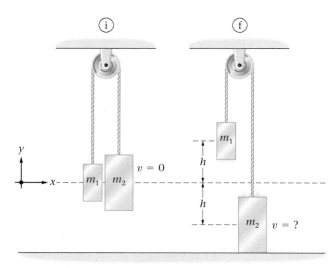

FIGURE 8-33 Problems 31, 32, and 63

32. ••• In the previous problem, suppose the masses have an initial speed of 0.20 m/s, and that m_2 is moving upward. How high does m_2 rise above its initial position before momentarily coming to rest, given that $m_1 = 3.7$ kg and $m_2 = 4.1$ kg?

SECTION 8-4 WORK DONE BY NONCONSERVATIVE FORCES

33. • **CE** Predict/Explain On reentry, the Space Shuttle's protective heat tiles became extremely hot. (a) Was the mechanical energy of the Shuttle–Earth system when the Shuttle landed greater than, less than, or the same as when it was in orbit? (b) Choose the *best explanation* from among the following:
   I. Dropping out of orbit increases the mechanical energy of the Shuttle.
   II. Gravity is a conservative force.
   III. A portion of the mechanical energy has been converted to heat energy.

34. • Catching a wave, a 77-kg surfer starts with a speed of 1.3 m/s, drops through a height of 1.65 m, and ends with a speed of 8.2 m/s. How much nonconservative work was done on the surfer?

35. • At a playground, a 19-kg child plays on a slide that drops through a height of 2.3 m. The child starts at rest at the top of the slide. On the way down, the slide does a nonconservative work of -361 J on the child. What is the child's speed at the bottom of the slide?

36. • Starting at rest at the edge of a swimming pool, a 72.0-kg athlete swims along the surface of the water and reaches a speed of 1.20 m/s by doing the work $W_{nc1} = +161$ J. Find the nonconservative work, W_{nc2}, done by the water on the athlete.

37. • A 22,000-kg airplane lands with a speed of 64 m/s on a stationary aircraft carrier deck that is 115 m long. Find the work done by nonconservative forces in stopping the plane.

38. •• A 78-kg skateboarder grinds down a hubba ledge that is 2.5 m long and inclined at 19° below the horizontal. Kinetic friction dissipates half of her initial potential energy to thermal and sound energies. What is the coefficient of kinetic friction between her skateboard and the ledge surface?

39. •• **CE** You ride your bicycle down a hill, maintaining a constant speed the entire time. (a) As you ride, does the gravitational potential energy of the you–bike–Earth system increase, decrease, or stay the same? Explain. (b) Does the kinetic energy of you and your bike increase, decrease, or stay the same? Explain. (c) Does the mechanical energy of the you–bike–Earth system increase, decrease, or stay the same? Explain.

40. •• A 111-kg seal at an amusement park slides from rest down a ramp into the pool below. The top of the ramp is 1.75 m higher than the surface of the water, and the ramp is inclined at an angle of 26.5° above the horizontal. If the seal reaches the water with a speed of 4.25 m/s, what are **(a)** the work done by kinetic friction and **(b)** the coefficient of kinetic friction between the seal and the ramp?

41. •• A 1.9-kg rock is released from rest at the surface of a pond 1.8 m deep. As the rock falls, a constant upward force of 4.6 N is exerted on it by water resistance. Calculate the nonconservative work, W_{nc}, done by water resistance on the rock, the gravitational potential energy of the system, U, the kinetic energy of the rock, K, and the total mechanical energy of the system, E, when the depth of the rock below the water's surface is **(a)** 0 m, **(b)** 0.50 m, and **(c)** 1.0 m. Let $y = 0$ be at the bottom of the pond.

42. •• A 1250-kg car drives up a hill that is 16.2 m high. During the drive, two nonconservative forces do work on the car: (i) the force of friction, and (ii) the force generated by the car's engine. The work done by friction is -3.11×10^5 J; the work done by the engine is $+6.44 \times 10^5$ J. Find the change in the car's kinetic energy from the bottom of the hill to the top of the hill.

43. ••• The Outlaw Run roller coaster in Branson, Missouri, features a track that is inclined at 81° below the horizontal and that spans a 49.4-m (162 ft) change in height. **(a)** If the coefficient of friction for the roller coaster is $\mu_k = 0.29$, what is the magnitude of the friction force on a 1700-kg coaster train? **(b)** How much work is done by friction during the time the coaster train travels along the incline? **(c)** What is the speed of the coaster train at the bottom of the incline, assuming it started from rest at the top?

44. ••• A 1.80-kg block slides on a rough horizontal surface. The block hits a spring with a speed of 2.00 m/s and compresses it a distance of 11.0 cm before coming to rest. If the coefficient of kinetic friction between the block and the surface is $\mu_k = 0.560$, what is the force constant of the spring?

SECTION 8-5 POTENTIAL ENERGY CURVES AND EQUIPOTENTIALS

45. • **FIGURE 8-34** shows a potential energy curve as a function of x. In qualitative terms, describe the subsequent motion of an object that starts at rest at point A.

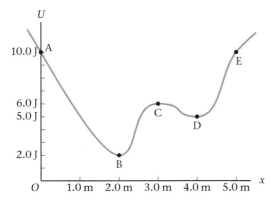

FIGURE 8-34 Problems 45, 46, 47, and 48

46. • An object moves along the x axis, subject to the potential energy shown in Figure 8-34. The object has a mass of 1.1 kg and starts at rest at point A. **(a)** What is the object's speed at point B? **(b)** At point C? **(c)** At point D? **(d)** What are the turning points for this object?

47. • A 1.34-kg object moves along the x axis, subject to the potential energy shown in Figure 8-34. If the object's speed at point C is 1.25 m/s, what are the approximate locations of its turning points?

48. •• The potential energy of a particle moving along the x axis is shown in Figure 8-34. When the particle is at $x = 1.0$ m it has 3.6 J of kinetic energy. Give approximate answers to the following questions. **(a)** What is the total mechanical energy of the system? **(b)** What is the smallest value of x the particle can reach? **(c)** What is the largest value of x the particle can reach?

49. •• A block of mass $m = 0.88$ kg is connected to a spring of force constant $k = 845$ N/m on a smooth, horizontal surface. **(a)** Plot the potential energy of the spring from $x = -5.00$ cm to $x = 5.00$ cm. **(b)** Determine the turning points of the block if its speed at $x = 0$ is 1.4 m/s.

50. •• A ball of mass $m = 0.75$ kg is thrown straight upward with an initial speed of 8.9 m/s. **(a)** Plot the gravitational potential energy of the ball from its launch height, $y = 0$, to the height $y = 5.0$ m. Let $U = 0$ correspond to $y = 0$. **(b)** Determine the turning point (maximum height) of the ball.

51. •• **FIGURE 8-35** depicts the potential energy of a 350-kg pallet that slides from rest down a frictionless roller conveyer ramp, then moves across a frictionless horizontal conveyer, and finally encounters a horizontal spring at $x = 4.0$ m. When the pallet encounters the spring, it momentarily comes to rest at $x = 6.0$ m. **(a)** What is the force constant of the spring? **(b)** At what height h above the horizontal conveyer did the pallet begin sliding from rest?

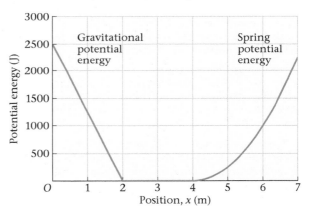

FIGURE 8-35 Problems 51 and 52

52. •• Figure 8-35 depicts the potential energy of a 655-kg pallet that slides from rest down a frictionless roller conveyer ramp, then moves across a frictionless horizontal conveyer, and finally encounters a horizontal spring at $x = 4.0$ m. **(a)** What is the angle of inclination of the conveyer ramp? **(b)** What will be the speed of the pallet at $x = 6.0$ m if it begins sliding from rest at $x = 0$?

GENERAL PROBLEMS

53. • **CE** You and a friend both solve a problem involving a skier going down a slope. When comparing solutions, you notice that your choice for the $y = 0$ level is different than the $y = 0$ level chosen by your friend. Will your answers agree or disagree on the following quantities: **(a)** the skier's potential energy; **(b)** the skier's change in potential energy; **(c)** the skier's kinetic energy?

54. • **CE** A particle moves under the influence of a conservative force. At point A the particle has a kinetic energy of 12 J; at point B the particle is momentarily at rest, and the potential energy of the system is 25 J; at point C the potential energy of the system is 5 J. **(a)** What is the potential energy of the system when the particle is at point A? **(b)** What is the kinetic energy of the particle at point C?

55. •• A sled slides without friction down a small, ice-covered hill. On its first run down the hill, the sled starts from rest at the top of the hill and its speed at the bottom is 7.50 m/s. On a second run, the sled starts with a speed of 1.50 m/s at the top. What is the speed of the sled at the bottom of the hill on the second run?

56. •• A 74-kg skier encounters a dip in the snow's surface that has a circular cross section with a radius of curvature of 12 m. If the skier's speed at point A in **FIGURE 8-36** is 7.1 m/s, what is the normal force exerted by the snow on the skier at point B? Ignore frictional forces.

FIGURE 8-36 Problem 56

57. •• **Running Shoes** The soles of a popular make of running shoe have a force constant of 2.0×10^5 N/m. Treat the soles as ideal springs for the following questions. **(a)** If a 62-kg person stands in a pair of these shoes, with her weight distributed equally on both feet, how much does she compress the soles? **(b)** How much energy is stored in the soles of her shoes when she's standing?

58. •• **Nasal Strips** The force required to flex a nasal strip and apply it to the nose is 0.25 N; the energy stored in the strip when flexed is 0.0022 J. Assume the strip to be an ideal spring for the following calculations. Find **(a)** the distance through which the strip is flexed and **(b)** the force constant of the strip.

59. •• The water slide shown in **FIGURE 8-37** ends at a height of 1.50 m above the pool. If the person starts from rest at point A and lands in the water at point B, what is the height h of the water slide? (Assume the water slide is frictionless.)

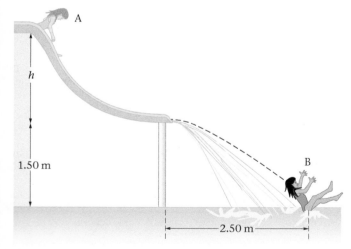

FIGURE 8-37 Problems 59

60. •• A skateboarder starts at point A in **FIGURE 8-38** and rises to a height of 2.64 m above the top of the ramp at point B. What was the skateboarder's initial speed at point A?

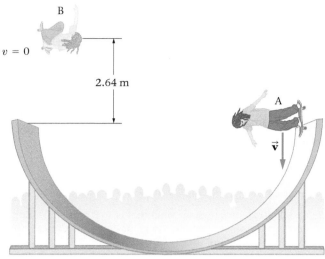

FIGURE 8-38 Problem 60

61. •• **The Crash of *Skylab*** NASA's *Skylab*, the largest spacecraft ever to fall back to the Earth, reentered the Earth's atmosphere on July 11, 1979, and broke into a myriad of pieces. One of the largest fragments was a 1770-kg, lead-lined film vault, which landed with an estimated speed of 120 m/s. If its speed at an altitude of 148 km was 7950 m/s, how much nonconservative work was done on the film vault during reentry?

62. •• **BIO Bird Tendons** Several studies indicate that the elastic properties of tendons can change in response to exercise. In one study, guinea fowl were divided into a group that ran for 30 minutes a day on a treadmill, and a control group that was kept in their cages. After 12 weeks, the researchers measured the mechanical properties of the gastrocnemius tendon in each bird. The results for one control bird and one treadmill bird are shown in **FIGURE 8-39**. **(a)** Treating the tendon as an ideal spring, what is the force constant for the control bird? **(b)** By how much does the control bird's tendon need to be stretched in order to have the same amount of stored energy as the tendon of the treadmill bird at a deformation of 1.0 mm?

FIGURE 8-39 Problem 62

63. •• In the Atwood's machine of Problem 31, the mass m_2 remains at rest once it hits the floor, but the mass m_1 continues moving upward. How much higher does m_1 go after m_2 has landed? Give your answer for the case $h = 1.2$ m, $m_1 = 3.7$ kg, and $m_2 = 4.1$ kg.

64. •• A 6.60-kg block slides with an initial speed of 2.24 m/s up a ramp inclined at an angle of 18.4° with the horizontal. The coefficient of kinetic friction between the block and the ramp is 0.44. Use energy conservation to find the distance the block slides before coming to rest.

65. •• Jeff of the Jungle swings on a 7.6-m vine that initially makes an angle of 37° with the vertical. If Jeff starts at rest and has a mass of 78 kg, what is the tension in the vine at the lowest point of the swing?

66. •• A 1.9-kg block slides down a frictionless ramp, as shown in **FIGURE 8-40**. The top of the ramp is 1.5 m above the ground; the bottom of the ramp is 0.25 m above the ground. The block leaves the ramp moving horizontally, and lands a horizontal distance d away. Find the distance d.

FIGURE 8-40 Problems 66 and 67

67. •• Suppose the ramp in Figure 8-40 is not frictionless. Find the distance d for the case in which friction on the ramp does $-9.7\,\text{J}$ of work on the block before it becomes airborne.

68. •• **BIO Compressing the Ground** A running track at Harvard University uses a surface with a force constant of $2.5 \times 10^5\,\text{N/m}$. This surface is compressed slightly every time a runner's foot lands on it. The force exerted by the foot, according to the Saucony shoe company, has a magnitude of 2700 N for a typical runner. Treating the track's surface as an ideal spring, find **(a)** the amount of compression caused by a foot hitting the track and **(b)** the energy stored briefly in the track every time a foot lands.

69. •• **BIO A Flea's Jump** The resilin in the body of a flea has a force constant of about 24 N/m, and when the flea cocks its jumping legs, the resilin associated with each leg is compressed by approximately 0.12 mm. Given that the flea has a mass of 0.56 mg, and that two legs are used in a jump, estimate the maximum height a flea can attain by using the energy stored in the resilin. (Assume the resilin to be an ideal spring.)

70. ••• Predict/Calculate **Tension at the Bottom** A ball of mass m is attached to a string of length L and released from rest at the point A in **FIGURE 8-41**. **(a)** Show that the tension in the string when the ball reaches point B is $3mg$, independent of the length l. **(b)** Give a detailed physical explanation for the fact that the tension at point B is independent of the length l.

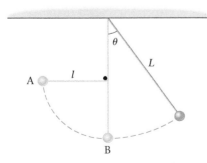

FIGURE 8-41 Problem 70

71. ••• An ice cube is placed on top of an overturned spherical bowl of radius r, as indicated in **FIGURE 8-42**. If the ice cube slides downward from rest at the top of the bowl, at what angle θ does it separate from the bowl? In other words, at what angle does the normal force between the ice cube and the bowl go to zero?

FIGURE 8-42 Problem 71

72. ••• Predict/Calculate The two blocks shown in **FIGURE 8-43** are moving with an initial speed v. **(a)** If the system is frictionless, find the distance d the blocks travel before coming to rest. (Let $U = 0$ correspond to the initial position of block 2.) **(b)** Is the work done on block 2 by the rope positive, negative, or zero? Explain. **(c)** Calculate the work done on block 2 by the rope.

FIGURE 8-43 Problem 72

73. ••• Predict/Calculate **Loop-the-Loop (a)** A block of mass m slides from rest on a frictionless loop-the-loop track, as shown in **FIGURE 8-44**. What is the minimum release height, h, required for the block to maintain contact with the track at all times? Give your answer in terms of the radius of the loop, r. **(b)** Explain why the release height obtained in part (a) is independent of the block's mass.

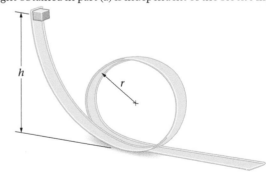

FIGURE 8-44 Problem 73

74. ••• **FIGURE 8-45** shows a 1.75-kg block at rest on a ramp of height h. When the block is released, it slides without friction to the bottom of the ramp, and then continues across a surface that is frictionless except for a rough patch of width 10.0 cm that has a coefficient of kinetic friction $\mu_k = 0.640$. Find h such that the block's speed after crossing the rough patch is 3.50 m/s.

FIGURE 8-45 Problem 74

75. ••• In **FIGURE 8-46** a 1.2-kg block is held at rest against a spring with a force constant $k = 730$ N/m. Initially, the spring is compressed a distance d. When the block is released, it slides across a surface that is frictionless except for a rough patch of width 5.0 cm that has a coefficient of kinetic friction $\mu_k = 0.44$. Find d such that the block's speed after crossing the rough patch is 2.3 m/s.

FIGURE 8-46 Problem 75

PASSAGE PROBLEMS

BIO The Flight of the Dragonflies

Of all the animals you're likely to see on a summer day, the most ancient is the dragonfly. In fact, the fossil record for dragonflies extends back over 250 million years, more than twice as long as for birds. Ancient dragonflies could be as large as a hawk, and were surely buzzing around the heads of both *T. Rex* and *Triceratops*.

Dragonflies belong to the order Odonata ("toothed jaws") and the suborder Anisoptera ("different wings"), a reference to the fact that their hindwings are wider front-to-back than their forewings. (Damselflies, in contrast, have forewings and hindwings that are the same width.) Although ancient in their lineage, dragonflies are among the fastest flying and most acrobatic of all insects; some of their maneuvers subject them to accelerations as great as 20*g*.

The properties of dragonfly wings, and how they account for such speed and mobility, have been of great interest to biologists. **FIGURE 8-47 (a)** shows an experimental setup designed to measure the force constant of Plexiglas models of wings, which are used in wind tunnel tests. A downward force is applied to the model wing at the tip (1 for hindwing, 2 for forewing) or at two-thirds the distance to the tip (3 for hindwing, 4 for forewing). As the force is varied in magnitude, the resulting deflection of the wing is measured. The results are shown in **FIGURE 8-47 (b)**. Notice that significant differences are seen between the hindwings and forewings, as one might expect from their different shapes.

76. • Treating the model wing as an ideal spring, what is the force constant of the hindwing when a force is applied to its tip?

   A. 94 N/m B. 130 N/m
   C. 290 N/m D. 330 N/m

77. • What is the force constant of the hindwing when a force is applied at two-thirds the distance from the base of the wing to the tip?

   A. 94 N/m B. 130 N/m
   C. 290 N/m D. 330 N/m

(a)

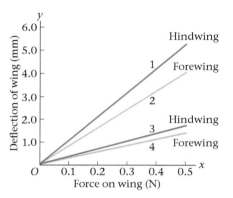

(b)

FIGURE 8-47 Problems 76, 77, 78, and 79

78. • Which of the wings is "stiffer"?

   A. The hindwing
   B. The forewing
   C. Depends on where the force is applied
   D. They are equally "stiff"

79. •• How much energy is stored in the forewing when a force at the tip deflects it by 3.5 mm?

   A. 0.766 mJ B. 49.0 mJ
   C. 0.219 J D. 1.70 kJ

80. •• Predict/Calculate **REFERRING TO EXAMPLE 8-12** Consider a spring with a force constant of 955 N/m. **(a)** Suppose the mass of the block is 1.70 kg, but its initial speed can be varied. What initial speed is required to give a maximum spring compression of 4.00 cm? **(b)** Suppose the initial speed of the block is 1.09 m/s, but its mass can be varied. What mass is required to give a maximum spring compression of 4.00 cm?

81. •• **REFERRING TO EXAMPLE 8-12** Suppose the block is released from rest with the spring compressed 5.00 cm. The mass of the block is 1.70 kg and the force constant of the spring is 955 N/m. **(a)** What is the speed of the block when the spring expands to a compression of only 2.50 cm? **(b)** What is the speed of the block after it leaves the spring?

82. •• **REFERRING TO EXAMPLE 8-17** Suppose we would like the landing speed of block 2 to be increased to 1.50 m/s. **(a)** Should the coefficient of kinetic friction between block 1 and the tabletop be increased or decreased? **(b)** Find the required coefficient of kinetic friction for a landing speed of 1.50 m/s. Note that $m_1 = 2.40$ kg, $m_2 = 1.80$ kg, and $d = 0.500$ m.

Linear Momentum and Collisions

Big Ideas

1 Momentum is mass times velocity.

2 Impulse is force multiplied by time.

3 Momentum is conserved when the net force is zero.

4 Collisions conserve momentum, but may or may not conserve mechanical energy.

5 Systems behave as if all their mass were concentrated at the center of mass.

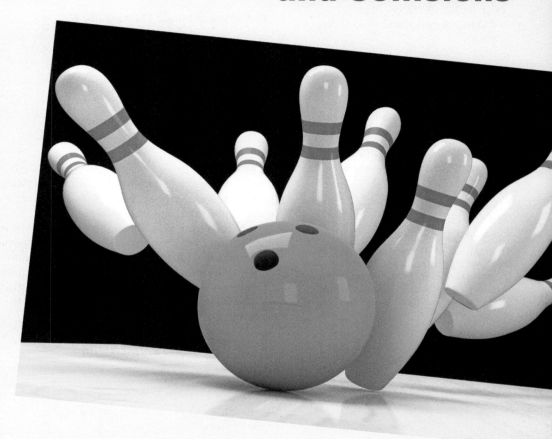

▲ A bowling ball hitting a group of pins is just one example of the many collisions that occur all around us everyday. As we shall see in this chapter, collisions transfer a new physical quantity—momentum—between the colliding objects, while keeping the total amount of momentum constant. In this chapter, we show how momentum is related to Newton's second law, and how it can be used to analyze a wide range of collisions.

C onservation laws play a central role in physics. In this chapter we introduce the concept of *momentum* and show that it, like energy, is a conserved quantity. Nothing we can do—in fact, nothing that can occur in nature—can change the total energy or total momentum of the universe.

9-1 Linear Momentum

Imagine standing at rest on a skateboard that rolls without friction on a horizontal surface. If you catch a heavy, slow-moving ball tossed to you by a friend, you begin to move. If, on the other hand, your friend tosses you a lightweight, yet fast-moving ball, the net effect may be the same—that is, catching a lightweight ball moving fast enough causes you to move with the same speed as when you caught the heavy ball.

In physics, these observations are made precise with a quantity called the **linear momentum**, $\vec{\mathbf{p}}$, which is defined as the product of the mass m and velocity $\vec{\mathbf{v}}$ of an object:

Definition of Linear Momentum, $\vec{\mathbf{p}}$

$$\vec{\mathbf{p}} = m\vec{\mathbf{v}} \qquad \qquad \text{9-1}$$

SI unit: $kg \cdot m/s$

Similarly, the magnitude of the linear momentum is simply the object's mass times its speed:

Magnitude of Linear Momentum, p

$$p = mv \qquad \qquad \text{9-2}$$

SI unit: $kg \cdot m/s$

Thus, if the heavy ball mentioned above has twice the mass of the light ball, but the light ball has twice the speed of the heavy ball, the momenta of the two balls are equal in magnitude; that is, $(2m)v = m(2v)$.

It follows from $\vec{\mathbf{p}} = m\vec{\mathbf{v}}$ and $p = mv$ that the units of linear momentum are those of mass times velocity—that is, $kg \cdot m/s$. No special shorthand name is given to this combination of units. In addition, notice that *linear momentum*—or just *momentum* for short—is the momentum of an object moving in a straight line. In Chapter 11 we introduce *angular momentum* for objects that are rotating.

The following Exercise gives a feeling for the magnitude of the momentum of everyday objects.

EXERCISE 9-1 MOMENTUM

(a) A 72.3-kg jogger runs with a speed of 3.13 m/s in the positive x direction. What is the jogger's momentum vector? **(b)** A 0.170-kg hockey puck has a momentum with a magnitude of $1.52\ kg \cdot m/s$. What is the speed of the puck? **(c)** The momentum of an airplane flying with a speed of 54.9 m/s has a magnitude of $3.93 \times 10^4\ kg \cdot m/s$. What is the mass of the airplane?

REASONING AND SOLUTION

We use the vector relationship, $\vec{\mathbf{p}} = m\vec{\mathbf{v}}$, when a direction is specified. In other cases, we work with magnitudes, and use the relationship $p = mv$.

a. Using $\vec{\mathbf{p}} = m\vec{\mathbf{v}}$, and noting that the direction of the velocity is $\hat{\mathbf{x}}$, we find

$$\vec{\mathbf{p}} = m\vec{\mathbf{v}} = (72.3\ \text{kg})(3.13\ \text{m/s})\hat{\mathbf{x}} = (226\ \text{kg} \cdot \text{m/s})\hat{\mathbf{x}}$$

b. Solving $p = mv$ for v yields

$$v = \frac{p}{m} = \frac{1.52\ \text{kg} \cdot \text{m/s}}{0.170\ \text{kg}} = 8.94\ \text{m/s}$$

c. Solving $p = mv$ for m yields

$$m = \frac{p}{v} = \frac{3.93 \times 10^4\ \text{kg} \cdot \text{m/s}}{54.9\ \text{m/s}} = 716\ \text{kg}$$

Calculating Changes in Momentum Because the velocity $\vec{\mathbf{v}}$ is a vector with both a magnitude and a direction, so too is the momentum, $\vec{\mathbf{p}} = m\vec{\mathbf{v}}$. In problems that are strictly one-dimensional, the vectors simplify to a single component. This makes the calculations much easier—we just have to be careful to use the correct sign of the component.

Big Idea 1 Momentum is a vector quantity equal to mass times velocity.

PHYSICS IN CONTEXT
Looking Back

Momentum is a vector quantity, and hence the vector tools introduced in Chapter 3 are important here as well. In particular, we use vector components in our analysis of many momentum problems in this chapter.

PHYSICS IN CONTEXT
Looking Ahead

The concept of momentum is used again in Chapter 11, when we study the dynamics of rotational motion. In particular, we introduce *angular momentum* as an extension of the linear momentum introduced in this chapter.

As an illustration of the vector nature of momentum, and its importance in analyzing real-world situations, consider the situations shown in **FIGURE 9-1**. In Figure 9-1 (a), a 0.10-kg beanbag bear is dropped to the floor, where it hits with a speed of 4.0 m/s and sticks. In Figure 9-1 (b), a 0.10-kg rubber ball also hits the floor with a speed of 4.0 m/s, but in this case the ball bounces upward off the floor. If we assume an ideal rubber ball, its initial upward speed is 4.0 m/s. Now the question in each case is, "What is the change in momentum?"

To approach the problem systematically, we introduce a coordinate system as shown in Figure 9-1. With this choice, it follows that neither object has momentum in the x direction; thus, the system is one-dimensional. We need only consider the y component of momentum, p_y, and be careful about its sign.

We begin with the beanbag. Just before hitting the floor, it moves downward (that is, in the negative y direction) with a speed of $v = 4.0$ m/s. Letting m stand for the mass of the beanbag, we find that the initial momentum is

$$p_{y,i} = m(-v)$$

After landing on the floor, the beanbag is at rest; hence, its final momentum is zero:

$$p_{y,f} = m(0) = 0$$

Therefore the change in momentum for the beanbag is

$$\Delta p_y = p_{y,f} - p_{y,i} = 0 - m(-v) = mv$$
$$= (0.10 \text{ kg})(4.0 \text{ m/s}) = 0.40 \text{ kg} \cdot \text{m/s}$$

Notice that the change in momentum is positive—that is, in the upward direction. This makes sense because, before the bag landed, it had a negative (downward) momentum in the y direction. In order to increase the momentum from a negative value to zero, it is necessary to add a positive (upward) momentum.

Next, consider the rubber ball in Figure 9-1 (b). Before bouncing, its momentum is

$$p_{y,i} = m(-v)$$

This is the same initial momentum as for the beanbag. After bouncing, the ball is moving in the upward (positive) direction and its momentum is

$$p_{y,f} = mv$$

As a result, the change in momentum for the rubber ball is

$$\Delta p_y = p_{y,f} - p_{y,i} = mv - m(-v) = 2mv$$
$$= 2(0.10 \text{ kg})(4.0 \text{ m/s}) = 0.80 \text{ kg} \cdot \text{m/s}$$

This is *twice* the change in momentum of the beanbag! The reason is that in this case the momentum in the y direction must first be increased from $-mv$ to 0, and then increased again from 0 to mv. For the beanbag, the change was simply from $-mv$ to 0.

Notice how important it is to be careful about the vector nature of the momentum and to use the correct sign for p_y. Otherwise, we might have concluded—erroneously—that the rubber ball had zero change in momentum, since the *magnitude* of its momentum was unchanged by the bounce. In fact, its momentum does change due to the change in its *direction* of motion.

Calculating the Momentum for a System of Objects Momentum is a vector, and hence the total momentum of a system of objects is the *vector* sum of the momenta of all the objects. That is,

$$\vec{\mathbf{P}}_{\text{total}} = \vec{\mathbf{P}}_1 + \vec{\mathbf{P}}_2 + \vec{\mathbf{P}}_3 + \cdots \qquad 9\text{-}3$$

This is illustrated for the case of three objects in the following Example.

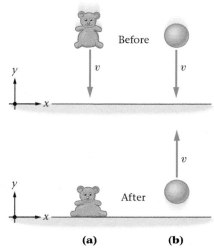

▲ **FIGURE 9-1 Change in momentum**
A beanbag bear and a rubber ball, with the same mass m and the same downward speed v, hit the floor. **(a)** The beanbag bear comes to rest on hitting the floor. Its change in momentum is mv upward. **(b)** The rubber ball bounces upward with a speed v. Its change in momentum is $2mv$ upward.

 Happy/Sad Pendulums

PROBLEM-SOLVING NOTE

Coordinate Systems

Be sure to draw a coordinate system for momentum problems, even if the problem is only one-dimensional. It is important to use the coordinate system to assign the correct sign to velocities and momenta in the system.

EXAMPLE 9-2   DUCK, DUCK, GOOSE

At a city park, a person throws some bread into a duck pond. Two 4.00-kg ducks and a 9.00-kg goose paddle rapidly toward the bread, as shown in our sketch. If the ducks swim at 1.10 m/s, and the goose swims with a speed of 1.30 m/s, find the magnitude and direction of the total momentum of the three birds.

CONTINUED

PICTURE THE PROBLEM
In our sketch we place the origin where the bread floats on the water. Notice that duck 1 swims in the positive x direction, duck 2 swims in the negative y direction, and the goose swims in the positive y direction. Therefore, $\vec{\mathbf{p}}_{d1} = m_d v_d \hat{\mathbf{x}}$, $\vec{\mathbf{p}}_{d2} = -m_d v_d \hat{\mathbf{y}}$, and $\vec{\mathbf{p}}_g = m_g v_g \hat{\mathbf{y}}$, where $v_d = 1.10$ m/s, $m_d = 4.00$ kg, $v_g = 1.30$ m/s, and $m_g = 9.00$ kg. The total momentum, $\vec{\mathbf{p}}_{total}$, points at an angle θ relative to the positive x axis, as indicated.

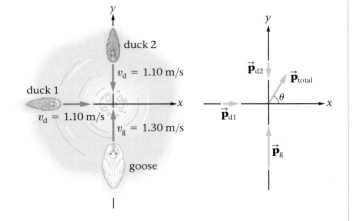

REASONING AND STRATEGY
We write the momentum of each bird as a vector, using unit vectors in the x and y directions. Next, we sum these vectors component by component to find the total momentum. Finally, we use the components of the total momentum to calculate its magnitude and direction.

Known   Mass of the ducks, $m_d = 4.00$ kg; mass of the goose, $m_g = 9.00$ kg; speed of the ducks, $v_d = 1.10$ m/s; speed of the goose, $v_g = 1.30$ m/s.
Unknown   Magnitude of total momentum, $p_{total} = $? Direction of total momentum, $\theta = $?

SOLUTION

1. Use x and y unit vectors to express the momentum of each bird in vector form:
$$\vec{\mathbf{p}}_{d1} = m_d v_d \hat{\mathbf{x}} = (4.00 \text{ kg})(1.10 \text{ m/s})\hat{\mathbf{x}}$$
$$= (4.40 \text{ kg} \cdot \text{m/s})\hat{\mathbf{x}}$$
$$\vec{\mathbf{p}}_{d2} = -m_d v_d \hat{\mathbf{y}} = -(4.00 \text{ kg})(1.10 \text{ m/s})\hat{\mathbf{y}}$$
$$= -(4.40 \text{ kg} \cdot \text{m/s})\hat{\mathbf{y}}$$
$$\vec{\mathbf{p}}_g = m_g v_g \hat{\mathbf{y}} = (9.00 \text{ kg})(1.30 \text{ m/s})\hat{\mathbf{y}}$$
$$= (11.7 \text{ kg} \cdot \text{m/s})\hat{\mathbf{y}}$$

2. Sum the momentum vectors to obtain the total momentum:
$$\vec{\mathbf{p}}_{total} = \vec{\mathbf{p}}_{d1} + \vec{\mathbf{p}}_{d2} + \vec{\mathbf{p}}_g$$
$$= (4.40 \text{ kg} \cdot \text{m/s})\hat{\mathbf{x}} + (-4.40 \text{ kg} \cdot \text{m/s} + 11.7 \text{ kg} \cdot \text{m/s})\hat{\mathbf{y}}$$
$$= (4.40 \text{ kg} \cdot \text{m/s})\hat{\mathbf{x}} + (7.30 \text{ kg} \cdot \text{m/s})\hat{\mathbf{y}}$$

3. Calculate the magnitude of the total momentum:
$$p_{total} = \sqrt{p_{total,x}^2 + p_{total,y}^2}$$
$$= \sqrt{(4.40 \text{ kg} \cdot \text{m/s})^2 + (7.30 \text{ kg} \cdot \text{m/s})^2}$$
$$= 8.52 \text{ kg} \cdot \text{m/s}$$

4. Calculate the direction of the total momentum:
$$\theta = \tan^{-1}\left(\frac{p_{total,y}}{p_{total,x}}\right) = \tan^{-1}\left(\frac{7.30 \text{ kg} \cdot \text{m/s}}{4.40 \text{ kg} \cdot \text{m/s}}\right) = 58.9°$$

INSIGHT
Notice that the momentum of each bird depends only on its mass and velocity; it is independent of the bird's location. In addition, we observe that the magnitude of the total momentum is less than the sum of the magnitudes of each individual bird's momentum. This is generally the case when dealing with vector addition—the only exception is when all vectors point in the same direction.

PRACTICE PROBLEM — PREDICT/CALCULATE
(a) Suppose we can change the speed of the goose. If the total momentum of the three birds is to have a y component equal to zero, should the speed of the goose be increased or decreased? Explain. (b) Verify your prediction by calculating the required speed.   [**Answer: (a)** The goose's speed must be decreased, because its current y component of momentum is greater in magnitude than that of duck 2. **(b)** Setting the magnitude of the goose's momentum equal to 4.40 kg · m/s, we find $v_g = 0.489$ m/s.]

Some related homework problems: Problem 1, Problem 3, Problem 4

PHYSICS IN CONTEXT
Looking Ahead

Linear momentum plays an important role in quantum physics. For example, the momentum of a particle is related to its de Broglie wavelength in Section 30-5, and to the uncertainty principle in Section 30-6.

Enhance Your Understanding   (Answers given at the end of the chapter)

1. A block of wood is struck by a bullet. **(a)** Is the block more likely to be knocked over if the bullet is metal and embeds itself in the wood, or if the bullet is rubber and bounces off the wood? **(b)** Choose the *best explanation* for your prediction from among the following:

   I. The change in momentum when a bullet rebounds is larger than when it is brought to rest.

CONTINUED

II. The metal bullet does more damage to the block.

III. The rubber bullet bounces off, and hence it has little effect.

Section Review

- The linear momentum of an object is $\vec{\mathbf{p}} = m\vec{\mathbf{v}}$.

- The magnitude of the linear momentum is $p = mv$.

- The total linear momentum for a system of objects is $\vec{\mathbf{p}}_{\text{total}} = \vec{\mathbf{p}}_1 + \vec{\mathbf{p}}_2 + \vec{\mathbf{p}}_3 + \cdots$.

9-2 Momentum and Newton's Second Law

In Chapter 5 we introduced Newton's second law:

$$\sum \vec{\mathbf{F}} = m\vec{\mathbf{a}}$$

We mentioned at the time that this expression is valid only for objects that have constant mass. Newton's original statement of the second law, which is valid even if the mass changes, is as follows:

Newton's Second Law

$$\sum \vec{\mathbf{F}} = \frac{\Delta \vec{\mathbf{p}}}{\Delta t} \qquad\qquad 9\text{-}4$$

That is, the net force acting on an object is equal to the rate of change of its momentum.

To show the connection between these two statements of the second law, consider the change in momentum, $\Delta \vec{\mathbf{p}}$. Recalling that $\vec{\mathbf{p}} = m\vec{\mathbf{v}}$, we have

$$\Delta \vec{\mathbf{p}} = \vec{\mathbf{p}}_f - \vec{\mathbf{p}}_i = m_f\vec{\mathbf{v}}_f - m_i\vec{\mathbf{v}}_i$$

If the mass is constant, so that $m_f = m_i = m$, it follows that the change in momentum is simply m times $\Delta \vec{\mathbf{v}}$:

$$\Delta \vec{\mathbf{p}} = m_f\vec{\mathbf{v}}_f - m_i\vec{\mathbf{v}}_i = m(\vec{\mathbf{v}}_f - \vec{\mathbf{v}}_i) = m\Delta \vec{\mathbf{v}}$$

As a result, Newton's second law for objects of constant mass is

$$\sum \vec{\mathbf{F}} = \frac{\Delta \vec{\mathbf{p}}}{\Delta t} = m\frac{\Delta \vec{\mathbf{v}}}{\Delta t}$$

If we recall that acceleration is the rate of change of velocity with time, $\vec{\mathbf{a}} = \Delta \vec{\mathbf{v}}/\Delta t$, the above equation becomes

$$\sum \vec{\mathbf{F}} = \frac{\Delta \vec{\mathbf{p}}}{\Delta t} = m\vec{\mathbf{a}} \qquad\qquad 9\text{-}5$$

Hence, the two statements of Newton's second law are equivalent if the mass is constant—as is often the case. The statement in terms of momentum is always valid, however. In the remainder of this chapter, we shall use the general form of Newton's second law to investigate the connections between forces and changes in momentum.

PHYSICS IN CONTEXT
Looking Ahead

In Chapter 11 we will see that the rate of change of angular momentum is related to the "rotational force," referred to as torque.

Enhance Your Understanding   (Answers given at the end of the chapter)

2. Object 1 has a mass of 1 kg and is acted on by a force of 100 N. Object 2 has a mass of 100 kg and is also acted on by a force of 100 N. **(a)** Is the change in momentum in one second greater for object 1, greater for object 2, or the same for the two objects? Explain. **(b)** Is the change in velocity in one second greater for object 1, greater for object 2, or the same for the two objects? Explain.

Section Review

- The net force acting on an object is equal to the rate of change of its momentum.

▶ **FIGURE 9-2 The average force during a collision** The force between two objects that collide, as when a bat hits a baseball, rises rapidly to very large values, then drops again to zero in a matter of milliseconds. Rather than try to describe the complex behavior of the force, we focus on its average value, F_{av}. Notice that the area under the F_{av} rectangle is the same as the area under the actual force curve.

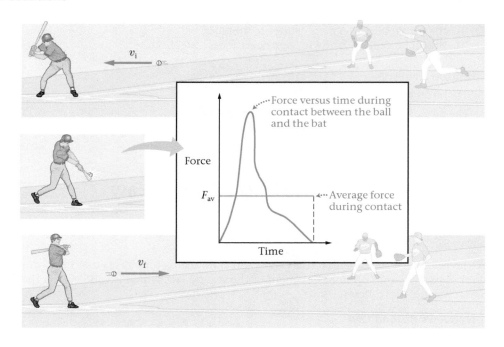

9-3 Impulse

The pitcher delivers a fastball, the batter takes a swing, and with a crack of the bat the ball that was approaching home plate at 95.0 mi/h is now heading toward the pitcher at 115 mi/h. In the language of physics, we say that the bat has delivered an **impulse**, $\vec{\mathbf{I}}$, to the ball.

During the brief time the ball and bat are in contact—perhaps as little as a thousandth of a second—the force between them rises rapidly to a large value, as shown in **FIGURE 9-2**, then falls back to zero as the ball takes flight. It would be almost impossible, of course, to describe every detail of the way the force varies with time. Instead, we focus on the average force exerted by the bat, $\vec{\mathbf{F}}_{av}$, which is also shown in Figure 9-2. The impulse delivered by the bat is defined to be $\vec{\mathbf{F}}_{av}$ times the amount of time, Δt, that the ball and bat are in contact:

Definition of Impulse, $\vec{\mathbf{I}}$

$$\vec{\mathbf{I}} = \vec{\mathbf{F}}_{av}\,\Delta t \qquad\qquad\qquad 9\text{-}6$$

SI unit: $N \cdot s = kg \cdot m/s$

The magnitude of the impulse is simply the area under the force-versus-time curve.

Notice that impulse is a vector and that it points in the same direction as the average force. In addition, its units are $N \cdot s = (kg \cdot m/s^2) \cdot s = kg \cdot m/s$, the same as the units of momentum. It's no accident that impulse and momentum have the same units. In fact, if we rearrange Newton's second law, $\sum \vec{\mathbf{F}} = \vec{\mathbf{F}}_{total} = \dfrac{\Delta \vec{\mathbf{p}}}{\Delta t}$, we can see that the total force times Δt is simply the change in momentum:

$$\vec{\mathbf{F}}_{total} = \frac{\Delta \vec{\mathbf{p}}}{\Delta t}$$

$$\vec{\mathbf{F}}_{total}\,\Delta t = \Delta \vec{\mathbf{p}}$$

PHYSICS IN CONTEXT
Looking Back

In this chapter we see that force multiplied by the *time* over which it acts is related to a change in momentum. In Chapters 7 and 8 we saw that force multiplied by the *distance* over which it acts leads to a change in energy.

The same applies to impulse if we replace $\vec{\mathbf{F}}_{total}$ with $\vec{\mathbf{F}}_{av}$. Thus, we find:

Momentum–Impulse Theorem

$$\vec{\mathbf{I}} = \vec{\mathbf{F}}_{av}\,\Delta t = \Delta \vec{\mathbf{p}} \qquad\qquad 9\text{-}7$$

For instance, if we know the impulse delivered to an object—that is, its change in momentum, $\Delta \vec{\mathbf{p}}$—and the time interval during which the change occurs, the average force that caused the impulse is simply $\Delta \vec{\mathbf{p}}/\Delta t$.

Impulse and Force   As an example of the connection between impulse and force, let's calculate the impulse given to the baseball considered at the beginning of this section. We will also calculate the average force between the ball and the bat.

First, we set up a coordinate system with the positive x axis pointing from home plate toward the pitcher's mound, as indicated in **FIGURE 9-3**. If the ball's mass is 0.145 kg, its initial momentum—which is in the negative x direction—is

$$\vec{\mathbf{p}}_i = -mv_i\hat{\mathbf{x}} = -(0.145\ \text{kg})(95.0\ \text{mi/h})\left(\frac{0.447\ \text{m/s}}{1\ \text{mi/h}}\right)\hat{\mathbf{x}} = -(6.16\ \text{kg}\cdot\text{m/s})\hat{\mathbf{x}}$$

Immediately after the hit, the ball's final momentum is in the positive x direction:

$$\vec{\mathbf{p}}_f = mv_f\hat{\mathbf{x}} = (0.145\ \text{kg})(115\ \text{mi/h})\left(\frac{0.447\ \text{m/s}}{1\ \text{mi/h}}\right)\hat{\mathbf{x}} = (7.45\ \text{kg}\cdot\text{m/s})\hat{\mathbf{x}}$$

The impulse, then, is

$$\vec{\mathbf{I}} = \Delta\vec{\mathbf{p}} = \vec{\mathbf{p}}_f - \vec{\mathbf{p}}_i = [7.45\ \text{kg}\cdot\text{m/s} - (-6.16\ \text{kg}\cdot\text{m/s})]\hat{\mathbf{x}} = (13.6\ \text{kg}\cdot\text{m/s})\hat{\mathbf{x}}$$

If the ball and bat are in contact for 1.20 ms = 1.20×10^{-3} s, a typical time, then the average force is

$$\vec{\mathbf{F}}_{av} = \frac{\Delta\vec{\mathbf{p}}}{\Delta t} = \frac{\vec{\mathbf{I}}}{\Delta t} = \frac{(13.6\ \text{kg}\cdot\text{m/s})\hat{\mathbf{x}}}{1.20 \times 10^{-3}\ \text{s}} = (1.13 \times 10^4\ \text{N})\hat{\mathbf{x}}.$$

RWP* The average force is in the positive x direction—that is, toward the pitcher, as expected. In addition, the magnitude of the average force is remarkably large, more than 2500 pounds! This explains why high-speed photographs show the ball flattening out during a hit, as in **FIGURE 9-4**. In addition, notice that the weight of the ball, which is only about 0.3 lb, is negligible compared to the forces involved during the hit. On the other hand, extending the time of impact can reduce the force, as in the cushion breaking the pole vaulter's fall in Figure 9-4.

Big Idea 2 Impulse is force multiplied by the time over which it acts. Impulse is equal to the change in momentum of a system.

▲ **FIGURE 9-3 Hitting a baseball** The impulse delivered to the ball by the bat changes the ball's momentum from $-p_i\hat{\mathbf{x}}$ to $p_f\hat{\mathbf{x}}$.

◀ **FIGURE 9-4 Visualizing Concepts Impulse** When a baseball is hit by a bat (left), an enormous force (thousands of newtons) acts for a very short period of time—perhaps only a few ms. During this time, the ball is dramatically deformed by the impact. To keep the same thing from happening to a pole vaulter (right), a deeply padded landing area is provided. The change in the pole vaulter's momentum as he is brought to a stop, $mv = F\Delta t$, is the same whether he lands on a mat or on concrete. However, the padding yields, thereby greatly prolonging the time Δt during which he is in contact with the mat. The corresponding force on the vaulter is thus markedly decreased.

QUICK EXAMPLE 9-3   FIND THE FINAL SPEED OF THE BALL

A 0.144-kg baseball is moving toward home plate with a speed of 43.0 m/s when it is bunted (hit softly). The bat exerts an average force of 6.50×10^3 N on the ball for 1.30 ms. The average force is directed toward the pitcher, which we take to be the positive x direction. What is the final speed of the ball?

REASONING AND SOLUTION
We know the force and the time over which it acts. This gives the impulse, which is the change in momentum. We also know the initial momentum of the ball, from its mass

CONTINUED

---
*Real World Physics applications are denoted by the acronym RWP.

and initial speed. Therefore, we can find the final momentum, and from that the final speed.

1. Use $\vec{\mathbf{I}} = \vec{\mathbf{F}}_{av}\,\Delta t = \Delta\vec{\mathbf{p}}$ to relate the change in momentum to the average force:

$$\vec{\mathbf{I}} = \vec{\mathbf{F}}_{av}\,\Delta t = \Delta\mathbf{p}$$
$$= \vec{\mathbf{p}}_f - \vec{\mathbf{p}}_i$$

2. Rearrange to solve for the final momentum:

$$\vec{\mathbf{p}}_f = \vec{\mathbf{I}} + \vec{\mathbf{p}}_i$$

3. Calculate the initial momentum, $\vec{\mathbf{p}}_i$. Recall that the ball's initial velocity is in the negative x direction:

$$\vec{\mathbf{p}}_i = m\vec{\mathbf{v}}_i$$
$$= (0.144\ \text{kg})[(-43.0\ \text{m/s})\hat{\mathbf{x}}]$$
$$= (-6.19\ \text{kg}\cdot\text{m/s})\hat{\mathbf{x}}$$

4. Calculate the impulse, $\vec{\mathbf{I}}$. Recall that the average force exerted by the bat is in the positive x direction:

$$\vec{\mathbf{I}} = \vec{\mathbf{F}}_{av}\,\Delta t$$
$$= [(6.50\times10^3\,\text{N})\hat{\mathbf{x}}](1.30\times10^{-3}\,\text{s})$$
$$= (8.45\ \text{kg}\cdot\text{m/s})\hat{\mathbf{x}}$$

5. Substitute the results for impulse and initial momentum into Step 2 to find the final momentum, $\vec{\mathbf{p}}_f$:

$$\vec{\mathbf{p}}_f = \vec{\mathbf{I}} + \vec{\mathbf{p}}_i$$
$$= (8.45\ \text{kg}\cdot\text{m/s})\hat{\mathbf{x}} + (-6.19\ \text{kg}\cdot\text{m/s})\hat{\mathbf{x}}$$
$$= (2.26\ \text{kg}\cdot\text{m/s})\hat{\mathbf{x}}$$

6. Divide the final momentum by the mass to find the final velocity, $\vec{\mathbf{v}}_f$. The final speed, v_f, is the magnitude of the final velocity:

$$\vec{\mathbf{v}}_f = \frac{\vec{\mathbf{p}}_f}{m} = \frac{(2.26\ \text{kg}\cdot\text{m/s})\hat{\mathbf{x}}}{0.144\ \text{kg}} = (15.7\ \text{m/s})\hat{\mathbf{x}}$$
$$v_f = 15.7\ \text{m/s}$$

With our choice of coordinate system, we see that the initial momentum of the ball was in the negative x direction. The impulse applied to the ball, however, resulted in a final momentum (and velocity) in the positive x direction.

Everyday Examples of Impulse and Force We saw in Section 9-1 that the change in momentum is different for an object that hits something and sticks compared with an object that hits and bounces off. This means that the impulse, and hence the force, is different in the two cases. We explore this in the following Conceptual Example.

CONCEPTUAL EXAMPLE 9-4   RAIN VERSUS HAIL   PREDICT/EXPLAIN

A person stands under an umbrella during a rain shower. A few minutes later the raindrops turn to hail—though the number of "drops" hitting the umbrella per time remains the same, as does their speed and mass. **(a)** Is the force required to hold the umbrella in the hail greater than, less than, or equal to the force required in the rain? **(b)** Which of the following is the *best explanation* for your prediction?

I. The force required is greater because the hail bounces off the umbrella, which causes a greater change in momentum and a greater impulse.

II. Less force is required because the hail bounces off the umbrella, and hence has little effect on it. In contrast, the raindrops hit the umbrella and come to rest.

III. The force is the same in either case because the same mass of water (liquid or ice) hits the umbrella with the same speed.

Rain

Hail

REASONING AND DISCUSSION

When raindrops strike the umbrella, they tend to splatter and run off; when hailstones hit the umbrella, they bounce back upward. As a result, the change in momentum is greater for the hail—just as the change in momentum is greater for a rubber ball bouncing off the floor than it is for a beanbag landing on the floor. Hence, the impulse and the force are greater with hail.

ANSWER

(a) The force is greater in the hail. **(b)** The best explanation is I.

We conclude this section with an additional calculation involving impulse.

EXAMPLE 9-5   JOGGING A THOUGHT

A force platform measures the horizontal force exerted on a person's foot by the ground. The sketch shows force-versus-time data taken by a force platform for a 71-kg person who is jogging with increasing speed. The first part of the data ($t = 0$ to 0.30 s) indicates a negative force as the person's foot comes to rest. The impulse (area under the curve) during this time period is $I_1 = -2.7$ kg · m/s. The second part of the data ($t = 0.30$ s $- 0.80$ s) shows a positive force as the person pushes off against the ground to gain speed. The impulse during this time period is $I_2 = 26.2$ kg · m/s. For the time period from $t = 0$ to $t = 0.80$ s, what are **(a)** the average horizontal force exerted on the jogger and **(b)** the jogger's change in speed?

PICTURE THE PROBLEM

The plot of data from the force platform shows the force exerted on a person's foot as it first comes to rest (from $t = 0$ to $t = 0.30$ s) and then pushes off against the ground (from $t = 0.30$ s to $t = 0.80$ s). The impulse is negative at first ($I_1 < 0$), but then becomes positive ($I_2 > 0$). The net impulse delivered to the person is positive, $I_{net} = I_1 + I_2 > 0$, indicating an increase in speed.

REASONING AND STRATEGY

a. The net impulse delivered to the jogger from $t = 0$ to $t = 0.80$ s is $I_{net} = I_1 + I_2$. The average horizontal force exerted during this time interval is the net impulse divided by the elapsed time, $F_{av} = I_{net}/\Delta t$.

b. The net impulse I_{net} delivered to the person from $t = 0$ to $t = 0.80$ s is equal to the change in momentum $\Delta p = m\Delta v$ during that time period. The corresponding change in speed is given by the change in momentum divided by the mass m of the jogger; that is, $\Delta v = \Delta p/m = I_{net}/m$.

Known   Mass of jogger, $m = 71$ kg; impulse from $t = 0$ to $t = 0.30$ s, $I_1 = -2.7$ kg · m/s; impulse from $t = 0.30$ s to $t = 0.80$ s, $I_2 = 26.2$ kg · m/s.

Unknown   **(a)** Average force, $F_{av} = ?$ **(b)** Change in speed, $\Delta v = ?$

SOLUTION

Part (a)

1. Calculate the net impulse in the horizontal direction, I_{net}:

$$I_{net} = I_1 + I_2$$
$$= -2.7 \text{ kg · m/s} + 26.2 \text{ kg · m/s} = 23.5 \text{ kg · m/s}$$

2. Divide the net impulse by the time period, $\Delta t = 0.80$ s, to find the average force, F_{av}:

$$F_{av} = \frac{I_{net}}{\Delta t} = \frac{23.5 \text{ kg · m/s}}{0.80 \text{ s}} = 29 \text{ N}$$

Part (b)

3. Divide the net impulse by the jogger's mass to find the change in speed:

$$\Delta v = \frac{\Delta p}{m} = \frac{I_{net}}{m} = \frac{23.5 \text{ kg · m/s}}{71 \text{ kg}} = 0.33 \text{ m/s}$$

INSIGHT

In this case, the floor delivers a *horizontal* impulse to the jogger, causing an increase in speed. In **FIGURE 9-5**, the ground delivers a *vertical* impulse to a vampire bat, propelling it into the air.

PRACTICE PROBLEM

For the time period from $t = 0$ to $t = 0.30$ s, when the jogger's foot is coming to rest on the floor, what is the jogger's change in speed?   [**Answer:** The change in speed is small, only -0.038 m/s.]

Some related homework problems: Problem 14, Problem 15, Problem 17

Enhance Your Understanding   (Answers given at the end of the chapter)

3. A steel ball and a rubber ball collide. Is the impulse received by the steel ball greater than, less than, or equal to the impulse received by the rubber ball? Explain.

CONTINUED

▲ **FIGURE 9-5** Most bats take off by simply dropping from their perch, but vampire bats like this one must leap from the ground to become airborne. They do so by rocking forward onto their front limbs and then pushing off, using their extremely strong pectoral muscles. As the bat pushes downward, the ground exerts an upward reaction force on the bat, with a corresponding impulse sufficient to propel it upward a considerable distance. In fact, a vampire bat can launch itself 1 m or more into the air in less than a tenth of a second.

Big Idea ❸ Momentum is conserved when the net force acting on a system is zero.

9-4 Conservation of Linear Momentum

In this section we turn to perhaps the most important aspect of linear momentum—the fact that it is a conserved quantity. In this respect, it plays a fundamental role in physics similar to that of energy. We shall also see that momentum conservation leads to simplifications in our calculations, making it of great practical significance.

First, recall that the net force acting on an object is equal to the rate of change of its momentum:

$$\sum \vec{\mathbf{F}} = \frac{\Delta \vec{\mathbf{p}}}{\Delta t}$$

Rearranging this expression, we find that the change in momentum during a time interval Δt is

$$\Delta \vec{\mathbf{p}} = \left(\sum \vec{\mathbf{F}}\right)\Delta t \qquad \text{9-8}$$

Clearly, then, if the net force acting on an object is zero, $\sum \vec{\mathbf{F}} = 0$, its change in momentum is also zero:

$$\Delta \vec{\mathbf{p}} = \left(\sum \vec{\mathbf{F}}\right)\Delta t = 0$$

Writing the change of momentum in terms of its initial and final values, we have

$$\Delta \vec{\mathbf{p}} = \vec{\mathbf{p}}_f - \vec{\mathbf{p}}_i = 0$$

This means that the initial and final momenta are equal:

$$\vec{\mathbf{p}}_f = \vec{\mathbf{p}}_i \qquad \text{9-9}$$

The momentum does not change in this case, and hence we say that it is **conserved**. To summarize:

Conservation of Momentum

If the net force acting on an object is zero, its momentum is conserved; that is,

$$\vec{\mathbf{p}}_f = \vec{\mathbf{p}}_i$$

In some cases, the force is zero in one direction and nonzero in another. For example, an object in free fall has no force in the x direction, $F_x = 0$, but has a nonzero component of force in the y direction, $F_y \neq 0$. As a result, the object's x component of momentum remains constant, while its y component of momentum changes with time. Therefore, in applying momentum conservation, we must remember that both the force and the momentum are vector quantities and that the momentum conservation principle applies separately to each coordinate direction.

Internal Versus External Forces

The net force acting on a system of objects is the sum of forces applied from outside the system (external forces, $\vec{\mathbf{F}}_{ext}$) and forces acting between objects within the system (internal forces, $\vec{\mathbf{F}}_{int}$). Thus, we can write

$$\vec{\mathbf{F}}_{net} = \sum \vec{\mathbf{F}} = \sum \vec{\mathbf{F}}_{ext} + \sum \vec{\mathbf{F}}_{int}$$

As we shall see, internal and external forces play very *different* roles in terms of how they affect the momentum of a system.

Internal forces always sum to zero . . .

$\vec{F}_1$ $\vec{F}_2$

1   2

. . . and hence they have no effect on the net momentum of the system.

◄ **FIGURE 9-6 Separating two canoes**
A system composed of two canoes and their occupants. The forces $\vec{F}_1$ and $\vec{F}_2$ are internal to the system, and hence they sum to zero.

To illustrate the distinction, consider the case of two canoes floating at rest next to one another on a lake, as described in Example 5-6 and shown in **FIGURE 9-6**. Let's consider the "system" to be the two canoes and the people inside them. When a person in canoe 1 pushes on canoe 2, a force $\vec{F}_2$ is exerted on canoe 2. By Newton's third law, an equal and opposite force, $\vec{F}_1 = -\vec{F}_2$, is exerted on the person in canoe 1. We say that $\vec{F}_1$ and $\vec{F}_2$ are internal forces, because they act between objects that are part of the same system. Notice, also, that these two forces sum to zero:

$$\vec{F}_1 + \vec{F}_2 = (-\vec{F}_2) + \vec{F}_2 = 0$$

This is a special case, of course, but it demonstrates the following general principles:

- Internal forces, like all forces, occur in action-reaction pairs. Both members of an action-reaction pair are within the system when the forces are internal.

- The forces in action-reaction pairs are equal and opposite—due to Newton's third law—and hence internal forces must *always* sum to zero:

$$\sum \vec{F}_{\text{int}} = 0$$

The fact that internal forces always cancel means that the net force acting on a system of objects is simply the sum of the *external* forces acting on it:

$$\vec{F}_{\text{net}} = \sum \vec{F}_{\text{ext}} + \sum \vec{F}_{\text{int}} = \sum \vec{F}_{\text{ext}} + 0 = \sum \vec{F}_{\text{ext}}$$

The external forces, on the other hand, may or may not sum to zero—it all depends on the particular situation. For example, if the system consists of the two canoes in Figure 9-6, the external forces are the weights of the people and the canoes acting downward, and the upward, normal force exerted by the water to keep the canoes afloat. These forces sum to zero, and there is no acceleration in the vertical direction. In other cases, like an object in free fall, the net external force is nonzero.

Finally, how do external and internal forces affect the momentum of a system? To see the connection, first note that Newton's second law gives the change in the net momentum for a given time interval Δt:

$$\Delta \vec{p}_{\text{net}} = \vec{F}_{\text{net}} \Delta t$$

Because the internal forces cancel, however, the change in the net momentum is directly related to the net *external* force:

$$\Delta \vec{p}_{\text{net}} = \left(\sum \vec{F}_{\text{ext}} \right) \Delta t \qquad \text{9-10}$$

Therefore, the key distinction between internal and external forces is the following:

Conservation of Momentum for a System of Objects

- *Internal* forces have absolutely no effect on the net momentum of a system.

- If the *net external* force acting on a system is zero, its net momentum is conserved. That is,

$$\vec{p}_{1,\text{f}} + \vec{p}_{2,\text{f}} + \vec{p}_{3,\text{f}} + \cdots = \vec{p}_{1,\text{i}} + \vec{p}_{2,\text{i}} + \vec{p}_{3,\text{i}} + \cdots$$

◀ **FIGURE 9-7** If the astronaut in this photo pushes on the satellite, the satellite exerts an equal but opposite force on him, in accordance with Newton's third law. If we are calculating the change in the astronaut's momentum, we must take this force into account. However, if we define the system to be the astronaut *and* the satellite, the forces between them are internal to the system. Whatever effect they may have on the astronaut or the satellite individually, they do not affect the momentum of the system as a whole. Therefore, whether a particular force counts as internal or external depends entirely on where we draw the boundaries of the system.

It's important to point out that these statements apply only to the *net* momentum of a system, and not to the momentum of each individual object. For example, suppose that a system consists of two objects, 1 and 2, and that the net external force acting on the system is zero. As a result, the net momentum must remain constant:

$$\vec{\mathbf{p}}_{net} = \vec{\mathbf{p}}_1 + \vec{\mathbf{p}}_2 = \text{constant}$$

This does not mean, however, that $\vec{\mathbf{p}}_1$ is constant or that $\vec{\mathbf{p}}_2$ is constant. All we can say is that the *sum* of $\vec{\mathbf{p}}_1$ and $\vec{\mathbf{p}}_2$ does not change. An example of this is the astronaut pushing off from a satellite in **FIGURE 9-7**.

PROBLEM-SOLVING NOTE

Internal Versus External Forces

It is important to keep in mind that internal forces cannot change the momentum of a system—only a net external force can do that.

Applying Momentum Conservation As a specific example of momentum conservation, consider the two canoes floating on a lake, as described previously. Initially the momentum of the system is zero, because the canoes are at rest. After a person pushes the canoes apart, they are both moving, and hence both have nonzero momentum. Thus, the momentum of each canoe has changed. On the other hand, because the net external force acting on the system is zero, the sum of the canoes' momenta must still vanish. We show this in the next Example.

EXAMPLE 9-6 | TIPPY CANOE

Two groups of canoeists meet in the middle of a lake. After a brief visit, a person in canoe 1 pushes on canoe 2 with a force of 46 N to separate the canoes. If the mass of canoe 1 and its occupants is 130 kg, and the mass of canoe 2 and its occupants is 250 kg, find the momentum of each canoe after 1.20 s of pushing.

PICTURE THE PROBLEM
We choose the positive x direction to point from canoe 1 to canoe 2. With this choice, the force exerted on canoe 2 is $\vec{\mathbf{F}}_2 = (46\,\text{N})\hat{\mathbf{x}}$ and the force exerted on canoe 1 is $\vec{\mathbf{F}}_1 = (-46\,\text{N})\hat{\mathbf{x}}$.

REASONING AND STRATEGY
First, we find the acceleration of each canoe using $a_x = F_x/m$. Next, we use kinematics, $v_x = v_{0x} + a_x t$, to find the velocity at time t. Notice that the canoes start at rest, and hence $v_{0x} = 0$. Finally, the momentum is calculated using $p_x = mv_x$.

Known Mass of canoe 1, $m_1 = 130\,\text{kg}$; mass of canoe 2, $m_2 = 250\,\text{kg}$; magnitude of pushing force, $F = 46\,\text{N}$; pushing time, $t = 1.20\,\text{s}$.

Unknown Final momentum of canoe 1, $\vec{\mathbf{p}}_{1,f} = ?$ Final momentum of canoe 2, $\vec{\mathbf{p}}_{2,f} = ?$

$F_1 = 46\,\text{N}$ $F_2 = 46\,\text{N}$

SOLUTION

1. Use Newton's second law to find the acceleration of canoe 1. Note that the acceleration of canoe 1 is in the negative x direction:

$$a_{1,x} = \frac{\sum F_{1,x}}{m_1} = \frac{-46\,\text{N}}{130\,\text{kg}} = -0.35\,\text{m/s}^2$$

2. Do the same calculation for canoe 2:

$$a_{2,x} = \frac{\sum F_{2,x}}{m_2} = \frac{46\,\text{N}}{250\,\text{kg}} = 0.18\,\text{m/s}^2$$

3. Calculate the velocity of each canoe at $t = 1.20\,\text{s}$:

$$v_{1,x} = a_{1,x}t = (-0.35\,\text{m/s}^2)(1.20\,\text{s}) = -0.42\,\text{m/s}$$
$$v_{2,x} = a_{2,x}t = (0.18\,\text{m/s}^2)(1.20\,\text{s}) = 0.22\,\text{m/s}$$

4. Calculate the momentum of each canoe at $t = 1.20\,\text{s}$:

$$p_{1,x} = m_1 v_{1,x} = (130\,\text{kg})(-0.42\,\text{m/s}) = -55\,\text{kg}\cdot\text{m/s}$$
$$p_{2,x} = m_2 v_{2,x} = (250\,\text{kg})(0.22\,\text{m/s}) = 55\,\text{kg}\cdot\text{m/s}$$

CONTINUED

INSIGHT

The sum of the momenta of the two moving canoes is zero. This is just what one would expect: The canoes start at rest with zero momentum, there is zero net external force acting on the system, and hence the final momentum must also be zero. The final velocities *do not* add to zero; it is momentum ($m\vec{\mathbf{v}}$) that is conserved, not velocity ($\vec{\mathbf{v}}$).

Finally, we solved this problem using one-dimensional kinematics so that we could clearly see the distinction between velocity and momentum. An alternative way to calculate the final momentum of each canoe is to use $\Delta\vec{\mathbf{p}} = \vec{\mathbf{p}}_f - \vec{\mathbf{p}}_i = \vec{\mathbf{F}}\Delta t$. For canoe 1 we have $\vec{\mathbf{p}}_{1,f} = \vec{\mathbf{F}}_1\Delta t + \vec{\mathbf{p}}_{1,i} = [(-46\text{ N})\hat{\mathbf{x}}](1.20\text{ s}) + 0 = (-55\text{ kg}\cdot\text{m/s})\hat{\mathbf{x}}$, in agreement with our results above. A similar calculation yields $\vec{\mathbf{p}}_{2,f} = (55\text{ kg}\cdot\text{m/s})\hat{\mathbf{x}}$ for canoe 2.

PRACTICE PROBLEM

What are the final momenta of the canoes if they are pushed apart with a force of 65 N? [**Answer:** $p_{1,x} = -78\text{ kg}\cdot\text{m/s}$; $p_{2,x} = 78\text{ kg}\cdot\text{m/s}$]

Some related homework problems: Problem 23, Problem 24, Problem 26

BIO The Physics of Recoil

In a situation like the one described in Example 9-6, the person in canoe 1 pushes canoe 2 away. At the same time, canoe 1 begins to move in the opposite direction. This is referred to as **recoil**. It is essentially the same as the recoil one experiences when firing a gun or when turning on a strong stream of water.

Participants in a number of athletic sports, including American football, soccer, hockey, and of course boxing, can experience a traumatic brain injury called a *concussion*. A concussion is an example of an injury caused by recoil. The human brain is surrounded by protective cerebrospinal fluid, which allows it to move a little bit inside the skull. Initially the brain and the skull are at rest relative to each other and the momentum of the system is zero. If the skull is suddenly accelerated forward, due to an impact at the back of the head as indicated in FIGURE 9-8, the brain remains at rest for a brief moment until it collides with the back of the skull. The brain then recoils from the skull and moves forward—in fact, it may even collide again, this time with the front of the skull. Such collision and recoil events cause the concussive brain injury.

Another interesting example of recoil involves the human body. Perhaps you've noticed, when resting quietly in a rocking or reclining chair, that the chair wobbles back and forth slightly about once a second. The reason for this movement is that each time your heart pumps blood in one direction (from the atria to the ventricles, then to the aorta and pulmonary arteries, and so on) your body recoils in the opposite direction. Because the recoil depends on the force exerted by your heart on the blood and the volume of blood expelled from the heart with each beat, it is possible to gain valuable medical information regarding the health of your heart by analyzing the recoil it produces.

The medical instrument that employs the physical principle of recoil is called the *ballistocardiograph*. It is a completely noninvasive technology that simply requires the patient to sit comfortably in a chair fitted with sensitive force sensors under the seat and behind the back. Sophisticated bathroom scales also utilize this technology. A ballistocardiographic (BCG) scale detects the recoil vibrations of the body as a person stands on the scale. This allows the BCG scale to display not only the person's body weight but his or her heart rate as well.

A more dramatic application of heartbeat recoil is currently being used at the Riverbend Maximum Security Institution in Tennessee. The only successful breakout from this prison occurred when four inmates hid in a secret compartment in a delivery truck that was leaving the facility. The institution now uses a heartbeat recoil detector that would have foiled this escape. Each vehicle leaving the prison must stop at a checkpoint where a small motion detector is attached to it with a suction cup. Any persons hidden in the vehicle will reveal their presence by the very beating of their hearts. These heartbeat detectors have proved to be 100% effective, even though the recoil of the heart may displace a large truck by only a few millionths of an inch. Similar systems are being used at other high-security installations and border crossings.

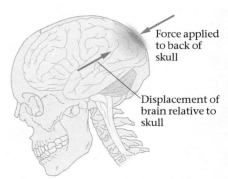

Force applied to back of skull

Displacement of brain relative to skull

▲ **FIGURE 9-8** A sharp force applied to the back of the head can accelerate the skull forward relative to the brain. The brain then collides with the back of the skull and rebounds off of it, which can cause a concussion.

CONCEPTUAL EXAMPLE 9-7 MOMENTUM VERSUS KINETIC ENERGY

The final momentum of the two canoes and their occupants in Example 9-6 is equal to their initial momentum. Is the final kinetic energy of the system (a) greater than, (b) less than, or (c) equal to its initial kinetic energy?

REASONING AND DISCUSSION

The final momentum of the two canoes is zero because one canoe has a positive momentum and the other has a negative momentum of the same magnitude. The two momenta, then, sum to zero. Kinetic energy, which is $\frac{1}{2}mv^2$, cannot be negative; hence no such cancellation is possible. Both canoes have positive kinetic energies, and therefore the final kinetic energy is greater than the initial kinetic energy, which is zero.

Where does the increase in kinetic energy come from? It comes from the muscular work done by the person who pushes the canoes apart.

ANSWER

(a) K_f is greater than K_i.

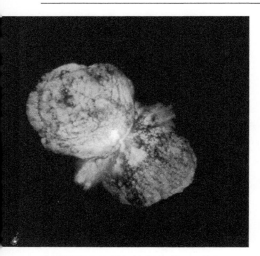

▲ **FIGURE 9-9** This Hubble Space Telescope photograph shows the aftermath of a violent explosion of the star Eta Carinae. The explosion produced two bright lobes of matter spewing outward in opposite directions. In this photograph, these lobes have expanded to about the size of our solar system. The momentum of the star before the explosion must be the same as the total momentum of the star and the bright lobes after the explosion. Since the lobes are roughly symmetrical and move in opposite directions, their net momentum is essentially zero. Thus, we conclude that the momentum of the star itself was virtually unchanged by the explosion.

RWP A special case of some interest is the universe. Since there is nothing external to the universe—by definition—it follows that the net external force acting on it is zero. Therefore, its net momentum is conserved. No matter what happens—a comet collides with the Earth, a star explodes and becomes a supernova, a black hole swallows part of a galaxy—the total momentum of the universe simply cannot change. A particularly vivid illustration of momentum conservation in our own galaxy is provided by the exploding star Eta Carinae. As can be seen in the Hubble Space Telescope photograph in **FIGURE 9-9**, jets of material are moving away from the star in opposite directions, just like the canoes moving apart from one another in Example 9-6.

Conservation of momentum also applies to the more down-to-Earth situation described in the next Quick Example.

QUICK EXAMPLE 9-8 THE SPEED OF A LOG

Predict/Calculate

A 61-kg lumberjack stands on a 320-kg log that floats on a pond. Both the log and the lumberjack are at rest initially. In addition, the log points directly toward the shore, which we take to be the positive x direction. (a) If the lumberjack trots along the log toward the shore with a speed v relative to the shore, the log moves away from the shore. Is the speed of the log relative to the shore greater than, less than, or equal to v? Explain. (b) If the lumberjack trots

with a velocity of 1.8 m/s relative to the shore, what is the velocity of the log relative to the shore? Ignore friction between the log and the water.

REASONING AND SOLUTION

Part (a) The speed of the log relative to the shore is less than v. To see this, notice that before the lumberjack begins trotting toward the shore, the system has zero momentum. Only internal forces are involved as the lumberjack trots, and so the total momentum remains equal to zero. Therefore, the momentum of the lumberjack moving toward the shore is balanced by the momentum of the log moving away from the shore. The log has the greater mass, and hence the speed of the log is less than v.

Part (b)

1. Set the total momentum of the lumberjack (lj) plus the log equal to zero: $\quad \vec{\mathbf{p}}_{lj} + \vec{\mathbf{p}}_{log} = 0$

CONTINUED

2. Solve for the momentum of the log in terms of the momentum of the lumberjack:

$$\vec{\mathbf{p}}_{\text{log}} = -\vec{\mathbf{p}}_{\text{lj}}$$

3. Calculate the momentum of the trotting lumberjack. Recall that the direction toward the shore is the positive x direction:

$$\vec{\mathbf{p}}_{\text{lj}} = m_{\text{lj}} v_{\text{lj}} \hat{\mathbf{x}} = (61 \text{ kg})(1.8 \text{ m/s})\hat{\mathbf{x}}$$
$$= (110 \text{ kg} \cdot \text{m/s})\hat{\mathbf{x}}$$

4. Calculate the momentum of the log:

$$\vec{\mathbf{p}}_{\text{log}} = -\vec{\mathbf{p}}_{\text{lj}} = -(110 \text{ kg} \cdot \text{m/s})\hat{\mathbf{x}}$$

5. Divide the log's momentum by its mass to find its velocity:

$$\vec{\mathbf{p}}_{\text{log}} = m_{\text{log}} \vec{\mathbf{v}}_{\text{log}}$$

$$\vec{\mathbf{v}}_{\text{log}} = \frac{\vec{\mathbf{p}}_{\text{log}}}{m_{\text{log}}} = \frac{-(110 \text{ kg} \cdot \text{m/s})\hat{\mathbf{x}}}{320 \text{ kg}} = -(0.34 \text{ m/s})\hat{\mathbf{x}}$$

As expected, the log moves away from the shore—in the minus x direction—and with a speed (0.34 m/s) that is less than the speed of the lumberjack (1.8 m/s). With these speeds, the magnitude of the log's momentum is equal to the magnitude of the lumberjack's momentum.

Enhance Your Understanding   (Answers given at the end of the chapter)

4. An object initially at rest breaks into two pieces as the result of an explosion. Piece 1 has twice the kinetic energy of piece 2. **(a)** If piece 1 has a momentum of magnitude p, what is the magnitude of the momentum of piece 2? Explain. **(b)** If the mass of piece 1 is m, what is the mass of piece 2? Explain.

Section Review

- Momentum is conserved when the net force acting on a system is zero.
- Internal forces have no effect on the net momentum of a system.

9-5 Inelastic Collisions

In this section we consider **collisions**. By a collision we mean a situation in which two objects strike one another, and in which the net external force is either zero or negligibly small. For example, if two train cars roll on a level track and hit one another, this is a collision. In this case, the net external force—the weight downward plus the normal force exerted by the tracks upward—is zero. As a result, the momentum of the two-car system is conserved.

Another example of a collision is a baseball that is struck by a bat. In this case, the external forces are not zero because the weight of the ball is not balanced by any other force. However, as we have seen in Section 9-3, the forces exerted during the hit are much larger than the weight of the ball or the bat. Hence, to a good approximation, we may neglect the external forces (the weight of the ball and bat) in this case, and say that the momentum of the ball–bat system is conserved.

Now it may seem surprising at first, but the fact that the momentum of a system is conserved during a collision does not necessarily mean that the system's kinetic energy is conserved. In fact most, or even all, of a system's kinetic energy may be converted to other forms during a collision. In general, collisions are categorized according to what happens to the kinetic energy of the system. There are two possibilities. After a collision, the final kinetic energy, K_f, is either equal to the initial kinetic energy, K_i, or it is not. If $K_f = K_i$, the collision is said to be **elastic.** We consider elastic collisions in the next section.

On the other hand, the kinetic energy may change during a collision. Usually it decreases due to losses associated with sound, heat, and deformation. Sometimes it increases—if the collision sets off an explosion, for instance. In any event, collisions in which the kinetic energy is not conserved are referred to as **inelastic:**

Big Idea ❹ Collisions conserve momentum, but may or may not conserve mechanical energy. Mechanical energy is conserved in elastic collisions, but not in inelastic collisions.

**PHYSICS
IN CONTEXT
Looking Back**

Kinetic energy (Chapter 7) plays a key role in analyzing collisions, leading to the distinction between elastic and inelastic collisions.

▲ **FIGURE 9-10 Visualizing Concepts
Collisions** In both elastic and inelastic collisions, momentum is conserved. The same is not true of kinetic energy, however. In the largely inelastic collision at top, much of the hockey players' initial kinetic energy is transformed into work—rearranging the players' anatomies and shattering the glass at the rink. In the highly elastic collision at bottom, the ball rebounds with very little diminution of its kinetic energy (though a little energy is lost as sound and heat).

Inelastic Collisions

In an inelastic collision, the momentum of a system is conserved:

$$\vec{\mathbf{p}}_f = \vec{\mathbf{p}}_i$$

The kinetic energy of the system is not conserved in an inelastic collision:

$$K_f \neq K_i$$

In the special case where objects stick together after the collision, we say that the collision is **completely inelastic**:

Completely Inelastic Collisions

When objects stick together after colliding, the collision is completely inelastic.

In a completely inelastic collision, the maximum amount of kinetic energy is lost. If the total momentum of the system is zero, this means that all of the kinetic energy is lost. For systems with nonzero total momentum, however, some kinetic energy remains after the collision. Examples of elastic and inelastic collisions are shown in **FIGURE 9-10**.

Inelastic Collisions in One Dimension

Consider a system consisting of two train cars of mass m_1 and m_2 on a smooth, level track. The cars have initial velocities $v_{1,i}$ and $v_{2,i}$, as shown in **FIGURE 9-11**. When the cars collide, the coupling mechanism latches, causing the cars to stick together and move as a unit. What is the speed of the cars after the collision?

To answer this question, we begin by noting that the system has zero net external force acting on it, and hence its momentum is conserved. The initial momentum is

$$p_i = m_1 v_{1,i} + m_2 v_{2,i}$$

After the collision, the objects move together with a common velocity v_f. Therefore, the final momentum is

$$p_f = (m_1 + m_2)v_f$$

Equating the initial and final momenta yields $m_1 v_{1,i} + m_2 v_{2,i} = (m_1 + m_2)v_f$, or

$$v_f = \frac{m_1 v_{1,i} + m_2 v_{2,i}}{m_1 + m_2}$$

9-11

Thus, the final velocity is equal to the initial momentum divided by the total mass. If the initial momentum is zero, as when two identical train cars approach one another with equal speed, the final velocity is zero and the system is at rest after the collision.

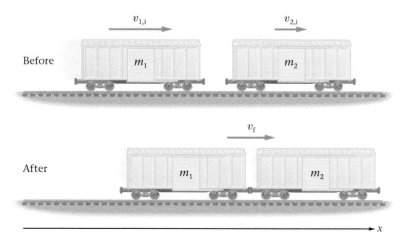

▲ **FIGURE 9-11 Railroad cars collide and stick together** After the collision, the cars stick together and move with the same speed.

EXERCISE 9-9 SPEED AFTER THE COLLISION

A 1400-kg car moving at 3.2 m/s is struck in the rear by a 2900-kg truck moving in the same direction at 5.3 m/s. If the vehicles stick together after the collision, what is their speed immediately after colliding? (Assume external forces may be ignored.)

REASONING AND SOLUTION
The vehicles stick together, and hence their collision is completely inelastic. As a result, we can find the final velocity with $v_f = (m_1 v_{1,i} + m_2 v_{2,i})/(m_1 + m_2)$ (Equation 9-11). In this case, $m_1 = 1400$ kg, $v_{1,i} = 3.2$ m/s, $m_2 = 2900$ kg, and $v_{2,i} = 5.3$ m/s; hence the final velocity is

$$v_f = \frac{m_1 v_{1,i} + m_2 v_{2,i}}{m_1 + m_2} = \frac{(1400 \text{ kg})(3.2 \text{ m/s}) + (2900 \text{ kg})(5.3 \text{ m/s})}{1400 \text{ kg} + 2900 \text{ kg}} = 4.6 \text{ m/s}$$

Let's apply $v_f = (m_1 v_{1,i} + m_2 v_{2,i})/(m_1 + m_2)$ to the simple case of two railroad cars of equal mass m, one of which is at rest before the collision. If the other car has an initial speed of v, what is the final speed of the two cars? Letting $m_1 = m_2 = m$, $v_{1,i} = v$, and $v_{2,i} = 0$, we find

$$v_f = \frac{mv + m \cdot 0}{m + m} = \frac{m}{2m}v = \frac{1}{2}v \qquad\qquad \text{9-12}$$

As you might have guessed, the final speed is one-half the initial speed.

No momentum is lost in this collision of railroad cars, but some of the initial kinetic energy is converted to other forms. Some propagates away as sound, some is converted to heat, and some creates permanent deformations in the metal of the latching mechanism. The precise amount of kinetic energy that is lost is addressed in the following Conceptual Example.

PROBLEM-SOLVING NOTE

Momentum Versus Energy Conservation

Be sure to distinguish between momentum conservation and energy conservation. A common error is to assume that kinetic energy is conserved just because the momentum is conserved.

CONCEPTUAL EXAMPLE 9-10 HOW MUCH KINETIC ENERGY IS LOST?

A railroad car of mass m and speed v collides with and sticks to an identical railroad car that is initially at rest. After the collision, is the kinetic energy of the system (a) 1/2, (b) 1/3, or (c) 1/4 of its initial kinetic energy?

REASONING AND DISCUSSION
Before the collision, the kinetic energy of the system is

$$K_i = \frac{1}{2}mv^2$$

After the collision, the mass is doubled and the speed is halved. Hence, the final kinetic energy is

$$K_f = \frac{1}{2}(2m)\left(\frac{v}{2}\right)^2 = \frac{1}{2}(2m)\left(\frac{v^2}{4}\right) = \frac{1}{2}\left(\frac{1}{2}mv^2\right) = \frac{1}{2}K_i$$

Therefore, one-half of the initial kinetic energy is converted to other forms of energy.

ANSWER
(a) The final kinetic energy is one-half the initial kinetic energy.

It's interesting that we know the precise amount of kinetic energy that was lost during the collision of the train cars, even though we don't know just how much went into sound, how much went into heat, and so on. It's not necessary to know all of the details to determine how much kinetic energy was lost.

We also know how much momentum was lost—none.

EXAMPLE 9-11   GOAL-LINE STAND

On a touchdown attempt, a 95.0-kg running back runs toward the end zone at 3.75 m/s. A 111-kg linebacker moving at 4.10 m/s meets the runner in a head-on collision. If the two players stick together, **(a)** what is their velocity immediately after the collision? **(b)** What are the initial and final kinetic energies of the system?

PICTURE THE PROBLEM

In our sketch, we let subscript 1 refer to the red-and-gray running back who carries the ball, and subscript 2 refers to the blue-and-gold linebacker who makes the tackle. The direction of the running back's initial motion is taken to be in the positive x direction. Therefore, the initial velocities of the players are $\vec{\mathbf{v}}_1 = (3.75 \text{ m/s})\hat{\mathbf{x}}$ and $\vec{\mathbf{v}}_2 = (-4.10 \text{ m/s})\hat{\mathbf{x}}$.

$\vec{\mathbf{v}}_1 = (3.75 \text{ m/s})\hat{\mathbf{x}}$ $\vec{\mathbf{v}}_2 = (-4.10 \text{ m/s})\hat{\mathbf{x}}$

REASONING AND STRATEGY

a. The final velocity can be found by applying momentum conservation to the system consisting of the two players. Initially, the players have momenta in opposite directions. After the collision, the players move together with a combined mass $m_1 + m_2$ and a velocity $\vec{\mathbf{v}}_f$.

b. The kinetic energies can be found by applying $\frac{1}{2}mv^2$ to the players individually to obtain the initial kinetic energy, and then to their combined motion for the final kinetic energy.

Known   Player 1: $m_1 = 95.0$ kg, $\vec{\mathbf{v}}_1 = (3.75 \text{ m/s})\hat{\mathbf{x}}$; Player 2: $m_2 = 111$ kg, $\vec{\mathbf{v}}_2 = (-4.10 \text{ m/s})\hat{\mathbf{x}}$.
Unknown   **(a)** $\vec{\mathbf{v}}_f = ?$ **(b)** $K_i = ? \; K_f = ?$

SOLUTION

Part (a)

1. Set the initial momenta of the two players equal to their final momentum:

$$m_1\vec{\mathbf{v}}_1 + m_2\vec{\mathbf{v}}_2 = (m_1 + m_2)\vec{\mathbf{v}}_f$$

2. Solve for the final velocity and substitute numerical values, being careful to use the appropriate signs:

$$\vec{\mathbf{v}}_f = \frac{m_1\vec{\mathbf{v}}_1 + m_2\vec{\mathbf{v}}_2}{m_1 + m_2}$$

$$= \frac{(95.0 \text{ kg})(3.75 \text{ m/s})\hat{\mathbf{x}} + (111 \text{ kg})(-4.10 \text{ m/s})\hat{\mathbf{x}}}{95.0 \text{ kg} + 111 \text{ kg}}$$

$$= (-0.480 \text{ m/s})\hat{\mathbf{x}}$$

Part (b)

3. Calculate the initial kinetic energy of the two players:

$$K_i = \tfrac{1}{2}m_1v_1{}^2 + \tfrac{1}{2}m_2v_2{}^2$$

$$= \tfrac{1}{2}(95.0 \text{ kg})(3.75 \text{ m/s})^2 + \tfrac{1}{2}(111 \text{ kg})(4.10 \text{ m/s})^2$$

$$= 1600 \text{ J}$$

4. Calculate the final kinetic energy of the players, noting that they both move with the same velocity after the collision:

$$K_f = \tfrac{1}{2}(m_1 + m_2)v_f{}^2$$

$$= \tfrac{1}{2}(95.0 \text{ kg} + 111 \text{ kg})(0.480 \text{ m/s})^2$$

$$= 23.7 \text{ J}$$

INSIGHT

After the collision, the two players are moving in the negative x direction—that is, away from the end zone—with a speed of 0.480 m/s. This is because the linebacker had more negative momentum than the running back had positive momentum. As for the kinetic energy, of the original 1600 J, only 23.7 J is left after the collision. This means that more than 98% of the original kinetic energy is converted to other forms. Even so, *none* of the momentum is lost.

PRACTICE PROBLEM — PREDICT/CALCULATE

(a) If the final speed of the two players is to be zero, should the speed of the running back (red and gray) be increased or decreased? Explain. **(b)** Check your prediction by calculating the required speed for the running back. [**Answer: (a)** The running back's speed should be increased, so that the magnitude of his momentum will match that of the linebacker. **(b)** The desired speed is 4.79 m/s.]

Some related homework problems: Problem 31, Problem 34

EXAMPLE 9-12 BALLISTIC PENDULUM

In a ballistic pendulum, an object of mass m is fired with an initial speed v_i at the bob of a pendulum. The bob has a mass M, and is suspended by a rod of negligible mass. After the collision, the object and the bob stick together and swing through an arc, eventually gaining a height h. Find the height h in terms of m, M, v_i, and g.

PICTURE THE PROBLEM

Our sketch shows the physical setup of a ballistic pendulum. Initially, only the object of mass m is moving, and it moves in the positive x direction with a speed v_i. Immediately after the collision, the bob and object move together with a new speed, v_f, which is determined by momentum conservation. Finally, the pendulum continues to swing to the right until its speed decreases to zero and it comes to rest at the height h.

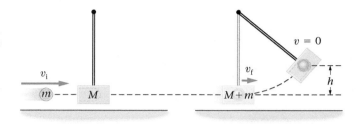

REASONING AND STRATEGY

There are two distinct physical processes at work in the ballistic pendulum. The first is a completely inelastic collision between the bob and the object. Momentum is conserved during this collision, but kinetic energy is not. After the collision, the remaining kinetic energy is converted to gravitational potential energy, which determines how high the bob and object will rise.

Known Mass of object, m; mass of bob, M; initial speed of object, v_i.
Unknown Height after collision, $h = ?$

SOLUTION

1. Set the momentum just before the bob–object collision equal to the momentum just after the collision. Let v_f be the speed just after the collision:

$$mv_i = (M + m)v_f$$

2. Solve for the speed just after the collision, v_f:

$$v_f = \left(\frac{m}{M + m}\right)v_i$$

3. Calculate the kinetic energy just after the collision:

$$K_f = \frac{1}{2}(M + m)v_f^2 = \frac{1}{2}(M + m)\left(\frac{m}{M + m}\right)^2 v_i^2$$

$$= \frac{1}{2}mv_i^2\left(\frac{m}{M + m}\right)$$

4. Set the kinetic energy after the collision equal to the gravitational potential energy of the bob–object combination at the height h:

$$\frac{1}{2}mv_i^2\left(\frac{m}{M + m}\right) = (M + m)gh$$

5. Solve for the height, h:

$$h = \left(\frac{m}{M + m}\right)^2\left(\frac{v_i^2}{2g}\right)$$

INSIGHT

A ballistic pendulum is often used to measure the speed of a rapidly moving object, such as a bullet. If a bullet were shot straight up, it would rise to the height $v_i^2/2g$, which can be thousands of feet. On the other hand, if a bullet of mass m is fired into a ballistic pendulum, in which M is much greater than m, the bullet will reach only a small fraction of this height. Thus, the ballistic pendulum makes for a more convenient and practical measurement.

PRACTICE PROBLEM

A 7.00-g bullet is fired into a ballistic pendulum whose bob has a mass of 0.950 kg. If the bob rises to a height of 0.220 m, what was the initial speed of the bullet? [**Answer:** $v_i = 284$ m/s. If this bullet were fired straight up, it would rise 4.11 km $\approx$ 13,000 ft in the absence of air resistance.]

Some related homework problems: Problem 32, Problem 33

Inelastic Collisions in Two Dimensions

Next we consider collisions in two dimensions, where we must conserve the momentum component by component. To do this, we set up a coordinate system and resolve the initial momentum into x and y components. Next, we demand that the final momentum have precisely the same x and y components as the initial momentum. That is,

$$p_{x,i} = p_{x,f} \quad \text{and} \quad p_{y,i} = p_{y,f}$$

PROBLEM-SOLVING NOTE

Sketch the System Before and After the Collision

In problems involving collisions, it is useful to draw the system before and after the collision. Be sure to label the relevant masses, velocities, and angles.

The following Example shows how to carry out such a calculation in a practical situation.

EXAMPLE 9-13 ANALYZING A TRAFFIC ACCIDENT

RWP A car with a mass of 950 kg and a speed of 16 m/s approaches an intersection, as shown in the sketch. A 1300-kg minivan traveling at 21 m/s is heading for the same intersection. The car and minivan collide and stick together. Find the direction and speed of the wrecked vehicles just after the collision, assuming external forces can be ignored.

PICTURE THE PROBLEM
In our sketch, we align the x and y axes with the crossing streets. With this choice, $\vec{\mathbf{v}}_1$ (the car's initial velocity) is in the positive x direction, and $\vec{\mathbf{v}}_2$ (the minivan's initial velocity) is in the positive y direction. In addition, the problem statement indicates that $m_1 = 950$ kg and $m_2 = 1300$ kg. After the collision, the two vehicles move together (as a unit) with a speed v_f in a direction θ with respect to the positive x axis.

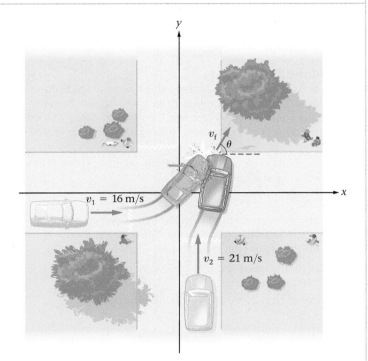

REASONING AND STRATEGY
Because external forces can be ignored, the total momentum of the system must be conserved during the collision. This is really two conditions: (i) The x component of momentum is conserved, and (ii) the y component of momentum is conserved. These two conditions determine the two unknowns: the final speed, v_f, and the final direction, θ.

Known   Mass of car, $m_1 = 950$ kg; initial velocity of car, $\vec{\mathbf{v}}_1 = (16\text{ m/s})\hat{\mathbf{x}}$; mass of minivan, $m_2 = 1300$ kg; initial velocity of minivan, $\vec{\mathbf{v}}_2 = (21\text{ m/s})\hat{\mathbf{y}}$.

Unknown   Final speed, $v_f = ?$ Final direction, $\theta = ?$

SOLUTION

1. Set the initial x component of momentum equal to the final x component of momentum. Note that the x component of $\vec{\mathbf{v}}_f$ is $v_f \cos\theta$:

$$m_1 v_1 = (m_1 + m_2)v_f \cos\theta$$

2. Set the initial y component of momentum equal to the final y component of momentum. Note that the y component of $\vec{\mathbf{v}}_f$ is $v_f \sin\theta$:

$$m_2 v_2 = (m_1 + m_2)v_f \sin\theta$$

3. Divide the y momentum equation by the x momentum equation. This eliminates v_f, giving an equation involving θ alone:

$$\frac{m_2 v_2}{m_1 v_1} = \frac{(m_1 + m_2)v_f \sin\theta}{(m_1 + m_2)v_f \cos\theta} = \frac{\sin\theta}{\cos\theta} = \tan\theta$$

4. Solve for θ:

$$\theta = \tan^{-1}\left(\frac{m_2 v_2}{m_1 v_1}\right) = \tan^{-1}\left[\frac{(1300\text{ kg})(21\text{ m/s})}{(950\text{ kg})(16\text{ m/s})}\right]$$
$$= \tan^{-1}(1.8) = 61°$$

5. The final speed can be found using either the x or the y momentum equation. Here we use the x equation:

$$v_f = \frac{m_1 v_1}{(m_1 + m_2)\cos\theta} = \frac{(950\text{ kg})(16\text{ m/s})}{(950\text{ kg} + 1300\text{ kg})\cos 61°}$$
$$= 14\text{ m/s}$$

INSIGHT
As a check, you should verify that the y momentum equation gives the same value for v_f.

When a collision occurs in the real world, a traffic-accident investigation team measures skid marks at the scene of the crash and uses this information—along with some basic physics—to determine the initial speeds and directions of the vehicles. This information is often presented in court, where it can lead to a clear identification of the driver at fault.

CONTINUED

Enhance Your Understanding   (Answers given at the end of the chapter)

5. Two objects travel in opposite directions along a straight line. They have momenta of equal magnitude, and they stick together when they collide. **(a)** What is the final speed of the two objects? Explain. **(b)** What percentage of the initial kinetic energy is converted to other forms as a result of the collision?

Section Review

• In an inelastic collision, the final kinetic energy of a system is not equal to the initial kinetic energy of the system.

• A completely inelastic collision is one in which objects hit and stick together.

• In a two-dimensional collision, momentum conservation must be applied to both the *x* and *y* directions.

9-6 Elastic Collisions

In this section we consider collisions in which both momentum and kinetic energy are conserved. As mentioned in the previous section, such collisions are referred to as elastic:

Elastic Collisions
In an elastic collision, momentum and kinetic energy are conserved. That is,

$$\vec{\mathbf{P}}_f = \vec{\mathbf{P}}_i$$

and

$$K_f = K_i$$

   Most collisions in everyday life are rather poor approximations to being elastic—usually a significant amount of energy is converted to other forms. However, the collision of objects that bounce off one another with little deformation—billiard balls, for example—provides a reasonably good approximation to an elastic collision. In the subatomic world, on the other hand, elastic collisions are common.

Elastic Collisions in One Dimension

Consider a head-on collision of two carts on an air track, as pictured in FIGURE 9-12. The carts are provided with bumpers that give an elastic bounce when the carts collide.

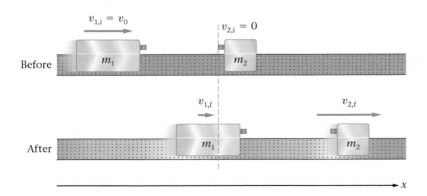

**PHYSICS
IN CONTEXT
Looking Ahead**

In Chapter 15 we shall see that elastic collisions between molecules and the wall of a container lead to the pressure exerted by a gas.

◀ **FIGURE 9-12 An elastic collision between two air carts** In the case pictured, $v_{1,f}$ is to the right (positive), which means that m_1 is greater than m_2. In fact, we have chosen $m_1 = 2m_2$ for this plot; therefore, $v_{1,f} = v_0/3$ and $v_{2,f} = 4v_0/3$ as given by Equations 9-13. If m_1 were less than m_2, cart 1 would bounce back toward the left, meaning that $v_{1,f}$ would be negative.

Let's suppose that initially cart 1 is moving to the right with a speed v_0, and cart 2 is at rest. If the masses of the carts are m_1 and m_2, respectively, then momentum conservation can be written as follows:

$$m_1 v_0 = m_1 v_{1,f} + m_2 v_{2,f}$$

In this expression, $v_{1,f}$ and $v_{2,f}$ are the final velocities of the two carts. Notice that we say velocities, not speeds, because it is possible for cart 1 to reverse direction, in which case $v_{1,f}$ would be negative.

Next, the fact that this is an elastic collision means the final velocities must also satisfy energy conservation:

$$\frac{1}{2}m_1 v_0^2 = \frac{1}{2}m_1 v_{1,f}^2 + \frac{1}{2}m_2 v_{2,f}^2$$

Thus, we now have two equations for the two unknowns, $v_{1,f}$ and $v_{2,f}$. Straightforward—though messy—algebra yields the following results:

$$v_{1,f} = \left(\frac{m_1 - m_2}{m_1 + m_2}\right)v_0$$

$$v_{2,f} = \left(\frac{2m_1}{m_1 + m_2}\right)v_0$$

9-13

Notice that the final velocity of cart 1 can be positive, negative, or zero, depending on whether m_1 is greater than, less than, or equal to m_2, respectively. The final velocity of cart 2, however, is always positive.

EXERCISE 9-14 BOUNCING BUMPER CARS

At an amusement park, a 98.5-kg bumper car moving with a speed of 1.18 m/s bounces elastically off a second 122-kg bumper car at rest. Find the final velocities of the two cars.

REASONING AND SOLUTION

The situation is just like the one described by Equations 9-13, where object 1 moves with an initial speed v_0 and collides elastically with object 2 at rest. Using $m_1 = 98.5$ kg, $m_2 = 122$ kg, and $v_0 = 1.18$ m/s, we can calculate the final velocities as follows:

$$v_{1,f} = \left(\frac{m_1 - m_2}{m_1 + m_2}\right)v_0 = \left(\frac{98.5 \text{ kg} - 122 \text{ kg}}{98.5 \text{ kg} + 122 \text{ kg}}\right)(1.18 \text{ m/s}) = -0.126 \text{ m/s}$$

$$v_{2,f} = \left(\frac{2m_1}{m_1 + m_2}\right)v_0 = \left(\frac{2(98.5 \text{ kg})}{98.5 \text{ kg} + 122 \text{ kg}}\right)(1.18 \text{ m/s}) = 1.05 \text{ m/s}$$

The direction of travel of car 1 has been reversed, as indicated by its negative velocity.

Let's check a few special cases of our results. First, consider the case where the two carts have equal masses: $m_1 = m_2 = m$. Substituting into Equations 9-13, we find

$$v_{1,f} = \left(\frac{m - m}{m + m}\right)v_0 = 0$$

and

$$v_{2,f} = \left(\frac{2m}{m + m}\right)v_0 = v_0$$

Thus, after the collision, the cart that was moving with velocity v_0 is now at rest, and the cart that was at rest is now moving with velocity v_0. In effect, the carts have "exchanged" velocities. This case is illustrated in **FIGURE 9-13 (a)**.

Next, suppose that m_2 is much greater than m_1 or, equivalently, that m_1 approaches zero. Returning to Equations 9-13, and setting $m_1 = 0$, we find

$$v_{1,f} = \left(\frac{0 - m_2}{0 + m_2}\right)v_0 = \left(\frac{-m_2}{m_2}\right)v_0 = -v_0$$

and

$$v_{2,f} = \frac{2 \cdot 0}{0 + m_2}v_0 = 0$$

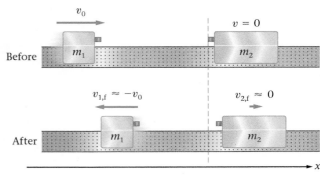

(a) $m_1 = m_2 = m$

(b) $m_1 << m_2$

▶ FIGURE 9-13 **Elastic collisions between air carts of various masses** **(a)** Carts of equal mass exchange velocities when they collide. **(b)** When a light cart collides with a stationary heavy cart, its direction of motion is reversed. Its speed is practically unchanged. **(c)** When a heavy cart collides with a stationary light cart, it continues to move in the same direction with essentially the same speed. The light cart moves off with a speed that is roughly twice the initial speed of the heavy cart.

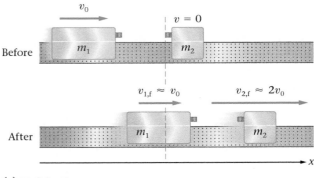

(c) $m_1 >> m_2$

Physically, we interpret these results as follows: A very light cart collides with a heavy cart that is at rest. The heavy cart hardly budges, but the light cart is reflected, heading *backward* (remember the minus sign in $-v_0$) with the same speed it had initially. For example, if you throw a ball against a wall, the wall is the very heavy object and the ball is the light object. The ball bounces back with the same speed it had initially (assuming an ideal elastic collision). We show a case in which m_1 is much less than m_2 in FIGURE 9-13 (b).

Finally, what happens when m_1 is much greater than m_2? To check this limit we can set m_2 equal to zero. We consider the results in the following Conceptual Example.

CONCEPTUAL EXAMPLE 9-15 **THE HOVERFLY AND THE ELEPHANT**

A hoverfly is happily maintaining a fixed position about 3 m above the ground when an elephant charges out of the bush and collides with it. The fly bounces elastically off the forehead of the elephant. If the initial speed of the elephant is v_0, is the speed v of the fly after the collision equal to **(a)** v_0, **(b)** $2v_0$, or **(c)** $3v_0$?

REASONING AND DISCUSSION
We can use Equations 9-13 to find the final speeds of the fly and the elephant. First, let m_1 be the mass of the elephant, and m_2 be the mass of the fly. Clearly, m_2 is vanishingly small compared with m_1, and hence we can evaluate Equations 9-13 in the limit $m_2 \to 0$. For the elephant, this yields

$$v_{1,f} = \left(\frac{m_1 - m_2}{m_1 + m_2}\right)v_0 \xrightarrow{m_2 \to 0} \left(\frac{m_1}{m_1}\right)v_0 = v_0$$

Similarly, for the fly we have

$$v_{2,f} = \left(\frac{2m_1}{m_1 + m_2}\right)v_0 \xrightarrow{m_2 \to 0} \left(\frac{2m_1}{m_1}\right)v_0 = 2v_0$$

As expected, the speed of the elephant is unaffected. The fly, however, rebounds with *twice* the speed of the elephant. FIGURE 9-13 (c) illustrates this case with air carts.

ANSWER
(b) The speed of the fly is $2v_0$.

▲ **FIGURE 9-14** This device, known as a Newton's cradle, illustrates some of the basic features of elastic collisions between objects of equal mass. When the end ball is pulled out to the side and then released, it creates a rapid succession of elastic collisions among the balls. In each collision, one ball comes to rest while the next one begins to move with the original speed, just as with the air carts in Figure 9-13 (a). When the collisions reach the other end of the apparatus, the last ball swings out to the same height from which the first ball was released.

If two balls are pulled out and released, two balls swing out at the other side, and so on. To see why this must be so, imagine that the two balls swing in with a speed v and a single ball swings out at the other side with a speed v'. What value must v' have (a) to conserve momentum and (b) to conserve kinetic energy? Since the required speed is $v' = 2v$ for (a) and $v' = \sqrt{2}v$ for (b), it follows that it is not possible to conserve both momentum and kinetic energy with two balls swinging in and one ball swinging out.

PROBLEM-SOLVING NOTE

Kinetic Energy in Elastic Collisions

Remember that in elastic collisions, by definition, the kinetic energy is conserved.

After the collision discussed above, the fly is separating from the elephant with the speed $2v_0 - v_0 = v_0$. Before the collision, the elephant was approaching the fly with the same speed, v_0. This is a special case of the following general result:

> The speed of separation after a head-on elastic collision is equal to the speed of approach before the collision.

An excellent example of elastic collisions is provided by the familiar apparatus shown in **FIGURE 9-14**. Another example with a rather surprising result is provided in **FIGURE 9-15**.

Elastic Collisions in Two Dimensions

In a two-dimensional elastic collision, if we are given the final speed and direction of one of the objects, we can find the speed and direction of the other object using energy conservation and momentum conservation. For example, consider the collision of two 7.00-kg curling stones, as depicted in **FIGURE 9-16**. One stone is at rest initially; the other approaches with a speed $v_{1,i} = 1.50$ m/s. The collision is not head-on, and after the collision, stone 1 moves with a speed of $v_{1,f} = 0.610$ m/s in a direction 66.0° away from the initial line of motion. What are the speed and direction of stone 2?

First, let's find the speed of stone 2. The easiest way to do this is to simply require that the final kinetic energy be equal to the initial kinetic energy. Initially, the kinetic energy is

$$K_i = \tfrac{1}{2}m_1 v_{1,i}{}^2 = \tfrac{1}{2}(7.00 \text{ kg})(1.50 \text{ m/s})^2 = 7.88 \text{ J}$$

After the collision stone 1 has a speed of 0.610 m/s and stone 2 has the speed $v_{2,f}$. Hence, the final kinetic energy is

$$K_f = \tfrac{1}{2}m_1 v_{1,f}{}^2 + \tfrac{1}{2}m_2 v_{2,f}{}^2 = \tfrac{1}{2}(7.00 \text{ kg})(0.610 \text{ m/s})^2 + \tfrac{1}{2}m_2 v_{2,f}{}^2$$
$$= 1.30 \text{ J} + \tfrac{1}{2}m_2 v_{2,f}{}^2 = K_i$$

Solving for the speed of stone 2, we find

$$v_{2,f} = 1.37 \text{ m/s}$$

Next, we can find the direction of motion of stone 2 by requiring that the momentum be conserved. For example, initially there is no momentum in the y direction. This must be true after the collision as well. Hence, we have the following condition:

$$0 = m_1 v_{1,f} \sin 66.0° - m_2 v_{2,f} \sin \theta$$

▲ **FIGURE 9-15 Visualizing Concepts**
Height Amplification In a collision between two objects of different mass, like the small and large balls in this photo, a significant amount of momentum can be transferred from the large object to the small object. Even though the total momentum is conserved, the small object can be given a speed that is significantly higher than any of the initial speeds. This is illustrated in the photo by the height to which the small ball bounces. A similar process occurs in the collapse of a star during a supernova explosion. The resulting collision can send jets of material racing away from the supernova at nearly the speed of light, just like the small ball that takes off with such a high speed in this collision.

◄ **FIGURE 9-16 Two curling stones undergo an elastic collision** The speed of curling stone 2 after this collision can be determined using energy conservation; its direction of motion can be found using momentum conservation in either the x or the y direction.

Solving for the angle θ we find

$$\theta = 24.0°$$

Conserving momentum in the x direction gives the same result.

EXAMPLE 9-16 | TWO FRUITS IN TWO DIMENSIONS: ANALYZING AN ELASTIC COLLISION

Two astronauts on opposite ends of a spaceship are comparing lunches. One has an apple, the other has an orange. They decide to trade. Astronaut 1 tosses the 0.130-kg apple toward astronaut 2 with a speed of 1.11 m/s. The 0.160-kg orange is tossed from astronaut 2 to astronaut 1 with a speed of 1.21 m/s. Unfortunately, the fruits collide, sending the orange off with a speed of 1.16 m/s at an angle of 42.0° with respect to its original direction of motion. Find the final speed and direction of the apple, assuming an elastic collision. Give the apple's direction relative to its original direction of motion.

PICTURE THE PROBLEM
In our sketch we label the apple as object 1 and the orange as object 2. We also choose the positive x direction to be in the initial direction of motion of the apple. We shall describe the "Before" and "After" sketches separately:

BEFORE
Initially, the apple moves in the positive x direction with a speed of 1.11 m/s, and the orange moves in the negative x direction with a speed of 1.21 m/s. There is no momentum in the y direction before the collision.

AFTER
After the collision, the orange moves with a speed of 1.16 m/s in a direction 42° below the negative x axis. As a result, the orange now has momentum in the negative y direction. To cancel this y momentum, the apple must move in a direction that is above the positive x axis, as indicated in the sketch.

REASONING AND STRATEGY
As described in the text, we first find the speed of the apple by demanding that the initial and final kinetic energies be the same. Next, we find the angle θ by conserving momentum in either the x or the y direction—the results are the same whichever direction is chosen.

CONTINUED

Known   Mass of apple, $m_1 = 0.130$ kg; mass of orange, $m_2 = 0.160$ kg; initial speed of apple, $v_{1,i} = 1.11$ m/s; initial speed of orange, $v_{2,i} = 1.21$ m/s; final speed of orange, $v_{2,f} = 1.16$ m/s; final direction of orange, 42.0°.

Unknown   Final speed of apple, $v_{1,f} = ?$ Final direction of apple, $\theta = ?$

SOLUTION

1. Calculate the initial kinetic energy of the system:

$$K_i = \tfrac{1}{2}m_1 v_{1,i}^2 + \tfrac{1}{2}m_2 v_{2,i}^2$$
$$= \tfrac{1}{2}(0.130 \text{ kg})(1.11 \text{ m/s})^2 + \tfrac{1}{2}(0.160 \text{ kg})(1.21 \text{ m/s})^2$$
$$= 0.197 \text{ J}$$

2. Calculate the final kinetic energy of the system in terms of $v_{1,f}$:

$$K_f = \tfrac{1}{2}m_1 v_{1,f}^2 + \tfrac{1}{2}m_2 v_{2,f}^2$$
$$= \tfrac{1}{2}(0.130 \text{ kg})v_{1,f}^2 + \tfrac{1}{2}(0.160 \text{ kg})(1.16 \text{ m/s})^2$$
$$= \tfrac{1}{2}(0.130 \text{ kg})v_{1,f}^2 + 0.108 \text{ J}$$

3. Set $K_f = K_i$ to find $v_{1,f}$:

$$v_{1,f} = \sqrt{\frac{2(0.197 \text{ J} - 0.108 \text{ J})}{0.130 \text{ kg}}} = 1.17 \text{ m/s}$$

4. Set the final y component of momentum equal to zero and solve for $\sin \theta$:

$$0 = m_1 v_{1,f} \sin \theta - m_2 v_{2,f} \sin 42.0°$$
$$\sin \theta = \frac{m_2 v_{2,f} \sin 42.0°}{m_1 v_{1,f}}$$

5. Substitute numerical values:

$$\sin \theta = \frac{(0.160 \text{ kg})(1.16 \text{ m/s}) \sin 42.0°}{(0.130 \text{ kg})(1.17 \text{ m/s})} = 0.817$$
$$\theta = \sin^{-1}(0.817) = 54.8°$$

INSIGHT

The x momentum equation gives the same value for θ, as expected.

PRACTICE PROBLEM

Suppose that after the collision the apple moves in the positive y direction with a speed of 1.27 m/s. What are the final speed and direction of the orange in this case?   [**Answer:** The orange moves with a speed of 1.07 m/s in a direction 74.6° below the negative x axis.]

Some related homework problems: Problem 35, Problem 40

Enhance Your Understanding   (Answers given at the end of the chapter)

6. The 1-kg block in **FIGURE 9-17** slides to the right on a frictionless surface and undergoes an elastic collision with a 2-kg block that is initially at rest. **(a)** What is the direction of travel of the 1-kg block after the collision? **(b)** Which block has the greater speed after the collision?

▲ **FIGURE 9-17** Enhance Your Understanding 6.

Section Review

- In elastic collisions, the final kinetic energy of a system is equal to the initial kinetic energy of the system.

- In a one-dimensional collision between a mass m_1 with an initial velocity of v_0, and a mass m_2 at rest, the final velocities are

$$v_{1,f} = \left(\frac{m_1 - m_2}{m_1 + m_2}\right)v_0 \quad \text{and} \quad v_{2,f} = \left(\frac{2m_1}{m_1 + m_2}\right)v_0.$$

- In a two-dimensional collision, kinetic energy is conserved, and so are the x and y components of momentum.

9-7 Center of Mass

In this section we introduce the concept of the center of mass. We begin by defining its location for a given system of masses. Next we consider the motion of the center of mass, and show how it is related to the net external force acting on the system.

Location of the Center of Mass

There is one point in any system of objects that has special significance—the **center of mass (CM)**. One of the reasons the center of mass is so special is that, in many ways, a system behaves as if all of its mass were concentrated there. As a result, a system can be balanced at its center of mass:

> The center of mass of a system of masses is the point where the system can be balanced in a uniform gravitational field.

For example, suppose you are making a mobile. At one stage in its construction, you want to balance a light rod with objects of mass m_1 and m_2 connected to either end, as indicated in FIGURE 9-18. To make the rod balance, you should attach a string to the center of mass of the system, just as if all its mass were concentrated at that point. In a sense, you can think of the center of mass as the "average" location of the system's mass. An elaborate example of a balanced mobile is shown in FIGURE 9-19.

To be more specific, suppose the two objects connected to the rod have the same mass. In this case the center of mass is at the midpoint of the rod, because this is where it balances. On the other hand, if one object has more mass than the other, the center of mass is closer to the heavier object, as indicated in FIGURE 9-20. In general, if a mass m_1 is on the x axis at the position x_1, and a mass m_2 is at the position x_2, as in Figure 9-18, then the location of the center of mass, X_{cm}, is defined as the "weighted" average of the two positions:

Center of Mass for Two Objects

$$X_{cm} = \frac{m_1 x_1 + m_2 x_2}{m_1 + m_2} = \frac{m_1 x_1 + m_2 x_2}{M} \qquad 9\text{-}14$$

Notice that we have used $M = m_1 + m_2$ for the total mass of the two objects, and that the two positions, x_1 and x_2, are multiplied—or weighted—by their respective masses.

To see that this definition of X_{cm} agrees with our expectations, consider first the case where the masses are equal: $m_1 = m_2 = m$. In this case, $M = m_1 + m_2 = 2m$ and $X_{cm} = (mx_1 + mx_2)/2m = \frac{1}{2}(x_1 + x_2)$. Thus, as expected, if two masses are equal, their center of mass is halfway between them. On the other hand, if one mass is larger than the other, the center of mass is closer to the larger mass.

EXERCISE 9-17 CENTER OF MASS

Suppose that the masses in Figure 9-18 are separated by 0.800 m, and that $m_1 = 0.325$ kg and $m_2 = 0.166$ kg. What is the distance from m_1 to the center of mass of the system?

REASONING AND SOLUTION

Let $x_1 = 0$ and $x_2 = 0.800$ m in Figure 9-18. This yields

$$X_{cm} = \frac{m_1 x_1 + m_2 x_2}{m_1 + m_2} = \frac{(0.325 \text{ kg}) \cdot 0 + (0.166 \text{ kg})(0.800 \text{ m})}{0.325 \text{ kg} + 0.166 \text{ kg}} = 0.270 \text{ m}$$

Thus, the center of mass is 0.270 m from x_1, the larger of the two masses. Because the midpoint between the masses is 0.400 m from x_1, it follows that the center of mass is closer to the larger mass, as expected.

Center of Mass for Multiple or Continuous Objects in Two Dimensions   To extend the definition of X_{cm} to more general situations, first consider a system that contains many objects, not just two. In that case, X_{cm} is the sum of m times x for each object,

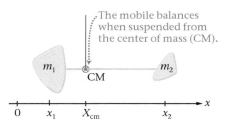

▲ FIGURE 9-18 **Balancing a mobile** Consider a portion of a mobile with masses m_1 and m_2 at the locations x_1 and x_2, respectively. The object balances when a string is attached at the center of mass. Since the center of mass is closer to m_1 than to m_2, it follows that m_1 is greater than m_2.

▲ FIGURE 9-19 This mobile by Alexander Calder illustrates the concept of center of mass with artistic flair. Each arm of the mobile is in balance because it is suspended at its center of mass.

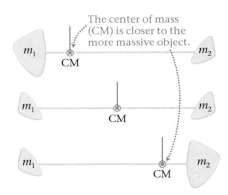

▲ FIGURE 9-20 **The center of mass of two objects** The center of mass is closer to the larger mass, or equidistant between the masses if they are equal.

The center of mass is at the geometric center of a uniform object . . .

. . . even if there is no mass at that location.

▲ **FIGURE 9-21 Locating the center of mass** In an object of continuous, uniform mass distribution, the center of mass is located at the geometric center of the object. In some cases, this means that the center of mass is not located within the object.

divided by the total mass of the system, M. If, in addition, the objects in the system are not in a line, but are distributed in two dimensions, the center of mass has both an x coordinate, X_{cm}, and a y coordinate, Y_{cm}. As one would expect, Y_{cm} is simply the sum of m times y for each object, divided by M. Thus, the x coordinate of the center of mass is

X Coordinate of the Center of Mass

$$X_{cm} = \frac{m_1 x_1 + m_2 x_2 + \cdots}{m_1 + m_2 + \cdots} = \frac{\sum mx}{M}$$ 9-15

Similarly, the y coordinate of the center of mass is

Y Coordinate of the Center of Mass

$$Y_{cm} = \frac{m_1 y_1 + m_2 y_2 + \cdots}{m_1 + m_2 + \cdots} = \frac{\sum my}{M}$$ 9-16

In systems with a continuous, uniform distribution of mass, the center of mass is at the geometric center of the object, as illustrated in **FIGURE 9-21**. Notice that it is common for the center of mass to be located in a position where no mass exists, as in a life preserver, where the center of mass is precisely in the center of the hole.

EXAMPLE 9-18   CENTER OF MASS OF THE ARM

RWP A person's arm is held with the upper arm vertical, the lower arm and hand horizontal. **(a)** Find the center of mass of the arm in this configuration, given the following data: The upper arm has a mass of 2.5 kg and a center of mass 0.18 m above the elbow; the lower arm has a mass of 1.6 kg and a center of mass 0.12 m to the right of the elbow; the hand has a mass of 0.64 kg and a center of mass 0.40 m to the right of the elbow. **(b)** A 0.14-kg baseball is placed on the palm of the hand. If the radius of the ball is 3.7 cm, find the center of mass of the arm–ball system.

PICTURE THE PROBLEM
We place the origin at the elbow, with the x and y axes pointing along the lower and upper arms, respectively. In the sketch, the center of mass of each of the three parts of the arm is indicated by an x; the center of mass of the entire arm is at the point labeled CM. The inset shows the baseball on the palm of the hand.

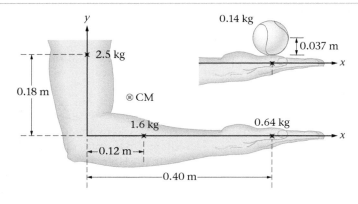

REASONING AND STRATEGY

a. Using the information given in the problem statement, we can treat the arm as a system of three point masses placed as follows: 2.5 kg at $(0, 0.18\ \text{m})$; 1.6 kg at $(0.12\ \text{m}, 0)$; 0.64 kg at $(0.40\ \text{m}, 0)$. We substitute these masses and locations into Equations 9-15 and 9-16 to find the x and y coordinates of the center of mass, respectively.

b. Treat the center of mass found in part (a) as a point particle with a mass 2.5 kg + 1.6 kg + 0.64 kg = 4.7 kg at the location (X_{cm}, Y_{cm}). The baseball can be treated as a point particle of mass 0.14 kg at the location $(0.40\ \text{m}, 0.037\ \text{m})$.

Known   Mass and location of parts of the arm and baseball given in drawing.
Unknown   **(a)** $X_{cm} = ?$ and $Y_{cm} = ?$ for arm **(b)** $X_{cm} = ?$ and $Y_{cm} = ?$ for arm–ball system

---

SOLUTION

Part (a)

1. Calculate the x coordinate of the center of mass:

$$X_{cm} = \frac{(2.5\ \text{kg})(0) + (1.6\ \text{kg})(0.12\ \text{m}) + (0.64\ \text{kg})(0.40\ \text{m})}{2.5\ \text{kg} + 1.6\ \text{kg} + 0.64\ \text{kg}}$$

$$= 0.095\ \text{m}$$

2. Do the same calculation for the y coordinate of the center of mass:

$$Y_{cm} = \frac{(2.5\ \text{kg})(0.18\ \text{m}) + (1.6\ \text{kg})(0) + (0.64\ \text{kg})(0)}{2.5\ \text{kg} + 1.6\ \text{kg} + 0.64\ \text{kg}}$$

$$= 0.095\ \text{m}$$

CONTINUED

Part (b)

3. Calculate the new x coordinate of the center of mass:

$$X_{cm} = \frac{(4.7 \text{ kg})(0.095 \text{ m}) + (0.14 \text{ kg})(0.40 \text{ m})}{4.7 \text{ kg} + 0.14 \text{ kg}}$$

$$= 0.10 \text{ m}$$

4. Calculate the new y coordinate of the center of mass:

$$Y_{cm} = \frac{(4.7 \text{ kg})(0.095 \text{ m}) + (0.14 \text{ kg})(0.037 \text{ m})}{4.7 \text{ kg} + 0.14 \text{ kg}}$$

$$= 0.093 \text{ m}$$

INSIGHT

As is often the case, the center of mass of an arm held in this position is in a location where no mass exists—you might say the center of mass is having an out-of-body experience. This effect can sometimes be put to good use, as when the center of mass of a high jumper passes beneath the horizontal bar while the body passes above it. (See Conceptual Question 16 for a photo of this technique in action, in the famous "Fosbury flop.")

PRACTICE PROBLEM — PREDICT/CALCULATE

Suppose the mass of the baseball is increased to 0.25 kg. **(a)** Does X_{cm} increase, decrease, or stay the same? **(b)** Does Y_{cm} increase, decrease, or stay the same? **(c)** Check your answers to parts (a) and (b) by finding the center of mass of the arm–ball system in this case.   [**Answer: (a)** The ball adds mass at a large value of x, and hence X_{cm} increases. **(b)** The ball adds mass at a small value of y, and hence Y_{cm} decreases. **(c)** $X_{cm} = 0.11$ m, $Y_{cm} = 0.092$ m]

Some related homework problems: Problem 47, Problem 48, Problem 50

Motion of the Center of Mass

Another reason the center of mass is important is that its motion often displays a re-markable simplicity when compared with the motion of other parts of a system. To analyze this motion, we consider both the velocity and acceleration of the center of mass. Each of these quantities is defined in analogy with the definition of the center of mass itself.

For example, to find the velocity of the center of mass, we first multiply the mass of each object in a system, m, by its velocity, $\vec{v}$, to give $m_1\vec{v}_1, m_2\vec{v}_2$, and so on. Next, we add all these products together, $m_1\vec{v}_1 + m_2\vec{v}_2 + \cdots$, and divide by the total mass, $M = m_1 + m_2 + \cdots$. The result, by definition, is the velocity of the center of mass, $\vec{V}_{cm}$:

Velocity of the Center of Mass

$$\vec{V}_{cm} = \frac{m_1\vec{v}_1 + m_2\vec{v}_2 + \cdots}{m_1 + m_2 + \cdots} = \frac{\sum m\vec{v}}{M} \qquad \text{9-17}$$

Comparing with Equation 9-15, we see that $\vec{V}_{cm}$ is the same as X_{cm} with each position x replaced with a velocity vector $\vec{v}$. In addition, note that the total mass of the system, M, times the velocity of the center of mass, $\vec{V}_{cm}$, is simply the total momentum of the system:

$$M\vec{V}_{cm} = m_1\vec{v}_1 + m_2\vec{v}_2 + \cdots = \vec{p}_1 + \vec{p}_2 + \cdots = \vec{p}_{total}$$

To gain more information on how the center of mass moves, we next consider its acceleration, $\vec{A}_{cm}$. As expected by analogy with $\vec{V}_{cm}$, the acceleration of the center of mass is defined as follows:

Acceleration of the Center of Mass

$$\vec{A}_{cm} = \frac{m_1\vec{a}_1 + m_2\vec{a}_2 + \cdots}{m_1 + m_2 + \cdots} = \frac{\sum m\vec{a}}{M} \qquad \text{9-18}$$

The vector $\vec{A}_{cm}$ contains the terms $m_1\vec{a}_1, m_2\vec{a}_2$, and so on, for each object in the sys-tem. From Newton's second law, however, we know that $m_1\vec{a}_1$ is simply $\vec{F}_1$, the net force acting on mass 1. The same conclusion applies to each of the masses. Therefore, we find that the total mass of the system, M, times the acceleration of the center of mass, $\vec{\mathbf{A}}_{cm}$, is simply the total force acting on the system:

$$M\vec{\mathbf{A}}_{cm} = m_1\vec{\mathbf{a}}_1 + m_2\vec{\mathbf{a}}_2 + \cdots = \vec{\mathbf{F}}_1 + \vec{\mathbf{F}}_2 + \cdots = \vec{\mathbf{F}}_{total}$$

Recall, however, that the total force acting on a system is the same as the net external force, $\vec{\mathbf{F}}_{net,ext}$, because the internal forces cancel. Therefore, $M\vec{\mathbf{A}}_{cm}$ is the net external force acting on the system:

Newton's Second Law for a System of Particles

$$M\vec{\mathbf{A}}_{cm} = \vec{\mathbf{F}}_{net,ext}$$

9-19

Zero Net External Force For systems in which $\vec{\mathbf{F}}_{net,ext}$ is zero, it follows that the acceleration of the center of mass is zero. Hence, if the center of mass is initially at rest, it remains at rest. Similarly, if the center of mass is moving initially, it continues to move with the same velocity. For example, in a collision between two air-track carts, the velocity of each cart changes as a result of the collision. The velocity of the center of mass of the two carts, however, is the same before and after the collision. We explore this situation in the following Example.

EXAMPLE 9-19   CRASH OF THE AIR CARTS

An air cart with a mass m and speed v_0 moves toward a second, identical air cart that is at rest. When the carts collide they stick together and move as one. Find the velocity of the center of mass of this system **(a)** before and **(b)** after the carts collide.

PICTURE THE PROBLEM
We choose the positive x direction to be the direction of motion of the incoming cart, whose initial speed is v_0. Notice that the carts have wads of putty on their bumpers; this ensures that they stick together when they collide and thereafter move as a unit. Their final speed is v_f.

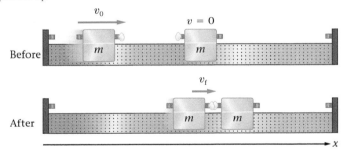

REASONING AND STRATEGY

a. We can find the velocity of the center of mass by applying Equation 9-17 to the case of just two masses; $\vec{\mathbf{V}}_{cm} = (m_1\vec{\mathbf{v}}_1 + m_2\vec{\mathbf{v}}_2)/M$. In this case, $\vec{\mathbf{v}}_1 = v_0\hat{\mathbf{x}}$, $\vec{\mathbf{v}}_2 = 0$, and $m_1 = m_2 = m$.

b. After the collision the two masses move with the same velocity, $\vec{\mathbf{v}}_f = v_f\hat{\mathbf{x}}$, which is given by momentum conservation (Equation 9-11, $v_f = (m_1v_{1,i} + m_2v_{2,i})/(m_1 + m_2))$. Hence, $\vec{\mathbf{V}}_{cm} = (m_1\vec{\mathbf{v}}_f + m_2\vec{\mathbf{v}}_f)/M$.

Known   Mass of air carts, m; initial speed of one air cart, v_0; initial speed of other air cart, $v = 0$.
Unknown   Velocity of center of mass, $\vec{\mathbf{V}}_{cm} = ?$ **(a)** before and **(b)** after collision

SOLUTION

Part (a)

1. Use $\vec{\mathbf{V}}_{cm} = (m_1\vec{\mathbf{v}}_1 + m_2\vec{\mathbf{v}}_2)/M$ to find the velocity of the center of mass before the collision:

$$\vec{\mathbf{V}}_{cm} = \frac{(m_1\vec{\mathbf{v}}_1 + m_2\vec{\mathbf{v}}_2)}{m_1 + m_2} = \frac{(mv_0\hat{\mathbf{x}} + m \cdot 0)}{m + m} = \tfrac{1}{2}v_0\hat{\mathbf{x}}$$

Part (b)

2. Use momentum conservation in the x direction to find the speed of the carts after the collision:

$$mv_0 = mv_f + mv_f$$
$$v_f = \tfrac{1}{2}v_0$$

3. Calculate the velocity of the center of mass of the two carts after the collision:

$$\vec{\mathbf{V}}_{cm} = \frac{(m_1\vec{\mathbf{v}}_1 + m_2\vec{\mathbf{v}}_2)}{m_1 + m_2} = \frac{(mv_f\hat{\mathbf{x}} + mv_f\hat{\mathbf{x}})}{m + m} = v_f\hat{\mathbf{x}} = \tfrac{1}{2}v_0\hat{\mathbf{x}}$$

CONTINUED

INSIGHT

As expected, the velocity of each cart changes when they collide: The velocity of cart 1 changes from v_0 to $\frac{1}{2}v_0$, and the velocity of cart 2 changes from 0 to $\frac{1}{2}v_0$. On the other hand, the velocity of the center of mass is completely unaffected by the collision.

This finding is illustrated in the figure, where we show a sequence of equal-time snapshots of the system just before and just after the collision. First, we note that the incoming cart moves two distance units for every time interval until it collides with the second cart. From that point on, the two carts are locked together, and move one distance unit per time interval. In contrast, the center of mass (CM), which is centered between the two equal-mass carts, progresses uniformly throughout the sequence, advancing one unit of distance for each time interval.

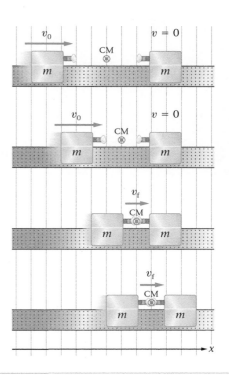

PRACTICE PROBLEM — PREDICT/CALCULATE

(a) If the mass of the cart that is moving initially is doubled to $2m$, does the velocity of the center of mass increase, decrease, or stay the same? Explain.
(b) Verify your answer to part (a) by calculating the velocity of the center of mass in this case.   [**Answer: (a)** More mass is moving, and hence V_{cm} increases. **(b)** The velocity of the center of mass is $\vec{V}_{cm} = (2v_0/3)\hat{x}$, both before and after the collision.]

Some related homework problems: Problem 53, Problem 64, Problem 73

Nonzero Net External Force   Recall that Newton's second law, as applied to the center of mass, states that the acceleration of the center of mass is related to the net external force as follows:

$$M\vec{A}_{cm} = \vec{F}_{net,ext}$$

This is completely analogous to the relationship between the acceleration of an object of mass m and the net force $\vec{F}_{net}$ applied to it:

$$m\vec{a} = \vec{F}_{net}$$

Therefore, when $\vec{F}_{net,ext}$ is nonzero, we can conclude the following:

> The center of mass of a system accelerates precisely as if it were a point particle of mass M acted on by the force $\vec{F}_{net,ext}$.

For this reason, the motion of the center of mass can be quite simple compared to the motion of its constituent parts. For example, a hammer tossed into the air with a rotation is shown in FIGURE 9-22. The motion of one part of the hammer—the yellow tip of the handle, let's say—follows a complicated path in space. On the other hand, the path of the center of mass (red dot) is a simple parabola, precisely the same path that a point mass would follow.

RWP Similarly, consider a fireworks rocket launched into the sky, as illustrated in FIGURE 9-23. The center of mass of the rocket follows a parabolic path, if air resistance is ignored. At some point in its path it explodes into numerous individual pieces. The explosion is due to internal forces, however, which must therefore sum to zero. Hence, the net external force acting on the pieces of the rocket is the same before, during, and after the explosion. As a result, the center of mass has a constant downward

Big Idea 5 The center of mass of a system behaves as if all the mass were concentrated there.

▶ **FIGURE 9-22 Simple Motion of the Center of Mass** As this hammer flies through the air, its motion is quite complex. Some parts of the hammer follow wild trajectories with strange loops and turns (yellow path on the top). There is one point on the hammer, however, that travels on a smooth, simple parabolic path—the center of mass. The center of mass (red path on the bottom) travels as if all the mass of the hammer were concentrated there.

▲ FIGURE 9-23 **Center of mass of an exploding rocket** A fireworks rocket follows a parabolic path, ignoring air resistance, until it explodes. After exploding, its center of mass continues on the same parabolic path until some of the fragments start to land.

acceleration and continues to follow the original parabolic path. It is only when an additional external force acts on the system, as when one of the pieces of the rocket hits the ground, that the path of the center of mass changes.

Enhance Your Understanding   (Answers given at the end of the chapter)

Predict/Explain

7. A piece of sheet metal of mass M is cut into the shape of a right triangle, as shown in **FIGURE 9-24**.

▲ FIGURE 9-24 Enhance Your Understanding 7.

A vertical dashed line is drawn on the sheet at the point where the mass on either side of the line is $M/2$. The sheet is now placed on a fulcrum just under the dashed line and released from rest. **(a)** Does the metal sheet remain level, rotate counterclockwise (left end down), or rotate clockwise (right end down)? **(b)** Choose the *best explanation* for your prediction from among the following:

   I. Equal mass on either side will keep the metal sheet level.
   II. The metal sheet extends for a greater distance to the left, which shifts the center of mass to the left of the dashed line.
   III. The center of mass is to the right of the dashed line because the sheet of metal is thicker there.

Section Review

- The location of the center of mass is given by $X_{cm} = \dfrac{m_1 x_1 + m_2 x_2 + \cdots}{m_1 + m_2 + \cdots} = \dfrac{\sum mx}{M}$
  and $Y_{cm} = \dfrac{m_1 y_1 + m_2 y_2 + \cdots}{m_1 + m_2 + \cdots} = \dfrac{\sum my}{M}$.

- The velocity of the center of mass is defined as $\vec{\mathbf{V}}_{cm} = \dfrac{m_1 \vec{\mathbf{v}}_1 + m_2 \vec{\mathbf{v}}_2 + \cdots}{m_1 + m_2 + \cdots} = \dfrac{\sum m\vec{\mathbf{v}}}{M}$.

- The acceleration of the center of mass is defined as $\vec{\mathbf{A}}_{cm} = \dfrac{m_1 \vec{\mathbf{a}}_1 + m_2 \vec{\mathbf{a}}_2 + \cdots}{m_1 + m_2 + \cdots} = \dfrac{\sum m\vec{\mathbf{a}}}{M}$. It is related to the net external force as follows: $M\vec{\mathbf{A}}_{cm} = \vec{\mathbf{F}}_{net,ext}$.

9-8 Systems with Changing Mass: Rocket Propulsion

We close this chapter with systems whose mass does change. The mass of a rocket, for example, changes as its engines operate because it ejects part of the fuel as it burns. The burning process is produced by internal forces; hence the total momentum of the rocket and its fuel remains constant.

Consider, then, a rocket at rest in outer space, far from any large, massive objects. When the rocket's engine is fired, it expels a certain mass of fuel out the back with a speed v. If the mass of the ejected fuel is Δm, then the momentum of the ejected fuel has a magnitude equal to $(\Delta m)v$. The total momentum of the system must still be zero, and therefore the rocket acquires an equivalent amount of momentum in the forward direction. Hence, the momentum of the rocket increases by the amount

$$\Delta p = (\Delta m)v$$

If this increase in momentum occurs in the time Δt, the force exerted on the rocket is the change in its momentum divided by the time interval (Equation 9-4); that is,

$$F = \frac{\Delta p}{\Delta t} = \left(\frac{\Delta m}{\Delta t}\right)v$$

The force exerted on the rocket by the ejected fuel is referred to as **thrust**:

Thrust

$$\text{thrust} = \left(\frac{\Delta m}{\Delta t}\right)v \qquad\qquad 9\text{-}20$$

SI unit: newton, N

RWP By $\Delta m/\Delta t$, we simply mean the amount of mass per time coming out of the rocket. For example, on the Delta IV Heavy rocket, among the most powerful rockets in operation today, the main engines eject fuel at the rate of 2510 kg/s with a speed of 3710 m/s. As a result, the thrust produced by these engines is

$$\text{thrust} = \left(\frac{\Delta m}{\Delta t}\right)v = (2510 \text{ kg/s})(3710 \text{ m/s}) = 9.31 \times 10^6 \text{ N}$$

This is equivalent to 2.09 million pounds, a small amount compared to the 7.60 million pounds generated by the Saturn V rocket that was used for the manned missions to the Moon. However, because the Delta IV Heavy rocket and its cargo weigh only 1.68 million pounds, there is sufficient thrust to give the rocket an upward acceleration. The acceleration increases as the rocket's mass decreases due to the expelled fuel. Real-world examples of rocket propulsion are shown in FIGURE 9-25.

EXERCISE 9-20 LEAVING THE MOON

The ascent stage of the lunar lander was designed to produce 15,600 N of thrust at liftoff. If the fuel was burned at the rate of 6.40 kg/s, what was the speed of the ejected fuel?

REASONING AND SOLUTION
The thrust, fuel consumption rate ($\Delta m/\Delta t$), and speed of ejected fuel v, are related by thrust $= (\Delta m/\Delta t)v$. We can rearrange this equation to solve for the speed:

$$v = \frac{\text{thrust}}{(\Delta m/\Delta t)} = \frac{15,600 \text{ N}}{6.40 \text{ kg/s}} = 2440 \text{ m/s}$$

This is about seven times greater than the speed of sound on Earth.

A common question regarding rockets is, "How can a rocket accelerate in outer space when it has nothing to push against?" The answer is that rockets, in effect, push against their own fuel. The situation is similar to firing a gun. When a bullet is ejected by the internal combustion of the gunpowder, the person firing the gun feels a recoil. If the person were in space, or standing on a frictionless surface, the recoil would give him or her a speed in the direction opposite to the bullet. The burning of a rocket engine provides a continuous recoil, almost as if the rocket were firing a steady stream of bullets out the back.

Enhance Your Understanding   (Answers given at the end of the chapter)

8. You place a bucket on a scale and pour water into it at a constant rate. When the weight of the bucket and the water in it is 20 N, and you're still pouring water into the bucket, is the reading on the scale greater than, less than, or equal to 20 N? Explain.

Section Review

- The thrust (force) produced by a rocket engine is thrust $= (\Delta m/\Delta t)v$.

Water Rocket

▲ **FIGURE 9-25 Visualizing Concepts**
Rocket Propulsion A Delta Heavy IV rocket (top) makes use of the principle of conservation of momentum: Mass (the products of explosive burning of fuel) is ejected at high speed in one direction, causing the rocket to move in the opposite direction. The same method of propulsion has evolved in octopi (bottom) and some other animals. When danger threatens and a quick escape is needed, powerful muscles contract to create a jet of water that propels the animal to safety. (A smokescreen of ink provides additional security.)

CHAPTER 9 REVIEW

CHAPTER SUMMARY

9-1 LINEAR MOMENTUM

The linear momentum of an object of mass m moving with velocity $\vec{v}$ is

$$\vec{p} = m\vec{v} \qquad \text{9-1}$$

Momentum of a System of Objects

In a system of several objects, the total linear momentum is the vector sum of the individual momenta:

$$\vec{p}_{total} = \vec{p}_1 + \vec{p}_2 + \vec{p}_3 + \cdots \qquad \text{9-3}$$

9-2 MOMENTUM AND NEWTON'S SECOND LAW

In terms of momentum, Newton's second law is

$$\sum \vec{F} = \frac{\Delta \vec{p}}{\Delta t} \qquad \text{9-4}$$

9-3 IMPULSE

The impulse delivered to an object by an average force $\vec{F}_{av}$ acting for a time Δt is

$$\vec{I} = \vec{F}_{av} \Delta t \qquad \text{9-6}$$

9-4 CONSERVATION OF LINEAR MOMENTUM

The momentum of an object is conserved (remains constant) if the net force acting on it is zero.

9-5 INELASTIC COLLISIONS

In an inelastic collision, the final kinetic energy is different from the initial kinetic energy. The kinetic energy is usually less after a collision, but it can also be more than the initial kinetic energy.

9-6 ELASTIC COLLISIONS

In an elastic collision, the final kinetic energy is equal to the initial kinetic energy.

9-7 CENTER OF MASS

The location of the center of mass of a two-dimensional system of objects is defined as follows:

$$X_{cm} = \frac{m_1 x_1 + m_2 x_2 + \cdots}{m_1 + m_2 + \cdots} = \frac{\sum mx}{M} \qquad \text{9-15}$$

and

$$Y_{cm} = \frac{m_1 y_1 + m_2 y_2 + \cdots}{m_1 + m_2 + \cdots} = \frac{\sum my}{M} \qquad \text{9-16}$$

Motion of the Center of Mass

The center of mass accelerates as if the net external force acted on a single object of mass $M = m_1 + m_2 + \cdots$.

9-8   SYSTEMS WITH CHANGING MASS: ROCKET PROPULSION

The mass of a rocket changes because its engines expel fuel when they are fired. If fuel is expelled with the speed v and at the rate $\Delta m/\Delta t$, the thrust experienced by the rocket is

$$\text{thrust} = \left(\frac{\Delta m}{\Delta t}\right)v \qquad \text{9-20}$$

ANSWERS TO ENHANCE YOUR UNDERSTANDING QUESTIONS

1. **(a)** The rubber bullet is more likely to topple the block.
   (b) The best explanation is I.

2. **(a)** The change in momentum is the same for the two objects because the force (momentum change per time) acting on them is the same. **(b)** The change in velocity is greater for the less massive object 1 because momentum is the product of mass and velocity.

3. The impulse is the same for the two balls because the force each experiences has the same magnitude (due to Newton's third law) and acts for the same amount of time.

4. **(a)** The total momentum must be zero after the explosion, and hence the momentum of piece 2 must have a magnitude p and point in a direction opposite to the momentum of piece 1. **(b)** The mass of piece 2 is $2m$, which, with $v_2 = \frac{1}{2}v_1$, satisfies both $p_1 = p_2$ and $K_1 = 2K_2$.

5. **(a)** The objects are at rest ($v = 0$) after the collision, because the total momentum of the system is zero. **(b)** 100% of the initial kinetic energy is converted to other forms, because the final kinetic energy is zero.

6. **(a)** The 1-kg mass rebounds from the 2-kg mass and travels to the left after the collision. **(b)** After the collision, the 2-kg mass has a greater speed ($2v/3$) than the 1-kg mass ($v/3$).

7. **(a)** The metal sheet will rotate counterclockwise (left end down). **(b)** The best explanation is II.

8. The reading is greater than 20 N because an additional upward force is needed to stop the water as it comes to rest in the bucket.

CONCEPTUAL QUESTIONS

For instructor-assigned homework, go to www.masteringphysics.com

(Answers to odd-numbered Conceptual Questions can be found in the back of the book.)

1. If you drop your keys, their momentum increases as they fall. Why is the momentum of the keys not conserved? Does this mean that the momentum of the universe increases as the keys fall? Explain.

2. By what factor does an object's kinetic energy change if its speed is doubled? By what factor does its momentum change?

3. A system of particles is known to have zero kinetic energy. What can you say about the momentum of the system?

4. A system of particles is known to have zero momentum. Does it follow that the kinetic energy of the system is also zero? Explain.

5. On a calm day you connect an electric fan to a battery on your sailboat and generate a breeze. Can the wind produced by the fan be used to power the sailboat? Explain.

6. Crash statistics show that it is safer to be riding in a heavy car in an accident than in a light car. Explain in terms of physical principles.

7. **(a)** As you approach a stoplight, you apply the brakes and bring your car to rest. What happened to your car's initial momentum? **(b)** When the light turns green, you accelerate until you reach cruising speed. What force was responsible for increasing your car's momentum?

8. An object at rest on a frictionless surface is struck by a second object. Is it possible for both objects to be at rest after the collision? Explain.

9. **(a)** Can two objects on a horizontal frictionless surface have a collision in which all the initial kinetic energy of the system is lost? Explain, and give a specific example if your answer is yes. **(b)** Can two such objects have a collision in which all the initial momentum of the system is lost? Explain, and give a specific example if your answer is yes.

10. Two cars collide at an intersection. If the cars do not stick together, can we conclude that their collision was elastic? Explain.

11. At the instant a bullet is fired from a gun, the bullet and the gun have equal and opposite momenta. Which object—the bullet or the gun—has the greater kinetic energy? Explain. How does your answer apply to the observation that it is safe to hold a gun while it is fired, whereas the bullet is deadly?

12. An hourglass is turned over, and the sand is allowed to pour from the upper half of the glass to the lower half. If the hourglass is resting on a scale, and the total mass of the hourglass and sand is M, describe the reading on the scale as the sand runs to the bottom.

13. In the classic movie *The Spirit of St. Louis*, Jimmy Stewart portrays Charles Lindbergh on his history-making transatlantic flight. Lindbergh is concerned about the weight of his fuel-laden airplane. As he flies over Newfoundland he notices a fly on the dashboard. Speaking to the fly, he wonders aloud, "Does the plane weigh less if you fly inside it as it's flying? Now that's an interesting question." What do you think?

14. A tall, slender drinking glass with a thin base is initially empty. **(a)** Where is the center of mass of the glass? **(b)** Suppose the glass is now filled slowly with water until it is completely full. Describe the position and motion of the center of mass during the filling process.

15. Lifting one foot into the air, you balance on the other foot. What can you say about the location of your center of mass?

16. In the "Fosbury flop" method of high jumping, named for the track and field star Dick Fosbury, an athlete's center of mass may pass under the bar while the athlete's body passes over the bar. Explain how this is possible.

The "Fosbury flop." (Conceptual Question 16)

PROBLEMS AND CONCEPTUAL EXERCISES

Answers to odd-numbered Problems and Conceptual Exercises can be found in the back of the book. **BIO** *identifies problems of biological or medical interest;* **CE** *indicates a conceptual exercise.* Predict/Explain *problems ask for two responses: (a) your prediction of a physical outcome, and (b) the best explanation among three provided; and* Predict/Calculate *problems ask for a prediction of a physical outcome, based on fundamental physics concepts, and follow that with a numerical calculation to verify the prediction. On all problems, bullets (•, ••, •••) indicate the level of difficulty.*

SECTION 9-1 LINEAR MOMENTUM

1. • What is the mass of a mallard duck whose speed is 8.9 m/s and whose momentum has a magnitude of 11 kg·m/s?

2. • **(a)** What is the magnitude of the momentum of a 0.0053-kg marble whose speed is 0.65 m/s? **(b)** What is the speed of a 0.132-kg baseball whose momentum has a magnitude of 3.28 kg·m/s?

3. • A 54-kg person walks due north with a speed of 1.2 m/s, and her 6.9-kg dog runs directly toward her, moving due south, with a speed of 1.7 m/s. What is the magnitude of the total momentum of this system?

4. •• A 26.2-kg dog is running northward at 2.70 m/s, while a 5.30-kg cat is running eastward at 3.04 m/s. Their 74.0-kg owner has the same momentum as the two pets taken together. Find the direction and magnitude of the owner's velocity.

5. •• Predict/Calculate Two air-track carts move toward one another on an air track. Cart 1 has a mass of 0.28 kg and a speed of 0.97 m/s. Cart 2 has a mass of 0.64 kg. **(a)** What speed must cart 2 have if the total momentum of the system is to be zero? **(b)** Since the momentum of the system is zero, does it follow that the kinetic energy of the system is also zero? **(c)** Verify your answer to part (b) by calculating the system's kinetic energy.

6. •• A 0.145-kg baseball is dropped from rest. If the magnitude of the baseball's momentum is 1.55 kg·m/s just before it lands on the ground, from what height was it dropped?

7. •• A 285-g ball falls vertically downward, hitting the floor with a speed of 2.5 m/s and rebounding upward with a speed of 2.0 m/s. **(a)** Find the magnitude of the change in the ball's momentum. **(b)** Find the change in the magnitude of the ball's momentum. **(c)** Which of the two quantities calculated in parts (a) and (b) is more directly related to the net force acting on the ball during its collision with the floor? Explain.

8. ••• Object 1 has a mass m_1 and a velocity $\vec{v}_1 = (2.80 \text{ m/s})\hat{x}$. Object 2 has a mass m_2 and a velocity $\vec{v}_2 = (3.10 \text{ m/s})\hat{y}$. The total momentum of these two objects has a magnitude of 17.6 kg·m/s and points in a direction 66.5° above the positive x axis. Find m_1 and m_2.

SECTION 9-3 IMPULSE

9. • **CE** Your car rolls slowly in a parking lot and bangs into the metal base of a light pole. In terms of safety, is it better for your collision with the light pole to be elastic, inelastic, or is the safety risk the same for either case? Explain.

10. • **CE** Predict/Explain A net force of 200 N acts on a 100-kg boulder, and a force of the same magnitude acts on a 100-g pebble. **(a)** Is the change of the boulder's momentum in one second greater than, less than, or equal to the change of the pebble's momentum in the same time period? **(b)** Choose the *best explanation* from among the following:
   I. The large mass of the boulder gives it the greater momentum.
   II. The force causes a much greater speed in the 100-g pebble, resulting in more momentum.
   III. Equal force means equal change in momentum for a given time.

11. • **CE** Predict/Explain Referring to the previous question, **(a)** is the change in the boulder's speed in one second greater than, less than, or equal to the change in speed of the pebble in the same

time period? **(b)** Choose the *best explanation* from among the following:
   I. The large mass of the boulder results in a small acceleration.
   II. The same force results in the same change in speed for a given time.
   III. Once the boulder gets moving it is harder to stop than the pebble.

12. • **CE** Predict/Explain Two identical cars, each traveling at 16 m/s, slam into a concrete wall and come to rest. In car A the air bag does not deploy and the driver hits the steering wheel; in car B the driver contacts the deployed air bag. **(a)** Is the impulse delivered by the steering wheel to driver A greater than, less than, or equal to the impulse delivered by the air bag to driver B? **(b)** Choose the *best explanation* from among the following:
   I. The larger force delivered by the steering wheel delivers a greater impulse.
   II. The longer interaction time with the air bag delivers a greater impulse.
   III. The change in momentum of each driver is the same.

13. • **CE** Force A has a magnitude F and acts for the time Δt, force B has a magnitude $2F$ and acts for the time $\Delta t/3$, force C has a magnitude $5F$ and acts for the time $\Delta t/10$, and force D has a magnitude $10F$ and acts for the time $\Delta t/100$. Rank these forces in order of increasing impulse. Indicate ties where appropriate.

14. • Find the magnitude of the impulse delivered to a soccer ball when a player kicks it with a force of 1250 N. Assume that the player's foot is in contact with the ball for 5.95×10^{-3} s.

15. • A 0.45-kg croquet ball is initially at rest on the grass. When the ball is struck by a mallet, the average force exerted on it is 270 N. If the ball's speed after being struck is 5.4 m/s, for what amount of time was the mallet in contact with the ball?

16. • When spiking a volleyball, a player changes the velocity of the ball from 4.2 m/s to −24 m/s along a certain direction. If the impulse delivered to the ball by the player is −9.3 kg·m/s, what is the mass of the volleyball?

17. •• **Force Platform** A force platform measures the horizontal force exerted on a person's foot by the ground. In **FIGURE 9-26** we see force-versus-time data taken by a force platform for a person who starts jogging at time $t = 0$. The impulse imparted to the jogger from $t = 0$ to $t = 0.6$ s is 198 N·s. **(a)** What is the average force exerted on the jogger from $t = 0$ to $t = 0.6$ s? **(b)** If the jogger's change in speed during this time period is 2.69 m/s, what is the jogger's mass?

FIGURE 9-26 Problem 17

18. •• **Air Bag Safety** If a driver makes contact with a steering wheel during a 35-mph crash, she comes to rest in about $\Delta t = 15$ ms. If, during an identical crash, the driver makes contact with an air bag, she comes to rest in about $\Delta t = 110$ ms. What is the ratio F_{sw}/F_{ab} of the force exerted by the steering wheel to the force exerted by the air bag on the driver?

19. •• To make a bounce pass, a player throws a 0.60-kg basketball toward the floor. The ball hits the floor with a speed of 5.4 m/s at an angle of 65° to the vertical. If the ball rebounds with the same speed and angle, what was the impulse delivered to it by the floor?

20. •• **BIO Concussion Impulse** One study suggests that a brain concussion can occur if a person's head is subjected to an acceleration as small as $60g$, where g is the acceleration of gravity. **(a)** If the mass of the person's head is 4.8 kg, what is the magnitude of the force? **(b)** If the impulse delivered to the person's head is 18 kg · m/s, over what time interval is the force applied?

21. •• Predict/Calculate A 0.14-kg baseball moves toward home plate with a velocity $\vec{v}_i = (-36 \text{ m/s})\hat{x}$. After striking the bat, the ball moves vertically upward with a velocity $\vec{v}_f = (18 \text{ m/s})\hat{y}$. **(a)** Find the direction and magnitude of the impulse delivered to the ball by the bat. Assume that the ball and bat are in contact for 1.5 ms. **(b)** How would your answer to part (a) change if the mass of the ball were doubled? **(c)** How would your answer to part (a) change if the mass of the bat were doubled instead?

22. •• A player bounces a 0.43-kg soccer ball off her head, changing the velocity of the ball from $\vec{v}_i = (8.8 \text{ m/s})\hat{x} + (-2.3 \text{ m/s})\hat{y}$ to $\vec{v}_f = (5.2 \text{ m/s})\hat{x} + (3.7 \text{ m/s})\hat{y}$. If the ball is in contact with the player's head for 6.7 ms, what are **(a)** the direction and **(b)** the magnitude of the impulse delivered to the ball?

SECTION 9-4 CONSERVATION OF LINEAR MOMENTUM

23. • Two ice skaters stand at rest in the center of an ice rink. When they push off against one another the 45-kg skater acquires a speed of 0.62 m/s. If the speed of the other skater is 0.89 m/s, what is this skater's mass?

24. • A 0.042-kg pet lab mouse sits on a 0.35-kg air-track cart, as shown in FIGURE 9-27. The cart is at rest, as is a second cart with a mass of 0.25 kg. The lab mouse now jumps to the second cart. After the jump, the 0.35-kg cart has a speed of $v_1 = 0.88$ m/s. What is the speed v_2 of the mouse and the 0.25-kg cart?

(a) Before the jump

(b) After the jump

FIGURE 9-27 Problem 24

25. •• An object initially at rest breaks into two pieces as the result of an explosion. One piece has twice the kinetic energy of the other piece. What is the ratio of the masses of the two pieces? Which piece has the larger mass?

26. •• A 92-kg astronaut and a 1200-kg satellite are at rest relative to the space shuttle. The astronaut pushes on the satellite, giving it a speed of 0.14 m/s directly away from the shuttle. Seven and a half seconds later the astronaut comes into contact with the shuttle. What was the initial distance from the shuttle to the astronaut?

27. •• The recoil of a shotgun can be significant. Suppose a 3.3-kg shotgun is held tightly by an arm and shoulder with a combined mass of 15.0 kg. When the gun fires a projectile with a mass of 0.038 kg and a speed of 385 m/s, what is the recoil velocity of the shotgun and arm–shoulder combination?

28. ••• A plate drops onto a smooth floor and shatters into three pieces of equal mass. Two of the pieces go off with equal speeds v at right angles to one another. Find the speed and direction of the third piece.

SECTION 9-5 INELASTIC COLLISIONS

29. • Suppose the car in Example 9-13 has an initial speed of 20.0 m/s and that the direction of the wreckage after the collision is 40.0° above the x axis. Find the initial speed of the minivan and the final speed of the wreckage.

30. • Two 78.5-kg hockey players skating at 4.47 m/s collide and stick together. If the angle between their initial directions was 105°, what is their speed after the collision?

31. • An air-track cart with mass $m_1 = 0.32$ kg and initial speed $v_0 = 0.85$ m/s collides with and sticks to a second cart that is at rest initially. If the mass of the second cart is $m_2 = 0.44$ kg, how much kinetic energy is lost as a result of the collision?

32. •• Predict/Calculate A bullet with a mass of 4.0 g and a speed of 650 m/s is fired at a block of wood with a mass of 0.095 kg. The block rests on a frictionless surface, and is thin enough that the bullet passes completely through it. Immediately after the bullet exits the block, the speed of the block is 23 m/s. **(a)** What is the speed of the bullet when it exits the block? **(b)** Is the final kinetic energy of this system equal to, less than, or greater than the initial kinetic energy? Explain. **(c)** Verify your answer to part (b) by calculating the initial and final kinetic energies of the system.

33. •• **BIO Concussion Recoil** The human head can be considered as a 3.3-kg cranium protecting a 1.5-kg brain, with a small amount of cerebrospinal fluid that allows the brain to move a little bit inside the cranium. Suppose a cranium at rest is subjected to a force of 2800 N for 6.5 ms in the forward direction. **(a)** What is the final speed of the cranium? **(b)** The back of the cranium then collides with the back of the brain, which is still at rest, and the two move together. What is their final speed? **(c)** The cranium now hits an external object and suddenly comes to rest, but the brain continues to move forward. If the front of the brain interacts with the front of the cranium over a period of 15 ms before coming to rest, what average force is exerted on the brain by the cranium?

34. ••• Two objects moving with a speed v travel in opposite directions in a straight line. The objects stick together when they collide, and move with a speed of $v/4$ after the collision. **(a)** What is the ratio of the final kinetic energy of the system to the initial kinetic energy? **(b)** What is the ratio of the mass of the more massive object to the mass of the less massive object?

SECTION 9-6 ELASTIC COLLISIONS

35. • In the apple-orange collision in Example 9-16, suppose the final velocity of the orange is 1.03 m/s in the negative y direction. What are the final speed and direction of the apple in this case?

36. • A 732-kg car stopped at an intersection is rear-ended by a 1720-kg truck moving with a speed of 15.5 m/s. If the car was in neutral and its brakes were off, so that the collision is approximately elastic, find the final speed of both vehicles after the collision.

37. •• The collision between a hammer and a nail can be considered to be approximately elastic. Calculate the kinetic energy acquired by a 7.2-g nail when it is struck by a 450-g hammer moving with an initial speed of 8.8 m/s.

38. •• Predict/Calculate A charging bull elephant with a mass of 5240 kg comes directly toward you with a speed of 4.55 m/s. You toss a 0.150-kg rubber ball at the elephant with a speed of 7.81 m/s. **(a)** When the ball bounces back toward you, what is its speed? **(b)** How do you account for the fact that the ball's kinetic energy has increased?

39. •• **Moderating a Neutron** In a nuclear reactor, neutrons released by nuclear fission must be slowed down before they can trigger additional reactions in other nuclei. To see what sort of material is most effective in slowing (or moderating) a neutron, calculate the ratio of a neutron's final kinetic energy to its initial kinetic energy, K_f/K_i, for a head-on elastic collision with each of the following stationary target particles. (*Note:* The mass of a neutron is $m = 1.009$ u, where the atomic mass unit, u, is defined as follows: 1 u = 1.66×10^{-27} kg.) **(a)** An electron ($M = 5.49 \times 10^{-4}$ u). **(b)** A proton ($M = 1.007$ u). **(c)** The nucleus of a lead atom ($M = 207.2$ u).

40. •• The three air carts shown in **FIGURE 9-28** have masses, reading from left to right, of $4m$, $2m$, and m, respectively.

FIGURE 9-28 Problem 40

The most massive cart has an initial speed of v_0; the other two carts are at rest initially. All carts are equipped with spring bumpers that give elastic collisions. **(a)** Find the final speed of each cart. **(b)** Verify that the final kinetic energy of the system is equal to the initial kinetic energy. (Assume the air track is long enough to accommodate all collisions.)

41. •• An air-track cart with mass $m = 0.25$ kg and speed $v_0 = 1.2$ m/s approaches two other carts that are at rest and have masses $2m$ and $3m$, as indicated in **FIGURE 9-29**. The carts have bumpers that make all the collisions elastic. Find the final speed of each of the carts, assuming the air track extends indefinitely in either direction.

FIGURE 9-29 Problem 41

SECTION 9-7 CENTER OF MASS

42. • **CE** Predict/Explain A stalactite in a cave has drops of water falling from it to the cave floor below. The drops are equally spaced in time and come in rapid succession, so that at any given moment there are many drops in midair. **(a)** Is the center of mass of the midair drops higher than, lower than, or equal to the halfway distance between the tip of the stalactite and the cave floor? **(b)** Choose the *best explanation* from among the following:
  I. The drops bunch up as they near the floor of the cave.
  II. The drops are equally spaced as they fall, since they are released at equal times.
  III. Though equally spaced in time, the drops are closer together higher up.

43. • **CE** A baseball bat balances when placed on a fulcrum in the position shown in **FIGURE 9-30**. Is the mass of the bat to the left of the fulcrum greater than, less than, or equal to the mass of the bat to the right of the fulcrum? Explain.

FIGURE 9-30 Problem 43

44. • Find the x coordinate of the center of mass of the bricks shown in **FIGURE 9-31**.

FIGURE 9-31 Problem 44

45. • You are holding a shopping basket at the grocery store with two 0.62-kg cartons of cereal at the left end of the basket. The basket is 0.61 m long. Where should you place a 1.9-kg half gallon of milk, relative to the left end of the basket, so that the center of mass of your groceries is at the center of the basket?

46. •• **CE** A pencil standing upright on its eraser end falls over and lands on a table. As the pencil falls, its eraser does not slip. The following questions refer to the contact force exerted on the pencil by the table. Let the positive x direction be in the direction the pencil falls, and the positive y direction be vertically upward. **(a)** During the pencil's fall, is the x component of the contact force positive, negative, or zero? Explain. **(b)** Is the y component of the contact force greater than, less than, or equal to the weight of the pencil? Explain.

47. •• **BIO Human Center of Mass** Suppose the human body can be reduced to individual point masses as shown in **FIGURE 9-32**. If $m_L = 14.0$ kg, $y_L = 0.65$ m, $m_A = 9.00$ kg, $y_A = 1.20$ m, $m_T = 24.5$ kg, $y_T = 1.25$ m, $m_H = 6.50$ kg, and $y_H = 1.65$ m, what is the vertical height of the center of mass?

48. •• The location of the center of mass of the partially eaten, 12-inch-diameter pizza shown in **FIGURE 9-33** is $X_{cm} = -1.4$ in. and $Y_{cm} = -1.4$ in. Assuming each quadrant of the pizza to be the same, find the center of mass of the uneaten pizza above the x axis (that is, the portion of the pizza in the second quadrant).

FIGURE 9-32 Problem 47

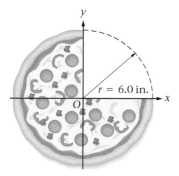

FIGURE 9-33 Problem 48

49. •• **The Center of Mass of Sulfur Dioxide** Sulfur dioxide (SO_2) consists of two oxygen atoms (each of mass 16 u, where u is defined in Problem 39) and a single sulfur atom (of mass 32 u). The center-to-center distance between the sulfur atom and either of the oxygen atoms is 0.143 nm, and the angle formed by the three atoms is 120°, as shown in **FIGURE 9-34**. Find the x and y coordinates of the center of mass of this molecule.

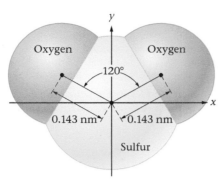

FIGURE 9-34 Problem 49

50. •• Predict/Calculate Three uniform metersticks, each of mass m, are placed on the floor as follows: stick 1 lies along the y axis from $y = 0$ to $y = 1.0$ m, stick 2 lies along the x axis from $x = 0$ to $x = 1.0$ m, stick 3 lies along the x axis from $x = 1.0$ m to $x = 2.0$ m. **(a)** Find the location of the center of mass of the metersticks. **(b)** How would the location of the center of mass be affected if the mass of the metersticks were doubled?

51. •• A 0.726-kg rope 2.00 meters long lies on a floor. You grasp one end of the rope and begin lifting it upward with a constant speed of 0.710 m/s. Find the position and velocity of the rope's center of mass from the time you begin lifting the rope to the time the last piece of rope lifts off the floor. Plot your results. (Assume the rope occupies negligible volume directly below the point where it is being lifted.)

52. •• Consider the system shown in **FIGURE 9-35**. Assume that after the string breaks the ball falls through the liquid with constant speed. If the mass of the bucket *and* the liquid is 1.20 kg, and the mass of the ball is 0.150 kg, what is the reading on the scale **(a)** before and **(b)** after the string breaks?

FIGURE 9-35 Problem 52

53. ••• A metal block of mass m is attached to the ceiling by a spring. Connected to the bottom of this block is a string that supports a second block of the same mass m, as shown in **FIGURE 9-36**. The string connecting the two blocks is now cut. **(a)** What is the net force acting on the two-block system immediately after the string is cut? **(b)** What is the acceleration of the center of mass of the two-block system immediately after the string is cut?

FIGURE 9-36 Problem 53

SECTION 9-8 SYSTEMS WITH CHANGING MASS: ROCKET PROPULSION

54. • **Helicopter Thrust** During a rescue operation, a 5300-kg helicopter hovers above a fixed point. The helicopter blades send air downward with a speed of 62 m/s. What mass of air must pass through the blades every second to produce enough thrust for the helicopter to hover?

55. • **Rocks for a Rocket Engine** A child sits in a wagon with a pile of 0.34-kg rocks. If she can throw each rock with a speed of 9.5 m/s relative to the ground, causing the wagon to move, how many rocks must she throw per minute to maintain a constant average speed against a 3.4-N force of friction?

56. • A 57.8-kg person holding two 0.880-kg bricks stands on a 2.10-kg skateboard. Initially, the skateboard and the person are at rest. The person now throws the two bricks at the same time so that their speed relative to the person is 17.0 m/s. What is the recoil speed of the person and the skateboard relative to the ground, assuming the skateboard moves without friction?

57. • A fire hose can expel water at a rate of 9.5 kg/s (150 gallons/minute) with a speed of 24 m/s. How much force must the firefighters exert on the hose in order to hold it steady?

58. •• A 0.540-kg bucket rests on a scale. Into this bucket you pour sand at the constant rate of 56.0 g/s. If the sand lands in the bucket with a speed of 3.20 m/s, **(a)** what is the reading of the scale when there is 0.750 kg of sand in the bucket? **(b)** What is the weight of the bucket and the 0.750 kg of sand?

59. •• Predict/Calculate Holding a long rope by its upper end, you lower it onto a scale. The rope has a mass of 0.13 kg per meter of length, and is lowered onto the scale at the constant rate of 1.4 m/s. **(a)** Calculate the thrust exerted by the rope as it lands on the scale. **(b)** At the instant when the amount of rope at rest on the scale has a weight of 2.5 N, does the scale read 2.5 N, more than 2.5 N, or less than 2.5 N? Explain. **(c)** Check your answer to part (b) by calculating the reading on the scale at this time.

GENERAL PROBLEMS

60. • **CE** Object A has a mass m, object B has a mass $2m$, and object C has a mass $m/2$. Rank these objects in order of increasing kinetic energy, given that they all have the same momentum. Indicate ties where appropriate.

61. • **CE** Object A has a mass m, object B has a mass $4m$, and object C has a mass $m/4$. Rank these objects in order of increasing momentum, given that they all have the same kinetic energy. Indicate ties where appropriate.

62. • **CE** A juggler performs a series of tricks with three bowling balls while standing on a bathroom scale. Is the average reading of the scale greater than, less than, or equal to the weight of the juggler plus the weight of the three balls? Explain.

63. • A golfer attempts a birdie putt, sending the 0.045-kg ball toward the hole with a speed of 1.2 m/s. Unfortunately, the ball does a "horseshoe" and goes halfway around the rim of the hole. The ball ends up heading back toward the golfer with a speed of 1.1 m/s. If we take the positive direction to be from the golfer to the hole, what is the impulse delivered to the ball by the hole?

64. • Predict/Calculate Two trucks drive directly away from one another on a straight line, as indicated in **FIGURE 9-37**. The trucks drive with their speeds adjusted precisely so that the center of mass (CM) of the two trucks remains stationary—that is, the velocity of the center of mass is zero. The mass and speed of truck 1 are $m_1 = 8400$ kg and $v_1 = 12$ m/s, respectively. The mass of truck 2 is $m_2 = 10,200$ kg. **(a)** Is the speed of truck 2 greater than, less than, or equal to the speed of truck 1? Explain. **(b)** Find the speed of truck 2.

FIGURE 9-37 Problem 64

65. •• **CE** Predict/Explain **FIGURE 9-38** shows a block of mass $2m$ at rest on a horizontal, frictionless table. Attached to this block by a string that passes over a pulley is a second block, with a mass m. The initial position of the center of mass of the blocks is indicated by the point i.

FIGURE 9-38 Problem 65

The blocks are now released and allowed to accelerate; a short time later their center of mass is at the point f. **(a)** Did the center of mass follow the red path, the green path, or the blue path? **(b)** Choose the *best explanation* from among the following:

I. The center of mass must always be closer to the $2m$ block than to the m block.

II. The center of mass starts at rest, and moves in a straight line in the direction of the net force.

III. The masses are accelerating, which implies parabolic motion.

66. •• A 1.35-kg block of wood sits at the edge of a table, 0.782 m above the floor. A 0.0105-kg bullet moving horizontally with a speed of 715 m/s embeds itself within the block. What horizontal distance does the block cover before hitting the ground?

67. •• In a stunt, three people jump off a platform and fall 9.9 m onto a large air bag. A fourth person at the other end of the air bag, the "flier," is launched 17 m vertically into the air. Assume all four people have a mass of 75 kg. **(a)** What fraction of the mechanical energy of the three jumpers is transferred to the mechanical energy of the flier? **(b)** What is the total momentum of the three jumpers just before they land on the air bag? Let upward be the positive direction. **(c)** What is the momentum of the flier just after launch? **(d)** Notice that momentum is not conserved during this stunt, unless you include the Earth. What impulse is delivered to the jumpers–flier system by the Earth at the instant of the launch?

68. •• Predict/Calculate The carton of eggs shown in **FIGURE 9-39** is filled with a dozen eggs, each of mass m. Initially, the center of mass of the eggs is at the center of the carton. **(a)** Does the location of the center of mass of the eggs change more if egg 1 is removed or if egg 2 is removed? Explain. **(b)** Find the center of mass of the eggs when egg 1 is removed. **(c)** Find the center of mass of the eggs if egg 2 is removed instead.

FIGURE 9-39 Problem 68

69. •• **The Force of a Storm** During a severe storm in Palm Beach, FL, on January 2, 1999, 31 inches of rain fell in a period of nine hours. Assuming that the raindrops hit the ground with a speed of 10 m/s, estimate the average upward force exerted by one square meter of ground to stop the falling raindrops during the storm. (*Note:* One cubic meter of water has a mass of 1000 kg.)

70. •• An experiment is performed in which two air carts collide on an air track. Position-versus-time data from a video of the experiment is shown in **FIGURE 9-40**. **(a)** Is the collision elastic or inelastic? Explain. **(b)** What is the speed of the center of mass of the two carts?

FIGURE 9-40 Problems 70 and 71

71. •• Figure 9-40 shows position-versus-time data from an experiment in which two air carts collide on an air track. If the mass of the air cart that is moving initially is 0.24 kg, what is the mass of the other cart?

72. •• To balance a 35.5-kg automobile tire and wheel, a mechanic must place a 50.2-g lead weight 25.0 cm from the center of the wheel. When the wheel is balanced, its center of mass is exactly at the center of the wheel. How far from the center of the wheel was its center of mass before the lead weight was added?

73. •• A hoop of mass M and radius R rests on a smooth, level surface. The inside of the hoop has ridges on either side, so that it forms a track on which a ball can roll, as indicated in **FIGURE 9-41**. If a ball of mass $2M$ and radius $r = R/4$ is released as shown, the system rocks back and forth until it comes to rest with the ball at the bottom of the hoop. When the ball comes to rest, what is the x coordinate of its center?

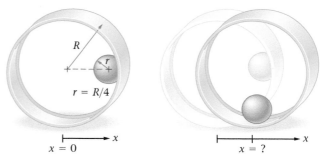

FIGURE 9-41 Problem 73

74. •• Predict/Calculate A 63-kg canoeist stands in the middle of her 22-kg canoe. The canoe is 3.0 m long, and the end that is closest to land is 2.5 m from the shore. The canoeist now walks toward the shore until she comes to the end of the canoe. **(a)** When the canoeist stops at the end of her canoe, is her distance from the shore equal to, greater than, or less than 2.5 m? Explain. **(b)** Verify your answer to part (a) by calculating the distance from the canoeist to shore.

75. •• A honeybee with a mass of 0.150 g lands on one end of a float-ing 4.75-g popsicle stick, as shown in FIGURE 9-42. After sitting at rest for a moment, it runs toward the other end with a velocity $\vec{v}_b$ relative to the still water. The stick moves in the opposite direction with a speed of 0.120 cm/s. What is the velocity of the bee? (Let the direction of the bee's motion be the positive x direction.)

FIGURE 9-42 Problem 75

76. •• A young hockey player stands at rest on the ice holding a 1.3-kg helmet. The player tosses the helmet with a speed of 6.5 m/s in a direction 11° above the horizontal, and recoils with a speed of 0.25 m/s. Find the mass of the hockey player.

77. •• Predict/Calculate A 0.341-kg block of wood hangs from the ceiling by a string, and a 0.0567-kg wad of putty is thrown straight upward, striking the bottom of the block with a speed of 8.05 m/s. The wad of putty sticks to the block. **(a)** Is the mechanical energy of this system conserved? **(b)** How high does the putty–block sys-tem rise above the original position of the block?

78. •• A 0.454-kg block is attached to a horizontal spring that is at its equilibrium length, and whose force constant is 25.0 N/m. The block rests on a frictionless surface. A 0.0500-kg wad of putty is thrown horizontally at the block, hitting it with a speed of 8.94 m/s and sticking. How far does the putty–block system com-press the spring?

79. •• **BIO Escaping Octopus** The giant Pacific octopus (*Enterocto-pus dofleini*) has a typical mass of 15 kg and can propel itself by means of a water jet when frightened. **(a)** If it expels water at a mass rate of 2.4 kg/s and with a speed of 27 m/s, what thrust can it generate? **(b)** If the seawater exerts a resistive force of 18 N on the octopus, what acceleration can it achieve? **(c)** If the octopus maintains its jet propulsion for 3.0 s, what maximum velocity can it attain?

80. •• **The Center of Mass of Water** Find the center of mass of a water molecule, referring to FIGURE 9-43 for the relevant angles and dis-tances. The mass of a hydrogen atom is 1.0 u, and the mass of an oxygen atom is 16 u, where u is the atomic mass unit (see Prob-lem 39). Use the center of the oxygen atom as the origin of your coordinate system.

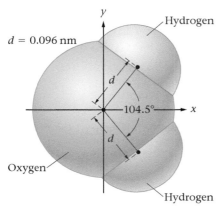

FIGURE 9-43 Problem 80

81. •• The three air carts shown in FIGURE 9-44 have masses, reading from left to right, of m, $2m$, and $4m$, respectively. Initially, the cart on the right is at rest, whereas the other two carts are moving to the right with a speed v_0. All carts are equipped with putty bum-pers that give completely inelastic collisions. **(a)** Find the final speed of the carts. **(b)** Calculate the ratio of the final kinetic energy of the system to the initial kinetic energy.

FIGURE 9-44 Problem 81

82. •• **Unlimited Overhang** Four identical textbooks, each of length L, are stacked near the edge of a table, as shown in FIGURE 9-45. The books are stacked in such a way that the distance they overhang the edge of the table, d, is maximized. Find the maximum overhang dis-tance d in terms of L. In particular, show that $d > L$; that is, the top book is completely to the right of the table edge. (In principle, the overhang distance d can be made as large as desired simply by increasing the number of books in the stack.)

FIGURE 9-45 Problem 82

83. ••• Consider a one-dimensional, head-on elastic collision. One object has a mass m_1 and an initial velocity v_1; the other has a mass m_2 and an initial velocity v_2. Use momentum conservation and energy conservation to show that the final velocities of the two masses are

$$v_{1,f} = \left(\frac{m_1 - m_2}{m_1 + m_2}\right)v_1 + \left(\frac{2m_2}{m_1 + m_2}\right)v_2$$

$$v_{2,f} = \left(\frac{2m_1}{m_1 + m_2}\right)v_1 + \left(\frac{m_2 - m_1}{m_1 + m_2}\right)v_2$$

84. ••• Two air carts of mass $m_1 = 0.84$ kg and $m_2 = 0.42$ kg are placed on a frictionless track. Cart 1 is at rest initially, and has a spring bumper with a force constant of 690 N/m. Cart 2 has a flat metal surface for a bumper, and moves toward the bumper of the stationary cart with an initial speed $v = 0.68$ m/s. **(a)** What is the speed of the two carts at the moment when their speeds are equal? **(b)** How much energy is stored in the spring bumper when the carts have the same speed? **(c)** What is the final speed of the carts after the collision?

85. ••• **Golden Earrings and the Golden Ratio** A popular earring design fea-tures a circular piece of gold of diameter D with a circular cutout of diameter d, as shown in FIGURE 9-46. If this earring is to balance at the point P, show that the diameters must satisfy the condi-tion $D = \phi d$, where $\phi = \left(1 + \sqrt{5}\right)/2 = 1.61803$ is the famous "golden ratio."

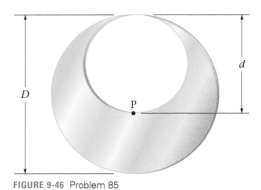

FIGURE 9-46 Problem 85

86. ••• **Amplified Rebound Height** Two small rubber balls are dropped from rest at a height h above a hard floor. When the balls are released, the lighter ball (with mass m) is directly above the heavier ball (with mass M). Assume the heavier ball reaches the floor first and bounces elastically; thus, when the balls collide, the ball of mass M is moving upward with a speed v and the ball of mass m is moving downward with essentially the same speed. In terms of h, find the height to which the ball of mass m rises after the collision. (Use the results given in Problem 83, and assume the balls collide at ground level.)

87. ••• Predict/Calculate **Weighing a Block on an Incline** A wedge of mass m_1 is firmly attached to the top of a scale, as shown in **FIGURE 9-47**. The inclined surface of the wedge makes an angle θ with the horizontal. Now, a block of mass m_2 is placed on the inclined surface of the wedge and allowed to accelerate without friction down the slope. **(a)** Show that the reading on the scale while the block slides is

$$(m_1 + m_2 \cos^2\theta)g$$

(b) Explain why the reading on the scale is less than $(m_1 + m_2)g$. **(c)** Show that the expression in part (a) gives the expected results for $\theta = 0$ and $\theta = 90°$.

FIGURE 9-47 Problem 87

88. ••• Predict/Calculate A uniform rope of length L and mass M rests on a table. **(a)** If you lift one end of the rope upward with a constant *speed, v*, show that the rope's center of mass moves upward with constant *acceleration.* **(b)** Next, suppose you hold the rope suspended in air, with its lower end just touching the table. If you now lower the rope with a constant speed, v, onto the table, is the acceleration of the rope's center of mass upward or downward? Explain your answer. **(c)** Find the magnitude and direction of the acceleration of the rope's center of mass for the case described in part (b). Compare with part (a).

PASSAGE PROBLEMS

BIO North Atlantic right whales (*Eubalaena glacialis*) are one of the largest and rarest marine mammals (adults can weigh up to 70 tons). Right whales move slowly near the surface to feed, and their thick blubber gives them enough buoyancy to float at the surface after they're killed, which explains why whalers called them the "right" whales to hunt. Given this, it's not surprising that centuries of whaling, which ended only in 1949, drove the species nearly to extinction. It's estimated that currently only about 400 individuals remain.

One of the reasons the population has been slow to recover is that right whales spend much of their time in shallow coastal waters, where they are susceptible to collisions with ships and entanglement in fishing gear. The United States has taken several steps to protect right whales from shipping-related injuries. For instance, ships weighing more than 300 tons are required to report when they enter specific regions off the U.S. coast that are important right-whale habitats. In addition, ships longer than 65 feet are required to slow down to 10 knots or less in certain critical areas, such as whale calving grounds. (Ship speeds are measured in knots, or nautical miles per hour; one nautical mile is 1.151 miles.)

In order to guide whale conservation policies, researchers are working to understand what factors affect the severity of injury to whales in a collision. Unfortunately, it's not always easy to gather information about the extent of whale injuries as a result of a particular collision. One approach is to mathematically model the effects of collisions. In one model of a collision between a ship and a whale, the duration of the impact decreases from 0.50 s for a ship moving at 5.0 knots to 0.38 s at 15 knots. Because the collision is this short, forces on the whale from water drag during the collision can be neglected.

89. •• The lightest ship that's subject to the mandatory reporting regulations, traveling at 5.0 knots, collides with a large stationary adult right whale of maximum size. As sometimes happens, the whale remains stuck on the prow of the ship after the collision. By how much does the speed of the ship change as a result of the collision? Give your answer in knots.

A. 0.95   B. 1.1   C. 2.5   D. 4.1

90. •• By how much does the whale's speed change (in knots) as a result of this collision?

A. 0.95   B. 1.1   C. 2.5   D. 4.1

91. •• What is the magnitude of the average force (in kN) the ship exerts on the whale during the collision? (*Note:* 1 ton = 907 kg)

A. 62   B. 270   C. 520   D. 620

92. •• If the ship was traveling at 5.0 knots but was 10 times more massive, how would the force of the collision on the whale compare?

A. It would be 100 times larger.
B. It would be 10 times larger.
C. It would be larger, but not as much as 10 times larger.
D. It would be the same.

93. •• **REFERRING TO EXAMPLE 9-12** Suppose a bullet of mass $m = 6.75$ g is fired into a ballistic pendulum whose bob has a mass of $M = 0.675$ kg. **(a)** If the bob rises to a height of 0.128 m, what was the initial speed of the bullet? **(b)** What was the speed of the bullet–bob combination immediately after the collision took place?

94. •• **REFERRING TO EXAMPLE 9-12** A bullet with a mass $m = 8.10$ g and an initial speed $v_i = 320$ m/s is fired into a ballistic pendulum. What mass must the bob have if the bullet–bob combination is to rise to a maximum height of 0.125 m after the collision?

95. •• **REFERRING TO EXAMPLE 9-19** Suppose that cart 1 has a mass of 3.00 kg and an initial speed of 0.250 m/s. Cart 2 has a mass of 1.00 kg and is at rest initially. **(a)** What is the final speed of the carts? **(b)** How much kinetic energy is lost as a result of the collision?

96. •• **REFERRING TO EXAMPLE 9-19** Suppose the two carts have equal masses and are both moving to the right before the collision. The initial speed of cart 1 (on the left) is v_0, and the initial speed of cart 2 (on the right) is $v_0/2$. **(a)** What is the speed of the center of mass of this system? **(b)** What percentage of the initial kinetic energy is lost as a result of the collision? **(c)** Suppose the collision is elastic. What are the final speeds of the two carts in this case?

Big Ideas

1 Rotational quantities are analogous to the corresponding linear quantities.

2 Rolling motion is a combination of linear and rotational motions.

3 A rotating object has kinetic energy, just like an object in linear motion.

Rotational Kinematics and Energy

▲ Because of the earth's rotation about its axis, the stars in this long-exposure photograph appear to follow circular paths across the sky—with Polaris, the north star, very near the axis of rotation. Notice that each star moves through the same angle in the course of the exposure. However, the farther a star is from the axis of rotation, the longer the arc it traces out. This is just one example of the connections among angle, radius, and arc length that we explore in this chapter.

Rotational motion is a part of everyday life, from the spinning planet on which we live to the spinning wheels of an automobile. In this chapter we present the basic concepts of rotational motion, and show the close connection between rotational and linear motions.

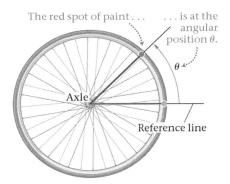

▲ FIGURE 10-1 **Angular position** The angular position, θ, of a spot of paint on a bicycle wheel. The reference line, where $\theta = 0$, is drawn horizontal here but any direction can be chosen.

10-1 Angular Position, Velocity, and Acceleration

To describe the motion of an object moving in a straight line, it's useful to establish a coordinate system with a definite origin and positive direction. In terms of this coordinate system we can measure the object's position, velocity, and acceleration.

Similarly, to describe rotational motion, we define "angular" quantities that are analogous to the linear position, velocity, and acceleration. These angular quantities, which are defined in terms of an angular "coordinate system," form the basis of our study of rotation. We begin by defining the most basic angular quantity—the angular position.

Angular Position, θ

Consider a bicycle wheel that is free to rotate about its axle, as shown in FIGURE 10-1. We say that the axle is the **axis of rotation** for the wheel. As the wheel rotates, each and every point on it moves in a circular path centered on the axis of rotation.

Now, suppose there is a small spot of red paint on the tire, and we want to describe the rotational motion of the spot. The **angular position** of the spot is defined to be the angle, θ, that a line from the axle to the spot makes with a reference line, as indicated in Figure 10-1.

Definition of Angular Position, θ

θ = angle measured from reference line   10-1

SI unit: radian (rad), which is dimensionless

The reference line simply defines $\theta = 0$; it is analogous to the origin in a linear coordinate system. The reference line begins at the axis of rotation, and may be chosen to point in any direction—just as an origin may be placed anywhere along a coordinate axis. Once chosen, however, the reference line must be used consistently.

The spot of paint in Figure 10-1 is rotated counterclockwise from the reference line by the angle θ. By convention, we say that this angle is positive. Clockwise rotations correspond to negative angles.

Sign Convention for Angular Position

By convention:

$\theta > 0$ counterclockwise rotation from reference line
$\theta < 0$ clockwise rotation from reference line

Examples of rotational motion are shown in FIGURE 10-2.

Units of Angular Position   Now that we have established a reference line (for $\theta = 0$) and a positive direction of rotation (counterclockwise), we must choose units in which to measure angles. Common units are degrees (°) and revolutions (rev), where one revolution—that is, going completely around a circle—corresponds to 360°:

$$1 \text{ rev} = 360°$$

▶ FIGURE 10-2 **Visualizing Concepts Rotational Motion** Rotational motion is everywhere in our universe, on every scale of length and time. A galaxy like the one at left may take millions of years to complete a single rotation about its center, while the skater in the middle photo spins several times in a second. The bacterium at right moves in a corkscrew path by rapidly twirling its flagella (the fine projections at either end of the cell) like whips.

The most convenient units for scientific calculations, however, are radians. A **radian** (rad) is defined as follows:

> A radian is the angle for which the arc length on a circle of radius r is equal to the radius of the circle.

This definition is useful because it establishes a particularly simple relationship between an angle measured in radians and the corresponding arc length, as illustrated in FIGURE 10-3. For example, it follows from our definition that for an angle of one radian, the arc length s is equal to the radius: $s = r$. Similarly, an angle of two radians corresponds to an arc length of two radii, $s = 2r$, and so on. Thus, the arc length s for an arbitrary angle θ measured in radians is given by the following relationship:

$$s = r\theta \qquad \text{10-2}$$

This simple and straightforward relationship does not hold for degrees or revolutions—additional conversion factors are needed in those cases. An example of the connection among arc length, radius, and angle is shown in FIGURE 10-4.

In one complete revolution, the arc length is the circumference of a circle, $C = 2\pi r$. Comparing with $s = r\theta$, we see that a complete revolution corresponds to 2π radians:

$$1 \text{ rev} = 360° = 2\pi \text{ rad}$$

Equivalently,

$$1 \text{ rad} = \frac{360°}{2\pi} = 57.3°$$

One final note on the units for angles: Radians, as well as degrees and revolutions, are dimensionless. In the relationship $s = r\theta$, for example, the arc length and the radius both have SI units of meters. For the equation to be dimensionally consistent, it is necessary that θ have no dimensions. Still, if an angle θ is, let's say, three radians, we will write it as $\theta = 3$ rad to remind us of the angular units being used.

Angular Velocity, ω

As the bicycle wheel in Figure 10-1 rotates, the angular position of the spot of red paint changes. This is illustrated in FIGURE 10-5. The **angular displacement** of the spot, $\Delta\theta$, is

$$\Delta\theta = \theta_f - \theta_i$$

If we divide the angular displacement by the time, Δt, during which the displacement occurs, the result is the **average angular velocity**, ω_{av}.

Definition of Average Angular Velocity, ω_{av}

$$\omega_{av} = \frac{\Delta\theta}{\Delta t} \qquad \text{10-3}$$

SI unit: radian per second (rad/s) $= \text{s}^{-1}$

This is analogous to the definition of the average linear velocity: $v_{av} = \Delta x/\Delta t$. Notice that the units of linear velocity are m/s, whereas the units of angular velocity are rad/s.

In addition to the average angular velocity, we can define an **instantaneous angular velocity** as the limit of ω_{av} as the time interval, Δt, approaches zero. The instantaneous angular velocity, then, is:

Definition of Instantaneous Angular Velocity, ω

$$\omega = \lim_{\Delta t \to 0} \frac{\Delta\theta}{\Delta t} \qquad \text{10-4}$$

SI unit: rad/s $= \text{s}^{-1}$

Generally, we shall refer to the instantaneous angular velocity simply as the angular velocity.

Notice that we call ω the angular velocity, not the angular speed. The reason is that ω can be positive or negative, depending on the sense of rotation. For example, if the

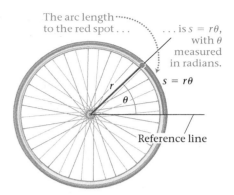

The arc length to the red spot . . . . . . is $s = r\theta$, with θ measured in radians.

$s = r\theta$

Reference line

▲ **FIGURE 10-3 Arc length** The arc length, s, from the reference line to the spot of paint is given by $s = r\theta$ if the angular position θ is measured in radians.

PROBLEM-SOLVING NOTE

Radians

Remember to measure angles in radians when using the relationship $s = r\theta$.

▲ **FIGURE 10-4** The rotating carnival ride in this time exposure illustrates the connection between radius and arc length. As the wheel rotates through a given angle, the lights on the spokes trace out luminous "wedges" that are wider the greater the distance from the axis of rotation—as one would expect from $s = r\theta$.

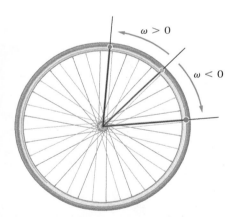

Angular displacement $\Delta\theta$

Reference line

▲ **FIGURE 10-5 Angular displacement** As the wheel rotates, the spot of paint undergoes an angular displacement, $\Delta\theta = \theta_f - \theta_i$.

red paint spot rotates in the counterclockwise sense, the angular position, θ, increases. As a result, $\Delta\theta$ is positive and, therefore, so is ω. Similarly, clockwise rotation corresponds to a negative $\Delta\theta$ and hence a negative ω.

Sign Convention for Angular Velocity
By convention:

$$\omega > 0 \text{ counterclockwise rotation}$$
$$\omega < 0 \text{ clockwise rotation}$$

The sign convention for angular velocity is illustrated in **FIGURE 10-6**. In analogy with linear motion, the sign of ω indicates the *direction* of the angular velocity *vector*, as we shall see in detail in Chapter 11. Similarly, the magnitude of the angular velocity is the **angular speed,** just as in the one-dimensional case.

In the following Exercise we utilize the definitions and conversion factors presented so far in this section.

EXERCISE 10-1 ANGULAR VELOCITY AND SPEED

(a) A wind turbine for generating electricity rotates clockwise at the rate of 17.0 rpm (revolutions per minute). What is its angular velocity in rad/s? **(b)** A CD rotates through an angle of 106° in 0.0860 s. What is the angular speed of the CD in rad/s?

REASONING AND SOLUTION
A rotation rate in revolutions per minute can be converted to rad/s by noting that 1 revolution $= 2\pi$ rad. In addition, knowing the angle $\Delta\theta$ through which an object rotates in a time Δt allows us to calculate the angular velocity with $\theta = \Delta\theta/\Delta t$. Recall that clockwise rotation corresponds to the negative direction, and that 2π rad $= 360°$.

a. Convert from rpm to rad/s. Use the fact that clockwise rotation corresponds to a negative angular velocity:

$$\omega = -17.0 \text{ rpm} = \left(-17.0\frac{\text{rev}}{\text{min}}\right)\left(\frac{2\pi \text{ rad}}{1 \text{ rev}}\right)\left(\frac{1 \text{ min}}{60 \text{ s}}\right) = -1.78 \text{ rad/s}$$

b. Convert $\Delta\theta$ from degrees (°) to radians (rad) in the relationship $\omega = \Delta\theta/\Delta t$. This gives the following result for the angular speed:

$$\omega = \frac{\Delta\theta}{\Delta t} = \frac{106°\left(\dfrac{2\pi \text{ rad}}{360°}\right)}{0.0860 \text{ s}} = 21.5 \text{ rad/s}$$

The same symbol, ω, is used for both angular velocity and angular speed in Exercise 10-1. Which quantity is meant in a given situation will be clear from the context in which it is used.

Angular Velocity and Period As a simple application of angular velocity, consider the following question: If an object rotates with a constant angular velocity, ω, how much time, T, is required for it to complete one full revolution?

To solve this problem, first note that because ω is constant, the instantaneous angular velocity is equal to the average angular velocity. That is,

$$\omega = \omega_{av} = \frac{\Delta\theta}{\Delta t}$$

In one revolution, we know that $\Delta\theta = 2\pi$ and $\Delta t = T$. Therefore,

$$\omega = \frac{\Delta\theta}{\Delta t} = \frac{2\pi}{T}$$

Finally, solving for T we find

$$T = \frac{2\pi}{\omega}$$

The time to complete one revolution, T, is referred to as the **period.**

▲ **FIGURE 10-6 Angular speed and velocity** Counterclockwise rotation is defined to correspond to a positive angular velocity, ω. Clockwise rotation corresponds to a negative angular velocity. The magnitude of the angular velocity is referred to as the angular speed.

PHYSICS IN CONTEXT
Looking Ahead

Just as linear speed is related to linear momentum (Chapter 9), angular speed is related to angular momentum. This is discussed in detail in Chapter 11.

Definition of Period, T

$$T = \frac{2\pi}{\omega}$$ 10-5

SI unit: second, s

The angular speed in Equation 10-5 must be measured in rad/s for the relationship to be valid.

EXERCISE 10-2 **PERIOD OF ROTATION**

Find the period of a microwave tray that is rotating with an angular speed of 5.0 rpm.

REASONING AND SOLUTION
To apply $T = 2\pi/\omega$ we must first express ω in terms of rad/s:

$$\omega = 5.0 \text{ rpm} = \left(5.0\frac{\text{rev}}{\text{min}}\right)\left(\frac{2\pi \text{ rad}}{1 \text{ rev}}\right)\left(\frac{1 \text{ min}}{60 \text{ s}}\right) = 0.52 \text{ rad/s}$$

Now we can calculate the period:

$$T = \frac{2\pi}{\omega} = \frac{2\pi \text{ rad}}{0.52 \text{ rad/s}} = 12 \text{ s}$$

Angular Acceleration, α

If the angular velocity of the rotating bicycle wheel we've been considering increases or decreases with time, we say that the wheel experiences an **angular acceleration,** α. The average angular acceleration is the change in angular velocity, $\Delta\omega$, in a given interval of time, Δt:

Definition of Average Angular Acceleration, α_{av}

$$\alpha_{\text{av}} = \frac{\Delta\omega}{\Delta t}$$ 10-6

SI unit: radian per second per second $(\text{rad/s}^2) = \text{s}^{-2}$

The SI units of α are rad/s^2, which, since rad is dimensionless, is simply s^{-2}. If the angular acceleration α is constant, which is often the case, then the average angular acceleration is simply equal to the angular acceleration itself, $\alpha_{\text{av}} = \alpha$.

As expected, the instantaneous angular acceleration is the limit of α_{av} as the time interval, Δt, approaches zero:

Definition of Instantaneous Angular Acceleration, α

$$\alpha = \lim_{\Delta t \to 0} \frac{\Delta\omega}{\Delta t}$$ 10-7

SI unit: $\text{rad/s}^2 = \text{s}^{-2}$

When referring to the instantaneous angular acceleration, we will usually just say angular acceleration.

The sign of the angular acceleration is determined by whether the change in the angular velocity is positive or negative. For example, if ω is becoming more positive, so that ω_f is greater than ω_i, it follows that α is positive. Similarly, if ω is becoming more negative, so that ω_f is less than ω_i, it follows that α is negative. Therefore, if ω and α have the same sign, the speed of rotation is increasing. If ω and α have opposite signs, the speed of rotation is decreasing. This is illustrated in **FIGURE 10-7**.

EXERCISE 10-3 **ANGULAR ACCELERATION**

A ceiling fan that was rotating with an angular velocity of 12.6 rad/s begins to slow down when it is turned off. How much time does it take for the fan to come to rest if it has a constant angular acceleration of -0.168 rad/s^2?

CONTINUED

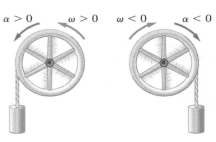

(a) Angular speed increases   **(b)** Angular speed increases

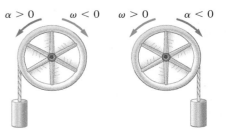

(c) Angular speed decreases   **(d)** Angular speed decreases

▲ **FIGURE 10-7 Angular acceleration and angular speed** When angular velocity and acceleration have the same sign, as in **(a)** and **(b)**, the angular speed increases. When angular velocity and angular acceleration have opposite signs, as in **(c)** and **(d)**, the angular speed decreases.

PHYSICS IN CONTEXT
Looking Back

Notice that our definitions of position, velocity, and acceleration from Chapter 2 are generalized here to apply to rotational motion.

PHYSICS IN CONTEXT
Looking Ahead

In Chapter 11, we relate force to angular acceleration in much the same way that force and acceleration are related in linear motion. This results in the concept of torque.

$\omega = 0$ $\omega < 0$ $\alpha < 0$

$v = 0$

v

$t = 0$ $t > 0$

▲ FIGURE 10-8 **A pulley with constant angular acceleration** A mass is attached to a string wrapped around a pulley. As the mass falls, it causes the pulley to increase its angular speed with a constant angular acceleration.

REASONING AND SOLUTION

We know the initial and final angular velocities, as well as the angular acceleration. The fact that the angular acceleration is constant means that $\alpha_{\mathrm{av}} = \alpha = -0.168 \text{ rad/s}^2$. We can then rearrange Equation 10-6 ($\alpha_{\mathrm{av}} = \Delta\omega/\Delta t$) to find the elapsed time, Δt:

$$\Delta t = \frac{\Delta\omega}{\alpha_{\mathrm{av}}} = \frac{\omega_{\mathrm{f}} - \omega_{\mathrm{i}}}{\alpha} = \frac{0 - 12.6 \text{ rad/s}}{-0.168 \text{ rad/s}^2} = 75.0 \text{ s}$$

Enhance Your Understanding (Answers given at the end of the chapter)

1. Rank the following systems (A, B, C, D) in order of increasing **(a)** angular speed and **(b)** angular velocity. Indicate ties where appropriate. **A:** 2 rad/s clockwise; **B:** 3 rpm counterclockwise; **C:** 100 rpm counterclockwise; **D:** 10 rad/s clockwise.

Section Review

- Angular position is the angle relative to a reference line, angular velocity is the rate of change of angular position, and angular acceleration is the rate of change of angular velocity.

- Rotation in the counterclockwise direction is positive; rotation in the clockwise direction is negative.

10-2 Rotational Kinematics

Just as the kinematics of Chapter 2 described linear motion, rotational kinematics describes rotational motion. In this section, as in Chapter 2, we concentrate on the important special case of constant acceleration.

Angular Velocity with Constant Angular Acceleration As an example of a system with constant angular acceleration, consider the pulley shown in **FIGURE 10-8**. Wrapped around the circumference of the pulley is a string, with a mass attached to its free end. When the mass is released, the pulley begins to rotate—slowly at first, but then faster and faster. As we will see in Chapter 11, the pulley is accelerating with constant angular acceleration.

Because α is constant, it follows that the average and instantaneous angular accelerations are equal. Hence,

$$\alpha = \alpha_{\mathrm{av}} = \frac{\Delta\omega}{\Delta t}$$

Suppose that the pulley starts with the initial angular velocity ω_0 at time $t = 0$, and that at the later time t its angular velocity is ω. Substituting these values into the preceding expression for α yields

$$\alpha = \frac{\Delta\omega}{\Delta t} = \frac{\omega - \omega_0}{t - 0} = \frac{\omega - \omega_0}{t}$$

Rearranging, we see that the angular velocity, ω, varies with time as follows:

$$\omega = \omega_0 + \alpha t \qquad\qquad 10\text{-}8$$

EXERCISE 10-4 ANGULAR KINEMATICS

If the angular velocity of the pulley in Figure 10-8 is -6.43 rad/s at a given time, and its constant angular acceleration is -0.491 rad/s^2, what is the angular velocity of the pulley 1.25 s later?

REASONING AND SOLUTION

The angular velocity, ω, is found by applying $\omega = \omega_0 + \alpha t$:

$$\omega = \omega_0 + \alpha t = -6.43 \text{ rad/s} + (-0.491 \text{ rad/s}^2)(1.25 \text{ s}) = -7.04 \text{ rad/s}$$

The angular speed has *increased*, as expected, because ω and α have the *same sign*.

PROBLEM-SOLVING NOTE

Sign of the Angular Acceleration

The sign of the angular acceleration (by itself) doesn't determine whether a rotating system speeds up or slows down. What really matters is whether the signs of the angular acceleration and the angular velocity are the same (speeds up) or different (slows down).

Rotational and Linear Analogies   Notice the close analogy between $\omega = \omega_0 + \alpha t$ for angular velocity and the corresponding relationship for linear velocity, Equation 2-7:

$$v = v_0 + at$$

Clearly, the equation for angular velocity can be obtained from our previous equation for linear velocity by replacing v with ω and replacing a with α. This type of analogy between linear and angular quantities can be most useful both in deriving angular equations—by starting with linear equations and using analogies—and in obtaining a better physical understanding of angular systems. Several linear-angular analogs are listed in the accompanying table.

Using these analogies, we can rewrite all the kinematic equations in Chapter 2 in angular form. The following table gives both the linear kinematic equations and their angular counterparts.

Linear Quantity	Angular Quantity
x	θ
v	ω
a	α

Big Idea 1 Rotational quantities, like angular position, angular velocity, and angular acceleration, are analogous to the corresponding linear quantities. In addition, the kinematics of rotational motion is essentially the same as the kinematics of linear motion.

Linear Equation (a = constant)		Angular Equation (α = constant)	
$v = v_0 + at$	2-7	$\omega = \omega_0 + \alpha t$	10-8
$x = x_0 + \frac{1}{2}(v_0 + v)t$	2-10	$\theta = \theta_0 + \frac{1}{2}(\omega_0 + \omega)t$	10-9
$x = x_0 + v_0 t + \frac{1}{2}at^2$	2-11	$\theta = \theta_0 + \omega_0 t + \frac{1}{2}\alpha t^2$	10-10
$v^2 = v_0^2 + 2a(x - x_0)$	2-12	$\omega^2 = \omega_0^2 + 2\alpha(\theta - \theta_0)$	10-11

In solving kinematic problems involving rotation, we apply these angular equations in the same way that the linear equations were applied in Chapter 2. In a sense, then, this material is a review—after all, the mathematics is essentially the same. The only difference comes in the physical interpretation of the results. We will emphasize the rotational interpretations throughout the chapter.

PROBLEM-SOLVING NOTE

Rotational Kinematics

Using analogies between linear and angular quantities often helps when solving problems involving rotational kinematics.

EXAMPLE 10-5 | THROWN FOR A CURVE

To throw a curve ball, a pitcher gives the ball an initial angular speed of 36.0 rad/s. Just before the catcher gloves the ball 0.595 s later, its angular speed has decreased (due to air resistance) to 34.2 rad/s. **(a)** What is the ball's angular acceleration, assuming it to be constant? **(b)** How many revolutions does the ball make before being caught?

PICTURE THE PROBLEM
We choose the ball's initial direction of rotation to be positive. As a result, the angular acceleration (which decreases the angular speed) will be negative. We can also identify the initial angular velocity to be $\omega_0 = 36.0$ rad/s, and the final angular velocity to be $\omega = 34.2$ rad/s.

REASONING AND STRATEGY
The problem states that the angular acceleration of the ball is constant. It follows that the equations of angular kinematics (Equations 10-8 to 10-11) apply.

a. To relate angular velocity to time, we use $\omega = \omega_0 + \alpha t$. This can be solved for α.

b. To relate angle to time we use $\theta = \theta_0 + \omega_0 t + \frac{1}{2}\alpha t^2$. The angular displacement of the ball is $\theta - \theta_0$.

Known   Initial angular speed, $\omega_0 = 36.0$ rad/s; final angular speed, $\omega = 34.2$ rad/s; elapsed time, $t = 0.595$ s.
Unknown   **(a)** Angular acceleration, $\alpha = ?$ **(b)** Number of revolutions, $\theta - \theta_0 = ?$

SOLUTION

Part (a)

1. Solve $\omega = \omega_0 + \alpha t$ for the angular acceleration, α:

$$\omega = \omega_0 + \alpha t$$
$$\alpha = \frac{\omega - \omega_0}{t}$$

CONTINUED

2. Substitute numerical values to find α:

$$\alpha = \frac{\omega - \omega_0}{t}$$

$$= \frac{34.2 \text{ rad/s} - 36.0 \text{ rad/s}}{0.595 \text{ s}} = -3.03 \text{ rad/s}^2$$

Part (b)

3. Use $\theta = \theta_0 + \omega_0 t + \frac{1}{2}\alpha t^2$ to calculate the angular displacement of the ball:

$$\theta - \theta_0 = \omega_0 t + \frac{1}{2}\alpha t^2$$
$$= (36.0 \text{ rad/s})(0.595 \text{ s}) + \frac{1}{2}(-3.03 \text{ rad/s}^2)(0.595 \text{ s})^2$$
$$= 20.9 \text{ rad}$$

4. Convert the angular displacement to revolutions:

$$\theta - \theta_0 = 20.9 \text{ rad} = 20.9 \text{ rad}\left(\frac{1 \text{ rev}}{2\pi \text{ rad}}\right) = 3.33 \text{ rev}$$

INSIGHT
The ball rotates through three-and-one-third revolutions during its time in flight.

An alternative method of solution is to use the kinematic relationship given in Equation 10-9. This procedure yields $\theta - \theta_0 = \frac{1}{2}(\omega_0 + \omega)t = 20.9$ rad, in agreement with our previous result.

PRACTICE PROBLEM
(a) What is the angular velocity of the ball 0.500 s after it is thrown? **(b)** What is the ball's angular velocity after it completes its first full revolution?   [**Answer: (a)** Use $\omega = \omega_0 + \alpha t$ to find $\omega = 34.5$ rad/s. **(b)** Use $\omega^2 = \omega_0^2 + 2\alpha(\theta - \theta_0)$ to find $\omega = 35.5$ rad/s.]

Some related homework problems: Problem 16, Problem 19

EXAMPLE 10-6 | WHEEL OF MISFORTUNE

On a certain game show, contestants spin a wheel when it is their turn. One contestant gives the wheel an initial angular speed of 3.40 rad/s. It then rotates through one-and-one-quarter revolutions and comes to rest on the BANKRUPT space. **(a)** Find the angular acceleration of the wheel, assuming it to be constant. **(b)** How much time does it take for the wheel to come to rest?

PICTURE THE PROBLEM
We choose the initial angular velocity to be positive, $\omega_0 = +3.40$ rad/s, and indicate it with a counterclockwise rotation in our sketch. Because the wheel is slowing down, its angular acceleration must have the opposite sign of its angular velocity, which means it is negative (clockwise) in this case. After a rotation of 1.25 rev the wheel will read BANKRUPT.

REASONING AND STRATEGY
As in Example 10-5, we can use the kinematic equations for constant angular acceleration, Equations 10-8 to 10-11.

a. To begin, we are given the initial angular velocity, $\omega_0 = +3.40$ rad/s, the final angular velocity, $\omega = 0$ (the wheel comes to rest), and the angular displacement, $\theta - \theta_0 = 1.25$ rev. We can find the angular acceleration using $\omega^2 = \omega_0^2 + 2\alpha(\theta - \theta_0)$.

b. Knowing the initial angular velocity and acceleration, we can find the time to stop rotating with $\omega = \omega_0 + \alpha t$.

Known   Initial angular speed, $\omega_0 = +3.40$ rad/s; final angular speed, $\omega = 0$; angle of rotation, $\theta - \theta_0 = 1.25$ rev.
Unknown   **(a)** Angular acceleration, $\alpha = ?$ **(b)** Time to stop, $t = ?$

SOLUTION

Part (a)

1. Solve $\omega^2 = \omega_0^2 + 2\alpha(\theta - \theta_0)$ for the angular acceleration, α:

$$\omega^2 = \omega_0^2 + 2\alpha(\theta - \theta_0)$$
$$\alpha = \frac{\omega^2 - \omega_0^2}{2(\theta - \theta_0)}$$

2. Convert $\theta - \theta_0 = 1.25$ rev to radians:

$$\theta - \theta_0 = 1.25 \text{ rev} = 1.25 \text{ rev}\left(\frac{2\pi \text{ rad}}{1 \text{ rev}}\right) = 7.85 \text{ rad}$$

CONTINUED

3. Substitute numerical values to find α:

$$\alpha = \frac{\omega^2 - \omega_0{}^2}{2(\theta - \theta_0)} = \frac{0 - (3.40 \text{ rad/s})^2}{2(7.85 \text{ rad})} = -0.736 \text{ rad/s}^2$$

Part (b)

4. Solve $\omega = \omega_0 + \alpha t$ for the time, t:

$$\omega = \omega_0 + \alpha t$$
$$t = \frac{\omega - \omega_0}{\alpha}$$

5. Substitute numerical values to find t:

$$t = \frac{\omega - \omega_0}{\alpha} = \frac{0 - 3.40 \text{ rad/s}}{(-0.736 \text{ rad/s}^2)} = 4.62 \text{ s}$$

INSIGHT

Notice that we converted the angular displacement in revolutions, 1.25 rev, to radians, 7.85 rad, in Step 2. This ensures that all angular quantities are measured with the same angular units. In addition, notice that it is not necessary to define a reference line—that is, a direction for $\theta = 0$. All we need to know is the angular displacement, $\theta - \theta_0$, not the individual angles θ and θ_0.

PRACTICE PROBLEM

What is the angular speed of the wheel after one complete revolution? [**Answer:** $\omega = 1.52$ rad/s]

Some related homework problems: Problem 15, Problem 17

Enhance Your Understanding   (Answers given at the end of the chapter)

2. An object at rest begins to rotate at $t = 0$ with constant angular acceleration. After a time t the object has rotated through an angle θ and has an angular velocity ω. **(a)** Through what angle has the object rotated in the time $2t$? **(b)** What is the angular velocity of the object at time $2t$?

Section Review

- The basic analogies between linear and angular quantities are as follows: position, x analogous to θ; velocity, v analogous to ω; acceleration, a analogous to α.

- Using the linear-angular analogies, we can convert the kinematic equations of Chapter 2 to kinematic equations for angular motion.

10-3 Connections Between Linear and Rotational Quantities

At a local county fair a child rides on a merry-go-round. The ride completes one revolution, 2π radians, every $T = 7.50$ s. Therefore, the angular speed of the child, from Equation 10-5 ($\omega = 2\pi/T$), is

$$\omega = \frac{2\pi}{T} = \frac{2\pi}{7.50 \text{ s}} = 0.838 \text{ rad/s}$$

The path followed by the child is circular, with the center of the circle at the axis of rotation of the merry-go-round. In addition, at any instant of time the child is moving in a direction that is *tangential* to the circular path, as **FIGURE 10-9** shows. What is the tangential speed, v_t, of the child? In other words, what is the speed of the wind in the child's face?

We can find the child's tangential speed by dividing the distance traveled—the circumference of the circular path, $2\pi r$—by the time required to complete one revolution, T. That is,

$$v_t = \frac{2\pi r}{T} = r\left(\frac{2\pi}{T}\right)$$

Because $2\pi/T$ is simply the angular speed, ω, we can express the tangential speed as follows:

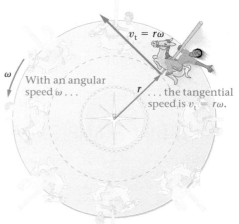

▲ **FIGURE 10-9 Angular and linear speed** Overhead view of a child riding on a merry-go-round. The child's path is a circle centered on the merry-go-round's axis of rotation. At any given time the child is moving tangential to the circular path with a speed $v_t = r\omega$.

▲ **FIGURE 10-10** **Visualizing Concepts**
Tangential Speed In the top photo, the letters on the label are blurred by the rotational motion of the record. Notice that the blurring is greater the greater the distance from the axis of rotation— showing that the linear speed of the record increases with radius, as expected. Similarly, the boy in the bottom photo is moving faster than the girl, because his path is at a greater distance from the axis of rotation.

PHYSICS
IN CONTEXT
Looking Ahead

Though it is a bit surprising at first, rotational motion is directly related to the motion of a pendulum swinging back and forth, and to the motion of a mass oscillating up and down on a spring. These connections are established in Section 13-3.

Tangential Speed of a Rotating Object

$$v_t = r\omega$$

SI unit: m/s   10-12

It is important to note that ω must be given in rad/s for this relationship to be valid.

In the case of the merry-go-round, if the radius of the child's circular path is $r = 4.25$ m, the tangential speed is $v_t = r\omega = (4.25\ \text{m})(0.838\ \text{rad/s}) = 3.56$ m/s. Examples of the dependence of linear speed on radius are shown in **FIGURE 10-10**. When it is clear that we are referring to the tangential speed, we will often drop the subscript t, and simply write $v = r\omega$.

An interesting application of the relationship between linear and angular speeds is provided in the operation of a compact disk (CD). As you may know, a CD is played by shining a laser beam onto the disk, and then converting the pattern of reflected light into a pattern of sound waves. For proper operation, however, the linear speed of the disk where the laser beam shines on it must be maintained at the constant value of 1.25 m/s. As the CD is played, the laser beam scans the disk in a spiral track from near the center outward to the rim. In order to maintain the required linear speed, the angular speed of the disk must decrease as the beam scans outward to larger distances, r. The required angular speeds are determined in the next Exercise.

EXERCISE 10-7   ANGULAR SPEED OF A CD

Find the angular speed a CD must have to give a linear speed of 1.25 m/s when the laser beam shines on the disk **(a)** 2.50 cm and **(b)** 6.00 cm from its center.

REASONING AND SOLUTION

    a. Using $v = 1.25$ m/s and $r = 0.0250$ m in $v_t = r\omega$, we find

$$\omega = \frac{v}{r} = \frac{1.25\ \text{m/s}}{0.0250\ \text{m}} = 50.0\ \text{rad/s} = 477\ \text{rpm}$$

    b. Similarly, with $r = 0.0600$ m we find

$$\omega = \frac{v}{r} = \frac{1.25\ \text{m/s}}{0.0600\ \text{m}} = 20.8\ \text{rad/s} = 199\ \text{rpm}$$

Thus, a CD slows from about 500 rpm to roughly 200 rpm as it plays.

    RWP* How do the angular and tangential speeds of an object vary from one point to another? We explore this question in the following Conceptual Example.

---

*Real World Physics applications are denoted by the acronym RWP.

CONCEPTUAL EXAMPLE 10-8   COMPARE THE SPEEDS   PREDICT/EXPLAIN

Two children ride on a merry-go-round, with child 1 at a greater distance from the axis of rotation than child 2, as shown in our sketch. **(a)** Is the angular speed of child 1 greater than, less than, or the same as the angular speed of child 2? **(b)** Which of the following is the *best explanation* for your prediction?

   I. Child 1 is moving faster than child 2, and hence has the greater angular speed.

  II. Child 2 is closer to the axis of rotation, which results in a greater angular speed.

 III. Both children complete a revolution in the same time, so their angular speeds are the same.

REASONING AND DISCUSSION

At any given time, the angle θ for child 1 is the same as the angle for child 2, as shown. Therefore, when the angle for child 1 has gone through 2π, for example, so has the angle for child 2. As a result, they have the same angular speed. In fact, *each and every point on the merry-go-round has exactly the same angular speed.*

CONTINUED

The tangential speeds are different, however. Child 1 has the greater tangential speed because he travels around a larger circle in the same time that child 2 travels around a smaller circle. This is in agreement with the relationship $v = r\omega$, because the radius to child 1 is greater than the radius to child 2. That is, $v_1 = r_1\omega > v_2 = r_2\omega$.

ANSWER
(a) The angular speeds are the same for the two children. **(b)** The best explanation is III.

Centripetal Acceleration   Because the children on the merry-go-round move in a circular path, they experience a centripetal acceleration, a_{cp} (Section 6-5). The centripetal acceleration is always directed toward the axis of rotation and has a magnitude given by

$$a_{cp} = \frac{v^2}{r}$$

The speed v in this expression is the tangential speed, $v = v_t = r\omega$, and therefore the centripetal acceleration in terms of ω is

$$a_{cp} = \frac{(r\omega)^2}{r}$$

Canceling one power of r, we have:

Centripetal Acceleration of a Rotating Object

$$a_{cp} = r\omega^2 \qquad\qquad \text{10-13}$$

SI unit: m/s^2

If the radius of a child's circular path on the merry-go-round is 4.25 m, and the angular speed of the ride is 0.838 rad/s, the centripetal acceleration of the child is $a_{cp} = r\omega^2 = (4.25\text{ m})(0.838\text{ rad/s})^2 = 2.98\text{ m/s}^2$.

RWP Although the centripetal acceleration of a merry-go-round is typically only a fraction of the acceleration of gravity, rotating devices called **centrifuges** can produce extremely large centripetal accelerations. Two examples of centrifuges of different sizes are shown in **FIGURE 10-11**. The world's most powerful research centrifuge, operated by the U.S. Army Corps of Engineers, can subject 2.2-ton payloads to accelerations as high as 350g (350 times greater than the acceleration of gravity). This centrifuge is used to study earthquake engineering and dam erosion. The Air Force uses centrifuges to subject prospective jet pilots to the accelerations they will experience during rapid flight maneuvers, and in the future NASA may even use a human-powered centrifuge for gravity studies aboard the International Space Station.

The centrifuges most commonly encountered in everyday life are those found in virtually every medical laboratory in the world. These devices, which can produce centripetal accelerations in excess of 13,000g, are used to separate blood cells from blood plasma. They do this by speeding up the natural tendency of cells to settle out of plasma from days to minutes. The ratio of the packed cell volume to the total blood volume gives the *hematocrit value*, which is a useful clinical indicator of blood quality. In the next Example we consider the operation of a *microhematocrit centrifuge*, which measures the hematocrit value of a small (micro) sample of blood.

▲ **FIGURE 10-11 Visualizing Concepts Centripetal Acceleration** The large centrifuge (top), at the Gagarin Cosmonaut Training Center, is used to train Russian cosmonauts for space missions. This device, which rotates at 36 rpm, can produce a centripetal acceleration of over 290 m/s^2, 30 times the acceleration of gravity. The microcentrifuge with Eppendorf tubes (bottom) is used to separate blood cells from plasma. The volume of red blood cells in a given quantity of whole blood is a major factor in determining the oxygen-carrying capacity of the blood.

11,500 rpm

9.07 cm

QUICK EXAMPLE 10-9 THE MICROHEMATOCRIT

BIO In a microhematocrit centrifuge, small samples of blood are placed in heparinized capillary tubes (heparin is an anticoagulant). The tubes are rotated at 11,500 rpm, with the bottoms of the tubes 9.07 cm from the axis of rotation. **(a)** Find the linear speed of the bottom of the tubes. **(b)** What is the centripetal acceleration at the bottom of the tubes?

REASONING AND SOLUTION
Our sketch shows a top view of the centrifuge, with the capillary tubes rotating at 11,500 rpm. Notice that the bottoms of the tubes move in a circular path of radius 9.07 cm.

CONTINUED

a. Linear and angular speeds are related by $v = r\omega$. Once we convert the angular speed to rad/s, we can use this relationship to determine the linear speed v.

b. The centripetal acceleration is $a_{cp} = r\omega^2$. Using ω from part (a) yields the desired result.

Part (a)

1. Convert the angular speed, ω, to radians per second:

$$\omega = (11{,}500 \text{ rev/min})\left(\frac{2\pi \text{ rad}}{1 \text{ rev}}\right)\left(\frac{1 \text{ min}}{60 \text{ s}}\right)$$
$$= 1.20 \times 10^3 \text{ rad/s}$$

2. Use $v = r\omega$ to calculate the linear speed:

$$v = r\omega = (0.0907 \text{ m})(1.20 \times 10^3 \text{ rad/s})$$
$$= 109 \text{ m/s}$$

Part (b)

3. Calculate the centripetal acceleration using $a_{cp} = r\omega^2$:

$$a_{cp} = r\omega^2 = (0.0907 \text{ m})(1.20 \times 10^3 \text{ rad/s})^2$$
$$= 131{,}000 \text{ m/s}^2$$

4. As a check, calculate the centripetal acceleration using $a_{cp} = v^2/r$:

$$a_{cp} = \frac{v^2}{r} = \frac{(109 \text{ m/s})^2}{0.0907 \text{ m}} = 131{,}000 \text{ m/s}^2$$

Every point on a centrifuge tube has the same angular speed. As a result, points near the top of a tube have smaller linear speeds and centripetal accelerations than do points near the bottom of a tube. In this case, the bottoms of the tubes experience a centripetal acceleration about 13,400 times greater than the acceleration of gravity on the surface of the Earth; that is, $a_{cp} = 131{,}000 \text{ m/s}^2 = 13{,}400g$. This means that the force exerted on the bottom of a tube by a 1-ounce sample would be about 800 lb.

Tangential Acceleration If the angular speed of the merry-go-round in Conceptual Example 10-8 changes, the tangential speed of the children changes as well. It follows, then, that the children will experience a tangential acceleration, a_t. We can determine a_t by considering the relationship $v_t = r\omega$. If ω changes by the amount $\Delta\omega$, with r remaining constant, the corresponding change in the tangential speed is

$$\Delta v_t = r\Delta\omega$$

If this change in ω occurs in the time Δt, the tangential acceleration is

$$a_t = \frac{\Delta v_t}{\Delta t} = r\frac{\Delta\omega}{\Delta t}$$

Recalling that $\Delta\omega/\Delta t$ is the angular acceleration, α, we find that

Tangential Acceleration of a Rotating Object

$$a_t = r\alpha \qquad \qquad \text{10-14}$$

SI unit: m/s^2

As with the tangential speed, we will often drop the subscript t in a_t when no confusion will arise.

In general, the children on the merry-go-round may experience both tangential and centripetal accelerations at the same time. Recall that a_t is due to a changing tangential speed, and that a_{cp} is caused by a changing direction of motion, even if the tangential speed remains constant. To summarize:

Tangential Versus Centripetal Acceleration

$$a_t = r\alpha \qquad \text{due to changing angular speed}$$
$$a_{cp} = r\omega^2 \qquad \text{due to changing direction of motion}$$

As the names suggest, the tangential acceleration is always tangential to an object's path; the centripetal acceleration is always perpendicular to its path.

In cases in which both the centripetal and tangential accelerations are present, the total acceleration is the vector sum of the two, as indicated in **FIGURE 10-12**. Notice that

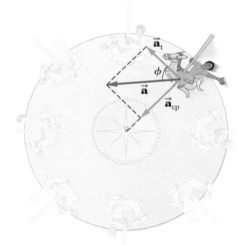

▲ **FIGURE 10-12 Centripetal and tangential accelerations** If the angular speed of the merry-go-round is increased, the child will experience two accelerations: (i) a tangential acceleration, $\vec{a}_t$, and (ii) a centripetal acceleration, $\vec{a}_{cp}$. The child's total acceleration, $\vec{a}$, is the vector sum of $\vec{a}_t$ and $\vec{a}_{cp}$.

$\vec{\mathbf{a}}_t$ and $\vec{\mathbf{a}}_{cp}$ are at right angles to one another, and hence the magnitude of the total acceleration is given by the Pythagorean theorem:

$$a = \sqrt{a_t^2 + a_{cp}^2}$$

The direction of the total acceleration, measured relative to the tangential direction, is

$$\phi = \tan^{-1}\left(\frac{a_{cp}}{a_t}\right)$$

This angle is shown in Figure 10-12.

Enhance Your Understanding   (Answers given at the end of the chapter)

3. Disk 1 has a radius r and rotates with an angular velocity ω. A point on the rim of disk 1 has a tangential speed of 1 m/s and a centripetal acceleration of 3 m/s². Disk 2 has a radius $2r$ and rotates with an angular velocity 2ω. What are **(a)** the tangential speed and **(b)** the centripetal acceleration of a point on the rim of disk 2?

Section Review

- The tangential speed of a point a distance r from the axis of a rotating object is $v_t = r\omega$.

- The magnitude of the centripetal acceleration in terms of angular speed is $a_{cp} = r\omega^2$, and the magnitude of the tangential acceleration in terms of angular acceleration is $a_t = r\alpha$.

10-4 Rolling Motion

We began this chapter with a bicycle wheel rotating about its axle. In that case, the axle was at rest and every point on the wheel, such as the spot of red paint, moved in a circular path about the axle. We would like to consider a different situation now. Suppose the bicycle wheel is rolling freely, as indicated in FIGURE 10-13, with no slipping between the tire and the ground. The wheel still rotates about the axle, but the axle itself is moving in a straight line. As a result, the motion of the wheel is a combination of both rotational motion and linear (or **translational**) motion.

Connecting Rotational and Tangential Motions in a Rolling Wheel To see the connection between the wheel's rotational and translational motions, we show one full rotation of the wheel in Figure 10-13. During this rotation, the axle translates (moves) forward through a distance equal to the circumference of the wheel, $2\pi r$. Because the time required for one rotation is the period, T, the translational speed of the axle is

$$v = \frac{2\pi r}{T}$$

Recalling that $\omega = 2\pi/T$, we find

$$v = r\omega = v_t \qquad\qquad \text{10-15}$$

To roll without slipping . . .

ω

v

. . . the linear (v) and angular (ω) speeds of the wheel must be related by $v = 2\pi r/T = r\omega$.

r

$2\pi r$

◄ FIGURE 10-13 **Rolling without slipping** A wheel of radius r rolling without slipping. During one complete revolution, the center of the wheel moves forward through a distance $2\pi r$.

(a) Pure rotational motion   **(b)** Pure translational motion

▲ **FIGURE 10-14 Rotational and translational motions of a wheel (a)** In pure rotational motion, the velocities at the top and bottom of the wheel are in opposite directions. **(b)** In pure translational motion, each point on the wheel moves with the same speed in the same direction.

▲ **FIGURE 10-15 Velocities in rolling motion** In a wheel that rolls without slipping, the point in contact with the ground is instantaneously at rest. The center of the wheel moves forward with the speed $v = r\omega$, and the top of the wheel moves forward with twice that speed, $v = 2r\omega$.

Hence, the translational speed of the axle is equal to the tangential speed of a point on the rim of a wheel spinning with angular speed ω.

A rolling object, then, combines rotational motion with angular speed ω, and translational motion with linear speed $v = r\omega$, where r is the radius of the object. Let's consider these two motions one at a time. First, in **FIGURE 10-14 (a)** we show pure rotational motion with angular speed ω. In this case, the axle is at rest, and points at the top and bottom of the wheel have tangential velocities that are equal in magnitude, $v = r\omega$, but point in opposite directions.

Next, we consider translational motion with speed $v = r\omega$. This is illustrated in **FIGURE 10-14 (b)**, where we see that each point on the wheel moves in the same direction with the same velocity. If this were the only motion the wheel had, it would be skidding across the ground, instead of rolling without slipping.

Finally, we combine these two motions by simply adding the velocity vectors in Figures 10-14 (a) and (b). The result is shown in **FIGURE 10-15**. At the top of the wheel the two velocity vectors are in the same direction, so they sum to give a speed of $2r\omega$. At the axle, the velocity vectors sum to give a speed $r\omega$. Finally, at the bottom of the wheel, the velocity vectors from rotation and translation have equal magnitude, but are in opposite directions. As a result, these velocities cancel, giving a speed of zero where the wheel is in contact with the ground. An example of a bicycle wheel rolling without slipping is shown in **FIGURE 10-16**. Notice how the spokes are blurry at the top of the wheel but sharp at the bottom.

The fact that the bottom of the wheel is instantaneously at rest, so that it is in static contact with the ground, is precisely what is meant by "rolling without slipping." Thus, a wheel that rolls without slipping is just like the situation when you are walking—even though your body as a whole moves forward, the soles of your shoes are momentarily at rest every time you place them on the ground. This point was discussed in detail in Conceptual Example 6-5.

EXERCISE 10-10   ROLLING TIRES

A car with tires of radius 31 cm drives on the highway at 65 mph. **(a)** What is the angular speed of the tires? **(b)** What is the linear speed of the tops of the tires?

REASONING AND SOLUTION

a. We know the linear speed and the radius of the tires. Thus, we can rearrange $v = r\omega$ to solve for the angular speed, ω. (A useful conversion factor here is 1 mph $= 0.447$ m/s.)

$$\omega = \frac{v}{r} = \frac{(65 \text{ mph})\left(\dfrac{0.447 \text{ m/s}}{1 \text{ mph}}\right)}{0.31 \text{ m}} = 94 \text{ rad/s}$$

This is about 15 revolutions per second.

b. The tops of the tires have a speed of $2v = 130$ mph.

◄ **FIGURE 10-16** This photograph of a rolling wheel gives a visual indication of the speed of its various parts. The bottom of the wheel is at rest at any instant, so the image there is sharp. The top of the wheel has the greatest speed, and the image there shows the most blurring. (Compare Figure 10-15.)

Enhance Your Understanding

(Answers given at the end of the chapter)

4. In a popular pirate movie, two pirates find themselves on a runaway water wheel, as shown in **FIGURE 10-17**. The wheel rolls without slipping to the right with a linear speed v. Pirate 1 at the bottom of the wheel runs to the right with a speed v_1, and pirate 2 on top of the wheel runs to the left with a speed v_2. Each pirate runs with the speed necessary to maintain a fixed position relative to the axle of the wheel as it rolls. What are the running speeds of **(a)** pirate 1 and **(b)** pirate 2 in terms of v? **(c)** What are the speeds of the two pirates relative to the ground?

Running with speed v_2 to the left

v

Running with speed v_1 to the right

▲ **FIGURE 10-17** Enhance Your Understanding 4.

Section Review

* A wheel that rolls without slipping has zero speed at the bottom of the wheel, speed v at the axle, and speed $2v$ at the top of the wheel.

10-5 Rotational Kinetic Energy and the Moment of Inertia

An object in motion has kinetic energy, whether that motion is translational, rotational, or a combination of the two. In translational motion, for example, the kinetic energy of a mass m moving with a speed v is $K = \frac{1}{2}mv^2$. We cannot use this expression for a rotating object, however, because the speed v of each particle within a rotating object varies with its distance r from the axis of rotation. Thus, there is no unique value of v for an entire rotating object. On the other hand, there *is* a unique value of ω, the angular speed, and it applies to all particles in the object.

Kinetic Energy in Terms of Angular Speed To see how the kinetic energy of a rotating object depends on its angular speed, we start with a particularly simple system consisting of a rod of negligible mass and length r rotating about one end with an angular speed ω. Attached to the other end of the rod is a point mass m, as **FIGURE 10-18** shows. To find the kinetic energy of the mass, recall that its linear speed is $v = r\omega$ (Equation 10-12). Therefore, the translational kinetic energy of the mass m is

$$K = \tfrac{1}{2}mv^2 = \tfrac{1}{2}m(r\omega)^2 = \tfrac{1}{2}(mr^2)\omega^2 \qquad \text{10-16}$$

Notice that the kinetic energy of the mass depends not only on the angular speed squared (analogous to the way the translational kinetic energy depends on the linear speed squared), but also on the radius squared—that is, the kinetic energy depends on the *distribution* of mass in the rotating object. To be specific, mass near the axis of rotation contributes little to the kinetic energy because its speed ($v = r\omega$) is small. On the other hand, the farther a mass is from the axis of rotation, the greater its speed v for a given angular velocity, and thus the greater its kinetic energy.

You've probably noticed that the kinetic energy in Equation 10-16 is similar in form to the translational kinetic energy. Instead of $\frac{1}{2}(m)v^2$, we now have $\frac{1}{2}(mr^2)\omega^2$. Clearly, then, the quantity mr^2 plays the role of the mass for the rotating object. This "rotational mass" is given a special name in physics: the **moment of inertia**, I. Thus, in general, the kinetic energy of an object rotating with an angular speed ω can be written as:

Rotational Kinetic Energy

$$K = \tfrac{1}{2}I\omega^2 \qquad \text{10-17}$$

SI unit: J

Big Idea 2 Rolling motion is a combination of linear and rotational motions. When a wheel rolls without slipping with a linear speed v, the rim of the wheel has an instantaneous speed of zero where it touches the ground and an instantaneous speed of $2v$ at the top of the wheel.

Big Idea 3 A rotating object has kinetic energy, just like an object in linear motion. The kinetic energy of rotation can be used in energy conservation in the same way as linear kinetic energy.

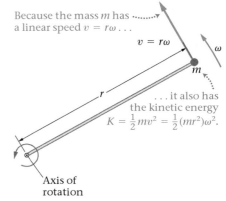

Because the mass m has a linear speed $v = r\omega$ . . .

$v = r\omega$

ω

m

r

. . . it also has the kinetic energy $K = \frac{1}{2}mv^2 = \frac{1}{2}(mr^2)\omega^2$.

Axis of rotation

▲ **FIGURE 10-18 Kinetic energy of a rotating object** As this rod rotates about the axis of rotation with an angular speed ω, the mass has a speed of $v = r\omega$. It follows that the kinetic energy of the mass is $K = \frac{1}{2}mv^2 = \frac{1}{2}(mr^2)\omega^2$.

▲ FIGURE 10-19 **Kinetic energy of a rotating object of arbitrary shape** To calculate the kinetic energy of an object of arbitrary shape as it rotates about an axis with angular speed ω, imagine dividing it into small mass elements, m_i. The total kinetic energy of the object is the sum of the kinetic energies of all the mass elements.

PHYSICS IN CONTEXT
Looking Back

The kinetic energy, first defined in Chapter 7, plays a key role in defining the moment of inertia.

The greater the moment of inertia, the greater an object's rotational kinetic energy. As we have just seen, in the special case of a point mass m a distance r from the axis of rotation, the moment of inertia is simply $I = mr^2$.

Determining the Moment of Inertia We now show how to find the moment of inertia for an object of arbitrary, fixed shape, as in FIGURE 10-19. Suppose, for example, that this object rotates about the axis indicated in the figure with an angular speed ω. To calculate the kinetic energy of the object, we first imagine dividing it into a collection of small mass elements, m_i. We then calculate the kinetic energy of each element and sum over all elements.

Following this plan, we find that the total kinetic energy of an arbitrary rotating object is

$$K = \sum \left(\tfrac{1}{2} m_i v_i^2 \right)$$

In this expression, m_i is the mass of one of the small mass elements and v_i is its speed. If m_i is at the radius r_i from the axis of rotation, as indicated in Figure 10-19, its speed is $v_i = r_i \omega$. Notice that it is not necessary to write a separate angular speed, ω_i, for each element, because *all* mass elements of the object have exactly the same angular speed, ω. Therefore,

$$K = \sum \left(\tfrac{1}{2} m_i r_i^2 \omega^2 \right) = \tfrac{1}{2} \left(\sum m_i r_i^2 \right) \omega^2$$

Now, in analogy with our results for the single mass, we can define the moment of inertia, I, as follows:

Definition of Moment of Inertia, *I*

$$I = \sum m_i r_i^2$$

10-18

SI unit: kg · m²

The precise value of I for a given object depends on its distribution of mass. A simple example of this dependence is given in the following Exercise.

EXERCISE 10-11 MOMENT OF INERTIA

Use the general definition of the moment of inertia, $I = \sum m_i r_i^2$, to find the moment of inertia for the dumbbell-shaped object shown in FIGURE 10-20. Notice that the axis of rotation goes through the center of the object and points out of the page. In addition, assume that the masses may be treated as point masses.

REASONING AND SOLUTION
Referring to Figure 10-20, we see that $m_1 = m_2 = m$ and $r_1 = r_2 = r$. Therefore, the moment of inertia is

$$I = \sum m_i r_i^2 = m_1 r_1^2 + m_2 r_2^2 = mr^2 + mr^2 = 2mr^2$$

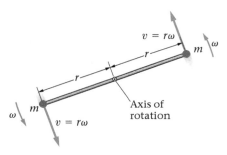

▲ FIGURE 10-20 **A dumbbell-shaped object rotating about its center**

The connection between rotational kinetic energy and the moment of inertia is explored in more detail in the following Example.

EXAMPLE 10-12 NOSE TO THE GRINDSTONE

A grindstone with a radius of 0.610 m is being used to sharpen an ax. **(a)** If the linear speed of the stone relative to the ax is 1.50 m/s, and the stone's rotational kinetic energy is 13.0 J, what is its moment of inertia? **(b)** If the linear speed is doubled to 3.00 m/s, what is the corresponding kinetic energy of the grindstone?

PICTURE THE PROBLEM
Our sketch shows the grindstone spinning with an angular speed ω, which is not given in the problem statement. We do know, however, that the linear speed of the grindstone at its rim is $v = 1.50$ m/s and that its radius is $r = 0.610$ m. At this rate of rotation, the stone has a kinetic energy of 13.0 J.

CONTINUED

REASONING AND STRATEGY

a. Recall that rotational kinetic energy and moment of inertia are related by $K = \frac{1}{2}I\omega^2$; thus $I = 2K/\omega^2$. We are not given ω, but we can find it from the connection between linear and angular speed, $v = r\omega$. Thus, we begin by finding ω. We then use ω, along with the kinetic energy K, to find I.

b. We can find the new angular speed with $\omega = v/r$. We can then use I from part (a), along with $K = \frac{1}{2}I\omega^2$, to find the new kinetic energy.

Known   Radius, $r = 0.610$ m; linear speed, $v = 1.50$ m/s; kinetic energy, $K = 13.0$ J.
Unknown   (a) Moment of inertia, $I = ?$ (b) $K = ?$ for $v = 3.00$ m/s

SOLUTION

Part (a)

1. Find the angular speed of the grindstone:

$$\omega = \frac{v}{r} = \frac{1.50 \text{ m/s}}{0.610 \text{ m}} = 2.46 \text{ rad/s}$$

2. Solve for the moment of inertia in terms of kinetic energy:

$$K = \frac{1}{2}I\omega^2 \text{ or } I = \frac{2K}{\omega^2}$$

3. Substitute numerical values for K and ω:

$$I = \frac{2K}{\omega^2} = \frac{2(13.0 \text{ J})}{(2.46 \text{ rad/s})^2} = 4.30 \text{ J} \cdot \text{s}^2 = 4.30 \text{ kg} \cdot \text{m}^2$$

Part (b)

4. Find the angular speed of the grindstone corresponding to $v = 3.00$ m/s:

$$\omega = \frac{v}{r} = \frac{3.00 \text{ m/s}}{0.610 \text{ m}} = 4.92 \text{ rad/s}$$

5. Determine the kinetic energy, K, using the moment of inertia, I, from part (a):

$$K = \frac{1}{2}I\omega^2 = \frac{1}{2}(4.30 \text{ kg} \cdot \text{m}^2)(4.92 \text{ rad/s})^2 = 52.0 \text{ J}$$

INSIGHT

(a) We found I by relating it to the rotational kinetic energy of the grindstone. Later in this section we show how to calculate the moment of inertia of a disk directly, given its radius and mass. (b) Doubling the linear speed, v, results in a doubling of the angular speed, ω. The kinetic energy K depends on ω^2; therefore doubling ω increases K by a factor of 4—from 13.0 J to 4(13.0 J) = 52.0 J.

PRACTICE PROBLEM — PREDICT/CALCULATE

When the ax is pressed firmly against the grindstone for sharpening, the angular speed of the grindstone decreases. (a) If the rotational kinetic energy of the grindstone is cut in half to 6.50 J, is the corresponding angular speed greater than, less than, or equal to one-half the original angular speed of 2.46 rad/s? Explain. (b) Find the new angular speed. [**Answer:** (a) The rotational kinetic energy depends on the square of the angular speed. Therefore, reducing the kinetic energy by a factor of 2 corresponds to reducing the angular speed by a factor of $\sqrt{2}$. Thus, the new angular speed is greater than ½ the initial angular speed. (b) The new angular speed (using $K = 6.50$ J) is $\omega = \sqrt{2K/I} = 1.74$ rad/s, which is greater than ½(2.46 rad/s) = 1.23 rad/s.]

Some related homework problems: Problem 47, Problem 48, Problem 51

The Moment of Inertia for Various Shapes   We return now to the dependence of the moment of inertia on the particular shape, or mass distribution, of an object. Suppose, for example, that a mass M is formed into the shape of a *hoop* of radius R. In addition, consider the case where the axis of rotation is perpendicular to the plane of the hoop and passes through its center, as shown in **FIGURE 10-21**. This is similar to a bicycle wheel rotating about its axle, if one ignores the spokes. In terms of small mass elements, we can write the moment of inertia as

$$I = \sum m_i r_i^2$$

Each mass element of the hoop, however, is at the same radius R from the axis of rotation; that is, $r_i = R$. Hence, the moment of inertia in this case is

$$I = \sum m_i r_i^2 = \sum m_i R^2 = \left(\sum m_i\right)R^2$$

Clearly, the sum of all the elementary masses is simply the total mass of the hoop, $\sum m_i = M$. Therefore, the moment of inertia of a hoop of mass M and radius R is

$$I = MR^2 \text{ (hoop)}$$

▲ **FIGURE 10-21 The moment of inertia of a hoop** Consider a hoop of mass M and radius R. Each small mass element is at the same distance, R, from the center of the hoop. The moment of inertia in this case is $I = MR^2$.

▲ **FIGURE 10-22 The moment of inertia of a disk** Consider a disk of mass M and radius R. Mass elements for the disk are at distances from the center ranging from 0 to R. The moment of inertia in this case is $I = \frac{1}{2}MR^2$.

In contrast, if the same mass, M, is formed into a uniform *disk* of the same radius, R, the moment of inertia is different. To see this, notice that it is no longer true that $r_i = R$ for all mass elements. In fact, most of the mass elements are closer to the axis of rotation than was the case for the hoop, as indicated in FIGURE 10-22. Thus, most of the r_i are less than R, and hence the moment of inertia will be smaller for the disk than for the hoop. A detailed calculation, summing over all mass elements, yields the following result:

$$I = \tfrac{1}{2}MR^2 \text{ (disk)}$$

As expected, I is less for the disk than for the hoop.

EXERCISE 10-13 UNIFORM DISK

If the grindstone in Example 10-12 is a uniform disk, what is its mass?

REASONING AND SOLUTION
Solving $I = \frac{1}{2}MR^2$ for the mass yields

$$M = \frac{2I}{R^2} = \frac{2(4.30\,\text{kg} \cdot \text{m}^2)}{(0.610\,\text{m})^2} = 23.1\,\text{kg}$$

Thus, the grindstone has a weight of roughly 51 lb.

Table 10-1 collects moments of inertia for a variety of objects. Notice that in all cases the moment of inertia is of the form $I = (\text{constant})MR^2$. It is only the numerical constant in front of MR^2 that changes from one object to another.

Note also that objects of the same general shape but with different mass distributions—such as solid and hollow spheres—have different moments of inertia. In particular, a hollow sphere has a larger I than a solid sphere of the same mass, for the same reason that a hoop's moment of inertia is greater than a disk's: More of its mass is at a greater distance from the axis of rotation. Thus, I is a measure of both the shape *and* the mass distribution of an object.

RWP Moment of Inertia of the Earth For a good example of how the moment of inertia depends on the distribution of mass, consider the moment of inertia of the Earth. If the Earth were a uniform sphere of mass M_E and radius R_E, its moment of inertia

TABLE 10-1 Moments of Inertia for Uniform, Rigid Objects of Various Shapes and Total Mass M

Hoop or cylindrical shell
$I = MR^2$

Disk or solid cylinder
$I = \frac{1}{2}MR^2$

Disk or solid cylinder (axis at rim)
$I = \frac{3}{2}MR^2$

Long thin rod (axis through midpoint)
$I = \frac{1}{12}ML^2$

Long thin rod (axis at one end)
$I = \frac{1}{3}ML^2$

Hollow sphere
$I = \frac{2}{3}MR^2$

Solid sphere
$I = \frac{2}{5}MR^2$

Solid sphere (axis at rim)
$I = \frac{7}{5}MR^2$

Solid plate (axis through center, in plane of plate)
$I = \frac{1}{12}ML^2$

Solid plate (axis perpendicular to plane of plate)
$I = \frac{1}{12}M(L^2 + W^2)$

would be $\frac{2}{5}M_E R_E^2 = 0.4 M_E R_E^2$. In fact, the Earth's moment of inertia is only $0.331 M_E R_E^2$, considerably less than for a uniform sphere. This is due to the fact that the Earth is not homogeneous, but instead has a dense inner core surrounded by a less dense outer core and an even less dense mantle, as shown in FIGURE 10-23. The resulting concentration of mass near its axis of rotation gives the Earth a much smaller moment of inertia than it would have if its mass were uniformly distributed.

On the other hand, if the polar ice caps were to melt and release their water into the oceans, the Earth's moment of inertia would increase. This is because mass that had been near the axis of rotation (in the polar ice) would now be distributed more or less uniformly around the Earth (in the oceans). With more of the Earth's mass at greater distances from the axis of rotation, the moment of inertia would increase. If such an event were to occur, not only would the moment of inertia increase, but the length of the day would increase as well. We will discuss the reasons for this—which have to do with the conservation of angular momentum—in the next chapter.

The moment of inertia of an object also depends on the location and orientation of the axis of rotation. If the axis of rotation is moved, all of the r_i change, leading to a different result for I. This is investigated for a dumbbell system in the following Conceptual Example.

▲ **FIGURE 10-23** The distribution of mass in the Earth is not uniform. Dense materials, like iron and nickel, have concentrated near the center, while less dense materials, like silicon and aluminum, have risen to the surface. This concentration of mass near the axis of rotation lowers the Earth's moment of inertia.

CONCEPTUAL EXAMPLE 10-14　COMPARE THE MOMENTS OF INERTIA

If the dumbbell-shaped object in Figure 10-20 is rotated about one end, is its moment of inertia greater than, less than, or the same as the moment of inertia about its center? As before, assume that the masses can be treated as point masses.

REASONING AND DISCUSSION

As we saw in Exercise 10-11, the moment of inertia about the center of the dumbbell is $I = 2mR^2$. When the axis is at one end, that mass is at the radius $r = 0$, and the other mass is at $r = 2R$. Therefore, the moment of inertia is

$$I = \sum m_i r_i^2 = m \cdot 0 + m(2R)^2 = 4mR^2$$

Thus, the moment of inertia doubles when the axis of rotation is moved from the center to one end.

The reason I increases is that the moment of inertia depends on the radius squared. Hence, even small increases in r can cause significant increases in I. By moving the axis to one end, the radius to the other mass is increased to its greatest possible value. As a result, I increases.

ANSWER

The moment of inertia is greater about one end than about the center.

Extending Linear-Rotational Analogies to Mass and Kinetic Energy　In the accompanying table, we summarize the similarities between the translational kinetic energy, $K = \frac{1}{2}mv^2$, and the rotational kinetic energy, $K = \frac{1}{2}I\omega^2$. As expected, we see that the linear speed, v, has been replaced with the angular speed, ω. In addition, notice that the mass, m, has been replaced with the moment of inertia, I.

As suggested by these analogies, the moment of inertia, I, plays the same role in rotational motion that mass plays in translational motion. For example, the larger I, the more resistant an object is to any change in its angular velocity—an object with a large I is difficult to start rotating, and once it is rotating, it is difficult to stop. We shall see further applications of this analogy in the next chapter when we consider angular momentum.

RWP　Rotational kinetic energy can be used to improve the energy efficiency of an automobile. Whenever the car slows down, its kinetic energy can be temporarily stored by rotating a flywheel at a high angular speed. When the driver releases the brake, the energy stored in the flywheel's rotation is transferred to the rear wheels by a specially designed transmission. An example of this type of energy-recovery system is shown in FIGURE 10-24. Volvo Car Group has extensively tested the technology on public roads with an experimental version of their S60 car. The measurements indicate that the

Linear Quantity	Angular Quantity
v	ω
m	I
$\frac{1}{2}mv^2$	$\frac{1}{2}I\omega^2$

▲ FIGURE 10-24 **Flywheel technology for energy recovery** A rapidly spinning flywheel in the circular structure on the passenger side of the car stores energy from braking and uses it to help accelerate the car from a stop.

flywheel energy storage can reduce fuel consumption by up to 25%, especially in urban start-and-stop traffic.

Enhance Your Understanding
(Answers given at the end of the chapter)

5. Four masses m are connected by rods of negligible mass in a rectangular arrangement, as shown in **FIGURE 10-25**. The system can be rotated about one of four different axes of rotation (1, 2, 3, or 4). Rank the four axes of rotation in order of increasing moment of inertia of the rotating system. Indicate ties where appropriate.

▲ FIGURE 10-25 Enhance Your Understanding 5.

Section Review

- Rotational kinetic energy is $\frac{1}{2}I\omega^2$, in close analogy to linear kinetic energy $\frac{1}{2}mv^2$.
- The moment of inertia I is the rotational analog to the mass in a linear system. The greater the mass, and the greater its distance from the axis of rotation, the greater the moment of inertia.

10-6 Conservation of Energy

In this section, we consider the mechanical energy of objects that roll without slipping, and show how to apply energy conservation to such systems. In addition, we consider objects that rotate as a string or rope unwinds: for example, a pulley with a string wrapped around its circumference, or a yo-yo with a string wrapped around its axle. As long as the unwinding process and the rolling motion occur without slipping, the two situations are basically the same—at least as far as energy considerations are concerned.

Energy Conservation for Rolling Objects To apply energy conservation to rolling objects, we first need to determine the kinetic energy of rolling motion. In Section 10-4 we saw that rolling motion is a combination of rotation and translation. It follows, then, that the kinetic energy of a rolling object is simply the sum of its translational kinetic energy, $\frac{1}{2}mv^2$, and its rotational kinetic energy, $\frac{1}{2}I\omega^2$:

Kinetic Energy of Rolling Motion

$$K = \tfrac{1}{2}mv^2 + \tfrac{1}{2}I\omega^2 \qquad\qquad 10\text{-}19$$

The moment of inertia I in this expression is the moment of inertia about the center of the rolling object.

We can simplify the expression for the kinetic energy of a rolling object by using the fact that linear and angular speeds are related. In fact, recall that $v = r\omega$ (Equation 10-12), which can be rewritten as $\omega = v/r$. Substituting this into our expression for the rolling kinetic energy yields

Kinetic Energy of Rolling Motion: Alternative Form

$$K = \tfrac{1}{2}mv^2 + \tfrac{1}{2}I\left(\frac{v}{r}\right)^2 = \tfrac{1}{2}mv^2\left(1 + \frac{I}{mr^2}\right) \qquad 10\text{-}20$$

Noting that $I = (\text{constant})mr^2$, we see that the last term in Equation 10-20 is a constant that depends on the shape and mass distribution of the rolling object.

A special case of some interest is the point particle. In this case, by definition, all of the mass is at a single point. Therefore, $r = 0$, and hence $I = 0$. Substituting $I = 0$ in either Equation 10-19 or Equation 10-20 yields $K = \frac{1}{2}mv^2$, as expected.

Next, we apply Equations 10-19 and 10-20 to a hollow sphere (a basketball) that rolls without slipping.

PHYSICS IN CONTEXT Looking Back

Conservation of energy, first introduced in Chapter 8, is just as important in rotational motion as it is in linear motion.

EXAMPLE 10-15 LIKE A ROLLING BASKETBALL

A basketball is basically a hollow sphere with a mass of 0.650 kg and a radius of 12.1 cm. If a basketball rolls without slipping with a linear speed of 1.33 m/s, find **(a)** the translational kinetic energy, **(b)** the rotational kinetic energy, and **(c)** the total kinetic energy of the ball.

PICTURE THE PROBLEM

The ball rolls without slipping, and hence its angular speed and linear speed are related by $v = r\omega$. The linear speed is $v = 1.33$ m/s and the radius is $r = 12.1$ cm. We are also given that the mass of the ball is 0.650 kg.

REASONING AND STRATEGY

We calculate each contribution to the kinetic energy separately. The translational kinetic energy, of course, is simply $\frac{1}{2}mv^2$. For the rotational kinetic energy, $\frac{1}{2}I\omega^2$, we use the fact that the moment of inertia for a hollow sphere is $I = \frac{2}{3}mr^2$. Finally, the ball rolls without slipping, and hence its angular speed is $\omega = v/r$. The total kinetic energy is the sum of the translational and rotational kinetic energies.

Known   Mass of ball, $m = 0.650$ kg; radius, $r = 12.1$ cm; linear speed, $v = 1.33$ m/s; moment of inertia, $I = \frac{2}{3}mr^2$.
Unknown **(a)** Translational kinetic energy $= ?$ **(b)** Rotational kinetic energy $= ?$ **(c)** Total kinetic energy, $K = ?$

SOLUTION

Part (a)

1. Calculate the translational kinetic energy, $\frac{1}{2}mv^2$:

$$\frac{1}{2}mv^2 = \frac{1}{2}(0.650\text{ kg})(1.33\text{ m/s})^2 = 0.575\text{ J}$$

Part (b)

2. Calculate the rotational kinetic energy symbolically, using $I = \frac{2}{3}mr^2$ and $\omega = v/r$. Notice that the radius cancels:

$$\frac{1}{2}I\omega^2 = \frac{1}{2}\left(\frac{2}{3}mr^2\right)\left(\frac{v}{r}\right)^2 = \frac{2}{3}\left(\frac{1}{2}mv^2\right) = \frac{1}{3}mv^2$$

3. Substitute the numerical values:

$$\frac{1}{2}I\omega^2 = \frac{1}{3}mv^2$$
$$= \frac{1}{3}(0.650\text{ kg})(1.33\text{ m/s})^2 = 0.383\text{ J}$$

Part (c)

4. Sum the kinetic energies obtained in parts (a) and (b):

$$K = 0.575\text{ J} + 0.383\text{ J} = 0.958\text{ J}$$

5. The same result is obtained using $K = \frac{1}{2}mv^2\left(1 + \frac{I}{mr^2}\right)$:

$$K = \frac{1}{2}mv^2\left(1 + \frac{I}{mr^2}\right) = \frac{1}{2}mv^2\left(1 + \frac{2}{3}\right)$$
$$= \frac{5}{3}\left(\frac{1}{2}mv^2\right) = \frac{5}{3}(0.575\text{ J}) = 0.958\text{ J}$$

INSIGHT

The symbolic result in Step 2 shows that the rotational kinetic energy of a hollow sphere rolling without slipping is precisely two-thirds the sphere's translational kinetic energy. This result is independent of the sphere's radius, as we can see by the cancellation of the radius r in Step 2.

PRACTICE PROBLEM — PREDICT/CALCULATE

Suppose the hollow sphere in this Example is replaced with a uniform disk of the same mass and radius. **(a)** If the disk rolls without slipping with the same speed as the hollow sphere, is its rotational kinetic energy greater than, less than, or equal to the rotational kinetic energy of the sphere? Explain. **(b)** Find the rotational kinetic energy of the disk. [**Answer: (a)** The disk has a smaller moment of inertia than the hollow sphere, and hence its rotational kinetic energy is smaller as well. **(b)** The disk's rotational kinetic energy is 0.287 J, which is less than the hollow sphere's rotational kinetic energy of 0.383 J.]

Some related homework problems: Problem 54, Problem 68, Problem 71

CONCEPTUAL EXAMPLE 10-16 COMPARE KINETIC ENERGIES

A solid sphere and a hollow sphere of the same mass and radius roll without slipping at the same speed. Is the kinetic energy of the solid sphere greater than, less than, or the same as the kinetic energy of the hollow sphere?

CONTINUED

REASONING AND DISCUSSION
Both spheres have the same translational kinetic energy, since they have the same mass and speed. The rotational kinetic energy, however, is proportional to the moment of inertia. The hollow sphere has the greater moment of inertia, and hence it has the greater kinetic energy.

ANSWER
The solid sphere has less kinetic energy than the hollow sphere.

▲ FIGURE 10-26 **An object rolls down an incline** An object starts at rest at the top of an inclined plane and rolls without slipping to the bottom. The speed of the object at the bottom depends on its moment of inertia—a larger moment of inertia results in a lower speed.

Canned Food Race

Energy Conservation with Rolling Motion   Now that we can calculate the kinetic energy of rolling motion, we show how to apply it to energy conservation. For example, consider an object of mass m, radius r, and moment of inertia I at the top of a ramp, as shown in FIGURE 10-26. The object is released from rest and allowed to roll to the bottom, a vertical height h below the starting point. What is the object's speed on reaching the bottom?

The simplest way to solve this problem is to use energy conservation. To do so, we set the initial mechanical energy at the top (i) equal to the final mechanical energy at the bottom (f). That is,

$$K_i + U_i = K_f + U_f$$

Because we are dealing with rolling motion, the kinetic energy is

$$K = \tfrac{1}{2}mv^2\left(1 + \frac{I}{mr^2}\right)$$

The potential energy is simply that due to the uniform gravitational field. Therefore,

$$U = mgy$$

With $y = h$ at the top of the ramp and the object starting at rest, we have

$$K_i + U_i = 0 + mgh = mgh$$

Similarly, with $y = 0$ at the bottom of the ramp and the object rolling with a speed v, we find

$$K_f + U_f = \tfrac{1}{2}mv^2\left(1 + \frac{I}{mr^2}\right) + 0 = \tfrac{1}{2}mv^2\left(1 + \frac{I}{mr^2}\right)$$

Setting the initial and final energies equal yields

$$mgh = \tfrac{1}{2}mv^2\left(1 + \frac{I}{mr^2}\right)$$

Solving for v, we find

$$v = \sqrt{\frac{2gh}{1 + \dfrac{I}{mr^2}}}$$

Let's quickly check one special case—namely, $I = 0$. With this substitution, we find

$$v = \sqrt{2gh}$$

This is the speed an object would have after falling straight down with no rotation through a distance h. Thus, setting $I = 0$ means there is no rotational kinetic energy, and hence the result is the same as for a point particle. As I becomes larger, the speed at the bottom of the ramp is smaller.

CONCEPTUAL EXAMPLE 10-17   WHICH OBJECT WINS THE RACE?

A disk and a hoop of the same mass and radius are released at the same time at the top of an inclined plane. Does the disk reach the bottom of the plane before, after, or at the same time as the hoop?

CONTINUED

REASONING AND DISCUSSION

As we've just seen, the larger the moment of inertia, I, the smaller the speed, v. Hence the object with the larger moment of inertia (the hoop in this case) loses the race to the bottom, because its speed is less than the speed of the disk at any given height.

Another way to think about this is to recall that both objects have the same mechanical energy to begin with—namely, mgh. For the hoop, more of this initial potential energy goes into rotational kinetic energy, because the hoop has the larger moment of inertia; therefore, less energy is left for translational motion. As a result, the hoop moves more slowly and loses the race.

ANSWER

The disk wins the race by reaching the bottom before the hoop.

In the next Conceptual Example, we consider the effects of a surface that changes from nonslip to frictionless.

CONCEPTUAL EXAMPLE 10-18   COMPARE HEIGHTS

A ball is released from rest on a no-slip surface, as shown. After reaching its lowest point, the ball begins to rise again, this time on a frictionless surface. When the ball reaches its maximum height on the frictionless surface, is it at a greater height, at a lesser height, or at the same height as when it was released?

REASONING AND DISCUSSION

As the ball descends on the no-slip surface, it begins to rotate, increasing its angular speed until it reaches the lowest point of the surface. When it begins to rise again, there is no friction to slow the rotational motion; thus, the ball continues to rotate with the same angular speed it had at its lowest point. Therefore, some of the ball's initial gravitational potential energy remains in the form of rotational kinetic energy. As a result, less energy is available to be converted back into gravitational potential energy, and the height is less.

ANSWER

The height on the frictionless side is less.

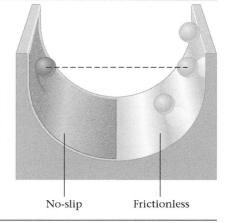

No-slip   Frictionless

Energy Conservation with Pulleys   We can also apply energy conservation to the case of a pulley, or similar object, with a string that winds or unwinds without slipping. In such cases, the relationship $v = r\omega$ is valid and we can follow the same methods applied to an object that rolls without slipping.

EXAMPLE 10-19   SPINNING WHEEL

A block of mass m is attached to a string that is wrapped around the circumference of a wheel of radius R and moment of inertia I. The wheel rotates freely about its axis and the string wraps around its circumference without slipping. Initially the wheel rotates with an angular speed ω, causing the block to rise with a linear speed v. To what height does the block rise before coming to rest? Give a symbolic answer.

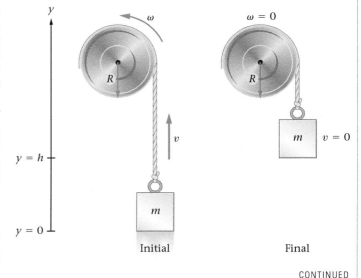

PICTURE THE PROBLEM

In our sketch, we choose the origin of the y axis to be at the initial height of the block. The positive y direction, as usual, is chosen to be upward. When the block comes to rest, then, it is at the height $y = h > 0$, where h is to be determined from the initial speed of the block and the properties of the wheel.

REASONING AND STRATEGY

The problem statement gives two key pieces of information. First, the string wraps onto the disk without slipping; therefore, $v = R\omega$. Second, the wheel rotates freely, which means that the

CONTINUED

mechanical energy of the system is conserved. Thus, at the height h the initial kinetic energy of the system has been converted to gravitational potential energy. This condition can be used to find h.

Before we continue, notice that the mechanical energy of the system includes the following contributions: (i) linear kinetic energy for the block, (ii) rotational kinetic energy for the wheel, and (iii) gravitational potential energy for the block. We do not include the gravitational potential energy of the wheel because its height does not change.

Known   Mass of block, m; radius of wheel, R; moment of inertia of wheel, I; initial angular speed, ω; initial linear speed, v.
Unknown   Height gained by the block, $h = ?$

SOLUTION

1. Write an expression for the initial mechanical energy of the system, E_i, including all three contributions mentioned in the Strategy:

$$E_i = \tfrac{1}{2}mv^2 + \tfrac{1}{2}I\omega^2 + mgy$$
$$= \tfrac{1}{2}mv^2 + \tfrac{1}{2}I\left(\frac{v}{R}\right)^2 + 0$$

2. Write an expression for the final mechanical energy of the system, E_f:

$$E_f = \tfrac{1}{2}mv^2 + \tfrac{1}{2}I\omega^2 + mgy$$
$$= 0 + 0 + mgh$$

3. Set the initial and final mechanical energies equal to one another, $E_i = E_f$:

$$E_i = \tfrac{1}{2}mv^2 + \tfrac{1}{2}I\left(\frac{v}{R}\right)^2 = \tfrac{1}{2}mv^2\left(1 + \frac{I}{mR^2}\right) = mgh = E_f$$

4. Solve for the height, h:

$$h = \left(\frac{v^2}{2g}\right)\left(1 + \frac{I}{mR^2}\right)$$

INSIGHT

If the block were moving upward with speed v on its own—not attached to anything—it would rise to the height $h = v^2/2g$. We recover this result if $I = 0$, because in that case it is as if the wheel were not there. If the wheel is present, and I is nonzero, the block rises to a height that is *greater* than $v^2/2g$. The reason is that the wheel has kinetic energy, in addition to the kinetic energy of the block, and the sum of these kinetic energies must be converted to gravitational potential energy before the block and the wheel stop moving.

PRACTICE PROBLEM

Suppose the wheel is a disk with a mass equal to the mass m of the block. Find an expression for the height h in this case. [**Answer:** The moment of inertia of the wheel is $I = \tfrac{1}{2}mR^2$. Therefore, $h = \left(\tfrac{3}{2}\right)(v^2/2g)$.]

Some related homework problems: Problem 57, Problem 59, Problem 62

EXAMPLE 10-20   FIND THE YO-YO'S SPEED

Yo-Yo man releases a yo-yo from rest and allows it to drop, as he keeps the top end of the string stationary. The mass of the yo-yo is 0.059 kg, its moment of inertia is 2.5×10^{-5} kg·m², and the radius r of the axle the string wraps around is 0.0068 m. What is the linear speed v of the yo-yo after it has dropped through a height of $h = 0.50$ m?

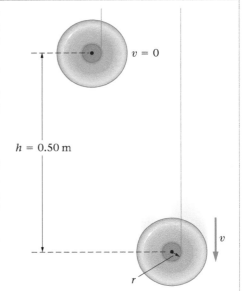

PICTURE THE PROBLEM

Our sketch shows the yo-yo released from rest and descending through a height of 0.50 m. As the yo-yo drops, it gains both linear and angular speed. The linear speed is related to the angular speed by $v = r\omega$, where r is the radius of the yo-yo's axle.

REASONING AND STRATEGY

The basic idea in this system is that the decrease in gravitational potential energy as the yo-yo drops is equal to the increase in the yo-yo's linear and rotational kinetic energies. That is, the final mechanical energy is equal to the initial mechanical energy. We start, then, by writing expressions for the initial and final mechanical energies. Setting these quantities equal yields a relationship that can be solved for the speed v.

Known   Mass of yo-yo, $m = 0.059$ kg; moment of inertia, $I = 2.5 \times 10^{-5}$ kg·m²; radius of axle, $r = 0.0068$ m; drop height, $h = 0.50$ m.
Unknown   Linear speed of yo-yo, $v = ?$

SOLUTION

1. Write the initial energy of the system, which is equal to its initial gravitational potential energy:

$$E_i = mgh$$

CONTINUED

2. Write the final energy of the system, which is equal to its final kinetic energy:

$$E_f = \tfrac{1}{2}mv^2\left(1 + \frac{I}{mr^2}\right)$$

3. Set the initial and final mechanical energies equal to one another, $E_i = E_f$, and solve for v:

$$v = \sqrt{\frac{2gh}{\left(1 + \dfrac{I}{mr^2}\right)}}$$

4. Substitute numerical values:

$$v = \sqrt{\frac{2(9.81 \text{ m/s}^2)(0.50 \text{ m})}{\left(1 + \dfrac{(2.5 \times 10^{-5} \text{ kg} \cdot \text{m}^2)}{(0.059 \text{ kg})(0.0068 \text{ m})^2}\right)}} = 0.98 \text{ m/s}$$

INSIGHT

The linear speed of the yo-yo is $v = r\omega$, where r is the radius of the axle from which the string unwraps without slipping. Therefore, the r in the term I/mr^2 is the radius of the axle. The outer radius of the yo-yo affects its moment of inertia, but I is given to us in the problem statement, and hence the outer radius is not needed.

PRACTICE PROBLEM — PREDICT/CALCULATE

(a) If the yo-yo's moment of inertia is increased, does the final linear speed increase, decrease, or stay the same? Explain. **(b)** Find the final linear speed for the case $I = 3.5 \times 10^{-5} \text{ kg} \cdot \text{m}^2$.   [**Answer: (a)** Increasing the moment of inertia increases the rotational kinetic energy, and leaves less energy for the translational kinetic energy. As a result, the linear speed decreases. **(b)** $v = 0.84 \text{ m/s}$]

Some related homework problems: Problem 62, Problem 82, Problem 84

Enhance Your Understanding   (Answers given at the end of the chapter)

6. A hoop, a disk, a solid sphere, and a hollow sphere are raced down an incline, as in Conceptual Example 10-17. The objects all have the same mass and radius. Rank the objects in the order in which they finish the race, from first to last. Indicate ties where appropriate.

Section Review

- To apply conservation of energy to a rotating system, it is necessary to include the kinetic energy of rotation.

CHAPTER 10 REVIEW

CHAPTER SUMMARY

10-1   ANGULAR POSITION, VELOCITY, AND ACCELERATION

To describe rotational motion, rotational analogs of position, velocity, and acceleration are defined.

Angular Position

Angular position, θ, is the angle measured from an arbitrary reference line:

$$\theta \text{ (in radians)} = \text{arc length/radius} = s/r \qquad \textbf{10-2}$$

Angular Velocity

Angular velocity, ω, is the rate of change of angular position. The average angular velocity is

$$\omega_{\text{av}} = \frac{\Delta\theta}{\Delta t} \qquad \textbf{10-3}$$

The instantaneous angular velocity is the limit of ω_{av} as Δt approaches zero:

$$\omega = \lim_{\Delta t \to 0} \frac{\Delta\theta}{\Delta t} \qquad \textbf{10-4}$$

Angular Acceleration

Angular acceleration, α, is the rate of change of angular velocity. The average angular acceleration is

$$\alpha_{av} = \frac{\Delta\omega}{\Delta t} \qquad \text{10-6}$$

The instantaneous angular acceleration is the limit of α_{av} as Δt approaches zero:

$$\alpha = \lim_{\Delta t \to 0} \frac{\Delta\omega}{\Delta t} \qquad \text{10-7}$$

Period of Rotation

The period, T, is the time required to complete one full rotation. If the angular velocity is constant, T is related to ω as follows:

$$T = \frac{2\pi}{\omega} \qquad \text{10-5}$$

Sign Convention

Counterclockwise rotations are positive; clockwise rotations are negative.

10-2 ROTATIONAL KINEMATICS

Rotational kinematics is the description of angular motion, in the same way that linear kinematics describes linear motion. In both cases, we assume constant acceleration.

Linear–Angular Analogs

Rotational kinematics is related to linear kinematics by the following linear–angular analogs:

Linear Quantity	Angular Quantity
x	θ
v	ω
a	α

10-3 CONNECTIONS BETWEEN LINEAR AND ROTATIONAL QUANTITIES

A point on a rotating object follows a circular path. At any instant of time, the point is moving in a direction tangential to the circle, with a linear speed and acceleration.

Tangential Speed

The tangential speed, v_t, of a point on a rotating object is

$$v_t = r\omega \qquad \text{10-12}$$

Centripetal Acceleration

The centripetal acceleration, a_{cp}, of a point on a rotating object is

$$a_{cp} = r\omega^2 \qquad \text{10-13}$$

Centripetal acceleration is due to a change in the direction of motion.

Tangential Acceleration

The tangential acceleration, a_t, of a point on a rotating object is

$$a_t = r\alpha \qquad \text{10-14}$$

Tangential acceleration is due to a change in speed.

Total Acceleration

The total acceleration of a point on a rotating object is the vector sum of its tangential and centripetal accelerations.

10-4 ROLLING MOTION

Rolling motion is a combination of translational and rotational motions. An object of radius r, rolling without slipping, translates with linear speed v and rotates with angular speed

$$\omega = v/r \qquad \text{10-15}$$

10-5 ROTATIONAL KINETIC ENERGY AND THE MOMENT OF INERTIA

Rotating objects have kinetic energy, just as objects in linear motion have kinetic energy.

Rotational Kinetic Energy

The kinetic energy of a rotating object is

$$K = \frac{1}{2}I\omega^2 \qquad \text{10-17}$$

The quantity I is the moment of inertia.

Moment of Inertia, Discrete Masses

The moment of inertia, I, of a collection of masses, m_i, at distances r_i from the axis of rotation is

$$I = \sum m_i r_i^2 \qquad \text{10-18}$$

10-6 CONSERVATION OF ENERGY

Energy conservation can be applied to a variety of rotational systems in the same way that it is applied to translational systems.

Kinetic Energy of Rolling Motion

The kinetic energy of an object that rolls without slipping is

$$K = \frac{1}{2}mv^2 + \frac{1}{2}I\omega^2 \qquad \text{10-19}$$

Since rolling without slipping implies that $\omega = v/r$, the kinetic energy can be written as follows:

$$K = \frac{1}{2}mv^2 + \frac{1}{2}I\left(\frac{v}{r}\right)^2 = \frac{1}{2}mv^2\left(1 + \frac{I}{mr^2}\right) \qquad \text{10-20}$$

ANSWERS TO ENHANCE YOUR UNDERSTANDING QUESTIONS

1. (a) B < A < D < C. (b) D < A < B < C.
2. (a) 4θ. (b) 2ω.
3. (a) 4 m/s. (b) 24 m/s².
4. (a) v. (b) v. (c) v.
5. $4 < 2 < 1 = 3$.
6. First = solid sphere; second = disk; third = hollow sphere; last = hoop.

CONCEPTUAL QUESTIONS

For instructor-assigned homework, go to www.masteringphysics.com MP

(Answers to odd-numbered Conceptual Questions can be found in the back of the book.)

1. A rigid object rotates about a fixed axis. Do all points on the object have the same angular speed? Do all points on the object have the same linear speed? Explain.

2. Can you drive your car in such a way that your tangential acceleration is zero while at the same time your centripetal acceleration is nonzero? Give an example if your answer is yes; state why not if your answer is no.

3. Can you drive your car in such a way that your tangential acceleration is nonzero while at the same time your centripetal acceleration is zero? Give an example if your answer is yes; state why not if your answer is no.

4. The fact that the Earth rotates gives people in New York a linear speed of about 750 mi/h. Where should you stand on the Earth to have the smallest possible linear speed?

5. At the local carnival you and a friend decide to take a ride on the Ferris wheel. As the wheel rotates with a constant angular speed, your friend poses the following questions: (a) Is my linear velocity constant? (b) Is my linear speed constant? (c) Is the magnitude of my centripetal acceleration constant? (d) Is the direction of my centripetal acceleration constant? What is your answer to each of these questions?

6. Why should changing the axis of rotation of an object change its moment of inertia, given that its shape and mass remain the same?

7. Give a common, everyday example for each of the following: (a) An object that has zero rotational kinetic energy but nonzero translational kinetic energy. (b) An object that has zero translational kinetic energy but nonzero rotational kinetic energy. (c) An object that has nonzero rotational and translational kinetic energies.

8. Two spheres have identical radii and masses. How might you tell which of these spheres is hollow and which is solid?

9. At the grocery store you pick up a can of beef broth and a can of chunky beef stew. The cans are identical in diameter and weight. Rolling both of them down the aisle with the same initial speed, you notice that the can of chunky stew rolls much farther than the can of broth. Why?

10. Suppose we change the race shown in Conceptual Example 10-17 so that a hoop of radius R and mass M races a hoop of radius R and mass $2M$. (a) Does the hoop with mass M finish before, after, or at the same time as the hoop with mass $2M$? Explain. (b) How would your answer to part (a) change if the hoops had different radii? Explain.

PROBLEMS AND CONCEPTUAL EXERCISES

Answers to odd-numbered Problems and Conceptual Exercises can be found in the back of the book. **BIO** *identifies problems of biological or medical interest;* **CE** *indicates a conceptual exercise.* Predict/Explain *problems ask for two responses: (a) your prediction of a physical outcome, and (b) the best explanation among three provided; and* Predict/Calculate *problems ask for a prediction of a physical outcome, based on fundamental physics concepts, and follow that with a numerical calculation to verify the prediction. On all problems, bullets (•, ••, •••) indicate the level of difficulty.*

SECTION 10-1 ANGULAR POSITION, VELOCITY, AND ACCELERATION

1. • The following angles are given in degrees. Convert them to radians: 30°, 45°, 90°, 180°.

2. • The following angles are given in radians. Convert them to degrees: $\pi/6, 0.70\pi, 1.5\pi, 5\pi$.

3. • Express the angular velocity of the second hand on a clock in the following units: **(a)** rev/h, **(b)** deg/min, and **(c)** rad/s.

4. • Rank the following in order of increasing angular speed: an automobile tire rotating at 2.00×10^3 deg/s, an electric drill rotating at 400.0 rev/min, and an airplane propeller rotating at 40.0 rad/s.

5. • A spot of paint on a bicycle tire moves in a circular path of radius 0.29 m. When the spot has traveled a linear distance of 2.73 m, through what angle has the tire rotated? Give your answer in radians.

6. • **The Crab Nebula** One of the most studied objects in the night sky is the Crab nebula, the remains of a supernova explosion observed by the Chinese in 1054. In 1968 it was discovered that a pulsar— a rapidly rotating neutron star that emits a pulse of radio waves with each revolution—lies near the center of the Crab nebula. The period of this pulsar is 33 ms. What is the angular speed (in rad/s) of the Crab nebula pulsar?

The photo is a true-color visible light image of the Crab nebula. In the false-color breakout, the pulsar can be seen as the left member of the pair of stars just above the center of the frame. (Problem 6)

7. • **BIO Hummingbird Flapping** The Ruby-throated Hummingbird (*Archilochus colubris*) can flap its wings with an angular speed of 346 rad/s. What is the period of its flapping motion?

8. • The following table gives the angular speed of a rotating fan blade at various times as it slows to a stop. Find the average angular acceleration for the times **(a)** $t = 0$ to $t = 2.0$ s, **(b)** $t = 0$ to $t = 4.0$ s, and **(c)** $t = 2.0$ s to $t = 4.0$ s.

Time (s)	Angular speed (rad/s)
0	5.0
2.0	4.1
4.0	3.0

9. • Four objects, a, b, c, and d, have the initial angular velocities and angular accelerations given in the following table. Determine the angular velocity of each object after an elapsed time of 3.5 s.

	Initial angular velocity (rad/s)	Angular acceleration (rad/s²)
(a)	15	4.6
(b)	16	−2.2
(c)	−17	3.6
(d)	−18	−7.2

10. •• Predict/Calculate A 3.5-inch floppy disk in a computer rotates with a period of 2.00×10^{-1} s. What are **(a)** the angular speed of the disk and **(b)** the linear speed of a point on the rim of the disk? **(c)** Does a point near the center of the disk have an angular speed that is greater than, less than, or the same as the angular speed found in part (a)? Explain. (*Note:* A 3.5-inch floppy disk is 3.5 inches in diameter.)

11. •• The angle an airplane propeller makes with the horizontal as a function of time is given by $\theta = (125 \text{ rad/s})t + (42.5 \text{ rad/s}^2)t^2$. **(a)** Estimate the instantaneous angular velocity at $t = 0.00$ by calculating the average angular velocity from $t = 0.00$ to $t = 0.010$ s. **(b)** Estimate the instantaneous angular velocity at $t = 1.000$ s by calculating the average angular velocity from $t = 1.000$ s to $t = 1.010$ s. **(c)** Estimate the instantaneous angular velocity at $t = 2.000$ s by calculating the average angular velocity from $t = 2.000$ s to $t = 2.010$ s. **(d)** Based on your results from parts (a), (b), and (c), is the angular acceleration of the propeller positive, negative, or zero? Explain. **(e)** Calculate the average angular acceleration from $t = 0.00$ to $t = 1.00$ s and from $t = 1.00$ s to $t = 2.00$ s.

SECTION 10-2 ROTATIONAL KINEMATICS

12. • **CE** An object at rest begins to rotate with a constant angular acceleration. If this object rotates through an angle θ in the time t, through what angle did it rotate in the time $t/2$?

13. • **CE** An object at rest begins to rotate with a constant angular acceleration. If the angular speed of the object is ω after the time t, what was its angular speed at the time $t/2$?

14. • The angular speed of a propeller on a boat increases with constant acceleration from 11 rad/s to 39 rad/s in 3.0 revolutions. What is the angular acceleration of the propeller?

15. •• After fixing a flat tire on a bicycle you give the wheel a spin. **(a)** If its initial angular speed was 6.35 rad/s and it rotated 14.2 revolutions before coming to rest, what was its average angular acceleration? **(b)** For what length of time did the wheel rotate?

16. •• Predict/Calculate A ceiling fan is rotating at 0.96 rev/s. When turned off, it slows uniformly to a stop in 2.4 min. **(a)** How many revolutions does the fan make in this time? **(b)** Using the result from part (a), find the number of revolutions the fan must make for its speed to decrease from 0.96 rev/s to 0.48 rev/s.

17. •• A discus thrower starts from rest and begins to rotate with a constant angular acceleration of 2.2 rad/s². **(a)** How many revolutions does it take for the discus thrower's angular speed to reach 6.3 rad/s? **(b)** How much time does this take?

18. •• **Half Time** At 3:00 the hour hand and the minute hand of a clock point in directions that are 90.0° apart. What is the first time after

3:00 that the angle between the two hands has decreased by half to 45.0°?

When the little hand is on the 3 and the big hand is on the 12 . . . . (Problem 18)

19. •• **BIO** A centrifuge is a common laboratory instrument that separates components of differing densities in solution. This is accomplished by spinning a sample around in a circle with a large angular speed. Suppose that after a centrifuge in a medical laboratory is turned off, it continues to rotate with a constant angular deceleration for 10.2 s before coming to rest. **(a)** If its initial angular speed was 3850 rpm, what is the magnitude of its angular acceleration? **(b)** How many revolutions did the centrifuge complete after being turned off?

20. •• **The Slowing Earth** The Earth's rate of rotation is constantly decreasing, causing the day to increase in duration. In the year 2006 the Earth took about 0.840 s longer to complete 365 revolutions than it did in the year 1906. What was the average angular acceleration, in rad/s², of the Earth during this time?

21. •• When a carpenter shuts off his circular saw, the 10.0-inch-diameter blade slows from 3984 rpm to 0 in 1.25 s. **(a)** What is the angular acceleration of the blade? **(b)** What is the distance traveled by a point on the rim of the blade during the deceleration? **(c)** What is the magnitude of the net displacement of a point on the rim of the blade during the deceleration?

22. •• **The World's Fastest Turbine** The drill used by most dentists today is powered by a small air turbine that can operate at angular speeds of 350,000 rpm. These drills, along with ultrasonic dental drills, are the fastest turbines in the world—far exceeding the angular speeds of jet engines. Suppose a drill starts at rest and comes up to operating speed in 2.1 s. **(a)** Find the angular acceleration produced by the drill, assuming it to be constant. **(b)** How many revolutions does the drill bit make as it comes up to speed?

SECTION 10-3 CONNECTIONS BETWEEN LINEAR AND ROTATIONAL QUANTITIES

23. • **CE** Predict/Explain Two children, Jason and Betsy, ride on the same merry-go-round. Jason is a distance R from the axis of rotation; Betsy is a distance $2R$ from the axis. **(a)** Is the rotational period of Jason greater than, less than, or equal to the rotational period of Betsy? **(b)** Choose the *best explanation* from among the following:
   I. The period is greater for Jason because he moves more slowly than Betsy.
   II. The period is greater for Betsy since she must go around a circle with a larger circumference.
   III. It takes the same amount of time for the merry-go-round to complete a revolution for all points on the merry-go-round.

24. • The hour hand on a certain clock is 8.2 cm long. Find the tangential speed of the tip of this hand.

25. • The outer edge of a rotating Frisbee with a diameter of 29 cm has a linear speed of 3.7 m/s. What is the angular speed of the Frisbee?

26. • A carousel at the local carnival rotates once every 45 seconds. **(a)** What is the linear speed of an outer horse on the carousel, which is 2.75 m from the axis of rotation? **(b)** What is the linear speed of an inner horse that is 1.75 m from the axis of rotation?

27. • A chainsaw is shown in **FIGURE 10-27**. When the saw is in operation, the chain moves with a linear speed of $v = 5.5$ m/s. At the end of the saw, the chain follows a semicircular path with a radius of $r = 0.044$ m. **(a)** What is the angular speed of the chain as it goes around the end of the saw? **(b)** What is the centripetal acceleration of the chain at the end of the saw?

FIGURE 10-27 Problem 27

28. •• Predict/Calculate Jeff of the Jungle swings on a vine that is 7.20 m long (**FIGURE 10-28**). At the bottom of the swing, just before hitting the tree, Jeff's linear speed is 8.50 m/s. **(a)** Find Jeff's angular speed at this time. **(b)** What centripetal acceleration does Jeff experience at the bottom of his swing? **(c)** What exerts the force that is responsible for Jeff's centripetal acceleration?

FIGURE 10-28 Problem 28

29. •• A compact disk, which has a diameter of 12.0 cm, speeds up uniformly from 0.00 to 4.00 rev/s in 3.00 s. What is the tangential acceleration of a point on the outer rim of the disk at the moment when its angular speed is **(a)** 2.00 rev/s and **(b)** 3.00 rev/s?

30. •• Predict/Calculate When a compact disk with a 12.0-cm diameter is rotating at 34.6 rad/s, what are **(a)** the linear speed and **(b)** the centripetal acceleration of a point on its outer rim? **(c)** Consider a point on the CD that is halfway between its center and its outer rim. Without repeating all of the calculations required for parts (a) and (b), determine the linear speed and the centripetal acceleration of this point.

31. •• Predict/Calculate As Tony the fisherman reels in a "big one," he turns the spool on his fishing reel at the rate of 2.0 complete revolutions every second (FIGURE 10-29). **(a)** If the radius of the reel is 2.3 cm, what is the linear speed of the fishing line as it is reeled in? **(b)** How would your answer to part (a) change if the radius of the reel were doubled?

FIGURE 10-29 Problem 31

32. •• A Ferris wheel with a radius of 9.5 m rotates at a constant rate, completing one revolution every 36 s. Find the direction and magnitude of a passenger's acceleration when **(a)** at the top and **(b)** at the bottom of the wheel.

33. •• Suppose the Ferris wheel in the previous problem begins to decelerate at the rate of 0.22 rad/s² when the passenger is at the top of the wheel. Find the direction and magnitude of the passenger's acceleration at that time.

34. •• To polish a filling, a dentist attaches a sanding disk with a radius of 3.20 mm to the drill. **(a)** When the drill is operated at 2.15×10^4 rad/s, what is the tangential speed of the rim of the disk? **(b)** What period of rotation must the disk have if the tangential speed of its rim is to be 275 m/s?

35. •• **The Bohr Atom** The Bohr model of the hydrogen atom pictures the electron as a tiny particle moving in a circular orbit about a stationary proton. In the lowest-energy orbit the distance from the proton to the electron is 5.29×10^{-11} m, and the linear speed of the electron is 2.18×10^6 m/s. **(a)** What is the angular speed of the electron? **(b)** How many orbits about the proton does it make each second? **(c)** What is the electron's centripetal acceleration?

36. ••• A wheel of radius R starts from rest and accelerates with a constant angular acceleration α about a fixed axis. At what time t will the centripetal and tangential accelerations of a point on the rim have the same magnitude?

SECTION 10-4 ROLLING MOTION

37. • **CE Microwave Tray** Most microwave ovens have a glass tray that sits on top of a circular ring with three small wheels, as shown in FIGURE 10-30. A small motor in the base of the microwave rotates the glass tray to provide uniform heating. When the glass tray completes one revolution, does the circular ring below it rotate through 0.5 rev, 1 rev, 2 rev, or 3 rev? Explain.

FIGURE 10-30 Problem 37

38. • The tires on a car have a radius of 31 cm. What is the angular speed of these tires when the car is driven at 15 m/s?

39. • A child pedals a tricycle, giving the driving wheel an angular speed of $\omega = 0.875$ rev/s (FIGURE 10-31). If the radius of the wheel is $r = 0.260$ m, what is the child's linear speed, v?

FIGURE 10-31 Problem 39

40. • A soccer ball, which has a circumference of 70.0 cm, rolls 14.0 yards in 3.35 s. What was the average angular speed of the ball during this time?

41. •• You drive down the road at 31 m/s (70 mi/h) in a car whose tires have a radius of 34 cm. **(a)** What is the period of rotation of the tires? **(b)** Through what angle does a tire rotate in one second?

42. •• The wheels of a skateboard roll without slipping as it accelerates at 0.35 m/s² down an 85-m-long hill. If the skateboarder travels at 1.8 m/s at the top of the hill, what is the average angular speed of the 2.6-cm-radius wheels during the entire trip down the hill?

43. •• The page feed roller of a computer printer grips each 11-inch-long sheet of paper and feeds it through the print mechanism. If the roller has a radius of 6.4 mm and the drive motor has a maximum angular speed of 480 rpm, what is the maximum number of pages that the printer can print each minute?

44. •• As you drive down the road at 17 m/s, you press on the gas pedal and speed up with a uniform acceleration of 1.12 m/s² for 0.65 s. If the tires on your car have a radius of 33 cm, what is their angular displacement during this period of acceleration?

SECTION 10-5 ROTATIONAL KINETIC ENERGY AND THE MOMENT OF INERTIA

45. • **CE** Predict/Explain The minute and hour hands of a clock have a common axis of rotation and equal mass. The minute hand is long, thin, and uniform; the hour hand is short, thick, and uniform. **(a)** Is the moment of inertia of the minute hand greater than, less than, or equal to the moment of inertia of the hour hand? **(b)** Choose the *best explanation* from among the following:

   I. The hands have equal mass, and hence equal moments of inertia.

   II. Having mass farther from the axis of rotation results in a greater moment of inertia.

   III. The more compact hour hand concentrates its mass and has the greater moment of inertia.

46. CE • Predict/Explain Suppose a bicycle wheel is rotated about an axis through its rim and parallel to its axle. **(a)** Is its moment of inertia about this axis greater than, less than, or equal to its moment of inertia about its axle? **(b)** Choose the *best explanation* from among the following:

   I. The moment of inertia is greatest when an object is rotated about its center.

   II. The mass and shape of the wheel remain the same.

   III. Mass is farther from the axis when the wheel is rotated about the rim.

47. • The moment of inertia of a 0.98-kg bicycle wheel rotating about its center is $0.13 \text{ kg} \cdot \text{m}^2$. What is the radius of this wheel, assuming the weight of the spokes can be ignored?

48. • An electric fan spinning with an angular speed of 13 rad/s has a kinetic energy of 4.6 J. What is the moment of inertia of the fan?

49. •• **BIO Spin-Dry Dragonflies** Some dragonflies splash down onto the surface of a lake to clean themselves. After this dunking, the dragonflies gain altitude, and then spin rapidly at about 1000 rpm to spray the water off their bodies. When the dragonflies do this "spin-dry," they tuck themselves into a "ball" with a moment of inertia of $2.6 \times 10^{-7} \text{ kg} \cdot \text{m}^2$. How much energy must the dragonfly generate to spin itself at this rate?

50. •• **CE** The L-shaped object in **FIGURE 10-32** can be rotated in one of the following three ways: case 1, rotation about the *x* axis; case 2, rotation about the *y* axis; and case 3, rotation about the *z* axis (which passes through the origin perpendicular to the plane of the figure). Rank these three cases in order of increasing moment of inertia. Indicate ties where appropriate.

FIGURE 10-32 Problem 50

51. •• Predict/Calculate A 12-g CD with a radius of 6.0 cm rotates with an angular speed of 34 rad/s. **(a)** What is its kinetic energy? **(b)** If its kinetic energy is to be quadrupled, should its angular speed be doubled, quadrupled, or increased by a factor of 16? **(c)** Calculate the angular speed required to quadruple the kinetic energy the CD had in part (a).

52. •• The engine of a model airplane must both spin a propeller and push air backward to propel the airplane forward. **(a)** Model the propeller as three 0.20-m-long thin rods of mass 0.040 kg each, with the rotation axis at one end. What is the moment of inertia of the propeller? **(b)** How much energy is required to rotate the propeller at 5500 rpm? Ignore the energy required to push the air.

53. •• **Flywheel Energy Storage** One way to store energy is in the rotational motion of a flywheel, and some have proposed using such technology to power automobiles. One unit is based on a 6.0-kg flywheel in the shape of a hoop of radius 0.10 m that spins as fast as 60,000 rpm. **(a)** How much kinetic energy is stored by the flywheel when it is rotating at its maximum rate? **(b)** If the vehicle requires an average power of 11 kW under normal driving conditions, for how much time can it operate with the energy stored in the flywheel?

54. •• When a pitcher throws a curve ball, the ball is given a fairly rapid spin. If a 0.15-kg baseball with a radius of 3.7 cm is thrown with a linear speed of 48 m/s and an angular speed of 42 rad/s, how much of its kinetic energy is translational and how much is rotational? Assume the ball is a uniform, solid sphere.

55. •• A lawn mower has a flat, rod-shaped steel blade that rotates about its center. The mass of the blade is 0.65 kg and its length is 0.55 m. **(a)** What is the rotational energy of the blade at its operating angular speed of 3500 rpm? **(b)** If all of the rotational kinetic energy of the blade could be converted to gravitational potential energy, to what height would the blade rise?

SECTION 10-6 CONSERVATION OF ENERGY

56. • **CE** Consider the physical situation shown in Conceptual Example 10-18. Suppose this time a ball is released from rest on the frictionless surface. When the ball comes to rest on the no-slip surface, is its height greater than, less than, or equal to the height from which it was released?

57. • Suppose the block in Example 10-19 has a mass of 2.1 kg and an initial upward speed of 0.33 m/s. Find the moment of inertia of the wheel if its radius is 8.0 cm and the block rises to a height of 7.4 cm before momentarily coming to rest.

58. •• Calculate the speeds of **(a)** the disk and **(b)** the hoop at the bottom of the inclined plane in Conceptual Example 10-17 if the height of the incline is 0.55 m.

59. •• Predict/Calculate **Atwood's Machine** The two masses ($m_1 = 5.0 \text{ kg}$ and $m_2 = 3.0 \text{ kg}$) in the Atwood's machine shown in **FIGURE 10-33** are released from rest, with m_1 at a height of 0.75 m above the floor. When m_1 hits the ground its speed is 1.8 m/s. Assuming that the pulley is a uniform disk with a radius of 12 cm, **(a)** outline a strategy that allows you to find the mass of the pulley. **(b)** Implement the strategy given in part (a) and determine the pulley's mass.

FIGURE 10-33 Problem 59

60. •• In Conceptual Example 10-18, assume the ball is a solid sphere of radius 2.9 cm and mass 0.14 kg. If the ball is released from rest at a height of 0.78 m above the bottom of the track on the no-slip side, **(a)** what is its angular speed when it is on the frictionless side of the track? **(b)** How high does the ball rise on the frictionless side?

61. •• Predict/Calculate After you pick up a spare, your bowling ball rolls without slipping back toward the ball rack with a linear speed of $v_i = 2.62$ m/s (**FIGURE 10-34**). To reach the rack, the ball rolls up a ramp that rises through a vertical height of $h = 0.47$ m. **(a)** What is the linear speed of the ball when it reaches the top of the ramp? **(b)** If the radius of the ball were increased, would the speed found in part (a) increase, decrease, or stay the same? Explain.

FIGURE 10-34 Problem 61

62. •• Predict/Calculate A 1.3-kg block is tied to a string that is wrapped around the rim of a pulley of radius 7.2 cm. The block is released from rest. **(a)** Assuming the pulley is a uniform disk with a mass of 0.31 kg, find the speed of the block after it has fallen through a height of 0.50 m. **(b)** If a small lead weight is attached near the rim of the pulley and this experiment is repeated, will the speed of the block increase, decrease, or stay the same? Explain.

63. •• After doing some exercises on the floor, you are lying on your back with one leg pointing straight up. If you allow your leg to fall freely until it hits the floor (**FIGURE 10-35**), what is the tangential speed of your foot just before it lands? Assume the leg can be treated as a uniform rod 0.95 m long that pivots freely about the hip.

FIGURE 10-35 Problem 63

64. ••• A 2.0-kg solid cylinder (radius = 0.10 m, length = 0.50 m) is released from rest at the top of a ramp and allowed to roll without slipping. The ramp is 0.75 m high and 5.0 m long. When the cylinder reaches the bottom of the ramp, what are **(a)** its total kinetic energy, **(b)** its rotational kinetic energy, and **(c)** its translational kinetic energy?

65. ••• A 2.5-kg solid sphere (radius = 0.10 m) is released from rest at the top of a ramp and allowed to roll without slipping. The ramp is 0.75 m high and 5.6 m long. When the sphere reaches the bottom of the ramp, what are **(a)** its total kinetic energy, **(b)** its rotational kinetic energy, and **(c)** its translational kinetic energy?

GENERAL PROBLEMS

66. • **CE** As you switch a fan setting from its slowest speed to its highest, its angular speed ω triples. **(a)** Does the period of rotation T of the fan blade increase, decrease, or stay the same? Explain. **(b)** What is the ratio of the new rotation period to the old value, T_{new}/T_{old}?

67. • **CE** When you stand on the observation deck of the Empire State Building in New York, is your linear speed due to the Earth's rotation greater than, less than, or the same as when you were waiting for the elevators on the ground floor?

68. • What linear speed must a 0.065-kg hula hoop have if its total kinetic energy is to be 0.12 J? Assume the hoop rolls on the ground without slipping.

69. • **BIO Losing Consciousness** A pilot performing a horizontal turn will lose consciousness if she experiences a centripetal acceleration greater than 7.00 times the acceleration of gravity. What is the minimum radius turn she can make without losing consciousness if her plane is flying with a constant speed of 245 m/s?

70. • The angular velocity of a rotating wheel as a function of time is shown in the graph in **FIGURE 10-36**. What is the angular displacement of the wheel, $\theta - \theta_0$, between the times $t = 0$ and $t = 5$ s?

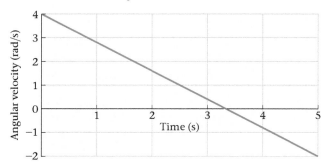

FIGURE 10-36 Problem 70

71. •• A 68-kg commuter rides on an 8.1-kg bicycle. Each bicycle wheel is a hoop of radius 33 cm and mass 1.4 kg. **(a)** What energy must the biker supply to move at 7.2 m/s? Include the kinetic energy of the biker, the kinetic energy of the bike frame, and the kinetic energy of the rotating wheels. **(b)** What percentage of the energy she expends goes into the rotation of the wheels? **(c)** What percentage goes into her own kinetic energy?

72. •• **CE** Place two quarters on a table with their rims touching, as shown in **FIGURE 10-37**. While holding one quarter fixed, roll the other one—without slipping—around the circumference of the fixed quarter until it has completed one round trip. How many revolutions has the rolling quarter made about its center?

FIGURE 10-37 Problem 72

73. •• **CE** The object shown in **FIGURE 10-38** can be rotated in three different ways: case 1, rotation about the *x* axis; case 2, rotation about the *y* axis; and case 3, rotation about the *z* axis. Rank these three cases in order of increasing moment of inertia. Indicate ties where appropriate.

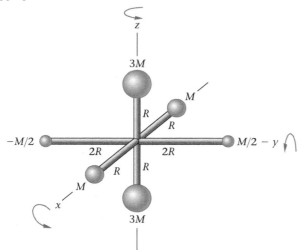

FIGURE 10-38 Problem 73

74. •• Predict/Calculate **When the Hands Align** A mathematically inclined friend e-mails you the following instructions: "Meet me in the cafeteria the first time after 2:00 P.M. today that the hands of a clock point in the same direction." **(a)** Is the desired meeting time before, after, or equal to 2:10 P.M.? Explain. **(b)** Is the desired meeting time before, after, or equal to 2:15 P.M.? Explain. **(c)** When should you meet your friend?

75. •• **BIO** Spinning Dragonflies Measurements show that when a dragonfly does a "spin-dry" (Problem 49), it can "spin up" from $\omega_0 = 0$ to $\omega = 104$ rad/s in 0.250 s. **(a)** Assuming constant angular acceleration, how many revolutions does the dragonfly make in this time? **(b)** How many revolutions has the dragonfly made when its angular speed is 51.2 rad/s?

76. •• Predict/Calculate A potter's wheel of radius 6.8 cm rotates with a period of 0.52 s. What are **(a)** the linear speed and **(b)** the centripetal acceleration of a small lump of clay on the rim of the wheel? **(c)** How do your answers to parts (a) and (b) change if the period of rotation is doubled?

77. •• Predict/Calculate **Playing a CD** The record in an old-fashioned record player always rotates at the same angular speed. With CDs, the situation is different. For a CD to play properly, the point on the CD where the laser beam shines must have a linear speed $v_t = 1.25$ m/s, as indicated in **FIGURE 10-39**. **(a)** As the CD plays from the center outward, does its angular speed increase, decrease, or stay the same? Explain. **(b)** Find the angular speed of a CD when the laser beam is 2.50 cm from its center. **(c)** Repeat part (b) for the laser beam 6.00 cm from the center. **(d)** If the CD plays for 66.5 min, and the laser beam moves from 2.50 cm to 6.00 cm during this time, what is the CD's average angular acceleration?

FIGURE 10-39 Problem 77

78. •• **BIO** **Roller Pigeons** Pigeons are bred to display a number of interesting characteristics. One breed of pigeon, the "roller," is remarkable for the fact that it does a number of backward somersaults as it drops straight down toward the ground. Suppose a roller pigeon drops from rest and free falls downward for a distance of 14 m. If the pigeon somersaults at the rate of 12 rad/s, how many revolutions has it completed by the end of its fall?

79. •• **BIO** **Dolphin Tricks** A bottlenose dolphin (*Tursiops truncatus*) is about 2.5 m long and has a mass of 250 kg. It can jump 3.5 m above the surface of the water while flipping nose-to-tail at 5.6 rad/s, fast enough to complete 1.5 rotations before landing in the water. **(a)** How much energy must the dolphin generate to jump 3.5 m above the surface of the water? **(b)** If the dolphin's moment of inertia about its rotation axis is 225 kg · m², how much energy must the dolphin generate to rotate its body in this way?

Jumping dolphin. (Problem 79)

80. •• As a marble with a diameter of 1.6 cm rolls down an incline, its center moves with a linear acceleration of 3.3 m/s². **(a)** What is the angular acceleration of the marble? **(b)** What is the angular speed of the marble after it rolls for 1.5 s from rest?

81. •• A rubber ball with a radius of 3.2 cm rolls along the horizontal surface of a table with a constant linear speed *v*. When the ball rolls off the edge of the table, it falls 0.66 m to the floor below. If the ball completes 0.37 revolution during its fall, what was its linear speed, *v*?

82. •• Predict/Calculate A yo-yo moves downward until it reaches the end of its string, where it "sleeps." As it sleeps—that is, spins in place—its angular speed decreases from 35 rad/s to 25 rad/s. During this time it completes 120 revolutions. **(a)** How much time did it take for the yo-yo to slow from 35 rad/s to 25 rad/s? **(b)** How much time does it take for the yo-yo to slow from 25 rad/s to 15 rad/s? Assume a constant angular acceleration as the yo-yo sleeps.

83. •• Predict/Calculate **(a)** An automobile with tires of radius 32 cm accelerates from 0 to 45 mph in 9.1 s. Find the angular acceleration of the tires. **(b)** How does your answer to part (a) change if the radius of the tires is halved?

84. •• **A Yo-Yo with a Brain** Yomega ("The yo-yo with a brain") is constructed with a clever clutch mechanism in its axle that allows it to rotate freely and "sleep" when its angular speed is greater than a certain critical value. When the yo-yo's angular speed falls below this value, the clutch engages, causing the yo-yo to climb

the string to the user's hand. If the moment of inertia of the yo-yo is 7.4×10^{-5} kg · m², its mass is 0.11 kg, and the string is 1.0 m long, what is the smallest angular speed that will allow the yo-yo to return to the user's hand?

85. •• The rotor in a centrifuge has an initial angular speed of 430 rad/s. After 8.2 s of constant angular acceleration, its angular speed has increased to 550 rad/s. During this time, what were (a) the angular acceleration of the rotor and (b) the angle through which it turned?

86. •• The Sun, with Earth in tow, orbits about the center of the Milky Way galaxy at a speed of 137 miles per second, completing one revolution every 240 million years. (a) Find the angular speed of the Sun relative to the center of the Milky Way. (b) Find the distance from the Sun to the center of the Milky Way.

87. •• A person walks into a room and switches on the ceiling fan. The fan accelerates with constant angular acceleration for 15 s until it reaches its operating angular speed of 1.9 rotations/s—after that its speed remains constant as long as the switch is "on." The person stays in the room for a short time; then, 5.5 minutes after turning the fan on, she switches it off again and leaves the room. The fan now decelerates with constant angular acceleration, taking 2.4 minutes to come to rest. What is the total number of revolutions made by the fan, from the time it was turned on until the time it stopped?

88. •• **BIO Preventing Bone Loss in Space** When astronauts return from prolonged space flights, they often suffer from bone loss, resulting in brittle bones that may take weeks for their bodies to rebuild. One solution may be to expose astronauts to periods of substantial "g forces" in a centrifuge carried aboard their spaceship. To test this approach, NASA conducted a study in which four people spent 22 hours each in a compartment attached to the end of a 28-foot arm that rotated with an angular speed of 10.0 rpm. (a) What centripetal acceleration, in terms of g, did these volunteers experience? (b) What was their linear speed?

89. ••• A thin, uniform rod of length L and mass M is pivoted about one end, as shown in FIGURE 10-40. The rod is released from rest in a horizontal position, and allowed to swing downward without friction or air resistance. When the rod is vertical, what are (a) its angular speed ω and (b) the tangential speed v_t of its free end?

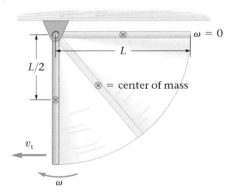

FIGURE 10-40 Problem 89

90. ••• **Center of Percussion** In the previous problem, suppose a small metal ball of mass $m = 2M$ is attached to the rod a distance d from the pivot. The rod and ball are released from rest in the horizontal position. (a) Show that when the rod reaches the vertical position, the speed of its tip is

$$v_t = \sqrt{3gL}\sqrt{\frac{1 + 4(d/L)}{1 + 6(d/L)^2}}$$

(b) At what finite value of d/L is the speed of the rod the same as it is for $d = 0$? (This value of d/L is the **center of percussion**, or "sweet spot," of the rod.)

91. ••• A wooden plank rests on two soup cans laid on their sides. Each can has a diameter of 6.5 cm, and the plank is 3.0 m long. Initially, one can is placed 1.0 m inward from either end of the plank, as FIGURE 10-41 shows. The plank is now pulled 1.0 m to the right, and the cans roll without slipping. (a) How far does the center of each can move? (b) How many rotations does each can make?

FIGURE 10-41 Problem 91

92. ••• A person rides on a 12-m-diameter Ferris wheel that rotates at the constant rate of 8.1 rpm. Calculate the magnitude and direction of the force that the seat exerts on a 65-kg person when he is (a) at the top of the wheel, (b) at the bottom of the wheel, and (c) halfway up the wheel.

93. ••• Predict/Calculate A solid sphere with a diameter of 0.17 m is released from rest; it then rolls without slipping down a ramp, dropping through a vertical height of 0.61 m. The ball leaves the bottom of the ramp, which is 1.22 m above the floor, moving horizontally (FIGURE 10-42). (a) Through what horizontal distance d does the ball move before landing? (b) How many revolutions does the ball make during its fall? (c) If the ramp were to be made frictionless, would the distance d increase, decrease, or stay the same? Explain.

FIGURE 10-42 Problem 93

PASSAGE PROBLEMS

BIO **Human-Powered Centrifuge**

Space travel is fraught with hazards, not the least of which are the many side effects of prolonged weightlessness, including weakened muscles, bone loss, decreased coordination, and unsteady balance. If you are fortunate enough to go on a trip to Mars, which could take more than a year each way, you might be a bit "weak in the knees" by the time you arrive. This could lead to problems when you try to take your first "small step" on the surface.

To counteract these effects, NASA is looking into ways to provide astronauts with "portable gravity" on long space flights. One method under consideration is the human-powered centrifuge, which not only subjects the astronauts to artificial gravity, but also gives them aerobic exercise. The device is basically a rotating, circular platform on which two astronauts lie supine along a diameter, head-to-head at the center, with their feet at opposite rims, as shown in the accompanying photo. The radius of the platform in this test model is 6.25 ft. As one astronaut pedals to rotate the platform, the astronaut facing the other direction can exercise in the artificial gravity. Alternatively, a third astronaut on a stationary bicycle can provide the rotation for the other two. While the astronauts' feet are at the outer rim of the platform, their heads are near the center of the platform, and their hearts are 4.50 feet from the rim, which means that different parts of the astronauts' bodies will experience different "gravitational" accelerations.

Human-powered centrifuge.

FIGURE 10-43 shows the centripetal acceleration (in g) produced by a rotating platform at four different radii. Notice that the acceleration increases as the square of the angular speed. Also indicated in Figure 10-43 are acceleration levels corresponding to 1, 3, and 5 gs. It is thought that enhanced gravitational effects may be desirable because the astronauts will experience the artificial gravity for only relatively brief periods of time during the flight.

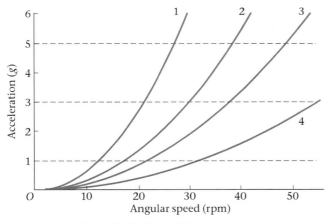

FIGURE 10-43 Problems 94, 95, 96, and 97

94. • Rank the four curves shown in Figure 10-43 in order of increasing radius. Indicate ties where appropriate.

95. • A particular experimental protocol calls for measuring the physiological parameters during exercise with 2.00g at the feet. What angular speed in rpm must the platform in this test model have in order to produce this centripetal acceleration at the feet?

    A. 3.20 rpm   **B.** 26.5 rpm
    C. 30.6 rpm   **D.** 57.8 rpm

96. • If the acceleration is 2.00g at the feet, what is the acceleration at the heart?

    A. 0   **B.** 0.560g
    C. 1.40g **D.** 2.00g

97. • Which of the curves shown in Figure 10-43 corresponds to the location of the feet in this test model?

    A. 1   **B.** 2
    C. 3   **D.** 4

98. •• **REFERRING TO CONCEPTUAL EXAMPLE 10-17** Suppose we race a disk and a hollow spherical shell, like a basketball. The spherical shell has a mass M and a radius R; the disk has a mass 2M and a radius 2R. **(a)** Which object wins the race? If the two objects are released at rest, and the height of the ramp is $h = 0.75$ m, find the speed of **(b)** the disk and **(c)** the spherical shell when they reach the bottom of the ramp.

99. •• **REFERRING TO CONCEPTUAL EXAMPLE 10-17** Consider a race between the following three objects: object 1, a disk; object 2, a solid sphere; and object 3, a hollow spherical shell. All objects have the same mass and radius. **(a)** Rank the three objects in the order in which they finish the race. Indicate a tie where appropriate. **(b)** Rank the objects in order of increasing kinetic energy at the bottom of the ramp. Indicate a tie where appropriate.

11

Rotational Dynamics and Static Equilibrium

Big Ideas

1 Torque is the rotational analog of force in a linear system.

2 A constant torque causes a constant angular acceleration.

3 A system is in static equilibrium if the net force and net torque are equal to zero.

4 Angular momentum is the rotational analog of linear momentum.

5 Angular momentum is conserved in systems that have zero net torque.

▲ Equilibrium, and the sense of serenity that comes with it, requires more than just forces that add to zero. To keep from falling, for example, this acrobat's hands must exert forces that add up to his total weight. But the total weight must be shared between his two hands in just the right way, or else his body will rotate and the pose will be lost. To ensure equilibrium, a new physical quantity—the torque—must also be zero. In this chapter we introduce the torque and show that equilibrium occurs only when both the net force and the net torque are zero. We will also consider the consequences of nonzero torque.

In this chapter we develop the concept of torque, and show that it is the rotational equivalent of force. We also introduce the notion of *angular momentum*, and show that it is conserved when the net torque acting on a system is zero.

11-1 Torque

Suppose you want to loosen a nut by rotating it counterclockwise with a wrench. If you have ever used a wrench in this way, you probably know that the nut is more likely to turn if you apply your force as far from the nut as possible, as indicated in FIGURE 11-1 and FIGURE 11-2 (a). Applying a force near the nut would not be very effective—you could still get the nut to turn, but it would require considerably more effort! Similarly, it is much easier to open a revolving door if you push far from the axis of rotation, as indicated in FIGURE 11-2 (b). Clearly, then, the tendency for a force to cause a rotation increases with the distance, r, from the axis of rotation to the force. As a result, it's useful to define a quantity called the **torque**, τ, that takes into account both the magnitude of the force, F, *and* the distance from the axis of rotation, r:

Definition of Torque, τ, for a Tangential Force

$$\tau = rF \qquad\qquad 11\text{-}1$$

SI unit: $\text{N} \cdot \text{m}$

With this definition, we see that torque increases with both the force and the distance.

▲ **FIGURE 11-1** The long handle of this wrench enables the user to produce a large torque without having to exert a very great force.

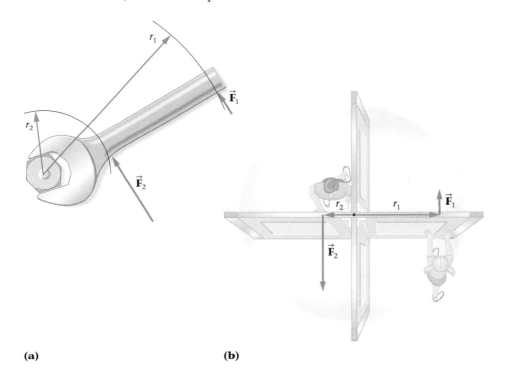

(a)　　　　　　　　　**(b)**

◄ **FIGURE 11-2 Applying a torque**
(a) When a wrench is used to loosen a nut, less force is required if it is applied far from the nut. **(b)** Similarly, less force is required to open a revolving door if it is applied far from the axis of rotation.

Equation 11-1 is valid only when the applied force is *tangential* to a circle of radius r centered on the axis of rotation, as in Figure 11-2. The more general case is considered later in this section. First, we use $\tau = rF$ to determine how much force is needed to open a swinging door, depending on where we apply the force.

EXERCISE 11-1 TORQUE AND FORCE

To open the door in Figure 11-2 (b) a tangential force F is applied at a distance r from the axis of rotation. Suppose the minimum torque required to open the door is $3.4 \ \text{N} \cdot \text{m}$. **(a)** What force must be applied to open the door if r is 0.91 m? **(b)** At what minimum distance must a force of 6.9 N be applied to open the door?

REASONING AND SOLUTION

a. Setting $\tau = rF = 3.4 \ \text{N} \cdot \text{m}$, we can solve for the required force:

$$F = \frac{\tau}{r} = \frac{3.4 \ \text{N} \cdot \text{m}}{0.91 \ \text{m}} = 3.7 \ \text{N}$$

CONTINUED

PROBLEM-SOLVING NOTE

The Units of Torque

Notice that the units of torque are $\text{N} \cdot \text{m}$, the same as the units of work. Though their units are the same, torque, τ, and work, W, represent different physical quantities and should not be confused with one another.

▲ **FIGURE 11-3 Starfish attack** A starfish attempts to open a clam by exerting a torque on the two halves of its shell. The clam uses its adductor muscles to exert an opposite torque on the shell.

▶ **FIGURE 11-4 Only the tangential component of a force causes a torque** **(a)** A radial force causes no rotation. In this case, the force $\vec{F}$ is opposed by an equal and opposite force exerted by the axle of the merry-go-round. The merry-go-round does not rotate. **(b)** A force applied at an angle θ with respect to the radial direction. The radial component of this force, $F \cos \theta$, causes no rotation; the tangential component, $F \sin \theta$, can cause a rotation.

Big Idea 1 Torque is the rotational analog of force in a linear system. Torque depends on the direction and magnitude of the force applied to an object, and also on the distance from the point of application of the force to the axis of rotation.

PHYSICS IN CONTEXT
Looking Back

The concept of force (Chapters 5 and 6) has been extended to torque, its rotational equivalent. We apply Newton's laws to rotational motion in Sections 11-2 and 11-5, just as was done for linear motion in Chapters 5 and 6.

PHYSICS IN CONTEXT
Looking Ahead

Torque arises in the discussion of magnetic fields and the forces they exert in Chapter 22. The torques due to magnetic fields are also the key element in the operation of electric motors, as we show in Chapter 23.

b. This time, we solve $\tau = rF = 3.4 \, \text{N} \cdot \text{m}$ for the required distance:

$$r = \frac{\tau}{F} = \frac{3.4 \, \text{N} \cdot \text{m}}{6.9 \, \text{N}} = 0.49 \, \text{m}$$

As expected, the required distance is smaller when the applied force is greater.

BIO Many human inventions make use of torque and rotation, but there are also examples in nature—from the hinged jaws of the Venus flytrap to those of the alligator. One classic example of a struggle between torques is the clam and the starfish, as shown in **FIGURE 11-3**. The hard clam (*Mercenaria mercenaria*) keeps its two halves partly open for filter feeding, but it also has very strong adductor muscles to hold the shell closed when it senses danger. A starfish such as *Asterias forbesi* feeds by prying open the two halves of the clamshell and then inserting its stomach into the clam in order to digest it *in situ*. The starfish uses hydraulic suction in its tube feet to exert a large torque on the clamshell. If the starfish can exert sufficient torque, and for a sufficient length of time, it can wear out the clam's adductor muscles and enjoy a meal.

(a) A radial force produces zero torque   **(b)** Only the tangential component of force causes a torque

Torque with Nontangential Forces So far we have considered tangential forces only. What happens if you exert a force in a direction that is not tangential? Suppose, for example, that you pull on a playground merry-go-round in a direction that is radial—that is, along a line that extends through the axis of rotation—as in **FIGURE 11-4 (a)**. In this case, your force has no tendency to cause a rotation. Instead, the axle of the merry-go-round simply exerts an equal and opposite force, and the merry-go-round remains at rest. Similarly, if you were to push or pull in a radial direction on a swinging door it would not rotate. We conclude that a *radial force produces zero torque*.

On the other hand, suppose your force is at an angle θ relative to a radial line, as shown in **FIGURE 11-4 (b)**. What is the torque now? To analyze this case, we first resolve the force vector $\vec{F}$ into radial and tangential components. Referring to the figure, we see that the radial component has a magnitude of $F \cos \theta$, and the tangential component has a magnitude of $F \sin \theta$. Because it is the tangential component alone that causes rotation, we define the torque to have a magnitude of $r(F \sin \theta)$. That is:

General Definition of the Magnitude of Torque, τ

$$\tau = r(F \sin \theta) \qquad \qquad \text{11-2}$$

SI units: N · m

(More generally, the torque can be defined as the **cross product** of the vectors $\vec{r}$ and $\vec{F}$; that is, $\vec{\tau} = \vec{r} \times \vec{F}$. The cross product is discussed in detail in Appendix A.)

As a quick check, notice that a radial force corresponds to $\theta = 0$. In this case, $\tau = r(F \sin 0) = 0$, as expected. If the force is tangential, however, it follows that $\theta = \pi/2$. This gives $\tau = r(F \sin \pi/2) = rF$, in agreement with our original definition in Equation 11-1.

Torque and the Moment Arm   An equivalent way to define torque is in terms of the **moment arm**, $r_\perp$. The idea here is to extend a line through the force vector, as in FIGURE 11-5, and then draw a second line from the axis of rotation perpendicular to the line of the force. The perpendicular distance from the axis of rotation to the line of the force is defined to be $r_\perp$. From the figure, we see that

$$r_\perp = r \sin \theta$$

In addition, we note that a simple rearrangement of the torque expression in Equation 11-2 yields

$$\tau = r(F \sin \theta) = (r \sin \theta)F$$

Thus, the torque can be written as the moment arm times the force:

$$\tau = r_\perp F \qquad\qquad 11\text{-}3$$

Sign Convention for Torque   Just as a force applied to an object gives rise to a linear acceleration, a torque applied to an object gives rise to an angular acceleration. Specifically, if a torque acts on an object at rest, the object will begin to rotate; if a torque acts on a rotating object, the object's angular velocity will change. In fact, the greater the torque applied to an object, the greater its angular acceleration, as we shall see in the next section. For this reason, the sign of the torque is determined by the same convention used previously in Section 10-1 for angular acceleration:

Sign Convention for Torque

By convention, if a torque τ acts alone, then

$\tau > 0$ if the torque causes a counterclockwise angular acceleration

$\tau < 0$ if the torque causes a clockwise angular acceleration

In a system with more than one torque, as in FIGURE 11-6, the sign of each torque is determined by the type of angular acceleration *it alone* would produce. The net torque acting on the system, then, is the sum of the individual torques, taking into account the proper signs. This is illustrated in the following Example.

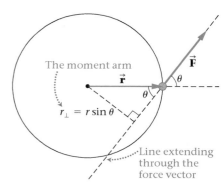

▲ **FIGURE 11-5 The moment arm** To find the moment arm, $r_\perp$, for a given force, first extend a line through the force vector. This line is sometimes referred to as the "line of action." Next, drop a perpendicular line from the axis of rotation to the line of the force. The perpendicular distance is $r_\perp = r \sin \theta$.

PROBLEM-SOLVING NOTE

The Sign of Torques

The sign of a torque is determined by the direction of rotation it would cause if it were the only torque acting in the system.

▲ **FIGURE 11-6** The net torque on the wheel of this ship is the sum of the torques exerted by the two helmsmen.

EXAMPLE 11-2   TORQUES TO THE LEFT AND TORQUES TO THE RIGHT

Two helmsmen, in disagreement about which way to turn a ship, exert the forces shown in the sketch of a ship's wheel. The wheel has a radius of 0.74 m, and the two forces have the magnitudes $F_1 = 72$ N and $F_2 = 58$ N. Find **(a)** the torque caused by $\vec{F}_1$ and **(b)** the torque caused by $\vec{F}_2$. **(c)** In which direction does the wheel turn as a result of these two forces?

PICTURE THE PROBLEM
Our sketch shows that both forces are applied at the distance $r = 0.74$ m from the axis of rotation. However, $F_1 = 72$ N is at an angle of 50.0° relative to the radial direction, whereas $F_2 = 58$ N is tangential, which means that its angle relative to the radial direction is 90.0°.

REASONING AND STRATEGY
For each force, we find the magnitude of the corresponding torque, using $\tau = rF \sin \theta$. As for the signs of the torques, we must consider the angular acceleration each force alone would cause. $\vec{F}_1$ acting alone would cause the wheel to accelerate counterclockwise, and

CONTINUED

hence its torque is positive. $\vec{\mathbf{F}}_2$ would accelerate the wheel clockwise if it acted alone, and hence its torque is negative. If the sum of the two torques is positive, the wheel accelerates counterclockwise; if the sum of the two torques is negative, the wheel accelerates clockwise.

Known   Radius of wheel, $r = 0.74$ m; force 1, $F_1 = 72$ N; force 2, $F_2 = 58$ N; direction of force 1, 50.0°; direction of force 2, 90.0°.

Unknown   (a) Torque caused by $\vec{\mathbf{F}}_1$, $\tau_1 = ?$ (b) Torque caused by $\vec{\mathbf{F}}_2$, $\tau_2 = ?$ (c) Direction of rotation?

SOLUTION

Part (a)

1. Use $\tau = rF \sin\theta$ to calculate the torque due to $\vec{\mathbf{F}}_1$. Recall that this torque is positive:

$$\tau_1 = rF_1 \sin 50.0° = (0.74\ \text{m})(72\ \text{N}) \sin 50.0° = 41\ \text{N} \cdot \text{m}$$

Part (b)

2. Similarly, calculate the torque due to $\vec{\mathbf{F}}_2$. Recall that this torque is negative:

$$\tau_2 = -rF_2 \sin 90.0° = -(0.74\ \text{m})(58\ \text{N}) = -43\ \text{N} \cdot \text{m}$$

Part (c)

3. Sum the torques from parts (a) and (b) to find the net torque:

$$\tau_{\text{net}} = \tau_1 + \tau_2 = 41\ \text{N} \cdot \text{m} - 43\ \text{N} \cdot \text{m} = -2\ \text{N} \cdot \text{m}$$

INSIGHT

Because the net torque is negative, the wheel accelerates clockwise. Thus, even though $\vec{\mathbf{F}}_2$ is the smaller force, it has the greater effect in determining the wheel's direction of acceleration. This is because $\vec{\mathbf{F}}_2$ is applied tangentially, whereas $\vec{\mathbf{F}}_1$ is applied in a direction that is partially radial.

PRACTICE PROBLEM

What magnitude of $\vec{\mathbf{F}}_2$ would yield zero net torque on the wheel, assuming everything else remains the same?   [**Answer:** $F_2 = 55$ N]

Some related homework problems: Problem 1, Problem 2

Enhance Your Understanding   (Answers given at the end of the chapter)

1. A bicycle wheel is mounted on an axle, as shown in **FIGURE 11-7**. Rank the four forces of equal magnitude (F_1, F_2, F_3, F_4) in order of the torque they produce, from most negative to most positive. Indicate ties where appropriate.

▲ **FIGURE 11-7** Enhance Your Understanding 1.

Section Review

- Torque has a magnitude given by $\tau = rF \sin\theta$, where F is the force, r is the distance to the axis of rotation, and θ is the angle between the force and the radial direction.

- A torque is positive if it alone would cause a counterclockwise rotation, and negative if it alone would cause a clockwise rotation.

11-2 Torque and Angular Acceleration

In the previous section, we indicated that a torque causes a change in the rotational motion of an object. To be more precise, a single torque, τ, acting on an object causes the object to have an angular acceleration, α. In this section we develop the specific relationship between τ and α.

Torque and Angular Acceleration Consider a small object of mass m connected to an axis of rotation by a light rod of length r, as in **FIGURE 11-8**. If a tangential force of magnitude F is applied to the mass, it will move with an acceleration given by Newton's second law:

$$a = \frac{F}{m}$$

From Equation 10-14, we know that the linear and angular accelerations are related by

$$\alpha = \frac{a}{r}$$

Combining these results yields the following expression for the angular acceleration:

$$\alpha = \frac{a}{r} = \frac{(F/m)}{r} = \frac{F}{mr}$$

Multiplying both numerator and denominator by r gives

$$\alpha = \left(\frac{r}{r}\right)\frac{F}{mr} = \frac{rF}{mr^2}$$

Now this last result is rather interesting, because the numerator and denominator have simple interpretations. First, the numerator is the torque, $\tau = rF$, for the case of a tangential force (Equation 11-1). Second, the denominator is the moment of inertia of a single mass m rotating at a radius r; that is, $I = mr^2$. Therefore, we find that

$$\alpha = \frac{rF}{mr^2} = \frac{\tau}{I}$$

Thus, the angular acceleration is directly proportional to the torque, and inversely proportional to the moment of inertia. Rearranging this slightly, we find

$$\tau = I\alpha$$

Notice the similarity between this result and Newton's second law for linear systems, $F = ma$.

Newton's Second Law for Rotational Motion The relationship $\tau = I\alpha$ was derived for the special case of a tangential force and a single mass rotating at a radius r. However, the result is completely general. For example, in a system with more than one torque, the relationship $\tau = I\alpha$ is replaced with $\tau_{\text{net}} = \Sigma\tau = I\alpha$, where τ_{net} is the net torque acting on the system. This gives the *rotational* version of Newton's second law:

Newton's Second Law for Rotational Motion

$$\Sigma\tau = I\alpha \qquad\qquad\qquad 11\text{-}4$$

If only a single torque acts on a system, we simply write $\tau = I\alpha$.

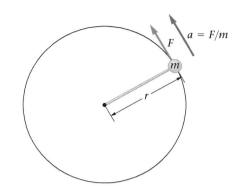

▲ **FIGURE 11-8 Torque and angular acceleration** A tangential force F applied to a mass m gives it a linear acceleration of magnitude $a = F/m$. The corresponding angular acceleration is $\alpha = \tau/I$, where $\tau = rF$ and $I = mr^2$.

QUICK EXAMPLE 11-3 ANGULAR ACCELERATION

A light rope wrapped around a disk-shaped pulley is pulled tangentially with a force of 0.51 N. Find the angular acceleration of the pulley, given that its mass is 2.4 kg and its radius is 0.13 m.

REASONING AND SOLUTION
To find the angular acceleration of the pulley, we first calculate the torque exerted on it by the rope ($\tau = rF$ for a tangential force). Next, we calculate the pulley's moment of inertia $\left(I = \frac{1}{2}mr^2 \text{ for a disk}\right)$. The angular acceleration of the pulley is the torque exerted on it divided by its moment of inertia.

1. Calculate the torque applied to the disk: $\qquad \tau = rF = (0.13\,\text{m})(0.51\,\text{N}) = 0.066\,\text{N}\cdot\text{m}$

CONTINUED

2. Determine the moment of inertia of the disk-shaped pulley: $\quad I = \frac{1}{2}mr^2 = \frac{1}{2}(2.4\text{ kg})(0.13\text{ m})^2 = 0.020\text{ kg}\cdot\text{m}^2$

3. Divide the torque by the moment of inertia to find the angular acceleration of the pulley: $\quad \alpha = \dfrac{\tau}{I} = \dfrac{0.066\text{ N}\cdot\text{m}}{0.020\text{ kg}\cdot\text{m}^2} = 3.3\text{ rad/s}^2$

How would the angular acceleration change if the radius of the pulley were doubled, with everything else remaining the same? Well, doubling the radius doubles the torque (see Step 1), but quadruples the moment of inertia (see Step 2). Hence, the angular acceleration of the pulley is halved if the radius of the pulley is doubled.

Big Idea 2 A constant torque applied to an object causes a constant angular acceleration, just as a constant force causes a constant linear acceleration.

PHYSICS IN CONTEXT
Looking Back

The connection between rotational and linear quantities (Chapter 10) is used here to relate torque to angular acceleration.

It's easy to remember the rotational version of Newton's second law, $\sum\tau = I\alpha$, by using analogies between rotational and linear quantities. We have already seen that I is the analog of m, and that α is the analog of a. Similarly, τ, which causes an angular acceleration, is the analog of F, which causes a linear acceleration. To summarize:

Linear Quantity	Angular Quantity
m	I
a	α
F	τ

Thus, just as $\sum F = ma$ describes linear motion, $\sum\tau = I\alpha$ describes rotational motion.

EXAMPLE 11-4 — GETTING WINDED

A wind turbine has a radius of 37 m, a moment of inertia of $8.4 \times 10^6\text{ kg}\cdot\text{m}^2$, and rotates with an initial angular speed of 1.1 rad/s. When the wind speed increases, the turbine experiences a net torque of $1.6 \times 10^5\text{ N}\cdot\text{m}$. **(a)** How many rotations does it take for the angular speed to increase to 1.8 rad/s? **(b)** What is the linear speed of the tips of the turbine blades 27 s after the acceleration begins?

PICTURE THE PROBLEM
Our sketch shows the wind turbine rotating with an initial angular speed of $\omega_0 = 1.1$ rad/s, and accelerating with an angular acceleration α in the same direction as the angular velocity. The result will be an increase in the angular speed.

REASONING AND STRATEGY
This is basically an angular kinematics problem, as in Chapter 10, but in this case we must first calculate the angular acceleration using $\alpha = \tau/I$. **(a)** Once α is known, we can find the angular displacement, $\Delta\theta$, using $\omega^2 = \omega_0^2 + 2\alpha\,\Delta\theta$. We divide $\Delta\theta$ by 2π to convert from radians to revolutions. **(b)** We can find the angular speed at $t = 27$ s by using $\omega = \omega_0 + \alpha t$. The linear speed at the tip of the blades is then $v = r\omega$.

$\omega_0 = 1.1$ rad/s

Known Radius of turbine, $r = 37$ m; moment of inertia, $I = 8.4 \times 10^6\text{ kg}\cdot\text{m}^2$; initial angular speed, $\omega_0 = 1.1$ rad/s; net torque, $\tau = 1.6 \times 10^5\text{ N}\cdot\text{m}$.

Unknown **(a)** Angular displacement, $\Delta\theta = ?$, when $\omega = 1.8$ rad/s
(b) Linear speed of blade tips, $v = ?$, at $t = 27$ s

SOLUTION

1. Use the torque and moment of inertia to calculate the angular acceleration of the turbine: $\quad \alpha = \dfrac{\tau}{I} = \dfrac{1.6 \times 10^5\text{ N}\cdot\text{m}}{8.4 \times 10^6\text{ kg}\cdot\text{m}^2} = 0.019\text{ rad/s}^2$

Part (a)

2. Rearrange $\omega^2 = \omega_0^2 + 2\alpha\,\Delta\theta$ to calculate the angular displacement $\Delta\theta$:

$$\Delta\theta = \frac{(\omega^2 - \omega_0^2)}{2\alpha}$$
$$= \frac{(1.8\text{ rad/s})^2 - (1.1\text{ rad/s})^2}{2(0.019\text{ rad/s}^2)} = 53\text{ rad}$$

CONTINUED

3. Convert from radians to revolutions:

$$\Delta\theta = (53 \text{ rad})\left(\frac{1 \text{ rev}}{2\pi \text{ rad}}\right) = 8.4 \text{ rev}$$

Part (b)

4. Use $\omega = \omega_0 + \alpha t$ to find the final angular speed at $t = 27$ s:

$$\omega = \omega_0 + \alpha t$$
$$= (1.1 \text{ rad/s}) + (0.019 \text{ rad/s}^2)(27 \text{ s}) = 1.6 \text{ rad/s}$$

5. Calculate the corresponding linear speed with $v = r\omega$:

$$v = r\omega = (37 \text{ m})(1.6 \text{ rad/s}) = 59 \text{ m/s}$$

INSIGHT

The speed of the blade tips at $t = 27$ s is about 132 mph. This speed is an important consideration in wind turbine design because the greater the tip speed, the more noise the turbine makes.

PRACTICE PROBLEM

What is the linear acceleration of the blade tips at $t = 27$ s? [**Answer:** $a = r\alpha = (37 \text{ m})(0.019 \text{ rad/s}^2) = 0.70 \text{ m/s}^2$. This result is independent of the time, as long as the angular acceleration remains constant.]

Some related homework problems: Problem 10, Problem 11, Problem 15

CONCEPTUAL EXAMPLE 11-5   WHICH BLOCK LANDS FIRST?   PREDICT/EXPLAIN

The rotating systems shown in the sketch differ only in that the two spherical movable masses are positioned either far from the axis of rotation (left) or near the axis of rotation (right). **(a)** If the hanging blocks are released simultaneously from rest, is it observed that the block on the left lands first, the block on the right lands first, or both blocks land at the same time? **(b)** Which of the following is the *best explanation* for your prediction?

I. The torque is the same in both cases, and therefore both blocks land at the same time.

II. The moment of inertia is less for the system on the right, and hence it has the greater angular acceleration. As a result the block on the right lands first.

III. The moment of inertia is greater for the system on the left, and hence it has more kinetic energy as it rotates. As a result the block on the left lands first.

REASONING AND DISCUSSION

The net external torque, which is supplied by the hanging blocks, is the same for each of these systems. However, the moment of inertia of the system on the right is less than that of the system on the left, because the movable masses are closer to the axis of rotation. The angular acceleration is inversely proportional to the moment of inertia ($\alpha = \tau/I$), and hence the system on the right has the greater angular acceleration, and it wins the race.

ANSWER

(a) The block on the right lands first. **(b)** The best explanation is II.

EXAMPLE 11-6   DROP IT

A person holds his outstretched arm at rest in a horizontal position. The mass of the arm is m and its length is 0.740 m. When the person releases his arm, allowing it to drop freely, it begins to rotate about the shoulder joint. Find **(a)** the initial angular acceleration of the arm, and **(b)** the initial linear acceleration of the man's hand. (*Hint*: In calculating the torque, assume the mass of the arm is concentrated at its midpoint. In calculating the angular acceleration, use the moment of inertia of a uniform rod of length L about one end; $I = \frac{1}{3}mL^2$.)

PICTURE THE PROBLEM

The arm is initially horizontal and at rest. When released, it rotates downward about the shoulder joint. The force of gravity, mg, acts at a distance of $(0.740 \text{ m})/2 = 0.370$ m from the shoulder.

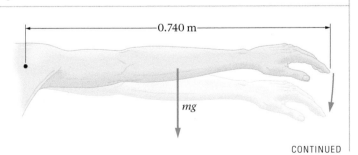

REASONING AND STRATEGY

The angular acceleration, α, can be found using $\tau = I\alpha$. In this case, the initial torque is $\tau = mg(L/2)$, where $L = 0.740$ m, and the moment of inertia is $I = \frac{1}{3}mL^2$.

CONTINUED

Once the initial angular acceleration is found, the corresponding linear acceleration is obtained from $a = r\alpha$.

Known Mass of arm, m; length of arm, $L = 0.740$ m.
Unknown (a) Initial angular acceleration of arm, $\alpha = ?$ (b) Initial linear acceleration, $a = ?$

SOLUTION

Part (a)

1. Use $\tau = I\alpha$ to find the angular acceleration, α:

$$\alpha = \frac{\tau}{I}$$

2. Write expressions for the initial torque, τ, and the moment of inertia, I:

$$\tau = mg\frac{L}{2}$$

$$I = \frac{1}{3}mL^2$$

3. Substitute τ and I into the expression for the angular acceleration. Notice that the mass of the arm cancels:

$$\alpha = \frac{\tau}{I} = \frac{mgL/2}{mL^2/3} = \frac{3g}{2L}$$

4. Substitute numerical values:

$$\alpha = \frac{3g}{2L} = \frac{3(9.81 \text{ m/s}^2)}{2(0.740 \text{ m})} = 19.9 \text{ rad/s}^2$$

Part (b)

5. Use $a = r\alpha$ to calculate the linear acceleration at the man's hand, a distance $r = L$ from the shoulder:

$$a = L\alpha = L\left(\frac{3g}{2L}\right) = \frac{3}{2}g = 14.7 \text{ m/s}^2$$

INSIGHT

The linear acceleration of the hand is 1.50 times greater than the acceleration of gravity, regardless of the mass of the arm. This can be demonstrated with the following simple experiment: Hold your arm straight out with a pen resting on your hand. Now, relax your deltoid muscles, and let your arm rotate freely downward about your shoulder joint. Notice that as your arm falls downward, your hand moves more rapidly than the pen, which appears to "lift off" your hand. The pen drops with the acceleration of gravity, which is clearly less than the acceleration of the hand. This effect can be seen in **FIGURE 11-9**.

PRACTICE PROBLEM

At what distance from the shoulder is the initial linear acceleration of the arm equal to the acceleration of gravity? [**Answer:** Set $a = r\alpha$ equal to g. This gives $r = 2L/3 = 0.493$ m.]

Some related homework problems: Problem 14, Problem 16, Problem 70

▶ **FIGURE 11-9** As a rod of length L rotates freely about one end, points farther from the axle than $2L/3$ have an acceleration greater than g (see the Practice Problem for Example 11-6). Thus, the rod falls out from under the last two dice.

Enhance Your Understanding   (Answers given at the end of the chapter)

2. Consider two objects with the following characteristics: Object 1 is a hoop with mass M and radius R; object 2 is a disk with mass $2M$ and radius $2R$. Suppose a torque τ is applied to object 1, and a torque 2τ is applied to object 2. Is the angular acceleration of object 1 greater than, less than, or equal to the angular acceleration of object 2? Explain.

Section Review

• The angular acceleration of an object is directly proportional to the net torque applied to it, and inversely proportional to the object's moment of inertia.

• In equation form that is similar to $F_{\text{net}} = ma$, we can write Newton's second law for rotational motion as $\tau_{\text{net}} = I\alpha$.

11-3 Zero Torque and Static Equilibrium

The parents of a young boy are supporting him on a long, lightweight plank, as illustrated in **FIGURE 11-10**. If the mass of the child is m, the upward forces exerted by the parents must sum to mg; that is,

$$F_1 + F_2 = mg$$

This condition ensures that the net force acting on the plank is zero. It *does not*, however, guarantee that the plank remains at rest.

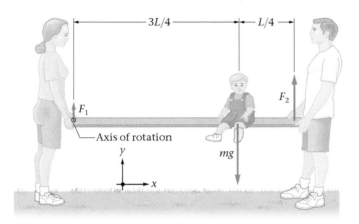

◄ **FIGURE 11-10** **Forces required for static equilibrium** Two parents support a child on a lightweight plank of length L. For the calculation described in the text, we choose the axis of rotation to be the left end of the plank.

To see why, imagine for a moment that the parent on the right lets go of the plank and that the parent on the left increases her force until it is equal to the weight of the child. In this case, $F_1 = mg$ and $F_2 = 0$, which clearly satisfies the force equation we have just written. The right end of the plank is no longer supported, however, and it drops toward the ground while the left end rises. In other words, the plank rotates in a clockwise sense. For the plank to remain completely at rest, with no translation or rotation, we must impose the following *two* conditions:

1. The net force acting on the plank must be zero, so that there is no translational acceleration.

2. The net torque acting on the plank must be zero, so that there is no rotational acceleration.

If both of these conditions are met, an extended object, like the plank, will remain at rest if it starts at rest. To summarize:

Conditions for Static Equilibrium
For an extended object to be in static equilibrium, the following two conditions must be met:

(i) The net force acting on the object must be zero:

$$\sum F_x = 0, \qquad \sum F_y = 0 \qquad\qquad \text{11-5}$$

(ii) The net torque acting on the object must be zero:

$$\sum \tau = 0 \qquad\qquad \text{11-6}$$

These conditions are independent of one another; that is, satisfying one does *not* guarantee that the other is satisfied.

Big Idea **3** A system is in static equilibrium if both the net force and the net torque acting on it are equal to zero.

Applying the Conditions for Equilibrium Let's apply these conditions to the plank that supports the child. First, we consider the forces acting on the plank, with upward chosen as the positive direction, as in Figure 11-10. Setting the net force equal to zero yields

$$F_1 + F_2 - mg = 0$$

Clearly, this agrees with the force equation we wrote earlier.

Next, we apply the torque condition. To do so, we must first choose an axis of rotation. For example, we might take the left end of the plank to be the axis, as in Figure 11-10. With this choice, we see that the force F_1 exerts zero torque, because it acts directly through the axis of rotation. On the other hand, F_2 acts at the far end of the plank, a distance L from the axis. In addition, F_2 would cause a counterclockwise (positive) rotation if it acted alone, as we can see in Figure 11-10. Therefore, the torque due to F_2 is positive and has the magnitude

$$\tau_2 = F_2 L$$

Finally, the weight of the child, mg, acts at a distance of $3L/4$ from the axis, and would cause a clockwise (negative) rotation if it acted alone. Hence, its torque is negative:

$$\tau_{mg} = -mg\left(\tfrac{3}{4}L\right)$$

Setting the net torque equal to zero, then, yields the following condition:

$$\tau_{net} = F_2 L - mg\left(\tfrac{3}{4}L\right) = 0$$

This torque condition, along with the force condition in $F_1 + F_2 - mg = 0$, can be used to determine the two unknowns, F_1 and F_2. For example, we can begin by canceling L in the torque equation to find F_2:

$$F_2 = \tfrac{3}{4}mg$$

Substituting this result into the force condition gives

$$F_1 + \tfrac{3}{4}mg - mg = 0$$

Therefore, F_1 is

$$F_1 = \tfrac{1}{4}mg$$

These two forces support the plank, *and* keep it from rotating. As one might expect, the force nearer the child is greater.

PROBLEM-SOLVING NOTE

Axis of Rotation

Any point in a system may be used as the axis of rotation when calculating torque. It is generally best, however, to choose an axis that gives zero torque for at least one of the unknown forces in the system. Such a choice simplifies the algebra needed to solve for the forces.

Any Location Can Be the Axis of Rotation   Our choice of the left end of the plank as the axis of rotation was completely arbitrary. In fact, if an object is in static equilibrium, the net torque acting on it is zero, regardless of the location of the axis of rotation. Hence, we are free to choose an axis of rotation that is most convenient for a given problem. In general, it is useful to pick the axis to be at the location of one of the unknown forces. This eliminates that force from the torque condition and simplifies the remaining algebra. In the following Example, we choose an axis of rotation that eliminates the unknown force F_2 from the torque equation.

EXAMPLE 11-7   FIND THE FORCES: AXIS ON THE RIGHT

A child of mass m is supported on a light plank by his parents, who exert the forces F_1 and F_2 as indicated in the sketch. Find the forces required to keep the plank in static equilibrium. Use the right end of the plank as the axis of rotation.

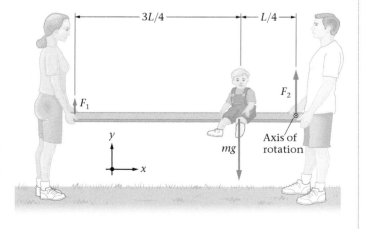

PICTURE THE PROBLEM
Our sketch shows the same physical system as in Figure 11-10, only this time the axis of rotation is at the right end of the plank.

REASONING AND STRATEGY
We set both the net force and the net torque equal to zero. The expression for the net torque is different this time, due to the new location for the axis of rotation, but the final result for the forces will be the same.

Known   Mass of child, m; distance for force 1, L; distance for force 2, 0; distance for weight (mg) of child, $L/4$.
Unknown $F_1 = ?$ $F_2 = ?$

CONTINUED

SOLUTION

1. Set the net force acting on the plank equal to zero:

$$F_1 + F_2 - mg = 0$$

2. Set the net torque acting on the plank equal to zero. [Notice that the force F_1 rotates the plank clockwise (negative torque), and the weight mg rotates it counterclockwise (positive torque) about the axis.]:

$$-F_1(L) + mg\left(\tfrac{1}{4}L\right) = 0$$

3. Solve the torque condition for the force F_1:

$$F_1 = \tfrac{1}{4}mg$$

4. Substitute F_1 into the force condition to find F_2:

$$F_2 = mg - \tfrac{1}{4}mg = \tfrac{3}{4}mg$$

INSIGHT

As expected, the final results for the forces are identical to those obtained previously. Notice, however, that the torque produced by the child would cause a counterclockwise rotation in this case, and hence it is positive. Thus, the magnitude *and* sign of the torque produced by a given force depend on the location chosen for the axis of rotation.

PRACTICE PROBLEM — PREDICT/CALCULATE

Suppose the child moves to a new position, with the result that the force exerted by the father is reduced from 0.75mg to 0.60mg. **(a)** Did the child move to the left or to the right? Explain. **(b)** Determine the distance and direction of the child's displacement. [**Answer: (a)** The child moved to the left, making for a more even distribution of the system's mass. **(b)** The new position of the child is 0.40L from the father. This means the child moved left a distance of 0.15L.]

Some related homework problems: Problem 23, Problem 24

In the next Example, we show that the forces supporting a person or other object sometimes act in different directions. To emphasize the direction of the forces, we solve the Example in terms of the components of the relevant forces.

EXAMPLE 11-8 TAKING THE PLUNGE

A 5.00-m-long diving board of negligible mass is supported by two pillars. One pillar is at the left end of the diving board, as shown in the sketch; the other is 1.50 m away. Find the forces exerted by the pillars when a 90.0-kg diver stands at the far end of the board.

PICTURE THE PROBLEM

We choose upward to be the positive direction for the forces. When calculating torques, we use the left end of the diving board as the axis of rotation, which means that $\vec{\mathbf{F}}_1$ causes zero torque. In addition, $\vec{\mathbf{F}}_2$ would cause a counterclockwise rotation if it acted alone, so its torque is positive. On the other hand, $m\vec{\mathbf{g}}$ would cause a clockwise rotation, so its torque is negative. Finally, $\vec{\mathbf{F}}_2$ acts at a distance d from the axis of rotation, and $m\vec{\mathbf{g}}$ acts at a distance L.

REASONING AND STRATEGY

As usual in static equilibrium problems, we use the conditions of (i) zero net force and (ii) zero net torque to determine the unknown forces, $\vec{\mathbf{F}}_1$ and $\vec{\mathbf{F}}_2$. In this system, all forces act in the positive or negative y direction; thus we need only set the net y component of force equal to zero.

Known   Length of diving board, $L = 5.00$ m; distance of second pillar, $d = 1.50$ m; mass of diver, $m = 90.0$ kg.

Unknown   Force exerted by pillar 1, $F_{1,y} = ?$; force exerted by pillar 2, $F_{2,y} = ?$

SOLUTION

1. Set the net y component of force acting on the diving board equal to zero:

$$\sum F_y = F_{1,y} + F_{2,y} - mg = 0$$

2. Calculate the torque due to each force, using the left end of the board as the axis of rotation. Notice that each force is at right angles to the radius and that $\vec{\mathbf{F}}_1$ goes directly through the axis of rotation:

$$\tau_1 = F_{1,y}(0) = 0$$
$$\tau_2 = F_{2,y}(d)$$
$$\tau_3 = -mg(L)$$

CONTINUED

3. Set the net torque acting on the diving board equal to zero:

$$\sum \tau = F_{1,y}(0) + F_{2,y}(d) - mg(L) = 0$$

4. Solve the torque equation for the force $F_{2,y}$:

$$F_{2,y} = mg(L/d)$$
$$= (90.0\ \text{kg})(9.81\ \text{m/s}^2)(5.00\ \text{m}/1.50\ \text{m}) = 2940\ \text{N}$$

5. Use the force equation to determine $F_{1,y}$:

$$F_{1,y} = mg - F_{2,y}$$
$$= (90.0\ \text{kg})(9.81\ \text{m/s}^2) - 2940\ \text{N} = -2060\ \text{N}$$

INSIGHT

The first point to notice about our solution is that $F_{1,y}$ is negative, which means that $\vec{F}_1$ is actually directed *downward*, as shown to the right. To see why, imagine for a moment that the board is no longer connected to the first pillar. In this case, the board would rotate clockwise about the second pillar, and the left end of the board would move upward. Thus, a downward force is required on the left end of the board to hold it in place.

The second point is that both pillars exert forces with magnitudes that are considerably larger than the diver's weight, $mg = 883\ \text{N}$. In particular, the first pillar must pull downward with a force of $2.33mg$, while the second pillar pushes upward with a force of $2.33mg + mg = 3.33mg$. This is not unusual. In fact, it is common for the forces in a structure, such as a bridge, a building, or the human body, to be much greater than the weight it supports.

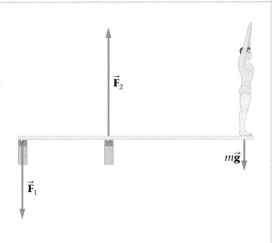

PRACTICE PROBLEM

Find the forces exerted by the pillars when the diver is 1.00 m to the left of the right end.   [**Answer:** $F_{1,y} = -1470\ \text{N}, F_{2,y} = 2350\ \text{N}$]

Some related homework problems: Problem 24, Problem 25, Problem 29

RWP* To this point we have ignored the mass of the plank holding the child and the diving board holding the swimmer, since they were described as lightweight. If we want to consider the torque exerted by the force of gravity on an extended object of finite mass, however, we can simply treat it as if all its mass were concentrated at its center of mass, as was done in similar situations in Section 9-7. We consider such a system in the next Example.

Walking the plank

EXAMPLE 11-9   WALKING THE PLANK: FIND THE MASS

A cat walks along a plank with mass $M = 6.00\ \text{kg}$. The plank is supported by two sawhorses. The center of mass of the plank is a distance $d_1 = 0.850\ \text{m}$ to the left of sawhorse B. When the cat is a distance $d_2 = 1.11\ \text{m}$ to the right of sawhorse B, the plank just begins to tip. What is the mass of the cat, m?

PICTURE THE PROBLEM

In our sketch, the cat is at the point where the plank just begins to tip. As a result, the force exerted by sawhorse A is zero, $F_A = 0$. The distance from sawhorse B to the plank's center of mass is $d_1 = 0.850\ \text{m}$, and the distance from sawhorse B to the cat is $d_2 = 1.11\ \text{m}$.

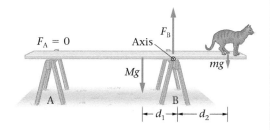

REASONING AND STRATEGY

We use sawhorse B as the axis of rotation, and we consider all of the mass of the plank to be concentrated at its center of mass. It follows that just two forces cause torques on the plank: (1) the weight of the cat, and (2) the weight of the plank. Setting the net torque caused by these two forces equal to zero determines the weight—and hence the mass—of the cat.

Known   Mass of plank, $M = 6.00\ \text{kg}$; distance 1, $d_1 = 0.850\ \text{m}$; distance 2, $d_2 = 1.11\ \text{m}$.
Unknown Mass of cat, $m = ?$

SOLUTION

1. Calculate the net torque about sawhorse B and set it equal to zero. Recall that clockwise rotation is negative, and counterclockwise rotation is positive:

$$\tau_{\text{net}} = -mg(d_2) + Mg(d_1) = 0$$

CONTINUED

---

*Real World Physics applications are denoted by the acronym RWP.

2. Solve for the mass of the cat and substitute numerical values:

$$mg(d_2) = Mg(d_1)$$

$$m = M\left(\frac{d_1}{d_2}\right) = (6.00 \text{ kg})\left(\frac{0.850 \text{ m}}{1.11 \text{ m}}\right) = 4.59 \text{ kg}$$

INSIGHT

We did not include a torque for sawhorse A because F_A is zero. The nonzero force F_B from sawhorse B also produces zero torque. This is because F_B goes through the axis of rotation. The magnitude of F_B is equal to $mg + Mg = 104$ N in order to make the net force acting on the system equal to zero.

PRACTICE PROBLEM

Write both the zero force and zero torque conditions for the case where the axis of rotation is at the center of mass of the plank.   [**Answer:** Zero force: $F_B - mg - Mg = 0$. Zero torque: $F_B(d_1) - mg(d_1 + d_2) = 0$. Combining these relationships gives $m = 4.59$ kg, as before.]

Some related homework problems: Problem 27, Problem 36

The same basic physics applies to a car or SUV that hits the brakes in a sudden stop, as illustrated in **FIGURE 11–11**. As the vehicle brakes, the frictional forces with the road tend to make it rotate in the clockwise sense. This is countered by a larger normal force on the front tires, which exerts a torque in the counterclockwise sense. The result is the vehicle goes into a "nose down" position during the braking.

Forces with Both Vertical and Horizontal Components

All of the previous examples dealt with forces that point either directly upward or directly downward. We now consider a more general situation, where forces may have both vertical and horizontal components. For example, consider the wall-mounted lamp (sconce) shown in **FIGURE 11–12**. The sconce consists of a light curved rod that is bolted to the wall at its lower end. Suspended from the upper end of the rod, a horizontal distance H from the wall, is the lamp of mass m. The rod is also connected to the wall by a horizontal wire a vertical distance V above the bottom of the rod.

Now, suppose we are designing this sconce to be placed in the lobby of a building on campus. To ensure its structural stability, we would like to know the tension T the wire must exert, as well as the vertical and horizontal components of the force $\vec{\mathbf{f}}$ that must be exerted by the bolt on the rod. This information will be important in deciding on the type of wire and bolt to be used in the construction.

To find these forces, we apply the same conditions as before: The net force and the net torque must be zero. In this case, however, forces may have both horizontal and vertical components. Thus, the condition of zero net force is really two separate conditions: (i) zero net force in the horizontal direction, and (ii) zero net force in the vertical direction. These two conditions plus (iii), zero net torque, allow for a full solution of the problem.

Zero Net Torque   We begin with the torque condition. A convenient choice for the axis of rotation is the bottom end of the rod, because this eliminates one of the unknown forces ($\vec{\mathbf{f}}$). With this choice we can readily calculate the net torque acting on the rod. We find

$$\sum \tau = T(V) - mg(H) = 0$$

This relationship can be solved immediately for the tension, giving

$$T = mg(H/V)$$

Notice that the tension is increased if the wire is connected closer to the bottom of the rod—that is, if V is reduced.

Zero Net Force   Next, we apply the force conditions. First, we sum the y components of all the forces and set the sum equal to zero:

$$\sum F_y = f_y - mg = 0$$

▲ **FIGURE 11-11** As the brakes are applied on this SUV, rotational equilibrium demands that the normal forces exerted on the front tires be greater than the normal forces exerted on the rear tires—which is why braking cars are "nose down" during a rapid stop. For this reason, many cars use disk brakes for the front wheels and less powerful drum brakes for the rear wheels. As the disk brakes wear, they tend to coat the front wheels with dust from the brake pads, which gives the front wheels a characteristic "dirty" look.

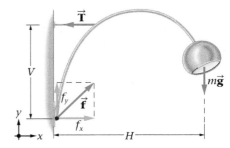

▲ **FIGURE 11-12 A lamp in static equilibrium** A wall-mounted lamp of mass m is suspended from a light curved rod. The bottom of the rod is bolted to the wall. The rod is also connected to the wall by a horizontal wire a vertical distance V above the bottom of the rod.

Thus, the vertical component of the force exerted by the bolt simply supports the weight of the lamp:

$$f_y = mg$$

Next, we sum the x components of the forces, and set that sum equal to zero as well:

$$\sum F_x = f_x - T = 0$$

Clearly, the x component of the force exerted by the bolt has the same magnitude as the tension, but it points in the opposite direction:

$$f_x = T = mg(H/V)$$

The bolt, then, pushes upward on the rod to support the lamp, and at the same time it pushes to the right to keep the rod from rotating.

For example, suppose the lamp in Figure 11-12 has a mass of 2.00 kg, and that $V = 12.0$ cm and $H = 15.0$ cm. In this case, we find the following forces:

$$T = mg(H/V) = (2.00 \text{ kg})(9.81 \text{ m/s}^2)(15.0 \text{ cm})/(12.0 \text{ cm}) = 24.5 \text{ N}$$
$$f_x = T = 24.5 \text{ N}$$
$$f_y = mg = (2.00 \text{ kg})(9.81 \text{ m/s}^2) = 19.6 \text{ N}$$

▲ **FIGURE 11-13** The chains that support this sign maintain it in a state of translational and rotational equilibrium. The forces in the chains are most easily analyzed by resolving them into vertical and horizontal components and applying the conditions for equilibrium. In particular, the net vertical force, the net horizontal force, and the net torque must all be zero.

Notice that f_x and T are greater than the weight of the lamp, $mg = 19.6$ N. Just as we found with the diving board in Example 11-8, the forces required of structural elements can be greater than the weight of the object to be supported—an important consideration when designing a structure like a bridge, or a hanging sign like the one in **FIGURE 11-13**. The same effect occurs in the human body. We find in Problem 23, for example, that the force exerted by the biceps to support a baseball in the hand is several times larger than the baseball's weight. Similar conclusions apply to muscles throughout the body.

BIO In the next Example we consider another system in which forces have both vertical and horizontal components.

EXAMPLE 11-10   ARM IN A SLING

A hiker who has broken his forearm rigs a temporary sling using a cord stretching from his shoulder to his hand. The cord holds the forearm level and makes an angle of 40.0° with the horizontal where it attaches to the hand. Considering the forearm and hand to be uniform, with a total mass of 1.30 kg and a length of 0.300 m, find **(a)** the tension in the cord and **(b)** the horizontal and vertical components of the force, $\vec{\mathbf{f}}$, exerted by the humerus (the bone of the upper arm) on the radius and ulna (the bones of the forearm).

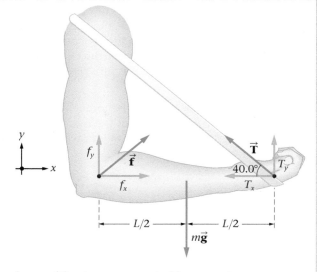

PICTURE THE PROBLEM
In our sketch, we use the typical conventions for the positive x and y directions. In addition, the forearm and hand are assumed to be a uniform object, and hence we indicate the weight mg as acting at its center. The length of the forearm and hand is $L = 0.300$ m. Finally, two other forces act on the forearm: (i) the tension in the cord, $\vec{\mathbf{T}}$, at an angle of 40.0° above the negative x axis, and (ii) the force $\vec{\mathbf{f}}$ exerted at the elbow joint.

REASONING AND STRATEGY
In this system there are three unknowns: T, f_x, and f_y. These unknowns can be determined using the following three conditions: (i) net torque equals zero, (ii) net x component of force equals zero, and (iii) net y component of force equals zero.

    We start with the torque condition, using the elbow joint as the axis of rotation. As we shall see, this choice of axis eliminates the force f, and gives a direct solution for the tension T. Next, we use T and the two force conditions to determine f_x and f_y.

CONTINUED

Known   Mass of arm and hand, $m = 1.30$ kg; length of arm and hand, $L = 0.300$ m; angle of cord, 40.0°.
Unknown   (a) Tension in cord, $T = ?$ (b) Forces at elbow, $f_x = ?$ and $f_y = ?$

SOLUTION

Part (a)

1. Calculate the torque about the elbow joint. Notice that f causes zero torque, mg causes a negative torque, and the vertical component of T causes a positive torque. The horizontal component of T produces no torque, because it is on a line with the axis:

$$\sum \tau = (T \sin 40.0°)L - mg(L/2) = 0$$

2. Solve the torque condition for the tension, T:

$$T = \frac{mg}{2 \sin 40.0°} = \frac{(1.30 \text{ kg})(9.81 \text{ m/s}^2)}{2 \sin 40.0°} = 9.92 \text{ N}$$

Part (b)

3. Set the sum of the x components of force equal to zero, and solve for f_x:

$$\sum F_x = f_x - T \cos 40.0° = 0$$
$$f_x = T \cos 40.0° = (9.92 \text{ N}) \cos 40.0° = 7.60 \text{ N}$$

4. Set the sum of the y components of force equal to zero, and solve for f_y:

$$\sum F_y = f_y - mg + T \sin 40.0° = 0$$
$$f_y = mg - T \sin 40.0°$$
$$= (1.30 \text{ kg})(9.81 \text{ m/s}^2) - (9.92 \text{ N}) \sin 40.0° = 6.38 \text{ N}$$

INSIGHT

It is not necessary to determine T_x and T_y separately, because we know the direction of the cord. In particular, it is clear from our sketch that the components of $\vec{T}$ are $T_x = -T \cos 40.0° = -7.60$ N and $T_y = T \sin 40.0° = 6.38$ N.

Did you notice that $\vec{f}$ is at an angle of 40.0° with respect to the positive x axis, the same angle that $\vec{T}$ makes with the negative x axis? The reason for this symmetry, of course, is that mg acts at the center of the forearm. If mg were to act closer to the elbow, for example, $\vec{f}$ would make a larger angle with the horizontal, as we see in the following Practice Problem.

PRACTICE PROBLEM

Suppose that the forearm and hand are nonuniform, and that the center of mass is located at a distance of $L/4$ from the elbow joint. What are T, f_x, and f_y in this case?   [**Answer:** $T = 4.96$ N, $f_x = 3.80$ N, $f_y = 9.56$ N. In this case, $\vec{f}$ makes an angle of 68.3° with the horizontal.]

Some related homework problems: Problem 30, Problem 76, Problem 80

EXAMPLE 11-11   DON'T WALK UNDER THE LADDER: FIND THE FORCES

An 85-kg person stands on a lightweight ladder, as shown in the sketch. The floor is rough; hence, it exerts both a normal force, f_1, and a frictional force, f_2, on the ladder. The wall, on the other hand, is frictionless; it exerts only a normal force, f_3. Using the dimensions given in the figure, find the magnitudes of f_1, f_2, and f_3.

PICTURE THE PROBLEM

Our sketch shows the positive x and y directions, as well as the location of the axis of rotation, which is at the bottom of the ladder.

REASONING AND STRATEGY

As usual, we set both the net force and the net torque equal to zero. In this case, the net force must be set equal to zero in both the x and y directions.

Known   Mass of person, $m = 85$ kg; vertical distance, $a = 3.8$ m; horizontal distance, $b = 0.70$ m.
Unknown   Forces exerted on ladder: $f_1 = ?; f_2 = ?; f_3 = ?$

f_3

$a = 3.8$ m

f_1

mg

y

Axis

x

f_2

$b = 0.70$ m

SOLUTION

1. Set the net torque acting on the ladder equal to zero. Use the bottom of the ladder as the axis:

$$\tau_{net} = f_3(a) - mg(b) = 0$$

2. Solve for f_3:

$$f_3 = mg(b/a) = 150 \text{ N}$$

3. Sum the x components of force and set equal to zero:

$$f_2 - f_3 = 0$$

CONTINUED

4. Solve for f_2: $\qquad\qquad\qquad\qquad\qquad\qquad$ $f_2 = f_3 = 150\,\text{N}$

5. Sum the y components of force and set equal to zero: $\quad$ $f_1 - mg = 0$

6. Solve for f_1: $\qquad\qquad\qquad\qquad\qquad\qquad\quad$ $f_1 = mg = 830\,\text{N}$

INSIGHT

If the floor is quite smooth, the ladder might slip—it depends on whether the coefficient of static friction is great enough to provide the needed force $f_2 = 150\,\text{N}$. In this system, the normal force exerted by the floor is $N = f_1 = 830\,\text{N}$. Therefore, if the coefficient of static friction is greater than 0.18 [noting that $0.18(830\,\text{N}) = 150\,\text{N}$], the ladder will stay put. Ladders often have rubberized pads on the bottom in order to increase the static friction, and hence increase the safety of the ladder.

PRACTICE PROBLEM

Write the zero-torque condition, and also the two zero-force conditions, for the case where the axis of rotation is at the top of the ladder. [**Answer:** Zero horizontal force: $f_2 - f_3 = 0$; zero vertical force, $f_1 - mg = 0$; zero torque, $-f_1(b + c) + f_2(a) + mg(c) = 0$, where c is the horizontal distance from the wall to the line through the force mg. These relationships give the same results for f_1, f_2, and f_3, as expected, independent of the value of c.]

Some related homework problems: Problem 30, Problem 31

Enhance Your Understanding $\qquad$ (Answers given at the end of the chapter)

3. A "Physics" sign is supported symmetrically by two ropes, as shown in FIGURE 11-14. Is the magnitude of the tension in each rope greater than, less than, or equal to half the weight of the sign? Explain.

▲ **FIGURE 11-14** Enhance Your Understanding 3.

Section Review

- A system is in equilibrium when both the net force and the net torque acting on it are zero. The condition of zero net force applies equally to the x and y directions.

11-4 Center of Mass and Balance

Suppose you decide to construct a mobile. To begin, you tie a thread to a light rod, as in FIGURE 11-15. Notice that the rod extends a distance x_1 to the left of the thread and a distance x_2 to the right. At the left end of the rod you attach an object of mass m_1. What mass, m_2, should be attached to the right end if the rod is to be balanced?

From the discussions in the previous sections, it is clear that if the rod is to be in static equilibrium (balanced), the net torque acting on it must be zero. Taking the point where the thread is tied to the rod as the axis of rotation, we can write this zero-torque condition as

$$m_1 g(x_1) - m_2 g(x_2) = 0$$

Canceling g and rearranging, we find

$$m_1 x_1 = m_2 x_2 \qquad\qquad 11\text{-}7$$

This gives the following result for m_2:

$$m_2 = m_1(x_1/x_2)$$

For example, if $x_2 = 2x_1$, it follows that m_2 should be one-half of m_1.

Let's now consider a slightly different question: Where is the center of mass of m_1 and m_2? Choosing the origin of the x axis to be at the location of the thread, as indicated in Figure 11-15, we can use the definition of the center of mass, Equation 9-14, to find x_{cm}:

$$x_{cm} = \frac{m_1(-x_1) + m_2(x_2)}{m_1 + m_2} = -\left(\frac{m_1 x_1 - m_2 x_2}{m_1 + m_2}\right)$$

▲ **FIGURE 11-15 Zero torque and balance** One section of a mobile. The rod is balanced when the net torque acting on it is zero. This is equivalent to having the center of mass directly under the suspension point.

Referring to the zero-torque condition in Equation 11-7, we see that $m_1x_1 - m_2x_2 = 0$; hence the center of mass is at the origin:

$$x_{cm} = 0$$

This is precisely where the string is attached. We conclude, then, that the rod balances when the center of mass is directly below the point from which the rod is suspended. This is a general result.

Let's apply this result to the case of the mobile shown in the next Example.

EXAMPLE 11-12   A WELL-BALANCED MEAL

As a grade-school project, students construct a mobile representing some of the major food groups. Their completed artwork is shown in the sketch. Find the masses m_1, m_2, and m_3 that are required for a perfectly balanced mobile. Assume the strings and the horizontal rods have negligible mass.

PICTURE THE PROBLEM
The dimensions of the horizontal rods and the values of the given masses are indicated in our sketch. Notice that each rod is balanced at its suspension point.

REASONING AND STRATEGY
We can find all three unknown masses by repeatedly applying the condition for balance, $m_1x_1 = m_2x_2$.

First, we apply the balance condition to m_1 and m_2, with the distances $x_1 = 12$ cm and $x_2 = 18$ cm. This gives a relationship between m_1 and m_2.

To get a second relationship between m_1 and m_2, we apply the balance condition again at the next higher level of the mobile. That is, the mass $(m_1 + m_2)$ at the distance 6.0 cm must balance the mass 0.30 kg at the distance 24 cm. These two conditions determine m_1 and m_2.

To find m_3 we again apply the balance condition, this time with the mass $(m_1 + m_2 + 0.30\text{ kg} + 0.20\text{ kg})$ at the distance 18 cm, and the mass m_3 at the distance 31 cm.

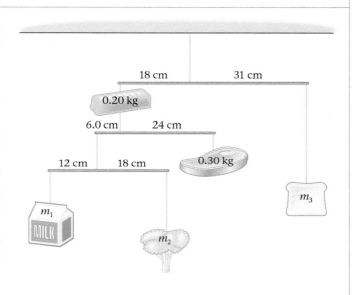

Known   Masses and distances are given in the sketch.
Unknown   Masses $m_1 = ?$, $m_2 = ?$, and $m_3 = ?$

SOLUTION

1. Apply the balance condition to m_1 and m_2:
$$m_1(12\text{ cm}) = m_2(18\text{ cm})$$
$$m_1 = (1.5)m_2$$

2. Apply the balance condition to the next level up in the mobile. Solve for the sum, $m_1 + m_2$:
$$(m_1 + m_2)(6.0\text{ cm}) = (0.30\text{ kg})(24\text{ cm})$$
$$m_1 + m_2 = \frac{(0.30\text{ kg})(24\text{ cm})}{6.0\text{ cm}} = 1.2\text{ kg}$$

3. Substitute $m_1 = (1.5)m_2$ into $m_1 + m_2 = 1.2$ kg to find m_2:
$$(1.5)m_2 + m_2 = (2.5)m_2 = 1.2\text{ kg}$$
$$m_2 = 1.2\text{ kg}/2.5 = 0.48\text{ kg}$$

4. Use $m_1 = (1.5)m_2$ to find m_1:
$$m_1 = (1.5)m_2 = (1.5)0.48\text{ kg} = 0.72\text{ kg}$$

5. Apply the balance condition to the top level of the mobile:
$$(0.72\text{ kg} + 0.48\text{ kg} + 0.30\text{ kg} + 0.20\text{ kg})(18\text{ cm}) = m_3(31\text{ cm})$$

6. Solve for m_3:
$$m_3 = \frac{(1.70\text{ kg})(18\text{ cm})}{31\text{ cm}} = 0.99\text{ kg}$$

INSIGHT
With the values for m_1, m_2, and m_3 found above, the mobile balances at every level. In fact, the center of mass of the *entire* mobile is directly below the point where the uppermost string attaches to the ceiling.

PRACTICE PROBLEM
Find m_1, m_2, and m_3 if the 0.30-kg mass is replaced with a 0.40-kg mass.   [**Answer:** $m_1 = 0.96$ kg, $m_2 = 0.64$ kg, $m_3 = 1.3$ kg]

Some related homework problems: Problem 37, Problem 39

(a) Zero torque   **(b)** Nonzero torque

▲ FIGURE 11-16 **Equilibrium of a suspended object** **(a)** If an object's center of mass is directly below the suspension point, its weight creates zero torque and the object is in equilibrium. **(b)** When an object is rotated, so that the center of mass is no longer directly below the suspension point, the object's weight creates a torque. The torque tends to rotate the object to bring the center of mass under the suspension point.

Center of Mass for a Suspended Object   In general, if you allow an arbitrarily shaped object to hang freely, its center of mass is directly below the suspension point. To see why, notice that when the center of mass is directly below the suspension point, the torque due to gravity is zero, because the force of gravity extends right through the axis of rotation. This is shown in FIGURE 11-16 (a). If the object is rotated slightly, as in FIGURE 11-16 (b), the force of gravity is not in line with the axis of rotation—hence gravity produces a torque. This torque tends to rotate the object, bringing the center of mass back under the suspension point.

For example, suppose you cut a piece of wood into the shape of the continental United States, as shown in FIGURE 11-17. Next, drill a small hole in it, and hang it from the point *A*. The result is that the center of mass lies somewhere on the line *aa'*. Similarly, if a second hole is drilled at point *B*, we find that the center of mass lies somewhere on the line *bb'*. The only point that is on both the line *aa'* *and* the line *bb'* is the point CM, near Smith Center, Kansas, which marks the location of the center of mass.

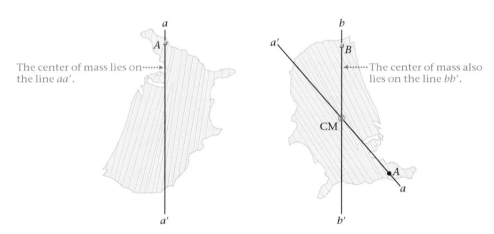

▲ FIGURE 11-17 **The geometric center of the United States** To find the center of mass of an irregularly shaped object, such as a wooden model of the continental United States, suspend it from two or more points. The center of mass lies on a vertical line extending downward from the suspension point. The intersection of these vertical lines gives the precise location of the center of mass.

Balancing a Meter Stick

CONCEPTUAL EXAMPLE 11-13   **COMPARE THE MASSES**   PREDICT/EXPLAIN

A croquet mallet balances when suspended from its center of mass, as indicated in the left-hand drawing to the right. You now cut the mallet into two pieces at its center of mass, as in the right-hand drawing. **(a)** Is the mass of piece 1 greater than, less than, or equal to the mass of piece 2? **(b)** Which of the following is the *best explanation* for your prediction?

  I. Piece 1 has more mass because it balances piece 2, even though it extends a smaller distance from the balance point.

 II. Piece 1 has less mass because it is physically smaller than piece 2.

III. The masses are equal because the mallet balanced at the cut point, and it would balance there only if there were equal mass on either side of that point.

REASONING AND DISCUSSION
The mallet balances because the *torques* due to the two pieces are of equal magnitude. The piece with the head of the mallet (piece 1) extends a smaller distance from the point of suspension than does the other piece, and hence its mass must be greater; that is, a large mass at a small distance creates the same torque as a small mass at a large distance.

ANSWER
(a) Piece 1 has the greater mass. **(b)** The best explanation is I.

▲ **FIGURE 11-18** Visualizing Concepts **Balance and Center of Mass** (Left) Although her knowledge may be based more on practical experience than on physics, this woman knows exactly what she must do to keep from falling. By extending one leg backward as she leans forward, she keeps her center of mass safely positioned over the foot that supports her. (Center) A kangaroo has a long, heavy tail, which helps it to maintain its balance, both on solid ground and in flight between hops. (Right) Although it looks precarious, this rock in Arches National Park, Utah, has probably been balancing above the desert for many thousands of years. It will remain secure on its perch as long as its center of mass lies above its base of support.

Center of Mass for a Standing Object Similar considerations apply to an object standing at rest on a surface, as opposed to being suspended from a point. In such a case, the object is in equilibrium as long as its center of mass is vertically above the base on which it is supported. For example, when you stand upright with normal posture, your feet provide a base of support, and your center of mass is above a point roughly halfway between your feet. If you lift your left foot from the floor—without changing your posture—you will begin to lose your balance and tip over. The reason is that your center of mass is no longer above the base of support, which is now your right foot. To balance on your right foot, you must lean slightly in that direction—and perhaps extend your left leg as well—so as to position your center of mass directly above the foot. This principle applies to everything from a person standing on a stool, to a kangaroo using its tail to balance, to one of the "balancing rocks" that are a familiar sight in the desert Southwest. Examples are shown in **FIGURE 11-18**. In Problem 38 we apply this condition for stability to a stack of books on the edge of a table.

Enhance Your Understanding (Answers given at the end of the chapter)

4. A mobile made from three piggy banks (A, B, C) is shown in **FIGURE 11-19**. The banks are identical, except that they contain different numbers of coins and hence have different masses. Rank the banks in order of increasing mass. Indicate ties where appropriate. (Which bank would you rather have?)

▲ **FIGURE 11-19** Enhance Your Understanding 4.

Section Review

- Extended objects balance when suspended from their center of mass.

(a) Physical picture

(b) Free-body diagram for mass

(c) Torque producing force acting on pulley

▲ FIGURE 11-20 **A mass suspended from a pulley** A mass m hangs from a string wrapped around the circumference of a disk-shaped pulley of radius R and mass M. When the mass is released, it accelerates downward. Positive directions of motion for the system are shown in parts (b) and (c). In part (c), the weight of the pulley acts downward at its center, and the axle exerts an upward force equal in magnitude to the weight of the pulley plus the tension in the string. Of the three forces acting on the pulley, only the tension in the string produces a torque about the axle.

11-5 Dynamic Applications of Torque

In this section we focus on applications of Newton's second law for rotational motion. To begin, consider a disk-shaped pulley of radius R and mass M with a string wrapped around its circumference, as in FIGURE 11-20 (a). Hanging from the string is a mass m. When the mass is released, it accelerates downward and the pulley begins to rotate. If the pulley rotates without friction, and the string unwraps without slipping, what are the acceleration of the mass and the tension in the string?

At first it may seem that because the pulley rotates freely, the mass will simply fall with the acceleration of gravity. But remember, the pulley has a nonzero moment of inertia, $I > 0$, which means that it resists any change in its rotational motion. In order for the pulley to rotate, the string must pull downward on it with a tension T. This means that the string also pulls upward on the mass m with a tension T. As a result, the net downward force on m is less than mg, and hence its acceleration is less than g.

Applying Newton's Second Law to Both Linear and Rotational Motions To solve for the acceleration of the mass, we must apply Newton's second law to both the linear motion of the mass *and* the rotational motion of the pulley. The first step is to define a consistent choice of positive directions for the two motions. In Figure 11-20 (a) we note that when the pulley rotates counterclockwise, the mass moves upward. Thus, we choose counterclockwise to be positive for the pulley and upward to be positive for the mass.

With our positive directions established, we proceed to apply Newton's second law. Referring to the free-body diagram for the mass, shown in FIGURE 11-20 (b), we see that

$$T - mg = ma \qquad \text{11-8}$$

Similarly, the free-body diagram for the pulley is shown in FIGURE 11-20 (c). Notice that the tension in the string, T, exerts a tangential force on the pulley at a distance R from the axis of rotation. This produces a torque of magnitude TR. The tension tends to cause a clockwise rotation, and hence it follows that the torque is negative; thus, $\tau = -TR$. As a result, Newton's second law for the pulley gives

$$-TR = I\alpha \qquad \text{11-9}$$

Now, these two statements of Newton's second law are related by the fact that the string unwraps without slipping. As was discussed in Chapter 10, when a string unwraps without slipping, the angular and linear accelerations are related by

$$\alpha = \frac{a}{R}$$

Using this relationship in Equation 11-9, we have

$$-TR = I\frac{a}{R}$$

Dividing by R yields

$$T = -I\frac{a}{R^2}$$

Substituting this result into $T - mg = ma$, we find

$$-I\frac{a}{R^2} - mg = ma$$

Finally, dividing by m and rearranging yields the acceleration, a:

$$a = -\frac{g}{\left(1 + \dfrac{I}{mR^2}\right)} \qquad \text{11-10}$$

Let's briefly check our solution for a. First, note that a is negative. This is to be expected because the mass accelerates downward, which is the negative direction. Second, if the moment of inertia were zero, $I = 0$, or if the mass m were infinite, $m \to \infty$, the mass would fall with the acceleration of gravity, $a = -g$. When I is greater than zero and m is finite, however, the acceleration of the mass has a magnitude less than g. In fact, in the limit of an infinite moment of inertia, $I \to \infty$, the acceleration vanishes—the mass is simply unable to cause the pulley to rotate in this case.

The next Example presents another system in which Newton's laws are used to relate linear and rotational motions.

EXAMPLE 11-14 THE PULLEY MATTERS

A 0.31-kg cart on a horizontal air track is attached to a string. The string passes over a disk-shaped pulley of mass 0.080 kg and radius 0.012 m and is pulled vertically downward with a constant force of 1.1 N. Find **(a)** the tension in the string between the pulley and the cart and **(b)** the acceleration of the cart.

PICTURE THE PROBLEM
The system is shown in the sketch. We label the mass of the cart with M, the mass of the pulley with m, and the radius of the pulley with r. The applied downward force creates a tension $T_1 = 1.1$ N in the vertical portion of the string. The horizontal portion of the string, from the pulley to the cart, has a tension T_2. If the pulley had zero mass, these two tensions would be equal. In this case, however, T_2 will have a different value than T_1.

We also show the relevant forces acting on the pulley and the cart. The positive direction of rotation is counterclockwise, and the corresponding positive direction of motion for the cart is to the left.

Physical picture   **Relevant forces**

REASONING AND STRATEGY
The two unknowns, T_2 and a, can be found by applying Newton's second law to both the pulley and the cart. This gives two equations for two unknowns.

In applying Newton's second law to the pulley, we note that the pulley is a disk and hence $I = \frac{1}{2}mr^2$. Also, because the string is not said to slip as it rotates the pulley, we can assume that the angular and linear accelerations are related by $\alpha = a/r$.

Known   Mass of cart, $M = 0.31$ kg; mass of pulley, $m = 0.080$ kg; radius of pulley, $r = 0.012$ m; downward force exerted on string, $T_1 = 1.1$ N.
Unknown **(a)** Horizontal tension in string, $T_2 = $? **(b)** Acceleration of the cart, $a = $?

SOLUTION

Part (a)

1. Apply Newton's second law to the cart:

$$T_2 = Ma$$

2. Apply Newton's second law to the pulley. Notice that T_1 causes a positive torque, and T_2 causes a negative torque. In addition, use the relationship $\alpha = a/r$:

$$\sum \tau = I\alpha$$
$$rT_1 - rT_2 = \left(\tfrac{1}{2}mr^2\right)\left(\frac{a}{r}\right) = \tfrac{1}{2}mra$$

3. Use the cart equation, $T_2 = Ma$, to eliminate the acceleration a in the pulley equation:

$$a = \frac{T_2}{M}$$
$$rT_1 - rT_2 = \tfrac{1}{2}mr\left(\frac{T_2}{M}\right)$$

4. Cancel r and solve for T_2:

$$T_2 = \frac{T_1}{1 + m/2M} = \frac{1.1 \text{ N}}{1 + 0.080 \text{ kg}/[2(0.31 \text{ kg})]} = 0.97 \text{ N}$$

Part (b)

5. Use $T_2 = Ma$ to find the acceleration of the cart:

$$a = \frac{T_2}{M} = \frac{0.97 \text{ N}}{0.31 \text{ kg}} = 3.1 \text{ m/s}^2$$

INSIGHT
We find that T_2 is less than T_1. As a result, the net torque acting on the pulley is in the counterclockwise direction, causing a rotation in that direction, as expected. If the mass of the pulley were zero ($m = 0$), the two tensions would be equal (as can be seen in Step 4), and the acceleration of the cart would be $T_1/M = 3.5$ m/s^2.

CONTINUED

PRACTICE PROBLEM

What applied force is necessary to give the cart an acceleration of 2.2 m/s²?

[**Answer:** $T_1 = T_2(1 + m/2M) = (Ma)(1 + m/2M) = 0.77$ N. Only 0.68 N would be required if the pulley had zero mass.]

Some related homework problems: Problem 40, Problem 42

Enhance Your Understanding   (Answers given at the end of the chapter)

5. Refer to Figure 11-20 (a) for the following questions: **(a)** Is the tension in the rope greater than, less than, or equal to *mg*? Explain. **(b)** Is the tension in the rope greater than, less than, or equal to 0? Explain.

Section Review

- In systems containing both linear and rotational motions, the two must be analyzed together. This is done by applying Newton's second law to both motions, and taking advantage of any connections that exist between the motions.

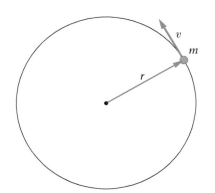

▲ **FIGURE 11-21 The angular momentum of circular motion** A particle of mass *m*, moving in a circle of radius *r* with a speed *v*. This particle has an angular momentum of magnitude $L = rmv$.

11-6 Angular Momentum

When an object of mass *m* moves with a speed *v* in a straight line, we say that it has a linear momentum, $p = mv$. When the same object moves with an angular speed ω along the circumference of a circle of radius *r*, as in **FIGURE 11-21**, we say that it has an **angular momentum**, *L*. The magnitude of *L* is given by replacing *m* and *v* in the expression for *p* with their angular analogs *I* and ω (Section 10-5). Thus, we define the angular momentum as follows:

Definition of Angular Momentum, *L*

$$L = I\omega \qquad \qquad \text{11-11}$$

SI unit: $\text{kg} \cdot \text{m}^2/\text{s}$

This expression applies to any object undergoing angular motion, whether it is a point mass moving in a circle, as in Figure 11-21, or a rotating hoop, disk, or other object.

For a point mass *m* moving in a circle of radius *r*, the moment of inertia in this case is $I = mr^2$ (Equation 10-18). In addition, the linear speed of the mass is $v = r\omega$ (Equation 10-12). Combining these results, we find

$$L = I\omega = (mr^2)(v/r) = rmv$$

Noting that *mv* is the linear momentum *p*, we find that the angular momentum of a point mass can be written in the following form:

$$L = rmv = rp \qquad \qquad \text{11-12}$$

It's important to recall that this expression applies specifically to a point particle moving along the circumference of a circle.

More generally, a point particle may be moving at an angle θ with respect to a radial line, as indicated in **FIGURE 11-22 (a)**. In this case, only the tangential component of the momentum, $p \sin \theta = mv \sin \theta$, contributes to the angular momentum, just as the tangential component of the force, $F \sin \theta$, is all that contributes to the torque. Thus, the magnitude of the angular momentum for a point particle is defined as:

(a)

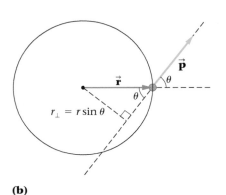

(b)

◀ **FIGURE 11-22 The angular momentum of nontangential motion (a)** When a particle moves at an angle θ with respect to the radial direction, only the tangential component of velocity, $v \sin \theta$, contributes to the angular momentum. In the case shown here, the particle's angular momentum has a magnitude given by $L = rmv \sin \theta$. **(b)** The angular momentum of an object can also be defined in terms of the moment arm, $r_\perp$. Since $r_\perp = r \sin \theta$, it follows that $L = rmv \sin \theta = r_\perp mv$. Note the similarity between this figure and Figure 11-5.

Angular Momentum, *L*, for a Point Particle

$$L = rp \sin \theta = rmv \sin \theta \qquad\qquad 11\text{-}13$$

SI unit: $\text{kg} \cdot \text{m}^2/\text{s}$

If the particle moves in a circular path, the angle θ is 90° and the angular momentum is $L = rmv$, in agreement with Equation 11-12. On the other hand, if the object moves radially, so that $\theta = 0$, the angular momentum is zero; $L = rmv \sin 0 = 0$.

EXERCISE 11-15   ANGULAR MOMENTUM

Find the angular momentum of **(a)** a 16.0-g DVD (considered to be a uniform disk of radius 6.00 cm) spinning with an angular speed of 125 rad/s, and **(b)** a 125-kg motorcycle driving with a speed of 8.92 m/s on a circular track of radius 35.7 m.

REASONING AND SOLUTION

a. Recalling that $L = I\omega$, and that $I = \frac{1}{2}mR^2$ for a uniform disk (Table 10-1), we find

$$L = I\omega = \left(\tfrac{1}{2}mR^2\right)\omega = \tfrac{1}{2}(0.0160 \text{ kg})(0.0600 \text{ m})^2(125 \text{ rad/s}) = 0.00360 \text{ kg} \cdot \text{m}^2/\text{s}$$

b. Treating the motorcycle as a particle of mass m, we find

$$L = rmv = (35.7 \text{ m})(125 \text{ kg})(8.92 \text{ m/s}) = 39{,}800 \text{ kg} \cdot \text{m}^2/\text{s}$$

Angular Momentum and the Moment Arm   An alternative definition of the angular momentum uses the moment arm, $r_\perp$, as was done for the torque in Equation 11-3. To apply this definition, start by extending a line through the momentum vector, $\vec{\mathbf{p}}$, as in FIGURE 11-22 (b). Next, draw a line from the axis of rotation perpendicular to the line through $\vec{\mathbf{p}}$. The perpendicular distance from the axis of rotation to the line of $\vec{\mathbf{p}}$ is the moment arm. From the figure we see that $r_\perp = r \sin \theta$. Hence, from Equation 11-13, the angular momentum is

$$L = r_\perp p = r_\perp mv$$

If an object moves in a circle of radius r, the moment arm is $r_\perp = r$ and the angular momentum reduces to our earlier result, $L = rp$.

CONCEPTUAL EXAMPLE 11-16   ANGULAR MOMENTUM?

Does an object moving in a straight line have nonzero angular momentum always, sometimes, or never?

REASONING AND DISCUSSION

The answer is sometimes, because it depends on the choice of the axis of rotation. If the axis of rotation is not on the line drawn through the momentum vector, as in the left-hand sketch, the moment arm is nonzero, and therefore $L = r_\perp p$ is also nonzero. If the axis of rotation is on the line of motion, as in the right-hand sketch, the moment arm is zero; hence the linear momentum is radial and L vanishes.

ANSWER

An object moving in a straight line may or may not have nonzero angular momentum, depending on the location of the axis of rotation.

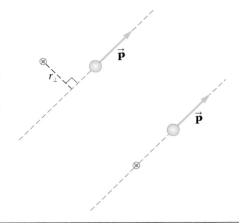

> **Big Idea** ❹ Angular momentum is the rotational analog of linear momentum. Angular momentum depends on the magnitude and direction of the linear momentum, and also on the distance from the axis of rotation.

PHYSICS IN CONTEXT
Looking Back

Angular momentum plays the same role in rotational systems as linear momentum does in linear systems in Chapter 9.

PHYSICS IN CONTEXT
Looking Ahead

Angular momentum is quantized (given discrete values) in the Bohr model of the hydrogen atom in Chapter 31.

An object moving with a momentum p in a straight line that does not go through the axis of rotation has an *angular* position that changes with time. This is illustrated in FIGURE 11-23 (a). It is for this reason that such an object is said to have an *angular* momentum.

Sign of the Angular Momentum   The sign of L is determined by whether the angle to a given object is increasing or decreasing with time. For example, the object moving counterclockwise in a circular path in FIGURE 11-23 (b) has a positive angular momentum because θ is increasing with time. Similarly, the object in Figure 11-23 (a) has an angle θ that increases with time, and hence its angular momentum is positive as well. On the other hand, if these objects were to have their direction of motion reversed, they would have angles that decrease with time and their angular momenta would be negative.

▶ **FIGURE 11-23 Angular momentum in linear and circular motions** An object moving in **(a)** a straight line and **(b)** a circular path. In both cases, the angular position increases with time; hence, the angular momentum is positive.

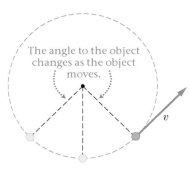

(a) Linear motion   **(b)** Circular motion

EXAMPLE 11-17   HOCKEY SHOT

A 0.170-kg hockey puck moves straight toward the goalie with a speed of $v = 4.42$ m/s, as shown in the sketch. The puck is also observed by a defender. **(a)** Find the angular momentum as measured by the defender, given that $r = 1.45$ m and $\theta = 72.5°$. Use the expression $L = rmv \sin\theta$. **(b)** Find the moment arm, $r_\perp$, for the defender. **(c)** Check that the angular momentum is also given by $L = r_\perp mv$.

PICTURE THE PROBLEM
The hockey puck moves directly toward the goalie with a linear momentum $p = mv$. For the defender, the line of motion is displaced by a distance equal to the moment arm, $r_\perp = r \sin\theta$. Hence, the defender observes a nonzero angular momentum, while the goalie does not.

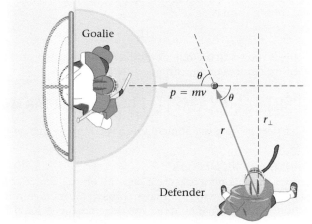

REASONING AND STRATEGY

a. The angular momentum of the puck can be found by applying $L = rmv \sin\theta$. In this case, the distance is $r = 1.45$ m and the angle between the line of sight and the puck's momentum is $\theta = 72.5°$. In addition, $m = 0.170$ kg and $v = 4.42$ m/s.

b. Notice that the moment arm, $r_\perp$, is the side of the right triangle opposite to the angle of $\theta = 72.5°$. It follows that $r_\perp = r \sin 72.5°$.

c. Use the moment arm from part **(b)** to calculate $L = r_\perp mv$.

Known   Mass of puck, $m = 0.170$ kg; linear speed of puck, $v = 4.42$ m/s; distance to puck, $r = 1.45$ m; angle between $\vec{r}$ and $\vec{p}$, $\theta = 72.5°$.
Unknown   **(a)** Angular momentum of puck, $L = ?$ **(b)** Moment arm, $r_\perp = ?$ **(c)** Angular momentum of puck, $L = ?$

SOLUTION

Part (a)

1. Evaluate $L = rmv \sin\theta$:

$$L = rmv \sin\theta$$
$$= (1.45 \text{ m})(0.170 \text{ kg})(4.42 \text{ m/s}) \sin 72.5° = 1.04 \text{ kg} \cdot \text{m}^2/\text{s}$$

Part (b)

2. Calculate the moment arm, $r_\perp$:

$$r_\perp = r \sin\theta = (1.45 \text{ m}) \sin 72.5° = 1.38 \text{ m}$$

Part (c)

3. Evaluate $L = r_\perp mv$:

$$L = r_\perp mv = (1.38 \text{ m})(0.170 \text{ kg})(4.42 \text{ m/s}) = 1.04 \text{ kg} \cdot \text{m}^2/\text{s}$$

CONTINUED

INSIGHT

The defender observes a nonzero angular momentum for the puck because she has to turn her head to follow its motion. The goalie always sees the puck at the same angle, and hence observes an angular momentum of zero.

PRACTICE PROBLEM — PREDICT/CALCULATE

As the puck gets closer to the goalie, the distance r from the defender increases and the angle θ decreases. (a) Does the puck's angular momentum, as observed by the defender, increase, decrease, or stay the same as it approaches the goalie? Explain. (b) Find the angular momentum relative to the defender when $r = 1.58$ m and $\theta = 61.0°$.   [Answer: (a) The distance r increases, and the angle θ decreases, but the moment arm $r_\perp = r \sin \theta$ stays the same. Hence the angular momentum also stays the same. (b) We find $r_\perp = 1.38$ m and $L = 1.04$ kg · m^2/s, as before.]

Some related homework problems: Problem 46, Problem 47

The Rate of Change of Angular Momentum   To conclude this section, we consider the rate of change of angular momentum with time. Assuming the moment of inertia is a constant—which is true as long as the mass and shape of the object remain unchanged—the change in L in a time interval Δt is

$$\frac{\Delta L}{\Delta t} = I \frac{\Delta \omega}{\Delta t}$$

Recall, however, that $\Delta \omega / \Delta t$ is the angular acceleration, α. Therefore, we have

$$\frac{\Delta L}{\Delta t} = I\alpha$$

Noting that $I\alpha$ is the torque, it follows that Newton's second law for rotational motion can be written as:

Newton's Second Law for Rotational Motion

$$\sum \tau = I\alpha = \frac{\Delta L}{\Delta t} \qquad \text{11-14}$$

Clearly, this is the rotational analog of $\sum F_x = ma_x = \Delta p_x / \Delta t$. Just as force can be expressed as the change in *linear* momentum in a given time interval, torque can be expressed as the change in *angular* momentum in a given time interval.

EXERCISE 11-18   TORQUE AND ANGULAR MOMENTUM

A ceiling fan experiences a constant torque of 1.9 N · m. If the fan is initially at rest, what is its angular momentum 2.6 s later?

REASONING AND SOLUTION

Solve $\sum \tau = \Delta L / \Delta t$ (Equation 11-14) for the change in angular momentum due to a single torque τ:

$$\Delta L = L_f - L_i = \left(\sum \tau \right) \Delta t = \tau \Delta t$$

Noting that the initial angular momentum of the fan is zero ($L_i = 0$), we find its final angular momentum is

$$L_f = \tau \Delta t = (1.9 \, \text{N} \cdot \text{m})(2.6 \, \text{s}) = 4.9 \, \text{kg} \cdot \text{m}^2/\text{s}$$

Enhance Your Understanding   (Answers given at the end of the chapter)

6. Consider two objects with the following characteristics: Object 1 is a hoop with mass $2M$ and radius R; object 2 is a disk with mass M and radius $2R$. If the angular speed of object 1 is ω, and the angular speed of object 2 is 2ω, which object has the greater angular momentum? Explain.

CONTINUED

▲ **FIGURE 11-24** Once she has launched herself into space, this diver is essentially a projectile. However, the principle of conservation of angular momentum allows her to control the rotational part of her motion. By curling her body up into a tight "tuck," she decreases her moment of inertia, thereby increasing the speed of her spin. To slow down for an elegant entry into the water, she will extend her body, thus increasing her moment of inertia.

Spinning Person Drops Weights

Big Idea $\mathbf{5}$ In systems with zero net torque, the angular momentum is conserved. It follows that the total angular momentum of the universe is conserved, just like the total linear momentum and the total energy.

**PHYSICS
IN CONTEXT
Looking Ahead**

The conservation of angular momentum plays a key role in the study of gravity. See, in particular, the discussion of Kepler's second law in Chapter 12.

Section Review

• Angular momentum plays the same role in rotational systems as linear momentum plays in linear systems.

• Angular momentum is given by $L = I\omega$ for a rotating system, or $L = rmv \sin \theta$ for a point particle.

• The rate of change of angular momentum in a system is equal to the net torque applied to the system.

11-7 Conservation of Angular Momentum

When an ice skater goes into a spin and pulls his arms inward to speed up, he probably doesn't think about angular momentum. Neither does a diver who springs into the air and folds her body to speed her rotation, as in **FIGURE 11-24**. Most people, in fact, are not aware that the actions of these athletes are governed by the same basic laws of physics that cause a collapsing star to spin faster as it becomes a rapidly rotating pulsar. Yet in all these cases, as we shall see, **conservation of angular momentum** is at work.

The Origin of Angular Momentum Conservation   To see how angular momentum conservation comes about, consider an object with an initial angular momentum L_i acted on by a single torque τ. After a period of time, Δt, the object's angular momentum changes in accordance with Newton's second law:

$$\tau = \frac{\Delta L}{\Delta t}$$

Solving for ΔL, we find

$$\Delta L = L_f - L_i = \tau \Delta t$$

Thus, the final angular momentum of the object is

$$L_f = L_i + \tau \Delta t$$

If the torque acting on the object is zero ($\tau = 0$), it follows that the initial and final angular momenta are equal—that is, the angular momentum is conserved:

$$L_f = L_i \quad (\text{if } \tau = 0)$$

Angular momentum is also conserved in systems acted on by more than one torque, provided that the *net external torque* is zero. The reason internal torques can be ignored is that, just as internal forces come in equal and opposite pairs that cancel, so too do internal torques. As a result, the internal torques in a system sum to zero, and the net torque acting on it is simply the net external torque. Thus, for a general system, angular momentum is conserved if $\tau_{\text{net, ext}}$ is zero:

Conservation of Angular Momentum

$$L_f = L_i \quad (\text{if } \tau_{\text{net, ext}} = 0)$$ 11-15

As an illustration of angular momentum conservation, we consider the case of a student rotating on a piano stool in the next Example. Notice how a change in the moment of inertia results in a change in angular speed.

EXAMPLE 11-19   GOING FOR A SPIN

For a classroom demonstration, a student sits on a piano stool holding a sizable mass in each hand. Initially, the student holds his arms outstretched and spins about the axis of the stool with an angular speed of 3.72 rad/s. The moment of inertia in this case is 5.33 kg·m². While still spinning, the student pulls his arms in to his chest, reducing the moment of inertia to 1.60 kg·m². **(a)** What is the student's angular speed now? **(b)** Find the initial and final angular momenta of the student.

CONTINUED

PICTURE THE PROBLEM

The initial and final configurations of the student are shown in our sketch. Clearly, the mass distribution in the final configuration, with the masses held closer to the axis of rotation, results in a smaller moment of inertia.

REASONING AND STRATEGY

Ignoring friction in the axis of the stool, we conclude that no external torques act on the system. As a result, the angular momentum is conserved. Therefore, setting the initial angular momentum, $L_i = I_i\omega_i$, equal to the final angular momentum, $L_f = I_f\omega_f$, yields the final angular speed.

Known   Initial angular speed, $\omega_i = 3.72$ rad/s; initial moment of inertia, $I_i = 5.33$ kg · m²; final moment of inertia, $I_f = 1.60$ kg · m².

Unknown (a) Final angular speed, $\omega_f = ?$ (b) Initial and final angular momenta of student, $L_i = ?, L_f = ?$

SOLUTION

Part (a)

1. Apply angular momentum conservation to this system:

$$L_i = L_f$$
$$I_i\omega_i = I_f\omega_f$$

2. Solve for the final angular speed, ω_f:

$$\omega_f = \left(\frac{I_i}{I_f}\right)\omega_i$$

3. Substitute numerical values:

$$\omega_f = \left(\frac{5.33\ \text{kg} \cdot \text{m}^2}{1.60\ \text{kg} \cdot \text{m}^2}\right)(3.72\ \text{rad/s}) = 12.4\ \text{rad/s}$$

Part (b)

4. Use $L = I\omega$ to calculate the angular momentum. Substitute both initial and final values as a check:

$$L_i = I_i\omega_i = (5.33\ \text{kg} \cdot \text{m}^2)(3.72\ \text{rad/s}) = 19.8\ \text{kg} \cdot \text{m}^2/\text{s}$$
$$L_f = I_f\omega_f = (1.60\ \text{kg} \cdot \text{m}^2)(12.4\ \text{rad/s}) = 19.8\ \text{kg} \cdot \text{m}^2/\text{s}$$

INSIGHT

Initially the student completes one revolution roughly every two seconds. After pulling the weights in, the student's rotation rate has increased to almost two revolutions a second—quite a dizzying pace. The same physics applies to a rotating diver or a spinning ice skater.

PRACTICE PROBLEM

What moment of inertia would be required to give a final spin rate of 10.0 rad/s?   [**Answer:** $I_f = (\omega_i/\omega_f)I_i = 1.98$ kg · m²]

Some related homework problems: Problem 54, Problem 58

Conservation of Angular Momentum in Nature An increasing angular speed, as experienced by the student in Example 11-19, can be observed in nature as well. Examples are shown in **FIGURE 11-25**. For example, a hurricane draws circulating air in at ground level toward its "eye," where it then rises to an altitude of 10 miles or more. As air moves inward toward the axis of rotation, its angular speed increases, just as the masses held by the student speed up when they are pulled inward. For example, if the wind has a speed of only 3.0 mph at a distance of 300 miles from the center of the hurricane, it would speed up to 150 mph when it comes to within 6.0 miles of the center. Of course, this analysis ignores friction, which would certainly decrease the wind speed. Still, the basic principle—that a decreasing distance from the axis of rotation implies an increasing speed—applies to both the student and the hurricane. Similar behavior is observed in tornadoes and waterspouts.

RWP Another example of conservation of angular momentum occurs in stellar explosions. On occasion a star explodes, sending a portion of its material out into space. After the explosion, the star collapses to a fraction of its original size, speeding up its rotation in the process. If the mass of the star is greater than 1.44 times the mass of

▲ **FIGURE 11-25 Visualizing Concepts Conservation of Angular Momentum** This 1992 satellite photo of Hurricane Andrew (left), one of the most powerful hurricanes in recent decades, clearly suggests the rotating structure of the storm. The violence of the hurricane winds can be attributed in large part to conservation of angular momentum: As air is pushed inward toward the low pressure near the eye of the storm, its rotational velocity increases. The same principle, operating on a smaller scale, explains the tremendous destructive power of tornadoes (right).

▲ **FIGURE 11-26** Among the fastest rotating objects known in nature are pulsars—stars that have collapsed to a tiny fraction of their original size. Since all the angular momentum of a star must be conserved when it collapses, the dramatic decrease in radius is accompanied by a correspondingly great increase in rotational speed. The Crab nebula pulsar spins at about 30 rev/s. This X-ray photograph shows rings and jets of high-energy particles flying outward from the whirling neutron star at the center.

the Sun, the collapse can continue until a *neutron star* is formed, with a radius of only about 10 to 20 km. Neutron stars have incredibly high densities; in fact, if you could bring a teaspoonful of neutron star material to the Earth, it would weigh about one billion tons! In addition, neutron stars produce powerful beams of X-rays and other radiation that sweep across the sky like a gigantic lighthouse beam as the star rotates. On the Earth we see pulses of radiation from these rotating beams, one for each revolution of the star. These "pulsating stars," or *pulsars*, typically have periods ranging from about 2 ms to nearly 1 s. The Crab nebula, shown in **FIGURE 11-26** (see Problem 6 in Chapter 10), is a famous example of such a system. The dependence of angular speed on radius for a collapsing star is considered in the next Quick Example.

QUICK EXAMPLE 11-20 A STELLAR PERFORMANCE: FIND THE ANGULAR SPEED

A star of radius $R = 2.3 \times 10^8$ m rotates with an angular speed $\omega = 2.4 \times 10^{-6}$ rad/s. If this star collapses to a radius of 20.0 km, find its final angular speed. (Treat the star as if it were a uniform sphere, and assume that no mass is lost as the star collapses.)

REASONING AND SOLUTION

Because the star is an isolated system, its angular momentum remains the same as it collapses. Thus, we set the initial angular momentum $(I_i\omega_i)$ equal to the final angular momentum $(I_f\omega_f)$ and solve for the final angular speed (ω_f).

1. Apply conservation of angular momentum:

$$I_i\omega_i = I_f\omega_f$$

2. Write expressions for the initial and final moments of inertia:

$$I_i = \tfrac{2}{5}MR_i^2 \quad \text{and} \quad I_f = \tfrac{2}{5}MR_f^2$$

3. Solve for the final angular speed:

$$\omega_f = (I_i/I_f)\omega_i = (R_i^2/R_f^2)\omega_i$$

4. Substitute numerical values:

$$\omega_f = \left(\frac{2.3 \times 10^8 \text{ m}}{20.0 \times 10^3 \text{ m}}\right)^2 (2.4 \times 10^{-6} \text{ rad/s})$$
$$= 320 \text{ rad/s}$$

RWP The final angular speed corresponds to a period of about 20 ms, a typical period for pulsars. Given that 320 rad/s is roughly 3000 rpm, it follows that a pulsar, which has the mass of a star, rotates as fast as the engine in a racing car.

Angular Momentum and the Length of the Day If the student in Example 11-19 were to stretch his arms back out again, the resulting *increase* in the moment of inertia would cause a *decrease* in his angular speed. The same effect might apply to the Earth one day. For example, a melting of the polar ice caps would lead to an increase in the Earth's moment of inertia (as we saw in Chapter 10) and thus, by angular momentum conservation, the angular speed of the Earth would decrease. This would mean that more time would be required for the Earth to complete a revolution about its axis of rotation; that is, the day would lengthen.

Angular Momentum and Kinetic Energy   Angular momentum is conserved in the systems we've studied so far, and hence it's natural to ask whether the energy is conserved as well. We consider this question in the next Conceptual Example.

CONCEPTUAL EXAMPLE 11-21   COMPARE KINETIC ENERGIES

A skater pulls in her arms, decreasing her moment of inertia by a factor of two, and doubling her angular speed. Is her final kinetic energy greater than, less than, or equal to her initial kinetic energy?

REASONING AND DISCUSSION

Let's calculate the initial and final kinetic energies, and compare them. The initial kinetic energy is

$$K_i = \tfrac{1}{2}I_i\omega_i{}^2$$

After pulling in her arms, the skater has half the moment of inertia and twice the angular speed. Hence, her final kinetic energy is

$$K_f = \tfrac{1}{2}I_f\omega_f{}^2 = \tfrac{1}{2}(I_i/2)(2\omega_i)^2 = 2\left(\tfrac{1}{2}I_i\omega_i{}^2\right) = 2K_i$$

Thus, the fact that K depends on the square of ω leads to an increase in the kinetic energy. The source of this additional energy is the work done by the muscles in the skater's arms as she pulls them in to her body.

ANSWER

The skater's kinetic energy increases.

Rotational Collisions

In the not-too-distant past, a person would play music by placing a record on a rotating turntable. Suppose, for example, that a turntable with a moment of inertia I_t is rotating freely with an initial angular speed ω_0. A record, with a moment of inertia I_r and initially at rest, is dropped straight down onto the rotating turntable, as in **FIGURE 11-27**. When the record lands, frictional forces between it and the turntable cause the record to speed up and the turntable to slow down, until they both have the same angular speed. Only internal forces are involved during this process, and hence it follows that the system's angular momentum is conserved. We can think of this event, then, as a "rotational collision."

Before the collision, the angular momentum of the system is

$$L_i = I_t\omega_0$$

After the collision, when both the record and the turntable are rotating with the angular speed ω_f, the system's angular momentum is

$$L_f = I_t\omega_f + I_r\omega_f$$

Setting $L_f = L_i$ yields the final angular speed:

$$\omega_f = \left(\frac{I_t}{I_t + I_r}\right)\omega_0 \qquad \text{11-16}$$

Because this collision is completely inelastic, we expect the final kinetic energy to be less than the initial kinetic energy.

We conclude this section with a somewhat different example of a rotational collision. The physical principles involved are precisely the same, however.

▲ **FIGURE 11-27 A rotational collision** A nonrotating record dropped onto a rotating turntable is an example of a "rotational collision." Since only internal forces are involved during the collision, the final angular momentum is equal to the initial angular momentum.

Off-Center Collision

QUICK EXAMPLE 11-22 CONSERVE ANGULAR MOMENTUM: FIND THE ANGULAR SPEED

A 34.0-kg child runs with a speed of 2.80 m/s tangential to the rim of a stationary merry-go-round. The merry-go-round has a moment of inertia of 512 kg·m^2 and a radius of 2.31 m. When the child jumps onto the merry-go-round, the entire system begins to rotate. What is the angular speed of the system? Ignore friction and any other type of external torque.

Before   After

Our sketch shows the system before and after the collision. Before the collision the merry-go-round is at rest and the child is running. After the collision the child is riding on the rotating merry-go-round.

REASONING AND SOLUTION

Angular momentum is conserved in this system, so we set the initial angular momentum equal to the final angular momentum to find the final angular speed. The initial angular momentum is the angular momentum of the running child. The final angular momentum is due to the rotation of the merry-go-round *and* the child about the axis of rotation.

1. Write the initial angular momentum of the child:

$$L_i = rmv$$

2. Write the final angular momentum of the system, noting that the moment of inertia of the merry-go-round is I and the moment of inertia of the child is mr^2:

$$L_f = (I + mr^2)\omega$$

3. Set $L_f = L_i$ and solve for the angular speed:

$$\omega = rmv/(I + mr^2)$$

4. Substitute numerical values:

$$\omega = \frac{(2.31\ \text{m})(34.0\ \text{kg})(2.80\ \text{m/s})}{512\ \text{kg·m}^2 + (34.0\ \text{kg})(2.31\ \text{m})^2}$$
$$= 0.317\ \text{rad/s}$$

If the moment of inertia of the merry-go-round had been zero, $I = 0$, the angular speed of the merry-go-round would be $\omega = v/r$. This, in turn, would mean that the linear speed of the child ($v = r\omega$) is unchanged. If $I > 0$, however, the linear speed of the child is decreased. In this particular case, the child's linear speed after the collision is only $v = r\omega = 0.732$ m/s, compared with her initial linear speed of 2.80 m/s.

Enhance Your Understanding   (Answers given at the end of the chapter)

7. Consider the system described in Quick Example 11-22, where a child runs and then jumps onto a merry-go-round. Do you expect the initial kinetic energy of this system to be greater than, less than, or equal to the final kinetic energy? Explain.

Section Review

• Angular momentum is conserved in a system if the net torque exerted on it is zero.

11-8 Rotational Work and Power

Just as a force acting through a distance does work on an object, so too does a torque acting through an angular displacement. To see this, consider the case of fishing line pulled from a reel. If the line is pulled with a force F for a distance Δx, as in FIGURE 11-28, the work done on the reel is

$$W = F\Delta x$$

Now, the line is unwinding without slipping, and hence it follows that the linear displacement of the line, Δx, is related to the angular displacement of the reel, $\Delta\theta$, by the following relationship:

$$\Delta x = R\Delta\theta$$

In this equation, R is the radius of the reel, and $\Delta\theta$ is measured in radians. Thus, the work can be written as

$$W = F\Delta x = FR\Delta\theta$$

Finally, the torque exerted on the reel by the line is $\tau = RF$, from which it follows that the work done on the reel is simply torque times angular displacement:

Work Done by Torque

$$W = \tau\Delta\theta \qquad\qquad\qquad \text{11-17}$$

Notice again the analogies between angular and linear quantities in $W = F\Delta x$ and $W = \tau\Delta\theta$. As usual, τ is the analog of F, and θ is the analog of x.

As we saw in Chapter 7, the net work done on an object is equal to the change in its kinetic energy. This is the work–energy theorem:

$$W = \Delta K = K_\mathrm{f} - K_\mathrm{i} \qquad\qquad\qquad \text{11-18}$$

The work–energy theorem applies regardless of whether the work is done by a force acting through a distance or a torque acting through an angle.

Similarly, power is the amount of work done in a given time, regardless of whether the work is done by a force or a torque. In the case of a torque, we have $W = \tau\Delta\theta$, and hence:

Power Produced by a Torque

$$P = \frac{W}{\Delta t} = \tau\frac{\Delta\theta}{\Delta t} = \tau\omega \qquad\qquad\qquad \text{11-19}$$

Again, the analogy is clear between $P = Fv$ for the linear case, and $P = \tau\omega$ for the rotational case.

▲ FIGURE 11-28 **Rotational work** A force F pulling a length of line Δx from a fishing reel does the work $W = F\Delta x$. In terms of torque and angular displacement, the work can be expressed as $W = \tau\Delta\theta$.

EXERCISE 11-23 ROTATIONAL POWER

It takes a good deal of effort to make homemade ice cream. **(a)** If the torque required to turn the handle on an ice cream maker is 4.5 N · m, how much work is expended on each complete revolution of the handle? **(b)** How much power is required to turn the handle if each revolution is completed in 1.2 s?

REASONING AND SOLUTION

a. Applying $W = \tau\Delta\theta$ with $\tau = 4.5$ N · m and $\Delta\theta = 2\pi$ yields the work done in one revolution:

$$W = \tau\Delta\theta = (4.5\,\text{N·m})(2\pi\,\text{rad}) = 28\,\text{J}$$

b. Power is the work per time; that is,

$$P = W/\Delta t = (28\,\text{J})/(1.2\,\text{s}) = 23\,\text{W}$$

Equivalently, the angular speed of the handle is $\omega = (2\pi)/T = (2\pi)/(1.2\,\text{s}) = 5.2$ rad/s, and therefore $P = \tau\omega$ (Equation 11-19) yields $P = \tau\omega = (4.5\,\text{N·m})(5.2\,\text{rad/s}) = 23$ W, in agreement with our previous result.

8. In system 1, a torque of $20 \, \text{N} \cdot \text{m}$ acts through an angular displacement of $45°$. In system 2, a torque of $10 \, \text{N} \cdot \text{m}$ acts through an angular displacement of $\pi/2$. Is the work done in system 1 greater than, less than, or equal to the work done in system 2? Explain.

Section Review

- The work done in a rotational system is equal to torque times angular displacement.

Curl fingers in direction of rotation...

$v = r\omega$

$\vec{\boldsymbol{\omega}}$

$\vec{\mathbf{L}} = I\vec{\boldsymbol{\omega}}$

...and thumb points in direction of angular velocity.

$v = r\omega$

▲ **FIGURE 11-29 The right-hand rule for angular velocity** The angular velocity, $\vec{\boldsymbol{\omega}}$, of a rotating wheel points along the axis of rotation. Its direction is given by the right-hand rule.

11-9 The Vector Nature of Rotational Motion

We've mentioned many times that the angular velocity is a vector, and that we must be careful to use the proper sign for ω. But if the angular velocity is a vector, what is its direction?

To address this question, consider the rotating wheel shown in **FIGURE 11-29**. Each point on the rim of this wheel has a velocity vector pointing in a different direction in the plane of rotation. But if different parts of the wheel move in different directions, how can we assign a single direction to the angular velocity vector, $\vec{\boldsymbol{\omega}}$? The answer is that there is only one direction that remains fixed as the wheel rotates—the direction of the axis of rotation. By definition, then, the angular velocity vector, $\vec{\boldsymbol{\omega}}$, is taken to point along the axis of rotation.

Given that $\vec{\boldsymbol{\omega}}$ points along the axis of rotation, we must still decide whether it points to the left or to the right in Figure 11-29. The convention we use for assigning the direction of $\vec{\boldsymbol{\omega}}$ is referred to as the right-hand rule:

Right-Hand Rule for Angular Velocity, $\vec{\boldsymbol{\omega}}$

Curl the fingers of the right hand in the direction of rotation.

The thumb now points in the direction of the angular velocity, $\vec{\boldsymbol{\omega}}$.

The right-hand rule (RHR) for $\vec{\boldsymbol{\omega}}$ is illustrated in Figure 11-29. Additional examples of rotating objects are given in **FIGURE 11-30**.

▲ **FIGURE 11-30** Children have always been fascinated by tops—but not only children. The physicists in the photo at right, Wolfgang Pauli and Niels Bohr, seem as delighted by a spinning top as any child. Their contributions to modern physics, discussed in Chapter 30, helped to show that subatomic particles, the ultimate constituents of matter, have a property (now referred to as "spin") that is in some ways analogous to the rotation of a top or a gyroscope.

The same convention for direction applies to the angular momentum vector. First, recall that the angular momentum has a magnitude given by $L = I\omega$. Hence, we choose the direction of $\vec{\mathbf{L}}$ to be the same as the direction of $\vec{\boldsymbol{\omega}}$. That is,

$$\vec{\mathbf{L}} = I\vec{\boldsymbol{\omega}} \qquad \text{11-20}$$

The angular momentum vector is also illustrated in Figure 11-29.

Similarly, torque is a vector, and it too is defined to point along the axis of rotation. The right-hand rule for torque is similar to that for angular velocity:

Right-Hand Rule for Torque, $\vec{\tau}$

Curl the fingers of the right hand in the direction of rotation that this torque would cause if it acted alone.

The thumb now points in the direction of the torque vector, $\vec{\tau}$.

Two examples of torque vectors are shown in **FIGURE 11-31**.

As an example of torque and angular momentum vectors, consider the spinning bicycle wheel shown in **FIGURE 11-32**. The angular momentum vector for the wheel points to the left, along the axis of rotation. If a person pushes on the rim of the wheel in the direction indicated, the resulting torque is also to the left, as shown in the figure. If this torque lasts for a time Δt, the angular momentum changes by the amount

$$\Delta \vec{L} = \vec{\tau} \Delta t$$

Adding $\Delta \vec{L}$ to the original angular momentum $\vec{L}_i$ yields the final angular momentum $\vec{L}_f$, shown in Figure 11-32. Since $\vec{L}_f$ is in the same direction as $\vec{L}_i$, but with a greater magnitude, it follows that the wheel is spinning in the same direction as before, only faster. This is to be expected, considering the direction of the person's push on the wheel.

On the other hand, if a person pushes on the wheel in the opposite direction, the torque vector points to the right. As a result, $\Delta \vec{L}$ points to the right as well. When we add $\Delta \vec{L}$ and $\vec{L}_i$ to obtain the final angular momentum $\vec{L}_f$, we find that it has the same direction as $\vec{L}_i$, but a smaller magnitude. Hence, we conclude that the wheel spins more slowly, as one would expect.

Finally, a case of considerable interest is when the torque and angular momentum vectors are at right angles to one another. The classic example of such a system is the **gyroscope**. To begin, consider a gyroscope whose axis of rotation is horizontal, as in **FIGURE 11-33**. If the gyroscope were to be released with no spin it would simply fall, rotating counterclockwise downward about its point of support. Curling the fingers of the right hand in the counterclockwise direction, we see that the thumb, and hence the torque due to gravity, points out of the page.

Next, imagine the gyroscope to be spinning rapidly—as would normally be the case—with its angular momentum pointing to the left in Figure 11-33. If the gyroscope is released now, it doesn't fall as before, even though the torque is the same. To see what happens instead, consider the change in angular momentum, $\Delta \vec{L}$, caused by the torque, $\vec{\tau}$, acting for a small interval of time. As shown in **FIGURE 11-34**, the small

▲ **FIGURE 11-31 The right-hand rule for torque** Examples of torque vectors obtained using the right-hand rule.

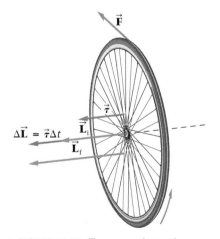

▲ **FIGURE 11-32 Torque and angular momentum vectors** A tangential push on the spinning wheel in the direction shown causes a torque to the left. As a result, the angular momentum increases. Hence, the wheel spins faster, as expected.

▲ **FIGURE 11-33 The torque exerted on a gyroscope** A spinning gyroscope has an initial angular momentum to the left. The torque due to gravity is out of the page.

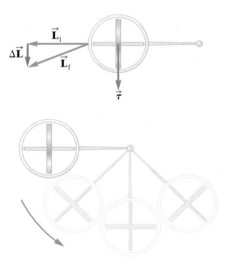

Top view

▲ **FIGURE 11-34 Precession of a gyroscope** The gyroscope as viewed from above. In a time Δt the angular momentum changes by the amount $\Delta \vec{L} = \vec{\tau} \Delta t$. This causes the angular momentum vector, and hence the gyroscope as a whole, to rotate in a counterclockwise direction.

▲ FIGURE 11-35

change, $\Delta\vec{L}$, is at right angles to $\vec{L}_i$; hence the final angular momentum, $\vec{L}_f$, is essentially the same length as $\vec{L}_i$, but pointing in a direction slightly out of the page. With each small interval of time, the angular momentum vector continues to change in direction so that, viewed from above as in Figure 11-34, the gyroscope as a whole rotates in a counterclockwise sense around its support point. This type of motion, where the axis of rotation changes direction with time, is referred to as **precession**. The gyroscope in FIGURE 11-35 is designed to use precession to test Einstein's theory of relativity.

RWP Because of its spinning motion about its rotational axis, the Earth may be considered as one rather large gyroscope. Gravitational forces exerted on the Earth by the Sun and the Moon subject it to external torques that cause its rotational axis to precess. At the moment, the rotational axis of the Earth points toward Polaris, the North Star, which remains almost fixed in position in time-lapse photographs while the other stars move in circular paths about it. In a few hundred years, however, Polaris will also move in a circular path in the sky because the Earth's axis of rotation will point in a different direction. After 26,000 years the Earth will complete one full cycle of precession, and Polaris will again be the pole star.

Enhance Your Understanding   (Answers given at the end of the chapter)

9. The angular velocity of the spinning bicycle wheel in FIGURE 11-36 points out of the page, toward the reader. Does the wheel spin clockwise or counterclockwise?

▲ FIGURE 11-36 Enhance Your Understanding 9.

Section Review

* The directions of the angular velocity, angular momentum, and torque are given by right-hand rules.

CHAPTER 11 REVIEW

CHAPTER SUMMARY

11-1 TORQUE

A force applied so as to cause an angular acceleration is said to exert a torque, τ.

Torque Due to a Tangential Force

A tangential force F applied at a distance r from the axis of rotation produces a torque

$$\tau = rF \qquad \text{11-1}$$

Torque for a General Force

A force exerted at an angle θ with respect to the radial direction, and applied at a distance r from the axis of rotation, produces the torque

$$\tau = rF \sin\theta \qquad \text{11-2}$$

11-2 TORQUE AND ANGULAR ACCELERATION

A single torque applied to an object gives it an angular acceleration.

Newton's Second Law for Rotation

The connection between torque and angular acceleration is

$$\sum \tau = I\alpha \qquad \text{11-4}$$

Rotational/Translational Analogies

Torque is analogous to force, the moment of inertia is analogous to mass, and angular acceleration is analogous to linear acceleration. Therefore, the rotational analog of $F = ma$ is $\tau = I\alpha$.

11-3 ZERO TORQUE AND STATIC EQUILIBRIUM

The conditions for an object to be in static equilibrium are that the total force and the total torque acting on the object must be zero:

$$\sum F_x = 0, \quad \sum F_y = 0, \quad \sum \tau = 0$$

11-4 CENTER OF MASS AND BALANCE

An object balances when it is supported at its center of mass.

11-5 DYNAMIC APPLICATIONS OF TORQUE

Newton's second law can be applied to rotational systems in a way that is completely analogous to its application to linear systems.

11-6 ANGULAR MOMENTUM

A moving object has angular momentum as long as its direction of motion does not extend through the axis of rotation.

Angular Momentum and Angular Speed

Angular momentum can be expressed in terms of angular speed and the moment of inertia as follows:

$$L = I\omega \qquad \text{11-11}$$

This is the rotational analog of $p = mv$.

Angular Momentum and Linear Speed

If an object of mass m is a distance r from the axis of rotation and moves with a linear speed v at an angle θ with respect to the radial direction, its angular momentum is

$$L = rmv \sin \theta \qquad \text{11-13}$$

Newton's Second Law

Newton's second law can be expressed in terms of the rate of change of the angular momentum:

$$\sum \tau = I\alpha = \frac{\Delta L}{\Delta t} \qquad \text{11-14}$$

This is the rotational analog of $\sum F = \Delta p / \Delta t$.

11-7 CONSERVATION OF ANGULAR MOMENTUM

If the net external torque acting on a system is zero, its angular momentum is conserved:

$$L_f = L_i$$

11-8 ROTATIONAL WORK AND POWER

A torque acting through an angle does work, just as a force acting through a distance does work.

Work Done by a Torque

A torque τ acting through an angle $\Delta\theta$ does work W given by

$$W = \tau\Delta\theta \qquad \text{11-17}$$

Work–Energy Theorem

The work–energy theorem is

$$W = \Delta K = K_f - K_i \qquad \text{11-18}$$

11-9 THE VECTOR NATURE OF ROTATIONAL MOTION

Rotational quantities have directions that point along the axis of rotation. The precise direction is given by the right-hand rule.

ANSWERS TO ENHANCE YOUR UNDERSTANDING QUESTIONS

1. Torque produced by $F_3 <$ torque produced by $F_2 =$ torque produced by $F_4 <$ torque produced by F_1.

2. The angular acceleration of object 1 is greater (by a factor of 2) than the angular acceleration of object 2.

3. Greater than half the weight, because the ropes have both vertical and horizontal components.

4. Mass of B $<$ mass of C $<$ mass of A.

5. **(a)** The tension is less than mg, because the mass accelerates downward and hence the net force is downward. **(b)** The tension is greater than zero, because it is accelerating the pulley.

6. Object 2 has the greater angular momentum, because both objects have the same moment of inertia and object 2 has the greater angular speed.

7. The initial kinetic energy is greater than the final kinetic energy because the inelastic collision between the child and the merry-go-round dissipates energy.

8. The work done in system 1 is equal to the work done in system 2 because torque times angular displacement is the same for the two systems.

9. The wheel spins counterclockwise, as can be verified with the RHR for angular velocity.

CONCEPTUAL QUESTIONS

For instructor-assigned homework, go to www.masteringphysics.com.

(Answers to odd-numbered Conceptual Questions can be found in the back of the book.)

1. Two forces produce the same torque. Does it follow that they have the same magnitude? Explain.

2. A car pitches down in front when the brakes are applied sharply. Explain this observation in terms of torques.

3. A tightrope walker uses a long pole to aid in balancing. Why?

4. When a motorcycle accelerates rapidly from a stop it sometimes "pops a wheelie"; that is, its front wheel may lift off the ground. Explain this behavior in terms of torques.

5. Give an example of a system in which the net torque is zero but the net force is nonzero.

6. Give an example of a system in which the net force is zero but the net torque is nonzero.

7. Is the normal force exerted by the ground the same for all four tires on your car? Explain.

8. Give two everyday examples of objects that are not in static equilibrium.

9. Give two everyday examples of objects that are in static equilibrium.

10. Can an object have zero translational acceleration and, at the same time, have nonzero angular acceleration? If your answer is no, explain why not. If your answer is yes, give a specific example.

11. Stars form when a large rotating cloud of gas collapses. What happens to the angular speed of the gas cloud as it collapses?

12. What purpose does the tail rotor on a helicopter serve?

13. Is it possible to change the angular momentum of an object without changing its linear momentum? If your answer is no, explain why not. If your answer is yes, give a specific example.

14. Suppose a diver springs into the air with no initial angular velocity. Can the diver begin to rotate by folding into a tucked position? Explain.

PROBLEMS AND CONCEPTUAL EXERCISES

Answers to odd-numbered Problems and Conceptual Exercises can be found in the back of the book. **BIO** *identifies problems of biological or medical interest;* **CE** *indicates a conceptual exercise.* Predict/Explain *problems ask for two responses: (a) your prediction of a physical outcome, and (b) the best explanation among three provided; and* Predict/Calculate *problems ask for a prediction of a physical outcome, based on fundamental physics concepts, and follow that with a numerical calculation to verify the prediction. On all problems, bullets (•, ••, •••) indicate the level of difficulty.*

SECTION 11-1 TORQUE

1. • To tighten a spark plug, it is recommended that a torque of 15 N·m be applied. If a mechanic tightens the spark plug with a wrench that is 25 cm long, what is the minimum force necessary to create the desired torque?

2. • **Pulling a Weed** The gardening tool shown in FIGURE 11-37 is used to pull weeds. If a 1.23-N·m torque is required to pull a given weed, what force did the weed exert on the tool?

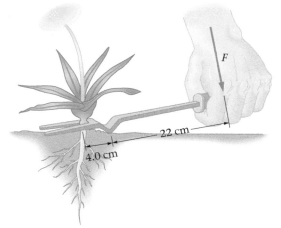

FIGURE 11-37 Problem 2

3. • A person slowly lowers a 3.6-kg crab trap over the side of a dock, as shown in FIGURE 11-38. What torque does the trap exert about the person's shoulder?

FIGURE 11-38 Problem 3

4. •• A squirrel-proof bird feeder has a lever that closes to protect the seeds when a 0.42-kg squirrel sits on it, but not when a 0.077-kg bird perches there. **(a)** If the lever radius is 9.6 cm, what torques do the squirrel and the bird each exert on it? **(b)** The torque produced by the animal is balanced by a spring that applies a perpendicular force a distance of 3.5 cm from the axis of rotation. If the squirrel must stretch the spring 2.5 cm in order to close the lever and protect the seeds, what should be the force constant of the spring?

5. •• At one position during its cycle, the foot pushes straight down with a 440-N force on a bicycle pedal arm that is rotated an angle $\phi = 65°$ from the vertical, as shown in FIGURE 11-39. If the force is applied $r = 0.17$ m from the axis of rotation, what torque is produced?

FIGURE 11-39 Problem 5

6. •• **BIO** Predict/Calculate **Force to Hold a Baseball** A person holds a 1.42-N baseball in his hand, a distance of 34.0 cm from the elbow joint, as shown in FIGURE 11-40. The biceps, attached at a distance of 2.75 cm from the elbow, exerts an upward force of 12.6 N on the forearm. Consider the forearm and hand to be a uniform rod with a mass of 1.20 kg. **(a)** Calculate the net torque acting on the forearm and hand. Use the elbow joint as the axis of rotation. **(b)** If the net torque obtained in part (a) is nonzero, in which direction will the forearm and hand rotate? **(c)** Would the torque exerted on the forearm by the biceps increase or decrease if the biceps were attached farther from the elbow joint?

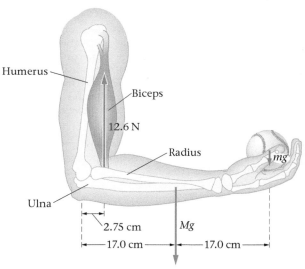

FIGURE 11-40 Problems 6 and 23

7. •• At the local playground, a 21-kg child sits on the right end of a horizontal teeter-totter, 1.8 m from the pivot point. On the left side of the pivot an adult pushes straight down on the teeter-totter with a force of 151 N. In which direction (clockwise or counterclockwise) does the teeter-totter rotate if the adult applies the force at a distance of **(a)** 3.0 m, **(b)** 2.5 m, or **(c)** 2.0 m from the pivot?

SECTION 11-2 TORQUE AND ANGULAR ACCELERATION

8. • **CE** Predict/Explain Consider the pulley–block systems shown in Conceptual Example 11-5. **(a)** Is the tension in the string on the left-hand rotating system (the one with the moveable masses at the ends of the rod) greater than, less than, or equal to the weight of the mass attached to that string? **(b)** Choose the *best explanation* from among the following:
   I. The mass is in free fall once it is released.
   II. The string rotates the pulley in addition to supporting the mass.
   III. The mass accelerates downward.

9. • **CE** Suppose a torque rotates your body about one of three different axes of rotation: case A, an axis through your spine; case B, an axis through your hips; and case C, an axis through your ankles. Rank these three axes of rotation in increasing order of the angular acceleration produced by the torque. Indicate ties where appropriate.

10. • A torque of 0.97 N · m is applied to a bicycle wheel of radius 35 cm and mass 0.75 kg. Treating the wheel as a hoop, find its angular acceleration.

11. • When a ceiling fan rotating with an angular speed of 3.66 rad/s is turned off, a frictional torque of 0.125 N · m slows it to a stop in 31.8 s. What is the moment of inertia of the fan?

12. • When the play button is pressed, a CD accelerates uniformly from rest to 450 rev/min in 3.0 revolutions. If the CD has a radius of 6.0 cm and a mass of 17 g, what is the torque exerted on it?

13. •• A person holds a ladder horizontally at its center. Treating the ladder as a uniform rod of length 3.66 m and mass 10.9 kg, find the torque the person must exert on the ladder to give it an angular acceleration of 0.333 rad/s².

14. •• A 0.180-kg wooden rod is 1.25 m long and pivots at one end. It is held horizontally and then released. **(a)** What is the angular acceleration of the rod after it is released? **(b)** What is the linear acceleration of a spot on the rod that is 0.95 m from the axis of rotation? **(c)** At what location along the rod should a die be placed (Figure 11-9) so that the die just begins to separate from the rod as it falls?

15. •• Predict/Calculate A wheel on a game show is given an initial angular speed of 1.22 rad/s. It comes to rest after rotating through 0.75 of a turn. **(a)** Find the average torque exerted on the wheel given that it is a disk of radius 0.71 m and mass 6.4 kg. **(b)** If the mass of the wheel is doubled and its radius is halved, will the angle through which it rotates before coming to rest increase, decrease, or stay the same? Explain. (Assume that the average torque exerted on the wheel is unchanged.)

16. •• The L-shaped object in FIGURE 11-41 consists of three masses connected by light rods. What torque must be applied to this object to give it an angular acceleration of 1.20 rad/s² if it is rotated about **(a)** the x axis, **(b)** the y axis, or **(c)** the z axis (which is through the origin and perpendicular to the page)?

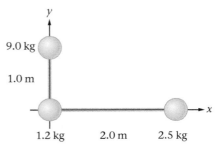

FIGURE 11-41 Problems 16, 17, and 66

17. •• **CE** The L-shaped object described in the previous problem can be rotated in one of the following three ways: case A, about the x axis; case B, about the y axis; and case C, about the z axis (which passes through the origin perpendicular to the plane of the figure). If the same torque τ is applied in each of these cases, rank them in increasing order of the resulting angular acceleration. Indicate ties where appropriate.

18. •• **CE** A motorcycle accelerates from rest, and both the front and rear tires roll without slipping. **(a)** Is the force exerted by the ground on the rear tire in the forward or in the backward direction? Explain. **(b)** Is the force exerted by the ground on the front tire in the forward or in the backward direction? Explain. **(c)** If the moment of inertia of the front tire is increased, will the motorcycle's acceleration increase, decrease, or stay the same? Explain.

19. •• Predict/Calculate A torque of 13 N · m is applied to the rectangular object shown in FIGURE 11-42. The torque can act about the x axis, the y axis, or the z axis, which passes through the origin and points out of the page. **(a)** In which case does the object experience the greatest angular acceleration? The least angular acceleration? Explain. Find the angular acceleration when the torque acts about **(b)** the x axis, **(c)** the y axis, and **(d)** the z axis.

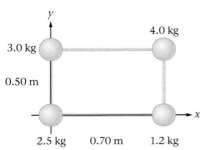

FIGURE 11-42 Problems 19 and 67

SECTION 11-3 ZERO TORQUE AND STATIC EQUILIBRIUM

20. • **CE** Predict/Explain Suppose the person in Example 11-11 climbs higher on the ladder. **(a)** As a result, is the ladder more likely, less likely, or equally likely to slip? **(b)** Choose the *best explanation* from among the following:
   I. The forces are the same regardless of the person's position.
   II. The magnitude of f_2 must increase as the person moves upward.
   III. When the person is higher, the ladder presses down harder on the floor.

21. • A string that passes over a pulley has a 0.321-kg mass attached to one end and a 0.635-kg mass attached to the other end. The pulley, which is a disk of radius 9.40 cm, has friction in its axle. What is the magnitude of the frictional torque that must be exerted by the axle if the system is to be in static equilibrium?

22. • To loosen the lid on a jar of jam 7.6 cm in diameter, a torque of 11 N · m must be applied to the circumference of the lid. If a jar wrench whose handle extends 15 cm from the center of the jar is attached to the lid, what is the minimum force required to open the jar?

23. •• **BIO** Predict/Calculate Referring to the person holding a baseball in Problem 6, suppose the biceps exert just enough upward force to keep the system in static equilibrium. **(a)** Is the force exerted by the biceps more than, less than, or equal to the combined weight of the forearm, hand, and baseball? Explain. **(b)** Determine the force exerted by the biceps.

24. •• **BIO** Predict/Calculate **A Person's Center of Mass** To determine the location of her center of mass, a physics student lies on a

lightweight plank supported by two scales 2.50 m apart, as indicated in **FIGURE 11-43**. If the left scale reads 435 N, and the right scale reads 183 N, find **(a)** the student's mass and **(b)** the distance from the student's head to her center of mass.

FIGURE 11-43 Problem 24

25. •• **Triceratops** A set of fossilized triceratops footprints discovered in Texas show that the front and rear feet were 3.2 m apart. The rear footprints were observed to be twice as deep as the front footprints. Assuming that the rear feet pressed down on the ground with twice the force exerted by the front feet, find the horizontal distance from the rear feet to the triceratops's center of mass.

26. •• Predict/Calculate A schoolyard teeter-totter with a total length of 6.4 m and a mass of 41 kg is pivoted at its center. A 21-kg child sits on one end of the teeter-totter. **(a)** Where should a parent push vertically downward with a force of 210 N in order to hold the teeter-totter level? **(b)** Where should the parent push with a force of 310 N? **(c)** How would your answers to parts (a) and (b) change if the mass of the teeter-totter were doubled? Explain.

27. •• A 0.122-kg remote control 23.0 cm long rests on a table, as shown in **FIGURE 11-44**, with a length L overhanging its edge. To operate the power button on this remote requires a force of 0.365 N. How far can the remote control extend beyond the edge of the table and still not tip over when you press the power button? Assume the mass of the remote is distributed uniformly, and that the power button is 1.41 cm from the overhanging end of the remote.

FIGURE 11-44 Problem 27

28. •• Predict/Calculate A 0.16-kg meterstick is held perpendicular to a vertical wall by a 2.5-m string going from the wall to the far end of the stick. **(a)** Find the tension in the string. **(b)** If a shorter string is used, will its tension be greater than, less than, or the same as that found in part (a)? **(c)** Find the tension in a 2.0-m string.

29. •• Babe Ruth steps to the plate and casually points to left center field to indicate the location of his next home run. The mighty Babe holds his bat across his shoulder, with one hand holding the small end of the bat. The bat is horizontal, and the distance from the small end of the bat to the shoulder is 22.5 cm. If the bat has a mass of 1.10 kg and has a center of mass that is 67.0 cm from the small end of the bat, find the magnitude and direction of the force exerted by **(a)** the hand and **(b)** the shoulder.

30. •• A uniform metal rod, with a mass of 2.0 kg and a length of 1.5 m, is attached to a wall by a hinge at its base. A horizontal wire bolted to the wall 0.70 m above the base of the rod holds the rod at an angle of 28° above the horizontal. The wire is attached to the top of the rod. **(a)** Find the tension in the wire. Find **(b)** the horizontal and **(c)** the vertical components of the force exerted on the rod by the hinge.

31. •• A rigid, vertical rod of negligible mass is connected to the floor by a bolt through its lower end, as shown in **FIGURE 11-45**. The rod also has a wire connected between its top end and the floor. If a horizontal force F is applied at the midpoint of the rod, find **(a)** the tension in the wire, and **(b)** the horizontal and **(c)** the vertical components of force exerted by the bolt on the rod.

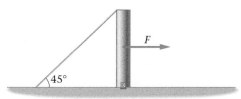

FIGURE 11-45 Problem 31

32. •• In **FIGURE 11-46** two acrobats perform a balancing maneuver. The arms of the top acrobat (in yellow), together with the weight of his head and shoulders, provide a downward force at a distance of 28 cm from the pivot point to balance the torque produced by his lower body. If the lower body of the top acrobat is modeled as a 54-kg mass at a distance of 39 cm from the pivot point, how much force must the arms and upper body produce?

FIGURE 11-46 Problem 32

33. ••• **BIO Forces in the Foot** In **FIGURE 11-47** we see the forces acting on a sprinter's foot just before she takes off at the start of the race. Find the magnitude of the force exerted on the heel by the Achilles tendon, F_H, and the magnitude of the force exerted on the foot at the ankle joint, F_J.

FIGURE 11-47 Problem 33

34. ••• A stick with a mass of 0.214 kg and a length of 0.436 m rests in contact with a bowling ball and a rough floor, as shown in **FIGURE 11-48**. The bowling ball has a diameter of 21.6 cm, and the angle the stick makes with the horizontal is 30.0°. You may assume there is no friction

FIGURE 11-48 Problem 34

between the stick and the bowling ball, though friction with the floor must be taken into account. **(a)** Find the magnitude of the force exerted on the stick by the bowling ball. **(b)** Find the horizontal component of the force exerted on the stick by the floor. **(c)** Repeat part (b) for the vertical component of the force.

SECTION 11-4 CENTER OF MASS AND BALANCE

35. • A hand-held shopping basket 62.0 cm long has a 1.81-kg carton of milk at one end, and a 0.722-kg box of cereal at the other end. Where should a 1.80-kg container of orange juice be placed so that the basket balances at its center?

36. • If the cat in Example 11-9 has a mass of 3.9 kg, how far to the right of sawhorse B can it walk before the plank begins to tip?

37. •• Predict/Calculate A 0.34-kg meterstick balances at its center. If a necklace is suspended from one end of the stick, the balance point moves 9.5 cm toward that end. **(a)** Is the mass of the necklace more than, less than, or the same as that of the meterstick? Explain. **(b)** Find the mass of the necklace.

38. •• **Maximum Overhang** Three identical, uniform books of length L are stacked one on top the other. Find the maximum overhang distance d in **FIGURE 11-49** such that the books do not fall over.

FIGURE 11-49 Problem 38

39. •• A baseball bat balances 71.1 cm from one end. If a 0.560-kg glove is attached to that end, the balance point moves 24.7 cm toward the glove. Find the mass of the bat.

SECTION 11-5 DYNAMIC APPLICATIONS OF TORQUE

40. •• A 2.85-kg bucket is attached to a rope wrapped around a disk-shaped pulley of radius 0.121 m and mass 0.742 kg. If the bucket is allowed to fall, **(a)** what is its linear acceleration? **(b)** What is the angular acceleration of the pulley? **(c)** How far does the bucket drop in 1.50 s?

41. •• A child exerts a tangential 53.4-N force on the rim of a disk-shaped merry-go-round with a radius of 2.29 m. If the merry-go-round starts at rest and acquires an angular speed of 0.125 rev/s in 3.50 s, what is its mass?

42. •• Predict/Calculate You pull downward with a force of 28 N on a rope that passes over a disk-shaped pulley of mass 1.2 kg and radius 0.075 m. The other end of the rope is attached to a 0.67-kg mass. **(a)** Is the tension in the rope the same on both sides of the pulley? If not, which side has the greater tension? **(b)** Find the tension in the rope on both sides of the pulley.

43. ••• One elevator arrangement includes the passenger car, a counterweight, and two large pulleys, as shown in **FIGURE 11-50**. Each pulley has a radius of 1.2 m and a moment of inertia of 380 kg · m². The top pulley is driven by a motor. The elevator car plus passengers has a mass of 3100 kg, and the counterweight has a mass of 2700 kg. If the motor is to accelerate the elevator car upward at 1.8 m/s², how much torque must it generate? *Hint*: The two pulleys move together, so you can model them as a single pulley with the sum of the moments of inertia.

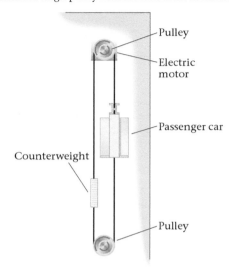

FIGURE 11-50 Problem 43

44. ••• **Atwood's Machine** An Atwood's machine consists of two masses, m_1 and m_2, connected by a string that passes over a pulley. If the pulley is a disk of radius R and mass M, find the acceleration of the masses.

SECTION 11-6 ANGULAR MOMENTUM

45. • A 1.4-kg bicycle tire with a radius of 33 cm rotates with an angular speed of 155 rpm. Find the angular momentum of the tire, assuming it can be modeled as a hoop.

46. • Jogger 1 in **FIGURE 11-51** has a mass of 65.3 kg and runs in a straight line with a speed of 3.35 m/s. **(a)** What is the magnitude of the jogger's linear momentum? **(b)** What is the magnitude of the jogger's angular momentum with respect to the origin, O?

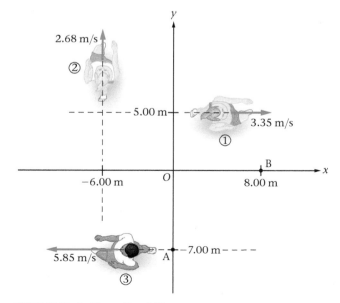

FIGURE 11-51 Problems 46 and 47

47. •• Predict/Calculate Suppose jogger 3 in Figure 11-51 has a mass of 62.2 kg and a speed of 5.85 m/s. **(a)** Is the magnitude of the jogger's angular momentum greater with respect to point A or point B? Explain. **(b)** Is the magnitude of the jogger's angular momentum with respect to point B greater than, less than, or the same as it is with respect to the origin, O? Explain. **(c)** Calculate the magnitude of the jogger's angular momentum with respect to points A, B, and O.

48. •• A torque of 0.12 N · m is applied to an egg beater. **(a)** If the egg beater starts at rest, what is its angular momentum after 0.65 s? **(b)** If the moment of inertia of the egg beater is 2.5×10^{-3} kg · m², what is its angular speed after 0.65 s?

49. •• A windmill has an initial angular momentum of 8500 kg · m²/s. The wind picks up, and 5.86 s later the windmill's angular momentum is 9700 kg · m²/s. What was the torque acting on the windmill, assuming it was constant during this time?

50. •• Two gerbils run in place with a linear speed of 0.55 m/s on an exercise wheel that is shaped like a hoop. Find the angular momentum of the system if each gerbil has a mass of 0.22 kg and the exercise wheel has a radius of 9.5 cm and a mass of 5.0 g.

SECTION 11-7 CONSERVATION OF ANGULAR MOMENTUM

51. • **CE** Predict/Explain A student rotates on a frictionless piano stool with his arms outstretched, a heavy weight in each hand. Suddenly he lets go of the weights, and they fall to the floor. As a result, does the student's angular speed increase, decrease, or stay the same? **(b)** Choose the *best explanation* from among the following:
  I. The loss of angular momentum when the weights are dropped causes the student to rotate more slowly.
  II. The student's moment of inertia is decreased by dropping the weights.
  III. Dropping the weights exerts no torque on the student, but the floor exerts a torque on the weights when they land.

52. • **CE** A puck on a horizontal, frictionless surface is attached to a string that passes through a hole in the surface, as shown in FIGURE 11-52. As the puck rotates about the hole, the string is pulled downward, bringing the puck closer to the hole. During this process, do the puck's **(a)** linear speed, **(b)** angular speed, and **(c)** angular momentum increase, decrease, or stay the same?

FIGURE 11-52 Problems 52 and 75

53. • **CE** A puck on a horizontal, frictionless surface is attached to a string that wraps around a pole of finite radius, as shown in FIGURE 11-53. **(a)** As the puck moves along the spiral path, does its speed increase, decrease, or stay the same? Explain. **(b)** Does its angular momentum increase, decrease, or stay the same? Explain.

FIGURE 11-53 Problem 53

54. • As an ice skater begins a spin, his angular speed is 3.04 rad/s. After pulling in his arms, his angular speed increases to 5.24 rad/s. Find the ratio of the skater's final moment of inertia to his initial moment of inertia.

55. •• A disk-shaped merry-go-round of radius 2.63 m and mass 155 kg rotates freely with an angular speed of 0.641 rev/s. A 59.4-kg person running tangential to the rim of the merry-go-round at 3.41 m/s jumps onto its rim and holds on. Before jumping on the merry-go-round, the person was moving in the same direction as the merry-go-round's rim. What is the final angular speed of the merry-go-round?

56. •• A student sits at rest on a piano stool that can rotate without friction. The moment of inertia of the student-stool system is 4.1 kg · m². A second student tosses a 1.5-kg mass with a speed of 2.7 m/s to the student on the stool, who catches it at a distance of 0.40 m from the axis of rotation. What is the resulting angular speed of the student and the stool?

57. •• Predict/Calculate A turntable with a moment of inertia of 5.4×10^{-3} kg · m² rotates freely with an angular speed of $33\frac{1}{3}$ rpm. Riding on the rim of the turntable, 15 cm from the center, is a cute, 32-g mouse. **(a)** If the mouse walks to the center of the turntable, will the turntable rotate faster, slower, or at the same rate? Explain. **(b)** Calculate the angular speed of the turntable when the mouse reaches the center.

58. •• A student on a piano stool rotates freely with an angular speed of 2.95 rev/s. The student holds a 1.25-kg mass in each outstretched arm, 0.759 m from the axis of rotation. The combined moment of inertia of the student and the stool, ignoring the two masses, is 5.43 kg · m², a value that remains constant. **(a)** As the student pulls his arms inward, his angular speed increases to 3.54 rev/s. How far are the masses from the axis of rotation at this time, considering the masses to be points? **(b)** Calculate the initial and final kinetic energies of the system.

59. ••• **Walking on a Merry-Go-Round** A child of mass m stands at rest near the rim of a stationary merry-go-round of radius R and moment of inertia I. The child now begins to walk around the circumference of the merry-go-round with a tangential speed v with respect to the merry-go-round's surface. **(a)** What is the child's speed with respect to the ground? Check your result in the limits **(b)** $I \to 0$ and **(c)** $I \to \infty$.

SECTION 11-8 ROTATIONAL WORK AND POWER

60. • **CE** Predict/Explain Two spheres of equal mass and radius are rolling across the floor with the same speed. Sphere 1 is a uniform solid; sphere 2 is hollow. Is the work required to stop sphere 1 greater than, less than, or equal to the work required to stop sphere 2? **(b)** Choose the *best explanation* from among the following:
  I. Sphere 2 has the greater moment of inertia and hence the greater rotational kinetic energy.
  II. The spheres have equal mass and speed; therefore, they have the same kinetic energy.
  III. The hollow sphere has less kinetic energy.

61. • Turning a doorknob through 0.25 of a revolution requires 0.14 J of work. What is the torque required to turn the doorknob?

62. • A person exerts a tangential force of 36.1 N on the rim of a disk-shaped merry-go-round of radius 2.74 m and mass 167 kg. If the merry-go-round starts at rest, what is its angular speed after the person has rotated it through an angle of 32.5°?

63. • To prepare homemade ice cream, a crank must be turned with a torque of 3.95 N · m. How much work is required for each complete turn of the crank?

64. • **Power of a Dental Drill** A popular make of dental drill can operate at a speed of 42,500 rpm while producing a torque of 3.68 oz · in. What is the power output of this drill? Give your answer in watts.

65. •• For a home repair job you must turn the handle of a screwdriver 27 times. **(a)** If you apply an average force of 13 N tangentially to the 3.0-cm-diameter handle, how much work have you done? **(b)** If you complete the job in 24 seconds, what was your average power output?

66. •• The L-shaped object in Figure 11-40 consists of three masses connected by light rods. Find the work that must be done on this object to accelerate it from rest to an angular speed of 2.35 rad/s about **(a)** the *x* axis, **(b)** the *y* axis, and **(c)** the *z* axis (which is through the origin and perpendicular to the page).

67. •• The rectangular object in Figure 11-41 consists of four masses connected by light rods. What power must be applied to this object to accelerate it from rest to an angular speed of 2.5 rad/s in 6.4 s about **(a)** the *x* axis, **(b)** the *y* axis, and **(c)** the *z* axis (which is through the origin and perpendicular to the page)?

68. •• **Predict/Calculate** A circular saw blade accelerates from rest to an angular speed of 3620 rpm in 6.30 revolutions. **(a)** Find the torque exerted on the saw blade, assuming it is a disk of radius 15.2 cm and mass 0.755 kg. **(b)** Is the angular speed of the saw blade after 3.15 revolutions greater than, less than, or equal to 1810 rpm? Explain. **(c)** Find the angular speed of the blade after 3.15 revolutions.

GENERAL PROBLEMS

69. • **CE** A uniform disk stands upright on its edge, and rests on a sheet of paper placed on a tabletop. If the paper is pulled horizontally to the right, as in **FIGURE 11-54**, **(a)** does the disk rotate clockwise or counterclockwise about its center? Explain. **(b)** Does the center of the disk move to the right, move to the left, or stay in the same location? Explain.

FIGURE 11-54 Problem 69

70. • **CE** Consider the two rotating systems shown in **FIGURE 11-55**, each consisting of a mass *m* attached to a rod of negligible mass pivoted at one end. On the left, the mass is attached at the midpoint of the rod; to the right, it is attached at the free end of the rod. The rods are released from rest in the horizontal position at the same time. When the rod to the left reaches the vertical position, is the rod to the right not yet vertical (location A), vertical (location B), or past vertical (location C)? Explain.

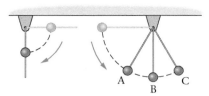

FIGURE 11-55 Problem 70

71. • **CE** Predict/Explain A disk and a hoop (bicycle wheel) of equal radius and mass each have a string wrapped around their circumferences. Hanging from the strings, halfway between the disk and the hoop, is a block of mass *m*, as shown in **FIGURE 11-56**. The disk and the hoop are free to rotate about their centers. When the block is allowed to fall, does it stay on the center line, move toward the right, or move toward the left? **(b)** Choose the *best explanation* from among the following:

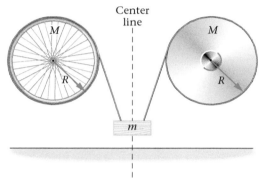

FIGURE 11-56 Problem 71

I. The disk is harder to rotate, and hence its angular acceleration is less than that of the wheel.

II. The wheel has the greater moment of inertia and unwinds more slowly than the disk.

III. The system is symmetric, with equal mass and radius on either side.

72. • **CE** A beetle sits at the rim of a turntable that is at rest but is free to rotate about a vertical axis. Suppose the beetle now begins to walk around the perimeter of the turntable. Does the beetle move forward, backward, or does it remain in the same location relative to the ground? Answer for two different cases, **(a)** the turntable is much more massive than the beetle and **(b)** the turntable is massless.

73. • After getting a drink of water, a hamster jumps onto an exercise wheel for a run. A few seconds later the hamster is running in place with a speed of 1.3 m/s. Find the work done by the hamster to get the exercise wheel moving, assuming it is a hoop of radius 0.13 m and mass 6.5 g.

74. •• A 47.0-kg uniform rod 4.25 m long is attached to a wall with a hinge at one end. The rod is held in a horizontal position by a wire attached to its other end. The wire makes an angle of 30.0° with the horizontal, and is bolted to the wall directly above the hinge. If the wire can withstand a maximum tension of 1450 N before breaking, how far from the wall can a 68.0-kg person sit without breaking the wire?

75. •• Predict/Calculate A puck attached to a string moves in a circular path on a frictionless surface, as shown in Figure 11-52. Initially, the speed of the puck is *v* and the radius of the circle is *r*. If the string passes through a hole in the surface, and is pulled downward until the radius of the circular path is $r/2$, **(a)** does the speed of the puck increase, decrease, or stay the same? **(b)** Calculate the final speed of the puck.

76. •• **BIO The Masseter Muscle** The masseter muscle, the principal muscle for chewing, is one of the strongest muscles for its size in the human body. It originates on the lower edge of the zygomatic arch (cheekbone) and inserts in the angle of the mandible. Referring to the lower diagram in **FIGURE 11-57**, where $d = 7.60$ cm, $h = 3.15$ cm and $D = 10.85$ cm, **(a)** find the torque produced about the axis of rotation by the masseter muscle. The force exerted by the masseter muscle is $F_M = 455$ N. **(b)** Find the biting force, F_B, exerted on the mandible by the upper teeth. Find **(c)** the horizontal and **(d)** the vertical component of the force F_J exerted on the mandible at the joint where it attaches to the skull. Assume that the mandible is in static equilibrium, and that upward is the positive vertical direction.

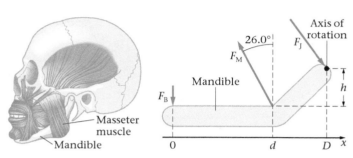

FIGURE 11-57 Problem 76

77. •• **Exercising the Biceps** You are designing exercise equipment to operate as shown in **FIGURE 11-58**, where a person pulls upward on an elastic cord. The cord behaves like an ideal spring and has an unstretched length of 31 cm. If you would like the torque about the elbow joint to be 81 N·m in the position shown, what force constant, k, is required for the cord?

FIGURE 11-58 Problem 77

78. •• **Balancing a T. rex** Paleontologists believe that *Tyrannosaurus rex* stood and walked with its spine almost horizontal, as indicated in

FIGURE 11-59, and that its tail was held off the ground to balance its upper body about the hip joint. Given that the total mass of *T. rex* was 5400 kg, and that the placement of the center of mass of the tail and the upper body was as shown in Figure 11-59, find the mass of the tail required for balance.

FIGURE 11-59 Problem 78

79. •• In Example 11-11, suppose the ladder is uniform, 4.0 m long, and weighs 60.0 N. Find the forces exerted on the ladder when the person is **(a)** halfway up the ladder and **(b)** three-fourths of the way up the ladder.

80. •• When you arrive at Duke's Dude Ranch, you are greeted by the large wooden sign shown in **FIGURE 11-60**. The left end of the sign is held in place by a bolt, the right end is tied to a rope that makes an angle of 20.0° with the horizontal. If the sign is uniform, 3.20 m long, and has a mass of 16.0 kg, what are **(a)** the tension in the rope, and **(b)** the horizontal and vertical components of the force, $\vec{F}$, exerted by the bolt?

FIGURE 11-60 Problem 80

81. •• A 67.0-kg person stands on a lightweight diving board supported by two pillars, one at the end of the board, the other 1.10 m away. The pillar at the end of the board exerts a downward force of 828 N. **(a)** How far from that pillar is the person standing? **(b)** Find the force exerted by the second pillar.

82. •• **Flats Versus Heels** A woman might wear a pair of flat shoes to work during the day, as in **FIGURE 11-61 (a)**, but a pair of high heels, **FIGURE 11-61 (b)**, when going out for the evening. Assume that each foot supports half her weight, $w = W/2 = 279$ N, and that the forces exerted by the floor on her feet occur at the points A and

B in both figures. Find the forces F_A (point A) and F_B (point B) for **(a)** flat shoes and **(b)** high heels. **(c)** How have the high heels changed the weight distribution between the woman's heels and toes?

(a)   **(b)**

FIGURE 11-61 Problem 82

83. •• **BIO** A young girl sits at the edge of a dock by the bay, dipping her feet in the water. At the instant shown in **FIGURE 11-62**, she holds her lower leg stationary with her quadriceps muscle at an angle of 39° with respect to the horizontal. Use the information given in the figure, plus the fact that her lower leg has a mass of 3.4 kg, to determine the magnitude of the force, F_Q, exerted on the lower leg by the quadriceps.

FIGURE 11-62 Problem 83

84. •• **BIO Deltoid Muscle** A crossing guard holds a STOP sign at arm's length, as shown in **FIGURE 11-63**. Her arm is horizontal, and we assume that the deltoid muscle is the only muscle supporting her arm. The weight of her upper arm is $W_u = 18$ N, the weight of her lower arm is $W_l = 11$ N, the weight of her hand is $W_h = 4.0$ N, and the weight of the sign is $W_s = 8.9$ N. The location where each of these forces acts on the arm is indicated in the figure. A force of magnitude f_d is exerted on the humerus by the deltoid, and the shoulder joint exerts a force on the humerus with horizontal and vertical components given by f_x and f_y, respectively. **(a)** Is the magnitude of f_d greater than, less than, or equal to the magnitude of f_x? Explain. Find **(b)** f_d, **(c)** f_x, and **(d)** f_y. (The weights in Figure 11-63 are drawn to scale; the unknown forces are to be determined. If a force is found to be negative, its direction is opposite to that shown.)

Free-Body Diagram of the Arm

FIGURE 11-63 Problem 84

85. •• **BIO Triceps** To determine the force a person's triceps muscle can exert, a doctor uses the procedure shown in **FIGURE 11-64**, where the patient pushes down with the palm of his hand on a force meter. Given that the weight of the lower arm is $Mg = 15.6$ N, and that the force meter reads $F = 89.0$ N, what is the force F_T exerted vertically upward by the triceps?

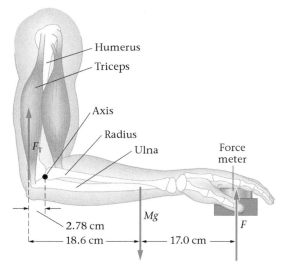

FIGURE 11-64 Problem 85

86. •• Predict/Calculate Suppose partial melting of the polar ice caps increases the moment of inertia of the Earth from $0.331\,M_E R_E^2$ to $0.332\,M_E R_E^2$. **(a)** Would the length of a day (the time required for the Earth to complete one revolution about its axis) increase or decrease? Explain. **(b)** Calculate the change in the length of a day. Give your answer in seconds.

87. ••• A bicycle wheel of radius R and mass M is at rest against a step of height $3R/4$, as illustrated in FIGURE 11-65. Find the minimum horizontal force F that must be applied to the axle to make the wheel start to rise up over the step.

FIGURE 11-65 Problem 87

88. ••• A 0.101-kg yo-yo has an outer radius R that is 5.60 times greater than the radius r of its axle. The yo-yo is in equilibrium if a mass m is suspended from its outer edge, as shown in FIGURE 11-66. Find the tension in the two strings, T_1 and T_2, and the mass m.

FIGURE 11-66 Problem 88

89. ••• **BIO Peak Pedaling Torque** The downward force produced by the quadriceps muscles during the power stroke of bicycle-pedaling motion is shown in FIGURE 11-67 as a function of the crank angle ϕ (see Figure 11-39). The force from these muscles decreases linearly, but the torque depends on the crank angle according to $\tau = rF\sin(180° - \phi)$. Use the information in the graph, together with a computer spreadsheet, to find the angle ϕ at which the pedaling torque produced by the quadriceps muscle is a maximum. (Note that the actual torque applied to the crank is a result of the action of many muscles in addition to the quadriceps.)

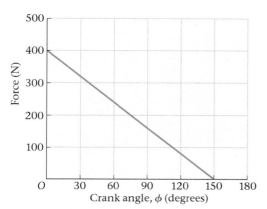

FIGURE 11-67 Problem 89

90. ••• A cylinder of mass m and radius r has a string wrapped around its circumference. The upper end of the string is held fixed, and the cylinder is allowed to fall. Show that its linear acceleration is $(2/3)g$.

91. ••• **Bricks in Equilibrium** Consider a system of four uniform bricks of length L stacked as shown in FIGURE 11-68. What is the maximum distance, x, that the middle bricks can be displaced outward before they begin to tip?

FIGURE 11-68 Problem 91

PASSAGE PROBLEMS

BIO Correcting Torsiversion

Torsiversion is a medical condition in which a tooth is rotated away from its normal position about the long axis of the root. Studies show that about 2 percent of the population suffer from this condition to some degree. For those who do, the improper alignment of the tooth can lead to tooth-to-tooth collisions during eating, as well as other problems. Typical patients display a rotation ranging from 20° to 60°, with an average around 30°.

An example is shown in FIGURE 11-69 (a), where the first premolar is not only displaced slightly from its proper location in the negative y direction, but also rotated clockwise from its normal orientation. To correct this condition, an orthodontist might use an archwire and a bracket to apply both a force and a torque to the tooth. In the simplest case, two forces are applied to the tooth in different locations, as indicated by F_1 and F_2 in Figure 11-69 (a). These two forces, if chosen properly, can reposition the tooth by exerting a net force in the positive y direction, and also reorient it by applying a torque in the counterclockwise direction.

(a)

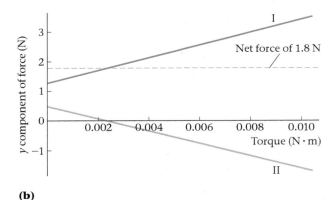

(b)

FIGURE 11-69 Problems 92, 93, 94, and 95

In a typical case, it may be desired to have a net force in the positive y direction of 1.8 N. In addition, the distances in Figure 11-69 (a) can be taken to be $d = 3.2$ mm and $D = 4.5$ mm. Given these conditions, a range of torques is possible for various values of the y components of the forces, F_{1y} and F_{2y}. For example, **FIGURE 11-69 (b)** shows the values of F_{1y} and F_{2y} necessary to produce a given torque, where the torque is measured about the center of the tooth (which is also the origin of the coordinate system). Notice that the two forces always add to 1.8 N in the positive y direction, though one of the forces changes sign as the torque is increased.

92. •• The two solid lines in Figure 11-69 (b) represent the two forces applied to the tooth. Which of the following statements about the two lines is correct?

    A. Line I $= F_{1y}$, line II $= F_{2y}$.
    B. The force indicated by line I changes sign.
    C. Line I $= F_{2y}$, line II $= F_{1y}$.
    D. The magnitude of the force indicated by line II is greater than the magnitude of the force indicated by line I.

93. • What is the value of the torque that corresponds to one of the forces being equal to zero?

    A. 0.0023 N·m   **B.** 0.0058 N·m
    C. 0.0081 N·m   **D.** 0.017 N·m

94. •• Find the values of F_{1y} and F_{2y} required to give zero net torque.

    A. $F_{1y} = -1.2$ N, $F_{2y} = 3.0$ N
    B. $F_{1y} = 1.1$ N, $F_{2y} = 0.75$ N
    C. $F_{1y} = -0.73$ N, $F_{2y} = 2.5$ N
    D. $F_{1y} = 0.52$ N, $F_{2y} = 1.3$ N

95. •• Find the values of F_{1y} and F_{2y} required to give a net torque of 0.0099 N·m. This is a torque that would be effective at rotating the tooth.

    A. $F_{1y} = -1.7$ N, $F_{2y} = 3.5$ N
    B. $F_{1y} = -3.8$ N, $F_{2y} = 5.6$ N
    C. $F_{1y} = -0.23$ N, $F_{2y} = 2.0$ N
    D. $F_{1y} = 4.0$ N, $F_{2y} = -2.2$ N

96. •• **REFERRING TO EXAMPLE 11-14** Suppose the mass of the pulley is doubled, to 0.160 kg, and that everything else in the system remains the same. **(a)** Do you expect the value of T_2 to increase, decrease, or stay the same? Explain. **(b)** Calculate the value of T_2 for this case.

97. •• **REFERRING TO EXAMPLE 11-14** Suppose the mass of the cart is doubled, to 0.62 kg, and that everything else in the system remains the same. **(a)** Do you expect the value of T_2 to increase, decrease, or stay the same? Explain. **(b)** Calculate the value of T_2 for this case.

98. •• **REFERRING TO QUICK EXAMPLE 11-22** Suppose the child runs with a different initial speed, but that everything else in the system remains the same. What initial speed does the child have if the angular speed of the system after the collision is 0.425 rad/s?

99. •• **REFERRING TO QUICK EXAMPLE 11-22** Suppose everything in the system is as described in Quick Example 11-22 except that the child approaches the merry-go-round in a direction that is not tangential. Find the angle θ between the direction of motion and the outward radial direction that is required if the final angular speed of the system is to be 0.272 rad/s.

Gravity

Big Ideas

1 Newton's law of universal gravitation states that all objects exert gravitational forces on all other objects.

2 The gravitational force of a spherical mass is the same as if all the mass were concentrated at the center of the sphere.

3 The orbits of planets, moons, comets, and asteroids are governed by Kepler's three laws.

4 Gravitational potential energy is a form of energy that must be taken into account when we apply energy conservation.

▲ Visitors from outer space have had significant impact on the Earth. For example, the Pingualuit crater in Ontario shows the result of a violent collision between the Earth and a rogue meteor or comet traveling many times faster than a bullet. But why do meteorites always seem to hit the Earth with such great speed? In this chapter we show that energy conservation, when applied to planetary bodies like the Earth, always implies a large impact speed, even when the alien object starts at rest far from the Earth.

The study of gravity has always been a central theme in physics—from Galileo's early experiments on free fall, to Einstein's general theory of relativity and Stephen Hawking's work on black holes. Perhaps the grandest milestone in this endeavor, however, was the discovery by Newton of the **law of universal gravitation.** With just one simple equation to describe the force of gravity, Newton was able to determine the orbits of planets, moons, and comets, and to explain such earthly phenomena as the tides and the fall of an apple.

12-1 Newton's Law of Universal Gravitation

It's ironic, but the first fundamental force of nature to be recognized as such, **gravity,** is also the weakest of the fundamental forces. Still, it is the force most apparent to us in our everyday lives, and it is the force responsible for the motions of the Moon, the Earth, and the planets. Yet the connection between objects falling on Earth and planets moving in their orbits was not known before Newton.

Newton's Insight The flash of insight that came to Newton—whether it was due to seeing an apple fall to the ground or not—is simply this:

> The force that causes an apple to accelerate downward is the same force that causes the Moon to move in a circular path around the Earth.

To put it another way, Newton was the first to realize that the Moon is *constantly falling* toward the Earth, without getting closer to it, and that the Moon falls for the same reason that an apple falls. This is illustrated in a classic drawing by Newton, shown in **FIGURE 12-1.**

To be specific, in the case of the apple the motion is linear as it accelerates downward toward the center of the Earth. In the case of the Moon the motion is circular with constant speed. As discussed in Section 6-5, an object in uniform circular motion accelerates toward the center of the circle. It follows, therefore, that the Moon *also* accelerates toward the center of the Earth. In fact, the force responsible for the Moon's centripetal acceleration is the Earth's gravitational attraction, the same force responsible for the fall of the apple.

To describe the force of gravity, Newton proposed the following simple law:

Newton's Law of Universal Gravitation
The force of gravity between any two point objects of mass m_1 and m_2 is attractive and of magnitude

$$F = G\frac{m_1 m_2}{r^2} \qquad \text{12-1}$$

In this expression, r is the distance between the masses, and G is a constant referred to as the **universal gravitation constant.** Its value is

$$G = 6.67 \times 10^{-11}\,\mathrm{N \cdot m^2/kg^2} \qquad \text{12-2}$$

The force is directed along the line connecting the masses, as indicated in **FIGURE 12-2.**

Notice that each mass in Figure 12-2 experiences a force of the same magnitude, $F = Gm_1 m_2/r^2$, but acting in opposite directions. That is, the force of gravity between two objects forms an action-reaction pair.

Universal Gravitation According to Newton's law, all objects in the universe attract all other objects in the universe by way of the gravitational interaction. It is in this sense that the force law is called "universal." Thus, the net gravitational force acting on you is due not only to the planet on which you stand, which is certainly responsible for the majority of the net force, but also to people nearby, planets, and even stars in far-off galaxies. In short, everything in the universe "feels" everything else, thanks to gravity.

The fact that G is such a small number means that the force of gravity between everyday objects is imperceptibly small. This is shown in the following Exercise.

EXERCISE 12-1 THE FORCE OF GRAVITY

Two apples of mass $m_1 = 0.110\,\mathrm{kg}$ and $m_2 = 0.220\,\mathrm{kg}$ sit on a dining room table. Treating the apples as point objects, find the force of gravity between them when their separation is **(a)** 13.2 cm and **(b)** 132 cm.

REASONING AND SOLUTION
The force in each case can be found with Newton's law of universal gravitation (Equation 12-1): $F = Gm_1 m_2/r^2$.

CONTINUED

▲ **FIGURE 12-1** In this illustration from his great work, the *Principia*, published in 1687, Newton presents a "thought experiment" to show the connection between free fall and orbital motion. Imagine throwing a projectile horizontally from the top of a mountain. The greater the initial speed of the projectile, the farther it travels in free fall before it strikes the ground. In the absence of air resistance, a great enough initial speed could result in the projectile circling the Earth and returning to its starting point. Thus, an object orbiting the Earth is actually in free fall—it simply has a large horizontal speed.

Big Idea **1** Newton's law of universal gravitation states that all objects in the universe exert gravitational forces on all other objects. The magnitude of the force between two objects is proportional to the product of their masses, and inversely proportional to the square of the distance between them. The force is always attractive, and along the line connecting the two objects.

▲ **FIGURE 12-2 Gravitational force between point masses** Two point masses, m_1 and m_2, separated by a distance r exert equal and opposite attractive forces on one another. The magnitude of the forces, F, is given by Equation 12-1.

a. Substituting $m_1 = 0.110$ kg, $m_2 = 0.220$ kg, and $r = 0.132$ m into $F = Gm_1m_2/r^2$ yields

$$F = G\frac{m_1m_2}{r^2} = (6.67 \times 10^{-11}\,\text{N}\cdot\text{m}^2/\text{kg}^2)\frac{(0.110\,\text{kg})(0.220\,\text{kg})}{(0.132\,\text{m})^2} = 9.26 \times 10^{-11}\,\text{N}$$

b. Repeating the calculation for $r = 1.32$ m gives

$$F = G\frac{m_1m_2}{r^2} = (6.67 \times 10^{-11}\,\text{N}\cdot\text{m}^2/\text{kg}^2)\frac{(0.110\,\text{kg})(0.220\,\text{kg})}{(1.32\,\text{m})^2} = 9.26 \times 10^{-13}\,\text{N}$$

The forces found in Exercise 12-1 are imperceptibly small—hence we can pick up one apple and not feel a tug on it from the other apple. In general, gravitational forces are significant only when large masses, such as the Earth or the Moon, are involved. That's why the weight of each apple, which is the gravitational force between it and the Earth, is so much greater than the attractive force between the apples themselves. In addition, notice that the force experienced by each apple has exactly the same magnitude, even though one has twice the mass of the other.

Exercise 12-1 also illustrates how rapidly the force of gravity decreases with distance. In particular, because F varies as $1/r^2$, it is said to have an **inverse square dependence** on distance. Thus, for example, an increase in distance by a factor of 10 results in a decrease in the force by a factor of $10^2 = 100$, as the Example illustrates. A plot of the force of gravity versus distance is given in **FIGURE 12-3**. Notice that even though the force diminishes rapidly with distance, it never completely vanishes; thus, we say that gravity is a force of infinite range.

Vector Addition and Gravity If a given mass is acted on by gravitational interactions with a number of other masses, the net force acting on it is simply the vector sum of the individual forces. This property of gravity is referred to as **superposition.** As an example, superposition implies that the net gravitational force exerted on you at this moment is the vector sum of the force exerted by the Earth, plus the force exerted by the Moon, plus the force exerted by the Sun, and so on. The following Example illustrates superposition.

▶ **FIGURE 12-3 Dependence of the gravitational force on the separation distance, *r*** The $1/r^2$ dependence of the gravitational force means that it decreases rapidly with distance. Still, it never completely vanishes. For this reason, we say that gravity is a force of infinite range; that is, every mass in the universe experiences a nonzero force from every other mass in the universe, no matter how far away.

PROBLEM-SOLVING NOTE

Net Gravitational Force

To find the net gravitational force acting on an object, you should (i) resolve each of the forces acting on the object into components and (ii) add the forces component by component.

**PHYSICS
IN CONTEXT
Looking Back**

The force of gravity is a vector quantity. Therefore, the vector calculations introduced in Chapter 3 are important here as well.

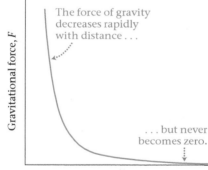

The force of gravity decreases rapidly with distance . . .

Gravitational force, F

. . . but never becomes zero.

Separation distance, r

EXAMPLE 12-2   HOW MUCH FORCE IS WITH YOU?

As part of a daring rescue attempt, the *Millennium Eagle* passes between a pair of twin asteroids, as shown in the sketch. If the mass of the spaceship is 2.50×10^7 kg and the mass of each asteroid is 3.50×10^{11} kg, find the net gravitational force exerted on the *Millennium Eagle* **(a)** when it is at location A and **(b)** when it is at location B. Treat the spaceship and the asteroids as if they were point objects.

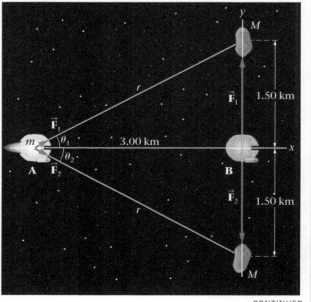

PICTURE THE PROBLEM
Our sketch shows the spaceship as it follows a path between the twin asteroids. The relevant distances and masses are indicated, as are the two points of interest, A and B. At location A the force $\vec{F}_1$ points above the x axis at the angle θ_1 (to be determined); the force $\vec{F}_2$ points below the x axis at the angle $\theta_2 = -\theta_1$, as can be seen by symmetry. At location B, the two forces act in opposite directions.

REASONING AND STRATEGY
To find the net gravitational force exerted on the spaceship, we first determine the magnitude of the force exerted on it by each asteroid. This is done by using $F = Gm_1m_2/r^2$ (Equation 12-1) and the distances given in our sketch. Next, we resolve these forces into x and y components. Finally, we sum the force components to find the net force.

CONTINUED

Known Mass of spaceship, $m = 2.50 \times 10^7$ kg; mass of each asteroid, $M = 3.50 \times 10^{11}$ kg; distances given in the sketch.

Unknown (a) Net force at location A, $\vec{\mathbf{F}}_{net} = ?$ (b) Net force at location B, $\vec{\mathbf{F}}_{net} = ?$

SOLUTION

Part (a)

1. Use the Pythagorean theorem to find the distance r from point A to each asteroid. Also, refer to the sketch to find the angle between $\vec{\mathbf{F}}_1$ and the x axis. The angle between $\vec{\mathbf{F}}_2$ and the x axis has the same magnitude but the opposite sign:

$$r = \sqrt{(3.00 \times 10^3 \text{ m})^2 + (1.50 \times 10^3 \text{ m})^2} = 3350 \text{ m}$$
$$\theta_1 = \tan^{-1}\left(\frac{1.50 \times 10^3 \text{ m}}{3.00 \times 10^3 \text{ m}}\right) = \tan^{-1}(0.500) = 26.6°$$
$$\theta_2 = -\theta_1 = -26.6°$$

2. Use r and $F = Gm_1m_2/r^2$ to calculate the magnitude of the forces $\vec{\mathbf{F}}_1$ and $\vec{\mathbf{F}}_2$ at point A:

$$F_1 = F_2 = G\frac{mM}{r^2}$$
$$= (6.67 \times 10^{-11} \text{ N} \cdot \text{m}^2/\text{kg}^2)\frac{(2.50 \times 10^7 \text{ kg})(3.50 \times 10^{11} \text{ kg})}{(3350 \text{ m})^2}$$
$$= 52.0 \text{ N}$$

3. Use the values of θ_1 and θ_2 found in Step 1 to calculate the x and y components of $\vec{\mathbf{F}}_1$ and $\vec{\mathbf{F}}_2$:

$$F_{1,x} = F_1 \cos\theta_1 = (52.0 \text{ N}) \cos 26.6° = 46.5 \text{ N}$$
$$F_{1,y} = F_1 \sin\theta_1 = (52.0 \text{ N}) \sin 26.6° = 23.3 \text{ N}$$
$$F_{2,x} = F_2 \cos\theta_2 = (52.0 \text{ N}) \cos(-26.6°) = 46.5 \text{ N}$$
$$F_{2,y} = F_2 \sin\theta_2 = (52.0 \text{ N}) \sin(-26.6°) = -23.3 \text{ N}$$

4. Add the components of $\vec{\mathbf{F}}_1$ and $\vec{\mathbf{F}}_2$ to find the components of the net force. Write $\vec{\mathbf{F}}_{net}$ in vector form:

$$F_x = F_{1,x} + F_{2,x} = 93.0 \text{ N}$$
$$F_y = F_{1,y} + F_{2,y} = 0$$
$$\vec{\mathbf{F}}_{net} = (93.0 \text{ N})\hat{\mathbf{x}}$$

Part (b)

5. Use $F = Gm_1m_2/r^2$ to find the magnitude of the forces exerted on the spaceship by the asteroids at location B:

$$F_1 = F_2 = G\frac{mM}{r^2}$$
$$= (6.67 \times 10^{-11} \text{ N} \cdot \text{m}^2/\text{kg}^2)\frac{(2.50 \times 10^7 \text{ kg})(3.50 \times 10^{11} \text{ kg})}{(1.50 \times 10^3 \text{ m})^2}$$
$$= 259 \text{ N}$$

6. Use the fact that $\vec{\mathbf{F}}_1$ and $\vec{\mathbf{F}}_2$ have equal magnitudes and point in opposite directions to determine the net force, $\vec{\mathbf{F}}_{net}$, acting on the spaceship:

$$\vec{\mathbf{F}}_{net} = \vec{\mathbf{F}}_1 + \vec{\mathbf{F}}_2 = 0$$

INSIGHT

The net force at location A is in the positive x direction, as one would expect by symmetry. At location B, where the force exerted by each asteroid is about 5 times greater than it is at location A, the *net* force is zero because the attractive forces exerted by the two asteroids are equal and opposite, and thus cancel. Notice that the forces in our sketch have been drawn in correct proportion.

PRACTICE PROBLEM

Find the net gravitational force acting on the spaceship when it is at the location $x = 5.00 \times 10^3$ m and $y = 0$.
[**Answer:** The net force is 41.0 N in the negative x direction; that is, $\vec{\mathbf{F}}_{net} = (-41.0 \text{ N})\hat{\mathbf{x}}$.]

Some related homework problems: Problem 8, Problem 10, Problem 11

BIO Although gravity is the weakest of the four fundamental forces of nature (the weak nuclear force, the electromagnetic force, and the strong nuclear force are discussed later in this book and summarized in Chapter 32), nearly every living organism on Earth has structures or features that help it adapt to an environment of uniform gravitation. For example, as plants grow, they tend to align their stems and roots along the direction of the gravitational field. This tendency, which is implemented by cellular elongation in response to differential concentrations of plant hormones known as auxins, is referred to as *gravitropism*. **FIGURE 12-4** illustrates the gravitropic response of a growing plant.

◄ **FIGURE 12-4** Plant stems grow vertically upward, opposite the force of gravity, because a concentration gradient of growth hormones called auxins develops in the stems when they are oriented away from the vertical.

Enhance Your Understanding

(Answers given at the end of the chapter)

1. Rank the four systems shown in FIGURE 12-5 in order of increasing gravitational force exerted on each of the masses. Indicate ties where appropriate.

▲ **FIGURE 12-5** Enhance Your Understanding 1.

Section Review

- The magnitude of the force of gravity between two masses m_1 and m_2 separated by a distance r is $F = Gm_1m_2/r^2$. In this expression, the universal gravitation constant has the value $G = 6.67 \times 10^{-11}\ \mathrm{N \cdot m^2/kg^2}$.

- The force of gravity is always attractive, and points along the line connecting two masses. The net force due to multiple masses is given by vector summation.

PHYSICS IN CONTEXT
Looking Ahead

In Chapter 19, we shall see that the force between two electric charges has exactly the same mathematical form as the force of gravity between two masses. The equation describing the electric force is referred to as Coulomb's law.

12-2 Gravitational Attraction of Spherical Bodies

Newton's law of gravity applies to point objects. How, then, do we calculate the force of gravity for an object of finite size? In general, the approach is to divide the finite object into a collection of small mass elements, then use superposition and the methods of calculus to determine the net gravitational force. For an arbitrary shape, this calculation can be quite involved. For objects with a uniform spherical shape, however, the final result is remarkably simple, as was shown by Newton.

Uniform Sphere

Consider a uniform sphere of radius R and mass M, as in FIGURE 12-6. A point object of mass m is brought near the sphere, though still outside it at a distance r from its center. The object experiences a relatively strong attraction from mass near the point A, and a weaker attraction from mass near point B. In both cases the force is along the line connecting the mass m and the center of the sphere—that is, along the x axis. In addition, masses at the points C and D exert a net force that is also along the x axis—just as in the case of the twin asteroids in Example 12-2. Thus, the symmetry of the sphere guarantees that the net force it exerts on m is directed toward the sphere's center. The magnitude of the force exerted by the sphere must be calculated with the methods of calculus—which Newton invented and then applied to this problem. As a result of his calculations, Newton was able to show that **the net force exerted by the sphere on the mass m is the same as if all the mass of the sphere were concentrated at its center.** That is, the force between the mass m and the sphere of mass M has a magnitude F that is simply

$$F = G\frac{mM}{r^2} \qquad \text{12-3}$$

The Force of Gravity on the Surface of the Earth   Let's apply this result to the case of a mass m on the surface of the Earth—assuming the Earth to be a uniform sphere of mass. If the mass of the Earth is M_E and its radius is R_E, it follows that the force exerted on m by the Earth is

$$F = G\frac{mM_E}{R_E^2} = m\left(\frac{GM_E}{R_E^2}\right)$$

We also know, however, that the gravitational force experienced by a mass m on the Earth's surface is simply $F = mg$, where g is the acceleration due to gravity. Therefore, we see that

$$m\left(\frac{GM_E}{R_E^2}\right) = mg$$

▲ **FIGURE 12-6 Gravitational force between a point mass and a sphere** The force is the same as if all the mass of the sphere were concentrated at its center.

Big Idea **2** The gravitational force exerted by a spherical mass distribution is the same as if all the mass of the sphere were concentrated at its center.

Canceling the mass m, we find

$$g = \frac{GM_E}{R_E^2} = \frac{(6.67 \times 10^{-11}\,\text{N}\cdot\text{m}^2/\text{kg}^2)(5.97 \times 10^{24}\,\text{kg})}{(6.37 \times 10^6\,\text{m})^2} = 9.81\,\text{m/s}^2 \qquad \text{12-4}$$

This result can be extended to objects above the Earth's surface, and hence farther from the center of the Earth, as we show in the next Example.

EXAMPLE 12-3 THE DEPENDENCE OF GRAVITY ON ALTITUDE

*RWP If you climb to the top of Mt. Everest, you will be about 5.50 mi above sea level. What is the acceleration due to gravity at this altitude?

PICTURE THE PROBLEM
At the top of the mountain, your distance from the center of the Earth is $r = R_E + h$, where $h = 5.50$ mi is the altitude.

REASONING AND STRATEGY
First, use $F = GmM_E/r^2$ to find the force due to gravity on the mountaintop. Then, set $F = mg_h$ to find the acceleration g_h at the height h.

Known   Height of Mt. Everest, $h = 5.50$ mi.
Unknown   Acceleration due to gravity at height h, $g_h = ?$

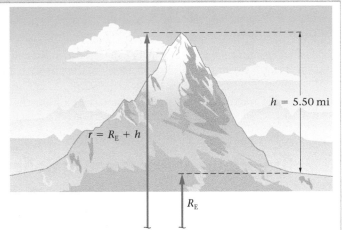

$h = 5.50$ mi

$r = R_E + h$

R_E

SOLUTION

1. Calculate the force F due to gravity at a height h above the Earth's surface:

$$F = G\frac{mM_E}{(R_E + h)^2}$$

2. Set F equal to mg_h and solve for g_h:

$$F = G\frac{mM_E}{(R_E + h)^2} = mg_h$$

$$g_h = G\frac{M_E}{(R_E + h)^2}$$

3. Factor out R_E^2 from the denominator, and use the fact that $GM_E/R_E^2 = g$:

$$g_h = \left(\frac{GM_E}{R_E^2}\right)\frac{1}{\left(1 + \dfrac{h}{R_E}\right)^2} = \frac{g}{\left(1 + \dfrac{h}{R_E}\right)^2}$$

4. Substitute numerical values, with $h = 5.50$ mi $= (5.50\,\text{mi})(1609\,\text{m/mi}) = 8850$ m and $R_E = 6.37 \times 10^6$ m:

$$g_h = \frac{g}{\left(1 + \dfrac{h}{R_E}\right)^2} = \frac{9.81\,\text{m/s}^2}{\left(1 + \dfrac{8850\,\text{m}}{6.37 \times 10^6\,\text{m}}\right)^2} = 9.78\,\text{m/s}^2$$

INSIGHT
As expected, the acceleration due to gravity is less as one moves farther from the center of the Earth. Thus, if you were to climb to the top of Mt. Everest, you would lose weight—not only because of the physical exertion required for the climb, but also because of the reduced gravity. In particular, a person with a mass of 60 kg (about 130 lb) would lose about half a pound of weight just by standing on the summit of the mountain.

A plot of g_h as a function of h is shown in **FIGURE 12-7 (a)**. The plot indicates the altitude of Mt. Everest and the orbit of the International Space Station. **FIGURE 12-7 (b)** shows g_h out to the orbits of communications and weather satellites, which are at an altitude of roughly 22,200 mi.

PRACTICE PROBLEM
Find the acceleration due to gravity at the altitude of the International Space Station's orbit, 322 km above the Earth's surface.   [**Answer:** $g_h = 8.89$ m/s^2, a reduction of 9.38% compared to the acceleration of gravity on the surface of the Earth]

Some related homework problems: Problem 14, Problem 16

CONTINUED

---

*Real World Physics applications are denoted by the acronym RWP.

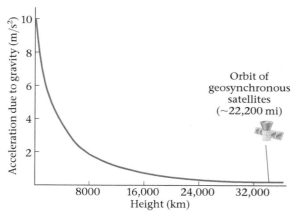

(a) Acceleration due to gravity near the Earth's surface

(b) Acceleration due to gravity far from the Earth

▲ **FIGURE 12-7 The acceleration due to gravity at a height *h* above the Earth's surface** (a) In this plot, the peak of Mt. Everest is at about $h = 5.50$ mi, and the International Space Station orbit is at roughly $h = 322$ km. (b) This plot shows the decrease in the acceleration of gravity from the surface of the Earth to an altitude of about 25,000 mi. The orbit of geosynchronous satellites—ones that orbit above a fixed point on the Earth—is at roughly $h = 22,200$ mi.

◄ **FIGURE 12-8** This global model of the Earth's gravitational strength was constructed from a combination of surface gravity measurements and satellite tracking data. It shows how the acceleration due to gravity varies from the value at an idealized "sea level" that takes into account the Earth's nonspherical shape. (The Earth is somewhat flattened at the poles—its radius is greatest at the equator.) Gravity is strongest in the red areas and weakest in the dark blue areas.

Of course, the Earth isn't actually a perfectly uniform sphere, though this is certainly a good first approximation. **FIGURE 12-8** shows the actual acceleration due to gravity on the Earth's surface. Variations are caused by the Earth's slightly nonspherical shape and nonuniform mass distribution.

RWP The local value of *g* can be measured with a *gravimeter*, the most sensitive of which can detect changes as small as $10^{-12}g$. A portable gravimeter is shown in **FIGURE 12-9 (a)**. Unusually dense materials nearby will slightly increase the value of *g* measured by a gravimeter, whereas a nearby deposit of low-density material will slightly decrease *g*. It follows that small variations in the acceleration of gravity can be used by those who search for buried oil and minerals, as well as those who study the motion of the Earth's crust. An orbiting gravimeter, the *Gravity Field and Steady-State Ocean Circulation Explorer* shown in **FIGURE 12-9 (b)**, recently completed a successful mission where it used small variations in *g* to map ocean currents, detect earthquakes, discover mantle plumes, and find remnants of ancient oceans and subduction zones.

Applications in the Solar System The result $g = GM_E/R_E^2$ can be used to calculate the acceleration due to gravity on other objects in the solar system besides the Earth.

(a)

(b)

▶ **FIGURE 12-9** (a) A portable gravimeter. (b) An artist's conception of the European Space Agency's *Gravity Field and Steady-State Ocean Circulation Explorer* (GOCE) satellite, which has mapped Earth's gravitational field in unprecedented detail.

▲ **FIGURE 12-10 Visualizing Concepts**
Gravity in the Solar System (Top) The weak lunar gravity permits astronauts, even encumbered by their massive space suits, to bound over the Moon's surface. The low gravitational pull, only about one-sixth that of Earth, is a consequence not only of the Moon's smaller size, but also of its lower average density. (Bottom) The force of gravity on the surface of Mars is only about 38% of its strength on Earth. This was an important factor in designing NASA's Phoenix Mars Lander, shown here lifting a scoop of dirt on its 16th Martian day after landing in May 2008.

For example, to calculate the acceleration due to gravity on the Moon, g_m, we simply use the mass and radius of the Moon in the equation. Once g_m is known, the weight of an object of mass m on the Moon is found by using $W_m = mg_m$. Similar comments apply to other astronomical bodies in the solar system, and everywhere else in the universe as well. **FIGURE 12-10** illustrates the concept for objects on both the Moon and Mars.

EXERCISE 12-4 GRAVITY IN THE SOLAR SYSTEM

a. Find the acceleration due to gravity on the surface of the Moon. (*Note:* The mass of the Moon is $M_m = 7.34 \times 10^{22}$ kg and its radius is $R_m = 1.74 \times 10^6$ m.)

b. The lunar rover had a mass of 225 kg. What was its weight on the Earth and on the Moon?

REASONING AND SOLUTION

a. To find the acceleration due to gravity on the surface of the Moon, we use $g = GM_E/R_E^2$ but with M_E replaced by $M_m = 7.34 \times 10^{22}$ kg, and R_E replaced with $R_m = 1.74 \times 10^6$ m. We find

$$g_m = \frac{GM_m}{R_m^{\;2}} = \frac{(6.67 \times 10^{-11}\,\text{N} \cdot \text{m}^2/\text{kg}^2)(7.34 \times 10^{22}\,\text{kg})}{(1.74 \times 10^6\,\text{m})^2} = 1.62\,\text{m/s}^2$$

This is about one-sixth the acceleration due to gravity on the Earth.

b. An object's weight is $W = mg$, where g is the acceleration due to gravity at the location of the object. Therefore, the lunar rover's weight on Earth was

$$W = mg = (225\,\text{kg})(9.81\,\text{m/s}^2) = 2210\,\text{N}$$

On the Moon, its weight was

$$W_m = mg_m = (225\,\text{kg})(1.62\,\text{m/s}^2) = 365\,\text{N}$$

As expected, this is roughly one-sixth its Earth weight.

The replacement of a sphere with a point mass at its center can be applied to many physical systems. For example, the force of gravity between two spheres of finite size is the same as if *both* were replaced by point masses. Thus, the gravitational force between the Earth, with mass M_E, and the Moon, with mass M_m, is

$$F = G\frac{M_E M_m}{r^2}$$

The distance r in this expression is the center-to-center distance between the Earth and the Moon, as shown in **FIGURE 12-11**. It follows, then, that in many calculations involving the solar system, moons and planets can be treated as point objects.

Weighing the Earth

The British physicist Henry Cavendish performed an experiment in 1798 that is often referred to as "weighing the Earth." What he did, in fact, was measure the value of the universal gravitation constant, G, that appears in Newton's law of gravity. As we've pointed out, G is a very small number; hence a sensitive experiment is needed for its measurement. It is because of this experimental difficulty that G was not measured until more than 100 years after Newton published the law of gravitation.

The Cavendish Experiment In the Cavendish experiment, illustrated in **FIGURE 12-12**, two masses m are connected by a thin rod and suspended from a thin thread. Near each suspended mass is a large stationary mass M, as shown. Each suspended mass is attracted by the force of gravity toward the large mass near it; hence the rod holding the suspended masses tends to rotate and twist the thread. The angle through which the thread twists can be measured by bouncing a beam of light from a mirror attached to the thread. If the force required to twist the thread through a given angle is known (from previous experiments), a measurement of the twist angle gives the magnitude

▲ **FIGURE 12-11 Gravitational force between the Earth and the Moon** The force is the same as if both the Earth and the Moon were point masses. (The sizes of the Earth and Moon are in correct proportion in this figure, but the separation between the two should be much greater than that shown here. In reality, it is about 30 times the diameter of the Earth, and so would be about 2 ft on this scale.)

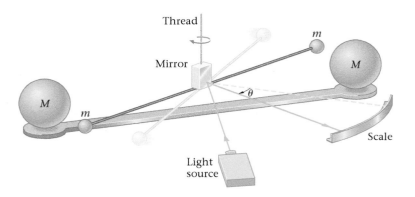

Thread

Mirror

M

m

m

M

θ

Light source

Scale

of the force of gravity. Finally, knowing the masses, $m_1 = m$ and $m_2 = M$, and the distance between their centers, r, we can use $F = Gm_1m_2/r^2$ to solve for G. Cavendish found $6.754 \times 10^{-11}\,\text{N} \cdot \text{m}^2/\text{kg}^2$, in good agreement with the currently accepted value of $G = 6.67 \times 10^{-11}\,\text{N} \cdot \text{m}^2/\text{kg}^2$.

To see why Cavendish is said to have weighed the Earth, recall that the force of gravity on the surface of the Earth, mg, can be written as follows:

$$mg = G\frac{mM_E}{R_E{}^2}$$

Canceling m and solving for M_E yields

$$M_E = \frac{gR_E{}^2}{G} \qquad\qquad 12\text{-}5$$

Before the Cavendish experiment, the quantities g and R_E were known from direct measurement, but G had yet to be determined. When Cavendish measured G, he didn't actually "weigh" the Earth, of course. Instead, he calculated its mass, M_E.

EXERCISE 12-5 MASS OF THE EARTH

Use $M_E = gR_E{}^2/G$ to calculate the mass of the Earth.

REASONING AND SOLUTION
Substituting numerical values into $M_E = gR_E{}^2/G$, we find

$$M_E = \frac{gR_E{}^2}{G} = \frac{(9.81\,\text{m/s}^2)(6.37 \times 10^6\,\text{m})^2}{6.67 \times 10^{-11}\,\text{N} \cdot \text{m}^2/\text{kg}^2} = 5.97 \times 10^{24}\,\text{kg}$$

Connections with Geology As soon as Cavendish determined the mass of the Earth, geologists were able to use the result to calculate its average density; that is, its average mass per volume. If we assume a spherical Earth of radius R_E, its total volume is

$$V_E = \tfrac{4}{3}\pi R_E{}^3 = \tfrac{4}{3}\pi(6.37 \times 10^6\,\text{m})^3 = 1.08 \times 10^{21}\,\text{m}^3$$

Dividing the total volume into the total mass yields the average density, ρ:

$$\rho = \frac{M_E}{V_E} = \frac{5.97 \times 10^{24}\,\text{kg}}{1.08 \times 10^{21}\,\text{m}^3} = 5530\,\text{kg/m}^3 = 5.53\,\text{g/cm}^3$$

This is an interesting result because typical rocks found near the surface of the Earth, such as granite, have a density of only about $3.00\,\text{g/cm}^3$. We conclude, then, that the interior of the Earth must have a greater density than its surface. In fact, by analyzing the propagation of seismic waves around the world, we now know that the Earth has a rather complex interior structure, including a solid inner core with a density of about $15.0\,\text{g/cm}^3$ (see Section 10-5).

A similar calculation for the Moon yields an average density of about $3.33\,\text{g/cm}^3$, essentially the same as the density of the lunar rocks brought back during the Apollo program. Hence, the Moon probably does not have an internal structure similar to that of the Earth.

Because G is a universal constant—with the same value everywhere in the universe—it can be used to calculate the mass of other bodies in the solar system as well. This is illustrated in the following Example.

EXAMPLE 12-6   MARS ATTRACTS!

After landing on Mars, an astronaut performs a simple experiment by dropping a rock. A quick calculation using the drop height and the time of fall yields a value of 3.73 m/s^2 for the rock's acceleration. **(a)** Find the mass of Mars, given that its radius is $R_M = 3.39 \times 10^6$ m. **(b)** What is the acceleration due to gravity at a distance $r = 2R_M$ from the center of Mars?

PICTURE THE PROBLEM
Our sketch shows an astronaut dropping a rock to the ground on the surface of Mars. When the acceleration of the rock is measured, we find $g_M = 3.73 \text{ m/s}^2$, where the subscript M refers to Mars. In addition, we indicate the radius of Mars in our sketch, where $R_M = 3.39 \times 10^6$ m.

REASONING AND STRATEGY

a. The acceleration due to gravity is g_M on the surface of Mars, and hence it follows that the Martian force of gravity on an object of mass m is $F = mg_M$. This force is also given by Newton's law of gravity—that is, $F = GmM_M/R_M^2$. Setting these expressions for the force equal to one another yields the mass of Mars, M_M.

b. We set $F = ma$ equal to $F = GmM_M/(2R_M)^2$ and solve for the acceleration, a.

Known   Acceleration due to gravity on Mars, $g_M = 3.73 \text{ m/s}^2$; radius of Mars, $R_M = 3.39 \times 10^6$ m.
Unknown **(a)** Mass of Mars, $M_M = ?$ **(b)** Acceleration due to gravity at $r = 2R_M$, $a = ?$

SOLUTION

Part (a)

1. Set mg_M equal to GmM_M/R_M^2:
$$mg_M = G\frac{mM_M}{R_M^2}$$

2. Cancel m and solve for the mass of Mars:
$$M_M = \frac{g_M R_M^2}{G}$$

3. Substitute numerical values:
$$M_M = \frac{(3.73 \text{ m/s}^2)(3.39 \times 10^6 \text{ m})^2}{6.67 \times 10^{-11} \text{ N} \cdot \text{m}^2/\text{kg}^2} = 6.43 \times 10^{23} \text{ kg}$$

Part (b)

4. Apply Newton's law of gravity with $r = 2R_M$. Use the fact that $g_M = GM_M/R_M^2$ from Step 1, and the given numerical value, $g_M = 3.73 \text{ m/s}^2$, to simplify the calculation:
$$ma = G\frac{mM_M}{(2R_M)^2} \quad \text{or}$$
$$a = G\frac{M_M}{(2R_M)^2} = \frac{1}{4}\left(G\frac{M_M}{R_M^2}\right) = \frac{1}{4}(g_M) = \frac{1}{4}(3.73 \text{ m/s}^2) = 0.933 \text{ m/s}^2$$

INSIGHT
The important point in this Example is that the universal gravitation constant, G, applies as well on Mars as on Earth, or any other object. Therefore, knowing the size and acceleration of gravity on the surface of an astronomical body is sufficient to determine its mass.

PRACTICE PROBLEM — PREDICT/CALCULATE
(a) If the radius of Mars were reduced to 3.00×10^6 m, with its mass remaining the same, would the acceleration due to gravity on the surface of Mars increase, decrease, or stay the same? Explain. **(b)** Check your answer by calculating the acceleration of gravity for this case.   [**Answer: (a)** The force of gravity increases as the radius is decreased, and hence the acceleration due to gravity would increase. **(b)** The acceleration due to gravity increases from 3.73 m/s^2 to 4.77 m/s^2.]

Some related homework problems: Problem 19, Problem 21

Enhance Your Understanding (Answers given at the end of the chapter)

2. Suppose the Sun suddenly collapsed to half its current radius, with its mass remaining the same. Would the gravitational force exerted by the Sun on the Earth increase, decrease, or stay the same? Explain.

Section Review

- The force of gravity between a spherical mass M and a mass m outside the sphere is the same as if all the mass of the sphere were concentrated at its center.

12-3 Kepler's Laws of Orbital Motion

If you go outside each clear night and observe the position of Mars with respect to the stars, you will find that its apparent motion across the sky is rather complex. Instead of moving on a simple curved path, it occasionally reverses direction (this is known as *retrograde motion*). A few months later it reverses direction yet again and resumes its original direction of motion. Other planets exhibit similar odd behavior.

The Danish astronomer Tycho Brahe (1546–1601) followed the paths of the planets, and Mars in particular, for many years, even though the telescope had not yet been invented. He used, instead, an elaborate sighting device to plot the precise positions of the planets. Brahe was joined in his work by Johannes Kepler (1571–1630) in 1600, and after Brahe's death, Kepler inherited his astronomical observations.

Kepler made good use of Brahe's life work, extracting from his carefully collected data the three laws of orbital motion we know today as Kepler's laws. These laws make it clear that the Sun and the planets do not orbit the Earth, as Ptolemy—the ancient Greek astronomer—claimed, but rather that the Earth, along with the other planets, orbits the Sun, as proposed by Copernicus (1473–1543).

Why the planets obey Kepler's laws no one knew—not even Kepler—until Newton considered the problem decades after Kepler's death. Newton was able to show that each of Kepler's laws follows as a direct consequence of the law of universal gravitation. In the remainder of this section we consider Kepler's three laws one at a time, and point out the connection between them and the law of gravitation.

Kepler's First Law

Kepler tried long and hard to find a circular orbit around the Sun that would match Brahe's observations of Mars. After all, up to that time everyone from Ptolemy to Copernicus believed that celestial objects moved in circular paths of one sort or another. Though the orbit of Mars was exasperatingly close to being circular, the small differences between a circular path and the experimental observations just could not be ignored. Eventually, after a great deal of hard work and disappointment over the loss of circular orbits, Kepler discovered that Mars followed an orbit that was elliptical rather than circular. The same applied to the other planets. This observation became Kepler's first law:

> Planets follow elliptical orbits, with the Sun at one focus of the ellipse.

This is a fine example of the scientific method in action. Though Kepler expected, and wanted, to find circular orbits, he would not allow himself to ignore the data. If Brahe's observations had not been so accurate, Kepler probably would have chalked up the small differences between the data and a circular orbit to error. As it was, he had to discard a treasured—but incorrect—theory and move on to an unexpected, but ultimately correct, view of nature.

Kepler's first law is illustrated in **FIGURE 12-13**, along with a definition of an ellipse in terms of its two foci. In the case where the two foci merge, as in **FIGURE 12-14**, the ellipse reduces to a circle. Thus, a circular orbit *is* allowed by Kepler's first law, but only as a special case.

Newton was able to show that, because the force of gravity decreases with distance as $1/r^2$, closed orbits must have the form of ellipses or circles, as stated in Kepler's first

Big Idea 3 The orbits of planets, moons, comets, and asteroids are governed by Kepler's three laws, all of which can be derived directly from Newton's law of universal gravitation.

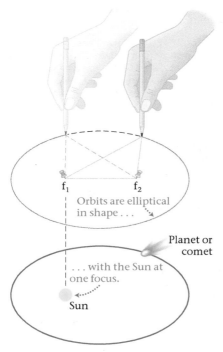

▲ **FIGURE 12-13 Drawing an ellipse** To draw an ellipse, put two tacks in a piece of cardboard. The tacks define the "foci" of the ellipse. Now connect a length of string to the two tacks, and use a pencil and the string to sketch out a smooth, closed curve, as shown. This closed curve is an ellipse. In a planetary orbit a planet follows an elliptical path, with the Sun at one focus. Nothing is at the other focus.

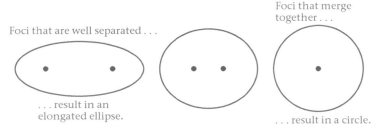

▲ **FIGURE 12-14 The circle as a special case of the ellipse** As the two foci of an ellipse approach one another, the ellipse becomes more circular. In the limit that the foci merge, the ellipse becomes a circle.

PHYSICS IN CONTEXT
Looking Back

Conservation of angular momentum, introduced in Chapter 11, plays a key role in gravity. In fact, it is the basis for Kepler's second law.

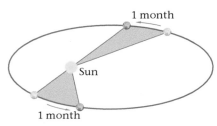

(a) Equal areas in equal times

(b) Equal areas in equal times for highly elliptical orbit

▲ **FIGURE 12-15 Kepler's second law**
(a) The second law states that a planet sweeps out equal areas in equal times.
(b) In a highly elliptical orbit, the long, thin area is equal to the broad, fan-shaped area.

law. He also showed that orbits that are not closed—say, the orbit of a comet that passes by the Sun once and then leaves the solar system—are either parabolic or hyperbolic in shape.

Kepler's Second Law

When Kepler plotted the position of a planet on its elliptical orbit, indicating at each position the time the planet was there, he made an interesting observation. First, draw a line from the Sun to a planet at a given time. Then a certain time later—perhaps a month—draw a line again from the Sun to the new position of the planet. The result is that the planet has "swept out" a wedge-shaped area, as indicated in **FIGURE 12-15 (a)**. If this procedure is repeated when the planet is on a different part of its orbit, another wedge-shaped area is generated. Kepler's observation was that the areas of these two wedges are equal:

> As a planet moves in its orbit, it sweeps out an equal amount of area in an equal amount of time.

Kepler's second law follows from the fact that the force of gravity on a planet is directly toward the Sun. As a result, gravity exerts zero torque about the Sun, which means that the angular momentum of a planet in its orbit must be conserved. As Newton showed, conservation of angular momentum is equivalent to the equal-area law stated by Kepler.

CONCEPTUAL EXAMPLE 12-7   COMPARE SPEEDS   PREDICT/EXPLAIN

The Earth's orbit is slightly elliptical. In fact, the Earth is closer to the Sun during the northern hemisphere's winter than it is during the summer. **(a)** Is the orbital speed of the Earth during winter greater than, less than, or equal to its orbital speed during summer? **(b)** Which of the following is the *best explanation* for your prediction?

  I. The orbital speed of the Earth is conserved, and so its value is the same at all times.

 II. The Earth sweeps out equal areas in equal times, and hence it moves faster in winter when its distance from the Sun is smaller.

III. The orbital speed of the Earth is smaller in winter, because a smaller distance from the Sun implies a smaller speed.

REASONING AND DISCUSSION
According to Kepler's second law, the area swept out by the Earth per month is the same in winter as it is in summer. In winter, however, the radius from the Sun to the Earth is smaller than it is in summer. Therefore, if this smaller radius is to sweep out the same area, the Earth must move more rapidly.

ANSWER
(a) The speed of the Earth is greater during the winter. **(b)** The best explanation is II.

Though we have stated the first two laws in terms of planets, they apply equally well to any object orbiting the Sun. For example, a comet might follow a highly elliptical orbit, as in **FIGURE 12-15 (b)**. When it is near the Sun, it moves very quickly, for the reason discussed in Conceptual Example 12-7, sweeping out a broad, wedge-shaped area in a month's time. Later in its orbit, the comet is far from the Sun and moving slowly. In this case, the area it sweeps out in a month is a long, thin wedge. Still, the two wedges have equal areas.

Kepler's Third Law

Finally, Kepler studied the relationship between the mean distance of a planet from the Sun, r, and its period—that is, the time, T, it takes for the planet to complete one orbit. **FIGURE 12-16** shows a plot of period versus distance for the planets of the solar system. Kepler tried to "fit" these results to a simple dependence between T and r. If he tried a linear fit—that is, T proportional to r (the bottom curve in Figure 12-16)—he found that the period did not increase rapidly enough with distance. On the other hand, if he tried T proportional to r^2 (the top curve in Figure 12-16), the period increased too rapidly. Splitting the difference—trying T proportional to $r^{3/2}$—yields a good fit (the middle curve in Figure 12-16). This is Kepler's third law:

> The period, T, of a planet increases as its mean distance from the Sun, r, raised to the 3/2 power. That is,

$$T = (\text{constant})r^{3/2}$$ 12-6

Deriving Kepler's Third Law It is straightforward to derive Kepler's third law for the special case of a circular orbit. Consider a planet orbiting the Sun at a distance r, as in **FIGURE 12-17**. The planet moves in a circular path, and hence a centripetal force must act on it, as we saw in Section 6-5. In addition, this force must be directed toward the center of the circle—that is, toward the Sun. It is as if you were to swing a ball on the end of a string in a circle above your head, as in Figure 6-22. In order for the ball to move in a circular path, you have to exert a force on the ball toward the center of the circular path. This force is exerted through the string. In the case of a planet orbiting the Sun, the centripetal force is provided by the force of gravity between the Sun and the planet.

If the planet has a mass m, and the Sun has a mass M_S, the force of gravity between them is

$$F = G\frac{mM_S}{r^2}$$

Now, this force creates the centripetal acceleration of the planet, a_{cp}, which, according to Equation 6-15, is

$$a_{cp} = \frac{v^2}{r}$$

Thus, the centripetal force necessary for the planet to orbit is ma_{cp}:

$$F = ma_{cp} = m\frac{v^2}{r}$$

Noting that the speed of the planet, v, is the circumference of the orbit, $2\pi r$, divided by the time to complete an orbit, T, we have

$$F = m\frac{v^2}{r} = m\frac{(2\pi r/T)^2}{r} = \frac{4\pi^2 rm}{T^2}$$

Setting the centripetal force equal to the force of gravity yields

$$\frac{4\pi^2 rm}{T^2} = G\frac{mM_S}{r^2}$$

Eliminating m and rearranging, we find

$$T^2 = \frac{4\pi^2}{GM_S}r^3$$

Taking the square root of both sides of the equation yields

$$T = \left(\frac{2\pi}{\sqrt{GM_S}}\right)r^{3/2} = (\text{constant})r^{3/2}$$ 12-7

As predicted by Kepler, T is proportional to $r^{3/2}$.

▲ **FIGURE 12-16 Kepler's third law and some near misses** These plots represent three possible mathematical relationships between period of revolution, T (in years), and mean distance from the Sun, r (in kilometers). The lower curve shows $T = (\text{constant})r$; the upper curve is $T = (\text{constant})r^2$. The middle curve, which fits the data, is $T = (\text{constant})r^{3/2}$. This is Kepler's third law.

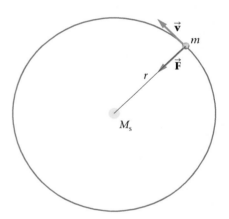

▲ **FIGURE 12-17 Centripetal force on a planet in orbit** As a planet revolves about the Sun in a circular orbit of radius r, the force of gravity between it and the Sun, $F = GmM_S/r^2$, provides the required centripetal force.

PHYSICS
IN CONTEXT
Looking Ahead

The analysis used to derive Kepler's third law is used again when we explore the Bohr model of the hydrogen atom in Chapter 31. The calculation is the same, but in hydrogen the Coulomb force between electric charges (Equation 19-5) is responsible for the orbital motion.

PROBLEM-SOLVING NOTE

The Mass in Kepler's Third Law

When applying Kepler's third law, recall that the mass in Equation 12-7, M_S, refers to the mass of the object being orbited. Thus, the third law can be applied to satellites of any object, as long as M_S is replaced by the orbited mass.

Applying Kepler's Third Law Deriving Kepler's third law by using Newton's law of gravitation has allowed us to calculate the constant that multiplies $r^{3/2}$. Notice that the constant depends on the mass of the Sun; that is, *T depends on the mass being orbited.* It does not depend on the mass of the planet orbiting the Sun, however, as long as the planet's mass is much less than the mass of the Sun. As a result, Equation 12-7 applies equally to all the planets.

This result can also be applied to the case of a moon or a satellite (an artificial moon) orbiting a planet. To do so, we simply note that it is the planet that is being orbited, not the Sun. Hence, to apply Equation 12-7, we replace the mass of the Sun, M_S, with the mass of the appropriate planet. This concept is illustrated in FIGURE 12-18, where we show that it also applies to moons orbiting planets and asteroids.

As an example, let's calculate the mass of Jupiter. One of the four moons of Jupiter discovered by Galileo is Io, which completes one orbit every 42 h 27 min = 1.53×10^5 s. Given that the average distance from the center of Jupiter to Io is 4.22×10^8 m, we can find the mass of Jupiter as follows:

$$T = \left(\frac{2\pi}{\sqrt{GM_J}}\right)r^{3/2}$$

$$M_J = \frac{4\pi^2 r^3}{GT^2} = \frac{4\pi^2(4.22 \times 10^8 \, \text{m})^3}{(6.67 \times 10^{-11} \, \text{N} \cdot \text{m}^2/\text{kg}^2)(1.53 \times 10^5 \, \text{s})^2} = 1.90 \times 10^{27} \, \text{kg}$$

This is roughly 300 times the mass of the Earth.

▶ FIGURE 12-18 Visualizing Concepts
Orbital Motion Kepler's laws of orbital motion apply to planetary satellites as well as planets. Jupiter, the largest planet in the solar system, has at least 16 moons, all of which travel in elliptical orbits that obey Kepler's laws. (The moons in the photo at left, passing in front of Jupiter, are Europa and Io) Even some asteroids have been found to have their own satellites. The large cratered object in the photo at right is 243 Ida, an asteroid some 56 km long; its miniature companion at the top of the photo is Dactyl, about 1.5 km in diameter. Like all gravitationally bound bodies, Ida and Dactyl orbit their common center of mass.

EXAMPLE 12-8   THE SUN AND MERCURY

The Earth revolves around the Sun once a year at an average distance of 1.50×10^{11} m. **(a)** Use this information to calculate the mass of the Sun. **(b)** Find the period of revolution for the planet Mercury, whose average distance from the Sun is 5.79×10^{10} m.

PICTURE THE PROBLEM
Our sketch shows the orbits of Mercury, Venus, and the Earth in correct proportion. In addition, each of these orbits is slightly elliptical, though the deviation from circularity is too small for the eye to see. Finally, we indicate that the orbital radius for Mercury is 5.79×10^{10} m and the orbital radius for the Earth is 1.50×10^{11} m.

REASONING AND STRATEGY

a. To find the mass of the Sun, we solve Equation 12-7 for M_S. Notice that the period $T = 1$ y must be converted to seconds before we evaluate the formula.

b. The period of Mercury is found by substituting $r = 5.79 \times 10^{10}$ m in Equation 12-7.

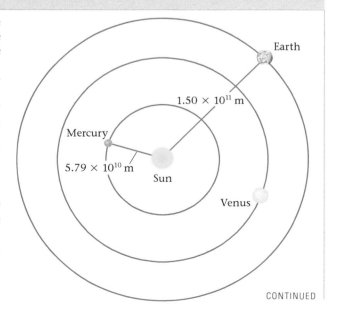

CONTINUED

Known   Orbital radius of Earth, $r = 1.50 \times 10^{11}$ m; orbital radius of Mercury, $r = 5.79 \times 10^{10}$ m.
Unknown   (a) Mass of the Sun, $M_S = ?$ (b) Period of Mercury, $T = ?$

SOLUTION

Part (a)

1. Solve Equation 12-7 for the mass of the Sun:

$$T = \left(\frac{2\pi}{\sqrt{GM_S}}\right)r^{3/2}$$

$$M_S = \frac{4\pi^2 r^3}{GT^2}$$

2. Calculate the period of the Earth in seconds:

$$T = 1\,\text{y}\left(\frac{365.24\,\text{days}}{1\,\text{y}}\right)\left(\frac{24\,\text{h}}{1\,\text{day}}\right)\left(\frac{3600\,\text{s}}{1\,\text{h}}\right) = 3.16 \times 10^7\,\text{s}$$

3. Substitute numerical values in the expression for the mass of the Sun obtained in Step 1:

$$M_S = \frac{4\pi^2 r^3}{GT^2}$$
$$= \frac{4\pi^2(1.50 \times 10^{11}\,\text{m})^3}{(6.67 \times 10^{-11}\,\text{N}\cdot\text{m}^2/\text{kg}^2)(3.16 \times 10^7\,\text{s})^2}$$
$$= 2.00 \times 10^{30}\,\text{kg}$$

Part (b)

4. Substitute $r = 5.79 \times 10^{10}$ m into Equation 12-7. In addition, use the mass of the Sun obtained in part (a):

$$T = \left(\frac{2\pi}{\sqrt{GM_S}}\right)r^{3/2}$$
$$= \left(\frac{2\pi}{\sqrt{(6.67 \times 10^{-11}\,\text{N}\cdot\text{m}^2/\text{kg}^2)(2.00 \times 10^{30}\,\text{kg})}}\right) \times (5.79 \times 10^{10}\,\text{m})^{3/2}$$
$$= 7.58 \times 10^6\,\text{s} = 0.240\,\text{y} = 87.7\,\text{days}$$

INSIGHT

In part (a), we find that the mass of the Sun is almost a million times greater than the mass of the Earth. In fact, the Sun accounts for 99.9% of all the mass in the solar system. In part (b) we see that Mercury, with its smaller orbital radius, has a shorter year than the Earth.

PRACTICE PROBLEM

Venus orbits the Sun with a period of 1.94×10^7 s. What is its average distance from the Sun?
[**Answer:** $r = 1.08 \times 10^{11}$ m]

Some related homework problems: Problem 25, Problem 28

RWP The 24 satellites of the Global Positioning System (GPS) are in relatively low orbits about the Earth. These satellites are used to provide a precise determination of an observer's position anywhere on Earth. The operating principle of the GPS is illustrated in **FIGURE 12-19**. Imagine, for example, that satellite 2 emits a radio signal at a particular time (all GPS satellites carry atomic clocks on board). This signal travels away from the satellite with the speed of light (see Chapter 25) and is detected a short time later by an observer's GPS receiver. Multiplying the time delay by the speed of light gives the distance of the receiver from satellite 2. Thus, in our example, the observer must lie somewhere on the red circle in Figure 12-19. Similar time-delay measurements for signals from satellite 11 show that the observer is also somewhere on the green circle; hence the observer is either at the point shown in Figure 12-19, or at the second intersection of the red and green circles on the other side of the planet. Measurements from satellite 6 can resolve the ambiguity and place the observer at the point shown in the figure. Measurements from additional satellites can even determine the observer's altitude. GPS receivers, which are used by hikers, boaters, and car navigation systems, typically use signals from as many as 12 satellites. As currently operated, the GPS gives positions with a typical accuracy of 2 m to 10 m. The altitude of GPS orbits is determined in the next Quick Example.

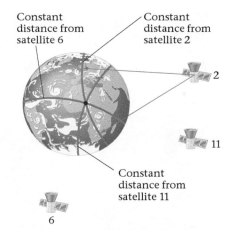

▲ **FIGURE 12-19 The Global Positioning System** A system of 24 satellites in orbit about the Earth makes it possible to determine a person's location with great accuracy. Measuring the distance of a person from satellite 2 places the person somewhere on the red circle. Similar measurements using satellite 11 place the person's position somewhere on the green circle, and further measurements can pinpoint the person's location.

QUICK EXAMPLE 12-9   FIND THE ALTITUDE OF A GPS SATELLITE

Find the altitude above the Earth's surface where a GPS satellite orbits with a period of 12 h. (Note that $R_E = 6.37 \times 10^6$ m, $M_E = 5.97 \times 10^{24}$ kg, and $T = 12\,\text{h} = 4.32 \times 10^4$ s.)

CONTINUED

REASONING AND SOLUTION

We find the period of an orbit around the Earth with $T = (2\pi/\sqrt{GM_E})r^{3/2}$, which is simply Equation 12-7 with the mass of the Earth, M_E, replacing the mass of the Sun, M_S.

1. Write the expression for the period of an object orbiting the Earth:

$$T = \left(\frac{2\pi}{\sqrt{GM_E}}\right)r^{3/2}$$

2. Solve for the radius, r:

$$r = \left(\frac{T}{2\pi}\right)^{2/3}(GM_E)^{1/3}$$

3. Substitute numerical values:

$$r = \left(\frac{4.32 \times 10^4\,\text{s}}{2\pi}\right)^{2/3}((6.67 \times 10^{-11}\,\text{N}\cdot\text{m}^2/\text{kg}^2)$$
$$\times\,(5.97 \times 10^{24}\,\text{kg}))^{1/3}$$
$$= 2.66 \times 10^7\,\text{m}$$

4. Subtract the radius of the Earth to find the altitude:

$$r - R_E = 2.66 \times 10^7\,\text{m} - 6.37 \times 10^6\,\text{m}$$
$$= 2.02 \times 10^7\,\text{m}$$

Thus, GPS satellites orbit at 2.02×10^7 m ~ 12,600 mi above our heads.

Not all spacecraft are placed in GPS orbits, of course. The International Space Station, for example, orbits at an altitude of only about 320 km (200 mi). At that altitude, it takes about an hour and a half to complete one orbit. Some of the higher satellites in everyday use are considered next.

RWP A *geosynchronous satellite* is one that orbits above the equator with a period equal to one day. From the Earth, such a satellite appears to be in the same location in the sky at all times, making it particularly useful for applications such as communications and weather forecasting. From Kepler's third law, we know that a satellite has a period of one day only if its orbital radius has a particular value. This value can be found in the same way that the orbital altitude of GPS satellites was determined in Quick Example 12-9. The result is that geosynchronous satellites orbit at an altitude of 22,200 mi above the Earth. A satellite that was boosted from a low Earth orbit to a geosynchronous orbit is shown in **FIGURE 12-20**.

▲ **FIGURE 12-20 Visualizing Concepts Satellites** Many weather and communications satellites are placed in geosynchronous orbits that allow them to remain "stationary" in the sky—that is, fixed over one point on the Earth's equator. Because the Earth rotates, the period of such a satellite must exactly match that of the Earth. The altitude needed for such an orbit is about 36,000 km (22,200 mi). Other satellites, such as those used in the Global Positioning System (GPS), the Hubble Space Telescope, and the International Space Station, operate at much lower altitudes—typically just a few hundred miles. The photo at left shows the communications satellite Intelsat VI just prior to its capture by astronauts of the space shuttle *Endeavour*. A launch failure had left the satellite stranded in a low orbit. The astronauts snared the satellite (right) and fitted it with a new engine that boosted it to its geosynchronous orbit.

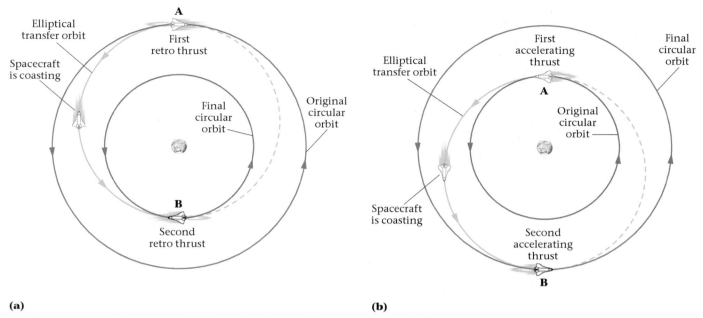

▲ **FIGURE 12-21 Orbital maneuvers (a)** The radius of a satellite's orbit can be decreased by firing the decelerating rockets once at point A and again at point B. Between firings the satellite follows an elliptical orbit. The satellite speeds up as it falls inward toward the Earth during this maneuver. For this reason its final speed in the new circular orbit is greater than its speed in the original orbit, even though the decelerating rockets have slowed it down twice. **(b)** The radius of a satellite's orbit can be increased by firing the accelerating rockets once at point A and again at point B. Between firings the satellite follows an elliptical orbit. The satellite slows down as it moves farther from the Earth during this maneuver. For this reason its final speed in the new circular orbit is less than its speed in the original orbit, even though the accelerating rockets have sped it up twice.

Orbital Maneuvers

RWP We now show how Kepler's laws can give insight into maneuvering a satellite in orbit. Suppose, for example, that you are piloting a spacecraft in a circular orbit, and you would like to move to a lower circular orbit. As you might expect, you should begin by using your rockets to decrease your speed—that is, fire the rockets that point in the forward direction so that their thrust (Section 9-8) is opposite to your direction of motion. The result of firing the decelerating rockets at a given point A in your original orbit is shown in **FIGURE 12-21 (a)**. Notice that your new orbit is not a circle, as desired, but rather an ellipse. To produce a circular orbit you can simply fire the decelerating rockets once again at point B, on the opposite side of the Earth from point A. The net result of these two firings is that you now move in a circular orbit of smaller radius.

Similarly, to move to a larger orbit, you must fire your accelerating rockets twice. The first firing puts you into an elliptical orbit that moves farther from the Earth, as **FIGURE 12-21 (b)** shows. After the second firing you are again in a circular orbit. This simplest type of orbital transfer, requiring just two rocket burns, is referred to as a *Hohmann transfer*. The Hohmann transfer is the basic maneuver used to send spacecraft such as the Mars lander from Earth's orbit about the Sun to the orbit of Mars.

CONCEPTUAL EXAMPLE 12-10   WHICH ROCKETS TO USE?

As you pilot your spacecraft in a circular orbit about the Earth, you notice the space station you want to dock with several miles ahead in the same orbit. To catch up with the space station, should you fire your accelerating rockets or fire your decelerating rockets?

REASONING AND DISCUSSION

You want to catch up with something miles ahead, and hence you must accelerate, right? Well, not in this case. Accelerating moves you into an elliptical orbit, as in Figure 12-21 (b), and with a second acceleration you can make your new orbit circular with a larger radius. Recall from Kepler's third law, however, that the larger the radius of an orbit, the greater the period.

CONTINUED

Thus, on your new higher path you take longer to complete an orbit, so you fall farther behind the space station. The same is true even if you fire your rockets only once and stay on the elliptical orbit—it also has a longer period than the original orbit.

On the other hand, two decelerating burns will put you into a circular orbit of smaller radius, and thus a smaller period. As a result, you complete an orbit in less time than before and catch up with the space station. After catching up, you can perform two accelerating burns to move you back into the original orbit to dock.

ANSWER
You should fire your decelerating rockets.

Enhance Your Understanding
(Answers given at the end of the chapter)

3. Satellite A orbits the Earth at the radius r; satellite B orbits the Earth at the radius $2r$. (a) Is the orbital period of satellite A greater than, less than, or equal to the orbital period of satellite B? Explain. (b) Is the orbital speed of satellite A greater than, less than, or equal to the orbital speed of satellite B? Explain.

Section Review

- Kepler's three laws describe the orbital motions of planets, comets, satellites, and other objects in space. All three laws follow directly from Newton's law of universal gravitation.

12-4 Gravitational Potential Energy

In Chapter 8 we saw that the principle of conservation of energy can be used to solve a number of problems that would be difficult to handle with a straightforward application of Newton's laws of mechanics. Before we can apply energy conservation to astronomical situations, however, we must know the gravitational potential energy for a spherical object such as the Earth. Now you may be wondering: Don't we already know the potential energy of gravity? Well, in fact, in Chapter 8 we said that the gravitational potential energy a distance h above the Earth's surface is $U = mgh$. As was mentioned at the time, however, this result is valid only near the Earth's surface, where we can say that the acceleration due to gravity, g, is approximately constant.

General Result for Gravitational Potential Energy As the distance from the Earth increases, we know that g decreases, as was shown in Example 12-3. It follows that $U = mgh$ cannot be valid for arbitrary h. Indeed, it can be shown that the general result for the gravitational potential energy of a system consisting of a mass m a distance r from the center of the Earth is

$$U = -G\frac{mM_E}{r}$$

12-8

A plot of $U = -GmM_E/r$ is presented in **FIGURE 12-22**. Notice that U approaches zero as r approaches infinity. This is a common convention in astronomical systems. In fact, because only *differences* in potential energy matter, as was mentioned in Chapter 8, the

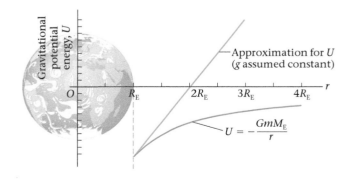

▶ **FIGURE 12-22 Gravitational potential energy as a function of distance r from the center of the Earth** The lower curve in this plot shows the gravitational potential energy, $U = -GmM_E/r$, for r greater than R_E. Near the Earth's surface, U is approximately linear, corresponding to the result $U = mgh$ given in Chapter 8.

choice of the reference point ($U = 0$) is completely arbitrary. When we considered systems that were near the Earth's surface, it was natural to let $U = 0$ at ground level. When we consider, instead, distances of astronomical scale, it is generally more convenient to choose the potential energy to be zero when objects are separated by an infinite distance.

EXERCISE 12-11   GRAVITATIONAL POTENTIAL ENERGY

Use $U = -GmM_E/r$ to find the gravitational potential energy of a 976-kg meteorite when it is **(a)** ten Earth radii from the center of the Earth and **(b)** on the surface of the Earth.

REASONING AND SOLUTION

a. In this case, the distance from the center of the Earth is $r = 10R_E$. It follows that the gravitational potential energy is

$$U = -G\frac{mM_E}{10R_E}$$

$$= -(6.67 \times 10^{-11}\,\text{N} \cdot \text{m}^2/\text{kg}^2)\frac{(976\,\text{kg})(5.97 \times 10^{24}\,\text{kg})}{10(6.37 \times 10^6\,\text{m})} = -6.10 \times 10^9\,\text{J}$$

b. Now, the distance from the center of the Earth is R_E, and therefore the gravitational potential energy is

$$U = -G\frac{mM_E}{R_E}$$

$$= -(6.67 \times 10^{-11}\,\text{N} \cdot \text{m}^2/\text{kg}^2)\frac{(976\,\text{kg})(5.97 \times 10^{24}\,\text{kg})}{(6.37 \times 10^6\,\text{m})} = -6.10 \times 10^{10}\,\text{J}$$

The potential energy in part (b) is ten times greater in magnitude than it is in part (a), because the distance from the center of the Earth to the meteorite is ten times smaller.

PHYSICS IN CONTEXT
Looking Ahead

The electric potential energy (Chapter 20) has the same mathematical form as the gravitational potential energy introduced in this chapter.

Connecting the General Gravitational Potential Energy with the Result Near the Surface of the Earth   At first glance, $U = -GmM_E/r$ doesn't seem to bear any similarity to $U = mgh$, which we know to be valid near the surface of the Earth. Even so, there is a direct connection between these two expressions. Recall that when we say that the potential energy at a height h is mgh, what we mean is that when a mass m is raised from the ground to a height h, the potential energy of the system increases by the amount mgh. Let's calculate the corresponding difference in potential energy using $U = -GmM_E/r$.

First, at a height h above the surface of the Earth we have $r = R_E + h$; hence the potential energy there is

$$U = -G\frac{mM_E}{R_E + h}$$

On the surface of the Earth, where $r = R_E$, we have

$$U = -G\frac{mM_E}{R_E}$$

The corresponding difference in potential energy is

$$\Delta U = \left(-G\frac{mM_E}{R_E + h}\right) - \left(-G\frac{mM_E}{R_E}\right)$$

$$= \left(-G\frac{mM_E}{R_E}\right)\left(\frac{1}{1 + h/R_E}\right) - \left(-G\frac{mM_E}{R_E}\right)$$

If h is much smaller than the radius of the Earth, it follows that h/R_E is a small number. In this case, we can apply the useful approximation $1/(1 + x) \approx 1 - x$ [see Figure A-5 (b) in Appendix A] to write $1/(1 + h/R_E) \approx 1 - h/R_E$. As a result, we have

$$\Delta U = \left(-G\frac{mM_E}{R_E}\right)(1 - h/R_E) - \left(-G\frac{mM_E}{R_E}\right) = m\left[\frac{GM_E}{R_E^2}\right]h$$

The term in square brackets should look familiar—according to Equation 12-4 it is simply g. Hence, the increase in potential energy at the height h is

$$\Delta U = mgh$$

This result is valid for heights that are small compared with the radius of the Earth.

The straight line in Figure 12-22 corresponds to the potential energy mgh. Near the Earth's surface, it is clear that mgh and $-GmM_E/r$ are in close agreement. For larger r, however, the fact that gravity is getting weaker means that the potential energy does not continue rising as rapidly as it would if gravity were of constant strength.

Gravitational Potential Energy Is a Scalar An important distinction between the potential energy, U, and the gravitational force, $\vec{\mathbf{F}}$, is that the force is a vector, whereas the potential energy is a scalar—that is, U is simply a number. As a result:

> The total gravitational potential energy of a system of objects is the sum of the gravitational potential energies of each pair of objects separately.

Because U is not a vector, there are no x or y components to consider, as would be the case with a vector. Finally, the potential energy $U = -GmM_E/r$ applies to a mass m a distance r from the center of the Earth, with mass M_E. More generally, if two point masses, m_1 and m_2, are separated by a distance r, their gravitational potential energy is:

Gravitational Potential Energy, U

$$U = -G\frac{m_1 m_2}{r} \qquad \text{12-9}$$

SI unit: joule, J

In the next Example we use this result, and the fact that U is a scalar, to find the total gravitational potential energy for a system of three point masses.

PHYSICS IN CONTEXT
Looking Back

Just as the force of gravity is generalized in this chapter from the simple result $F = mg$ to $F = Gm_1m_2/r^2$, so too is the gravitational potential energy. Thus, the expression $U = mgh$ (Chapter 8) is generalized to $U = -Gm_1m_2/r$.

EXAMPLE 12-12   SIMPLE ADDITION

The sketch shows the spaceship *Millennium Condor* as it approaches a pair of twin asteroids. Assuming all three objects can be treated as point masses, find the total gravitational potential energy of the system, given that the mass of the spaceship is $m = 3.75 \times 10^7$ kg and the mass of each asteroid is $M = 1.54 \times 10^9$ kg. The relevant distances are given in the sketch.

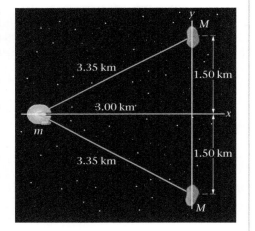

PICTURE THE PROBLEM
The positions of the spaceship and the two asteroids are shown in the sketch. Also shown are the distances between the various masses.

REASONING AND STRATEGY
The gravitational potential energy associated with each pair of masses is given by $U = -Gm_1m_2/r$ (Equation 12-9). The total gravitational potential energy of the system is the sum of the potential energies for each of the three pairs of masses.

Known   Mass of spaceship, $m = 3.75 \times 10^7$ kg; mass of asteroid, $M = 1.54 \times 10^9$ kg; distances given in the sketch.

Unknown   Total gravitational potential energy, $U_{\text{total}} = ?$

SOLUTION

1. Calculate the gravitational potential energy for the two asteroids, each with mass M:

$$U_{MM} = -G\frac{MM}{r_{MM}}$$

$$= -(6.67 \times 10^{-11}\,\text{N} \cdot \text{m}^2/\text{kg}^2)\frac{(1.54 \times 10^9\,\text{kg})^2}{2(1.50 \times 10^3\,\text{m})}$$

$$= -5.27 \times 10^4\,\text{J}$$

2. Calculate the gravitational potential energy for one of the asteroids (mass M) and the spaceship (mass m). The same calculation applies to the other asteroid as well:

$$U_{Mm} = -G\frac{Mm}{r_{Mm}}$$

$$= -(6.67 \times 10^{-11}\,\text{N} \cdot \text{m}^2/\text{kg}^2)\frac{(1.54 \times 10^9\,\text{kg})(3.75 \times 10^7\,\text{kg})}{(3.35 \times 10^3\,\text{m})}$$

$$= -1.15 \times 10^3\,\text{J}$$

CONTINUED

3. The total gravitational potential energy is the sum of the potential energies for the three pairs of objects:

$$U_{total} = U_{MM} + U_{Mm} + U_{Mm}$$
$$= -5.27 \times 10^4 \, \text{J} - 2(1.15 \times 10^3 \, \text{J}) = -5.50 \times 10^4 \, \text{J}$$

INSIGHT

The total gravitational potential energy of this system is negative, which is less than it would be if the separation of the three objects were to approach infinity, in which case $U_{total} = 0$. The implications of this change in potential energy are considered in the next section, when we explore the consequences of energy conservation.

PRACTICE PROBLEM — PREDICT/CALCULATE

(a) If the spaceship were moved to the origin of the coordinate system, halfway between the asteroids, would the potential energy of the system increase, decrease, or stay the same? Explain. **(b)** Verify your answer by calculating the potential energy in this case. **[Answer: (a)** The potential energy would decrease—that is, become more negative—because the distance from each asteroid to the spaceship would decrease. **(b)** $U_{total} = -5.78 \times 10^4 \, \text{J}]$

Some related homework problems: Problem 39, Problem 40

Enhance Your Understanding   (Answers given at the end of the chapter)

4. Consider the two systems shown in **FIGURE 12-23**. Is the total gravitational potential energy of system A greater than, less than, or equal to the total gravitational potential energy of system B? Explain.

▲ **FIGURE 12-23** Enhance Your Understanding 4.

Section Review

• The gravitational potential energy of a system consisting of point masses m_1 and m_2 separated by a distance r is $U = -Gm_1m_2/r$.

• The gravitational potential energy of a system consisting of a mass m a distance r from the center of the Earth is $U = -GmM_E/r$. Near the surface of the Earth, this expression simplifies to $U = mgh$.

12-5 Energy Conservation

Now that we know the gravitational potential energy, U, at an arbitrary distance from a spherical object, we can apply energy conservation to astronomical situations in the same way we applied it to systems near the Earth's surface in Chapter 8. To be specific, the mechanical energy, E, of an object of mass m a distance r from the center of the Earth is

$$E = K + U = \tfrac{1}{2}mv^2 - G\frac{mM_E}{r} \qquad \text{12-10}$$

This more general expression replaces $E = K + U = \tfrac{1}{2}mv^2 + mgh$, which applies for objects near the Earth's surface.

Speed of an Impacting Asteroid   If we set the initial mechanical energy of a system equal to its final mechanical energy, we can answer questions such as the following: Suppose that an asteroid has zero speed infinitely far from the Earth. If this asteroid were to fall directly toward the Earth, what speed would it have when it strikes the Earth's surface?

Big Idea **4** Gravitational potential energy is a form of energy that must be taken into account when we apply energy conservation. The general expression for gravitational potential energy is $U = -Gm_1m_2/r$, which simplifies to $U = mgh$ for objects near the Earth's surface.

PROBLEM-SOLVING NOTE

Energy Conservation in Astronomical Systems

To apply conservation of energy to an object that moves far from the surface of a planet, one must use $U = -GmM/r$, where r is the distance from the center of the planet.

▲ **FIGURE 12-24 Visualizing Concepts**
Impacts with Earth Bodies from space
have struck the Earth countless times in
the past and continue to do so on a regular basis. Most such objects are relatively
small, but larger objects, including the occasional comet or asteroid, also cross our
path from time to time. The crater on top,
in Rotorua, New Zealand, must be of relatively recent origin (thousands rather than
millions of years old), because erosion
has not yet erased this scar on the Earth's
surface. The center image is a false-color
gravity anomaly map of the Chicxulub
impact crater in Mexico. The object that
struck there some 65 million years ago
may have produced such far-reaching
climatic disruption that the dinosaurs and
many other species became extinct as a
result. The photo on the bottom shows the
12,500-metric-ton meteorite that blazed
across the sky over Chelyabinsk, Russia,
on February 15, 2013, injuring hundreds of
people with its powerful shock wave.

RWP As you probably know, this is not an entirely academic question. Asteroids
and comets, both large and small, have struck the Earth innumerable times during
its history. Three such impacts are shown in **FIGURE 12-24**. A particularly large object
appears to have struck the Earth on the Yucatan Peninsula in Mexico, near the town
of Chicxulub, some 65 million years ago. Evidence suggests that this impact may
have led to the mass extinctions at the end of the Cretaceous period, during which
the dinosaurs disappeared from the Earth. Unfortunately, such events are not limited to the distant past. For example, as recently as 50,000 years ago, an iron asteroid tens of meters in diameter and shining 10,000 times brighter than the Sun (from
atmospheric heating) slammed into the ground near Winslow, Arizona, forming the
1.2-km-wide Barringer Meteor Crater. More recently yet, at sunrise on June 30, 1908,
a relatively small stony asteroid streaked through the atmosphere and exploded at an
altitude of several kilometers near the Tunguska River in Siberia. The energy released
by the explosion was comparable to that of an H-bomb, and it flattened the forest for
kilometers in all directions. And on February 15, 2013, another meteorite struck in
Siberia, causing extensive damage and injuring hundreds.

Applying Energy Conservation Returning to our original question, we can use
energy conservation to determine the speed such an asteroid or comet might have
when it hits the Earth. To begin, we assume the asteroid starts at rest, and hence its
initial kinetic energy is zero, $K_i = 0$. In addition, the initial potential energy of the
system, U_i, is also zero, because $U = -GmM_E/r$ approaches zero as r approaches infinity. As a result, the total initial mechanical energy of the asteroid-Earth system is
zero: $E_i = K_i + U_i = 0$. Because gravity is a conservative force (as discussed in Section
8-1), the total mechanical energy remains constant as the asteroid falls toward the
Earth. Thus, as the asteroid moves closer to the Earth and U becomes increasingly
negative, the kinetic energy K must become increasingly positive so that their sum,
$U + K$, is always zero.

We now set the initial energy equal to the final energy to determine the final speed,
v_f. Recalling that the final distance r is the radius of the Earth, R_E, we have

$$E_i = E_f$$

$$0 = \tfrac{1}{2}mv_f^2 - G\frac{mM_E}{R_E}$$

Solving for the final speed yields

$$v_f = \sqrt{\frac{2GM_E}{R_E}}$$

12-11

Substituting numerical values into this expression gives

$$v_f = \sqrt{\frac{2GM_E}{R_E}} = \sqrt{\frac{2(6.67 \times 10^{-11}\,\text{N} \cdot \text{m}^2/\text{kg}^2)(5.97 \times 10^{24}\,\text{kg})}{6.37 \times 10^6\,\text{m}}}$$

$$= 11,200\,\text{m/s}\ (25,000\,\text{mi/h})$$

12-12

Thus, a typical asteroid hits the Earth moving at about 7.0 mi/s—about 16 times faster
than a rifle bullet! This result is independent of the asteroid's mass.

Visualizing Energy Conservation To help visualize energy conservation in this system, we plot the gravitational potential energy U in **FIGURE 12-25**. Also indicated in the
plot is the total energy, $E = 0$. Noting that $U + K$ must always equal zero, we see that
the value of K goes up as the value of U goes down. This is illustrated graphically in the
figure with the help of histogram bars.

A somewhat more elaborate plot showing the same physics is presented in **FIGURE 12-26**.
The two-dimensional surface in this case represents the potential energy function U as
one moves away from the Earth in any direction. In particular, the dependence of U on
distance r along any radial line in Figure 12-26 is the same as the shape of U versus r in
Figure 12-25. Because the potential energy drops downward in such a plot, this type of
situation is often referred to as a "potential well." If a marble is allowed to roll on such
a surface, its motion is similar in many ways to the motion of an object near the Earth.

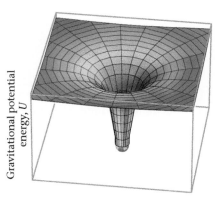

▲ **FIGURE 12-26 A gravitational potential "well"** A three-dimensional plot of the gravitational potential energy near an object such as the Earth. An object approaching the Earth speeds up as it "falls" into the gravitational potential well.

▲ **FIGURE 12-25 Potential and kinetic energies of an object falling toward Earth** As an object with zero total energy moves closer to the Earth, its gravitational potential energy, U, becomes increasingly negative. In order for the total energy to remain zero, $E = U + K = 0$, it is necessary for the kinetic energy to become increasingly positive.

In fact, if the marble is started with the right initial velocity, it will roll in a circular or elliptical "orbit" for a long time before falling into the center of the well. (Eventually, of course, the well does swallow up the marble. Though the retarding force of rolling friction is quite small, it still causes the marble to descend into a lower and lower orbit—just as air resistance causes a satellite to descend lower and lower into the Earth's atmosphere until it finally burns up.)

EXAMPLE 12-13 ARMAGEDDON RENDEZVOUS

In the movie *Armageddon*, a crew of hard-boiled oil drillers rendezvous with a menacing asteroid just as it passes the orbit of the Moon on its way toward Earth. Assuming the asteroid starts at rest infinitely far from the Earth, as in the previous discussion—and hence has zero initial energy—find its speed when it passes the Moon's orbit. Assume that the Moon orbits at a distance of $60R_E$ from the center of the Earth and that its gravitational influence may be ignored.

PICTURE THE PROBLEM
Our sketch shows the Earth, the Moon, and the asteroid. The initial position of the asteroid is at infinity, where its speed is zero. For the purposes of this problem, its final position is at the Moon's orbit, where its speed is v_f. At this point, the asteroid is heading directly for the Earth.

REASONING AND STRATEGY
The basic strategy is the same as that used above to obtain the speed of an asteroid when it hits the Earth; namely, we set the initial energy equal to the final energy and solve for the final speed v_f. In this case, the final radius is $r_f = 60R_E$. As before, the initial energy is zero; $E_i = 0$.

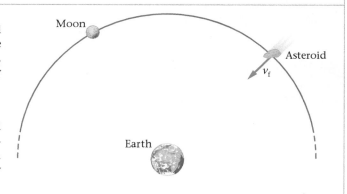

Known Initial mechanical energy, $E_i = 0$; final radius, $r_f = 60R_E$.
Unknown Speed at Moon's orbit, $v_f = ?$

SOLUTION

1. Set the initial energy of the system equal to its final energy:

$$E_i = E_f$$
$$0 = \tfrac{1}{2}mv_f^2 - G\frac{mM_E}{60R_E}$$

CONTINUED

2. Solve for the final speed, v_f:

$$v_f = \sqrt{\frac{2GM_E}{60R_E}} = \frac{1}{\sqrt{60}}\left(\sqrt{\frac{2GM_E}{R_E}}\right)$$

3. Substitute the numerical value given in Equation 12-12 for the quantity in parentheses:

$$v_f = \frac{1}{\sqrt{60}}(11,200\,\text{m/s}) = 1450\,\text{m/s} \sim 3200\,\text{mi/h}$$

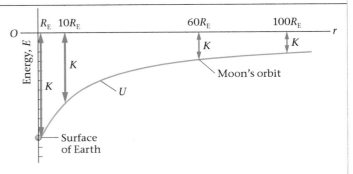

INSIGHT

The asteroid's speed increases from 0 at infinity to 1450 m/s at the Moon's orbit, and then to 11,200 m/s at the Earth's surface. The majority of this total increase in speed occurs after the asteroid passes the Moon. The reason for this can be seen in the accompanying plot of the gravitational potential energy, U.

This plot shows that U drops downward more and more rapidly as the Earth is approached. Thus, while there is relatively little increase in K from infinite distance to $r = 60R_E$, there is a substantially larger increase in K from $r = 60R_E$ to $r = R_E$.

PRACTICE PROBLEM

At what distance from the center of the Earth is the asteroid's speed equal to 3535 m/s?

[**Answer:** $r = 6.37 \times 10^7\,\text{m} = 10R_E$]

Some related homework problems: Problem 49, Problem 52

QUICK EXAMPLE 12-14   FIND THE DISTANCE TO A SATELLITE

A satellite in an elliptical orbit has a speed of 9.00 km/s when it is at its closest approach to the Earth (perigee). The satellite is 7.00×10^6 m from the center of the Earth at this time. When the satellite is at its greatest distance from the center of the Earth (apogee), its speed is 3.66 km/s. How far is the satellite from the center of the Earth at apogee? ($R_E = 6.37 \times 10^6$ m, $M_E = 5.97 \times 10^{24}$ kg)

REASONING AND SOLUTION

The mechanical energy (kinetic plus potential) of the satellite is conserved as it orbits the Earth. Setting the mechanical energy at perigee equal to the mechanical energy at apogee gives a relationship that can be solved for the distance at apogee.

1. Set the mechanical energy at perigee (speed v_1 at the distance r_1) equal to the energy at apogee (speed v_2 at the distance r_2):

$$\frac{1}{2}mv_1^2 - \frac{GmM_E}{r_1} = \frac{1}{2}mv_2^2 - \frac{GmM_E}{r_2}$$

2. Solve for $\frac{1}{r_2}$:

$$\frac{1}{r_2} = \frac{1}{r_1} + \frac{(v_2^2 - v_1^2)}{(2GM_E)}$$

3. Substitute numerical values:

$$\frac{1}{r_2} = 5.80 \times 10^{-8}\,\text{m}^{-1}$$

4. Invert the fraction to obtain the distance at apogee, r_2:

$$r_2 = 1.72 \times 10^7\,\text{m}$$

In this case, apogee is about 2.5 times farther from the center of the Earth than is perigee.

▲ **FIGURE 12-27** Comet Hale-Bopp, one of the largest and brightest comets to visit our celestial neighborhood in recent decades, photographed in April 1997. While most of the planets and planetary satellites in the solar system have roughly circular orbits, the orbits of many comets are highly elliptical. In accordance with Kepler's second law, these objects spend most of their time moving slowly through cold, distant regions of the solar system (often far beyond the orbit of Pluto). Their visits to the inner solar system are infrequent and relatively brief.

The concepts in Quick Example 12-14 apply to comets as well. For example, the comet shown in **FIGURE 12-27** moves much faster when it passes close to the Sun than when its orbit takes it far from the Sun.

Escape Speed

Resisting the pull of Earth's gravity has always held a fascination for the human species, from Daedalus and Icarus with their wings of feathers and wax, to Leonardo da Vinci and his flying machine, to the Montgolfier brothers and their hot-air balloons. In his 1865 novel, *From the Earth to the Moon*, Jules Verne imagined launching a spacecraft to the Moon by firing it straight upward from a cannon. Not a bad idea—if you could survive the initial blast. Today, rockets are fired into space using the same basic idea, though they smooth out the initial blast by burning their engines over a period of several minutes.

Suppose, then, that you would like to launch a rocket of mass m with an initial speed sufficient not only to reach the Moon, but to allow it to escape the Earth altogether. If we refer to this speed as the **escape speed** for the Earth, v_e, the initial energy of the rocket is

$$E_i = K_i + U_i = \frac{1}{2}mv_e^2 - G\frac{mM_E}{R_E}$$

If the rocket just barely escapes the Earth, its speed will decrease to zero as its distance from the Earth approaches infinity. Therefore, the rocket's final kinetic energy is zero, as is the potential energy of the system, because $U = -GmM_E/r$ goes to zero as $r \to \infty$. It follows that

$$E_f = K_f + U_f = 0 - 0 = 0$$

Equating these energies yields

$$\frac{1}{2}mv_e^2 - G\frac{mM_E}{R_E} = 0$$

Therefore, the escape speed from the Earth is

$$v_e = \sqrt{\frac{2GM_E}{R_E}} = 11{,}200 \text{ m/s} \approx 25{,}000 \text{ mi/h} \qquad \text{12-13}$$

Notice that the escape speed is precisely the same as the speed of the asteroid calculated in Equation 12-12. This is not surprising when you consider that an object launched from the Earth to infinity is just the reverse of an object falling from infinity to the Earth.

The result given in Equation 12-13 can be applied to other astronomical objects as well by simply replacing M_E and R_E with the appropriate mass and radius for another object.

EXERCISE 12-15   ESCAPE SPEED

Calculate the escape speed for an object launched from the Moon.

REASONING AND SOLUTION
For the Moon we use $M_m = 7.34 \times 10^{22}$ kg and $R_m = 1.74 \times 10^6$ m. With these values, the escape speed is

$$v_e = \sqrt{\frac{2GM_m}{R_m}} = \sqrt{\frac{2(6.67 \times 10^{-11} \text{ N} \cdot \text{m}^2/\text{kg}^2)(7.34 \times 10^{22} \text{ kg})}{1.74 \times 10^6 \text{ m}}}$$
$$= 2370 \text{ m/s (5300 mi/h)}$$

The relatively low escape speed of the Moon means that it is much easier to launch a rocket into space from the Moon than from the Earth. For example, the tiny Lunar Module that blasted off from the Moon to return the Apollo astronauts to Earth could not have come close to escaping from the Earth.

RWP The Moon's low escape speed has another important consequence—it is the reason the Moon has no atmosphere. Even if you could magically supply the Moon with an atmosphere, it would soon evaporate into space because the individual molecules in the air move with speeds great enough to escape. On Earth, however, where the escape speed is much higher, gravity can prevent the rapidly moving molecules from moving off into space. Even so, light molecules, like hydrogen and helium, move faster at a given temperature than the heavier molecules like nitrogen and oxygen, as we shall see in Chapter 17. For this reason, the Earth's atmosphere contains virtually no hydrogen or helium. (In fact, helium was first discovered on the Sun, as we point out in Chapter 31—hence its name, which derives from the Greek word for the Sun, "helios.") Because a stable atmosphere is a likely requirement for the development of life, it follows that the escape speed is an important factor when considering the possibility of life on other planets.

CONCEPTUAL EXAMPLE 12-16 COMPARE ESCAPE SPEEDS

PREDICT/EXPLAIN

(a) Is the escape speed for a 10,000-kg rocket greater than, less than, or equal to the escape speed for a 1000-kg rocket?
(b) Which of the following is the *best explanation* for your prediction?

 I. The greater the mass of a rocket, the greater the energy needed to make it escape, and hence the greater the escape speed.

 II. A rocket with more mass has more kinetic energy; hence less speed is needed to make it escape.

III. The escape speed depends on the mass of the Earth, but not on the mass of the rocket.

REASONING AND DISCUSSION
The derivation of the escape speed, $v_e = \sqrt{2GM_E/R_E}$ (Equation 12-13), shows that the mass of the rocket, m, cancels. Hence, the escape speed is the same for all objects, regardless of their mass. On the other hand, the kinetic energy required to give the 10,000-kg rocket the escape speed is 10 times greater than the kinetic energy required for the 1000-kg rocket.

ANSWER
(a) The escape speed is the same for both rockets. (b) The best explanation is III.

EXAMPLE 12-17 HALF ESCAPE

Suppose Jules Verne's cannon launches a rocket straight upward with an initial speed equal to one-half the escape speed. How far from the center of the Earth does this rocket travel before momentarily coming to rest? (Ignore air resistance in the Earth's atmosphere.)

PICTURE THE PROBLEM
Our sketch shows the rocket launched vertically from the Earth's surface with an initial speed equal to half the escape speed, $v_0 = \frac{1}{2}v_e$. The rocket moves radially away from the Earth until it comes to rest, $v = 0$, at a distance r from the center of the Earth.

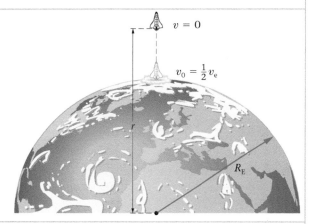

REASONING AND STRATEGY
We ignore air resistance in this system, and hence the final energy of the rocket, E_f, must be equal to its initial energy, E_0. Setting these energies equal determines the point where the rocket comes to rest.

Known Initial speed, $v_0 = v_e/2$; initial radius, R_E.
Unknown Final radius, $r = ?$

SOLUTION

1. The initial speed, v_0, is one-half the escape speed. Use $v_e = \sqrt{2GM_E/R_E}$ (Equation 12-13) to write an expression for v_0:

$$v_0 = \frac{1}{2}v_e = \frac{1}{2}\sqrt{2GM_E/R_E} = \sqrt{GM_E/2R_E}$$

2. Write the initial energy of the rocket, E_0. Substitute the result for v_0 from Step 1 and simplify:

$$E_0 = K_0 + U_0 = \frac{1}{2}mv_0{}^2 - \frac{GmM_E}{R_E}$$

$$= \frac{1}{2}m\left(\sqrt{\frac{GM_E}{2R_E}}\right)^2 - \frac{GmM_E}{R_E} = -\frac{3}{4}\frac{GmM_E}{R_E}$$

3. Write the final energy of the rocket. Recall that the rocket is a distance r from the center of the Earth when it comes to rest:

$$E_f = K_f + U_f = 0 - \frac{GmM_E}{r} = -\frac{GmM_E}{r}$$

4. Equate the initial and final energies:

$$-\frac{3}{4}\frac{GmM_E}{R_E} = -\frac{GmM_E}{r}$$

5. Solve for the final radius, r:

$$r = \frac{4}{3}R_E$$

INSIGHT
An initial speed of v_e allows the rocket to go to infinity before stopping. If the rocket is launched with half that initial speed, however, it can rise to a height of only $4R_E/3 - R_E = R_E/3$ above the Earth's surface. Quite a dramatic difference.

PRACTICE PROBLEM
Find the rocket's maximum distance from the center of the Earth, r, if its launch speed is $3v_e/4$.
[**Answer:** $r = 16R_E/7 = 2.29R_E$]

Some related homework problems: Problem 48, Problem 53, Problem 55

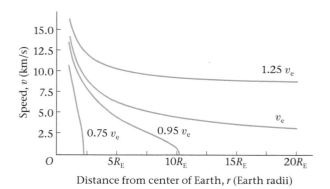

◀ **FIGURE 12-28 Speed of a rocket as a function of distance from the center of the Earth, *r*, for various vertical launch speeds** The lower two curves show launch speeds that are less than the escape speed, v_e. In these cases the rocket comes to rest momentarily at a finite height above the Earth. The next higher curve shows the speed of a rocket launched with the escape speed, v_e. In this case, the rocket slows to zero speed as the distance approaches infinity. The top curve corresponds to a launch speed greater than the escape speed—this rocket has a finite speed even at infinite distance.

A plot of the speed of a rocket as a function of its distance from the center of the Earth is presented in **FIGURE 12-28** for a variety of initial speeds. Notice that when the initial speed is less than the escape speed, the rocket comes to rest momentarily at a finite distance, *r*. In particular, if the launch speed is $0.75v_e$, as in the Practice Problem of Example 12-17, the rocket's maximum distance from the center of the Earth is $2.29R_E$.

Black Holes and Gravitational Lensing

As we can see from $v_e = \sqrt{2GM_E/R_E}$, the escape speed of an object increases with increasing mass and decreasing radius. Thus, for example, if a massive star were to collapse to a relatively small size, its escape speed would become very large. According to Einstein's theory of general relativity, the escape speed of a compressed, massive star could even exceed the speed of light. In this case nothing—not even light—could escape from the star. For this reason, such objects are referred to as *black holes*. Anything entering a black hole would be making a one-way trip to an unknown destiny.

RWP Just as a black hole can bend a beam of light back on itself and prevent it from escaping, any massive object can bend light—at least a little. For example, light from distant stars is deflected as it passes by the Sun by 1.75 seconds of arc (the angular size of a quarter at a distance of 1.8 miles). Light passing by an entire galaxy of stars, or even a cluster of galaxies, can be bent by more significant amounts, however, as **FIGURE 12-29** indicates. This effect is referred to as *gravitational lensing*, because the galaxies act much like the lenses we will study in Chapter 26. As a result of gravitational lensing, the images of very distant galaxies or quasars in deep-space astronomical photographs sometimes appear in duplicate, in quadruplicate, or even spread out into circular arcs.

Enhance Your Understanding   (Answers given at the end of the chapter)

5. Consider a distant planet that has twice the mass and twice the radius of the Earth. Is the escape speed on this planet greater than, less than, or equal to the escape speed on the Earth? Explain.

Section Review

- The total mechanical energy of a gravitational system is $E = K + U = \frac{1}{2}mv^2 - GmM/r$.

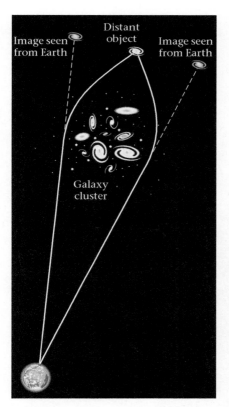

▲ **FIGURE 12-29 Gravitational lensing** Astronomers often find that very distant objects seem to produce multiple images in their photographs. The cause is the gravitational attraction of intervening galaxies or clusters of galaxies, which are so massive that they can significantly bend the light from remote objects as it passes by them on its way to Earth.

12-6 Tides

The reason for the ocean tides that rise and fall twice a day was a perplexing and enduring mystery until Newton introduced his law of universal gravitation. Even Galileo, who made so many advances in physics and astronomy, could not explain the tides. However, with the understanding that a force is required to cause an object to move in a circular path, and that the force of gravity becomes weaker with distance, it is possible to describe the tides in detail. In this section we show how it can be done. In addition, we extend the basic idea of tides to several related phenomena.

▶ **FIGURE 12-30 The reason for two tides a day (a)** Tides are caused by a disparity between the gravitational force exerted at various points on a finite-sized object (dark red arrows) and the centripetal force needed for circular motion (light red arrows). Note that the gravitational force decreases with distance, as expected. On the other hand, the centripetal force required to keep an object moving in a circular path *increases* with distance. On the near side, therefore, the gravitational force is stronger than required, and the object is stretched inward. On the far side, the gravitational force is weaker than required, and the object stretches outward. **(b)** On the Earth, the water in the oceans responds more to the deforming effects of tides than do the solid rocks of the land. The result is two high tides and two low tides daily on opposite sides of the Earth.

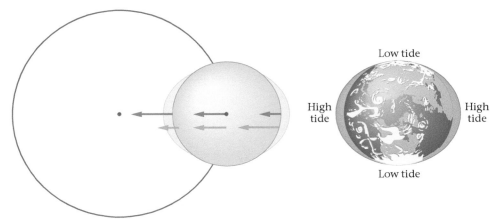

(a) The mechanism responsible for tides

(b) Tidal deformations on Earth

▲ **FIGURE 12-31** Tides on Earth are caused chiefly by the Moon's gravitational pull, though at full and new moon, when the Moon and Sun are aligned, the Sun's gravity can enhance the effect. In some places on Earth, such as the Bay of Fundy between Maine and Nova Scotia, local topographic conditions produce abnormally large tides.

Variation of Force To begin, consider the idealized situation shown in **FIGURE 12-30 (a)**. Here we see an object of finite size (a moon or a planet, for example) orbiting a point mass. If all the mass of the object were concentrated at its center, the gravitational force exerted on it by the central mass would be precisely the amount needed to cause it to move in its circular path. Because the object has a finite size, however, the force exerted on various parts of it has different magnitudes. For example, points closer to the central mass experience a greater force than points farther away.

To see the effect of this variation in force, we use a dark red vector in Figure 12-30 (a) to indicate the force exerted by the central mass at three different points on the object. In addition, we use a light red vector to show the force that is required at each of these three points to cause a mass at that distance to orbit the central mass. Comparing these vectors, we see that the forces are identical at the center of the object—as expected. On the near side of the object, however, the force exerted by the central mass is larger than the force needed to hold the object in orbit, and on the far side the force due to the central mass is less than the force needed to hold the object in orbit. The result is that the near side of the object is pulled closer to the central mass and the far side tends to move farther from the central mass. This causes an egg-shaped deformation of the object, as indicated in Figure 12-30 (a).

Any two objects orbiting one another cause deformations of this type. For example, the Earth causes a deformation in the Moon, and the Moon causes a similar deformation in the Earth. In **FIGURE 12-30 (b)** we show the Earth and the waters of its oceans deformed into an egg shape. Because the waters in the oceans can flow, they deform much more than the underlying rocky surface of the Earth. As a result, the water level relative to the surface of the Earth is higher at the *tidal bulges* shown in the figure. As the Earth rotates about its axis, a person at a given location will observe two high tides and two low tides each day. This is the basic mechanism of the tides on Earth, but the actual situation is complicated by the shape of the coastline at different locations and by the additional tidal effects due to the Sun. An example of a coastline that produces greater than normal tidal ranges is shown in **FIGURE 12-31**.

Tidal Locking The Moon has no oceans, of course, but the tidal bulges produced in it by the Earth are the reason we see only one side of the Moon. Specifically, the Earth exerts gravitational forces on the tidal bulges of the Moon, causing them to point directly toward the Earth. If the Moon were to rotate slightly away from this alignment, the forces exerted by the Earth would cause a torque that would return the Moon to the original alignment. The net result is that the Moon's period of rotation about its axis is equal to its period of revolution about the Earth. This effect, known as **tidal locking**, is common among the various moons in the solar system.

RWP A particularly interesting example of tidal locking is provided by Jupiter's moon Io, a site of intense volcanism. Io follows an elliptical orbit around Jupiter, and its tidal deformation is larger when it is closer to Jupiter than when it is farther away. As a result, each time Io orbits Jupiter it is squeezed into a greater deformation and then released. This continual flexing of Io causes its internal temperature to rise, just as a rubber ball gets warm if you squeeze and release it in your hand over and over. It is this mechanism that is largely responsible for Io's ongoing volcanic activity.

RWP In extreme cases, tidal deformation can become so large that an object is literally torn apart. Because tidal deformation increases as a moon moves closer to the planet it orbits, there is a limiting orbital radius—known as the **Roche limit**—inside of which the tidal force pulling the moon apart exceeds the gravitational attraction holding it together. A most spectacular example of this effect can be seen in the rings of Saturn, all of which exist well within the Roche limit. The small chunks of ice and other materials that make up the rings may be the remains of a moon that moved too close to Saturn and was destroyed by tidal forces. On the other hand, they may represent material that tidal forces prevented from aggregating to form a moon in the first place. In either case, this dramatic debris field will never coalesce to form a moon—tidal effects will not allow such a process to occur. Similar remarks apply to the smaller, much fainter rings that spacecraft have observed around Jupiter, Uranus, and Neptune.

Enhance Your Understanding   (Answers given at the end of the chapter)

6. If the radius of the Moon's orbit around the Earth were increased, would tides on the Earth be higher than, lower than, or the same as they are now? Explain. (The radius of the Moon's orbit is actually increasing at the rate of about 3.8 cm per year.)

Section Review

- Tides are caused by the difference in strength of the gravitational force from one side of an object to the other side.

CHAPTER 12 REVIEW

CHAPTER SUMMARY

12-1   NEWTON'S LAW OF UNIVERSAL GRAVITATION

The force of gravity between two point masses, m_1 and m_2, separated by a distance r is attractive and of magnitude

$$F = G\frac{m_1 m_2}{r^2}$$ 12-1

G is the universal gravitation constant:

$$G = 6.67 \times 10^{-11}\,\text{N} \cdot \text{m}^2/\text{kg}^2$$ 12-2

Inverse Square Dependence
The force of gravity decreases with distance, r, as $1/r^2$. This is referred to as an inverse square dependence.

Superposition
If more than one mass exerts a gravitational force on a given object, the net force is simply the vector sum of the individual forces.

12-2   GRAVITATIONAL ATTRACTION OF SPHERICAL BODIES

In calculations of gravitational forces, spherical objects can be replaced by point masses.

Uniform Sphere
If a mass m is outside a uniform sphere of mass M, the gravitational force between m and the sphere is equivalent to the force exerted by a point mass M located at the center of the sphere.

Acceleration of Gravity

Replacing the Earth with a point mass at its center, we find that the acceleration of gravity on the surface of the Earth is

$$g = \frac{GM_E}{R_E^2} \qquad \text{12-4}$$

12-3 KEPLER'S LAWS OF ORBITAL MOTION

Kepler determined three laws that describe the motion of the planets in our solar system. Newton showed that Kepler's laws are a direct consequence of his law of universal gravitation.

Kepler's First Law
The orbits of the planets are ellipses, with the Sun at one focus.

Kepler's Second Law
Planets sweep out equal areas in equal times.

Kepler's Third Law
The period of a planet's orbit, T, is proportional to the 3/2 power of its average distance from the Sun, r:

$$T = \left(\frac{2\pi}{\sqrt{GM_S}}\right)r^{3/2} = (\text{constant})r^{3/2} \qquad \text{12-7}$$

12-4 GRAVITATIONAL POTENTIAL ENERGY

The gravitational potential energy, U, between two point masses, m_1 and m_2, separated by a distance r is

$$U = -G\frac{m_1 m_2}{r} \qquad \text{12-9}$$

Zero Level
The zero level of the gravitational potential energy between two point masses is chosen to be at infinite separation of the two masses.

12-5 ENERGY CONSERVATION

With the gravitational potential energy given in Section 12-4, energy conservation can be applied to astronomical situations.

Total Mechanical Energy
An object with mass m, speed v, and at a distance r from the center of the Earth has a total mechanical energy given by

$$E = K + U = \frac{1}{2}mv^2 - \frac{GmM_E}{r} \qquad \text{12-10}$$

12-6 TIDES

Tides result from the variation of the gravitational force from one side of an astronomical object to the other side.

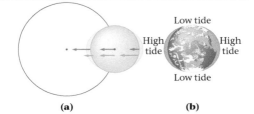

(a)   (b)

ANSWERS TO ENHANCE YOUR UNDERSTANDING QUESTIONS

1. B < A = D < C.

2. The gravitational force would stay the same because the masses and center-to-center distance did not change.

3. (a) The orbital period of satellite A is smaller because period increases with orbital radius. (b) The orbital speed of satellite A is greater. This follows because doubling the radius of satellite B doubles the circumference, which more than doubles the period, resulting in a lower speed (distance/time).

4. The total gravitational potential energy of system A is less (more negative) than that of system B. In general, a more "compact" system, like system A with the large masses close together, has a lower gravitational potential energy.

5. The escape speed on the distant planet is the same as the escape speed on Earth because the two factors of 2 (doubling the mass and doubling the radius) cancel.

6. The tides on Earth would be lower, because the gravitational forces exerted by the Moon on the Earth would be smaller than they are now.

CONCEPTUAL QUESTIONS

For instructor-assigned homework, go to www.masteringphysics.com.

(Answers to odd-numbered Conceptual Questions can be found in the back of the book.)

1. It is often said that astronauts in orbit experience weightlessness because they are beyond the pull of Earth's gravity. Is this statement correct? Explain.

2. When a person passes you on the street, you do not feel a gravitational tug. Explain.

3. Imagine bringing the tips of your index fingers together. Each finger contains a certain finite mass, and the distance between them goes to zero as they come into contact. From the force law $F = Gm_1m_2/r^2$ one might conclude that the attractive force between the fingers is infinite, and, therefore, that your fingers must remain forever stuck together. What is wrong with this argument?

4. Does the radius vector of Mars sweep out the same amount of area per time as that of the Earth? Why or why not?

5. When a communications satellite is placed in a geosynchronous orbit above the equator, it remains fixed over a given point on the ground. Is it possible to put a satellite into an orbit so that it remains fixed above the North Pole? Explain.

6. **The Mass of Pluto** On June 22, 1978, James Christy made the first observation of a moon orbiting Pluto. Until that time the mass of Pluto was not known, but with the discovery of its moon, Charon, its mass could be calculated with some accuracy. Explain.

7. Rockets are launched into space from Cape Canaveral in an easterly direction. Is there an advantage to launching to the east versus launching to the west? Explain.

8. One day in the future you may take a pleasure cruise to the Moon. While there you might climb a lunar mountain and throw a rock horizontally from its summit. If, in principle, you could throw the rock fast enough, it might end up hitting you in the back. Explain.

9. Apollo astronauts orbiting the Moon at low altitude noticed occasional changes in their orbit that they attributed to localized concentrations of mass below the lunar surface. Just what effect would such "mascons" have on their orbit?

10. If you light a candle on the International Space Station—which would not be a good idea—would it burn the same as on the Earth? Explain.

11. The force exerted by the Sun on the Moon is more than twice the force exerted by the Earth on the Moon. Should the Moon be thought of as orbiting the Earth or the Sun? Explain.

12. **The Path of the Moon** The Earth and Moon exert gravitational forces on one another as they orbit the Sun. As a result, the path they follow is not the simple circular orbit you would expect if either one orbited the Sun alone. Occasionally you will see a suggestion that the Moon follows a path like a sine wave centered on a circular path, as in the upper part of **FIGURE 12-32**. This is *incorrect*. The Moon's path is qualitatively like that shown in the lower part of Figure 12-32. Explain. (Refer to Conceptual Question 11.)

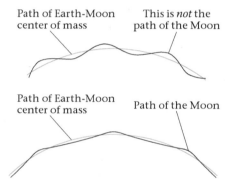

FIGURE 12-32 Conceptual Question 12

PROBLEMS AND CONCEPTUAL EXERCISES

Answers to odd-numbered Problems and Conceptual Exercises can be found in the back of the book. **BIO** *identifies problems of biological or medical interest;* **CE** *indicates a conceptual exercise.* Predict/Explain *problems ask for two responses: (a) your prediction of a physical outcome, and (b) the best explanation among three provided; and* Predict/Calculate *problems ask for a prediction of a physical outcome, based on fundamental physics concepts, and follow that with a numerical calculation to verify the prediction. On all problems, bullets (•, ••, •••) indicate the level of difficulty. Planetary data can be found in Appendix C.*

SECTION 12-1 NEWTON'S LAW OF UNIVERSAL GRAVITATION

1. • **CE** System A has masses m and m separated by a distance r; system B has masses m and $2m$ separated by a distance $2r$; system C has masses $2m$ and $3m$ separated by a distance $2r$; and system D has masses $4m$ and $5m$ separated by a distance $3r$. Rank these systems in order of increasing gravitational force. Indicate ties where appropriate.

2. • A 6.3-kg bowling ball and a 7.1-kg bowling ball rest on a rack 0.85 m apart. **(a)** What is the force of gravity exerted on each of the balls by the other ball? **(b)** At what separation is the force of gravity between the balls equal to 2.0×10^{-9} N?

3. • A communications satellite with a mass of 520 kg is in a circular orbit about the Earth. The radius of the orbit is 35,000 km as measured from the center of the Earth. Calculate **(a)** the weight of the satellite on the surface of the Earth and **(b)** the gravitational force exerted on the orbiting satellite by the Earth.

4. • **The Attraction of Ceres** Ceres, the largest asteroid known, has a mass of roughly 8.7×10^{20} kg. If Ceres passes within 14,000 km of the spaceship in which you are traveling, what force does it exert on you? (Use an approximate value for your mass, and treat yourself and the asteroid as point objects.)

5. • In one hand you hold a 0.13-kg apple, in the other hand a 0.22-kg orange. The apple and orange are separated by 0.75 m. What is the magnitude of the force of gravity that **(a)** the orange exerts on the apple and **(b)** the apple exerts on the orange?

6. •• Predict/Calculate A spaceship of mass m travels from the Earth to the Moon along a line that passes through the center of the Earth and the center of the Moon. **(a)** At what distance from the center of the Earth is the force due to the Earth twice the magnitude of the force due to the Moon? **(b)** How does your answer to part (a) depend on the mass of the spaceship? Explain.

7. •• At new moon, the Earth, Moon, and Sun are in a line, as indicated in **FIGURE 12-33**. Find the direction and magnitude of the net

gravitational force exerted on **(a)** the Earth, **(b)** the Moon, and **(c)** the Sun.

FIGURE 12-33 Problem 7

8. •• When the Earth, Moon, and Sun form a right triangle, with the Moon located at the right angle, as shown in **FIGURE 12-34**, the Moon is in its third-quarter phase. (The Earth is viewed here from above its North Pole.) Find the magnitude and direction of the net force exerted on the Moon. Give the direction relative to the line connecting the Moon and the Sun.

FIGURE 12-34 Problems 8, 9, and 69

9. •• Repeat the previous problem, this time finding the magnitude and direction of the net force acting on the Sun. Give the direction relative to the line connecting the Sun and the Moon.

10. •• Predict/Calculate Three 7.25-kg masses are at the corners of an equilateral triangle and located in space far from any other masses. **(a)** If the sides of the triangle are 0.610 m long, find the magnitude of the net force exerted on each of the three masses. **(b)** How does your answer to part (a) change if the sides of the triangle are doubled in length?

11. •• Predict/Calculate Four masses are positioned at the corners of a rectangle, as indicated in **FIGURE 12-35**. **(a)** Find the magnitude and direction of the net force acting on the 2.0-kg mass. **(b)** How do your answers to part (a) change (if at all) if all sides of the rectangle are doubled in length?

1.0 kg ———— 0.20 m ———— 2.0 kg

0.10 m

4.0 kg   3.0 kg

FIGURE 12-35 Problems 11 and 39

12. ••• Suppose that three astronomical objects (1, 2, and 3) are observed to lie on a line, and that the distance from object 1 to object 3 is D. Given that object 1 has four times the mass of object 3 and seven times the mass of object 2, find the distance between objects 1 and 2 for which the net force on object 2 is zero.

SECTION 12-2 GRAVITATIONAL ATTRACTION OF SPHERICAL BODIES

13. • Find the acceleration due to gravity on the surface of **(a)** Mercury and **(b)** Venus.

14. • At what altitude above the Earth's surface is the acceleration due to gravity equal to $g/4$?

15. • Two 6.4-kg bowling balls, each with a radius of 0.11 m, are in contact with one another. What is the gravitational attraction between the bowling balls?

16. • What is the acceleration due to Earth's gravity at a distance from the center of the Earth equal to the orbital radius of the Moon?

17. • **Extrasolar Planet Gravity** Kepler-62e is an exoplanet that orbits within the habitable zone around its parent star. The planet has a mass that is 3.57 times larger than Earth's and a radius that is 1.61 times larger than Earth's. Calculate the acceleration of gravity on the surface of Kepler-62e.

18. •• Predict/Calculate At a certain distance from the center of the Earth, a 5.0-kg object has a weight of 3.6 N. **(a)** Find this distance. **(b)** If the object is released at this location and allowed to fall toward the Earth, what is its initial acceleration? **(c)** If the object is now moved twice as far from the Earth, by what factor does its weight change? Explain. **(d)** By what factor does its initial acceleration change? Explain.

19. •• The acceleration due to gravity on the Moon's surface is known to be about one-sixth the acceleration due to gravity on the Earth. Given that the radius of the Moon is roughly one-quarter that of the Earth, find the mass of the Moon in terms of the mass of the Earth.

20. •• **Gravitational Tug of War** At some point along the direct path from the center of the Earth to the center of the Moon, the gravitational force of attraction on a spacecraft from the Moon becomes greater than the force from the Earth. **(a)** How far from the center of the Earth does this occur? **(b)** At this location, how far is the spacecraft from the surface of the Moon? How far is it from the surface of the Earth?

21. •• Predict/Calculate **An Extraterrestrial Volcano** Several volcanoes have been observed erupting on the surface of Jupiter's closest Galilean moon, Io. Suppose that material ejected from one of these volcanoes reaches a height of 5.00 km after being projected straight upward with an initial speed of 134 m/s. Given that the radius of Io is 1820 km, **(a)** outline a strategy that allows you to calculate the mass of Io. **(b)** Use your strategy to calculate Io's mass.

22. ••• Consider an asteroid with a radius of 19 km and a mass of 3.35×10^{15} kg. Assume the asteroid is roughly spherical. **(a)** What is the acceleration due to gravity on the surface of the asteroid? **(b)** Suppose the asteroid spins about an axis through its center, like the Earth, with a rotational period T. What is the smallest value T can have before loose rocks on the asteroid's equator begin to fly off the surface?

SECTION 12-3 KEPLER'S LAWS OF ORBITAL MOTION

23. • **CE** A satellite orbits the Earth in a circular orbit of radius r. At some point its rocket engine is fired so that its speed increases rapidly by a small amount. As a result, do the **(a)** apogee distance and **(b)** perigee distance increase, decrease, or stay the same?

24. • **CE** Predict/Explain **The Earth-Moon Distance Is Increasing** Laser reflectors left on the surface of the Moon by the Apollo astronauts show that the average distance from the Earth to the Moon is increasing at the rate of 3.8 cm per year. **(a)** As a result, will the length of the month increase, decrease, or remain the same? **(b)** Choose the *best explanation* from among the following:

   I. The greater the radius of an orbit, the greater the period, which implies a longer month.

   II. The length of the month will remain the same due to conservation of angular momentum.

   III. The speed of the Moon is greater with increasing radius; therefore, the length of the month will decrease.

25. • **Apollo Missions** On Apollo missions to the Moon, the command module orbited at an altitude of 110 km above the lunar surface. How much time did it take for the command module to complete one orbit?

26. • Find the orbital speed of a satellite in a geosynchronous circular orbit 3.58×10^7 m above the surface of the Earth.

27. • **An Extrasolar Planet** In July of 1999 a planet was reported to be orbiting the Sun-like star Iota Horologii with a period of 320 days. Find the radius of the planet's orbit, assuming that Iota Horologii has the same mass as the Sun. (This planet is presumably similar to Jupiter, but it may have large, rocky moons that enjoy a relatively pleasant climate.)

28. • Phobos, one of the moons of Mars, orbits at a distance of 9378 km from the center of the red planet. What is the orbital period of Phobos?

29. •• Predict/Calculate **An Asteroid with Its Own Moon** The asteroid 243 Ida has its own small moon, Dactyl. (See the photo on p. 394.) Another such system, asteroid 624 Hektor, has a mass of 7.9×10^{18} kg, with its moon orbiting at a radius of 623.5 km and a period of 2.965 days. (a) Given that the orbital radius of Dactyl is 108 km, and its period is 1.54 days, is the mass of 243 Ida greater than, less than, or equal to the mass of 624 Hektor? (b) Calculate the mass of 243 Ida.

30. •• **GPS Satellites** GPS (Global Positioning System) satellites orbit at an altitude of 2.0×10^7 m. Find (a) the orbital period, and (b) the orbital speed of such a satellite.

31. •• Predict/Calculate Two satellites orbit the Earth, with satellite 1 at a higher altitude than satellite 2. (a) Which satellite has the greater orbital speed? Explain. (b) Calculate the orbital speed of a satellite that orbits at an altitude of one Earth radius above the surface of the Earth. (c) Calculate the orbital speed of a satellite that orbits at an altitude of two Earth radii above the surface of the Earth.

32. •• Predict/Calculate Satellite A has a mass of 1000 kg and satellite B has a mass of 2000 kg. (a) When each satellite orbits at one Earth radius above the surface of the Earth, will the period of satellite B be longer, shorter, or the same as the period of satellite A? (b) Calculate the orbital periods of satellites A and B.

33. •• Predict/Calculate The Martian moon Deimos has an orbital period that is greater than the other Martian moon, Phobos. Both moons have approximately circular orbits. (a) Is Deimos closer to or farther from Mars than Phobos? Explain. (b) Calculate the distance from the center of Mars to Deimos given that its orbital period is 1.10×10^5 s.

34. •• Predict/Calculate (a) Calculate the orbital period of a satellite that orbits at two Earth radii above the surface of the Earth. (b) If the Earth suddenly became more massive while the satellite remains in circular orbit at the same altitude, will the orbital period increase, decrease, or remain the same? (c) Calculate the orbital period of the satellite if the Earth's mass were to double to 1.194×10^{25} kg and everything else were to remain the same.

35. ••• **Binary Stars** Alpha Centauri A and Alpha Centauri B are binary stars with a separation of 3.45×10^{12} m and an orbital period of 2.52×10^9 s. Assuming the two stars are equally massive (which is approximately the case), determine their mass.

SECTION 12-4 GRAVITATIONAL POTENTIAL ENERGY

36. • **Sputnik** The first artificial satellite to orbit the Earth was Sputnik I, launched October 4, 1957. The mass of Sputnik I was 83.5 kg, and its distances from the center of the Earth at apogee and perigee were 7330 km and 6610 km, respectively. Find the difference in gravitational potential energy for Sputnik I as it moved from apogee to perigee.

37. • How much gravitational potential energy is required to lift a 9380-kg *Progress* spacecraft to the altitude of the International Space Station, 422 km above the surface of the Earth?

38. •• CE Predict/Explain (a) Is the amount of energy required to get a spacecraft from the Earth to the Moon greater than, less than, or equal to the energy required to get the same spacecraft from the Moon to the Earth? (b) Choose the *best explanation* from among the following:
    I. The escape speed of the Moon is less than that of the Earth; therefore, less energy is required to leave the Moon.
    II. The situation is symmetric, and hence the same amount of energy is required to travel in either direction.
    III. It takes more energy to go from the Moon to the Earth because the Moon is orbiting the Earth.

39. •• Predict/Calculate Consider the four masses shown in Figure 12-35. (a) Find the total gravitational potential energy of this system. (b) How does your answer to part (a) change if all the masses in the system are doubled? (c) How does your answer to part (a) change if, instead, all the sides of the rectangle are halved in length?

40. •• Calculate the gravitational potential energy of a 9.50-kg mass (a) on the surface of the Earth and (b) at an altitude of 325 km. (c) Take the difference between the results for parts (b) and (a), and compare with mgh, where $h = 325$ km.

41. •• Three bowling balls form an equilateral triangle, as shown in **FIGURE 12-36**. Each ball has a radius of 10.8 cm and a mass of 7.26 kg. What is the total gravitational potential energy of this system?

42. •• Two 0.59-kg basketballs, each with a radius of 12 cm, are just touching. How much energy is required to change the separation between the centers of the basketballs to (a) 1.0 m and (b) 10.0 m? (Ignore any other gravitational interactions.)

FIGURE 12-36

43. •• Find the minimum kinetic energy needed for a 32,000-kg rocket to escape (a) the Moon or (b) the Earth.

SECTION 12-5 ENERGY CONSERVATION

44. • CE Predict/Explain Suppose the Earth were to suddenly shrink to half its current diameter, with its mass remaining constant. (a) Would the escape speed of the Earth increase, decrease, or stay the same? (b) Choose the best explanation from among the following:
    I. Since the radius of the Earth would be smaller, the escape speed would also be smaller.
    II. The Earth would have the same amount of mass, and hence its escape speed would be unchanged.
    III. The force of gravity would be much stronger on the surface of the compressed Earth, leading to a greater escape speed.

45. • CE Is the energy required to launch a rocket vertically to a height h greater than, less than, or equal to the energy required to put the same rocket into orbit at the height h? Explain.

46. • Suppose one of the Global Positioning System satellites has a speed of 4.46 km/s at perigee and a speed of 3.64 km/s at apogee. If the distance from the center of the Earth to the satellite at perigee is 2.00×10^4 km, what is the corresponding distance at apogee?

47. • **Meteorites from Mars** Several meteorites found in Antarctica are believed to have come from Mars, including the famous ALH84001

meteorite that was once thought to contain fossils of ancient life on Mars. Meteorites from Mars are thought to get to Earth by being blasted off the Martian surface when a large object (such as an asteroid or a comet) crashes into the planet. What speed must a rock have to escape Mars?

48. • What is the launch speed of a projectile that rises vertically above the Earth to an altitude equal to one Earth radius before coming to rest momentarily?

49. • A projectile launched vertically from the surface of the Moon rises to an altitude of 425 km. What was the projectile's initial speed?

50. •• (a) How much gravitational potential energy must a 3130-kg satellite acquire in order to attain a geosynchronous orbit? (b) How much kinetic energy must it gain? Note that because of the rotation of the Earth on its axis, the satellite had a velocity of 463 m/s relative to the center of the Earth just before launch.

51. •• Predict/Calculate **Halley's Comet** Halley's comet, which passes around the Sun every 76 years, has an elliptical orbit. When closest to the Sun (perihelion) it is at a distance of 8.823×10^{10} m and moves with a speed of 54.6 km/s. The greatest distance between Halley's comet and the Sun (aphelion) is 6.152×10^{12} m. (a) Is the speed of Halley's comet greater than or less than 54.6 km/s when it is at aphelion? Explain. (b) Calculate its speed at aphelion.

52. •• **The End of the Lunar Module** On Apollo Moon missions, the lunar module would blast off from the Moon's surface and dock with the command module in lunar orbit. After docking, the lunar module would be jettisoned and allowed to crash back onto the lunar surface. Seismometers placed on the Moon's surface by the astronauts would then pick up the resulting seismic waves. Find the impact speed of the lunar module, given that it is jettisoned from an orbit 110 km above the lunar surface moving with a speed of 1630 m/s.

53. •• If a projectile is launched vertically from the Earth with a speed equal to the escape speed, how high above the Earth's surface is it when its speed is one-third the escape speed?

54. •• Suppose a planet is discovered orbiting a distant star. If the mass of the planet is 10 times the mass of the Earth, and its radius is one-tenth the Earth's radius, how does the escape speed of this planet compare with that of the Earth?

55. •• A projectile is launched vertically from the surface of the Moon with an initial speed of 1010 m/s. At what altitude is the projectile's speed one-half its initial value?

56. •• To what radius would the Sun have to be contracted for its escape speed to equal the speed of light?

57. •• Predict/Calculate Two baseballs, each with a mass of 0.148 kg, are separated by a distance of 395 m in outer space, far from any other objects. (a) If the balls are released from rest, what speed do they have when their separation has decreased to 145 m? (b) Suppose the mass of the balls is doubled. Would the speed found in part (a) increase, decrease, or stay the same? Explain.

58. ••• On Earth, a person can jump vertically and rise to a height h. What is the radius of the largest spherical asteroid from which this person could escape by jumping straight upward? Assume that each cubic meter of the asteroid has a mass of 3500 kg.

SECTION 12-6 TIDES

59. •• The magnitude of the tidal force exerted on a linear object of mass m and length L is approximately $2GmML/r^3$. In this expression, M is the mass of the body causing the tidal force and r is the distance from the center of m to the center of M. Suppose you are 1 million miles (1.6×10^9 m) away from a black hole whose mass

is 1.99×10^{36} kg (one million times that of the Sun). (a) Estimate the tidal force exerted on your body ($L = 1.8$ m) by the black hole. (b) At what distance will the tidal force be approximately 10 times greater than your weight?

60. •• The magnitude of the tidal force between the International Space Station (ISS) and a nearby astronaut on a spacewalk is approximately $2GmMa/r^3$. In this expression, M is the mass of the Earth, $r = 6.79 \times 10^6$ m is the distance from the center of the Earth to the orbit of the ISS, $m = 125$ kg is the mass of the astronaut, and $a = 10$ m is the distance from the astronaut to the center of mass of the ISS. (a) Calculate the magnitude of the tidal force for this astronaut. This force tends to separate the astronaut from the ISS if the astronaut is located along the line that connects the center of the Earth with the center of mass of the ISS. (b) Calculate the force of gravitational attraction between the astronaut and the ISS if they are 10 m apart and the ISS has a mass of 420,000 kg. (c) Is the ISS orbit inside or outside the Roche limit?

61. ••• A dumbbell has a mass m on either end of a rod of length $2a$. The center of the dumbbell is a distance r from the center of the Earth, and the dumbbell is aligned radially. If $r \gg a$, show that the difference in the gravitational force exerted on the two masses by the Earth is approximately $4GmM_Ea/r^3$. (*Note:* The difference in force causes a tension in the rod connecting the masses. We refer to this as a *tidal force.*) [*Hint:* Use the fact that $1/(r-a)^2 - 1/(r+a)^2 \sim 4a/r^3$ for $r \gg a$.]

62. ••• Referring to the previous problem, suppose the rod connecting the two masses m is removed. In this case, the only force between the two masses is their mutual gravitational attraction. In addition, suppose the masses are spheres of radius a and mass $m = \frac{4}{3}\pi a^3\rho$ that touch each other. (The Greek letter ρ stands for the density of the masses.) (a) Write an expression for the gravitational force between the masses m. (b) Find the distance from the center of the Earth, r, for which the gravitational force found in part (a) is equal to the tidal force found in the previous problem. This distance is known as the *Roche limit.* (c) Calculate the Roche limit for Saturn, assuming $\rho = 3330$ kg/m³. (The famous rings of Saturn are within the Roche limit for that planet. Thus, the innumerable small objects, composed mostly of ice, that make up the rings will never coalesce to form a moon.)

GENERAL PROBLEMS

63. • CE You weigh yourself on a scale inside an airplane flying due east above the equator. If the airplane now turns around and heads due west with the same speed, will the reading on the scale increase, decrease, or stay the same? Explain.

64. • CE Rank objects A, B, and C in FIGURE 12-37 in order of increasing net gravitational force experienced by the object. Indicate ties where appropriate.

FIGURE 12-37 Problems 64 and 65

65. • CE Referring to Figure 12-37, rank objects A, B, and C in order of increasing initial acceleration each would experience if it alone were allowed to move. Indicate ties where appropriate.

66. • CE **The Crash of Skylab** Skylab, the largest spacecraft ever to fall back to the Earth, met its fiery end on July 11, 1979, after flying directly over Everett, WA, on its last orbit. On the *CBS Evening News* the night before the crash, anchorman Walter Cronkite, in his rich baritone voice, made the following statement: "NASA says there is a little chance that Skylab will land in a populated area."

After the commercial, he immediately corrected himself by saying, "I meant to say '*there is little chance*' Skylab will hit a populated area." In fact, it landed primarily in the Indian Ocean off the west coast of Australia, though several pieces were recovered near the town of Esperance, Australia, which later sent the U.S. State Department a $400 bill for littering. The cause of Skylab's crash was the friction it experienced in the upper reaches of the Earth's atmosphere. As the radius of Skylab's orbit decreased, did its speed increase, decrease, or stay the same? Explain.

67. • Consider a system consisting of three masses on the x axis. Mass $m_1 = 1.00$ kg is at $x = 1.00$ m; mass $m_2 = 2.00$ kg is at $x = 2.00$ m; and mass $m_3 = 3.00$ kg is at $x = 3.00$ m. What is the total gravitational potential energy of this system?

68. •• An astronaut exploring a distant solar system lands on an unnamed planet with a radius of 3520 km. When the astronaut jumps upward with an initial speed of 2.91 m/s, she rises to a height of 0.525 m. What is the mass of the planet?

69. •• Predict/Calculate When the Moon is in its third-quarter phase, the Earth, Moon, and Sun form a right triangle, as shown in Figure 12-34. Calculate the magnitude of the force exerted on the Moon by **(a)** the Earth and **(b)** the Sun. **(c)** Does it make more sense to think of the Moon as orbiting the Sun, with a small effect due to the Earth, or as orbiting the Earth, with a small effect due to the Sun?

70. •• An equilateral triangle 10.0 m on a side has a 1.00-kg mass at one corner, a 2.00-kg mass at another corner, and a 3.00-kg mass at the third corner (**FIGURE 12-38**). Find the magnitude and direction of the net force acting on the 2.00-kg mass.

1.00 kg

10.0 m   10.0 m

3.00 kg   10.0 m   2.00 kg

FIGURE 12-38 Problems 70 and 71

71. •• Suppose that each of the three masses in Figure 12-38 is replaced by a mass of 5.95 kg and radius 0.0714 m. If the masses are released from rest, what speed will they have when they collide at the center of the triangle? Ignore gravitational effects from any other objects.

72. •• **A Near Miss!** In the early morning hours of June 14, 2002, the Earth had a remarkably close encounter with an asteroid the size of a small city. The previously unknown asteroid, now designated 2002 MN, remained undetected until three days after it had passed the Earth. At its closest approach, the asteroid was 73,600 miles from the center of the Earth—about a third of the distance to the Moon. **(a)** Find the speed of the asteroid at closest approach, assuming its speed at infinite distance to be zero and considering only its interaction with the Earth. **(b)** Observations indicate the asteroid to have a diameter of about 0.730 km. Estimate the kinetic energy of the asteroid at closest approach, assuming it has an average density of 3.33 g/cm³. (For comparison, a 1-megaton nuclear weapon releases about 4.2×10^{15} J of energy.)

73. •• Predict/Calculate Suppose a planet is discovered that has the same total mass as the Earth, but half its radius. **(a)** Is the acceleration due to gravity on this planet more than, less than, or the

same as the acceleration due to gravity on the Earth? Explain. **(b)** Calculate the acceleration due to gravity on this planet.

74. •• Show that the speed of a satellite in a circular orbit a height h above the surface of the Earth is

$$v = \sqrt{\frac{GM_E}{R_E + h}}$$

75. •• **Walking into Orbit** A spherical asteroid of average density would have a mass of 8.7×10^{13} kg if its radius were 2.0 km. **(a)** If you and your spacesuit had a mass of 125 kg, how much would you weigh when standing on the surface of this asteroid? **(b)** If you could walk on the surface of this asteroid, what minimum speed would you need to launch yourself into an orbit just above the surface of the asteroid?

76. •• In a binary star system, two stars orbit about their common center of mass, as shown in **FIGURE 12-39**. If $r_2 = 2r_1$, what is the ratio of the masses m_2/m_1 of the two stars?

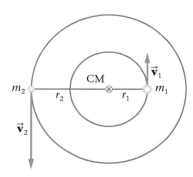

FIGURE 12-39 Problems 76 and 77

77. •• Find the orbital period of the binary star system described in the previous problem.

78. •• **Exploring Mars** In the not-too-distant future astronauts will travel to Mars to carry out scientific explorations. As part of their mission, it is likely that a "geosynchronous" satellite will be placed above a given point on the Martian equator to facilitate communications. At what altitude above the surface of Mars should such a satellite orbit? (*Note:* The Martian "day" is 24.6229 hours. Other relevant information can be found in Appendix C.)

79. •• **Comet Wild 2** In 2004, a NASA spacecraft named *Stardust* flew within 147 miles of Comet Wild 2 (pronounced "Vilt 2"), zooming by it at 6200 m/s, about six times the speed of a rifle bullet. Photos taken by *Stardust* show that the comet is roughly spherical, as shown in **FIGURE 12-40**, with a radius of 2.7 km. It has also been determined that the acceleration due to gravity on the surface of Wild 2 is 0.00010g. What is the minimum speed needed for an object to escape from the surface of Wild 2?

FIGURE 12-40 Comet Wild 2 and some of its surface features, including the Walker basin, the site of unusual jets of outward-flowing dust and rocks. (Problem 79)

80. •• **Predict/Calculate** **(a)** If you want to launch a probe that orbits the Sun with a period of 2.00 years, what should be its orbital radius? Give your answer in astronomical units. The astronomical unit AU is defined as the mean distance from the Sun to the Earth (1 AU = 1.50×10^{11} m). **(b)** Is the orbital speed of this probe greater than, less than, or equal to the orbital speed of the Earth? Explain. **(c)** What is the orbital speed of this probe?

81. •• **Predict/Calculate** A satellite is placed in Earth orbit at an altitude of 23,300 mi, which is 1100 miles higher than the altitude of a geosynchronous satellite. **(a)** Is the period of this satellite greater than or less than 24 hours? **(b)** As viewed from the surface of the Earth, does the satellite move eastward or westward? Explain. **(c)** Find the orbital period of this satellite.

82. •• Show that the force of gravity between the Moon and the Sun is always greater than the force of gravity between the Moon and the Earth.

83. •• The astronomical unit AU is defined as the mean distance from the Sun to the Earth (1 AU = 1.50×10^{11} m). Apply Kepler's third law (Equation 12-7) to the solar system, and show that it can be written as

$$T = Cr^{3/2}$$

In this expression, the period T is measured in years, the distance r is measured in astronomical units, and the constant C has a magnitude that you must determine.

84. •• **(a)** Find the kinetic energy of a 1940-kg satellite in a circular orbit about the Earth, given that the radius of the orbit is 12,600 miles. **(b)** How much energy is required to move this satellite to a circular orbit with a radius of 25,200 miles?

85. •• **Predict/Calculate** **Space Station Orbit** The International Space Station ($m = 4.5 \times 10^5$ kg) orbits at an altitude of 415 km above the Earth's surface. **(a)** Does the orbital speed of the station depend on its mass? Explain. **(b)** Find the speed of the station in its orbit. **(c)** How much time does it take for the station to complete one orbit of the Earth?

86. •• **Approaching the ISS** A Russian *Soyuz* module, with three astronauts and a full load of cargo, has a mass of 7500 kg. The International Space Station (ISS) has a mass of 420,000 kg. When *Soyuz* docks at the ISS, the two centers of mass are separated by 9.5 m. **(a)** Find the average of the gravitational force of attraction between the two spacecraft when *Soyuz* is docked and when it is 110 m from the ISS. **(b)** If *Soyuz* approached the ISS from rest at a distance of 110 m, a thruster would have to counteract the average force of gravitational attraction between it and the ISS. If this thruster has an exhaust velocity of 590 m/s, at what average rate must it burn fuel (in kg/s) to counteract the pull of the ISS? (See Section 9-8 for a discussion of thrust.) **(c)** If the approach requires 330 s, how much fuel will the thruster burn in order to counteract the gravitational attraction?

87. ••• **Predict/Calculate** Consider an object of mass m orbiting the Earth at a radius r. **(a)** Find the speed of the object. **(b)** Show that the total mechanical energy of this object is equal to (-1) times its kinetic energy. **(c)** Does the result of part (b) apply to an object orbiting the Sun? Explain.

88. ••• The Earth's orbit around the Sun is slightly elliptical (see Conceptual Example 12-7). At Earth's closest approach to the Sun (perihelion) the orbital radius is 1.471×10^{11} m, and at its farthest distance (aphelion) the orbital radius is 1.521×10^{11} m. Given

that the mass of the Earth is 5.972×10^{24} kg, and that of the Sun is 1.989×10^{30} kg, **(a)** find the difference in gravitational potential energy between when the Earth is at its aphelion and perihelion radii. **(b)** If the orbital speed of the Earth is 29,290 m/s at aphelion, what is its orbital speed at perihelion?

89. ••• Three identical stars, at the vertices of an equilateral triangle, orbit about their common center of mass (**FIGURE 12-41**). Find the period of this orbital motion in terms of the orbital radius, R, and the mass of each star, M.

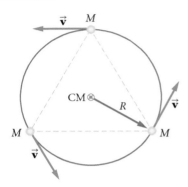

FIGURE 12-41 Problem 89

90. ••• Find an expression for the kinetic energy of a satellite of mass m in an orbit of radius r about a planet of mass M.

PASSAGE PROBLEMS

If life exists elsewhere in our solar system, it may not have developed independently from life on Earth. Instead, it's possible that microbes from Earth may have colonized other planets or moons by hitching a ride on a rock blasted from Earth's surface by a meteor impact. If the impact gives the rock enough energy to escape into space (while at the same time not raising its temperature so high as to "cook" the microbes), the rock may eventually reach another body in the solar system. In fact, rocks from Mars are known to have reached Earth in just this way, although none are currently known to have contained microbes. Computer modeling can be used to estimate the probability that a rock ejected from the surface of the Earth with a speed greater than the escape speed will reach another planet. These computer models indicate that under the influence of gravitational fields from the other objects in the solar system, an ejected rock can take millions of years to travel from one planet to another. During this time any life "aboard" is continually exposed to the high radiation levels of space. Some researchers have calculated that a 3.0-m-diameter rock at a typical rock density of 3.0 g/cm³ is sufficient to shield some types of microbes from the hostile environment of space for several million years of travel.

The accompanying plot shows the residual speed of an ejected object—that is, the speed the object would have when infinitely far from the Earth—as a function of its speed at the surface of the Earth (its original ejection speed). By simulating the motion of rocks ejected from the Earth with a variety of speeds, researchers conclude that 0.03% of the rocks ejected such that they have a residual speed of 2.5 km/s will have reached Mars 2.0 million years later. Although this doesn't seem like a high probability, there have been so many meteor impacts over the long history of the Earth that many ejected rocks must have reached Mars—though whether they carried microbes, and if they did, whether the microbes would have survived, are open questions.

91. • What is the minimum kinetic energy an impact must give to a 3.0-m-diameter rock for the rock to be able to escape from the Earth?

    A. 5.3×10^3 J   **B.** 1.3×10^6 J
    C. 2.7×10^{12} J   **D.** 2.1×10^{13} J

92. • What is the speed a rock needs to be given at the surface of the Earth in order for it to have a residual speed of 2.5 km/s?

    A. 8.7 km/s   **B.** 10.9 km/s
    C. 11.5 km/s   **D.** 13.7 km/s

93. • Would increasing the ejection speed from 12 km/s to 13 km/s change the residual speed by more than, less than, or the same amount as increasing the ejection speed from 15 km/s to 16 km/s?

94. •• Consider a similar plot for rocks ejected from Mars. Where would this plot intercept the *x* axis?

A. The plot for Mars would intercept the *x* axis at 5.0 km/s.
B. The plot for Mars would intercept the *x* axis at 16.2 km/s.
C. The plot for Mars would intercept the *x* axis at 11.2 km/s.
D. The plot for Mars would not intercept the *x* axis.

95. •• **REFERRING TO EXAMPLE 12-8** Find the orbital radius that corresponds to a "year" of 150 days.

96. •• Predict/Calculate **REFERRING TO EXAMPLE 12-8** Suppose the mass of the Sun is suddenly doubled, but the Earth's orbital radius remains the same. **(a)** Would the length of an Earth year increase, decrease, or stay the same? **(b)** Find the length of a year for the case of a Sun with twice the mass. **(c)** Suppose the Sun retains its present mass, but the mass of the Earth is doubled instead. Would the length of the year increase, decrease, or stay the same?

97. •• Predict/Calculate **REFERRING TO EXAMPLE 12-17** **(a)** If the mass of the Earth were doubled, would the escape speed of a rocket increase, decrease, or stay the same? **(b)** Calculate the escape speed of a rocket for the case of an Earth with twice its present mass. **(c)** If the mass of the Earth retains its present value, but the mass of the rocket is doubled, does the escape speed increase, decrease, or stay the same?

98. •• Predict/Calculate **REFERRING TO EXAMPLE 12-17** Suppose the Earth is suddenly shrunk to half its present radius without losing any of its mass. **(a)** Would the escape speed of a rocket increase, decrease, or stay the same? **(b)** Find the escape speed for an Earth with half its present radius.

13

Oscillations About Equilibrium

▲ A struck tuning fork and a plucked guitar string are just two examples of objects that oscillate about their equilibrium positions. In this chapter we explore the behavior of a variety of oscillating systems, including the classic case of a swinging pendulum, and lay the foundations for understanding many other natural phenomena as well.

Big Ideas

1 Simple harmonic motion occurs when the restoring force is proportional to the displacement.

2 Simple harmonic motion is closely related to uniform circular motion.

3 The period of a mass on a spring depends on the mass and the force constant.

4 The period of a pendulum is proportional to the square root of its length.

5 An oscillator driven at its resonance frequency has a large amplitude of oscillation.

A system in equilibrium is seldom left undisturbed for very long, it seems, before it's given a bump, a kick, or a nudge that displaces it from its equilibrium position. When this happens, the system **oscillates** back and forth from one side of the equilibrium position to the other. This chapter explores the basic physics of all oscillating systems, from a pendulum in a grandfather clock, to water molecules in a microwave oven, to the universe oscillating between "big bangs" and "big crunches."

13-1 Periodic Motion

A motion that repeats itself over and over is referred to as **periodic motion**. The beating of your heart, as illustrated in FIGURE 13-1, the ticking of a clock, and the movement of a child on a swing are familiar examples.

Period One of the key characteristics of a periodic system is the time required for the completion of one cycle of its repetitive motion. For example, the pendulum in a grandfather clock might take one second to swing from maximum displacement in one direction to maximum displacement in the opposite direction, and another second to swing back to its original position. Thus, the **period**, T, of this pendulum—the time required to complete one oscillation—is 2 s.

Definition of Period, T

T = time required for one cycle of a periodic motion

SI unit: seconds/cycle = s

Notice that a cycle (that is, an oscillation) is dimensionless.

Frequency Closely related to the period is another common measure of periodic motion, the **frequency**, f. The frequency of an oscillation is simply the number of oscillations per unit of time. Thus, f tells us how frequently, or rapidly, an oscillation takes place—the higher the frequency, the more rapid the oscillations. By definition, the frequency is simply the inverse of the period, T:

Definition of Frequency, f

$$f = \frac{1}{T}$$ 13-1

SI unit: cycle/second = 1/s = s^{-1}

If the period of an oscillation, T, is very small, resulting in rapid oscillations, the corresponding frequency, $f = 1/T$, is large, as expected.

A special unit has been introduced for the measurement of frequency. It is the **hertz** (Hz), named for the German physicist Heinrich Hertz (1857–1894), in honor of his pioneering studies of radio waves. By definition, one Hz is one cycle per second:

$$1 \text{ Hz} = 1 \text{ cycle/second}$$

High frequencies are often measured in kilohertz (kHz), where $1 \text{ kHz} = 10^3 \text{ Hz}$, or megahertz (MHz), where $1 \text{ MHz} = 10^6 \text{ Hz}$.

EXERCISE 13-1 COMPUTER PERIOD

The processing "speed" of a computer refers to the number of binary operations it can perform in 1 second, so it is really a frequency. If the processor of a certain computer operates at 3.70 GHz, how much time is required for one processing cycle?

REASONING AND SOLUTION
The frequency of the computer's processor is given as f = 3.70 GHz. Therefore, we can rearrange $f = 1/T$ to solve for the processing period:

$$T = \frac{1}{f} = \frac{1}{3.70 \text{ GHz}} = \frac{1}{3.70 \times 10^9 \text{ cycles/s}} = 2.70 \times 10^{-10} \text{ s}$$

The high frequency of a processor corresponds to a very small period—less than a billionth of a second to complete one operation. We next consider a situation in which the frequency is considerably smaller.

EXERCISE 13-2 PLAYING CATCH

Two baseball players warm up with a game of catch. If it takes 6.6 s for the ball to go from one player to the other and then back to the first, what is the frequency of the ball's motion?

CONTINUED

▲ **FIGURE 13-1** Periodic phenomena are found everywhere in nature, from the movements of the heavenly bodies to the vibration of individual atoms. The trace of an electrocardiogram (ECG or EKG), as shown here, records the rhythmic electrical activity that accompanies the beating of our hearts.

PHYSICS IN CONTEXT
Looking Ahead

Period, T, and frequency, f, play key roles in the study of sound waves, as shown in Chapter 14. They reappear when we study electromagnetic waves in Chapter 25.

PHYSICS IN CONTEXT
Looking Ahead

In Chapter 30 we study the beginnings of modern physics, and blackbody radiation in particular. As we shall see, the frequency f of light is directly related to the energy of a "particle of light," referred to as a photon.

REASONING AND SOLUTION

The period of this motion is the time for the ball to complete one round trip from one player to the other and back. Therefore, the period is $T = 6.6$ s, and the corresponding frequency is the inverse of the period:

$$f = \frac{1}{T} = \frac{1}{6.6 \text{ s}} = 0.15 \text{ Hz}$$

The period and frequency of periodic motion can vary over a remarkably large range. Table 13-1 gives a sampling of typical values of T and f.

TABLE 13-1 Typical Periods and Frequencies

System	Period (s)	Frequency (Hz)
Precession of the Earth	8.2×10^{11} (26,000 y)	1.2×10^{-12}
Hour hand of a clock	43,200 (12 h)	2.3×10^{-5}
Minute hand of a clock	3600	2.8×10^{-4}
Second hand of clock	60	0.017
Pendulum in grandfather clock	2.0	0.50
Human heartbeat	1.0	1.0
Lower range of human hearing	5.0×10^{-2}	20
Wing beat of housefly	5.0×10^{-3}	200
Upper range of human hearing	5.0×10^{-5}	20,000
Computer processor	5.6×10^{-10}	1.8×10^9

BIO An example of rapid periodic motion in nature is the drumming of a Hairy Woodpecker (*Picoides villosus*) (**FIGURE 13-2**), which can hammer its beak into a tree trunk as many as 25 times a second. It accomplishes this feat by using very strong thigh muscles to rotate its entire body, and then transferring the rotational motion to its head and beak. While the motion of its body can be roughly described by a sine or cosine function, the speed of its head is described by a more complex sawtooth function, achieving its maximum speed just before striking the tree. The rapid, periodic impacts allow the birds to drill for hidden bugs, construct hollowed-out nests, and make the characteristic woodpecker drumming sound, which itself is useful for declaring its territory and attracting a mate.

▲ **FIGURE 13-2** The Hairy Woodpecker can hammer a tree trunk as many as 25 times a second, and perform as many as 12,000 hits in a day. The impacts are severe enough (1000g or more) to render a person unconscious, but are just part of the daily routine for the woodpecker.

Enhance Your Understanding   (Answers given at the end of the chapter)

1. If the frequency of an oscillator is halved, by what factor does the period change?

Section Review

- Period, T, is the amount of time required to complete one cycle of an oscillation.
- Frequency, f, is the number of cycles per second. In terms of the period, $f = 1/T$.

13-2 Simple Harmonic Motion

Periodic motion can take many forms, such as a tennis ball going back and forth between players, or an expectant father pacing up and down a hospital hallway. There is one type of periodic motion, however, that is of particular importance. It is referred to as **simple harmonic motion**.

Oscillations of a Mass on a Spring A classic example of simple harmonic motion is the oscillation of a mass attached to a spring. (See Section 6-2 for a discussion of ideal springs and the forces they exert.) To be specific, consider an air-track cart of mass m

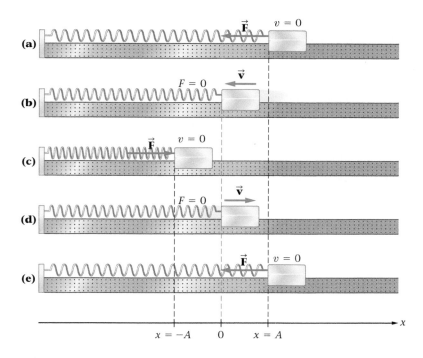

attached to a spring of force constant k, as in **FIGURE 13-3**. When the spring is at its equilibrium length—neither stretched nor compressed—the cart is at the position $x = 0$, where it will remain at rest if left undisturbed. If the cart is displaced from equilibrium by a distance x, however, the spring exerts a restoring force given by Hooke's law, $F_x = -kx$. In words:

> A spring exerts a restoring force whose magnitude is proportional to the distance it is displaced from equilibrium.

This direct proportionality between distance from equilibrium and force is the key feature of a mass–spring system that leads to simple harmonic motion. As for the direction of the spring force:

> The force exerted by a spring is opposite in direction to its displacement from equilibrium; this accounts for the minus sign in $F_x = -kx$. In general, a restoring force is one that *always* points toward the equilibrium position.

Now, suppose the cart is released from rest at the location $x = A$. As indicated in Figure 13-3, the spring exerts a force on the cart to the left, causing the cart to accelerate toward the equilibrium position. When the cart reaches $x = 0$, the net force acting on it is zero. Its speed is not zero at this point, however, and so it continues to move to the left. As the cart compresses the spring, it experiences a force to the right, causing it to decelerate and finally come to rest at $x = -A$. The spring continues to exert a force to the right; thus, the cart immediately begins to move to the right until it comes to rest again at $x = A$, completing one oscillation in the time T.

Position as a Function of Time   If a pen is attached to the cart, it can trace its motion on a strip of paper moving with constant speed, as indicated in **FIGURE 13-4**. On this "strip chart" we obtain a record of the cart's motion as a function of time. As we see in Figure 13-4, the motion of the cart looks like a sine or a cosine function.

Mathematical analysis, using the methods of calculus, shows that this is indeed the case; that is, the position of the cart as a function of time can be represented by a sine or a cosine function. The reason that either function works can be seen by considering **FIGURE 13-5**. If we take $t = 0$ to be at point 1, for example, the position as a function of time starts at zero, just like a sine function; if we choose $t = 0$ to be at point 2, however, the position versus time starts at its maximum value, just like a cosine function. It's really the same mathematical function, differing only in the choice of starting point.

Big Idea **1** Simple harmonic motion occurs when a system's restoring force is proportional to the displacement from equilibrium.

▶ **FIGURE 13-4 Displaying position versus time for simple harmonic motion** As an air-track cart oscillates about its equilibrium position, a pen attached to it traces its motion onto a moving sheet of paper. This produces a "strip chart," showing that the cart's motion has the shape of a sine or a cosine function.

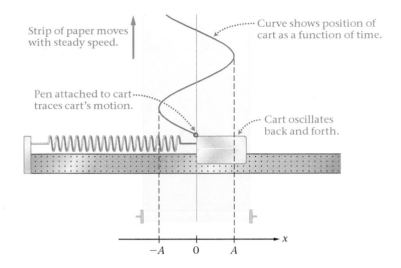

Strip of paper moves with steady speed.

Curve shows position of cart as a function of time.

Pen attached to cart traces cart's motion.

Cart oscillates back and forth.

Returning to Figure 13-3, notice that the position of the mass oscillates between $x = +A$ and $x = -A$. Because A represents the extreme displacement of the cart on either side of equilibrium, we refer to it as the **amplitude** of the motion. It follows that the amplitude is one-half the total range of motion. In addition, recall that the cart's motion repeats with a period T. As a result, the position of the cart is the same at the time $t + T$ as it is at the time t, as shown in Figure 13-5. Combining all these observations results in the following mathematical description of position versus time:

Position Versus Time in Simple Harmonic Motion

$$x = A \cos\left(\frac{2\pi}{T}t\right)$$

13-2

SI unit: m

PROBLEM-SOLVING NOTE

Using Radians

When evaluating Equation 13-2, be sure you have your calculator set to "radians" mode.

This type of dependence on time—as a sine or a cosine—is characteristic of simple harmonic motion. With this particular choice, the position at $t = 0$ is $x = A \cos(0) = A$; thus, Equation 13-2 describes an object that has its maximum displacement at $t = 0$, like point 2 in Figure 13-5.

To see how Equation 13-2 works, recall that the cosine oscillates between $+1$ and -1. Therefore, $x = A \cos(2\pi t/T)$ will oscillate between $+A$ and $-A$, just as in the strip chart. Next, consider what happens if we replace the time t with the time $t + T$. This gives

$$x = A \cos\left(\frac{2\pi}{T}(t + T)\right)$$
$$= A \cos\left(\frac{2\pi}{T}t + \frac{2\pi}{T}T\right) = A \cos\left(\frac{2\pi}{T}t + 2\pi\right)$$

▶ **FIGURE 13-5 Simple harmonic motion as a sine or a cosine** The strip chart from Figure 13-4. The cart oscillates back and forth from $x = +A$ to $x = -A$, completing one cycle in the time T. The function traced by the pen can be represented as a sine function if $t = 0$ is taken to be at point 1, where the function is equal to zero. The function can be represented by a cosine if $t = 0$ is taken to be at point 2, where the function has its maximum value.

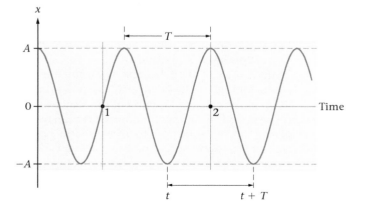

Using the fact that $\cos(\theta + 2\pi) = \cos\theta$ for any angle θ, we can rewrite the last expression as follows:

$$x = A\cos\left(\frac{2\pi}{T}t\right)$$

Therefore, as expected, the position at time $t + T$ is precisely the same as the position at time t.

EXAMPLE 13-3   SPRING TIME

An air-track cart attached to a spring completes one oscillation every 2.4 s. At $t = 0$ the cart is released from rest at a distance of 0.10 m from its equilibrium position. What is the position of the cart at (a) 0.30 s, (b) 0.60 s, (c) 2.7 s, and (d) 3.0 s?

PICTURE THE PROBLEM

In our sketch, we place the origin of the x-axis at the equilibrium position of the cart and choose the positive direction to point to the right. The cart is released from rest at $x = 0.10$ m, which means that its amplitude is $A = 0.10$ m. After it is released, the cart oscillates back and forth between $x = 0.10$ m and $x = -0.10$ m.

$x = -0.10$ m 0 $x = 0.10$ m

REASONING AND STRATEGY

Notice that the period of oscillation, $T = 2.4$ s, is given in the problem statement. Thus, we can find the position of the cart by evaluating $x = A\cos(2\pi t/T)$ at the desired times, using $A = 0.10$ m. (*Note*: Remember to have your calculator set to radian mode when evaluating these cosine functions.)

Known   Period of cart, $T = 2.4$ s; release point, $x = 0.10$ m.
Unknown   Position of cart at indicated times, $x = ?$

SOLUTION

Part (a)
1. Calculate x at the time $t = 0.30$ s:

$$x = A\cos\left(\frac{2\pi}{T}t\right) = (0.10\,\text{m})\cos\left[\left(\frac{2\pi}{2.4\,\text{s}}\right)(0.30\,\text{s})\right]$$
$$= (0.10\,\text{m})\cos(\pi/4) = 7.1\,\text{cm}$$

Part (b)
2. Now, substitute $t = 0.60$ s:

$$x = A\cos\left(\frac{2\pi}{T}t\right) = (0.10\,\text{m})\cos\left[\left(\frac{2\pi}{2.4\,\text{s}}\right)(0.60\,\text{s})\right]$$
$$= (0.10\,\text{m})\cos(\pi/2) = 0$$

Part (c)
3. Repeat with $t = 2.7$ s:

$$x = A\cos\left(\frac{2\pi}{T}t\right) = (0.10\,\text{m})\cos\left[\left(\frac{2\pi}{2.4\,\text{s}}\right)(2.7\,\text{s})\right]$$
$$= (0.10\,\text{m})\cos(9\pi/4) = 7.1\,\text{cm}$$

Part (d)
4. Repeat with $t = 3.0$ s:

$$x = A\cos\left(\frac{2\pi}{T}t\right) = (0.10\,\text{m})\cos\left[\left(\frac{2\pi}{2.4\,\text{s}}\right)(3.0\,\text{s})\right]$$
$$= (0.10\,\text{m})\cos(5\pi/2) = 0$$

INSIGHT

The results for parts (c) and (d) are the same as for parts (a) and (b), respectively. This is because the times in (c) and (d) are greater than the corresponding times in (a) and (b) by one period; that is, 2.7 s = 0.30 s + 2.4 s and 3.0 s = 0.60 s + 2.4 s.

PRACTICE PROBLEM

What is the first time the cart is at the position $x = -5.0$ cm?   [**Answer:** $t = T/3 = 0.80$ s]

Some related homework problems: Problem 10, Problem 17, Problem 18

Enhance Your Understanding   (Answers given at the end of the chapter)

2. Which of the following equations corresponds to the position-versus-time graph shown in **FIGURE 13-6**? In each equation, x is measured in meters and t is measured in

CONTINUED

seconds. **(a)** $x = (0.1\,\text{m})\cos(2\pi t)$, **(b)** $x = (0.2\,\text{m})\cos(2\pi t)$, **(c)** $x = (0.1\,\text{m})\cos(\pi t)$, **(d)** $x = (0.2\,\text{m})\cos(\pi t)$.

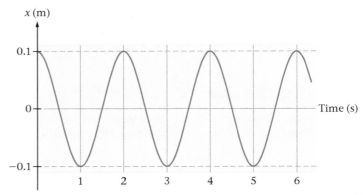

▲ **FIGURE 13-6** Enhance Your Understanding 2.

Section Review

- Simple harmonic motion can be represented mathematically by either a sine or a cosine function. In terms of the cosine function, we have $x = A\cos(2\pi t/T)$, where x is the position at time t, A is the amplitude (maximum displacement), and T is the period.

▲ **FIGURE 13-7 The relationship between uniform circular motion and simple harmonic motion** A peg is placed at the rim of a turntable that rotates with constant angular velocity. Viewed from above, the peg exhibits uniform circular motion. If the peg is viewed from the side, however, it appears to move back and forth in a straight line, as we can see by shining a light to cast a shadow of the peg onto a screen. The shadow moves with simple harmonic motion. If we compare this motion with the behavior of a mass on a spring, moving with the same period as the turntable and having an amplitude of motion equal to the radius of the turntable, we find that the mass and the shadow of the peg move together in simple harmonic motion.

13-3 Connections Between Uniform Circular Motion and Simple Harmonic Motion

Though the motion of the air-track cart is strictly one-dimensional, it bears a close relationship to uniform circular motion. In this section, we explore this connection between simple harmonic motion and uniform circular motion in detail.

Uniform Circular Motion   Imagine a turntable that rotates with a constant angular speed $\omega = 2\pi/T$, taking the time T to complete a revolution. At the rim of the turntable we place a small peg, as indicated in **FIGURE 13-7**. If we view the turntable from above, we see the peg undergoing uniform circular motion.

On the other hand, suppose we view the turntable from the side, so that the peg appears to move back and forth. Perhaps the easiest way to view this motion is to shine a light that casts a shadow of the peg on a screen, as shown in Figure 13-7. While the peg itself moves on a circular path, its shadow moves back and forth in a straight line.

To be specific, let the radius of the turntable be $r = A$, so that the shadow moves from $x = +A$ to $x = -A$. When the shadow is at $x = +A$, release a mass on a spring that is also at $x = +A$, so that the mass and the shadow start together. If we adjust the period of the turntable so that it completes one revolution in the same time T that the mass completes one oscillation, we find that the mass and the shadow move as one for all times. This is also illustrated in Figure 13-7. Recalling that the mass undergoes simple harmonic motion, it follows that the shadow does so as well.

We now use this connection, plus our knowledge of circular motion, to obtain detailed results for the position, velocity, and acceleration of a particle undergoing simple harmonic motion.

Position

In **FIGURE 13-8** we show the peg at the angular position θ, where θ is measured relative to the x axis. If the peg starts at $\theta = 0$ at $t = 0$, and the turntable rotates with a constant angular speed ω, we know from Equation 10-10 that the angular position of the peg is simply

$$\theta = \omega t \qquad \text{13-3}$$

That is, the angular position increases linearly with time.

Now, imagine drawing a radius vector of length A to the position of the peg, as indicated in Figure 13-8. When we project the shadow of the peg onto the screen, the shadow is at the location $x = A \cos \theta$, which is the x component of the radius vector. Therefore, the position of the shadow as a function of time is:

Position of Shadow as a Function of Time

$$x = A \cos \theta = A \cos(\omega t) = A \cos\left(\frac{2\pi}{T}t\right) \qquad \text{13-4}$$

SI unit: m

Notice that we have used $\theta = \omega t$ (Equation 13-3) to express θ in terms of the time, t. Equations 13-4 and 13-2 are identical, so the shadow does indeed exhibit simple harmonic motion, just like a mass on a spring.

For notational simplicity, we often write the position of a mass on a spring in the form $x = A \cos(\omega t)$, which is more compact than $x = A \cos(2\pi t/T)$. When referring to a rotating turntable, ω is called the angular speed; when referring to simple harmonic motion, or other periodic motion, we have a slightly different name for ω. In these situations, ω is called the **angular frequency**:

Definition of Angular Frequency, ω

$$\omega = 2\pi f = \frac{2\pi}{T} \qquad \text{13-5}$$

SI unit: rad/s = s^{-1}

Velocity

We can find the velocity of the shadow in the same way that we determined its position; first find the velocity of the peg, then take its x component. The result of this calculation is the velocity as a function of time for simple harmonic motion.

To begin, recall that the velocity of an object in uniform circular motion of radius r has a magnitude equal to

$$v = r\omega$$

In addition, the velocity is tangential to the object's circular path, as indicated in FIGURE 13-9 (a). Therefore, referring to the figure, we see that when the peg is at the angular position θ, the velocity vector makes an angle θ with the vertical. As a result, the x component of the velocity is $-v \sin \theta$. Combining these results, we find that the velocity of the peg, along the x axis, is

$$v_x = -v \sin \theta = -r\omega \sin \theta$$

In what follows we shall drop the x subscript, since we know that the shadow and a mass on a spring move only along the x axis. Recalling that $r = A$ and $\theta = \omega t$, we have:

Velocity in Simple Harmonic Motion

$$v = -A\omega \sin(\omega t) \qquad \text{13-6}$$

SI unit: m/s

We plot x and v for simple harmonic motion in FIGURE 13-9 (b). Notice that when the displacement from equilibrium is a maximum, the velocity is zero. This is to be expected, because at $x = +A$ and $x = -A$ the object is momentarily at rest as it turns around. Appropriately, then, these points are referred to as **turning points** of the motion.

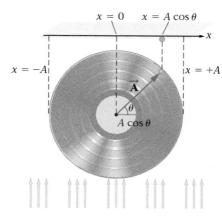

▲ FIGURE 13-8 **Position versus time in simple harmonic motion** A peg rotates on the rim of a turntable of radius A. When the peg is at the angular position θ, its shadow is at $x = A \cos \theta$. Note that $A \cos \theta$ is also the x component of the radius vector $\vec{A}$ from the center of the turntable to the peg.

Big Idea 2 Simple harmonic motion is closely related to uniform circular motion. In fact, the x or y component of an object in uniform circular motion has the same time dependence as an object in simple harmonic motion.

PHYSICS IN CONTEXT
Looking Back

Uniform circular motion (Chapter 10) is used to better understand simple harmonic motion.

▶ FIGURE 13-9 **Velocity versus time in simple harmonic motion** (a) The velocity of a peg rotating on the rim of a turntable is tangential to its circular path. As a result, when the peg is at the angle θ, its velocity makes an angle of θ with the vertical. The x component of the velocity, then, is $-v \sin \theta$. (b) Position, x, and velocity, v, as a function of time for simple harmonic motion. The speed is greatest when the object passes through equilibrium, $x = 0$. On the other hand, the speed is zero when the position is greatest—that is, at the turning points. Finally, note that as x moves in the negative direction, the velocity is negative. Similar remarks apply to the positive direction.

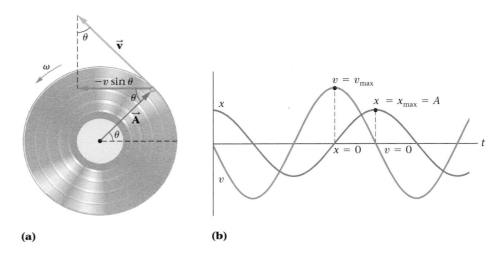

(a) **(b)**

On the other hand, the speed is a maximum when the displacement from equilibrium is zero. Similarly, a mass on a spring is moving with its greatest speed as it goes through $x = 0$. From the expression $v = -A\omega \sin(\omega t)$, and the fact that the largest value of $\sin \theta$ is 1, we see that the maximum speed of the mass is

$$v_{max} = A\omega \qquad \text{13-7}$$

After the mass passes $x = 0$ it begins either to compress or to stretch the spring, and hence it slows down.

Acceleration

The acceleration of an object in uniform circular motion has a magnitude given by

$$a_{cp} = r\omega^2$$

In addition, the direction of the acceleration is toward the center of the circular path, as indicated in FIGURE 13-10 (a). Thus, when the angular position of the peg is θ, the acceleration vector is at an angle θ below the x axis, and its x component is $-a_{cp} \cos \theta$. Once again setting $r = A$ and $\theta = \omega t$, we find:

PROBLEM-SOLVING NOTE

Remember to Use Radians

Note that Equations 13-4, 13-6, and 13-8 must all be evaluated in terms of radians.

Acceleration in Simple Harmonic Motion

$$a = -A\omega^2 \cos(\omega t) \qquad \text{13-8}$$

SI unit: m/s^2

The Connection Between Acceleration and Position   The position and acceleration for simple harmonic motion are plotted in FIGURE 13-10 (b). If you look closely, you'll see that the acceleration and position vary with time in the same way but with opposite signs. That is, when the position has its maximum *positive* value, the acceleration has its maximum *negative* value, and so on. After all, the restoring force of the spring is

▶ FIGURE 13-10 **Acceleration versus time in simple harmonic motion** (a) The acceleration of a peg on the rim of a uniformly rotating turntable is directed toward the center of the turntable. Hence, when the peg is at the angle θ, the acceleration makes an angle θ with the horizontal. The x component of the acceleration is $-a \cos \theta$. (b) Position, x, and acceleration, a, as a function of time for simple harmonic motion. Note that when the position has its greatest positive value, the acceleration has its greatest negative value.

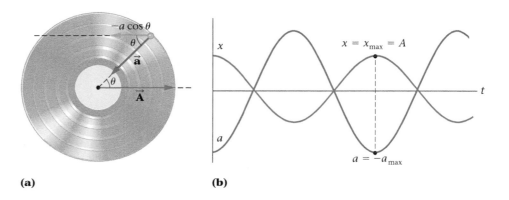

(a) **(b)**

opposite to the position, and hence the acceleration, $a = F/m$, must also be opposite to the position. In fact, comparing $x = A\cos(\omega t)$ (Equation 13-4) and $a = -A\omega^2\cos(\omega t)$ (Equation 13-8), we see that the acceleration can be written as

$$a = -\omega^2 x$$

This expression is valid at all times. In addition, noting that the largest value of x is the amplitude A, we see that the maximum acceleration has the following magnitude:

$$a_{max} = A\omega^2 \qquad\qquad \text{13-9}$$

We conclude this section with a few examples using position, velocity, and acceleration in simple harmonic motion.

EXAMPLE 13-4 VELOCITY AND ACCELERATION

As in Example 13-3, an air-track cart attached to a spring completes one oscillation every 2.4 s. At $t = 0$ the cart is released from rest with the spring stretched 0.10 m from its equilibrium position. What are the velocity and acceleration of the cart at **(a)** 0.30 s and **(b)** 0.60 s?

PICTURE THE PROBLEM

Once again, we place the origin of the x axis at the equilibrium position of the cart, with the positive direction pointing to the right. In addition, the cart is released from rest at $x = 0.10$ m, which means that the amplitude of its motion is $A = 0.10$ m. It follows that the cart has zero speed at $x = 0.10$ m and $x = -0.10$ m. Its speed will be a maximum at $x = 0$, however, which is also the point where the acceleration is zero.

REASONING AND STRATEGY

After calculating the angular frequency, $\omega = 2\pi/T$, we simply substitute $t = 0.30$ s and $t = 0.60$ s into $v = -A\omega\sin(\omega t)$ and $a = -A\omega^2\cos(\omega t)$. Remember to set your calculators to radian mode.

Known Period of cart, $T = 2.4$ s; release point, $x = 0.10$ m.
Unknown Velocity and acceleration ($v = ?, a = ?$) at indicated times

SOLUTION

1. Calculate the angular frequency for this motion:

$$\omega = \frac{2\pi}{T} = \frac{2\pi}{(2.4\text{ s})} = 2.6\text{ rad/s}$$

Part (a)

2. Calculate v at the time $t = 0.30$ s. Express ω in terms of π—that is, $\omega = 2\pi/(2.4\text{ s})$. This isn't necessary, but it makes it easier to evaluate the sine and cosine functions (see the discussion following the Example for additional details):

$$v = -A\omega\sin(\omega t)$$
$$= -(0.10\text{ m})(2.6\text{ rad/s})\sin\left[\left(\frac{2\pi}{2.4\text{ s}}\right)(0.30\text{ s})\right]$$
$$= -(26\text{ cm/s})\sin(\pi/4) = -18\text{ cm/s}$$

3. Similarly, calculate a at $t = 0.30$ s:

$$a = -A\omega^2\cos(\omega t)$$
$$= -(0.10\text{ m})(2.6\text{ rad/s})^2\cos\left[\left(\frac{2\pi}{2.4\text{ s}}\right)(0.30\text{ s})\right]$$
$$= -(68\text{ cm/s}^2)\cos\left(\frac{\pi}{4}\right) = -48\text{ cm/s}^2$$

Part (b)

4. Calculate v at the time $t = 0.60$ s:

$$v = -A\omega\sin(\omega t)$$
$$= -(0.10\text{ m})(2.6\text{ rad/s})\sin\left[\left(\frac{2\pi}{2.4\text{ s}}\right)(0.60\text{ s})\right]$$
$$= -(26\text{ cm/s})\sin\left(\frac{\pi}{2}\right) = -26\text{ cm/s}$$

5. Similarly, calculate a at $t = 0.60$ s:

$$a = -A\omega^2\cos(\omega t)$$
$$= -(0.10\text{ m})(2.6\text{ rad/s})^2\cos\left[\left(\frac{2\pi}{2.4\text{ s}}\right)(0.60\text{ s})\right]$$
$$= -(68\text{ cm/s}^2)\cos\left(\frac{\pi}{2}\right) = 0$$

CONTINUED

INSIGHT
Did you notice that the cart speeds up from $t = 0.30$ s to $t = 0.60$ s? In fact, the maximum speed of the cart, $v_{max} = A\omega = 26$ cm/s, occurs at $t = 0.60$ s $= T/4$. Referring to Example 13-3, we can see that this is precisely the time when the cart is at the equilibrium position, $x = 0$. As expected, the acceleration is zero at this time.

PRACTICE PROBLEM
What is the first time the velocity of the cart is $+26$ cm/s? [**Answer:** $t = 3T/4 = 1.8$ s]

Some related homework problems: Problem 21, Problem 22

PROBLEM-SOLVING NOTE

Expressing Time in Terms of the Period

When evaluating the expression $x = A\cos(2\pi t/T)$ at the time t, it is often helpful to express t in terms of the period T.

In problems like the preceding Example, it's often useful to express the time t in terms of the period, T. For example, if the period is $T = 2.4$ s it follows that $t = 0.60$ s is $T/4$. Thus, the angular frequency multiplied by the time is

$$\omega t = \left(\frac{2\pi}{T}\right)\left(\frac{T}{4}\right) = \frac{\pi}{2}$$

This result was used in Steps 4 and 5 in Example 13-4. Using $\omega t = \pi/2$ we find that the position of the cart is

$$x = A\cos(\omega t) = A\cos\left(\frac{\pi}{2}\right) = 0$$

Similarly, the velocity of the cart is

$$v = -A\omega\sin(\omega t) = -A\omega\sin\left(\frac{\pi}{2}\right) = -A\omega$$

The acceleration of the cart is

$$a = -A\omega^2\cos(\omega t) = -A\omega^2\cos\left(\frac{\pi}{2}\right) = 0$$

When expressed in this way, it's clear why x and a are zero and why v has its maximum negative value.

EXAMPLE 13-5   TURBULENCE!

On December 29, 1997, a United Airlines flight from Tokyo to Honolulu was hit with severe turbulence 31 minutes after takeoff. Data from the airplane's "black box" indicated that the 747 moved up and down with an amplitude of 30.0 m and a maximum acceleration of 1.8g. Treating the up-and-down motion of the plane as simple harmonic motion, find **(a)** the time required for one complete oscillation and **(b)** the plane's maximum vertical speed.

PICTURE THE PROBLEM
Our sketch shows the 747 airliner moving up and down with an amplitude of $A = 30.0$ m relative to its normal horizontal flight path. The period of motion, T, is the time required for one complete cycle of this up-and-down motion.

REASONING AND STRATEGY
We are given the maximum acceleration and the amplitude of motion. With these quantities, and the basic equations of simple harmonic motion, we can determine the other characteristics of the motion.

a. We know that the maximum acceleration of simple harmonic motion is $a_{max} = A\omega^2$ (Equation 13-9). This relationship can be solved for ω in terms of the known quantities a_{max} and A. We can then rearrange $\omega = 2\pi/T$ to solve for the period of motion, T.

b. The maximum vertical speed is found using $v_{max} = A\omega$ (Equation 13-7).

Known   Amplitude of motion, $A = 30.0$ m; maximum vertical acceleration, $a_{max} = 1.8g$.
Unknown   **(a)** Period of motion, $T = ?$ **(b)** Maximum vertical speed, $v_{max} = ?$

CONTINUED

SOLUTION

Part (a)

1. Relate a_{max} to the angular frequency, ω:

$$a_{max} = A\omega^2$$

2. Solve for ω, and express in terms of T:

$$\omega = \sqrt{a_{max}/A} = 2\pi/T$$

3. Solve for T and substitute numerical values for g and A:

$$T = \frac{2\pi}{\sqrt{a_{max}/A}}$$

$$= \frac{2\pi}{\sqrt{1.8g/A}} = \frac{2\pi}{\sqrt{1.8(9.81 \text{ m/s}^2)/(30.0 \text{ m})}} = 8.2 \text{ s}$$

Part (b)

4. Calculate the maximum vertical speed using $v_{max} = A\omega = 2\pi A/T$:

$$v_{max} = A\omega = \frac{2\pi A}{T} = \frac{2\pi(30.0 \text{ m})}{8.2 \text{ s}} = 23 \text{ m/s}$$

INSIGHT

We don't expect the up-and-down motion of the plane to be exactly simple harmonic motion—after all, it's unlikely that the plane's path has precisely the shape of a cosine function. Still, the approximation is reasonable. In fact, many systems are well approximated by simple harmonic motion, and hence the equations derived in this section are used widely in physics.

Notice that the maximum vertical speed of the passengers (23 m/s) is roughly 50 mi/h, and that this speed is first in the upward direction and then 4.1 s later in the downward direction.

PRACTICE PROBLEM

What amplitude of motion would result in a maximum acceleration of 0.50g, with everything else remaining the same?
[**Answer:** $A = 0.50g/\omega^2 = 0.50gT^2/4\pi^2 = 8.4$ m]

Some related homework problems: Problem 21, Problem 23

EXAMPLE 13-6   A BOBBING DRAGONFLY

When dragonflies hover, they tend to "bob" up and down with a motion that is a very good approximation to simple harmonic motion. Suppose that a dragonfly is at $x = A$ at time $t = 0$, and that its bobbing motion has an amplitude of $A = 0.40$ cm and a period of $T = 0.24$ s. Find **(a)** the position, **(b)** the velocity, and **(c)** the acceleration of the dragonfly at the time $t = 0.10$ s.

PICTURE THE PROBLEM

Our sketch shows a hovering dragonfly bobbing up and down along the x axis between $x = +A$ and $x = -A$. The positive x direction is chosen to be upward, and the dragonfly is at $x = +A$ at $t = 0$.

Position of dragonfly at $t = 0$

x

$x = +A$

0

$x = -A$

REASONING AND STRATEGY

The dragonfly bobs up and down in simple harmonic motion with an amplitude $A = 0.40$ cm and a period $T = 0.24$ s. It follows that the position of the dragonfly as a function of time is $x = A\cos(\omega t)$, where the angular frequency is $\omega = 2\pi/T$. (Notice that $x = A$ at $t = 0$, in agreement with the problem statement.) Similarly, the velocity is $v = -A\omega\sin(\omega t)$ and the acceleration is $a = -A\omega^2\cos(\omega t)$. These expressions are to be evaluated at $t = 0.10$ s.

Known   Amplitude of motion, $A = 0.40$ cm; period of bobbing, $T = 0.24$ s; time, $t = 0.10$ s.
Unknown   **(a)** Position, $x = ?$ **(b)** Velocity, $v = ?$ **(c)** Acceleration, $a = ?$

SOLUTION

1. Evaluate ωt at the time $t = 0.10$ s. This quantity will be used in evaluating x, v, and a:

$$\omega t = \left(\frac{2\pi}{T}\right)t = \left(\frac{2\pi}{0.24 \text{ s}}\right)(0.10 \text{ s}) = 2.6 \text{ rad}$$

Part (a)

2. Evaluate $x = A\cos(\omega t)$ with $\omega t = 2.6$ rad, which corresponds to $t = 0.10$ s:

$$x = A\cos(\omega t)$$
$$= (0.40 \text{ cm})\cos(2.6 \text{ rad}) = -0.34 \text{ cm}$$

Part (b)

3. Evaluate $v = -A\omega\sin(\omega t)$ with $\omega t = 2.6$ rad:

$$v = -A\omega\sin(\omega t)$$
$$= -(0.40 \text{ cm})\left(\frac{2\pi}{0.24 \text{ s}}\right)\sin(2.6 \text{ rad}) = -5.4 \text{ cm/s}$$

CONTINUED

Part (c)

4. Evaluate $a = -A\omega^2 \cos(\omega t)$ with $\omega t = 2.6$ rad:

$$a = -A\omega^2 \cos(\omega t)$$
$$= -(0.40 \text{ cm})\left(\frac{2\pi}{0.24 \text{ s}}\right)^2 \cos(2.6 \text{ rad}) = 230 \text{ cm/s}^2 = 2.3 \text{ m/s}^2$$

INSIGHT
At $t = 0.10$ s the dragonfly is almost at the bottom of its harmonic motion, which occurs at $t = T/2 = 0.12$ s. Hence, it makes sense that its position is negative, its velocity is small and negative, and its acceleration is positive.

PRACTICE PROBLEM
What are x, v, and a at the time $t = T/2 = 0.12$ s?   [**Answer:** $x = -0.40$ cm, $v = 0$, and $a = 2.7$ m/s^2]

Some related homework problems: Problem 21, Problem 23, Problem 27

Enhance Your Understanding (Answers given at the end of the chapter)

3. An object moves with simple harmonic motion about $x = 0$. At one instant of time its position is $x = -0.1$ m and its velocity is 0. Is the acceleration of the object at this time positive, negative, or zero? Explain.

Section Review

- The position, velocity, and acceleration in simple harmonic motion are given by $x = A \cos(\omega t)$, $v = -A\omega \sin(\omega t)$, and $a = -A\omega^2 \cos(\omega t)$, respectively, where A is the amplitude and $\omega = 2\pi/T$.

13-4 The Period of a Mass on a Spring

In this section, we show how the period of a mass oscillating on a spring is related to the mass, m, the force constant of the spring, k, and the amplitude of motion, A. As a first step, we note that the net force acting on the mass at the position x is

$$F = -kx$$

Next, we apply Newton's second law, $F = ma$, to find the following:

$$ma = -kx$$

Substituting the time dependence of x and a, as given in Equations 13-4 and 13-8 in the previous section, we find

$$m[-A\omega^2 \cos(\omega t)] = -k[A \cos(\omega t)]$$

Canceling $-A \cos(\omega t)$ from each side of the equation yields

$$\omega^2 = \frac{k}{m}$$

Taking the square root on both sides of the equation yields

$$\omega = \sqrt{\frac{k}{m}} \qquad \text{13-10}$$

Big Idea 3 The period of a mass oscillating on an ideal spring is proportional to the square root of the mass, and inversely protional to the square root of the force constant. The period is independent of the amplitude of motion.

Finally, recalling that $\omega = 2\pi/T$, it follows that the period of a mass oscillating on a spring is:

Period of a Mass on a Spring

$$T = 2\pi \sqrt{\frac{m}{k}} \qquad \text{13-11}$$

SI unit: s

PHYSICS IN CONTEXT
Looking Back

Newton's second law, $F = ma$ (Chapter 5), played a key role in our discussion of a mass on a spring. We also used the force law for a spring, $F = -kx$ (Chapter 6).

Notice that the period is independent of the amplitude of motion, A. This is a key feature of simple harmonic motion.

EXERCISE 13-7   MASS, FORCE CONSTANT, AND PERIOD

A horizontal spring with a force constant of 11 N/m is connected to an air-track cart. When the cart is set in motion, the spring causes it to oscillate back and forth with a period of 1.2 s. What is the mass of the cart?

REASONING AND SOLUTION

The period of an object attached to a spring is related to the mass of the object, m, and the force constant of the spring, k, by the equation $T = 2\pi\sqrt{m/k}$. In this case we know k and T, and hence we can rearrange the equation to solve for the mass:

$$m = \frac{T^2 k}{4\pi^2} = \frac{(1.2\ \text{s})^2(11\ \text{N/m})}{4\pi^2} = 0.40\ \text{kg}$$

▲ **FIGURE 13-11** Because astronauts are in free fall when orbiting the Earth, they behave as if they were "weightless." It is nevertheless possible to determine the mass of an astronaut by exploiting the properties of oscillatory motion. The Body Mass Measurement Device (BMMD), into which astronaut Tamara Jernigan is strapped, is attached to a spring. If the force constant of the spring is known, her mass can be determined simply by measuring the period with which she rocks back and forth.

As one might expect, the period increases with the mass and decreases with the spring's force constant. For example, a larger mass has greater inertia, and hence it takes longer for the mass to move back and forth through an oscillation. On the other hand, a larger value of the force constant, k, indicates a stiffer spring. Clearly, a mass on a stiff spring completes an oscillation in less time than one on a soft, squishy spring.

RWP* The relationship between mass and period given by $T = 2\pi\sqrt{m/k}$ is used by NASA to measure the mass of astronauts in orbit. Recall that astronauts are in free fall as they orbit, as was discussed in Chapter 12, and therefore they are "weightless." As a result, they cannot simply step onto a bathroom scale to determine their mass. Thus, NASA has developed a device, known as the Body Mass Measurement Device (BMMD), to get around this problem. The BMMD is basically a spring attached to a chair, into which an astronaut is strapped, as shown in **FIGURE 13-11**. As the astronaut oscillates back and forth, the period of oscillation is measured. Since the force constant, k, and the period of oscillation, T, are known, the astronaut's mass can be determined in the same way we determined the mass of the air-track cart in Exercise 13-7.

Period and Amplitude   The period, $T = 2\pi\sqrt{m/k}$, is independent of the amplitude, A, which canceled in the derivation of this equation. This might seem counterintuitive at first: Shouldn't it take more time for a mass to cover the greater distance implied by a larger amplitude? While it's true that a mass will cover a greater distance when the amplitude is increased, it's also true that a larger amplitude implies a larger force exerted by the spring. With a greater force acting on it, the mass moves more rapidly; in fact, the speed of the mass is increased just enough that it covers the greater distance in precisely the same time.

PHYSICS IN CONTEXT
Looking Ahead

Though a mass on a spring may seem far removed from an electric circuit, we show in Chapter 24 that there is in fact a deep connection. In particular, the motion of a mass on a spring is directly analogous to the current in a specific type of circuit referred to as an *RLC* circuit.

EXAMPLE 13-8   SPRING INTO MOTION

When a 0.120-kg air-track cart is attached to a horizontal spring it oscillates with an amplitude of 0.0750 m and a maximum speed of 0.524 m/s. Find **(a)** the force constant and **(b)** the period of motion.

PICTURE THE PROBLEM

Our sketch shows a mass oscillating about the equilibrium position of a spring, which we place at $x = 0$. The amplitude of the oscillations is 0.0750 m, and therefore the mass moves back and forth between $x = 0.0750$ m and $x = -0.0750$ m. The maximum speed of the mass, which occurs at $x = 0$, is $v_{\text{max}} = 0.524$ m/s.

$x = -0.0750\ \text{m} \qquad 0 \qquad x = 0.0750\ \text{m}$

REASONING AND STRATEGY

a. To find the force constant, we first use the maximum speed, $v_{\text{max}} = A\omega$ (Equation 13-7), to determine the angular frequency ω. Once we know ω, we can obtain the force constant k with $\omega = \sqrt{k/m}$ (Equation 13-10).

b. We can find the period from the angular frequency, using $\omega = 2\pi/T$. Alternatively, we can use the force constant and the mass in $T = 2\pi\sqrt{m/k}$ (Equation 13-11).

Known   Mass of cart, $m = 0.120$ kg; amplitude of motion, $A = 0.0750$ m; maximum speed of cart, $v_{\text{max}} = 0.524$ m/s.
Unknown   **(a)** Force constant, $k = ?$ **(b)** Period of motion, $T = ?$

CONTINUED

---

*Real World Physics applications are denoted by the acronym RWP.

SOLUTION

Part (a)

1. Calculate the angular frequency in terms of the maximum speed:

$$v_{max} = A\omega$$

$$\omega = \frac{v_{max}}{A} = \frac{0.524 \text{ m/s}}{0.0750 \text{ m}} = 6.99 \text{ rad/s}$$

2. Solve $\omega = \sqrt{k/m}$ for the force constant:

$$\omega = \sqrt{\frac{k}{m}}$$

$$k = m\omega^2 = (0.120 \text{ kg})(6.99 \text{ rad/s})^2 = 5.86 \text{ N/m}$$

Part (b)

3. Use $\omega = 2\pi/T$ to find the period:

$$\omega = \frac{2\pi}{T}$$

$$T = \frac{2\pi}{\omega} = \frac{2\pi}{6.99 \text{ rad/s}} = 0.899 \text{ s}$$

4. Alternatively, use $T = 2\pi\sqrt{m/k}$ to find the period:

$$T = 2\pi\sqrt{\frac{m}{k}} = 2\pi\sqrt{\frac{0.120 \text{ kg}}{5.86 \text{ N/m}}} = 0.899 \text{ s}$$

INSIGHT

What would happen if we were to attach a greater mass to this same spring and release it with the same amplitude? The answer is that the period would increase. As a result, the angular frequency ω would decrease, and the maximum speed, $v_{max} = A\omega$, would decrease as well.

PRACTICE PROBLEM

What is the maximum acceleration of the mass described in this Example?   [**Answer:** $a_{max} = A\omega^2 = 3.66 \text{ m/s}^2$]

Some related homework problems: Problem 36, Problem 39

These relationships between the motion of a mass on a spring and the mass, the force constant, and the amplitude are summarized in **FIGURE 13-12**.

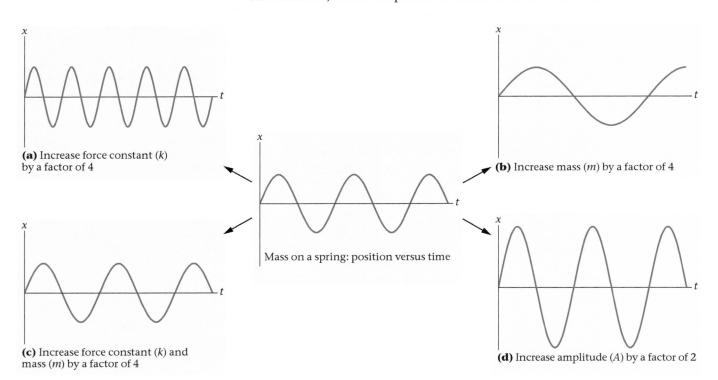

(a) Increase force constant (k) by a factor of 4

(b) Increase mass (m) by a factor of 4

Mass on a spring: position versus time

(c) Increase force constant (k) and mass (m) by a factor of 4

(d) Increase amplitude (A) by a factor of 2

▲ **FIGURE 13-12 Factors affecting the motion of a mass on a spring** The motion of a mass on a spring is determined by the force constant of the spring, k, the mass, m, and the amplitude, A. **(a)** Increasing the force constant causes the mass to oscillate with a greater frequency. **(b)** Increasing the mass lowers the frequency of oscillation. **(c)** If the force constant and the mass are both increased by the same factor, the effects described in parts (a) and (b) cancel, resulting in no change in the motion. **(d)** An increase in amplitude has no effect on the oscillation frequency. However, it increases the maximum speed and maximum acceleration of the mass.

QUICK EXAMPLE 13-9   MASS ON A SPRING: FIND THE FORCE CONSTANT AND THE MASS

When a 0.420-kg air-track cart is attached to a horizontal spring, it oscillates with a period of 0.350 s. If, instead, a different cart with mass m_2 is attached to the same spring, it oscillates with a period of 0.700 s. Find **(a)** the force constant of the spring and **(b)** the mass m_2.

REASONING AND STRATEGY

The information about the first cart allows us to determine the force constant of the spring. Once we know that, we can use the equation for the period of a mass on a spring to find the second mass.

Part (a)

1. Let the initial mass and period be m_1 and T_1, respectively:

$$m_1 = 0.420 \text{ kg}$$
$$T_1 = 0.350 \text{ s}$$

2. Use $T = 2\pi\sqrt{\dfrac{m}{k}}$ to write an expression for T_1:

$$T_1 = 2\pi\sqrt{\dfrac{m_1}{k}}$$

3. Solve this expression for the force constant, k:

$$k = \dfrac{4\pi^2 m_1}{T_1{}^2} = 135 \text{ N/m}$$

Part (b)

4. Write an expression for T_2:

$$T_2 = 2\pi\sqrt{\dfrac{m_2}{k}}$$

5. Solve this expression for m_2:

$$m_2 = \dfrac{kT_2{}^2}{4\pi^2} = 1.68 \text{ kg}$$

To double the period of a mass on a given spring—as in this case where it went from 0.350 s to 0.700 s—the mass must be increased by a factor of 4, in accordance with $T = 2\pi\sqrt{m/k}$. Therefore, an alternative way to find the second mass is $m_2 = 4(0.420 \text{ kg}) = 1.68 \text{ kg}$, in agreement with our result in Step 5.

A Vertical Spring

To this point we've considered only springs that are horizontal, and that are therefore unstretched at their equilibrium position. In many cases, however, we may wish to consider a vertical spring, as in **FIGURE 13-13**.

Now, when a mass m is attached to a vertical spring, it causes the spring to stretch. In fact, the vertical spring is in equilibrium when it exerts an upward force equal to the weight of the mass. That is, the spring stretches by an amount y_0 given by

$$ky_0 = mg$$

Rearranging, we find

$$y_0 = \dfrac{mg}{k} \qquad\qquad \text{13-12}$$

Thus, a mass on a vertical spring oscillates about the equilibrium point $y = -y_0$. In all other respects the oscillations are the same as for a horizontal spring. In particular, the motion is simple harmonic motion, and the period is given by $T = 2\pi\sqrt{m/k}$. Examples of a mass on a vertical spring are shown in **FIGURE 13-14**.

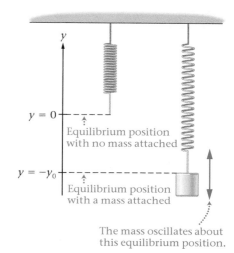

▲ **FIGURE 13-13 A mass on a vertical spring** A mass stretches a vertical spring from its initial equilibrium at $y = 0$ to a new equilibrium at $y = -y_0 = -mg/k$. The mass executes simple harmonic motion about this new equilibrium.

◄ **FIGURE 13-14 Visualizing Concepts Mass on a Vertical Spring** A ball attached to a vertical spring oscillates up and down with simple harmonic motion. If successive images taken at equal time intervals are displaced laterally, as in the sequence of photos on the left, the ball appears to trace out a sinusoidal pattern (compare with Figure 13-5). The bungee jumper on the right will oscillate in a similar fashion, though friction and air resistance will reduce the amplitude of his bounces.

EXAMPLE 13-10 IT'S A STRETCH

A 0.260-kg mass is attached to a vertical spring. When the mass is put into motion, its period is 1.12 s. **(a)** How much does the mass stretch the spring when it is at rest in its equilibrium position? **(b)** Suppose this experiment is repeated on a planet where the acceleration due to gravity is twice what it is on Earth. By what multiplicative factors do the period and equilibrium stretch change?

PICTURE THE PROBLEM
We choose the vertical axis to have its origin at the unstretched position of the spring. Once the mass is attached, the spring stretches to the position $y = -y_0$. The mass oscillates about this point with a period of 1.12 s.

REASONING AND STRATEGY
a. In order to find the stretch of the spring, $y_0 = mg/k$, we need to know the force constant, k. We can find k from the period of oscillation—that is, from $T = 2\pi\sqrt{m/k}$.

b. We replace g with $2g$ in both $T = 2\pi\sqrt{m/k}$ and $y_0 = mg/k$. Notice that g does not occur in the expression for the period T.

Known Mass, $m = 0.260$ kg; period, $T = 1.12$ s.
Unknown **(a)** Equilibrium stretch, $y_0 = ?$ **(b)** Period and equilibrium stretch ($T = ?$, $y_0 = ?$) when $g \rightarrow 2g$

SOLUTION

Part (a)

1. Use the period $T = 2\pi\sqrt{m/k}$ to solve for the force constant:

$$T = 2\pi\sqrt{\frac{m}{k}}$$

$$k = \frac{4\pi^2 m}{T^2} = \frac{4\pi^2(0.260\text{ kg})}{(1.12\text{ s})^2} = 8.18\text{ kg/s}^2 = 8.18\text{ N/m}$$

2. Set the magnitude of the spring force, ky_0, equal to mg to solve for y_0:

$$ky_0 = mg$$

$$y_0 = \frac{mg}{k} = \frac{(0.260\text{ kg})(9.81\text{ m/s}^2)}{8.18\text{ N/m}} = 0.312\text{ m}$$

Part (b)

3. Use $2g$ in place of g in the expressions for the period T and the magnitude of the equilibrium stretch, y_0:

$$T = 2\pi\sqrt{m/k} \xrightarrow{g \rightarrow 2g} T$$
$$y_0 = mg/k \xrightarrow{g \rightarrow 2g} 2y_0$$

INSIGHT
We see that doubling the force of gravity has no effect on the period (changes it by a factor of 1), but doubles the equilibrium stretch. Therefore, one would observe the same period of oscillation on the Moon or Mars—or even in orbit, as in the case of the Body Mass Measurement Device mentioned earlier in this section.

PRACTICE PROBLEM
A 0.170-kg mass stretches a vertical spring 0.250 m when at rest. What is its period when set into vertical motion?
[**Answer:** $T = 1.00$ s]

Some related homework problems: Problem 40, Problem 44

CONCEPTUAL EXAMPLE 13-11 COMPARE PERIODS

PREDICT/EXPLAIN

When a mass m is attached to a vertical spring with a force constant k, it oscillates with a period T. **(a)** If the spring is cut in half and the same mass is attached to it, is the period of oscillation greater than, less than, or equal to T? **(b)** Which of the following is the *best explanation* for your prediction?

 I. Cutting the spring in half effectively makes it stiffer, and hence the period is less than T.

 II. The spring is still made of the same material, and hence the period stays the same.

III. A shorter spring is easier to stretch. As a result, the period is greater than T.

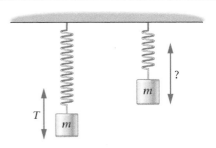

REASONING AND DISCUSSION
The downward force exerted by the mass is of magnitude mg, as is the upward force exerted by the spring. *Each coil* of the spring experiences the *same force*, just as each point in a string

CONTINUED

experiences the same tension. Therefore, each coil elongates by the same amount, regardless of how many coils there are in a given spring. It follows, then, that the total elongation of the spring with half the number of coils is half the total elongation of the longer spring. Half the elongation for the same applied force means a greater force constant, and hence the half-spring has a larger value of k—it is stiffer. As a result, its period of oscillation is less than the period of the full spring.

ANSWER
(a) The period for the half-spring is less than for the full spring. (b) The best explanation is I.

Enhance Your Understanding   (Answers given at the end of the chapter)

4. Rank the four mass–spring systems in **FIGURE 13-15** in order of increasing period. Indicate ties where appropriate.

▲ **FIGURE 13-15** Enhance Your Understanding 4.

Section Review

- The period of a mass on a spring is $T = 2\pi\sqrt{m/k}$, where m is the mass and k is the force constant.

13-5 Energy Conservation in Oscillatory Motion

In an ideal system, with no friction or other nonconservative forces, the total energy is conserved. For example, the total mechanical energy E of a mass on a horizontal spring is the sum of its kinetic energy, $K = \frac{1}{2}mv^2$, and its potential energy, $U = \frac{1}{2}kx^2$. Therefore, in the absence of nonconservative forces we have

$$E = K + U = \frac{1}{2}mv^2 + \frac{1}{2}kx^2 = \text{constant} \qquad 13\text{-}13$$

Because E remains the same throughout the motion, it follows that there is a continual tradeoff between kinetic and potential energy.

This energy tradeoff is illustrated in **FIGURE 13-16**, where the horizontal line represents the total energy of the system, E, and the parabolic curve is the spring's potential

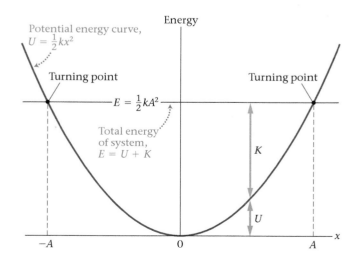

◄ **FIGURE 13-16 Energy as a function of position in simple harmonic motion** The parabola represents the potential energy of the spring, $U = \frac{1}{2}kx^2$. The horizontal line shows the total energy of the system, $E = U + K$, which is constant. It follows that the distance from the parabola to the total-energy line is the kinetic energy, K. Note that the kinetic energy vanishes at the turning points, $x = A$ and $x = -A$. At these points the energy is purely potential, and thus the total energy of the system is $E = \frac{1}{2}kA^2$.

PHYSICS
IN CONTEXT
Looking Back

The concepts of kinetic energy (Chapter 7) and potential energy (Chapter 8) play key roles in understanding oscillatory motion.

energy, U. At any given value of x, the sum of U and K must equal E; therefore, because U is the amount of energy from the axis to the parabola, K is the amount of energy from the parabola to the horizontal line. We can see, then, that the kinetic energy vanishes at the turning points, $x = +A$ and $x = -A$, as expected. On the other hand, the kinetic energy is greatest at $x = 0$, where the potential energy vanishes.

Potential and Kinetic Energy as a Function of Time As a mass on a spring oscillates back and forth with time, its kinetic and potential energies also change with time. For example, the potential energy is

$$U = \tfrac{1}{2} k x^2$$

Letting $x = A \cos(\omega t)$, we have

$$U = \tfrac{1}{2} k A^2 \cos^2(\omega t) \qquad \text{13-14}$$

Clearly, the maximum value of U is

$$U_{\text{max}} = \tfrac{1}{2} k A^2$$

From Figure 13-16, we see that the maximum value of U is simply the total energy of the system, E. Therefore

$$E = U_{\text{max}} = \tfrac{1}{2} k A^2 \qquad \text{13-15}$$

This result leads to the following conclusion:

> In simple harmonic motion, the total energy is proportional to the square of the amplitude of motion.

Similarly, the kinetic energy of the mass is

$$K = \tfrac{1}{2} m v^2$$

Letting $v = -A\omega \sin(\omega t)$ yields

$$K = \tfrac{1}{2} m A^2 \omega^2 \sin^2(\omega t)$$

It follows that the maximum kinetic energy is

$$K_{\text{max}} = \tfrac{1}{2} m A^2 \omega^2 \qquad \text{13-16}$$

As noted, the maximum kinetic energy occurs when the potential energy is zero, and hence the maximum kinetic energy must equal the total energy, E. That is,

$$E = U + K = U_{\text{max}} + 0 = 0 + K_{\text{max}}$$

At first glance, Equation 13-16 $\left(E = K_{\text{max}} = \tfrac{1}{2} m A^2 \omega^2 \right)$ doesn't seem to be the same as Equation 13-15 $\left(E = U_{\text{max}} = \tfrac{1}{2} k A^2 \right)$. However, if we recall that $\omega^2 = k/m$ (Equation 13-10), we see that

$$K_{\text{max}} = \tfrac{1}{2} m A^2 \omega^2 = \tfrac{1}{2} m A^2 \left(\frac{k}{m} \right) = \tfrac{1}{2} k A^2$$

Therefore, $K_{\text{max}} = U_{\text{max}} = E$, as expected, and the kinetic energy as a function of time is

$$K = \tfrac{1}{2} k A^2 \sin^2(\omega t) \qquad \text{13-17}$$

K and U are plotted as functions of time in **FIGURE 13-17**. The horizontal line at the top is E, the sum of U and K at all times. This shows quite graphically the back-and-forth tradeoff of energy between kinetic and potential. Mathematically, we can see that the total energy E is constant as follows:

$$E = U + K = \tfrac{1}{2} k A^2 \cos^2(\omega t) + \tfrac{1}{2} k A^2 \sin^2(\omega t)$$
$$= \tfrac{1}{2} k A^2 [\cos^2(\omega t) + \sin^2(\omega t)] = \tfrac{1}{2} k A^2$$

The last step follows from the fact that $\cos^2 \theta + \sin^2 \theta = 1$ for all θ.

PROBLEM-SOLVING NOTE

Maximum Potential and Kinetic Energy

The maximum potential energy of a mass–spring system is the same as the maximum kinetic energy of the mass. When the system has its maximum potential energy, the kinetic energy of the mass is zero; when the mass has its maximum kinetic energy, the potential energy of the system is zero.

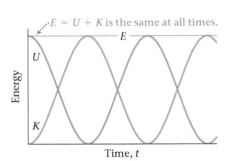

▲ **FIGURE 13-17 Energy as a function of time in simple harmonic motion** The sum of the potential energy, U, and the kinetic energy, K, is equal to the (constant) total energy E at all times. Note that when one energy (U or K) has its maximum value, the other energy is zero.

EXAMPLE 13-12 STOP THE BLOCK

A 0.980-kg block slides on a frictionless, horizontal surface with a speed of 1.32 m/s. The block encounters an unstretched spring with a force constant of 245 N/m, as shown in the sketch. **(a)** How far is the spring compressed before the block comes to rest? **(b)** For what amount of time is the block in contact with the spring before it comes to rest?

PICTURE THE PROBLEM
As our sketch shows, the initial energy of the system is entirely kinetic—namely, the kinetic energy of the block with mass $m = 0.980$ kg and speed $v_0 = 1.32$ m/s. When the block momentarily comes to rest after compressing the spring by the amount A, its kinetic energy has been converted into the potential energy of the spring.

REASONING AND STRATEGY

a. We can find the compression, A, by using energy conservation. We set the initial kinetic energy of the block, $\frac{1}{2}mv_0^2$, equal to the spring potential energy, $\frac{1}{2}kA^2$, and solve for A.

b. If the mass were attached to the spring, it would complete one oscillation in the time $T = 2\pi\sqrt{m/k}$. In moving from the equilibrium position of the spring to maximum compression, the mass has undergone one-quarter of a cycle; thus the time is $T/4$.

Known   Mass of block, $m = 0.980$ kg; initial speed of block, $v_0 = 1.32$ m/s; force constant, $k = 245$ N/m.
Unknown **(a)** Maximum compression, $A = ?$ **(b)** Time to come to rest, $t = ?$

SOLUTION

Part (a)

1. Set the initial kinetic energy of the block equal to the spring potential energy:
$$\tfrac{1}{2}mv_0^2 = \tfrac{1}{2}kA^2$$

2. Solve for A, the maximum compression:
$$A = v_0\sqrt{\frac{m}{k}} = (1.32\,\text{m/s})\sqrt{\frac{0.980\,\text{kg}}{245\,\text{N/m}}} = 0.0835\,\text{m}$$

Part (b)

3. Calculate the period of one oscillation:
$$T = 2\pi\sqrt{\frac{m}{k}} = 2\pi\sqrt{\frac{0.980\,\text{kg}}{245\,\text{N/m}}} = 0.397\,\text{s}$$

4. The block has been in contact with the spring for one-quarter of an oscillation, and hence $t = T/4$:
$$t = \tfrac{1}{4}T = \tfrac{1}{4}(0.397\,\text{s}) = 0.0993\,\text{s}$$

INSIGHT
If the horizontal surface had been rough, some of the block's initial kinetic energy would have been converted to thermal energy. In this case, the maximum compression of the spring would be less than that just calculated.

PRACTICE PROBLEM — PREDICT/CALCULATE
(a) If the initial speed of the mass in this Example is increased, does the time required to bring it to rest increase, decrease, or stay the same? Explain. **(b)** Check your answer by calculating the time for an initial speed of 1.50 m/s. [**Answer: (a)** Increasing v increases the amplitude, A. The period is independent of amplitude, however. Thus, the time is the same. **(b)** The time to bring the mass to rest is $t = 0.0993$ s, as before.]

Some related homework problems: Problem 53, Problem 83

In the next Example, we consider a bullet striking a block that is attached to a spring. As the bullet embeds itself in the block, the completely inelastic bullet–block collision dissipates some of the initial kinetic energy into thermal energy. Only the kinetic energy remaining after the collision is available for compressing the spring.

EXAMPLE 13-13 BULLET–BLOCK COLLISION: FIND THE COMPRESSION

A bullet of mass m embeds itself in a block of mass M, which is attached to a spring of force constant k. If the initial speed of the bullet is v_0, find the maximum compression of the spring.

PICTURE THE PROBLEM
Our sketch shows the bullet with mass m and speed v_0 moving toward the block of mass M that is initially at rest. After the collision, the bullet and block move as if they were a single object until they reach the maximum compression of the spring, A, at which time the system is momentarily at rest before the spring expands again.

CONTINUED

REASONING AND STRATEGY

We start by applying momentum conservation to the collision. This determines the speed of the bullet–block system after the collision. The kinetic energy of the bullet–block system after the collision is converted to the potential energy of the spring as it is compressed. The maximum compression occurs when all of the bullet–block system's kinetic energy has been converted to potential energy.

Known   Mass of bullet, m; mass of block, M; force constant, k; initial speed of bullet, v_0.

Unknown Maximum compression of spring, A = ?

SOLUTION

1. Use momentum conservation to find the final speed, v, of the bullet-block system just after the collision:

$$mv_0 = (m + M)v$$
$$v = \frac{mv_0}{(m + M)}$$

2. Find the kinetic energy of the bullet-block system just after the collision:

$$\tfrac{1}{2}(m + M)v^2 = \tfrac{1}{2}\frac{m^2 v_0^2}{(m + M)}$$

3. Set this kinetic energy equal to $\tfrac{1}{2}kA^2$ to find the maximum compression of the spring, A:

$$\tfrac{1}{2}\frac{m^2 v_0^2}{(m + M)} = \tfrac{1}{2}kA^2$$
$$A = \frac{mv_0}{\sqrt{k(m + M)}}$$

INSIGHT

The maximum compression depends directly on the initial speed of the bullet. Thus, a measurement of the compression can be used to determine the bullet's speed.

PRACTICE PROBLEM — PREDICT/CALCULATE

Suppose the force constant of the spring is quadrupled. **(a)** Does the maximum compression, A, increase, decrease, or stay the same? Explain. **(b)** By what factor does the maximum compression, A, change?   [**Answer: (a)** A stiffer spring compresses less for a given amount of energy, and hence the maximum compression decreases. **(b)** Substituting $4k$ for k in Step 3 shows that the maximum compression is halved.]

Some related homework problems: Problem 53, Problem 54

Enhance Your Understanding   (Answers given at the end of the chapter)

5. The total mechanical energy of an ideal oscillating mass–spring system is 15 J. **(a)** What is the potential energy of the system when the kinetic energy of the mass is 5 J? **(b)** What is the kinetic energy of the mass when the system's potential energy is 12 J?

Section Review

- The total mechanical energy of an oscillating mass–spring system is given by $E = K + U = \tfrac{1}{2}mv^2 + \tfrac{1}{2}kx^2$, assuming an absence of nonconservative forces in the system.

13-6 The Pendulum

One Sunday in 1583, as Galileo Galilei attended services in a cathedral in Pisa, Italy, he suddenly realized something interesting about the chandeliers hanging from the ceiling. Air currents circulating through the cathedral had set them in motion with small oscillations, and Galileo noticed that chandeliers of equal length oscillated with equal periods, even if their amplitudes were different. Indeed, as any given chandelier oscillated with decreasing amplitude, its period remained constant. He verified this observation by timing the oscillations with his pulse!

RWP Galileo was much struck by this observation, and after rushing home, he experimented with pendula constructed from different lengths of string and different weights. Continuing to use his pulse as a stopwatch, he observed that the period of a pendulum varies with its length, but is independent of the weight attached to the string. Thus, in one exhilarating afternoon, the young medical student discovered the key characteristics of a pendulum and launched himself on a new career in science. Later, he would go on to construct the first crude pendulum clock and a medical device, known as the pulsilogium, to measure a patient's pulse rate.

In modern terms, we would say that the chandeliers observed by Galileo were undergoing simple harmonic motion, as expected for small oscillations. As we know, the period of simple harmonic motion is independent of amplitude. The fact that the period is also independent of the mass is a special property of the pendulum, as we shall see next.

The Simple Pendulum

A simple pendulum consists of a mass m suspended by a light string or rod of length L. The pendulum has a stable equilibrium when the mass is directly below the suspension point, and oscillates about this position if displaced from it.

To understand the behavior of the pendulum, let's begin by considering the potential energy of the system. As shown in **FIGURE 13-18**, when the pendulum is at an angle θ with respect to the vertical, the mass m is above its lowest point by a vertical height $L(1 - \cos\theta)$. If we let the potential energy be zero at $\theta = 0$, the potential energy for general θ is

$$U = mgL(1 - \cos\theta) \qquad \text{13-18}$$

This function is plotted in **FIGURE 13-19**.

Notice that the stable equilibrium of the pendulum corresponds to a minimum of the potential energy, as expected. Near this minimum, the shape of the potential energy curve is approximately the same as for a mass on a spring, as indicated in Figure 13-19. As a result, when a pendulum oscillates with small displacements from the vertical, its motion is virtually the same as the motion of a mass on a spring; that is, the pendulum exhibits simple harmonic motion.

Next, we consider the forces acting on the mass m. In **FIGURE 13-20** we show the force of gravity, $m\vec{g}$, and the tension force in the supporting string, $\vec{T}$. The tension acts in the radial direction and supplies the force needed to keep the mass moving along its circular path. The net tangential force acting on m, then, is simply the tangential component of its weight:

$$F = mg\sin\theta$$

The direction of the net tangential force is always toward the equilibrium point. Thus F is a restoring force, as expected.

Now, for small angles θ (measured in *radians*), the sine of θ is approximately equal to the angle itself. That is,

$$\sin\theta \approx \theta$$

▲ **FIGURE 13-18 Motion of a pendulum** **(a)** As a pendulum swings away from its equilibrium position, it rises a vertical distance $L - L\cos\theta = L(1 - \cos\theta)$. **(b)** As a pendulum bob swings back and forth, it leaks a trail of sand onto a moving sheet of paper, creating a "strip chart" similar to that produced by a mass on a spring in Figure 13-4. In particular, the angle of the pendulum with the vertical varies with time like a sine or a cosine.

◀ **FIGURE 13-19 The potential energy of a simple pendulum** As a simple pendulum swings away from the vertical by an angle θ, its potential energy increases, as indicated by the solid curve. Near $\theta = 0$ the potential energy of the pendulum is essentially the same as that of a mass on a spring (dashed curve). Therefore, when a pendulum oscillates with small displacements from the vertical, it exhibits simple harmonic motion—the same as a mass on a spring.

Chin Basher?

Potential energy, U

Spring potential energy

Pendulum potential energy

$-180°$ 0 $180°$ θ

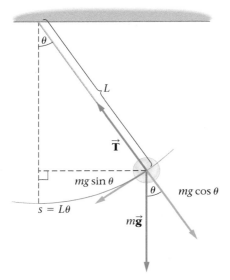

▲ **FIGURE 13-20 Forces acting on a pendulum bob** When the bob is displaced by an angle θ from the vertical, the restoring force is the tangential component of the weight, $mg \sin \theta$.

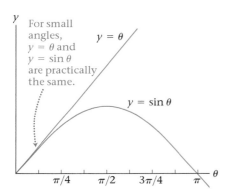

▲ **FIGURE 13-21 Relationship between sin θ and θ** For small angles measured in radians, $\sin \theta$ is approximately equal to θ. Thus, when considering small oscillations of a pendulum, we can replace $\sin \theta$ with θ. (See also Appendix A.)

Big Idea 4 The period of a pendulum is proportional to the square root of the length of the pendulum, and is independent of the mass of the pendulum. The period is also independent of the amplitude of oscillation, for small amplitudes.

This is illustrated in **FIGURE 13-21**. Notice that there is little difference between $\sin \theta$ and θ for angles smaller than about $\pi/8$ rad = 22.5 degrees. In addition, we see from Figure 13-20 that the arc length displacement of the mass from equilibrium is

$$s = L\theta$$

Equivalently,

$$\theta = \frac{s}{L}$$

Therefore, if the mass m is displaced from equilibrium by a small arc length s, the force it experiences is restoring and of magnitude

$$F = mg \sin \theta \approx mg\theta = \left(\frac{mg}{L}\right)s \qquad \text{13-19}$$

The restoring force is proportional to the displacement, s, just as expected for simple harmonic motion.

Let's compare the pendulum to a mass on a spring. In the latter case, the restoring force has a magnitude given by

$$F = kx$$

The restoring force acting on the pendulum has precisely the same form if we let $x = s$ and

$$k = \frac{mg}{L}$$

Therefore, the period of a pendulum is simply the period of a mass on a spring, $T = 2\pi\sqrt{m/k}$, with k replaced by mg/L:

$$T = 2\pi\sqrt{\frac{m}{k}} = 2\pi\sqrt{\frac{m}{(mg/L)}}$$

Canceling the mass m, we find:

Period of a Pendulum (small amplitude)

$$T = 2\pi\sqrt{\frac{L}{g}} \qquad \text{13-20}$$

SI unit: s

This is the classic formula for the period of a pendulum. Notice that T depends on the length of the pendulum, L, and on the acceleration of gravity, g. It is independent, however, of the mass, m, and the amplitude, A, as noted by Galileo.

The mass does not appear in the expression for the period of a pendulum, for the same reason that different masses free-fall with the same acceleration. In particular, a large mass tends to move more slowly because of its large inertia; on the other hand, the larger a mass, the greater the gravitational force acting on it. These two effects cancel in free fall, as well as in a pendulum.

EXERCISE 13-14 PERIOD AND LENGTH

The pendulum in a grandfather clock is designed to take 1.00 second to swing in each direction—that is, 2.00 seconds for a complete period. Find the length of a pendulum that has a period of 2.00 seconds.

REASONING AND SOLUTION
We can solve $T = 2\pi\sqrt{L/g}$ (Equation 13-20) for L and substitute numerical values. The result is

$$L = \frac{gT^2}{4\pi^2} = \frac{(9.81 \text{ m/s}^2)(2.00 \text{ s})^2}{4\pi^2} = 0.994 \text{ m}$$

CONCEPTUAL EXAMPLE 13-15 RAISE OR LOWER THE WEIGHT?

RWP If you look carefully at a grandfather clock, you will notice that the weight at the bottom of the pendulum can be moved up or down by turning a small screw. Suppose you have a grandfather clock at home that runs slow. Should you turn the adjusting screw so as to **(a)** raise the weight or **(b)** lower the weight?

REASONING AND DISCUSSION

To make the clock run faster, we want it to go from *tick* to *tick* more rapidly; in other words, we want the period of the pendulum to be decreased. From $T = 2\pi\sqrt{L/g}$ we can see that shortening the pendulum—that is, decreasing L—decreases the period. Hence, the weight should be raised, which effectively shortens the pendulum.

ANSWER

(a) The weight should be raised. This shortens the period and makes the clock run faster.

EXAMPLE 13-16 DROP TIME

A pendulum is constructed from a string 0.627 m long attached to a mass of 0.250 kg. When set in motion, the pendulum completes one oscillation every 1.59 s. If the pendulum is held at rest and the string is cut, how much time will it take for the mass to fall through a distance of 1.00 m?

PICTURE THE PROBLEM

The pendulum has a length $L = 0.627$ m and a period of oscillation $T = 1.59$ s. When the pendulum is held at rest in a vertical position, a pair of scissors is used to cut the string. The mass then falls straight downward with an acceleration g through a distance $y = 1.00$ m.

REASONING AND STRATEGY

At first it might seem that the period of oscillation and the time of fall are unrelated. Recall, however, that the period of a pendulum depends on both its length *and* the acceleration of gravity g *at the location of the pendulum.*

To solve this problem, then, we first use the period T to find the acceleration of gravity g. Once g is known, the time of fall is a straightforward kinematics problem (see Example 2-18).

Known Length of pendulum, $L = 0.627$ m; mass of bob, $m = 0.250$ kg; period of oscillation, $T = 1.59$ s; distance of drop, $y = 1.00$ m.

Unknown Time to fall through drop distance, $t = ?$

SOLUTION

1. Use the formula for the period of a pendulum to solve for the acceleration of gravity:

$$T = 2\pi\sqrt{\frac{L}{g}} \quad \text{or} \quad g = \frac{4\pi^2 L}{T^2}$$

2. Substitute numerical values to find g:

$$g = \frac{4\pi^2 L}{T^2} = \frac{4\pi^2(0.627 \text{ m})}{(1.59 \text{ s})^2} = 9.79 \text{ m/s}^2$$

3. Use kinematics to solve for the time to drop from rest through a distance y:

$$y = \tfrac{1}{2}gt^2 \quad \text{or} \quad t = \sqrt{\frac{2y}{g}}$$

4. Substitute $y = 1.00$ m and the value of g just found to find the time:

$$t = \sqrt{\frac{2y}{g}} = \sqrt{\frac{2(1.00 \text{ m})}{9.79 \text{ m/s}^2}} = 0.452 \text{ s}$$

INSIGHT

The fact that the acceleration of gravity g varies from place to place on the Earth was mentioned in Chapter 2, and again in Chapter 12. In fact, "gravity maps" are valuable tools for geologists attempting to understand the underground properties of a given region. One instrument geologists use to make gravity maps is basically a very precise pendulum whose period can be accurately measured. Slight changes in the period from one location to another correspond to slight changes in g. More common in recent decades is an electronic "gravimeter" that uses a mass on a spring and relates the force of gravity to the stretch of the spring. These gravimeters can measure g with an accuracy of one part in 1000 million.

CONTINUED

PRACTICE PROBLEM
If a mass falls 1.00 m in a time of 0.451 s, what are **(a)** the acceleration of gravity and **(b)** the period of a pendulum of length 0.500 m at that location?
[**Answer: (a)** 9.83 m/s^2, **(b)** 1.42 s]

Some related homework problems: Problem 58, Problem 59

▲ **FIGURE 13-22 Examples of physical pendulums** In each case, an object of definite size and shape oscillates about a given pivot point. The period of oscillation depends in detail on the location of the pivot point as well as on the distance ℓ from it to the center of mass, CM.

The Physical Pendulum

In the ideal version of a simple pendulum, a bob of mass m swings back and forth a distance L from the suspension point. All of the pendulum's mass is assumed to be concentrated in the bob, which is treated as a point mass. On the other hand, a **physical pendulum** is one in which the mass is not concentrated at a point, but instead is distributed over a finite volume. Examples are shown in **FIGURE 13-22**. Detailed mathematical analysis shows that if the moment of inertia (Chapter 10) of a physical pendulum about its axis of rotation is I, and the distance from the axis to the center of mass is ℓ, the period of the pendulum is given by the following:

Period of a Physical Pendulum

$$T = 2\pi\sqrt{\frac{\ell}{g}}\left(\sqrt{\frac{I}{m\ell^2}}\right)$$

13-21

SI unit: s

Notice that the first part of the expression, $2\pi\sqrt{\ell/g}$, is the period of a simple pendulum with all its mass concentrated at the center of mass. The second factor, $\sqrt{I/m\ell^2}$, is a correction that takes into account the size and shape of the physical pendulum. Thus, writing the period in this form, rather than canceling one power of ℓ, makes for a convenient comparison with the simple pendulum.

To see how Equation 13-21 works in practice, we first apply it to a simple pendulum of mass m and length L. In this case, the moment of inertia is $I = mL^2$, and the distance to the center of mass is $\ell = L$. As a result, the period is

$$T = 2\pi\sqrt{\frac{\ell}{g}}\left(\sqrt{\frac{I}{m\ell^2}}\right) = 2\pi\sqrt{\frac{L}{g}}\left(\sqrt{\frac{mL^2}{mL^2}}\right) = 2\pi\sqrt{\frac{L}{g}}$$

Thus, as expected, Equation 13-21 also applies to a simple pendulum.

Walking Next, we apply Equation 13-21 to a nontrivial physical pendulum—namely, your leg. When you walk, your leg rotates about the hip joint much like a uniform rod pivoted about one end, as indicated in **FIGURE 13-23**. Thus, if we approximate your leg as a uniform rod of length L, its period can be found using Equation 13-21. Recall from Chapter 10 (see Table 10-1) that the moment of inertia of a rod of length L about one end is

$$I = \tfrac{1}{3}mL^2$$

Similarly, the center of mass of your leg is essentially at the center of your leg; thus,

$$\ell = \tfrac{1}{2}L$$

Combining these results, we find that the period of your leg is roughly

$$T = 2\pi\sqrt{\frac{\ell}{g}}\left(\sqrt{\frac{I}{m\ell^2}}\right) = 2\pi\sqrt{\frac{\tfrac{1}{2}L}{g}}\left(\sqrt{\frac{\tfrac{1}{3}mL^2}{m\left(\tfrac{1}{2}L\right)^2}}\right) = 2\pi\sqrt{\frac{L}{g}}\left(\sqrt{\frac{2}{3}}\right)$$

▲ **FIGURE 13-23 The leg as a physical pendulum** As a person walks, each leg swings much like a physical pendulum. A reasonable approximation is to treat the leg as a uniform rod about 1 m in length.

PROBLEM-SOLVING NOTE

The Period of a Physical Pendulum

When finding the period of a physical pendulum, recall that I is the moment of inertia about the pivot point and ℓ is the distance from the pivot point to the center of mass.

Given that a typical human leg is about a meter long ($L = 1.0\,\mathrm{m}$), we find that its period of oscillation about the hip is approximately $T = 1.6\,\mathrm{s}$.

RWP The significance of this result is that the natural walking pace of humans and other animals is largely controlled by the swinging motion of their legs as physical pendula. This is illustrated in **FIGURE 13-24**, where we compare the walking motions of Beatles and elephants. In fact, a great deal of research in animal locomotion has focused on precisely this type of analysis. In the case just considered, suppose that the leg in Figure 13-23 swings through an angle of roughly 1.0 radian, as the foot moves from behind the hip to in front of the hip for the next step. The arc length through which the foot moves is $s = r\theta \approx (1.0\,\text{m})(1.0\,\text{rad}) = 1.0\,\text{m}$. Because it takes half a period ($T/2 = 0.80\,\text{s}$) for the foot to move from behind the hip to in front of it, the average speed of the foot is roughly

$$v = \frac{d}{t} \approx \frac{1.0\,\text{m}}{0.80\,\text{s}} \approx 1.3\,\text{m/s}$$

This is the typical speed of a person taking a brisk walk.

Returning to $T = 2\pi\sqrt{\ell/g}(\sqrt{I/m\ell^2})$, we can see that the smaller the moment of inertia I, the smaller the period. After all, an object with a small moment of inertia rotates quickly and easily. It therefore completes an oscillation in less time than an object with a larger moment of inertia. In the case of the human leg, the fact that its mass is distributed uniformly along its length—rather than concentrated at the far end—means that its moment of inertia and period are less than for a simple pendulum 1 m long.

QUICK EXAMPLE 13-17   PERIOD OF OSCILLATION

A Hula Hoop with mass $m = 0.68$ kg and radius $R = 0.51$ m hangs from a peg in a garage. When someone bumps the hoop, it begins to swing back and forth. What is the hoop's period of oscillation? (*Note*: The moment of inertia of a hoop about a point on its rim is $I = 2mR^2$.)

▲ **FIGURE 13-24 Visualizing Concepts**
Physical Pendulums and Walking When animals walk, the swinging movement of their legs can be approximated fairly well by treating the legs as physical pendula. Such analysis has proved useful in analyzing the gaits of various creatures, from Beatles to elephants.

REASONING AND STRATEGY

The period is given by $T = 2\pi\sqrt{\ell/g}\left(\sqrt{I/(m\ell^2)}\right)$. In this case, the distance to the center of mass is $\ell = R$, and the moment of inertia is $I = 2mR^2$.

1. Substitute $I = 2mR^2$ and $\ell = R$ into $T = 2\pi\sqrt{\ell/g}\left(\sqrt{I/(m\ell^2)}\right)$:
$$T = 2\pi\sqrt{\frac{R}{g}}\left(\sqrt{\frac{2mR^2}{mR^2}}\right) = 2\pi\sqrt{\frac{2R}{g}}$$

2. Evaluate with $R = 0.51$ m and $g = 9.81$ m/s²:
$$T = 2\pi\sqrt{\frac{2(0.51\,\text{m})}{9.81\,\text{m/s}^2}} = 2.0\,\text{s}$$

This period is about 1.4 times greater than the period of a simple pendulum of length 0.51 m. This makes sense when you consider that much of the hoop's mass is farther from the axis of rotation than 0.51 m, resulting in a larger moment of inertia than is the case for a simple pendulum.

Enhance Your Understanding   (Answers given at the end of the chapter)

6. Rank the four pendulum systems in **FIGURE 13-25** in order of increasing period. Indicate ties where appropriate.

▲ **FIGURE 13-25** Enhance Your Understanding 6.   CONTINUED

Section Review

- The period of a simple pendulum of length L is $T = 2\pi\sqrt{L/g}$.
- The period of a physical pendulum is $T = 2\pi\sqrt{\ell/g}\left(\sqrt{I/m\ell^2}\right)$.

13-7 Damped Oscillations

To this point we have restricted our considerations to oscillating systems in which no mechanical energy is gained or lost. In most physical systems, however, there is some loss of mechanical energy to friction, air resistance, or other nonconservative forces. As the mechanical energy of a system decreases, its amplitude of oscillation decreases as well, as expected from $E = \frac{1}{2}kA^2$ (Equation 13-15). This type of motion is referred to as a **damped oscillation**.

In a typical situation, an oscillating mass may lose its mechanical energy to a force such as air resistance that is proportional to the speed of the mass and opposite in direction. The force in such a case can be written as

$$\vec{\mathbf{F}} = -b\vec{\mathbf{v}}$$

The constant b is referred to as the **damping constant**; it is a measure of the strength of the damping force, and its SI units are kg/s.

If the damping constant is small, the system will continue to oscillate, but with a continuously decreasing amplitude. This type of motion, referred to as **underdamped**, is illustrated in **FIGURE 13-26 (a)**. In such cases, the amplitude decreases exponentially with time. Thus, if A_0 is the initial amplitude of an oscillating mass m, the amplitude at the time t is

$$A = A_0 e^{-bt/2m}$$

The exponential dependence of the amplitude is indicated by the dashed curve in the figure.

As the damping is increased, a point is reached where the system no longer oscillates, but instead simply relaxes back to the equilibrium position, as shown in **FIGURE 13-26 (b)**. A system with this type of behavior is said to be **critically damped**. If the damping is increased beyond this point, the system is said to be **overdamped**. In this case, the system still returns to equilibrium without oscillating, but the time required is greater. This is also illustrated in Figure 13-26 (b).

Some mechanical systems are designed to be near the condition for critical damping. If such a system is displaced from equilibrium, it will return to equilibrium, without oscillating, in the shortest possible time. For example, shock absorbers are designed so that a car that has just hit a bump will return to equilibrium quickly, without a lot of up-and-down oscillations.

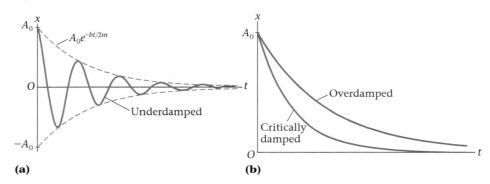

(a) **(b)**

▲ **FIGURE 13-26 Damped oscillations (a)** In underdamped oscillations, the position continues to oscillate as a function of time, but the amplitude of oscillation decreases exponentially. **(b)** In critically damped and overdamped motion, no oscillations occur. Instead, an object simply settles back to its equilibrium position without overshooting. Equilibrium is reached most rapidly in the critically damped case.

Enhance Your Understanding
(Answers given at the end of the chapter)

7. The amplitude of a damped oscillation decreases from A at $t = 0$ to $\frac{3}{4}A$ at $t = T$. What is the amplitude of the system at $t = 2T$? Explain.

Section Review

- The amplitude of a damped oscillating system decreases with time as $A = A_0 e^{-bt/2m}$, where b is the damping constant.

13-8 Driven Oscillations and Resonance

In the previous section we considered the effects of removing energy from an oscillating system. It is also possible to increase the energy of a system, or to replace the energy lost to various forms of friction. This can be done by applying an external force that does positive work on the system.

Suppose, for example, that you hold the end of a string from which a small weight is suspended, as in **FIGURE 13-27**. If the weight is set in motion and you hold your hand still, it will soon stop oscillating. If you move your hand back and forth in a horizontal direction, however, you can keep the weight oscillating indefinitely. The motion of your hand is said to be "driving" the weight, leading to **driven oscillations**.

The response of the weight in this example depends on the frequency of your hand's back-and-forth motion, as you can readily verify for yourself. For instance, if you move your hand very slowly, the weight will simply track the motion of your hand. Similarly, if you oscillate your hand very rapidly, the weight will exhibit only small oscillations. Oscillating your hand at an intermediate frequency, however, can result in large amplitude oscillations for the weight.

Just what is an appropriate intermediate frequency? Well, to achieve a large response, your hand should drive the weight at the frequency at which it oscillates when not being driven—that is, when it oscillates by itself. This is referred to as the **natural frequency**, f_0, of the system. For example, the natural frequency of a pendulum of length L is simply the inverse of its period:

$$f_0 = \frac{1}{T} = \frac{1}{2\pi}\sqrt{\frac{g}{L}}$$

Similarly, the natural frequency of a mass on a spring is

$$f_0 = \frac{1}{T} = \frac{1}{2\pi}\sqrt{\frac{k}{m}}$$

In general, driving any system at a frequency near its natural frequency results in large oscillations.

As an example, **FIGURE 13-28** shows a plot of amplitude, A, versus driving frequency, f, for a mass on a spring. Notice the large amplitude for frequencies near f_0. This type of large response, due to frequency matching, is known as **resonance**, and the curves shown in Figure 13-28 are referred to as **resonance curves**. The five curves shown in Figure 13-28 correspond to different amounts of damping. As can be seen, systems with small damping have a high, narrow peak on their resonance curve. This means that resonance is a large effect in these systems, and that it is very sensitive to frequency. On the other hand, systems with large damping have resonance peaks that are broad and low.

RWP Resonance plays an important role in a variety of physical systems, from a pendulum to atoms in a laser to a tuner in a radio or TV. As we shall see in Chapter 24, for example, adjusting the tuning knob in a radio changes the resonance frequency of the electric circuit in the tuner. When its resonance frequency matches the frequency being broadcast by a station, that station is picked up. To change stations, we simply

Damping reduces the amplitude of motion.

Adding energy to the system . . .

. . . can offset the damping.

▲ **FIGURE 13-27 Driven oscillations** If the support point of a pendulum is held still, its oscillations quickly die away due to damping. If the support point is oscillated back and forth, however, the pendulum will continue swinging. This is called "driving" the pendulum. If the driving frequency is close to the natural frequency of the pendulum—the frequency at which it oscillates when the support point is held still—its amplitude of motion can become quite large.

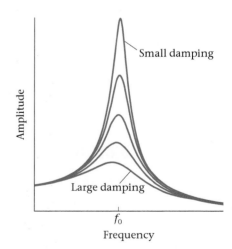

▲ **FIGURE 13-28 Resonance curves for various amounts of damping** When the damping is small, the amplitude of oscillation can become very large for frequencies close to the natural frequency, f_0. When the damping is large, the amplitude has only a low, broad peak near the natural frequency.

(a)

(b)

 Vibrating Rods

Big Idea 5 An oscillator driven at its resonance frequency responds with a large amplitude of oscillation. The smaller the damping, the larger the resonant amplitude.

◄ **FIGURE 13-29 Visualizing Concepts Resonance (a)** Anyone who has ever pushed a child on a swing knows that the timing of the pushes is critical. If they are synchronized with the natural frequency of the swing, the amplitude can increase rapidly. **(b)** The phenomenon of resonance can have dangerous consequences. In 1940, the Tacoma Narrows Bridge, completed only four months earlier, collapsed when high winds set the bridge swaying at one of its resonant frequencies.

change the resonance frequency of the tuner to the frequency of another station. A good tuner will have little damping, so stations that are even slightly "off resonance" will have small response, and hence will not be heard.

Mechanical examples of resonance are all around us as well. In fact, you might want to try the following experiment the next time you notice a spider in its web. Move close to the web and hum, starting with a low pitch. Slowly increase the pitch of your humming, and soon you will notice the spider react excitedly. You have hit the resonance frequency of its web, causing it to vibrate and making the spider think he has snagged a lunch. If you continue to increase your humming frequency, the spider quiets down again, because you have gone past the resonance. Each individual spiderweb you encounter will resonate at a different, and specific, frequency.

RWP Man-made structures can show resonance effects as well. A playful, everyday example is shown in **FIGURE 13-29 (a)**. On a more serious note, one of the most dramatic and famous examples of mechanical resonance is the collapse of Washington's Tacoma Narrows Bridge in 1940. High winds through the narrows had often set the bridge into a gentle swaying motion, resulting in its being known by the affectionate nickname "Galloping Girdie." During one particular windstorm, however, the bridge experienced a resonance-like effect, and the amplitude of its swaying motion began to increase. Alarmed officials closed the bridge to traffic, and a short time later the swaying motion became so great that the bridge broke apart and fell into the waters below, as shown in **FIGURE 13-29 (b)**. Needless to say, bridges built since that time have been designed to prevent such catastrophic oscillations.

Enhance Your Understanding   (Answers given at the end of the chapter)

8. When you drive a pendulum at a frequency f_1, you observe an amplitude of 2 cm. When you gradually increase the driving frequency to $f_2 > f_1$, you observe that the amplitude decreases to 1 cm. Is the resonance frequency of the pendulum less than f_1, between f_1 and f_2, or greater than f_2? Explain.

Section Review

- The amplitude of a driven oscillating system is greatest when it is driven at its natural frequency.

CHAPTER 13 REVIEW

CHAPTER SUMMARY

13-1 PERIODIC MOTION

Periodic motion repeats after a definite length of time.

Period
The period, T, is the time required for a motion to repeat:
T = time required for one cycle of a periodic motion

Frequency
Frequency, f, is the number of oscillations per unit time.
Equivalently, f is the inverse of the period:

$$f = \frac{1}{T}$$

13-1

13-2 SIMPLE HARMONIC MOTION

A particular type of periodic motion is simple harmonic motion.

Restoring Force

Simple harmonic motion occurs when the restoring force is proportional to the displacement from equilibrium.

Amplitude

The maximum displacement from equilibrium is the amplitude, A.

Position Versus Time

The position, x, of an object in simple harmonic motion is

$$x = A \cos\left(\frac{2\pi}{T}t\right) = A \cos(\omega t) \qquad \text{13-2, 13-4}$$

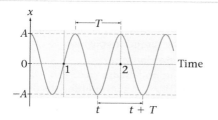

13-3 CONNECTIONS BETWEEN UNIFORM CIRCULAR MOTION AND SIMPLE HARMONIC MOTION

Circular motion viewed from the side—by projecting a shadow onto a screen, for example—is simple harmonic motion.

Angular Frequency

Angular frequency, ω, is 2π times the frequency:

$$\omega = 2\pi f = \frac{2\pi}{T} \qquad \text{13-5}$$

Velocity

Velocity as a function of time in simple harmonic motion is

$$v = -A\omega \sin(\omega t) \qquad \text{13-6}$$

Acceleration

Acceleration as a function of time in simple harmonic motion is

$$a = -A\omega^2 \cos(\omega t) \qquad \text{13-8}$$

13-4 THE PERIOD OF A MASS ON A SPRING

An important special case of simple harmonic motion is a mass on a spring.

Period

The period of a mass m attached to a spring of force constant k is

$$T = 2\pi \sqrt{\frac{m}{k}} \qquad \text{13-11}$$

13-5 ENERGY CONSERVATION IN OSCILLATORY MOTION

In an ideal oscillatory system, the total energy remains constant.

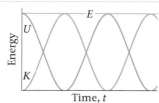

13-6 THE PENDULUM

A pendulum oscillating with small amplitude exhibits simple harmonic motion.

Period of a Simple Pendulum

The period of a simple pendulum of length L is

$$T = 2\pi \sqrt{\frac{L}{g}} \qquad \text{13-20}$$

Period of a Physical Pendulum

The period of a physical pendulum is

$$T = 2\pi \sqrt{\frac{\ell}{g}}\left(\sqrt{\frac{I}{m\ell^2}}\right) \qquad \text{13-21}$$

In this expression, I is the moment of inertia about the pivot point, and ℓ is the distance from the pivot point to the center of mass.

13-7 DAMPED OSCILLATIONS

Systems in which mechanical energy is lost to other forms eventually come to rest at the equilibrium position.

Underdamping
In an underdamped case, a system of mass m and damping constant b continues to oscillate as its amplitude steadily decreases with time. The decrease in amplitude is exponential:

$$A = A_0 e^{-bt/2m}$$

Critical Damping
In a system with critical damping, the system relaxes back to equilibrium with no oscillations in the least possible time.

Overdamping
An overdamped system also relaxes back to equilibrium with no oscillations, but more slowly than in critical damping.

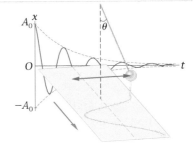

13-8 DRIVEN OSCILLATIONS AND RESONANCE

If an oscillating system is driven by an external force, it is possible for energy to be added to the system.

Resonance
Resonance is the response of an oscillating system to a driving force of the appropriate frequency.

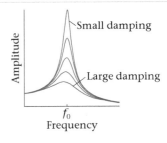

ANSWERS TO ENHANCE YOUR UNDERSTANDING QUESTIONS

1. The period, which is the inverse of the frequency, doubles.

2. (c)

3. The acceleration is positive because the object, which just came to rest momentarily, will begin to gain speed in the positive direction.

4. B < A = C < D

5. (a) 10 J. (b) 3 J.

6. A < B = C < D.

7. Each time interval of T reduces the amplitude by a factor of $\frac{3}{4}$. Thus, at $t = 2T$ the amplitude is $\left(\frac{3}{4}\right)^2 A = \left(\frac{9}{13}\right)A$.

8. The amplitude decreases as the frequency is increased beyond the value f_1. Therefore, the resonance frequency is less than f_1.

CONCEPTUAL QUESTIONS

For instructor-assigned homework, go to www.masteringphysics.com. (MP)

(Answers to odd-numbered Conceptual Questions can be found in the back of the book.)

1. A basketball player dribbles a ball with a steady period of T seconds. Is the motion of the ball periodic? Is it simple harmonic motion? Explain.

2. A person rides on a Ferris wheel that rotates with constant angular speed. If the Sun is directly overhead, does the person's shadow on the ground undergo periodic motion? Does it undergo simple harmonic motion? Explain.

3. An air-track cart bounces back and forth between the two ends of an air track. Is this motion periodic? Is it simple harmonic motion? Explain.

4. If a mass m and a mass $2m$ oscillate on identical springs with identical amplitudes, they both have the same maximum kinetic energy. How can this be? Shouldn't the larger mass have more kinetic energy? Explain.

5. An object oscillating with simple harmonic motion completes a cycle in a time T. If the object's amplitude is doubled, the time required for one cycle is still T, even though the object covers twice the distance. How can this be? Explain.

6. The position of an object undergoing simple harmonic motion is given by $x = A \cos(Bt)$. Explain the physical significance of the constants A and B. What is the frequency of this object's motion?

7. The pendulum bob in Figure 13-18 leaks sand onto the strip chart. What effect does this loss of sand have on the period of the pendulum? Explain.

PROBLEMS AND CONCEPTUAL EXERCISES

Answers to odd-numbered Problems and Conceptual Exercises can be found in the back of the book. **BIO** *identifies problems of biological or medical interest;* **CE** *indicates a conceptual exercise.* Predict/Explain *problems ask for two responses: (a) your prediction of a physical outcome, and (b) the best explanation among three provided; and* Predict/Calculate *problems ask for a prediction of a physical outcome, based on fundamental physics concepts, and follow that with a numerical calculation to verify the prediction. On all problems, bullets (•, ••, •••) indicate the level of difficulty.*

SECTION 13-1 PERIODIC MOTION

1. • A person in a rocking chair completes 12 cycles in 21 s. What are the period and frequency of the rocking?

2. • While fishing for catfish, a fisherman suddenly notices that the bobber (a floating device) attached to his line is bobbing up and down with a frequency of 1.8 Hz. What is the period of the bobber's motion?

3. • If you dribble a basketball with a frequency of 1.9 Hz, how much time does it take for you to complete 12 dribbles?

4. • You take your pulse and observe 74 heartbeats in a minute. What are the period and frequency of your heartbeat?

5. • **BIO Slow-Motion Dragonfly** A frame-by-frame analysis of a slow-motion video shows that a hovering dragonfly takes 7 frames to complete one wing beat. If the video is shot at 240 frames per second, **(a)** what is the period of the wing beat, and **(b)** how many wing beats does the dragonfly perform per second?

6. •• Predict/Calculate **(a)** Your heart beats with a frequency of 1.45 Hz. How many beats occur in a minute? **(b)** If the frequency of your heartbeat increases, will the number of beats in a minute increase, decrease, or stay the same? **(c)** How many beats occur in a minute if the frequency increases to 1.55 Hz?

7. •• You rev your car's engine to 3300 rpm (rev/min). **(a)** What are the period and frequency of the engine? **(b)** If you change the period of the engine to 0.033 s, how many rpms is it doing?

SECTION 13-2 SIMPLE HARMONIC MOTION

8. • **CE** A mass moves back and forth in simple harmonic motion with amplitude A and period T. **(a)** In terms of A, through what distance does the mass move in the time T? **(b)** Through what distance does it move in the time $5T/2$?

9. • **CE** A mass moves back and forth in simple harmonic motion with amplitude A and period T. **(a)** In terms of T, how much time does it take for the mass to move through a total distance of $2A$? **(b)** How much time does it take for the mass to move through a total distance of $3A$?

10. • The position of a mass oscillating on a spring is given by $x = (3.9 \text{ cm}) \cos[2\pi t/(0.38 \text{ s})]$. **(a)** What is the period of this motion? **(b)** What is the first time the mass is at the position $x = 0$?

11. • The position of a mass oscillating on a spring is given by $x = (7.8 \text{ cm}) \cos[2\pi t/(0.68 \text{ s})]$. **(a)** What is the frequency of this motion? **(b)** When is the mass first at the position $x = -7.8 \text{ cm}$?

12. •• **CE** A position-versus-time plot for an object undergoing simple harmonic motion is given in **FIGURE 13-30**. Rank the six points indicated in the figure in order of increasing **(a)** speed, **(b)** velocity, and **(c)** acceleration. Indicate ties where necessary.

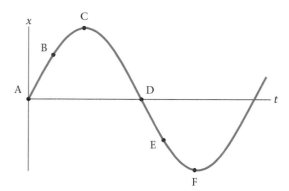

FIGURE 13-30 Problem 12

13. •• **CE** A mass on a spring oscillates with simple harmonic motion of amplitude A about the equilibrium position $x = 0$. Its maximum speed is v_{max} and its maximum acceleration is a_{max}. **(a)** What is the speed of the mass at $x = 0$? **(b)** What is the acceleration of the mass at $x = 0$? **(c)** What is the speed of the mass at $x = A$? **(d)** What is the acceleration of the mass at $x = A$?

14. •• A mass oscillates on a spring with a period of 0.63 s and an amplitude of 4.4 cm. Write an equation giving x as a function of time, assuming the mass starts at $x = A$ at time $t = 0$.

15. •• Predict/Calculate **Molecular Oscillations** An atom in a molecule oscillates about its equilibrium position with a frequency of 2.00×10^{14} Hz and a maximum displacement of 3.50 nm. **(a)** Write an expression giving x as a function of time for this atom, assuming that $x = A$ at $t = 0$. **(b)** If, instead, we assume that $x = 0$ at $t = 0$, would your expression for position versus time use a sine function or a cosine function? Explain.

16. •• A mass oscillates on a spring with a period T and an amplitude 0.48 cm. The mass is at the equilibrium position $x = 0$ at $t = 0$, and is moving in the positive direction. Where is the mass at the times **(a)** $t = T/8$, **(b)** $t = T/4$, **(c)** $t = T/2$, and **(d)** $t = 3T/4$? **(e)** Plot your results for parts (a) through (d) with the vertical axis representing position and the horizontal axis representing time.

17. •• The position of a mass on a spring is given by $x = (3.8 \text{ cm}) \cos[2\pi t/(0.88 \text{ s})]$. **(a)** What is the period, T, of this motion? **(b)** Where is the mass at $t = 0.55$ s? **(c)** Show that the mass is at the same location at 0.55 s $+ T$ seconds as it is at 0.55 s.

18. •• Predict/Calculate A mass attached to a spring oscillates with a period of 3.35 s. **(a)** If the mass starts from rest at $x = 0.0440$ m and time $t = 0$, where is it at time $t = 6.37$ s? **(b)** Is the mass moving in the positive or negative x direction at $t = 6.37$ s? Explain.

19. ••• A lawn sprinkler oscillates with simple harmonic motion of period $T = 52.0$ s, and sprays water with an angle to the vertical $(\theta = 0)$ that varies from $\theta = -45°$ to $\theta = 45°$, as shown in **FIGURE 13-31**. Water from the sprinkler lands in a nearby garden patch when the angle the sprinkler makes with the vertical is greater than $\theta = 36°$. For what amount of time is the sprinkler watering the garden during one complete cycle of its oscillation?

FIGURE 13-31 Problem 19

SECTION 13-3 CONNECTIONS BETWEEN UNIFORM CIRCULAR MOTION AND SIMPLE HARMONIC MOTION

20. • A ball rolls on a circular track of radius 0.62 m with a constant angular speed of 1.3 rad/s in the counterclockwise direction. If the angular position of the ball at $t = 0$ is $\theta = 0$, find the x component of the ball's position at the times 2.5 s, 5.0 s, and 7.5 s. Let $\theta = 0$ correspond to the positive x direction.

21. • An object executing simple harmonic motion has a maximum speed of 4.1 m/s and a maximum acceleration of 0.85 m/s². Find **(a)** the amplitude and **(b)** the period of this motion.

22. • A child rocks back and forth on a porch swing with an amplitude of 0.204 m and a period of 2.80 s. Assuming the motion is approximately simple harmonic, find the child's maximum speed.

23. •• Predict/Calculate A 30.0-g goldfinch lands on a slender branch, where it oscillates up and down with simple harmonic motion of amplitude 0.0335 m and period 1.65 s. **(a)** What is the maximum acceleration of the finch? Express your answer as a fraction of the acceleration of gravity, g. **(b)** What is the maximum speed of the goldfinch? **(c)** At the time when the goldfinch experiences its maximum acceleration, is its speed a maximum or a minimum? Explain.

24. •• **BIO Tuning Forks in Neurology** Tuning forks are used in the diagnosis of nervous afflictions known as large-fiber polyneuropathies, which are often manifested in the form of reduced sensitivity to vibrations. Disorders that can result in this type of pathology include diabetes and nerve damage from exposure to heavy metals. The tuning fork in FIGURE 13-32 has a frequency of 165 Hz. If the tips of the fork move with an amplitude of 1.10 mm, find **(a)** their maximum speed and **(b)** their maximum acceleration. Give your answer to part (b) as a multiple of g.

FIGURE 13-32 A Buck neurological hammer with tuning fork and Wartenburg pinwheel. (Problem 24)

25. •• A vibrating structural beam in a spacecraft can cause problems if the frequency of vibration is fairly high. Even if the amplitude of vibration is only a fraction of a millimeter, the acceleration of the beam can be several times greater than the acceleration due to gravity. As an example, find the maximum acceleration of a beam that vibrates with an amplitude of 0.25 mm at the rate of 110 vibrations per second. Give your answer as a multiple of g.

26. •• A peg on a turntable moves with a constant tangential speed of 0.82 m/s in a circle of radius 0.33 m. The peg casts a shadow on a wall. Find the following quantities related to the motion of the shadow: **(a)** the period, **(b)** the amplitude, **(c)** the maximum speed, and **(d)** the maximum magnitude of the acceleration.

27. •• The pistons in an internal combustion engine undergo a motion that is approximately simple harmonic motion. If the amplitude of the motion is 3.5 cm, and the engine runs at 1700 rev/min, find **(a)** the maximum acceleration of the pistons and **(b)** their maximum speed.

28. •• **Vomit Comet** NASA trains astronauts to deal with weightlessness (and its associated nausea) by flying them in the "vomit comet," a modified KC-135 airplane that flies in an oscillating path with a period of 72 seconds and a maximum acceleration of magnitude 9.81 m/s². What is the amplitude of the airplane's oscillation, assuming it undergoes simple harmonic motion?

29. •• A 0.84-kg air cart is attached to a spring and allowed to oscillate. If the displacement of the air cart from equilibrium is $x = (10.0 \text{ cm}) \cos[(2.00 \text{ s}^{-1})t + \pi]$, find **(a)** the maximum kinetic energy of the cart and **(b)** the maximum force exerted on it by the spring.

30. •• Predict/Calculate A person rides on a mechanical bucking horse (see FIGURE 13-33) that oscillates up and down with simple harmonic motion. The period of the bucking is 0.74 s and the amplitude is slowly increasing. At a certain amplitude the rider must hang on to prevent separating from the mechanical horse. **(a)** Give a strategy that will allow you to calculate this amplitude. **(b)** Carry out your strategy and find the desired amplitude.

FIGURE 13-33 Problem 30

31. ••• An object moves with simple harmonic motion of period T and amplitude A. During one complete cycle, for what length of time is the speed of the object greater than $v_{max}/2$?

32. ••• An object executing simple harmonic motion has a maximum speed v_{max} and a maximum acceleration a_{max}. Find **(a)** the amplitude and **(b)** the period of this motion. Express your answers in terms of v_{max} and a_{max}.

SECTION 13-4 THE PERIOD OF A MASS ON A SPRING

33. • CE Predict/Explain If a mass m is attached to a given spring, its period of oscillation is T. If two such springs are connected end to end and the same mass m is attached, **(a)** is the resulting period of oscillation greater than, less than, or equal to T? **(b)** Choose the *best explanation* from among the following:
 I. Connecting two springs together makes the spring stiffer, which means that less time is required for an oscillation.
 II. The period of oscillation does not depend on the length of a spring, only on its force constant and the mass attached to it.
 III. The longer spring stretches more easily, and hence takes longer to complete an oscillation.

34. • CE Predict/Explain An old car with worn-out shock absorbers oscillates with a given frequency when it hits a speed bump. If the driver adds a couple of passengers to the car and hits another speed bump, **(a)** is the car's frequency of oscillation greater than, less than, or equal to what it was before? **(b)** Choose the *best explanation* from among the following:
 I. Increasing the mass on a spring increases its period, and hence decreases its frequency.
 II. The frequency depends on the force constant of the spring but is independent of the mass.
 III. Adding mass makes the spring oscillate more rapidly, which increases the frequency.

35. • CE Predict/Explain The two blocks in FIGURE 13-34 have the same mass, m. All the springs have the same force constant, k, and are at their equilibrium length. When the blocks are set into oscillation, **(a)** is the period of block 1 greater than, less than, or equal to the period of block 2? **(b)** Choose the *best explanation* from among the following:

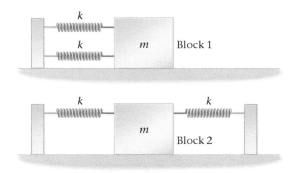

FIGURE 13-34 Problems 35 and 39

I. Springs in parallel are stiffer than springs in series; therefore the period of block 1 is smaller than the period of block 2.

II. The two blocks experience the same restoring force for a given displacement from equilibrium, and hence they have equal periods of oscillation.

III. The force of the two springs on block 2 partially cancel one another, leading to a longer period of oscillation.

36. • A 0.49-kg mass attached to a spring undergoes simple harmonic motion with a period of 0.67 s. What is the force constant of the spring?

37. • A freshly caught catfish is placed on a spring scale, and it oscillates up and down with a period of 0.204 s. If the spring constant of the scale is 2160 N/m, what is the mass of the catfish?

38. •• **CE** System A consists of a mass m attached to a spring with a force constant k; system B has a mass $2m$ attached to a spring with a force constant k; system C has a mass $3m$ attached to a spring with a force constant $6k$; and system D has a mass m attached to a spring with a force constant $4k$. Rank these systems in order of increasing period of oscillation.

39. •• Find the periods of block 1 and block 2 in Figure 13-34, given that $k = 49.2$ N/m and $m = 1.25$ kg.

40. •• When a 0.62-kg mass is attached to a vertical spring, the spring stretches by 12 cm. How much mass must be attached to the spring to result in a 0.67-s period of oscillation?

41. •• A spring with a force constant of 82 N/m is attached to a 0.47-kg mass. Assuming that the amplitude of motion is 3.4 cm, determine the following quantities for this system: **(a)** ω, **(b)** v_{max}, **(c)** T.

42. •• A bunch of grapes is placed in a spring scale at a supermarket. The grapes oscillate up and down with a period of 0.48 s, and the spring in the scale has a force constant of 650 N/m. What are **(a)** the mass and **(b)** the weight of the grapes?

43. •• Two people with a combined mass of 125 kg hop into an old car with worn-out shock absorbers. This causes the springs to compress by 8.00 cm. When the car hits a bump in the road, it oscillates up and down with a period of 1.65 s. Find **(a)** the total load supported by the springs and **(b)** the mass of the car.

44. •• A 0.95-kg mass attached to a vertical spring of force constant 130 N/m oscillates with a maximum speed of 0.45 m/s. Find the following quantities related to the motion of the mass: **(a)** the period, **(b)** the amplitude, **(c)** the maximum magnitude of the acceleration.

45. •• When a 0.184-kg mass is attached to a vertical spring, it causes the spring to stretch a distance d. If the mass is now displaced slightly from equilibrium, it is found to make 25.0 oscillations in 12.6 s. Find the stretch distance, d.

46. •• Predict/Calculate The springs of a 511-kg motorcycle have an effective force constant of 9130 N/m. **(a)** If a person sits on the motorcycle, does its period of oscillation increase, decrease, or stay the same? **(b)** By what percent and in what direction does the period of oscillation change when a 112-kg person rides the motorcycle?

47. ••• Predict/Calculate If a mass m is attached to a given spring, its period of oscillation is T. If two such springs are connected end to end, and the same mass m is attached, **(a)** is its period greater than, less than, or the same as with a single spring? **(b)** Verify your answer to part (a) by calculating the new period, T', in terms of the old period T.

SECTION 13-5 ENERGY CONSERVATION IN OSCILLATORY MOTION

48. • A 0.285-kg mass is attached to a spring with a force constant of 53.4 N/m. If the mass is displaced 0.180 m from equilibrium and released, what is its speed when it is 0.135 m from equilibrium?

49. • A 1.6-kg mass attached to a spring oscillates with an amplitude of 7.3 cm and a frequency of 2.8 Hz. What is its energy of motion?

50. •• Predict/Calculate A 0.40-kg mass is attached to a spring with a force constant of 26 N/m and released from rest a distance of 3.2 cm from the equilibrium position of the spring. **(a)** Give a strategy that allows you to find the speed of the mass when it is halfway to the equilibrium position. **(b)** Use your strategy to find this speed.

51. •• **(a)** What is the maximum speed of the mass in the previous problem? **(b)** How far is the mass from the equilibrium position when its speed is half the maximum speed?

52. •• **BIO Astronaut Mass** An astronaut uses a Body Mass Measurement Device to measure her mass. If the force constant of the spring is 2700 N/m, her mass is 75 kg, and the amplitude of her oscillation is 1.9 cm, what is her maximum speed during the measurement?

53. •• Predict/Calculate A 0.505-kg block slides on a frictionless horizontal surface with a speed of 1.18 m/s. The block encounters an unstretched spring and compresses it 23.2 cm before coming to rest. **(a)** What is the force constant of this spring? **(b)** For what length of time is the block in contact with the spring before it comes to rest? **(c)** If the force constant of the spring is increased, does the time required to stop the block increase, decrease, or stay the same? Explain.

54. •• A 3.55-g bullet embeds itself in a 1.47-kg block, which is attached to a spring of force constant 825 N/m. If the maximum compression of the spring is 5.88 cm, find **(a)** the initial speed of the bullet and **(b)** the time for the bullet–block system to come to rest.

SECTION 13-6 THE PENDULUM

55. • **CE** Metronomes, such as the penguin shown in FIGURE 13-35, are useful devices for music students. If it is desired to have the metronome tick with a greater frequency, should the penguin's bow tie be moved upward or downward?

56. • **CE** Predict/Explain A grandfather clock keeps correct time at sea level. If the clock is taken to the top of a nearby mountain, **(a)** would you expect it to keep correct time, run slow, or run fast? **(b)** Choose the *best explanation* from among the following:

FIGURE 13-35 How do you like my tie? (Problem 55)

I. Gravity is weaker at the top of the mountain, leading to a greater period of oscillation.

II. The length of the pendulum is unchanged, and therefore its period remains the same.

III. The extra gravity from the mountain causes the period to decrease.

57. • An observant fan at a baseball game notices that the radio commentators have lowered a microphone from their booth to just a few inches above the ground, as shown in FIGURE 13-36. The microphone is used to pick up sound from the field and from the fans. The fan also notices that the microphone is slowly swinging back and forth like a simple pendulum. Using her digital watch, she finds that 10 complete oscillations take 60.0 s. How high above the field is the radio booth? (Assume the microphone and its cord can be treated as a simple pendulum.)

FIGURE 13-36 Problem 57

58. • A simple pendulum of length 2.3 m makes 5.0 complete swings in 37 s. What is the acceleration of gravity at the location of the pendulum?

59. • **United Nations Pendulum** A large pendulum with a 200-lb gold-plated bob 12 inches in diameter is on display in the lobby of the United Nations building. The pendulum has a length of 75 ft. It is used to show the rotation of the Earth—for this reason it is referred to as a Foucault pendulum. What is the least amount of time it takes for the bob to swing from a position of maximum displacement to the equilibrium position of the pendulum? (Assume that the acceleration due to gravity is $g = 9.81 \text{ m/s}^2$ at the UN building.)

60. •• Predict/Calculate If the pendulum in the previous problem were to be taken to the Moon, where the acceleration of gravity is $g/6$, (a) would its period increase, decrease, or stay the same? (b) Check your result in part (a) by calculating the period of the pendulum on the Moon.

61. •• A Hula Hoop hangs from a peg. Find the period of the hoop as it gently rocks back and forth on the peg. (For a hoop with axis at the rim $I = 2mR^2$, where R is the radius of the hoop.)

62. •• A fireman tosses his 0.98-kg hat onto a peg, where it oscillates as a physical pendulum (FIGURE 13-37). If the center of mass of the hat is 8.4 cm from the pivot point, and its period of oscillation is 0.73 s, what is the moment of inertia of the hat about the pivot point?

FIGURE 13-37 Problem 62

63. •• Predict/Calculate Consider a meterstick that oscillates back and forth about a pivot point at one of its ends. (a) Is the period of a simple pendulum of length $L = 1.00$ m greater than, less than, or the same as the period of the meterstick? Explain. (b) Find the length L of a simple pendulum that has a period equal to the period of the meterstick.

64. •• On the construction site for a new skyscraper, a uniform beam of steel is suspended from one end. If the beam swings back and forth with a period of 2.00 s, what is its length?

65. •• BIO (a) Find the period of a child's leg as it swings about the hip joint. Assume the leg is 0.55 m long and can be treated as a uniform rod. (b) Estimate the child's walking speed.

66. ••• Suspended from the ceiling of an elevator is a simple pendulum of length L. What is the period of this pendulum if the elevator (a) accelerates upward with an acceleration a, or (b) accelerates downward with an acceleration whose magnitude is greater than zero but less than g? Give your answer in terms of L, g, and a.

GENERAL PROBLEMS

67. • CE An object undergoes simple harmonic motion with a period T. In the time $3T/2$ the object moves through a total distance of $12D$. In terms of D, what is the object's amplitude of motion?

68. • CE If the amplitude of a simple harmonic oscillator is doubled, by what multiplicative factor do the following quantities change: (a) angular frequency, (b) frequency, (c) period, (d) maximum speed, (e) maximum acceleration, (f) total mechanical energy?

69. • CE A mass m is suspended from the ceiling of an elevator by a spring of force constant k. When the elevator is at rest, the period of the mass is T. Does the period increase, decrease, or remain the same when the elevator (a) moves upward with constant speed or (b) moves upward with constant acceleration?

70. • CE A pendulum of length L is suspended from the ceiling of an elevator. When the elevator is at rest, the period of the pendulum is T. Does the period increase, decrease, or remain the same when the elevator (a) moves upward with constant speed or (b) moves upward with constant acceleration?

71. • A 1.3-kg mass is attached to a spring with a force constant of 52 N/m. If the mass is released with a speed of 0.28 m/s at a distance of 8.1 cm from the equilibrium position of the spring, what is its speed when it is halfway to the equilibrium position?

72. • BIO **Measuring an Astronaut's Mass** An astronaut uses a Body Mass Measurement Device (BMMD) to determine her mass. What is the astronaut's mass, given that the force constant of the BMMD is 2600 N/m and the period of oscillation is 0.85 s? (See the discussion on page 431 for more details on the BMMD.)

73. • **Sunspot Observations** Sunspots vary in number as a function of time, exhibiting an approximately 11-year cycle. Galileo made the first European observations of sunspots in 1610, and daily observations were begun in Zurich in 1749. At the present time we are well into the 24th observed cycle. What is the frequency of the sunspot cycle? Give your answer in Hz.

74. • BIO **Weighing a Bacterium** Scientists are using tiny, nanoscale cantilevers 4 micrometers long and 500 nanometers wide—essentially miniature diving boards—as a sensitive way to measure mass. An example is shown in FIGURE 13-38. The cantilevers oscillate up and down with a frequency that depends on the mass placed near the tip, and a laser beam is used to measure the frequency. A single E. coli bacterium was measured to have a mass of 665 femtograms = 6.65×10^{-16} kg with this device, as the cantilever oscillated with a frequency of 14.5 MHz. Treating the cantilever as an ideal, massless spring, find its effective force constant.

FIGURE 13-38 A silicon and silicon nitride cantilever with a 50-nanometer gold dot near its tip. (Problem 74)

75. •• **CE** An object undergoing simple harmonic motion with a period T is at the position $x = 0$ at the time $t = 0$. At the time $t = 0.25T$ the position of the object is positive. State whether x is positive, negative, or zero at the following times: **(a)** $t = 1.5T$, **(b)** $t = 2T$, **(c)** $t = 2.25T$, and **(d)** $t = 6.75T$.

76. •• The maximum speed of a 4.1-kg mass attached to a spring is 0.78 m/s, and the maximum force exerted on the mass is 13 N. **(a)** What is the amplitude of motion for this mass? **(b)** What is the force constant of the spring? **(c)** What is the frequency of this system?

77. •• The acceleration of a block attached to a spring is given by $a = -(0.302\ \text{m/s}^2)\cos([2.41\ \text{rad/s}]t)$. **(a)** What is the frequency of the block's motion? **(b)** What is the maximum speed of the block? **(c)** What is the amplitude of the block's motion?

78. •• **Helioseismology** In 1962, physicists at Cal Tech discovered that the surface of the Sun vibrates due to the violent nuclear reactions that roil within its core. This has led to a new field of solar science known as helioseismology. A typical vibration of the Sun is shown in **FIGURE 13-39**; it has a period of 5.7 minutes. The blue patches in Figure 13-39 are moving outward; the red patches are moving inward. **(a)** Find the angular frequency of this vibration. **(b)** The maximum speed at which a patch of the surface moves during a vibration is 4.5 m/s. What is the amplitude of the vibration, assuming it to be simple harmonic motion?

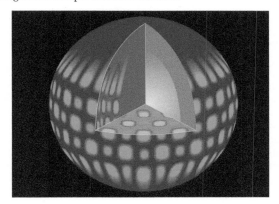

FIGURE 13-39 A typical vibration pattern of the Sun. (Problem 78)

79. •• Predict/Calculate A 9.50-g bullet, moving horizontally with an initial speed v_0, embeds itself in a 1.45-kg pendulum bob that is initially at rest. The length of the pendulum is $L = 0.745$ m. After the collision, the pendulum swings to one side and comes to rest when it has gained a vertical height of 12.4 cm. **(a)** Is the kinetic energy of the bullet–bob system immediately after the collision greater than, less than, or the same as the kinetic energy of the system just before the collision? Explain. **(b)** Find the initial speed of the bullet. **(c)** How much time does it take for the bullet–bob system to come to rest for the first time?

80. •• **BIO Spiderweb Oscillations** A 1.44-g spider oscillates on its web, which has a damping constant of 3.30×10^{-5} kg/s. How much time does it take for the spider's amplitude of oscillation to decrease by 10.0 percent?

81. •• A service dog tag (**FIGURE 13-40**) is a circular disk of radius 1.9 cm and mass 0.013 kg that can pivot about a small hole near its rim. Refer to Table 10-1 to determine the moment of inertia about the perpendicular axis that passes through the pivot hole, and find the period of oscillation of the dog tag.

FIGURE 13-40 Problem 81

82. •• Calculate the ratio of the kinetic energy to the potential energy of a simple harmonic oscillator when its displacement is half its amplitude.

83. •• A 0.340-kg mass slides on a frictionless floor with a speed of 1.34 m/s. The mass strikes and compresses a spring with a force constant of 53.4 N/m. **(a)** How far does the mass travel after contacting the spring before it comes to rest? **(b)** How much time does it take for the spring to stop the mass?

84. •• A shock absorber is designed to quickly damp out the oscillations that a car would otherwise make because it is suspended on springs. **(a)** Find the period of oscillation of a 1610-kg car that is suspended by springs that make an effective force constant of 5.75×10^4 N/m. **(b)** Find the damping constant b that will reduce the amplitude of oscillations of this car by a factor of 5.00 within a time equal to half the period of oscillation.

85. •• Predict/Calculate **FIGURE 13-41** shows a displacement-versus-time graph of the periodic motion of a 3.8-kg mass on a spring. **(a)** Referring to the figure, do you expect the maximum speed of the mass to be greater than, less than, or equal to 0.50 m/s? Explain. **(b)** Calculate the maximum speed of the mass. **(c)** How much energy is stored in this system?

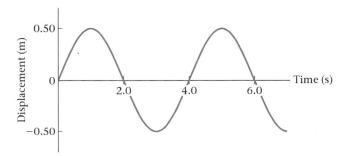

FIGURE 13-41 Problems 85 and 86

86. •• Predict/Calculate A 3.2-kg mass on a spring oscillates as shown in the displacement-versus-time graph in Figure 13-41.

(a) Referring to the graph, at what times between $t = 0$ and $t = 6.0$ s does the mass experience a force of maximum magnitude? Explain. **(b)** Calculate the magnitude of the maximum force exerted on the mass. **(c)** At what times shown in the graph does the mass experience zero force? Explain. **(d)** How much force is exerted on the mass at the time $t = 0.50$ s?

87. •• A 0.45-kg crow lands on a slender branch and bobs up and down with a period of 1.5 s. An eagle flies up to the same branch, scaring the crow away, and lands. The eagle now bobs up and down with a period of 4.8 s. Treating the branch as an ideal spring, find **(a)** the effective force constant of the branch and **(b)** the mass of the eagle.

88. ••• A mass m is connected to the bottom of a vertical spring whose force constant is k. Attached to the bottom of the mass is a string that is connected to a second mass m, as shown in **FIGURE 13-42**. Both masses are undergoing simple harmonic vertical motion of amplitude A. At the instant when the acceleration of the masses is a maximum in the upward direction the string breaks, allowing the lower mass to drop to the floor. Find the resulting amplitude of motion of the remaining mass.

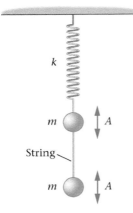

FIGURE 13-42 Problem 88

89. ••• Predict/Calculate Consider the pendulum shown in **FIGURE 13-43**. Note that the pendulum's string is stopped by a peg when the bob swings to the left, but moves freely when the bob swings to the right. **(a)** Is the period of this pendulum greater than, less than, or the same as the period of the same pendulum without the peg? **(b)** Calculate the period of this pendulum in terms of L and ℓ. **(c)** Evaluate your result for $L = 1.0$ m and $\ell = 0.25$ m.

FIGURE 13-43 Problem 89

90. ••• An object undergoes simple harmonic motion of amplitude A and angular frequency ω about the equilibrium point $x = 0$. Use energy conservation to show that the speed of the object at the general position x is given by the following expression:

$$v = \omega\sqrt{A^2 - x^2}$$

91. ••• A physical pendulum consists of a light rod of length L suspended in the middle. A large mass m_1 is attached to one end of the rod, and a lighter mass m_2 is attached to the other end, as illustrated in **FIGURE 13-44**. Find the period of oscillation for this pendulum.

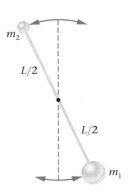

FIGURE 13-44 Problem 91

92. ••• Predict/Calculate A vertical hollow tube is connected to a speaker, which vibrates vertically with simple harmonic motion (**FIGURE 13-45**). The speaker operates with constant amplitude, A, but variable frequency, f. A slender pencil is placed inside the tube. **(a)** At low frequencies the pencil stays in contact with the speaker at all times; at higher frequencies the pencil begins to rattle. Explain the reason for this behavior. **(b)** Find an expression for the frequency at which rattling begins.

FIGURE 13-45 Problem 92

PASSAGE PROBLEMS

BIO A Cricket Thermometer, by Jiminy

Insects are ectothermic, which means their body temperature is largely determined by the temperature of their surroundings. This can have a number of interesting consequences. For example, the wing coloration in some butterfly species is determined by the ambient temperature, as is the body color of several species of dragonfly. In addition, the wing beat frequency of beetles taking flight varies with temperature due to changes in the resonant frequency of their thorax.

The origin of such temperature effects can be traced back to the fact that molecules have higher speeds and greater energy as temperature is increased (see Chapters 16 and 17). Thus, for example, molecules that collide and react as part of the metabolic process will do so more rapidly when the reactions are occurring at a higher temperature. As a result, development rates, heart rates, wing beats, and other processes all occur more rapidly.

One of the most interesting thermal effects is the temperature dependence of chirp rate in certain insects. This behavior has been observed in cone-headed grasshoppers, as well as several types of cricket. A particularly accurate connection between chirp rate and temperature is found in the snowy tree cricket (*Oecanthus fultoni* Walker), which chirps at a rate that follows the expression $N = T - 39$, where N is the number of chirps in 13 seconds, and T is the numerical value of the temperature in degrees Fahrenheit. This formula, which is known as Dolbear's law, is plotted in FIGURE 13-46 (green line) along with data points (blue dots) for the snowy tree cricket.

FIGURE 13-46 Problems 93, 94, 95, and 96

93. • If the temperature is increased by 10 degrees Fahrenheit, how many additional chirps are heard in a 13-s interval?

A. 5   B. 10   C. 13   D. 39

94. • What is the temperature in degrees Fahrenheit if a cricket is observed to give 35 chirps in 13 s?

A. 13 °F   B. 35 °F   C. 74 °F   D. 90 °F

95. • What is the frequency of the cricket's chirping (in Hz) when the temperature is 68 °F?

A. 0.45 Hz   B. 2.2 Hz   C. 5.2 Hz   D. 29 Hz

96. •• Suppose the temperature decreases uniformly from 75 °F to 63 °F in 12 minutes. How many chirps does the cricket produce during this time?

A. 28   B. 1700   C. 3800   D. 22,000

97. •• Predict/Calculate REFERRING TO EXAMPLE 13-5 Suppose we can change the plane's period of oscillation, while keeping its amplitude of motion equal to 30.0 m. (a) If we want to reduce the maximum acceleration of the plane, should we increase or decrease the period? Explain. (b) Find the period that results in a maximum acceleration of 1.0g.

98. •• Predict/Calculate REFERRING TO EXAMPLE 13-12 Suppose the force constant of the spring is doubled, but the mass and speed of the block are still 0.980 kg and 1.32 m/s, respectively. (a) By what multiplicative factor do you expect the maximum compression of the spring to change? Explain. (b) Find the new maximum compression of the spring. (c) Find the time required for the mass to come to rest after contacting the spring.

99. •• Predict/Calculate REFERRING TO EXAMPLE 13-12 (a) If the block's initial speed is increased, does the total time the block is in contact with the spring increase, decrease, or stay the same? (b) Find the total time of contact for $v_0 = 1.65$ m/s, $m = 0.980$ kg, and $k = 245$ N/m.

14

Waves and Sound

Big Ideas

1 Waves can be transverse, longitudinal, or a combination of the two.

2 Sound waves travel through the air with a speed of 343 m/s.

3 The intensity of a sound wave is its power divided by the area through which it passses.

4 The frequency of a sound wave depends on the motion of both the source and the observer.

5 Overlapping waves add algebraically to either increase or decrease amplitude.

6 Beats are the result of two sound waves of different frequency adding together.

▲ Have you ever wondered why the instruments in a string quartet differ in size? It's not just tradition—there's also a physical reason, having to do with the way vibrating strings produce sound. But to understand this and other aspects of sound, it is first necessary to learn about waves in general. This chapter considers a variety of waves—including sound—and explores their characteristic properties and behavior.

In this chapter, we extend the study of a single oscillator to the behavior of a series of oscillators that are connected to one another. Connecting the oscillators leads to an assortment of new phenomena, including waves on a string, water waves, and sound.

14-1 Types of Waves

Consider a group of swings in a playground swing set. We know that each swing by itself behaves like a simple pendulum—that is, like an oscillator. Now, let's connect the swings to one another. To be specific, suppose we tie a rope from the seat of the first swing to its neighbor, and then another rope from the second swing to the third swing, and so on. When the swings are at rest—in equilibrium—the connecting ropes have no effect. If you now sit in the first swing and begin oscillating—thus "disturbing" the equilibrium—the connecting ropes cause the other swings along the line to start oscillating as well. You have created a traveling disturbance.

In general, a disturbance that propagates from one place to another is referred to as a **wave**. Waves propagate with well-defined speeds determined by the properties of the material through which they travel. In addition, waves carry energy. For example, part of the energy you put into sound waves when you speak is carried to the ears of others, where some of the sound energy is converted into electrical energy carried by nerve impulses to the brain, which, in turn, creates the sensation of hearing.

It's important to distinguish between the motion of the wave itself and the motion of the individual particles that make up the wave. Common examples include the waves that propagate through a field of wheat. The individual wheat stalks sway back and forth as a wave passes, but they do not change their location. Similarly, surfers waiting for a big wave simply rise and fall in place as a small wave passes beneath them, as shown in FIGURE 14-1 (a). In addition, a "wave" at a ball game may propagate around the stadium more quickly than a person can run, as shown in FIGURE 14-1 (b), but the individual people making up the wave simply stand and sit in one place. From these simple examples it's clear that waves can come in a variety of types. We discuss some of the more common types in this section. In addition, we show how the speed of a wave is related to some of its basic properties.

Transverse Waves

Perhaps the easiest type of wave to visualize is a wave on a string, as illustrated in FIGURE 14-2. To generate such a wave, start by tying one end of a long string or rope to a wall. Pull on the free end with your hand, producing a tension in the string, and then move your hand up and down. As you do so, a wave will travel along the string toward the wall. In fact, if your hand moves up and down with simple harmonic motion, the wave on the string will have the shape of a sine or a cosine function; we refer to such a wave as a **harmonic wave**.

(a)

(b)

▲ **FIGURE 14-1 Visualizing Concepts Waves** A wave can be viewed as a disturbance that travels from one location to another. Although the wave itself moves steadily in one direction, the particles in the wave (people in the two cases shown here) do not share in this motion. Instead, they oscillate back and forth about their equilibrium positions as the wave moves past them.

◄ **FIGURE 14-2 A wave on a string** Vibrating one end of a string with an up-and-down motion generates a wave that travels away from its point of origin.

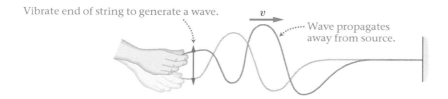

Vibrate end of string to generate a wave.

v

Wave propagates away from source.

Notice that the wave in Figure 14-2 travels in the horizontal direction, even though your hand oscillates vertically about one spot. In fact, if you look at any point on the string, it too moves vertically up and down, with no horizontal motion at all. This is shown in FIGURE 14-3, where we see that an individual point on a string oscillates up and down with an amplitude A as a wave travels past. Notice, in particular, that the displacement of particles in a string is at *right angles* to the direction of propagation of the wave. A wave with this property is called a **transverse wave**:

> In a transverse wave, the displacement of individual particles is at right angles to the direction of propagation of the wave.

Other examples of transverse waves include light and radio waves. These will be discussed in detail in Chapter 25.

PHYSICS IN CONTEXT Looking Ahead

We next encounter waves when we study electricity and magnetism. In fact, we shall see in Chapter 25 that light is an electromagnetic wave, with both the electric and magnetic fields propagating much like a wave on a string.

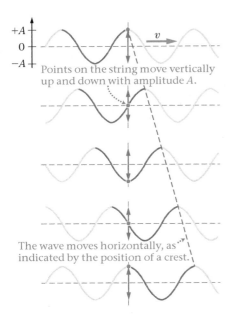

▲ **FIGURE 14-3 The motion of a wave on a string** As a wave on a string moves horizontally, all points on the string vibrate in the vertical direction, as indicated by the blue arrow.

As the wave propagates outward . . .

. . . the crests and troughs form concentric circles.

▲ **FIGURE 14-5 Water waves from a disturbance** An isolated disturbance in a pool of water, caused by a pebble dropped into the water, creates waves that propagate symmetrically away from the disturbance. The crests and troughs form concentric circles on the surface of the water as they move outward.

Big Idea 1 Waves can be transverse, longitudinal, or a combination of the two.

Longitudinal Waves

Longitudinal waves differ from transverse waves in the way that particles in the wave move. In particular, a longitudinal wave is defined as follows:

> In a longitudinal wave, the displacement of individual particles is *parallel* to the direction of propagation of the wave.

The classic example of a longitudinal wave is sound. When you speak, the vibrations in your vocal cords create a series of compressions and expansions (rarefactions) in the air. The same kind of situation occurs with a loudspeaker, as illustrated in FIGURE 14-4. Here we see a speaker diaphragm vibrating horizontally with simple harmonic motion. As it moves to the right it compresses the air momentarily; as it moves to the left it rarefies the air. A series of compressions and rarefactions then travel horizontally away from the loudspeaker with the speed of sound.

Figure 14-4 also indicates the motion of an individual particle in the air as a sound wave passes. Notice that the particle moves back and forth horizontally; that is, in the same direction as the propagation of the wave. The particle does not travel with the wave—each individual particle simply oscillates about a given position in space.

Water Waves

If a pebble is dropped into a pool of water, a series of concentric waves move away from the drop point. This is illustrated in FIGURE 14-5. To visualize the movement of the water as a wave travels by a given point, place a small piece of cork into the water. As a wave passes, the motion of the cork traces out the motion of the water itself, as indicated in FIGURE 14-6.

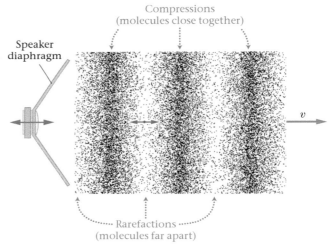

▲ **FIGURE 14-4 Sound produced by a speaker** As the diaphragm of a speaker vibrates back and forth, it alternately compresses and rarefies the surrounding air. These regions of high and low density propagate away from the speaker with the speed of sound. Individual particles in the air oscillate back and forth about a given position, as indicated by the blue arrow.

Notice that the cork moves in a roughly circular path, returning to approximately its starting point. Thus, each element of water moves both vertically and horizontally as the wave propagates by in the horizontal direction. In this sense, a water wave is a combination of both transverse and longitudinal waves. This makes the water wave more difficult to analyze. Hence, most of our results will refer to the simpler cases of purely transverse and purely longitudinal waves.

Wavelength, Frequency, and Speed

A simple wave can be thought of as a regular, rhythmic disturbance that propagates from one point to another, repeating itself both in *space* and in *time*. We now show that the repeat length and the repeat time of a wave are directly related to its speed of propagation.

▶ FIGURE 14-6 **The motion of a water wave** As a water wave passes a given point, a molecule (or a small piece of cork) moves in a roughly circular path. This means that the water molecules move both vertically and horizontally. In this sense, the water wave has characteristics of both transverse and longitudinal waves.

We begin by considering the snapshots of a wave shown in FIGURE 14-7. Points on the wave corresponding to maximum upward displacement are referred to as **crests**; points corresponding to maximum downward displacement are called **troughs**. The distance from one crest to the next, or from one trough to the next, is the repeat length—or **wavelength,** λ—of the wave.

Definition of Wavelength, λ

λ = distance over which a wave repeats

SI unit: m

Similarly, the repeat time—or **period,** T—of a wave is the time required for one wavelength to pass a given point, as illustrated in Figure 14-7. Closely related to the period of a wave is its **frequency,** f, which, as with oscillations, is defined as the inverse of the period: $f = 1/T$.

Combining these observations, we see that a wave travels a distance λ in the time T. If we apply the definition of speed—distance divided by time—it follows that the speed of a wave is:

Speed of a Wave

$$v = \frac{\text{distance}}{\text{time}} = \frac{\lambda}{T} = \lambda f \qquad\qquad 14\text{-}1$$

SI unit: m/s

This result applies to all waves.

EXERCISE 14-1 FREQUENCY AND WAVELENGTH

Sound waves travel in air with a speed of 343 m/s. **(a)** Middle C on a piano produces sound waves with a wavelength of 1.31 m. What is the frequency of middle C? **(b)** Dogs can hear sounds with frequencies as high as 45,000 Hz. What is the wavelength of such sounds?

REASONING AND SOLUTION

Wave speed, wavelength, and frequency are related by $v = \lambda f$. This relationship can be rearranged to solve for any one quantity in terms of the other two. **(a)** We solve $v = \lambda f$ for the frequency, and then substitute $v = 343$ m/s and $\lambda = 1.31$ m:

$$f = \frac{v}{\lambda} = \frac{343 \text{ m/s}}{1.31 \text{ m}} = 262 \text{ Hz}$$

(b) We solve $v = \lambda f$ for the wavelength, and then substitute $v = 343$ m/s and $f = 45,000$ Hz:

$$\lambda = \frac{v}{f} = \frac{343 \text{ m/s}}{45,000 \text{ Hz}} = 7.62 \text{ mm}$$

Enhance Your Understanding (Answers given at the end of the chapter)

1. Rank the following systems in order of increasing speed of the corresponding wave. Indicate ties where appropriate. **A,** $\lambda = 1$ m, $T = 1$ s; **B,** $\lambda = 4$ m, $T = 2$ s; **C,** $\lambda = 10$ m, $f = 1$ Hz; **D,** $\lambda = 1$ m, $f = 2$ Hz.

Section Review

- Transverse waves have particles that oscillate at right angles to the motion of the wave; longitudinal waves have particles that move parallel to the motion of the wave. Some waves combine characteristics of both transverse and longitudinal waves.

CONTINUED

$t = 0$

$t = T/4$

Crests

$t = T/2$

Troughs

$t = 3T/4$

$t = T$

▲ FIGURE 14-7 **The speed of a wave** A wave repeats over a distance equal to the wavelength, λ. The time necessary for a wave to move one wavelength is the period, T. Thus, the speed of a wave is $v = \lambda/T = \lambda f$.

**PHYSICS
IN CONTEXT
Looking Back**

The concept of frequency, first introduced in terms of oscillations (Chapter 13), is applied here to waves and sound.

- The wavelength of a wave, λ, is the distance over which the wave repeats. The period of a wave is the time T for one wavelength to pass a given point, and the frequency f is the inverse of the period. The speed v of a wave is given by $v = \lambda f$.

14-2 Waves on a String

In this section, we consider some of the basic properties of waves traveling on a string. Our results apply equally well to waves on a rope, a flexible wire, or any similar linear medium.

The Speed of a Wave on a String

The speed of a wave is determined by the properties of the medium through which it propagates. In the case of a string of length L, there are two basic characteristics that determine the speed of a wave:

1. The tension in the string
2. The mass of the string

Let's begin with the tension, which is the force F transmitted through the string (we will use F for the tension rather than T, to avoid confusion between the tension and the period). Clearly, there must be a tension in a string in order for it to propagate a wave. Imagine, for example, that a string lies on a smooth floor with both ends free. If you shake one end of it, the portions of the string near your hand will oscillate slightly, but no wave will travel to the other end of the string. If someone else takes hold of the other end of the string and pulls enough to set up a tension, then any movement you make on your end will propagate to the other end. In fact, if the tension is increased, the waves will travel through the string more rapidly.

Next, let's consider the mass m of the string. A heavy string responds slowly to a given disturbance because of its inertia. Thus, if you try sending a wave through a kite string or a thick rope, both under the same tension, you will find that the wave in the rope travels more slowly. In general, the heavier a rope or string, the slower the speed of waves in it. Of course, the total mass of a string doesn't really matter; a longer string has more mass, but its other properties are basically the same. What is important is the mass of the string per length. We give this quantity the label μ:

Definition of Mass per Length, μ

$$\mu = \text{mass per length} = \frac{m}{L}$$

SI unit: kg/m

Applying Dimensional Analysis From the above discussion, we expect the speed v to increase with the tension F and decrease with the mass per length, μ. Assuming these are the only factors determining the speed of a wave on a string, we can obtain the dependence of v on F and μ using dimensional analysis (see Section 1-3). First, we identify the dimensions of v, F, and μ:

$$[v] = \text{m/s}$$
$$[F] = \text{N} = \text{kg} \cdot \text{m/s}^2$$
$$[\mu] = \text{kg/m}$$

Next, we seek a combination of F and μ that has the dimensions of v; namely, m/s. Suppose, for example, that v depends on F to the power a and μ to the power b. Then, we have

$$v = F^a \mu^b$$

In terms of dimensions, this equation is

$$\text{m/s} = (\text{kg} \cdot \text{m/s}^2)^a (\text{kg/m})^b = \text{kg}^{a+b} \text{m}^{a-b} \text{s}^{-2a}$$

Comparing dimensions, we see that kg does not appear on the left side of the equation; therefore, we conclude that $a + b = 0$ so that kg does not appear on the right side of the equation. Hence, $a = -b$. Looking at the time dimension, s, we see that on the left we have s^{-1}; thus on the right side we must have $-2a = -1$, or $a = \frac{1}{2}$. It follows that $b = -a = -\frac{1}{2}$. This gives the following result:

Speed of a Wave on a String, v

$$v = \sqrt{\frac{F}{\mu}}$$ 14-2

SI unit: m/s

As expected, the speed increases with F and decreases with μ.

Dimensional analysis does not guarantee that this is the complete, final result; there could be a dimensionless factor like $\frac{1}{2}$ or 2π left unaccounted for. It turns out, however, that a complete analysis based on Newton's laws gives precisely the same result.

PROBLEM-SOLVING NOTE

Mass Versus Mass-per-Length

To find the mass of a string, multiply its mass per length, μ, by its length L. That is, $m = \mu L$.

EXERCISE 14-2 FIND THE SPEED

A typical guitar string has a mass per length of 0.401×10^{-3} kg/m and is tightened to a tension of 71.4 N. What is the speed of waves on the string?

REASONING AND SOLUTION

The speed of waves on the guitar string is given by $v = \sqrt{F/\mu}$. We substitute $F = 71.4$ N and $\mu = 0.401 \times 10^{-3}$ kg/m to find the speed:

$$v = \sqrt{\frac{F}{\mu}} = \sqrt{\frac{71.4 \text{ N}}{0.401 \times 10^{-3} \text{ kg/m}}} = 422 \text{ m/s}$$

EXAMPLE 14-3 A WAVE ON A ROPE

A 12-m rope is pulled tight with a tension of 49 N. When one end is given a quick "flick," a wave is generated that takes 0.54 s to travel to the other end of the rope. What is the mass of the rope?

PICTURE THE PROBLEM

Our sketch shows a wave pulse traveling with a speed v from one end of the rope to the other, a distance of 12 m. The tension in the rope is 49 N, and the travel time of the pulse is 0.54 s.

REASONING AND STRATEGY

We know that the speed of waves on a rope is determined by the tension and the mass per length. Thus, we first calculate the speed of the wave from the distance and time information given in the problem statement. Next, we solve for the mass per length in terms of speed and tension. Finally, we multiply by the length to get the mass.

Known Length of rope, $L = 12$ m; force of tension in rope, $F = 49$ N; travel time, $t = 0.54$ s.
Unknown Mass of rope, $m = ?$

SOLUTION

1. Calculate the speed of the wave:

$$v = \frac{d}{t} = \frac{12 \text{ m}}{0.54 \text{ s}} = 22 \text{ m/s}$$

2. Use $v = \sqrt{F/\mu}$ to solve for the mass per length:

$$\mu = \frac{F}{v^2}$$

3. Substitute numerical values for F and v:

$$\mu = \frac{F}{v^2} = \frac{49 \text{ N}}{(22 \text{ m/s})^2} = 0.10 \text{ kg/m}$$

4. Multiply μ by $L = 12$ m to find the mass:

$$m = \mu L = (0.10 \text{ kg/m})(12 \text{ m}) = 1.2 \text{ kg}$$

INSIGHT

The speed of a wave on this rope (almost 50 mi/h) is comparable to the speed of a car on a highway.

CONTINUED

PRACTICE PROBLEM — PREDICT/CALCULATE
(a) If the tension in the rope is increased, will the time for the wave to travel from one end to the other increase, decrease, or stay the same? Explain. **(b)** If the tension in the rope is doubled, how much time will it take for the wave to travel from one end to the other? **[Answer: (a)** The time will decrease, because increasing the tension in the rope increases the speed of waves. **(b)** The wave speed is increased to $v = 31$ m/s, and the time decreases to $t = 0.39$ s.]

Some related homework problems: Problem 14, Problem 15, Problem 16

In the following Conceptual Example, we consider the speed of a wave on a vertical rope of finite mass.

CONCEPTUAL EXAMPLE 14-4   SPEED OF A WAVE   PREDICT/EXPLAIN

A rope of length L and mass M hangs from a ceiling. If the bottom of the rope is given a gentle wiggle, a wave will travel to the top of the rope. **(a)** As the wave travels upward, does its speed increase, decrease, or stay the same? **(b)** Which of the following is the *best explanation* for your prediction?

  I. The tension in the rope increases as the wave approaches the top, and hence the speed of the wave increases as well.

  II. The speed of the wave decreases as it moves upward because of the downward acceleration of gravity.

 III. The speed of the wave stays the same because it's the same rope all the way to the top.

REASONING AND DISCUSSION

The speed of the wave is determined by the tension in the rope and its mass per length. The mass per length is the same from bottom to top, but not the tension. In particular, the tension at any point in the rope is equal to the weight of the rope below that point. Thus, the tension increases from almost zero near the bottom to essentially Mg near the top. Because the tension increases with height, so too does the speed, according to the relationship $v = \sqrt{F/\mu}$.

ANSWER

(a) The speed increases. **(b)** The best explanation is I.

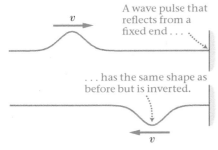

▲ **FIGURE 14-8 A reflected wave pulse: fixed end** A wave pulse on a string is inverted when it reflects from an end that is tied down.

Reflections

Thus far we have discussed only the situation in which a wave travels along a string, but at some point the wave must reach the end of the string. What happens then? Clearly, we expect the wave to be reflected, but the precise way in which the reflection occurs needs to be considered.

    Suppose, for example, that the far end of a string is anchored firmly into a wall, as shown in **FIGURE 14-8**. If you give a flick to your end of the string, you set up a wave "pulse" that travels toward the far end. When it reaches the end, it exerts an upward force on the wall, trying to pull it up into the pulse. The end is tied down, however, and hence the wall exerts an equal and opposite downward force to keep the end at rest. Thus, the wall exerts a downward force on the string that is just the *opposite* of the upward force you exerted when you created the pulse. As a result, the reflection is an inverted, or upside-down, pulse, as indicated in Figure 14-8. We shall encounter this same type of inversion under reflection when we consider the reflection of light in Chapter 28.

    Another way to tie off the end of the string is shown in **FIGURE 14-9**. In this case, the string is tied to a small ring that slides vertically with little friction on a vertical pole. In this way, the string still has a tension in it, since it pulls on the ring, but it is also free to move up and down.

    Consider a pulse moving along such a string, as in Figure 14-9. When the pulse reaches the end, it lifts the ring upward and then lowers it back down. In fact, the pulse flicks the far end of the string in the *same* way that you flicked it when you created the pulse. Therefore, the far end of the string simply creates a new pulse, identical to the first, only traveling in the opposite direction. This is illustrated in the figure.

Thus, when waves reflect, they may or may not be inverted, depending on how the reflection occurs.

Enhance Your Understanding   (Answers given at the end of the chapter)

2. Suppose the tension in a string is doubled, its mass remains the same, and its length is doubled. By what factor does the speed of waves on the string change?

Section Review

- The speed v of a wave on a string depends on the tension in the string, F, and the mass per unit length of the string, μ. The precise dependence is given by the equation $v = \sqrt{F/\mu}$.

- A wave pulse on a string is inverted when it reflects from a fixed end, but is not inverted when it reflects from an end that is free to move up and down.

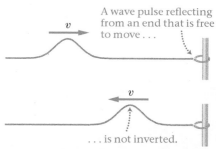

▲ **FIGURE 14-9 A reflected wave pulse: free end** A wave pulse on a string whose end is free to move is reflected without inversion.

14-3 Harmonic Wave Functions

If a wave is generated by oscillating one end of a string with simple harmonic motion, the waves will have the shape of a sine or a cosine function. This is shown in FIGURE 14-10, where the y direction denotes the vertical displacement of the string, and $y = 0$ corresponds to the flat string with no wave present. In what follows, we consider the mathematical formula that describes y as a function of time, t, and position, x, for such a harmonic wave.

First, notice that the harmonic wave in Figure 14-10 repeats when x increases by an amount equal to the wavelength, λ. Thus, the dependence of the wave on x must be of the form

$$y(x) = A \cos\left(\frac{2\pi}{\lambda}x\right)$$ 14-3

In this expression, A is the amplitude of the wave, and hence the vertical displacement ranges in value from $+A$ to $-A$.

To see that this is the correct dependence, notice that replacing x with $x + \lambda$ gives the same value for y:

$$y(x + \lambda) = A \cos\left[\frac{2\pi}{\lambda}(x + \lambda)\right] = A \cos\left(\frac{2\pi}{\lambda}x + 2\pi\right) = A \cos\left(\frac{2\pi}{\lambda}x\right) = y(x)$$

It follows that Equation 14-3 describes a vertical displacement that repeats with a wavelength λ, as desired for a wave.

This is only part of the "wave function," however, because we have not yet described the way the wave changes with time. This is illustrated in Figure 14-10, where we see a harmonic wave at times $t = 0$, $t = T/4$, $t = T/2$, $t = 3T/4$, and $t = T$. Notice that the peak in the wave that was originally at $x = 0$ at $t = 0$ moves to $x = \lambda/4$, $x = \lambda/2$, $x = 3\lambda/4$, and $x = \lambda$ for the times just given. Thus, the position x of this peak can be written as follows:

$$x = \lambda \frac{t}{T}$$

Equivalently, we can say that the peak that was at $x = 0$ is now at the location given by

$$x - \lambda \frac{t}{T} = 0$$

Similarly, the peak that was originally at $x = \lambda$ at $t = 0$ is at the following position at the general time t:

$$x - \lambda \frac{t}{T} = \lambda$$

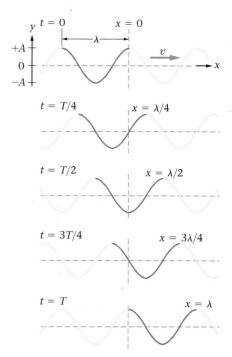

▲ **FIGURE 14-10 A harmonic wave moving to the right** As a wave moves, the peak that was at $x = 0$ at time $t = 0$ moves to the position $x = \lambda t/T$ at time t.

In general, if the position of a given point on a wave at $t = 0$ is $x(0)$, and its position at the time t is $x(t)$, the relationship between these positions is $x(t) - \lambda t/T = x(0)$. Therefore, to take into account the time dependence of a wave, we replace $x = x(0)$ in Equation 14-3 with $x(0) = x - \lambda t/T$. This yields the harmonic wave function:

$$y(x, t) = A \cos\left[\frac{2\pi}{\lambda}\left(x - \lambda\frac{t}{T}\right)\right] = A \cos\left(\frac{2\pi}{\lambda}x - \frac{2\pi}{T}t\right) \qquad \text{14-4}$$

The wave function, $y(x, t)$, depends on both time and position, and repeats whenever position increases by the wavelength, λ, or time increases by the period, T, as desired. An example of a harmonic wave produced by a tuning fork is shown in **FIGURE 14-11**.

▲ **FIGURE 14-11** An oscilloscope connected to a microphone can be used to display the waveform of a pure tone, created here by a tuning fork. The trace on the screen shows that the wave has the shape of a sine or cosine function, as expected for a harmonic wave function.

Enhance Your Understanding (Answers given at the end of the chapter)

3. A particular harmonic wave is described by the following equation: $y(x, t) = (0.12 \text{ m}) \cos(\pi x/4 - 8\pi t)$. In this expression, x is measured in meters, and t is measured in seconds. What are the **(a)** amplitude, **(b)** wavelength, and **(c)** period for this wave?

Section Review

- The mathematical function that describes a harmonic wave is $y(x, t) = A\cos\left(\dfrac{2\pi}{\lambda}x - \dfrac{2\pi}{T}t\right)$, where A is the amplitude of the wave, λ is the wavelength, and T is the period.

Oscillating one end of a Slinky . . .   . . . generates a longitudinal wave.

▲ **FIGURE 14-12 A wave on a Slinky** If one end of a Slinky is oscillated back and forth, a series of longitudinal waves are produced. These Slinky waves are analogous to sound waves.

(a)

(b)

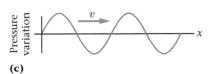

(c)

14-4 Sound Waves

The first thing we do when we come into this world is make a sound. It's many years later before we realize that sound is a wave propagating through the air at a speed of about 770 mi/h. More years are required to gain an understanding of the physics of a sound wave.

A useful mechanical model of a sound wave is provided by a Slinky. If we oscillate one end of a Slinky back and forth horizontally, as in **FIGURE 14-12**, we send out a longitudinal wave that also travels in the horizontal direction. The wave consists of regions where the coils of the Slinky are compressed alternating with regions where the coils are more widely spaced.

In close analogy with the Slinky model, a loudspeaker produces sound waves by oscillating a diaphragm back and forth horizontally, as we saw in Figure 14-4. Just as with the Slinky, a wave travels away from the source horizontally. The wave consists of compressed regions alternating with rarefied regions.

At first glance, the sound wave seems very different from the wave on a string. In particular, the sound wave doesn't seem to have the nice sinusoidal shape of a wave. Certainly, Figure 14-4 gives no hint of such a wavelike shape.

If we plot the appropriate quantities, however, the classic wave shape emerges. For example, in **FIGURE 14-13 (a)** we plot the rarefactions and compressions of a typical sound wave, while in **FIGURE 14-13 (b)** we plot the fluctuations in the density of the air versus x. Clearly, the density oscillates in a wavelike fashion. Similarly, **FIGURE 14-13 (c)** shows a plot of the fluctuations in the pressure of the air as a function of x. In regions where the density is high, the pressure is also high; where the density is low, the pressure is low. Thus, pressure versus position again shows that a sound wave has the usual wavelike properties.

The Speed of Sound The speed of sound is determined by the properties of the medium through which it propagates, just as with a wave on a string. In air, under normal atmospheric pressure and temperature, the speed of sound is approximately the following:

◄ **FIGURE 14-13 Wave properties of sound** A sound wave moving through the air **(a)** produces a wavelike disturbance in the **(b)** density and **(c)** pressure of the air.

Speed of Sound in Air (at room temperature, 20 °C)

$v = 343 \, \text{m/s} \approx 770 \, \text{mi/h}$

SI unit: m/s

When we refer to the speed of sound in this text we will always assume the value is 343 m/s, unless stated specifically otherwise.

As we shall see in Chapter 17, where we study the kinetic theory of gases, the speed of sound in air is directly related to the speed of the molecules themselves. Did you know, for example, that the air molecules colliding with your body at this moment have speeds that are essentially the speed of sound? As the air is heated the molecules will move faster, and hence the speed of sound also increases with temperature.

In a solid, the speed of sound is determined in part by how stiff the material is. The stiffer the material, the faster the sound wave, just as having more tension in a string causes a faster wave. Thus the speed of sound in plastic is rather high (2680 m/s), and in steel it is higher still (5960 m/s). Both speeds are much higher than the speed in air (343 m/s), which is certainly a "squishy" material in comparison. Table 14-1 gives a sampling of sound speed in a range of different materials.

TABLE 14-1 Speed of Sound in Various Materials

Material	Speed (m/s)
Aluminum	6420
Granite	6000
Steel	5960
Pyrex glass	5640
Copper	5010
Plastic	2680
Fresh water (20 °C)	1482
Fresh water (0 °C)	1402
Hydrogen (0 °C)	1284
Helium (0 °C)	965
Air (20 °C)	343
Air (0 °C)	331

CONCEPTUAL EXAMPLE 14-5   HOW FAR TO THE LIGHTNING?

Five seconds after a brilliant flash of lightning, thunder shakes the house. Was the lightning about a mile away, much closer than a mile, or much farther away than a mile?

REASONING AND DISCUSSION
As mentioned, the speed of sound is 343 m/s, which is just over 1000 ft/s. Thus, in five seconds sound travels slightly more than one mile. This gives rise to the following rule of thumb: The distance to a lightning strike (in miles) is the time for the thunder to arrive (in seconds) divided by 5.

Notice that we have ignored the travel time of light in our discussion. This is because light propagates with such a high speed (approximately 186,000 mi/s) that its travel time is about a million times less than that of sound.

ANSWER
The lightning was about a mile away.

EXAMPLE 14-6   WISHING WELL

You drop a stone from rest into a well that is 7.35 m deep. How much time elapses before you hear the splash?

PICTURE THE PROBLEM
Our sketch shows the well into which the stone is dropped. Notice that the depth of the well is $d = 7.35$ m. After the stone falls a distance d, the sound from the splash rises the same distance d before it is heard.

REASONING AND STRATEGY
The time until the splash is heard is the sum of (i) the time, t_1, for the stone to drop a distance d, and (ii) the time, t_2, for sound to travel a distance d.

For the time of the drop, we use one-dimensional kinematics with an initial velocity $v = 0$, since the stone is dropped from rest, and an acceleration g. Therefore, the relationship between distance and time for the stone is $d = \frac{1}{2}gt_1^2$, with $g = 9.81$ m/s^2.

For the sound wave, we use $d = vt_2$, with $v = 343$ m/s.

Known   Depth of well, $d = 7.35$ m.
Unknown Time to hear splash, $t = ?$

$d = 7.35$ m

CONTINUED

SOLUTION

1. Calculate the time for the stone to drop:

$$d = \frac{1}{2}gt_1^2$$

$$t_1 = \sqrt{\frac{2d}{g}} = \sqrt{\frac{2(7.35 \text{ m})}{9.81 \text{ m/s}^2}} = 1.22 \text{ s}$$

2. Calculate the time for the sound to travel a distance d:

$$d = vt_2$$

$$t_2 = \frac{d}{v} = \frac{7.35 \text{ m}}{343 \text{ m/s}} = 0.0214 \text{ s}$$

3. Sum the times found above:

$$t = t_1 + t_2 = 1.22 \text{ s} + 0.0214 \text{ s} = 1.24 \text{ s}$$

INSIGHT

The time of travel for the sound is quite small, only a couple hundredths of a second. It is still nonzero, however, and must be taken into account to obtain the correct total time.

In addition, notice that we use the same speed for a sound wave whether it is traveling horizontally or vertically—its speed is independent of its direction of motion. As a result, the waves emanating from a source of sound propagate outward in a spherical pattern, with the wave crests forming concentric spheres around the source.

PRACTICE PROBLEM

You drop a stone into a well and hear the splash 1.47 s later. How deep is the well? [**Answer:** 10.2 m]

Some related homework problems: Problem 32, Problem 33

The Frequency of a Sound Wave

Big Idea 2 Sound waves travel through the air with a speed of 343 m/s. The frequency of a sound wave is referred to as its pitch.

When we hear a sound, its frequency makes a great impression on us; in fact, the frequency determines the **pitch** of a sound. For example, the keys on a piano produce sounds with frequencies ranging from 55 Hz for the key farthest to the left to 4187 Hz for the rightmost key. Similarly, as you hum a song you change the shape and size of your vocal chords slightly to change the frequency of the sound you produce.

The frequency range of human hearing extends well beyond the range of a piano, however. As a rule of thumb, humans can hear sounds between 20 Hz on the low-frequency end and 20,000 Hz on the high-frequency end. Sounds with frequencies above this range are referred to as **ultrasonic,** while those with frequencies lower than 20 Hz are classified as **infrasonic.** Though we are unable to hear ultrasound and infrasound, these frequencies occur commonly in nature and are used in many technological applications as well.

BIO Bats and dolphins produce ultrasound almost continuously as they go about their daily lives. By listening to the echoes of their calls—that is, by using *echolocation*—they are able to navigate about their environment and detect their prey. As a defense mechanism, some of the insects that are preyed upon by bats have the ability to hear the ultrasonic frequency of a hunting bat and take evasive action. For instance, certain moths fold their wings in flight and drop into a precipitous dive toward the ground when they hear a bat on the prowl, as shown in **FIGURE 14-14 (a)**. In addition, the praying mantis has

(a)

(b)

(c)

▲ **FIGURE 14-14 Visualizing Concepts Ultrasound** Sound waves with frequencies that are too high for human ears to detect are used by many animal species, and in technological applications as well. **(a)** Bats navigate in the dark and locate their prey by means of a system of biological sonar. They emit a continuous stream of ultrasonic sounds and detect the echoes from objects around them. **(b)** Ultrasound scans, or sonograms, are created by beaming ultrasonic pulses into the body and measuring the time required for the echoes to return. This technique is commonly used to evaluate heart function (echocardiograms) and to visualize the fetus in the uterus. **(c)** In shock wave lithotripsy, pulses of high-frequency sound waves are used to shatter kidney stones into fragments that can be passed in the urine.

a specialized ultrasound receptor on its abdomen that allows it to "duck and cover" in response to an approaching bat.

BIO Medical applications of ultrasound are also common. Perhaps the most familiar is the ultrasound scan that is used to image a fetus in the womb, as in FIGURE 14-14 (b). By sending bursts of ultrasound into the body and measuring the time delay of the resulting echoes—the technological equivalent of echolocation—it is possible to map out the location of structures that lie hidden beneath the skin. In addition to imaging the interior of a body, ultrasound can also produce changes within the body that would otherwise require surgery. For example, in a technique called *shock wave lithotripsy* (SWL), an intense beam of ultrasound is concentrated onto a kidney stone that must be removed. Such a procedure is shown in FIGURE 14-14 (c). After being hit with as many as 1000 to 3000 pulses of sound (at 23 joules per pulse), the stone is fractured into small pieces that the body can then eliminate on its own.

RWP As for infrasound, it has been discovered in recent years that elephants (FIGURE 14-15 (a)) can communicate with one another using sounds with frequencies as low as 5 Hz. In fact, it may be that *most* elephant communication is infrasonic. These sounds, which humans feel as vibration rather than hear as sound, can carry over an area of about 30 square kilometers on the dry African savanna. And elephants are not alone in this ability. Whales, such as the blue whales shown in FIGURE 14-15 (b), produce powerful infrasonic calls as well. Because sound generally travels farther in water than in air, the whale calls can be heard by others of their species over distances of thousands of kilometers.

RWP Not all sources of infrasound are biological, however. On February 15, 2013, a 12,000-ton asteroid crashed into the atmosphere near Chelyabinsk, Russia (FIGURE 14-15 (c)). It exploded with a flash of light bright enough to temporarily blind some people, and created a shock wave that shattered windows and injured hundreds. The event was observed not just visually, but with infrasound as well. A worldwide array of special microphones—originally designed to listen for clandestine nuclear weapons tests—heard the infrasonic boom created by the meteor. It was the largest infrasound wave ever recorded by the United Nations' monitoring system. In fact, the wave was strong enough to be detected after traveling 85,000 km, more than twice around the globe! Infrasound signals can be analyzed to discern the trajectories of meteors much smaller than the one that hit Chelyabinsk, as well as to study volcanic eruptions, earthquakes, and tsunamis.

It should be noted, in light of the wide range of frequencies observed in sound, that the speed of sound in air is the same for all frequencies. Thus, the speed v is a constant in the relationship

$$v = \lambda f$$

For example, if the frequency of a wave is doubled, its wavelength is halved, so that the speed v stays the same. The fact that different frequencies travel with the same speed is evident when we listen to an orchestra in a large room. Different instruments are producing sounds of different frequencies, but we hear the sounds at the same time. Otherwise, listening to music from a distance would be quite a different and inharmonious experience.

(a)

(b)

(c)

▲ **FIGURE 14-15 Visualizing Concepts Infrasound** Sound waves with frequencies that are too low to be heard with human ears occur in many natural settings. **(a)** Elephants use infrasound to communicate with one another with sounds that humans can't hear, but instead feel as "rumbles" in the ground. **(b)** Blue whales communicate over great distances by means of infrasonic sounds of about 10 Hz. **(c)** Even nonbiological systems, like a meteor streaking across the sky, can produce infrasonic sound waves.

Enhance Your Understanding   (Answers given at the end of the chapter)

4. Which is faster: wave 1 in medium 1 with a wavelength of 2 m and a frequency of 4 Hz, or wave 2 in medium 2 with a wavelength of 0.5 m and a frequency of 50 Hz?

Section Review

- Sound is a longitudinal wave that travels through the air at about 343 m/s. It also travels through liquids and solids, at speeds determined by the particular medium.

- The frequency of a sound is its pitch. Sounds with frequencies too high to be heard are referred to as ultrasonic; sounds with frequencies too low to be heard are referred to as infrasonic.

---

*Real World Physics applications are denoted by the acronym RWP.

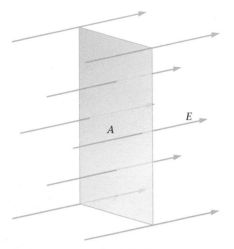

▲ FIGURE 14-16 **Intensity of a wave** If a wave carries an energy E through an area A in the time t, the corresponding intensity is $I = E/At = P/A$, where $P = E/t$ is the power.

PHYSICS IN CONTEXT
Looking Back

We use the concept of power, first developed in Chapter 7, in our definition of the intensity of a wave.

TABLE 14-2 Sound Intensities (W/m²)

Loudest sound produced in laboratory	10^9
Saturn V rocket at 50 m	10^8
Rupture of the eardrum	10^4
Jet engine at 50 m	10^2
Threshold of pain	1
Rock concert	10^{-1}
Jackhammer at 1 m	10^{-3}
Heavy street traffic	10^{-5}
Conversation at 1 m	10^{-6}
Classroom	10^{-7}
Whisper at 1 m	10^{-10}
Normal breathing	10^{-11}
Threshold of hearing	10^{-12}

PROBLEM-SOLVING NOTE

Intensity Variation with Distance

Suppose the intensity of a point source is I_1 at a distance r_1. This is enough information to find its intensity at any other distance. For example, to find the intensity I_2 at a distance r_2 we use the relationship $I_2 = (r_1/r_2)^2 I_1$.

14-5 Sound Intensity

The noise made by a jackhammer is much louder than the song of a sparrow. On this we can all agree. But how do we express such an observation physically? What physical quantity determines the loudness of a sound? We address these questions in this section, and we also present a quantitative scale by which loudness may be measured.

Intensity

Suppose you spread some jam on a piece of toast. The intensity of the taste is greater, the greater the amount of jam you use. On the other hand, the intensity of taste is smaller if you spread the same amount of jam over a larger area.

Similarly, the loudness of a sound is determined by its **intensity**—that is, by the amount of energy that passes through a given area in a given time. This is illustrated in FIGURE 14-16. If the energy E passes through the area A in the time t, the intensity I of the wave carrying the energy is

$$I = \frac{E}{At}$$

Recalling that power is energy per time, $P = E/t$, we can express the intensity as follows:

Definition of Intensity, I

$$I = \frac{P}{A}$$

14-5

SI unit: W/m²

The units are those of power (watts, W) divided by area (meters squared, m²).

Though we have introduced the concept of intensity in terms of sound, it applies to all types of waves. For example, the intensity of light from the Sun as it reaches the Earth's upper atmosphere is about 1380 W/m². If this intensity could be heard as sound, it would be painfully loud—roughly the equivalent of four jet airplanes taking off simultaneously. By comparison, the intensity of microwaves in a microwave oven is even greater, about 6000 W/m², whereas the intensity of a whisper is an incredibly tiny 10^{-10} W/m². A selection of representative intensities is given in Table 14-2.

EXERCISE 14-7 INTENSITY

Sounds from the street below come through an open window measuring 0.75 m by 0.88 m. If the power of the sound is 1.2×10^{-6} W, what is its intensity?

REASONING AND SOLUTION
Applying $I = P/A$ with $P = 1.2 \times 10^{-6}$ W and $A = (0.75 \text{ m})(0.88 \text{ m})$, we find

$$I = \frac{P}{A} = \frac{1.2 \times 10^{-6} \text{ W}}{(0.75 \text{ m})(0.88 \text{ m})} = 1.8 \times 10^{-6} \text{ W/m}^2$$

This is an intensity similar to that of normal conversation.

Dependence of Intensity on Distance When we listen to a source of sound, such as a person speaking or a radio playing a song, we notice that the loudness of the sound decreases as we move away from the source. This means that the intensity of the sound is also decreasing. The reason for this reduction in intensity is simply that the energy emitted per time by the source spreads out over a larger area—just as spreading a certain amount of jam over a larger piece of toast reduces the intensity of the taste.

In FIGURE 14-17 we show a source of sound (a bat) and two observers (moths) listening at the distances r_1 and r_2. The waves emanating from the bat propagate outward spherically, with the wave crests forming a series of concentric spheres. If we assume no reflections of sound, and a power output by the bat equal to P, the intensity detected by the first moth is

$$I_1 = \frac{P}{4\pi r_1^2}$$

In writing this expression, we have used the fact that the area of a sphere of radius r is $A = 4\pi r^2$. Similarly, the second moth hears the same sound with an intensity of

$$I_2 = \frac{P}{4\pi r_2^{\,2}}$$

The power P is the same in each case—it simply represents the amount of sound emitted by the bat. Solving for the intensity at moth 2 in terms of the intensity at moth 1, we find

$$I_2 = \left(\frac{r_1}{r_2}\right)^2 I_1 \qquad\qquad 14\text{-}6$$

In words, the intensity falls off with the square of the distance; doubling the distance reduces the intensity by a factor of four.

To summarize, the intensity a distance r from a point source of power P is:

Intensity with Distance from a Point Source

$$I = \frac{P}{4\pi r^2} \qquad\qquad 14\text{-}7$$

SI unit: W/m^2

This result assumes that no sound is reflected—which could increase the amount of energy passing through a given area—that no sound is absorbed, and that the sound propagates outward spherically. These assumptions are applied in the next Example.

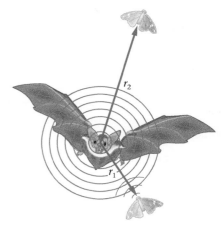

▲ **FIGURE 14-17 Echolocation** Two moths, at distances r_1 and r_2, hear the sonar signals sent out by a bat. The intensity of the signal decreases with the square of the distance from the bat. The bat, in turn, hears the echoes sent back by the moths. It can then use the direction and intensity of the returning echoes to locate its prey.

EXAMPLE 14-8   THE POWER OF SONG

Two people relaxing on a deck listen to a songbird sing. One person, only 1.66 m from the bird, hears the sound with an intensity of $2.86 \times 10^{-6}\,\text{W/m}^2$. **(a)** What intensity is heard by the second person, who is 6.71 m from the bird? Assume that no reflected sound is heard by either person. **(b)** What is the power output of the bird's song?

PICTURE THE PROBLEM

Our sketch shows the two observers, one at a distance of $r_1 = 1.66\,\text{m}$ from the bird, and the other at a distance of $r_2 = 6.71\,\text{m}$. The sound emitted by the bird is assumed to spread out spherically, with no reflections.

REASONING AND STRATEGY

a. The two intensities are related by $I_2 = (r_1/r_2)^2 I_1$ (Equation 14-6), with $r_1 = 1.66\,\text{m}$ and $r_2 = 6.71\,\text{m}$.

b. The power output can be obtained from the definition of intensity, $I = P/A$. We can calculate P for either observer, noting that $A = 4\pi r^2$.

Known   Distance to bird for person 1, $r_1 = 1.66\,\text{m}$; distance for person 2, $r_2 = 6.71\,\text{m}$; intensity for person 1, $I_1 = 2.86 \times 10^{-6}\,\text{W/m}^2$.

Unknown   **(a)** Intensity for person 2, $I_2 = ?$ **(b)** Power output of bird, $P = ?$

SOLUTION

Part (a)

1. Substitute numerical values into $I_2 = (r_1/r_2)^2 I_1$:

$$I_2 = \left(\frac{r_1}{r_2}\right)^2 I_1 = \left(\frac{1.66\,\text{m}}{6.71\,\text{m}}\right)^2 (2.86 \times 10^{-6}\,\text{W/m}^2)$$
$$= 1.75 \times 10^{-7}\,\text{W/m}^2$$

Part (b)

2. Solve $I = P/A$ for the power, P, using data for observer 1:

$$I_1 = \frac{P}{A_1}$$
$$P = I_1 A_1 = (2.86 \times 10^{-6}\,\text{W/m}^2)\,[4\pi(1.66\,\text{m})^2]$$
$$= 9.90 \times 10^{-5}\,\text{W}$$

CONTINUED

3. As a check, repeat the calculation for observer 2:

$$I_2 = \frac{P}{A_2}$$
$$P = I_2 A_2 = (1.75 \times 10^{-7} \, \text{W/m}^2)[4\pi(6.71 \, \text{m})^2]$$
$$= 9.90 \times 10^{-5} \, \text{W}$$

INSIGHT

The intensity at observer 1 is about 16.3 times the intensity at observer 2. Even so, the bird doesn't *seem* to be 16.3 times *louder*, as perceived by observer 2; in fact, the bird seems to be only a little more than twice as loud. The connection between intensity and perceived (subjective) loudness is discussed in detail later in this section.

PRACTICE PROBLEM

If the intensity at a third observer is $7.43 \times 10^{-8} \, \text{W/m}^2$, how far is that observer from the bird?   [**Answer:** $r_3 = 10.3 \, \text{m}$]

Some related homework problems: Problem 38, Problem 39

Human Perception of Sound

BIO Hearing, like most of our senses, is incredibly versatile and sensitive. We can detect sounds that are about a million times fainter than a typical conversation, and listen to sounds that are a million times louder before experiencing pain. In addition, we are able to hear sounds over a wide range of frequencies, from 20 Hz to 20,000 Hz.

When detecting the faintest of sounds, our hearing is more sensitive than one would ever guess. For example, a faint sound, with an intensity of about $10^{-11} \, \text{W/m}^2$, causes a displacement of molecules in the air of about $10^{-10} \, \text{m}$. This displacement is roughly the diameter of an atom!

Equally interesting is the way we perceive the loudness of a sound. As an example, suppose you hear a sound of intensity I_1. Next, you listen to a second sound of intensity I_2, and this sound seems to be "twice as loud" as the first. If the two intensities are measured, it turns out that I_2 is about ten times greater than I_1. Similarly, a third sound, twice as loud as I_2, has an intensity I_3 that is ten times greater than I_2. Thus, $I_2 = 10 \, I_1$ and $I_3 = 10 \, I_2 = 100 \, I_1$.

Our perception of sound, then, is such that a uniform increase in loudness corresponds to intensities that increase by multiplicative factors. For this reason, a convenient scale to measure loudness depends on the logarithm of intensity, as we discuss next.

Intensity Level and Decibels

In the study of sound, loudness is measured by the **intensity level** of a wave. Notice that we use the new term *intensity level*, which is different from *intensity*. In fact, the intensity level, designated by the symbol β, is defined as follows:

Definition of Intensity Level, β

$$\beta = (10 \, \text{dB}) \log\left(\frac{I}{I_0}\right) \qquad \text{14-8}$$

SI unit: decibel (dB), which is dimensionless

In this expression, log indicates the logarithm to the base 10, and I_0 is the intensity of the faintest sounds that can be heard. Experiments show this lowest detectable intensity to be

$$I_0 = 10^{-12} \, \text{W/m}^2$$

As is clear from its definition, the intensity level β is dimensionless; the only dimensions that enter into the definition are those of intensity, and they cancel in the logarithm. Still, just as with radians, it's convenient to label the values of intensity level with a name. The name we use—the *bel*—honors the work of Alexander Graham Bell (1847–1922), the inventor of the telephone. Because the bel is a fairly large unit, it's more common to measure β in units that are one-tenth of a bel. This unit is referred to as the **decibel**, and its abbreviation is dB.

PROBLEM-SOLVING NOTE

Intensity Versus Intensity Level

When reading a problem statement, be sure to note carefully whether it refers to the intensity, I, or to the intensity level, β. These two quantities have similar names but completely different meanings and units, as indicated in the following table:

Physical quantity	Physical meaning	Units
Intensity, I	Energy per time per area	W/m²
Intensity level, β	A measure of relative loudness	dB

PROBLEM-SOLVING NOTE

Calculating the Intensity Level

When determining the intensity level β, be sure to use the base-10 logarithm (log), as opposed to the "natural," or base e, logarithm (ln).

To get a feeling for the decibel scale, let's start with the faintest sounds. If a sound has an intensity $I = I_0$, the corresponding intensity level is

$$\beta = (10\,\text{dB})\log\!\left(\frac{I_0}{I_0}\right) = (10\,\text{dB})\log(1) = 0$$

Increasing the intensity by a factor of ten makes the sound seem twice as loud. In terms of decibels, we have

$$\beta = (10\,\text{dB})\log(10I_0/I_0) = (10\,\text{dB})\log(10) = 10\,\text{dB}$$

Going up in intensity by another factor of ten doubles the loudness of the sound again, and yields

$$\beta = (10\,\text{dB})\log(100I_0/I_0) = (10\,\text{dB})\log(100) = 20\,\text{dB}$$

Thus, *the loudness of a sound doubles with each increase in intensity level of 10 dB*. The *smallest* increase in intensity level that can be detected by the human ear is about 1 dB.

The intensity of a number of independent sound sources is simply the sum of the individual intensities. We use this fact in the following Example.

Big Idea **3** The intensity of a sound wave is the power of the wave divided by the area through which it passses. The logarithm of a sound's intensity is related to its loudness, or intensity level. Increasing the intensity of a sound by a factor of ten doubles the perceived loudness.

EXAMPLE 14-9 PASS THE PACIFIER

A crying child emits sound with an intensity of $8.0 \times 10^{-6}\,\text{W/m}^2$. Find **(a)** the intensity level in decibels for the child's sounds, and **(b)** the intensity level for this child and its twin, both crying with identical intensities.

I $2I$

(a) (b)

PICTURE THE PROBLEM
We consider the crying sounds of either one or two children. Each child emits sound with an intensity $I = 8.0 \times 10^{-6}\,\text{W/m}^2$. If two children are crying together, the total intensity of their sound is $2I$.

REASONING AND STRATEGY
The intensity level, β, is obtained by applying $\beta = (10\,\text{dB})\log(I/I_0)$ (Equation 14-8).

Known   Intensity of one crying child, $I = 8.0 \times 10^{-6}\,\text{W/m}^2$; intensity of two crying children, $2I$.
Unknown **(a)** Intensity level for crying child, $\beta = ?$ **(b)** Intensity level for two crying children, $\beta = ?$

SOLUTION

Part (a)
1. Calculate β for $I = 8.0 \times 10^{-6}\,\text{W/m}^2$:

$$\beta = (10\,\text{dB})\log\!\left(\frac{I}{I_0}\right)$$
$$= (10\,\text{dB})\log\!\left(\frac{8.0 \times 10^{-6}\,\text{W/m}^2}{10^{-12}\,\text{W/m}^2}\right) = (10\,\text{dB})\log(8.0 \times 10^6)$$
$$= (10\,\text{dB})\log(8.0) + (10\,\text{dB})\log(10^6) = 69\,\text{dB}$$

Part (b)
2. Repeat the calculation with I replaced by $2I$:

$$\beta = (10\,\text{dB})\log\!\left(\frac{2I}{I_0}\right)$$
$$= (10\,\text{dB})\log(2) + (10\,\text{dB})\log\!\left(\frac{I}{I_0}\right)$$
$$= 3.0\,\text{dB} + 69\,\text{dB} = 72\,\text{dB}$$

INSIGHT
We see that when the intensity is doubled, as in this Example, the intensity level is increased by $(10\,\text{dB})\log(2) = 3\,\text{dB}$. Similarly, when the intensity is *halved*, β *decreases* by 3 dB.

PRACTICE PROBLEM
What is the intensity level of four identically crying quadruplets?   **[Answer:** $\beta = 75\,\text{dB}$**]**

Some related homework problems: Problem 40, Problem 41

Decibels

Eardrum ruptures — 160
— 150
— 140
Jet taking off — 130
Pain — 120
Loud rock band — 110
Subway — 100
— 90
— 80
Heavy traffic — 70
Conversation — 60
Classroom — 50
— 40
— 30
Whisper — 20 } 10-dB increment
Rustling leaves — 10 } = 2 × loudness
Threshold of hearing — 0

▲ **FIGURE 14-18 Representative intensity levels for common sounds**

Even though a change of 3 dB is relatively small—after all, a change of 10 dB is required to make a sound seem twice as loud—it still requires changing the intensity by a factor of two. For example, suppose a large nursery in a hospital has so many crying babies that the intensity level is 6 dB above the safe value, as determined by OSHA (Occupational Safety and Health Administration). To reduce the level by 6 dB it would be necessary to remove three-quarters of the children, leaving only one-quarter of the original number. To our ears, however, the nursery will *sound* only 40 percent quieter!

FIGURE 14-18 shows the decibel scale with representative values indicated for a variety of common sounds.

Enhance Your Understanding   (Answers given at the end of the chapter)

5. A sound has an intensity of I and a corresponding intensity level of β. If the intensity is increased to $10I$, what is the new intensity level? **(a)** 10β, **(b)** $\beta + 3$ dB, **(c)** $\beta + 10$ dB, **(d)** $\beta/2$.

Section Review

*   The intensity of a sound, I, is given by $I = P/A$, where P is the power and A is the area through which the sound passes.
*   The intensity level of a sound is $\beta = (10\text{ dB})\log(I/I_0)$, where $I_0 = 10^{-12}$ W/m^2.

14-6 The Doppler Effect

One of the most common physical phenomena involving sound is the change in pitch of a train whistle or a car horn as the vehicle moves past us. This change in pitch, due to the relative motion between a source of sound and the receiver, is called the **Doppler effect,** after the Austrian physicist Christian Doppler (1803–1853). If you listen carefully to the Doppler effect, you will notice that the pitch increases when the observer and the source are moving closer together, and decreases when the observer and the source are separating.

One of the more fascinating aspects of the Doppler effect is that it applies to all wave phenomena, not just to sound. In particular, the frequency of light is also Doppler-shifted when there is relative motion between the source and the receiver. For light, this change in frequency means a change in color. In fact, most distant galaxies are observed to be red-shifted in the color of their light, which means they are moving away from the Earth. Some galaxies, however, are moving toward us, and their light shows a blue shift.

In the remainder of this section, we focus on the Doppler effect in sound waves. We show that the effect is different depending on whether the observer or the source is moving. Finally, both observer and source may be in motion, and we present results for such cases as well.

Moving Observer

In **FIGURE 14-19** we see a stationary source of sound in still air. The radiated sound is represented by the circular patterns of compressions moving away from the source with a speed v. The distance between the compressions is the wavelength, λ, and the frequency of the sound is f. As for any wave, these quantities are related by

$$v = \lambda f$$

▶ **FIGURE 14-19 The Doppler effect: a moving observer** Sound waves from a stationary source form concentric circles moving outward with a speed v. To the observer, who moves toward the source with a speed u, the waves are moving with a speed $v + u$.

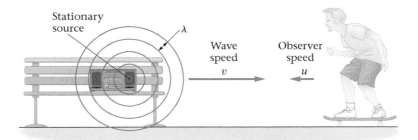

Stationary source λ Wave speed v Observer speed u

For an observer moving toward the source with a speed u, as in Figure 14-19, the sound *appears* to have a higher speed, $v + u$ (though, of course, the speed of sound relative to the air is always the same). As a result, more compressions move past the observer in a given time than if the observer had been at rest. To the observer, then, the sound has a frequency, f', that is higher than the frequency of the source, f.

We can find the frequency f' by first noting that the wavelength of the sound does not change—it is still λ. The speed, however, has increased to $v' = v + u$. Thus, we can solve $v' = \lambda f'$ for the frequency, which yields

$$f' = \frac{v'}{\lambda} = \frac{v + u}{\lambda}$$

Finally, we recall that $v = \lambda f$ can be rearranged to give $\lambda = v/f$, and hence

$$f' = \frac{v + u}{(v/f)} = \left(\frac{v + u}{v}\right)f = (1 + u/v)f$$

Notice that f' is greater than f for this observer. This is the Doppler effect.

If the observer had been moving away from the source with a speed u, the sound would *appear* to the observer to have the reduced speed $v' = v - u$. Repeating the calculation just given, we find that

$$f' = \frac{v'}{\lambda} = \frac{v - u}{\lambda} = (1 - u/v)f$$

In this case the Doppler effect results in f' being less than f.

Combining these results, we have:

Doppler Effect for Moving Observer

$f' = (1 \pm u/v)f$ 14-9

SI unit: $1/s = s^{-1} = Hz$

In this expression u and v are speeds, with u referring to the observer and v referring to sound, and hence they are always positive. The appropriate signs are obtained by using the *plus* sign when the observer moves *toward* the source, and the *minus* sign when the observer moves *away from* the source.

PROBLEM-SOLVING NOTE

Using the Correct Sign for the Doppler Effect

When an observer approaches a source, the frequency heard by the observer is greater than the frequency of the source; that is, $f' > f$. This means that we must use the plus sign in $f' = (1 \pm u/v)f$. Similarly, we must use the minus sign when the observer moves away from the source.

EXAMPLE 14-10   A MOVING OBSERVER

A street musician sounds the A string of his violin, producing a tone of 440 Hz. What frequency does a bicyclist hear as she **(a)** approaches and **(b)** recedes from the musician with a speed of 11.0 m/s?

PICTURE THE PROBLEM
The sketch shows a stationary source of sound and a moving observer. In part (a) the observer approaches the source with a speed $u = 11.0$ m/s; in part (b) the observer has passed the source and is moving away with the same speed.

REASONING AND STRATEGY
The frequency heard by the observer is given by $f' = (1 \pm u/v)f$ (Equation 14-9), with the plus sign for part (a) and the minus sign for part (b).

Known   Frequency of sound, $f = 440$ Hz; speed of bicyclist, $u = 11.0$ m/s; speed of sound, $v = 343$ m/s.

Unknown **(a)** Frequency heard by cyclist when approaching, $f' = ?$ **(b)** Frequency heard by cyclist when receding, $f' = ?$

(a)

(b)

CONTINUED

SOLUTION

Part (a)

1. Apply $f' = (1 \pm u/v)f$ with the plus sign and $u = 11.0$ m/s:
$$f' = (1 + u/v)f = \left(1 + \frac{11.0 \text{ m/s}}{343 \text{ m/s}}\right)(440 \text{ Hz}) = 454 \text{ Hz}$$

Part (b)

2. Now use the minus sign in $f' = (1 \pm u/v)f$:
$$f' = (1 - u/v)f = \left(1 - \frac{11.0 \text{ m/s}}{343 \text{ m/s}}\right)(440 \text{ Hz}) = 426 \text{ Hz}$$

INSIGHT

As the bicyclist passes the musician, the observed frequency of sound decreases, giving a "wow" effect where the frequency of the sound changes rapidly. The difference in frequency is about 1 semitone, the frequency difference between adjacent notes on the piano. See Table 14-3 on page 485 for the frequencies of notes in the vicinity of middle C.

PRACTICE PROBLEM

If the bicyclist hears a frequency of 451 Hz when approaching the musician, what is her speed?   [**Answer:** $u = 8.58$ m/s]

Some related homework problems: Problem 45, Problem 46

Moving Source

In the case of a moving source, the Doppler effect is not due to the sound wave appearing to have a higher or lower speed, as when the observer moves. On the contrary, the speed of a wave is determined solely by the properties of the medium through which it propagates. Thus, once the source emits a sound wave, it travels through the medium with its characteristic speed v regardless of what the source is doing.

By way of analogy, consider a water wave. The speed of such waves is the same whether they are created by a rock dropped into the water or by a stick moved rapidly through the water. To take an extreme case, the waves coming to the beach from a slow-moving tugboat move with the same speed as the waves produced by a 100-mph speed boat. The same is true of sound waves.

▶ **FIGURE 14-20 The Doppler effect: a moving source** Sound waves from a moving source are bunched up in the forward direction, causing a shorter wavelength and a higher frequency.

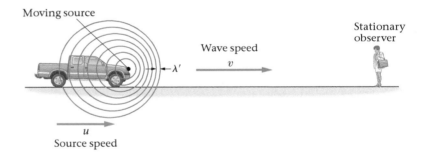

Consider, then, a source moving toward an observer with a speed u, as shown in **FIGURE 14-20**. If the frequency of the source is f, it emits one compression every T seconds, where $T = 1/f$. Therefore, during one cycle of the wave a compression travels a distance vT while the source moves a distance uT. As a result, the next compression is emitted a distance $vT - uT$ behind the previous compression, as illustrated in **FIGURE 14-21**. This means that the wavelength in the forward direction is

$$\lambda' = vT - uT = (v - u)T$$

Now, the speed of the wave is still v, as mentioned above, and hence

$$v = \lambda' f'$$

Solving for the new frequency, f', we find

$$f' = \frac{v}{\lambda'} = \frac{v}{(v - u)T}$$

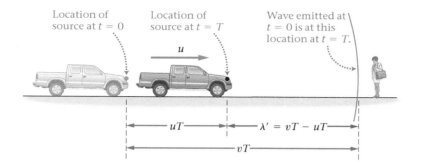

Location of source at $t = 0$

Location of source at $t = T$

Wave emitted at $t = 0$ is at this location at $t = T$.

u

uT

$\lambda' = vT - uT$

vT

◄ **FIGURE 14-21 The Doppler-shifted wavelength** During one period, T, the wave emitted at $t = 0$ moves through a distance vT. In the same time, the source moves toward the observer through the distance uT. At the time $t = T$ the next wave is emitted from the source; hence, the distance between the waves (the wavelength) is $\lambda' = vT - uT$.

Finally, recalling that $T = 1/f$, we have

$$f' = \frac{v}{(v-u)(1/f)} = \frac{v}{v-u}f = \left(\frac{1}{1-u/v}\right)f$$

Notice that f' is greater than f, as expected.

In the reverse direction, the wavelength is increased by the amount uT. Thus,

$$\lambda' = vT + uT = (v+u)T$$

It follows that the Doppler-shifted frequency is

$$f' = \frac{v}{(v+u)T} = \left(\frac{1}{1+u/v}\right)f$$

This is less than the source frequency, f.

Finally, we can combine these results to yield:

Doppler Effect for Moving Source

$$f' = \left(\frac{1}{1 \mp u/v}\right)f \qquad\qquad 14\text{-}10$$

SI unit: $1/s = s^{-1} = Hz$

As before, u and v are positive quantities. The *minus* sign is used when the source moves *toward* the observer, and the *plus* sign when the source moves *away from* the observer. The speed v refers to the speed of sound, and the speed u refers to the speed of the source of the sound.

PHYSICS IN CONTEXT
Looking Back

Basic concepts from kinematics (Chapter 2) are used to derive the Doppler effect.

PROBLEM-SOLVING NOTE

Using Correct Signs

When a source approaches an observer, the frequency heard by the observer is greater than the frequency of the source; that is, $f' > f$. This means that we must use the minus sign in Equation 14-10, $f' = f/(1 - u/v)$, since this makes the denominator less than one. Similarly, we use the plus sign when the source moves away from the observer.

EXAMPLE 14-11   WHISTLE STOP

A train sounds its whistle as it approaches a tunnel in a cliff. The whistle produces a tone of 650.0 Hz, and the train travels with a speed of 21.2 m/s. **(a)** Find the frequency heard by an observer standing near the tunnel entrance. **(b)** The sound from the whistle reflects from the cliff back to the engineer in the train. What frequency does the engineer hear?

PICTURE THE PROBLEM
The train moves with a speed $u = 21.2$ m/s and emits sound with a frequency $f = 650.0$ Hz. The observer near the tunnel hears the Doppler-shifted frequency f', and the engineer on the train hears the reflected sound at an even higher frequency f''.

REASONING AND STRATEGY
Two Doppler shifts are involved in this problem. The first is due to the motion of the train toward the cliff. This shift causes the observer at the cliff to hear sound with a higher frequency f', given by $f' = (1/(1 \mp u/v))f$ (Equation 14-10) with the minus sign. The reflected sound has the same frequency, f'.

The second shift is due to the engineer moving toward the reflected sound. Thus, the engineer hears a frequency f'' that is greater than f'. We find f'' using $f' = (1 \pm u/v)f$

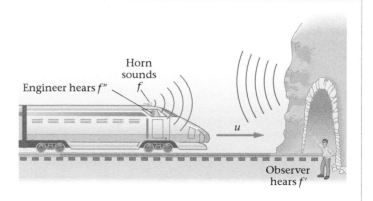

Engineer hears f''

Horn sounds f

u

Observer hears f'

CONTINUED

(Equation 14-9) with the plus sign. We use this Doppler shift equation because in this case the source of the sound (the cliff) is stationary, while the observer (the engineer) is moving.

Known Frequency of train whistle, $f = 650.0$ Hz; speed of train, $u = 21.2$ m/s; speed of sound, $v = 343$ m/s.

Unknown (a) Frequency heard by observer at rest, $f' = ?$ (b) Frequency of echo heard by engineer, $f'' = ?$

SOLUTION

Part (a)

1. Use $f' = \left(\dfrac{1}{1 \mp u/v}\right)f$, with the minus sign, to Doppler-shift from f to f':

$$f' = \left(\frac{1}{1 - u/v}\right)f = \left(\frac{1}{1 - \dfrac{21.2\text{ m/s}}{343\text{ m/s}}}\right)(650.0\text{ Hz})$$

$$= \left(\frac{1}{1 - 0.0618}\right)(650.0\text{ Hz}) = 693\text{ Hz}$$

Part (b)

2. Now use $f' = (1 \pm u/v)f$, with the plus sign, to Doppler-shift from f' to f'':

$$f'' = (1 + u/v)f' = \left(1 + \frac{21.2\text{ m/s}}{343\text{ m/s}}\right)(693\text{ Hz})$$

$$= (1 + 0.0618)(693\text{ Hz}) = 736\text{ Hz}$$

INSIGHT

The reflected sound has the same frequency f' heard by the stationary observer, because all the cliff does is reverse the direction of motion of the sound heard at the cliff. Therefore, the cliff acts as a stationary source of sound at the frequency f'. The engineer's motion toward this stationary source results in the Doppler shift from f' to f''.

PRACTICE PROBLEM

If the stationary observer hears a frequency of 700.0 Hz, what are (a) the speed of the train and (b) the frequency heard by the engineer? [**Answer:** (a) $u = 24.5$ m/s, (b) $f'' = 750$ Hz]

Some related homework problems: Problem 47, Problem 96

Comparing Doppler Shifts Now that we've obtained the Doppler effect for both moving observers and moving sources, it's interesting to compare the results. FIGURE 14-22 shows the Doppler-shifted frequency versus speed for a 400-Hz source of sound. The upper curve corresponds to a moving source, the lower curve to a moving observer. Notice that while the two cases give similar results for low speed, the high-speed behavior is quite different. In fact, the Doppler frequency for the moving source grows without limit for speeds near the speed of sound, while the Doppler frequency for the moving observer is relatively small.

These results can be understood both in terms of mathematics—by simply comparing Equations 14-9 and 14-10—and physically. In physical terms, recall that a moving observer encounters wave crests separated by the wavelength, as indicated in Figure 14-19. Doubling the speed of the observer simply reduces the time required to move from one crest to the next by a factor of two, which doubles the frequency. Thus, in general, the frequency is proportional to the speed, as we see in the lower curve in Figure 14-22. In contrast, when the source moves, as in Figure 14-20, the wave crests become "bunched up" in the forward direction, because the source is almost keeping up with the propagating waves. As the speed of the source approaches the speed of sound, the separation between the crests approaches zero. Consequently, the frequency with which the crests pass a stationary observer approaches infinity, as indicated by the upper curve in Figure 14-22.

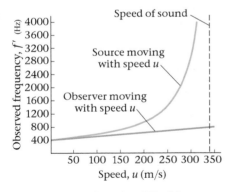

▲ **FIGURE 14-22 Doppler-shifted frequency versus speed for a 400-Hz sound source** The upper curve corresponds to a moving source, the lower curve to a moving observer. Notice that while the two cases give similar results for low speed, the high-speed behavior is quite different. In fact, the Doppler frequency for the moving source increases without limit for speeds near the speed of sound, while the Doppler frequency for the moving observer is relatively low. If a source moves faster than the speed of sound, the sound it produces is perceived not as a pure tone, but as a shock wave, commonly referred to as a sonic boom.

General Case

The results derived earlier in this section can be combined to give the Doppler effect for situations in which both observer and source move. Letting u_s be the speed of the source, and u_o be the speed of the observer, we have:

Doppler Effect for Moving Source and Observer

$$f' = \left(\frac{1 \pm u_o/v}{1 \mp u_s/v}\right)f \qquad\qquad 14\text{-}11$$

SI unit: $1/s = s^{-1} = $ Hz

As in the previous cases, u_s, u_o, and v are positive quantities. In the numerator, the plus sign corresponds to the case in which the observer moves in the direction of the source, whereas the minus sign indicates motion in the opposite direction. In the denominator, the minus sign corresponds to the case in which the source moves in the direction of the observer, whereas the plus sign indicates motion in the opposite direction.

EXERCISE 14-12   GENERAL DOPPLER SHIFT

A car traveling north at 21 m/s sounds its 480-Hz horn. A bicyclist travels south on the same road with a speed of 6.7 m/s, heading directly toward the car—though at a safe distance. What frequency does the bicyclist hear?

REASONING AND SOLUTION

We use $f' = ((1 \pm u_o/v)/(1 \mp u_s/v))f$ (Equation 14-11) to find the Doppler-shifted frequency, because both the observer and the source are in motion. In this case, $u_s = 21$ m/s and $u_o = 6.7$ m/s. Because the source and the observer are approaching, we use the plus sign in the numerator (which raises the frequency) and the minus sign in the denominator (which also raises the frequency). The resulting frequency is

$$f' = \left(\frac{1 + u_o/v}{1 - u_s/v}\right)f = \left(\frac{1 + 6.7/343}{1 - 21/343}\right)(480\,\text{Hz}) = 520\,\text{Hz}$$

RWP The Doppler effect is used in an amazing variety of technological applications. Perhaps one of the most familiar is the "radar gun," which is used to measure the speed of a pitched baseball or a car that is breaking the speed limit. Although the radar gun uses radio waves rather than sound waves, the basic physical principle is the same: By measuring the Doppler-shifted frequency of waves reflected from an object, it is possible to determine its speed. Doppler radar, used in weather forecasting, applies the same technology to tracking the motion of precipitation inside storm clouds. An example showing the formation of a tornado is presented in FIGURE 14-23 (a).

BIO In medicine, the Doppler shift is used to measure the speed of blood flow in an artery, as in FIGURE 14-23 (b), or in the heart itself. In this application, a beam of ultrasound is directed toward an artery in a patient. Some of the sound is reflected back by red blood cells moving through the artery. The reflected sound is detected, and its frequency is used to determine the speed of blood flow. If this information is color coded, with different colors indicating different speeds and directions of flow, an impressive image of blood flow can be constructed.

RWP Finally, the Doppler effect applies to the light of distant galaxies as well, as in FIGURE 14-23 (c). For example, if a galaxy moves away from us—as most do—the light we observe from that galaxy has a lower frequency than if the galaxy were at rest relative to us. Because red light has the lowest frequency of visible light, we refer to this reduction in frequency as a "red shift." Thus, by measuring the red shift of a galaxy, we can determine its speed relative to us. Such measurements show that the more distant a galaxy, the greater its speed relative to us—a result known as Hubble's law. This is just what one would expect from an explosion, or "Big Bang," in which rapidly moving pieces travel farther in a given amount of time. Thus the Doppler effect, and red-shift measurements, provide strong evidence for the Big Bang and an expanding universe.

Big Idea 4 The frequency of a sound wave depends on the motion of both the source of the sound and the observer. This is the Doppler effect. When the source and the observer approach one another, the observed frequency is higher; when the source and the observer move away from one another, the observed frequency decreases.

(a)

(b)

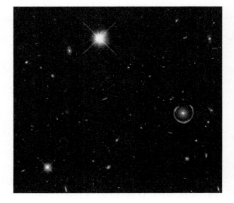

(c)

▶ **FIGURE 14-23 Visualizing Concepts The Doppler Effect** Many familiar and not-so-familiar devices utilize the Doppler effect. **(a)** Doppler radar, now widely used at airports and for weather forecasting, makes it possible for a forecaster to determine the speed and direction of winds in a distant storm by measuring the Doppler shift they produce. This image is the Doppler radar scan of a severe thunderstorm that struck Pawnee County, Oklahoma, on May 19, 2010. The hook-shaped echo is characteristic of tornadoes in the making. **(b)** A doppler echocardiogram of the human heart indicates the direction of blood flow with different colors. **(c)** Most distant galaxies are moving away from us, producing a "red shift" (lower frequency) in their light. The galaxies in this Hubble Deep Field photo have extreme red shifts—their velocities relative to us are near the speed of light.

Enhance Your Understanding　　　　(Answers given at the end of the chapter)

6. Observer 1 approaches a stationary 1000-Hz source of sound with a speed of 25 m/s. Observer 2 is at rest and an identical 1000-Hz source approaches her with a speed of 25 m/s. Is the frequency heard by observer 1 greater than, less than, or equal to the frequency heard by observer 2?

Section Review

- The observed frequency of sound depends on the relative motion of the source and the observer. The observed frequency is higher when the source and observer approach one another, and lower when they separate from one another. This is referred to as the Doppler effect.

- The observed frequency f' from a source of sound with a frequency f is $f' = \left(\dfrac{1 \pm u_\mathrm{o}/v}{1 \mp u_\mathrm{s}/v} \right) f$, where u_o is the speed of the observer and u_s is the speed of the source. The plus and minus signs are chosen to give a higher frequency when the observer and the source are approaching one another, and to give a lower frequency when they are separating.

y_1

y_2

$y = y_1 + y_2$

▲ **FIGURE 14-24 Wave superposition** Waves of small amplitude superpose (that is, combine) by simple addition.

Big Idea **5** Waves that overlap one another add together algebraically, which can result in either an increase or decrease in amplitude. This is referred to as interference.

PHYSICS IN CONTEXT
Looking Ahead

We return to the wave properties of light in Chapter 28, where we see that superposition and interference play a similar role for light as they do for sound.

PHYSICS IN CONTEXT
Looking Ahead

Another type of wave behavior is known as diffraction. This concept is developed in Chapter 28.

14-7 Superposition and Interference

So far we've considered only a single wave at a time. In this section we turn our attention to the way waves combine when more than one is present. As we shall see, the behavior of waves is quite simple in this respect.

Superposition

The combination of two or more waves to form a resultant wave is referred to as **superposition**. When waves are of small amplitude, they superpose in the simplest of ways—they just add. For example, consider two waves on a string, as in **FIGURE 14-24**, described by the wave functions y_1 and y_2. If these two waves are present on the same string at the same time, the result is a wave given by

$$y = y_1 + y_2$$

The sum $y = y_1 + y_2$ is referred to as an algebraic sum, because y_1 and y_2 can each be either positive or negative.

To see how superposition works as a function of time, consider a string with two wave pulses on it, one traveling in each direction as shown in **FIGURE 14-25 (a)**. When the pulses arrive in the same region, they add, as stated. This is illustrated in Figure 14-25(a). The question is: What do the pulses look like after they have passed through one another? Does their interaction change them in any way?

The answer is that the waves are completely unaffected by their interaction. This is also shown in Figure 14-25 (a). After the wave pulses pass through one another they continue on as if nothing had happened. It's like listening to an orchestra, where many different instruments are playing simultaneously, and their sounds are combining throughout the concert hall. Even so, you can still hear individual instruments, each making its own sound as if the others were not present.

Interference

As simple as the principle of superposition is, it still leads to interesting consequences. For example, consider the wave pulses on a string shown in Figure 14-25 (a). When they combine, the resulting pulse has an amplitude equal to the sum of the amplitudes of the individual pulses. This is referred to as **constructive interference**.

On the other hand, two pulses like those in **FIGURE 14-25 (b)** may combine. When this happens, the positive displacement of one wave adds to the negative displacement

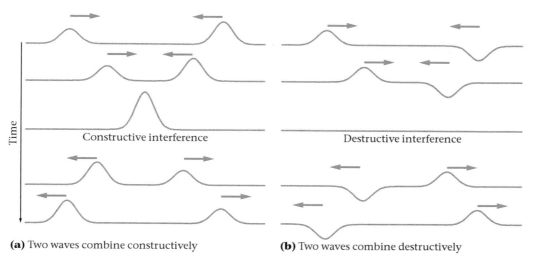

Constructive interference

Destructive interference

(a) Two waves combine constructively

(b) Two waves combine destructively

◄ **FIGURE 14-25 Interference** Wave pulses superpose as they pass through one another. Afterward, the pulses continue on unchanged. In **(a)**, an example of *constructive interference*, two positive pulses combine to give a larger amplitude. In **(b)**, an example of *destructive interference*, a positive pulse and a symmetrical negative pulse momentarily cancel each other.

of the other to create a net displacement of zero. That is, the pulses momentarily cancel one another. This is **destructive interference**.

It's important to note that the waves don't simply disappear when they experience destructive interference. For example, in Figure 14-25 (b) the wave pulses continue on unchanged after they interact. This makes sense from an energy point of view—after all, each wave carries energy, and hence the waves, along with their energy, cannot simply vanish. In fact, when the string is flat in Figure 14-25(b) it has its greatest transverse speed—just as a swing has its highest speed when it is in its equilibrium position. Therefore, the energy of the wave is still present at this instant of time—it is just in the form of the kinetic energy of the string.

Interference Patterns It should also be noted that interference is not limited to waves on a string; all waves exhibit interference effects. In fact, you could say that interference is one of the key characteristics that define waves. In general, when waves combine, they form **interference patterns** that include regions of both constructive and destructive interference. An example is shown in **FIGURE 14-26**, where two circular waves are interfering. Notice the regions of constructive interference separated by regions of destructive interference.

To understand the formation of such patterns, consider a system of two identical sources, as in **FIGURE 14-27**. Each source sends out waves consisting of alternating crests and troughs. We set up the system so that when one source emits a crest, the other emits a crest as well. Sources that are synchronized like this are said to be **in phase**.

Now, at a point like A, the distance to each source is the same. Thus, if the wave from one source produces a crest at point A, so too does the wave from the other source. As a result, with crest combining with crest, the interference at A is constructive.

Next consider point B. At this location the wave from source 1 must travel a greater distance than the wave from source 2. If the extra distance is half a wavelength, it follows that when the wave from source 2 produces a crest at B the wave from source 1 produces a trough. As a result, the waves combine to give destructive interference at B. At point C, on the other hand, the distance from source 1 is one wavelength greater than the distance from source 2. Hence the waves are in phase again at C, with crest meeting crest for constructive interference.

In general, then, we can say that constructive and destructive interference occur under the following conditions for two sources that are in phase:

Constructive interference occurs when the path length from two in-phase sources differs by $0, \lambda, 2\lambda, 3\lambda, \dots$.

Destructive interference occurs when the path length from two in-phase sources differs by $\lambda/2, 3\lambda/2, 5\lambda/2, \dots$.

Constructive interference along this line, where crest meets crest

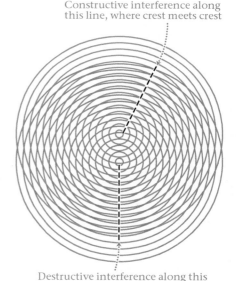

Destructive interference along this line, where crest meets trough

▲ **FIGURE 14-26 Interference of circular waves** Interference pattern formed by the superposition of two sets of circular waves. The light radial "rays" are regions where crest meets crest and trough meets trough (constructive interference). The dark areas in between the light rays are regions where the crest of one wave overlaps the trough of another wave (destructive interference).

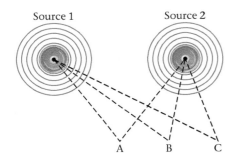

Source 1   Source 2

A   B   C

◄ **FIGURE 14-27 Interference with two sources** Suppose the two sources emit waves in phase. At point A the distance to each source is the same; hence, crest meets crest. The result is constructive interference. At B the distance from source 1 is greater than the distance from source 2 by half a wavelength. The result is crest meeting trough and destructive interference. Finally, at C the distance from source 1 is one wavelength greater than the distance from source 2. Hence, we find constructive interference at C. If the sources had been opposite in phase, then A and C would be points of destructive interference, and B would be a point of constructive interference.

Other path length differences result in intermediate degrees of interference, between the extremes of destructive and constructive interference.

A specific example of interference patterns is provided by sound, using speakers that emit sound in phase with the same frequency. This situation is analogous to the two sources in Figure 14-27. As a result, constructive or destructive interference is to be expected, depending on the path length from each speaker. This is illustrated in the next Example.

EXAMPLE 14-13 SOUND OFF

Two speakers separated by a distance of 4.30 m emit sound of frequency 221 Hz. The speakers are in phase with one another. A person listens from a location 2.80 m directly in front of one of the speakers. Does the person hear constructive or destructive interference?

PICTURE THE PROBLEM
Our sketch shows the two speakers emitting sound in phase at the frequency $f = 221$ Hz. The speakers are separated by the distance $D = 4.30$ m, and the observer is a distance $d_1 = 2.80$ m directly in front of one of the speakers.

REASONING AND STRATEGY
The type of interference depends on whether the difference in path length, $d_2 - d_1$, is one or more wavelengths or an odd multiple of half a wavelength. Thus, we begin by calculating the wavelength, λ. Next, we find d_2, and compare the difference in path length to λ.

Known Separation of speakers, $D = 4.30$ m; frequency of speakers, $f = 221$ Hz; distance from speaker 1 to person, $d_1 = 2.80$ m.
Unknown Type of interference = ?

SOLUTION

1. Calculate the wavelength of this sound, using $v = \lambda f$. As usual, let $v = 343$ m/s be the speed of sound:

$$\lambda = \frac{v}{f} = \frac{343 \text{ m/s}}{221 \text{ Hz}} = 1.55 \text{ m}$$

2. Find the path length d_2:

$$d_2 = \sqrt{D^2 + d_1{}^2} = \sqrt{(4.30 \text{ m})^2 + (2.80 \text{ m})^2} = 5.13 \text{ m}$$

3. Determine the difference in path length, $d_2 - d_1$:

$$d_2 - d_1 = 5.13 \text{ m} - 2.80 \text{ m} = 2.33 \text{ m}$$

4. Divide λ into $d_2 - d_1$ to find the number of wavelengths that fit into the path difference:

$$\frac{d_2 - d_1}{\lambda} = \frac{2.33 \text{ m}}{1.55 \text{ m}} = 1.50$$

INSIGHT
The path difference is $3\lambda/2$, and hence we expect destructive interference. In the ideal case, the person would hear no sound. As a practical matter, some sound will be reflected from objects in the vicinity, resulting in a finite sound intensity.

PRACTICE PROBLEM
We know that 221 Hz gives destructive interference. What is the lowest frequency (corresponding to the largest wavelength) that gives constructive interference for the case described in this Example? [**Answer:** Set $\lambda = d_2 - d_1 = 2.33$ m, the largest wavelength that gives constructive interference. This yields $f = 147$ Hz.]

Some related homework problems: Problem 57, Problem 59

RWP If the two speakers in Figure 14-27 have **opposite phase** (by connecting one of them with its wires reversed), the conditions for constructive and destructive interference are changed, as are the interference patterns. For example, at point A, where the distances from the two speakers are the same, the wave from one speaker is a compression when the wave from the other speaker is a rarefaction. Thus, point A is now a point of destructive interference rather than constructive interference. In general, then, the conditions for constructive and destructive interference are simply reversed—a path difference of $0, \lambda, 2\lambda, \ldots$ results in destructive interference, a path difference of $\lambda/2, 3\lambda/2, \ldots$ results in constructive interference.

Out-of-Phase Speakers

QUICK EXAMPLE 14-14 OPPOSITE PHASE INTERFERENCE

The speakers shown in our sketch have opposite phase. They are separated by a distance of 5.20 m and emit sound with a frequency of 104 Hz. A person stands 3.00 m in front of the speakers and 1.30 m to one side of the center line between them. What type of interference occurs at the person's location?

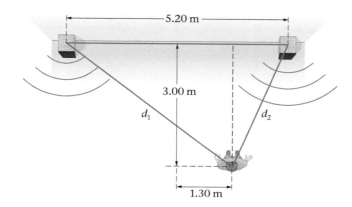

REASONING AND SOLUTION

We start by finding the wavelength of the sound and the distance to the person from each speaker. If the difference in distances is a multiple of the wavelength of the sound, the interference is destructive (because the speakers are opposite in phase); if the difference in distances is an odd multiple of half a wavelength, the interference is constructive.

1. Calculate the wavelength of the sound: $\qquad$ $\lambda = 3.30\,\text{m}$

2. Find the path length d_1: $\qquad$ $d_1 = 4.92\,\text{m}$

3. Find the path length d_2: $\qquad$ $d_2 = 3.27\,\text{m}$

4. Calculate the path length difference, $d_1 - d_2$: $\qquad$ $d_1 - d_2 = 1.65\,\text{m}$

5. Divide the path length difference by the wavelength: $\qquad$ $\dfrac{d_1 - d_2}{\lambda} = 0.500$

The path length difference is half a wavelength, and the speakers have opposite phase. As a result, the person experiences constructive interference.

RWP Destructive interference can be used to reduce the intensity of noise in a variety of situations, such as a factory, a busy office, or even the cabin of an airplane. The process, referred to as active noise reduction (ANR), begins with a microphone that picks up the noise to be reduced. The signal from the microphone is then reversed in phase and sent to a speaker. As a result, the speaker emits sound that is opposite in phase to the incoming noise—in effect, the speaker produces "anti-noise." In this way, the noise is *actively* canceled by destructive interference, rather than simply reduced by absorption. The effect when wearing a pair of ANR headphones can be as much as a 30-dB reduction in the intensity level of noise.

▲ **FIGURE 14-28** Enhance Your Understanding 7.

7. Two in-phase speakers emit sound with a wavelength of 2 m. An observer is a distance d_1 from one speaker and a distance d_2 from the other speaker, as shown schematically in **FIGURE 14-28**. For each of the following cases, state whether the observer experiences constructive or destructive interference: **(a)** $d_1 = 2$ m, $d_2 = 1$ m; **(b)** $d_1 = 4$ m, $d_2 = 2$ m; **(c)** $d_1 = 4$ m, $d_2 = 1$ m; **(d)** $d_1 = 6$ m, $d_2 = 1$ m.

Section Review

• Waves add by simple addition.

• If the crest of one wave meets the crest of another wave, the result is an increased amplitude—referred to as constructive interference. If the crest of one wave meets the trough of another wave, the result is a decreased amplitude—referred to as destructive interference.

14-8 Standing Waves

If you've ever plucked a guitar string, or blown across the mouth of a pop bottle to create a tone, you've generated **standing waves.** In general, a standing wave is one that oscillates with time, but remains fixed in its location. It's in this sense that the wave is said to be "standing."

In some respects, a standing wave can be considered as resulting from constructive interference of a wave with itself. As one might expect, then, standing waves occur only if specific conditions are satisfied. We explore these conditions in this section for two cases: (i) waves on a string and (ii) sound waves in a hollow, cylindrical structure.

Waves on a String

We begin by considering a string of length L that is tied down at both ends, as in **FIGURE 14-29 (a)**. If you pluck this string in the middle it vibrates as shown in **FIGURE 14-29 (b)**. This is referred to as the **fundamental mode** of vibration for this string or, also, as the **first harmonic.** Clearly, the string assumes a wavelike shape, but because of the boundary conditions—the ends tied down—the wave stays in place.

As is clear from **FIGURE 14-29 (c)**, the fundamental mode corresponds to half a wavelength of a usual wave on a string. One can think of the fundamental as being formed by this wave reflecting back and forth between the walls holding the string. If the frequency is just right, the reflections combine to give constructive interference and the fundamental is formed; if the frequency differs from the fundamental frequency, the reflections result in destructive interference and a standing wave does not result.

We can find the frequency of the fundamental, or first harmonic, as follows: First we use the fact that the wavelength of the first harmonic is twice the distance between the walls. Thus,

$$\lambda_1 = 2L$$

If the speed of waves on the string is v, it follows that the frequency of the first harmonic, f_1, is determined by the speed relationship $v = \lambda f$. In this case, $v = \lambda_1 f_1 = (2L)f_1$. Solving for the frequency yields

$$f_1 = \frac{v}{\lambda_1} = \frac{v}{2L}$$

Notice that the frequency of the first harmonic increases with the speed of the waves, and decreases as the string is lengthened.

(a)

(b)

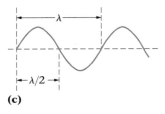

(c)

▲ **FIGURE 14-29 A standing wave (a)** A string is tied down at both ends. **(b)** If the string is plucked in the middle, a standing wave results. This is the fundamental mode of oscillation of the string. **(c)** The fundamental consists of one-half $\left(\frac{1}{2}\right)$ a wavelength between the two ends of the string. Hence, its wavelength is $2L$.

(a) First harmonic (fundamental) **(b) Second harmonic** **(c) Third harmonic**

▲ **FIGURE 14-30 Harmonics** The first three harmonics for a string tied down at both ends. Note that an extra half wavelength is added to go from one harmonic to the next. **(a)** $\lambda/2 = L$, $\lambda = 2L$; **(b)** $\lambda = L$; **(c)** $3\lambda/2 = L$, $\lambda = 2L/3$.

Higher Harmonics The first harmonic is not the only standing wave that can exist on a string. In fact, there are an infinite number of standing wave modes—or **harmonics**—for any given string, with frequencies that are integer multiples of the first harmonic. To find the higher harmonics, first note that the two ends of the string must remain fixed. Points on a standing wave that stay fixed are referred to as **nodes**. Halfway between any two nodes is a point on the wave that has a maximum displacement, as indicated in FIGURE 14-30 and in the photo in FIGURE 14-31. Such a point is called an **antinode**. Referring to Figure 14-30 (a), then, we see that the first harmonic consists of two nodes (N) and one antinode (A); the sequence is N-A-N.

The *second* harmonic can be constructed by including one more half wavelength in the standing wave, as shown in Figure 14-30 (b). This mode has the sequence N-A-N-A-N, and has one complete wavelength between the walls; that is, $\lambda_2 = L$. Therefore, the frequency of the second harmonic, f_2, is

$$f_2 = \frac{v}{\lambda_2} = \frac{v}{L} = 2f_1$$

Similarly, the *third* harmonic again includes one more half wavelength, as in Figure 14-30 (c). Now there are one and a half wavelengths in the length L, and hence $(3/2)\lambda_3 = L$, or $\lambda_3 = 2L/3$. The corresponding third-harmonic frequency, f_3, is

$$f_3 = \frac{v}{\lambda_3} = \frac{v}{\frac{2}{3}L} = 3\frac{v}{2L} = 3f_1$$

Notice that the frequencies of the harmonics are increasing in integer steps; that is, each harmonic has a frequency that is an integer multiple of the first-harmonic frequency. Clearly, then, the sequence of standing waves can be characterized as follows:

Standing Waves on a String

The first harmonic (fundamental) has the following frequency and wavelength:

$$f_1 = \frac{v}{2L}$$
$$\lambda_1 = 2L \qquad\qquad\qquad 14\text{-}12$$

In general, the nth harmonic, with $n = 1, 2, 3, \ldots$, has the following frequency and wavelength:

$$f_n = nf_1 = n\frac{v}{2L} \qquad n = 1, 2, 3, \ldots$$
$$\lambda_n = \lambda_1/n = \frac{2L}{n} \qquad n = 1, 2, 3, \ldots \qquad 14\text{-}13$$

The integer n in these expressions represents the number of half wavelengths in the standing waves. In addition, notice that the difference in frequency between any two successive harmonics is equal to the first-harmonic frequency, f_1.

Waves on a string provide a one-dimensional example of standing waves. Standing waves can also exist in two- and three-dimensional systems. An example of a standing wave in two dimensions is shown in FIGURE 14-32.

(a)

(b)

(c)

▲ **FIGURE 14-31** The string in these multiflash photographs vibrates in one of three different standing wave patterns, each with its own characteristic frequency. **(a)** The lowest frequency standing wave—also referred to as the fundamental, or first harmonic—is shown here. In this case, there is one antinode. **(b)** The second harmonic has two antinodes. **(c)** The third harmonic has three antinodes.

▲ **FIGURE 14-32** The photos show a time sequence (from left to right) as a square metal plate that is initially at rest is vibrated vertically about its center. Initially the plate is covered with a uniform coating of salt crystals. As the plate is vibrated, however, a standing wave develops. The salt makes the wave pattern visible by collecting at the nodes, where the plate is at rest. Clearly, standing waves on a two-dimensional plate can be much more complex than the standing waves on a string.

EXAMPLE 14-15   IT'S FUNDAMENTAL

One of the harmonics on a string 1.30 m long has a frequency of 15.60 Hz. The next higher harmonic has a frequency of 23.40 Hz. Find **(a)** the fundamental frequency and **(b)** the speed of waves on this string.

PICTURE THE PROBLEM
The problem statement does not tell us directly which two harmonics have the given frequencies. We do know, however, that they are *successive* harmonics of the string. For example, if one harmonic has one node between the two ends of the string, the next harmonic has two nodes. Our sketch illustrates this case, which turns out to be appropriate for this problem.

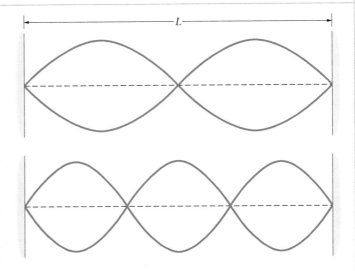

REASONING AND STRATEGY

a. We know from $f_n = nf_1 = n(v/2L)$ (Equation 14-13) that the frequencies of successive harmonics increase by f_1. That is, $f_2 = f_1 + f_1 = 2f_1$, $f_3 = f_2 + f_1 = 3f_1$, $f_4 = f_3 + f_1 = 4f_1, \ldots$. Therefore, we can find the fundamental frequency, f_1, by taking the difference between the given frequencies.

b. Once the fundamental frequency is determined, we can find the speed of waves in the string from the relationship $f_1 = v/2L$.

Known   Length of string, $L = 1.30$ m; one harmonic of the string $= 15.60$ Hz; next higher harmonic of the string $= 23.40$ Hz.

Unknown   **(a)** Fundamental frequency of the string, $f_1 = ?$ **(b)** Speed of waves on string, $v = ?$

SOLUTION

Part (a)

1. The fundamental frequency is the difference between the two given frequencies:

$$f_1 = 23.40 \text{ Hz} - 15.60 \text{ Hz} = 7.80 \text{ Hz}$$

Part (b)

2. Solve $f_1 = v/2L$ for the speed, v:

$$f_1 = \frac{v}{2L}$$
$$v = 2Lf_1$$

3. Substitute numerical values:

$$v = 2Lf_1 = 2(1.30 \text{ m})(7.80 \text{ Hz}) = 20.3 \text{ m/s}$$

INSIGHT
Now that we know the fundamental frequency, we can identify the harmonics given in the problem statement. First, 15.6 Hz $= 2(7.80$ Hz), so this is the second harmonic. The next mode, 23.4 Hz $= 3(7.80$ Hz), is the third harmonic, as expected.

PRACTICE PROBLEM
Suppose the tension in this string is increased until the speed of the waves is 22.0 m/s. What are the frequencies of the first three harmonics in this case?   [**Answer:** $f_1 = 8.46$ Hz, $f_2 = 16.9$ Hz, $f_3 = 25.4$ Hz]

Some related homework problems: Problem 69, Problem 71

Musical Instruments When a guitar string is plucked or a piano string is struck, it vibrates primarily in its fundamental mode, with smaller contributions coming from the higher harmonics. It follows that notes of different pitch can be produced by using strings of different length. Recalling that the fundamental frequency for a string of length L is $f_1 = v/2L$, we see that long strings produce low frequencies and short strings produce high frequencies—all other variables remaining the same.

RWP This fact accounts for the general shape of a piano. Notice that the strings shorten toward the right side of the piano, where the notes are of higher frequency. Similarly, a double bass is a larger instrument with longer strings than a violin, as one would expect by the different frequencies the instruments produce. To tune a stringed instrument, the tension in the strings is adjusted—since changing the length of the instrument is impractical. This in turn varies the speed v of waves on the string, and hence the fundamental frequency $f_1 = v/2L$ can be adjusted as desired. Once the string is tuned, a performer can produce different notes by pressing the string against a fret to change its effective length—that is, the length that is free to vibrate.

RWP The human ear perceives frequencies as being equally spaced when they are related to one another by a multiplicative factor. For instance, middle C on the piano has a frequency of 261.7 Hz. If we move up one octave to the next C, the frequency is 523.3 Hz; up one more octave, the next C is 1047 Hz. Notice that with each octave the frequency doubles; that is, it goes up by a multiplicative factor of two. Since there are 12 semitones in one octave of the chromatic scale, it follows that the frequency increase from one semitone to the next is $(2)^{1/12}$. The frequencies for a full chromatic octave are given in Table 14-3.

TABLE 14-3 Chromatic Musical Scale

Note	Frequency (Hz)
Middle C	261.7
C♯ (C-sharp)	
D♭ (D-flat)	277.2
D	293.7
D♯, E♭	311.2
E	329.7
F	349.2
F♯, G♭	370.0
G	392.0
G♯, A♭	415.3
A	440.0
A♯, B♭	466.2
B	493.9
C	523.3

▲ **FIGURE 14-33** Three factors determine the pitch of a vibrating string: mass per unit length, μ; tension, F; and length, L. In a guitar, the first of these factors is fixed once the strings are put on the guitar. The strings vary in thickness; other things being equal, the heavier the string, the lower the pitch. The second factor, the tension, can be varied by twisting the pegs at the top of the neck (left), adjusting the pitch of each "open" string to its correct value. The third factor, the length of the string, is controlled by the musician pressing a string against one of the frets (right), changing its effective length—the length of string that is free to vibrate—and thus the note that is produced.

RWP On a guitar, a performer can produce two full octaves and more on a single string by pressing the string down against frets to change its effective length. The separation between frets is not uniform, however, as can be seen in **FIGURE 14-33**. In particular, suppose the unfretted string has a fundamental frequency of 250 Hz. Noting that one octave up on the scale would be twice the frequency, 500 Hz, the length of the string must be halved to produce that note. To go to the next octave, and double the frequency again to 1000 Hz, the string must be shortened by a factor of two again, to one-quarter of its original length. This is illustrated in **FIGURE 14-34**. Because the distance between successive octaves is decreasing—in this case from $L/2$ to $L/4$—but the number of notes in an octave stays the same, it follows that the spacing between frets must decrease as one goes to higher notes. As a result, the frets on a guitar are always more closely spaced as one moves toward the base of the neck.

▲ **FIGURE 14-34 Frets on a guitar** To go up one octave from the fundamental, the effective length of a guitar string must be halved. To increase one more octave, it is necessary to halve the length of the string again. Thus, the distance between frets is not uniform; they are more closely spaced near the base of the neck.

Vibrating Columns of Air

If you blow across the open end of a pop bottle, as in **FIGURE 14-35**, you hear a tone of a certain frequency. If you pour some water into the bottle and repeat the experiment,

▲ **FIGURE 14-35 Exciting a standing wave** When air is blown across the open top of a soda pop bottle, the turbulent air flow can cause an audible standing wave. The standing wave will have an antinode, A, at the top (where the air is moving) and a node, N, at the bottom (where the air cannot move).

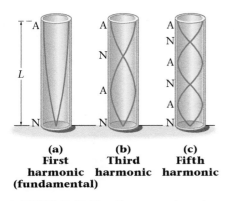

(a)	(b)	(c)
First	**Third**	**Fifth**
harmonic	**harmonic**	**harmonic**
(fundamental)		

▲ **FIGURE 14-36 Standing waves in a pipe that is open at one end** The first three harmonics for waves in a column of air of length L that is open at one end: **(a)** $\lambda/4 = L$, $\lambda = 4L$; **(b)** $3\lambda/4 = L$, $\lambda = 4L/3$; **(c)** $5\lambda/4 = L$, $\lambda = 4L/5$.

the sound you hear has a higher frequency. In both cases you have excited the fundamental mode of the column of air within the bottle. When water was added to the bottle, however, the column of air was shortened, leading to a higher frequency—in the same way that a shortened string has a higher frequency.

Columns Closed at One End Let's examine the situation of an empty bottle more carefully. When you blow across the opening in the bottle, the result is a swirling movement of air that excites rarefactions and compressions, as illustrated in Figure 14-35. For this reason, the opening is an antinode (A) for sound waves. In contrast, the bottom of the bottle is closed, preventing movement of the air; hence, it must be a node (N). Any standing wave in the bottle must have a node at the bottom and an antinode at the top.

The lowest frequency standing wave that is consistent with these conditions is shown in **FIGURE 14-36 (a)**. If we plot the density variation of the air for this wave, we see that one-quarter of a wavelength fits into the column of air in the bottle. Thus, if the length of the bottle is L, the first harmonic (fundamental) has a wavelength satisfying the following relationship:

$$\tfrac{1}{4}\lambda = L$$
$$\lambda = 4L$$

The first-harmonic frequency, f_1, is given by

$$v = \lambda f_1$$

Solving for f_1 we find

$$f_1 = \frac{v}{\lambda} = \frac{v}{4L}$$

This is half of the corresponding fundamental frequency for a wave on a string.

Higher Harmonics The next harmonic is produced by adding half a wavelength, just as in the case of the string. Thus, if the fundamental is represented by N-A, the second harmonic can be written as N-A-N-A. Because the distance from a node to an antinode is a quarter of a wavelength, we see that three-quarters $\left(\tfrac{3}{4}\right)$ of a wavelength fits into the bottle for this mode. This is shown in FIGURE 14-36 (b). Therefore, $3\lambda/4 = L$, and hence

$$\lambda = \tfrac{4}{3}L$$

As a result, the frequency is

$$\frac{v}{\lambda} = \frac{v}{\frac{4}{3}L} = 3\frac{v}{4L} = 3f_1$$

Notice that this is the *third* harmonic of the pipe; that is, its frequency is three times f_1.

Similarly, the next-higher harmonic is written as N-A-N-A-N-A, as indicated in FIGURE 14-36 (c). In this case, $5\lambda/4 = L$, and the frequency is

$$\frac{v}{\lambda} = \frac{v}{\frac{4}{5}L} = 5\frac{v}{4L} = 5f_1$$

This is the *fifth* harmonic of the pipe.

Clearly, the progression of harmonics for a column of air that is closed at one end and open at the other end is described by the following frequencies and wavelengths:

Standing Waves in a Column of Air Closed at One End

$$f_1 = \frac{v}{4L}$$
$$f_n = nf_1 = n\frac{v}{4L} \quad n = 1, 3, 5, \dots$$
$$\lambda_n = \frac{\lambda_1}{n} = \frac{4L}{n} \quad n = 1, 3, 5, \dots$$

14-14

Only odd harmonics are present in this case, as opposed to waves on a string, in which all integer harmonics occur.

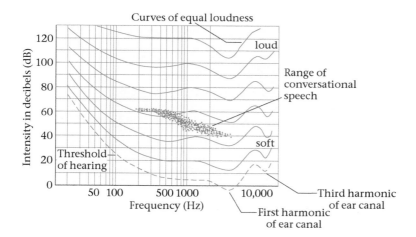

Curves of equal loudness

loud

Range of conversational speech

soft

Threshold of hearing

Third harmonic of ear canal

First harmonic of ear canal

◄ **FIGURE 14-37 Human response to sound** The human ear is more sensitive to some frequencies of sound than to others. For example, every point on a "curve of equal loudness" seems just as loud to us as any other point, even though the corresponding physical intensities may be quite different. To illustrate, note that the threshold of hearing is not equal to 0 dB for all frequencies. In fact, it is approximately 25 dB at 100 Hz, about 5 dB at 1000 Hz, and even slightly negative near 3500 Hz. Thus, regions where the curves dip downward correspond to increased sensitivity of the ear—in fact, near 3500 Hz we can hear sounds that are a thousand times less intense than sounds at 100 Hz. The two most prominent dips occur near 3500 Hz and 11,000 Hz, corresponding to standing waves in the ear canal analogous to those shown in Figure 14-36 (a) and (b), respectively. (See Problem 68.)

RWP The human ear canal is an example of a column of air that is closed at one end (the eardrum) and open at the other end. Standing waves in the ear canal can lead to an increased sensitivity of hearing. This is illustrated in **FIGURE 14-37**, which shows "curves of equal loudness" as a function of frequency. Where these curves dip downward, sounds of lower intensity seem just as loud as sounds of higher intensity at other frequencies. The two prominent dips near 3500 Hz and 11,000 Hz are due to standing waves in the ear canal corresponding to Figures 14-36 (a) and (b), respectively.

EXAMPLE 14-16   POP MUSIC

An empty soda pop bottle is to be used as a musical instrument in a band. In order to be tuned properly, the fundamental frequency of the bottle must be 440.0 Hz. **(a)** If the bottle is 26.0 cm tall, how high should it be filled with water to produce the desired frequency? Treat the bottle as a pipe that is closed at one end (the surface of the water) and open at the other end. **(b)** What is the frequency of the next higher harmonic for this bottle?

PICTURE THE PROBLEM

In our sketch, we label the height of the bottle with $H = 26.0$ cm, and the unknown height of water with h. Clearly, then, the length of the vibrating column of air is $L = H - h$.

REASONING AND STRATEGY

a. Given the frequency of the fundamental ($f_1 = 440.0$ Hz) and the speed of sound in air ($v = 343$ m/s), we can use $f_1 = v/4L$ to solve for the length L of the air column. The height of water is then $h = H - L$.

b. The next higher harmonic for a pipe open at one end is the third harmonic ($n = 3$ in $f_n = n(v/4L)$). Thus, the next higher harmonic frequency for this bottle is $f_3 = 3f_1$.

Known   Fundamental frequency of bottle, $f_1 = 440.0$ Hz; height of bottle, $H = 0.260$ m.

Unknown   **(a)** Height of water in bottle, $h = ?$ **(b)** Frequency of next higher harmonic, $f_3 = ?$

SOLUTION

Part (a)

1. Solve $f_1 = v/4L$ for the length L:
$$f_1 = \frac{v}{4L} \quad \text{or} \quad L = \frac{v}{4f_1}$$

2. Substitute numerical values:
$$L = \frac{v}{4f_1} = \frac{343 \text{ m/s}}{4(440.0 \text{ Hz})} = 0.195 \text{ m}$$

3. Use $h = H - L$ to find the height of the water:
$$h = H - L = 0.260 \text{ m} - 0.195 \text{ m} = 0.065 \text{ m} = 6.5 \text{ cm}$$

Part (b)

4. Calculate the frequency of the third harmonic (the next higher) with $f_3 = 3f_1$:
$$f_3 = 3f_1 = 3(440.0 \text{ Hz}) = 1320 \text{ Hz}$$

CONTINUED

INSIGHT
Filling a series of bottles with different levels of water can produce a nice musical instrument that sounds similar to a xylophone.

PRACTICE PROBLEM — PREDICT/CALCULATE
(a) If the water level in the bottle is raised, does the fundamental frequency increase, decrease, or stay the same? Explain. **(b)** Find the fundamental frequency if the water level is increased to 7.00 cm. [**Answer: (a)** Raising the water level reduces the length of the air column. As a result, the fundamental wavelength decreases, and the fundamental frequency increases. **(b)** The fundamental frequency increases to $f_1 = 451$ Hz.]

Some related homework problems: Problem 66, Problem 68, Problem 72

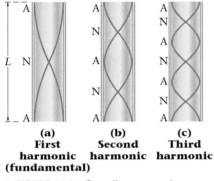

▲ **FIGURE 14-38 Standing waves in a pipe that is open at both ends** The first three harmonics for waves in a column of air of length L that is open at both ends: **(a)** $\lambda/2 = L$, $\lambda = 2L$; **(b)** $\lambda = L$; **(c)** $3\lambda/2 = L$, $\lambda = 2L/3$.

Columns Open at Both Ends Standing waves can also be produced in columns of air that are open at both ends, as illustrated in **FIGURE 14-38**. In this case there is an anti-node at each end of the column. Hence, the first harmonic, or fundamental, is A-N-A, as shown in Figure 14-38 (a). In this case, half a wavelength fits into the pipe. As a result,

$$f_1 = \frac{v}{2L}$$

This is the same as the corresponding result for a wave on a string.

The next harmonic is A-N-A-N-A, which fits one complete wavelength in the pipe. This harmonic is shown in Figure 14-38 (b), and has the frequency

$$f_2 = \frac{v}{L} = 2f_1$$

This is the *second* harmonic of the pipe. The rest of the harmonics continue in exactly the same manner as for waves on a string, with all integer harmonics present. Thus, the frequencies and wavelengths in a column of air open at both ends are as follows:

Standing Waves in a Column of Air Open at Both Ends

$$f_1 = \frac{v}{2L}$$

$$f_n = nf_1 = n\frac{v}{2L} \quad n = 1, 2, 3, \ldots$$

$$\lambda_n = \frac{\lambda_1}{n} = \frac{2L}{n} \quad n = 1, 2, 3, \ldots$$

14-15

CONCEPTUAL EXAMPLE 14-17   TALKING WITH HELIUM

If you fill your lungs with helium and speak, you sound something like Donald Duck. From this observation, can we conclude that the speed of sound in helium is greater than, less than, or equal to the speed of sound in air?

REASONING AND DISCUSSION
When we speak with helium, our words are higher pitched. Looking at $f_n = n(v/2L)$ (Equation 14-15), we see that for the frequency to increase while the length of the vocal chords remains the same, the speed of sound must be higher.

ANSWER
The speed of sound is greater in helium than in air.

RWP A pipe organ uses a variety of pipes of different length, with some open at both ends and others open at one end only. When a key is pressed on the console of the organ, air is forced through a given pipe. By accurately adjusting the length of the pipe, we can create the desired tone. The principle is the same as playing a note on a pop bottle, as in **FIGURE 14-39**. In addition, because open and closed pipes have different harmonic frequencies, they sound distinctly different to the ear, even if they have the same fundamental frequency. Thus, by judiciously choosing both the length and the type of a pipe, an organist can play a range of different sounds, allowing the organ to mimic a trumpet, a trombone, a clarinet, and so on.

Standing waves have also been observed in the Sun. Like an enormous, low-frequency musical instrument, the Sun vibrates once roughly every five minutes, a result of the roiling nuclear reactions that take place within its core. One of the goals of SOHO, the Solar and Heliospheric Observatory, is to study these solar vibrations in detail. By observing the variety of standing waves produced in the Sun, we can learn more about its internal structure and dynamics.

Enhance Your Understanding   (Answers given at the end of the chapter)

8. When a string oscillates with the standing wave shown in **FIGURE 14-40** (a) its frequency is 100 Hz. What is the frequency when the same string oscillates with the standing wave shown in **FIGURE 14-40 (b)**?

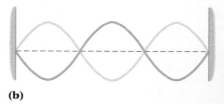

(a)   (b)

▲ **FIGURE 14-40** Enhance Your Understanding 8.

Section Review

- Standing waves oscillate in place. The standing wave with the largest wavelength and lowest frequency is the fundamental, or first harmonic. Higher harmonics have higher frequencies.

- The standing waves on a string tied down at both ends have frequencies $f_n = n(v/2L)$, where n is the harmonic number, L is the length of the string, and v is the speed of waves on the string.

- The standing waves in a column of air of length L and closed at one end have frequencies $f_n = n(v/4L)$. In this expression, v is the speed of sound in air. The standing waves in a column of air of length L and open at both ends have frequencies $f_n = n(v/2L)$. This is the same expression as for the standing waves on a string.

▲ **FIGURE 14-39 Visualizing Concepts**
Waves in a Column of Air Blowing across the mouth of a bottle sets the air column within the bottle vibrating, producing a tone. This principle is put to use on a large scale in a pipe organ. A typical pipe organ may contain hundreds of pipes of different lengths, some open at both ends and some at only one, affording the performer great control over the tonal quality of the sound produced, as well as its pitch.

14-9 Beats

An interference pattern, such as that shown in Figure 14-26, is a snapshot at a given time, showing locations where constructive and destructive interference occur. It is an interference pattern in space. **Beats,** on the other hand, can be thought of as an interference pattern in time.

To be specific, imagine plucking two guitar strings that have slightly different frequencies. If you listen carefully, you notice that the sound produced by the strings is not constant in time. In fact, the intensity increases and decreases with a definite period. These fluctuations in intensity are the beats, and the frequency of successive maximum intensities is the **beat frequency.**

As an example, suppose two waves, with frequencies $f_1 = 1/T_1$ and $f_2 = 1/T_2$, interfere at a given, fixed location. At this location, each wave moves up and down with simple harmonic motion, as described by Equation 13-2. Applying this result to the vertical position, y, of each wave yields the following:

$$y_1 = A\cos\left(\frac{2\pi}{T_1}t\right) = A\cos(2\pi f_1 t)$$

$$y_2 = A\cos\left(\frac{2\pi}{T_2}t\right) = A\cos(2\pi f_2 t)$$

14-16

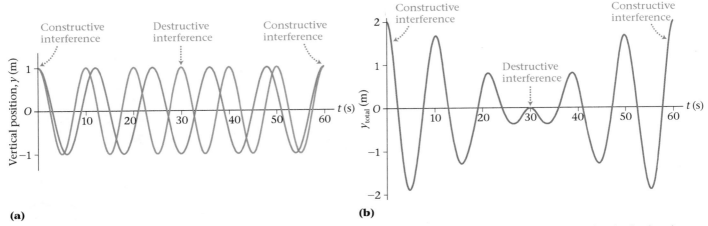

(a)

(b)

▲ **FIGURE 14-41 Interference of two waves with slightly different frequencies (a)** A plot of the two waves, y_1 (blue) and y_2 (red), given in Equations 14-16. **(b)** The resultant wave y_{total} for the two waves shown in part (a). Note the alternately constructive and destructive interference leading to beats.

These equations are plotted in **FIGURE 14-41 (a)**, with $A = 1$ m, and their superposition, $y_{total} = y_1 + y_2$, is shown in **FIGURE 14-41 (b)**.

Notice that at the time $t = 0$, both y_1 and y_2 are equal to A; thus their superposition gives $2A$. Because the waves have different frequencies, however, they do not stay in phase. At a later time, t_1, we find that $y_1 = A$ and $y_2 = -A$; their superposition gives zero at this time. At a still later time, $t_2 = 2t_1$, the waves are again in phase and add to give $2A$. Thus, a person listening to these two waves hears a sound whose amplitude and loudness vary with time; that is, the person hears beats.

Superposing these waves mathematically, we find

$$
\begin{aligned}
y_{total} &= y_1 + y_2 \\
&= A\cos(2\pi f_1 t) + A\cos(2\pi f_2 t) \\
&= 2A\cos\left(2\pi\frac{f_1 - f_2}{2}t\right)\cos\left(2\pi\frac{f_1 + f_2}{2}t\right)
\end{aligned}
$$

14-17

> **Big Idea 6** When two sound waves of different frequency superpose, the result is sound with a loudness that varies with time. The frequency of the loudness variation is known as the beat frequency.

The final step in the expression follows from the trigonometric identities given in Appendix A. The first part of y_{total} is

$$
2A\cos\left(2\pi\frac{f_1 - f_2}{2}t\right)
$$

This gives the slowly varying amplitude of the beats, as indicated in **FIGURE 14-42**. A loud sound is heard whenever this term is equal to $-2A$ or $2A$—which happens twice during any given oscillation—and hence the beat frequency is:

▶ **FIGURE 14-42 Beats** Beats can be understood as oscillations at the average frequency, modulated by a slowly varying amplitude.

Amplitude: oscillates with beat frequency, $|f_1 - f_2|$

y_{total}: oscillates with average frequency, $\frac{f_1 + f_2}{2}$

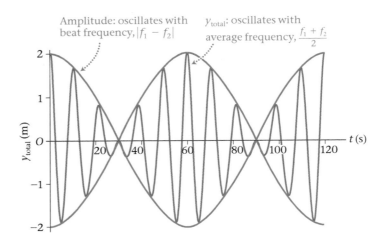

Definition of Beat Frequency

$$f_{\text{beat}} = |f_1 - f_2|$$ 14-18

SI unit: $1/s = s^{-1} = \text{Hz}$

Finally, the rapid oscillations within each beat are due to the second part of y_{total}:

$$\cos\left(2\pi\frac{f_1 + f_2}{2}t\right)$$

These oscillations have a frequency that is the average of the two input frequencies.

These results apply to any type of wave. In particular, if two sound waves produce beats, your ear will hear the average frequency with a loudness that varies with the beat frequency. For example, suppose the two guitar strings mentioned at the beginning of this section have the frequencies 438 Hz and 442 Hz. If you sound them simultaneously, you will hear the average frequency, 440 Hz, increasing and decreasing in loudness with a beat frequency of 4 Hz. This means that you will hear maximum loudness four times a second. If the frequencies are brought closer together, the beat frequency will be reduced, and fewer maxima will be heard each second.

Clearly, beats can be used to tune a musical instrument to a desired frequency. To tune a guitar string to 440 Hz, for example, the string can be played simultaneously with a 440-Hz tuning fork. Listening to the beats, the person tuning the guitar increases or decreases the tension in the string until the beat frequency becomes vanishingly small. This technique applies only to frequencies that are reasonably close to begin with, because the maximum beat frequencies the ear can detect are about 15 Hz to 20 Hz.

PROBLEM-SOLVING NOTE

Calculating the Beat Frequency

The beat frequency of two waves is the *magnitude* of the difference in their frequencies. Thus, the beat frequency is always positive.

EXAMPLE 14-18 | GETTING A TUNE-UP

A guitarist tunes one of her strings to a 440-Hz tuning fork. Initially, a beat frequency of 5 Hz is heard when the string and tuning fork are sounded. When the guitarist tightens the string slowly and steadily, the beat frequency decreases to 4 Hz. What was the initial frequency of the string?

PICTURE THE PROBLEM
A guitarist is tightening a string at the same time that a tuning fork is giving off (emitting) sound with a frequency of 440 Hz. A beat frequency is heard, which decreases as the string is tightened.

Known Frequency of tuning fork, $f = 440$ Hz; initial beat frequency = 5 Hz; final beat frequency = 4 Hz.

Unknown Initial frequency of guitar string = ?

Beat frequency = 5 Hz — tightening string — Beat frequency = 4 Hz

440 Hz

REASONING AND SOLUTION
The fact that the initial beat frequency is 5 Hz means the initial frequency of the string is either 435 Hz or 445 Hz. Now, tightening the string increases its frequency. Why? Well, tightening the string means the tension is increased, which in turn means the speed of waves in the string is increased (recall $v = \sqrt{F/\mu}$ from Equation 14-2). As a result, the fundamental frequency is also increased (recall $f_1 = v/2L$ from Equation 14-12). The fact that increasing the frequency of the string decreased the beat frequency—bringing it closer to 440 Hz—means the initial frequency was 435 Hz.

INSIGHT
In this case, the initial frequency was too low (435 Hz). If tightening the string had increased the beat frequency, then the initial frequency would have been too high (445 Hz) and the guitarist would have to loosen the string.

CONTINUED

PRACTICE PROBLEM

Suppose that the initial beat frequency was 4 Hz, and that tightening the string caused the beat frequency to increase steadily to 6 Hz. What was the initial frequency of the string?   [**Answer:** 444 Hz]

Some related homework problems: Problem 78, Problem 79

Enhance Your Understanding   (Answers given at the end of the chapter)

9. Rank the following systems in order of increasing beat frequency. Indicate ties where appropriate. **A**, $f_1 = 100\,\text{Hz}$, $f_2 = 110\,\text{Hz}$; **B**, $f_1 = 90\,\text{Hz}$, $f_2 = 95\,\text{Hz}$; **C**, $f_1 = 110\,\text{Hz}$, $f_2 = 100\,\text{Hz}$; **D**, $f_1 = 303\,\text{Hz}$, $f_2 = 300\,\text{Hz}$.

Section Review

- Playing two sounds with different frequencies f_1 and f_2 simultaneously produces beats, which can be heard as an alternation in the loudness of the sound. The frequency of the beats is given by $f_{\text{beat}} = |f_1 - f_2|$.

CHAPTER 14 REVIEW

CHAPTER SUMMARY

14-1 TYPES OF WAVES

A wave is a propagating disturbance.

Transverse Waves and Longitudinal Waves
In a transverse wave individual particles move at right angles to the direction of wave propagation. In a longitudinal wave individual particles move in the same direction as the wave propagation.

Wavelength, Frequency, and Speed
The wavelength, λ, frequency, f, and speed, v, of a wave are related by

$$v = \lambda f \qquad \textbf{14-1}$$

14-2 WAVES ON A STRING

Transverse waves can propagate on a string held taut with a tension force, F.

Speed of a Wave on a String
The speed of a wave on a string with a tension force F and a mass per length μ is

$$v = \sqrt{\frac{F}{\mu}} \qquad \textbf{14-2}$$

14-3 HARMONIC WAVE FUNCTIONS

A harmonic wave has the shape of a sine or a cosine function.

Wave Function
A harmonic wave of wavelength λ and period T is described by the following expression:

$$y(x, t) = A \cos\!\left(\frac{2\pi}{\lambda}x - \frac{2\pi}{T}t\right) \qquad \textbf{14-4}$$

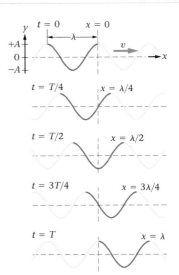

14-4   SOUND WAVES

A sound wave is a longitudinal wave of compressions and rarefactions that can travel through the air, as well as through other gases, liquids, and solids.

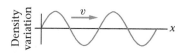

14-5   SOUND INTENSITY

The loudness of a sound is determined by its intensity.

Intensity

Intensity, I, is a measure of the amount of energy per time that passes through a given area. Since energy per time is power, P, the intensity of a wave is

$$I = \frac{P}{A} \qquad \text{14-5}$$

Intensity Level and Decibels

The intensity level, β, of a sound gives an indication of how loud it sounds to our ears. The intensity level is defined as

$$\beta = (10\,\text{dB}) \log\!\left(\frac{I}{I_0}\right) \qquad \text{14-8}$$

The value of β is given in decibels.

14-6   THE DOPPLER EFFECT

The change in frequency due to the relative motion between a source and a receiver is called the Doppler effect.

General Case

If the observer moves with a speed u_o and the source moves with a speed u_s, the Doppler effect gives

$$f' = \left(\frac{1 \pm u_o/v}{1 \mp u_s/v}\right) f \qquad \text{14-11}$$

The plus and minus signs are chosen to give a higher frequency when the observer and the source are approaching one another, and to give a lower frequency when they are separating.

14-7   SUPERPOSITION AND INTERFERENCE

Waves can combine to give a variety of effects.

Constructive Interference

Waves that add to give a larger amplitude exhibit constructive interference.

Destructive Interference

Waves that add to give a smaller amplitude exhibit destructive interference.

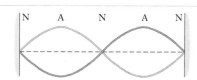

14-8   STANDING WAVES

Standing waves oscillate in a fixed location.

Waves on a String

The fundamental, or first harmonic, corresponds to half a wavelength fitting into the length of the string. The fundamental for waves of speed v on a string of length L is

$$f_1 = \frac{v}{2L}$$
$$\lambda_1 = 2L \qquad \text{14-12}$$

The higher harmonics, with $n = 1, 2, 3, \ldots$, are described by

$$f_n = nf_1 = n(v/2L)$$
$$\lambda_n = \lambda_1/n = 2L/n$$

14-13

Vibrating Columns of Air

The harmonics for a column of air closed at one end are

$$f_n = nf_1 = n(v/4L) \quad n = 1, 3, 5, \ldots$$
$$\lambda_n = \lambda_1/n = 4L/n$$

14-14

The harmonics for a column of air open at both ends are

$$f_n = nf_1 = n(v/2L) \quad n = 1, 2, 3, \ldots$$
$$\lambda_n = \lambda_1/n = 2L/n$$

14-15

In all four of these expressions the speed of sound is v and the length of the column is L.

14-9 BEATS

Beats occur when waves of slightly different frequencies interfere.

Beat Frequency

If waves of frequencies f_1 and f_2 interfere, the beat frequency is

$$f_{\text{beat}} = |f_1 - f_2|$$

14-18

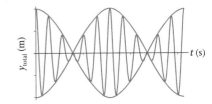

ANSWERS TO ENHANCE YOUR UNDERSTANDING QUESTIONS

1. $A < B = D < C$.

2. The speed doubles.

3. (a) $A = 0.12$ m. (b) $\lambda = 8$ m. (c) $T = 0.25$ s.

4. Wave 2 has the greater speed (25 m/s compared to 8 m/s).

5. (c)

6. Observer 1 hears a frequency that is less than the frequency heard by observer 2.

7. (a) Destructive. (b) Constructive. (c) Destructive. (d) Destructive.

8. 150 Hz.

9. $D < B < A = C$.

CONCEPTUAL QUESTIONS

For instructor-assigned homework, go to www.masteringphysics.com. MP

(Answers to odd-numbered Conceptual Questions can be found in the back of the book.)

1. A long nail has been driven halfway into the side of a barn. How should you hit the nail with a hammer to generate a longitudinal wave? How should you hit it to generate a transverse wave?

2. What type of wave is exhibited by "amber waves of grain"?

3. In a classic TV commercial, a group of cats feed from bowls of cat food that are lined up side by side. Initially there is one cat for each bowl. When an additional cat is added to the scene, it runs to a bowl at the end of the line and begins to eat. The cat that was there originally moves to the next bowl, displacing that cat, which moves to the next bowl, and so on down the line. What type of wave have the cats created? Explain.

4. Describe how the sound of a symphony played by an orchestra would be altered if the speed of sound depended on the frequency of sound.

5. A "radar gun" is often used to measure the speed of a major-league pitch by reflecting a beam of radio waves off a moving ball. Describe how the Doppler effect can give the speed of the ball from a measurement of the frequency of the reflected beam.

6. When you drive a nail into a piece of wood, you hear a tone with each blow of the hammer. In fact, the tone increases in pitch as the nail is driven farther into the wood. Explain.

7. Explain the function of the sliding part of a trombone.

8. On a guitar, some strings are single wires, others are wrapped with another wire to increase the mass per length. Which type of string would you expect to be used for a low-frequency note? Explain.

9. To play a C major chord on the piano, you hit the C, E, and G keys simultaneously. When you do so, you hear no beats. Why? (Refer to Table 14-3.)

PROBLEMS AND CONCEPTUAL EXERCISES

Answers to odd-numbered Problems and Conceptual Exercises can be found in the back of the book. **BIO** *identifies problems of biological or medical interest;* **CE** *indicates a conceptual exercise.* Predict/Explain *problems ask for two responses: (a) your prediction of a physical outcome, and (b) the best explanation among three provided; and* Predict/Calculate *problems ask for a prediction of a physical outcome, based on fundamental physics concepts, and follow that with a numerical calculation to verify the prediction. On all problems, bullets (•, ••, •••) indicate the level of difficulty. Planetary data can be found in Appendix C. Assume the speed of sound in air is 343 m/s unless stated otherwise.*

SECTION 14-1 TYPES OF WAVES

1. • A wave travels along a stretched horizontal rope. The vertical distance from crest to trough for this wave is 16 cm and the horizontal distance from crest to trough is 22 cm. What are (a) the wavelength and (b) the amplitude of this wave?

2. • A surfer floating beyond the breakers notes 17 waves per minute passing her position. If the wavelength of these waves is 38 m, what is their speed?

3. • The speed of surface waves in water decreases as the water becomes shallower. Suppose waves travel across the surface of a lake with a speed of 2.0 m/s and a wavelength of 1.5 m. When these waves move into a shallower part of the lake, their speed decreases to 1.6 m/s, though their frequency remains the same. Find the wavelength of the waves in the shallower water.

4. • **Tsunami** A tsunami traveling across deep water can have a speed of 750 km/h and a wavelength of 310 km. What is the frequency of such a wave?

5. •• A stationary boat bobs up and down with a period of 2.3 s when it encounters the waves from a moving boat. **(a)** What is the frequency of the waves? **(b)** If the crests of the waves are 8.5 m apart, what is their speed?

6. •• Predict/Calculate A 4.5-Hz wave with an amplitude of 12 cm and a wavelength of 27 cm travels along a stretched horizontal string. **(a)** How far does a given peak on the wave travel in a time interval of 0.50 s? **(b)** How far does a knot on the string travel in the same time interval? **(c)** How would your answers to parts (a) and (b) change if the amplitude of the wave were halved? Explain.

7. •• **Deepwater Waves** The speed of a deepwater wave with a wavelength λ is given approximately by $v = \sqrt{g\lambda/2\pi}$. Find **(a)** the speed and **(b)** the frequency of a deepwater wave with a wavelength of 4.5 m.

8. •• **Shallow-Water Waves** In shallow water of depth d the speed of waves is approximately $v = \sqrt{gd}$. Find **(a)** the speed and **(b)** the period of a wave with a wavelength of 2.6 cm in water that is 0.75 cm deep.

SECTION 14-2 WAVES ON A STRING

9. • CE Consider a wave on a string with constant tension. If the frequency of the wave is doubled, by what multiplicative factor do **(a)** the speed and **(b)** the wavelength change?

10. • CE Suppose you would like to double the speed of a wave on a string. By what multiplicative factor must you increase the tension?

11. • CE Predict/Explain Two strings are made of the same material and have equal tensions. String 1 is thick; string 2 is thin. **(a)** Is the speed of waves on string 1 greater than, less than, or equal to the speed of waves on string 2? **(b)** Choose the *best explanation* from among the following:
  I. Since the strings are made of the same material, the wave speeds will also be the same.
  II. A thick string implies a large mass per length and a slow wave speed.
  III. A thick string has a greater force constant, and therefore a greater wave speed.

12. • CE Predict/Explain Two strings are made of the same material and have waves of equal speed. String 1 is thick; string 2 is thin. **(a)** Is the tension in string 1 greater than, less than, or equal to the tension in string 2? **(b)** Choose the *best explanation* from among the following:
  I. String 1 must have a greater tension to compensate for its greater mass per length.
  II. String 2 will have a greater tension because it is thinner than string 1.
  III. Equal wave speeds implies equal tensions.

13. • CE The three waves, A, B and C, shown in FIGURE 14-43 propagate on strings with equal tensions and equal mass per length. Rank the waves in order of increasing **(a)** frequency, **(b)** wavelength, and **(c)** speed. Indicate ties where appropriate.

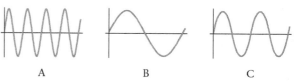

A   B   C

FIGURE 14-43 Problem 13

14. •• A brother and sister try to communicate with a string tied between two tin cans (FIGURE 14-44). If the string is 7.6 m long, has a mass of 29 g, and is pulled taut with a tension of 18 N, how much time does it take for a wave to travel from one end of the string to the other?

|← 7.6 m →|

FIGURE 14-44 Problems 14 and 15

15. •• Predict/Calculate **(a)** Suppose the tension is increased in the previous problem. Does a wave take more, less, or the same time to travel from one end to the other? Calculate the time of travel for tensions of **(b)** 22 N and **(c)** 32 N.

16. •• Predict/Calculate A 5.7-m wire with a mass of 81 g is attached to the mast of a sailboat. If the wire is given a "thunk" at one end, it takes 0.084 s for the resulting wave to reach the other end. **(a)** What is the tension in the wire? **(b)** Would the tension found in part (a) be larger or smaller if the mass of the wire is greater than 81 g? **(c)** Calculate the tension for a 91-g wire.

17. •• A 4.5-m-long rope of mass 1.8 kg hangs from a ceiling. **(a)** What is the tension in the rope at the bottom end? **(b)** What is the wave speed in the rope at the bottom end? **(c)** What is the tension in the rope at the top end, where it is attached to the ceiling? **(d)** What is the wave speed in the rope at the top end? **(e)** It can be shown that the average wave speed in the rope is $\frac{1}{2}\sqrt{gL}$, where L is the length of the rope. Use the average wave speed to calculate the time required for a pulse produced at the bottom of the rope to travel to the top.

18. ••• Two steel guitar strings have the same length. String A has a diameter of 0.50 mm and is under 410.0 N of tension. String B has a diameter of 1.0 mm and is under a tension of 820.0 N. Find the ratio of the wave speeds, v_A/v_B, in these two strings.

19. ••• Use dimensional analysis to show how the speed v of a wave on a string of circular cross section depends on the tension in the string, F, the radius of the string, R, and its mass per volume, ρ.

SECTION 14-3 HARMONIC WAVE FUNCTIONS

20. • Write an expression for a harmonic wave with an amplitude of 0.19 m, a wavelength of 2.6 m, and a period of 1.2 s. The wave is transverse, travels to the right, and has a displacement of 0.19 m at $t = 0$ and $x = 0$.

21. • Write an expression for a transverse harmonic wave that has a wavelength of 2.6 m and propagates to the right with a speed of 14.3 m/s. The amplitude of the wave is 0.11 m, and its displacement at $t = 0$ and $x = 0$ is 0.11 m.

22. •• **CE** The vertical displacement of a wave on a string is described by the equation $y(x, t) = A \sin(Bx - Ct)$, in which A, B, and C are positive constants. (a) Does this wave propagate in the positive or negative x direction? (b) What is the wavelength of this wave? (c) What is the frequency of this wave? (d) What is the smallest positive value of x where the displacement of this wave is zero at $t = 0$?

23. •• As it travels through a crystal, a light wave is described by the function $E(x,t) = A \cos[(1.54 \times 10^7)x - (2.98 \times 10^{15})t]$. In this expression, x is measured in meters and t is measured in seconds. What is the speed of the light wave?

24. •• Predict/Calculate A wave on a string is described by the following equation:

$$y = (15 \text{ cm}) \cos\left(\frac{\pi}{5.0 \text{ cm}}x - \frac{\pi}{12 \text{ s}}t\right)$$

(a) What is the amplitude of this wave? (b) What is its wavelength? (c) What is its period? (d) What is its speed? (e) In which direction does the wave travel?

25. •• Consider a harmonic wave with the following wave function:

$$y = (12 \text{ cm}) \cos\left(\frac{\pi}{4.5 \text{ cm}}x - \frac{\pi}{16 \text{ s}}t\right)$$

Sketch this wave from $x = 0$ to $x = 9.0$ cm for the following times: (a) $t = 0$; (b) $t = 4.0$ s; (c) $t = 8.0$ s. (d) What is the least amount of time required for a given point on this wave to move from $y = 0$ to $y = 12$ cm?

26. •• Predict/Calculate Four waves are described by the following equations, in which all distances are measured in centimeters and all times are measured in seconds:

$$y_A = 10 \cos(3x - 4t)$$
$$y_B = 10 \cos(5x + 4t)$$
$$y_C = 20 \cos(-10x + 60t)$$
$$y_D = 20 \cos(-4x - 20t)$$

(a) Which of these waves travel in the $+x$ direction? (b) Which of these waves travel in the $-x$ direction? (c) Which wave has the highest frequency? (d) Which wave has the greatest wavelength? (e) Which wave has the greatest speed?

SECTION 14-4 SOUND WAVES

27. • At Zion National Park a loud shout produces an echo 1.55 s later from a colorful sandstone cliff. How far away is the cliff?

28. • **BIO Dolphin Ultrasound** Dolphins of the open ocean are classified as Type II Odontocetes (toothed whales). These animals use ultrasonic "clicks" with a frequency of about 55 kHz to navigate and find prey. (a) Suppose a dolphin sends out a series of clicks that are reflected back from the bottom of the ocean 75 m below. How much time elapses before the dolphin hears the echoes of the clicks? (The speed of sound in seawater is approximately 1530 m/s.) (b) What is the wavelength of 55-kHz sound in the ocean?

29. • The lowest note on a piano is A, four octaves below the A given in Table 14-3. The highest note on a piano is C, four octaves above middle C. Find the frequencies and wavelengths (in air) of these notes.

30. •• **Meteor Infrasound** A meteor that explodes in the atmosphere creates infrasound waves that can travel multiple times around the globe. On February 15, 2013, the Chelyabinsk meteor produced infrasound waves with a wavelength of 6.4 km that propagated 8840 km in 8.53 h. What was the frequency of these waves?

31. •• Predict/Calculate A sound wave in air has a frequency of 405 Hz. (a) What is its wavelength? (b) If the frequency of the sound is increased, does its wavelength increase, decrease, or stay the same? Explain. (c) Calculate the wavelength for a sound wave with a frequency of 485 Hz.

32. •• Predict/Calculate When you drop a rock into a well, you hear the splash 1.5 seconds later. (a) How deep is the well? (b) If the depth of the well were doubled, would the time required to hear the splash be greater than, less than, or equal to 3.0 seconds? Explain.

33. ••• A rock is thrown downward into a well that is 7.62 m deep. If the splash is heard 1.17 seconds later, what was the initial speed of the rock?

SECTION 14-5 SOUND INTENSITY

34. • **CE** If the distance to a point source of sound is doubled, by what multiplicative factor does the intensity change?

35. • The intensity level of sound in a truck is 88 dB. What is the intensity of this sound?

36. • The distance to a point source is decreased by a factor of three. (a) By what multiplicative factor does the intensity increase? (b) By what additive amount does the intensity level increase?

37. • Sound 1 has an intensity of 48.0 W/m². Sound 2 has an intensity level that is 3.5 dB greater than the intensity level of sound 1. What is the intensity of sound 2?

38. •• A bird-watcher is hoping to add the white-throated sparrow to her "life list" of species. How far could she be from the bird described in Example 14-8 and still be able to hear it? Assume no reflections or absorption of the sparrow's sound.

39. •• Residents of Hawaii are warned of the approach of a tsunami by sirens mounted on the tops of towers. Suppose a siren produces a sound that has an intensity level of 120 dB at a distance of 2.0 m. Treating the siren as a point source of sound, and ignoring reflections and absorption, find the intensity level heard by an observer at a distance of (a) 12 m and (b) 21 m from the siren. (c) How far away can the siren be heard?

40. •• In a pig-calling contest, a caller produces a sound with an intensity level of 107 dB. How many such callers would be required to reach the pain level of 120 dB?

41. •• Predict/Calculate Twenty violins playing simultaneously with the same intensity combine to give an intensity level of 82.5 dB. (a) What is the intensity level of each violin? (b) If the number of violins is increased to 40, will the combined intensity level be more than, less than, or equal to 165 dB? Explain.

42. •• **BIO The Human Eardrum** The radius of a typical human eardrum is about 4.0 mm. Find the energy per second received by an eardrum when it listens to sound that is (a) at the threshold of hearing and (b) at the threshold of pain.

SECTION 14-6 THE DOPPLER EFFECT

43. • **CE** Predict/Explain A horn produces sound with frequency f_0. Let the frequency you hear when you are at rest and the horn moves toward you with a speed u be f_1; let the frequency you hear when the horn is at rest and you move toward it with a speed u be

f_2. **(a)** Is f_1 greater than, less than, or equal to f_2? **(b)** Choose the *best explanation* from among the following:

 I. A moving observer encounters wave crests more often than a stationary observer, leading to a higher frequency.

 II. The relative speeds are the same in either case. Therefore, the frequencies will be the same as well.

 III. A moving source causes the wave crests to "bunch up," leading to a higher frequency than for a moving observer.

44. • **CE** You are heading toward an island in your speedboat when you see a friend standing on shore at the base of a cliff. You sound the boat's horn to get your friend's attention. Let the wavelength of the sound produced by the horn be λ_1, the wavelength as heard by your friend be λ_2, and the wavelength of the echo as heard on the boat be λ_3. Rank these wavelengths in order of increasing length. Indicate ties where appropriate.

45. • When the bell in a clock tower rings with a sound of 475 Hz, a pigeon roosting in the belfry flies directly away from the bell. If the pigeon hears a frequency of 448 Hz, what is its speed?

46. • A car approaches a train station with a speed of 24 m/s. A stationary train at the station sounds its 166-Hz horn. What frequency is heard by the driver of the car?

47. • **BIO** A bat moving with a speed of 3.25 m/s and emitting sound of 35.0 kHz approaches a moth at rest on a tree trunk. **(a)** What frequency is heard by the moth? **(b)** If the speed of the bat is increased, is the frequency heard by the moth higher or lower? **(c)** Calculate the frequency heard by the moth when the speed of the bat is 4.25 m/s.

48. • A motorcycle and a police car are moving toward one another. The police car emits sound with a frequency of 523 Hz and has a speed of 32.2 m/s. The motorcycle has a speed of 14.8 m/s. What frequency does the motorcyclist hear?

49. •• Hearing the siren of an approaching fire truck, you pull over to the side of the road and stop. As the truck approaches, you hear a tone of 460 Hz; as the truck recedes, you hear a tone of 410 Hz. How much time will it take for the truck to get from your position to the fire 5.0 km away, assuming it maintains a constant speed?

50. •• Predict/Calculate For a science experiment, you would like to use the Doppler effect to double the perceived frequency of a harmonica, from the 261.7-Hz tone (middle C) that it emits, to 523.3 Hz (C_5 note). **(a)** Which strategy will accomplish your goal more easily: move the harmonica, or move the observer? **(b)** Calculate the speed the harmonica must have in order for a stationary observer to hear 523.3 Hz. **(c)** Calculate the speed an observer must have in order to perceive a 523.3-Hz sound from a stationary harmonica that emits sound at 261.7 Hz.

51. •• Predict/Calculate Two bicycles approach one another, each traveling with a speed of 8.50 m/s. **(a)** If bicyclist A beeps a 315-Hz horn, what frequency is heard by bicyclist B? **(b)** Which of the following would cause the greater increase in the frequency heard by bicyclist B: (i) bicyclist A speeds up by 1.50 m/s, or (ii) bicyclist B speeds up by 1.50 m/s? Explain.

52. •• A train on one track moves in the same direction as a second train on the adjacent track. The first train, which is ahead of the second train and moves with a speed of 36.8 m/s, blows a horn whose frequency is 124 Hz. If the frequency heard on the second train is 135 Hz, what is its speed?

53. •• Two cars traveling with the same speed move directly away from one another. One car sounds a horn whose frequency is 422 Hz

and a person in the other car hears a frequency of 385 Hz. What is the speed of the cars?

54. ••• **The Bullet Train** The Shinkansen, the Japanese "bullet" train, runs at high speed from Tokyo to Nagoya. Riding on the Shinkansen, you notice that the frequency of a crossing signal changes markedly as you pass the crossing. As you approach the crossing, the frequency you hear is f; as you recede from the crossing the frequency you hear is $2f/3$. What is the speed of the train?

SECTION 14-7 SUPERPOSITION AND INTERFERENCE

55. • Pulses A, B, C, and D all travel at 10 m/s on the same string but in opposite directions. The string is depicted at time $t = 0$ in FIGURE 14-45. The small pulses have an amplitude of 2.0 cm, and the large pulses have an amplitude of 4.0 cm. Find the displacement of point P at times **(a)** $t = 0.10$ s and **(b)** $t = 0.20$ s.

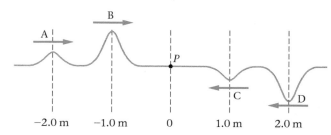

FIGURE 14-45 Problem 55

56. • Pulses A, B, C, and D all travel at 10 m/s on the same string but in opposite directions. The string is depicted at time $t = 0$ in FIGURE 14-46. The small pulses have an amplitude of 2.0 cm, and the large pulse has an amplitude of 4.0 cm. Find the displacement of point P at times **(a)** $t = 0.10$ s and **(b)** $t = 0.20$ s.

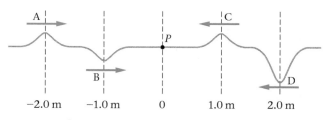

FIGURE 14-46 Problem 56

57. •• A pair of in-phase stereo speakers is placed side by side, 0.914 m apart. You stand directly in front of one of the speakers, 2.44 m from the speaker. What is the lowest frequency that will produce constructive interference at your location?

58. •• Predict/Calculate Two violinists, one directly behind the other, play for a listener directly in front of them. Both violinists sound concert A (440 Hz). **(a)** What is the smallest separation between the violinists that will produce destructive interference for the listener? **(b)** Does this smallest separation increase or decrease if the violinists produce a note with a higher frequency? **(c)** Repeat part (a) for violinists who produce sounds of 540 Hz.

59. •• Two loudspeakers are placed at either end of a gymnasium, both pointing toward the center of the gym and equidistant from it. The speakers emit 266-Hz sound that is in phase. An observer at the center of the gym experiences constructive interference. How far toward either speaker must the observer walk to first experience destructive interference?

60. •• Two speakers with opposite phase are positioned 3.5 m apart, both pointing toward a wall 5.0 m in front of them (**FIGURE 14-47**). An observer standing against the wall midway between the speakers hears destructive interference. If the observer hears constructive interference after moving 0.84 m to one side along the wall, what is the frequency of the sound emitted by the speakers?

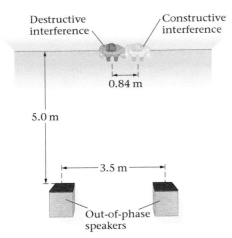

FIGURE 14-47 Problem 60

61. •• Suppose, in Example 14-13, that the speakers have opposite phase. What is the lowest frequency that gives destructive interference in this case?

SECTION 14-8 STANDING WAVES

62. • **CE** Predict/Explain When you blow across the opening of a partially filled 2-L soda pop bottle you hear a tone. **(a)** If you take a sip of the pop and blow across the opening again, does the tone you hear have a higher frequency, a lower frequency, or the same frequency as before? **(b)** Choose the *best explanation* from among the following:
  I. The same pop bottle will give the same frequency regardless of the amount of pop it contains.
  II. The greater distance from the top of the bottle to the level of the pop results in a higher frequency.
  III. A lower level of pop results in a longer column of air, and hence a lower frequency.

63. • An organ pipe that is open at both ends is 3.5 m long. What is its fundamental frequency?

64. • A string 2.5 m long with a mass of 3.6 g is stretched between two fixed points with a tension of 91 N. Find the frequency of the fundamental on this string.

65. • **BIO Human Vocal Resonance** The unique sound of a person's voice arises from the frequencies of the standing waves that occur in a person's chest, trachea, larynx, mouth, nose, and sinuses. Suppose a person's voice has a strong resonant frequency at 890 Hz when she is breathing air. What will be the frequency of the same resonance when the person breathes helium? (*Note:* The speed of sound in helium is 1007 m/s.)

66. •• The fundamental wavelength for standing sound waves in an empty pop bottle is 0.88 m. **(a)** What is the length of the bottle? **(b)** What is the frequency of the third harmonic of this bottle?

67. •• **CE** A string is tied down at both ends. Some of the standing waves on this string have the following frequencies: 100 Hz, 200 Hz, 250 Hz, and 300 Hz. It is also known that there are no standing waves with frequencies between 250 Hz and 300 Hz. **(a)** What is the fundamental frequency of this string? **(b)** What is the frequency of the third harmonic?

68. •• **BIO** Predict/Calculate **Standing Waves in the Human Ear** The human ear canal is much like an organ pipe that is closed at one end (at the tympanic membrane or eardrum) and open at the other (**FIGURE 14-48**). A typical ear canal has a length of about 2.4 cm. **(a)** What are the fundamental frequency and wavelength of the ear canal? **(b)** Find the frequency and wavelength of the ear canal's third harmonic. (Recall that the third harmonic in this case is the standing wave with the second-lowest frequency.) **(c)** Suppose a person has an ear canal that is shorter than 2.4 cm. Is the fundamental frequency of that person's ear canal greater than, less than, or the same as the value found in part (a)? Explain. [Note that the frequencies found in parts (a) and (b) correspond closely to the frequencies of enhanced sensitivity in Figure 14-37.]

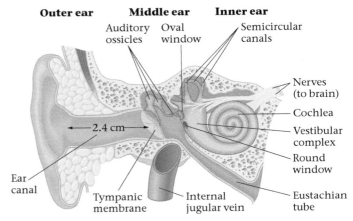

FIGURE 14-48 Problem 68

69. •• A guitar string 66 cm long vibrates with a standing wave that has three antinodes. **(a)** Which harmonic is this? **(b)** What is the wavelength of this wave?

70. •• Predict/Calculate A guitar string has a mass per length of 2.33×10^{-3} kg/m and a fundamental frequency of 146.8 Hz when it is under a tension of 82.4 N. The string breaks, and its owner has only a spare string of mass per length 6.79×10^{-3} kg/m. **(a)** In order to use the new string to play the same note, should the string tension be greater than, less than, or equal to 82.4 N? Explain. **(b)** What tension should the 6.79×10^{-3} kg/m string have so that its fundamental frequency is 146.8 Hz?

71. •• Predict/Calculate A 0.323-kg clothesline is stretched with a tension of 53.4 N between two poles 6.10 m apart. What is the frequency of **(a)** the fundamental and **(b)** the second harmonic? **(c)** If the tension in the clothesline is increased, do the frequencies in parts (a) and (b) increase, decrease, or stay the same? Explain.

72. •• The organ pipe in **FIGURE 14-49** is 2.75 m long. **(a)** What is the frequency of the standing wave shown in the pipe? **(b)** What is the fundamental frequency of this pipe?

FIGURE 14-49 Problem 72

73. •• The frequency of the standing wave shown in FIGURE 14-50 is 185 Hz. **(a)** What is the fundamental frequency of this pipe? **(b)** What is the length of the pipe?

FIGURE 14-50 Problem 73

74. ••• An organ pipe open at both ends has a harmonic with a frequency of 440 Hz. The next higher harmonic in the pipe has a frequency of 495 Hz. Find **(a)** the frequency of the fundamental and **(b)** the length of the pipe.

SECTION 14-9 BEATS

75. • **CE** When guitar strings A and B are plucked at the same time, a beat frequency of 2 Hz is heard. If string A is tightened, the beat frequency increases to 3 Hz. Which of the two strings had the lower frequency initially?

76. • **CE** Predict/Explain **(a)** Is the beat frequency produced when a 245-Hz tone and a 240-Hz tone are played together greater than, less than, or equal to the beat frequency produced when a 140-Hz tone and a 145-Hz tone are played together? **(b)** Choose the *best explanation* from among the following:

   I. The beat frequency is determined by the difference in frequencies and is independent of their actual values.

   II. The higher frequencies will produce a higher beat frequency.

   III. The percentage change in frequency for 240 and 245 Hz is less than for 140 and 145 Hz, resulting in a lower beat frequency.

77. • You have three tuning forks with frequencies of 252 Hz, 256 Hz, and 259 Hz. What beat frequencies are possible with these tuning forks?

78. • **Tuning a Piano** To tune middle C on a piano, a tuner hits the key and at the same time sounds a 261-Hz tuning fork. If the tuner hears 3 beats per second, what are the possible frequencies of the piano key?

79. •• Two musicians are comparing their clarinets. The first clarinet produces a tone that is known to be 466 Hz. When the two clarinets play together they produce eight beats every 3.00 seconds. If the second clarinet produces a higher pitched tone than the first clarinet, what is the second clarinet's frequency?

80. •• Predict/Calculate Two strings that are fixed at each end are identical, except that one is 0.560 cm longer than the other. Waves on these strings propagate with a speed of 34.2 m/s, and the fundamental frequency of the shorter string is 212 Hz. **(a)** What beat frequency is produced if each string is vibrating with its fundamental frequency? **(b)** Does the beat frequency in part (a) increase or decrease if the longer string is increased in length? **(c)** Repeat part (a), assuming that the longer string is 0.761 cm longer than the shorter string.

81. •• Identical cellos are being tested. One is producing a fundamental frequency of 120.9 Hz on a string that is 1.25 m long and has a mass of 109 g. On the second cello an identical string is fingered to reduce the length that can vibrate. If the beat frequency produced by these two strings is 5.25 Hz, what is the vibrating length of the second string?

82. ••• A friend in another city tells you that she has two organ pipes of different lengths, one open at both ends, the other open at one end only. In addition, she has determined that the beat frequency caused by the second-lowest frequency of each pipe is equal to the beat frequency caused by the third-lowest frequency of each pipe. Her challenge to you is to calculate the length of the organ pipe that is open at both ends, given that the length of the other pipe is 1.00 m.

GENERAL PROBLEMS

83. • **CE** FIGURE 14-51 shows a wave on a string moving to the right. For each of the points indicated on the string, A-F, state whether it is (**I**, moving upward; **II**, moving downward; or **III**, instantaneously at rest) at the instant pictured.

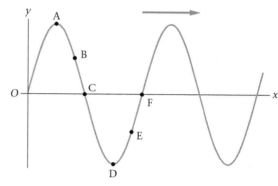

FIGURE 14-51 Problem 83

84. • The fundamental of an organ pipe that is closed at one end and open at the other end is 261.6 Hz (middle C). The second harmonic of an organ pipe that is open at both ends has the same frequency. What are the lengths of these two pipes?

85. • **The Loudest Animal** The loudest sound produced by a living organism on Earth is made by the bowhead whale (*Balaena mysticetus*). These whales can produce a sound that, if heard in air at a distance of 3.00 m, would have an intensity level of 127 dB. This is roughly the equivalent of 5000 trumpeting elephants. How far away can you be from a 127-dB sound and still just barely hear it? (Assume a point source, and ignore reflections and absorption.)

86. • **Hearing a Good Hit** Physicist Robert Adair, once appointed the "official physicist to the National League" by the commissioner of baseball, believes that the "crack of the bat" can tell an outfielder how well the ball has been hit. According to Adair, a good hit makes a sound of 510 Hz, while a poor hit produces a sound of 170 Hz. What is the difference in wavelength of these sounds?

87. • A standing wave of 603 Hz is produced on a string that is 1.33 m long and fixed on both ends. If the speed of waves on this string is 402 m/s, how many antinodes are there in the standing wave?

88. •• **Playing Harmonics** When a 63-cm-long guitar string is plucked, all possible standing waves simultaneously exist on the string. When you place your finger on the string, any standing waves with nonzero amplitude at that location are stopped, but any standing waves with a node at that location continue vibrating. If you wish to stop all standing waves with frequencies lower than the third harmonic on this string, how far from the end of the string should you place your finger?

89. •• **BIO Measuring Hearing Loss** To determine the amount of temporary hearing loss loud music can cause in humans, researchers studied a group of 20 adult females who were exposed to 110-dB music for 60 minutes. Eleven of the 20 subjects showed a 20.0-dB reduction in hearing sensitivity at 4000 Hz. What is the intensity corresponding to the threshold of hearing for these subjects?

90. •• **BIO Hearing a Pin Drop** The ability to hear a "pin drop" is the sign of sensitive hearing. Suppose a 0.48-g pin is dropped from a height of 31 cm, and that the pin emits sound for 1.2 s when it lands. Assuming all of the mechanical energy of the pin is converted to sound energy, and that the sound radiates uniformly in all directions, find the maximum distance from which a person can hear the pin drop. (This is the ideal maximum distance, but atmospheric absorption and other factors will make the actual maximum distance considerably smaller.)

91. •• A cannon 105 m away from you shoots a cannonball straight up into the air with an initial speed of 46 m/s. What is the cannonball's speed when you hear the blast from the cannon? Ignore air resistance.

92. •• A machine shop has 120 equally noisy machines that together produce an intensity level of 92 dB. If the intensity level must be reduced to 82 dB, how many machines must be turned off?

93. •• Predict/Calculate A bottle has a standing wave with a frequency of 375 Hz. The next higher-frequency standing wave in the bottle is 625 Hz. **(a)** Is 375 Hz the fundamental frequency for the bottle? Explain. **(b)** What is the next-higher frequency standing wave after 625 Hz?

94. •• **Speed of a Tsunami** Tsunamis can have wavelengths between 100 and 400 km. Since this is much greater than the average depth of the oceans (about 4.3 km), the ocean can be considered as shallow water for these waves. Using the speed of waves in shallow water of depth d given in Problem 8, find the typical speed for a tsunami. (*Note*: In the open ocean, tsunamis generally have an amplitude of less than a meter, allowing them to pass ships unnoticed. As they approach shore, however, the water depth decreases and the waves slow down. This can result in an increase of amplitude to as much as 37 m or more.)

95. •• Two trains with 124-Hz horns approach one another. The slower of the two trains has a speed of 26 m/s. What is the speed of the fast train if an observer standing near the tracks between the trains hears a beat frequency of 4.4 Hz?

96. •• Predict/Calculate Jim is speeding toward James Island with a speed of 24 m/s when he sees Betsy standing on shore at the base of a cliff (**FIGURE 14-52**). Jim sounds his 330-Hz horn. **(a)** What frequency does Betsy hear? **(b)** Jim can hear the echo of his horn reflected back to him by the cliff. Is the frequency of this echo greater than or equal to the frequency heard by Betsy? Explain. **(c)** Calculate the frequency Jim hears in the echo from the cliff.

330 Hz

24 m/s

FIGURE 14-52 Problem 96

97. •• Two ships in a heavy fog are blowing their horns, both of which produce sound with a frequency of 175.0 Hz (**FIGURE 14-53**). One ship is at rest; the other moves on a straight line that passes through the one at rest. If people on the stationary ship hear a beat frequency of 3.5 Hz, what are the two possible speeds and directions of motion of the moving ship?

$v = 0$

$v = ?$

FIGURE 14-53 Problem 97

98. •• **BIO Cracking Your Knuckles** When you "crack" a knuckle, you cause the knuckle cavity to widen rapidly. This, in turn, allows the synovial fluid to expand into a larger volume. If this expansion is sufficiently rapid, it causes a gas bubble to form in the fluid in a process known as *cavitation*. This is the mechanism responsible for the cracking sound. (Cavitation can also cause pits in rapidly rotating ship's propellers.) If a "crack" produces a sound with an intensity level of 57 dB at your ear, which is 18 cm from the knuckle, how far from your knuckle can the "crack" be heard? Assume the sound propagates uniformly in all directions, with no reflections or absorption.

99. •• A steel guitar string has a tension F, length L, and diameter D. Give the multiplicative factor by which the fundamental frequency of the string changes under the following conditions: **(a)** The tension in the string is increased by a factor of 4. The diameter is D and the length is L. **(b)** The diameter of the string is increased by a factor of 3. The tension is F and the length is L. **(c)** The length of the string is halved. The tension is F and the diameter is D.

100. •• A Slinky has a mass of 0.28 kg and negligible length. When it is stretched 1.5 m, it is found that transverse waves travel the length of the Slinky in 0.75 s. **(a)** What is the force constant, k, of the Slinky? **(b)** If the Slinky is stretched farther, will the time required for a wave to travel the length of the Slinky increase, decrease, or stay the same? Explain. **(c)** If the Slinky is stretched 3.0 m, how much time does it take a wave to travel the length of the Slinky? (The Slinky stretches like an ideal spring, and propagates transverse waves like a rope with variable tension.)

101. •• **BIO** Predict/Calculate **OSHA Noise Standards** OSHA, the Occupational Safety and Health Administration, has established standards for workplace exposure to noise. According to OSHA's Hearing Conservation Standard, the permissible noise exposure per day is 95.0 dB for 4 hours or 105 dB for 1 hour. Assuming the eardrum is 9.5 mm in diameter, find the energy absorbed by the eardrum **(a)** with 95.0 dB for 4 hours and **(b)** with 105 dB for 1 hour. **(c)** Is OSHA's safety standard simply a measure of the amount of energy absorbed by the ear? Explain.

102. •• An organ pipe 3.4 m long is open at one end and closed at the other end. What is the linear distance between a node and the adjacent antinode for the third harmonic in this pipe?

103. •• Two identical strings with the same tension vibrate at 631 Hz. If the tension in one of the strings is increased by 2.25%, what is the resulting beat frequency?

104. •• **BIO The Love Song of the Midshipman Fish** When the plainfin midshipman fish (*Porichthys notatus*) migrates from deep Pacific waters to the west coast of North America each summer, the males begin to sing their "love song," which some describe as sounding like a low-pitched motorboat. Houseboat residents and shore dwellers are kept awake for nights on end by the amorous fish. The love song consists of a single note, the second A flat below middle C. **(a)** If the speed of sound in seawater is 1531 m/s,

what is the wavelength of the midshipman's song? **(b)** What is the wavelength of the sound after it emerges into the air? (Information on the musical scale is given in Table 14-3.)

105. ••• Predict/Calculate A rope of length L and mass M hangs vertically from a ceiling. The tension in the rope is only that due to its own weight. **(a)** Suppose a wave starts near the bottom of the rope and propagates upward. Does the speed of the wave increase, decrease, or stay the same as it moves up the rope? Explain. **(b)** Show that the speed of waves a height y above the bottom of the rope is $v = \sqrt{gy}$.

106. ••• **Beats and Standing Waves** In Problem 59, suppose the observer walks toward one speaker with a speed of 1.35 m/s. **(a)** What frequency does the observer hear from each speaker? **(b)** What beat frequency does the observer hear? **(c)** How far must the observer walk to go from one point of constructive interference to the next? **(d)** How many times per second does the observer hear maximum loudness from the speakers? Compare your result with the beat frequency from part (b).

PASSAGE PROBLEMS

BIO The Sound of a Dinosaur

Modern-day animals make extensive use of sounds in their interactions with others. Some sounds are meant primarily for members of the same species, like the cooing calls of a pair of doves, the long-range infrasound communication between elephants, or the songs of the humpback whale. Other sounds may be used as a threat to other species, such as the rattle of a rattlesnake or the roar of a lion.

There is little doubt that extinct animals used sounds in much the same ways. But how can we ever hear the call of a long-vanished animal like a dinosaur when sounds don't fossilize? In some cases, basic physics may have the answer.

Consider, for example, the long-crested, duck-billed dinosaur *Parasaurolophus walkeri*, which roamed the Earth 75 million years ago. This dinosaur, shown in **FIGURE 14-54 (a)**, possessed the largest crest of any duck bill—so long, in fact, that there was a notch in *P. walkeri*'s spine to make room for the crest when its head was tilted backward. Many paleontologists believe the air passages in the dinosaur's crest acted like bent organ pipes open at both ends, and that they produced sounds *P. walkeri* used to communicate with others of its kind. As air was forced through the passages, the predominant sound they produced would be the fundamental standing wave, with a small admixture of higher harmonics as well. The frequencies of these standing waves can be determined with basic physical principles. **FIGURE 14-54 (b)** presents a plot of the lowest ten harmonics of a pipe that is open at both ends as a function of the length of the pipe.

107. • Suppose the air passages in a certain *P. walkeri* crest produce a bent tube 2.7 m long. What is the fundamental frequency of this tube, assuming the bend has no effect on the frequency? (For comparison, a typical human hearing range is 20 Hz to 20 kHz.)

 A. 0.0039 Hz   B. 32 Hz   C. 64 Hz   D. 130 Hz

108. • Paleontologists believe the crest of a female *P. walkeri* was probably shorter than the crest of a male. If this was the case, would the fundamental frequency of a female be greater than, less than, or equal to the fundamental frequency of a male?

109. • Suppose the fundamental frequency of a particular female was 74 Hz. What was the length of the air passages in this female's crest?

 A. 1.2 m   B. 2.3 m   C. 2.7 m   D. 4.6 m

110. • As a young *P. walkeri* matured, the air passages in its crest might increase in length from 1.5 to 2.7 m, causing a decrease in the standing wave frequencies. Referring to Figure 14-54, do you expect the change in the fundamental frequency to be greater than, less than, or equal to the change in the second harmonic frequency?

111. •• Predict/Calculate **REFERRING TO EXAMPLE 14-11** Suppose the engineer adjusts the speed of the train until the sound he hears reflected from the cliff is 775 Hz. The train's whistle still produces a tone of 650.0 Hz. **(a)** Is the new speed of the train greater than, less than, or equal to 21.2 m/s? Explain. **(b)** Find the new speed of the train.

112. •• **REFERRING TO EXAMPLE 14-11** Suppose the train is backing away from the cliff with a speed of 18.5 m/s and is sounding its 650.0-Hz whistle. **(a)** What is the frequency heard by the observer standing near the tunnel entrance? **(b)** What is the frequency heard by the engineer?

113. •• Predict/Calculate **REFERRING TO EXAMPLE 14-16** Suppose we add more water to the soda pop bottle. **(a)** Does the fundamental frequency increase, decrease, or stay the same? Explain. **(b)** Find the fundamental frequency if the height of water in the bottle is increased to 7.5 cm. The height of the bottle is still 26.0 cm.

114. •• Predict/Calculate **REFERRING TO EXAMPLE 14-16** The speed of sound increases slightly with temperature. **(a)** Does the fundamental frequency of the bottle increase, decrease, or stay the same as the air heats up on a warm day? Explain. **(b)** Find the fundamental frequency if the speed of sound in air increases to 348 m/s. Assume the bottle is 26.0 cm tall, and that it contains water to a depth of 6.5 cm.

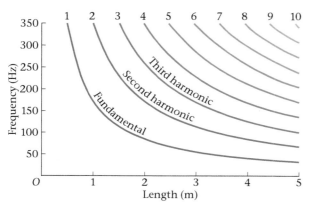

FIGURE 14-54 Standing wave frequencies as a function of length for a pipe open at both ends. The first ten harmonics ($n = 1 - 10$) are shown. (Problems 107, 108, 109, and 110)

15

Fluids

Big
Ideas

1 Pressure is the force exerted by a fluid on an area.

2 Pressure increases with depth in a fluid.

3 Archimedes' principle states that the buoyant force is equal to the weight of the displaced fluid.

4 Bernoulli's equation is a statement of energy conservation applied to fluids.

5 Surface tension tends to make the surface of a fluid as small as possible.

▲ How do doughnuts cooking in oil illustrate physics in action? Well, did you notice that each one has a white ring around the middle? Why is that? When you think about it, you realize that as the doughnuts float in the cooking oil, more than half their volume is out of the oil—only about 45 percent is being cooked. When they're flipped over to cook the other side, another 45 percent is cooked, leaving 10 percent as a white uncooked ring around the middle. Archimedes' principle, one of the key topics in this chapter, has never had a tastier application.

W hen we speak of *fluids*, we are referring to substances like gases and liquids that can readily flow from place to place, and that take on the shape of a container rather than retain a shape of their own. In this chapter, we examine some of the fundamental principles that apply to fluids. As we shall see, all of these principles derive from the basic physics we have learned to this point.

15-1 Density

The properties of a fluid can be hard to pin down, given that it can flow, change shape, and either split into smaller portions or combine into a larger system. Thus, one of the best ways to quantify a fluid is in terms of its **density.** Specifically, the density, ρ, of a material (fluid or not) is defined as the mass, M, per volume, V:

Definition of Density, ρ

$$\rho = M/V \qquad \qquad 15\text{-}1$$

SI unit: kg/m^3

The denser a material, the more mass it has in any given volume.

To get a feel for densities in common substances, we start with water. For example, to fill a cubic container 1 meter on a side would take more than 2000 pounds of water. More precisely, water has the following density:

$$\rho_w = \text{density of water} = 1000 \text{ kg/m}^3$$

A gallon (1 gallon = 3.79 L = 3.79×10^{-3} m^3) of water, then, has a mass of

$$M = \rho V = (1000 \text{ kg/m}^3)(3.79 \times 10^{-3} \text{ m}^3) = 3.79 \text{ kg}$$

As a rule of thumb, a gallon of water weighs just over 8 pounds.

In comparison, the density of the air in your room is roughly 1.29 kg/m^3, and the helium in a helium-filled balloon has a density of only about 0.179 kg/m^3. On the higher end of the density scale, solid gold "weighs in" with a hefty 19,300 kg/m^3. The densities of some other common materials are listed in Table 15-1.

TABLE 15-1 Densities of Common Substances

Substance	Density (kg/m^3)
Gold	19,300
Mercury	13,600
Lead	11,300
Silver	10,500
Iron	7860
Aluminum	2700
Ebony (wood)	1220
Ethylene glycol (antifreeze)	1114
Whole blood (37 °C)	1060
Seawater	1025
Freshwater	1000
Olive oil	920
Ice	917
Ethyl alcohol	806
Cherry (wood)	800
Balsa (wood)	120
Styrofoam	100
Oxygen	1.43
Air	1.29
Helium	0.179

CONCEPTUAL EXAMPLE 15-1 THE WEIGHT OF AIR

One day you look in your refrigerator and find nothing but a dozen eggs (45 g each). A quick measurement shows that the inside of the refrigerator is 1.0 m by 0.61 m by 0.75 m. Is the weight of the *air* in your refrigerator much less than, about the same as, or much more than the weight of the *eggs*?

REASONING AND DISCUSSION

At first it might seem that the "thin air" in the refrigerator weighs practically nothing compared with a carton full of eggs. A brief calculation shows this is not the case. For the eggs, we have

$$m_{eggs} = 12(45 \text{ g}) = 0.54 \text{ kg}$$

For the air, with a density of 1.29 kg/m^3, the mass is

$$m_{air} = \rho V = (1.29 \text{ kg/m}^3)(1.0 \text{ m} \times 0.61 \text{ m} \times 0.75 \text{ m}) = 0.59 \text{ kg}$$

Thus, the air, which has a mass of 0.59 kg (1.3 lb), actually weighs slightly more than the eggs, which have a mass of 0.54 kg (1.2 lb)!

ANSWER

The weight of the air is about the same as the weight of the eggs.

Enhance Your Understanding (Answers given at the end of the chapter)

1. System 1 consists of 10 kg of water; system 2 consists of 100 kg of water. **(a)** Is the volume of system 1 greater than, less than, or equal to the volume of system 2? Explain. **(b)** Is the density of system 1 greater than, less than, or equal to the density of system 2? Explain.

Section Review

- Density, ρ, is the amount of mass per volume: $\rho = M/V$.

PHYSICS IN CONTEXT
Looking Back

Pressure can be thought of as an application of the concept of force (Chapters 5 and 6) to fluids.

PHYSICS IN CONTEXT
Looking Ahead

The pressure of an ideal gas, and its connection with temperature and volume, will be explored in detail in Chapter 17 when we study thermodynamics.

15-2 Pressure

If you've ever pushed a button, or pressed a key on a keyboard, you've applied pressure. Now, you might object that you simply exerted a force on the button or the key, which is correct. That force is spread out over an area, however. For example, when you press a button, the tip of your finger contacts the button over a small but finite area. **Pressure,** P, is a measure of the amount of force, F, per area, A:

Definition of Pressure, P

$$P = F/A \qquad \qquad 15\text{-}2$$

SI unit: N/m^2

The pressure is increased if the force applied to a given area is increased, or if a given force is applied to a smaller area. For example, if you press your finger against a balloon, not much happens—your finger causes a small indentation. On the other hand, if you push a needle against the balloon with the same force, you get an explosive pop. The difference is that the same force applied to the small area of a needle tip causes a large enough pressure to rupture the balloon.

EXAMPLE 15-2 POPPING A BALLOON

Find the pressure exerted on the surface of a balloon if you press with a force of 2.6 N using **(a)** your finger or **(b)** a needle. Assume the area of your fingertip is $1.5 \times 10^{-4}\ m^2$, and the area of the needle tip is $2.3 \times 10^{-7}\ m^2$. **(c)** Find the minimum force necessary to pop the balloon with the needle, given that the balloon pops with a pressure of $3.7 \times 10^5\ N/m^2$.

PICTURE THE PROBLEM
Our sketch shows a balloon deformed by the press of a finger and of a needle. The difference is not the amount of force that is applied, but the area over which it is applied. In the case of the needle, the force may be sufficient to pop the balloon.

REASONING AND STRATEGY

a, b. The force F and area A are given in the problem statement. We can find the pressure by applying the definition: $P = F/A$.

c. Now we rearrange $P = F/A$ to find the force corresponding to a given pressure and area.

Known Force, $F = 2.6$ N; area of fingertip, $A_f = 1.5 \times 10^{-4}\ m^2$; area of needle tip, $A_n = 2.3 \times 10^{-7}\ m^2$; pressure to pop balloon, $P = 3.7 \times 10^5\ N/m^2$.

Unknown **(a)** Pressure, $P = ?$ **(b)** Pressure, $P = ?$ **(c)** Force to pop balloon, $F = ?$

SOLUTION

Part (a)

1. Calculate the pressure exerted by the finger:
$$P = \frac{F}{A_f} = \frac{2.6\ N}{1.5 \times 10^{-4}\ m^2} = 1.7 \times 10^4\ N/m^2$$

Part (b)

2. Calculate the pressure exerted by the needle:
$$P = \frac{F}{A_n} = \frac{2.6\ N}{2.3 \times 10^{-7}\ m^2} = 1.1 \times 10^7\ N/m^2$$

Part (c)

3. Solve $P = F/A$ for the force:
$$F = PA_n$$

4. Substitute numerical values:
$$F = (3.7 \times 10^5\ N/m^2)(2.3 \times 10^{-7}\ m^2) = 0.085\ N$$

INSIGHT
The pressure exerted by the needle in part (b) is about 650 times greater than the pressure due to the finger in part (a). This increase in pressure with decreasing area accounts for the sharp tips found on a variety of objects, including nails, pens, pencils, and syringes.

PRACTICE PROBLEM
Find the area that a force of 2.6 N would have to act on to produce a pressure of $3.7 \times 10^5\ N/m^2$.
[**Answer:** $A = 7.0 \times 10^{-6}\ m^2$]

Some related homework problems: Problem 7, Problem 8

RWP* An interesting example of force, area, and pressure in nature is provided by the family of small aquatic birds referred to as rails, and in particular by the gallinule. This bird has exceptionally long toes that are spread out over a large area, as can be seen in FIGURE 15-1. The result is that the weight of the bird causes only a relatively small pressure as it walks on the soft, muddy shorelines encountered in its habitat. In some species, the pressure exerted while walking is so small that the birds can actually walk across lily pads without sinking into the water.

Atmospheric Pressure and Gauge Pressure

Atmospheric Pressure   We're all used to working under pressure—about 14.7 pounds per square inch to be precise. This is **atmospheric pressure,** P_{at}, a direct result of the weight of the air above us. In SI units, atmospheric pressure has the following value:

Atmospheric Pressure, P_{at}

$$P_{at} = 1.01 \times 10^5 \, \text{N/m}^2 \qquad\qquad \text{15-3}$$

SI unit: N/m^2

A shorthand unit for N/m^2 is the **pascal** (Pa):

$$1 \, \text{Pa} = 1 \, \text{N/m}^2 \qquad\qquad \text{15-4}$$

The pascal honors the pioneering studies of fluids by the French scientist Blaise Pascal (1623–1662). Thus, atmospheric pressure can be written as

$$P_{at} = 101 \, \text{kPa}$$

In British units, pressure is measured in pounds per square inch, and

$$P_{at} = 14.7 \, \text{lb/in}^2$$

Finally, a common unit for atmospheric pressure in weather forecasting is the **bar,** defined as follows:

$$1 \, \text{bar} = 10^5 \, \text{Pa} \approx 1 \, P_{at}$$

EXERCISE 15-3   THE FORCE OF ATMOSPHERIC PRESSURE

Find the force exerted on the palm of your hand by atmospheric pressure. Assume your palm measures 0.077 m by 0.094 m.

REASONING AND SOLUTION
Rearranging $P = F/A$ to solve for the force, and using $P_{at} = 1.01 \times 10^5 \, \text{N/m}^2$, we find

$$F = P_{at}A = (1.01 \times 10^5 \, \text{Pa})(0.077 \, \text{m})(0.094 \, \text{m}) = 730 \, \text{N}$$

Thus, the atmosphere pushes on the palm of your hand with a force of approximately 160 pounds! Of course, it also pushes on the back of your hand with essentially the same force—but in the opposite direction.

FIGURE 15-2 illustrates the forces exerted on your hand by atmospheric pressure. If your hand is vertical, atmospheric pressure pushes to the right and to the left equally, so your hand feels zero net force. If your hand is horizontal, atmospheric pressure exerts upward and downward forces on your hand that are essentially the same in magnitude, again giving zero net force. This cancellation of forces occurs no matter what the orientation of your hand; thus, we can conclude the following:

> The pressure in a fluid acts equally in all directions, and acts at right angles to any surface.

A vivid illustration of the strength of atmospheric pressure is shown in FIGURE 15-3 (a). When air is removed from the inside of the can, atmospheric pressure causes it to collapse. A method of overcoming atmospheric pressure to protect a broken arm is shown in FIGURE 15-3 (b).

---

*Real World Physics applications are denoted by the acronym RWP.

PHYSICS IN CONTEXT
Looking Ahead

Pressure appears again in Chapter 18 when we consider thermal processes, and the back-and-forth conversion of thermal and mechanical energy.

PROBLEM-SOLVING NOTE

Pressure Is Force Per Area

Remember that pressure is proportional to the applied force and *inversely* proportional to the area over which it acts.

▲ **FIGURE 15-1** This bird exerts only a small pressure on the lily pad on which it walks because its weight is spread out over a large area by its long toes. Since the pressure is not enough to sink a lily pad, the bird can seemingly "walk on water."

▲ **FIGURE 15-2 Pressure is the same in all directions** The forces exerted on the two sides of a hand cancel, regardless of the hand's orientation. Hence, pressure acts equally in all directions.

▶ **FIGURE 15-3 Visualizing Concepts At-mospheric Pressure (a)** When the air is pumped out of a sealed can, atmospheric pressure produces an inward force that is unopposed. The resulting collapse of the can vividly illustrates the pressure that is all around us. **(b)** An air splint utilizes the same principle of unequal pressure. A plastic sleeve is placed around an injured limb and inflated to a pressure greater than that of the atmosphere—and thus of the body's internal pressure. The increased external pressure retards bleeding from the injured area, and also tends to immobilize the limb in case it might be fractured.

(a)

(b)

PROBLEM-SOLVING NOTE

Gauge Pressure

If a problem gives you the gauge pressure, recall that the actual pressure is the gauge pressure *plus* atmospheric pressure.

Gauge Pressure In many cases we're interested in the difference between a given pressure and atmospheric pressure. For example, a flat tire doesn't have zero pressure in it; the pressure in the tire is atmospheric pressure. To inflate the tire to 241 kPa (35 lb/in^2), the pressure inside the tire must be greater than atmospheric pressure by this amount; that is, $P = 241$ kPa $+ P_{at} = 241$ kPa $+ 101$ kPa $= 342$ kPa.

To deal with such situations, we introduce the **gauge pressure**, P_g, defined as the amount of pressure that is in excess of atmospheric pressure:

$$P_g = P - P_{at} \qquad \text{15-5}$$

It is the gauge pressure that is determined by a tire gauge. Many problems in this chapter refer to the gauge pressure. Hence it must be remembered that the actual pressure in these cases is greater by the amount P_{at}; that is, $P = P_g + P_{at}$.

EXAMPLE 15-4 PRESSURING THE BALL: ESTIMATE THE GAUGE PRESSURE

Estimate the gauge pressure in a basketball by pushing down on it and noting the area of contact it makes with the surface on which it rests.

PICTURE THE PROBLEM
Our sketch shows the basketball both in its original state, and when a force F pushes downward on it. In the latter case, the ball flattens out on the bottom, forming a circular area of contact with the floor of diameter d.

REASONING AND STRATEGY
To solve this problem, we have to make reasonable estimates of the force applied to the ball and the area of contact.

   Suppose, for example, that we push down with a moderate force of 22 N (about 5 lb). The circular area of contact will probably have a diameter of about 2.0 centimeters. This can be verified by carrying out the experiment. Thus, given $F = 22$ N and $A = \pi(d/2)^2$, we can find the gauge pressure.

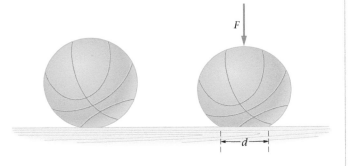

Known Force of push, $F = 22$ N; diameter of contact with floor, $d = 2.0$ cm.

Unknown Gauge pressure of basketball, $P_g = ?$

SOLUTION

1. Using these estimates, calculate the gauge pressure, P_g: $P_g = \dfrac{F}{A} = \dfrac{22 \text{ N}}{\pi\left(\dfrac{0.020 \text{ m}}{2}\right)^2} = 7.0 \times 10^4$ Pa

INSIGHT
Given that a pressure of 1 atmosphere ($P_{at} = 101$ kPa $= 1.01 \times 10^5$ Pa) corresponds to 14.7 lb/in^2, it follows that the gauge pressure of the ball ($P_g = 7.0 \times 10^4$ Pa) corresponds to a pressure of roughly 10 lb/in^2. Thus, a basketball will typically have a gauge pressure in the neighborhood of 10 lb/in^2, and hence a total pressure inside the ball of about 25 lb/in^2.

CONTINUED

PRACTICE PROBLEM

What is the diameter of the circular area of contact if a basketball with a 12 lb/in² gauge pressure is pushed down with a force of 44 N (about 10 lb)?   [**Answer:** $d = 2.6$ cm]

Some related homework problems: Problem 9, Problem 11

Enhance Your Understanding   (Answers given at the end of the chapter)

2. A force F acts on a circular area of radius r. Rank the following systems in order of increasing pressure. Indicate ties where appropriate. **A**, $F = 10$ N, $r = 1$ cm; **B**, $F = 100$ N, $r = 10$ cm; **C**, $F = 20$ N, $r = 2$ cm; **D**, $F = 1$ N, $r = 0.1$ cm.

Section Review

-   Pressure, P, is force per area: $P = F/A$.
-   Atmospheric pressure is $P_{at} = 1.01 \times 10^5$ N/m² $= 101$ kPa, and gauge pressure is $P_g = P - P_{at}$.

15-3 Static Equilibrium in Fluids: Pressure and Depth

Pressure is exerted equally in all directions, but it is not equal at all *locations* within a fluid. Just think of the low air pressure on a mountaintop compared with the higher pressure at sea level. We explore this aspect of pressure in this section.

Pressure with Depth

Countless war movies have educated us on the perils of taking a submarine too deep. The hull creaks and groans, rivets start to pop, water begins to spray into the ship, and the captain keeps a close eye on the depth gauge. But what causes the pressure to increase as a submarine dives, and how much does it go up for a given increase in depth?

The answer to the first question is that the increased pressure is due to the added weight of water pressing on the submarine as it goes deeper. To see how this works, consider a cylindrical container filled to a height h with a fluid of density ρ, as in **FIGURE 15-4**. The top surface of the fluid is open to the atmosphere, with a pressure P_{at}. If the cross-sectional area of the container is A, the downward force exerted on the top surface by the atmosphere is

$$F_{top} = P_{at}A$$

Now, at the bottom of the container, the downward force is F_{top} *plus* the weight of the fluid. Recalling that $M = \rho V$, and that $V = hA$ for a cylinder of height h and area A, this weight is

$$W = Mg = \rho Vg = \rho(hA)g$$

Hence, we have

$$F_{bottom} = F_{top} + W = P_{at}A + \rho(hA)g$$

Finally, the pressure at the bottom is obtained by dividing F_{bottom} by the area A:

$$P_{bottom} = \frac{F_{bottom}}{A} = \frac{P_{at}A + \rho(hA)g}{A} = P_{at} + \rho gh$$

Of course, this relationship holds not only for the bottom of the container, but for *any depth h below the surface.* Thus, the answer to the second question is that if the depth increases by the amount h, the pressure increases by the amount ρgh. At a depth h below the surface of a fluid, then, the pressure P is given by

$$P = P_{at} + \rho gh \qquad\qquad 15\text{-}6$$

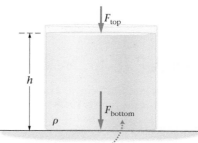

The force on the bottom is equal to the force on the top plus the weight of fluid in the flask.

▲ **FIGURE 15-4 Pressure and the weight of a fluid** The force pushing down on the bottom of the flask is greater than the force pushing down on the surface of the fluid. The difference in the force is the weight of the fluid in the flask.

Big Idea 2 Pressure increases with depth in a fluid. This is true whether the fluid is the air in the atmosphere or water in an ocean.

Pressure in Water and Alcohol

Pressure Depends Only on Depth

The pressure in a static fluid depends only on the depth of the fluid. It is independent of the shape of the container.

▲ FIGURE 15-5 **Pressure variation with depth** If point 2 is deeper than point 1 by the amount h, its pressure is greater by the amount $\rho g h$.

This expression holds for any fluid with constant density ρ and a pressure P_{at} at its upper surface.

EXERCISE 15-5 TITANIC PRESSURE

RWP The *Titanic* was found in 1985 lying on the bottom of the North Atlantic at a depth of about 2.35 miles. What is the pressure at this depth?

REASONING AND SOLUTION

Applying $P = P_{at} + \rho g h$, with $\rho = 1025$ kg/m³ for seawater, we have

$$P = P_{at} + \rho g h = 1.01 \times 10^5 \text{ Pa} + (1025 \text{ kg/m}^3)(9.81 \text{ m/s}^2)\left(\frac{1609 \text{ m}}{1 \text{ mi}}\right)(2.35 \text{ mi})$$

$$= 3.81 \times 10^7 \text{ Pa}$$

This is about 377 atmospheres.

The relationship $P = P_{at} + \rho g h$ can be applied to any two points in a fluid. For example, if the pressure at one point is P_1, the pressure P_2 at a depth h below that point is the following:

Dependence of Pressure on Depth

$$P_2 = P_1 + \rho g h \qquad\qquad 15\text{-}7$$

This relationship is illustrated in FIGURE 15-5 and utilized in the next Example.

EXAMPLE 15-6 PRESSURE AND DEPTH

A cubical box 20.00 cm on a side is completely immersed in a fluid. At the top of the box the pressure is 105.0 kPa; at the bottom the pressure is 106.8 kPa. What is the density of the fluid?

PICTURE THE PROBLEM

Our sketch shows the box at an unknown depth d below the surface of the fluid. The important dimension for this problem is the height of the box, which is 20.00 cm. We are also given that the pressures at the top and bottom of the box are $P_1 = 105.0$ kPa and $P_2 = 106.8$ kPa, respectively.

REASONING AND STRATEGY

The pressures at the top and bottom of the box are related by $P_2 = P_1 + \rho g h$. The pressures and the height of the box are given, and hence this relationship can be solved for the unknown density, ρ.

Known Vertical height of cubical box, $h = 20.00$ cm; pressure at top of box, $P_1 = 105.0$ kPa; pressure at bottom of box, $P_2 = 106.8$ kPa.

Unknown Density of fluid, $\rho = ?$

SOLUTION

1. Solve $P_2 = P_1 + \rho g h$ for the density:

$$\rho = \frac{P_2 - P_1}{gh}$$

2. Substitute numerical values:

$$\rho = \frac{(1.068 \times 10^5 \text{ Pa}) - (1.050 \times 10^5 \text{ Pa})}{(9.81 \text{ m/s}^2)(0.2000 \text{ m})} = 920 \text{ kg/m}^3$$

INSIGHT

Comparing this density with the values in Table 15-1, we see that the fluid in question is probably olive oil. If the box had been immersed in water instead, with its greater density, the difference in pressure between the top and bottom of the box would have been greater.

PRACTICE PROBLEM

Given the density obtained above, what is the depth of the fluid, d, at the top of the box? [**Answer:** $d = 0.44$ m]

Some related homework problems: Problem 14, Problem 19

CONCEPTUAL EXAMPLE 15-7 THE SIZE OF BUBBLES

One day, while swimming below the surface of the ocean, you let out a small bubble of air from your mouth. As the bubble rises toward the surface, does its diameter increase, decrease, or stay the same?

REASONING AND DISCUSSION

As the bubble rises, the pressure in the surrounding water decreases. This allows the air in the bubble to expand and occupy a larger volume.

ANSWER

The diameter of the bubble increases.

RWP An interesting application of the variation of pressure with depth is the **barometer,** which can be used to measure atmospheric pressure. We consider here the simplest type of barometer, first proposed by Evangelista Torricelli (1608–1647) in 1643. First, take a long glass tube—open at one end and closed at the other end—and fill it with a fluid of density ρ. Next, invert the tube and place its open end below the surface of the same fluid in a bowl, as shown in **FIGURE 15-6**. Some of the fluid in the tube will flow into the bowl, leaving an empty space (vacuum) at the top. Enough will remain, however, to create a difference in level, h, between the fluid in the bowl and the fluid in the tube.

The basic idea of the barometer is that this height difference is directly related to the atmospheric pressure that pushes down on the fluid in the bowl. To see how this works, first note that the pressure in the vacuum at the top of the tube is zero. Hence, the pressure in the tube at a depth h below the vacuum is $0 + \rho gh = \rho gh$. Now, at the level of the fluid in the bowl we know that the pressure is 1 atmosphere, P_{at}. Therefore, it follows that

$$P_{at} = \rho gh$$

If these pressures were not the same, there would be a pressure difference between the fluid in the tube and the fluid in the bowl, resulting in a net force and a flow of fluid. Thus, a measurement of h immediately gives atmospheric pressure.

A fluid that is often used in such a barometer is mercury (Hg), with a density of $\rho = 1.3595 \times 10^4 \text{ kg/m}^3$. Because of mercury's relatively high density, the corresponding height for a column of mercury is reasonably small. We find the following:

$$h = \frac{P_{at}}{\rho g} = \frac{1.013 \times 10^5 \text{ Pa}}{(1.3595 \times 10^4 \text{ kg/m}^3)(9.81 \text{ m/s}^2)} = 760 \text{ mm}$$

In fact, atmospheric pressure is *defined* in terms of millimeters of mercury (mmHg):

$$1 \text{ atmosphere} = P_{at} = 760 \text{ mmHg}$$

Table 15-2 lists the various expressions we have developed for atmospheric pressure.

▲ **FIGURE 15-6 A simple barometer** Atmospheric pressure, P_{at}, is related to the height of fluid in the tube by $P_{at} = \rho gh$.

TABLE 15-2 Atmospheric Pressure

1 atmosphere = P_{at}	
	= 760 mmHg (definition)
	= 14.7 lb/in²
	= 101 kPa
	= 101 kN/m²
	~ 1 bar = 100 kPa

Water Seeks Its Own Level

We're all familiar with the aphorism that water seeks its own level. In order for this to hold true, however, it's necessary that the pressure at the surface of the water (or other fluid) be the same everywhere on the surface. This was not the case for the barometer just discussed, where the pressure was P_{at} on one portion of the surface and zero on another. Let's take a moment, then, to consider the level assumed by a fluid with constant pressure on its surface. In doing so, we'll apply considerations involving force, pressure, and energy.

Balancing Forces First, we consider the force-pressure point of view. In **FIGURE 15-7 (a)** we show a U-shaped tube containing a quantity of fluid of density ρ. The fluid rises to the same level in each arm of the U, where it is open to the atmosphere. Therefore, the

(a) A fluid is in equilibrium when the levels are equal

(b) Unequal levels result in unequal pressures and fluid flow

▲ **FIGURE 15-7 Fluids seek their own level (a)** When the levels are equal, the pressure is the same at the base of each arm of the U tube. As a result, the fluid in the horizontal section of the U is in equilibrium. **(b)** With unequal heights, the pressures are different. In this case, the pressure is greater at the base of the right arm; hence fluid will flow toward the left and the levels will equalize.

To create different levels, fluid must be moved from one side . . .

. . . to the other . . .

. . . resulting in an increase in height and greater potential energy.

▲ **FIGURE 15-8 Gravitational potential energy of a fluid** In order to create unequal levels in the two arms of the U tube, an element of fluid must be raised by the height Δh. This increases the gravitational potential energy of the system. The lowest potential energy corresponds to equal levels.

Water Level in Pascal's Vases

pressure at the base of each arm is the same: $P_{at} + \rho g h$. Thus, the fluid in the horizontal section of the U is pushed with equal force from each side, giving zero net force. As a result, the fluid remains at rest.

In contrast, in **FIGURE 15-7 (b)** the two arms of the U are filled to different levels. Therefore, the pressure at the base of the two arms is different, with the greater pressure at the base of the right arm. It follows that the fluid in the horizontal section experiences a net force to the left, causing it to move in that direction. This will tend to equalize the fluid levels of the two arms.

Minimizing Energy We can arrive at the same conclusion on the basis of energy minimization. Consider a U tube that is initially filled to the same level in both arms, as in Figure 15-7 (a). Now, consider moving a small element of fluid from one arm to the other, to create different levels, as in **FIGURE 15-8**. To move this fluid element to the other arm, it is necessary to lift it upward. This, in turn, causes an increase in the gravitational potential energy of the system. Nothing else in the system has changed its position, and therefore the only change in potential energy is the increase associated with the fluid element. Thus, we conclude that the system has a minimum energy when the fluid levels are the same and a higher energy when the levels are different. Just as a ball rolls to the bottom of a hill, where its energy is minimized, the fluid seeks its own level and a minimum energy. An illustration of a fluid seeking its own level is shown in **FIGURE 15-9**.

If two different liquids, with different densities, are combined in the same U tube, the levels in the arms are not the same. Still, the pressures at the base of each arm must be equal, as before. This is discussed in the next Example.

◀ **FIGURE 15-9** The containers shown here are connected at the bottom by a hollow tube, which allows fluid to flow freely between them. As a result the fluid level is the same in each container regardless of its shape and size.

EXAMPLE 15-8   OIL AND WATER DON'T MIX

A U-shaped tube is filled mostly with water, but a small amount of vegetable oil has been added to one side, as shown in the sketch. The density of the water is $1.00 \times 10^3 \text{ kg/m}^3$, and the density of the vegetable oil is $9.20 \times 10^2 \text{ kg/m}^3$. If the depth of the oil is 5.00 cm, what is the difference in level h between the top of the oil on one side of the U tube and the top of the water on the other side?

PICTURE THE PROBLEM
The U-shaped tube and the relevant dimensions are shown in our sketch. In particular, notice that the depth of the oil is 5.00 cm, and that the oil rises to a greater height on its side of the U tube than the water does on its side. This is due to the oil having the lower density. Both sides of the U tube are open to the atmosphere, so the pressure at the top surfaces is P_{at}.

CONTINUED

REASONING AND STRATEGY

For the system to be in equilibrium, it is necessary that the pressure be the same at the bottom of each side of the U tube—that is, at points C and D. If the pressure is the same at C and D, it will remain equal as one moves up through the water to points A and B. Above this point the pressures will differ because of the presence of the oil.

Therefore, setting the pressure at point A equal to the pressure at point B determines the depth h_1 in terms of the known depth h_2. It follows that the difference in level between the two sides of the U tube is $h = h_2 - h_1$.

Known Density of water, $\rho_{water} = 1.00 \times 10^3$ kg/m³; density of vegetable oil, $\rho_{oil} = 9.20 \times 10^2$ kg/m³; depth of oil, $h_2 = 5.00$ cm.

Unknown Height difference between the sides of the U tube, $h = ?$

SOLUTION

1. Find the pressure at point A, where the depth of the water is h_1:

$$P_A = P_{at} + \rho_{water}\, gh_1$$

2. Find the pressure at point B, where the depth of the oil is $h_2 = 5.00$ cm:

$$P_B = P_{at} + \rho_{oil}\, gh_2$$

3. Set P_A equal to P_B:

$$P_{at} + \rho_{water}\, gh_1 = P_{at} + \rho_{oil}\, gh_2$$

4. Solve the equation in Step 3 for the depth of the water, h_1, and substitute numerical values:

$$h_1 = h_2\left(\frac{\rho_{oil}}{\rho_{water}}\right)$$

$$= (5.00\text{ cm})\left(\frac{9.20 \times 10^2\text{ kg/m}^3}{1.00 \times 10^3\text{ kg/m}^3}\right) = 4.60\text{ cm}$$

5. Calculate the difference in levels between the water and oil sides of the U tube:

$$h = h_2 - h_1 = 5.00\text{ cm} - 4.60\text{ cm} = 0.40\text{ cm}$$

INSIGHT

The weight of the height h_1 of water is equal to the weight of the height h_2 of oil. This is a special case, however, and is due to the fact that our U tube has equal diameters on its two sides. If the oil side of the U tube had been wider, for example, the weight of the oil would have been greater, though the difference in height still would have been $h = 0.40$ cm.

PRACTICE PROBLEM

Find the pressure at points A and B. [**Answer:** $P_A = P_B = P_{at} + 451$ Pa]

Some related homework problems: Problem 17, Problem 23, Problem 24

Pascal's Principle

Recall that if the surface of a fluid of density ρ is exposed to the atmosphere with a pressure P_{at}, the pressure at a depth h below the surface is

$$P = P_{at} + \rho gh$$

Suppose, now, that atmospheric pressure is increased from P_{at} to $P_{at} + \Delta P$. As a result, the pressure at the depth h is also increased:

$$P = P_{at} + \Delta P + \rho gh = (P_{at} + \rho gh) + \Delta P$$

Thus, by increasing the pressure at the top of the fluid by the amount ΔP, we have increased the pressure by the same amount everywhere in the fluid. This is **Pascal's principle:**

> An external pressure applied to an enclosed fluid is transmitted unchanged to every point within the fluid.

This principle applies no matter the size or shape of the container.

RWP A classic example of Pascal's principle at work is the **hydraulic lift,** which is shown schematically in **FIGURE 15-10.** Here we see two cylinders, one of cross-sectional area A_1, the other of cross-sectional area $A_2 > A_1$. The cylinders, each of which is fitted with a piston, are connected by a tube and filled with a fluid. Initially the pistons are at the same level and exposed to the atmosphere.

▲ **FIGURE 15-10 A hydraulic lift** A force F_1 exerted on the small piston causes a much larger force, F_2, to act on the large piston.

Now, suppose we push down on piston 1 with the force F_1. This increases the pressure in that cylinder by the amount

$$\Delta P = \frac{F_1}{A_1}$$

By Pascal's principle, the pressure in cylinder 2 increases by the *same* amount. Therefore, the increased upward force on piston 2 due to the fluid is

$$F_2 = (\Delta P)A_2$$

Substituting the increase in pressure, $\Delta P = F_1/A_1$, we find

$$F_2 = \left(\frac{F_1}{A_1}\right)A_2 = F_1\left(\frac{A_2}{A_1}\right) > F_1 \qquad \text{15-8}$$

To be specific, let's assume that A_2 is 100 times greater than A_1. Then, by pushing down on piston 1 with a force F_1, we push upward on piston 2 with a force of $100F_1$. Our force has been magnified 100 times!

If this sounds too good to be true, rest assured that we are not getting something for nothing. Just as with a lever, there is a tradeoff between the distance through which a force must be applied and the force magnification. This is illustrated in Figure 15-10, where we show piston 1 being pushed down through a distance d_1. This displaces a volume of fluid equal to A_1d_1. The same volume flows into cylinder 2, where it causes piston 2 to rise through a distance d_2. Equating the two volumes, we have

$$A_1d_1 = A_2d_2$$

Solving for d_2 gives

$$d_2 = d_1\left(\frac{A_1}{A_2}\right)$$

Thus, in the example just given, if we move piston 1 down a distance d_1, piston 2 rises a distance $d_2 = d_1/100$. Our force at piston 2 has been magnified 100 times, but the distance it moves has been reduced 100 times.

EXERCISE 15-9 HYDRAULIC LIFT

To inspect a 21,000-N truck, a mechanic raises it with a hydraulic lift. If the radius of the small piston in Figure 15-10 is 4.5 cm, and the radius of the large piston is 18 cm, find the force that must be exerted on the small piston to lift the truck.

REASONING AND SOLUTION

Solving $F_2 = (F_1/A_1)A_2$ for F_1, and noting that the area is πr^2, we find

$$F_1 = F_2\left(\frac{A_1}{A_2}\right) = (21{,}000 \text{ N})\left[\frac{\pi(0.045 \text{ m})^2}{\pi(0.18 \text{ m})^2}\right] = 1300 \text{ N}$$

Thus, the 4700-lb truck can be raised with a force of 290 lb.

Enhance Your Understanding   (Answers given at the end of the chapter)

3. Is the increase in pressure from the surface of the ocean to a depth of 10 m greater than, less than, or equal to the increase in pressure from a depth of 100 m in the ocean to a depth of 110 m? Explain.

Section Review

- Pressure in a fluid with density ρ increases with depth h as follows: $P = P_{at} + \rho gh$.
- Pascal's principle states that an external pressure applied to a fluid is transmitted unchanged to every point within the fluid.

15-4 Archimedes' Principle and Buoyancy

The fact that a fluid's pressure increases with depth leads to many interesting consequences. Among them is the fact that a fluid exerts a net upward force on any object it surrounds. This is referred to as a **buoyant force.**

The Origin of Buoyancy   To see how buoyancy comes about, consider a cubical block immersed in a fluid of density ρ, as shown in **FIGURE 15-11**. The surrounding fluid exerts normal forces on all of its faces. Clearly, the horizontal forces pushing to the right and to the left are equal; hence they cancel and have no effect on the block.

   The situation is quite different for the vertical forces, however. Notice, for example, that the downward force exerted on the top face is less than the upward force exerted on the lower face, because the pressure at the lower face is greater. This difference in forces gives rise to a net upward force—the buoyant force.

   Let's calculate the buoyant force acting on the block. First, we assume that the cubical block is of length L on a side, and that the pressure on the top surface is P_1. The downward force on the block, then, is

$$F_1 = P_1 A = P_1 L^2$$

Note that we have used the fact that the area of a square face of side L is L^2. Next, we consider the bottom face. The pressure there is given by $P_2 = P_1 + \rho g h$, with a difference in depth of $h = L$:

$$P_2 = P_1 + \rho g L$$

Therefore, the upward force exerted on the bottom face of the cube is

$$F_2 = P_2 A = (P_1 + \rho g L)L^2 = P_1 L^2 + \rho g L^3 = F_1 + \rho g L^3$$

If we take upward as the positive direction, the net vertical force exerted by the fluid on the block—that is, the buoyant force, F_b—is

$$F_b = F_2 - F_1 = \rho g L^3$$

As expected, the block experiences a net upward force from the surrounding fluid.

Archimedes' Principle   The precise value of the buoyant force is of some significance, as we now show. First, recall that the volume of a cube is L^3. It follows that $\rho g L^3$ is the weight of *fluid* of density ρ that would occupy the same volume as the cube. Therefore, the buoyant force ($F_b = \rho g L^3$) is equal to the weight of fluid that is displaced by the cube. This is **Archimedes' principle:**

> An object immersed in a fluid experiences an upward buoyant force equal to the weight of the fluid displaced by the object.

This principle applies to an object of any size and shape, and to any type of fluid. Stated mathematically, if a volume V of an object is immersed in a fluid of density ρ_{fluid}, the buoyant force F_b is the following:

Buoyant Force When a Volume *V* Is Submerged in a Fluid of Density ρ_{fluid}

$$F_b = \rho_{\text{fluid}} g V \qquad \qquad 15\text{-}9$$

SI unit: N

The volume V may be the total volume of the object, or any fraction of the total volume that is immersed in the fluid.

   To verify that Archimedes' principle is completely general, consider the submerged object shown in **FIGURE 15-12 (a)**. If we were to replace this object with an equivalent volume of fluid, as in **FIGURE 15-12 (b)**, the container would hold nothing but fluid and

▲ **FIGURE 15-11 Buoyant force due to a fluid**   A fluid surrounding an object exerts a buoyant force in the upward direction. This is due to the fact that pressure increases with depth, and hence the upward force on the object, F_2, is greater than the downward force, F_1. Forces acting to the left and to the right cancel.

PROBLEM-SOLVING NOTE

The Buoyant Force

Note that the buoyant force is equal to the weight of displaced fluid. It does not depend on the weight of the object that displaces the fluid.

(a) The forces acting on an object surrounded by fluid

(b) The same forces act when the object is replaced by fluid

▶ **FIGURE 15-12 Buoyant force equals the weight of displaced fluid**   The buoyant force acting on the object in **(a)** is equal to the weight of the "fluid object" (with the same size and shape) in **(b)**. This is because the fluid in (b) is at rest; hence the buoyant force acting on the fluid object must cancel its weight. The same forces act on the original object in (a), however, and therefore the buoyant force it experiences is also equal to the weight of the fluid object.

Big Idea 3 Archimedes' principle states that the buoyant force exerted by a fluid on an object is equal to the weight of the *fluid* that is displaced by the object.

would be in static equilibrium. As a result, we conclude that the net buoyant force acting on this "fluid object" must be upward and equal in magnitude to its weight. Now here's the key idea: Because the original object and the "fluid object" occupy the same position, the forces acting on their surfaces are identical, and hence the net buoyant force is the same for both objects. Therefore, the original object experiences a buoyant force equal to the weight of fluid that it displaces—that is, equal to the weight of the "fluid object."

CONCEPTUAL EXAMPLE 15-10   HOW IS THE SCALE READING AFFECTED?   PREDICT/EXPLAIN

A partially filled flask of water rests on a scale. **(a)** If you dip your finger into the water, without touching the flask, does the reading on the scale increase, decrease, or stay the same? **(b)** Which of the following is the *best explanation* for your prediction?

  I. The upward buoyant force on your finger results in a downward force on the flask, which increases the reading on the scale.

  II. The reading on the scale decreases because your finger has displaced some water.

  III. There is no change in the reading of the scale because your finger doesn't actually touch the flask, and hence you can't exert a downward force on it.

REASONING AND DISCUSSION
Your finger experiences an upward buoyant force when it is dipped into the water. By Newton's third law, the water experiences an equal and opposite reaction force acting downward. This downward force is transmitted to the scale, which in turn gives a higher reading.

Another way to look at this result is to note that when you dip your finger into the water, its depth increases. This results in a greater pressure at the bottom of the flask, and hence a greater downward force on the flask. The scale reads this increased downward force.

ANSWER
(a) The reading on the scale increases. **(b)** The best explanation is I.

Enhance Your Understanding   (Answers given at the end of the chapter)

  4. Is the buoyant force exerted on a cubical block of aluminum 10 cm on a side greater than, less than, or equal to the buoyant force exerted on a cubical block of lead 10 cm on a side, when both are completely submerged in water? Explain.

Section Review

  • Archimedes' principle is as follows: An object immersed in a fluid experiences an upward buoyant force equal to the weight of the fluid displaced by the object.

  • The mathematical statement of Archimedes' principle is $F_b = \rho_{fluid} g V$.

15-5 Applications of Archimedes' Principle

In this section we consider a variety of applications of Archimedes' principle. We begin with situations in which an object is fully immersed. Later we consider systems in which an object floats.

Complete Submersion

An interesting application of complete submersion can be found in an apparatus commonly used in determining a person's percentage of body fat. We consider the basic physics of the apparatus and the measurement procedure in the next Example. After the Example, we derive the relationship between overall body density and the body-fat percentage.

Weighing Weights in Water

EXAMPLE 15-11   MEASURING THE BODY'S DENSITY

RWP A person who weighs 720.0 N in air is lowered into a tank of water to about chin level. He sits in a harness of negligible mass suspended from a scale that reads his apparent weight. He now exhales as much air as possible and dunks his head underwater, submerging his entire body. If his apparent weight while submerged is 34.3 N, find **(a)** his volume and **(b)** his density.

PICTURE THE PROBLEM

The scale, the tank, and the person are shown in our sketch. We also show the free-body diagram for the person. The weight of the person in air is designated by $W = mg = 720.0$ N, and the apparent weight in water, which is the upward force exerted by the scale on the person, is designated by $W_a = 34.3$ N. The buoyant force exerted by the water on the person is F_b.

Physical picture   Free-body diagram

REASONING AND STRATEGY

To find the volume, V_p, and density, ρ_p, of the person, we must use two separate conditions. They are as follows:

a. When the person is submerged, the surrounding water exerts an upward buoyant force given by Archimedes' principle: $F_b = \rho_{water}V_p g$. This relationship, and Newton's second law, can be used to determine V_p.

b. The weight of the person in air is $W = mg = \rho_p V_p g$. Combining this relationship with the volume, V_p, found in part (a) allows us to determine the density, ρ_p.

Known   Weight of person, $W = 720.0$ N; apparent weight, $W_a = 34.3$ N.
Unknown **(a)** Volume of person, $V_p = ?$ **(b)** Density of person, $\rho_p = ?$

SOLUTION

Part (a)

1. Apply Newton's second law to the person. The person remains at rest, and therefore the net force acting on him is zero:

$$W_a + F_b - W = 0$$

2. Substitute $F_b = \rho_{water}V_p g$ for the buoyant force, and solve for the volume of the person, V_p:

$$W_a + \rho_{water}V_p g - W = 0$$
$$V_p = \frac{W - W_a}{\rho_{water}g} = \frac{720.0\text{ N} - 34.3\text{ N}}{(1.00 \times 10^3 \text{ kg/m}^3)(9.81 \text{ m/s}^2)}$$
$$= 6.99 \times 10^{-2} \text{ m}^3$$

Part (b)

3. Use $W = \rho_p V_p g$ to solve for the density of the person, ρ_p:

$$W = \rho_p V_p g$$
$$\rho_p = \frac{W}{V_p g} = \frac{720.0\text{ N}}{(6.99 \times 10^{-2} \text{ m}^3)(9.81 \text{ m/s}^2)}$$
$$= 1050 \text{ kg/m}^3$$

INSIGHT

As in Conceptual Example 15-10, the water exerts an upward buoyant force on the person, and an equal and opposite reaction force acts downward on the tank and water, making it press against the floor with a greater force.

   In addition, notice that the density of the person (1050 kg/m³) is only slightly greater than the density of seawater (1025 kg/m³), as given in Table 15-1.

PRACTICE PROBLEM

The person can float in water if his lungs are partially filled with air, which increases the volume of his body. What volume must his body have to just float?   [**Answer:** $V_p = 7.34 \times 10^{-2}$ m³]

Some related homework problems: Problem 43, Problem 44

BIO Once the overall density of the body is determined, the percentage of body fat can be obtained by noting that body fat has a density of $\rho_f = 9.00 \times 10^2$ kg/m³, whereas the lean body mass (muscles and bone) has a density of $\rho_l = 1.10 \times 10^3$ kg/m³. Suppose, for example, that a fraction x_f of the total body mass M is fat mass, and a fraction $(1 - x_f)$ is lean mass; that is, the fat mass is $m_f = x_f M$ and the lean mass is $m_l = (1 - x_f)M$. The total volume of the body is $V = V_f + V_l$. Using the fact that $V = m/\rho$, we can write the total volume as $V = m_f/\rho_f + m_l/\rho_l = x_f M/\rho_f + (1 - x_f)M/\rho_l$. Combining these results, we find that the overall density of a person's body is

$$\rho_p = \frac{M}{V} = \frac{1}{\dfrac{x_f}{\rho_f} + \dfrac{(1 - x_f)}{\rho_l}}$$

Solving for the body-fat fraction, x_f, yields

$$x_f = \frac{1}{\rho_p}\left(\frac{\rho_l \rho_f}{\rho_l - \rho_f}\right) - \frac{\rho_f}{\rho_l - \rho_f}$$

Finally, substituting the values for ρ_f and ρ_l, we find

$$x_f = \frac{(4950 \text{ kg/m}^3)}{\rho_p} - 4.50$$

This result is known in medical circles as *Siri's formula*. For example, if $\rho_p = 900 \text{ kg/m}^3$ (all fat), we find $x_f = 1$; if $\rho_p = 1100 \text{ kg/m}^3$ (no fat), we find $x_f = 0$. In the case of Example 15-11, where $\rho_p = 1050 \text{ kg/m}^3$, we find that this person's body-fat fraction is $x_f = 0.214$, for a percentage of 21.4%. This is a reasonable value for a healthy adult male.

A recent refinement to the measurement of body-fat percentage is the Bod Pod (FIGURE 15-13), an egg-shaped, airtight chamber in which a person sits comfortably—high and dry, surrounded only by air. This device works on the same physical principle as submerging a person in water, only it uses air instead of water. Since air is about a thousand times less dense than water, the measurements of apparent weight must be roughly a thousand times more sensitive. Fortunately, this is possible with today's technology, allowing for a much more convenient means of measurement.

Next we consider a low-density object, such as a piece of wood, held down below the surface of a denser fluid.

▲ **FIGURE 15-13** This device, known as the Bod Pod, measures the body-fat percentage of a person inside it by varying the air pressure in the chamber and measuring the corresponding changes in the person's apparent weight. Archimedes' principle is at work here, just as it is in Example 15-11.

EXAMPLE 15-12   FIND THE TENSION IN THE STRING

A piece of wood with a density $\rho_{\text{wood}} = 706 \text{ kg/m}^3$ is tied with a string to the bottom of a water-filled flask. The wood is completely immersed and has a volume $V_{\text{wood}} = 8.00 \times 10^{-6} \text{ m}^3$. What is the tension in the string? (Recall that the density of water is $\rho_{\text{water}} = 1.00 \times 10^3 \text{ kg/m}^3$.)

PICTURE THE PROBLEM
Our sketch shows the piece of wood held in place underwater by a string. Also shown is the free-body diagram for the wood, indicating the three forces acting on it.

REASONING AND STRATEGY
To find the tension in the string we apply Newton's second law to the piece of wood, noting that the wood is at rest and hence has zero acceleration. The weight of the wood has a magnitude of $\rho_{\text{wood}} g V_{\text{wood}}$, and the buoyant force has a magnitude of $\rho_{\text{water}} g V_{\text{wood}}$.

Known   Density of wood, $\rho_{\text{wood}} = 706 \text{ kg/m}^3$; volume of wood, $V_{\text{wood}} = 8.00 \times 10^{-6} \text{ m}^3$; density of water, $\rho_{\text{water}} = 1.00 \times 10^3 \text{ kg/m}^3$.
Unknown   Tension in the string, $T = ?$

SOLUTION

1. Apply Newton's second law to the piece of wood: $F_b - T - mg = 0$

2. Solve for the tension, T: $T = F_b - mg$

3. Calculate the buoyant force: $F_b = \rho_{\text{water}} g V_{\text{wood}}$
$= (1.00 \times 10^3 \text{ kg/m}^3)(9.81 \text{ m/s}^2)(8.00 \times 10^{-6} \text{ m}^3)$
$= 0.0785 \text{ N}$

4. Calculate the weight of the wood: $mg = \rho_{\text{wood}} g V_{\text{wood}} = (706 \text{ kg/m}^3)(9.81 \text{ m/s}^2)(8.00 \times 10^{-6} \text{ m}^3)$
$= 0.0554 \text{ N}$

5. Subtract to obtain the tension: $T = F_b - mg$
$= 0.0785 \text{ N} - 0.0554 \text{ N} = 0.0231 \text{ N}$

CONTINUED

INSIGHT

The wood is less dense than water, and hence its buoyant force when completely immersed (0.0785 N) is greater than its weight (0.0554 N).

PRACTICE PROBLEM — PREDICT/CALCULATE

(a) If the density of the wood is increased, does the tension in the string increase, decrease, or stay the same? Explain. **(b)** What is the tension in the string if the piece of wood has a density of 822 kg/m³? [**Answer: (a)** The tension decreases because the weight of the wood is closer to the value of the buoyant force. **(b)** The tension is 0.0140 N, which is less than that found above.]

Some related homework problems: Problem 40, Problem 93

► **FIGURE 15-14 Floatation (a)** The block of wood displaces some water, but not enough to equal its weight. Thus, the block would not float at this position. **(b)** The weight of displaced water equals the weight of the block in this case. The block floats now.

Floatation

When an object floats, the buoyant force acting on it equals its weight. For example, suppose we slowly lower a block of wood into a flask of water. At first, as in **FIGURE 15-14 (a)**, only a small amount of water is displaced and the buoyant force is a fraction of the block's weight. If we were to release the block now, it would drop farther into the water. As we continue to lower the block, more water is displaced, increasing the buoyant force.

Eventually, we reach the situation pictured in **FIGURE 15-14 (b)**, where the block begins to float. In this case, the buoyant force equals the weight of the wood. This, in turn, means that the weight of the displaced water is equal to the weight of the wood. In general:

> An object floats when it displaces an amount of fluid whose weight is equal to the weight of the object.

This is illustrated in **FIGURE 15-15 (a)**, where we show the volume of water equal to the weight of a block of wood. The wood displaces just this amount of water when it floats. A similar case involving human floatation is shown in **FIGURE 15-16**.

On the other hand, in **FIGURE 15-15 (b)** we show the amount of water necessary to have the same weight as a block of metal. Clearly, if the metal is completely submerged, the buoyant force is only a fraction of its weight, and so it sinks. On the other hand, if the metal is formed into the shape of a bowl, as in **FIGURE 15-15 (c)**, it can displace a

(a) Some water is displaced, but not enough to float the wood

(b) More water is displaced, and now the wood floats

(a) Block of wood and water of equal weight

(b) Block of metal and water of equal weight

(c) Same block of metal as in (b), but shaped like a bowl to displace more water

▲ **FIGURE 15-15 Floating an object that is more dense than water (a)** A wood block and the volume of water that has the same weight. Because the wood has a larger volume than this, it floats. **(b)** A metal block and the volume of water that has the same weight. Since the metal displaces less water than this, it sinks. **(c)** If the metal in (b) is shaped like a bowl, it can displace more water than the volume of the metal itself. In fact, it can displace enough water to float.

▲ **FIGURE 15-16** The water of the Dead Sea is unusually dense because of its high salt content. As a result, swimmers can float higher in the water than they are accustomed and engage in recreational activities that we don't ordinarily associate with a dip in the ocean.

► **FIGURE 15-17 Visualizing Concepts Floatation (a)** Although steel is denser than water, a ship's bowl-like hull displaces enough water to allow it to float. (The boundary between the red and black areas of the hull is the Plimsoll line mentioned in Conceptual Example 15-14.) **(b)** For hot-air balloons, the key to buoyancy is the low density of the heated air in its envelope.

(a)   **(b)**

Squeezing sides of flask increases the water pressure . . .

. . . and shrinks the air bubble, causing the diver to drop.

▲ **FIGURE 15-18 A Cartesian diver** A Cartesian diver floats because of the bubble of air trapped within it. When the bottle is squeezed, which increases the pressure in the water, the bubble is reduced in size and the diver sinks.

▲ **FIGURE 15-19** Felix Baumgartner prepares to jump from an altitude of 38,969 m (24.2 mi). He was carried to that altitude by a helium-filled balloon that operates on Archimedes' principle.

volume of water equal to its weight and float. A similar example is shown in **FIGURE 15-17 (a)**.

Another way to change the buoyancy of an object is to alter its overall density. A good example is provided by the hot-air balloons in **FIGURE 15-17 (b)**, which use low-density heated air to produce an overall density less than that of air. Similar considerations apply to the *Cartesian diver* shown in **FIGURE 15-18**. As illustrated, the diver is simply a small glass tube, similar in shape to an eyedropper, with an air bubble trapped inside. Initially, the overall density of the tube and the air bubble is less than the density of water, and the diver floats. When the bottle containing the diver is squeezed, however, the pressure in the water rises, and the air bubble is compressed to a smaller volume. Now, the overall density of the tube and air bubble is greater than that of water, and the diver descends. By adjusting the pressure on the bottle, the diver can be made to float at any depth in the bottle.

BIO The principle behind the Cartesian diver also applies to the swim bladder of ray-finned bony fishes. The swim bladder is basically an air sac whose volume can be controlled by the fish. By adjusting the size of the swim bladder, the fish can give itself "neutral buoyancy"—that is, the fish can float without effort at a given depth, just like the Cartesian diver. Similar considerations apply to certain diving sea mammals, such as the bottlenose dolphin, Weddell seal, elephant seal, and blue whale. All of these animals are capable of diving to great depths—in fact, some of the seals have been observed at depths of nearly 400 m. In order to conserve energy on their long dives, they take advantage of the fact that the pressure of the surrounding water compresses their bodies and flattens the air sacs in their lungs. Just as with the Cartesian diver, this decreases their buoyancy to the point where they begin to sink. As a result, they can glide effortlessly to the desired depth, saving energy for the swim back to the surface.

RWP Just as fish and dolphins are surrounded by an ocean of liquid water, balloons are surrounded by an ocean of air. The warm air inside a hot-air balloon is less dense than the surrounding cold air, so that the buoyant force on the balloon (from the displacement of the cold air) is larger than its weight, and the balloon rises into the sky. Hydrogen or helium-filled balloons work on the same principle, and are frequently used to make weather measurements. The behemoth high-altitude balloons, some as large as 30 m in diameter, rise into the atmosphere until the average density of the balloon matches the density of the surrounding atmosphere, often at an altitude of 30 km (100,000 ft) or higher. On October 14, 2012, Austrian skydiver Felix Baumgartner rode a helium-filled balloon to an altitude of 38,969 m (127,852 ft, or 24.2 mi) and jumped (**FIGURE 15-19**), setting a record for the highest skydive ever, and becoming the first human to break the sound barrier without any form of engine power.

QUICK EXAMPLE 15-13   FLOATING A BLOCK OF WOOD

How much water (density 1000 kg/m³) must be displaced to float a cubical block of wood (density 634 kg/m³) that is 13.3 cm on a side?

REASONING AND SOLUTION

We start by finding the volume of the wood block, V_{wood}, from its dimensions. Once we have this, we can find its mass ($m_{\text{wood}} = \rho_{\text{wood}} V_{\text{wood}}$) and weight ($W_{\text{wood}} = m_{\text{wood}} g$). Next,

CONTINUED

we set the weight of a volume of water ($W_{water} = \rho_{water} V_{water} g$) equal to the weight of the wood. This expression can be solved for the only unknown, the volume of the water.

1. Calculate the volume of the wood: $V_{wood} = (0.133 \text{ m})^3 = 0.00235 \text{ m}^3$

2. Find the weight of the wood:
$$W_{wood} = m_{wood} g = \rho_{wood} V_{wood} g$$
$$= (634 \text{ kg/m}^3)(0.00235 \text{ m}^3)(9.81 \text{ m/s}^2)$$
$$= 14.6 \text{ N}$$

3. Write an expression for the weight of an arbitrary volume of water, V_{water}:
$$W_{water} = m_{water} g = \rho_{water} V_{water} g$$

4. Set the weight of water equal to the weight of the wood:
$$W_{water} = \rho_{water} V_{water} g$$
$$= W_{wood} = 14.6 \text{ N}$$

5. Solve for the volume of water:
$$V_{water} = \frac{W_{wood}}{\rho_{water} g} = \frac{14.6 \text{ N}}{(1000 \text{ kg/m}^3)(9.81 \text{ m/s}^2)}$$
$$= 0.00149 \text{ m}^3 = 0.634 V_{wood}$$

As expected, only a fraction of the wood (63.4%) must be submerged for it to float.

CONCEPTUAL EXAMPLE 15-14 THE PLIMSOLL MARK

On the side of a cargo ship you may see a horizontal line indicating "maximum load." (The line is sometimes known as the "Plimsoll mark," after the nineteenth-century British legislator who caused it to be adopted.) When a ship is loaded to capacity, the maximum load line is at water level. The ship shown here has two maximum load lines, one for freshwater and one for salt water. Which line should be marked "maximum load for salt water": (a) the top line or (b) the bottom line?

REASONING AND DISCUSSION
If a ship sails from freshwater into salt water it floats higher, just as it is easier for you to float in an ocean than in a lake. The reason is that salt water is denser than freshwater; hence less of it needs to be displaced to provide a given buoyant force. Because the ship floats higher in salt water, the bottom line should be used to indicate maximum load.

ANSWER
(b) The bottom line should be used in salt water.

Tip of the Iceberg

As we've seen, an object floats when its weight is equal to the weight of the fluid it displaces. Let's use this condition to determine just how much of a floating object is submerged. We will then apply our result to the classic case of an iceberg, as shown in **FIGURE 15-20**.

Consider a solid of density ρ_s floating in a fluid of density ρ_f, as in **FIGURE 15-21**. If the solid has a volume V_s, its total weight is

$$W_s = \rho_s V_s g$$

Similarly, the weight of a volume V_f of displaced fluid is

$$W_f = \rho_f V_f g$$

Equating these weights, we find the following:

$$W_s = W_f$$
$$\rho_s V_s g = \rho_f V_f g$$

Canceling g, and solving for the volume of displaced fluid, we have

$$V_f = V_s(\rho_s/\rho_f)$$

Noting that the volume of displaced fluid, V_f, is the same as the volume of the solid that is submerged, V_{sub}, we find:

▲ **FIGURE 15-20** Most people know that the bulk of an iceberg lies below the surface of the water. But, as with ships and swimmers, the actual proportion that is submerged depends on whether the water is fresh or salt (see Example 15-15).

◀ **FIGURE 15-21 Submerged volume of a floating object** A solid, of volume V_s and density ρ_s, floats in a fluid of density ρ_f. The volume of the solid that is submerged is $V_{sub} = V_s(\rho_s/\rho_f)$.

Submerged Volume V_{sub} for a Solid of Volume V_s and Density ρ_s Floating in a Fluid of Density ρ_f

$$V_{sub} = V_s(\rho_s/\rho_f)$$

15-10

SI unit: m^3

As a check, notice that this result agrees with our calculations in Quick Example 15-13. We now apply this result to the classic case of ice floating in water.

EXAMPLE 15-15 THE TIP OF THE ICEBERG

RWP A chunk of ice floats in fresh water. What percentage of the ice projects above the level of the water? Assume a density of 917 kg/m³ for the ice and 1.00×10^3 kg/m³ for the water.

PICTURE THE PROBLEM

Our sketch shows a chunk of ice floating in a glass of water. In this case the solid object is ice, with a density $\rho_s = \rho_{ice} = 917$ kg/m³, and the fluid is water, with a density $\rho_f = \rho_{water} = 1.00 \times 10^3$ kg/m³. Thus, in this system, we make the following identifications: s = solid → ice, f = fluid → water.

REASONING AND STRATEGY

We can use Equation 15-10, $V_{sub} = V_s(\rho_s/\rho_f)$, to find the volume of ice that is below the water (that is, *sub*merged). It follows that the fraction of the total volume of the ice, V_s, that is submerged is $V_{sub}/V_s = \rho_s/\rho_f$. The fraction of the ice that is above the water, then, is $1 - V_{sub}/V_s$. Multiplying this fraction by 100 yields the percentage p above water.

Known Density of ice, $\rho_s = \rho_{ice} = 917$ kg/m³; density of water, $\rho_f = \rho_{water} = 1.00 \times 10^3$ kg/m³.

Unknown Percentage of ice above water level, $p = 100(1 - V_{sub}/V_s) = ?$

SOLUTION

1. Calculate the fraction of the total volume of the ice that is submerged:
$$\frac{V_{sub}}{V_s} = \frac{\rho_s}{\rho_f} = \frac{917 \text{ kg/m}^3}{1.00 \times 10^3 \text{ kg/m}^3} = 0.917$$

2. Calculate the fraction of the ice that is above the water level:
$$1 - \frac{V_{sub}}{V_s} = 1 - 0.917 = 0.083$$

3. Multiply by 100 to obtain a percentage:
$$p = 100\left(1 - \frac{V_{sub}}{V_s}\right) = 100(0.083) = 8.3\%$$

INSIGHT

Because we seek a percentage, it's not necessary to know the total volume of the ice. Thus, our result that 8.3% of the ice is above the water level applies whether we are talking about an ice cube in a drinking glass, or an iceberg floating in a freshwater lake. If an iceberg floats in the ocean, however, it will float higher due to the higher density of seawater. We consider this case in the following Practice Problem.

PRACTICE PROBLEM

What percentage of an ice chunk is above the water level if it floats in seawater? (The density of seawater can be found in Table 15-1.) **[Answer: 10.5%]**

Some related homework problems: Problem 46, Problem 47, Problem 87

CONCEPTUAL EXAMPLE 15-16 THE NEW WATER LEVEL I

A cup is filled to the brim with water and a floating ice cube. When the ice melts, which of the following occurs: **(a)** water overflows the cup, **(b)** the water level is lower, or **(c)** the water level remains the same?

CONTINUED

REASONING AND DISCUSSION

The ice cube floats, and hence it displaces a volume of water equal to its weight. But when the ice cube melts, it becomes water, and its weight is the same. Hence, the melted water fills exactly the same volume that the ice cube displaced when floating. As a result, the water level is unchanged.

ANSWER

(c) The water level remains the same.

CONCEPTUAL EXAMPLE 15-17   THE NEW WATER LEVEL II

A cup is filled to the brim with water and a floating ice cube. Resting on top of the ice cube is a small pebble. When the ice melts, which of the following occurs: **(a)** water overflows the cup, **(b)** the water level is lower, or **(c)** the water level remains the same?

REASONING AND DISCUSSION

We know from the previous Conceptual Example that the ice itself makes no difference to the water level. As for the pebble, when it floats on the ice it displaces an amount of water equal to its *weight*. When the ice melts, the pebble drops to the bottom of the cup, where it displaces a volume of water equal to its own *volume*. Because the volume of the pebble is less than the volume of water with the same weight, we conclude that less water is displaced after the ice melts. Hence, the water level is lower.

ANSWER

(b) The water level is lower.

Enhance Your Understanding   (Answers given at the end of the chapter)

5. A cup is filled to the brim with water. Floating in the water is an ice cube, on top of which rests a small block of wood, as shown in **FIGURE 15-22**. When the ice melts, and the block of wood floats in the water, is the level of the water higher (causing the cup to overflow), is it lower, or does it stay the same? Explain.

Section Review

- The condition for an object to float in a fluid is as follows: An object floats when it displaces an amount of fluid whose weight is equal to the weight of the object.

▲ **FIGURE 15-22** Enhance Your Understanding 5.

15-6 Fluid Flow and Continuity

Suppose you want to water the yard, but you don't have a spray nozzle for the end of the hose. Without a nozzle the water flows rather slowly from the hose and hits the ground within half a meter. But if you place your thumb over the end of the hose, narrowing the opening to a fraction of its original size, the water sprays out with a high speed and a large range. Why does decreasing the size of the opening have this effect?

Equation of Continuity To answer this question, we begin by considering a simple system that shows the same behavior. Imagine a fluid that flows with a speed v_1 through a cylindrical pipe of cross-sectional area A_1, as in the left-hand portion of **FIGURE 15-23**. If the pipe narrows to a cross-sectional area A_2, as in the right-hand portion of Figure 15-23, the fluid will flow with a new speed, v_2.

We can find the speed in the narrow section of the pipe by assuming that any amount of fluid that passes point 1 in a given time, Δt, must also flow past point 2 in the same time. If this were not the case, the system would be gaining or losing fluid. To

▲ **FIGURE 15-23 Fluid flow through a pipe of varying diameter** As a fluid flows from a large pipe to a small pipe, the same mass of fluid passes a given point in a given amount of time. Thus, the speed in the small pipe is greater than it is in the large pipe.

PHYSICS
IN CONTEXT
Looking Ahead

When considering fluid flow, we introduced the idea of the amount of material passing through a given area in a given time. This is referred to as flux. We generalize the concept of flux to electric fields in Chapter 19 and to magnetic fields in Chapter 23 .

PROBLEM-SOLVING NOTE

Continuity of Flow

The speed of an incompressible fluid is inversely proportional to the area through which it flows.

▲ FIGURE 15-24 Narrowing the opening in a hose with a nozzle (or thumb) increases the velocity of flow, as one would expect from the equation of continuity.

find the mass of fluid passing point 1 in the time Δt, note that the fluid moves through a distance $v_1 \Delta t$ in this time. As a result, the volume of fluid going past point 1 is

$$\Delta V_1 = A_1 v_1 \Delta t$$

Hence, the mass of fluid passing point 1 is

$$\Delta m_1 = \rho_1 \Delta V_1 = \rho_1 A_1 v_1 \Delta t$$

Similarly, the mass passing point 2 in the time Δt is

$$\Delta m_2 = \rho_2 \Delta V_2 = \rho_2 A_2 v_2 \Delta t$$

Notice that we have allowed for the possibility of the fluid having different densities at points 1 and 2.

Finally, equating these two masses yields the relationship between v_1 and v_2:

$$\Delta m_1 = \Delta m_2$$
$$\rho_1 A_1 v_1 \Delta t = \rho_2 A_2 v_2 \Delta t$$

Canceling Δt we find:

Equation of Continuity

$$\rho_1 A_1 v_1 = \rho_2 A_2 v_2 \qquad \text{15-11}$$

This relationship is referred to as the **equation of continuity.**

Incompressible Fluids Most gases are readily compressed, which means that their densities can change. In contrast, most liquids are practically incompressible, so their densities are essentially constant. Unless stated otherwise, we will assume that all liquids discussed in this text are perfectly incompressible. Thus, for liquids, ρ_1 and ρ_2 are the same in Equation 15-11, and the equation of continuity reduces to the following:

Equation of Continuity for an Incompressible Fluid

$$A_1 v_1 = A_2 v_2 \qquad \text{15-12}$$

Reducing the area of flow increases the speed, as we see in FIGURE 15-24.

We next apply this relationship to the case of water flowing through the nozzle of a fire hose.

EXAMPLE 15-18   SPRAY I

Water travels through a 9.8-cm-diameter fire hose with a speed of 1.5 m/s. At the end of the hose, the water flows out through a nozzle whose diameter is 2.6 cm. What is the speed of the water coming out of the nozzle?

PICTURE THE PROBLEM
In our sketch, we label the speed of the water in the hose with v_1 and the speed of the water coming out the nozzle with v_2. We are given that $v_1 = 1.5$ m/s. We also know that the diameter of the hose is $d_1 = 9.8$ cm and the diameter of the nozzle is $d_2 = 2.6$ cm.

REASONING AND STRATEGY
We can find the water speed in the nozzle by applying $A_1 v_1 = A_2 v_2$ (Equation 15-12). In addition, we assume that the hose and nozzle are circular in cross section; hence, their areas are given by $A = \pi d^2/4$, where d is the diameter.

Known Diameter of fire hose, $d_1 = 9.8$ cm; speed of water in fire hose, $v_1 = 1.5$ m/s; diameter of nozzle, $d_2 = 2.6$ cm.
Unknown Speed of water in nozzle, $v_2 = ?$

SOLUTION

1. Solve $A_1 v_1 = A_2 v_2$ for v_2, the speed of water in the nozzle: $\quad v_2 = v_1(A_1/A_2)$

2. Replace the areas with $A = \pi d^2/4$: $\quad v_2 = v_1\left(\dfrac{\pi d_1^2/4}{\pi d_2^2/4}\right) = v_1\left(\dfrac{d_1^2}{d_2^2}\right)$

3. Substitute numerical values: $\quad v_2 = v_1\left(\dfrac{d_1^2}{d_2^2}\right) = (1.5\text{ m/s})\left(\dfrac{9.8\text{ cm}}{2.6\text{ cm}}\right)^2 = 21\text{ m/s}$

CONTINUED

Enhance Your Understanding   (Answers given at the end of the chapter)

6. Water flows with a speed v through a pipe. If the diameter of the pipe is decreased to half its initial value, what is the new speed of the water? Explain.

Section Review

- The equation of continuity for a flowing fluid is $\rho_1 A_1 v_1 = \rho_2 A_2 v_2$. For an incompressible fluid, the density is constant, and hence the equation of continuity simplifies to $A_1 v_1 = A_2 v_2$.

15-7 Bernoulli's Equation

In this section, we apply the work-energy theorem to fluids. The result is a relationship among the pressure of a fluid, its speed, and its height. This relationship is known as **Bernoulli's equation.**

Change in Speed

We begin by considering a system in which the speed of the fluid changes, but its height remains the same. To be specific, the system of interest is the same as that shown in Figure 15-23. We have already shown that the speed of the fluid increases as it flows from region 1 to region 2; we now investigate the corresponding change in pressure.

Our plan of attack is to first calculate the total work done on the fluid as it moves from one region to the next. This result will depend on the pressure in the fluid. Once the total work is obtained, the work-energy theorem allows us to equate it to the change in kinetic energy of the fluid. This will give the pressure-speed relationship we desire.

Consider an element of fluid of length Δx_1, as shown in **FIGURE 15-25**. This element is pushed in the direction of motion by the pressure P_1. Thus, P_1 does positive work, ΔW_1, on the fluid element. Since the force exerted on the element is $F_1 = P_1 A_1$, and work is force times distance, the work done on the element by P_1 is

$$\Delta W_1 = F_1 \Delta x_1 = P_1 A_1 \Delta x_1$$

The volume of the fluid element is $\Delta V_1 = A_1 \Delta x_1$, so the work done by P_1 is

$$\Delta W_1 = P_1 \Delta V_1$$

Next, when the fluid element emerges into region 2, it experiences a force opposite to its direction of motion due to the pressure P_2. Thus, P_2 does negative work on the element. Following the same steps given previously, we can write the work done by P_2 as

$$\Delta W_2 = -P_2 \Delta V_2$$

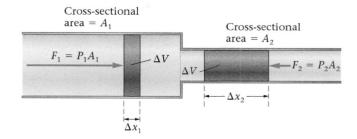

◀ **FIGURE 15-25 Work done on a fluid element** As an incompressible fluid element of volume ΔV moves from pipe 1 to pipe 2, the pressure P_1 does a positive work $P_1 \Delta V$ and the pressure P_2 does a negative work $P_2 \Delta V$. Since P_1 is greater than P_2, the net result is that positive work is done, and the fluid element speeds up.

Now, for an incompressible fluid, the volume of the element does not change as it goes from region 1 to region 2. Therefore,

$$\Delta V_1 = \Delta V_2 = \Delta V$$

Using this result, we can write the total work done on the fluid element as follows:

$$\Delta W_{total} = \Delta W_1 + \Delta W_2 = P_1 \Delta V - P_2 \Delta V = (P_1 - P_2)\Delta V$$

The final step is to equate the total work to the change in kinetic energy:

$$\Delta W_{total} = (P_1 - P_2)\Delta V = K_{final} - K_{initial} = K_2 - K_1 \qquad \text{15-13}$$

What is the kinetic energy of the fluid element? Well, the mass of the element is

$$\Delta m = \rho \Delta V$$

Thus, its kinetic energy is simply

$$K = \tfrac{1}{2}(\Delta m)v^2 = \tfrac{1}{2}(\rho \Delta V)v^2$$

Using this expression in Equation 15-13, we have

$$\Delta W_{total} = (P_1 - P_2)\Delta V = \left(\tfrac{1}{2}\rho v_2{}^2 - \tfrac{1}{2}\rho v_1{}^2\right)\Delta V$$

Canceling the common factor ΔV and rearranging, we find the following:

Bernoulli's Equation (constant height)

$$P_1 + \tfrac{1}{2}\rho v_1{}^2 = P_2 + \tfrac{1}{2}\rho v_2{}^2 \qquad \text{15-14}$$

Notice that $P_1 + \tfrac{1}{2}\rho v_1{}^2 = P_2 + \tfrac{1}{2}\rho v_2{}^2$ is equivalent to saying that $P + \tfrac{1}{2}\rho v^2$ is constant. Thus, there is a tradeoff between the pressure in a fluid and its speed—as the fluid speeds up, its pressure decreases. If this seems odd, recall that P_1 acts to increase the speed of the fluid element and P_2 acts to decrease its speed. The element will speed up, then, only if P_2 is less than P_1. The fluid speeds up as it enters a region of lower pressure.

PHYSICS IN CONTEXT
Looking Back

The connection between a force acting over a distance and kinetic energy (Chapter 7) is applied to a fluid in the derivation of Bernoulli's equation.

EXAMPLE 15-19   SPRAY II

Predict/Calculate

Suppose the pressure in the fire hose in Example 15-18 is 350 kPa. **(a)** Is the pressure in the nozzle greater than, less than, or equal to 350 kPa? **(b)** Find the pressure in the nozzle.

PICTURE THE PROBLEM
In our sketch, we use the same numbering system as in Example 15-18; that is, 1 refers to the hose, 2 refers to the nozzle. Therefore, $P_1 = 350$ kPa, and P_2 is to be determined.

REASONING AND STRATEGY

a. The pressure and speed of the fluid at constant height are related by $P_1 + \tfrac{1}{2}\rho v_1{}^2 = P_2 + \tfrac{1}{2}\rho v_2{}^2$. This result can be used to compare P_1 to P_2.

b. In Example 15-18, we used the equation of continuity $(A_1 v_1 = A_2 v_2)$ to determine the speed of water in the nozzle, v_2. We now use this result, along with $P_1 + \tfrac{1}{2}\rho v_1{}^2 = P_2 + \tfrac{1}{2}\rho v_2{}^2$, to determine P_2.

Known   Pressure in fire hose, $P_1 = 350$ kPa; speed in fire hose, $v_1 = 1.5$ m/s; speed in nozzle, $v_2 = 21$ m/s.
Unknown   Pressure in nozzle, $P_2 = ?$

SOLUTION

Part (a)

1. Apply $P_1 + \tfrac{1}{2}\rho v_1{}^2 = P_2 + \tfrac{1}{2}\rho v_2{}^2$, noting that v_2 is greater than v_1, as determined in Example 15-18:

     The pressure and speed are related by $P + \tfrac{1}{2}\rho v^2 =$ constant. Thus, an increase in speed (v) can occur only when there is a corresponding decrease in pressure (P). **Answer:** P_2 (nozzle) is less than P_1 (hose); that is, $P_2 < 350$ kPa.

Part (b)

2. Solve $P_1 + \tfrac{1}{2}\rho v_1{}^2 = P_2 + \tfrac{1}{2}\rho v_2{}^2$ for the pressure in the nozzle, P_2:

     $P_2 = P_1 + \tfrac{1}{2}\rho(v_1{}^2 - v_2{}^2)$

CONTINUED

3. Substitute numerical values, including $v_1 = 1.5$ m/s and $v_2 = 21$ m/s from Example 15-18:

$$P_2 = 350 \text{ kPa} + \tfrac{1}{2}(1.00 \times 10^3 \text{ kg/m}^3)[(1.5 \text{ m/s})^2 - (21 \text{ m/s})^2]$$
$$= 131 \text{ kPa}$$

INSIGHT

The pressure in the nozzle is less than the pressure in the hose by roughly a factor of 2.7. This is because part of the energy associated with the high pressure in the hose has been converted to kinetic energy as the water passes into the nozzle. The connection between pressure and energy will be explored in more detail later in this section.

PRACTICE PROBLEM

What nozzle speed would be required to give a nozzle pressure of 161 kPa? [**Answer:** $v_2 = 19.5$ m/s]

Some related homework problems: Problem 60, Problem 62

Change in Height

If a fluid flows through the pipe shown in **FIGURE 15-26**, its height increases from y_1 to y_2 as it goes from one section of the pipe to the next. Since the cross-sectional area of the pipe is constant, however, the speed of the fluid is unchanged, according to $A_1 v_1 = A_2 v_2$. Thus, the change in kinetic energy of the fluid element shown in Figure 15-26 is zero.

The total work done on the fluid element is the sum of the works done by the pressure in each of the two regions, plus the work done by gravity. As before, the work done by pressure is

$$\Delta W_{\text{pressure}} = \Delta W_1 + \Delta W_2 = (P_1 - P_2)\Delta V$$

As the fluid element rises, gravity does negative work on it. Recall that the mass of the element is

$$\Delta m = \rho \Delta V$$

Therefore, the work done by gravity is

$$\Delta W_{\text{gravity}} = -\Delta m g(y_2 - y_1) = -\rho \Delta V g(y_2 - y_1)$$

Setting the total work equal to zero (since $\Delta K = 0$) yields

$$\Delta W_{\text{total}} = \Delta W_{\text{pressure}} + \Delta W_{\text{gravity}}$$
$$= (P_1 - P_2)\Delta V - \rho g(y_2 - y_1)\Delta V = 0$$

Canceling ΔV and rearranging gives:

Bernoulli's Equation (constant speed)

$$P_1 + \rho g y_1 = P_2 + \rho g y_2 \qquad\qquad \textbf{15-15}$$

In this case, it is $P + \rho g y$ that is constant—hence, pressure decreases as the height within a fluid increases. In fact, $P_1 + \rho g y_1 = P_2 + \rho g y_2$ (Equation 15-15) is precisely the same as the pressure–depth relationship, Equation 15-7 ($P_2 = P_1 + \rho g h$), which was obtained using force considerations. Here we obtained the same result using energy considerations and the work-energy theorem.

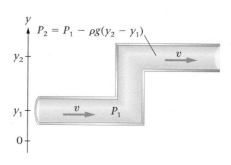

▲ **FIGURE 15-26 Fluid pressure in a pipe of varying elevation** Fluid of density ρ flows in a pipe of uniform cross-sectional area from height y_1 to height y_2. As it does so, its pressure decreases by the amount $\rho g(y_2 - y_1)$.

EXERCISE 15-20   PRESSURE AT THE TOP

Water flows with constant speed through a garden hose that goes up a step 27.5 cm high, as in Figure 15-26. If the water pressure is 132 kPa at the bottom of the step, what is its pressure at the top of the step?

REASONING AND SOLUTION

We apply $P_1 + \rho g y_1 = P_2 + \rho g y_2$, letting subscript 1 refer to the bottom of the step and subscript 2 refer to the top of the step. Rearranging and solving for P_2, we have

$$P_2 = P_1 + \rho g(y_1 - y_2)$$
$$= 132 \text{ kPa} + (1.00 \times 10^3 \text{ kg/m}^3)(9.81 \text{ m/s}^2)(0 - 0.275 \text{ m}) = 129 \text{ kPa}$$

This is precisely the pressure difference that would be observed if the water had been at rest.

General Case

In a more general case, both the height of a fluid and its speed may change. Combining the results obtained in Equations 15-14 and 15-15 yields the general form of Bernoulli's equation:

Bernoulli's Equation (general case)

$$P_1 + \tfrac{1}{2}\rho v_1^2 + \rho g y_1 = P_2 + \tfrac{1}{2}\rho v_2^2 + \rho g y_2$$

15-16

Thus, in general, the quantity $P + \tfrac{1}{2}\rho v^2 + \rho g y$ is a constant within a fluid. This is basically a statement of energy conservation. For example, recalling the definition of density in Equation 15-1, we find that $\tfrac{1}{2}\rho v^2$ is $\tfrac{1}{2}(M/V)v^2 = \left(\tfrac{1}{2}Mv^2\right)/V$. Clearly, this term represents the kinetic energy per volume of the fluid. Similarly, the term $\rho g y$ can be written as $(M/V)g y = (Mg y)/V$, which is the gravitational potential energy per volume.

Finally, the first term in Bernoulli's equation—the pressure—can also be thought of as an energy per volume. Recall that $P = F/A$. If we multiply numerator and denominator by a distance, d, we have $P = Fd/Ad$. But Fd is the work done by the force F as it acts through the distance d, and Ad is the volume swept out by an area A moved through a distance d. Therefore, the pressure can be thought of as work (energy) per volume: $P = W/V$.

As a result, Bernoulli's equation is simply a restatement of the work-energy theorem in terms of energy per volume. Of course, this relationship holds only as long as we can ignore frictional losses, which would lead to heating. We will consider the energy aspects of heat in Chapter 16.

> **Big Idea 4** Bernoulli's equation is a statement of energy conservation applied to fluids. For example, P refers to the energy stored in a fluid due to pressure, $\tfrac{1}{2}\rho v^2$ refers to the kinetic energy per volume of a flowing fluid, and $\rho g y$ refers to the gravitational potential energy per volume of a fluid.

QUICK EXAMPLE 15-21 FIND THE PRESSURE

A hose goes up a step of height 26.4 cm. The speed of water in the hose at the bottom of the step is 1.31 m/s, and the cross-sectional area of the hose on top of the step is half the cross-sectional area at the bottom of the step. If the pressure in the water at the bottom of the step is 147 kPa, what is the pressure in the water on top of the step?

REASONING AND SOLUTION

First, we use the equation of continuity to find the speed of water on the top of the step. Combining this with the known height at the top of the step allows us to find the pressure there. (Let subscript 1 refer to the bottom of the step, and subscript 2 refer to the top of the step.)

1. Use the continuity equation to find the water's speed on the top of the step:

$$v_2 = \left(\frac{A_1}{A_2}\right)v_1 = 2v_1 = 2(1.31 \text{ m/s}) = 2.62 \text{ m/s}$$

2. Use Bernoulli's equation for the general case to solve for P_2:

$$P_2 = P_1 - \rho g(y_2 - y_1) - \tfrac{1}{2}\rho(v_2^2 - v_1^2)$$

3. Substitute numerical values:

$$P_2 = 147 \text{ kPa} - (1000 \text{ kg/m}^3)(9.81 \text{ m/s}^2)(0.264 \text{ m})$$
$$- \tfrac{1}{2}(1000 \text{ kg/m}^3)[(2.62 \text{ m/s})^2 - (1.31 \text{ m/s})^2]$$
$$= 142 \text{ kPa}$$

The pressure on the top of the step is less than the pressure at the bottom of the step for two reasons: (i) The top of the step is higher, and (ii) the water has a greater speed at the top of the step.

Enhance Your Understanding   (Answers given at the end of the chapter)

7. Water flows through a pipe with a varying diameter, as shown in FIGURE 15-27. Small vertical tubes open to the air indicate the pressure in each segment of the pipe. **(a)** Is the height of water in tube B greater than, less than, or equal to the height of water in tube A? Explain. **(b)** Is the height of water in tube C greater than, less than, or equal to the height of water in tube A? Explain. **(c)** Is the height of water in tube C greater than, less than, or equal to the height of water in tube B? Explain.

▲ **FIGURE 15-27** Enhance Your Understanding 7.

Section Review

- Bernoulli's equation for the general case is $P_1 + \frac{1}{2}\rho v_1^2 + \rho g y_1 = P_2 + \frac{1}{2}\rho v_2^2 + \rho g y_2$. This is basically a statement of energy conservation per volume for a fluid.

15-8 Applications of Bernoulli's Equation

We now consider a variety of real-world examples that illustrate various applications of Bernoulli's equation.

Pressure and Speed

As mentioned, it often seems counterintuitive that a fast-moving fluid should have less pressure than a slow-moving one. Remember, however, that pressure can be thought of as a form of energy. From this point of view, there is an energy tradeoff between pressure and kinetic energy.

Perhaps the easiest way to demonstrate the dependence of pressure on speed is to blow across the top of a piece of paper. If you hold the paper as shown in FIGURE 15-28, then blow over the top surface, the paper will lift upward. The reason is that there is a difference in air speed between the top and the bottom of the paper, with the higher speed on top. As a result, the pressure above the paper is lower. This pressure difference, in turn, results in a net upward force, referred to as **lift**, and the paper rises.

A similar example of pressure and speed is provided by the airplane wing. A cross section of a typical wing is shown in FIGURE 15-29. The shape of the wing is designed so that air flows more rapidly over the top surface than the lower surface. As a result, the pressure is less on top. As with the piece of paper, the pressure difference results in a net upward force (lift) on the wing. (The wing also deflects air downward, which results in an upward force on the wing due to Newton's third law.) It follows that aerodynamic lift is a dynamic effect; it requires a flow of air. The greater the speed difference, the greater the upward force on the wing.

▲ **FIGURE 15-28 The bernoulli effect on a sheet of paper** If you hold a piece of paper by its end, it will bend downward. Blowing across the top of the paper reduces the pressure there, resulting in a net upward force that lifts the paper to a nearly horizontal position.

 Air Jet Blows between Bowling Balls

▶ **FIGURE 15-29 Air flow and lift in an airplane wing** This figure shows the cross section of an airplane wing with air flowing past it. The wing is shaped so that air flows more rapidly over the top of the wing than along the bottom. As a result, the pressure on the top of the wing is reduced, and a net upward force (lift) is generated. The wing also deflects moving air downward, which also results in an upward force on the wing.

(a)

(b)

◀ **FIGURE 15-30 Visualizing Concepts Bernoulli's Equation (a)** We often say that a hurricane or tornado "blew the roof off a house." Actually, the high-speed wind passing over the roof creates a region of reduced pressure, and normal atmospheric pressure inside the house then "blows" the roof off. **(b)** Prairie dogs use the Bernoulli effect to ventilate their burrows. One end of a burrow is always at a greater height than the other. This means that the wind speed is different at the two ends, and the resulting pressure difference produces a flow of air through the burrow.

EXERCISE 15-22 PRESSURE DIFFERENCE

During a hurricane, a 42.6 m/s wind blows across the flat roof of a small house, as shown in the sketch. Find the difference in pressure between the air inside the home and the air just above the roof. (The density of air is 1.29 kg/m^3.)

REASONING AND SOLUTION

Use Bernoulli's equation with point 1 just under the roof and point 2 just above the roof. The air speed is zero at point 1, but is equal to the wind speed at point 2. There is little difference in elevation between these points, however, and hence $y_1 = y_2 = y$. Thus, Bernoulli's general equation reduces to

$$P_1 + 0 + \rho g y = P_2 + \tfrac{1}{2}\rho v_2{}^2 + \rho g y$$

Solving this expression for the pressure difference, $P_1 - P_2$, yields the following:

$$P_1 - P_2 = \tfrac{1}{2}\rho v_2{}^2 = \tfrac{1}{2}(1.29 \text{ kg/m}^3)(42.6 \text{ m/s})^2 = 1.17 \text{ kPa}$$

▲ **FIGURE 15-31 Air circulation in a prairie dog burrow** A prairie dog burrow typically has a high mound on one end and a low mound on the other. Since the wind speed increases with height above the ground, the pressure is smaller at the high-mound end of the burrow. The result is a very convenient circulation of fresh air through the burrow.

A difference in pressure of 1.17 kPa might seem rather small, considering that atmospheric pressure is 101 kPa. However, it can still cause a significant force on a relatively large area, such as a roof. If a typical roof has an area of about 130 m^2, for example, a pressure difference of 1.17 kPa results in an upward force of over 34,000 pounds! This is why roofs are often torn from houses during severe windstorms, as illustrated in **FIGURE 15-30 (a)**.

On a lighter note, the prairie dogs in **FIGURE 15-30 (b)** *benefit* from the effects of Bernoulli's equation. A schematic prairie dog burrow is shown in **FIGURE 15-31**. Notice that one of the entrance/exit mounds is higher than the other. This is significant because the speed of air flow due to the incessant prairie winds varies with height; the speed goes to zero right at ground level, and increases to its maximum value within a few feet above the surface. As a result, the speed of air over the higher mound is greater than that over the lower mound. This causes the pressure to be less over the higher mound. With a pressure difference between the two mounds, air is drawn through the burrow, giving a form of natural air conditioning.

Similar effects are seen in an atomizer, which sprays perfume in a fine mist. As the bulb shoots a gust of air, as in **FIGURE 15-32**, it passes through a narrow orifice, which causes the air speed to increase. The pressure decreases as a result, and perfume is drawn up by the pressure difference into the stream of air.

◀ **FIGURE 15-32 An atomizer** The operation of an "atomizer" can be understood in terms of Bernoulli's equation. The high-speed jet of air made by squeezing the bulb creates a low pressure at the top of the vertical tube. This causes fluid to be drawn up the tube and expelled with the air jet as a fine spray.

CONCEPTUAL EXAMPLE 15-23 A RAGTOP ROOF

A small ranger vehicle has a soft, ragtop roof. When the car is at rest, the roof is flat. When the car is cruising at highway speeds with its windows rolled up, does the roof bow upward, remain flat, or bow downward?

REASONING AND DISCUSSION

When the car is in motion, air flows over the top of the roof, while the air inside the car is at rest—since the windows are closed. Thus, there is less pressure over the roof than under it. As a result, the roof bows upward.

ANSWER

The roof bows upward.

Torricelli's Law

Our final example of Bernoulli's equation deals with the speed of a fluid as it flows through a hole in a container. Consider, for example, the tank of water shown in **FIGURE 15-33**. If a hole is poked through the side of the tank at a depth h below the surface, what is the speed of the water as it emerges?

To answer this question, we apply Bernoulli's equation to the two points shown in the figure. First, at point 1 we notice that the water is open to the atmosphere; thus $P_1 = P_{at}$. Next, with the origin at the level of the hole, the height of the water surface is $y_1 = h$. Finally, if the hole is relatively small and the tank is large, the top surface of the water will have essentially zero speed; thus we can set $v_1 = 0$. Collecting these results in Bernoulli's general equation, we have the following for point 1:

$$P_1 + \tfrac{1}{2}\rho v_1{}^2 + \rho g y_1 = P_{at} + 0 + \rho g h$$

Now for point 2. At this point the height is $y_2 = 0$, by definition of the origin, and the speed of the escaping water is the unknown, v_2. *Now here is the key step*: The pressure P_2 is *atmospheric pressure*, because the hole opens the water to the atmosphere. Thus, for point 2 we have the following:

$$P_2 + \tfrac{1}{2}\rho v_2{}^2 + \rho g y_2 = P_{at} + \tfrac{1}{2}\rho v_2{}^2 + 0$$

Equating the results for points 1 and 2 yields

$$P_1 + \tfrac{1}{2}\rho v_1{}^2 + \rho g y_1 = P_2 + \tfrac{1}{2}\rho v_2{}^2 + \rho g y_2$$
$$P_{at} + \rho g h = P_{at} + \tfrac{1}{2}\rho v_2{}^2$$

Eliminating P_{at} and ρ we find

$$v_2 = \sqrt{2gh} \qquad\qquad \text{15-17}$$

This result is known as **Torricelli's law.**

This expression for v_2 should look familiar; it is the speed of an object that falls freely through a distance h. That is, the water emerges from the tank with the same speed as if it had fallen from the surface of the water to the hole. Similarly, if the emerging stream of water were to be directed upward, as in **FIGURE 15-34**, it would have just enough speed to rise through a height h—right back to the water's surface. This is precisely what one would expect on the basis of energy conservation.

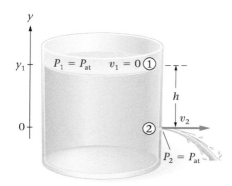

▲ **FIGURE 15-33 Fluid emerging from a hole in a container** Since the fluid exiting the hole is in contact with the atmosphere, the pressure there is just as it is on the top surface of the fluid.

▲ **FIGURE15-34 Maximum height of a stream of water** If the fluid emerging from a hole in a container is directed upward, it has just enough speed to reach the surface level of the fluid. This is an example of energy conservation.

EXAMPLE 15-24   A WATER FOUNTAIN

In designing a backyard water fountain, a gardener wants a stream of water to exit from the bottom of one tub (tub 1) and land in a second one (tub 2), as shown in the sketch. The top of tub 2 is 0.775 m below the hole in tub 1. In addition, tub 1 is filled with water to a depth of 0.218 m. How far to the right of tub 1 must the center of tub 2 be placed to catch the stream of water?

PICTURE THE PROBLEM

Our sketch labels the pertinent quantities for this problem. We know that $h = 0.218$ m and $H = 0.775$ m. The distance D is to be found. We have also chosen an appropriate coordinate system, with the y direction vertical and the x direction horizontal. The $y = 0$ level is at the top of tub 2.

REASONING AND STRATEGY

This problem combines Torricelli's law and kinematics. First, we find the speed v of the stream of water as it leaves tub 1 using Equation 15-17 ($v = \sqrt{2gh}$). Next, we find the time t required for the stream to fall freely through a distance H to tub 2 using kinematics (Chapter 2). Finally, because the stream moves with constant speed in the x direction, the distance D is given by $D = vt$.

Known   Water depth in tub 1, $h = 0.218$ m; vertical drop to tub 2, $H = 0.775$ m.

Unknown   Horizontal distance to tub 2, $D = ?$

SOLUTION

1. Find the speed v of the stream when it leaves tub 1:
$$v = \sqrt{2gh} = \sqrt{2(9.81 \text{ m/s}^2)(0.218 \text{ m})} = 2.07 \text{ m/s}$$

2. Find the time t for free fall through a height H:
$$y = H - \tfrac{1}{2}gt^2 = 0$$
$$t = \sqrt{\frac{2H}{g}} = \sqrt{\frac{2(0.775 \text{ m})}{9.81 \text{ m/s}^2}} = 0.397 \text{ s}$$

3. Multiply v times t to find the distance D:
$$x = vt = (2.07 \text{ m/s})(0.397 \text{ s}) = 0.822 \text{ m} = D$$

INSIGHT

Our solution for x can also be written symbolically as $x = vt = \sqrt{2gh}(\sqrt{2H/g}) = 2\sqrt{hH}$. Thus, if the values of h and H are interchanged, the distance $x = D$ remains the same. Also, notice that x is independent of the acceleration due to gravity; therefore, the fountain would work just as well on the Moon with its reduced gravity.

PRACTICE PROBLEM

Find the distance D for $h = 0.775$ m and $H = 0.218$ m.   [**Answer:** $D = 2\sqrt{hH} = 0.822$ m, as expected.]

Some related homework problems: Problem 65, Problem 106

Enhance Your Understanding   (Answers given at the end of the chapter)

8. A rock is dropped into the container shown in Figure 15-33 before the hole is made at the $y = 0$ level. When the hole is made, does the water emerge with a speed that is greater than, less than, or equal to the speed without the rock? Explain.

Section Review

- When fluid emerges from a hole in a container at a depth h below the surface of the fluid, its speed is $v = \sqrt{2gh}$. This is known as Torricelli's law.

15-9 Viscosity and Surface Tension

To this point we have considered only "ideal" fluids. In particular, we've assumed that fluids flow without frictional losses and that the molecules in a fluid have no interaction with one another. In this section, we consider the consequences that follow from relaxing these assumptions.

▶ **FIGURE 15-35 Fluid flow through a tube (a)** An ideal fluid flows through a tube with a speed that is the same everywhere in the fluid. **(b)** In a fluid with finite viscosity, the speed of the fluid goes to zero on the walls of the tube and reaches its maximum value in the center of the tube. The average speed of the fluid depends on the pressure difference between the ends of the tube, $P_1 - P_2$; the length of the tube, L; the cross-sectional area of the tube, A; and the coefficient of viscosity of the liquid, η.

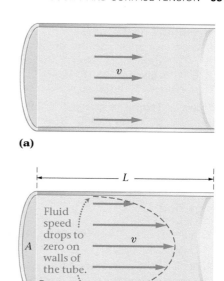

(a)

(b)

Viscosity

When a block slides across a rough floor, it experiences a frictional force opposing the motion. Similarly, a fluid flowing past a stationary surface experiences a force opposing the flow. This tendency to resist flow is referred to as the **viscosity** of a fluid. Fluids like air have low viscosities, thicker fluids like water are more viscous, and fluids like honey and motor oil are characterized by high viscosity.

To be specific, consider a situation of great practical importance—the flow of a fluid through a tube. Examples of this type of system include water flowing through a metal pipe in a house and blood flowing through an artery or a vein. If the fluid were ideal, with zero viscosity, it would flow through the tube with a speed that is the same throughout the fluid, as indicated in **FIGURE 15-35 (a)**. Real fluids with finite viscosity are found to have flow patterns like the one shown in **FIGURE 15-35 (b)**. In this case, the fluid is at rest next to the walls of the tube and flows with its greatest speed in the center of the tube. Because adjacent portions of the fluid flow past one another with different speeds, a force must be exerted on the fluid to maintain the flow, just as a force is required to keep a block sliding across a rough surface.

The force causing a viscous fluid to flow is provided by the pressure difference, $P_1 - P_2$, across a given length, L, of tube. Experiments show that the required pressure difference is proportional to the length of the tube and to the average speed, v, of the fluid. In addition, it is inversely proportional to the cross-sectional area, A, of the tube. Combining these observations, we can write the pressure difference in the following form:

$$P_1 - P_2 \propto \frac{vL}{A}$$

The constant of proportionality between the pressure difference and vL/A is related to the **coefficient of viscosity, η,** of a fluid. In fact, the viscosity is *defined* in such a way that the pressure difference is given by the following expression:

$$P_1 - P_2 = 8\pi\eta\frac{vL}{A} \qquad \text{15-18}$$

From this equation we can see that the dimensions of the coefficient of viscosity are $N \cdot s/m^2$. A common unit in the study of viscous fluids is the **poise,** named for the French physiologist Jean Leonard Marie Poiseuille (1797–1869) and defined as

$$1 \text{ poise} = 1 \text{ dyne} \cdot s/cm^2 = 0.1 \text{ N} \cdot s/m^2$$

For example, the viscosity of water at room temperature is $1.0016 \times 10^{-3} \text{ N} \cdot s/m^2$ and the viscosity of blood at 37 °C is $2.72 \times 10^{-3} \text{ N} \cdot s/m^2$. Some additional viscosities are given in Table 15-3.

TABLE 15-3 Viscosities (η) of Various Fluids ($N \cdot s/m^2$)

Honey	10
Glycerine (20 °C)	1.50
10-wt motor oil (30 °C)	0.250
Whole blood (37 °C)	2.72×10^{-3}
Water (0 °C)	1.79×10^{-3}
Water (20 °C)	1.0055×10^{-3}
Water (100 °C)	2.82×10^{-4}
Air (20 °C)	1.82×10^{-5}

EXAMPLE 15-25   BLOOD SPEED IN THE PULMONARY ARTERY

BIO The pulmonary artery, which connects the heart to the lungs, is 8.9 cm long and has a pressure difference over this length of 450 Pa. If the inside radius of the artery is 2.3 mm, what is the average speed of blood in the pulmonary artery?

PICTURE THE PROBLEM

Our sketch shows a schematic representation of the pulmonary artery, not drawn to scale. The length (8.9 cm) and radius (2.3 mm) of the artery are indicated. In addition, we note that the pressure difference between the two ends of the artery is 450 Pa.

CONTINUED

REASONING AND STRATEGY

The average speed of the blood can be found by using $P_1 - P_2 = 8\pi\eta(vL/A)$. Notice that the pressure difference, $P_1 - P_2$, is given as $450\,\text{Pa} = 450\,\text{N/m}^2$, and that the cross-sectional area of the blood vessel is $A = \pi r^2$.

Known Length of artery, $L = 8.9$ cm; pressure difference, $P_1 - P_2 = 450$ N/m²; inside radius of artery, $r = 2.3$ mm.

Unknown Speed of blood in artery, $v = ?$

P_1 P_2

$$P_1 - P_2 = 450 \text{ Pa}$$

SOLUTION

1. Solve $P_1 - P_2 = 8\pi\eta\,(vL/A)$ for the average speed, v:

$$v = \frac{(P_1 - P_2)A}{8\pi\eta L}$$

2. Replace the cross-sectional area A with πr^2 and cancel π from the numerator and denominator:

$$v = \frac{(P_1 - P_2)r^2}{8\eta L}$$

3. Substitute numerical values:

$$v = \frac{(450\,\text{Pa})(0.0023\,\text{m})^2}{8(0.00272\,\text{N}\cdot\text{s/m}^2)(0.089\,\text{m})} = 1.2\,\text{m/s}$$

INSIGHT

The viscosity of blood increases rapidly with its hematocrit value—that is, with the concentration of red blood cells in the whole blood (see Chapter 10 for more information on the hematocrit value of blood). Thus, thick blood, with a high hematocrit value, requires a significantly larger pressure difference for a given rate of blood flow. This higher pressure must be provided by the heart, which consequently works harder with each beat.

PRACTICE PROBLEM

What pressure difference is required to give the blood in this pulmonary artery an average speed of 1.5 m/s?
[**Answer:** 550 Pa]

Some related homework problems: Problem 99, Problem 100

Volume Flow Rate A convenient way to characterize the flow of a fluid is in terms of its volume flow rate—the volume of fluid that passes a given point in a given amount of time. Referring to Section 15-6, we see that the volume flow rate of a fluid is simply vA, where v is the average speed of the fluid and A is the cross-sectional area of the tube through which it flows. Solving $P_1 - P_2 = 8\pi\eta(vL/A)$ (Equation 15-18) for the average speed gives $v = (P_1 - P_2)A/8\pi\eta L$. Multiplying this result by the cross-sectional area of the tube yields the volume flow rate:

$$\text{volume flow rate} = \frac{\Delta V}{\Delta t} = vA = \frac{(P_1 - P_2)A^2}{8\pi\eta L}$$

Using the fact that the cross-sectional area of a cylindrical tube is $A = \pi r^2$, where r is its radius, we obtain the result known as **Poiseuille's equation:**

$$\frac{\Delta V}{\Delta t} = \frac{(P_1 - P_2)\pi r^4}{8\eta L} \qquad \text{15-19}$$

Notice that the volume flow rate varies with the *fourth power* of the tube's radius; thus a small change in radius corresponds to a large change in volume flow rate.

To see the significance of the r^4 dependence, consider an artery that branches into an arteriole with half the artery's radius. Letting r go to $r/2$ in Poiseuille's equation, and solving for the pressure difference, we find

$$P_1 - P_2 = \frac{8\eta L}{\pi(r/2)^4}\left(\frac{\Delta V}{\Delta t}\right) = 16\left[\frac{8\eta L}{\pi r^4}\left(\frac{\Delta V}{\Delta t}\right)\right]$$

Thus, the pressure difference across a given length of arteriole is 16 times what it is across the same length of artery. In fact, in the human body the pressure drop along an artery is small compared to the rather large pressure drop observed in the arterioles. This is a direct consequence of the increased viscous drag of the blood as it flows through narrower blood vessels.

Similarly, a narrowing, or *stenosis*, of an artery can produce significant increases in blood pressure. For example, a reduction in radius of only 20%, from r to $0.8r$, causes an increase in pressure by a factor of $(1/0.8)^4 \sim 2.4$. Thus, even a small narrowing of an artery can lead to an increased risk for heart disease and stroke.

Surface Tension

A small insect or spider resting on the surface of a pond or a lake is a common sight in the summertime. If you look carefully, you can see that the insect creates tiny dimples in the water's surface (**FIGURE 15-36**), almost as if it were supported by a thin sheet of rubber. In fact, the surface of water and other fluids behaves in many respects as if it were an elastic membrane. This effect is known as **surface tension.**

To understand the origin of surface tension, we start by noting that the molecules in a fluid exert attractive forces on one another. Thus, a molecule deep within a fluid experiences forces in all directions, as indicated in **FIGURE 15-37**, due to the molecules that surround it on all sides. The net force on such a molecule is zero. As a molecule nears the surface, however, it experiences a net force away from the surface, because there are no fluid molecules on the other side of the surface to attract it in that direction. It follows that work must be done on a molecule to move it from within a fluid to the surface, and that the *energy* of a fluid is increased for every molecule on its surface.

In general, physical systems tend toward configurations of minimum energy. For example, a ball on a slope rolls downhill, lowering its gravitational potential energy. If the energy of a droplet of liquid is to be minimized, it must have the smallest surface area possible for a given volume; that is, it must have the shape of a sphere. This is the reason small drops of dew are always spherical. Larger drops of water may be distorted by the downward pull of Earth's gravity, but in orbit drops of all sizes are spherical.

Energy is required to increase the area of a liquid surface, and hence the situation is similar to the energy required to stretch a spring or to stretch a sheet of rubber. Thus, the surface of a liquid behaves as if it were elastic, resisting tendencies to increase its area. For example, if a drop of dew is distorted into an ellipsoid, it quickly returns to its original spherical shape. Similarly, when an insect alights on the surface of a pond, it creates dimples that increase the surface area. The water resists this distortion with a force sufficient to support the weight of the insect—if the insect is not too large. In fact, even a needle or a razor blade can be supported on the surface of water if it is put into place gently, even though it has a density significantly greater than the density of water.

BIO Surface tension is important in many biological systems, and it plays a particularly important role in human breathing. For example, the crucial exchange of oxygen and carbon dioxide between inspired air and the blood occurs across the membranes of small balloonlike structures called *alveoli*. During inhalation, the alveoli expand from a radius of about 0.050 mm to 0.10 mm as they draw in fresh air. The walls of the alveoli are coated with a thin film of water, however, and in order to expand they must push outward against the water's surface tension—like trying to inflate a balloon against the surface tension of the rubber. To reduce the rather large surface tension of water, and make breathing easier, the lungs produce a substance called a *surfactant* that mixes with the water. This surfactant is produced rather late in the development of a fetus, however, and therefore premature infants may experience respiratory distress and even death as a result of too much surface tension in their alveoli.

Enhance Your Understanding (Answers given at the end of the chapter)

9. Which pipe requires a greater pressure difference to make a given fluid flow through it with a speed v: pipe 1 with $L = 1$ m and a diameter of 0.1 m, or pipe 2 with $L = 2$ m and a diameter of 0.2 m?

Section Review

- The pressure difference required to push a fluid with a speed v through a pipe of length L and cross-sectional area A is $P_1 - P_2 = 8\pi\eta(vL/A)$. In this expression, η is the coefficient of viscosity.

▲ **FIGURE 15-36** Surface tension causes the surface of a liquid to behave like an elastic skin or membrane. When a small force is applied to the liquid surface, it tends to stretch, resisting penetration. This phenomenon enables a fishing spider to launch itself vertically upward as high as 4 cm from the water surface, as if from a trampoline. Such an acrobatic maneuver can save the spider from the attack of a predatory fish.

Big Idea **5** Surface tension tends to make the surface area of a fluid as small as possible. Because of this tendency, the surface of a fluid acts much like an elastic membrane, resisting forces that try to stretch it and increase its area.

A molecule on the surface is pulled toward the interior.

Drop of liquid

An interior molecule is pulled by its neighbors equally in all directions.

▲ **FIGURE 15-37 The origin of surface tension** A molecule in the interior of a fluid experiences attractive forces of equal magnitude in all directions, giving a net force of zero. A molecule near the surface of the fluid experiences a net attractive force toward the interior of the fluid. This causes the surface to be pulled inward, resulting in a surface of minimum area.

CHAPTER 15 REVIEW

CHAPTER SUMMARY

15-1 DENSITY

The density, ρ, of a material is its mass M per volume V:

$$\rho = M/V \qquad \text{15-1}$$

15-2 PRESSURE

Pressure, P, is force F per area A:

$$P = F/A \qquad \text{15-2}$$

Atmospheric Pressure
The pressure exerted by the atmosphere is $P_{at} = 1.01 \times 10^5 \, \text{N/m}^2 \approx 14.7 \, \text{lb/in}^2$.

Pascals
Pressure is often given in terms of the pascal (Pa), where $1 \, \text{Pa} = 1 \, \text{N/m}^2$.

Gauge Pressure
Gauge pressure is the difference between the actual pressure and atmospheric pressure:

$$P_g = P - P_{at} \qquad \text{15-5}$$

15-3 STATIC EQUILIBRIUM IN FLUIDS: PRESSURE AND DEPTH

The pressure of a fluid in static equilibrium increases with depth.

Pressure with Depth
If the pressure at one point in a fluid is P_1, the pressure at a depth h below that point is

$$P_2 = P_1 + \rho g h \qquad \text{15-7}$$

Pascal's Principle
An external pressure applied to an enclosed fluid is transmitted unchanged to every point within the fluid.

15-4 ARCHIMEDES' PRINCIPLE AND BUOYANCY

The fact that the pressure in a fluid increases with depth leads to a net upward force on any object that is immersed in the fluid. This upward force is referred to as a buoyant force. The magnitude of the buoyant force is equal in magnitude to the weight of fluid displaced by the object.

$$F_b = \rho_{fluid} g V \qquad \text{15-9}$$

15-5 APPLICATIONS OF ARCHIMEDES' PRINCIPLE

Archimedes' principle applies equally well to objects that are completely immersed, partially immersed, or floating.

Floatation
An object floats when it displaces an amount of fluid equal to its weight.

15-6 FLUID FLOW AND CONTINUITY

The speed of a fluid changes as the cross-sectional area of the pipe through which it flows changes.

Equation of Continuity for an Incompressible Fluid

$$A_1 v_1 = A_2 v_2 \qquad \text{15-12}$$

15-7 BERNOULLI'S EQUATION

Bernoulli's equation can be thought of as energy conservation per volume for a fluid.

General Case
The pressure, speed, and height of a fluid at two different points are related by

$$P_1 + \tfrac{1}{2}\rho v_1^2 + \rho g y_1 = P_2 + \tfrac{1}{2}\rho v_2^2 + \rho g y_2 \qquad \text{15-16}$$

15-8 APPLICATIONS OF BERNOULLI'S EQUATION

Bernoulli's equation applies to a wide range of everyday situations, including airplane wings and prairie dog burrows.

Torricelli's Law

According to Torricelli's law, if a hole is poked in a container at a depth h below the surface, the fluid exits with the speed

$$v = \sqrt{2gh} \qquad \text{15-17}$$

15-9 VISCOSITY AND SURFACE TENSION

Viscosity in a fluid is similar to friction between two solid surfaces.

$$P_1 - P_2 = 8\pi\eta\frac{vL}{A} \qquad \text{15-18}$$

Surface Tension

A fluid tends to pull inward on its surface, resulting in a surface of minimum area. The surface of the fluid behaves much like an elastic membrane enclosing the fluid.

ANSWERS TO ENHANCE YOUR UNDERSTANDING QUESTIONS

1. **(a)** The volume of system 1 is less than the volume of system 2 because it contains less of the same substance (water). **(b)** The density of system 1 is equal to the density of system 2 because they consist of the same substance.

2. B < C < A < D.

3. The increase in pressure from 0 to 10 m is the same as the increase in pressure from 100 m to 110 m because the density of water is the same at 0 and 100 m.

4. The buoyant force exerted on the aluminum block is the same as the buoyant force on the lead block because they displace the same amount of water.

5. The water level stays the same because the wood displaces the same amount of water whether it rests on the ice cube or floats by itself.

6. The new speed is $4v$ because the area has decreased by a factor of $2^2 = 4$.

7. **(a)** The height in tube B is less than in tube A because the speed is greater at B. **(b)** The height in tube C is less than in tube A because the speed is greater at C. **(c)** The height in tube C is greater than in tube B because the speed is smaller at C.

8. The speed is greater because the rock raises the water level.

9. A greater pressure difference is required for pipe 1.

CONCEPTUAL QUESTIONS

For instructor-assigned homework, go to www.masteringphysics.com. (MP)

(Answers to odd-numbered Conceptual Questions can be found in the back of the book.)

1. Suppose you drink a liquid through a straw. Explain why the liquid moves upward, against gravity, into your mouth.

2. Considering your answer to the previous question, is it possible to sip liquid through a straw on the surface of the Moon? Explain.

3. Water towers on the roofs of buildings have metal bands wrapped around them for support. The spacing between bands is smaller near the base of a tower than near its top. Explain.

4. What holds a suction cup in place?

5. Suppose a force of 400 N is required to push the top off a wine barrel. In a famous experiment, Blaise Pascal attached a tall, thin tube to the top of a filled wine barrel, as shown in FIGURE 15-38. Water was slowly added to the tube until the barrel burst. The puzzling result found by Pascal was that the barrel broke when the weight of water in the tube was much less than 400 N. Explain Pascal's observation.

FIGURE 15-38 Conceptual Question 5 and Problem 18

6. Why is it more practical to use mercury in the barometer shown in Figure 15-6 than water?

7. An object's density can be determined by first weighing it in air, then in water (provided the density of the object is greater than the density of water, so that it is totally submerged when placed in water). Explain how these two measurements can give the desired result.

8. How does a balloonist control the vertical motion of a hot-air balloon?

9. Why is it possible for people to float without effort in Utah's Great Salt Lake?

10. **Physics in the Movies** In the movie *Voyage to the Bottom of the Sea*, the Earth is experiencing a rapid warming. In one scene, large icebergs break up into small, car-size chunks that drop downward through the water and bounce off the hull of the submarine *Seaview*. Is this an example of good, bad, or ugly physics? Explain.

11. One day, while snorkeling near the surface of a crystal-clear ocean, it occurs to you that you could go considerably deeper by simply lengthening the snorkel tube. Unfortunately, this does not work well at all. Why?

12. Since metal is more dense than water, how is it possible for a metal boat to float?

13. A sheet of water passing over a waterfall is thicker near the top than near the bottom. Similarly, a stream of water emerging from a water faucet becomes narrower as it falls. Explain.

14. It is a common observation that smoke rises more rapidly through a chimney when there is a wind blowing outside. Explain.

15. Is it best for an airplane to take off against the wind or with the wind? Explain.

16. If you have a hair dryer and a Ping Pong ball at home, try this demonstration. Direct the air from the dryer in a direction just above horizontal. Next, place the Ping Pong ball in the stream of air. If done just right, the ball will remain suspended in midair. Use the Bernoulli effect to explain this behavior.

17. Suppose a pitcher wants to throw a baseball so that it rises as it approaches the batter. How should the ball be spinning to accomplish this feat? Explain.

PROBLEMS AND CONCEPTUAL EXERCISES

Answers to odd-numbered Problems and Conceptual Exercises can be found in the back of the book. **BIO** *identifies problems of biological or medical interest;* **CE** *indicates a conceptual exercise.* Predict/Explain *problems ask for two responses: (a) your prediction of a physical outcome, and (b) the best explanation among three provided; and* Predict/Calculate *problems ask for a prediction of a physical outcome, based on fundamental physics concepts, and follow that with a numerical calculation to verify the prediction. On all problems, bullets (•, ••, •••) indicate the level of difficulty.*

SECTION 15-1 DENSITY

1. • Estimate the weight of the air in your physics classroom.

2. • What weight of water is required to fill a 25-gallon aquarium?

3. • You buy a "gold" ring at a pawn shop. The ring has a mass of 4.7 g and a volume of 0.55 cm³. Is the ring solid gold?

4. • A cube of metal has a mass of 0.347 kg and measures 3.21 cm on a side. Calculate the density and identify the metal.

SECTION 15-2 PRESSURE

5. • What is the downward force exerted by the atmosphere on a soccer field whose dimensions are 105 m by 55 m?

6. • **BIO Bioluminescence** Some species of dinoflagellate (a type of unicellular plankton) can produce light as the result of biochemical reactions within the cell. This light is an example of bioluminescence. It is found that bioluminescence in dinoflagellates can be triggered by deformation of the cell surface with a pressure as low as one dyne $(10^{-5}\,\text{N})$ per square centimeter. What is this pressure in **(a)** pascals and **(b)** atmospheres?

7. • A 71-kg person sits on a 3.9-kg chair. Each leg of the chair makes contact with the floor in a circle that is 1.1 cm in diameter. Find the pressure exerted on the floor by each leg of the chair, assuming the weight is evenly distributed.

8. • To prevent damage to floors (and to increase friction), a crutch will often have a rubber tip attached to its end. If the end of the crutch is a circle of radius 0.95 cm without the tip, and the tip is a circle of radius 2.0 cm, by what factor does the tip reduce the pressure exerted by the crutch?

9. •• Suppose that when you ride on your 7.85-kg bike the weight of you and the bike is supported equally by the two tires. If the gauge pressure in the tires is 70.2 lb/in² and the area of contact between each tire and the road is 7.03 cm², what is your weight?

10. •• **Shock Wave Pressure** On February 15, 2013, a 12,000,000-kg asteroid exploded in the atmosphere above Chelyabinsk, Russia, generating a shock wave with a pressure amplitude of 750 Pa. What force (in N) would this pressure difference exert on a window with dimensions of 0.86 m × 1.17 m? (The shock wave shattered many glass windows, resulting in a number of injuries.)

11. •• Predict/Calculate The weight of your 1420-kg car is supported equally by its four tires, each inflated to a gauge pressure of 35.0 lb/in². **(a)** What is the area of contact each tire makes with the road? **(b)** If the gauge pressure is increased, does the area of contact increase, decrease, or stay the same? **(c)** What gauge pressure is required to give an area of contact of 116 cm² for each tire?

SECTION 15-3 STATIC EQUILIBRIUM IN FLUIDS: PRESSURE AND DEPTH

12. • CE Two drinking glasses, 1 and 2, are filled with water to the same depth. Glass 1 has twice the diameter of glass 2. **(a)** Is the weight of the water in glass 1 greater than, less than, or equal to the weight of the water in glass 2? **(b)** Is the pressure at the bottom of glass 1 greater than, less than, or equal to the pressure at the bottom of glass 2?

13. • CE FIGURE 15-39 shows four containers, each filled with water to the same level. Rank the containers in order of increasing pressure at the depth *h*. Indicate ties where appropriate.

FIGURE 15-39 Problem 13

14. • Water in the lake behind Hoover Dam is 221 m deep. What is the water pressure at the base of the dam?

15. • In a classroom demonstration, the pressure inside a soft drink can is suddenly reduced to essentially zero. Assuming the can to be a cylinder with a height of 12 cm and a diameter of 6.5 cm, find the net inward force exerted on the vertical sides of the can due to atmospheric pressure.

16. •• As a storm front moves in, you notice that the column of mercury in a barometer rises to only 736 mm. **(a)** What is the air pressure? **(b)** If the mercury in this barometer is replaced with water, to what height does the column of water rise? Assume the same air pressure found in part (a).

17. •• In the hydraulic system shown in FIGURE 15-40, the piston on the left has a diameter of 5.1 cm and a mass of 2.3 kg. The piston on the right has a diameter of 13 cm and a mass of 3.6 kg. If the density of the fluid is 750 kg/m³, what is the height difference *h* between the two pistons?

FIGURE 15-40 Problem 17

18. •• A circular wine barrel 75 cm in diameter will burst if the net upward force exerted on the top of the barrel is 643 N. A tube 1.0 cm in diameter extends into the barrel through a hole in the top, as indicated in Figure 15-38. Initially, the barrel is filled to the top and the tube is empty above that level. What weight of water must be poured into the tube in order to burst the barrel?

19. •• A cylindrical container with a cross-sectional area of 65.2 cm² holds a fluid of density 806 kg/m³. At the bottom of the container the pressure is 126 kPa. **(a)** What is the depth of the fluid? **(b)** Find the pressure at the bottom of the container after an additional 2.05×10^{-3} m³ of this fluid is added to the container. Assume that no fluid spills out of the container.

20. •• Predict/Calculate **Tourist Submarine** A submarine called the *Deep View 66* is being developed to take 66 tourists at a time on sight-seeing trips to tropical coral reefs. According to guidelines of the American Society of Mechanical Engineers (ASME), to be safe for human occupancy the *Deep View 66* must be able to withstand a pressure of 10.0 N per square millimeter. **(a)** To what depth can the *Deep View 66* safely descend in seawater? **(b)** If the submarine is used in freshwater instead, is its maximum safe depth greater than, less than, or the same as in seawater? Explain.

21. •• Predict/Calculate A water storage tower is filled with freshwater to a depth of 6.4 m. What is the pressure at **(a)** 4.5 m and **(b)** 5.5 m below the surface of the water? **(c)** Why are the metal bands on such towers more closely spaced near the base of the tower?

22. •• Predict/Calculate You step into an elevator holding a glass of water filled to a depth of 7.9 cm. After a moment, the elevator moves upward with constant acceleration, increasing its speed from 0 to 2.4 m/s in 2.9 s. **(a)** During the period of acceleration, is the pressure exerted on the bottom of the glass greater than, less than, or the same as before the elevator began to move? Explain. **(b)** Find the change in the pressure exerted on the bottom of the glass as the elevator accelerates.

23. •• Suppose you pour water into a container until it reaches a depth of 14 cm. Next, you carefully pour in a 7.5-cm thickness of olive oil so that it floats on top of the water. What is the pressure at the bottom of the container?

24. •• Referring to Example 15-8, suppose that some vegetable oil has been added to both sides of the U tube. On the right side of the tube, the depth of oil is 5.00 cm, as before. On the left side of the tube, the depth of the oil is 3.00 cm. Find the difference in fluid level between the two sides of the tube.

25. •• Predict/Calculate As a stunt, you want to sip some water through a very long, vertical straw. **(a)** First, explain why the liquid moves upward, against gravity, into your mouth when you sip. **(b)** What is the tallest straw that you could, in principle, drink from in this way?

26. •• **BIO** Predict/Calculate The patient in FIGURE 15-41 is to receive an intravenous injection of medication. In order to work properly, the pressure of fluid containing the medication must be 109 kPa at the injection point. **(a)** If the fluid has a density of 1020 kg/m³, find the height at which the bag of fluid must be suspended above the patient. Assume that the pressure inside the bag is one atmosphere. **(b)** If a less dense fluid is used instead, must the height of suspension be increased or decreased? Explain.

FIGURE 15-41 Problem 26

27. ••• A cylindrical container 1.0 m tall contains mercury to a certain depth, *d*. The rest of the cylinder is filled with water. If the pressure at the bottom of the cylinder is two atmospheres, what is the depth *d*?

SECTION 15-4 ARCHIMEDES' PRINCIPLE AND BUOYANCY

28. • **CE** Predict/Explain **Beebe and Barton** On Wednesday, August 15, 1934, William Beebe and Otis Barton made history by descending in the Bathysphere—basically a steel sphere 4.75 ft in diameter—3028 ft below the surface of the ocean, deeper than anyone had been before. **(a)** As the Bathysphere was lowered, was the buoyant force exerted on it at a depth of 10 ft greater than, less than, or equal to the buoyant force exerted on it at a depth of 50 ft? **(b)** Choose the *best explanation* from among the following:
  I. The buoyant force depends on the density of the water, which is essentially the same at 10 ft and 50 ft.
  II. The pressure increases with depth, and this increases the buoyant force.
  III. The buoyant force decreases as an object sinks below the surface of the water.

29. • **CE** Lead is more dense than aluminum. **(a)** Is the buoyant force on a solid lead sphere greater than, less than, or equal to the buoyant force on a solid aluminum sphere of the same diameter? **(b)** Does your answer to part (a) depend on the fluid that is causing the buoyant force?

30. • A fish adjusts its buoyancy to hover in one place in a small bowl as it drops a pebble it was holding in its mouth. When the pebble is released, does the water level in the bowl rise, fall, or stay the same?

31. • A raft is 3.7 m wide and 6.1 m long. When a horse is loaded onto the raft, it sinks 3.8 cm deeper into the water. What is the weight of the horse?

32. •• **BIO** Predict/Calculate **Swim Bladder** Many ray-finned fish adjust their buoyancy by changing the volume of gas in a *swim bladder* located just below their spine. **(a)** If the volume of the fish increases while the mass remains the same, does the buoyant force on the fish increase, decrease, or remain the same? **(b)** Calculate the change in the buoyant force on a fish if its volume increases by 11 cm³, assuming it swims in seawater.

33. •• Predict/Calculate **Submarine Buoyancy** Submarines adjust their buoyancy by changing the amount of water held in rigid containers called *ballast tanks*. **(a)** If the mass of a submerged submarine increases while the volume remains the same, does the buoyant force on the submarine increase, decrease, or remain the same? **(b)** Calculate the change in the buoyant force on a submerged submarine if it pumps 0.85 m³ of seawater into its ballast tank.

34. •• A 3.2-kg balloon is filled with helium (density = 0.179 kg/m³). If the balloon is a sphere with a radius of 4.9 m, what is the maximum weight it can lift?

35. •• A hot-air balloon plus cargo has a mass of 312 kg and a volume of 2210 m³ on a day when the outside air density is 1.22 kg/m³. The balloon is floating at a constant height of 9.14 m above the ground. What is the density of the hot air in the balloon?

36. ••• In the lab you place a beaker that is half full of water (density ρ_w) on a scale. You now use a light string to suspend a piece of metal of volume *V* in the water. The metal is completely submerged, and none of the water spills out of the beaker. Give a symbolic expression for the change in reading of the scale.

SECTION 15-5 APPLICATIONS OF ARCHIMEDES' PRINCIPLE

37. • **CE** Predict/Explain A block of wood has a steel ball glued to one surface. The block can be floated with the ball "high and dry" on its top surface. **(a)** When the block is inverted, and the ball is immersed in water, does the volume of wood that is submerged increase, decrease, or stay the same? **(b)** Choose the *best explanation* from among the following:

  I. When the block is inverted the ball pulls it downward, causing more of the block to be submerged.

  II. The same amount of mass is supported in either case, therefore the amount of the block that is submerged is the same.

  III. When the block is inverted the ball experiences a buoyant force, which reduces the buoyant force that must be provided by the wood.

38. • **CE** Predict/Explain In the preceding problem, suppose the block of wood with the ball "high and dry" is floating in a tank of water. **(a)** When the block is inverted, does the water level in the tank increase, decrease, or stay the same? **(b)** Choose the *best explanation* from among the following:

  I. Inverting the block makes the block float higher in the water, which lowers the water level in the tank.

  II. The same mass is supported by the water in either case, and therefore the amount of displaced water is the same.

  III. The inverted block floats lower in the water, which displaces more water and raises the level in the tank.

39. • **CE Measuring Density with a Hydrometer** A hydrometer, a device for measuring fluid density, is constructed as shown in **FIGURE 15-42**. If the hydrometer samples fluid 1, the small float inside the tube is submerged to level 1. When fluid 2 is sampled, the float is submerged to level 2. Is the density of fluid 1 greater than, less than, or equal to the density of fluid 2? (This is how mechanics test your antifreeze level. Since antifreeze [ethylene glycol] is more dense than water, the higher the density of coolant in your radiator the more antifreeze protection you have.)

FIGURE 15-42 Problem 39

40. • **CE** Predict/Explain Referring to Example 15-12, suppose the flask with the wood tied to the bottom is placed on a scale. At some point the string breaks and the wood rises to the surface where it floats. **(a)** When the wood is floating, is the reading on the scale greater than, less than, or equal to its previous reading? **(b)** Choose the *best explanation* from among the following:

  I. The same mass is supported by the scale before and after the string breaks, and therefore the reading on the scale remains the same.

  II. When the block is floating the water level drops, and this reduces the reading on the scale.

  III. When the block is floating it no longer pulls upward on the flask; therefore, the reading on the scale increases.

41. • **CE** On a planet in a different solar system the acceleration of gravity is greater than it is on Earth. If you float in a pool of water on this planet, do you float higher than, lower than, or at the same level as when you float in water on Earth?

42. • An air mattress is 2.3 m long, 0.66 m wide, and 14 cm deep. If the air mattress itself has a mass of 0.22 kg, what is the maximum mass it can support in freshwater?

43. •• A solid block is attached to a spring scale. When the block is suspended in air, the scale reads 21.2 N; when it is completely immersed in water, the scale reads 18.2 N. What are **(a)** the volume and **(b)** the density of the block?

44. •• As in the previous problem, a solid block is suspended from a spring scale. If the reading on the scale when the block is completely immersed in water is 25.0 N, and the reading when it is completely immersed in alcohol of density 806 kg/m³ is 25.7 N, what are **(a)** the block's volume and **(b)** its density?

45. •• **BIO** A person weighs 756 N in air and has a body-fat percentage of 28.1%. **(a)** What is the overall density of this person's body? **(b)** What is the volume of this person's body? **(c)** Find the apparent weight of this person when completely submerged in water.

46. •• Predict/Calculate A log floats in a river with one-fourth of its volume above the water. **(a)** What is the density of the log? **(b)** If the river carries the log into the ocean, does the portion of the log above the water increase, decrease, or stay the same? Explain.

47. •• A person with a mass of 78 kg and a volume of 0.086 m³ floats quietly in water. **(a)** What is the volume of the person that is above water? **(b)** If an upward force F is applied to the person by a friend, the volume of the person above water increases by 0.0032 m³. Find the force F.

48. •• Predict/Calculate A block of wood floats on water. A layer of oil is now poured on top of the water to a depth that more than covers the block, as shown in **FIGURE 15-43**. **(a)** Is the volume of wood submerged in water greater than, less than, or the same as before? **(b)** If 90% of the wood is submerged in water before the oil is added, find the fraction submerged when oil with a density of 875 kg/m³ covers the block.

FIGURE 15-43 Problem 48

49. •• A piece of lead has the shape of a hockey puck, with a diameter of 7.5 cm and a height of 2.5 cm. If the puck is placed in a mercury bath, it floats. How deep below the surface of the mercury is the bottom of the lead puck?

50. ••• Predict/Calculate A lead weight with a volume of 0.82×10^{-5} m³ is lowered on a fishing line into a lake to a depth of 1.0 m. **(a)** What tension is required in the fishing line to give the weight an upward acceleration of 2.1 m/s²? **(b)** If the initial depth of the weight is increased to 2.0 m, does the tension found in part (a) increase, decrease, or stay the same? Explain. **(c)** What acceleration will the weight have if the tension in the fishing line is 1.2 N? Give both direction and magnitude.

SECTION 15-6 FLUID FLOW AND CONTINUITY

51. • To water the yard, you use a hose with a diameter of 3.6 cm. Water flows from the hose with a speed of 1.3 m/s. If you partially block the end of the hose so the effective diameter is now 0.52 cm, with what speed does water spray from the hose?

52. • Water flows through a pipe with a speed of 2.4 m/s. Find the flow rate in kg/s if the diameter of the pipe is 3.1 cm.

53. • To fill a child's inflatable wading pool, you use a garden hose with a diameter of 2.9 cm. Water flows from this hose with a speed of 1.3 m/s. How much time will it take to fill the pool to a depth of 26 cm if the pool is circular and has a diameter of 2.0 m?

54. • **BIO Heart Pump Rate** When at rest, your heart pumps blood at the rate of 5.00 liters per minute (L/min). What are the volume and mass of blood pumped by your heart in one day?

55. •• **BIO Blood Speed in an Arteriole** A typical arteriole has a diameter of 0.080 mm and carries blood at the rate of 9.6×10^{-5} cm³/s. **(a)** What is the speed of the blood in an arteriole? **(b)** Suppose an arteriole branches into 8800 capillaries, each with a diameter of

6.0×10^{-6} m. What is the blood speed in the capillaries? (The low speed in capillaries is beneficial; it promotes the diffusion of materials to and from the blood.)

56. •• Predict/Calculate Water flows at the rate of 3.11 kg/s through a hose with a diameter of 3.22 cm. **(a)** What is the speed of water in this hose? **(b)** If the hose is attached to a nozzle with a diameter of 0.732 cm, what is the speed of water in the nozzle? **(c)** Is the number of kilograms per second flowing through the nozzle greater than, less than, or equal to 3.11 kg/s? Explain.

57. •• A river narrows at a rapids from a width of 12 m to a width of only 5.8 m. The depth of the river before the rapids is 2.7 m; the depth in the rapids is 0.85 m. Find the speed of water flowing in the rapids, given that its speed before the rapids is 1.2 m/s. Assume the river has a rectangular cross section.

58. •• **BIO How Many Capillaries?** The aorta has an inside diameter of approximately 2.1 cm, compared to that of a capillary, which is about 1.0×10^{-5} m ($10 \, \mu m$). In addition, the average speed of flow is approximately 1.0 m/s in the aorta and 0.30 mm/s in a capillary. Assuming that all the blood that flows through the aorta also flows through the capillaries, how many capillaries does the circulatory system have?

SECTION 15-7 BERNOULLI'S EQUATION

59. • **BIO Plaque in an Artery** The buildup of plaque on the walls of an artery may decrease its diameter from 1.1 cm to 0.75 cm. If the speed of blood flow was 15 cm/s before reaching the region of plaque buildup, find **(a)** the speed of blood flow and **(b)** the pressure drop within the plaque region.

60. • A horizontal pipe contains water at a pressure of 120 kPa flowing with a speed of 1.9 m/s. When the pipe narrows to one-half its original diameter, what are **(a)** the speed and **(b)** the pressure of the water?

61. •• Unfiltered olive oil must flow at a minimum speed of 3.0 m/s to prevent settling of debris in a pipe. The oil leaves a pump at a pressure of 88 kPa through a pipe of radius 9.5 mm. It then enters a horizontal pipe at atmospheric pressure. Ignore the effects of viscosity. **(a)** What is the speed of the oil as it leaves the pump if it flows at 3.0 m/s in the horizontal pipe? **(b)** What is the radius of the horizontal pipe?

62. •• **BIO A Blowhard** Tests of lung capacity show that adults are able to exhale 1.5 liters of air through their mouths in as little as 1.0 second. **(a)** If a person blows air at this rate through a drinking straw with a diameter of 0.60 cm, what is the speed of air in the straw? **(b)** If the air from the straw in part (a) is directed horizontally across the upper end of a second straw that is vertical, as shown in FIGURE 15-44, to what height does water rise in the vertical straw?

FIGURE 15-44 Problem 62

63. •• Predict/Calculate Water flows through a horizontal tube of diameter 2.5 cm that is joined to a second horizontal tube of diameter 1.2 cm. The pressure difference between the tubes is 7.3 kPa. **(a)** Which tube has the higher pressure? **(b)** Which tube has the higher speed of flow? **(c)** Find the speed of flow in the first tube.

64. •• A garden hose is attached to a water faucet on one end and a spray nozzle on the other end. The water faucet is turned on, but the nozzle is turned off so that no water flows through the hose. The hose lies horizontally on the ground, and a stream of water sprays vertically out of a small leak to a height of 0.68 m. What is the pressure inside the hose?

SECTION 15-8 APPLICATIONS OF BERNOULLI'S EQUATION

65. • A water tank springs a leak. Find the speed of water emerging from the hole if the leak is 2.9 m below the surface of the water, which is open to the atmosphere.

66. •• **(a)** Find the pressure difference on an airplane wing if air flows over the upper surface with a speed of 125 m/s, and along the bottom surface with a speed of 109 m/s. **(b)** If the area of the wing is 32 m^2, what is the net upward force exerted on the wing?

67. •• On a vacation flight, you look out the window of the jet and wonder about the forces exerted on the window. Suppose the air outside the window moves with a speed of approximately 170 m/s shortly after takeoff, and that the air inside the plane is at atmospheric pressure. **(a)** Find the pressure difference between the inside and outside of the window. **(b)** If the window is 25 cm by 42 cm, find the force exerted on the window by air pressure.

68. •• **Pitot Tube** The pitot tube is commonly used to measure the air speed of an aircraft. Air flows into a small opening at the end of a tube that is closed at the other end, bringing the air to rest and allowing the measurement of the pressure difference between air at rest inside the tube and air moving rapidly just outside the tube. If the high-altitude air density is 0.364 kg/m^3, and the pressure difference between inside and outside the tube is 9460 Pa, what is the airplane's speed relative to the air?

69. •• Predict/Calculate During a thunderstorm, winds with a speed of 47.7 m/s blow across a flat roof with an area of 668 m^2. **(a)** Find the magnitude of the force exerted on the roof as a result of this wind. **(b)** Is the force exerted on the roof in the upward or downward direction? Explain.

70. •• A garden hose with a diameter of 1.6 cm has water flowing in it with a speed of 1.3 m/s and a pressure of 1.5 atmospheres. At the end of the hose is a nozzle with a diameter of 0.64 cm. Find **(a)** the speed of water in the nozzle and **(b)** the pressure in the nozzle.

71. ••• Predict/Calculate Water flows in a cylindrical, horizontal pipe. As the pipe narrows to half its initial diameter, the pressure in the pipe changes. **(a)** Is the pressure in the narrow region greater than, less than, or the same as the initial pressure? Explain. **(b)** Calculate the change in pressure between the wide and narrow regions of the pipe. Give your answer symbolically in terms of the density of the water, ρ, and its initial speed v.

SECTION 15-9 VISCOSITY AND SURFACE TENSION

72. • **BIO Vasodilation** When the body requires an increased blood flow rate in a particular organ or muscle, it can accomplish this by increasing the diameter of arterioles in that area. This is referred to as vasodilation. What percentage increase in the diameter of an arteriole is required to double the volume flow rate of blood, all other factors remaining the same?

73. • **BIO (a)** Find the volume of blood that flows per second through the pulmonary artery described in Example 15-25. **(b)** If the radius of the artery is reduced by 18%, by what factor is the blood flow rate reduced? Assume that all other properties of the artery remain unchanged.

74. • **BIO An Occlusion in an Artery** Suppose an occlusion in an artery reduces its diameter by 15%, but the volume flow rate of blood in the artery remains the same. By what factor has the pressure drop across the length of this artery increased?

75. •• **Motor Oil** The viscosity of 5W-30 motor oil changes from $6.4 \, \mathrm{N \cdot s/m^2}$ at $-30 \, °C$ to $0.0090 \, \mathrm{N \cdot s/m^2}$ at $100 \, °C$. Model the oil circulation system of a car engine as a tube of radius 5.0 mm and length 2.1 m, driven by a pump that produces a pressure difference of 110 kPa across the ends of the tube. **(a)** What volume flow rate does the pump produce when the oil is at $-30 \, °C$? **(b)** What is the volume flow rate when the oil is at the operating temperature of $100 \, °C$?

76. •• **Predict/Calculate** Water at $20 \, °C$ flows through a horizontal garden hose at the rate of $5.0 \times 10^{-4} \, \mathrm{m^3/s}$. The diameter of the garden hose is 2.5 cm. **(a)** What is the water speed in the hose? **(b)** What is the pressure drop across a 15-m length of hose? Suppose the cross-sectional area of the hose is halved, but the length and pressure drop remain the same. **(c)** By what factor does the water speed change? **(d)** By what factor does the volume flow rate change? Explain.

GENERAL PROBLEMS

77. • **CE Reading a Weather Glass** A weather glass, as shown in **FIGURE 15-45**, is used to give an indication of a change in the weather. Does the water level in the neck of the weather glass move up or move down when a low-pressure system approaches?

FIGURE 15-45 Problem 77

78. • **CE Predict/Explain** A person floats in a boat in a small backyard swimming pool. Inside the boat with the person are some bricks. **(a)** If the person drops the bricks overboard to the bottom of the pool, does the water level in the pool increase, decrease, or stay the same? **(b)** Choose the *best explanation* from among the following:
   I. When the bricks sink they displace less water than when they were floating in the boat; hence, the water level decreases.
   II. The same mass (boat + bricks + person) is in the pool in either case, and therefore the water level remains the same.
   III. The bricks displace more water when they sink to the bottom than they did when they were above the water in the boat; therefore the water level increases.

79. • **CE** A person floats in a boat in a small backyard swimming pool. Inside the boat with the person are several blocks of wood. Suppose the person now throws the blocks of wood into the pool, where they float. **(a)** Does the boat float higher, lower, or at the same level relative to the water? **(b)** Does the water level in the pool increase, decrease, or stay the same?

80. • **CE** The three identical containers in **FIGURE 15-46** are open to the air and filled with water to the same level. A block of wood floats in container A; an identical block of wood floats in container B, supporting a small lead weight; container C holds only water. **(a)** Rank the three containers in order of increasing weight of water they contain. Indicate ties where appropriate. **(b)** Rank the three containers in order of increasing weight of the container plus its contents. Indicate ties where appropriate.

FIGURE 15-46 Problem 80

81. • **CE BIO Sphygmomanometer** When a person's blood pressure is taken with a device known as a sphygmomanometer, it is measured on the arm, at approximately the same level as the heart. If the measurement were to be taken on a standing patient's leg instead, would the reading on the sphygmomanometer be greater than, less than, or the same as when the measurement is made on the arm?

82. • A water main broke on Lake Shore Drive in Chicago on November 8, 2002, shooting water straight upward to a height of 8.0 ft. What was the pressure in the pipe?

83. • **BIO Measuring Pain Threshold** A useful instrument for evaluating fibromyalgia and trigger-point tenderness is the doloriometer or algorimeter. This device consists of a force meter attached to a circular probe that is pressed against the skin until pain is experienced. If the reading of the force meter is 3.25 lb, and the diameter of the circular

A doloriometer/algorimeter used to evaluate pain threshold. (Problem 83)

probe is 1.39 cm, what is the pressure applied to the skin? Give your answer in pascals.

84. •• **BIO Power Output of the Heart** The power output of the heart is given by the product of the average blood pressure, $1.33 \, \mathrm{N/cm^2}$, and the flow rate, $105 \, \mathrm{cm^3/s}$. **(a)** Find the power of the heart. Give your answer in watts. **(b)** How much energy does the heart expend in a day? **(c)** Suppose the energy found in part (b) is used to lift a 72-kg person vertically to a height h. Find h, in meters.

85. •• A solid block is suspended from a spring scale. When the block is in air, the scale reads 35.0 N, when immersed in water the scale reads 31.1 N, and when immersed in oil the scale reads 31.8 N. **(a)** What is the density of the block? **(b)** What is the density of the oil?

86. •• A wooden block with a density of $710 \, \mathrm{kg/m^3}$ and a volume of $0.012 \, \mathrm{m^3}$ is attached to the top of a vertical spring whose force constant is $k = 540 \, \mathrm{N/m}$. Find the amount by which the spring is stretched or compressed if it and the wooden block are **(a)** in air or **(b)** completely immersed in water. [The density of air may be neglected in part (a).]

87. •• **Predict/Calculate Floating a Ball and Block** A 1.35-kg wooden block has an iron ball of radius 1.22 cm glued to one side. **(a)** If the block floats in water with the iron ball "high and dry," what is the volume of wood that is submerged? **(b)** If the block is now inverted, so that the iron ball is completely immersed, does the volume of wood that is submerged in water increase, decrease, or remain the same? Explain. **(c)** Calculate the volume of wood that is submerged when the block is in the inverted position.

88. •• **The Depth of the Atmosphere** Evangelista Torricelli (1608–1647) was the first to put forward the idea that we live at the bottom of

an ocean of air. **(a)** Given the value of atmospheric pressure at the surface of the Earth, and the fact that there is zero pressure in the vacuum of space, determine the depth of the atmosphere, assuming that the density of air and the acceleration of gravity are constant. **(b)** According to this model, what is the atmospheric pressure at the summit of Mt. Everest, 29,029 ft above sea level. (In fact, the density of air and the acceleration of gravity decrease with altitude, so the result obtained here is less than the actual depth of the atmosphere. Still, this is a reasonable first estimate.)

89. •• **The Hydrostatic Paradox I** Consider the lightweight containers shown in **FIGURE 15-47**. Both containers have bases of area $A_{\text{base}} = 24 \text{ cm}^2$ and depths of water equal to 18 cm. As a result, the downward force on the base of container 1 is equal to the downward force on container 2, even though the containers clearly hold different weights of water. This is referred to as the hydrostatic paradox. **(a)** Given that container 2 has an annular (ring-shaped) region of area $A_{\text{ring}} = 72 \text{ cm}^2$, determine the *net* downward force exerted on the container by the water. **(b)** Show that your result from part (a) is equal to the weight of the water in container 2.

Container 1   **Container 2**

FIGURE 15-47 Problem 89

90. •• **The Hydrostatic Paradox II** Consider the two lightweight containers shown in **FIGURE 15-48**. As in the previous problem, these containers have equal forces on their bases but contain different weights of water. This is another version of the hydrostatic paradox. **(a)** Determine the *net* downward force exerted by the water on container 2. Note that the bases of the containers have an area $A_{\text{base}} = 24 \text{ cm}^2$, the annular region has an area $A_{\text{ring}} = 18 \text{ cm}^2$, and the depth of the water is 18 cm. **(b)** Show that your result from part (a) is equal to the weight of the water in container 2. **(c)** If a hole is poked in the annular region of container 2, how fast will water exit the hole? **(d)** How high above the hole will the stream of water rise?

Container 1   **Container 2**

FIGURE 15-48 Problem 90

91. •• Predict/Calculate A backyard swimming pool is circular in shape and contains water to a uniform depth of 42 cm. It is 2.4 m in diameter and is not completely filled. **(a)** What is the pressure at the bottom of the pool? **(b)** If a person gets into the pool and floats peacefully, does the pressure at the bottom of the pool increase, decrease, or stay the same? **(c)** Calculate the pressure due to the

water and the person at the bottom of the pool if the floating person has a mass of 73 kg.

92. •• A prospector finds a solid rock composed of granite ($\rho = 2650 \text{ kg/m}^3$) and gold. If the volume of the rock is $3.84 \times 10^{-4} \text{ m}^3$, and its mass is 3.40 kg, **(a)** what mass of gold is contained in the rock? What percentage of the rock is gold by **(b)** volume and **(c)** mass?

93. •• Predict/Calculate **(a)** If the tension in the string in Example 15-12 is 0.89 N, what is the volume of the wood? Assume that everything else remains the same. **(b)** If the string breaks and the wood floats on the surface, does the water level in the flask rise, drop, or stay the same? Explain. **(c)** Assuming the flask is cylindrical with a cross-sectional area of 62 cm^2, find the change in water level after the string breaks.

94. •• Predict/Calculate **A Siphon for Irrigation** A siphon is a device that allows water to flow from one level to another. The siphon shown in **FIGURE 15-49** delivers water from an irrigation canal to a field of crops. To operate the siphon, water is first drawn through the length of the tube. After the flow is started in this way it continues on its own. **(a)** Using points 1 and 3 in Figure 15-49, find the speed v of the water leaving the siphon at its lower end. Give a symbolic answer. **(b)** Is the speed of the water at point 2 greater than, less than, or the same as its speed at point 3? Explain.

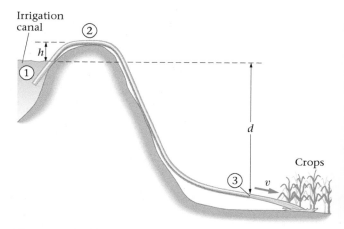

FIGURE 15-49 Problem 94

95. •• **Milking Cows** The density of raw milk at body temperature is 1023 kg/m^3. After being extracted from the cow, the milk is pumped through a pipe to the bottom of a cooling tank filled to a depth of $h = 1.8$ m (**FIGURE 15-50**). **(a)** If the pressure at location 2 in Figure 15-50 is atmospheric pressure, what minimum speed v_1 must the milk have in order to enter the cooling tank? Assume the milk at location 2 is essentially at rest. **(b)** What is the pressure at the bottom of the cooling tank?

FIGURE 15-50 Problem 95

96. •• A tin can is filled with water to a depth of 39 cm. A hole 11 cm above the bottom of the can produces a stream of water that is directed at an angle of 36° above the horizontal. Find **(a)** the range and **(b)** the maximum height of this stream of water.

97. •• **BIO** A person weighs 685 N in air but only 497 N when standing in water up to the hips. Find **(a)** the volume of each of the person's legs and **(b)** the mass of each leg, assuming they have a density that is 1.05 times the density of water.

98. •• **Thunderstorm Outflow** Rain-cooled air near the core of a thunderstorm sinks and then spreads out in front of the storm in a *forward flank downdraft gust front* (**FIGURE 15-51**). These gusts can vary from a cool breeze to a violent and damaging wind. Thunderstorms are extremely complex, but modeling air as an incompressible fluid can offer some insight. **(a)** Suppose 1.0 m³ of rain-cooled air has a density of 0.835 kg/m³, while the warmer air surrounding it has a density of 0.819 kg/m³. Taking into account the buoyant force on the parcel of air, find its downward acceleration. **(b)** If the parcel maintains the same acceleration from rest at an altitude of 4000 m, what is its speed when it arrives at the surface? **(c)** Now model the air as an incompressible fluid of constant density 1.02 kg/m³ that is at rest and has a pressure of 61.6 kPa at $h = 4000$ m altitude, but is moving and has a pressure of 101.3 kPa at $h = 0$. What is the speed of the air at the surface?

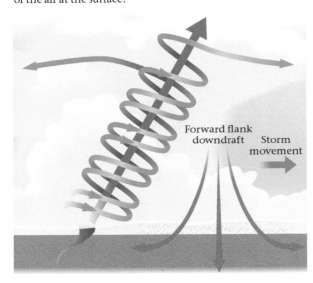

FIGURE 15-51 Problem 98

99. •• A horizontal pipe carries oil whose coefficient of viscosity is 0.00012 N · s/m². The diameter of the pipe is 5.2 cm, and its length is 55 m. **(a)** What pressure difference is required between the ends of this pipe if the oil is to flow with an average speed of 1.2 m/s? **(b)** What is the volume flow rate in this case?

100. •• **BIO** A patient is given an injection with a hypodermic needle 3.3 cm long and 0.26 mm in diameter. Assuming the solution being injected has the same density and viscosity as water at 20 °C, find the pressure difference needed to inject the solution at the rate of 1.5 g/s.

101. •• **Going Over Like a *Mythbuster* Lead Balloon** On one episode of *Mythbusters*, Jamie and Adam try to make a lead balloon that will float when filled with helium. The balloon they constructed was approximately cubical in shape, and 10 feet on a side. They used

a thin lead foil, which gave the finished balloon a mass of 11 kg. **(a)** What was the thickness of the foil? **(b)** Would the lead balloon float if filled with helium? **(c)** If the balloon does float, what would be the most mass it could lift in addition to its own mass?

102. ••• A round wooden log with a diameter of 73 cm floats with one-half of its radius out of the water. What is the log's density?

103. ••• The hollow, spherical glass shell shown in **FIGURE 15-52** has an inner radius R and an outer radius $1.2R$. The density of the glass is ρ_g. What fraction of the shell is submerged when it floats in a liquid of density $\rho = 1.5\rho_g$? (Assume the interior of the shell is a vacuum.)

FIGURE 15-52 Problem 103

104. ••• A geode is a hollow rock with a solid shell and an air-filled interior. Suppose a particular geode weighs twice as much in air as it does when completely submerged in water. If the density of the solid part of the geode is 2500 kg/m³, what fraction of the geode's volume is hollow?

105. ••• A tank of water filled to a depth d has a hole in its side a height h above the table on which it rests. Show that water emerging from the hole hits the table at a horizontal distance of $2\sqrt{(d - h)h}$ from the base of the tank.

106. ••• The water tank in **FIGURE 15-53** is open to the atmosphere and has two holes in it, one 0.80 m and one 3.6 m above the floor on which the tank rests. If the two streams of water strike the floor in the same place, what is the depth of water in the tank?

FIGURE 15-53 Problem 106

PASSAGE PROBLEMS

Cooking Doughnuts

Doughnuts are cooked by dropping the dough into hot vegetable oil until it changes from white to a rich, golden brown. One popular brand of doughnut automates this process; the doughnuts are made on an assembly line that customers can view in operation as they wait to order. Watching the doughnuts cook gives the customers time to develop an appetite as they ponder the physics of the process.

First, the uncooked doughnut is dropped into hot vegetable oil, whose density is $\rho = 919$ kg/m³. There it browns on one side as it floats on the oil. After cooking for the proper amount of time, a mechanical lever flips the doughnut over so it can cook on the other side. The doughnut floats fairly high in the oil, with less than half of its volume submerged. As a result, the final product has a characteristic white stripe around the middle where the dough is always out of the oil, as shown in **FIGURE 15-54**.

After cooking on one side, the doughnut is flipped over . . .

. . . to continue cooking on the other side.

Cooking oil

A white, uncooked stripe remains around the middle.

FIGURE 15-54 **Steps in cooking a doughnut**

The relationship between the density of the doughnut and the height of the white stripe is graphed in **FIGURE 15-55**. On the x axis we plot the density of the doughnut as a fraction of the density of the vegetable cooking oil; the y axis shows the height of the white stripe as a fraction of the total height of the doughnut. Notice that the height of the white stripe is plotted for both positive and negative values.

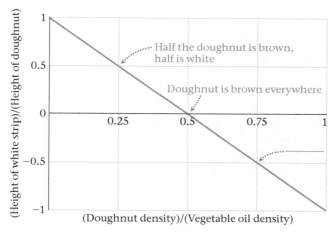

Half the doughnut is brown, half is white

Doughnut is brown everywhere

(Height of white strip)/(Height of doughnut)

(Doughnut density)/(Vegetable oil density)

FIGURE 15-55 **Cooking a floating doughnut** A finished doughnut has a stripe around the middle whose height is related to the density of the doughnut by the straight line shown here. (Problems 107, 108, 109, and 110)

107. • Assuming the doughnut has a cylindrical shape of height H and diameter D, and that the height of the white stripe is 0.22H, what is the density of the doughnut?

 A. 260 kg/m^3 B. 360 kg/m^3
 C. 720 kg/m^3 D. 820 kg/m^3

108. • Figure 15-57 has comments where the "height" of the white stripe is 0.5H and 0. The comment for −0.5H has been left blank, however. Which of the following comments is most appropriate for this case?

 A. The doughnut sinks.
 B. The white stripe is still present, but has a negative height.
 C. Half the doughnut is light brown, half is dark brown.
 D. The top and bottom of the doughnut are white, the middle one-half is brown.

109. •• A new doughnut is being planned whose density will be 330 kg/m^3. If the height of the doughnut is H, what will be the height of the white stripe?

 A. 0.14H B. 0.24H
 C. 0.28H D. 0.64H

110. •• Suppose the density of a doughnut is 550 kg/m^3. In terms of the height of the doughnut, H, what is the height of the portion of the doughnut that is out of the oil?

 A. 0.20H B. 0.40H
 C. 0.67H D. 0.80H

111. •• Predict/Calculate **REFERRING TO EXAMPLE 15-8** Suppose we use a different vegetable oil that has a higher density than the one in Example 15-8. **(a)** If everything else remains the same, will the height difference, h, increase, decrease, or remain the same? Explain. **(b)** Find the height difference for an oil that has a density of 9.60×10^2 kg/m^3.

112. •• **REFERRING TO EXAMPLE 15-8** Find the height difference, h, if the depth of the oil is increased to 7.50 cm. Assume everything else in the problem remains the same.

113. •• **REFERRING TO EXAMPLE 15-24** **(a)** Find the height H required to make $D = 0.655$ m. Assume everything else in the problem remains the same. **(b)** Find the depth h required to make $D = 0.455$ m. Assume everything else in the problem remains the same.

114. •• **REFERRING TO EXAMPLE 15-24** Suppose both h and H are increased by a factor of two. By what factor is the distance D increased?

16

Temperature and Heat

Big Ideas

1 Heat is energy that is transferred due to a difference in temperature.

2 Most substances expand when their temperature is increased.

3 Mechanical work can be converted into an equivalent amount of heat.

4 Different substances require different amounts of heat for a given increase in temperature.

5 Thermal energy can be transferred by conduction, convection, or radiation.

▲ Penguins are gregarious animals, but it's more than a desire to be close to one another that drives these birds to huddle together in tight groups. By packing in closely, the penguins reduce their surface area in contact with the frigid temperatures of the Antarctic. As a result, the transfer of thermal energy from their warm bodies to the surrounding air is minimized. This chapter explores these and other issues involving temperature, heat, and thermal energy.

In our study of physics so far, every calculation has been in terms of mass, length, and time, or some combination of the three. We now add a fourth physical quantity—*temperature*.

With the introduction of temperature, we broaden the scope of physics, allowing the study of a wide variety of new physical situations that mechanics alone cannot address.

16-1 Temperature and the Zeroth Law of Thermodynamics

Even as children we learn to avoid objects that are "hot." We also discover early on that if we forget to wear our coats outside we can become "cold." Later, we associate high values of something called "temperature" with hot objects, and low values with cold objects. These basic notions about temperature carry over into physics, though with a bit more precision.

Heat and the Flow of Energy   When we put a pan of cool water on a hot stove burner, we say that "heat flows" from the hot burner to the cool water. To be more precise, we define **heat** as follows:

> Heat is the energy that is *transferred* between objects because of a temperature difference.

Therefore, when we say there is a "transfer of heat" or a "heat flow" from object A to object B, what we mean is that the total energy of object A decreases and the total energy of object B increases. The total amount of energy in a substance—the sum of all of its kinetic and potential energy—is referred to as its *internal energy*, or **thermal energy**. Thus, an object's thermal energy refers both to the random motion of its particles (kinetic energy) and to the separation and orientation of the particles relative to one another (potential energy). Adding thermal energy to a system is known as *heating* and removing thermal energy is known as *cooling*. To summarize, an object does not "contain" heat—it has a certain energy content, and the energy it *exchanges* with other objects due to temperature differences is called heat.

Now, objects are said to be in **thermal contact** if heat can flow between them. In general, when a hot object is brought into thermal contact with a cold object, heat will be exchanged. The result is that the hot object cools off—and its molecules move more slowly—while the cold object warms up—and its molecules move more rapidly. After some time in thermal contact, the transfer of heat ceases. At this point, we say that the objects are in **thermal equilibrium**. The study of physical processes involving the transfer of heat is the area of physics referred to as **thermodynamics**. Because thermodynamics deals with the flow of energy within and between objects, it has great practical importance in all areas of the life sciences, physical sciences, and engineering.

Zeroth Law of Thermodynamics

We now introduce perhaps the most fundamental law obeyed by thermodynamic systems—referred to, appropriately enough, as the zeroth law of thermodynamics. The zeroth law spells out the basic properties of temperature. Its name reflects not only its basic importance, but also its subtlety, in that it was not recognized as a separate law until after the other three laws of thermodynamics had been accepted. Later, in Chapter 18, we introduce the remaining three laws of thermodynamics. These laws enable us to analyze the behaviors of engines and refrigerators, and to show, among other things, that perpetual motion machines are not possible.

The basic idea of the zeroth law of thermodynamics is that thermal equilibrium is determined by a single physical quantity—the **temperature**. For example, two objects in thermal contact are in equilibrium when they have the *same* temperature. If one object has a higher temperature, heat flows from that object to the other object until their temperatures are equal.

This may seem almost too obvious to mention, at least until you give it more thought. Suppose, for example, that you have a piece of metal and a pool of water, and you want to know if heat will flow between them when you put the metal in the pool. You measure the temperature of each, and if they are the same, you can conclude that no heat will flow. If the temperatures are different, however, it follows that there will be a flow of heat. Nothing else matters—not the type of metal, its mass, its shape, the amount of water, whether the water is fresh or salt, and so on—all that matters is one number, the temperature.

No heat flows between A and B . . .

A   B

. . . which means A and B are in equilibrium.

Now place B in contact with C.

B   C

Again, no heat flows between the objects. Therefore, C is also in equilibrium with B.

A   C

If A and C are now placed in contact, we find that no heat flows, so they too are in equilibrium.

▲ **FIGURE 16-1 An illustration of the zeroth law of thermodynamics** If A and C are each in thermal equilibrium with B, then A will be in thermal equilibrium with C when they are brought into thermal contact.

Big Idea 1 Heat is energy that is transferred from one physical system to another due to a difference in temperature. If two systems have the same temperature, there will be no transfer of energy when they are brought into thermal contact.

PHYSICS IN CONTEXT Looking Ahead

We make extensive use of the concept of temperature throughout Chapters 17 and 18.

To summarize, the **zeroth law of thermodynamics** can be stated as follows:

If object A is in thermal equilibrium with object B, and object C is also in thermal equilibrium with object B, then objects A and C will be in thermal equilibrium if brought into thermal contact.

This law is illustrated in **FIGURE 16-1**. We begin with objects A and B in thermal contact and in equilibrium. Next, object B is separated from A, and placed in contact with C. Objects C and B are also found to be in equilibrium. Hence, by the zeroth law, we are assured that when A and C are placed in contact they also will be in equilibrium.

To apply this principle to our example, let object A be the piece of metal and object C be the pool of water. Object B, then, can be a thermometer, used to measure the temperature of the metal and the water. If A and C are each separately in equilibrium with B—which means that they have the same temperature—they will be in equilibrium with one another.

Enhance Your Understanding   (Answers given at the end of the chapter)

1. When bricks A and B are placed in thermal contact, energy flows from brick B to brick A. Which brick had the higher initial temperature?

Section Review

- Heat is the energy transferred between objects due to a difference in temperature. If objects have the same temperature, they are said to be in thermal equilibrium.

- Two objects in thermal equilibrium with a third object will be in thermal equilibrium with one another if placed in thermal contact.

16-2 Temperature Scales

A variety of temperature scales are commonly used both in everyday situations and in physics. Some are related to familiar reference points, such as the temperature of boiling or freezing water. Others have more complex, historical rationales for their values. Here we consider three of the more frequently used temperature scales. We also examine the connections between them.

The Celsius Scale

Perhaps the easiest temperature scale to remember is the Celsius scale, named in honor of the Swedish astronomer Anders Celsius (1701–1744). Originally, Celsius assigned zero degrees to boiling water and 100 degrees to freezing water. These values were later reversed by the biologist Carolus Linnaeus (1707–1778). Thus, today we say that water freezes at zero degrees Celsius, which we abbreviate as 0 °C, and boils at 100 °C.

The choice of zero level for a temperature scale is quite arbitrary, as is the number of degrees between any two reference points. In the Celsius scale, as in others, there is no upper limit to the value a temperature may have. There is a lower limit, however. For the Celsius scale, the lowest possible temperature is −273.15 °C, as we shall see later in this section.

One bit of notation should be pointed out before we continue. When we write a Celsius temperature, we give a numerical value followed by the degree symbol, °, and the capital letter C. For example, 5 °C is the temperature five degrees Celsius. On the other hand, if the temperature of an object is *changed* by a given amount, we use the notation C°. Thus, if we increase the temperature by five degrees on the Celsius scale, we say that the change in temperature is 5 C°—that is, five Celsius degrees. This is summarized below:

A temperature T of five degrees is 5 °C (five degrees Celsius).

A temperature change ΔT of five degrees is 5 C° (five Celsius degrees).

The Fahrenheit Scale

The Fahrenheit scale was developed by Daniel Gabriel Fahrenheit (1686–1736), who chose zero to be the lowest temperature he was able to achieve in his laboratory. He also chose 96 degrees to be body temperature, though why he made this choice is not known. In the modern version of the Fahrenheit scale, body temperature is 98.6 °F; in addition, water freezes at 32 °F and boils at 212 °F. Lastly, using the same convention as for °C and C°, we say that an increase of 180 F° is required to bring water from freezing to boiling.

The Fahrenheit scale not only has a different zero than the Celsius scale, it also has a different "size" for its degree. As just noted, 180 Fahrenheit degrees are required for the same change in temperature as 100 Celsius degrees. Hence, the Fahrenheit degrees are smaller by a factor of $100/180 = 5/9$.

To convert between a Fahrenheit temperature, T_F, and a Celsius temperature, T_C, we start by writing a linear relationship between them. Thus, let

$$T_F = aT_C + b$$

We would like to determine the constants a and b. This requires two independent pieces of information, which we have in the freezing and boiling points of water. Using the freezing point, we find

$$32\,°F = a(0\,°C) + b = b$$

Thus, b is 32 °F. Next, the boiling point gives

$$212\,°F = a(100\,°C) + 32\,°F$$

Solving for the constant a, we find

$$a = \frac{(212\,°F - 32\,°F)}{(100\,°C)} = \frac{180\,F°}{100\,C°} = \frac{9}{5}\,F°/C°$$

Combining our results gives the following conversion relationship:

Conversion Between Degrees Celsius and Degrees Fahrenheit

$$T_F = \left(\tfrac{9}{5}\,F°/C°\right)T_C + 32\,°F \qquad\qquad \text{16-1}$$

Similarly, this relationship can be rearranged to convert from Fahrenheit to Celsius:

Conversion Between Degrees Fahrenheit and Degrees Celsius

$$T_C = \left(\tfrac{5}{9}\,C°/F°\right)(T_F - 32\,°F) \qquad\qquad \text{16-2}$$

Because conversion factors like $\tfrac{9}{5}\,F°/C°$ are a bit clumsy and tend to clutter up an equation, we will generally drop the degree symbols until the final result. For example, to convert 10 °C to degrees Fahrenheit, we write

$$T_F = \tfrac{9}{5}\,T_C + 32 = \tfrac{9}{5}(10) + 32 = 50\,°F$$

Similar remarks apply in converting from Fahrenheit to Celsius.

EXAMPLE 16-1   TEMPERATURE CONVERSIONS

(a) On a fine spring day you notice that the temperature is 75 °F. What is the corresponding temperature on the Celsius scale? **(b)** If the temperature on a brisk winter morning is −2.0 °C, what is the corresponding Fahrenheit temperature?

PICTURE THE PROBLEM
Our sketch shows a circular thermometer with both a Fahrenheit and a Celsius scale. The range of the scales extends well beyond the temperatures 75 °F and −2.0 °C needed for the problem.

CONTINUED

REASONING AND STRATEGY

The conversions asked for in this problem are straightforward applications of the relationships between T_F and T_C. In particular, for **(a)** we use $T_C = \frac{5}{9}(T_F - 32)$, and for **(b)** we use $T_F = \frac{9}{5}T_C + 32$.

Known   **(a)** Temperature is $T_F = 75\,°F$. **(b)** Temperature is $T_C = -2.0\,°C$.

Unknown   **(a)** Temperature in Celsius, $T_C = ?$ **(b)** Temperature in Fahrenheit, $T_F = ?$

SOLUTION

Part (a)

1. Substitute $T_F = 75\,°F$ in $T_C = \frac{5}{9}(T_F - 32)$: $T_C = \frac{5}{9}(75 - 32) = 24\,°C$

Part (b)

2. Substitute $T_C = -2.0\,°C$ in $T_F = \frac{9}{5}T_C + 32$: $T_F = \frac{9}{5}(-2.0) + 32 = 28\,°F$

INSIGHT

The results given here agree with the scales shown in the sketch.

PRACTICE PROBLEM

Find the Celsius temperature that corresponds to 110 °F.   [**Answer:** $T_C = 43\,°C$]

Some related homework problems: Problem 2, Problem 3

QUICK EXAMPLE 16-2   SAME TEMPERATURE

What temperature is the same on both the Celsius and Fahrenheit scales?

REASONING AND SOLUTION

We can use either $T_F = \left(\frac{9}{5}\,F°/C°\right)T_C + 32\,°F$ or $T_C = \left(\frac{5}{9}\,C°/F°\right)(T_F - 32\,°F)$ to set the two temperatures equal to one another: $T_F = T_C = t$. The algebra is a bit simpler with $T_F = \left(\frac{9}{5}\,F°/C°\right)T_C + 32\,°F$, so we'll use that relationship.

1. Set $T_F = T_C = t$ in $T_F = \frac{9}{5}T_C + 32$: $t = 9t/5 + 32$

2. Move all terms involving t to the left side of the equation: $-4t/5 = 32$

3. Solve for t: $t = -40$

4. As a check, substitute $T_F = -40\,°F$ in $T_C = \frac{5}{9}(T_F - 32)$: $T_C = \frac{5}{9}(-40 - 32) = -40\,°C$

Thus, $-40\,°F$ is the same as $-40\,°C$. This is consistent with the sketch shown in Example 16-1.

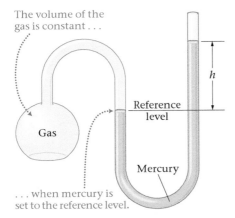

The volume of the gas is constant . . .

Gas

Reference level

Mercury

. . . when mercury is set to the reference level.

▲ **FIGURE 16-2 A constant-volume gas thermometer** By adjusting the height of mercury in the right-hand tube, we can set the level in the left-hand tube at the reference level. This assures that the gas occupies a constant volume.

Absolute Zero

Experiments show conclusively that there is a lowest temperature below which it is impossible to cool an object. This is referred to as **absolute zero**. Though absolute zero can be approached from above arbitrarily closely, it can never be attained.

    To give an idea of just where absolute zero is on the Celsius scale, we start with the following observation: If a given volume V of air—say, the air in a balloon—is cooled from 100 °C to 0 °C, its volume decreases by roughly $V/4$. Imagine this trend continuing uninterrupted. In this case, cooling from 0 °C to −100 °C would reduce the volume by another $V/4$, from −100 °C to −200 °C by another $V/4$, and finally, from −200 °C to −300 °C by another $V/4$, which brings the volume down to zero. Clearly, it doesn't make sense for the volume to be less than zero, and hence absolute zero must be roughly −300 °C.

    This result, though crude, is in the right ballpark. A precise determination of absolute zero can be made with a device known as a **constant-volume gas thermometer** (FIGURE 16-2). The basic idea is that by adjusting the level of mercury in the right-hand tube, we can set the level of mercury in the left-hand tube to a fixed reference level. With the mercury so adjusted, the gas occupies a constant volume and its pressure is simply $P_{gas} = P_{at} + \rho g h$ (Equation 15-7), where ρ is the density of mercury.

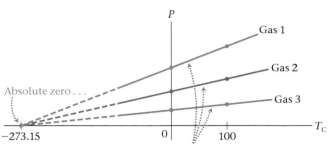

... is the temperature at which the pressure of any gas would become zero.

As the temperature of the gas is changed, the mercury level in the right-hand tube can be readjusted as described. The gas pressure can be determined again, and the process repeated. The results of a series of such measurements are shown in **FIGURE 16-3**.

Notice that as a gas is cooled, its pressure decreases, as one would expect. In fact, the decrease in pressure is approximately linear. At low enough temperatures the gas eventually liquefies, and its behavior changes, but if we extrapolate the straight line obtained before liquefaction, we see that it reaches zero pressure (the lowest pressure possible) at −273.15 °C.

What is remarkable about this result is that it is independent of the gas we use in the thermometer. For example, gas 2 and gas 3 in Figure 16-3 have pressures that are different from one another, and from gas 1. Yet all three gases extrapolate to zero pressure at precisely the same temperature. Thus, we conclude that there is indeed a *unique* value of absolute zero, below which further cooling is not possible.

EXAMPLE 16-3 | IT'S A GAS

The gas in a constant-volume gas thermometer has a pressure of 97.6 kPa at 0.00 °C. Assuming ideal behavior, as in Figure 16-3, what is the pressure of this gas at 105 °C?

PICTURE THE PROBLEM
Our sketch plots the pressure of the gas as a function of temperature. At $T_C = 0.00\,°C$ the pressure is 97.6 kPa, and at $T_C = -273.15\,°C$ the pressure extrapolates linearly to zero. We also indicate the point corresponding to 105 °C on the graph.

REASONING AND STRATEGY
We assume that the pressure lies on a straight line, as in Figure 16-3. To find the pressure at $T_C = 105\,°C$ we simply extend the straight line.

Thus, we start with the information that the pressure increases linearly from 0 to 97.6 kPa when the temperature increases from −273.15 °C to 0.00 °C. This rate of increase in pressure must also apply to an increase in temperature from −273.15 °C to 105 °C. Using this rate we can find the desired pressure.

Known $P = 0$ at $T_C = -273.15$, $P = 97.6$ kPa at $T_C = 0.00\,°C$.
Unknown $P = ?$ at $T_C = 105\,°C$.

SOLUTION

1. Calculate the rate at which pressure increases with temperature for this gas:

$$\text{rate} = \frac{97.6\text{ kPa}}{273.15\text{ C}°} = 0.357\text{ kPa/C}°$$

2. Multiply this rate by the temperature change from −273.15 °C to 105 °C (a change of 378.15 C°):

$$(0.357\text{ kPa/C}°)(378.15\text{ C}°) = 135\text{ kPa}$$

INSIGHT
The pressure of this gas increases from slightly less than one atmosphere at 0.00 °C to roughly 34% more than one atmosphere at 105 °C.

PRACTICE PROBLEM
Find the temperature at which the pressure of the gas is 74 kPa.   [**Answer:** $T_C = -66\,°C$]

Some related homework problems: Problem 7, Problem 8

The Kelvin Scale

The Kelvin temperature scale, named for the Scottish physicist William Thomson, Lord Kelvin (1824–1907), is based on the existence of absolute zero. In fact, the zero of the Kelvin scale, abbreviated 0 K, is set exactly at absolute zero. Thus, in this scale there are no negative equilibrium temperatures. The Kelvin scale is also chosen to have the same degree size as the Celsius scale.

As mentioned, absolute zero occurs at $-273.15\,°C$; hence the conversion between a Kelvin-scale temperature, T, and a Celsius temperature, T_C, is as follows:

Conversion Between a Celsius Temperature and a Kelvin Temperature

$$T = T_C + 273.15 \qquad\qquad 16\text{-}3$$

The difference between the Celsius and Kelvin scales is simply a difference in the zero level.

The notation for the Kelvin scale differs somewhat from that for the Celsius and Fahrenheit scales. In particular, by international agreement, the degree terminology and the degree symbol, °, are not used in the Kelvin scale. Instead, a temperature of 5 K is read simply as 5 kelvin. In addition, a change in temperature of 5 kelvin is written 5 K, the same as for a temperature of 5 kelvin.

Though the Celsius and Fahrenheit scales are the ones most commonly used in everyday situations, the Kelvin scale is used more than any other in physics. This stems from the fact that the Kelvin scale incorporates the significant concept of absolute zero. As a result, the thermal energy of a system depends in a very simple way on the Kelvin temperature. This will be discussed in detail in the next chapter.

EXERCISE 16-4   KELVIN SCALE

You visit Palm Springs on a nice warm day, when the temperature is 113 °F. Convert this temperature to the Kelvin temperature scale.

REASONING AND SOLUTION
We will do two temperature conversions to go from Fahrenheit to Kelvin. First, we convert from °F to °C using $T_C = \left(\frac{5}{9}\,C°/F°\right)\left(T_F - 32\,°F\right)$:

$$T_C = \frac{5}{9}(113 - 32) = 45\,°C$$

Next, we convert °C to K with $T = T_C + 273.15$:

$$T = 45 + 273.15 = 318\,K$$

°F	°C	K
Boiling point of water 212	100	373
Freezing point of water 32	0	273
Freezing point of dry ice (CO_2) −109	−78	195
Boiling point of nitrogen −321	−196	77
Absolute zero −460	−273	0

▲ **FIGURE 16-4 Temperature scales** A comparison of the Fahrenheit, Celsius, and Kelvin temperature scales. Physically significant temperatures, such as the freezing and boiling points of water, are indicated for each scale.

The three temperature scales presented in this section are shown side by side in **FIGURE 16-4**. Temperatures of particular interest are indicated as well. This permits a useful visual comparison between the scales.

Enhance Your Understanding   (Answers given at the end of the chapter)

2. Is the "size" of a degree in the Fahrenheit scale greater than, less than, or equal to the size of a degree in the Kelvin scale?

Section Review

- Three temperature scales are in common usage. The conversions between these scales are as follows: Celsius to Fahrenheit, $T_F = \left(\frac{9}{5}\,F°/C°\right)T_C + 32\,°F$; Fahrenheit to Celsius, $T_C = \left(\frac{5}{9}\,C°/F°\right)(T_F - 32\,°F)$; Celsius to Kelvin, $T = T_C + 273.15$.

16-3 Thermal Expansion

Most substances expand when heated. For example, power lines on a hot summer day hang low compared to their appearance on a cold day in winter. In fact, thermal expansion is the basis for many thermometers, including the familiar thermometers used for measuring a fever. The expansion of a liquid, such as mercury or alcohol, results in a column of liquid of variable height within the glass neck of the thermometer. The height is read against markings on the glass, which gives the temperature.

You may wonder what bizarre substance could possibly be an exception to this common response to heating. The most important exception occurs in a substance you drink every day—water. This is just one of the many special properties that sets water apart from most other substances.

In this section we consider the physics of thermal expansion, including linear, area, and volume expansion. We also discuss briefly the unusual thermal behavior of water, and some of its more significant consequences.

> **Big Idea** **2** Most substances expand when their temperature is increased. A notable exception is water, which expands as its temperature is lowered from 4 °C to 0 °C.

Linear Expansion

Consider a rod whose length is L_0 at the temperature T_0. Experiments show that when the rod is heated or cooled, its length changes in direct proportion to the temperature change. Thus, if the change in temperature is ΔT, the change in length of the rod, ΔL, is

$$\Delta L = (\text{constant})\Delta T$$

The constant of proportionality depends, among other things, on the substance from which the rod is made.

CONCEPTUAL EXAMPLE 16-5 COMPARE EXPANSIONS

When rod 1 is heated by an amount ΔT, its length increases by ΔL. If rod 2, which is twice as long as rod 1 and made of the same material, is heated by the same amount, does its length increase by **(a)** ΔL, **(b)** $2\Delta L$, or **(c)** $\Delta L/2$?

REASONING AND DISCUSSION

We can imagine rod 2 to be composed of two copies of rod 1 placed end to end, as shown.

When the temperature is increased by ΔT, each copy of rod 1 expands by ΔL. Hence, the total expansion of the two copies is $2\Delta L$, as is the total expansion of rod 2.

ANSWER

(b) The rod that is twice as long expands twice as much, by $2\Delta L$.

We conclude, then, on the basis of the preceding Conceptual Example, that the change in length is proportional to *both* the initial length, L_0, and the temperature change, ΔT. The constant of proportionality is referred to as α, the **coefficient of linear expansion**, and is defined as follows:

Definition of Coefficient of Linear Expansion, α

$$\Delta L = \alpha L_0 \Delta T \qquad \qquad 16\text{-}4$$

SI unit for α: $\text{K}^{-1} = (\text{C}°)^{-1}$

Table 16-1 gives values of α for a variety of substances.

EXERCISE 16-6 THERMAL EXPANSION

The Eiffel Tower (**FIGURE 16-5**), constructed in 1889 by Alexandre Eiffel, is an impressive lat-ticework structure made of iron. If the tower is 324 m high on a 0.0 °C day, by how much does its height increase on a day when the temperature is 29 °C?

CONTINUED

▲ **FIGURE 16-5** The Eiffel Tower in Paris gains about a sixth of an inch in height for each Celsius degree that the temperature rises.

REASONING AND SOLUTION

We can calculate the change in height with $\Delta L = \alpha L_0 \Delta T$. Table 16-1 shows that the coefficient of linear expansion for iron is $\alpha = 12 \times 10^{-6} \text{ K}^{-1}$, and we know that the initial length (height in this case) is $L_0 = 324$ m. The change in temperature is $\Delta T = 29 \text{ C}° = 29 \text{ K}$, and the corresponding change in height is

$$\Delta L = \alpha L_0 \Delta T = (12 \times 10^{-6} \text{ K}^{-1})(324 \text{ m})(29 \text{ K}) = 11 \text{ cm}$$

TABLE 16-1 Coefficients of Thermal Expansion near 20°C

Substance	Coefficient of linear expansion, $\alpha(\text{K}^{-1})$	Substance	Coefficient of volume expansion, $\beta(\text{K}^{-1})$
Lead	29×10^{-6}	Ether	1.51×10^{-3}
Aluminum	24×10^{-6}	Carbon tetrachloride	1.18×10^{-3}
Brass	19×10^{-6}		
Copper	17×10^{-6}	Alcohol	1.01×10^{-3}
Iron (steel)	12×10^{-6}	Gasoline	0.95×10^{-3}
Concrete	12×10^{-6}	Olive oil	0.68×10^{-3}
Window glass	11×10^{-6}	Water	0.21×10^{-3}
Pyrex glass	3.3×10^{-6}	Mercury	0.18×10^{-3}
Quartz	0.50×10^{-6}		

RWP* An interesting application of thermal expansion is in the behavior of a bimetallic strip. As the name suggests, a bimetallic strip consists of two metals bonded together to form a linear strip of metal. This is illustrated in **FIGURE 16-6**. Because two different metals will, in general, have different coefficients of linear expansion, α, the two sides of the strip will change lengths by different amounts when heated or cooled.

For example, suppose metal B in Figure 16-6 (a) has the larger coefficient of linear expansion. This means that its length will change by greater amounts than metal A for the same temperature change. Hence, if this strip is cooled, the B side will shrink more than the A side, resulting in the strip bending toward the B side, as in Figure 16-6 (b). On the other hand, if the strip is heated, the B side expands by a greater amount than the A side, and the strip curves toward the A side, as in Figure 16-6 (c). Thus, the shape of the bimetallic strip depends sensitively on temperature.

Because of this property, bimetallic strips are used in a variety of thermal applications. For example, a bimetallic strip can be used as a thermometer; as the strip

(a) A bimetallic strip

(b) Chilling the strip

(c) Heating the strip

(d) A bimetallic strip in a thermostat

▲ **FIGURE 16-6 A bimetallic strip** **(a)** A bimetallic strip composed of metals A and B. **(b)** If metal B has a larger coefficient of linear expansion than metal A, it will shrink more when cooled and **(c)** expand more when heated. **(d)** A bimetallic strip can be used to construct a thermostat. If the temperature falls, the strip bends downward and closes the electric circuit, which then operates a heater. When the temperature rises, the strip deflects in the opposite direction, breaking the circuit and turning off the heater.

*Real World Physics applications are denoted by the acronym RWP.

changes its shape, it can move a needle to indicate the temperature. Similarly, many thermostats have a bimetallic strip to turn on or shut off a heater. This is shown in Figure 16-6 (d). As the temperature of the room changes, the bimetallic strip deflects in one direction or the other, which either closes or breaks the electric circuit connected to the heater.

RWP Another common use of thermal expansion is the *antiscalding device* for water faucets. An antiscalding device is simply a valve inside a water faucet that is attached to a spring. When the water temperature is at a safe level, the valve permits water to flow freely through the faucet. If the temperature of the water reaches a dangerous level, however, the thermal expansion of the spring is enough to close the valve and stop the flow of water—thus preventing inadvertent scalding. When the water cools down again, the valve reopens and the flow of water resumes.

RWP Finally, thermal expansion can have undesirable effects in some cases. For example, you may have noticed that bridges often have gaps between different sections of the structure, as seen in **FIGURE 16-7 (a)**. When the air temperature rises in the summer, the sections of the bridge can expand freely into these gaps. If the gaps were not present, the expansion of the different sections could cause the bridge to buckle and warp. Thus, these gaps, referred to as *expansion joints*, are a way of avoiding this type of heat-related damage. Expansion joints can also be found in railroad tracks and oil pipelines (**FIGURE 16-7 (b)**), to name just two other examples.

Area Expansion

The length of an object changes with temperature, and hence it follows that its area changes as well. To see precisely how the area changes, consider a square piece of metal of length L on a side. The initial area of the square is $A = L^2$. If the temperature of the square is increased by ΔT, the length of each side increases from L to $L + \Delta L = L + \alpha L \Delta T$. As a result, the square has an increased area, A', given by

$$A' = (L + \Delta L)^2 = (L + \alpha L \Delta T)^2$$
$$= L^2 + 2\alpha L^2 \Delta T + \alpha^2 L^2 \Delta T^2$$

Now, if $\alpha \Delta T$ is much less than one—which is certainly the case for typical changes in temperature, ΔT—then $\alpha^2 \Delta T^2$ is even smaller. Hence, if we ignore this small contribution, we find

$$A' \approx L^2 + 2\alpha L^2 \Delta T = A + 2\alpha A \Delta T$$

As a result, the change in area, ΔA, is

$$\Delta A = A' - A \approx 2\alpha A \Delta T \qquad \text{16-5}$$

Notice the similarity between this relationship and Equation 16-4 ($\Delta L = \alpha L_0 \Delta T$); the length L has been replaced with the area A, and the expansion coefficient has been doubled to 2α.

Though this calculation was done for the simple case of a square, the result applies to an area of any shape. For example, the area of a circular disk of radius r and area $A = \pi r^2$ will increase by the amount $2\alpha A \Delta T$ with an increase in temperature of ΔT. What about a washer, however, which is a disk of metal with a circular hole cut out of its center? What happens to the area of the hole as the washer is heated? Does it expand along with everything else, or does the expanding washer "expand into the hole" to make it smaller? We consider this question in the next Conceptual Example.

(a)

(b)

▲ **FIGURE 16-7 Visualizing Concepts**
Thermal Expansion Thermal expansion, though small, is far from negligible in many everyday situations. This is especially true when long objects such as railroad tracks, bridges, or pipelines are involved. **(a)** Bridges and elevated highways must include expansion joints to prevent the roadway from buckling when it expands in hot weather. **(b)** Pipelines typically include loops that allow for expansion and contraction when the temperature changes.

PROBLEM-SOLVING NOTE

Expansion of a Hole

A hole in a material expands the same as if it were made of the material itself. Thus, to find the expansion of a hole in a steel plate, we use the coefficient of linear expansion for steel.

CONCEPTUAL EXAMPLE 16-7   **HEATING A HOLE**   PREDICT/EXPLAIN

A washer has a hole in the middle. **(a)** As the washer is heated, does the hole expand, shrink, or stay the same size? **(b)** Which of the following is the *best explanation* for you prediction?

   I. The hole shrinks because the metal of the washer expands inward into the hole.

  II. The hole expands along with everything else—every dimension of the washer expands.

 III. The size of the hole remains the same because it is empty space, rather than metal like the washer.

CONTINUED

REASONING AND DISCUSSION

To make a washer from a disk of metal, we can cut along a circular shown in the sketch, and remove the inner disk. If we now heat the both the washer and the inner disk expand. On the other hand, left the inner disk in place and heated the original disk, it would pand. Removing the *heated* inner disk would create an expanded with an expanded hole in the middle. We obtain the same result we remove the inner disk and then heat, or heat first and then re- the inner disk.

Thus heating the washer causes both it and its hole to expand, and they both expand with the same coefficient of linear expan- sically, the system behaves the same as if we had produced a photo- enlargement—everything expands.

ANSWER

(a) The hole expands along with everything else. **(b)** The best ex- tion is II.

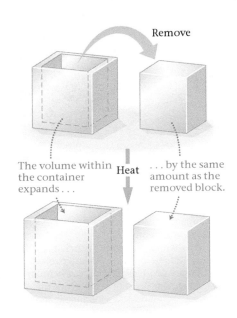

The volume within the container expands . . . Heat . . . by the same amount as the removed block.

◄ **FIGURE 16-8 Volume expansion** A portion of a cube is removed to create a container. When heated, the removed portion expands, just as the volume within the container expands.

Volume Expansion

Just as the hole in a washer increases in area with heating, so does the empty volume within a cup or other container. This is illustrated in **FIGURE 16-8**, where we show a block of material with a volume removed to convert it into a container. As the system is heated, there will be an expansion of the container, of the volume within it, and of the volume that was removed. As with the area of a hole, the volume within a con- tainer expands with the same coefficient of expansion as the container itself.

To calculate the change in volume, consider a cube of length L on a side. The ini- tial volume of the cube is $V = L^3$. Increasing the temperature results in an increased volume given by

$$V' = (L + \Delta L)^3 = (L + \alpha L \Delta T)^3$$
$$= L^3 + 3\alpha L^3 \Delta T + 3\alpha^2 L^3 \Delta T^2 + \alpha^3 L^3 \Delta T^3$$

Neglecting the smaller contributions, as we did with the area, we find

$$V' \approx L^3 + 3\alpha L^3 \Delta T = V + 3\alpha V \Delta T$$

Therefore, the change in volume, ΔV, is

$$\Delta V = V' - V \approx 3\alpha V \Delta T$$

PROBLEM-SOLVING NOTE

Expansion of a Volume

The empty volume inside a container ex- pands the same as if it were made of the same material as the container. For ex- ample, to find the increase in volume of a steel tank, we use the coefficient of linear expansion for steel, just as if the tank were actually filled with steel.

This expression, though calculated for a cube, applies to any volume.

In general, volume expansion is described in the same way as linear expansion, but with a **coefficient of volume expansion**, β, defined as follows:

Definition of Coefficient of Volume Expansion, β

$$\Delta V = \beta V \Delta T \qquad \text{16-6}$$

SI unit for β: $K^{-1} = (C°)^{-1}$

Values of β are given in Table 16-1 for a number of different liquids. If Table 16-1 lists a value of α for a given substance, but not for β, its change in volume is calculated as follows:

$$\Delta V = \beta V \Delta T \approx 3\alpha V \Delta T \qquad \text{16-7}$$

PROBLEM-SOLVING NOTE

Temperature Change

Remember that a change in temperature of 1 C° is the same as a change in tempera- ture of 1 K. Thus, when finding the thermal expansion of an object, we can express the change in temperature ΔT in terms of either the Celsius or the Kelvin temperature scale.

That is, we simply make the identification $\beta = 3\alpha$. For substances that have a specific value for β listed in Table 16-1, we simply use Equation 16-6. This is illustrated in the next Example.

EXAMPLE 16-8 OIL SPILL

A copper flask with a volume of 201 cm³ is filled to the brim with olive oil. If the temperature of the system is increased from 11 °C to 36 °C, how much oil spills from the flask?

Heat by 25 C°

PICTURE THE PROBLEM
Our sketch shows the flask filled to the top initially, then spilling over when heated by 25 C°. (*Note:* Because degrees have the same size on the Celsius and Kelvin scales, it follows that $\Delta T = 25$ C° = 25 K.)

REASONING AND STRATEGY
As the system is heated, both the flask and the olive oil expand. Thus, we start by calculating the expansion of the oil and the flask separately. A quick glance at Table 16-1 shows that olive oil, with $\beta_{oil} = 0.68 \times 10^{-3}$ K⁻¹, expands much more with temperature than does the copper flask, with $\beta_{copper} = 3\alpha_{copper} = 3(17 \times 10^{-6}$ K⁻¹). The difference in expansion volumes is what spills out.

Known Volume of copper flask, $V_{flask} = 201$ cm³; change in temperature, $\Delta T = 25$ K; coefficients of thermal expansion, $\beta_{oil} = 0.68 \times 10^{-3}$ K⁻¹ and $\beta_{copper} = 3\alpha_{copper} = 3(17 \times 10^{-6}$ K⁻¹).
Unknown Amount of oil that spills out of flask = $\Delta V_{oil} - \Delta V_{flask}$ = ?

SOLUTION

1. Calculate the change in volume of the olive oil:

$$\Delta V_{oil} = \beta_{oil} V \Delta T$$
$$= (0.68 \times 10^{-3} \text{ K}^{-1})(201 \text{ cm}^3)(25 \text{ K}) = 3.4 \text{ cm}^3$$

2. Calculate the change in volume of the copper flask:

$$\Delta V_{flask} = \beta_{copper} V \Delta T = 3\alpha_{copper} V \Delta T$$
$$= 3(17 \times 10^{-6} \text{ K}^{-1})(201 \text{ cm}^3)(25 \text{ K}) = 0.26 \text{ cm}^3$$

3. Find the difference in volume expansions. This is the volume of oil that spills out:

$$\Delta V_{oil} - \Delta V_{flask} = 3.4 \text{ cm}^3 - 0.26 \text{ cm}^3 = 3.1 \text{ cm}^3$$

INSIGHT
If the system were cooled, the oil would lose volume more rapidly than the flask. This would result in a drop in the level of the oil.

This Example illustrates why fuel tanks on cars are designed to stop filling before the gas reaches the top—otherwise, the tank could overflow on a hot day.

PRACTICE PROBLEM — PREDICT/CALCULATE
Suppose the copper flask is initially filled to the brim with gasoline rather than olive oil. (a) Referring to Table 16-1, do you expect the amount of liquid that spills with heating by 25 °C to be greater than, less than, or the same as in the case of olive oil? (b) Find the volume of spilled gasoline in this case. [**Answer: (a)** More liquid will spill because gasoline has a greater coefficient of thermal expansion. **(b)** $\Delta V = 4.5$ cm³]

Some related homework problems: Problem 22, Problem 27

Special Properties of Water

RWP As we've mentioned, water is a substance rich with unusual behavior. For example, in the last chapter we discussed the fact that the solid form of water (ice) is less dense than the liquid form. Hence, icebergs float. What is remarkable about icebergs floating is not that 90% is submerged, but that 10%, or any amount at all, is above the water. The solids of most substances are denser than their liquids; hence when they freeze, their solids immediately sink.

Here we consider the unusual *thermal* behavior of water. FIGURE 16-9 shows the density of water over a wide range of temperatures. Notice that the density is a maximum at about 4 °C. Thus, when you *heat* water from 0 °C to 4 °C, it actually *shrinks*, rather than expands, and becomes *more dense*. The reason is that water molecules that were once part of the rather open crystal structure of ice are now able to pack more closely together in the liquid. Above 4 °C water expands with heating, just like other substances. The implications of this expansion are quite evident in the El Niño effect, shown in FIGURE 16-10.

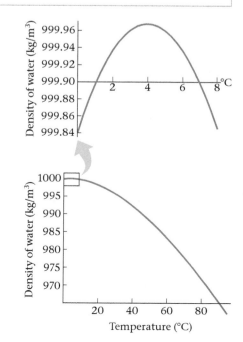

► FIGURE 16-9 **The unusual behavior of water near 4 °C** The density of water actually *increases* as the water is heated between 0 °C and 4 °C. Maximum density occurs near 4 °C.

▲ **FIGURE 16-10** Above 4 °C water expands with increasing temperature like any normal liquid. This photo shows the expansion of water on a grand scale. It depicts an El Niño event: the appearance of a mass of unusually warm water in the equatorial region of the eastern Pacific Ocean every few years. An El Niño causes worldwide changes in climate and weather patterns. In this satellite image, the warmest water temperatures are represented by red, the coolest by violet. The El Niño is the red spike west of the South American coast. In this region, sea levels are as much as 20 cm higher than their average values, due to thermal expansion of the water.

▲ **FIGURE 16-11** The Arctic permafrost landscape can be significantly altered by frost heaving, which often results in polygonal patterns.

RWP The expansion of water from 0 °C to 4 °C has significant consequences for the ecology of lakes in northern latitudes. When temperatures drop in the winter, the surface waters of a lake cool first and sink, allowing warmer water to rise to the surface to be cooled in turn. Eventually, a lake can fill with water at 4 °C. Further drops in temperature result in cooler, less dense water near the surface, where it floats until it freezes. Thus, lakes freeze on the top surface first, with the bottom water staying relatively warm at about 4 °C. In addition, the ice and snow on top of a lake act as thermal insulation, slowing the continued growth of ice.

On the other hand, if water had more ordinary behavior—like shrinking when cooled, and a solid form that is more dense than the liquid—a lake would freeze from the bottom up. There would be no insulating layer of ice on top, and if the winter were long enough, and cold enough, the lake could freeze solid. This, of course, would be disastrous for fish and other creatures that live in the water.

RWP The same physics that is responsible for floating icebergs and ice-covered lakes is to blame for water pipes that burst in the winter. Even a water pipe made of steel is not strong enough to keep from rupturing when the ice forming within it expands outward. Later, when the temperature rises above freezing again, the burst pipe will make itself known by springing a leak.

Freezing water can also transform the landscape. In Arctic regions, for instance, the thawing and freezing cycles of water-saturated soils leads to *frost heaving*, a phenomenon that can create fantastic shapes in the landscape, including a pattern of polygons, as shown in **FIGURE 16-11**.

Enhance Your Understanding   (Answers given at the end of the chapter)

3. The following systems consist of a metal rod with a given initial length. The rods are subjected to the indicated increases in temperature. Rank the systems in order of increasing change in length. Indicate ties where appropriate. (Refer to Table 16-1 for the coefficients of thermal expansion.)

System	A	B	C	D
Metal	aluminum	steel	steel	aluminum
Initial length	2 m	2 m	1 m	1 m
Temperature increase	40 C°	20 C°	30 C°	10 C°

Section Review

- Most substances expand with an increase in temperature, as described by the equation $\Delta L = \alpha L_0 \Delta T$. Water from 0 °C to 4 °C is an important exception.

16-4 Heat and Mechanical Work

As mentioned previously, heat is the energy *transferred* from one object to another. At one time it was thought—erroneously—that an object contained a certain amount of "fluid," or caloric, that could flow from one place to another. This idea was overturned by the observations of Benjamin Thompson (1753–1814), also known as Count Rumford, the American-born physicist, spy, and social reformer who at one point in his eclectic career supervised the boring of cannon barrels by large drills. He observed that as long as mechanical work was done to turn the drill bits, they continued to produce heat in unlimited quantities. Clearly, the unlimited heat observed in boring the cannons was not present initially in the metal, but instead was produced by continually turning the drill bit.

With this observation, it became clear that heat was simply another form of energy transfer that must be taken into account when applying conservation of energy. For example, if you rub sandpaper back and forth over a piece of wood, you do work. The energy associated with that work is not lost; instead, it produces an increase in temperature. Taking into account the energy associated with this temperature change, we find that energy is indeed conserved. In fact, no observation has ever indicated a situation in which energy is not conserved.

The Equivalence Between Work and Heat   The equivalence between work and heat was first explored quantitatively by James Prescott Joule (1818–1889), the British physicist. In one of his experiments, Joule observed the increase in temperature in a device similar to that shown in FIGURE 16-12. Here, a total mass $2m$ falls through a certain distance h, during which gravity does the mechanical work, $2mgh$. As the mass falls, it turns the paddles in the water, which results in a slight warming of the water. By measuring the mechanical work, $2mgh$, and the increase in the water's temperature, ΔT, Joule was able to show that energy was indeed conserved—it had been converted from gravitational potential energy into an increased thermal energy of the water, as indicated by a higher temperature. Joule's experiments established the precise amount of mechanical work that has the same effect as a given transfer of heat.

Before Joule's work, heat was measured in a unit called the calorie (cal). In particular, one kilocalorie (kcal) was defined as the amount of heat needed to raise the temperature of 1 kg of water from 14.5 °C to 15.5 °C. With his experiments, Joule was able to show that 1 kcal = 4186 J, or equivalently, that one calorie of heat transfer is the equivalent of 4.186 J of mechanical work. This is referred to as the **mechanical equivalent of heat**:

The Mechanical Equivalent of Heat

1 cal = 4.186 J   16-8

SI unit: J

Other Units of Heat   In studies of nutrition a different calorie is used. It is the Calorie, with a capital C, and it is simply a kilocalorie; that is, 1 C = 1 kcal. Perhaps this helps people to feel a little better about their calorie intake. After all, a 250-C candy bar sounds a lot better than a 250,000-cal candy bar. They are equivalent, however.

Another common unit for measuring heat is the British thermal unit (Btu). By definition, a Btu is the energy required to heat 1 lb of water from 63 °F to 64 °F. In terms of calories and joules, a Btu is as follows:

$$1 \text{ Btu} = 0.252 \text{ kcal} = 1055 \text{ J} \qquad \text{16-9}$$

Finally, we shall use the symbol Q to denote heat:

Heat, Q

Q = heat = energy transferred due to temperature differences   16-10

SI unit: J

Using the mechanical equivalent of heat as the conversion factor, we will typically give heat in either calories or joules.

▲ FIGURE 16-12 **The mechanical equivalent of heat** A device of this type was used by James Joule to measure the mechanical equivalent of heat.

Big Idea **3** Mechanical work can be converted into an equivalent amount of heat. For example, the work done by friction on a sliding object increases the temperature of the object.

PHYSICS IN CONTEXT
Looking Back

The concept of work (Chapter 7) plays a key role in this section, where we relate mechanical work to its equivalent in the form of heat.

PHYSICS IN CONTEXT
Looking Ahead

Heat plays an important role in Chapter 18, where we consider the heat and work exchanged during different types of thermal processes.

EXAMPLE 16-9   LIFT THAT BALE!

A farmhand lifts a 23.5-kg bale of hay to a height H = 1.1 m to get it into the back of a flatbed truck. How many bales of hay can the farmhand lift using the energy provided by a thick, rich, 485-C chocolate milkshake?

PICTURE THE PROBLEM
Our sketch shows the farmhand lifting a bale of hay of mass m from the ground to the height H to get it into the truck. Work must be done against gravity to lift the bale, and the energy for that work is provided by a milkshake.

REASONING AND STRATEGY
We know that the energy provided by the milkshake is equivalent to the heat Q = 485,000 cal. This energy can be converted to joules by using the relationship 1 cal = 4.186 J. Finally, we set the energy of the milkshake equal to the work done against gravity in lifting N bales of hay to a height H; that is, $Q = N(mgH)$. We can solve this for N.

CONTINUED

Known Energy provided by milkshake, $Q = 485,000$ cal; mass of hay bale, $m = 23.5$ kg; height of lift, $H = 1.1$ m.
Unknown Number of hay bales that can be lifted, $N = ?$

SOLUTION

1. Convert the energy of the milkshake to joules:

$$Q = 485,000 \text{ cal} = 485,000 \text{ cal}\left(\frac{4.186 \text{ J}}{1 \text{ cal}}\right) = 2.03 \times 10^6 \text{ J}$$

2. Equate the energy of the milkshake to the work done against gravity in lifting N hay bales to a height H:

$$Q = N(mgH)$$

3. Solve for the number of bales N and substitute numerical values:

$$N = \frac{Q}{mgH} = \frac{2.03 \times 10^6 \text{ J}}{(23.5 \text{ kg})(9.81 \text{ m/s}^2)(1.1 \text{ m})} = 8000$$

INSIGHT

8000 bales is a lot of hay! Even assuming a metabolic efficiency of 25%, the milkshake could give the energy to lift 2000 bales of hay. Of course, energy is also expended in walking from bale to bale, in bending over and standing up, and other physical activities. The point, however, is that the chemical energy stored in food is large compared with the mechanical energy you expend while doing everyday tasks.

PRACTICE PROBLEM

If the farmhand lifts 100 bales of hay into the truck, how many Calories are burned? Assume 25% metabolic efficiency. [**Answer: 24 C**]

Some related homework problems: Problem 29, Problem 31

Enhance Your Understanding   (Answers given at the end of the chapter)

4. Your friend claims that stirring a glass of water with a spoon, as in **FIGURE 16-13**, increases the temperature of the water. Is she correct? Explain.

Section Review

• Doing 4.186 J of mechanical work on a system, and converting that work to an increase in temperature, is equivalent to heating the system by adding 1 cal of thermal energy.

▲ **FIGURE 16-13** Enhance Your Understanding 4.

16-5 Specific Heats

In the previous section, we mentioned that it takes 4186 J of heat to raise the temperature of 1 kg of water by 1 C°. We have to be clear about the fact that we are heating water, however, because the heat required for a 1 C° increase varies considerably from one substance to another. For example, it takes only 128 J of heat to increase the temperature of 1 kg of lead by 1 C°. In general, the heat required for a given increase in temperature is given by the **heat capacity** of a substance.

Heat Capacity

Suppose we add the heat Q to a given object, and its temperature increases by the amount ΔT. The heat capacity, C, of this object is defined as follows:

Definition of Heat Capacity, *C*

$$C = \frac{Q}{\Delta T} \qquad\qquad 16\text{-}11$$

SI unit: J/K = J/C°

The units of heat capacity are joules per kelvin (J/K). Because the degree size is the same for the Kelvin and Celsius scales, C can also be expressed in units of joules per Celsius degree (J/C°). (Note that the heat capacity is written with an italic capital C, as opposed to the roman capital C used for degrees Celsius.)

Heating Water and Aluminum

The name "heat capacity" is perhaps a bit unfortunate. It derives from the mistaken idea of a "heat fluid," mentioned in the previous section. Objects were imagined to "contain" a certain amount of this nonexistent fluid. Today, we know that an object can readily gain or release heat when it is in thermal contact with other objects—objects cannot be thought of as holding a certain amount of heat.

Instead, the heat capacity should be viewed as the amount of heat necessary for a given temperature change. An object with a large heat capacity, like water, requires a large amount of heat for each increment in temperature. Just the opposite is true for an object with a small heat capacity, like a piece of lead.

To find the heat required for a given ΔT, we simply rearrange $C = Q/\Delta T$ to solve for Q. This yields

$$Q = C\Delta T \qquad\qquad \text{16-12}$$

It should be noted that the *heat capacity is always positive*—just like a speed. Thus, the relationship $Q = C\Delta T$ implies that the heat Q and the temperature change ΔT have the same sign. This observation leads to the following sign conventions for Q:

Q is positive if ΔT is positive—that is, if heat is *added* to a system.

Q is negative if ΔT is negative—that is, if heat is *removed* from a system.

EXERCISE 16-10 TEMPERATURE CHANGE

The heat capacity of 1.00 kg of iron is 448 J/K. **(a)** What is the temperature change of the iron if 493 J of heat is added to the system? **(b)** How much heat must be removed from the iron for its temperature to drop by 2.68 K?

REASONING AND SOLUTION
a. Heat, heat capacity, and temperature change are related by $Q = C\Delta T$. Solving for the change in temperature, we find

$$\Delta T = \frac{Q}{C} = \frac{493\,\text{J}}{448\,\text{J/K}} = 1.10\,\text{K}$$

b. Heat is removed in this case, and hence Q is negative. Substituting $\Delta T = -2.68$ K in $Q = C\Delta T$, we find

$$Q = C\Delta T = (448\,\text{J/K})(-2.68\,\text{K}) = -1200\,\text{J}$$

Big Idea **4** Different substances are characterized by their specific heats; that is, they require different amounts of heat for a given increase in temperature.

Specific Heat
It takes 4186 J to increase the temperature of one kilogram of water by one degree Celsius, and hence it takes twice that much to make the same temperature change in two kilograms of water, and so on. Thus, the heat capacity varies not only with the type of substance, but also with the mass of the substance.

It is useful, then, to define a new quantity—the **specific heat**, c—that depends only on the substance, and not on its mass. This definition is as follows:

Definition of Specific Heat, c

$$c = \frac{Q}{m\Delta T} \qquad\qquad \text{16-13}$$

SI unit: $J/(kg\cdot K) = J/(kg\cdot C°)$

Thus, for example, the specific heat of water is

$$c_{\text{water}} = 4186\,\text{J}/(\text{kg}\cdot\text{K})$$

Specific heats for common substances are listed in Table 16-2. Notice that the specific heat of water is by far the largest of any common material. This is just another of the many unusual properties of water. Having such a large specific heat means that water can give off or take in large quantities of heat with little change in temperature. It is for this reason that if you take a bite of a pie that is just out of the oven, you are much

TABLE 16-2 Specific Heats at Atmospheric Pressure

Substance	Specific heat, c [J/(kg·K)]
Water	4186
Ice	2090
Steam	2010
Beryllium	1820
Air	1004
Aluminum	900
Glass	837
Silicon	703
Iron (steel)	448
Copper	387
Silver	234
Gold	129
Lead	128

▲ **FIGURE 16-14** These two blocks of metal (aluminum on the left and lead on the right) have equal volumes. In addition, they were heated to equal temperatures before being placed on the block of paraffin wax. Notice, however, that the aluminum melted more wax—and hence gave off more heat—even though the lead block is about four times heavier than the aluminum block. The reason is that lead has a very small specific heat; in fact, lead's specific heat is about seven times smaller than the specific heat of aluminum, as we see in Table 16-2. As a result, even this relatively large mass of lead melts considerably less wax per degree of temperature change than the lightweight aluminum—a direct visual illustration of lead's low specific heat.

PROBLEM-SOLVING NOTE

Heat Flow and Thermal Equilibrium

To find the temperature of thermal equilibrium when two objects with different temperatures are brought into thermal contact, we simply use the idea that the heat that flows *out* of one of the objects flows *into* the other object.

 Water Balloon Held over
Candle Flame

PHYSICS IN CONTEXT
Looking Back

Energy and energy conservation (Chapter 8) are important in our understanding of specific heats.

more likely to burn your tongue on the fruit filling (which has a high water content) than on the much drier crust. An illustration of the difference in specific heats of aluminum and lead is shown in **FIGURE 16-14**.

RWP Water's unusually large specific heat also accounts for the moderate climates in regions near great bodies of water. In particular, the enormous volume and large heat capacity of an ocean serve to maintain a nearly constant temperature in the water, which in turn acts to even out the temperature of adjacent coastal areas. For example, the West Coast of the United States benefits from the moderating effect of the Pacific Ocean, aided by the prevailing breezes that come from the ocean onto the coastal regions. In the Midwest, on the other hand, temperature variations can be considerably greater as the land (with a relatively small specific heat) quickly heats up in the summer and cools off in the winter.

Calorimetry

Let's use the specific heat to solve a practical problem. Suppose a block of mass m_b, specific heat c_b, and initial temperature T_b is dropped into a **calorimeter** (basically, a lightweight, insulated flask) containing water. If the water has a mass m_w, a specific heat c_w, and an initial temperature T_w, find the final temperature of the block and the water. Assume that the calorimeter is light enough that it can be ignored, and that no heat is transferred from the calorimeter to its surroundings.

There are two basic ideas involved in solving this problem:

1. The total energy of the system is conserved.

2. The system is in thermal equilibrium at the end, and hence the final temperature of the block is equal to the final temperature of the water.

In particular, the first condition means that the amount of heat lost by the block is equal to the heat gained by the water—or vice versa if the water's initial temperature is higher.

Mathematically, we can write condition 1 as

$$Q_b + Q_w = 0 \qquad\qquad 16\text{-}14$$

This means that the heat flow from the block is equal and opposite to the heat flow from the water; in other words, energy is conserved. If we write the heats Q in terms of the specific heats and temperatures, and let the common final temperature be T, to satisfy condition 2, we have

$$m_b c_b(T - T_b) + m_w c_w(T - T_w) = 0$$

Notice that for each heat we write the change in temperature as T_{final} minus $T_{initial}$, as it should be. Solving for the final temperature, T, we find

$$T = \frac{m_b c_b T_b + m_w c_w T_w}{m_b c_b + m_w c_w} \qquad\qquad 16\text{-}15$$

This result need not be memorized, of course. Whenever solving a problem of this sort, one simply writes down energy conservation and solves for the desired unknown.

In some cases, we may wish to consider the influence of the container itself. This is illustrated in the following Quick Example.

QUICK EXAMPLE 16-11 FIND THE FINAL TEMPERATURE

Suppose 54 g of water at 35 °C are poured into a 65-g aluminum cup with an initial temperature of 11 °C. Find the final temperature of the system, assuming no heat is exchanged with the surroundings.

REASONING AND SOLUTION

We apply energy conservation—setting the heat flow out of the water equal to the heat flow into the aluminum—and solve for the common final temperature.

1. Write an expression for the heat flow out of the water:

$$Q_w = m_w c_w(T - T_w)$$

CONTINUED

2. Write an expression for the heat flow into the aluminum:

$$Q_a = m_a c_a (T - T_a)$$

3. Apply energy conservation:

$$Q_w + Q_a = 0$$

4. Solve for the final temperature, T:

$$T = \frac{m_a c_a T_a + m_w c_w T_w}{m_a c_a + m_w c_w} = 30\,°C$$

As one might expect from water's large specific heat (five times that of aluminum), the final common temperature T is much closer to the initial temperature of the water (T_w) than to that of the aluminum (T_a), even though the mass of the aluminum cup is greater than the mass of the water.

EXAMPLE 16-12 | COOLING OFF

A 0.50-kg block of metal with an initial temperature of 54.5 °C is dropped into a container holding 1.1 kg of water at 20.0 °C. If the final temperature of the block–water system is 21.4 °C, what is the specific heat of the metal block? Assume the container can be ignored, and that no heat is exchanged with the surroundings.

PICTURE THE PROBLEM

Initially, when the block is first dropped into the water, the temperatures of the block and water are $T_b = 54.5\,°C$ and $T_w = 20.0\,°C$, respectively. When thermal equilibrium is established, both the block and the water have the same temperature, $T = 21.4\,°C$.

REASONING AND STRATEGY

Heat flows from the block to the water. Setting the heat flow *out of the block* plus the heat flow *into the water* equal to zero (conservation of energy) yields the block's specific heat.

Initial conditions	Equilibrium

Known Mass of metal block, $m_b = 0.50$ kg; initial temperature of block, $T_b = 54.5\,°C$; mass of water, $m_w = 1.1$ kg; initial temperature of water, $T_w = 20.0\,°C$; final temperature of system, $T = 21.4\,°C$.

Unknown Specific heat of metal block, $c_b = ?$

SOLUTION

1. Write an expression for the heat flow out of the block. Notice that Q_{block} is negative, since T is less than T_b:

$$Q_{block} = m_b c_b (T - T_b)$$

2. Write an expression for the heat flow into the water. Notice that Q_{water} is positive, since T is greater than T_w:

$$Q_{water} = m_w c_w (T - T_w)$$

3. Set the sum of the heats equal to zero:

$$Q_{block} + Q_{water} = m_b c_b (T - T_b) + m_w c_w (T - T_w) = 0$$

4. Solve for the specific heat of the block, c_b

$$c_b = \frac{m_w c_w (T - T_w)}{m_b (T_b - T)}$$

5. Substitute numerical values:

$$c_b = \frac{(1.1\ \text{kg})[4186\ \text{J/(kg·K)}](21.4\,°C - 20.0\,°C)}{(0.50\ \text{kg})(54.5\,°C - 21.4\,°C)}$$

$$= 390\ \text{J/(kg·K)}$$

INSIGHT

From Table 16-2 we can see that the block is probably made of copper.

In addition, notice that the final temperature is much closer to the initial temperature of the water than to the initial temperature of the block. This is due in part to the fact that the mass of the water is about twice that of the block; more important, however, is the fact that the water's specific heat is more than ten times greater than that of the block. In the following Practice Problem we set the mass of the water equal to the mass of the block, so that we can see clearly the effect of the different specific heats.

PRACTICE PROBLEM

If the mass of the water is also 0.50 kg, what is the equilibrium temperature? [**Answer:** $T = 23\,°C$—still much closer to the water's initial temperature than to the block's]

Some related homework problems: Problem 40, Problem 41

Enhance Your Understanding   (Answers given at the end of the chapter)

5. Objects A and B have the same mass. When they receive the same amount of thermal energy, the temperature of object A increases more than the temperature of object B. Is the specific heat of object A greater than, less than, or equal to the specific heat of object B? Explain.

Section Review

- Heat capacity, $C = Q/\Delta T$, is the amount of heat per temperature change. Specific heat, $c = Q/(m\Delta T)$, is the amount of heat per mass per temperature change.
- Calorimetry uses energy conservation to determine the final temperature of a system.

16-6 Conduction, Convection, and Radiation

Big Idea 5 Thermal energy can be transferred in a number of ways, ranging from direct contact (conduction), to physical movement (convection), to transfer in the form of light and infrared waves (radiation).

Heat can be exchanged in a variety of ways. The Sun, for example, warms the Earth from across 93 million miles of empty space by a process known as radiation. As the sunlight strikes the ground and raises its temperature, the ground-level air gets warmer and begins to rise, producing a further exchange of heat by means of convection. Finally, if you walk across the ground in bare feet, you will feel the warming effect of heat entering your body by conduction. In this section we consider each of these three mechanisms of heat exchange in detail.

Conduction

Perhaps the most familiar form of heat exchange is **conduction,** which is the flow of heat directly through a physical material. For example, if you hold one end of a metal rod and put the other end into a fire, it doesn't take long before you begin to feel warmth on your end. The heat you feel is transported along the rod by conduction.

Let's consider this observation from a microscopic point of view. To begin, when you placed one end of the rod into the fire, the high temperature at that location caused the molecules to vibrate with an increased amplitude. These molecules in turn jostle their neighbors and cause them to vibrate with greater amplitude as well. Eventually, the effect travels from molecule to molecule across the length of the rod, resulting in the macroscopic phenomenon of conduction.

If you were to repeat the experiment, this time with a wooden rod, the hot end of the rod would heat up so much that it might even catch on fire, but your end would still be comfortably cool. Thus, conduction depends on the type of material involved. Some materials, called **conductors,** conduct heat very well, whereas other materials, called **insulators,** conduct heat poorly.

Just how much heat flows as a result of conduction? To answer this question we consider the simple system shown in FIGURE 16-15. Here we show a rod of length L and cross-sectional area A, with one end at the temperature T_1 and the other at the temperature $T_2 > T_1$. Experiments show that the amount of heat Q that flows through the rod has these properties:

- Q increases in proportion to the rod's cross-sectional area, A.
- Q increases in proportion to the temperature difference, $\Delta T = T_2 - T_1$.
- Q increases steadily with time, t.
- Q decreases with the length of the rod, L.

Combining these observations in a mathematical expression gives the following equation:

Heat Flow by Conduction

$$Q = kA\left(\frac{\Delta T}{L}\right)t$$

16-16

T_1 T_2

Cross-sectional area A

▲ **FIGURE 16-15 Heat conduction through a rod** The amount of heat that flows through a rod of length L and cross-sectional area A per time is proportional to $A(T_2 - T_1)L$.

PROBLEM-SOLVING NOTE

Area and Length in Heat Conduction

When applying Equation 16-16, note that A is the area through which the heat flows. Thus, the plane of the area A is perpendicular to the direction of heat flow. The length L is the distance from the high-temperature side of an object to its low-temperature side.

TABLE 16-3 Thermal Conductivities

Substance	Thermal conductivity, $k[W/(m \cdot K)]$
Silver	417
Copper	395
Gold	291
Aluminum	217
Steel, low carbon	66.9
Lead	34.3
Stainless steel—alloy 302	16.3
Ice	1.6
Concrete	1.3
Glass	0.84
Water	0.60
Asbestos	0.25
Wood	0.10
Wool	0.040
Air	0.0234

◄ **FIGURE 16-16 Visualizing Concepts**
Conduction Maintaining proper body temperature in an environment that is often too hot or too cold is a problem for many animals. When the sand is blazing hot, this lizard (top) keeps its contact with the ground to a minimum. By standing on two legs instead of four, it reduces conduction of heat from the ground to its body. Polar bears (bottom) have the opposite problem. The loss of precious body heat to their surroundings is retarded by their thick fur, which is actually made up of hollow fibers. Air trapped within these fibers provides enhanced insulation, just as it does in our thermal blankets and double-paned windows.

PHYSICS IN CONTEXT
Looking Ahead

The flux of heat transferred between two objects in thermal contact is a useful analogy in understanding the flux of electric fields in Chapter 19 and magnetic fields in Chapter 23.

The constant k is referred to as the **thermal conductivity** of the rod. It varies from material to material, as indicated in Table 16-3. An illustration of animals that deal with thermal conduction in different ways is shown in **FIGURE 16-16**.

CONCEPTUAL EXAMPLE 16-13   THE FEEL OF TILE IN THE MORNING

You get up in the morning and walk barefoot from the bedroom to the bathroom. In the bedroom you walk on carpet, but in the bathroom the floor is tile. Does the tile feel warmer, cooler, or the same temperature as the carpet?

REASONING AND DISCUSSION
Everything in the house is at the same temperature, so it might seem that the carpet and the tile would feel the same. As you probably know from experience, however, the tile feels cooler. The reason is that tile has a much larger thermal conductivity than carpet, which is actually a fairly good insulator. As a result, more heat flows from your skin to the tile than from your skin to the carpet. To your feet, then, it is as if the tile were much cooler than the carpet.

To get an idea of the thermal conductivities that would apply in this case, let's examine Table 16-3. For the tile, we might expect a thermal conductivity of roughly 0.84, the value for glass. For the carpet, the thermal conductivity might be as low as 0.04, the thermal conductivity of wool. Thus, the tile could have a thermal conductivity that is 20 times larger than that of the carpet.

ANSWER
The tile feels cooler.

Thermal conductivity is an important consideration when insulating a home. We consider some of these issues in the next Example.

EXAMPLE 16-14   WHAT A PANE!

One of the windows in a house has the shape of a square 1.0 m on a side. The glass in the window is 0.50 cm thick, and has a coefficient of thermal conductivity of $k = 0.84$ W/(m·K). **(a)** How much heat is lost through this window in one day if the temperature difference between the inside and outside of the window is 21 C°? **(b)** Suppose all the dimensions of the window—height, width, and thickness—are doubled. If everything else remains the same, by what factor does the heat flow change?

CONTINUED

PICTURE THE PROBLEM

The glass from the window is shown in our sketch, along with its relevant dimensions. Heat flows across a temperature difference of 21 C° through a length equal to the thickness of the glass, $L = 0.50$ cm.

REASONING AND STRATEGY

a. The heat flow is given by $Q = kA(\Delta T/L)t$ (Equation 16-16). The area is $A = (1.0 \text{ m})^2$ and the length over which heat is conducted—in this case the thickness of the glass—is $L = 0.0050$ m. The temperature difference is $\Delta T = 21$ C° = 21 K. Also, recall from Section 7-4 that 1 W = 1 J/s.

b. Doubling all dimensions increases the thickness by a factor of two and increases the area by a factor of four; that is, $L \rightarrow 2L$ and $A \rightarrow (2 \times \text{height}) \times (2 \times \text{width}) = 4A$. We can use these results in $Q = kA(\Delta T/L)t$.

Known Area of window, $A = (1.0 \text{ m})^2$; thickness of glass, $L = 0.50$ cm; temperature difference across window, $\Delta T = 21$ K; thermal conductivity of glass, $k = 0.84$ W/(m·K).

Unknown (a) Heat lost during one day, $Q = ?$ (b) Heat lost when dimensions are doubled, $Q = ?$

SOLUTION

Part (a)

1. Calculate the heat flow for a given time, t:

$$Q = kA\left(\frac{\Delta T}{L}\right)t$$

$$= [0.84 \text{ W/(m·K)}](1.0 \text{ m})^2\left(\frac{21 \text{ K}}{0.0050 \text{ m}}\right)t = (3500 \text{ W})t$$

2. Substitute the number of seconds in a day (86,400 s) for the time t:

$$Q = (3500 \text{ W})t = (3500 \text{ W})(86,400 \text{ s}) = 3.0 \times 10^8 \text{ J}$$

Part (b)

3. Replace L with $2L$ and A with $4A$ in Step 1. The result is a doubling of the heat flow, Q:

$$Q = kA\left(\frac{\Delta T}{L}\right)t \rightarrow k(4A)\left[\frac{\Delta T}{(2L)}\right]t \rightarrow 2\left[kA\left(\frac{\Delta T}{L}\right)t\right] = 2Q$$

INSIGHT

Q is a sizable amount of heat, roughly equivalent to the energy released in burning a gallon of gasoline. A considerable reduction in heat loss can be obtained by using a double-paned window, which has an insulating layer of air (actually argon or krypton) sandwiched between the two panes of glass. This is discussed in more detail later in this section, and is explored in Homework Problems 56 and 90.

PRACTICE PROBLEM

Suppose the window is replaced with a plate of solid silver. How thick must this plate be to have the same heat flow in a day as the glass? [**Answer:** $L = 2.5$ m]

Some related homework problems: Problem 53, Problem 55

Next, we consider the heat flow through a combination of two different materials with different thermal conductivities.

CONCEPTUAL EXAMPLE 16-15 COMPARE THE HEAT FLOW
PREDICT/EXPLAIN

Two metal rods are to be used to conduct heat from a region at 100 °C to a region at 0 °C. The rods can be placed in parallel, as shown on the left, or in series, as on the right. **(a)** Is the heat conducted in the parallel arrangement greater than, less than, or the same as the heat conducted with the rods in series? **(b)** Which of the following is the *best explanation* for your prediction?

I. The parallel arrangement conducts more heat because the cross-sectional area through which heat flows is greater.

II. More heat is conducted in series because the heat that flows through one rod also flows through the other rod.

III. The heat conducted is the same in both cases because the heat has to flow through the same two rods.

CONTINUED

REASONING AND DISCUSSION

The parallel arrangement conducts more heat for two reasons. First, the cross-sectional area available for heat flow is twice as great for the parallel rods. A greater cross-sectional area gives a greater heat flow—everything else being equal. Second, more heat flows through each rod in the parallel configuration because they both have the full temperature difference of 100 C° between their ends. In the series configuration, each rod has a smaller temperature difference between its ends, so less heat flows.

ANSWER

(a) More heat is conducted when the rods are in parallel. (b) The best explanation is I.

In the next Example, we consider the case of heat conduction when two rods are placed in parallel. In the homework we shall consider the corresponding case of the same two rods conducting heat in series.

EXAMPLE 16-16 PARALLEL RODS

Two 0.525-m rods, one lead and the other copper, are connected between metal plates held at 2.00 °C and 106 °C. The rods have a square cross section, 1.50 cm on a side. How much heat flows through the two rods in 1.00 s? Assume that no heat is exchanged between the rods and the surroundings, except at the ends.

PICTURE THE PROBLEM

The lead and the copper rods, each 0.525 m long, are shown in the sketch. Both rods have a temperature difference of 104 C° = 104 K between their ends.

REASONING AND STRATEGY

The heat flowing through each rod can be calculated using $Q = kA(\Delta T/L)t$ and the value of k given in Table 16-3. The total heat flow is simply the sum of that calculated for each rod.

Known   Length of rods, $L = 0.525$ m; metal in top rod = lead; metal in bottom rod = copper; temperature difference across rods, $\Delta T = 104$ K; cross-sectional area of each rod, $A = (0.0150\ \text{m})^2$; time during which heat flows, $t = 1.00$ s.

Unknown Total amount of heat that flows, $Q = ?$

SOLUTION

1. Calculate the heat flow in 1.00 second through the lead rod:

$$Q_{\text{lead}} = k_{\text{lead}}A\left(\frac{\Delta T}{L}\right)t$$

$$= [34.3\ \text{W/(m·K)}](0.0150\ \text{m})^2\left(\frac{104\ \text{K}}{0.525\ \text{m}}\right)(1.00\ \text{s}) = 1.53\ \text{J}$$

2. Calculate the heat flow in 1.00 second through the copper rod:

$$Q_{\text{copper}} = k_{\text{copper}}A\left(\frac{\Delta T}{L}\right)t$$

$$= [395\ \text{W/(m·K)}](0.0150\ \text{m})^2\left(\frac{104\ \text{K}}{0.525\ \text{m}}\right)(1.00\ \text{s}) = 17.6\ \text{J}$$

3. Sum the heats found in Steps 1 and 2 to get the total heat:

$$Q_{\text{total}} = Q_{\text{lead}} + Q_{\text{copper}} = 1.53\ \text{J} + 17.6\ \text{J} = 19.1\ \text{J}$$

INSIGHT

As our results show, the copper rod is by far the better conductor of heat. It is also a very good conductor of electricity, as we shall see in Chapter 21.

PRACTICE PROBLEM

What temperature difference would be required for the total heat flow in 1.00 second to be 15.0 J?

[**Answer:** $\Delta T = 81.5$ C° = 81.5 K]

Some related homework problems: Problem 57, Problem 58

As we shall see in Homework Problem 59, the heat conducted by the same two rods connected in series is only 1.41 J, which is less than either of the two rods conducts when placed in parallel. This verifies the conclusion reached in Conceptual Example 16-15.

▲ **FIGURE 16-17** Many wading birds use countercurrent exchange in their circulatory systems. This mechanism allows them to keep the temperature of their legs well below that of their bodies. In this way they reduce the conductive loss of body heat to the water.

Candle Chimneys

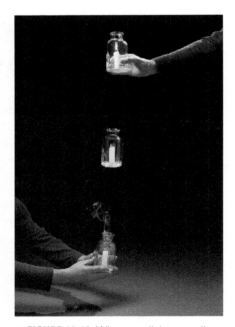

▲ **FIGURE 16-19** When you light a candle, it heats the air near the flame as it burns. Because hot air is less dense—and hence more buoyant—than cool air, a circulation pattern is established with hot air rising and being replaced from below by cool, oxygenated air. Thus, convection is necessary for a candle to continue burning. When the burning candle in this jar is dropped, it suddenly finds itself in free fall—an essentially "weightless" environment where buoyancy has no effect. As a result, convection ceases and the flame is quickly extinguished as it consumes all the oxygen in its immediate vicinity.

RWP An application of thermal conductivity in series is the *insulated window*. Most homes today have insulated windows as a means of increasing their energy efficiency. If you look closely at one of these windows, you will see that it is actually constructed from two panes of glass separated by a gas-filled gap—usually argon or krypton gas. Thus, heat flows through three different materials in series as it passes into or out of a home. The fact that the thermal conductivity of a gas is about 40 times smaller than that of glass means that the insulated window results in significantly less heat flow than would be experienced with a single pane of glass.

BIO As a final example of conduction, we note that many biological systems transfer heat by a mechanism known as *countercurrent exchange*. Consider, for example, a wading bird (**FIGURE 16-17**) standing in cool water all day. As warm blood flows through the arteries on its way to the legs and feet of the bird, it passes through constricted regions where the legs join to the body. Here, where the arteries and veins are packed closely together, the body-temperature arterial blood flowing into the legs transfers heat to the much cooler venous blood returning to the body. Thus, the counter-flowing streams of blood serve to maintain the core body temperature of the bird, while at the same time keeping the legs and feet at much cooler temperatures. The feet still receive the oxygen and nutrients carried by the blood, but they stay at a relatively low temperature to reduce the amount of heat lost to the water.

BIO Similar effects occur in humans. It is common, for example, to hear complaints that a person's hands or feet are cold. There is good reason for this, because they are in fact much cooler than the core body temperature. Just as with the wading birds, the warm arterial blood flowing to the hands and feet exchanges heat with the cool venous blood flowing in the opposite direction (**FIGURE 16-18**). This helps to reduce the heat loss to our surroundings, and to maintain the desired temperature in the core of the body.

Vein

Cold blood returning to trunk · 23 °C · 30 °C · 36 °C

Heat transfer

To capillaries · 24 °C · 31 °C · 37 °C · Warm blood from trunk

Artery

▶ **FIGURE 16-18 Countercurrent heat exchange in the human arm** Arteries bringing warm blood to the limbs lie close to veins returning cooler blood to the body. This arrangement assures that a temperature difference (gradient) is maintained over the entire length that the vessels run parallel to one another, maximizing heat exchange between the warm arterial blood and the cooler venous blood.

Convection

Suppose you want to heat a small room. To do so, you bring a portable electric heater into the room and turn it on. As the heating coils get red hot, they heat the air in their vicinity; as this air warms, it expands, becoming less dense. Because of its lower density, the warm air rises, to be replaced by cold, dense air descending from overhead. This sets up a circulating flow of air that transports heat from the heating coils to the air throughout the room. Heat exchange of this type is referred to as **convection**. Convection is also important in the burning of a candle, as illustrated in **FIGURE 16-19**, where we also show what happens when convection is "shut off."

In general, convection occurs when a fluid is unevenly heated. As with the room heater, the warm portions of the fluid rise because of their lower density and the cool portions sink because of their higher density. Thus, in convection, temperature differences result in a flow of fluid. It is this physical flow of matter that carries heat throughout the system.

Convection occurs on an enormous range of length scales. For example, the same type of uneven heating produced by an electric heater in a room can occur in the atmosphere of the Earth as well. The common seashore occurrence of sea breezes during the day and land breezes in the evening is one such example. (See FIGURE 16-20 for an illustration of this effect.) On a larger scale, the Sun causes greater warming near the equator of the Earth than near the poles; as a result, warm equatorial air rises, cool polar air descends, and global convection patterns are established. Similar convection patterns occur in ocean waters, and plate tectonics is believed to be caused, at least in part, by convection currents in the Earth's mantle. The Sun also has convection currents, due to the intense heating that occurs in its interior, and disturbances in these currents are often visible as sunspots.

Radiation

Convection and conduction occur primarily in specific situations, but *all* objects give off energy as a result of **radiation.** The energy radiated by an object is in the form of electromagnetic waves (Chapter 25), which include visible light as well as infrared and ultraviolet radiation. Thus, unlike convection and conduction, radiation has no need for a physical material to mediate the energy transfer, because electromagnetic waves can propagate through empty space—that is, through a vacuum. Therefore, the heat you feel radiated from a hot furnace would reach you even if the air were suddenly removed—just as radiant energy from the Sun reaches the Earth across 150 million kilometers of vacuum. Examples of infrared radiation from objects as diverse a house, a baby, and an oil pipeline are shown in FIGURE 16-21.

RWP Because radiation often includes visible light, it's sometimes possible to "see" the temperature of an object. This is the physical basis of the **optical pyrometer,** invented by Josiah Wedgwood (1730–1795), the renowned English potter and relative of Charles Darwin. For instance, when objects are about 800 °C, as in an oven's heating coil or in molten lava, they appear to be a dull "red hot," as shown in FIGURE 16-22 (a). Molten steel at 1500 °C glows yellow, and the 6000 °C plasma created by arc welding is bluish-white hot. Even stars have colors that indicate their surface temperature. For example, red stars like Betelgeuse in FIGURE 16-22 (b) (upper left) are cooler than blue stars like the nearby Rigel (lower right).

The energy radiated per time by an object—that is, the radiated power, P—is proportional to the surface area, A, over which the radiation occurs. It also depends on

▲ FIGURE 16-20 **Alternating land and sea breezes (a)** During the day, the sun warms the land more rapidly than the water. This is because the land, which is mostly rocks, has a lower specific heat than the water. The warm land heats the air above it, which becomes less dense and rises. Cooler air from over the water flows in to take its place, producing a "sea breeze." **(b)** At night, the land cools off more rapidly than the water—again because of its lower specific heat. Now it is the air above the relatively warm water that rises and is replaced by cooler air from over the land, producing a "land breeze."

(a)

(b)

(c)

▲ FIGURE 16-21 **Visualizing Concepts Infrared Radiation** Many everyday objects radiate energy in the infrared part of the electromagnetic spectrum (see Chapter 25), which is invisible to our eyes. **(a)** Careful measurements of the infrared radiation emitted by a house (depicted here in false color) can be used to locate thermal leaks. **(b)** A baby's temperature can be taken easily and unobtrusively with an infrared sensor. **(c)** Infrared radiation is given off by the finned "radiators" attached to the supports of the Alaska pipeline. These fins are designed to prevent melting of the environmentally sensitive permafrost over which the pipeline runs.

(a)

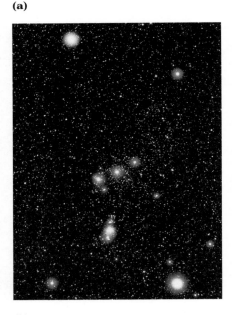

(b)

◄ **FIGURE 16-22** Visualizing Concepts **Visible Radiation (a)** Red-hot volcanic lava is just hot enough to radiate in the visible range. Even when it cools enough to stop glowing, it still emits energy, but most of it is in the form of invisible infrared radiation. **(b)** Stars have colors that are indicative of their surface temperatures. Thus, the red supergiant Betelgeuse (upper left) in the constellation Orion is much cooler than the blue supergiant Rigel (lower right).

the temperature of the object. In fact, the dependence is on the fourth power of the temperature, T^4, where T is the Kelvin-scale temperature. Thus, for instance, if T is doubled, the radiated power increases by a factor of 16. All this behavior is contained in the **Stefan–Boltzmann law:**

Stefan–Boltzmann Law for Radiated Power, *P*

$$P = e\sigma A T^4 \qquad\qquad 16\text{-}17$$

SI unit: W

The constant σ in this expression is a fundamental physical constant, the **Stefan–Boltzmann constant:**

$$\sigma = 5.67 \times 10^{-8}\,\text{W/(m}^2 \cdot \text{K}^4) \qquad\qquad 16\text{-}18$$

The other constant in the Stefan–Boltzmann law is the **emissivity, *e*.** The emissivity is a dimensionless number between 0 and 1 that indicates how effective an object is in radiating energy. A value of 1 means that the object is a perfect radiator. In general, a dark-colored object will have an emissivity near 1, and a light-colored object will have an emissivity closer to 0. People wearing garments with an emissivity near 1 are shown in **FIGURE 16-23.**

EXERCISE 16-17 RADIATED POWER

Calculate the power radiated by a cube with a side length of 7.70 cm at the temperature 339 K (about 151 °F). Assume the emissivity is unity.

REASONING AND SOLUTION
A cube has 6 faces, each with an area of L^2, where L is the length of a side. The total surface area of this cube, then, is $A = 6L^2$, with $L = 0.0770$ m. Substituting A and $T = 339$ K in $P = e\sigma A T^4$ yields

$$P = e\sigma A T^4 = (1)\left[5.67 \times 10^{-8}\,\text{W/(m}^2 \cdot \text{K}^4)\right]6(0.0770\,\text{m})^2(339\,\text{K})^4 = 26.6\,\text{W}$$

Experiments show that objects absorb radiation from their surroundings according to the same law, the Stefan–Boltzmann law, by which they emit radiation. Thus, if the temperature of an object is T, and its surroundings are at the temperature T_s, the *net* power radiated by the object is:

Net Radiated Power, P_{net}

$$P_{net} = e\sigma A(T^4 - T_s^4) \qquad\qquad 16\text{-}19$$

SI unit: W

If the object's temperature is greater than its surroundings, it radiates more energy than it absorbs and P_{net} is positive. On the other hand, if its temperature is lower than the surroundings, it absorbs more energy than it radiates and P_{net} is negative. When the object has the same temperature as its surroundings, it is in equilibrium and the net power is zero.

PROBLEM-SOLVING NOTE

Radiated Power

To correctly calculate the radiated power, Equations 16-17 and 16-19 the temperatures must be expressed in the Kelvin scale.

◄ **FIGURE 16-23** Wear black in the desert? It actually helps—by radiating heat away from the wearer more efficiently.

EXAMPLE 16-18   HUMAN POLAR BEARS

On New Year's Day, several human "polar bears" prepare for their annual dip into the icy waters of Narragansett Bay. One of these hardy souls has a surface area of 1.15 m^2 and a surface temperature of 303 K (~30 °C). Find the net radiated power from this person **(a)** in a dressing room, where the temperature is 293 K (~20 °C), and **(b)** outside, where the temperature is 273 K (~0 °C). Assume an emissivity of 0.900 for the person's skin.

PICTURE THE PROBLEM

Our sketch shows the person radiating power in a room where the surroundings are at 293 K, and outside where the temperature is 273 K. The person also absorbs radiation from the surroundings; hence the net radiated power is greater when the surroundings are cooler.

(a)   **(b)**

REASONING AND STRATEGY

A straightforward application of $P_{net} = e\sigma A(T^4 - T_s^4)$ can be used to find the net power for both parts (a) and (b). The only difference is the temperature of the surroundings, T_s.

Known   Surface area of person, $A = 1.15$ m^2; surface temperature of person, $T = 303$ K; temperature in room, $T_s = 293$ K; temperature outside, $T_s = 273$ K; emissivity of person's skin, $e = 0.900$.

Unknown **(a)** Net radiated power in room, $P_{net} = ?$ **(b)** Net radiated power outside, $P_{net} = ?$

SOLUTION

Part (a)

1. Calculate the net power using $P_{net} = e\sigma A(T^4 - T_s^4)$ (Equation 16-19) and $T_s = 293$ K:

$$P_{net} = e\sigma A(T^4 - T_s^4)$$
$$= (0.900)[5.67 \times 10^{-8}\ \text{W/(m}^2 \cdot \text{K}^4)](1.15\ \text{m}^2)$$
$$\times [(303\ \text{K})^4 - (293\ \text{K})^4]$$
$$= 62.1\ \text{W}$$

Part (b)

2. Calculate the net power using $P_{net} = e\sigma A(T^4 - T_s^4)$ and $T_s = 273$ K:

$$P_{net} = e\sigma A(T^4 - T_s^4)$$
$$= (0.900)[5.67 \times 10^{-8}\ \text{W/(m}^2 \cdot \text{K}^4)](1.15\ \text{m}^2)$$
$$\times [(303\ \text{K})^4 - (273\ \text{K})^4]$$
$$= 169\ \text{W}$$

INSIGHT

In the warm room the net radiated power is roughly that of a small lightbulb (about 60 W); outdoors, the net radiated power has more than doubled, and is comparable to that of a 150-W lightbulb.

PRACTICE PROBLEM

What temperature must the surroundings have if the net radiated power is to be 155 W?   [**Answer:** $T_s = 276$ K ≈ 3 °C]

Some related homework problems: Problem 54, Problem 62

The same emissivity e applies to both the emission and absorption of energy. Thus, a perfect emitter ($e = 1$) is also a perfect absorber. Such an object is referred to as a **blackbody**. As we shall see later, in Chapter 30, the study of blackbody radiation near the turn of the twentieth century ultimately led to one of the most fundamental revolutions in science—the introduction of quantum physics.

RWP The opposite of a blackbody is an ideal reflector, which absorbs *no* radiation ($e = 0$). It follows that an ideal reflector also radiates no energy. This is why the inside of a Thermos bottle is highly reflective. As an almost ideal reflector, the inside of the bottle radiates very little of the energy contained in the hot liquid that it holds. In addition to its shiny interior, a Thermos bottle has a vacuum between its inner and outer walls, as shown in **FIGURE 16-24**. Because convection and conduction cannot occur in a vacuum, this design vastly reduces heat exchange between the contents of the bottle and the surroundings. This type of double-walled insulating container was invented by Sir James Dewar (1842–1923), a Scottish physicist and chemist.

PHYSICS IN CONTEXT
Looking Back

Notice that we have used the concept of power (Chapter 7) in talking about the amount of energy radiated by an object in a given amount of time.

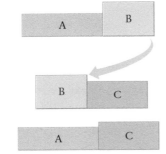

Outer glass wall   Silver film

Inner glass wall

Hot or cold liquid

Stopper

Vacuum

Gasket

▲ **FIGURE 16-24 The Thermos bottle** The hot or cold liquid stored in a Thermos bottle is separated from the outside world by a vacuum between the inner and outer glass walls. In addition, the inner glass wall has a reflective coating so that it is a good reflector and a poor radiator.

Enhance Your Understanding   (Answers given at the end of the chapter)

6. The following systems consist of a cylindrical metal rod with a given length and radius. Rank the systems in order of increasing heat conducted in 1 s, assuming each rod has the same temperature difference between its ends. Indicate ties where appropriate. (Refer to Table 16-3 for thermal conductivities.)

System	A	B	C	D
Metal	lead	copper	silver	aluminum
Length	20 cm	20 cm	10 cm	10 cm
Radius	3 cm	2 cm	1 cm	10 cm

Section Review

- The amount of heat conducted through a substance of length L and cross-sectional area A in the time t is $Q = kA(\Delta T/L)t$. The constant k is the thermal conductivity of the substance, and ΔT is the temperature difference across the substance.

- The net power radiated by an object with surface area A and temperature T in surroundings of temperature T_s is $P_{net} = e\sigma A(T^4 - T_s^4)$. The constant e is the emissivity (a number between 0 and 1), and σ is the Stefan–Boltzmann constant, $\sigma = 5.67 \times 10^{-8}\ \text{W}/(\text{m}^2 \cdot \text{K}^4)$.

CHAPTER 16 REVIEW

CHAPTER SUMMARY

16-1 TEMPERATURE AND THE ZEROTH LAW OF THERMODYNAMICS

This section defines several new terms dealing with heat and temperature.

Heat
Heat is the energy transferred between objects because of a temperature difference.

Thermal Contact
Objects are in thermal contact if heat can flow between them.

Thermal Equilibrium
Objects that are in thermal contact, but have no heat exchange between them, are said to be in thermal equilibrium.

Zeroth Law of Thermodynamics
If objects A and C are in thermal equilibrium with object B, then they are in thermal equilibrium with each other.

16-2 TEMPERATURE SCALES

Temperature is commonly measured in terms of three different scales.

Celsius Scale
In the Celsius scale, water freezes at 0 °C and boils at 100 °C.

Fahrenheit Scale
In the Fahrenheit scale, water freezes at 32 °F and boils at 212 °F.

Absolute Zero
The lowest temperature attainable is referred to as absolute zero. It is impossible to cool an object to a temperature lower than absolute zero, which is −273.15 °C.

Kelvin Scale
In the Kelvin scale, absolute zero is 0 K. In addition, water freezes at 273.15 K and boils at 373.15 K. The degree size is the same for the Kelvin and Celsius scales.

16-3 THERMAL EXPANSION

Most, though not all, substances expand when heated.

Linear Expansion

When an object of length L_0 is heated by the amount ΔT, its length increases by ΔL:

$$\Delta L = \alpha L_0 \Delta T \qquad \text{16-4}$$

The constant α is the coefficient of linear expansion (Table 16-1).

16-4 HEAT AND MECHANICAL WORK

An important step forward in the understanding of heat was the recognition that it is a form of energy.

Mechanical Equivalent of Heat

$$1 \text{ cal} = 4.186 \text{ J} \qquad \text{16-8}$$

16-5 SPECIFIC HEATS

Heat Capacity

The heat capacity of an object is the heat Q divided by the associated temperature change, ΔT:

$$C = \frac{Q}{\Delta T} \qquad \text{16-11}$$

Specific Heat

The specific heat is the heat capacity per unit mass. Thus, the specific heat is independent of the quantity of a material in a given object:

$$c = \frac{Q}{m \Delta T} \qquad \text{16-13}$$

16-6 CONDUCTION, CONVECTION, AND RADIATION

This section considers three common mechanisms of heat exchange.

Conduction

In conduction, heat flows through a material with no bulk motion. The heat flows as a result of the interactions of individual atoms with their neighbors.

If the thermal conductivity of a material is k, its cross-sectional area is A, and its length L, the heat exchanged in the time t is

$$Q = kA\left(\frac{\Delta T}{L}\right)t \qquad \text{16-16}$$

Convection

Convection is heat exchange due to the bulk motion of an unevenly heated fluid.

Radiation

Radiation is the heat exchange due to electromagnetic radiation, such as infrared rays and light.

The energy per time, or power P, radiated by an object with a surface area A at the Kelvin temperature T is

$$P = e\sigma A T^4 \qquad \text{16-17}$$

where e is the emissivity (a constant between 0 and 1) and σ is the Stefan–Boltzmann constant: $\sigma = 5.67 \times 10^{-8} \text{ W/(m}^2 \cdot \text{K}^4)$.

Since an object will also absorb radiation from its surroundings at the temperature T_s, the net radiated power is

$$P_{\text{net}} = e\sigma A(T^4 - T_s^4) \qquad \text{16-19}$$

ANSWERS TO ENHANCE YOUR UNDERSTANDING QUESTIONS

1. Heat flows from high temperature to low temperature, so brick B has the higher temperature.

2. Fahrenheit degrees are "smaller" than Kelvin degrees because there are 180 Fahrenheit degrees between freezing and boiling water, but only 100 Kelvin degrees.

3. D < C < B < A.

4. Yes, part of the mechanical work done in stirring the water goes into increasing its temperature.

5. The temperature increase is inversely proportional to the specific heat; thus the specific heat of object A is less than that of object B.

6. A < C < B < D.

CONCEPTUAL QUESTIONS

For instructor-assigned homework, go to www.masteringphysics.com

(Answers to odd-numbered Conceptual Questions can be found in the back of the book.)

1. A cup of hot coffee is placed on the table. Is it in thermal equilibrium? What condition determines when the coffee is in equilibrium?

2. To find the temperature at the core of the Sun, you consult some Web sites on the Internet. One site says the temperature is about 15 million °C, another says it is 15 million kelvin. Is this a serious discrepancy? Explain.

3. Is it valid to say that a hot object contains more heat than a cold object?

4. If the glass in a glass thermometer had the same coefficient of volume expansion as mercury, the thermometer would not be very useful. Explain.

5. When a mercury-in-glass thermometer is inserted into a hot liquid the mercury column first drops and then rises. Explain this behavior.

6. Sometimes the metal lid on a glass jar has been screwed on so tightly that it is very difficult to open. Explain why holding the lid under hot running water often loosens it enough for easy opening.

7. Two different objects receive the same amount of heat. Give at least two reasons why their temperature changes may not be the same.

8. The specific heat of concrete is greater than that of soil. Given this fact, would you expect a major-league baseball field or the parking lot that surrounds it to cool off more in the evening following a sunny day?

9. When you touch a piece of metal and a piece of wood that are both at room temperature the metal feels cooler. Why?

10. The rate of heat flow through a slab does *not* depend on which of the following? **(a)** The temperature difference between opposite faces of the slab. **(b)** The thermal conductivity of the slab. **(c)** The thickness of the slab. **(d)** The cross-sectional area of the slab. **(e)** The specific heat of the slab.

11. If a lighted match is held beneath a balloon inflated with air, the balloon quickly bursts. If, instead, the lighted match is held beneath a balloon filled with water, the balloon remains intact, even if the flame comes in contact with the balloon. Explain.

12. Updrafts of air allow hawks and eagles to glide effortlessly, all the while gaining altitude. What causes the updrafts?

13. **BIO** The fur of polar bears consists of hollow fibers. (Sometimes algae will grow in the hollow regions, giving the fur a green cast.) Explain why hollow hairs can be beneficial to the polar bears.

14. Object 2 has twice the emissivity of object 1, though they have the same size and shape. If the two objects radiate the same power, what is the ratio of their Kelvin temperatures?

PROBLEMS AND CONCEPTUAL EXERCISES

Answers to odd-numbered Problems and Conceptual Exercises can be found in the back of the book. **BIO** *identifies problems of biological or medical interest;* **CE** *indicates a conceptual exercise.* Predict/Explain *problems ask for two responses: (a) your prediction of a physical outcome, and (b) the best explanation among three provided; and* Predict/Calculate *problems ask for a prediction of a physical outcome, based on fundamental physics concepts, and follow that with a numerical calculation to verify the prediction. On all problems, bullets (•, ••, •••) indicate the level of difficulty.*

SECTION 16-2 TEMPERATURE SCALES

1. • **CE** Predict/Explain The temperature inside a freezer is 22 °F and the temperature outside is 42 °F. The temperature difference is 20 F°. **(a)** Is the temperature difference ΔT in degrees Celsius greater than, less than, or equal to 20 C°? **(b)** Choose the *best explanation* from among the following:

    I. The temperature difference is equal to 20 C° because temperature differences are the same in all temperature scales.

    II. The temperature difference is less than 20 C° because $\Delta T_C = \frac{5}{9}(20 \, ^\circ F) = 11 \, ^\circ C$.

    III. The temperature difference is greater than 20 C° because $\Delta T_C = \frac{9}{5}(20 \, ^\circ F) + 32 = 68 \, ^\circ C$.

2. • **Lowest Temperature on Earth** The official record for the lowest temperature ever recorded on Earth was set at Vostok, Antarctica, on July 21, 1983. The temperature on that day fell to −89.2 °C, well below the temperature of dry ice. What is this temperature in degrees Fahrenheit?

3. • Incandescent lightbulbs heat a tungsten filament to a temperature of about 4580 °F, which is almost half as hot as the surface of the Sun. What is this temperature in degrees Celsius?

4. • Normal body temperature for humans is 98.6 °F. What is the corresponding temperature in **(a)** degrees Celsius and **(b)** kelvins?

5. • The temperature at the surface of the Sun is about 6000 K. Convert this temperature to the **(a)** Celsius and **(b)** Fahrenheit scales.

6. • One day you notice that the outside temperature increased by 17 F° between your early morning jog and your lunch at noon. What is the corresponding change in temperature in the **(a)** Celsius and **(b)** Kelvin scales?

7. • The gas in a constant-volume gas thermometer has a pressure of 93.5 kPa at 105 °C. **(a)** What is the pressure of the gas at 47.5 °C? **(b)** At what temperature does the gas have a pressure of 105 kPa?

8. •• Predict/Calculate A constant-volume gas thermometer has a pressure of 80.3 kPa at −10.0 °C and a pressure of 86.4 kPa at

10.0 °C. **(a)** At what temperature does the pressure of this system extrapolate to zero? **(b)** What are the pressures of the gas at the freezing and boiling points of water? **(c)** In general terms, how would your answers to parts (a) and (b) change if a different constant-volume gas thermometer is used? Explain.

9. •• **Greatest Change in Temperature** A world record for the greatest change in temperature was set in Spearfish, SD, on January 22, 1943. At 7:30 A.M. the temperature was −4.0 °F; two minutes later the temperature was 45 °F. Find the average rate of temperature change during those two minutes in kelvins per second.

10. •• We know that −40 °C corresponds to −40 °F. What temperature has the same value in both the Fahrenheit and Kelvin scales?

11. •• At what temperature is the reading on a Fahrenheit scale twice the reading on a Celsius scale?

12. ••• When the bulb of a constant-volume gas thermometer is placed in a beaker of boiling water at 100 °C, the pressure of the gas is 227 mmHg. When the bulb is moved to an ice–salt mixture, the pressure of the gas drops to 162 mmHg. Assuming ideal behavior, as in Figure 16-3, what is the Celsius temperature of the ice–salt mixture?

SECTION 16-3 THERMAL EXPANSION

13. • **CE** Bimetallic strip A is made of copper and steel; bimetallic strip B is made of aluminum and steel. **(a)** Referring to Table 16-1, which strip bends more for a given change in temperature? **(b)** Which of the metals listed in Table 16-1 would give the greatest amount of bend when combined with steel in a bimetallic strip?

14. • **CE** Referring to Table 16-1, which would be more accurate for all-season outdoor use: a tape measure made of steel or one made of aluminum?

15. • **CE** Predict/Explain A brass plate has a circular hole whose diameter is slightly smaller than the diameter of an aluminum ball. If the ball and the plate are always kept at the same temperature, **(a)** should the temperature of the system be increased or decreased in order for the ball to fit through the hole? **(b)** Choose the *best explanation* from among the following:
   I. The aluminum ball changes its diameter more with temperature than the brass plate, and therefore the temperature should be decreased.
   II. Changing the temperature won't change the fact that the ball is larger than the hole.
   III. Heating the brass plate makes its hole larger, and that will allow the ball to pass through.

16. • **CE FIGURE 16-25** shows five metal plates, all at the same temperature and all made from the same material. They are all placed in an oven and heated by the same amount. Rank the plates in order of increasing expansion in **(a)** the vertical and **(b)** the horizontal direction. Indicate ties where appropriate.

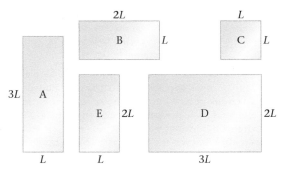

FIGURE 16-25 Problem 16

17. • **Longest Suspension Bridge** The world's longest suspension bridge is the Akashi Kaikyo Bridge in Japan. The bridge is 3910 m long and is constructed of steel. How much longer is the bridge on a warm summer day (30.0 °C) than on a cold winter day (−5.00 °C)?

18. •• A vinyl siding panel for a house is installed on a day when the temperature is 15.6 °C. If the coefficient of thermal expansion for vinyl siding is 55.8×10^{-6} K^{-1}, how much room (in mm) should the installer leave for expansion of a 3.66-m length if the sunlit temperature of the siding could reach 48.9 °C?

19. •• A cylinder bore in an aluminum engine block has a diameter of 96.00 mm at 20.00 °C. **(a)** What is the diameter of the bore when the engine operates at 119.0 °C? **(b)** At what temperature is the diameter of the hole equal to 95.85 mm?

20. •• Predict/Calculate It is desired to slip an aluminum ring over a steel bar (**FIGURE 16-26**). At 10.00 °C the inside diameter of the ring is 4.000 cm and the diameter of the rod is 4.040 cm. **(a)** In order for the ring to slip over the bar, should the ring be heated or cooled? Explain. **(b)** Find the temperature of the ring at which it fits over the bar. The bar remains at 10.0 °C.

Steel   Aluminum

FIGURE 16-26 Problems 20 and 77

21. •• At 18.75 °C a brass sleeve has an inside diameter of 2.21988 cm and a steel shaft has a diameter of 2.22258 cm. It is desired to shrink-fit the sleeve over the steel shaft. **(a)** To what temperature must the sleeve be heated in order for it to slip over the shaft? **(b)** Alternatively, to what temperature must the shaft be cooled before it is able to slip through the sleeve?

22. •• Early in the morning, when the temperature is 5.5 °C, gasoline is pumped into a car's 53-L steel gas tank until it is filled to the top. Later in the day the temperature rises to 27 °C. Since the volume of gasoline increases more for a given temperature increase than the volume of the steel tank, gasoline will spill out of the tank. How much gasoline spills out in this case?

23. •• Some cookware has a stainless steel interior ($\alpha = 17.3 \times 10^{-6}$ K^{-1}) and a copper bottom ($\alpha = 17.0 \times 10^{-6}$ K^{-1}) for better heat distribution. Suppose an 8.0-in. pot of this construction is heated to 610 °C on the stove. If the initial temperature of the pot is 22 °C, what is the difference in diameter change for the copper and the steel?

24. •• Predict/Calculate You construct two wire-frame cubes, one using copper wire, the other using aluminum wire. At 23 °C the cubes enclose equal volumes of 0.016 m^3. **(a)** If the temperature of the cubes is increased, which cube encloses the greater volume? **(b)** Find the difference in volume between the cubes when their temperature is 97 °C.

25. •• A metal ball that is 1.2 m in diameter expands by 2.2 mm when the temperature is increased by 95 °C. **(a)** What is the coefficient

of thermal expansion for the ball? **(b)** What substance is this ball likely made from? (Refer to Table 16-1.)

26. •• A copper ball with a radius of 1.7 cm is heated until its diameter has increased by 0.16 mm. Assuming an initial temperature of 21 °C, find the final temperature of the ball.

27. •• Predict/Calculate An aluminum saucepan with a diameter of 23 cm and a height of 6.0 cm is filled to the brim with water. The initial temperature of the pan and water is 19 °C. The pan is now placed on a stove burner and heated to 88 °C. **(a)** Will water overflow from the pan, or will the water level in the pan decrease? Explain. **(b)** Calculate the volume of water that overflows or the drop in water level in the pan, whichever is appropriate.

SECTION 16-4 HEAT AND MECHANICAL WORK

28. • BIO **Sleeping Metabolic Rate** When people sleep, their metabolic rate is about 2.6×10^{-4} C/(s·kg). How many Calories does an 81-kg person metabolize while getting a good night's sleep of 8.5 h?

29. •• BIO An exercise machine indicates that you have worked off 2.1 Calories in a minute-and-a-half of running in place. What was your power output during this time? Give your answer in both watts and horsepower.

30. •• BIO A certain sandwich cookie contains 53 C of nutritional energy. **(a)** For what amount of time must you swim in order to work off three cookies if swimming consumes 540 C/h? **(b)** Instead of swimming you decide to run for 45 min at a pace that works off 850 C/h. How many cookies have you worked off?

31. •• BIO During a workout, a person repeatedly lifts a 16-lb barbell through a distance of 1.1 ft. How many "reps" of this lift are required to work off 150 C?

32. •• Predict/Calculate Consider the apparatus that Joule used in his experiments on the mechanical equivalent of heat, shown in Figure 16-12. Suppose both blocks have a mass of 0.95 kg and that they fall through a distance of 0.48 m. **(a)** Find the expected rise in temperature of the water, given that 6200 J are needed for every 1.0 C° increase. Give your answer in Celsius degrees. **(b)** Would the temperature rise in Fahrenheit degrees be greater than or less than the result in part (a)? Explain. **(c)** Find the rise in temperature in Fahrenheit degrees.

33. •• BIO It was shown in Example 16-18 that a typical person radiates about 62 W of power at room temperature. Given this result, how much time does it take for a person to radiate away the energy acquired by consuming a 230-C doughnut?

SECTION 16-5 SPECIFIC HEATS

34. • CE Predict/Explain Two objects are made of the same material but have different temperatures. Object 1 has a mass m, and object 2 has a mass $2m$. If the objects are brought into thermal contact, **(a)** is the temperature change of object 1 greater than, less than, or equal to the temperature change of object 2? **(b)** Choose the *best explanation* from among the following:
   I. The larger object gives up more heat, and therefore its temperature change is greatest.
   II. The heat given up by one object is taken up by the other object. Since the objects have the same heat capacity, the temperature changes are the same.
   III. One object loses heat of magnitude Q, the other gains heat of magnitude Q. With the same magnitude of heat involved, the smaller object has the greater temperature change.

35. • CE Predict/Explain A certain amount of heat is transferred to 2 kg of aluminum, and the same amount of heat is transferred to 1 kg of ice. Referring to Table 16-2, **(a)** is the increase in temperature of the aluminum greater than, less than, or equal to the increase in temperature of the ice? **(b)** Choose the *best explanation* from among the following:
   I. Twice the specific heat of aluminum is less than the specific heat of ice, and hence the aluminum has the greater temperature change.
   II. The aluminum has the smaller temperature change since its mass is less than that of the ice.
   III. The same heat will cause the same change in temperature.

36. • Suppose 72.3 J of heat are added to a 101-g piece of aluminum at 20.5 °C. What is the final temperature of the aluminum?

37. • Estimate the heat required to heat a 0.15-kg apple from 12 °C to 36 °C. (Assume the apple is mostly water.)

38. • A 9.7-g lead bullet is fired into a fence post. The initial speed of the bullet is 720 m/s, and when it comes to rest, half its kinetic energy goes into heating the bullet. How much does the bullet's temperature increase?

39. •• Thermal energy is added to 150 g of water at the rate of 55 J/s for 2.5 min. How much does the temperature of the water increase?

40. •• Predict/Calculate Silver pellets with a mass of 1.0 g and a temperature of 85 °C are added to 220 g of water at 14 °C. **(a)** How many pellets must be added to increase the equilibrium temperature of the system to 25 °C? Assume no heat is exchanged with the surroundings. **(b)** If copper pellets are used instead, does the required number of pellets increase, decrease, or stay the same? Explain. **(c)** Find the number of copper pellets that are required.

41. •• A 225-g lead ball at a temperature of 81.2 °C is placed in a light calorimeter containing 155 g of water at 20.3 °C. Find the equilibrium temperature of the system.

42. •• If 2200 J of heat are added to a 190-g object, its temperature increases by 12 C°. **(a)** What is the heat capacity of this object? **(b)** What is the object's specific heat?

43. •• **Chips by the Ton** Tortilla chips are manufactured by submerging baked and partially cooled *masa* at 41 °C into corn oil at 182 °C. **(a)** If the specific heat of masa is 1850 J/(kg·K), how much thermal energy is required to warm 0.454 kg of masa to the temperature of the oil? **(b)** If the production line makes 227 kg of chips every hour, at what rate (in kW) must thermal energy be supplied to the corn oil, assuming that all of the energy goes into warming the masa?

44. •• A 117-g lead ball is dropped from rest from a height of 7.62 m. The collision between the ball and the ground is totally inelastic. Assuming all the ball's kinetic energy goes into heating the ball, find its change in temperature.

45. •• To determine the specific heat of an object, a student heats it to 100 °C in boiling water. She then places the 34.5-g object in a 151-g aluminum calorimeter containing 114 g of water. The aluminum and water are initially at a temperature of 20.0 °C, and are thermally insulated from their surroundings. If the final temperature is 23.6 °C, what is the specific heat of the object? Referring to Table 16-2, identify the material in the object.

46. •• Predict/Calculate A student drops a 0.33-kg piece of steel at 42 °C into a container of water at 22 °C. The student also drops a 0.51-kg chunk of lead into the same container at the same time. The temperature of the water remains the same. **(a)** Was the temperature of the lead greater than, less than, or equal to 22 °C? Explain. **(b)** What was the temperature of the lead?

47. ••• A ceramic coffee cup, with $m = 116$ g and $c = 1090$ J/(kg·K), is initially at room temperature (24.0 °C). If 225 g of 80.3 °C

coffee and 12.2 g of 5.00 °C cream are added to the cup, what is the equilibrium temperature of the system? Assume that no heat is exchanged with the surroundings, and that the specific heat of coffee and cream are the same as the specific heat of water.

SECTION 16-6 CONDUCTION, CONVECTION, AND RADIATION

48. • **CE** Predict/Explain In a popular lecture demonstration, a sheet of paper is wrapped around a rod that is made from wood on one half and metal on the other half. If held over a flame, the paper on one half of the rod is burned while the paper on the other half is unaffected. (a) Is the burned paper on the wooden half of the rod, or on the metal half of the rod? (b) Choose the *best explanation* from among the following:
    I. The metal will be hotter to the touch than the wood; therefore the metal side will be burnt.
    II. The metal conducts heat better than the wood, and hence the paper on the metal half is unaffected.
    III. The metal has the smaller specific heat; hence it heats up more and burns the paper on its half of the rod.

49. • **CE** FIGURE 16-27 shows a composite slab of three different materials with equal thickness but different thermal conductivities. The opposite sides of the composite slab are held at the fixed temperatures T_1 and T_2. Given that $k_B > k_A > k_C$, rank the materials in order of the temperature difference across them, starting with the smallest. Indicate ties where appropriate.

FIGURE 16-27 Problem 49

50. • **CE** Heat is transferred from an area where the temperature is 20 °C to an area where the temperature is 0 °C through a composite slab consisting of four different materials, each with the same thickness. The temperatures at the interface between each of the materials are given in FIGURE 16-28. Rank the four materials in order of increasing thermal conductivity. Indicate ties where appropriate.

FIGURE 16-28 Problem 50

51. • **CE** Predict/Explain Two identical bowls of casserole need to be kept warm. The cook covers bowl 1 with transparent plastic wrap and bowl 2 with shiny aluminum foil, in each case trapping a layer of air between the covering and the casserole. (a) After a few minutes, will the temperature of casserole 1 be greater than, less than, or equal to the temperature of casserole 2? (b) Choose the *best explanation* from among the following:
    I. The lower thermal conductivity of the plastic wrap over casserole 1 will suppress heat loss by conduction better than aluminum foil.
    II. In each case heat loss by convection is suppressed, so the final temperatures will be the same.
    III. The high reflectivity of the aluminum foil over casserole 2 will suppress heat loss by radiation, while the heat losses by conduction and convection for casseroles 1 and 2 will be similar.

52. • **CE** Two bowls of soup with identical temperatures are placed on a table. Bowl 1 has a metal spoon in it; bowl 2 does not. After a few minutes, is the temperature of the soup in bowl 1 greater than, less than, or equal to the temperature of the soup in bowl 2?

53. • A glass window 0.33 cm thick measures 81 cm by 39 cm. How much heat flows through this window per minute if the inside and outside temperatures differ by 13 C°?

54. • **BIO** Assuming your skin temperature is 37.2 °C and the temperature of your surroundings is 20.0 °C, determine the length of time required for you to radiate away the energy gained by eating a 278-C ice cream cone. Let the emissivity of your skin be 0.915 and its area be 1.78 m².

55. • Find the heat that flows in 1.0 s through a lead brick 25 cm long if the temperature difference between the ends of the brick is 8.5 C°. The cross-sectional area of the brick is 14 cm².

56. •• Consider a double-paned window consisting of two panes of glass, each with a thickness of 0.500 cm and an area of 0.725 m², separated by a layer of air with a thickness of 1.75 cm. The temperature on one side of the window is 0.00 °C; the temperature on the other side is 20.0 °C. In addition, note that the thermal conductivity of glass is roughly 36 times greater than that of air. (a) Approximate the heat transfer through this window by ignoring the glass. That is, calculate the heat flow per second through 1.75 cm of air with a temperature difference of 20.0 C°. (The exact result for the complete window is 19.1 J/s.) (b) Use the approximate heat flow found in part (a) to find an approximate temperature difference across each pane of glass. (The exact result is 0.157 C°.)

57. •• Predict/Calculate Two metal rods of equal length—one aluminum, the other stainless steel—are connected in parallel with a temperature of 20.0 °C at one end and 118 °C at the other end. Both rods have a circular cross section with a diameter of 3.50 cm. (a) Determine the length the rods must have if the combined rate of heat flow through them is to be 27.5 J per second. (b) If the length of the rods is doubled, by what factor does the rate of heat flow change?

58. •• Two cylindrical metal rods—one copper, the other lead—are connected in parallel with a temperature of 21.0 °C at one end and 112 °C at the other end. Both rods are 0.650 m in length, and the lead rod is 2.76 cm in diameter. If the combined rate of heat flow through the two rods is 33.2 J/s, what is the diameter of the copper rod?

59. •• Predict/Calculate Two metal rods—one lead, the other copper—are connected in series, as shown in FIGURE 16-29. These are the same two rods that were connected in parallel in Example 16-16. Note that each rod is 0.525 m in length and has a square cross section 1.50 cm on a side. The temperature at the lead end of the rods is 2.00 °C; the temperature at the copper end is 106 °C. (a) The average temperature of the two ends is 54.0 °C. Is the temperature in the middle, at the lead-copper interface, greater than, less than, or equal to 54.0 °C? Explain. (b) Given that the heat flow through each of these rods in 1.00 s is 1.41 J, find the temperature at the lead-copper interface.

FIGURE 16-29 Problem 59

60. •• Predict/Calculate Consider two cylindrical metal rods with equal cross section—one lead, the other aluminum—connected in series. The temperature at the lead end of the rods is 22.5 °C; the temperature at the aluminum end is 85.0 °C. **(a)** Given that the temperature at the lead-aluminum interface is 52.5 °C, and that the lead rod is 14.5 cm long, what condition can you use to find the length of the aluminum rod? **(b)** Find the length of the aluminum rod.

61. •• A copper rod 85 cm long is used to poke a fire. The hot end of the rod is maintained at 115 °C and the cool end has a constant temperature of 22 °C. What is the temperature of the rod 25 cm from the cool end?

62. •• Two identical objects are placed in a room at 24 °C. Object 1 has a temperature of 99 °C, and object 2 has a temperature of 27 °C. What is the ratio of the net power emitted by object 1 to that radiated by object 2?

63. ••• A block has the dimensions L, $2L$, and $3L$. When one of the $L \times 2L$ faces is maintained at the temperature T_1 and the other $L \times 2L$ face is held at the temperature T_2, the rate of heat conduction through the block is P. Answer the following questions in terms of P. **(a)** What is the rate of heat conduction in this block if one of the $L \times 3L$ faces is held at the temperature T_1 and the other $L \times 3L$ face is held at the temperature T_2? **(b)** What is the rate of heat conduction in this block if one of the $2L \times 3L$ faces is held at the temperature T_1 and the other $2L \times 3L$ face is held at the temperature T_2?

GENERAL PROBLEMS

64. • **CE** Predict/Explain A pendulum is made from an aluminum rod with a mass attached to its free end. If the pendulum is cooled, **(a)** does the pendulum's period increase, decrease, or stay the same? **(b)** Choose the *best explanation* from among the following:
    I. The period of a pendulum depends only on its length and the acceleration of gravity. It is independent of mass and temperature.
    II. Cooling makes everything move more slowly, and hence the period of the pendulum increases.
    III. Cooling shortens the aluminum rod, which decreases the period of the pendulum.

65. • **CE** A copper ring stands on edge with a metal rod placed inside it, as shown in **FIGURE 16-30**. As this system is heated, will the rod ever touch the top of the ring? Answer yes or no for the case of a rod that is made of **(a)** copper, **(b)** aluminum, and **(c)** steel.

FIGURE 16-30 Problems 65 and 66

66. • **CE** Referring to the copper ring in the previous problem, imagine that initially the ring is hotter than room temperature, and that an aluminum rod that is colder than room temperature fits snugly inside the ring. When this system reaches thermal equilibrium at room temperature, is the rod firmly wedged in the ring **(A)** or can it be removed easily **(B)**?

67. • **CE** Predict/Explain The specific heat of alcohol is about half that of water. Suppose you have 0.5 kg of alcohol at the temperature 20 °C in one container, and 0.5 kg of water at the temperature 30 °C in a second container. When these fluids are poured into the same container and allowed to come to thermal equilibrium, **(a)** is the final temperature greater than, less than, or equal to 25 °C? **(b)** Choose the *best explanation* from among the following:
    I. The low specific heat of alcohol pulls in more heat, giving a final temperature that is less than 25 °C.

II. More heat is required to change the temperature of water than to change the temperature of alcohol. Therefore, the final temperature will be greater than 25 °C.
    III. Equal masses are mixed together; therefore, the final temperature will be 25 °C, the average of the two initial temperatures.

68. • **Making Steel Sheets** In the continuous-caster process, steel sheets 25.4 cm thick, 2.03 m wide, and 10.0 m long are produced at a temperature of 872 °C. What are the dimensions of a steel sheet once it has cooled to 20.0 °C?

69. • **The Coldest Place in the Universe** The Boomerang nebula holds the distinction of having the lowest recorded temperature in the universe, a frigid −272 °C. What is this temperature in kelvins?

70. • **BIO The Hottest Living Things** From the surreal realm of deep-sea hydrothermal vents 200 miles offshore from Puget Sound, comes a newly discovered hyperthermophilic—or extreme heat-loving—microbe that holds the record for the hottest existence known to science. This microbe is tentatively known as Strain 121 for the temperature at which it thrives: 121 °C. (At sea level, water at this temperature would boil vigorously, but the extreme pressures at the ocean floor prevent boiling from occurring.) What is this temperature in degrees Fahrenheit?

71. •• Thermal energy is added to 180 g of water at a constant rate for 3.5 min, resulting in an increase in temperature of 12 C°. What is the heating rate, in joules per second?

72. •• You wish to chill 12 cans of soda at 25.0 °C down to 5.0 °C before serving them to guests. Each can has a mass of 0.354 kg, and the specific heat of soda is the same as that of water. **(a)** How much thermal energy must be removed in order to chill all 12 cans? **(b)** If the chilling process requires 4.0 h, what is the average rate (in W) at which thermal energy is removed from the soda?

73. •• **BIO Brain Power** As you read this problem, your brain is consuming about 22 W of power. **(a)** If your mass is 65 kg, how many steps with a height of 21 cm must you climb to expend a mechanical energy equivalent to one hour of brain operation? **(b)** A typical human brain, which is 77% water, has a mass of 1.4 kg. Assuming that the 22 W of brain power is converted to heat, what temperature rise would you estimate for the brain in one hour of operation? Ignore the significant heat transfer that occurs between a human head and its surroundings, as well as the 23% of the brain that is not water.

74. •• **BIO Brain Food** Your brain consumes about 22 W of power, and avocados have been shown to promote brain health. If each avocado contains 310 C, and your brain were powered entirely by the energy from the avocados, how many must you eat each day?

75. •• **BIO The Cricket Thermometer** The rate of chirping of the snowy tree cricket (*Oecanthus fultoni* Walker) varies with temperature in a predictable way. A linear relationship provides a good match to the chirp rate, but an even more accurate relationship is the following:

$$N = (5.63 \times 10^{10})e^{-(6290\,\text{K})/T}$$

In this expression, N is the number of chirps in 13.0 s and T is the temperature in kelvins. If a cricket is observed to chirp 185 times in 60.0 s, what is the temperature in degrees Fahrenheit?

76. •• Predict/Calculate A pendulum consists of a large weight suspended by a steel wire that is 0.9500 m long. **(a)** If the temperature increases, does the period of the pendulum increase, decrease, or stay the same? Explain. **(b)** Calculate the change in length of the pendulum if the temperature increase is 150.0 C°. **(c)** Calculate

the period of the pendulum before and after the temperature increase. (Assume that the coefficient of linear expansion for the wire is $12.00 \times 10^{-6}\,\text{K}^{-1}$, and that $g = 9.810\,\text{m/s}^2$ at the location of the pendulum.)

77. •• Predict/Calculate Once the aluminum ring in Problem 20 is slipped over the bar, the ring and bar are allowed to equilibrate at a temperature of 22 °C. The ring is now stuck on the bar. **(a)** If the temperatures of both the ring and the bar are changed together, should the system be heated or cooled to remove the ring? **(b)** Find the temperature at which the ring can be removed.

78. •• A 256-kg rock sits in full sunlight on the edge of a cliff 5.75 m high. The temperature of the rock is 33.2 °C. If the rock falls from the cliff into a pool containing $6.00\,\text{m}^3$ of water at 15.5 °C, what is the final temperature of the rock–water system? Assume that the specific heat of the rock is $1010\,\text{J/(kg}\cdot\text{K)}$.

79. •• Water going over Iguaçu Falls on the border of Argentina and Brazil drops through a height of about 72 m. Suppose that all the gravitational potential energy of the water goes into raising its temperature. Find the increase in water temperature at the bottom of the falls as compared with the top.

80. •• **Thermal Storage** Solar heating of a house is much more efficient if there is a way to store the thermal energy collected during the day to warm the house at night. Suppose one solar-heated home utilizes a concrete slab of area 12 m² and 25 cm thick. **(a)** If the density of concrete is $2400\,\text{kg/m}^3$, what is the mass of the slab? **(b)** The slab is exposed to sunlight and absorbs energy at a rate of $1.4 \times 10^7\,\text{J/h}$ for 10 h. If it begins the day at 22 °C and has a specific heat of $750\,\text{J/(kg}\cdot\text{K)}$, what is its temperature at sunset? **(c)** Model the concrete slab as being surrounded on both sides (contact area 24 m²) with a 2.0-m-thick layer of air in contact with a surface that is 5.0 °C cooler than the concrete. At sunset, what is the rate at which the concrete loses thermal energy by conduction through the air layer? **(d)** Model the concrete slab as having a surface area of 24 m² and surrounded by an environment 5.0 °C cooler than the concrete. If its emissivity is 0.94, what is the rate at which the concrete loses thermal energy by radiation at sunset?

81. •• **Pave It Over** Suppose city 1 leaves an entire block ($100\,\text{m} \times 100\,\text{m}$) as a park with trees and grass (emissivity 0.96) while city 2 paves the same area over with asphalt (emissivity 1.0). Sunlight heats each surface to 40.0 °C by sunset, and then the surface radiates its heat into a cube of air 100 m on a side and at 30.0 °C. **(a)** At what rate does the park in city 1 deliver energy to the air at sunset? **(b)** At what rate does the asphalt in city 2 deliver energy to the air at sunset? **(c)** If each city block maintains the same radiated power for 2.0 h and there are no other energy losses, what are the final temperatures of the cubes of air above each city block? The density of air at 30.0 °C is $1.16\,\text{kg/m}^3$. Although this example is oversimplified, a more sophisticated analysis recently showed that a city park can cool the air that passes through it by more than 4 °C.

82. •• **BIO** Suppose you could convert the 495 C in the cheeseburger you ate for lunch into mechanical energy with 100% efficiency. **(a)** How high could you throw a 0.141-kg baseball with the energy contained in the cheeseburger? **(b)** How fast would the ball be moving at the moment of release?

83. •• You turn a crank on a device similar to that shown in Figure 16-12 and produce a power of 0.16 hp. If the paddles are immersed in 0.75 kg of water, for what length of time must you turn the crank to increase the temperature of the water by 5.5 C°?

84. •• Predict/Calculate **BIO Heat Transport in the Human Body** The core temperature of the human body is 37.0 °C, and the skin, with a

surface area of 1.40 m², has a temperature of 34.0 °C. **(a)** Find the rate of heat transfer out of the body under the following assumptions: (i) The average thickness of tissue between the core and the skin is 1.20 cm; (ii) the thermal conductivity of the tissue is that of water. **(b)** Without repeating the calculation of part (a), what rate of heat transfer would you expect if the skin temperature were to fall to 31.0 °C? Explain.

85. •• **The Solar Constant** The surface of the Sun has a temperature of 5500 °C. **(a)** Treating the Sun as a perfect blackbody, with an emissivity of 1.0, find the power that it radiates into space. The radius of the Sun is $7.0 \times 10^8\,\text{m}$, and the temperature of space can be taken to be 3.0 K. **(b)** The *solar constant* is the number of watts of sunlight power falling on a square meter of the Earth's upper atmosphere. Use your result from part (a) to calculate the solar constant, given that the distance from the Sun to the Earth is $1.5 \times 10^{11}\,\text{m}$.

86. ••• Bars of two different metals are bolted together, as shown in FIGURE 16-31. Show that the distance D does not change with temperature if the lengths of the two bars have the following ratio: $L_A/L_B = \alpha_B/\alpha_A$.

FIGURE 16-31 Problem 86

87. ••• A grandfather clock has a simple brass pendulum of length L. One night, the temperature in the house is 25.0 °C and the period of the pendulum is 1.00 s. The clock keeps correct time at this temperature. If the temperature in the house quickly drops to 17.1 °C just after 10 P.M., and stays at that value, what is the actual time when the clock indicates that it is 10 A.M. the next morning?

88. ••• Predict/Calculate A sheet of aluminum has a circular hole with a diameter of 10.0 cm. A 9.99-cm-long steel rod is placed inside the hole, along a diameter of the circle, as shown in FIGURE 16-32. It is desired to change the temperature of this system until the steel rod just touches both sides of the circle. **(a)** Should the temperature of the system be increased or decreased? Explain. **(b)** By how much should the temperature be changed?

FIGURE 16-32 Problem 88

89. ••• A layer of ice has formed on a small pond. The air just above the ice is at −5.4 °C, the water-ice interface is at 0 °C, and the water at the bottom of the pond is at 4.0 °C. If the total depth from the top of the ice to the bottom of the pond is 1.4 m, how thick is the

layer of ice? *Note*: The thermal conductivity of ice is 1.6 W/(m·C°) and that of water is 0.60 W/(m·C°).

90. ••• **A Double-Paned Window** An energy-efficient double-paned window consists of two panes of glass, each with thickness L_1 and thermal conductivity k_1, separated by a layer of air of thickness L_2 and thermal conductivity k_2. Show that the equilibrium rate of heat flow through this window per unit area, A, is

$$\frac{Q}{At} = \frac{(T_2 - T_1)}{2L_1/k_1 + L_2/k_2}$$

In this expression, T_1 and T_2 are the temperatures on either side of the window.

PASSAGE PROBLEMS

Cool Medicine In situations in which the brain is deprived of oxygen, such as in a heart attack or traumatic brain injury, irreversible damage to the brain can happen very quickly. It's been proposed that cooling the body of a patient after cardiac arrest or traumatic brain injury may help protect the brain against long-term damage, perhaps by reducing the metabolic rate and thus the oxygen demand of the brain.

In order to achieve some protection for the brain without causing other undesired physiological effects such as disruptions to heart rhythms, the goal is to reduce the core body temperature from its normal value of 37 °C to between 30 °C and 35 °C. Animal experiments have shown some promising results, and human trials are being done. In one trial, 95% of patients had their core temperature reduced to 34 °C in 2.0 hours. To cool the body this quickly, each patient was placed between two cooled blankets, one above the patient and one below. Water and alcohol were sprayed on the patient, and any skin that wasn't covered by the top blanket was exposed to the environment.

91. • Which method of cooling the patient in this study relies primarily on conduction?

    A. Exposing skin to the environment
    B. Spraying water and alcohol on the patient
    C. Using cooled blankets above and below
    D. None of these

92. •• What total amount of heat must be removed to drop the whole body temperature of a typical 65-kg patient in this study by 1.0 °C?

Because the human body is mostly water, the average specific heat of the human body is relatively high, 3.5×10^3 J/(kg·K).

    A. 3.5×10^3 J   **B.** 2.3×10^5 J
    C. 5.4×10^6 J   **D.** 6.2×10^7 J

93. •• During the cooling of a typical 65-kg patient in this study, what is the average rate of heat loss, assuming the whole body's temperature changes by the same amount as the core temperature does?

    A. 95 W   **B.** 200 W
    C. 1100 W   **D.** 8700 W

94. •• For a patient in a typical hospital setting, the most important mechanism of heat loss is radiation. How does the power radiated by a patient change when the skin temperature drops from 34 °C to 33 °C, assuming no other changes?

    A. It's now 99% of what it originally was.
    B. It's now 97% of what it originally was.
    C. It's now 89% of what it originally was.
    D. It doesn't change.

95. •• **REFERRING TO EXAMPLE 16-12** Suppose the mass of the block is to be increased enough to make the final temperature of the system equal to 22.5 °C. What is the required mass? Everything else in Example 16-12 remains the same.

96. •• **REFERRING TO EXAMPLE 16-12** Suppose the initial temperature of the block is to be increased enough to make the final temperature of the system equal to 22.5 °C. What is the required initial temperature? Everything else remains the same as in Example 16-12.

97. •• Predict/Calculate **REFERRING TO EXAMPLE 16-16** Suppose the lead rod is replaced with a second copper rod. **(a)** Will the heat that flows in 1.00 s increase, decrease, or stay the same? Explain. **(b)** Find the heat that flows in 1.00 s with two copper rods. Everything else remains the same as in Example 16-16.

98. •• Predict/Calculate **REFERRING TO EXAMPLE 16-16** Suppose the temperature of the hot plate is to be changed to give a total heat flow of 25.2 J in 1.00 s. **(a)** Should the new temperature of the hot plate be greater than or less than 106 °C? Explain. **(b)** Find the required temperature of the hot plate. Everything else is the same as in Example 16-16.

Big Ideas

1 The pressure of an ideal gas depends on the number of molecules, the temperature, and the volume.

2 The pressure of a gas is due to molecules constantly hitting the wall of a container.

3 The average kinetic energy of a gas is proportional to its absolute temperature.

4 When two phases are in equilibrium, adding heat does not change the temperature.

5 The amount of heat needed to change one phase to another is the latent heat.

Phases and Phase Changes

▲ Though many people are surprised to learn that a liquid can boil and freeze at the same time, it is in fact true. At a unique combination of temperature and pressure called the triple point, all three phases of matter—solid, liquid, and gas—coexist in equilibrium. In this chapter we'll learn more about the three phases of matter, their differences and similarities, and what happens when a substance changes from one phase to another.

The matter we encounter in everyday life is generally a solid, a liquid, or a gas. The air we breathe is a gas, the water we drink is a liquid, and the salt on our popcorn is a solid. In this chapter we consider these three *phases* of matter in detail, as well as changes from one phase to another.

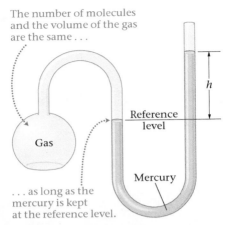

The number of molecules and the volume of the gas are the same . . .

Gas

Reference level

Mercury

h

. . . as long as the mercury is kept at the reference level.

▲ FIGURE 17-1 **A constant-volume gas thermometer** A device like this can be used as a thermometer. Note that both the volume of the gas and the number of molecules in the gas remain constant.

The pump forces more molecules into the ball . . .

. . . which increases the pressure.

▲ FIGURE 17-2 **Inflating a basketball** A hand pump can be used to increase the number of molecules inside a basketball. This raises the pressure of the gas in the ball, causing it to inflate.

PHYSICS IN CONTEXT
Looking Back

We use the concept of temperature (Chapter 16) throughout this chapter. We also make use of pressure (Chapter 15) in our discussion of ideal gases.

▲ FIGURE 17-3 **Increasing pressure by decreasing volume** Sitting on a basketball reduces its volume and increases the pressure of the gas it contains.

17-1 Ideal Gases

In Chapter 16 we discussed the thermal behavior of gases. In particular, we saw that the pressure of a constant volume of gas decreases linearly with decreasing temperature over a wide range of temperatures. Eventually, however, when the temperature is low enough, real gases change to liquid and then solid form, and their behavior changes. These changes are due to interactions between the molecules in a gas. Hence, the weaker these interactions, the wider the range of temperatures over which the simple, linear gas behavior persists. We now consider a simplified model of a gas, the **ideal gas,** in which intermolecular interactions are vanishingly small.

Though ideal gases do not actually exist in nature, the behavior of real gases is generally quite well approximated by that of an ideal gas, especially when the gas is dilute. By studying the simple "ideal" case, we can gain considerable insight into the workings of a real gas. This is similar to the kind of idealizations we made in mechanics. For example, we often considered a surface to be perfectly frictionless, when real surfaces have at least some friction, and we imagined springs to obey Hooke's law exactly, though real springs show some deviations. The ideal gas plays an analogous role in our study of thermodynamics.

Equation of State

We can describe the way the pressure, P, of an ideal gas depends on temperature, T, number of molecules, N, and volume, V, from just a few simple observations. First, imagine holding the number of molecules and the volume of a gas constant, as in the constant-volume gas thermometer shown in FIGURE 17-1. As we have already noted, the pressure of a gas under these conditions varies linearly with temperature. Therefore, we conclude the following:

$$P = (\text{constant})T$$
(*fixed volume, V; fixed number of molecules, N*)

In this expression, the constant multiplying the temperature depends on the number of molecules in the gas and its volume; the temperature T is measured on the Kelvin scale.

Second, imagine you have a basketball that is just slightly underinflated, as in FIGURE 17-2. The ball has the size and shape of a basketball, but is a bit too squishy. To increase the pressure you "pump" more molecules from the atmosphere into the ball. Thus, while the temperature and volume of the gas in the ball remain constant, its pressure increases as the number of molecules increases:

$$P = (\text{constant})N$$
(*fixed volume, V; fixed temperature, T*)

Our third and final observation concerns the volume dependence of pressure. Returning to the basketball, suppose that instead of pumping it up, you sit on it, as pictured in FIGURE 17-3. This deforms it a bit, *reducing* its volume. At the same time, the pressure of the gas in the ball *increases*. Thus, when the number of molecules and temperature remain constant, the pressure varies *inversely* with volume:

$$P = \frac{(\text{constant})}{V}$$
(*fixed number of molecules, N; fixed temperature, T*)

Combining these observations, we arrive at the following mathematical expression for the pressure of a gas:

$$P = k\frac{NT}{V}$$

The constant k in this expression is a fundamental constant of nature, the **Boltzmann constant,** named for the Austrian physicist Ludwig Boltzmann (1844-1906):

Boltzmann Constant, *k*

$$k = 1.38 \times 10^{-23}\,\text{J/K} \qquad \text{17-1}$$

SI unit: J/K

In general, a relationship between the thermal properties of a substance is referred to as an **equation of state.** Rearranging slightly, we can write the equation of state for an ideal gas:

Equation of State for an Ideal Gas

$$PV = NkT \qquad \text{17-2}$$

We apply this result to the gas contained in a person's lungs in the next Example.

Big Idea 1 The pressure of an ideal gas is proportional to the number of molecules and the absolute temperature, and inversely proportional to the volume.

PROBLEM-SOLVING NOTE

Ideal-Gas Law

When using the equation of state for an ideal gas, remember that the temperature must always be given in terms of the Kelvin scale.

EXAMPLE 17-1 TAKE A DEEP BREATH

Predict/Calculate

BIO A person's lungs can hold 6.0 L (1 L = $10^{-3}\,\text{m}^3$) of air at body temperature (310 K) and atmospheric pressure (101 kPa). **(a)** Given that air is 21% oxygen, find the number of oxygen molecules in the lungs. **(b)** If the person now climbs to the top of a mountain, where the air pressure is considerably less than 101 kPa, does the number of molecules in the lungs increase, decrease, or stay the same?

PICTURE THE PROBLEM
Our sketch shows a person's lungs, with their combined volume of $V = 6.0\,\text{L}$. In addition, we indicate that the pressure in the lungs is $P = 101\,\text{kPa}$ and the temperature is $T = 310\,\text{K}$.

REASONING AND STRATEGY

a. We will treat the air in the lungs as an ideal gas. Given the volume, temperature, and pressure of the gas, we can use the equation of state, $PV = NkT$, to solve for the number, N.
   Finally, only 21% of the molecules in the air are oxygen. Therefore, we multiply N by 0.21 to find the number of oxygen molecules.

b. We apply $PV = NkT$ again, but this time with a reduced pressure P. Notice that V and T remain the same, however, because they are determined by the shape and size of person's body.

Known Volume of lungs, $V = 6.0 \times 10^{-3}\,\text{m}^3$; temperature, $T = 310\,\text{K}$; pressure, $P = 101\,\text{kPa}$; fraction of molecules that are oxygen, $f = 0.21$.
Unknown **(a)** Number of oxygen molecules, $N_{\text{oxy}} = ?$ **(b)** Change in number of molecules?

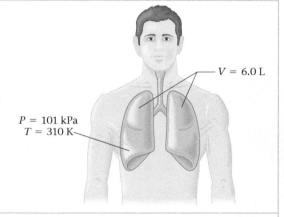

$V = 6.0\,\text{L}$

$P = 101\,\text{kPa}$
$T = 310\,\text{K}$

SOLUTION

Part (a)

1. Solve the equation of state for the number of molecules:

$$PV = NkT \quad \text{or} \quad N = PV/kT$$

2. Substitute numerical values to find the number of molecules in the lungs:

$$N = \frac{PV}{kT} = \frac{(1.01 \times 10^5\,\text{Pa})(6.0 \times 10^{-3}\,\text{m}^3)}{(1.38 \times 10^{-23}\,\text{J/K})(310\,\text{K})} = 1.4 \times 10^{23}$$

3. Multiply N by $f = 0.21$ to find the number of molecules that are oxygen, O_2:

$$N_{\text{oxy}} = fN = 0.21N = 0.21(1.4 \times 10^{23}) = 2.9 \times 10^{22}$$

Part (b)
From Step 1 we see that the number of molecules is $N = PV/kT$. As a result, if the volume V of a person's lungs remains the same, along with the body temperature T, it follows that the number of molecules N will decrease linearly with pressure. For example, at an altitude of 10,000 ft the air pressure is about 70% of its value at sea level, meaning that the number of oxygen molecules in the lungs is only 70% of its usual value. No wonder people have a hard time "catching their breath" at high altitudes.

INSIGHT
As one might expect, the number of molecules in even an average-size pair of lungs is enormous. In fact, a number this large is difficult to compare to anything we are familiar with. For example, the number of stars in the Milky Way galaxy is estimated to be "only" about 10^{11}. Thus, if each star in the Milky Way were itself a galaxy of 10^{11} stars, then all of these stars combined would just begin to approach the number of oxygen molecules in your lungs at this very moment!

CONTINUED

PRACTICE PROBLEM

If the person takes a particularly deep breath, so that the lungs hold a total of 1.5×10^{23} molecules, what is the new volume of the lungs? [**Answer:** $V = 6.4 \times 10^{-3}\,\text{m}^3 = 6.4\,\text{L}$]

Some related homework problems: Problem 4, Problem 5

CONCEPTUAL EXAMPLE 17-2   AIR PRESSURE

Feeling a bit cool, you turn up the thermostat in your living room. A short time later the air is warmer. Assuming the room is well sealed, is the pressure of the air greater than, less than, or the same as before you turned up the heat?

REASONING AND DISCUSSION

Because the room is well sealed, the number of molecules and the volume they occupy are approximately constant. Thus, increasing T, while holding N and V fixed, leads to an increased pressure, P.

ANSWER

The air pressure is greater.

PHYSICS IN CONTEXT
Looking Ahead

The equation of state for an ideal gas plays an important role in Chapter 18, especially when we consider thermal processes as well as specific heats at constant pressure and constant volume.

▲ **FIGURE 17-4 Moles of various substances** A mole of any substance contains 6.022×10^{23} entities (atoms or molecules). This number is referred to as Avogadro's number (N_A). At standard temperature and pressure, N_A molecules of a gas occupy a volume of 22.4 L. One mole of a solid or liquid substance, on the other hand, is a mass in grams that is equal to its atomic mass. Thus, the mole bridges the realm of atoms and molecules and the macroscopic world of observable masses and volumes. This photo shows molar amounts of four different substances. From left to right they are sugar, salt, carbon, hydrogen, and copper.

Describing a Gas in Terms of Moles Another common way to write the ideal-gas equation of state is in terms of the number of **moles** in a gas—as opposed to using the number of molecules, N, as in Equation 17-2. In the SI system of units, the mole (mol) is defined in terms of the most abundant isotope of carbon, which is referred to as carbon-12. The definition is as follows:

> A mole is the amount of a substance that contains as many elementary entities as there are atoms in 12 g of carbon-12.

The phrase "elementary entities" refers to molecules, which may contain more than one atom, as in water (H_2O), or just a single atom, as in carbon (C) and helium (He). Experiments show that the number of atoms in a mole of carbon-12 is 6.022×10^{23}. This number is known as **Avogadro's number**, N_A, named for the Italian physicist and chemist Amedeo Avogadro (1776–1856):

Avogadro's Number, N_A

$$N_A = 6.022 \times 10^{23}\ \text{molecules/mol} \qquad \text{17-3}$$

SI unit: mol^{-1}; the number of molecules is dimensionless

Examples of one mole of various substances are shown in **FIGURE 17-4**.

Now, if we let n denote the number of moles in a gas, the number of molecules it contains is

$$N = nN_A$$

Substituting this into the ideal-gas equation of state yields

$$PV = nN_A kT$$

The constants N_A and k are combined to form the **universal gas constant**, R, defined as follows:

Universal Gas Constant, R

$$R = N_A k = (6.022 \times 10^{23}\ \text{molecules/mol})(1.38 \times 10^{-23}\ \text{J/K})$$
$$= 8.31\ \text{J/(mol} \cdot \text{K)} \qquad \text{17-4}$$

SI unit: $\text{J/(mol} \cdot \text{K)}$

Thus, an alternative form of the equation of state is:

Equation of State for an Ideal Gas

$$PV = nRT \qquad \text{17-5}$$

This relationship is used in the following Quick Example.

QUICK EXAMPLE 17-3 THE AMOUNT OF AIR IN A BASKETBALL

How many moles of air are in an inflated basketball? Assume the pressure in the ball is 159 kPa, the temperature is 293 K, and the diameter of the ball is 23.0 cm.

REASONING AND SOLUTION

First, we solve the equation of state, $PV = nRT$, for the number of moles, n. Next, we substitute the numerical values $P = 159$ kPa and $T = 293$ K. The volume of the (spherical) basketball is $V = 4\pi r^3/3$, with $r = \frac{1}{2}(0.230\,\text{m}) = 0.115\,\text{m}$.

1. Solve $PV = nRT$ for the number of moles, n:
$$n = \frac{PV}{RT}$$

2. Calculate the volume of the ball:
$$V = \frac{4\pi r^3}{3} = \frac{4\pi(0.115\,\text{m})^3}{3} = 0.00637\,\text{m}^3$$

3. Substitute the numerical values in Step 1:
$$n = \frac{PV}{RT} = \frac{(159 \times 10^3\,\text{Pa})(0.00637\,\text{m}^3)}{(8.31\,\text{J/(mol}\cdot\text{K))}(293\,\text{K})} = 0.416\,\text{mol}$$

Thus, a typical basketball contains slightly less than one-half mole of air.

The Mass of a Mole As mentioned before, a mole of anything has precisely the same number (N_A) of particles. What differs from substance to substance is the mass of one mole. For example, one mole of helium atoms has a mass of 4.00260 g, and one mole of copper atoms has a mass of 63.546 g. In general, we define the **atomic or molecular mass,** M, of a substance to be the mass in grams of one mole of that substance. Thus, the atomic mass for helium is $M = 4.00260$ g/mol and for copper it is $M = 63.546$ g/mol. The periodic table in Appendix E gives the atomic masses for all elements.

The atomic mass provides a convenient bridge between the macroscopic world, where we measure the mass of a substance in grams, and the microscopic world, where the number of molecules is typically 10^{23} or greater. As we have seen, if you measure out a mass of copper equal to 63.546 g, you have, in effect, counted out $N_A = 6.022 \times 10^{23}$ atoms of copper. It follows that the mass of an individual copper atom, m, is the atomic mass of copper divided by Avogadro's number; that is,

$$m = \frac{M}{N_A} \qquad 17\text{-}6$$

We use this relationship to find the masses of an iron atom and a nitrogen molecule in the following Exercise.

EXERCISE 17-4 FIND THE MASS

Find the masses of **(a)** an atom of iron, Fe, and **(b)** a molecule of nitrogen, N_2. Atomic masses are listed in Appendix E.

REASONING AND SOLUTION

a. The atomic mass of iron is 55.85 g/mol (Appendix E). Thus, an iron atom has the mass

$$m_{\text{Fe}} = \frac{M}{N_A} = \frac{55.85\,\text{g/mol}}{6.022 \times 10^{23}\,\text{atoms/mol}} = 9.274 \times 10^{-23}\,\text{g/atom} = 9.274 \times 10^{-26}\,\text{kg/atom}$$

b. The atomic mass of nitrogen is 14.01 g/mol (Appendix E), and hence the molecular mass of N_2 is $2(14.01\,\text{g/mol}) = 28.02$ g/mol. Therefore, the mass of an N_2 molecule is

$$m_{N_2} = \frac{M}{N_A} = \frac{28.02\,\text{g/mol}}{6.022 \times 10^{23}\,\text{molecules/mol}} = 4.653 \times 10^{-23}\,\text{g/molecule}$$
$$= 4.653 \times 10^{-26}\,\text{kg/molecule}$$

Isotherms

Historically, the ideal-gas equation of state was arrived at piece by piece, as a result of the combined efforts of a number of researchers. For example, the English scientist Robert Boyle (1627–1691) established the fact that the pressure of a gas varies inversely

PROBLEM-SOLVING NOTE

Pressure and Volume in an Isotherm

If the temperature of an ideal gas is constant, the pressure and volume change in such a way that their product has a constant value.

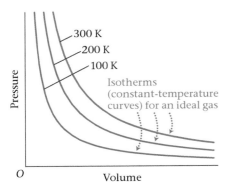

▲ FIGURE 17-5 **Ideal-gas isotherms**
Pressure-versus-volume isotherms for an ideal gas. Each isotherm is of the form $P = NkT/V = \text{constant}/V$. The three isotherms shown are for the temperatures 100 K, 200 K, and 300 K.

with volume—as long as temperature and the number of molecules are held constant. This is known as **Boyle's law:**

$$P_i V_i = P_f V_f \qquad 17\text{-}7$$

(fixed number of molecules, N; fixed temperature, T)

To see that Boyle's law is consistent with $PV = NkT$, notice that if N and T are constant, then so too is PV. Another way of saying that PV is a constant is to say that its initial value must be equal to its final value. This is Boyle's law.

In addition, when N and T are constant, the equation of state $PV = NkT$ implies that

$$P = \frac{NkT}{V} = \frac{\text{constant}}{V}$$

This result is plotted in **FIGURE 17-5**, where we show P as a function of V. The greater the temperature, T, the larger the constant in the numerator. Therefore, the curves farther from the origin correspond to higher temperatures, as indicated.

Each of the curves in Figure 17-5 corresponds to a different, fixed temperature. As a result, these curves are known as **isotherms,** which means, literally, "constant temperature." We shall return to the topic of isotherms in the next chapter, when we consider various thermal processes.

EXAMPLE 17-5 UNDER PRESSURE

A cylindrical flask of cross-sectional area A is fitted with an air-tight piston that is free to slide up and down. Contained within the flask is an ideal gas. Initially the pressure applied by the piston is 140 kPa and the height of the piston above the base of the flask is 22 cm. When additional mass is added to the piston, the pressure increases to 180 kPa. Assuming the system is always at the temperature 290 K, find the new height of the piston.

PICTURE THE PROBLEM
The physical system is shown in our sketch. The initial pressure in the flask is $P_1 = 140$ kPa, and the initial height of the piston is $h_1 = 22$ cm. When additional mass is placed on the piston, the pressure increases to $P_2 = 180$ kPa and the height decreases to h_2. The temperature remains constant.

REASONING AND STRATEGY
The temperature is held constant in this system, and hence it follows that $PV = NkT$ is also constant. As a result, $P_1V_1 = P_2V_2$, as in Boyle's law. This relationship can be used to find V_2 (since we are given P_1, P_2, and V_1).

Next, the cylindrical volume is related to the height by $V = Ah$. The area A is the same in both cases (and hence it will cancel), allowing us to solve for the height, h_2.

Known Initial pressure, $P_1 = 140$ kPa; initial height of piston, $h_1 = 22$ cm; final pressure, $P_2 = 180$ kPa; temperature, $T = 290$ K.

Unknown Final height of piston, $h_2 = ?$

SOLUTION

1. Set the initial and final values of PV equal to one another, then solve for V_2:

$$P_1 V_1 = P_2 V_2$$
$$V_2 = V_1 \left(\frac{P_1}{P_2} \right)$$

2. Substitute $V_1 = Ah_1$ and $V_2 = Ah_2$, and solve for h_2:

$$Ah_2 = Ah_1 \left(\frac{P_1}{P_2} \right)$$
$$h_2 = h_1 \left(\frac{P_1}{P_2} \right)$$

3. Substitute numerical values:

$$h_2 = h_1 \frac{P_1}{P_2} = (22 \text{ cm}) \frac{140 \text{ kPa}}{180 \text{ kPa}} = 17 \text{ cm}$$

CONTINUED

INSIGHT

In order to keep the temperature of this system constant, it is necessary to allow some heat to flow out of the cylindrical flask as the gas is compressed by the additional weight. Thus, the flask is not thermally insulated during this experiment.

PRACTICE PROBLEM

What pressure would be required to change the height of the piston to 28 cm?   [**Answer:** $P = 110$ kPa]

Some related homework problems: Problem 16, Problem 17

▶ **FIGURE 17-6** Charles's law states that the volume of a gas at constant pressure is proportional to its Kelvin temperature. For example, this balloon was fully inflated at room temperature (293 K) and atmospheric pressure. When the balloon was cooled by vapors from liquid nitrogen (77 K), however, the volume of the air inside the balloon decreased markedly, causing it to shrivel.

Finally, recall that in Conceptual Example 15-7 we considered the behavior of an air bubble rising from a swimmer in shallow water. At the time, we said that the diameter of the bubble increases as it rises because the pressure of the surrounding water is decreasing. This is certainly the case, but now we can be more precise. If we assume that the water temperature is constant, and that no gas molecules are added to or removed from the bubble, then the volume of the bubble has the following dependence:

$$V = \frac{NkT}{P} = \frac{\text{constant}}{P}$$

This is just our isotherm, again, and we see that the volume indeed increases as pressure decreases.

Constant Pressure

Another aspect of ideal-gas behavior was discovered by the French scientist Jacques Charles (1746–1823) and later studied in greater detail by fellow Frenchman Joseph Gay-Lussac (1778–1850). Known today as **Charles's law,** their result is that the volume of a gas divided by its temperature is constant, as long as the pressure and number of molecules are constant:

$$\frac{V_i}{T_i} = \frac{V_f}{T_f}$$

17-8

(fixed number of molecules, N; fixed pressure, P)

As with Boyle's law, this result follows immediately from the ideal-gas equation of state. Solving $PV = NkT$ for V/T, we have

$$\frac{V}{T} = \frac{Nk}{P}$$

If N and P are constant, then so too is the quantity V/T.

Charles's law can be rewritten as a linear relationship between volume and temperature:

$$V = (\text{constant})T$$

The constant in this expression is Nk/P. This result is illustrated by the shrunken balloon in **FIGURE 17-6** and also in **FIGURE 17-7**, where we see that the volume of an ideal gas vanishes as the temperature approaches absolute zero.

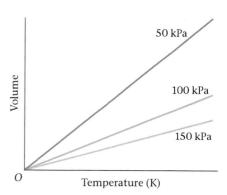

▲ **FIGURE 17-7 Volume versus temperature for an ideal gas at constant pressure** The temperature in this plot is the Kelvin temperature, T; hence $T = 0$ corresponds to absolute zero. At absolute zero the volume attains its lowest possible value—zero.

PROBLEM-SOLVING NOTE

Volume and Temperature in an Ideal Gas

If the pressure of an ideal gas is constant, the ratio of the volume and temperature remains constant. This is true no matter what the value of the pressure.

EXAMPLE 17-6   FIND THE VOLUME

A weather balloon is being prepared for launch. Initially it holds 6.86 m³ of hydrogen at 277 K. As pre-launch preparations continue in the morning sun, the temperature of the hydrogen rises to 288 K. What is the volume of the hydrogen at this new temperature? (The skin of the balloon is loose at launch, and has negligible effect on the pressure of the gas it contains.)

CONTINUED

PICTURE THE PROBLEM
Our sketch shows the weather balloon partially inflated before launch. (At high altitude, the balloon expands and eventually bursts.) The initial temperature of the gas in the balloon is T_1, and its initial volume is V_1.

REASONING AND STRATEGY
The number of molecules and the pressure in the enclosed gas remain the same, and hence the volume changes linearly with the absolute temperature. Another way of saying this is that V/T is constant. Therefore, we set the initial value of V/T equal to its final value. This allows us to solve for the final volume of the hydrogen in the balloon.

Known   Initial temperature, $T_1 = 277$ K; initial volume, $V_1 = 6.86$ m^3; final temperature, $T_2 = 288$ K.

Unknown   Final volume, $V_2 = ?$

SOLUTION

1. Set the initial value of V/T equal to the final value of V/T:
$$\frac{V_1}{T_1} = \frac{V_2}{T_2}$$

2. Solve for the final volume V_2:
$$V_2 = V_1\left(\frac{T_2}{T_1}\right)$$

3. Substitute numerical values in Step 2:
$$V_2 = V_1\left(\frac{T_2}{T_1}\right) = (6.86 \text{ m}^3)\left(\frac{288 \text{ K}}{277 \text{ K}}\right) = 7.13 \text{ m}^3$$

INSIGHT
The volume of the gas increased by the same factor that the absolute temperature increased.

PRACTICE PROBLEM
At what temperature is the volume of the gas equal to 6.97 m^3?   [**Answer:** 281 K]

Some related homework problems: Problem 8, Problem 10

Enhance Your Understanding   (Answers given at the end of the chapter)

1. Rank the following ideal-gas systems in order of increasing pressure. Indicate ties where appropriate.

System	A	B	C	D
n (moles)	1	2	10	4
T (K)	300	350	300	350
V (m^3)	1	5	5	10

Section Review

- The equation of state for an ideal gas is $PV = NkT$, where N is the number of molecules, T is the Kelvin temperature, and $k = 1.38 \times 10^{-23}$ J/K is Boltzmann's constant. Similarly, the equation of state can be written as $PV = nRT$, where n is the number of moles and $R = 8.31$ J/(mol·K) is the universal gas constant.

Big Idea **2** The pressure of a gas is due to the continuous bombardment of molecules against the wall of a container.

17-2 The Kinetic Theory of Gases

We can readily measure the pressure and temperature of a gas using a pressure gauge and a thermometer. These are *macroscopic* quantities that apply to the gas as a whole. It is not so easy, however, to measure *microscopic* quantities, such as the position or velocity of an individual molecule. Still, there must be some connection between what happens on the microscopic level and what we observe on the macroscopic level. This connection is described by the **kinetic theory of gases.**

In the kinetic theory, we imagine a gas to be made up of a collection of molecules moving about inside a container of volume V. In particular, we assume the following:

- The container holds a very large number N of identical molecules, each of mass m. The molecules themselves can be thought of as "point particles"—that is, the volume of the molecules is negligible compared to the volume of the container, and the size of the molecules is negligible compared to the distance between them. (This is why dilute real gases are the best approximation to the ideal case.)
- The molecules move about the container in a random manner. They obey Newton's laws of motion at all times.
- When molecules hit the walls of the container or collide with one another, they bounce elastically. Other than these collisions, the molecules have no interactions.

With these basic assumptions we can relate the pressure of a gas to the behavior of the molecules themselves.

The Origin of Pressure

As we shall see, the pressure exerted by a gas is due to the innumerable collisions between gas molecules and the walls of their container. Each collision results in a change of momentum for a given molecule, just like throwing a ball at a wall and having it bounce back. The total change in momentum of the molecules in a given time, divided by the time, is equal to the force a wall must exert on the gas to contain it (see Section 9-2). The average of this force over time, and over the area of a wall, is the pressure of the gas.

To be specific, imagine a container that is a cube of length L on a side. Its volume, then, is $V = L^3$. In addition, consider a given molecule that happens to be moving in the negative x direction toward a wall, as in **FIGURE 17-8**. If its speed is v_x, its initial momentum is $p_{i,x} = -mv_x$. After bouncing from the wall, it moves in the positive x direction with the same speed (since the collision is elastic), and hence its final momentum is $p_{f,x} = +mv_x$. As a result, the molecule's change in momentum is

$$\Delta p_x = p_{f,x} - p_{i,x} = mv_x - (-mv_x) = 2mv_x$$

The wall exerts a force on the molecule to cause this momentum change.

After the bounce, the molecule travels to the other side of the container and back before bouncing off the same wall again. The time required for this round trip of length $2L$ is

$$\Delta t = \frac{2L}{v_x}$$

Thus, by Newton's second law, the average force exerted by the wall on the molecule is

$$F = \frac{\Delta p_x}{\Delta t} = \frac{2mv_x}{2L/v_x} = \frac{mv_x^2}{L}$$

The average pressure exerted by this wall on a single molecule, then, is simply the force divided by the area. Because the area of the wall is $A = L^2$, we have

$$P_{\text{molecule}} = \frac{F}{A} = \frac{(mv_x^2/L)}{L^2} = \frac{mv_x^2}{L^3} = \frac{mv_x^2}{V} \qquad \text{17-9}$$

Notice that we have used the fact that the volume of the container is $V = L^3$.

In this calculation we assumed that the molecule moves in the x direction. This was merely to simplify the derivation. If, instead, the molecule moves at some angle to the x axis, the calculation applies to its x component of motion. The final conclusions are unchanged.

Speed Distribution of Molecules

In deriving Equation 17-9 ($P_{\text{molecule}} = mv_x^2/V$), we considered a single molecule with a particular speed. Other molecules will have different speeds, of course. In addition,

▲ **FIGURE 17-8 Force exerted by a molecule on the wall of a container**
A molecule bounces off a wall of a container, changing its momentum from $-mv_x$ to $+mv_x$; the change in momentum is $2mv_x$. A round trip will be completed in the time $\Delta t = 2L/v_x$, so the average force exerted on the molecule by the wall is $F = \Delta p/\Delta t = 2mv_x/(2L/v_x) = mv_x^2/L$.

▲ **FIGURE 17-9 The Maxwell speed distribution** The Maxwell distribution of molecular speeds for O_2 at the temperatures $T = 300$ K and $T = 1100$ K. Note that the most probable speed increases with increasing temperature.

the speed of any given molecule changes with time as it collides with other molecules in the gas. What remains constant, however, is the overall **distribution of speeds.**

This is illustrated in **FIGURE 17-9**, where we present the results obtained by the Scottish physicist James Clerk Maxwell (1831–1879). This plot shows the relative probability that an O_2 molecule will have a given speed. For example, on the curve labeled 300 K, the most probable speed is about 390 m/s. When the temperature is increased to 1100 K, the most probable speed increases to roughly 750 m/s. Other speeds occur as well, from speeds near zero to those that are very large, but these have much lower probabilities.

Thus, in the expression $P_{molecule} = mv_x^2/V$, the term v_x^2 should be replaced with the average of v_x^2 over all the molecules in the gas, to take into the account the different speeds of different molecules. Writing this average as $(v_x^2)_{av}$, we have

$$P_{molecule} = \frac{m(v_x^2)_{av}}{V}$$

But all N molecules in the gas follow the same distribution, and hence the pressure exerted by the gas as a whole is N times the average for a single molecule:

$$P = NP_{molecule} = N\frac{m(v_x^2)_{av}}{V} = \left(\frac{N}{V}\right)m(v_x^2)_{av} \qquad \text{17-10}$$

Of course, there's nothing special about the x direction—the result $P = (N/V)m(v_x^2)_{av}$ applies equally well with $(v_y^2)_{av}$ or $(v_z^2)_{av}$ in place of $(v_x^2)_{av}$. Thus, it would be preferable to express the pressure of the gas in terms of the overall speed of the molecules, rather than in terms of a single component. The speed squared of a molecule is

$$v^2 = v_x^2 + v_y^2 + v_z^2$$

Hence, the average of v^2 is

$$(v^2)_{av} = (v_x^2)_{av} + (v_y^2)_{av} + (v_z^2)_{av}$$

Noting that the x, y, and z directions are equivalent, it follows that

$$(v_x^2)_{av} = (v_y^2)_{av} = (v_z^2)_{av}$$

As a result,

$$(v^2)_{av} = (v_x^2)_{av} + (v_y^2)_{av} + (v_z^2)_{av} = 3(v_x^2)_{av}$$

Equivalently,

$$(v_x^2)_{av} = \tfrac{1}{3}(v^2)_{av}$$

Using this replacement in $P = (N/V)m(v_x^2)_{av}$ yields

$$P = \tfrac{1}{3}\left(\frac{N}{V}\right)m(v^2)_{av}$$

The last part of this expression, $m(v^2)_{av}$, is simply twice the average kinetic energy of a molecule. Thus, using K for the kinetic energy, as in Chapters 7 and 8, we can write the pressure as follows:

Pressure in the Kinetic Theory of Gases

$$P = \tfrac{1}{3}\left(\frac{N}{V}\right)2K_{av} = \tfrac{2}{3}\left(\frac{N}{V}\right)K_{av} = \tfrac{2}{3}\left(\frac{N}{V}\right)\left(\tfrac{1}{2}mv^2\right)_{av} \qquad \text{17-11}$$

To summarize, using the kinetic theory, we have shown that the pressure of a gas is proportional to the number of molecules and inversely proportional to the volume. We discussed both of these dependences before when considering the ideal gas. In addition, we can make the following observation:

> The pressure of an ideal gas is directly proportional to the average kinetic energy of its individual molecules.

This is the key connection between microscopic behavior and macroscopic observables.

Kinetic Energy and Temperature

If we compare the ideal-gas equation of state, $PV = NkT$, with the result from kinetic theory, $P = \frac{2}{3}(N/V)\left(\frac{1}{2}mv^2\right)_{av}$, we find

$$PV = NkT = \frac{2}{3}N\left(\frac{1}{2}mv^2\right)_{av}$$

As a result, it follows that

$$\frac{2}{3}\left(\frac{1}{2}mv^2\right)_{av} = kT$$

Equivalently:

Kinetic Energy and Temperature

$$\left(\frac{1}{2}mv^2\right)_{av} = K_{av} = \frac{3}{2}kT \qquad \text{17-12}$$

This is one of the most important results of kinetic theory. It says that the average kinetic energy of a gas molecule is directly proportional to the Kelvin temperature, T. Thus, when we heat a gas, what happens on the microscopic level is that the molecules move with speeds that are, on average, greater. Similarly, cooling a gas causes the molecules to slow. This result also shows the fundamental significance of the Kelvin temperature scale.

EXERCISE 17-7 AVERAGE KINETIC ENERGY

Find the average kinetic energy of oxygen molecules in the air. Assume the air is at a temperature of 23 °C.

REASONING AND SOLUTION
Using $K_{av} = \frac{3}{2}kT$, and the fact that 23 °C = 296 K, we find

$$K_{av} = \frac{3}{2}kT = \frac{3}{2}(1.38 \times 10^{-23}\,\text{J/K})(296\,\text{K}) = 6.13 \times 10^{-21}\,\text{J}$$

This is also the average kinetic energy of *nitrogen* molecules in the air. In fact, the type of molecule is not important in determining the average kinetic energy; all that matters is the temperature of the gas.

Returning to Equation 17-12, $\left(\frac{1}{2}mv^2\right)_{av} = \frac{3}{2}kT$, we find with a slight rearrangement that

$$\left(v^2\right)_{av} = \frac{3kT}{m}$$

Now, the square root of $\left(v^2\right)_{av}$ is given a special name—it is called the **root mean square** (rms) speed. Specifically, the root mean square speed is the square *root* of the *mean* of the *square* of the speed. For gas molecules, then, the rms speed, v_{rms}, is the following:

RMS Speed of a Gas Molecule

$$v_{rms} = \sqrt{\left(v^2\right)_{av}} = \sqrt{\frac{3kT}{m}} \qquad \text{17-13}$$

SI unit: m/s

Rewriting this in terms of the molecular mass, M, we have

$$v_{rms} = \sqrt{\frac{3kT}{m}} = \sqrt{\frac{3kT}{(M/N_A)}} = \sqrt{\frac{3N_A kT}{M}}$$

Finally, using $N_A k = R$ yields

$$v_{rms} = \sqrt{\frac{3RT}{M}} \qquad \text{17-14}$$

The rms speed is one of the characteristic speeds of the Maxwell speed distribution. As shown in **FIGURE 17-10**, v_{rms} is slightly greater than the most probable speed, v_{mp}, and the average speed, v_{av}.

**PHYSICS
IN CONTEXT
Looking Back**

Kinetic energy (Chapter 7) plays a central role in our understanding of an ideal gas. In particular, we can make a clear and specific connection between the mechanical concept of kinetic energy and the thermodynamic concept of temperature.

Big Idea 3 The average kinetic energy of a gas is proportional to the absolute temperature of the gas.

**PHYSICS
IN CONTEXT
Looking Ahead**

The rms value of a quantity is an important consideration in Chapter 24, when we study alternating-current circuits.

▲ **FIGURE 17-10 Most probable, average, and rms speeds** Characteristic speeds for O_2 at the temperature $T = 1100$ K. From left to right, the indicated speeds are the most probable speed, v_{mp}, the average speed, v_{av}, and the rms speed, v_{rms}.

EXAMPLE 17-8 · FRESH AIR

Predict/Calculate

The atmosphere is composed primarily of nitrogen N_2 (78%) and oxygen O_2 (21%). **(a)** Is the rms speed of N_2 (28.0 g/mol) greater than, less than, or the same as the rms speed of O_2 (32.0 g/mol)? **(b)** Find the rms speed of N_2 and O_2 at 293 K.

PICTURE THE PROBLEM
Our sketch shows molecules in the air bouncing off the walls of a room—they also bounce off a person in the room. Notice that the molecules move in random directions with a variety of speeds. Most of the molecules are N_2, but about 21% are O_2.

REASONING AND STRATEGY
The rms speeds are calculated by straightforward substitution into the equation $v_{rms} = \sqrt{3RT/M}$. The only point to be careful about is the molecular mass of each molecule. In particular, notice that for nitrogen the molecular mass is 28.0 g/mol = 0.0280 kg/mol, and that for oxygen the molecular mass is 32.0 g/mol = 0.0320 kg/mol.

Known Molecular mass of N_2, $M = 0.0280$ kg/mol; molecular mass of O_2, $M = 0.0320$ kg/mol; temperature, $T = 293$ K.
Unknown The rms speed of N_2, $v_{rms} = ?$ The rms speed of O_2, $v_{rms} = ?$

SOLUTION

Part (a)
Both molecules are at the same temperature, and hence they have the same kinetic energy. The nitrogen molecule has less mass, however; thus, if it is to have the same kinetic energy, it must have a higher speed. This is also what one would expect on the basis of $v_{rms} = \sqrt{3RT/M}$ because M is smaller for the nitrogen molecule.

Part (b)
1. To find the rms speed of N_2, substitute $M = 0.0280$ kg/mol in $v_{rms} = \sqrt{3RT/M}$:

$$v_{rms} = \sqrt{\frac{3[8.31\,\text{J/(mol}\cdot\text{K)}](293\text{ K})}{0.0280\text{ kg/mol}}} = 511\text{ m/s}$$

2. To find the rms speed of O_2, substitute $M = 0.0320$ kg/mol in $v_{rms} = \sqrt{3RT/M}$:

$$v_{rms} = \sqrt{\frac{3[8.31\,\text{J/(mol}\cdot\text{K)}](293\text{ K})}{0.0320\text{ kg/mol}}} = 478\text{ m/s}$$

INSIGHT
As expected, N_2 has the higher rms speed. For comparison, the speed of sound at 293 K is 343 m/s, which is just over 750 mi/h. Thus, the molecules in the air are bouncing off your skin with the speed of a supersonic jet.

The N_2 and O_2 molecules each contribute a *partial pressure* on your skin because of these collisions. The sum of these partial pressures, about 78 kPa for N_2 and 21 kPa for O_2, is the total pressure of the gas. This relationship is called *Dalton's law*. (See also Problems 43 and 44.)

PRACTICE PROBLEM
One of the substances found in air is monatomic and has an rms speed of 428 m/s, at 293 K. What is this substance?
[**Answer:** $M = 0.0399$ kg/mol; thus the substance is argon, Ar, which comprises about 0.94% of the atmosphere.]

Some related homework problems: Problem 21, Problem 22

RWP* The speed of molecules in the air determines the average rate at which one molecule will collide with another molecule. For air at room temperature, the collision rate is roughly 1.6 billion collisions per second. The speed of molecules also limits the speed of sound in a gas, because sound is a longitudinal wave that depends on those collisions in order to propagate (see Section 14-4). In fact, the speed of sound is proportional to the square root of the temperature, and thus sound travels about 10% faster in air on a hot summer day (33°C) than it does on a bitter cold winter day (−18°C). In addition, the rate at which molecules collide with each other determines the rate at which chemical reactions can occur. This is why a cricket chirps at a faster rate with increasing temperature, making it a novel type of thermometer. (See the Passage Problems in Chapters 1 and 13, and Problem 75 in Chapter 16.)

---

*Real World Physics applications are denoted by the acronym RWP.

CONCEPTUAL EXAMPLE 17-9   COMPARE SPEEDS

Each of two containers of equal volume holds an ideal gas. Container A has twice as many molecules as container B. If the gas pressure is the same in the two containers, is the rms speed of the molecules in container A greater than, less than, or the same as the rms speed of the molecules in container B?

REASONING AND DISCUSSION

Because P and V are the same, you might think the rms speeds are the same as well. Recall, however, that *the pressure of a gas is caused by the collisions of gas molecules with the walls of the container*. If more molecules occupy a given volume, and they bounce off the walls with the same speed as in a container with fewer molecules, the pressure will be greater. Therefore, in order for the pressure in the two containers to be the same, it is necessary that the rms speed of the molecules in container A be less than the rms speed in container B.

Mathematically, we note that $P_A = P_B$ and $V_A = V_B$; hence $P_A V_A = P_B V_B$. Using the ideal-gas equation, $PV = NkT$, we can write this condition as $N_A k T_A = N_B k T_B$. Thus, if $N_A = 2N_B$, it follows that $T_A = T_B/2$. Because the temperature is lower in container A, its molecules have the lower rms speed.

ANSWER

The rms speed is less in container A.

The Internal Energy of an Ideal Gas

The internal energy of a substance is the sum of all its potential and kinetic energies. In an ideal gas there are no interactions between molecules, other than perfectly elastic collisions; hence, there is no potential energy. As a result, the total energy of the system is the sum of the kinetic energy of each of its molecules. Thus, for an ideal gas of N pointlike molecules—that is, a monatomic gas—the internal energy is simply:

Internal Energy of a Monatomic Ideal Gas

$$U = \tfrac{3}{2} NkT \qquad\qquad\qquad \text{17-15}$$

SI unit: J

In terms of moles, this result is

$$U = \tfrac{3}{2} nRT \qquad\qquad\qquad \text{17-16}$$

We shall return to this result in the next chapter.

EXERCISE 17-10   INTERNAL ENERGY

A basketball at 290 K holds 0.95 mol of air molecules. What is the internal energy of the air in the ball?

REASONING AND SOLUTION

Applying the relationship $U = \tfrac{3}{2} nRT$, we find

$$U = \tfrac{3}{2} nRT = \tfrac{3}{2}(0.95 \text{ mol})[8.31 \text{ J/(mol} \cdot \text{K)}](290 \text{ K}) = 3400 \text{ J}$$

This is roughly the kinetic energy a basketball would have if you dropped it from a height of 700 m.

If an ideal gas is diatomic, there are additional contributions to the internal energy. For example, a diatomic molecule, which is shaped somewhat like a dumbbell, can have rotational kinetic energy. Diatomic molecules can also vibrate along the line joining the two atoms, which is yet another contribution to the total energy. Thus, the result $U = \tfrac{3}{2} nRT$ applies only to the simplest case, the ideal monatomic gas.

Enhance Your Understanding   (Answers given at the end of the chapter)

2. If the Kelvin temperature of a gas is doubled, by what factor does the root mean square speed of the molecules increase?

CONTINUED

Section Review

- In the kinetic theory of gases, the pressure of a gas is viewed as the result of innumerable collisions of molecules bouncing off the walls of a container. The average kinetic energy of a monatomic molecule in an ideal gas is $\frac{3}{2}kT$.

17-3 Solids and Elastic Deformation

The defining characteristic of a solid object is that it has a particular shape. For example, when Michelangelo carved the statue of David from a solid block of marble, he was confident it would retain its shape long after his work was done. Even so, the shape of a solid *can* be changed—though usually only slightly—if it is acted on by a force. In this section we consider various types of deformations that can occur in solids and the way these deformations are related to the forces that cause them.

Changing the Length of a Solid

A useful model for a solid is a lattice of small balls representing molecules connected to one another by springs representing the intermolecular forces. Pulling on a solid rod with a force F, for example, causes each "intermolecular spring" in the direction of the force to expand by an amount proportional to F. The net result is that the entire solid increases its length by an amount proportional to the force, $\Delta L \propto F$, as shown in **FIGURE 17-11**.

▶ **FIGURE 17-11 Stretching a rod**
Equal and opposite forces applied to the ends of a rod cause it to stretch. On the atomic level, the forces stretch the "intermolecular springs" in the solid, resulting in an overall increase in length. The stretch is proportional to the force F and the initial length L_0, and inversely proportional to the cross-sectional area A.

TABLE 17-1 Young's Modulus for Various Materials

Material	Young's modulus, $Y\,(\text{N/m}^2)$
Tungsten	36×10^{10}
Steel	20×10^{10}
Copper	11×10^{10}
Brass	9.0×10^{10}
Aluminum	6.9×10^{10}
Pyrex glass	6.2×10^{10}
Lead	1.6×10^{10}
Bone	
Tension	1.6×10^{10}
Compression	0.94×10^{10}
Nylon	0.37×10^{10}

The stretch ΔL is also proportional to the initial length of the rod, L_0. To see why, we first note that each intermolecular spring expands by the same amount, giving a total stretch that is proportional to the number of such springs. But the number of intermolecular springs is proportional to the total initial length of the rod. Thus, it follows that $\Delta L \propto FL_0$.

Finally, the amount of stretch for a given force F is inversely proportional to the cross-sectional area A of the rod. For example, a rod with a cross-sectional area $2A$ is like two rods of cross-sectional area A placed side by side. Thus, applying a force F to a rod of area $2A$ is equivalent to applying a force $F/2$ to two rods of area A. The result is half the stretch when the area is doubled; that is, $\Delta L \propto FL_0/A$. Solving this relationship for the amount of force F required to produce a given stretch ΔL, we find $F \propto (\Delta L/L_0)A$ or, as an equality,

$$F = Y\left(\frac{\Delta L}{L_0}\right)A \qquad\qquad 17\text{-}17$$

The proportionality constant Y in this expression is **Young's modulus,** named for the English physicist Thomas Young (1773–1829). Comparing the two sides of Equation 17-17, we see that Young's modulus has the units of force per area (N/m^2).

Typical values for Young's modulus are given in Table 17-1. Notice that the values vary from material to material, but are all rather large. This means that a large force is required to cause even a small stretch in a solid. Of course, Equation 17-17 applies equally well to a compression or a stretch. However, some materials have a slightly

---

PROBLEM-SOLVING NOTE

Area and Young's Modulus

When using Young's modulus, remember that A is the area that is at right angles to the applied force.

different Young's modulus for compression and stretching. For example, human bones under tension (stretching) have a Young's modulus of $1.6 \times 10^{10} \, \text{N/m}^2$, while bones under compression have a slightly smaller Young's modulus of $9.4 \times 10^9 \, \text{N/m}^2$.

EXAMPLE 17-11 STRETCHING A BONE

BIO At the local airport, a person carries a 21-kg duffel bag in one hand. Assuming the humerus (the upper arm bone) supports the entire weight of the bag, determine the amount by which the bone stretches. (The humerus may be assumed to be 33 cm in length and to have an effective cross-sectional area of $5.2 \times 10^{-4} \, \text{m}^2$.)

$A = 5.2 \times 10^{-4} \, \text{m}^2$

$L_0 = 0.33 \, \text{m}$

mg

PICTURE THE PROBLEM
The humerus is oriented vertically, and hence the weight of the duffel bag applies tension to the bone. The Young's modulus in this case is $1.6 \times 10^{10} \, \text{N/m}^2$. In addition, we note that the initial length of the bone is $L_0 = 0.33 \, \text{m}$ and its cross-sectional area is $A = 5.2 \times 10^{-4} \, \text{m}^2$.

REASONING AND STRATEGY
We can find the amount of stretch by solving the relationship $F = Y(\Delta L/L_0)A$ for the quantity ΔL. The force applied to the bone is simply the weight of the duffel bag, $F = mg$, with $m = 21 \, \text{kg}$. (We ignore the relatively small weight of the forearm and hand.)

Known Mass of duffel bag, $m = 21$ kg; length of humerus, $L_0 = 0.33$ m; cross-sectional area of humerus, $A = 5.2 \times 10^{-4} \, \text{m}^2$.
Unknown Change in length of humerus, $\Delta L = ?$

SOLUTION

1. Solve $F = Y(\Delta L/L_0)A$ for the amount of stretch, ΔL:
$$\Delta L = \frac{FL_0}{YA}$$

2. Calculate the force applied to the humerus:
$$F = mg = (21 \, \text{kg})(9.81 \, \text{m/s}^2) = 210 \, \text{N}$$

3. Substitute numerical values into the expression for ΔL:
$$\Delta L = \frac{FL_0}{YA} = \frac{(210 \, \text{N})(0.33 \, \text{m})}{(1.6 \times 10^{10} \, \text{N/m}^2)(5.2 \times 10^{-4} \, \text{m}^2)}$$
$$= 8.3 \times 10^{-6} \, \text{m}$$

INSIGHT
We find that the amount of stretch is imperceptibly small. The reason for this, of course, is that Young's modulus is such a large number. If the bone had been compressed rather than stretched, its change in length, though still minuscule, would have been greater by a factor of 16/9.4.

PRACTICE PROBLEM
Suppose the humerus is reduced uniformly in size by a factor of two. This means that both the length and diameter are halved. What is the amount of stretch in this case? **[Answer:** With $L_0 \rightarrow L_0/2$ and $A \rightarrow A/4$, it follows that the stretch is doubled. Thus, $\Delta L \rightarrow 2 \, \Delta L = 1.7 \times 10^{-5} \, \text{m}$.]

Some related homework problems: Problem 31, Problem 32

There is a straightforward connection between $F = Y(\Delta L/L_0)A$ (Equation 17-17) and Hooke's law for a spring, $F = kx$ (Equation 6-5). To see the connection, we rewrite Equation 17-17 as

$$F = \left(\frac{YA}{L_0}\right)\Delta L$$

Notice that the force required to cause a certain stretch is proportional to the stretch—just as in Hooke's law. In fact, if we identify ΔL with the displacement x of a spring from equilibrium and YA/L_0 with the force constant k, we have Hooke's law:

$$F = \left(\frac{YA}{L_0}\right)\Delta L = kx$$

Thus, we see that the force constant of a spring depends on the Young's modulus, Y, of the material from which it is made, the cross-sectional area A of the wire, and the length of the wire, L_0. Two structures with changes in length due to applied forces are shown in **FIGURE 17-12**.

PHYSICS IN CONTEXT
Looking Back

We gain a deeper understanding of Hooke's law (Chapter 6) in terms of the elastic deformation of solids.

▶ **FIGURE 17-12 Visualizing Concepts**
Changing the Length of a Solid **(a)** The cables in this suspension bridge have significant forces pulling on them from either end. As a result, they are under tension, just like the rod in Figure 17-11. It follows that the length of each cable increases by an amount that is proportional to its initial length. **(b)** This dam experiences forces that tend to compress it. The magnitude of its change in length can be described in terms of $F = Y(\Delta L/L_0)A$, with the appropriate value of Young's modulus.

(a)   **(b)**

CONCEPTUAL EXAMPLE 17-12   COMPARE FORCE CONSTANTS   PREDICT/EXPLAIN

Two identical springs are connected end to end. **(a)** Is the force constant of the resulting compound spring greater than, less than, or equal to the force constant of a single spring? **(b)** Which of the following is the *best explanation* for your prediction?

I. The resulting force constant is greater because two springs are acting together, doubling the force.

II. The two springs produce a longer compound spring—but the material is the same, and therefore the force constant is the same as well.

III. The force constant is less for the compound spring because it stretches twice as far for the same applied force.

REASONING AND DISCUSSION

It might seem that the force constant would be the same, because we simply have twice the length of the same spring. On the other hand, it might seem that the force constant is greater, because two springs are exerting a force rather than just one. In fact, the force constant decreases.

The reason for the reduced force constant is that if we apply a force F to the compound spring, each individual spring stretches by a certain amount. The compound spring, then, stretches by twice this amount. By Hooke's law, $F = kx$, a spring that stretches twice as far for the same applied force has half the force constant.

We can also obtain this result by recalling that the force constant is $k = YA/L_0$. Thus, if the length of a spring is doubled—as in the compound spring—the force constant is halved.

ANSWER

(a) The force constant of the compound spring is less, by a factor of two. **(b)** The best explanation is III.

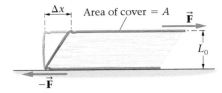

◀ **FIGURE 17-13 Shear deformation** Equal and opposite forces applied to the top and bottom of a book result in a shear deformation. The amount of deformation is proportional to the force F and the thickness of the book L_0, and inversely proportional to the area A.

Changing the Shape of a Solid

Another type of deformation, referred to as a **shear deformation**, changes the *shape* of a solid. Consider a book of thickness L_0 resting on a table, as shown in **FIGURE 17-13**. A force F is applied to the right on the top cover of the book, and static friction applies a force F to the left on the bottom cover of the book. The result is that the book remains at rest but becomes slanted by the amount Δx. The force required to cause a given amount of slant is proportional to Δx, inversely proportional to the thickness of the book L_0, and proportional to the surface area A of the book's cover; that is, $F \propto A\Delta x/L_0$. Writing this as an equality, we have

$$F = S\left(\frac{\Delta x}{L_0}\right)A \qquad\qquad 17\text{-}18$$

The constant of proportionality in this case is the **shear modulus,** S. Like Young's modulus, the shear modulus has the units N/m^2. Typical values of the shear modulus are collected in Table 17-2. As with Young's modulus, the shear modulus is large in magnitude, meaning that most solids require a large force to cause even a small amount of shear.

TABLE 17-2 Shear Modulus for Various Materials

Material	Shear modulus, S (N/m^2)
Tungsten	15×10^{10}
Steel	8.1×10^{10}
Bone	8.0×10^{10}
Copper	4.2×10^{10}
Brass	3.5×10^{10}
Aluminum	2.4×10^{10}
Lead	0.54×10^{10}

The expressions $F = Y(\Delta L/L_0)A$ (Equation 17-17) and $F = S(\Delta x/L_0)A$ (Equation 17-18) are similar in structure, but it's important to be aware of their differences as well. For example, the term L_0 in the Young's modulus equation refers to the length of a solid measured in the direction of the applied force. In contrast, L_0 in the shear modulus equation refers to the thickness of the solid as measured in a direction perpendicular to the applied force. Similarly, the area A in Equation 17-17 is the cross-sectional area of the solid, and this area is perpendicular to the applied force. In contrast, the area A in Equation 17-18 is the area of the solid in the plane of the applied force.

PROBLEM-SOLVING NOTE

Area and the Shear Modulus

When using the shear modulus, remember that A is the area that is parallel to the applied force.

EXAMPLE 17-13 | DEFORMING A STACK OF PANCAKES: FIND THE SHEAR MODULUS

A horizontal force of 1.3 N is applied to the top of a stack of pancakes 16 cm in diameter and 8.5 cm high. The result is a shear deformation of 2.4 cm. What is the shear modulus for these pancakes?

PICTURE THE PROBLEM
In our sketch, we see the stack of pancakes deformed by a force of magnitude $F = 1.3$ N to the right at the top of the stack and a force of equal magnitude to the left at the bottom of the stack. The result is a shear deformation of 2.4 cm.

REASONING AND STRATEGY
We can solve $F = S(\Delta x/L_0)A$ for the shear modulus, S. The other quantities in the equation are $F = 1.3$ N, $\Delta x = 2.4$ cm, $A = \pi d^2/4$ with $d = 16$ cm, and $L_0 = 8.5$ cm.

Known Horizontal force, $F = 1.3$ N; diameter of pancakes, $d = 0.16$ m; height of pancakes, $L_0 = 0.085$ m; shear of pancakes, $\Delta x = 0.024$ m.
Unknown Shear modulus of pancakes, $S = ?$

SOLUTION

1. Solve $F = S(\Delta x/L_0)A$ for the shear modulus:
$$S = \frac{FL_0}{A\Delta x}$$

2. Calculate the area of the pancakes:
$$A = \frac{\pi d^2}{4} = \frac{\pi(0.16\text{ m})^2}{4} = 0.020\text{ m}^2$$

3. Substitute numerical values in Step 1:
$$S = \frac{FL_0}{A\Delta x} = \frac{(1.3\text{ N})(0.085\text{ m})}{(0.020\text{ m}^2)(0.024\text{ m})} = 230\text{ N/m}^2$$

INSIGHT
Notice the small value of the pancakes' shear modulus, especially when compared to the shear modulus of a typical solid. This is a reflection of the fact that the pancake stack is easily deformed.

PRACTICE PROBLEM
Suppose the stack of pancakes is doubled in height, but everything else in the system remains the same. By what factor does the shear deformation change? [**Answer:** The shear deformation doubles.]

Some related homework problems: Problem 33, Problem 81

Changing the Volume of a Solid

If a piece of Styrofoam is taken deep into the ocean, the tremendous pressure of the water causes it to shrink to a fraction of its original volume, as can be seen in **FIGURE 17-14**. This is an extreme example of the volume change that occurs in all solids when the pressure of their surroundings is changed. The general situation is illustrated in **FIGURE 17-15**, where we show a spherical solid whose volume decreases by the amount

▶ **FIGURE 17-14** Styrofoam has a very small bulk modulus, which means that even a relatively small increase in pressure can cause a large decrease in volume. The Styrofoam cup at left was immersed in water to a depth of 1955 m, where the pressure was nearly 200 atm.

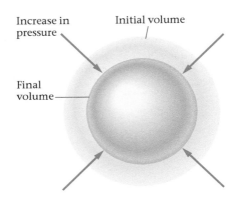

Increase in pressure

Initial volume

Final volume

◀ **FIGURE 17-15 Changing the volume of a solid** As the pressure surrounding an object increases, its volume decreases. The amount of volume change is proportional to the initial volume and to the change in pressure.

ΔV when the pressure acting on it increases by the amount ΔP. Experiments show that the pressure difference required to cause a given change in volume, ΔV, is proportional to ΔV and inversely proportional to the initial volume of the object, V_0. Therefore, we can write ΔP as follows:

$$\Delta P = -B\left(\frac{\Delta V}{V_0}\right)$$

17-19

The constant of proportionality in this case, B, is called the **bulk modulus.** Like Young's modulus and the shear modulus, the bulk modulus is a positive quantity with units of N/m^2.

Table 17-3 gives some representative values of the bulk modulus. Notice the large magnitudes of B, indicating that even small volume changes require large changes in pressure.

TABLE 17-3 Bulk Modulus for Various Materials

Material	Bulk modulus, B (N/m^2)
Gold	22×10^{10}
Tungsten	20×10^{10}
Steel	16×10^{10}
Copper	14×10^{10}
Aluminum	7.0×10^{10}
Brass	6.1×10^{10}
Ice	0.80×10^{10}
Water	0.22×10^{10}
Oil	0.17×10^{10}

Stress and Strain

In general, we refer to an applied force per area as a **stress** and the resulting deformation as a **strain.** If the stress applied to an object is not too large, a proportional relationship between the strain and stress is found to hold, and a plot of strain versus stress gives a straight line, as indicated in **FIGURE 17-16**. This straight-line relationship is simply a generalization of Hooke's law—extending the result for a spring to any solid object. As the stress becomes larger, the strain eventually begins to increase at a rate that is greater than the straight line.

For stresses less than the elastic limit, the deformation of an object is reversible; that is, the deformation vanishes as the stress that caused it is reduced to zero. This is just like a spring returning to its equilibrium position when the applied force vanishes. When a deformation is reversible, we say that it is an **elastic deformation.** For stresses greater than the *elastic limit*, on the other hand, the object becomes permanently deformed, and it will not return to its original size and shape even when the stress is removed. The elastic limit is indicated in Figure 17-16. Exceeding the elastic limit is like a spring that has been stretched too far, or a car fender that has been dented. If the stress on an object is increased even further, the object eventually tears apart or fractures.

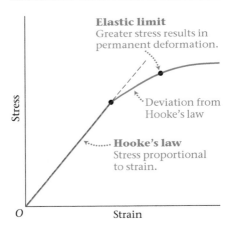

Elastic limit
Greater stress results in permanent deformation.

Deviation from Hooke's law

Hooke's law
Stress proportional to strain.

O   Strain

▲ **FIGURE 17-16 Stress Versus Strain** When stress and strain are small, they are proportional, as given by Hooke's law. Larger stresses can result in deviations from Hooke's law. Stresses greater than the elastic limit cause permanent deformation of the material.

Enhance Your Understanding

(Answers given at the end of the chapter)

3. A metal rod of a given initial length and cross-sectional area stretches when a force is applied to it. Rank the following rods in order of increasing amount of stretch. (All the rods are made of the same material.) Indicate ties where appropriate.

Rod	A	B	C	D
F (N)	5	10	50	100
L_0 (m)	1	0.5	2	1
A (10^{-4} m^2)	4	10	30	5

Section Review

- When a force is applied to a solid, the result is a change in its length, shape, or volume.

- A force applied to a solid causes a change in length given by $F = Y(\Delta L/L_0)A$. Similar expressions apply to changes in shape and volume.

17-4 Phase Equilibrium and Evaporation

To this point, we have studied the behavior of substances when they are in a single phase of matter. For example, we considered the properties of liquids in Chapter 15, and in this chapter we turned our attention to the characteristics of gases and solids. It's common, however, for a substance to exist in more than one phase at a time. In particular, if a substance has two or more phases that coexist in a stable fashion, we say that the substance exhibits **phase equilibrium.**

Equilibrium Vapor Pressure   To see what phase equilibrium means in a specific case, consider a closed container that is partially filled with a liquid, as in **FIGURE 17-17.** The container is kept at the constant temperature T_0, and initially the volume above the liquid is empty of molecules—it is a vacuum. Soon, however, some of the faster molecules in the liquid begin to escape the relatively strong intermolecular forces of their neighbors in the liquid and start to form a low-density gas. Occasionally a gas molecule collides with the liquid and reenters it, but initially, more molecules are entering the gas than returning to the liquid.

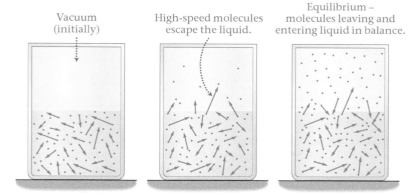

Vacuum (initially)

High-speed molecules escape the liquid.

Equilibrium – molecules leaving and entering liquid in balance.

◀ **FIGURE 17-17 A liquid in equilibrium with its vapor** Initially, a liquid is placed in a container, and the volume above it is a vacuum. High-speed molecules in the liquid are able to escape into the upper region of the container, forming a low-density gas. As the gas becomes more dense, the number of molecules leaving the liquid is balanced by the number returning to the liquid. At this point the system is in equilibrium.

This process continues until the gas is dense enough that the number of molecules returning to the liquid equals the number entering the gas. There is a constant "flow" of molecules in both directions, but when *phase equilibrium* is reached, these flows cancel, and the number of molecules in each phase remains constant. The pressure of the gas when equilibrium is established is referred to as the **equilibrium vapor pressure.** In **FIGURE 17-18** we plot the equilibrium vapor pressure of water.

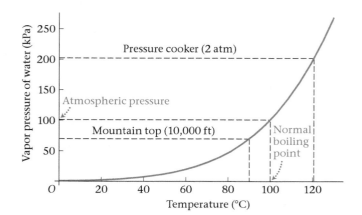

◀ **FIGURE 17-18 The vapor-pressure curve for water** The vapor pressure of water increases with increasing temperature. In particular, at $T = 100\,°C$ the vapor pressure is one atmosphere, 101 kPa.

What happens when we change the temperature? Well, if we increase the temperature, there will be more high-speed molecules in the liquid that can escape into the gas. Thus, to have an equal number of gas molecules returning to the liquid, it will be necessary for the pressure of the vapor to be greater. It follows that the equilibrium vapor pressure increases with temperature. This is also illustrated in Figure 17-18.

For each temperature there is just one equilibrium vapor pressure—that is, just one pressure where the precise balance between the phases is established. Thus, when we plot the equilibrium vapor pressure versus temperature, as in Figure 17-18, the result is a curve—the **vapor-pressure curve.** The significance of this curve is that it determines the boiling point of a liquid. In particular:

> A liquid boils at the temperature at which its vapor pressure equals the external pressure.

In Figure 17-18, we can see that the vapor pressure is equal to atmospheric pressure, $P_{at} = 101$ kPa, when the temperature of the water is 100 °C, as expected.

CONCEPTUAL EXAMPLE 17-14   BOILING TEMPERATURE

When water boils at the top of a mountain, is its temperature greater than, less than, or equal to 100 °C?

REASONING AND DISCUSSION
At the top of a mountain, air pressure is less than it is at sea level. Therefore, according to Figure 17-18, the boiling temperature will be lower as well.

For example, the atmospheric pressure at the top of a 10,000-ft mountain is roughly three-quarters what it is at sea level. The corresponding boiling point, as shown in Figure 17-18, is about 90 °C. Thus, cooking food in boiling water may be very difficult at such altitudes, because the water really isn't that hot.

ANSWER
The temperature of the water is less than 100 °C.

▲ **FIGURE 17-19** Old Faithful geyser in Yellowstone National Park erupts on schedule.

BIO If taken to an extreme, the principle discussed in Conceptual Example 17-14 implies that water in an open glass would begin boiling at room temperature when the atmospheric pressure is reduced to 2350 Pa (2.3% of P_{at}). You may have heard that this effect would cause an astronaut's blood to boil if he or she were exposed to the vacuum of space, but that would not occur because the heart can maintain a high pressure inside a person's closed circulatory system. Humans who have been exposed to very low pressures lose consciousness in less than 30 s due to a lack of oxygen, but they recover quickly with no ill effects once air pressure is restored.

If, instead, the atmospheric pressure were greatly increased, the boiling temperature of water would greatly increase as well. An *autoclave* (which is basically an elaborate pressure cooker) sterilizes surgical tools at temperatures well above 100 °C by using this principle. In a kitchen pressure cooker, this elevated temperature results in a reduced cooking time.

This principle is closely linked to how *geysers* work. A geyser, like the one shown in **FIGURE 17-19**, is a naturally heated spring of water that erupts periodically because of its underground structure. Below the geyser lies a tall column of water (with an average depth of 2000 m) that has at least one constriction. The narrow part suppresses thermal convection and allows the deepest water to become superheated by nearby magma. When the deepest water finally begins to boil, the bubbling action reduces the pressure, which causes the surrounding water to suddenly become far above the boiling-point temperature. The explosive boiling event that follows can hurl water as high as 60 m above the surface opening. After the eruption, cool water again fills the water column and begins heating to restart the cycle.

RWP The boiling temperature of water may be increased by adding a pinch of salt. The salt dissolves into sodium and chloride ions in the water. These ions have a strong interaction with the water molecules, making it harder for them to break free of the liquid and enter the vapor phase. As a result, the boiling temperature rises. This is why salt is often added to water when boiling eggs; the result is a higher temperature of boiling, which helps to solidify any material that might leak out of small cracks in the eggs.

Phase Diagram The vapor-pressure curve separates areas of liquid and gas on a graph of pressure versus temperature, as shown in **FIGURE 17-20**. A curve similar to the vapor-pressure curve indicates where the solid and liquid phases are in equilibrium.

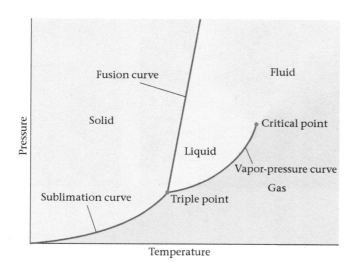

PHYSICS IN CONTEXT
Looking Ahead

The concept of a phase change is used in Chapter 21 when we discuss the fact that certain materials become superconducting (have zero electrical resistance) below a critical temperature.

PHYSICS IN CONTEXT
Looking Ahead

Phase changes appear again in Chapter 32 when we discuss the fundamental forces of nature and how they evolved as the universe cooled over time.

It is referred to as the **fusion curve.** Similarly, equilibrium between the solid and gas phases occurs along the **sublimation curve.** All three equilibrium curves are shown in Figure 17-20, on a plot that is referred to as a **phase diagram.**

Notice that the vapor-pressure curve comes to an end at a finite temperature and pressure. This end point is called the **critical point.** Beyond the critical point there is no longer a distinction between liquid and gas. They are one and the same, and are referred to, simply, as a fluid. Thus, as mentioned before, the liquid and gas phases are very similar—the liquid is just more dense than the gas and its molecules are somewhat less mobile.

The phase diagram also indicates that there is one particular temperature and pressure where all three phases are in equilibrium. This point is called the **triple point.** In water, the triple point occurs at the temperature $T = 273.16$ K and the pressure $P = 611.2$ Pa. At this temperature and pressure, ice, water, and steam are all in equilibrium with one another.

Finally, there is one feature of the fusion line that is of particular interest. In most substances, this line has a positive slope, as in Figure 17-20. This means that as the pressure is increased, the melting temperature of the substance also increases. This makes sense, because a solid is generally more dense than the corresponding liquid. Hence, if you apply pressure to a liquid—with the temperature held constant—the system will tend to become more dense and eventually solidify. This is indicated in FIGURE 17-21 (a), where we see that an increase in pressure at constant temperature (as from A to B) results in crossing the fusion curve into the solid region.

RWP There are exceptions to this rule, however, and it will probably come as no surprise that water is one such exception. This is related to the fact that ice is less dense than water. As a result, the fusion curve for water has a negative slope coming out of the triple point. This is illustrated in FIGURE 17-21 (b). What this means is that if the temperature is held constant, ice will melt when the pressure applied to it is increased by a large amount. This effect plays a role in how glaciers move: The tremendous pressure at the bottom of the glacier can melt the bottom layer of ice, providing a low-friction layer that helps the glacier to flow downhill. Similarly, the pressure exerted by an ice skate's blade can cause the ice beneath it to melt.

RWP The expansion of water as it freezes has important implications in geology as well. Suppose, for example, that water finds its way into cracks and fissures on a rocky cliff. If the temperature dips below freezing, the water in these cracks will begin to freeze and expand. This tends to split the rock farther apart in a process known as *frost wedging.* Over time, the repeated freezing and melting of water can break a "solid" rock cliff into a pile of debris, forming a talus slope at the base of the cliff. Similarly, repeated thermal expansion and contraction can result in fractured and cracked *patterned ground.* Examples on both Earth and Mars are shown in FIGURE 17-22.

(a)

(b)

▲ **FIGURE 17-21 The solid-liquid phase boundary** Fusion curves for **(a)** a typical substance and **(b)** water. For the typical substance, we note that an increase in pressure at constant temperature results in a liquid being converted to a solid. For water, increasing the pressure exerted on ice at constant temperature can result in the ice melting.

▶ **FIGURE 17-22 Visualizing Concepts**
Thermal Expansion The repeated expansion and contraction of water as it freezes and thaws over long periods of time can cause the ground to crack and buckle. Eventually, the resulting cracks may combine to form polygonal networks referred to as patterned ground (left), a common feature in Arctic regions of the Earth. The same type of patterned ground has been observed on the surface of Mars (right). This, along with recent observations of dark streaking on hillsides, gives strong evidence that water exists just below the surface of Mars in certain parts of the planet.

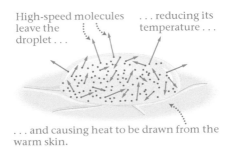
High-speed molecules leave the droplet . . .   . . . reducing its temperature . . .

. . . and causing heat to be drawn from the warm skin.

▲ **FIGURE 17-23 A droplet of sweat resting on the skin** High-speed molecules in a droplet of sweat are able to escape the droplet and become part of the atmosphere. The average speed of the molecules that remain in the droplet is reduced, so the temperature of the droplet is reduced as well.

▲ **FIGURE 17-24** Although the body temperature of this athlete is well below the boiling point of water, the water contained in his sweat is evaporating rapidly from his skin. (Since water vapor is invisible, the "steam" we see in this photo, though it indicates the presence of evaporation, is not the water vapor itself. Rather, it is a cloud of tiny droplets that form when the water vapor loses heat to the cold air around it and condenses back to the liquid state.)

BIO Finally, the expansion of ice can have devastating effects in living systems. If the blood cells of an organism were to freeze, for example, the resulting expansion could rupture the cells. To combat such effects, the crocodile icefish (*Chionodraco hamatus*), a small white-blooded fish found in southern polar seas, has a high concentration of glycoproteins in its bloodstream that serve as a biological antifreeze, inhibiting the formation and growth of ice crystals.

Evaporation

If you think about it for a moment, you realize that evaporation is a bit odd. After all, when you're hot and sweaty from physical exertion, steam rises from your skin. But how can this be, when water boils at 100 °C? How can a skin temperature of only 30 °C or 35 °C result in steam?

Recall that if a liquid is placed in a closed container with a vacuum above it, some of its fastest molecules will break loose and form a gas. When the gas becomes dense enough, its pressure rises to the equilibrium vapor pressure of the liquid, and the system attains equilibrium. But what if you open the container and let a breeze blow across it, removing much of the gas? In that case, the release of molecules from the liquid into the gas—that is, **evaporation**—continues without reaching equilibrium. As the molecules are continually removed, the liquid progressively evaporates until none is left. This is the basic mechanism of evaporation.

BIO Let's investigate how evaporation helps to cool us when we exercise or work up a sweat. First, consider a droplet of sweat on the skin, as illustrated in **FIGURE 17-23**. As mentioned before, the high-speed molecules in the drop are the ones that will escape from the liquid into the surrounding air. The breeze takes these molecules away as soon as they escape; hence, the chance of their reentering the drop of sweat is very small.

Now, what does this mean for the molecules that are left behind in the sweat droplet? Well, the droplet is preferentially losing high-speed molecules; therefore, the average kinetic energy of the remaining molecules must decrease. As we know from kinetic theory, this means that the temperature of the droplet must also decrease. Because the droplet is now cooler than its surroundings, including the skin on which it rests, it draws in heat from the body. This warms the droplet, increasing the speed of its molecules and continuing the evaporation process at more or less the same rate. Thus, sweat droplets are an effective means of drawing heat from the body and releasing it into the surrounding air in the form of high-speed water molecules. A dramatic example of evaporative cooling is shown in **FIGURE 17-24**, and two more examples are presented in **FIGURE 17-25**.

To look at this a bit more quantitatively, consider the Maxwell speed distribution for water molecules in a sweat droplet at 30 °C. This is shown in **FIGURE 17-26**. The rms speed at this temperature is 648 m/s. Also shown in Figure 17-26 is the speed distribution for water molecules at 100 °C. In this case the rms speed is 719 m/s, only slightly greater than that at 30 °C. Thus, it's clear that if water molecules at 100 °C have enough speed to escape into the gas phase, many molecules at 30 °C will be able to escape as well.

(a)

(b)

◀ **FIGURE 17-25 Visualizing Concepts**
Evaporative Cooling (a) Human beings keep cool by sweating. Because the most energetic molecules are the ones most likely to escape by evaporation, significant quantities of heat are removed from the body when perspiration evaporates from the skin. **(b)** Dogs, lacking sweat glands, nevertheless take advantage of the same mechanism to help regulate body temperature. In hot weather they pant to promote evaporation from the tongue. Of course, sitting in a cool pond can also help. (What mechanism is involved here?)

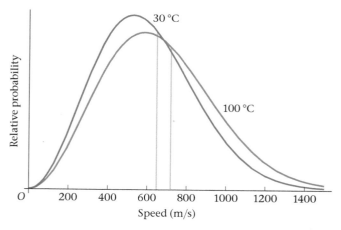

◀ **FIGURE 17-26 Speed distribution for water** The blue curve shows the speed distribution for water at 30 °C, and the red curve corresponds to 100 °C. Note that the rms speed for 100 °C is only slightly greater than that for 30 °C.

The Evaporating Atmosphere

RWP The atmosphere of a planet or a moon can evaporate in much the same way a drop of sweat evaporates from your forehead. In the case of an atmosphere, however, it is the force of gravity that must be overcome. If an astronomical body has a weak gravitational field, some molecules in its atmosphere may be moving rapidly enough to escape; that is, some molecules may have speeds in excess of the escape speed for that body.

Consider the Earth, for example. If a rocket is to escape the Earth, it must have a speed of at least 11,200 m/s. Recall from Chapter 12, however, that the escape speed is independent of the mass of the rocket. Thus, it applies equally well to molecules and rockets. As a result, molecules moving faster than 11,200 m/s may escape the Earth.

Now, let's compare the escape speed to the speeds of some of the molecules in the Earth's atmosphere. We have already seen in this chapter that the speeds of nitrogen and oxygen are on the order of the speed of sound—that is, several hundred meters per second. This is far below the Earth's escape speed. Thus, the odds against a nitrogen or oxygen molecule having enough speed to escape the Earth are truly astronomical. Good thing, too, because this is why these molecules have persisted in our atmosphere for billions of years.

On the other hand, consider a lightweight molecule like hydrogen, H_2. The fact that its average kinetic energy is the same as that of all the other molecules in the air (see Equation 17-12) means that its speed will be much greater than the speed of, say, oxygen or nitrogen. This is illustrated in **FIGURE 17-27**, where we show the speed distributions for both O_2 and H_2 at 20 °C. Clearly, it's very likely to find an H_2 molecule with a speed on the order of a couple thousand meters per second. Because of the higher speeds for H_2, the probability that an H_2 molecule will have enough speed to escape the Earth is about

▲ **FIGURE 17-27 Speed distribution for O_2 and H_2 at 20 °C** The typical speeds of an H_2 molecule (red curve) are much greater than those for an O_2 molecule (blue curve). In fact, some H_2 molecules move fast enough to escape from the Earth's atmosphere.

300 orders of magnitude greater than the corresponding probability for O_2! It is no surprise, then, that Earth's atmosphere contains essentially no hydrogen.

On Jupiter, however, gravity is more intense than on Earth and the temperature is lower. As a result, not even hydrogen moves fast enough to escape. In fact, Jupiter's atmosphere is composed of mostly H_2 and He.

At the other extreme, the Moon has a rather weak gravitational field. In fact, it is unable to maintain any atmosphere at all. Whatever atmosphere it may have had early in its history has long since evaporated. You might say that the Moon's atmosphere is "lost in space."

Enhance Your Understanding   (Answers given at the end of the chapter)

4. A portion of a substance's phase diagram is shown in FIGURE 17-28. **(a)** Is the curve in the figure the fusion curve, the vapor-pressure curve, or the sublimation curve? **(b)** Which of the processes (AB, BC, CD, DE) corresponds to heating the substance at constant pressure? **(c)** Which of the processes (AB, BC, CD, DE) corresponds to increasing the pressure of the substance at constant temperature?

Section Review

- When two phases of a substance are in equilibrium, the number of molecules in each phase remains constant.

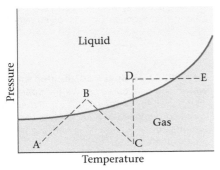

▲ **FIGURE 17-28** Enhance Your Understanding 4.

17-5 Latent Heats

When two phases coexist, something surprising happens—the temperature remains the same even when you add a small amount of heat. How can that be?

To understand this behavior, let's start by considering an ice cube initially at a temperature of $-10\,°C$. Adding an amount of heat Q results in a temperature increase ΔT given by the relationship $Q = mc_{ice}\Delta T$, as discussed in Chapter 16 . When the ice cube's temperature reaches $0\,°C$, however, adding more heat does not cause an additional increase in temperature. Instead, the heat goes into converting some of the ice into water. On a microscopic level, the added heat causes some of the molecules in the solid ice to vibrate with enough energy to break loose from neighboring molecules and become part of the liquid.

Thus, as long as any ice remains in a cup of water, *and* the water and ice are in equilibrium, you can be sure that both the ice and the water are at $0\,°C$. If heat is added to the system, the amount of ice decreases; if heat is removed, the amount of ice increases. The amount of heat required to completely convert 1 kg of ice to water is referred to as the **latent heat,** L. In general, the latent heat is defined as follows:

> The latent heat, L, is the heat that must be added to or removed from one kilogram of a substance to convert it from one phase to another.
>
> During the conversion process, the temperature of the system remains constant.

Just as with the specific heat, the latent heat is always a positive quantity.

In mathematical form, we can say that the heat Q required to convert a mass m from one phase to another is $Q = mL$. This gives the following definition for the latent heat:

Definition of Latent Heat, L

$$L = \frac{Q}{m}$$

17-20

SI unit: J/kg

A Variety of Latent Heats The value of the latent heat depends not only on which substance we're considering, but also on which phases are involved. For example, the latent heat to melt (or fuse) a substance is referred to as the **latent heat of fusion,**

Big Idea 4 When two phases of a substance are in equilibrium, the addition or removal of heat does not change the temperature of the substance. Instead, the amount of each phase in the system changes.

Big Idea 5 The amount of heat necessary to change one phase of a substance to another at constant temperature is the latent heat. The latent heat depends on the substance, and also on the two phases in question.

L_f. Similarly, the latent heat required to convert a liquid to a gas is the **latent heat of vaporization,** L_v, and the latent heat needed to convert a solid directly to a gas is the **latent heat of sublimation,** L_s. A novel example of a sublimating system on Mars is shown in FIGURE 17-29.

Typical latent heats for a variety of substances are given in Table 17-4.

▲ **FIGURE 17-29** This recently discovered ice lake on the surface of Mars lies on the floor of an impact crater. It grows or shrinks with the Martian seasons. Because atmospheric pressure is so low on Mars, however, the ice does not melt during the Martian summer—instead, it sublimates directly to the vapor phase. On Mars, water ice would have behavior similar to that of dry ice here on Earth.

TABLE 17-4 **Latent Heats for Various Materials**

Material	Latent heat of fusion, L_f (J/kg)	Latent heat of vaporization, L_v (J/kg)
Water	33.5×10^4	22.6×10^5
Ammonia	33.2×10^4	13.7×10^5
Copper	20.7×10^4	47.3×10^5
Benzene	12.6×10^4	3.94×10^5
Ethyl alcohol	10.8×10^4	8.55×10^5
Gold	6.28×10^4	17.2×10^5
Nitrogen	2.57×10^4	2.00×10^5
Lead	2.32×10^4	8.59×10^5
Oxygen	1.39×10^4	2.13×10^5

The relationship between the temperature of a substance (in this case water) and the heat added to it is illustrated in FIGURE 17-30. Initially 1 kg of H_2O is in the form of ice at $-20\,°C$. As heat is added to the ice, its temperature rises until it begins to melt at 0 °C. The temperature then remains constant until the latent heat of fusion is supplied to the system. When all the ice has melted to water at 0 °C, continued heating results in a renewed increase in temperature. When the temperature of the water rises to 100 °C, boiling begins and the temperature again remains constant—this time until an amount of heat equal to the latent heat of vaporization is added to the system. Finally, with the entire system converted to steam, continued heating again produces an increasing temperature.

▲ **FIGURE 17-30** **Temperature versus heat added or removed** The temperature of $m = 1.000$ kg of water as heat is added to or removed from the system. Note that the temperature stays the same—even as heat is added—when the system is changing from one phase to another. The points A, B, C, and D are referred to in Homework Problems 56 and 57.

CONCEPTUAL EXAMPLE 17-15 SEVERITY OF A BURN

Both water at 100 °C and steam at 100 °C can cause serious burns. Is a burn produced by steam likely to be more severe, less severe, or the same severity as a burn produced by water?

REASONING AND DISCUSSION

As the water or steam comes into contact with the skin, it cools from 100 °C to a skin temperature of something like 35 °C. For the case of water, this means that a certain amount of heat is transferred to the skin, which can cause a burn. The steam, on the other hand, must first give off the heat required for it to condense to water at 100 °C. After that, the condensed water cools to body temperature, as before. Thus, the heat transferred to the skin will be greater in the case of steam, resulting in a more serious burn.

ANSWER

The steam burn is more severe.

▶ **FIGURE 17-31 Heat required for a given change in temperature** The amount of heat required to raise 0.550 kg of H_2O from ice at −20.0 °C to water at 20.0 °C is the heat difference between points A and D. To calculate this heat we sum the following three heats: (1) the heat to warm the ice from −20.0 °C to 0 °C; (2) the heat to melt all the ice; (3) the heat to warm the water from 0 °C to 20.0 °C.

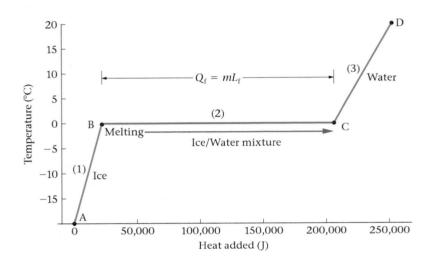

Practical Applications of Latent Heats and Specific Heats As a numerical example of using latent heat, let's calculate the heat energy required to raise the temperature of 0.550 kg of ice from −20.0 °C to water at 20.0 °C—that is, from point A to point D in **FIGURE 17-31**. The way to approach a problem like this is to take it one step at a time—that is, each phase, and each conversion from one phase to another, should be treated separately.

Thus, the first step (A to B in Figure 17-31) is to find the heat necessary to warm the ice from −20.0 °C to 0 °C. Using the specific heat of ice, $c_{ice} = 2090 \, J/(kg \cdot C°)$, we find

$$Q_1 = mc_{ice}\,\Delta T = (0.550 \text{ kg})[2090 \, J/(kg \cdot C°)](20.0 \, C°) = 23,000 \, J$$

The second step in the process (B to C) is to melt the ice at 0 °C. The heat required for this is found using the latent heat of fusion for water ($L_f = 33.5 \times 10^4 \, J/kg$):

$$Q_2 = mL_f = (0.550 \text{ kg})(33.5 \times 10^4 \, J/kg) = 184,000 \, J$$

Finally, the third step (C to D) is to heat the water at 0 °C to 20.0 °C. This time we use the specific heat for water, $c_{water} = 4186 \, J/(kg \cdot C°)$:

$$Q_3 = mc_{water}\,\Delta T = (0.550 \text{ kg})[4186 \, J/(kg \cdot C°)](20.0 \, C°) = 46,000 \, J$$

The total heat required for this process, then, is

$$Q_{total} = Q_1 + Q_2 + Q_3$$
$$= 23,000 \, J + 184,000 \, J + 46,000 \, J = 253,000 \, J$$

We consider a similar problem in the following Example.

PROBLEM-SOLVING NOTE

Specific Heats Versus Latent Heats

When you solve problems involving specific heats and latent heats, recall that specific heats give the heat related to a *change in temperature* in a given phase, and latent heats give the heat related to a *change in phase* at a given temperature.

EXAMPLE 17-16 STEAM HEAT

To make steam, you add 5.60×10^5 J of heat to 0.220 kg of water at an initial temperature of 50.0 °C. Find the final temperature of the steam.

PICTURE THE PROBLEM
Our sketch shows the temperature versus the heat added for water. The initial point for this system, placed at the origin, is at 50.0 °C. As we shall see, adding the given amount of heat, $Q = 5.60 \times 10^5$ J, raises the temperature to the point labeled "Final" in the plot.

REASONING AND STRATEGY
To find the final temperature, we first calculate the amount of heat that must be added to heat the water to 100 °C. If this is less than the total heat added, we continue by calculating the amount of heat needed to vaporize all the water. If the sum of these two heats is still less than the total heat added to the water, we calculate the increase in temperature when the remaining heat is added to the steam.

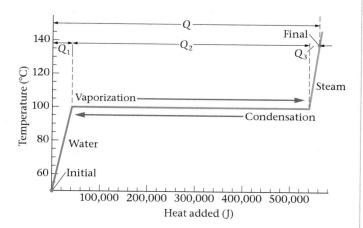

Known Heat added, $Q = 5.60 \times 10^5$ J; mass of water, $m = 0.220$ kg; initial temperature of water, $T_i = 50.0$ °C.
Unknown Final temperature of water, $T_f = ?$

SOLUTION

1. Calculate the heat that must be added to the water to heat it to 100 °C. Call the result Q_1:

$$Q_1 = mc_{water}\,\Delta T$$
$$= (0.220 \text{ kg})[4186 \text{ J}/(\text{kg} \cdot \text{C}°)](50.0 \text{ C}°)$$
$$= 4.60 \times 10^4 \text{ J}$$

2. Next, calculate the heat that must be added to the water to convert it to steam. Let this result be Q_2:

$$Q_2 = mL_f = (0.220 \text{ kg})(22.6 \times 10^5 \text{ J/kg})$$
$$= 4.97 \times 10^5 \text{ J}$$

3. Determine the heat that is still to be added to the system. Let this remaining heat be Q_3:

$$Q_3 = 5.60 \times 10^5 \text{ J} - Q_1 - Q_2$$
$$= 5.60 \times 10^5 \text{ J} - 4.60 \times 10^4 \text{ J} - 4.97 \times 10^5 \text{ J}$$
$$= 17{,}000 \text{ J}$$

4. Use Q_3 to find the increase in temperature of the steam:

$$Q_3 = mc_{steam}\,\Delta T$$
$$\Delta T = \frac{Q_3}{mc_{steam}} = \frac{17{,}000 \text{ J}}{(0.220 \text{ kg})[(2010 \text{ J})/(\text{kg} \cdot \text{C}°)]} = 38 \text{ C}°$$

5. The final temperature is 100 °C plus the increase in temperature found in Step 4:

$$T_f = 100 °C + 38 \text{ C}° = 138 °C$$

INSIGHT
Thus, the system ends up completely converted to steam at a temperature of 138 °C. If the amount of heat added to the system had been greater than Q_1 but less than $Q_1 + Q_2$, the final temperature of the system would have been 100 °C. In this case, the final state of the system is a mixture of liquid water and steam.

PRACTICE PROBLEM
Find the final temperature if the amount of heat added to the system is 3.40×10^5 J. [**Answer:** Only 0.130 kg of water vaporizes into steam. Hence the final temperature is 100 °C, with water and steam in equilibrium.]

Some related homework problems: Problem 49, Problem 54

RWP A pleasant application of latent heat is found in the making of homemade ice cream. As you may know, it is necessary to add salt to the ice-water mixture surrounding the container holding the ingredients for the ice cream. The dissolved salt molecules interact with water molecules in the liquid, impairing their ability to interact with one another and freeze. This means that ice and water are no longer in equilibrium at 0 °C; a lower temperature is required. The result is that ice begins to melt in the ice-water mixture, and, in the process of melting, the ice draws the required latent heat from its surroundings—which includes the ice cream. Thus, the salt together with the ice produces a temperature lower than the melting temperature of ice alone.

Enhance Your Understanding (Answers given at the end of the chapter)

5. Which requires more heat: melting 100 kg of copper, or boiling 10 kg of water?

Section Review

- Latent heat is the amount of heat needed to change one phase of a substance to another phase at constant temperature. Latent heat depends on the substance, and also on the phases in question.

17-6 Phase Changes and Energy Conservation

In the last section, we considered problems in which a given amount of heat is simply added to or removed from a system. We now turn to a more interesting type of problem involving energy conservation. In these problems, heat is exchanged within a system—that is, between its parts—but not with the external world. As a result, the total energy of the system is constant. Still, the heat flow within the system can cause changes in temperatures and phases.

The basic idea in solving energy-conservation problems is the following:

> Set the magnitude of the heat *lost* by one part of the system equal to the magnitude of the heat *gained* by another part of the system.

For example, consider the following problem: A 0.0420-kg ice cube at −10.0 °C is placed in a Styrofoam cup containing 0.350 kg of water at 0 °C. Assuming that the cup can be ignored, and that no heat is exchanged with the surroundings, find the mass of ice in the system when the equilibrium temperature of 0 °C is reached.

Applying Energy Conservation The initial setup for this problem is illustrated in **FIGURE 17-32 (a)**. Because the water is warmer than the ice, it follows that heat flows into the ice from the water, as indicated in **FIGURE 17-32 (b)**. The amount of heat will be just enough to raise the temperature of the ice from −10.0 °C to 0 °C. Thus,

$$Q = \text{heat gained by the ice}$$
$$= mc_{ice}\,\Delta T = (0.0420\text{ kg})[2090\text{ J}/(\text{kg}\cdot\text{C}°)](10.0\text{ C}°) = 878\text{ J}$$

By energy conservation, we can say that the heat lost by the water has the same magnitude as the heat gained by the ice. (In general, in these problems it is simplest to calculate the magnitude of each heat—so that all the heats are positive—and then set the "lost" and "gained" magnitudes equal.) In this case, then, we have

$$\text{heat lost by the water} = \text{heat gained by the ice} = 878\text{ J}$$

Thus, 878 J of heat are removed from the water.

PROBLEM-SOLVING NOTE

Determining Equilibrium

In problems involving two different phases—like solid and liquid—it may not be clear in advance whether both phases or only one is present in the equilibrium state. It may be necessary to assume one case or the other and proceed with the calculation based on that assumption. If the result obtained in this way is not physical, the assumption must be changed.

The heat Q that leaves the water . . .

. . . causes some of it to freeze, enlarging the ice cube.

▶ **FIGURE 17-32 Water freezing to form ice** An ice cube at −10.0 °C is placed in water at 0 °C. Since the ice is colder than the water, heat flows *from* the water *into* the ice. Heat that leaves the water results in the formation of additional ice in the system.

(a) Initial   **(b)** Heat goes from the warmer water into the colder ice   **(c)** Final

Now, noting that the water is already at 0 °C, removing heat from it does not lower its temperature—instead, it merely converts some of the water to ice. How much is converted? The amount is determined by the latent heat of fusion. In particular, we have

$$Q = \text{heat lost by the water}$$
$$= mL_f = 878\,\text{J}$$

In this expression, m is the mass of water that has been converted to ice. Solving for the mass yields

$$m = \frac{Q}{L_f} = \frac{878\,\text{J}}{33.5 \times 10^4\,\text{J/kg}} = 0.00262\,\text{kg}$$

Thus, the final amount of ice in the system is 0.0420 kg + 0.00262 kg = 0.0446 kg. This is illustrated in **FIGURE 17-32 (c)**.

Finally, in the next Example we consider a system in which a quantity of ice completely melts in warm water. The result is a container of liquid water at an intermediate temperature. Still, the basic idea is to apply energy conservation.

EXAMPLE 17-17 WARM PUNCH

A large punch bowl holds 3.95 kg of lemonade (which is essentially just water) at 20.0 °C. A 0.0450-kg ice cube at 0 °C is placed in the lemonade. What is the final temperature of the system, and how much ice (if any) remains when the system reaches equilibrium? Ignore any heat exchange with the bowl or the surroundings.

PICTURE THE PROBLEM
Our sketch shows the various stages imagined for this problem, starting with the initial condition in which the ice is at 0 °C and the lemonade (water) is at 20.0 °C. As one might guess from the large amount of water and the small amount of ice, all of the ice will melt. Therefore, the final condition is a container of liquid water at a final temperature, T_f.

| Initial condition | Calculate heat Q_w lost by water, 20 °C → 0 °C | Calculate heat Q_{ice} needed to melt ice cube | Calculate remaining heat $Q = Q_w - Q_{ice}$ | Calculate final temperature when heat Q is added to water and melted ice |

REASONING AND STRATEGY
To apply energy conservation to this problem, we first calculate the heat that would be lost by the water, Q_w, if we cooled it to 0 °C. We then imagine using part of this heat, Q_{ice}, to melt the ice.

As we shall see, a great deal of heat, $Q = Q_w - Q_{ice}$, is left after the ice is melted. We imagine adding this heat back into the system, which now contains 3.95 kg + 0.0450 kg of water at 0 °C. Calculating the increase in temperature caused by the heat Q gives us the final temperature of the system.

Known Initial mass of water, m_{water} = 3.95 kg; initial temperature of water, 20.0 °C; mass of ice cube, m_{ice} = 0.0450 kg; initial temperature of ice, 0 °C.
Unknown Final temperature of system, T_f = ?

SOLUTION

1. Find the heat lost by the water, Q_w, if it is cooled to 0 °C:

$$Q_w = m_{water}c_{water}\Delta T$$
$$= (3.95\,\text{kg})[4186\,\text{J/(kg·C°)}](20.0\,\text{C°})$$
$$= 3.31 \times 10^5\,\text{J}$$

2. Calculate the amount of heat, Q_{ice}, needed to melt all the ice:

$$Q_{ice} = m_{ice}L_f$$
$$= (0.0450\,\text{kg})(33.5 \times 10^4\,\text{J/kg}) = 1.51 \times 10^4\,\text{J}$$

3. Determine the amount of heat, Q, that is left:

$$Q = Q_w - Q_{ice}$$
$$= 3.31 \times 10^5\,\text{J} - 1.51 \times 10^4\,\text{J} = 3.16 \times 10^5\,\text{J}$$

CONTINUED

4. Use the remaining heat Q to warm the 3.95 kg + 0.0450 kg of water at 0 °C to its final temperature:

$$Q = (m_{water} + m_{ice})c_{water}\, \Delta T$$
$$\Delta T = \frac{Q}{(m_{water} + m_{ice})c_{water}}$$
$$= \frac{3.16 \times 10^5\, J}{(3.95\, kg + 0.0450\, kg)[4186\, J/(kg \cdot C°)]}$$
$$= 18.9\, C°$$

5. The final temperature is 0 °C plus the change in temperature found in Step 4:

$$T_f = 0\,°C + 18.9\, C° = 18.9\,°C$$

INSIGHT

As expected, the relatively small ice cube did not lower the temperature of the system very much. In fact, the temperature of the lemonade (water) decreased only from 20.0 °C to 18.9 °C.

PRACTICE PROBLEM

What would the final temperature of the system be if the ice cube's mass were 0.0750 kg? [**Answer:** $T_f = 18.2\,°C$]

Some related homework problems: Problem 60, Problem 64

Enhance Your Understanding   (Answers given at the end of the chapter)

6. An ice cube is placed in a cup of water. A few minutes later half the ice has melted, and the remainder is in equilibrium with the water. **(a)** Was the initial temperature of the water greater than, less than, or equal to 0 °C? Explain. **(b)** Is the final temperature of the water greater than, less than, or equal to 0 °C? Explain.

Section Review

• When heat is exchanged between various parts of a closed system, the amount of each phase may change, but the total energy of the system remains constant.

CHAPTER 17 REVIEW

CHAPTER SUMMARY

17-1   IDEAL GASES

An ideal gas is a simplified model of a real gas in which interactions between molecules are ignored.

Equation of State

The equation of state for an ideal gas is

$$PV = NkT \qquad \text{17-2}$$

In this expression, N is the number of molecules, T is the Kelvin temperature, and $k = 1.38 \times 10^{-23}$ J/K is Boltzmann's constant.

In terms of the universal gas constant, $R = 8.31$ J/(mol·K), and the number of moles in the gas, n, the ideal-gas equation of state is

$$PV = nRT \qquad \text{17-5}$$

Avogadro's Number and Moles

The number of molecules in a mole (mol) is Avogadro's number, $N_A = 6.022 \times 10^{23}$.

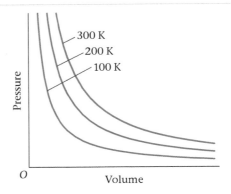

17-2   THE KINETIC THEORY OF GASES

In kinetic theory, a gas is imagined to be composed of a large number of pointlike molecules bouncing off the walls of a container.

Kinetic Energy and Temperature

Kinetic theory relates the average kinetic energy of the molecules in a gas to the Kelvin temperature of the gas, T:

$$\left(\tfrac{1}{2}mv^2\right)_{\text{av}} = K_{\text{av}} = \tfrac{3}{2}kT \qquad\qquad \text{17-12}$$

RMS Speed

The rms (root mean square) speed of the molecules in a gas at the Kelvin temperature T is

$$v_{\text{rms}} = \sqrt{\frac{3kT}{m}} = \sqrt{\frac{3RT}{M}} \qquad\qquad \text{17-13}$$

Internal Energy of an Ideal Gas

The internal energy of a monatomic ideal gas is

$$U = \tfrac{3}{2}NkT = \tfrac{3}{2}nRT \qquad\qquad \text{17-15}$$

17-3 SOLIDS AND ELASTIC DEFORMATION

When a force is applied to a solid, its size and shape may change.

Changing the Length of a Solid

The force required to change the length of a solid by the amount ΔL is

$$F = Y\left(\frac{\Delta L}{L_0}\right)A \qquad\qquad \text{17-17}$$

In this expression, Y is Young's modulus, L_0 is the initial length parallel to the applied force, and A is the cross-sectional area perpendicular to the applied force.

Shear Deformation

The force required to shear, or deform, a solid by the amount Δx is

$$F = S\left(\frac{\Delta x}{L_0}\right)A \qquad\qquad \text{17-18}$$

In this expression, S is the shear modulus, L_0 is the initial length perpendicular to the applied force, and A is the cross-sectional area parallel to the applied force.

Changing the Volume of a Solid

The change in pressure required to change the volume of a solid by the amount ΔV is

$$\Delta P = -B\left(\frac{\Delta V}{V_0}\right) \qquad\qquad \text{17-19}$$

In this expression, B is the bulk modulus and V_0 is the initial volume.

17-4 PHASE EQUILIBRIUM AND EVAPORATION

The three most common phases of matter are solid, liquid, and gas. Solids maintain a definite shape, whereas gases and liquids flow to take on the shape of their container.

Equilibrium Between Phases

When phases are in equilibrium, the number of molecules in each phase remains constant.

Equilibrium

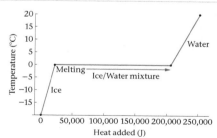

17-5 LATENT HEATS

The latent heat, L, is the amount of heat per unit mass that must be added to or removed from a substance to convert it from one phase to another.

Latent Heat of Fusion

The heat required for melting or freezing is called the latent heat of fusion, L_f.

Latent Heat of Vaporization

The heat required for vaporizing or condensing is the latent heat of vaporization, L_v.

Latent Heat of Sublimation

The heat required to sublime a solid directly to a gas, or to condense a gas to a solid, is the latent heat of sublimation, L_s.

17-6 PHASE CHANGES AND ENERGY CONSERVATION

When heat is exchanged within a system, with no exchanges with the surroundings, the energy of the system is conserved.

ANSWERS TO ENHANCE YOUR UNDERSTANDING QUESTIONS

1. B = D < A < C.

2. The root mean square speed increases by a factor of the square root of 2.

3. B < A < C < D.

4. **(a)** The plot shows the vapor-pressure curve. **(b)** DE corresponds to heating at constant pressure. **(c)** CD corresponds to increasing pressure at constant temperature.

5. More heat is required to boil the water.

6. **(a)** The initial temperature of the water was greater than 0 °C because heat flowed from the water to the ice as half of the ice melted. **(b)** The final temperature of the water is 0 °C because the system is in equilibrium.

CONCEPTUAL QUESTIONS

For instructor-assigned homework, go to www.masteringphysics.com.

(Answers to odd-numbered Conceptual Questions can be found in the back of the book.)

1. How is the air pressure in a tightly sealed house affected by operating the furnace? Explain.

2. The average speed of air molecules in your room is on the order of the speed of sound. What is their average velocity?

3. Is it possible to change both the pressure and the volume of an ideal gas without changing the average kinetic energy of its molecules? If your answer is no, explain why not. If your answer is yes, give a specific example.

4. **An Airport at Great Elevation** One of the highest airports in the world is located in La Paz, Bolivia. Pilots prefer to take off from this airport in the morning or the evening, when the air is quite cold. Explain.

5. A camping stove just barely boils water on a mountaintop. When the stove is used at sea level, will it be able to boil water? Explain your answer.

6. An autoclave is a device used to sterilize medical instruments. It is essentially a pressure cooker that heats the instruments in water under high pressure. This ensures that the sterilization process occurs at temperatures greater than the normal boiling point of water. Explain why the autoclave produces such high temperatures.

7. As the temperature of ice is increased, it changes first into a liquid and then into a vapor. On the other hand, dry ice, which is solid carbon dioxide, changes directly from a solid to a vapor as its temperature is increased. How might one produce liquid carbon dioxide?

8. **BIO** Isopropyl alcohol is sometimes rubbed onto a patient's arms and legs to lower their body temperature. Why is this effective?

9. A drop of water on a kitchen counter evaporates in a matter of minutes. However, only a relatively small fraction of the molecules in the drop move rapidly enough to escape through the drop's surface. Why, then, does the entire drop evaporate rather than just a small fraction of it?

PROBLEMS AND CONCEPTUAL EXERCISES

Answers to odd-numbered Problems and Conceptual Exercises can be found in the back of the book. **BIO** *identifies problems of biological or medical interest;* **CE** *indicates a conceptual exercise.* Predict/Explain *problems ask for two responses: (a) your prediction of a physical outcome, and (b) the best explanation among three provided; and* Predict/Calculate *problems ask for a prediction of a physical outcome, based on fundamental physics concepts, and follow that with a numerical calculation to verify the prediction. On all problems, bullets (•, ••, •••) indicate the level of difficulty*

SECTION 17-1 IDEAL GASES

1. • **CE** **(a)** Is the number of molecules in one mole of N_2 greater than, less than, or equal to the number of molecules in one mole of O_2? **(b)** Is the mass of one mole of N_2 greater than, less than, or equal to the mass of one mole of O_2?

2. • **CE** Predict/Explain If you put a helium-filled balloon in the refrigerator, **(a)** will its volume increase, decrease, or stay the same? **(b)** Choose the *best explanation* from among the following:
   I. Lowering the temperature of an ideal gas at constant pressure results in a reduced volume.
   II. The same amount of gas is in the balloon; therefore, its volume remains the same.
   III. The balloon can expand more in the cool air of the refrigerator, giving an increased volume.

3. • **CE** Two containers hold ideal gases at the same temperature. Container A has twice the volume and half the number of molecules as container B. What is the ratio P_A/P_B, where P_A is the pressure in container A and P_B is the pressure in container B?

4. • Standard temperature and pressure (STP) is defined as a temperature of 0 °C and a pressure of 101.3 kPa. What is the volume occupied by one mole of an ideal gas at STP?

5. • **BIO** After emptying her lungs, a person inhales 4.3 L of air at 0.0 °C and holds her breath. How much does the volume of the air increase as it warms to her body temperature of 35 °C?

6. • An automobile tire has a volume of 0.0185 m³. At a temperature of 294 K the absolute pressure in the tire is 212 kPa. How many moles of air must be pumped into the tire to increase its pressure to 252 kPa, given that the temperature and volume of the tire remain constant?

7. • **Amount of Helium in a Blimp** The Goodyear blimp *Spirit of Akron* is 62.6 m long and contains 7023 m³ of helium. When the temperature of the helium is 285 K, its absolute pressure is 112 kPa. Find the mass of the helium in the blimp.

8. • A compressed-air tank holds 0.500 m³ of air at a temperature of 295 K and a pressure of 820 kPa. What volume would the air occupy if it were released into the atmosphere, where the pressure is 101 kPa and the temperature is 303 K?

9. •• **CE** Four ideal gases have the following pressures, P, volumes, V, and mole numbers, n: gas A, P = 100 kPa, V = 1 m³, n = 10 mol; gas B, P = 200 kPa, V = 2 m³, n = 20 mol; gas C, P = 50 kPa, V = 1 m³, n = 50 mol; gas D, P = 50 kPa, V = 4 m³, n = 5 mol. Rank these gases in order of increasing temperature. Indicate ties where appropriate.

10. •• A balloon contains 3.9 liters of nitrogen gas at a temperature of 84 K and a pressure of 101 kPa. If the temperature of the gas is allowed to increase to 28 °C and the pressure remains constant, what volume will the gas occupy?

11. •• Predict/Calculate A balloon is filled with helium at a pressure of 2.4×10^5 Pa. The balloon is at a temperature of 18 °C and has a radius of 0.25 m. **(a)** How many helium atoms are contained in the balloon? **(b)** Suppose we double the number of helium atoms in the balloon, keeping the pressure and the temperature fixed. By what factor does the radius of the balloon increase? Explain.

12. •• Predict/Calculate A bicycle tire with a volume of 0.00212 m³ is filled to its recommended absolute pressure of 495 kPa on a cold winter day when the tire's temperature is −15 °C. The cyclist then brings his bicycle into a hot laundry room at 32 °C. **(a)** If the tire warms up while its volume remains constant, will the pressure increase be greater than, less than, or equal to the manufacturer's stated 10% overpressure limit? **(b)** Find the absolute pressure in the tire when it warms to 32 °C at constant volume.

13. •• A 515-cm³ flask contains 0.460 g of a gas at a pressure of 153 kPa and a temperature of 322 K. What is the molecular mass of this gas?

14. •• Predict/Calculate **The Atmosphere of Mars** On Mars, the average temperature is −64 °F and the average atmospheric pressure is 0.92 kPa. **(a)** What is the number of molecules per volume in the Martian atmosphere? **(b)** Is the number of molecules per volume on the Earth greater than, less than, or equal to the number per volume on Mars? Explain your reasoning. **(c)** Estimate the number of molecules per volume in Earth's atmosphere.

15. •• The air inside a hot-air balloon has an average temperature of 79.2 °C. The outside air has a temperature of 20.3 °C. What is the ratio of the density of air in the balloon to the density of air in the surrounding atmosphere?

16. •• A cylindrical flask is fitted with an airtight piston that is free to slide up and down, as shown in FIGURE 17-33. A mass rests on top of the piston. The initial temperature of the system is 313 K and the pressure of the gas is held constant at 137 kPa. The temperature is now increased until the height of the piston rises from 23.4 cm to 26.0 cm. What is the final temperature of the gas?

23.4 cm

FIGURE 17-33 Problems 16 and 17

17. •• Consider the system described in the previous problem. Contained within the flask is an ideal gas at a constant temperature of 313 K. Initially the pressure applied by the piston and the mass is 137 kPa and the height of the piston above the base of the flask is 23.4 cm. When additional mass is added to the piston, the height of the piston decreases to 20.0 cm. Find the new pressure applied by the piston.

18. ••• Predict/Calculate A gas has a temperature of 310 K and a pressure of 101 kPa. **(a)** Find the volume occupied by 1.25 mol of this gas, assuming it is ideal. **(b)** Assuming the gas molecules can be approximated as small spheres of diameter 2.5×10^{-10} m, determine the fraction of the volume found in part (a) that is occupied by the molecules. **(c)** In determining the properties of an ideal gas, we assume that molecules are points of zero volume. Discuss the validity of this assumption for the case considered here.

SECTION 17-2 THE KINETIC THEORY OF GASES

19. • CE Predict/Explain The air in your room is composed mostly of oxygen (O_2) and nitrogen (N_2) molecules. The oxygen molecules are more massive than the nitrogen molecules. **(a)** Is the rms speed of the O_2 molecules greater than, less than, or equal to the rms speed of the N_2 molecules? **(b)** Choose the *best explanation* from among the following:
I. The more massive oxygen molecules have greater momentum and therefore greater speed.
II. Equal temperatures for the oxygen and nitrogen molecules imply they have equal rms speeds.
III. The temperature is the same for both molecules, and hence their average kinetic energies are equal. As a result, the more massive oxygen molecules have lower speeds.

20. • CE If the translational speed of molecules in an ideal gas is doubled, by what factor does the Kelvin temperature change? Explain.

21. • At what temperature is the rms speed of H_2 equal to the rms speed that O_2 has at 303 K?

22. • Suppose a planet has an atmosphere of pure ammonia at 5.5 °C. What is the rms speed of the ammonia molecules? (The molecular weight of ammonia, NH_3, is 17.03 g/mol.)

23. •• Predict/Calculate Three moles of oxygen gas (that is, 3.0 mol of O_2) are placed in a portable container with a volume of 0.0035 m³. If the temperature of the gas is 295 °C, find **(a)** the pressure of the gas and **(b)** the average kinetic energy of an oxygen molecule. **(c)** Suppose the volume of the gas is doubled, while the temperature and number of moles are held constant. By what factor do your answers to parts (a) and (b) change? Explain.

24. •• Predict/Calculate The rms speed of O_2 is 1550 m/s at a given temperature. **(a)** Is the rms speed of H_2O at this temperature greater than, less than, or equal to 1550 m/s? Explain. **(b)** Find the rms speed of H_2O at this temperature.

25. •• Predict/Calculate An ideal gas is kept in a container of constant volume. The pressure of the gas is also kept constant. **(a)** If the number of molecules in the gas is doubled, does the rms speed increase, decrease, or stay the same? Explain. **(b)** If the initial rms speed is 1300 m/s, what is the final rms speed?

26. •• What is the temperature of a gas of CO_2 molecules whose rms speed is 309 m/s?

27. •• The rms speed of a sample of gas is increased by 1%. **(a)** What is the percent change in the temperature of the gas? **(b)** What is the percent change in the pressure of the gas, assuming its volume is held constant?

28. ••• **Enriching Uranium** In naturally occurring uranium atoms, 99.3% are ^{238}U (atomic mass = 238 u, where u = 1.6605×10^{-27} kg) and only 0.7% are ^{235}U (atomic mass = 235 u). Uranium-fueled reactors require an enhanced proportion of ^{235}U. Since both isotopes of uranium have identical chemical properties, they can be separated only by methods that depend on their differing masses. One such method is gaseous diffusion, in which uranium hexafluoride (UF_6) gas diffuses through a series of porous barriers. The lighter $^{235}UF_6$ molecules have a slightly higher rms speed at a given temperature than the heavier $^{238}UF_6$ molecules, and this allows the two isotopes to be separated. Find the ratio of the rms speeds of the two isotopes at 23.0 °C.

29. ••• A 380-mL spherical flask contains 0.065 mol of an ideal gas at a temperature of 283 K. What is the average force exerted on the walls of the flask by a single molecule?

SECTION 17-3 SOLIDS AND ELASTIC DEFORMATION

30. • **CE** Predict/Explain A hollow cylindrical rod (rod 1) and a solid cylindrical rod (rod 2) are made of the same material. The two rods have the same length and the same outer radius. If the same compressional force is applied to each rod, **(a)** is the change in length of rod 1 greater than, less than, or equal to the change in length of rod 2? **(b)** Choose the *best explanation* from among the following:

    I. The solid rod has the larger effective cross-sectional area, since the empty part of the hollow rod doesn't resist compression. Therefore, the solid rod has the smaller change in length.

    II. The rods have the same outer radius and hence the same cross-sectional area. As a result, their change in length is the same.

    III. The walls of the hollow rod are hard and resist compression more than the uniform material in the solid rod. Therefore the hollow rod has the smaller change in length.

31. • A rock climber hangs freely from a nylon rope that is 16 m long and has a diameter of 8.1 mm. If the rope stretches 4.2 cm, what is the mass of the climber?

32. • **BIO** To stretch a relaxed biceps muscle 2.5 cm requires a force of 25 N. Find the Young's modulus for the muscle tissue, assuming it to be a uniform cylinder of length 0.24 m and cross-sectional area 47 cm^2.

33. • A 22-kg chimpanzee hangs from the end of a horizontal, broken branch 1.1 m long, as shown in FIGURE 17-34. The branch is a uniform cylinder 4.6 cm in diameter, and the end of the branch supporting the chimp sags downward through a vertical distance of 13 cm. What is the shear modulus for this branch?

FIGURE 17-34 Problem 33

34. • **The Marianas Trench** The deepest place in all the oceans is the Marianas Trench, where the depth is 10.9 km and the pressure is 1.10×10^8 Pa. If a copper ball 15.0 cm in diameter is taken to the bottom of the trench, by how much does its volume decrease?

35. •• **CE** Four cylindrical rods with various cross-sectional areas and initial lengths are stretched by an applied force, as in Figure 17-11. The resulting change in length of each rod is given in the following table. Rank these rods in order of increasing Young's modulus. Indicate ties where appropriate.

Rod	Applied Force	Cross-Sectional Area	Initial Length	Change in Length
1	F	A	L	ΔL
2	F	$2A$	$2L$	ΔL
3	$2F$	$2A$	L	$2\Delta L$
4	$3F$	A	$L/2$	ΔL

36. •• Predict/Calculate A steel wire 4.1 m long stretches 0.13 cm when it is given a tension of 380 N. **(a)** What is the diameter of the wire? **(b)** If it is desired that the stretch be less than 0.13 cm, should its diameter be increased or decreased? Explain.

37. •• **BIO** **Spiderweb** An orb weaver spider with a mass of 0.26 g hangs vertically by one of its threads. The thread has a Young's modulus of 4.7×10^9 N/m^2 and a radius of 9.8×10^{-6} m. **(a)** What is the fractional increase in the thread's length caused by the spider? **(b)** Suppose a 76-kg person hangs vertically from a nylon rope. What radius must the rope have if its fractional increase in length is to be the same as that of the spider's thread?

38. •• Predict/Calculate Two rods of equal length (0.55 m) and diameter (1.7 cm) are placed end to end. One rod is aluminum, the other is brass. If a compressive force of 8400 N is applied to the rods, **(a)** how much does their combined length decrease? **(b)** Which of the rods changes its length by the greatest amount? Explain.

39. •• A piano wire 0.82 m long and 0.93 mm in diameter is fixed on one end. The other end is wrapped around a tuning peg 3.5 mm in diameter. Initially the wire, whose Young's modulus is 2.4×10^{10} N/m^2, has a tension of 14 N. Find the tension in the wire after the tuning peg has been turned through one complete revolution.

SECTION 17-4 PHASE EQUILIBRIUM AND EVAPORATION

40. • **CE** The formation of ice from water is accompanied by which of the following: **(a)** an absorption of heat by the water; **(b)** an increase in temperature; **(c)** a decrease in volume; **(d)** a removal of heat from the water; **(e)** a decrease in temperature?

41. • **Vapor Pressure for Water** FIGURE 17-35 shows a portion of the vapor-pressure curve for water. Referring to the figure, estimate the pressure that would be required for water to boil at 30 °C.

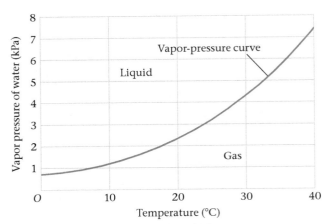

FIGURE 17-35 Problems 41, 42, 43, and 44

42. • Using the vapor-pressure curve given in Figure 17-35, find the temperature at which water boils when the pressure is 1.5 kPa.

43. •• **Dew-Point Temperature** One measure of the amount of water in the atmosphere is the *dew point*, the temperature at which the atmosphere's water content would be in equilibrium with liquid water and droplets of dew would begin to form. **(a)** On a warm summer day when the dew-point temperature is 68 °F, what is the pressure of water in the atmosphere? Refer to Figure 17-35 and note that the pressure of the water vapor (called the *partial pressure of water*) is equal to the vapor pressure at the dew-point temperature. **(b)** If the weather changes and the partial pressure of water is reduced to 1.7 kPa, what is the new dew-point temperature?

44. •• **Relative Humidity** One measure of the amount of water in the atmosphere is the *relative humidity*, the ratio of the actual partial pressure of water vapor to the maximum possible partial pressure of water. If the partial pressure of water vapor exceeds the vapor pressure (Figure 17-35) at that temperature, the vapor condenses to liquid and either fog or dew begins to form. In that case the relative humidity would be 100%. Referring to Figure 17-35, if the partial pressure of water in the atmosphere is 1.7 kPa when the air temperature is 25 °C, what is the relative humidity?

45. •• Predict/Calculate **The Vapor Pressure of CO₂** A portion of the vapor-pressure curve for carbon dioxide is given in FIGURE 17-36. **(a)** Estimate the pressure at which CO_2 boils at 0 °C. **(b)** If the temperature is increased, does the boiling pressure increase, decrease, or stay the same? Explain.

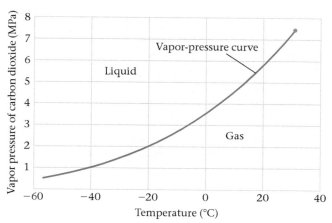

FIGURE 17-36 Problem 45

46. •• **Phase Diagram for Water** The phase diagram for water is shown in FIGURE 17-37. **(a)** What is the temperature T_1 on the phase diagram? **(b)** What is the temperature T_2 on the phase diagram? **(c)** What happens to the melting/freezing temperature of water if atmospheric pressure is *decreased*? Justify your answer by referring to the phase diagram. **(d)** What happens to the boiling/condensation temperature of water if atmospheric pressure is *increased*? Justify your answer by referring to the phase diagram.

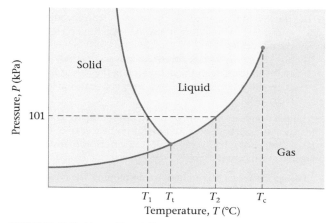

FIGURE 17-37 Problems 46 and 48

47. •• **Phase Diagram for CO₂** The phase diagram for CO_2 is shown in FIGURE 17-38. **(a)** What is the phase of CO_2 at $T = 20$ °C and $P = 500$ kPa? **(b)** What is the phase of CO_2 at $T = -80$ °C and $P = 120$ kPa? **(c)** For reasons of economy and convenience, bulk CO_2 is often transported in liquid form in pressurized tanks. Using the phase diagram, determine the minimum pressure required to keep CO_2 in the liquid phase at 20 °C.

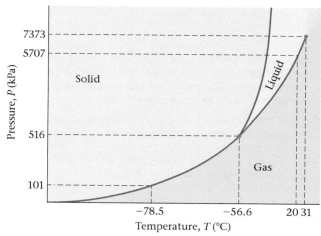

FIGURE 17-38 Problem 47

48. •• A sample of liquid water at atmospheric pressure has a temperature just above the freezing point. Refer to Figure 17-37 to answer the following questions. **(a)** What phase changes occur if the temperature of the system is increased while the pressure is held constant? **(b)** Suppose, instead, that the temperature of the system is held constant just above the freezing point while the pressure is decreased. What phase changes occur now?

SECTION 17-5 LATENT HEATS

49. • How much heat must be removed from 1.96 kg of water at 0 °C to make ice cubes at 0 °C?

50. • A heat transfer of 9.5×10^5 J is required to convert a block of ice at -15 °C to water at 15 °C. What was the mass of the block of ice?

51. • How much heat must be added to 2.55 kg of copper to change it from a solid at 1358 K to a liquid at 1358 K?

52. • An ammonia refrigeration cycle involves the conversion of 0.85 kg of liquid ammonia into vapor every minute at the boiling-point temperature. At what rate does the ammonia absorb energy?

53. •• **CE** Four liquids are at their freezing temperature. Heat is now removed from each of the liquids until it becomes completely solidified. The amount of heat that must be removed, Q, and the mass, m, of each of the liquids are as follows: liquid A, $Q = 33{,}500$ J, $m = 0.100$ kg; liquid B, $Q = 166{,}000$ J, $m = 0.500$ kg; liquid C, $Q = 31{,}500$ J, $m = 0.250$ kg; liquid D, $Q = 5400$ J, $m = 0.0500$ kg. Rank these liquids in order of increasing latent heat of fusion. Indicate ties where appropriate.

54. •• Predict/Calculate A 1.1-kg block of ice is initially at a temperature of -5.0 °C. **(a)** If 5.2×10^5 J of heat are added to the ice, what is the final temperature of the system? Find the amount of ice, if any, that remains. **(b)** Suppose the amount of heat added to the ice block is doubled. By what factor must the mass of the ice be increased if the system is to have the same final temperature? Explain.

55. •• Predict/Calculate Referring to the previous problem, suppose the amount of heat added to the block of ice is reduced by a factor of 2 to 2.6×10^5 J. Note that this amount of heat is still sufficient to melt at least some of the ice. **(a)** Do you expect the temperature increase in this case to be one-half that found in the previous problem? Explain. **(b)** What is the final temperature of the system in this case? Find the amount of ice, if any, that remains.

56. •• Figure 17-30 shows a temperature-versus-heat plot for 1.000 kg of water. **(a)** Calculate the heat corresponding to the points A, B, C, and D. **(b)** Calculate the slope of the line from point B to point C. Show that this slope is equal to $1/c$, where c is the specific heat of liquid water.

57. •• Predict/Calculate Suppose the 1.000 kg of water in Figure 17-30 starts at point A at time zero. Heat is added to this system at the rate of 12,250 J/s. How much time does it take for the system to reach **(a)** point B, **(b)** point C, and **(c)** point D? **(d)** Describe the physical state of the system at time $t = 63.00$ s.

58. •• **BIO** In Conceptual Example 17-15 we pointed out that steam can cause more serious burns than water at the same temperature. Here we examine this effect quantitatively, noting that flesh becomes badly damaged when its temperature reaches 50.0 °C. **(a)** Calculate the heat released as 12.5 g of liquid water at 100 °C is cooled to 50.0 °C. **(b)** Calculate the heat released when 12.5 g of steam at 100 °C is condensed and cooled to 50.0 °C. **(c)** Find the mass of flesh that can be heated from 37.0 °C (normal body temperature) to 50.0 °C for the cases considered in parts (a) and (b). (The average specific heat of flesh is 3500 J/kg · K.)

59. •• When you go out to your car one cold winter morning you discover a 0.58-cm-thick layer of ice on the windshield, which has an area of 1.6 m². If the temperature of the ice is −2.0 °C, and its density is 917 kg/m³, find the heat required to melt all the ice.

SECTION 17-6 PHASE CHANGES AND ENERGY CONSERVATION

60. •• A large punch bowl holds 3.99 kg of lemonade (which is essentially water) at 20.5 °C. A 0.0550-kg ice cube at −10.2 °C is placed in the lemonade. What are the final temperature of the system, and the amount of ice (if any) remaining? Ignore any heat exchange with the bowl or the surroundings.

61. •• A 155-g aluminum cylinder is removed from a liquid nitrogen bath, where it has been cooled to −196 °C. The cylinder is immediately placed in an insulated cup containing 80.0 g of water at 15.0 °C. What is the equilibrium temperature of this system? If your answer is 0 °C, determine the amount of water that has frozen. The average specific heat of aluminum over this temperature range is 653 J/(kg · K).

62. •• An 825-g iron block is heated to 352 °C and placed in an insulated container (of negligible heat capacity) containing 40.0 g of water at 20.0 °C. What is the equilibrium temperature of this system? If your answer is 100 °C, determine the amount of water that has vaporized. The average specific heat of iron over this temperature range is 560 J/(kg · K).

63. •• **Party Planning** You are expecting to serve 32 cups of soft drinks to your guests tonight. Each cup will hold 285 g of a soft drink that has a specific heat of 4186 J/(kg · K) and an initial temperature of 24 °C. If each guest would like to enjoy the drink at 3.0 °C, how much ice (in kg) should you buy? Assume the initial temperature of the ice is 0 °C, and ignore the heat exchange with the plastic cups and the surroundings.

64. •• Predict/Calculate A 35-g ice cube at 0.0 °C is added to 110 g of water in a 62-g aluminum cup. The cup and the water have an initial temperature of 23 °C. **(a)** Find the equilibrium temperature of the cup and its contents. **(b)** Suppose the aluminum cup is replaced with one of equal mass made from silver. Is the equilibrium temperature with the silver cup greater than, less than, or the same as with the aluminum cup? Explain.

65. •• A 48-g block of copper at −12 °C is added to 110 g of water in a 75-g aluminum cup. The cup and the water have an initial temperature of 4.1 °C. **(a)** Find the equilibrium temperature of the cup and its contents. **(b)** What mass of ice, if any, is present when the system reaches equilibrium?

66. •• A 0.075-kg ice cube at 0.0 °C is dropped into a Styrofoam cup holding 0.33 kg of water at 14 °C. **(a)** Find the final temperature of the system and the amount of ice (if any) remaining. Assume the cup and the surroundings can be ignored. **(b)** Find the initial temperature of the water that would be enough to *just barely* melt all of the ice.

67. •• To help keep her barn warm on cold days, a farmer stores 865 kg of warm water in the barn. How many hours would a 2.00-kilowatt electric heater have to operate to provide the same amount of heat as is given off by the water as it cools from 20.0 °C to 0 °C and then freezes at 0 °C?

GENERAL PROBLEMS

68. • **CE** As you go up in altitude, do you expect the ratio of oxygen to nitrogen in the atmosphere to increase, decrease; or stay the same? Explain.

69. • **CE** Predict/Explain Suppose the Celsius temperature of an ideal gas is doubled from 100 °C to 200 °C. **(a)** Does the average kinetic energy of the molecules in this gas increase by a factor that is greater than, less than, or equal to 2? **(b)** Choose the *best explanation* from among the following:
  I. Changing the temperature from 100 °C to 200 °C goes beyond the boiling point, which will increase the kinetic energy by more than a factor of 2.
  II. The average kinetic energy is directly proportional to the temperature, so doubling the temperature doubles the kinetic energy.
  III. Doubling the Celsius temperature from 100 °C to 200 °C changes the Kelvin temperature from 373.15 K to 473.15 K, which is an increase of less than a factor of 2.

70. • **CE** Predict/Explain Suppose the absolute temperature of an ideal gas is doubled from 100 K to 200 K. **(a)** Does the average speed of the molecules in this gas increase by a factor that is greater than, less than, or equal to 2? **(b)** Choose the *best explanation* from among the following:
  I. Doubling the Kelvin temperature doubles the average kinetic energy, but this implies an increase in the average speed by a factor of $\sqrt{2} = 1.414\ldots$, which is less than 2.
  II. The Kelvin temperature is the one we use in the ideal-gas law, and therefore doubling it also doubles the average speed of the molecules.
  III. The change in average speed depends on the mass of the molecules in the gas, and hence doubling the Kelvin temperature generally results in an increase in speed that is greater than a factor of 2.

71. • **Largest Raindrops** Atmospheric scientists studying clouds in the Marshall Islands have observed what they believe to be the world's largest raindrops, with a radius of 0.52 cm. How many molecules are in these monster drops?

72. • **Cooling Computers** Researchers are developing "heat exchangers" for laptop computers that take heat from the laptop—to keep it from being damaged by overheating—and use it to vaporize methanol. Given that 5100 J of heat is removed from the laptop when 4.6 g of methanol is vaporized, what is the latent heat of vaporization for methanol?

73. •• A 1.10-kg block of ice is initially at a temperature of −5.0 °C. **(a)** If 3.1×10^5 J of thermal energy are added to the ice, what is the amount of ice that remains? **(b)** How much additional thermal energy must be added to this system to convert it to 1.10 kg of water at 5.0 °C?

74. •• **Scuba Tanks** In scuba diving circles, "an 80" refers to a scuba tank that holds 80 cubic feet of air, a standard amount for recreational diving. Given that a scuba tank is a cylinder 2 feet long and half a foot in diameter, determine **(a)** the volume of a tank and **(b)** the pressure in a tank when 80 cubic feet of air is compressed into its relatively small volume. **(c)** What is the mass of air in a tank that holds 80 cubic feet of air? Assume the temperature is 21 °C and that the walls of the tank are of negligible thickness.

75. •• **Evaporating Atmosphere** Hydrogen gas evaporates into space even though its rms speed is less than one-fifth of the gravitational escape speed. This is because the distribution of molecular speeds at equilibrium (see Figure 17-27) shows that some of the molecules do have speeds that exceed the escape speed. To what temperature must the atmosphere be heated for the air around us to have an rms speed that is one-fifth of the gravitational escape speed? Treat the air as having a molecular mass of 0.029 kg/mol.

76. •• To make steam, you add 5.60×10^5 J of thermal energy to 0.220 kg of water at an initial temperature of 50.0 °C. Find the final temperature of the steam.

77. •• **A Boiling Geyser (a)** The column of water that forms as a geyser erupts is 2100 m tall. What is the pressure at the bottom of the column of water? **(b)** The water vapor-pressure curve (Figure 17-18) near the bottom of the column can be approximated by $P = (1.86 \times 10^5)T - 4.80 \times 10^7$, where P is the pressure in Pa and T is the Celsius temperature. What is the boiling-point temperature for water at the bottom of the column?

78. •• **A Melting Glacier (a)** A glacier is made of ice of density 850 kg/m³ and is 92 m thick. Treating the glacial ice as if it were a liquid, what is the pressure at the bottom of the ice? **(b)** The water solid-liquid curve (Figure 17-21 (b)) near the bottom of the glacier can be approximated by $P = (-1.31 \times 10^7)T + 1.01 \times 10^5$, where P is the pressure in Pa and T is the Celsius temperature. What is the melting-point temperature for water at the bottom of the glacier?

79. •• Peter catches a 4.2-kg striped bass on a fishing line 0.55 mm in diameter and begins to reel it in. He fishes from a pier well above the water, and his fish hangs vertically from the line out of the water. The fishing line has a Young's modulus of 5.1×10^9 N/m². **(a)** What is the fractional increase in length of the fishing line if the fish is at rest? **(b)** What is the fractional increase in the fishing line's length when the fish is pulled upward with a constant speed of 2.5 m/s? **(c)** What is the fractional increase in the fishing line's length when the fish is pulled upward with a constant acceleration of 2.5 m/s²?

80. •• A steel ball (density = 7860 kg/m³) with a diameter of 6.4 cm is tied to an aluminum wire 82 cm long and 2.5 mm in diameter. The ball is whirled about in a vertical circle with a tangential speed of 7.8 m/s at the top of the circle and 9.3 m/s at the bottom of the circle. Find the amount of stretch in the wire **(a)** at the top and **(b)** at the bottom of the circle.

81. •• A lead brick with the dimensions shown in **FIGURE 17-39** rests on a rough solid surface. A force of 2400 N is applied as indicated. Find **(a)** the change in height of the brick and **(b)** the amount of shear deformation.

FIGURE 17-39 Problem 81

82. •• **(a)** Find the amount of heat that must be extracted from 1.3 kg of steam at 120 °C to convert it to ice at 0.0 °C. **(b)** What speed would this 1.3-kg block of ice have if its translational kinetic energy were equal to the thermal energy calculated in part (a)?

83. ••• **Mighty Ice Lift** A tremendous force is generated when water freezes into ice and expands in volume by 9.0%. Suppose a 1.000-m³ cube of liquid water freezes into ice that is 1.000 m on a side by 1.090 m tall. How many 68-kg students would the ice be able to lift? Determine this by calculating the amount of force on the top 1.000-m² face that would be required to squeeze 1.090 m³ of ice back to 1.000 m³, assuming all of the volume change occurs along the vertical direction.

84. ••• **Orthopedic Implants** Metals such as titanium and stainless steel are frequently used for orthopedic implants such as artificial hip and knee joints. As with most metals, though, their elastic properties are significantly different from those of bone. Recently, metal "foams" made from aluminum and steel have been shown to have promising properties for use as implants. **(a)** A 0.5-m-long piece of bone with a certain cross-sectional area shortens by 0.10 mm under a given compressive force. By how much does a piece of steel with the same length and cross-sectional area shorten if the same force is applied? **(b)** In order to determine whether a material such as the aluminum-steel foam behaves similarly to bone, plots of measured stress (force per area) versus strain ($\Delta L/L_0$) such as the ones shown in **FIGURE 17-40** may be used. Which one of these plots corresponds to a material with elastic properties equal to those of bone in compression? **(c)** What is the value of Young's modulus for the material represented by curve D?

FIGURE 17-40 Problem 84

85. ••• Students on a spring break picnic bring a cooler that contains 5.1 kg of ice at 0.0 °C. The cooler has walls that are 3.8 cm thick and are made of Styrofoam, which has a thermal conductivity of 0.030 W/(m·C°). The surface area of the cooler is 1.5 m², and it rests in the shade where the air temperature is 21 °C. **(a)** Find the rate at which heat flows into the cooler. **(b)** How much time does it take for the ice in the cooler to melt?

86. ••• A 5.9-kg block of ice at −1.5 °C slides on a horizontal surface with a coefficient of kinetic friction equal to 0.069. The initial speed of the block is 7.1 m/s and its final speed is 5.3 m/s. Assuming that all the energy dissipated by kinetic friction goes into melting a small mass m of the ice, and that the rest of the ice block remains at −1.5 °C, determine the value of m.

87. ••• A cylindrical copper rod 37 cm long and 7.5 cm in diameter is placed upright on a hot plate held at a constant temperature of 120 °C, as indicated in **FIGURE 17-41**. A small depression on top of the rod holds a 25-g ice cube at an initial temperature of 0.0 °C. How much time does it take for the ice cube to melt? Assume there is no heat loss through the vertical surface of the rod, and that the thermal conductivity of copper is 390 W/(m · C°).

FIGURE 17-41 Problem 87

PASSAGE PROBLEMS

Diving in the Bathysphere

The American naturalist Charles William Beebe (1877–1962) set a world record in 1934 when he and Otis Barton (1899–1992) made a dive to a depth of 923 m below the surface of the ocean. The dive was made just 10 miles from Nonsuch Island, off the coast of Bermuda, in a device known as the bathysphere, designed and built by Barton. The bathysphere was basically a steel sphere 4.75 ft in diameter with three small ports made of fused quartz. Lowered into the ocean on a steel cable whose radius was 1.85 cm, the bathysphere also carried bottles of oxygen, chemicals to absorb carbon dioxide, and a telephone line to the surface.

Beebe was fascinated by the new forms of life he and Barton encountered on their numerous dives. At one point he saw a "creature, several feet long, dart toward the window, turn sideways and—explode. At the flash, which was so strong that it illumined my face . . . I saw the great red shrimp and the outpouring fluid of flame." No wonder he considered the ocean depths to be "a world as strange as that of Mars."

The dives were not without their risks, however. It was not uncommon, for example, to have the bathysphere return to the surface partially filled with water after a window seal failed. On one deep dive, water began to stream rapidly into the sphere. Beebe quickly called to the surface and asked—not to be raised quickly—but to be lowered more rapidly, in the hope that increasing water pressure would force the leaking window into its seals to stop the leak. It worked, showing that Beebe was not only an exceptional naturalist, but also a cool-headed scientist with a good knowledge of basic physics!

88. • What pressure did the bathysphere experience at its record depth?

A. 9.37 atm   B. 89.6 atm
C. 91.9 atm   D. 92.9 atm

89. • How many moles of air did the bathysphere contain when it was sealed at the surface, assuming a temperature of 297 K and ignoring the thickness of the metal shell? (*Note:* A resting person breathes roughly 0.5 mol of air per minute.)

A. 65.2 mol   B. 270 mol
C. 392 mol   D. 523 mol

90. •• How much did the volume of the bathysphere decrease as it was lowered to its record depth? (For simplicity, treat the bathysphere as a solid metal sphere.)

A. $9.0 \times 10^{-5}\,\mathrm{m^3}$ B. $9.2 \times 10^{-5}\,\mathrm{m^3}$
C. $1.1 \times 10^{-4}\,\mathrm{m^3}$ D. $3.8 \times 10^{-4}\,\mathrm{m^3}$

91. •• Suppose the bathysphere and its occupants had a combined mass of 12,700 kg. How much did the cable stretch when the bathysphere was at a depth of 923 m? (Neglect the weight of the cable itself, but include the effects of the bathysphere's buoyancy.)

A. 47 cm   B. 48 cm
C. 52 cm   D. 53 cm

92. •• **REFERRING TO EXAMPLE 17-17 (a)** Find the final temperature of the system if *two* 0.0450-kg ice cubes are added to the warm lemonade. The temperature of the ice is 0 °C; the temperature and mass of the warm lemonade are 20.0 °C and 3.95 kg, respectively. **(b)** How many 0.0450-kg ice cubes at 0 °C must be added to the original warm lemonade if the final temperature of the system is to be at least as cold as 15.0 °C?

93. •• **REFERRING TO EXAMPLE 17-17 (a)** Find the final temperature of the system if a single 0.045-kg ice cube at 0 °C is added to 2.00 kg of lemonade at 1.00 °C. **(b)** What initial temperature of the lemonade will be just high enough to melt all of the ice in a single ice cube and result in an equilibrium temperature of 0 °C? The mass of the lemonade is 2.00 kg and the temperature of the ice cube is 0 °C.

The Laws of Thermodynamics

Big Ideas

1 Energy conservation applies to thermodynamic systems, just as it does to mechanical systems.

2 Heat engines use a difference in temperature to convert thermal energy to mechanical energy.

3 Entropy is a measure of disorder in a physical system.

4 The lowest temperature possible is absolute zero.

▲ Did this plane just fly through a cloud? No, it created its own cloud. As the plane moves forward at high speed it creates a low-pressure region near its tail. When the surrounding air expands into this region, it cools by the process of adiabatic cooling. As a result, water vapor in the air condenses, creating the plane's own personal cloud that follows it around. This chapter explores various thermodynamic processes, including adiabatic cooling, and shows how these processes can be used to understand the laws of thermodynamics and the operation of heat engines.

I n this chapter we discuss the fundamental laws of nature that govern thermodynamic processes. Of particular interest is the second law of thermodynamics, which introduces the idea of directionality in the behavior of nature. Melting ice, cooling lava, and the crumbling ruins of the Parthenon all illustrate the second law in action.

18-1 The Zeroth Law of Thermodynamics

Although the zeroth law of thermodynamics has already been presented in Chapter 16, we repeat it here so that all the laws of thermodynamics can be collected together in one chapter. As you recall, the zeroth law states the conditions for thermal equilibrium between objects. To be precise:

Zeroth Law of Thermodynamics

If object A is in thermal equilibrium with object C, and object B is separately in thermal equilibrium with object C, then objects A and B will be in thermal equilibrium if they are placed in thermal contact.

The physical quantity that is equal when two objects are in thermal equilibrium is the temperature. In particular, if two objects have the same temperature, then we can be assured that *no* energy in the form of heat will flow when they are placed in thermal contact. On the other hand, if heat does flow between two objects, it follows that they are not in thermal equilibrium, and they do not have the same temperature.

Any temperature scale can be used to determine whether objects will be in thermal equilibrium. As we saw in the previous chapter, however, the Kelvin scale is particularly significant in physics. For example, the average kinetic energy of a gas molecule is directly proportional to the Kelvin temperature, as is the volume of an ideal gas. In this chapter we present additional illustrations of the special significance of the Kelvin scale.

Enhance Your Understanding   (Answers given at the end of the chapter)

1. System 1 is at 0 °C and system 2 is at 0 °F. If these systems are placed in thermal contact, which of the following statements is correct? **(A)** Heat flows from system 1 to system 2. **(B)** Heat flows from system 2 to system 1. **(C)** No heat flows.

Section Review

- Objects are in thermal equilibrium when their temperatures are the same.

Adding the heat Q to the system . . .   . . . increases the system's energy from U_i to $U_i + Q$.

▲ FIGURE 18-1 **The internal energy of a system** A system initially has the internal energy U_i (left). After the heat Q is added, the system's new internal energy is $U_f = U_i + Q$ (right). The system has rigid walls; hence, it can do no work on the external world.

PHYSICS IN CONTEXT
Looking Back

The concepts of heat, work, and temperature, from Chapters 16 and 17, are used throughout this chapter.

Big Idea 1 Energy conservation applies to thermodynamic systems, just as it does to mechanical systems. When we apply energy conservation, we must consider not only mechanical work, but energy transferred in the form of heat as well.

18-2 The First Law of Thermodynamics

The first law of thermodynamics is a statement of energy conservation that specifically includes energy transferred in the form of heat. For example, consider the system shown in FIGURE 18-1. The internal energy of this system—that is, the sum of all its potential and kinetic energies—has the initial value U_i. If an amount of heat Q flows into the system, its internal energy increases to the final value $U_f = U_i + Q$. Thus,

$$\Delta U = U_f - U_i = Q \qquad \text{18-1}$$

Of course, if heat is removed from the system, its internal energy decreases. We can take this into account by giving Q a *positive* value when the system *gains* heat, and a *negative* value when it *loses* heat. (In the following, for the sake of brevity, we will generally just say "heat" as opposed to a more complete statement like "energy flow in the form of heat.")

Similarly, suppose the system under consideration does a work W on the external world, as in FIGURE 18-2. If the system is insulated so that no heat can flow in or out, the energy to do the work must come from the internal energy of the system. Thus, if the initial internal energy is U_i, the final internal energy is $U_f = U_i - W$. Therefore,

$$\Delta U = U_f - U_i = -W \qquad \text{18-2}$$

On the other hand, if work is done *on* the system, its internal energy increases. Thus, we use the following sign conventions for the work, W:

> W is *positive* when a system *does work on the external world*.
>
> W is *negative* when *work is done on a system*.

These sign conventions are summarized in Table 18-1.

Doing work on the external world . . .

Insulation

W

. . . decreases the energy of the system.

U_i

$U_f = U_i - W$

◄ **FIGURE 18-2 Work and internal energy** A system initially has the internal energy U_i (left). After the system does the work W on the external world, its remaining internal energy is $U_f = U_i - W$ (right). Note that the insulation guarantees that no heat is gained or lost by the system.

Combining the results in Equations 18-1 and 18-2 yields the first law of thermodynamics:

First Law of Thermodynamics

The change in a system's internal energy, ΔU, is related to the heat Q and the work W as follows:

$$\Delta U = Q - W \qquad \qquad 18\text{-}3$$

If you apply the sign conventions given here, it is straightforward to verify that adding heat to a system, and/or doing work on it, increases the internal energy. On the other hand, if the system does work, and/or heat is removed, its internal energy decreases. Example 18-1 gives a specific numerical application of the first law.

TABLE 18-1 Signs of *Q* and *W*

Q positive	System *gains* heat
Q negative	System *loses* heat
W positive	Work done *by* system
W negative	Work done *on* system

PROBLEM-SOLVING NOTE

Proper Signs for *Q* and *W*

When applying the first law of thermodynamics, it is important to determine the proper signs for Q, W, and ΔU.

EXAMPLE 18-1   SHELL GAME

(a) Rowing your shell to work one day, you give off 4.8×10^5 J of heat and do 6.2×10^5 J of work. What is the change in your internal energy? **(b)** On your return trip, you give off 3.9×10^5 J of heat and your internal energy decreases by 5.1×10^5 J. How much work did you do on the return trip?

PICTURE THE PROBLEM

Our sketch shows a person rowing a racing shell. The fact that the person does work on the external world means that W is positive. As for the heat, the fact that heat is given off by the person means that Q is negative.

REASONING AND STRATEGY

The signs of W and Q have been determined in our sketch, and the magnitudes are given in the problem statement. To find ΔU for part (a), we simply use the first law of thermodynamics, $\Delta U = Q - W$. To find the work W for part (b), we solve the first law for W, which yields $W = Q - \Delta U$.

$Q < 0$

$W > 0$

U decreases

Known   **(a)** Heat, $Q = -4.8 \times 10^5$ J; work, $W = 6.2 \times 10^5$ J.
(b) Heat, $Q = -3.9 \times 10^5$ J; change in internal energy, $\Delta U = -5.1 \times 10^5$ J.

Unknown   **(a)** Change in internal energy, $\Delta U = ?$ **(b)** Work, $W = ?$

SOLUTION

Part (a)

1. Calculate ΔU, using $Q = -4.8 \times 10^5$ J and $W = 6.2 \times 10^5$ J:

$$\Delta U = Q - W$$
$$= (-4.8 \times 10^5 \text{ J}) - 6.2 \times 10^5 \text{ J} = -1.1 \times 10^6 \text{ J}$$

Part (b)

2. Solve $\Delta U = Q - W$ for the work, W:

$$W = Q - \Delta U$$

3. Substitute $Q = -3.9 \times 10^5$ J and $\Delta U = -5.1 \times 10^5$ J to find the work:

$$W = -3.9 \times 10^5 \text{ J} - (-5.1 \times 10^5 \text{ J}) = 1.2 \times 10^5 \text{ J}$$

CONTINUED

INSIGHT
Notice how important it is to use the correct signs for Q, W, and ΔU. When the proper signs are used, the first law is simply a way of keeping track of all the energy exchanges—including mechanical work and heat—that can occur in a given system.

PRACTICE PROBLEM
After coasting for a few minutes you begin to row again, doing 5.7×10^5 J of work and decreasing your internal energy by 7.6×10^5 J. How much energy did you give off in the form of heat?   [**Answer:** $Q = -1.9 \times 10^5$ J. The negative sign means you have given off heat, as expected.]

Some related homework problems: Problem 3, Problem 5

Comparing Heat, Work, and Internal Energy Just looking at the first law, $\Delta U = Q - W$, we can easily get the false impression that U, Q, and W are basically the same type of physical quantity. Certainly, they are all measured in the same units (J). In other respects, however, they are quite different. For example, the heat Q represents energy that flows through thermal contact. In contrast, the work W indicates a transfer of energy by the action of a force through a distance.

The most important distinction between these quantities, however, is the way they depend on the **state of a system,** which is determined by its temperature, pressure, and volume. The internal energy, for example, depends only on the state of a system, and not on how the system is brought to that state. A simple example is the ideal gas, where U depends only on the temperature T, and not on any previous values T may have had. Because U depends only on the *state* of a system—whether the system is an ideal gas or something more complicated—it is referred to as a **state function.**

On the other hand, Q and W are *not* state functions; they depend on the precise way—that is, on the process—by which a system is changed from one state, A, to another state, B. For example, one process connecting the states A and B may result in a heat $Q_1 = 19$ J and a work $W_1 = 32$ J. With a different process connecting the *same two states*, we may find a heat $Q_2 = -24$ J $\neq Q_1$ and a work $W_2 = -11$ J $\neq W_1$. Still, the difference in internal energy—which depends only on the initial and final states—must be the same for all processes connecting A and B. It follows that

$$\Delta U_{AB} = Q_1 - W_1 = Q_2 - W_2 = -13\,\text{J}$$

Therefore, the internal energy of this system always decreases by 13 J when it changes from state A to state B.

Enhance Your Understanding   (Answers given at the end of the chapter)

2. The following systems exchange energy with the external world in the form of heat Q and work W. Rank the systems in order of increasing change in internal energy, from most negative to most positive. Indicate ties where appropriate.

System	A	B	C	D
Q	10 J	− 10 J	50 J	− 50 J
W	10 J	10 J	− 20 J	− 20 J

Section Review

*   The first law of thermodynamics is a statement of energy conservation that combines both work and heat. It can be stated as $\Delta U = Q - W$, where U is the internal energy, Q is the heat, and W is the work. Q is positive when heat energy is added to the system, and W is positive when work is done on the system.

18-3 Thermal Processes

In this section we consider a variety of thermodynamic processes that can be used to change the state of a system. Among these are ones that take place at constant pressure, constant volume, or constant temperature. We also consider processes in which no heat is allowed to flow into or out of the system.

All of the processes discussed in this section are assumed to be **quasi-static,** which is a fancy way of saying they occur so slowly that at any given time the system and its surroundings are essentially in equilibrium. Thus, in a quasi-static process, the pressure and temperature are always uniform throughout the system. Furthermore, we assume that the system in question is free from friction or other dissipative forces.

These assumptions can be summarized by saying that the processes we consider are **reversible.** To be precise, a reversible process is defined by the following condition:

> For a process to be reversible, it must be possible to return both the system and its surroundings to exactly the same states they were in before the process began.

The fact that both the system and its surroundings must be returned to their initial states is the key element of this definition. For example, if there is friction between a piston and the cylinder in which it slides, a reversible process is not possible. Even if the piston is returned to its original location, the heat generated by friction will warm the cylinder and eventually flow into the surrounding air. Thus, it is not possible to "undo" the effects of friction, and such a process is said to be **irreversible.** Even without friction, a process can be irreversible if it occurs rapidly enough to cause effects such as turbulence, or if the system is far from equilibrium.

PHYSICS IN CONTEXT
Looking Ahead

Chapter 21 shows how an electric current can generate heat, much like the heating caused by friction as one object (like a piston) slides against another (like a cylinder wall).

Constant-temperature heat bath

(a) Compressing at constant temperature expels heat.   **(b)** Expanding at constant temperature draws in heat.

▲ **FIGURE 18-3 An idealized reversible process (a)** A piston is slowly moved downward, compressing a gas. In order for its temperature to remain constant, a heat Q goes from the gas into the constant-temperature heat bath, which may be nothing more than a large volume of water. **(b)** As the piston is allowed to slowly rise back to its initial position, it draws the same amount of heat Q from the heat bath that it gave to the bath in (a). Hence, both the system (gas) and the surroundings (heat bath) return to their initial states.

In practice, then, all real processes are irreversible to some extent. It is still possible, however, to have a process that closely approximates a perfectly reversible process, just as we can have systems that are practically free of friction. For example, in **FIGURE 18-3 (a)** we consider a "frictionless" piston that is slowly forced downward, while the gas in the cylinder is kept at constant temperature. An amount of heat, Q, goes from the gas to its surroundings in order to keep the temperature from rising. In **FIGURE 18-3 (b)** the piston is slowly moved back upward, drawing in the same heat Q from its surroundings to keep the temperature from dropping. In an "ideal" case like this, the process is reversible, because the system and its surroundings are left unchanged. We shall assume that all the processes described in this section are reversible.

Constant-Pressure Processes

We begin by considering a process that occurs at **constant pressure.** To be specific, suppose a gas with the pressure P_0 is held in a cylinder of cross-sectional area A, as in **FIGURE 18-4.** If the piston moves outward, so that the volume of the gas increases from an initial value V_i to a final value V_f, the process can be represented graphically as shown in **FIGURE 18-5.** Here we plot pressure P versus volume V in a PV plot; the process just described is the bold horizontal line segment with an arrow on it.

As the gas expands, it does work on the piston. First, the gas exerts a force on the piston equal to the pressure times the area:

$$F = P_0 A$$

▲ **FIGURE 18-4 Work done by an expanding gas** A gas in a cylinder of cross-sectional area A expands with a constant pressure of P_0 from an initial volume $V_i = Ax_i$ to a final volume $V_f = Ax_f$. As it expands, it does the work $W = P_0(V_f - V_i)$.

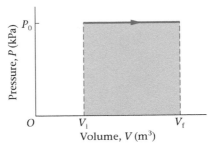

▲ **FIGURE 18-5 A constant-pressure process** A *PV* plot representing the constant-pressure process shown in Figure 18-4. The area of the shaded region, $P_0(V_f - V_i)$, is equal to the work done by the expanding gas in Figure 18-4.

PROBLEM-SOLVING NOTE

Work and the *PV* Diagram

When finding the work done by calculating the area on a *PV* diagram, recall that a pressure of 1 Pa times a volume of 1 m³ gives an energy equal to 1 J.

Second, the gas moves the piston from the position x_i to the position x_f, where $V_i = Ax_i$ and $V_f = Ax_f$, as indicated in Figure 18-4. Thus, the work done by the gas is the force times the distance through which the piston moves:

$$W = F(x_f - x_i) = P_0 A(x_f - x_i) = P_0(Ax_f - Ax_i) = P_0(V_f - V_i)$$

In general, if the pressure, *P*, is constant, and the volume changes by the amount ΔV, the work done by the gas is

$$W = P\Delta V \qquad (constant\ pressure) \qquad 18\text{-}4$$

The gas does positive work if the volume increases, because ΔV is positive.

EXERCISE 18-2   WORK DONE BY A GAS

A gas with a constant pressure of 126 kPa expands from a volume of 0.982 m³ to a volume of 1.19 m³. How much work does the gas do on its surroundings?

REASONING AND SOLUTION
The work done by the gas can be found with $W = P\Delta V$. Substituting numerical values, we find

$$W = P\Delta V = (126\ \text{kPa})(1.19\ \text{m}^3 - 0.982\ \text{m}^3) = 26,200\ \text{J}$$

For comparison, this is the energy required to raise the temperature of 1.0 kg of water by 6.26 C°.

Looking closely at the *PV* plot in Figure 18-5, we can see that the work done by the gas is equal to the area under the horizontal line representing the constant-pressure process. In particular, the shaded region is a rectangle of height P_0 and width $\Delta V = V_f - V_i$, and therefore its area is

$$\text{area} = P_0(V_f - V_i) = W$$

Though this result was obtained for the special case of constant pressure, it applies to any process:

> The work done by an expanding gas is equal to the area under the curve representing the process in a *PV* plot.

This result is applied in the next Example.

EXAMPLE 18-3   WORK AREA

A gas expands from an initial volume of 0.40 m³ to a final volume of 0.62 m³ as the pressure increases linearly from 110 kPa to 230 kPa. Find the work done by the gas.

PICTURE THE PROBLEM
The sketch shows the process for this problem. As the volume increases from $V_i = 0.40$ m³ to $V_f = 0.62$ m³, the pressure increases linearly from $P_i = 110$ kPa to $P_f = 230$ kPa.

REASONING AND STRATEGY
The work done by this gas is equal to the shaded area in the sketch. We can calculate this area as the sum of the area of a rectangle plus the area of a triangle.

In particular, the rectangle has a height P_i and a width $(V_f - V_i)$. Similarly, the triangle has a height $(P_f - P_i)$ and a base of $(V_f - V_i)$.

Known   Initial volume of gas, $V_i = 0.40$ m³; final volume of gas, $V_f = 0.62$ m³; initial pressure of gas, $P_i = 110$ kPa; final pressure of gas, $P_f = 230$ kPa.

Unknown   Work done by gas, $W = ?$

SOLUTION

1. Calculate the area of the rectangular portion of the total area:

$$A_{\text{rectangle}} = P_i(V_f - V_i)$$
$$= (110\ \text{kPa})(0.62\ \text{m}^3 - 0.40\ \text{m}^3) = 2.4 \times 10^4\ \text{J}$$

CONTINUED

2. Next, calculate the area of the triangular portion of the total area:

$$A_{triangle} = \tfrac{1}{2}(P_f - P_i)(V_f - V_i)$$
$$= \tfrac{1}{2}(230 \text{ kPa} - 110 \text{ kPa})(0.62 \text{ m}^3 - 0.40 \text{ m}^3)$$
$$= 1.3 \times 10^4 \text{ J}$$

3. Sum these areas to find the work done by the gas:

$$W = A_{rectangle} + A_{triangle}$$
$$= 2.4 \times 10^4 \text{ J} + 1.3 \times 10^4 \text{ J} = 3.7 \times 10^4 \text{ J}$$

INSIGHT

We could also have solved this problem by noting that the pressure varies linearly, and hence its average value is simply $P_{av} = \tfrac{1}{2}(P_f + P_i) = 170$ kPa. The work done by the gas, then, is $W = P_{av} \Delta V = (170 \text{ kPa})(0.22 \text{ m}^3) = 3.7 \times 10^4$ J, as before.

PRACTICE PROBLEM

Suppose the pressure varies linearly from 110 kPa to 290 kPa. How much work does the gas do in this case?
[**Answer:** $W = 4.4 \times 10^4$ J]

Some related homework problems: Problem 18, Problem 19

Constant-Volume Processes

Next, we consider a **constant-volume** process. Suppose, for example, that heat is added to a gas in a container of fixed volume, as in **FIGURE 18-6**, causing the pressure to increase. Noting that there is no displacement of any of the walls, we conclude that the force exerted by the gas does no work. Thus, for *any* constant-volume process, we have

$$W = 0 \qquad \qquad (constant\ volume)$$

From the first law of thermodynamics, $\Delta U = Q - W$, it follows that $\Delta U = Q$; that is, the change in internal energy is equal to the amount of heat that is added to or removed from the system. In addition, the conclusion that zero work is done in a constant-volume process is consistent with our earlier statement that work is equal to the area under the curve representing the process. For example, in **FIGURE 18-7** we show a constant-volume process in which the pressure is increased from P_i to P_f. This line is vertical, and hence the area under it is zero; that is, $W = 0$, as expected.

▲ **FIGURE 18-6 Adding heat to a system of constant volume** Heat is added to a system of constant volume, increasing its pressure from P_i to P_f. Since there is no displacement of the walls, there is no work done in this process. Therefore, $W = 0$ and the change in internal energy is simply equal to the heat added to or removed from the system, $\Delta U = Q$.

QUICK EXAMPLE 18-4   FIND THE TOTAL WORK

A gas undergoes the three-part process shown in the sketch, connecting the states A and B. The three parts of the process are labeled 1, 2, and 3. Find the total work done by the gas during this process.

▲ **FIGURE 18-7 A constant-volume process** The pressure increases from P_i to P_f, just as in Figure 18-6, while the volume remains constant at its initial value, V_i. The area under this process is zero, as is the work.

REASONING AND SOLUTION

As before, the work done by the gas is equal to the shaded area in the sketch. We can calculate this area as the sum of the areas of each part of the total process.

1. Calculate the work done during part 1 of the total process:

$$W_1 = \tfrac{1}{2}(140 \text{ kPa})(0.25 \text{ m}^3) = 17,500 \text{ J}$$

2. Calculate the work done during part 2 of the total process:

$$W_2 = 0$$

CONTINUED

3. Calculate the work done during part 3 of the total process:

$$W_3 = (210 \text{ kPa})(0.30 \text{ m}^3) = 63,000 \text{ J}$$

4. Sum the results to find the total work:

$$W = W_1 + W_2 + W_3 = 80,500 \text{ J}$$

The work done by the gas would be different if the process had been a straight-line (uniform) expansion from point A to point B. In that case, the total work done would be 57,750 J. Thus, the work W—which is *not* a state function—depends on the process, as expected.

**PHYSICS
IN CONTEXT
Looking Back**

Our understanding of ideal gases from Chapter 17 is put to good use here, in the study of common thermal processes.

Isothermal Processes

Another common process is one that takes place at constant temperature—that is, an **isothermal process.** For an ideal gas the isotherm has a relatively simple form. In particular, if T is constant, it follows that PV is a constant as well:

$$PV = NkT = \text{constant}$$

Thus, ideal-gas isotherms have the following pressure-volume relationship:

$$P = \frac{NkT}{V} = \frac{\text{constant}}{V}$$

This is illustrated in **FIGURE 18-8**, where we show several isotherms corresponding to different temperatures. (Recall that we've seen isotherms before, in Figure 17-5.)

▶ **FIGURE 18-8 Isotherms on a *PV* plot** These isotherms are for one mole of an ideal gas at the temperatures 300 K, 500 K, 700 K, and 900 K. Notice that each isotherm has the shape of a hyperbola. As the temperature is increased, however, the isotherms move farther from the origin. Thus, the pressure corresponding to a given volume increases with temperature, as one would expect.

▲ **FIGURE 18-9 An isothermal expansion** In an isothermal expansion from the volume V_i to the volume V_f, the work done is equal to the shaded area. For n moles of an ideal gas at the temperature T, the work done by the gas is $W = nRT \ln(V_f/V_i)$.

For any given isotherm, such as the one shown in **FIGURE 18-9**, the work done by an expanding gas is equal to the area under the curve, as usual. In particular, the work done in expanding from V_i to V_f in Figure 18-9 is equal to the shaded area. This area may be derived by using the methods of calculus. The result is found to be

$$W = NkT \ln\left(\frac{V_f}{V_i}\right) = nRT \ln\left(\frac{V_f}{V_i}\right) \qquad (\textit{constant temperature}) \qquad \text{18-5}$$

The symbol "ln" stands for the natural logarithm; that is, log to the base e. This result is utilized in the next Example.

EXAMPLE 18-5   ISOTHERMAL HEAT FLOW

A cylinder holds 0.54 mol of a monatomic ideal gas at a temperature of 310 K. As the gas expands isothermally from an initial volume of 4.2×10^{-3} m³ to a final volume of 6.8×10^{-3} m³, determine the amount of heat that must be added to the gas to maintain a constant temperature.

CONTINUED

PICTURE THE PROBLEM

The physical process is illustrated on the left in the sketch. Notice that heat flows into the gas as it expands in order to keep its temperature constant at $T = 310$ K. The graph on the right is a PV plot showing the same process. The work done by the expanding gas is equal to the shaded area from $V_i = 4.2 \times 10^{-3}$ m³ to $V_f = 6.8 \times 10^{-3}$ m³.

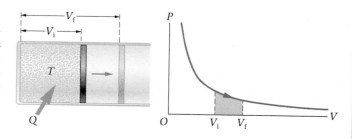

REASONING AND STRATEGY

We can use the first law of thermodynamics, $\Delta U = Q - W$, to find the heat Q in terms of W and ΔU; that is, $Q = \Delta U + W$. We can find W and ΔU as follows: (i) The work W is found using $W = nRT \ln(V_f/V_i)$ (Equation 18-5). (ii) To determine ΔU, recall that the internal energy of an ideal gas depends only on the temperature. Specifically, the internal energy of a monatomic ideal gas is $U = \frac{3}{2}nRT$. Because the temperature is constant in this process, there is no change in internal energy; that is, $\Delta U = 0$.

Known Moles of gas, $n = 0.54$ mol; constant temperature, $T = 310$ K; initial volume, $V_i = 4.2 \times 10^{-3}$ m³; final volume, $V_f = 6.8 \times 10^{-3}$ m³.

Unknown Heat added to gas, $Q = ?$

SOLUTION

1. Solve the first law of thermodynamics for the heat, Q:

$$\Delta U = Q - W$$
$$Q = \Delta U + W$$

2. Calculate the change in the gas's internal energy:

$$\Delta U = \tfrac{3}{2}nR(T_f - T_i) = 0$$

3. Calculate the work done by the expanding gas:

$$W = nRT \ln\left(\frac{V_f}{V_i}\right)$$
$$= (0.54 \text{ mol})[8.31 \text{ J/(mol} \cdot \text{K)}](310 \text{ K}) \ln\left(\frac{6.8 \times 10^{-3} \text{ m}^3}{4.2 \times 10^{-3} \text{ m}^3}\right)$$
$$= 670 \text{ J}$$

4. Substitute numerical values to find Q:

$$Q = \Delta U + W = 0 + 670 \text{ J} = 670 \text{ J}$$

INSIGHT

Thus, when an ideal gas undergoes an isothermal expansion, the heat gained by the gas is equal to the work it does; that is, $Q = W$. This is a direct consequence of the first law of thermodynamics, $\Delta U = Q - W$, and the fact that $\Delta U = 0$ for an ideal gas at constant temperature.

PRACTICE PROBLEM

Find the final volume of this gas when it has expanded enough to do 730 J of work.
[**Answer:** $V_f = 7.1 \times 10^{-3}$ m³]

Some related homework problems: Problem 20, Problem 21

Adiabatic Processes

The final process we consider is one in which no heat flows into or out of a system. Such a process, in which $Q = 0$, is said to be **adiabatic.** One way to produce an adiabatic process is illustrated in **FIGURE 18-10 (a)**. Here we see a cylinder that is insulated well enough that no heat can pass through the insulation (*adiabatic* means, literally, "not passable"). When the piston is pushed downward in the cylinder—decreasing the volume and *compressing* the gas—the gas heats up and its pressure increases. Similarly, in **FIGURE 18-10 (b)**, an adiabatic *expansion* causes the temperature of the gas to decrease, as does the pressure.

What does an adiabatic process look like on a PV plot? Certainly, its general shape must be similar to that of an isotherm; in particular, as the volume is decreased, the pressure increases. However, it can't be identical to an isotherm because, as we have pointed out, the temperature changes during an adiabatic process. The comparison between an adiabatic curve and an isothermal curve is the subject of the following Conceptual Example.

Insulation

P increases
T increases

P decreases
T decreases

(a) Compressing with no heat flow increases *P* and *T*.

(b) Expanding with no heat flow decreases *P* and *T*.

▲ **FIGURE 18-10 An adiabatic process** In adiabatic processes, no heat flows into or out of the system. In the cases shown in this figure, heat flow is prevented by insulation. **(a)** An adiabatic compression increases both the pressure and the temperature. **(b)** An adiabatic expansion results in a decrease in pressure and temperature.

CONCEPTUAL EXAMPLE 18-6 PRESSURE VERSUS VOLUME

A certain gas has an initial volume and pressure given by point A in the *PV* plot. If the gas is compressed isothermally, its pressure rises as indicated by the curve labeled "Isotherm." If, instead, the gas is compressed adiabatically from point A, does its pressure follow curve i, curve ii, or curve iii?

REASONING AND DISCUSSION
In an isothermal compression, some heat flows out of the system in order for its temperature to remain constant. No heat flows out of the system in the adiabatic process, however, and therefore its temperature rises. As a result, the pressure is greater for any given volume when the compression is adiabatic, as compared to isothermal. It follows that the adiabatic curve, or adiabat, is represented by curve iii.

ANSWER
Curve iii is the adiabat.

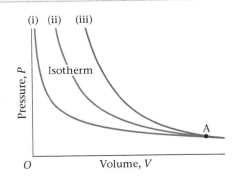

(i) (ii) (iii)

Isotherm

A

Pressure, *P*

O   Volume, *V*

As we've seen, an adiabatic curve is similar to an isotherm, only steeper. The precise mathematical relationship describing an adiabat is presented in the next section.

EXAMPLE 18-7 WORK INTO ENERGY

When a certain gas is compressed adiabatically, the amount of work done on it is 730 J. Find the change in internal energy of the gas.

PICTURE THE PROBLEM
Our sketch shows a piston being pushed downward, compressing a gas in an insulated cylinder. The insulation ensures that no heat can flow—as required in an adiabatic process.

REASONING AND STRATEGY
We know that 730 J of work is done on the gas, and we know that no heat is exchanged ($Q = 0$) in an adiabatic process. Thus, we can find ΔU by substituting Q and W into the first law of thermodynamics. One note of caution: Be careful to use the correct sign for the work. In particular, recall that the work done *on* a system is *negative*.

Insulation

Known Work, $W = -730$ J; process = adiabatic.
Unknown Change in internal energy, $\Delta U = ?$

SOLUTION

1. Identify the work and heat for this process:

$W = -730$ J
$Q = 0$

CONTINUED

2. Substitute Q and W into the first law of thermodynamics to find the change in internal energy, ΔU:

$$\Delta U = Q - W = 0 - (-730\,\text{J}) = 730\,\text{J}$$

INSIGHT

Because no energy can enter or leave the system in the form of heat, all the work done on the system goes into increasing its internal energy. This follows from the first law of thermodynamics, $\Delta U = Q - W$, which for $Q = 0$ reduces to $\Delta U = -W$. As a result, the temperature of the gas increases.

A familiar example of this type of effect is the heating that occurs when you pump air into a tire or a ball—the work done on the pump appears as an increased temperature. The effect occurs in the reverse direction as well. When air is let out of a tire, for example, it does work on the atmosphere as it expands, producing a cooling effect that can be quite noticeable. In extreme cases, the cooling can be great enough to create frost on the valve stem of the tire.

PRACTICE PROBLEM

If a system's internal energy decreases by 550 J in an adiabatic process, how much work was done by the system? [**Answer:** $W = +550\,\text{J}$]

Some related homework problems: Problem 17, Problem 24

▶ **FIGURE 18-11 Adiabatic heating** When a piston that fits snugly inside a cylinder is pushed downward rapidly, the temperature of the gas within the cylinder increases before there is time for heat to flow out of the system. Thus, the process is essentially adiabatic. As a result, the temperature of the gas can increase enough to ignite bits of paper in the cylinder. In a diesel engine, the same principle is used to ignite an air-gasoline mixture without a spark plug.

Examples of Adiabatic Processes An adiabatic process can occur when the system is thermally insulated, as in Figure 18-10, or in a system where the change in volume occurs rapidly. For example, if an expansion or compression happens quickly enough, there is no time for heat flow to occur. As a result, the process is adiabatic, even if there is no insulation.

RWP* An example of a rapid process is shown in **FIGURE 18-11**. Here, a piston is fitted into a cylinder that contains a certain volume of gas and a small piece of tissue paper. If the piston is driven downward rapidly—by a sharp impulsive blow, for example—the gas is compressed before heat has a chance to flow. As a result, the temperature of the gas rises rapidly. In fact, the rise in temperature can be enough for the paper to burst into flames.

The same principle applies to the operation of a diesel engine. As you may know, a diesel differs from a standard internal combustion engine in that it has no spark plugs. It doesn't need them. Instead of using a spark to ignite the fuel in a cylinder, it uses adiabatic heating. Fuel and air are admitted into the cylinder; then the piston rapidly compresses the air-fuel mixture. Just as with the piece of paper in Figure 18-11, the rising temperature is sufficient to ignite the fuel and run the engine.

The characteristics of constant-pressure, constant-volume, isothermal, and adiabatic processes are summarized in Table 18-2.

TABLE 18-2 Thermodynamic Processes and Their Characteristics

Constant pressure	$W = P\Delta V$	$Q = \Delta U + P\Delta V$
Constant volume	$W = 0$	$Q = \Delta U$
Isothermal (constant temperature)	$W = Q$	$\Delta U = 0$
Adiabatic (no heat flow)	$W = -\Delta U$	$Q = 0$

Enhance Your Understanding

(Answers given at the end of the chapter)

3. The pressure of a system doubles during an isothermal process. What happens to its volume?

Section Review

- An isothermal process is one that occurs with no change in temperature. An adiabatic process is one that occurs with no energy transfer in the form of heat.

*Real World Physics applications are denoted by the acronym RWP.

18-4 Specific Heats for an Ideal Gas: Constant Pressure, Constant Volume

Recall that the specific heat of a substance is the amount of heat needed to raise the temperature of 1 kg of the substance by 1 Celsius degree. As we know, however, the amount of heat depends on the type of process used to raise the temperature. Thus, we should specify, for example, whether a specific heat applies to a process at constant pressure or constant volume.

Constant Volume A constant-volume process is illustrated in FIGURE 18-12. Here we see an ideal gas of mass m in a container of fixed volume V. A heat Q flows into the container. As a result of this added heat, the temperature of the gas rises by the amount ΔT, and its pressure increases as well. Now, the specific heat at constant volume, c_v, is defined by the following relationship:

$$Q_v = mc_v\Delta T$$

In what follows, it will be more convenient to use the **molar specific heat,** denoted by a capital letter C, which is defined in terms of the number of moles rather than the mass of the substance. Thus, if a gas contains n moles, its molar specific heat at constant volume is given by

$$Q_v = nC_v\Delta T$$

$$C_v = Q_v/n\Delta T$$

▲ **FIGURE 18-12 Heating at constant volume** If the heat Q_v is added to n moles of gas, and the temperature rises by ΔT, the molar specific heat at constant volume is $C_v = Q_v/n\Delta T$. No mechanical work is done at constant volume.

$$C_p = Q_p/n\Delta T$$

▲ **FIGURE 18-13 Heating at constant pressure** If the heat Q_p is added to n moles of gas, and the temperature rises by ΔT, the molar specific heat at constant pressure is $C_p = Q_p/n\Delta T$. Note that the heat Q_p must increase the temperature *and* do mechanical work by lifting the piston.

Constant Pressure Similarly, a constant-pressure process is illustrated in FIGURE 18-13. In this case, the gas is held in a container with a movable piston that applies a constant pressure P. As a heat Q is added to the gas, its temperature increases, which causes the piston to rise—after all, if the piston didn't rise, the pressure of the gas would increase. If the temperature of the gas increases by the amount ΔT, the molar specific heat at constant pressure is given by

$$Q_p = nC_p\Delta T$$

We would now like to obtain a relationship between C_p and C_v.

Before we carry out the calculation, let's consider the qualitative relationship between these specific heats. This is addressed in the following Conceptual Example.

CONCEPTUAL EXAMPLE 18-8 COMPARING SPECIFIC HEATS

How does the molar specific heat at constant pressure, C_p, compare with the molar specific heat at constant volume, C_v, for a given system? **(a)** $C_p > C_v$; **(b)** $C_p = C_v$; **(c)** $C_p < C_v$.

REASONING AND DISCUSSION

In a constant-volume process, as in Figure 18-12, the heat that is added to a system goes entirely into increasing the temperature, since no work is done. On the other hand, at constant pressure, the heat added to a system increases the temperature

CONTINUED

and does mechanical work. This is illustrated in Figure 18-13, where we see that the heat must not only raise the temperature, but also supply enough energy to lift the piston. Thus, more heat is required in the constant-pressure process, and hence that specific heat is greater.

ANSWER

(a) $C_p > C_v$; the specific heat at constant pressure is greater than the specific heat at constant volume.

Comparing C_v and C_p We turn now to a detailed calculation of both C_v and C_p for a monatomic ideal gas. To begin, we rearrange the first law of thermodynamics, $\Delta U = Q - W$, to solve for the heat, Q:

$$Q = \Delta U + W$$

Recall from the previous section, however, that the work is zero, $W = 0$, for any constant-volume process. Hence, for constant volume we have

$$Q_v = \Delta U$$

Recalling that $U = \frac{3}{2} NkT = \frac{3}{2} nRT$ for a monatomic ideal gas, we find

$$Q_v = \Delta U = \frac{3}{2} nR\Delta T$$

Comparing with the definition of the molar specific heat, it follows that:

Molar Specific Heat for a Monatomic Ideal Gas at Constant Volume

$$C_v = \frac{3}{2} R \qquad\qquad\text{18-6}$$

Let's perform a similar calculation for constant pressure. In this case, we know that the work done by a gas is $W = P\Delta V$. In an ideal gas, for which $PV = nRT$, it follows that

$$W = P\Delta V = nR\Delta T$$

Combining this with the first law of thermodynamics yields

$$Q_p = \Delta U + W$$
$$= \frac{3}{2} nR\Delta T + nR\Delta T = \frac{5}{2} nR\Delta T$$

Applying the definition of molar specific heat yields:

Molar Specific Heat for a Monatomic Ideal Gas at Constant Pressure

$$C_p = \frac{5}{2} R \qquad\qquad\text{18-7}$$

As expected, the specific heat at constant pressure is larger than the specific heat at constant volume, and the difference is precisely the extra contribution due to the work done in lifting the piston in the constant-pressure case. In particular, we see that

$$C_p - C_v = R \qquad\qquad\text{18-8}$$

Though this relationship was derived for a monatomic ideal gas, it holds for all ideal gases, regardless of the structure of their molecules. It is also a good approximation for most real gases, as can be seen in Table 18-3.

TABLE 18-3 $C_p - C_v$ for Various Gases

Helium	$0.995R$
Nitrogen	$1.00R$
Oxygen	$1.00R$
Argon	$1.01R$
Carbon dioxide	$1.01R$
Methane	$1.01R$

EXERCISE 18-9 FIND THE HEAT

Find the heat required to raise the temperature of 0.209 mol of a monatomic ideal gas by 7.45 C° at **(a)** constant volume and **(b)** constant pressure.

REASONING AND SOLUTION

In this system, we know that $n = 0.209$ mol and $\Delta T = 7.45$ C°. The value of the specific heat depends on the type of process.

(a) To find the heat required at constant volume we use $Q_v = nC_v\Delta T$ with $C_v = \frac{3}{2} R$. This gives

$$Q_v = nC_v\Delta T = \frac{3}{2} nR\Delta T = \frac{3}{2}(0.209 \text{ mol})[8.31 \text{ J/(mol·K)}](7.45 \text{ K}) = 19.4 \text{ J}$$

CONTINUED

PROBLEM-SOLVING NOTE

Constant Volume Versus Constant Pressure

The heat required to increase the temperature of an ideal gas depends on whether the process is at constant pressure or constant volume. More heat is required when the process occurs at constant pressure.

Figure on left side:

(graph) Pressure, P (kPa) vs Volume, V (m³)

Adiabat

Adiabats rise more steeply than isotherms.

Isotherm at 1300 K

Isotherm at 700 K

▲ **FIGURE 18-14 A comparison between isotherms and adiabats** Two isotherms are shown, one for 700 K and one for 1300 K. An adiabat is also shown. Note that the adiabat is a steeper curve than the isotherms.

PHYSICS IN CONTEXT
Looking Ahead

The expansion of the universe is discussed in Chapter 32. During the expansion, the temperature of the universe has decreased, very much as it does for an ideal gas undergoing an adiabatic expansion.

(b) For constant pressure we use $Q_p = nC_p\Delta T$ with $C_p = \frac{5}{2}R$. This gives

$$Q_p = nC_p\Delta T = \frac{5}{2}nR\Delta T = \frac{5}{2}(0.209\ \text{mol})[8.31\ \text{J}/(\text{mol}\cdot\text{K})](7.45\ \text{K}) = 32.3\ \text{J}$$

As expected, the required heat is greater at constant pressure.

Adiabatic Processes

We return now briefly to a consideration of adiabatic processes. As we shall see, the relationship between C_p and C_v is important in determining the behavior of a system undergoing an adiabatic process.

FIGURE 18-14 shows an adiabatic curve and two isotherms. As mentioned before, the adiabatic curve is steeper and it cuts across the isotherms. For the isotherms, we recall that the curves are described by the equation

$$PV = \text{constant} \qquad\qquad (isothermal)$$

A similar equation applies to adiabats. In this case, calculus may be used to show that the appropriate equation is

$$PV^\gamma = \text{constant} \qquad\qquad (adiabatic) \qquad 18\text{-}9$$

In this expression, the constant γ is the ratio C_p/C_v:

$$\gamma = \frac{C_p}{C_v} = \frac{\frac{5}{2}R}{\frac{3}{2}R} = \frac{5}{3}$$

This value of γ applies to monatomic ideal gases—and is a good approximation for monatomic real gases as well. The value of γ is different, however, for gases that are diatomic, triatomic, and so on.

EXAMPLE 18-10   HOT AIR

A container with an initial volume of 0.0607 m³ holds 2.45 moles of a monatomic ideal gas at a temperature of 325 K. The gas is now compressed adiabatically to a volume of 0.0377 m³. Find **(a)** the final pressure and **(b)** the final temperature of the gas.

PICTURE THE PROBLEM

Our sketch shows a gas being compressed in an insulated container. The gas starts with a temperature of 325 K, but because no heat can flow outward through the insulation (adiabatic process), the work done on the gas results in an increase in temperature.

Insulation

REASONING AND STRATEGY

a. We can find the final pressure as follows: (i) Find the initial pressure using the ideal-gas equation of state, $P_iV_i = nRT_i$. (ii) Next, let $P_iV_i^\gamma = P_fV_f^\gamma$, since this is an adiabatic process. Solve this equation for the final pressure.

b. Use the final pressure and volume to find the final temperature, using the ideal-gas equation: $P_fV_f = nRT_f$.

Known   Initial volume, $V_i = 0.0607$ m³; final volume, $V_f = 0.0377$ m³; number of moles, $n = 2.45$ mol; initial temperature, $T_i = 325$ K; process = adiabatic.

Unknown   **(a)** Final pressure, $P_f = ?$ **(b)** Final temperature, $T_f = ?$

SOLUTION

Part (a)

1. Find the initial pressure, using $PV = nRT$:

$$P_i = \frac{nRT_i}{V_i}$$
$$= \frac{(2.45\ \text{mol})[8.31\ \text{J}/(\text{mol}\cdot\text{K})](325\ \text{K})}{0.0607\ \text{m}^3} = 109\ \text{kPa}$$

2. Use $PV^\gamma = \text{constant}$ to find a relationship for P_f:

$$P_iV_i^\gamma = P_fV_f^\gamma$$
$$P_f = P_i\left(\frac{V_i}{V_f}\right)^\gamma$$

CONTINUED

3. Substitute numerical values:

$$P_f = (109 \text{ kPa})\left(\frac{0.0607 \text{ m}^3}{0.0377 \text{ m}^3}\right)^{5/3} = 241 \text{ kPa}$$

Part (b)

4. Use $PV = nRT$ to solve for the final temperature:

$$T_f = \frac{P_f V_f}{nR}$$

$$= \frac{(241 \text{ kPa})(0.0377 \text{ m}^3)}{(2.45 \text{ mol})[8.31 \text{ J/(mol} \cdot \text{K)}]} = 446 \text{ K}$$

INSIGHT

Decreasing the volume of the gas increased its pressure from 109 kPa to 241 kPa, and at the same time increased its temperature from 325 K to 446 K. This is a specific example of adiabatic heating. Adiabatic cooling is the reverse effect, where the temperature of a gas decreases as its volume increases.

PRACTICE PROBLEM

In part (a), to what volume must the gas be compressed to yield a final pressure of 278 kPa?   [**Answer:** $V_f = 0.0346 \text{ m}^3$]

Some related homework problems: Problem 39, Problem 40

RWP Adiabatic heating and cooling can have important effects on the climate of a given region. For example, moisture-laden winds blowing from the Pacific Ocean into western Oregon are deflected upward when they encounter the Cascade Mountains. As the air rises, the atmospheric pressure decreases (see Chapter 15), allowing the air to expand and undergo adiabatic cooling. The result is that the moisture in the air condenses to form clouds and precipitation on the west side of the mountains. In some cases, where the air holds relatively little moisture, this mechanism may result in isolated, lens-shaped (lenticular) clouds just above the peak of a mountain. An example is shown in **FIGURE 18-15**. When the winds continue on to the east side of the mountains, they have little moisture remaining; thus, eastern Oregon is in the *rain shadow* of the Cascade Mountains. In addition, as the air descends on the east side of the mountains, it undergoes adiabatic heating. These are the primary reasons the summers in western Oregon are moist and mild, while the summers in eastern Oregon are hot and dry.

▲ **FIGURE 18-15** A spectacular lenticular (lens-shaped) cloud floats above a mountain in Tierra del Fuego, at the southern tip of Chile. Lenticular clouds are often seen "parked" above and just downwind of high mountain peaks, even when there are no other clouds in the sky. The reason is that as moisture-laden winds are deflected upward by the mountain, the moisture they contain cools due to adiabatic expansion and condenses to form a cloud.

Enhance Your Understanding   (Answers given at the end of the chapter)

4. Rank the following systems in order of the amount of heat needed to produce the indicated temperature change, from least to greatest. Indicate ties where appropriate.

System	A	B	C	D
Moles, n	1	2	4	10
ΔT (K)	2	10	10	1
Type of process	constant pressure	constant volume	constant pressure	constant volume

Section Review

- The molar specific heat for a monatomic ideal gas at constant volume is $C_v = \frac{3}{2}R$. The corresponding molar specific heat for constant pressure is $C_p = \frac{5}{2}R$.

18-5 The Second Law of Thermodynamics

Have you ever warmed your hands by pressing them against a block of ice? Probably not. But if you think about it, you might wonder why it doesn't work. After all, the first law of thermodynamics would be satisfied if energy simply flowed from the ice to your hands. The ice would get colder while your hands got warmer, and the energy of the universe would remain the same.

As we know, however, this sort of thing just doesn't happen—the spontaneous flow of heat is *always* from warmer objects to cooler objects, and never in the reverse direction. This simple observation, in fact, is one of many ways of expressing the **second law of thermodynamics:**

Second Law of Thermodynamics: Heat Flow

When objects of different temperatures are brought into thermal contact, the spontaneous flow of heat that results is always from the high-temperature object to the low-temperature object. Spontaneous heat flow never proceeds in the reverse direction.

Thus, the second law of thermodynamics is more restrictive than the first law; it says that of all the processes that conserve energy, only those that proceed in a certain direction actually occur. In a sense, the second law implies a definite "directionality" to the behavior of nature. For this reason, the second law is sometimes referred to as the "arrow of time."

For example, suppose you saw a movie that showed a snowflake landing on a person's hand and melting to a small drop of water. Nothing would seem particularly noteworthy about the scene from a physics point of view. But if the movie showed a drop of water on a person's hand suddenly freeze into the shape of a snowflake, and then lift off the person's hand into the air, it wouldn't take long to realize the film was running backward. It's clear in which direction time should "flow."

We study further consequences of the second law in the next few sections. As we do so, we'll find other more precise, but equivalent, ways of stating the second law.

Enhance Your Understanding   (Answers given at the end of the chapter)

5. Is any process that conserves energy allowed in a physical system? Give an example to support your answer.

Section Review

• The second law of thermodynamics states that heat always flows from a hot object to a cool object, and never in the reverse direction.

(a)

(b)

▲ **FIGURE 18-16 Visualizing Concepts**
Heat Engines (a) A modern version of Hero's engine, invented by the Greek mathematician and engineer Hero of Alexandria. In this simple heat engine, the steam that escapes from a heated vessel of water is directed tangentially, causing the vessel to rotate. This converts the thermal energy supplied to the water into mechanical energy, in the form of rotational motion.
(b) A steam engine of slightly more recent design hauls passengers up and down Mt. Washington in New Hampshire. Note that the locomotive is belching two clouds, one black and one white. Can you explain their origin?

18-6 Heat Engines and the Carnot Cycle

RWP A **heat engine,** simply put, is a device that converts heat into work. A simple, easy-to-understand example is shown in **FIGURE 18-16 (a)**, but the classic example is the steam engine, as shown in **FIGURE 18-16 (b)**. The basic elements of a steam engine are illustrated in **FIGURE 18-17**. First, some form of fuel (oil, wood, coal, etc.) is used to vaporize water in the boiler. The resulting steam is then allowed to enter the engine itself, where it expands against a piston, doing mechanical work. As the piston moves, it causes gears or wheels to rotate, which delivers the mechanical work to the external world. After leaving the engine, the steam proceeds to the condenser, where it gives off heat to the cool air in the atmosphere and condenses to liquid form.

All heat engines have the following features in common:

A high-temperature reservoir that supplies heat to the engine (the boiler in the steam engine)

A low-temperature reservoir where "waste" heat is released (the condenser in the steam engine)

An engine that converts thermal energy to mechanical work in a cyclic fashion

These features are illustrated schematically in **FIGURE 18-18**. In addition, though not shown in the figure, heat engines have a working substance (steam in the steam engine) that causes the engine to operate.

To begin our analysis of heat engines, we note that a certain amount of heat, Q_h, is supplied to the engine from the high temperature or "hot" reservoir during each cycle.

▲ FIGURE 18-17 **A schematic steam engine** The basic elements of a steam engine are a boiler—in which heat converts water to steam—and a piston that can be displaced by the expanding steam. In some engines, the steam is simply exhausted into the atmosphere after it has expanded against the piston. More sophisticated engines send the exhaust steam to a condenser, where it is cooled and condensed back to liquid water, then recycled to the boiler.

▲ FIGURE 18-18 **A schematic heat engine** The engine absorbs a heat Q_h from the hot reservoir, performs the work W, and gives off the heat Q_c to the cold reservoir. Energy conservation gives $Q_h = W + Q_c$, where Q_h and Q_c are the magnitudes of the hot- and cold-temperature heats.

Of this heat, a fraction appears as work, W, during each cycle, and the rest is given off as waste heat, Q_c, at a relatively low temperature to the "cold" reservoir. There is no change in energy for the engine, because it returns to its initial state at the completion of each cycle. To simplify notation, we shall let Q_h and Q_c denote magnitudes for the remainder of this chapter, so that both quantities are positive. With this convention, energy conservation can be written as follows:

$$\text{energy in } (Q_h) = \text{energy out } (W + Q_c)$$

Rearranging this relationship, we find that the work done during a cycle is

$$W = Q_h - Q_c \qquad \text{18-10}$$

As we shall see, the second law of thermodynamics requires that heat engines must *always* exhaust a finite amount of heat to a cold reservoir.

The Efficiency of a Heat Engine Of particular interest for any engine is its **efficiency**, e, which is simply the fraction of the heat supplied to the engine that appears as work. Thus, we define the efficiency to be

$$e = \frac{W}{Q_h} \qquad \text{18-11}$$

Using the energy-conservation result just derived in Equation 18-10 ($W = Q_h - Q_c$), we find:

Efficiency of a Heat Engine, e

$$e = \frac{W}{Q_h} = \frac{Q_h - Q_c}{Q_h} = 1 - \frac{Q_c}{Q_h} \qquad \text{18-12}$$

SI unit: dimensionless

For example, if $e = 0.20$, we say that the engine is 20% efficient. In this case, 20% of the input heat is converted to work, $W = 0.20Q_h$, and 80% goes to waste heat, $Q_c = 0.80Q_h$. Efficiency can be thought of as the ratio of how much you receive (work) to how much you have to pay to run the engine (input heat).

PHYSICS IN CONTEXT
Looking Ahead

In Chapter 23 we shall see how the mechanical work produced by a heat engine can be converted to electrical energy by a generator, in the form of an electric current.

EXAMPLE 18-11   HEAT INTO WORK

A heat engine with an efficiency of 24.0% performs 1250 J of work. Find **(a)** the heat absorbed from the hot reservoir and **(b)** the heat given off to the cold reservoir.

PICTURE THE PROBLEM
Our sketch shows a schematic of the heat engine. We know the amount of work that is done and the efficiency of the engine. We seek the heats Q_h and Q_c.
    Note that an efficiency of 24.0% means that $e = 0.240$.

REASONING AND STRATEGY

a. We can find the heat absorbed from the hot reservoir directly from the definition of efficiency, $e = W/Q_h$.

b. We can find Q_c by using energy conservation, $W = Q_h - Q_c$, or by using the expression for efficiency in terms of the heats, $e = 1 - Q_c/Q_h$.

Known   Efficiency, $e = 0.240$; work done, $W = 1250$ J.
Unknown   **(a)** Heat absorbed from hot reservoir, $Q_h = ?$ **(b)** Heat given off to cold reservoir, $Q_c = ?$

SOLUTION

Part (a)

1. Use $e = W/Q_h$ to solve for the heat Q_h:

$$e = \frac{W}{Q_h}$$
$$Q_h = \frac{W}{e} = \frac{1250\,\text{J}}{0.240} = 5210\,\text{J}$$

Part (b)

2. Use energy conservation to solve for Q_c:

$$W = Q_h - Q_c$$
$$Q_c = Q_h - W = 5210\,\text{J} - 1250\,\text{J} = 3960\,\text{J}$$

3. As a double check, use the efficiency $(e = 1 - Q_c/Q_h)$ to find Q_c:

$$e = 1 - \frac{Q_c}{Q_h}$$
$$Q_c = (1 - e)Q_h = (1 - 0.240)(5210\,\text{J}) = 3960\,\text{J}$$

INSIGHT
When the efficiency of a heat engine is less than one-half (50%), as in this case, the amount of heat given off as waste to the cold reservoir is greater than the amount of heat converted to work.

PRACTICE PROBLEM
What is the efficiency of a heat engine that does 1250 J of work and gives off 5250 J of heat to the cold reservoir?
[Answer: $e = 0.192$]

Some related homework problems: Problem 43, Problem 44

A temperature difference is essential to the operation of a heat engine. As heat flows from the hot to the cold reservoir in Figure 18-18, for example, the heat engine is able to tap into that flow and convert part of it to work—the greater the efficiency of the engine, the more heat that is converted to work. The second law of thermodynamics imposes limits, however, on the maximum efficiency a heat engine can have. We explore these limits next.

The Carnot Cycle and Maximum Efficiency

In 1824, the French engineer Sadi Carnot (1796–1832) published a book entitled *Reflections on the Motive Power of Fire* in which he considered the following question: Under what conditions will a heat engine have maximum efficiency? To address this question, let's consider a heat engine that operates between a single hot reservoir at the fixed temperature T_h and a single cold reservoir at the fixed temperature T_c. Carnot's result, known today as **Carnot's theorem,** can be expressed as follows:

Carnot's Theorem

If an engine operating between two constant-temperature reservoirs is to have maximum efficiency, it must be an engine in which all processes are reversible.

In addition, all reversible engines operating between the same two temperatures, T_c and T_h, have the same efficiency.

We should point out that no real engine can ever be perfectly reversible, just as no surface can be perfectly frictionless. Nonetheless, the concept of a reversible engine is a useful idealization.

Carnot's theorem is remarkable for a number of reasons. First, consider what the theorem says: No engine, no matter how sophisticated or technologically advanced, can exceed the efficiency of a reversible engine. We can strive to improve the technology of heat engines, but there is an upper limit to the efficiency that can never be exceeded. Second, the theorem is just as remarkable for what it does not say. It says nothing, for example, about the working substance that is used in the engine—it is as valid for a liquid or solid working substance as for one that is gaseous. Furthermore, it says nothing about the type of reversible engine that is used, what the engine is made of, or how it is constructed. Diesel engine, jet engine, rocket engine—none of these things matter. In fact, all that *does* matter are the two temperatures, T_c and T_h.

Recall that the efficiency of a heat engine can be written as follows:

$$e = 1 - \frac{Q_c}{Q_h}$$

Because the efficiency e depends only on the temperatures T_c and T_h, according to Carnot's theorem, it follows that Q_c/Q_h must also depend only on T_c and T_h. In fact, Lord Kelvin used this observation to propose that, instead of using a thermometer to measure temperature, we measure the efficiency of a heat engine and from this determine the temperature. Thus, he suggested that we *define* the ratio of the temperatures of two reservoirs, T_c/T_h, to be equal to the ratio of the heats Q_c/Q_h:

$$\frac{Q_c}{Q_h} = \frac{T_c}{T_h}$$

If we choose the size of a degree in this temperature scale to be equal to 1 C°, then we have, in fact, defined the Kelvin temperature scale discussed in Chapter 16. Thus, if T_h and T_c are given in kelvins, the maximum efficiency of a heat engine is:

Maximum Efficiency of a Heat Engine

$$e_{max} = 1 - \frac{T_c}{T_h} \qquad\qquad \text{18-13}$$

Limits to Efficiency Suppose for a moment that we *could* construct an ideal engine, perfectly reversible and free from all forms of friction. Would this ideal engine have 100% efficiency? No, it would not. From Equation 18-13 $[e_{max} = 1 - (T_c/T_h)]$ we can see that the only way the efficiency of a heat engine could be 100% (that is, $e_{max} = 1$) would be for T_c to be 0 K. As we shall see in the last section of this chapter, this is ruled out by the third law of thermodynamics. Hence, the maximum efficiency will always be less than 100%. No matter how perfect the engine, some of the input heat will always be wasted—given off as Q_c—rather than converted to work.

Since the efficiency is defined to be $e = W/Q_h$, it follows that the maximum work a heat engine can do with the input heat Q_h is

$$W_{max} = e_{max}Q_h = \left(1 - \frac{T_c}{T_h}\right)Q_h \qquad\qquad \text{18-14}$$

If the hot and cold reservoirs have the same temperature, so that $T_c = T_h$, it follows that the maximum efficiency is zero. As a result, the amount of work that such an engine can do is also zero. As mentioned before, a heat engine requires different temperatures

PROBLEM-SOLVING NOTE

Maximum Efficiency

The maximum efficiency a heat engine can have is determined solely by the temperature of the hot (T_h) and cold (T_c) reservoirs. The numerical value of this efficiency is $e = 1 - T_c/T_h$. Remember, however, that the temperatures must be expressed in the Kelvin scale for this expression to be valid.

Big Idea 2 Heat engines use a difference in temperature to convert thermal energy to mechanical energy. The maximum efficiency of the conversion is determined solely by the operating temperatures of the engine.

in order to operate. For example, for a fixed T_c, the higher the temperature of T_h, the greater the efficiency.

Finally, even though Carnot's theorem may seem quite different from the second law of thermodynamics, they are, in fact, equivalent. It can be shown, for example, that if Carnot's theorem were violated, it would be possible for heat to flow spontaneously from a cold object to a hot object.

CONCEPTUAL EXAMPLE 18-12   COMPARING EFFICIENCIES

Suppose you have a heat engine that can operate in one of two different modes. In mode 1, the temperatures of the two reservoirs are T_c = 200 K and T_h = 400 K; in mode 2, the temperatures are T_c = 400 K and T_h = 600 K. Is the maximum efficiency of mode 1 greater than, less than, or equal to the maximum efficiency of mode 2?

REASONING AND DISCUSSION
At first, you might think that because the temperature difference is the same in the two modes, the maximum efficiency is the same as well. This is not the case, however, because the maximum efficiency depends on the *ratio* of the two temperatures ($e = 1 - T_c/T_h$) rather than on their difference. In this case, the maximum efficiency of mode 1 is $e_1 = 1 - (200/400) = 1/2$ and the maximum efficiency of mode 2 is $e_2 = 1 - (400/600) = 1/3$. Thus, mode 1, even though it operates at the lower temperatures, is more efficient.

ANSWER
The efficiency of mode 1 is greater than the efficiency of mode 2.

QUICK EXAMPLE 18-13   FIND THE TEMPERATURE

If the heat engine in Example 18-11 is operating at its maximum efficiency of 24.0%, and its cold reservoir is at a temperature of 295 K, what is the temperature of the hot reservoir?

REASONING AND SOLUTION
Our sketch shows the schematic for this heat engine. We know that the maximum efficiency of the engine is 0.240, and that the temperature of the cold reservoir is T_c = 295 K.

We can solve the efficiency equation ($e = 1 - T_c/T_h$) for the temperature of the hot reservoir, T_h.

1. Write the efficiency, e, in terms of the hot and cold temperatures:

$$e = 1 - \frac{T_c}{T_h}$$

2. Solve for T_h:

$$T_h = T_c/(1 - e)$$

3. Substitute numerical values for T_c and e to find T_h:

$$T_h = \frac{295\ \text{K}}{(1 - 0.240)} = 388\ \text{K}$$

Although an efficiency of 24.0% may seem low, it is typical of many real engines.

We can summarize the conclusions of this section as follows: The first law of thermodynamics states that you cannot get something for nothing. To be specific, you cannot get more work out of a heat engine than the amount of heat you put in. The best you can do is break even. The second law of thermodynamics is more restrictive than the first law; it says that *you can't even break even*—some of the input heat must be wasted. It's a law of nature.

Enhance Your Understanding   (Answers given at the end of the chapter)

6. Heat engines A and B operate with different hot and cold reservoirs, as indicated in **FIGURE 18-19**. Is the maximum efficiency of engine A greater than, less than, or equal to the maximum efficiency of engine B? Explain.

CONTINUED

▲ **FIGURE 18-19** Enhance Your Understanding 6.

Section Review

- The maximum efficiency of a heat engine is $e_{max} = 1 - T_c/T_h$, where the temperatures are measured on the Kelvin scale.

18-7 Refrigerators, Air Conditioners, and Heat Pumps

When we stated the second law of thermodynamics in Section 18-5, we said that the spontaneous flow of heat is always from high temperature to low temperature. The key word here is *spontaneous*. It *is* possible for heat to flow "uphill," from a cold object to a hot one, but it doesn't happen spontaneously—work must be done on the system to make it happen, just as work must be done to pump water up from a well. Refrigerators, air conditioners, and heat pumps are devices that use work to transfer heat from a cold object to a hot object.

RWP Let's begin by comparing the operations of a heat engine, **FIGURE 18-20 (a)**, and a **refrigerator**, **FIGURE 18-20 (b)**. Notice that all the arrows are reversed in the refrigerator; in effect, *a refrigerator is a heat engine running backward*. In particular, the refrigerator uses a work W to remove a certain amount of heat, Q_c, from the cold reservoir (the interior of the refrigerator). It then exhausts an even larger amount of heat, Q_h, to the hot reservoir (the air in the kitchen). By energy conservation, it follows that

$$Q_h = Q_c + W$$

Thus, as a refrigerator operates, it warms the kitchen at the same time that it cools the food stored within it.

To design an effective refrigerator, you would like it to remove as much heat from its interior as possible for the smallest amount of work. After all, the work is supplied by electrical energy that must be paid for each month. Thus, we define the **coefficient of performance (COP) for a refrigerator** as an indicator of its effectiveness:

Coefficient of Performance for a Refrigerator, COP

$$COP = \frac{Q_c}{W} \qquad\qquad 18\text{-}15$$

SI unit: dimensionless

Typical values for the coefficient of performance range from 2 to 6.

▲ **FIGURE 18-20 Cooling and heating devices** Schematic comparison of a generalized heat engine **(a)** with a refrigerator, an air conditioner, and a heat pump. **(b)** In the refrigerator, a work W is done to remove a heat Q_c from the cold reservoir (the inside of the refrigerator). By energy conservation, the heat given off to the hot reservoir (the kitchen) is $Q_h = W + Q_c$. **(c)** An air conditioner removes heat from the cool air inside a house and exhausts it into the hot air of the atmosphere. **(d)** A heat pump moves heat from a cold reservoir to a hot reservoir, just like an air conditioner. The difference is that the reservoirs are switched, so that heat is pumped into the house rather than out to the atmosphere.

EXERCISE 18-14 WORK SUPPLIED

A refrigerator has a coefficient of performance of 3.25. How much work must be supplied to this refrigerator in order to remove 261 J of heat from its interior?

REASONING AND SOLUTION
Rearranging COP = Q_c/W to solve for the work yields

$$W = \frac{Q_c}{\text{COP}} = \frac{261 \text{ J}}{3.25} = 80.3 \text{ J}$$

Thus, 80.3 J of work removes 261 J of heat from the refrigerator, and exhausts 80.3 J + 261 J = 341 J of heat into the kitchen.

RWP An **air conditioner,** as illustrated in **FIGURE 18-20 (c)**, is basically a refrigerator in which the cold reservoir is the room that is being cooled. To be specific, the air conditioner uses electrical energy to "pump" heat from the cool room to the warmer air outside. As with the refrigerator, more heat is exhausted to the hot reservoir than is removed from the cold reservoir; that is, $Q_h = Q_c + W$, as before.

CONCEPTUAL EXAMPLE 18-15 ROOM TEMPERATURE   PREDICT/EXPLAIN

You haven't had time to install your new air conditioner in the window yet, so as a short-term measure you decide to place it on the dining-room table and turn it on to cool things off a bit. **(a)** As a result, does the air in the dining room get warmer, get cooler, or stay at the same temperature? **(b)** Which of the following is the *best explanation* for your prediction?

  I. The room gets warmer due to the combined effect of the heat exhausted by the air conditioner and the work done by its motor.

 II. The room gets cooler because the air conditioner cools the air.

III. The room temperature stays the same because the warm air in the exhaust of the air conditioner offsets the cool air it produces.

REASONING AND DISCUSSION
You might think the room stays at the same temperature, because the air conditioner draws heat from the room as usual, but then exhausts heat back into the room that would normally be sent outside. However, the motor of the air conditioner is doing work in order to draw heat from the room, and the heat that would normally be exhausted outdoors is equal to the

CONTINUED

heat drawn from the room *plus* the work done by the motor: $Q_h = Q_c + W$. Thus, the net effect is that the motor of the air conditioner is continually adding heat to the room, causing it to get warmer.

ANSWER
(a) The air in the dining room gets warmer. **(b)** The best explanation is I.

RWP A **heat pump** can be thought of as an air conditioner with the reservoirs switched. As we see in **FIGURE 18-20 (d)**, a heat pump does a work W to remove an amount of heat Q_c from the cold reservoir of outdoor air, then exhausts a heat Q_h into the hot reservoir of air in the room. Just as with the refrigerator and the air conditioner, the heat going to the hot reservoir is $Q_h = Q_c + W$.

In an **ideal**, reversible heat pump with only two operating temperatures, T_c and T_h, the Carnot relationship $Q_c/Q_h = T_c/T_h$ holds, just as it does for a heat engine. Thus, if you want to add a heat Q_h to a room, the work that must be done to accomplish this is

$$W = Q_h - Q_c = Q_h\left(1 - \frac{Q_c}{Q_h}\right) = Q_h\left(1 - \frac{T_c}{T_h}\right) \qquad \text{18-16}$$

We use this result in the next Example.

EXAMPLE 18-16   **PUMPING HEAT**

An ideal heat pump, one that satisfies the Carnot relationship $W = Q_h(1 - T_c/T_h)$, is used to heat a room that is at 293 K. If the pump does 275 J of work, how much heat does it supply to the room if the outdoor temperature is **(a)** 273 K or **(b)** 263 K?

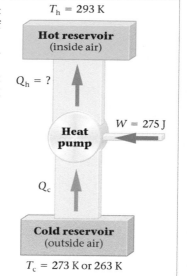

T_h = 293 K

Hot reservoir
(inside air)

Q_h = ?

W = 275 J

Heat
pump

Q_c

Cold reservoir
(outside air)

T_c = 273 K or 263 K

PICTURE THE PROBLEM
Our sketch shows the heat pump doing 275 J of mechanical work to transfer the heat Q_h to the hot reservoir at the temperature 293 K. The temperature of the cold reservoir is 273 K for part (a) and 263 K for part (b). We wish to determine Q_h for each of these cases.

REASONING AND STRATEGY
For an ideal heat pump, we know that $W = Q_h(1 - T_c/T_h)$. Therefore, given the hot and cold temperatures, as well as the mechanical work, it is straightforward to determine the heat delivered to the hot reservoir, Q_h.

Known   Temperature of room (hot reservoir), T_h = 293 K; work done by pump, W = 275 J; temperature of outdoors (cold reservoir), **(a)** T_c = 273 K, **(b)** T_c = 263 K.
Unknown   Heat supplied to room (hot reservoir), Q_h = ?

SOLUTION

1. Solve $W = Q_h(1 - T_c/T_h)$ for the heat Q_h: $\qquad Q_h = W/(1 - T_c/T_h)$

Part (a)

2. Substitute W = 275 J, T_c = 273 K, and T_h = 293 K into the expression for Q_h:

$$Q_h = \frac{W}{1 - \dfrac{T_c}{T_h}} = \frac{275\,\text{J}}{1 - \dfrac{273\,\text{K}}{293\,\text{K}}} = 4030\,\text{J}$$

Part (b)

3. Substitute W = 275 J, T_c = 263 K, and T_h = 293 K into the expression for Q_h:

$$Q_h = \frac{W}{1 - \dfrac{T_c}{T_h}} = \frac{275\,\text{J}}{1 - \dfrac{263\,\text{K}}{293\,\text{K}}} = 2690\,\text{J}$$

INSIGHT
As one might expect, the same amount of work provides less heat when the outside temperature is lower. That is, more work must be done on a colder day to provide the same heat to the inside air.

In addition, notice that if 275 J of heat is supplied to an electric heater, then 275 J of heat is given to the air in the room. When that same energy is used to run a heat pump, a good deal more than 275 J of heat is added to the room.

CONTINUED

PRACTICE PROBLEM
How much work must be done by this heat pump to supply 2550 J of heat on a day when the outside temperature is 253 K?
[**Answer:** $W = 348$ J]

Some related homework problems: Problem 58, Problem 59

The purpose of a heat pump is to add heat to a room; in fact, we would like to add as much heat (Q_h) as possible for the least work (W). As a result, the **coefficient of performance (COP) for a heat pump** is defined as follows:

Coefficient of Performance for a Heat Pump, COP

$$COP = \frac{Q_h}{W}$$

18-17

SI unit: dimensionless

The COP for a heat pump, which usually ranges from 3 to 4, depends on the inside and outside temperatures. We show how to use the COP in the next Exercise.

EXERCISE 18-17 WORK REQUIRED

A heat pump with a coefficient of performance equal to 3.65 supplies 3220 J of heat to a room. How much work is required?

REASONING AND SOLUTION
Work is related to heat and the coefficient of performance by COP $= Q_h/W$. Rearranging this equation, and solving for the work, W, we find

$$W = \frac{Q_h}{COP} = \frac{3220 \text{ J}}{3.65} = 882 \text{ J}$$

Enhance Your Understanding   (Answers given at the end of the chapter)

7. If the coefficient of performance (COP) of a particular refrigerator is increased, does the work required to remove a certain amount of heat from the cold reservoir increase, decrease, or remain the same? Explain.

Section Review

• A heat engine can be run in reverse to serve as a refrigerator, a device that uses work to "pump" heat from a cold reservoir to a hot reservoir.

18-8 Entropy

In this section we introduce a new quantity that is as fundamental to physics as energy or temperature. This quantity is referred to as the *entropy*, and it is related to the amount of disorder in a system. For example, a messy room has more entropy than a neat one, a pile of bricks has more entropy than a building constructed from the bricks, a freshly laid egg has more entropy than one that is just about to hatch, and a puddle of water has more entropy than the block of ice from which it melted. We begin by considering the connection between entropy and heat, and later develop more fully the connection between disorder and entropy.

When discussing heat engines, we saw that if an engine is reversible, it satisfies the following relationship:

$$\frac{Q_c}{Q_h} = \frac{T_c}{T_h}$$

Rearranging slightly, we can rewrite this as

$$\frac{Q_c}{T_c} = \frac{Q_h}{T_h}$$

Notice that the quantity Q/T is the same for both the hot and the cold reservoirs. This relationship prompted the German physicist Rudolf Clausius (1822–1888) to propose the following definition: The **entropy**, S, is a quantity whose change is given by the heat Q divided by the absolute temperature T:

Definition of Entropy Change, ΔS

$$\Delta S = \frac{Q}{T}$$
18-18

SI unit: J/K

For this definition to be valid, the heat Q must be transferred reversibly at the fixed Kelvin temperature T. Notice that if heat is added to a system ($Q > 0$), the entropy of the system increases; if heat is removed from a system ($Q < 0$), its entropy decreases.

Entropy is a state function, just like the internal energy, U. This means that the value of S depends only on the state of a system, and not on how the system gets to that state. It follows, then, that the *change* in entropy, ΔS, depends only on the initial and final states of a system. Thus, if a process is irreversible—so that Equation 18-18 ($\Delta S = Q/T$) *does not* hold—we can still calculate ΔS by using a reversible process to connect the same initial and final states.

PROBLEM-SOLVING NOTE

Calculating Entropy Change

When calculating the entropy change, $\Delta S = Q/T$, be sure to convert the temperature T to kelvins. It is *only* when this is done that the correct value for ΔS can be obtained.

EXAMPLE 18-18   MELTS IN YOUR HAND

(a) Calculate the change in entropy when a 0.125-kg chunk of ice melts at 0 °C. Assume the melting occurs reversibly.
(b) Suppose heat is now drawn reversibly from the 0 °C meltwater, causing a decrease in entropy of 112 J/K. How much ice freezes in the process?

PICTURE THE PROBLEM

In our sketch we show a 0.125-kg chunk of ice at the temperature 0 °C. As the ice absorbs the heat Q from its surroundings, it melts to water at 0 °C. Because the system absorbs heat, its entropy increases. When heat is drawn out of the meltwater, the entropy will decrease.

REASONING AND STRATEGY

a. The entropy change is $\Delta S = Q/T$, where $T = 0\,°C = 273$ K. To find the heat Q, we note that to melt the ice we must add to it the latent heat of fusion, L_f. Thus, the heat is $Q = mL_f$, where $L_f = 33.5 \times 10^4$ J/kg.

b. The magnitude of heat withdrawn from the meltwater is $Q = T|\Delta S|$, where $\Delta S = -112$ J/K. Using this heat, we can calculate the mass of re-frozen ice using $m = Q/L_f$.

Known   (a) Mass of ice, $m = 0.125$ kg; temperature, $T = 0\,°C = 273$ K. (b) Change in entropy, $\Delta S = -112$ J/K.
Unknown   (a) Change in entropy, $\Delta S = ?$ (b) Mass of ice that freezes, $m = ?$

SOLUTION

Part (a)

1. Find the heat that must be absorbed by the ice for it to melt: $Q = mL_f = (0.125\ \text{kg})(33.5 \times 10^4\ \text{J/kg}) = 4.19 \times 10^4\ \text{J}$

2. Calculate the change in entropy: $\Delta S = \dfrac{Q}{T} = \dfrac{4.19 \times 10^4\ \text{J}}{273\ \text{K}} = 153\ \text{J/K}$

Part (b)

3. Find the magnitude of the heat that is removed from the meltwater: $|Q| = T|\Delta S| = (273\ \text{K})(112\ \text{J/K}) = 3.06 \times 10^4\ \text{J}$

4. Use this heat to determine the amount of water that is re-frozen: $m = \dfrac{|Q|}{L_f} = \dfrac{(3.06 \times 10^4\ \text{J})}{(33.5 \times 10^4\ \text{J/kg})} = 0.0913\ \text{kg}$

INSIGHT

Notice that we were careful to convert the temperature of the system from 0 °C to 273 K before we applied $\Delta S = Q/T$. This must always be done when calculating the entropy change. If we had neglected to do the conversion in this case, we would have found an infinite increase in entropy—which is clearly not possible physically.

In addition, notice that the entropy change is positive in part (a) and negative in part (b). This illustrates the general rule that entropy increases when heat is added to a system, and decreases when it is removed.

CONTINUED

PRACTICE PROBLEM
Find the mass of ice that would be required to give an entropy change of 264 J/K.   [**Answer:** $m = 0.215$ kg]

Some related homework problems: Problem 65, Problem 66

Entropy and Heat Engines   Let's apply the definition of entropy change to the case of a reversible heat engine. First, a heat Q_h leaves the hot reservoir at the temperature T_h. Thus, the entropy of this reservoir decreases by the amount Q_h/T_h:

$$\Delta S_h = -\frac{Q_h}{T_h}$$

Recall that Q_h is the magnitude of the heat leaving the hot reservoir; hence, the minus sign is used to indicate a decrease in entropy. Similarly, heat is added to the cold reservoir; hence, its entropy increases by the amount Q_c/T_c:

$$\Delta S_c = \frac{Q_c}{T_c}$$

The total entropy change for this system is

$$\Delta S_{total} = \Delta S_h + \Delta S_c = -\frac{Q_h}{T_h} + \frac{Q_c}{T_c}$$

Recalling that $Q_h/T_h = Q_c/T_c$, as pointed out at the beginning of this section, we find that the total entropy change vanishes:

$$\Delta S_{total} = -\frac{Q_h}{T_h} + \frac{Q_c}{T_c} = 0$$

What is special about a reversible engine, then, is the fact that its entropy does not change.

On the other hand, a real engine will always have a lower efficiency than a reversible engine operating between the same temperatures. This means that in a real engine less of the heat from the hot reservoir is converted to work; hence, more heat is given off as waste heat to the cold reservoir. Thus, for a given value of Q_h, the heat Q_c is greater in an irreversible engine than in a reversible one. As a result, instead of $Q_c/T_c = Q_h/T_h$, we have

$$\frac{Q_c}{T_c} > \frac{Q_h}{T_h}$$

Therefore, if an engine is irreversible, the total entropy change is positive:

$$\Delta S_{total} = -\frac{Q_h}{T_h} + \frac{Q_c}{T_c} > 0$$

In general, *any irreversible process results in an increase of entropy.*

These results can be summarized in the following general statements:

Entropy in the Universe

The total entropy of the universe *increases* whenever an *irreversible* process occurs.

The total entropy of the universe is *unchanged* whenever a *reversible* process occurs.

Because all *real* processes are irreversible (with reversible processes being a useful idealization), the total entropy of the universe continually increases. Thus, in terms of the entropy, the universe moves in only one direction—toward an ever-increasing entropy. This is quite different from the behavior with regard to energy, which remains constant no matter what type of process occurs.

In fact, this statement about entropy in the universe is yet another way of expressing the second law of thermodynamics. Recall, for example, that our original statement of the second law said that heat flows spontaneously from a hot object to a cold object. During this flow of heat the entropy of the universe increases, as we show in the next Example. Hence, the direction in which heat flows is seen to be the result of the

PROBLEM-SOLVING NOTE

Entropy Change

Though it is tempting to treat entropy like energy, setting the final value equal to the initial value, this is not the case in general. Only in a reversible process is the entropy unchanged—otherwise, it increases. Still, the entropy of part of a system can decrease, as long as the entropy of other parts increases by the same amount or more.

Big Idea 3 Entropy is a measure of disorder in a physical system. All thermodynamic processes proceed in a direction that either increases entropy or leaves it unchanged. The entropy of the universe never decreases.

general principle of entropy increase in the universe. Again, we see a directionality in nature—the ever-present "arrow of time."

EXAMPLE 18-19 ENTROPY IS NOT CONSERVED!

A hot reservoir at the temperature 576 K transfers 1050 J of heat irreversibly to a cold reservoir at the temperature 305 K. Find the change in entropy of the universe.

PICTURE THE PROBLEM

The relevant physical situation is shown in our sketch. Notice that the heat $Q = 1050$ J is transferred from the hot reservoir at the temperature $T_h = 576$ K directly to the cold reservoir at the temperature $T_c = 305$ K.

REASONING AND STRATEGY

As the heat Q leaves the hot reservoir, its entropy *decreases* by Q/T_h. When the same heat Q enters the cold reservoir, its entropy *increases* by the amount Q/T_c. Summing these two contributions gives the entropy change of the universe.

Known   Temperature of hot reservoir, $T_h = 576$ K; temperature of cold reservoir, $T_c = 305$ K; heat, $Q = 1050$ J.

Unknown   Change in entropy of the universe, $\Delta S_{universe} = $?

SOLUTION

1. Calculate the entropy change of the hot reservoir. This entropy change is negative:

$$\Delta S_h = -\frac{Q}{T_h} = -\frac{1050\,\text{J}}{576\,\text{K}} = -1.82\,\text{J/K}$$

2. Calculate the entropy change of the cold reservoir. This entropy change is positive:

$$\Delta S_c = \frac{Q}{T_c} = \frac{1050\,\text{J}}{305\,\text{K}} = 3.44\,\text{J/K}$$

3. Sum these contributions to obtain the entropy change of the universe:

$$\Delta S_{universe} = \Delta S_h + \Delta S_c = -\frac{Q}{T_h} + \frac{Q}{T_c}$$
$$= -1.82\,\text{J/K} + 3.44\,\text{J/K} = 1.62\,\text{J/K}$$

INSIGHT

The decrease in entropy of the hot reservoir is more than made up for by the increase in entropy of the cold reservoir. This is a general result.

PRACTICE PROBLEM

What amount of heat must be transferred between these reservoirs for the entropy of the universe to increase by 1.50 J/K? [**Answer:** $Q = 972$ J]

Some related homework problems: Problem 67, Problem 71

When certain processes occur, it sometimes appears as if the entropy of the universe has decreased. On closer examination, however, it can always be shown that there is a larger increase in entropy elsewhere that results in an overall increase. This issue is addressed in the next Conceptual Example.

CONCEPTUAL EXAMPLE 18-20 ENTROPY CHANGE

You put a tray of water in the kitchen freezer, and some time later it has turned to ice. Has the entropy of the universe increased, decreased, or stayed the same?

REASONING AND DISCUSSION

It might seem that the entropy of the universe has decreased. After all, heat is removed from the water to freeze it, and, as we know, removing heat from an object lowers its entropy. On the other hand, we also know that the freezer does work to draw heat from the water; hence, it exhausts more heat into the kitchen than it absorbs from the water. As a result, the entropy increase of the heated air in the kitchen is greater than the magnitude of the entropy decrease of the water. Thus, the entropy of the universe increases—as it must for *any* real process.

ANSWER

The entropy of the universe has increased.

Useful Thermal Energy As the entropy in the universe increases, the amount of work that can be done is diminished. For example, the heat flow in Example 18-19 resulted in an increase in the entropy of the universe by the amount 1.62 J/K. If this same heat had been used in a reversible engine, however, it could have done work, and because the engine is reversible, the entropy of the universe would stay the same. In the next Example, we calculate the work that could be done with a reversible engine in this situation.

EXAMPLE 18-21   FIND THE WORK

Suppose a reversible heat engine operates between the two heat reservoirs described in Example 18-19. Find the amount of work done by such an engine when 1050 J of heat is drawn from the hot reservoir.

PICTURE THE PROBLEM
Our sketch shows a reversible heat engine operating between a hot reservoir at 576 K and a cold reservoir at 305 K.

REASONING AND STRATEGY
We first calculate the efficiency of this engine using $e = 1 - T_c/T_h$. We then use $W = eQ_h$ to find the work done.

Known   Temperature of hot reservoir, $T_h = 576$ K; temperature of cold reservoir, $T_c = 305$ K; heat, $Q = 1050$ J.

Unknown   Work done, $W = ?$

SOLUTION

1. Calculate the efficiency of this engine:
$$e = 1 - \frac{T_c}{T_h} = 1 - \left(\frac{305\text{ K}}{576\text{ K}}\right) = 0.470$$

2. Multiply the efficiency by Q_h to find the work done:
$$W = eQ_h = (0.470)(1050\text{ J}) = 494\text{ J}$$

INSIGHT
This engine is reversible, and hence its total entropy change must be zero. The decrease in entropy of the hot reservoir is $-(1050\text{ J})/576\text{ K} = -1.82$ J/K. It follows that the increase in entropy of the cold reservoir must have the same magnitude. The amount of heat that flows into the cold reservoir is $Q_c = Q_h - W = 1050\text{ J} - 494\text{ J} = 556\text{ J}$. This heat causes an entropy increase equal to $(556\text{ J})/(305\text{ K}) = +1.82$ J/K, as expected. If the engine were irreversible, it would exhaust more than 556 J of heat, and this would create a net increase in entropy and a reduction in the amount of work done. Clearly, then, a reversible engine produces the maximum amount of work.

PRACTICE PROBLEM
Suppose this engine is not reversible, and that only 457 J of work is done when 1050 J of heat is drawn from the hot reservoir. What is the entropy increase of the universe in this case? **[Answer: 0.12 J/K]**

Some related homework problems: Problem 71, Problem 86

The Quality of Energy When 1050 J of heat is transferred directly from the hot reservoir to the cold reservoir in Example 18-19, the entropy of the universe increases by 1.62 J/K. When this same heat is transferred reversibly, with an ideal engine, the entropy of the universe stays the same, but 494 J of work is done. The connection between the work done by a reversible engine, and the entropy increase of the irreversible process, is very simple:

$$W = T_c \Delta S_{\text{universe}} = (305\text{ K})(1.62\text{ J/K}) = 494\text{ J}$$

Thus, the flow of thermal energy can increase the entropy of the universe, or it can be tapped to produce useful work.

To see that this expression is valid in general, recall that in Example 18-19 the total change in entropy is $\Delta S_{\text{universe}} = Q/T_c - Q/T_h$. That is, the heat $Q_h = Q$ is withdrawn from the hot reservoir (lowering the entropy by the amount Q/T_h), and the same heat

$Q_c = Q$ is added to the cold reservoir (increasing the entropy by the larger amount, Q/T_c). If we multiply this increase in entropy by the temperature of the cold reservoir, T_c, we have $T_c \Delta S_{universe} = Q - QT_c/T_h = Q(1 - T_c/T_h)$. Recalling that the efficiency of an ideal engine is $e = 1 - T_c/T_h$, we see that $T_c \Delta S_{universe} = Qe$. Finally, the work done by an ideal engine is $W = eQ_h$, or in this case, $W = eQ$, since $Q_h = Q$. Therefore, we see that $W = T_c \Delta S_{universe}$, as expected.

In general, a process in which the entropy of the universe increases is one in which less work is done than if the process had been reversible. Thus, we lose forever the ability for that work to be done, because to restore the universe to its former state would mean lowering its entropy, which cannot be done. We conclude that with every increase in entropy, there is that much less useful work that can be done by the universe.

For this reason, entropy is sometimes referred to as a measure of the "quality" of energy. When an irreversible process occurs, and the entropy of the universe increases, we say that the energy of the universe has been "degraded" because less of it is available to do work. This process of increasing degradation of energy and increasing entropy in the universe is a continuing aspect of nature.

Enhance Your Understanding   (Answers given at the end of the chapter)

8. In system A, the heat Q is added to a cold reservoir with a temperature of 300 K. In system B, the same heat is added to a cold reservoir with a temperature of 350 K. Is the increase in entropy for system A greater than, less than, or equal to the increase in entropy for system B? Explain.

Section Review

- The change in entropy is defined as $\Delta S = Q/T$. Entropy is a state function, and depends only on the state of a system and not on how the state is produced.

- The total entropy of the universe *increases* whenever an *irreversible* process occurs. The total entropy of the universe is *unchanged* whenever a *reversible* process occurs.

18-9 Order, Disorder, and Entropy

In the previous section we considered entropy from the point of view of thermodynamics. We saw that as heat flows from a hot object to a cold object, the entropy of the universe increases. In this section we show that entropy can also be thought of as a measure of the amount of **disorder** in the universe.

We begin with the situation of heat flow from a hot to a cold object. In **FIGURE 18-21 (a)** we show two bricks, one hot and the other cold. As we know from kinetic theory, the molecules in the hot brick have more kinetic energy than the molecules in the cold brick. This means that the system is rather orderly, in that all the high-kinetic-energy molecules are grouped together in the hot brick, and all the low-kinetic-energy molecules are grouped together in the cold brick. There is a definite regularity, or order, to the distribution of the molecular speeds.

High temperature

Heat flow Q

Intermediate temperature

Low temperature

(a)   (b)   (c)

◀ **FIGURE 18-21 Heat flow and disorder**
(a) Initially, two bricks have different temperatures, and hence different average kinetic energies. **(b)** Heat flows from the hot brick to the cold brick. **(c)** The final result is that both bricks have the same intermediate temperature, and all the molecules have the same average kinetic energy. Thus, the initial orderly segregation of molecules by kinetic energy has been lost.

▲ **FIGURE 18-22** All processes that occur spontaneously increase the entropy of the universe. In this case, the movements of the water molecules become more random and chaotic when they reach the tumultuous, swirling pool at the bottom of the falls. In addition, some of their kinetic energy is converted to thermal energy—the most disordered and degraded form of energy.

Now bring the bricks into thermal contact, as in **FIGURE 18-21 (b)**. The result is a flow of heat from the hot brick to the cold brick until the temperatures become equal. The final result is indicated in **FIGURE 18-21 (c)**. During the heat transfer the entropy of the universe increases, as we know, and the system loses the nice orderly distribution it had in the beginning. In the final state, all the molecules have the same average kinetic energy; hence, the system is randomized, or disordered. We are led to the following conclusion:

> As the entropy of a system increases, its disorder increases as well; that is, an *increase* in entropy is the same as a *decrease* in order.

If heat had flowed in the opposite direction—from the cold brick to the hot brick—the ordered distribution of molecules would have been reinforced, rather than lost.

To take another example, consider the 0.125-kg chunk of ice discussed in Example 18-18. As we saw there, the entropy of the universe increases as the ice melts. Now let's consider what happens on the molecular level. To begin, the molecules are well ordered in their crystalline positions. As heat is absorbed, however, the molecules begin to free themselves from the ice and move about randomly in the growing puddle of water. Thus, the regular order of the solid is lost. Again, we see that as entropy increases, so too does the disorder of the molecules.

Thus, the second law of thermodynamics can be stated as the principle that the disorder of the universe is continually increasing. Everything that happens in the universe, such as the water pouring over a falls in **FIGURE 18-22**, is simply making it a more disorderly place. And there is nothing you can do to prevent it—nothing you can do will result in the universe being more ordered. Just as freezing a tray of water to make ice actually results in more entropy—and more disorder—in the universe, so does any action you take.

Heat Death

If one carries the previous discussion to its logical conclusion, it seems that the universe is "running down." That is, the disorder of the universe constantly increases, and as it does, the amount of energy available to do work decreases. If this process continues, might there come a day when no more work can be done? And, if that day does come, what then?

This is one possible scenario for the fate of the universe, and it is referred to as the "heat death" of the universe. In this scenario, heat continues to flow from hotter regions in space (like stars) to cooler regions (like planets) until, after many billions of years, all objects in the universe have the same temperature. With no temperature differences left, there can be no work done, and the universe would cease to do anything of particular interest. Not a pretty picture, but certainly a possibility. The universe may simply continue with its present expansion until the stars burn out and the galaxies fade away like the dying embers of a scattered campfire.

Living Systems

BIO So far we have focused on the rather gloomy prospect of the universe constantly evolving toward greater disorder. Is it possible, however, that life is an exception to this rule? After all, we know that an embryo utilizes simple raw materials to produce a complex, highly ordered living organism, as illustrated in **FIGURE 18-23**. Similarly, the well-known biological aphorism "ontogeny recapitulates phylogeny," while not strictly correct, expresses the fact that the development of an individual organism from embryo to adult often reflects certain aspects of the evolutionary development of the species as a whole. Thus, over time, species often evolve toward more complex forms. Finally, living systems are able to use disordered raw materials in the environment to

◀ **FIGURE 18-23** Many species, including humans, develop from a single fertilized egg into a complex multicellular organism. In the process they create large, intricately ordered molecules such as proteins and DNA from smaller, simpler precursors. In the metabolic processes of living things, however, heat is produced, which increases the entropy of the universe as a whole. Thus, the second law is not violated.

produce orderly structures in which to live. It seems, then, that there are many ways in which living systems produce increasing order.

This conclusion is flawed, however, because it fails to take into account the entropy of the environment in which the organism lives. It is similar to the conclusion that a freezer violates the second law of thermodynamics because it reduces the entropy of water as it freezes it into ice. This analysis neglects the fact that the freezer exhausts heat into the room, increasing the entropy of the air by an amount that is greater than the entropy decrease of the water. In the same way, living organisms constantly give off heat to the atmosphere as a by-product of their metabolism, increasing its entropy. Thus, if we build a house from a pile of bricks—decreasing the entropy of the bricks— the heat we give off during our exertions increases the entropy of the atmosphere more than enough to give a net increase in entropy.

Finally, all living organisms can be thought of as heat engines, tapping into the flow of energy from one place to another to produce mechanical work. Plants, for example, tap into the flow of energy from the high temperature of the Sun to the cold of deep space and use a small fraction of this energy to sustain themselves and reproduce. Animals consume plants and generate heat within their bodies as they metabolize their food. A fraction of the energy released by the metabolism is in turn converted to mechanical work. Living systems, then, obey the same laws of physics as steam engines and refrigerators—they simply produce different results as they move the universe toward greater disorder.

Enhance Your Understanding   (Answers given at the end of the chapter)

9. Is the entropy of a bag of popcorn greater than, less than, or equal to the entropy of the kernels of corn before popping? Explain.

Section Review

*   Entropy can be thought of as a measure of the disorder of a system. When the entropy of a system increases, its disorder increases as well.

18-10 The Third Law of Thermodynamics

Finally, we consider the **third law of thermodynamics,** which states that there is no temperature lower than absolute zero, and that absolute zero is unattainable. It is possible to cool an object to temperatures arbitrarily close to absolute zero—experiments have reached temperatures as low as 4.5×10^{-10} K—but no object can ever be cooled to precisely 0 K.

As an analogy to cooling toward absolute zero, imagine walking toward a wall, with each step half the distance between you and the wall. Even if you take an infinite number of steps, you will still not reach the wall. You can get arbitrarily close, of course, but you never get all the way there.

The same sort of thing happens when cooling. To cool an object, you can place it in thermal contact with an object that is colder. Heat transfer will occur, with your object ending up cooler and the other object ending up warmer. In particular, suppose you had a collection of objects at 0 K to use for cooling. You put your object in contact with one of the 0 K objects and your object cools, while the 0 K object warms slightly. You continue this process, each time throwing away the "warmed up" 0 K object and using a new one. Each time you do this your object gets closer to 0 K, but it ever actually gets there.

In light of this discussion, we can express the third law of thermodynamics as follows:

Third Law of Thermodynamics
It is impossible to lower the temperature of an object to absolute zero in a finite number of steps.

Big Idea **4** The lowest temperature possible is absolute zero. Systems can be cooled arbitrarily close to absolute zero, but can never attain it.

As with the second law of thermodynamics, this law can be expressed in a number of different but equivalent ways. The essential idea, however, is always the same: Absolute zero is the limiting temperature and, though it can be approached arbitrarily closely, it can never be attained.

Enhance Your Understanding   (Answers given at the end of the chapter)

10. Explain the difference between 0 °C, 0 °F, and 0 K.

Section Review

* No equilibrium temperature can be less than absolute zero, 0 K. A system can be cooled arbitrarily close to 0 K, but can never actually reach it.

CHAPTER 18 REVIEW

CHAPTER SUMMARY

18-1   THE ZEROTH LAW OF THERMODYNAMICS

When two objects have the same temperature, they are in thermal equilibrium.

18-2   THE FIRST LAW OF THERMODYNAMICS

The first law of thermodynamics is a statement of energy conservation that includes heat.

If U is the internal energy of an object, Q is the heat added to it, and W is the work done by the object, the first law of thermodynamics can be written as follows:

$$\Delta U = Q - W \qquad \text{18-3}$$

18-3   THERMAL PROCESSES

Reversible
In a reversible process it is possible to return the system and its surroundings to their initial states.

Irreversible
Irreversible processes cannot be "undone." When the system is returned to its initial state, the surroundings have been altered.

Work
In general, the work done during a process is equal to the area under the process curve in a PV plot.

Constant Pressure
In a PV plot, a constant-pressure process is represented by a horizontal line. The work done at constant pressure is $W = P\Delta V$.

Constant Volume
In a PV plot, a constant-volume process is represented by a vertical line. The work done at constant volume is zero; $W = 0$.

Isothermal Process
In a PV plot, an isothermal process is represented by $PV = $ constant. The work done in an isothermal expansion from V_i to V_f is $W = nRT \ln(V_f/V_i)$.

Adiabatic Process
An adiabatic process occurs with no heat transfer; that is, $Q = 0$.

Constant-temperature heat bath

18-4   SPECIFIC HEATS FOR AN IDEAL GAS: CONSTANT PRESSURE, CONSTANT VOLUME

Specific heats have different values depending on whether they apply to a process at constant pressure or a process at constant volume.

Molar Specific Heat

The molar specific heat, C, is defined by $Q = nC\Delta T$, where n is the number of moles.

Constant Volume

The molar specific heat for an ideal monatomic gas at constant volume is

$$C_v = \frac{3}{2}R \qquad \text{18-6}$$

Constant Pressure

The molar specific heat for an ideal monatomic gas at constant pressure is

$$C_p = \frac{5}{2}R \qquad \text{18-7}$$

Adiabatic Process

In a PV plot, an adiabatic process is represented by $PV^\gamma = \text{constant}$, where γ is the ratio C_p/C_v. For a monatomic, ideal gas $\gamma = \frac{5}{3}$.

$C_v = Q_v/n\Delta T \qquad C_p = Q_p/n\Delta T$

18-5 THE SECOND LAW OF THERMODYNAMICS

When objects of different temperatures are brought into thermal contact, the spontaneous flow of heat that results is always from the high-temperature object to the low-temperature object. Spontaneous heat flow never proceeds in the reverse direction.

18-6 HEAT ENGINES AND THE CARNOT CYCLE

A heat engine is a device that converts heat into work—for example, a steam engine.

Carnot's Theorem

If an engine operating between two constant-temperature reservoirs is to have maximum efficiency, it must be an engine in which all processes are reversible. In addition, all reversible engines operating between the same two temperatures have the same efficiency.

Maximum Efficiency

The maximum efficiency of a heat engine operating between the Kelvin temperatures T_h and T_c is

$$e_{\max} = 1 - \frac{T_c}{T_h} \qquad \text{18-13}$$

18-7 REFRIGERATORS, AIR CONDITIONERS, AND HEAT PUMPS

Refrigerators, air conditioners, and heat pumps are devices that use work to make heat flow from a cold region to a hot region.

Coefficient of Performance

The coefficient of performance of a refrigerator or air conditioner doing the work W to remove a heat Q_c from a cold reservoir is

$$\text{COP} = \frac{Q_c}{W} \qquad \text{18-15}$$

18-8 ENTROPY

Like the internal energy U, the entropy S is a state function.

Change in Entropy

The change in entropy during a reversible exchange of the heat Q at the Kelvin temperature T is

$$\Delta S = \frac{Q}{T} \qquad \text{18-18}$$

18-9 ORDER, DISORDER, AND ENTROPY

Entropy is a measure of the disorder of a system. As entropy increases, a system becomes more disordered.

Heat Death

A possible fate of the universe is heat death, in which everything is at the same temperature and no more work can be done.

18-10 THE THIRD LAW OF THERMODYNAMICS

It is impossible to lower the temperature of an object to absolute zero in a finite number of steps.

ANSWERS TO ENHANCE YOUR UNDERSTANDING QUESTIONS

1. **A**. Heat flows from system 1 (the warmer system) to system 2.

2. D < B < A < C.

3. The volume is halved, because pressure and volume are inversely related in an isotherm.

4. A < D < B < C.

5. No. A broken dish coming back together and flying up into your hand would conserve energy, but would not be allowed in a physical system.

6. The efficiency of heat engine A is equal to the efficiency of heat engine B because both have the same ratio of temperatures between the cold and hot reservoirs.

7. The required work decreases, because a higher COP implies that less heat is needed to remove a given amount of energy.

8. The increase in entropy is greater for system A, because it has the lower temperature.

9. The entropy of the popped popcorn is greater, because popping the corn is an irreversible process.

10. The temperature 0 K (absolute zero) is the lowest temperature possible. The temperatures 0 °F and 0 °C are arbitrary assignments of zero in these everyday temperature scales—they have no deeper significance.

CONCEPTUAL QUESTIONS

For instructor-assigned homework, go to www.masteringphysics.com.

(Answers to odd-numbered Conceptual Questions can be found in the back of the book.)

1. The temperature of a substance is held fixed. Is it possible for heat to flow **(a)** into or **(b)** out of this system? For each case, give an explanation if your answer is no. If your answer is yes, give a specific example.

2. Heat is added to a substance. Is it safe to conclude that the temperature of the substance will rise? Give an explanation if your answer is no. If your answer is yes, give a specific example.

3. Are there thermodynamic processes in which all the heat absorbed by an ideal gas goes completely into mechanical work? If so, give an example.

4. An ideal gas is held in an insulated container at the temperature T. All the gas is initially in one-half of the container, with a partition separating the gas from the other half of the container, which is a vacuum. If the partition ruptures, and the gas expands to fill the entire container, what is its final temperature?

5. Which of the following processes are approximately reversible? **(a)** Lighting a match. **(b)** Pushing a block up a frictionless inclined plane. **(c)** Frying an egg. **(d)** Swimming from one end of a pool to the other. **(e)** Stretching a spring by a small amount. **(f)** Writing a report for class.

6. Which law of thermodynamics would be violated if heat were to spontaneously flow between two objects of equal temperature?

7. Heat engines always give off a certain amount of heat to a low-temperature reservoir. Would it be possible to use this "waste" heat as the heat input to a second heat engine, and then use the "waste" heat of the second engine to run a third engine, and so on?

8. A heat pump uses 100 J of energy as it operates for a given time. Is it possible for the heat pump to deliver more than 100 J of heat to the inside of the house in this same time? Explain.

9. Which law of thermodynamics is most pertinent to the statement that "all the king's horses and all the king's men couldn't put Humpty Dumpty back together again"?

10. Which has more entropy: **(a)** popcorn kernels, or the resulting popcorn; **(b)** two eggs in a carton, or an omelet made from the eggs; **(c)** a pile of bricks, or the resulting house; **(d)** a piece of paper, or the piece of paper after it has been burned?

PROBLEMS AND CONCEPTUAL EXERCISES

Answers to odd-numbered Problems and Conceptual Exercises can be found in the back of the book. **BIO** *identifies problems of biological or medical interest;* **CE** *indicates a conceptual exercise.* Predict/Explain *problems ask for two responses: (a) your prediction of a physical outcome, and (b) the best explanation among three provided; and* Predict/Calculate *problems ask for a prediction of a physical outcome, based on fundamental physics concepts, and follow that with a numerical calculation to verify the prediction. On all problems, bullets (•, ••, •••) indicate the level of difficulty.*

SECTION 18-2 THE FIRST LAW OF THERMODYNAMICS

1. • **CE** Give the change in internal energy of a system if **(a)** $W = 50\,J$, $Q = 50\,J$; **(b)** $W = -50\,J$, $Q = -50\,J$; or **(c)** $W = 50\,J$, $Q = -50\,J$.

2. • **CE** A gas expands, doing 100 J of work. How much heat must be added to this system for its internal energy to increase by 200 J?

3. • A swimmer does $7.7 \times 10^5\,J$ of work and gives off $3.9 \times 10^5\,J$ of heat during a workout. Determine ΔU, W, and Q for the swimmer.

4. • When 1310 J of heat are added to one mole of an ideal monatomic gas, its temperature increases from 272 K to 276 K. Find the work done by the gas during this process.

5. • Three different processes act on a system. **(a)** In process A, 42 J of work are done on the system and 77 J of heat are added to the system. Find the change in the system's internal energy. **(b)** In process B, the system does 42 J of work and 77 J of heat are added to the system. What is the change in the system's internal energy? **(c)** In process C, the system's internal energy decreases by 120 J

while the system performs 120 J of work on its surroundings. How much heat was added to the system?

6. • An ideal gas is taken through the four processes shown in **FIGURE 18-24**. The changes in internal energy for three of these processes are as follows: $\Delta U_{AB} = +82\,J$; $\Delta U_{BC} = +15\,J$; $\Delta U_{DA} = -56\,J$. Find the change in internal energy for the process from C to D.

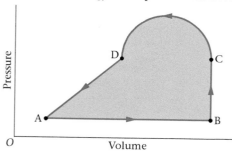

FIGURE 18-24 Problem 6

7. •• A basketball player does $4.81 \times 10^5\,J$ of work during her time in the game, and evaporates 0.151 kg of water. Assuming a latent heat of $2.26 \times 10^6\,J/kg$ for the perspiration (the same as for water), determine **(a)** the change in the player's internal energy and **(b)** the number of nutritional calories the player has converted to work and heat.

8. •• Predict/Calculate One mole of an ideal monatomic gas is initially at a temperature of 274 K. **(a)** Find the final temperature of the gas if 3260 J of heat are added to it and it does 712 J of work. **(b)** Suppose the amount of gas is doubled to two moles. Does the final temperature found in part (a) increase, decrease, or stay the same? Explain.

9. •• Predict/Calculate **Energy from Gasoline** Burning a gallon of gasoline releases $1.19 \times 10^8\,J$ of internal energy. If a certain car requires $5.20 \times 10^5\,J$ of work to drive one mile, **(a)** how much heat is given off to the atmosphere each mile, assuming the car gets 25.0 miles to the gallon? **(b)** If the miles per gallon of the car is increased, does the amount of heat released to the atmosphere increase, decrease, or stay the same? Explain.

10. •• A cylinder contains 4.0 moles of a monatomic gas at an initial temperature of 27 °C. The gas is compressed by doing 560 J of work on it, and its temperature increases by 130 °C. How much heat flows into or out of the gas?

11. •• An ideal gas is taken through the three processes shown in **FIGURE 18-25**. Fill in the missing entries in the following table:

	Q	W	ΔU
A → B	− 136 J	(a)	(b)
B → C	− 341 J	− 136 J	(c)
C → A	(e)	153 J	(d)

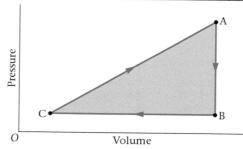

FIGURE 18-25 Problems 11 and 88

SECTION 18-3 THERMAL PROCESSES

12. • **CE FIGURE 18-26** shows three different multistep processes, labeled A, B, and C. Rank these processes in order of increasing work done by a gas that undergoes the process. Indicate ties where appropriate.

FIGURE 18-26 Problem 12

13. • The volume of a monatomic ideal gas triples in an isothermal expansion. By what factor does its pressure change?

14. • An ideal gas is compressed at constant pressure to one-half its initial volume. If the pressure of the gas is 120 kPa, and 790 J of work is done on it, find the initial volume of the gas.

15. • As an ideal gas expands at constant pressure from a volume of $0.74\,m^3$ to a volume of $2.3\,m^3$ it does 93 J of work. What is the gas pressure during this process?

16. •• A system consisting of an ideal gas at the constant pressure of 120 kPa gains 960 J of heat. Find the change in volume of the system if the internal energy of the gas increases by **(a)** 920 J or **(b)** 360 J.

17. •• Predict/Calculate **(a)** If the internal energy of a system increases as the result of an adiabatic process, is work done on the system or by the system? **(b)** Calculate the work done on or by the system in part (a) if its internal energy increases by 670 J.

18. •• **(a)** Find the work done by a monatomic ideal gas as it expands from point A to point C along the path shown in **FIGURE 18-27**. **(b)** If the temperature of the gas is 267 K at point A, what is its temperature at point C? **(c)** How much heat has been added to or removed from the gas during this process?

FIGURE 18-27 Problems 18 and 19

19. •• Predict/Calculate A fluid expands from point A to point B along the path shown in Figure 18-27. **(a)** How much work is done by the fluid during this expansion? **(b)** Does your answer to part (a) depend on whether the fluid is an ideal gas? Explain.

20. •• Predict/Calculate If 9.50 moles of a monatomic ideal gas at a temperature of 235 K are expanded isothermally from a volume of 1.12 L to a volume of 4.33 L, calculate **(a)** the work done and

(b) the heat flow into or out of the gas. **(c)** If the number of moles is doubled, by what factors do your answers to parts (a) and (b) change? Explain.

21. •• Suppose 118 moles of a monatomic ideal gas undergo an isothermal expansion from 1.00 m³ to 4.00 m³, as shown in FIGURE 18-28. **(a)** What is the temperature at the beginning and at the end of this process? **(b)** How much work is done by the gas during this expansion?

FIGURE 18-28 Problems 21 and 76

22. •• A weather balloon contains an ideal gas and has a volume of 2.5 m³ at launch, with a pressure of 101 kPa. As it rises slowly through the atmosphere, the gas temperature remains constant at 273 K. **(a)** High in the atmosphere the pressure drops to 19.3 kPa. What is the volume of the balloon at that point? **(b)** How much work is done by the gas in the balloon as it expands?

23. •• Predict/Calculate **(a)** A monatomic ideal gas expands at constant pressure. Is heat added to the system or taken from the system during this process? **(b)** Find the heat added to or taken from the gas in part (a) if it expands at a pressure of 130 kPa from a volume of 0.76 m³ to a volume of 0.93 m³.

24. •• During an adiabatic process, the temperature of 3.22 moles of a monatomic ideal gas drops from 475 °C to 215 °C. For this gas, find **(a)** the work it does, **(b)** the heat it exchanges with its surroundings, and **(c)** the change in its internal energy.

25. •• An ideal gas follows the three-part process shown in FIGURE 18-29. At the completion of one full cycle, find **(a)** the net work done by the system, **(b)** the net change in internal energy of the system, and **(c)** the net heat absorbed by the system.

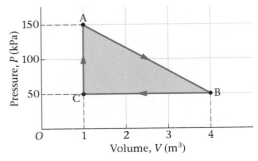

FIGURE 18-29 Problems 25 and 27

26. •• With the pressure held constant at 260 kPa, 43 mol of a monatomic ideal gas expands from an initial volume of 0.75 m³ to a final volume of 1.9 m³. **(a)** How much work was done by the gas during the expansion? **(b)** What were the initial and final temperatures of the gas? **(c)** What was the change in the internal energy of the gas? **(d)** How much heat was added to the gas?

27. •• Predict/Calculate Suppose 67.5 moles of an ideal monatomic gas undergo the series of processes shown in Figure 18-29. **(a)** Calculate the temperature at the points A, B, and C. **(b)** For each

process, A → B, B → C, and C → A, state whether heat enters or leaves the system. Explain in each case. **(c)** Calculate the heat exchanged with the gas during each of the three processes.

28. •• A system expands by 0.75 m³ at a constant pressure of 125 kPa. Find the heat that flows into or out of the system if its internal energy **(a)** increases by 65 kJ or **(b)** decreases by 33 kJ. In each case, give the direction of heat flow.

29. •• Predict/Calculate An ideal monatomic gas is held in a perfectly insulated cylinder fitted with a movable piston. The initial pressure of the gas is 110 kPa, and its initial temperature is 280 K. By pushing down on the piston, you are able to increase the pressure to 140 kPa. **(a)** During this process, did the temperature of the gas increase, decrease, or stay the same? Explain. **(b)** Find the final temperature of the gas.

30. ••• A certain amount of a monatomic ideal gas undergoes the process shown in FIGURE 18-30, in which its pressure doubles and its volume triples. In terms of the number of moles, n, the initial pressure, P_i, and the initial volume, V_i, determine **(a)** the work done by the gas W, **(b)** the change in internal energy of the gas U, and **(c)** the heat added to the gas Q.

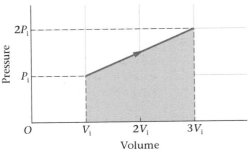

FIGURE 18-30 Problem 30

31. ••• An ideal gas doubles its volume in one of three different ways: (i) at constant pressure; (ii) at constant temperature; (iii) adiabatically. Explain your answers to each of the following questions: **(a)** In which expansion does the gas do the most work? **(b)** In which expansion does the gas do the least work? **(c)** Which expansion results in the highest final temperature? **(d)** Which expansion results in the lowest final temperature?

SECTION 18-4 SPECIFIC HEATS FOR AN IDEAL GAS: CONSTANT PRESSURE, CONSTANT VOLUME

32. * CE Predict/Explain You plan to add a certain amount of heat to a gas in order to raise its temperature. **(a)** If you add the heat at constant volume, is the increase in temperature greater than, less than, or equal to the increase in temperature if you add the heat at constant pressure? **(b)** Choose the *best explanation* from among the following:

   I. The same amount of heat increases the temperature by the same amount, regardless of whether the volume or the pressure is held constant.

   II. All the heat goes into raising the temperature when added at constant volume; none goes into mechanical work.

   III. Holding the pressure constant will cause a greater increase in temperature than simply having a fixed volume.

33. • Find the amount of heat needed to increase the temperature of 5.5 mol of an ideal monatomic gas by 27 K if **(a)** the pressure or **(b)** the volume is held constant.

34. • **(a)** If 585 J of heat are added to 49 moles of a monatomic gas at constant volume, how much does the temperature of the gas increase? **(b)** Repeat part (a), this time for a constant-pressure process.

35. • A system consists of 3.5 mol of an ideal monatomic gas at 315 K. How much heat must be added to the system to double its internal energy at **(a)** constant pressure or **(b)** constant volume?

36. • Find the change in temperature if 170 J of heat are added to 2.8 mol of an ideal monatomic gas at **(a)** constant pressure or **(b)** constant volume.

37. •• **Gasoline Ignition** Consider a short time span just before and after the spark plug in a gasoline engine ignites the fuel-air mixture and releases 1970 J of thermal energy into a volume of 47.6 cm³ that is at a gas pressure of 1.17×10^6 Pa. In this short period of time the volume of the gas can be considered constant. **(a)** Treating the fuel-air mixture as a monatomic ideal gas at an initial temperature of 325 K, what is its temperature after ignition? **(b)** What is the pressure in the gas after ignition?

38. •• Predict/Calculate A cylinder contains 18 moles of a monatomic ideal gas at a constant pressure of 160 kPa. **(a)** How much work does the gas do as it expands 3200 cm³, from 5400 cm³ to 8600 cm³? **(b)** If the gas expands by 3200 cm³ again, this time from 2200 cm³ to 5400 cm³, is the work it does greater than, less than, or equal to the work found in part **(a)**? Explain. **(c)** Calculate the work done as the gas expands from 2200 cm³ to 5400 cm³.

39. •• Predict/Calculate The volume of a monatomic ideal gas doubles in an adiabatic expansion. By what factor do **(a)** the pressure and **(b)** the temperature of the gas change? **(c)** Verify your answers to parts (a) and (b) by considering 135 moles of gas with an initial pressure of 330 kPa and an initial volume of 1.2 m³. Find the pressure and temperature of the gas after it expands adiabatically to a volume of 2.4 m³.

40. •• A monatomic ideal gas is held in a thermally insulated container with a volume of 0.0750 m³. The pressure of the gas is 101 kPa, and its temperature is 325 K. **(a)** To what volume must the gas be compressed to increase its pressure to 145 kPa? **(b)** At what volume will the gas have a temperature of 295 K?

41. ••• Consider the expansion of 60.0 moles of a monatomic ideal gas along processes 1 and 2 in FIGURE 18-31. In process 1 the gas is heated at constant volume from an initial pressure of 106 kPa to a final pressure of 212 kPa. In process 2 the gas expands at constant pressure from an initial volume of 1.00 m³ to a final volume of 3.00 m³. **(a)** How much heat is added to the gas during these two processes? **(b)** How much work does the gas do during this expansion? **(c)** What is the change in the internal energy of the gas?

FIGURE 18-31 Problems 41 and 78

SECTION 18-6 HEAT ENGINES AND THE CARNOT CYCLE

42. • **CE** A Carnot engine can be operated with one of the following four sets of reservoir temperatures: **A**, 400 K and 800 K; **B**, 400 K and 600 K; **C**, 800 K and 1200 K; and **D**, 800 K and 1000 K. Rank these reservoir temperatures in order of increasing efficiency of the Carnot engine. Indicate ties where appropriate.

43. • What is the efficiency of an engine that exhausts 840 J of heat in the process of doing 360 J of work?

44. • An engine receives 660 J of heat from a hot reservoir and gives off 420 J of heat to a cold reservoir. What are **(a)** the work done and **(b)** the efficiency of this engine?

45. • A Carnot engine operates between the temperatures 410 K and 290 K. **(a)** How much heat must be given to the engine to produce 2500 J of work? **(b)** How much heat is discarded to the cold reservoir as this work is done?

46. • A nuclear power plant has a reactor that produces heat at the rate of 838 MW. This heat is used to produce 253 MW of mechanical power to drive an electrical generator. **(a)** At what rate is heat discarded to the environment by this power plant? **(b)** What is the thermal efficiency of the plant?

47. • At a coal-burning power plant a steam turbine is operated with a power output of 548 MW. The thermal efficiency of the power plant is 32.0%. **(a)** At what rate is heat discarded to the environment by this power plant? **(b)** At what rate must heat be supplied to the power plant by burning coal?

48. •• Predict/Calculate A portable generator produces 2.5 kW of mechanical work with an efficiency of 0.20 to supply electricity at a work site. **(a)** If the efficiency is increased but the work output remains the same, will the rate of fuel consumption by the engine increase, decrease, or stay the same? Explain. **(b)** Will the rate of waste heat emission increase, decrease, or stay the same? Explain. **(c)** Calculate the rates of fuel consumption and waste heat emission for engine efficiencies of 0.20 and 0.30.

49. •• Predict/Calculate The efficiency of a particular Carnot engine is 0.280. **(a)** If the high-temperature reservoir is at a temperature of 565 K, what is the temperature of the low-temperature reservoir? **(b)** To increase the efficiency of this engine to 40.0%, must the temperature of the low-temperature reservoir be increased or decreased? Explain. **(c)** Find the temperature of the low-temperature reservoir that gives an efficiency of 0.400.

50. •• During each cycle a reversible engine absorbs 3100 J of heat from a high-temperature reservoir and performs 1900 J of work. **(a)** What is the efficiency of this engine? **(b)** How much heat is exhausted to the low-temperature reservoir during each cycle? **(c)** What is the ratio, T_h/T_c, of the two reservoir temperatures?

51. •• **Solar Thermal Power** The Ivanpah Solar Electric Generating System is the largest solar thermal power-tower facility in the world. It collects 1321 MW of solar thermal energy to heat special collectors to 566 °C. The thermal energy is then used to make steam to operate an electric generator. **(a)** What is the maximum efficiency of the power plant, taking the cold reservoir to be the surrounding environment at 12 °C. **(b)** How much mechanical work (electricity) could the facility generate if it operated at its maximum efficiency? **(c)** If the facility generates 392 MW of electricity, what is its actual efficiency?

52. ••• The operating temperatures for a Carnot engine are T_c and $T_h = T_c + 55$ K. The efficiency of the engine is 11%. Find T_c and T_h.

53. ••• A certain Carnot engine takes in the heat Q_h and exhausts the heat $Q_c = 2Q_h/3$, as indicated in FIGURE 18-32. **(a)** What is the efficiency of this engine? **(b)** Using the Kelvin temperature scale, find the ratio T_c/T_h.

FIGURE 18-32 Problem 53

SECTION 18-7 REFRIGERATORS, AIR CONDITIONERS, AND HEAT PUMPS

54. • **CE Predict/Explain** (a) If the temperature in the kitchen is decreased, is the cost (work needed) to freeze a dozen ice cubes greater than, less than, or equal to what it was before the kitchen was cooled? (b) Choose the *best explanation* from among the following:
    I. The difference in temperature between the inside and the outside of the refrigerator is decreased, and hence less work is required to freeze the ice.
    II. The same amount of ice is frozen in either case, which requires the same amount of heat to be removed and hence the same amount of work.
    III. Cooling the kitchen means that the refrigerator must do more work, both to freeze the ice cubes and to warm the kitchen.

55. • The refrigerator in your kitchen does 490 J of work to remove 95 J of heat from its interior. (a) How much heat does the refrigerator exhaust into the kitchen? (b) What is the refrigerator's coefficient of performance?

56. • A refrigerator with a coefficient of performance of 1.75 absorbs 3.45×10^4 J of heat from the low-temperature reservoir during each cycle. (a) How much mechanical work is required to operate the refrigerator for a cycle? (b) How much heat does the refrigerator discard to the high-temperature reservoir during each cycle?

57. •• Suppose you must remove an average of 3.4×10^8 J of thermal energy per day to keep your house cool during the summer. (a) If you upgrade from an old air conditioner with a COP of 2.0 to a new air conditioner with a COP of 6.0, by how many joules is the required mechanical work reduced each day? (b) If the cost of supplying your air conditioner with mechanical work is $0.12 per 3.6×10^6 J, how much money will you save during a 92-day cooling season?

58. •• To keep a room at a comfortable 20.5 °C, a Carnot heat pump does 315 J of work and supplies the room with 4180 J of heat. (a) How much heat is removed from the outside air by the heat pump? (b) What is the temperature of the outside air?

59. •• An air conditioner is used to keep the interior of a house at a temperature of 21 °C while the outside temperature is 32 °C. If heat leaks into the house at the rate of 11 kW, and the air conditioner has the efficiency of a Carnot engine, what is the mechanical power required to keep the house cool?

60. •• A reversible refrigerator has a coefficient of performance equal to 9.50. What is its efficiency?

61. •• A freezer has a coefficient of performance equal to 4.0. How much electrical energy must this freezer use to produce 1.5 kg of ice at −5.0 °C from water at 15 °C?

SECTION 18-8 ENTROPY

62. • **CE Predict/Explain** (a) If you rub your hands together, does the entropy of the universe increase, decrease, or stay the same? (b) Choose the *best explanation* from among the following:
    I. Rubbing hands together draws heat from the surroundings, and therefore lowers the entropy.
    II. No mechanical work is done by the rubbing, and hence the entropy does not change.
    III. The heat produced by rubbing raises the temperature of your hands and the air, which increases the entropy.

63. • **CE Predict/Explain** (a) An ideal gas is expanded slowly and isothermally. Does its entropy increase, decrease, or stay the same? (b) Choose the *best explanation* from among the following:
    I. Heat must be added to the gas to maintain a constant temperature, and this increases the entropy of the gas.
    II. The temperature of the gas remains constant, which means its entropy also remains constant.
    III. As the gas is expanded its temperature and entropy will decrease.

64. • **CE Predict/Explain** (a) A gas is expanded reversibly and adiabatically. Does its entropy increase, decrease, or stay the same? (b) Choose the *best explanation* from among the following:
    I. The process is reversible, and no heat is added to the gas. Therefore, the entropy of the gas remains the same.
    II. Expanding the gas gives it more volume to occupy, and this increases its entropy.
    III. The gas is expanded with no heat added to it, and hence its temperature will decrease. This, in turn, will lower its entropy.

65. • Find the change in entropy when 1.85 kg of water at 100 °C is boiled away to steam at 100 °C.

66. • Determine the change in entropy that occurs when 4.3 kg of water freezes at 0 °C.

67. •• **CE** You heat a pan of water on the stove. Rank the following temperature increases in order of increasing entropy change. Indicate ties where appropriate: **A,** 25 °C to 35 °C; **B,** 35 °C to 45 °C; **C,** 45 °C to 50 °C; and **D,** 50 °C to 55 °C.

68. •• On a cold winter's day heat leaks slowly out of a house at the rate of 20.0 kW. If the inside temperature is 22 °C, and the outside temperature is −14.5 °C, find the rate of entropy increase.

69. •• An 88-kg parachutist descends through a vertical height of 380 m with constant speed. Find the increase in entropy produced by the parachutist, assuming the air temperature is 21 °C.

70. •• **Predict/Calculate** Consider the air-conditioning system described in Problem 59. (a) Does the entropy of the universe increase, decrease, or stay the same as the air conditioner keeps the imperfectly insulated house cool? Explain. (b) What is the rate at which the entropy of the universe changes during this process?

71. •• A heat engine operates between a high-temperature reservoir at 610 K and a low-temperature reservoir at 320 K. In one cycle, the engine absorbs 6400 J of heat from the high-temperature reservoir and does 2200 J of work. What is the net change in entropy as a result of this cycle?

72. ••• It can be shown that as a mass m with specific heat c changes temperature from T_i to T_f its change in entropy is $\Delta S = mc \ln (T_f/T_i)$ if the temperatures are expressed in kelvin. Suppose you put 79 g of milk at 278 K into an insulated cup containing 296 g of coffee at 355 K, and that each has the specific heat of water. The system comes to an equilibrium temperature of 339 K. (a) What is the entropy change of the milk? (b) What is the entropy change of the coffee? (c) What is the entropy change of the universe due to adding the milk to the coffee?

GENERAL PROBLEMS

73. • **CE** Consider the three-process cycle shown in **FIGURE 18-33.** For each process in the cycle, (a) A→B, (b) B→C, and (c) C→A, state whether the work done by the system is positive, negative, or zero.

FIGURE 18-33 Problem 73

74. • **CE** An ideal gas has the pressure and volume indicated by point I in **FIGURE 18-34.** At this point its temperature is T_1. The temperature of the gas can be increased to T_2 by using the constant-volume process, I→II, or the constant-pressure process, I→III. Is the entropy change for the process I→II greater than, less than, or equal to the entropy change on the process I→III? Explain.

FIGURE 18-34 Problem 74

75. •• The heat that goes into a particular Carnot engine is 4.00 times greater than the work it performs. What is the engine's efficiency?

76. •• Predict/Calculate Consider 132 moles of a monatomic gas undergoing the isothermal expansion shown in Figure 18-28. **(a)** What is the temperature T of this expansion? **(b)** Does the entropy of the gas increase, decrease, or stay the same during the process? Explain. **(c)** Calculate the change in entropy for the gas, ΔS, if it is nonzero. **(d)** Calculate the work done by the gas during this process, and compare to $T\Delta S$.

77. •• **Inflating a Hot-Air Balloon** A hot-air balloon is inflated by heating air at constant pressure. Assume the number of air molecules inside the balloon remains constant and that it expands at atmospheric pressure. **(a)** If the initial temperature and volume of the air inside the balloon are 18 °C and 1600 m³, respectively, what is the final volume of the balloon when its temperature is increased to 65 °C? **(b)** How much work does the balloon do to expand against atmospheric pressure?

78. •• Predict/Calculate Referring to Figure 18-31, suppose 60.0 moles of a monatomic ideal gas are expanded along process 5. **(a)** How much work does the gas do during this expansion? **(b)** What is the change in the internal energy of the gas? **(c)** How much heat is added to the gas during this process?

79. •• Predict/Calculate Engine A has an efficiency of 66%. Engine B absorbs the same amount of heat from the hot reservoir and exhausts twice as much heat to the cold reservoir. **(a)** Which engine has the greater efficiency? Explain. **(b)** What is the efficiency of engine B?

80. •• **Nuclear Versus Natural Gas Energy** Because of safety concerns and other constraints, nuclear power plants must be run at lower efficiencies than natural gas-fired power plants. The efficiency of a particular gas-fired power plant is 44.8% and that of a nuclear plant is 32.6%. **(a)** For every 1.00 kJ of mechanical work (to generate electricity) produced by the gas-fired power plant, how much energy is exhausted as waste heat? **(b)** How much waste heat is generated for every 1.00 kJ of work produced by the nuclear power plant? **(c)** What is the percentage increase of waste heat generated by the nuclear plant over the natural gas plant as a result of the lower efficiency?

81. •• A freezer with a coefficient of performance of 3.88 is used to convert 1.75 kg of water to ice in one hour. The water starts at a temperature of 20.0 °C, and the ice that is produced is cooled to a temperature of −5.00 °C. **(a)** How much heat must be removed from the water for this process to occur? **(b)** How much electrical energy does the freezer use during this hour of operation? **(c)** How much heat is discarded into the room that houses the freezer?

82. •• **Entropy and the Sun** The surface of the Sun has a temperature of 5500 °C and the temperature of deep space is 3.0 K. **(a)** Find the entropy increase produced by the Sun in one day, given that it radiates heat at the rate of 3.80×10^{26} W. **(b)** How much work could have been done if this heat had been used to run an ideal heat engine?

83. •• The following table lists results for various processes involving n moles of a monatomic ideal gas. Fill in the missing entries.

	Q	W	ΔU
Constant pressure	$\frac{5}{2}nR\Delta T$	(a)	$\frac{3}{2}nR\Delta T$
Adiabatic	(b)	$-\frac{3}{2}nR\Delta T$	(c)
Constant volume	$\frac{3}{2}nR\Delta T$	(d)	(e)
Isothermal	(f)	$nRT\ln(V_f/V_i)$	(g)

84. •• A cylinder with a movable piston holds 2.95 mol of argon at a constant temperature of 235 K. As the gas is compressed isothermally, its pressure increases from 101 kPa to 121 kPa. Find **(a)** the final volume of the gas, **(b)** the work done by the gas, and **(c)** the heat added to the gas.

85. •• **Making Ice** You place 0.410 kg of cold water inside a freezer that has a constant temperature of 0 °C. The water eventually freezes and becomes ice at 0 °C. **(a)** What is the change in entropy of the water as it turns into ice? (See Table 17-4 for the appropriate latent heat.) **(b)** If the coefficient of performance of the freezer is 4.50, how much heat does the freezer exhaust into your kitchen as a result of freezing the water? **(c)** If your kitchen temperature is 22 °C, what is the increase in entropy of your kitchen? **(d)** What is the change in entropy of the universe during the freezing process?

86. •• An inventor claims a new cyclic engine that uses organic grape juice as its working material. According to the claims, the engine absorbs 1250 J of heat from a 1010 K reservoir and performs 1120 J of work each cycle. The waste heat is exhausted to the atmosphere at a temperature of 302 K. **(a)** What is the efficiency that is implied by these claims? **(b)** What is the efficiency of a reversible engine operating between the same high and low temperatures used by this engine? (Should you invest in this invention?)

87. •• Predict/Calculate A small dish containing 530 g of water is placed outside for the birds. During the night the outside temperature drops to −5.0 °C and stays at that value for several hours. **(a)** When the water in the dish freezes at 0 °C, does its entropy increase, decrease, or stay the same? Explain. **(b)** Calculate the change in entropy that occurs as the water freezes. **(c)** When the water freezes, by how much does the entropy of the environment increase? **(d)** What is the change in entropy of the universe as a result of this process?

88. •• Predict/Calculate An ideal gas is taken through the three processes shown in Figure 18-25. Fill in the missing entries in the following table.

	Q	W	ΔU
A → B	(a)	(b)	−95.3 J
B → C	(c)	−90.8 J	−136 J
C → A	331 J	(d)	(e)

89. ••• One mole of an ideal monatomic gas follows the three-part cycle shown in **FIGURE 18-35**. **(a)** Fill in the table below. **(b)** What is the efficiency of this cycle?

	Q	W	ΔU
A → B			
B → C			
C → A			

FIGURE 18-35 Problem 89

90. ••• When a heat Q is added to a monatomic ideal gas at constant pressure, the gas does a work W. Find the ratio, W/Q.

91. ••• **The Carnot Cycle** FIGURE 18-36 shows an example of a Carnot cycle. The cycle consists of the following four processes: (1) an isothermal expansion from V_1 to V_2 at the temperature T_h; (2) an adiabatic expansion from V_2 to V_3 during which the temperature drops from T_h to T_c; (3) an isothermal compression from V_3 to V_4 at the temperature T_c; and (4) an adiabatic compression from V_4 to V_1 during which the temperature increases from T_c to T_h. Show that the efficiency of this cycle is $e = 1 - T_c/T_h$, as expected.

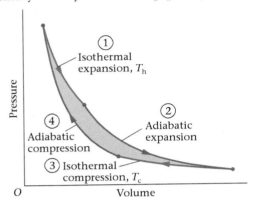

FIGURE 18-36 Problem 91

92. ••• A Carnot engine and a Carnot refrigerator operate between the same two temperatures. Show that the coefficient of performance, COP, for the refrigerator is related to the efficiency, e, of the engine by the following expression: COP = $(1 - e)/e$.

PASSAGE PROBLEMS

Energy from the Ocean

Whenever two objects are at different temperatures, thermal energy can be extracted with a heat engine. A case in point is the ocean, where one "object" is the warm water near the surface, and the other is the cold water at considerable depth. Tropical seas, in particular, can have significant temperature differences between the sun-warmed surface waters, and the cold, dark water 1000 m or more below the surface. A typical oceanic "temperature profile" is shown in FIGURE 18-37, where we see a rapid change in temperature—a thermocline—between depths of approximately 400 m and 900 m.

FIGURE 18-37 Temperature versus depth for ocean waters in the tropics (Problems 93, 94, 95, and 96)

The idea of tapping this potential source of energy has been around for a long time. In 1870, for example, Captain Nemo in Jules Verne's *Twenty Thousand Leagues Under the Sea*, said, "I owe all to the ocean; it produces electricity, and electricity gives heat, light, motion, and, in a word, life to the Nautilus." Just 11 years later, the French physicist Jacques Arsene d'Arsonval proposed a practical system referred to as Ocean Thermal Energy Conversion (OTEC), and in 1930 Georges Claude, one of d'Arsonval's students, built and operated the first experimental OTEC system off the coast of Cuba.

OTEC systems, which are potentially low-cost and carbon neutral, can provide not only electricity, but also desalinated water as part of the process. In fact, an OTEC plant generating 2 MW of electricity is expected to produce over 14,000 cubic feet of desalinated water a day. The governments of Hawaii, Japan, and Australia are actively pursuing plans for OTEC systems.

93. • Suppose an OTEC system operates with surface water at 22 °C and deep water at 4.0 °C. What is the maximum efficiency this system could have?

    A. 6.10%   **B.** 8.20%   **C.** 9.40%   **D.** 18.0%

94. • If we could go deeper for colder water, say to a depth of 4500 m, what is the maximum efficiency that can be achieved?

    A. 6.64%   **B.** 7.52%   **C.** 14.0%   **D.** 25.5%

95. • After they are used in the OTEC plant, the warm surface water and cold deep water (which is richer in inorganic nutrients) may be mixed together and discharged back into the ocean. In this process, an environmental concern is to not affect the local temperature profile so as to reduce impacts on sea life. If 22 °C and 5.5 °C water are mixed in equal volumes, at about what depth should the mixture be discharged so that the discharged mixture is at the same temperature as the surroundings?

    A. 0 m   **B.** 100 m   **C.** 500 m   **D.** 1000 m

96. •• A commercial OTEC system may take in 1500 kg of water per second at 22 °C and cool it to 4.0 °C. How much energy is released in one second by this system? (For comparison, the energy released in burning a gallon of gasoline is 1.3×10^8 J.)

    A. 2.5×10^7 J   **B.** 1.1×10^8 J
    C. 1.4×10^8 J   **C.** 1.6×10^8 J

97. •• Predict/Calculate REFERRING TO EXAMPLE 18-21 Suppose we lower the temperature of the cold reservoir to 295 K; the temperature of the hot reservoir is still 576 K. **(a)** Is the new efficiency of the engine greater than, less than, or equal to 0.470? Explain. **(b)** What is the new efficiency? **(c)** Find the work done by this engine when 1050 J of heat is drawn from the hot reservoir.

98. •• Predict/Calculate REFERRING TO EXAMPLE 18-21 Suppose the temperature of the hot reservoir is increased by 16 K, from 576 K to 592 K, *and* that the temperature of the cold reservoir is also increased by 16 K, from 305 K to 321 K. **(a)** Is the new efficiency greater than, less than, or equal to 0.470? Explain. **(b)** What is the new efficiency? **(c)** What is the change in entropy of the hot reservoir when 1050 J of heat is drawn from it? **(d)** What is the change in entropy of the cold reservoir?

APPENDIX A BASIC MATHEMATICAL TOOLS

This text is designed for students with a working knowledge of basic algebra and trigonometry. Even so, it is useful to review some of the mathematical tools that are of particular importance in the study of physics. In this Appendix we cover a number of topics related to mathematical notation, trigonometry, algebra, mathematical expansions, and vector multiplication.

Mathematical Notation
COMMON MATHEMATICAL SYMBOLS

In Table A-1 we present some of the more common mathematical symbols, along with a translation into English. Though these symbols are probably completely familiar, it is worthwhile to be sure we all interpret them in the same way.

TABLE A-1 Mathematical Symbols

$=$	is equal to
$\neq$	is not equal to
$\approx$	is approximately equal to
$\propto$	is proportional to
$>$	is greater than
$\geq$	is greater than or equal to
$\gg$	is much greater than
$<$	is less than
$\leq$	is less than or equal to
$\ll$	is much less than
$\pm$	plus or minus
$\mp$	minus or plus
x_{av} or $\bar{x}$	average value of x
Δx	change in $x (x_f - x_i)$
$|x|$	absolute value of x
Σ	sum of
$\rightarrow 0$	approaches 0
∞	infinity

A couple of the symbols in Table A-1 warrant further discussion. First, Δx, which means "change in x," is used frequently, and in many different contexts. Pronounced "delta x," it is defined as the final value of x, x_f, minus the initial value of x, x_i:

$$\Delta x = x_f - x_i \qquad \text{A-1}$$

Thus, Δx is not Δ times x; it is a shorthand way of writing $x_f - x_i$. The same delta notation can be applied to any quantity—it does not have to be x. In general, we can say that

$$\Delta(anything) = (anything)_f - (anything)_i$$

For example, $\Delta t = t_f - t_i$ is the change in time, $\Delta \vec{v} = \vec{v}_f - \vec{v}_i$ is the change in velocity, and so on. Throughout this text, we use the delta notation whenever we want to indicate the change in a given quantity.

Second, the Greek letter Σ (capital sigma) is also encountered frequently. In general, Σ is shorthand for "sum." For example, suppose we have a system comprised of nine masses, m_1 through m_9. The total mass of the system, M, is simply

$$M = m_1 + m_2 + m_3 + m_4 + m_5 + m_6 + m_7 + m_8 + m_9$$

This is a rather tedious way to write M, however, and would be even more so if the number of masses were larger. To simplify our equation, we use the Σ notation:

$$M = \sum_{i=1}^{9} m_i \qquad \text{A-2}$$

With this notation we could sum over any number of masses, simply by changing the upper limit of the sum.

In addition, Σ is often used to designate a general summation, where the number of terms in the sum may not be known, or may vary from one system to another. In a case like this we would simply write Σ without specific upper and lower limits. Thus, a general way of writing the total mass of a system is as follows:

$$M = \sum m \qquad \text{A-3}$$

VECTOR NOTATION

When we draw a vector to represent a physical quantity, we typically use an arrow whose length is proportional to the magnitude of the quantity, and whose direction is the direction of the quantity. (This and other aspects of vector notation are discussed in Chapter 3.) A slight problem arises, however, when a physical quantity points into or out of the page. In such a case, we use the conventions illustrated in **Figure A-1**.

Figure A-1 (a) shows a vector pointing out of the page. Note that we see only the tip. Below, we show the corresponding convention, which is a dot set off by a circle. The dot represents the point of the vector's arrow coming out of the page toward you.

A similar convention is employed in Figure A-1 (b) for a vector pointing into the page. In this case, the arrow moves directly away from you, giving a view of its "tail feathers." The feathers are placed in an X-shaped pattern, so we represent the vector as an X set off by a circle.

These conventions are used in Chapter 22 to represent the magnetic field vector, $\vec{B}$, and in other locations in the text as well.

(a)   **(b)**

▲ FIGURE A-1 **Vectors pointing out of and into the page** **(a)** A vector pointing out of the page is represented by a dot in a circle. The dot indicates the tip of the vector's arrow. **(b)** A vector pointing into the page is represented by an X in a circle. The X indicates the "tail feathers" of the vector's arrow.

SCIENTIFIC NOTATION

In physics, the numerical value of a physical quantity can cover an enormous range, from the astronomically large to the microscopically small. For example, the mass of the Earth is roughly

$$M_E = 5970000000000000000000000 \text{ kg}$$

In contrast, the mass of a hydrogen atom is approximately

$$M_{hydrogen} = 0.00000000000000000000000000167 \text{ kg}$$

Clearly, representing such large and small numbers with a long string of zeros is clumsy and prone to error.

The preferred method for handling such numbers is to replace the zeros with the appropriate power of ten. For example, the mass of the Earth can be written as follows:

$$M_E = 5.97 \times 10^{24} \text{ kg}$$

The factor of 10^{24} simply means that the decimal point for the mass of the Earth is 24 places to the right of its location in 5.97. Similarly, the mass of a hydrogen atom is

$$M_{hydrogen} = 1.67 \times 10^{-27} \text{ kg}$$

In this case, the correct location of the decimal point is 27 places to the left of its location in 1.67. This type of representation, using powers of ten, is referred to as **scientific notation**.

Scientific notation also simplifies various mathematical operations, such as multiplication and division. For example, the product of the mass of the Earth and the mass of a hydrogen atom is

$$
\begin{aligned}
M_E M_{hydrogen} &= (5.97 \times 10^{24} \text{ kg})(1.67 \times 10^{-27} \text{ kg}) \\
&= (5.97 \times 1.67)(10^{24} \times 10^{-27}) \text{ kg}^2 \\
&= 9.99 \times 10^{24-27} \text{ kg}^2 \\
&= 9.99 \times 10^{-3} \text{ kg}^2
\end{aligned}
$$

Similarly, the mass of a hydrogen atom divided by the mass of the Earth is

$$
\begin{aligned}
\frac{M_{hydrogen}}{M_E} &= \frac{1.67 \times 10^{-27} \text{ kg}}{5.97 \times 10^{24} \text{ kg}} = \frac{1.67}{5.97} \times \frac{10^{-27}}{10^{24}} \\
&= 0.280 \times 10^{-27-24} \\
&= 0.280 \times 10^{-51} = 2.80 \times 10^{-52}
\end{aligned}
$$

Note the change in location of the decimal point in the last two expressions, and the corresponding change in the power of ten.

Exponents and their manipulation are discussed in greater detail later in this Appendix.

Trigonometry

DEGREES AND RADIANS

We all know the definition of a degree; there are 360 degrees in a circle. The definition of a radian is somewhat less well known; there are 2π radians in a circle. An equivalent definition of the radian is the following:

A radian is the angle for which the corresponding arc length is equal to the radius.

To visualize this definition, consider a pie with a piece cut out, as shown in Figure A-2 (a). Note that a piece of pie has three sides—two radial lines from the center, and an arc of crust. If a piece of pie is cut with an angle of one radian, all three sides are equal in

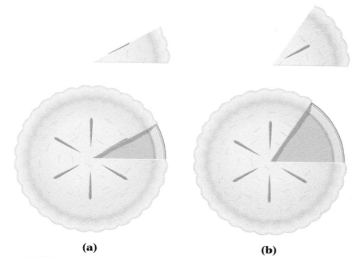

(a)   (b)

▲ **FIGURE A-2 The definition of a radian (a)** This piece of pie is cut with an angle less than a radian. Thus, the two radial sides (coming out from the center) are longer than the arc of crust. **(b)** The angle for this piece of pie is equal to one radian (about 57.3°). Thus, all three sides of the piece are of equal length.

length, as shown in **Figure A-2 (b)**. Since a radian is about 57.3°, this amounts to a fairly good-sized piece of pie. Thus, if you want a healthy helping of pie, just tell the server, "One radian, please."

Now, radians are particularly convenient when we are interested in the length of an arc. In **Figure A-3** we show a circular arc corresponding to the radius r and the angle θ. If the angle θ is measured in radians, the length of the arc, s, is given by

$$s = r\theta \qquad \text{A-4}$$

Note that this simple relation is *not valid* when θ is measured in degrees. For a full circle, in which case $\theta = 2\pi$, the length of the arc (which is the circumference of the circle) is $2\pi r$, as expected.

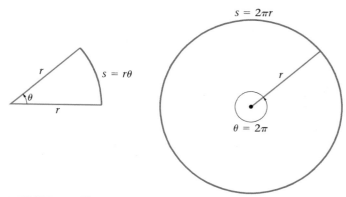

▲ **FIGURE A-3 The Length of an arc** The arc created by a radius r rotated through an angle θ has a length $s = r\theta$. If the radius is rotated through a full circle, the angle is $\theta = 2\pi$, and the arc length is the circumference of a circle, $s = 2\pi r$.

TRIGONOMETRIC FUNCTIONS AND THE PYTHAGOREAN THEOREM

Next, we consider some of the more important and frequently used results from trigonometry. We start with the right triangle, shown in **Figure A-4**, and the basic **trigonometric functions**, sin θ (sine theta), cos θ (cosine theta), and tan θ (tangent theta). The cosine of an angle θ is defined to be the side adjacent to the angle divided by the hypotenuse; $\cos \theta = x/r$. Similarly, the sine is defined to be the opposite side divided by the hypotenuse, $\sin \theta = y/r$, and the tangent is the opposite side divided by the adjacent side, $\tan \theta = y/x$. These relations are summarized in the following equations:

$$\cos \theta = \frac{x}{r}$$
$$\sin \theta = \frac{y}{r}$$ A-5
$$\tan \theta = \frac{y}{x} = \frac{\sin \theta}{\cos \theta}$$

Note that each of the trigonometric functions is the ratio of two lengths, and hence is dimensionless.

According to the **Pythagorean theorem,** the sides of the right triangle in Figure A-4 are related as follows:

$$x^2 + y^2 = r^2$$ A-6

Dividing by r^2 yields

$$\frac{x^2}{r^2} + \frac{y^2}{r^2} = 1$$

This can be re-written in terms of sine and cosine to give

$$\sin^2 \theta + \cos^2 \theta = 1$$

Figure A-4 also shows how sine and cosine are used in a typical calculation. In many cases, the hypotenuse of a triangle, r, and one of its angles, θ, are given. To find the short sides of the triangle we rearrange the relations given in Equation A-5. For example, in Figure A-4 we see that $x = r \cos \theta$ is the length of the short side adjacent to the angle, θ, and $y = r \sin \theta$ is the length of the short side opposite the angle. The following Example applies this type of calculation to the case of an inclined roadway.

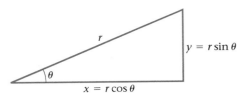

▲ FIGURE A-4 **Relating the short sides of a right triangle to its hypotenuse** The trigonometric functions sin θ and cos θ and the Pythagorean theorem are useful in relating the lengths of the short sides of a right triangle to the length of its hypotenuse.

EXAMPLE A–1 HIGHWAY TO HEAVEN

You are driving on a long straight road that slopes uphill at an angle of 6.4° above the horizontal. At one point you notice a sign that reads, "Elevation 1500 feet." What is your elevation after you have driven another 1.0 mi?

PICTURE THE PROBLEM
From our sketch, we see that the car is moving along the hypotenuse of a right triangle. The length of the hypotenuse is one mile.

REASONING AND STRATEGY
The elevation gain is the vertical side of the triangle, y. We find y by multiplying the hypotenuse, r, by the sine of theta. That is, since $\sin \theta = y/r$ it follows that $y = r \sin \theta$.

Known   Angle of slope, $q = 6.4°$; driving distance, $r = 10$ mi; initial elevation = 1500 ft.
Unknown Gain in elevation, $y = ?$; final elevation = initial elevation + change in elevation = ?

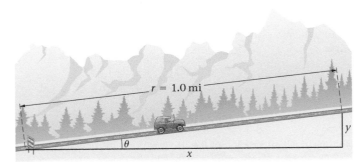

SOLUTION

1. Calculate the elevation gain, y:

$y = r \sin \theta$
$\quad = (1.0 \text{ mi}) \sin 6.4° = (1.0 \text{ mi})(0.11) = 0.11 \text{ mi}$

2. Convert y from miles to feet:

$y = (0.11 \text{ mi})\left(\dfrac{5280 \text{ ft}}{1 \text{ mi}}\right) = 580 \text{ ft}$

3. Add the elevation gain to the original elevation to obtain the new elevation:

elevation = 1500 ft + 580 ft = 2100 ft

INSIGHT
As surprising as it may seem, the horizontal distance covered by the car is $r \cos \theta = (5280 \text{ ft}) \cos 6.4° = 5200$ ft, only about 80 ft less than the total distance driven by the car. At the same time, the car rises a distance of 580 ft.

PRACTICE PROBLEM
How far up the road from the first sign should the road crew put another sign reading "Elevation 3500 ft"?
[**Answer:** 18,000 ft = 3.4 mi]

In some problems, the sides of a triangle (x and y) are given and it is desired to find the corresponding hypotenuse, r, and angle, θ. For example, suppose that $x = 5.0$ m and $y = 2.0$ m. Using the Pythagorean theorem, we find $r = \sqrt{x^2 + y^2} = \sqrt{(5.0 \text{ m})^2 + (2.0 \text{ m})^2} = 5.4$ m. Similarly, to find the angle we use the definition of tangent: $\tan \theta = y/x$. The inverse of this relation is $\theta = \tan^{-1}(y/x) = \tan^{-1}(2.0 \text{ m}/5.0 \text{ m}) = \tan^{-1}(0.40)$. Note that the expression $\tan^{-1}$ is the *inverse tangent function*—it does not mean 1 divided by tangent, but rather "the angle whose tangent is——." Your calculator should have a button on it labeled $\tan^{-1}$. If you enter 0.40 and then press $\tan^{-1}$, you should get 22° (to two significant figures), which means that $\tan 22° = 0.40$. Inverse sine and cosine functions work in the same way.

TRIGONOMETRIC IDENTITIES

In addition to the basic definitions of sine, cosine, and tangent just given, there are a number of useful relationships involving these functions referred to as **trigonometric identities**. First, consider changing the sign of an angle. This corresponds to flipping the triangle in Figure A-4 upside-down, which changes the sign of y but leaves x unaffected. The result is that sine changes its sign, but cosine does not. Specifically, for a general angle A we find the following:

$$\sin(-A) = -\sin A$$
$$\cos(-A) = \cos A \qquad \text{A-7}$$

Next, we consider trigonometric identities relating to the sum or difference of two angles. For example, consider two general angles A and B. The sine and cosine of the sum of these angles, $A + B$, are given below:

$$\sin(A + B) = \sin A \cos B + \sin B \cos A$$
$$\cos(A + B) = \cos A \cos B - \sin A \sin B \qquad \text{A-8}$$

By changing the sign of B, and using the results given in Equation A-7, we obtain the corresponding results for the difference between two angles:

$$\sin(A - B) = \sin A \cos B - \sin B \cos A$$
$$\cos(A - B) = \cos A \cos B + \sin A \sin B \qquad \text{A-9}$$

Applications of these relations can be found in Chapters 4, 14, 23, and 24.

To see how one might use a relation like $\sin(A + B) = \sin A \cos B + \sin B \cos A$, consider the case where $A = B = \theta$. With this substitution we find

$$\sin(\theta + \theta) = \sin \theta \cos \theta + \sin \theta \cos \theta$$

Simplifying somewhat yields the commonly used double-angle formula

$$\sin 2\theta = 2 \sin \theta \cos \theta \qquad \text{A-10}$$

This expression is used in deriving Equation 4-12.

As a final example of using trigonometric identities, let $A = 90°$ and $B = \theta$. Making these substitutions in Equations A-9 yields

$$\sin(90° - \theta) = \sin 90° \cos \theta - \sin \theta \cos 90° = \cos \theta$$
$$\cos(90° - \theta) = \cos 90° \cos \theta + \sin 90° \sin \theta = \sin \theta \qquad \text{A-11}$$

Algebra
THE QUADRATIC EQUATION

A well-known result that finds many uses in physics is the solution to the **quadratic equation**

$$ax^2 + bx + c = 0 \qquad \text{A-12}$$

In this equation, a, b, and c are constants and x is a variable. When we refer to the solution of the quadratic equation, we mean the values of x that satisfy Equation A-12. These values are given by the following expression:

Solutions to the Quadratic Equation

$$x = \frac{-b \pm \sqrt{b^2 - 4ac}}{2a} \qquad \text{A-13}$$

Note that there are two solutions to the quadratic equation, in general, corresponding to the plus and minus sign in front of the square root. In the special case that the quantity under the square root vanishes, there will be only a single solution. If the quantity under the square root is negative the result for x is not physical, which means a mistake has probably been made in the calculation.

To illustrate the use of the quadratic equation and its solution, we consider a standard one-dimensional kinematics problem, such as one might encounter in Chapter 2:

A ball is thrown straight upward with an initial speed of 11 m/s. How long does it take for the ball to first reach a height of 4.5 m above its launch point?

The first step in solving this problem is to write the equation giving the height of the ball, y, as a function of time. Referring to Equation 2-11, and replacing x with y, we have

$$y = y_0 + v_0 t - \tfrac{1}{2}gt^2$$

To make this look more like a quadratic equation, we move all the terms onto the left-hand side, which yields

$$\tfrac{1}{2}gt^2 - v_0 t + y - y_0 = 0$$

This is the same as Equation A-12 if we make the following identifications: $x = t$; $a = \tfrac{1}{2}g$; $b = -v_0$; $c = y - y_0$. The desired solution, then, is given by making these substitutions in Equation A-13:

$$t = \frac{v_0 \pm \sqrt{v_0^2 - 2g(y - y_0)}}{g}$$

The final step is to use the appropriate numerical values; $g = 9.81$ m/s^2, $v_0 = 11$ m/s, $y - y_0 = 4.5$ m. Straightforward calculation gives $t = 0.54$ s and $t = 1.7$ s. Therefore, the time it takes to first reach a height of 4.5 m is 0.54 s; the second solution is the time when the ball is again at a height of 4.5 m, this time on its way down.

TWO EQUATIONS IN TWO UNKNOWNS

In some problems, two unknown quantities are determined by two interlinked equations. In such cases it often seems at first that you have not been given enough information to obtain a solution. By patiently writing out what is known, however, you can generally use straightforward algebra to solve the problem.

As an example, consider the following problem: A father and daughter share the same birthday. On one birthday the father announces to his daughter, "Today I am four times older than you, but in 5 years I will be only three times older." How old are the father and daughter now?

You might be able to solve this problem by guessing, but here's how to approach it systematically. First, write what is given in the form of equations. Letting F be the father's age in years, and D the daughter's age in years, we know that on this birthday

$$F = 4D \qquad \text{A-14}$$

In 5 years, the father's age will be $F + 5$, the daughter's age will be $D + 5$, and the following will be true:

$$F + 5 = 3(D + 5)$$

Multiplying through the parenthesis gives

$$F + 5 = 3D + 15 \qquad \text{A-15}$$

Now if we subtract Equation A-15 from Equation A-14 we can eliminate one of the unknowns, F:

$$\begin{aligned} F &= 4D \\ -(F + 5 &= 3D + 15) \\ \hline -5 &= D - 15 \end{aligned}$$

The solution to this new equation shows that the daughter's age now is $D = 10$, and hence the father's age now is $F = 4D = 40$.

The following Example investigates a similar problem. In this case, we use the fact that if you drive with a speed v for a time t the distance covered is $d = vt$.

EXAMPLE A–2 HIT THE ROAD

It takes 1.50 h to drive with a speed v from home to a nearby town, a distance d away. Later, on the way back, the traffic is lighter, and you are able to increase your speed by 15 mi/h. With this higher speed, you get home in just 1.00 h. Find your initial speed v, and the distance to the town, d.

PICTURE THE PROBLEM
Our sketch shows home and the town, separated by a distance d. Going to town the speed is v, returning home the speed is $v + 15$ mi/h.

REASONING AND STRATEGY
To determine the two unknowns, v and d, we need two separate equations. One equation corresponds to what we know about the trip to the town, the second equation corresponds to what we know about the return trip.

Known Drive time outbound, $t = 1.50$ h; increase in speed = 15 mi/h; drive time going home = 1.00 h.
Unknown Initial speed, $v = ?$; distance to town, $d = ?$

SOLUTION

1. Write an equation for the trip to the town. Recall that this trip takes one and a half hours:

$$d = vt = v(1.50 \text{ h})$$

2. Write an equation for the trip home. This trip takes one hour, and covers the same distance d:

$$d = (v + 15 \text{ mi/h})t = (v + 15 \text{ mi/h})(1.00 \text{ h})$$

3. Subtract these two equations to eliminate d:

$$\begin{aligned} d &= v(1.50 \text{ h}) \\ -d &= (v + 15 \text{ mi/h})(1.00 \text{ h}) \\ \hline 0 &= v(1.50 \text{ h}) - v(1.00 \text{ h}) - (15 \text{ mi/h})(1.00 \text{ h}) \end{aligned}$$

4. Solve this new equation for v:

$$0 = v(1.50 \text{ h}) - v(1.00 \text{ h}) - (15 \text{ mi/h})(1.00 \text{ h})$$
$$0 = v(0.50 \text{ h}) - (15 \text{ mi/h})(1.00 \text{ h})$$
$$v = \frac{(15 \text{ mi/h})(1.00 \text{ h})}{(0.50 \text{ h})} = 30 \text{ mi/h}$$

5. Use the first equation to solve for d:

$$d = vt = (30 \text{ mi/h})(1.50 \text{ h}) = 45 \text{ mi}$$

INSIGHT
We could also use the second equation to solve for d. The algebra is a bit messier, but it provides a good double check on our results: $d = (30 \text{ mi/h} + 15 \text{ mi/h})(1.00 \text{ h}) = (45 \text{ mi/h})(1.00 \text{ h}) = 45 \text{ mi}$.

PRACTICE PROBLEM
Suppose the speed going home is 12 mi/h faster than the speed on the way to town, but the times are the same as above. What are v and d in this case? [**Answer:** $v = 24$ mi/h, $d = 36$ mi]

EXPONENTS AND LOGARITHMS

An **exponent** is the power to which a number is raised. For example, in the expression 10^3, we say that the exponent of 10 is 3. To evaluate 10^3 we simply multiply 10 by itself three times:

$$10^3 = 10 \times 10 \times 10 = 1000$$

Similarly, a negative exponent implies an inverse, as in the relation $10^{-1} = 1/10$. Thus, to evaluate a number like 10^{-4}, for example, we multiply 1/10 by itself four times:

$$10^{-4} = \frac{1}{10} \times \frac{1}{10} \times \frac{1}{10} \times \frac{1}{10} = \frac{1}{10,000} = 0.0001$$

The relations just given apply not just to powers of 10, of course, but to any number at all. Thus, x^4 is

$$x^4 = x \times x \times x \times x$$

and x^{-3} is

$$x^{-3} = \frac{1}{x} \times \frac{1}{x} \times \frac{1}{x} = \frac{1}{x^3}$$

Using these basic rules, it follows that exponents add when two or more numbers are multiplied together:

$$x^2 x^3 = (x \times x)(x \times x \times x)$$
$$= x \times x \times x \times x \times x = x^5 = x^{2+3}$$

On the other hand, exponents multiply when a number is raised to a power:

$$(x^2)^3 = (x \times x) \times (x \times x) \times (x \times x)$$
$$= x \times x \times x \times x \times x \times x = x^6 = x^{2 \times 3}$$

In general, the rules obeyed by exponents can be summarized as follows:

$$x^n x^m = x^{n+m}$$
$$x^{-n} = \frac{1}{x^n}$$
$$\frac{x^n}{x^m} = x^{n-m} \qquad \text{A-16}$$
$$(xy)^n = x^n y^n$$
$$(x^n)^m = x^{nm}$$

Fractional exponents, such as $1/n$, indicate the nth root of a number. Specifically, the square root of x is written as

$$\sqrt{x} = x^{1/2}$$

For n greater than 2 we write the nth root in the following form:

$$\sqrt[n]{x} = x^{1/n} \qquad \text{A-17}$$

Thus, the nth root of a number, x, is the value that gives x when multiplied by itself n times: $(x^{1/n})^n = x^{n/n} = x^1 = x$.

A general method for calculating the exponent of a number is provided by the **logarithm**. For example, suppose x is equal to 10 raised to the power n:

$$x = 10^n$$

In this expression, 10 is referred to as the base. The exponent, n, is equal to the logarithm (log) of x:

$$n = \log x$$

The notation "log" is known as the *common logarithm,* and it refers specifically to base 10.

As an example, suppose that $x = 1000 = 10^n$. Clearly, we can write x as 10^3, which means that the exponent of x is 3:

$$\log x = \log 1000 = \log 10^3 = 3$$

When dealing with a number this simple, the exponent can be determined without a calculator. Suppose, however, that $x = 1205 = 10^n$. To find the exponent for this value of x we use the "log" button on a calculator. The result is

$$n = \log 1205 = 3.081$$

Thus, 10 raised to the 3.081 power gives 1205.

Another base that is frequently used for calculating exponents is $e = 2.718\ldots$. To represent $x = 1205$ in this base we write

$$x = 1205 = e^m$$

The logarithm to base e is known as the *natural logarithm,* and it is represented by the notation "ln." Using the "ln" button on a calculator, we find

$$m = \ln 1205 = 7.094$$

Thus, e raised to the 7.094 power gives 1205. The connection between the common and natural logarithms is as follows:

$$\ln x = 2.3026 \log x \qquad \text{A-18}$$

In the example just given, we have $\ln 1205 = 7.094 = 2.3026 \log 1205 = 2.3026(3.081)$.

The basic rules obeyed by logarithms follow directly from the rules given for exponents in Equation A-16. In particular,

$$\ln(xy) = \ln x + \ln y$$
$$\ln\left(\frac{x}{y}\right) = \ln x - \ln y \qquad \text{A-19}$$
$$\ln x^n = n \ln x$$

Though these rules are stated in terms of natural logarithms, they are satisfied by logarithms with any base.

Mathematical Expansions

We conclude with a brief consideration of small quantities in mathematics. Consider the following equation:

$$(1 + x)^3 = 1 + 3x + 3x^2 + x^3$$

This expression is valid for all values of x. However, if x is much smaller than one, $x \ll 1$, we can say to a good approximation that

$$(1 + x)^3 \approx 1 + 3x$$

Now, just how good is this approximation? After all, it ignores two terms that would need to be included to produce an equality. In the case $x = 0.001$, for example, the two terms that are neglected, $3x^2$ and x^3, have a combined contribution of only about 3 ten-thousandths of a percent! Clearly, then, little error is made in the approximation $(1 + 0.001)^3 \sim 1 + 3(0.001) = 1.003$. This can be seen visually in **Figure A-5 (a)**, where we plot $(1 + x)^3$ and $1 + 3x$ for x ranging from 0 to 1. Note that there is little difference in the two expressions for x less than about 0.1.

This is just one example of a general result in mathematics that can be derived from the **binomial expansion.** In general, we can say that the following approximation is valid for $x \ll 1$:

$$(1 + x)^n \approx 1 + nx \qquad \text{A-20}$$

This result holds for arbitrary n, not just for the case of $n = 3$. For example, if $n = -1$ we have

$$(1 + x)^{-1} = \frac{1}{1 + x} \approx 1 - x$$

We plot $(1 + x)^{-1}$ and $1 - x$ in **Figure A-5 (b)**, and again we see that the results are in good agreement for x less than about 0.1.

An example of an expansion that arises in the study of relativity concerns the following quotient:

$$\frac{1}{\sqrt{1 - \dfrac{v^2}{c^2}}}$$

In this expression v is the speed of an object and c is the speed of light. Since objects we encounter generally have speeds much less than the speed of light, the ratio v/c is much less than one, and v^2/c^2 is even smaller than v/c. Thus, if we let $x = v^2/c^2$ we have

$$\frac{1}{\sqrt{1 - x}}$$

We can apply the binomial expansion to this result if we replace n with $-1/2$ and x with $-x$ in Equation A-20. This yields

$$\frac{1}{\sqrt{1 - \dfrac{v^2}{c^2}}} \approx 1 + \frac{1}{2}\frac{v^2}{c^2}$$

The two sides of this approximate equality are plotted in **Figure A-5 (c)**, showing the accuracy of the approximation for small v/c.

Another type of mathematical expansion leads to the following useful results:

$$\sin \theta \approx \theta$$
$$\cos \theta \approx 1 - \frac{1}{2}\theta^2 \qquad \text{A-21}$$

These expansions are valid for small angles θ measured in radians. Note that the result $\sin \theta \approx \theta$ is used to derive Equations 6-14 and 13-19. (See Table 6-2, p. 170, and Figure 13-21, p. 435, for more details on this expansion.)

Vector Multiplication

There are two distinct ways to multiply vectors, referred to as the **dot product** and the **cross product.** The difference between these two types of multiplication is that the dot product yields a scalar (a number) as its result, whereas the cross product results in a vector. Both types of product have important applications in physics. In what follows, we present the basic techniques associated with dot and cross products, and point out places in the text where they are used.

THE DOT PRODUCT

Consider two vectors, $\vec{A}$ and $\vec{B}$, as shown in **Figure A-6 (a)**. The magnitudes of these vectors are A and B, respectively, and the angle

(a)

(b)

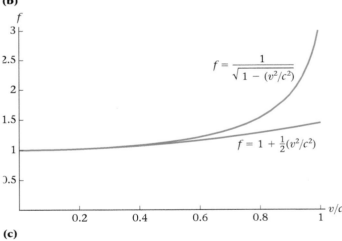

(c)

▲ **FIGURE A-5 Examples of mathematical expansions (a)** A comparison between $(1 + x)^3$ and the result obtained from the binomial expansion, $1 + 3x$. **(b)** A comparison between $1/(1 + x)$ and the result obtained from the binomial expansion, $1 - x$. **(c)** A comparison between $1/\sqrt{1 - (v^2/c^2)}$ and the result obtained from the binomial expansion, $1 + \frac{1}{2}(v^2/c^2)$.

between them is θ. We define the dot product of $\vec{A}$ and $\vec{B}$ as follows:

$$\vec{A} \cdot \vec{B} = AB \cos \theta \qquad \text{A-22}$$

(a)

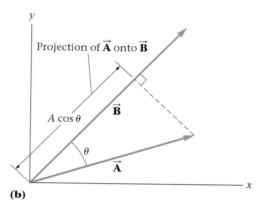

(b)

▲ FIGURE A-6 **The dot product between vectors $\vec{\mathbf{A}}$ and $\vec{\mathbf{B}}$.**

In words, the dot product of two vectors is a scalar equal to the magnitude of one vector times the magnitude of the second vector times the cosine of the angle between them.

A geometric interpretation of the dot product is presented in **Figure A-6 (b)**. We begin by projecting the vector $\vec{\mathbf{A}}$ onto the direction of vector $\vec{\mathbf{B}}$. This is done by dropping a perpendicular from the tip of $\vec{\mathbf{A}}$ onto the line that passes through $\vec{\mathbf{B}}$, as shown in **Figure A-6 (b)**. Note that the projection of $\vec{\mathbf{A}}$ on the direction of $\vec{\mathbf{B}}$ has a length given by $A \cos \theta$. It follows that the dot product is simply the projection of $\vec{\mathbf{A}}$ onto $\vec{\mathbf{B}}$ times the magnitude of $\vec{\mathbf{B}}$; that is, $(A \cos \theta)B = AB \cos \theta = \vec{\mathbf{A}} \cdot \vec{\mathbf{B}}$. Equivalently, the dot product can be thought of as the projection of $\vec{\mathbf{B}}$ onto $\vec{\mathbf{A}}$ times the magnitude of $\vec{\mathbf{A}}$.

A few special cases will help to clarify the dot product. Suppose, for example, that $\vec{\mathbf{A}}$ and $\vec{\mathbf{B}}$ are parallel. In this case, $\theta = 0$ and $\vec{\mathbf{A}} \cdot \vec{\mathbf{B}} = AB$. Therefore, when vectors are parallel, the dot product is simply the product of their magnitudes. On the other hand, suppose $\vec{\mathbf{A}}$ and $\vec{\mathbf{B}}$ point in opposite directions. Now we have $\theta = 180$, and therefore $\vec{\mathbf{A}} \cdot \vec{\mathbf{B}} = -AB$. In general, the sign of $\vec{\mathbf{A}} \cdot \vec{\mathbf{B}}$ is positive if the angle between $\vec{\mathbf{A}}$ and $\vec{\mathbf{B}}$ is less than 90°, and is negative if the angle between them is greater than 90°. Finally, if $\vec{\mathbf{A}}$ and $\vec{\mathbf{B}}$ are perpendicular to one another—that is, if $\theta = 90°$—we see that $\vec{\mathbf{A}} \cdot \vec{\mathbf{B}} = AB \cos 90° = 0$. In this case, neither vector has a nonzero projection onto the other vector.

Dot products have a particularly simple form when applied to unit vectors. Recall, for example, that $\hat{\mathbf{x}}$ and $\hat{\mathbf{y}}$ have unit magnitude and are perpendicular to one another. It follows that

$$\hat{\mathbf{x}} \cdot \hat{\mathbf{x}} = 1, \quad \hat{\mathbf{x}} \cdot \hat{\mathbf{y}} = \hat{\mathbf{y}} \cdot \hat{\mathbf{x}} = 0, \quad \hat{\mathbf{y}} \cdot \hat{\mathbf{y}} = 1 \qquad \text{A-23}$$

These results can be applied to the general two-dimensional vectors $\vec{\mathbf{A}} = A_x \hat{\mathbf{x}} + A_y \hat{\mathbf{y}}$ and $\vec{\mathbf{B}} = B_x \hat{\mathbf{x}} + B_y \hat{\mathbf{y}}$ to give

$$\begin{aligned}
\vec{\mathbf{A}} \cdot \vec{\mathbf{B}} &= (A_x \hat{\mathbf{x}} + A_y \hat{\mathbf{y}}) \cdot (B_x \hat{\mathbf{x}} + B_y \hat{\mathbf{y}}) \\
&= A_x B_x \hat{\mathbf{x}} \cdot \hat{\mathbf{x}} + A_x B_y \hat{\mathbf{x}} \cdot \hat{\mathbf{y}} + A_y B_x \hat{\mathbf{y}} \cdot \hat{\mathbf{x}} + A_y B_y \hat{\mathbf{y}} \cdot \hat{\mathbf{y}} \qquad \text{A-24} \\
&= A_x B_x + A_y B_y
\end{aligned}$$

Thus, the dot product of tpwo-dimensional vectors is simply the product of their x components plus the product of their y components.

At first glance the result $\vec{\mathbf{A}} \cdot \vec{\mathbf{B}} = AB \cos \theta$ looks quite different from the result $\vec{\mathbf{A}} \cdot \vec{\mathbf{B}} = A_x B_x + A_y B_y$. They are identical, however, as we now show. Suppose that $\vec{\mathbf{A}}$ is at an angle θ_A to the positive x axis, and that $\vec{\mathbf{B}}$ is at an angle $\theta_B > \theta_A$ to the x axis, from which it follows that the angle between $\vec{\mathbf{A}}$ and $\vec{\mathbf{B}}$ is $\theta = \theta_B - \theta_A$. Noting that $A_x = A \cos \theta_A$ and $A_y = A \sin \theta_A$, and similarly for B_x and B_y, we have

$$\vec{\mathbf{A}} \cdot \vec{\mathbf{B}} = AB(\cos \theta_A \cos \theta_B + \sin \theta_A \sin \theta_B)$$

The second trigonometric identity in Equation A-9 can be applied to the quantity in brackets, with the result that $\vec{\mathbf{A}} \cdot \vec{\mathbf{B}} = A_x B_x + A_y B_y = AB \cos(\theta_B - \theta_A) = AB \cos \theta$, as desired.

The most prominent application of dot products in this text is in Chapter 7, where in Equation 7-3 we define the work to be $W = Fd \cos \theta$, with θ the angle between $\vec{\mathbf{F}}$ and $\vec{\mathbf{d}}$. Clearly, this is simply a statement that work is the dot product of force and displacement:

$$W = \vec{\mathbf{F}} \cdot \vec{\mathbf{d}} = Fd \cos \theta$$

Later in the text, in Equation 19-11, we define the electric flux to be $\Phi = EA \cos \theta$. If we let $\vec{\mathbf{A}}$ represent a vector that has a magnitude equal to the area, A, and points in the direction of the normal to the area, we can write the electric flux as a dot product:

$$\Phi = \vec{\mathbf{E}} \cdot \vec{\mathbf{A}} = EA \cos \theta$$

Similar remarks apply to the magnetic flux, defined in Equation 23-1.

THE CROSS PRODUCT

When two vectors are multiplied with the cross product, the result is a third vector that is perpendicular to both original vectors. An example is shown in **Figure A-7**, where we see a vector $\vec{\mathbf{A}}$, a vector $\vec{\mathbf{B}}$, and their cross product, $\vec{\mathbf{C}}$:

$$\vec{\mathbf{C}} = \vec{\mathbf{A}} \times \vec{\mathbf{B}} \qquad \text{A-25}$$

Notice that $\vec{\mathbf{C}}$ is perpendicular to the plane formed by the vectors $\vec{\mathbf{A}}$ and $\vec{\mathbf{B}}$. In addition, the direction of $\vec{\mathbf{C}}$ is given by the following right-hand rule:

To find the direction of $\vec{\mathbf{C}} = \vec{\mathbf{A}} \times \vec{\mathbf{B}}$, point the fingers of your right hand in the direction of $\vec{\mathbf{A}}$ and curl them toward $\vec{\mathbf{B}}$. Your thumb is now pointing in the direction of $\vec{\mathbf{C}}$.

It is clear from this rule that if $\vec{\mathbf{A}} \times \vec{\mathbf{B}} = \vec{\mathbf{C}}$, then $\vec{\mathbf{B}} \times \vec{\mathbf{A}} = -\vec{\mathbf{C}}$.

The magnitude of $\vec{\mathbf{C}} = \vec{\mathbf{A}} \times \vec{\mathbf{B}}$ depends on the magnitudes of the vectors $\vec{\mathbf{A}}$ and $\vec{\mathbf{B}}$, and on the angle θ between them. In particular,

$$C = AB \sin \theta \qquad \text{A-26}$$

Comparing with Equation A-22, we see that the cross product involves a sin θ, whereas the dot product depends on cos θ. As a result, it follows that the cross product has zero magnitude when $\vec{\mathbf{A}}$ and

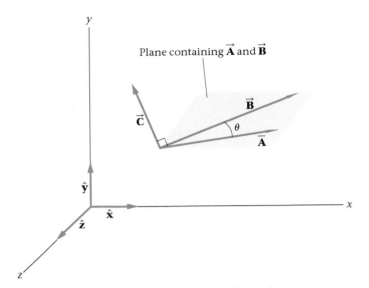

▲ FIGURE A-7 **The vector cross product of $\vec{A}$ and $\vec{B}$.**

$\vec{A} = A_x\hat{\mathbf{x}} + A_y\hat{\mathbf{y}}$ and $\vec{B} = B_x\hat{\mathbf{x}} + B_y\hat{\mathbf{y}}$. Straightforward application of Equation A-27 yields

$$
\begin{aligned}
\vec{C} = \vec{A} \times \vec{B} &= (A_x\hat{\mathbf{x}} + A_y\hat{\mathbf{y}}) \times (B_x\hat{\mathbf{x}} + B_y\hat{\mathbf{y}}) \\
&= A_xB_x(\hat{\mathbf{x}} \times \hat{\mathbf{x}}) + A_xB_y(\hat{\mathbf{x}} \times \hat{\mathbf{y}}) \\
&\quad + A_yB_x(\hat{\mathbf{y}} \times \hat{\mathbf{x}}) + A_yB_y(\hat{\mathbf{y}} \times \hat{\mathbf{y}}) \\
&= (A_xB_y - A_yB_x)\hat{\mathbf{z}}
\end{aligned}
\qquad \text{A-28}
$$

Notice that $\vec{C}$ is perpendicular to both $\vec{A}$ and $\vec{B}$, as required for a cross product. In addition, the magnitude of $\vec{C}$ is

$$A_xB_y - A_yB_x = AB(\cos\theta_A \sin\theta_B - \sin\theta_A \cos\theta_B)$$

If we now apply the first trigonometric identity in Equation A-9, we recover the result given in Equation A-26:

$$
\begin{aligned}
C &= AB(\cos\theta_A \sin\theta_B - \sin\theta_A \cos\theta_B) \\
&= AB\sin(\theta_B - \theta_A) = AB\sin\theta
\end{aligned}
$$

The first application of cross products in this text is torque, which is discussed in Chapter 11. In fact, Equation 11-2 defines the magnitude of the torque, τ, as follows: $\tau = r(F\sin\theta)$. As one might expect by referring to Equations A-25 and A-26, the torque vector, $\vec{\tau}$, can be written as the following cross product:

$$\vec{\tau} = \vec{r} \times \vec{F}$$

Similarly, the angular momentum vector, $\vec{L}$, whose magnitude is given in Equation 11-13, is simply

$$\vec{L} = \vec{r} \times \vec{p}$$

Finally, the cross product appears again in magnetism. In fact, the magnitude of the magnetic force on a charge q with a velocity $\vec{v}$ in a magnetic field $\vec{B}$ is $F = qvB\sin\theta$, as given in Equation 22-1. As one might guess, the vector form of this force is

$$\vec{F} = q\vec{v} \times \vec{B}$$

The advantage of a cross product expression like this is that it contains both the direction and magnitude of a vector in one compact equation. In fact, we can now see the origin of the right-hand rule for magnetic forces given in Section 22-2.

$\vec{B}$ point in the same direction ($\theta = 0°$) or in opposite directions ($\theta = 180°$). On the other hand, the cross product has its greatest magnitude, $C = AB$, when $\vec{A}$ and $\vec{B}$ are perpendicular to one another ($\theta = 90°$).

When we apply these rules to unit vectors, which are at right angles to one another and of unit magnitude, the results are particularly simple. For example, consider the cross product $\hat{\mathbf{x}} \times \hat{\mathbf{y}}$. Referring to Figure A-7, we see that this cross product points in the positive z direction. In addition, the magnitude of $\hat{\mathbf{x}} \times \hat{\mathbf{y}}$ is $(1)(1)\sin 90° = 1$. It follows, therefore, that $\hat{\mathbf{x}} \times \hat{\mathbf{y}} = \hat{\mathbf{z}}$. On the other hand, $\hat{\mathbf{x}} \times \hat{\mathbf{x}} = \hat{\mathbf{y}} \times \hat{\mathbf{y}} = \hat{\mathbf{z}} \times \hat{\mathbf{z}} = 0$ because $\theta = 0°$ in each of these cases. To summarize:

$$
\begin{array}{lll}
\hat{\mathbf{x}} \times \hat{\mathbf{y}} = \hat{\mathbf{z}}, & \hat{\mathbf{y}} \times \hat{\mathbf{z}} = \hat{\mathbf{x}}, & \hat{\mathbf{z}} \times \hat{\mathbf{x}} = \hat{\mathbf{y}} \\
\hat{\mathbf{y}} \times \hat{\mathbf{x}} = -\hat{\mathbf{z}}, & \hat{\mathbf{z}} \times \hat{\mathbf{y}} = -\hat{\mathbf{x}}, & \hat{\mathbf{x}} \times \hat{\mathbf{z}} = -\hat{\mathbf{y}} \\
\hat{\mathbf{x}} \times \hat{\mathbf{x}} = 0, & \hat{\mathbf{y}} \times \hat{\mathbf{y}} = 0, & \hat{\mathbf{z}} \times \hat{\mathbf{z}} = 0
\end{array}
\qquad \text{A-27}
$$

As an example of how to use these unit-vector results, consider the cross product of the two-dimensional vectors shown in Figure A-6,

APPENDIX B TYPICAL VALUES

Mass

Sun	2.00×10^{30} kg
Earth	5.97×10^{24} kg
Moon	7.35×10^{22} kg
747 airliner (maximum takeoff weight)	3.5×10^{5} kg
blue whale	178,000 kg = 197 tons
elephant	5400 kg
mountain gorilla	180 kg
human	70 kg
bowling ball	7 kg
half gallon of milk	1.81 kg = 4 lbs
baseball	0.141–0.148 kg
golf ball	0.045 kg
female calliope hummingbird (smallest bird in North America)	3.5×10^{-3} kg = $\frac{1}{8}$ oz
raindrop	3×10^{-5} kg
antibody molecule (IgG)	2.5×10^{-22} kg
hydrogen atom	1.67×10^{-27} kg

Length

orbital radius of Earth (around Sun)	1.5×10^{8} km
orbital radius of Moon (around Earth)	3.8×10^{5} km
altitude of geosynchronous satellite	35,800 km = 22,300 mi
radius of Earth	6370 km
altitude of Earth's ozone layer	50 km
height of Mt. Everest	8848 m
height of Washington Monument	169 m = 555 ft
pitcher's mound to home plate	18.44 m
baseball bat	1.067 m
CD (diameter)	120 mm
aorta (diameter)	18 mm
period in sentence (diameter)	0.5 mm
red blood cell	7.8 μm = $\frac{1}{3300}$ in.
typical bacterium (*E. coli*)	2 μm
wavelength of green light	550 nm
virus	20–300 nm
large protein molecule	25 nm
diameter of DNA molecule	2.0 nm
radius of hydrogen atom	5.29×10^{-11} m

Time

estimated age of Earth	approx. 4.6 billion y $\approx 10^{17}$ s
estimated age of human species	approx. 150,000 y $\approx 5 \times 10^{12}$ s
half life of carbon-14	5730 y = 1.81×10^{11} s
period of Halley's comet	76 y = 2.40×10^{9} s
half life of technetium-99	6 h = 2.16×10^{4} s
time for driver of car to apply brakes	0.46 s
human reaction time	60–180 ms
air bag deployment time	10 ms

period of middle C sound wave	3.9 ms
collision time for batted ball	2 ms
decay of excited atomic state	10^{-8} s
period of green light wave	1.8×10^{-15} s

Speed

light	3×10^8 m/s
meteor	35–95 km/s
space shuttle (orbital velocity)	8.5 km/s = 19,000 mi/h
rifle bullet	700–750 m/s
sound in air (STP)	340 m/s
fastest human nerve impulses	140 m/s
747 at takeoff	80.5 m/s
kangaroo	18.1 m/s = 40.5 mi/h
200-m dash (Olympic record)	10.1 m/s
butterfly	1 m/s
blood speed in aorta	0.35 m/s
giant tortoise	0.076 m/s = 0.170 mi/h
Mer de Glace glacier (French Alps)	4×10^{-6} m/s

Acceleration

protons in particle accelerator	9×10^{13} m/s^2
ultracentrifuge	3×10^6 m/s^2
meteor impact	10^5 m/s^2
baseball struck by bat	3×10^4 m/s^2
loss of consciousness	$7.14g$ = 70 m/s^2
acceleration of gravity on Earth (g)	9.81 m/s^2
braking auto	8 m/s^2
acceleration of gravity on the moon	1.62 m/s^2
rotation of Earth at equator	3.4×10^{-2} m/s^2

APPENDIX C   PLANETARY DATA

Name	Equatorial Radius (km)	Mass (Relative to Earth's)*	Mean Density (kg / m³)	Surface Gravity (Relative to Earth's)	Orbital Semimajor Axis × 10⁶ km	Orbital Semimajor Axis A. U.	Escape Speed (km/s)	Orbital Period (Years)	Orbital Eccentricity
Mercury	2440	0.0553	5430	0.38	57.9	0.387	4.2	0.240	0.206
Venus	6052	0.816	5240	0.91	108.2	0.723	10.4	0.615	0.007
Earth	6370	1	5510	1	149.6	1	11.2	1.000	0.017
Mars	3394	0.108	3930	0.38	227.9	1.523	5.0	1.881	0.093
Jupiter	71,492	318	1360	2.53	778.4	5.203	60	11.86	0.048
Saturn	60,268	95.1	690	1.07	1427.0	9.539	36	29.42	0.054
Uranus	25,559	14.5	1270	0.91	2871.0	19.19	21	83.75	0.047
Neptune	24,776	17.1	1640	1.14	4497.1	30.06	24	163.7	0.009
Pluto	1137	0.0021	2060	0.07	5906	39.84	1.2	248.0	0.249

*Mass of Earth = 5.97×10^{24} kg

APPENDIX D ELEMENTS OF ELECTRICAL CIRCUITS

Circuit Element	Symbol	Physical Characteristics
resistor		Resists the flow of electric current. Converts electrical energy to thermal energy.
capacitor		Stores electrical energy in the form of an electric field.
inductor		Stores electrical energy in the form of a magnetic field.
incandescent lightbulb		A device containing a resistor that gets hot enough to give off visible light.
battery		A device that produces a constant difference in electrical potential between its terminals.
ac generator		A device that produces a potential difference between its terminals that oscillates with time.
switches (open and closed)		Devices to control whether electric current is allowed to flow through a portion of a circuit.
ground		Sets the electric potential at a point in a circuit equal to a constant value usually taken to be $V = 0$.

APPENDIX E   PERIODIC TABLE OF THE ELEMENTS

Key:

Atomic number →	26 **Fe** ← Symbol
	55.85 ← Atomic mass
Outer electron configuration →	$3d^6 4s^2$

s block — Transition elements (d) — p block — f block

GROUP I
- 1 **H** 1.01 $1s^1$
- 3 **Li** 6.94 $2s^1$
- 11 **Na** 22.99 $3s^1$
- 19 **K** 39.10 $4s^1$
- 37 **Rb** 85.47 $5s^1$
- 55 **Cs** 132.91 $6s^1$
- 87 **Fr** (223) $7s^1$

GROUP II
- 4 **Be** 9.01 $2s^2$
- 12 **Mg** 24.31 $3s^2$
- 20 **Ca** 40.08 $4s^2$
- 38 **Sr** 87.62 $5s^2$
- 56 **Ba** 137.33 $6s^2$
- 88 **Ra** 226.03 $7s^2$

Transition elements (d)

Period 4:
- 21 **Sc** 44.96 $3d^1 4s^2$
- 22 **Ti** 47.88 $3d^2 4s^2$
- 23 **V** 50.94 $3d^3 4s^2$
- 24 **Cr** 52.00 $3d^5 4s^1$
- 25 **Mn** 54.94 $3d^5 4s^2$
- 26 **Fe** 55.85 $3d^6 4s^2$
- 27 **Co** 58.93 $3d^7 4s^2$
- 28 **Ni** 58.69 $3d^8 4s^2$
- 29 **Cu** 63.55 $3d^{10} 4s^1$
- 30 **Zn** 65.39 $3d^{10} 4s^2$

Period 5:
- 39 **Y** 88.96 $4d^1 5s^2$
- 40 **Zr** 91.22 $4d^2 5s^2$
- 41 **Nb** 92.91 $4d^4 5s^1$
- 42 **Mo** 95.94 $4d^5 5s^1$
- 43 **Tc** (98) $4d^5 5s^2$
- 44 **Ru** 101.07 $4d^7 5s^1$
- 45 **Rh** 102.91 $4d^8 5s^1$
- 46 **Pd** 106.42 $4d^{10} 5s^6$
- 47 **Ag** 107.87 $4d^{10} 5s^1$
- 48 **Cd** 112.41 $4d^{10} 5s^2$

Period 6:
- 57 **La** 138.91 $5d^1 6s^2$
- 72 **Hf** 178.49 $5d^2 6s^2$
- 73 **Ta** 180.95 $5d^3 6s^2$
- 74 **W** 183.85 $5d^4 6s^2$
- 75 **Re** 186.21 $5d^5 6s^2$
- 76 **Os** 190.2 $5d^6 6s^2$
- 77 **Ir** 192.22 $5d^7 6s^2$
- 78 **Pt** 195.08 $5d^9 6s^1$
- 79 **Au** 196.97 $5d^{10} 6s^1$
- 80 **Hg** 200.59 $5d^{10} 6s^2$

Period 7:
- 89 **Ac** 227.03 $6d^1 7s^2$
- 104 **Rf** (261) $6d^2 7s^2$
- 105 **Db** (262) $6d^3 7s^2$
- 106 **Sg** (266) $6d^4 7s^2$
- 107 **Bh** (264) $6d^5 7s^2$
- 108 **Hs** (269) $6d^6 7s^2$
- 109 **Mt** (268) $6d^7 7s^2$
- 110 (271)
- 111 (272)
- 112 (277)

GROUP III
- 5 **B** 10.81 $2p^1$
- 13 **Al** 26.98 $3p^1$
- 31 **Ga** 69.72 $4p^1$
- 49 **In** 114.82 $5p^1$
- 81 **Tl** 204.36 $6p^1$

GROUP IV
- 6 **C** 12.01 $2p^2$
- 14 **Si** 28.09 $3p^2$
- 32 **Ge** 72.61 $4p^2$
- 50 **Sn** 118.71 $5p^2$
- 82 **Pb** 207.2 $6p^2$
- 114 (289)

GROUP V
- 7 **N** 14.01 $2p^3$
- 15 **P** 30.97 $3p^3$
- 33 **As** 74.92 $4p^3$
- 51 **Sb** 121.76 $5p^3$
- 83 **Bi** 208.98 $6p^3$

GROUP VI
- 8 **O** 16.00 $2p^4$
- 16 **S** 32.07 $3p^4$
- 34 **Se** 78.96 $4p^4$
- 52 **Te** 127.60 $5p^4$
- 84 **Po** (209) $6p^4$
- 116 (289)

GROUP VII
- 9 **F** 19.00 $2p^5$
- 17 **Cl** 35.45 $3p^5$
- 35 **Br** 79.90 $4p^5$
- 53 **I** 126.90 $5p^5$
- 85 **At** (210) $6p^5$

GROUP VIII
- 2 **He** 4.00 $1s^2$
- 10 **Ne** 20.18 $2p^6$
- 18 **Ar** 39.95 $3p^6$
- 36 **Kr** 83.80 $4p^6$
- 54 **Xe** 131.29 $5p^6$
- 86 **Rn** (222) $6p^6$
- 118 (293)

Lanthanides (f) (*)
- 58 **Ce** 140.12 $5d^1 4f^1 6s^2$
- 59 **Pr** 140.91 $4f^3 6s^2$
- 60 **Nd** 144.24 $4f^4 6s^2$
- 61 **Pm** (145) $4f^5 6s^2$
- 62 **Sm** 150.36 $4f^6 6s^2$
- 63 **Eu** 151.96 $4f^7 6s^2$
- 64 **Gd** 157.25 $5d^1 4f^7 6s^2$
- 65 **Tb** 158.93 $4f^9 6s^2$
- 66 **Dy** 162.50 $4f^{10} 6s^2$
- 67 **Ho** 164.93 $4f^{11} 6s^2$
- 68 **Er** 167.26 $4f^{12} 6s^2$
- 69 **Tm** 168.93 $4f^{13} 6s^2$
- 70 **Yb** 173.04 $4f^{14} 6s^2$
- 71 **Lu** 174.97 $5d^1 4f^{14} 6s^2$

Actinides (†)
- 90 **Th** 232.04 $6d^2 7s^2$
- 91 **Pa** 231.04 $5f^2 6d^1 7s^2$
- 92 **U** 238.03 $5f^3 6d^1 7s^2$
- 93 **Np** 237.05 $5f^4 6d^1 7s^2$
- 94 **Pu** (244) $5f^6 6d^0 7s^2$
- 95 **Am** (243) $5f^7 6d^0 7s^2$
- 96 **Cm** (247) $5f^7 6d^1 7s^2$
- 97 **Bk** (247) $5f^8 6d^1 7s^2$
- 98 **Cf** (251) $5f^{10} 6d^0 7s^2$
- 99 **Es** (252) $5f^{11} 6d^0 7s^2$
- 100 **Fm** (257) $5f^{12} 6d^0 7s^2$
- 101 **Md** (258) $5f^{13} 6d^0 7s^2$
- 102 **No** (259) $5f^{14} 6d^0 7s^2$
- 103 **Lr** (262) $5f^{14} 6d^1 7s^2$

PERIODS: 1, 2, 3, 4, 5, 6, 7

APPENDIX F   PROPERTIES OF SELECTED ISOTOPES

Atomic Number (Z)	Element	Symbol	Mass Number (A)	Atomic Mass*	Abundance (%) or Decay Mode† (if radioactive)	Half-Life (if radioactive)
0	(Neutron)	n	1	1.008665	β^-	10.6 min
1	Hydrogen	H	1	1.007825	99.985	
	Deuterium	D	2	2.014102	0.015	
	Tritium	T	3	3.016049	β^-	12.33 y
2	Helium	He	3	3.016029	0.00014	
			4	4.002603	≈ 100	
3	Lithium	Li	6	6.015123	7.5	
			7	7.016003	92.5	
4	Beryllium	Be	7	7.016930	EC, γ	53.3 d
			8	8.005305	2α	6.7×10^{-17} s
			9	9.012183	100	
5	Boron	B	10	10.012938	19.9	
			11	11.009305	80.1	
			12	12.014353	β^-	20.2 ms
6	Carbon	C	11	11.011433	β^+, EC	20.3 min
			12	12.000000	98.89	
			13	13.003355	1.11	
			14	14.003242	β^-	5730 y
7	Nitrogen	N	13	13.005739	β^-	9.96 min
			14	14.003074	99.63	
			15	15.000109	0.37	
8	Oxygen	O	15	15.003065	β^+, EC	122 s
			16	15.994915	99.76	
			18	17.999159	0.204	
9	Fluorine	F	19	18.998403	100	
			18	18.000938	EC	109.77 min
10	Neon	Ne	20	19.992439	90.51	
			22	21.991384	9.22	
11	Sodium	Na	22	21.994435	β^+, EC, γ	2.602 y
			23	22.989770	100	
			24	23.990964	β^-, γ	15.0 h
12	Magnesium	Mg	24	23.985045	78.99	
13	Aluminum	Al	27	26.981541	100	
14	Silicon	Si	28	27.976928	92.23	
			31	30.975364	β^-, γ	2.62 h
15	Phosphorus	P	31	30.973763	100	
			32	31.973908	β^-	14.28 d
16	Sulfur	S	32	31.972072	95.0	
			35	34.969033	β^-	87.4 d
17	Chlorine	Cl	35	34.968853	75.77	
			37	36.965903	24.23	
18	Argon	Ar	40	39.962383	99.60	
19	Potassium	K	39	38.963708	93.26	
			40	39.964000	β^-, EC, γ, β^+	1.28×10^9 y

Atomic Number (Z)	Element	Symbol	Mass Number (A)	Atomic Mass*	Abundance (%) or Decay Mode[†] (if radioactive)	Half-Life (if radioactive)
20	Calcium	Ca	30	39.962591	96.94	
24	Chromium	Cr	52	51.940510	83.79	
25	Manganese	Mn	55	54.938046	100	
26	Iron	Fe	56	55.934939	91.8	
27	Cobalt	Co	59	58.933198	100	
			60	59.933820	β^-, γ	5.271 y
28	Nickel	Ni	58	57.935347	68.3	
			60	59.930789	26.1	
			64	63.927968	0.91	
29	Copper	Cu	63	62.929599	69.2	
			64	63.929766	β^-, β^+	12.7 h
			65	64.927792	30.8	
30	Zinc	Zn	64	63.929145	48.6	
			66	65.926035	27.9	
33	Arsenic	As	75	74.921596	100	
35	Bromine	Br	79	78.918336	50.69	
36	Krypton	Kr	84	83.911506	57.0	
			89	88.917563	β^-	3.2 min
			92	91.926153	β^-	1.84 s
38	Strontium	Sr	86	85.909273	9.8	
			88	87.905625	82.6	
			90	89.907746	β^-	28.8 y
39	Yttrium	Y	89	89.905856	100	
41	Niobium	Nb	98	97.910331	β^-	2.86 s
43	Technetium	Tc	98	97.907210	β^-, γ	4.2×10^6 y
47	Silver	Ag	107	106.905095	51.83	
			109	108.904754	48.17	
48	Cadmium	Cd	114	113.903361	28.7	
49	Indium	In	115	114.90388	95.7; β^-	5.1×10^{14} y
50	Tin	Sn	120	119.902199	32.4	
51	Antimony	Sb	133	132.915237	β^-	2.5 min
53	Iodine	I	127	126.904477	100	
			131	130.906118	β^-, γ	8.04 d
54	Xenon	Xe	132	131.90415	26.9	
			136	135.90722	8.9	
55	Cesium	Cs	133	132.90543	100	
56	Barium	Ba	137	136.90582	11.2	
			138	137.90524	71.7	
			141	140.914406	β^-	18.27 min
			144	143.92273	β^-	11.9 s
61	Promethium	Pm	145	144.91275	EC, α, γ	17.7 y
74	Tungsten (Wolfram)	W	184	183.95095	30.7	
76	Osmium	Os	191	190.96094	β^-, γ	15.4 d
			192	191.96149	41.0	
78	Platinum	Pt	195	194.96479	33.8	
79	Gold	Au	197	196.96656	100	

Atomic Number (Z)	Element	Symbol	Mass Number (A)	Atomic Mass*	Abundance (%) or Decay Mode† (if radioactive)	Half-Life (if radioactive)
81	Thallium	Tl	205	204.97441	70.5	
			210	209.990069	β^-	1.3 min
82	Lead	Pb	204	203.973044	β^-, 1.48	1.4×10^{17} y
			206	205.97446	24.1	
			207	206.97589	22.1	
			208	207.97664	52.3	
			210	209.98418	α, β^-, γ	22.3 y
			211	210.98874	β^-, γ	36.1 min
			212	211.99188	β^-, γ	10.64 h
			214	213.99980	β^-, γ	26.8 min
83	Bismuth	Bi	209	208.98039	100	
			211	210.98726	α, β^-, γ	2.15 min
			212	211.991272	α	60.55 min
84	Polonium	Po	210	209.98286	α, γ	138.38 d
			212	211.988852	α	0.299 μs
			214	213.99519	α, γ	164 μs
86	Radon	Rn	222	222.017574	α, β	3.8235 d
87	Francium	Fr	223	223.019734	α, β^-, γ	21.8 min
88	Radium	Ra	226	226.025406	α, γ	1.60×10^3 y
			228	228.031069	β^-	5.76 y
89	Actinium	Ac	227	227.027751	α, β^-, γ	21.773 y
90	Thorium	Th	228	228.02873	α, γ	1.9131 y
			231	231.036297	α, β^-	25.52 h
			232	232.038054	100; α, γ	1.41×10^{10} y
			234	234.043596	β^-	24.10 d
91	Protactium	Pa	234	234.043302	β^-	6.70 h
92	Uranium	U	232	232.03714	α, γ	72 y
			233	233.039629	α, γ	1.592×10^5 y
			235	235.043925	0.72; α, γ	7.038×10^8 y
			236	236.045563	α, γ	2.342×10^7 y
			238	238.050786	99.275; α, γ	4.468×10^9 y
			239	239.054291	β^-, γ	23.5 min
93	Neptunium	Np	239	239.052932	β^-, γ	2.35 d
94	Plutonium	Pu	239	239.052158	α, γ	2.41×10^4 y
95	Americium	Am	243	243.061374	α, γ	7.37×10^3 y
96	Curium	Cm	245	245.065487	α, γ	8.5×10^3 y
97	Berkelium	Bk	247	247.07003	α, γ	1.4×10^3 y
98	Californium	Cf	249	249.074849	α, γ	351 y
99	Einsteinium	Es	254	254.08802	α, γ, β^-	276 d
100	Fermium	Fm	253	253.08518	EC, α, γ	3.0 d
101	Mendelevium	Md	255	255.0911	EC, α	27 min
102	Nobelium	No	255	255.0933	EC, α	3.1 min
103	Lawrencium	Lr	257	257.0998	α	≈ 35 s

*The masses given throughout this table are those for the neutral atom, including the Z electrons.
†EC stands for electron capture.

CHAPTER 1

1. (a) 0.152 gigadollars (b) 1.52×10^{-4} teradollars
3. 3×10^8 m/s
5. (a) Yes (b) No (a) Yes
7. (a) Yes (b) Yes (c) No (d) Yes
9. $p = 1$
11. $p = \frac{1}{2}$
13. $[M]/[T]^2$
15. 330.7 m
17. (a) three (b) four
19. (a) 3.14 (b) 3.1416 (c) 3.141593
21. 18 ft^3
23. 0.9788 km
25. 32.9 m
27. 322 ft/s^2
29. (a) 75 ft/s (b) 51 mi/h
31. (a) 67 mutchkin (b) 0.031 gal
33. 32.2 ft/s^2
35. 10^5 seats
37. (a) 10^{10} gal/y (b) 10^9 lb/y
39. (a) Yes (b) No (c) No (d) Yes
41. (a) 6.75×10^{-4} mm (b) 2.66×10^{-5} in
43. 1.2×10^3 lb
45. 55 shakes/s
47. (a) 4.7×10^{-4} mm/s (b) 0.94 ft/week
49. (a) greater than (b) 46 ft/s^2
51. (a) 310 mi/h (b) 0.70 m
53. (a) 51 rev (b) 8.7 ft/rev
55. $p = 1$; $q = -1$
57. 43 mi/h
59. D. 52 s
61. C. 4.78×10^9

CHAPTER 2

1. (a) 1.95 mi (b) 0.75 mi
3. (a) distance = 5 m, $\Delta x = 5$ m
   (b) distance = 2 m, $\Delta x = -2$ m
5. (a) distance = 130 m, $\Delta x = 100$ m
   (b) distance = 260 m, $\Delta x = 0$
7. (a) less than (b) I
9. 23.32 mi/h
11. (a) 0.098 m/s (b) 0.22 mi/h
13. 2.2 km
15. 5.5 m/s
17. 8.5 m
19. (a) less than (b) 0.75 m/s
21. (a) positive (b) zero (c) positive
   (d) negative (e) 2.0 m/s (f) 0 m/s
   (g) 1.0 m/s (h) -1.5 m/s
23. (a) x (m)

(b) 4.0 m/s (c) 4.0 m/s
25. 35 mi/h

27. (a)

(b) 0.55 m/s (c) 0.56 m/s
   (d) closer to 0.56 m/s
29. (a) greater than (b) I
31. (a) 3.8 m/s (b) 9.9 m/s
33. (a) 32.8 m/s north (b) 15.4 m/s north
35. (a) 10 m (b) 20 m (c) 40 m
37. (a) factor of two (b) 4.3 s (c) 8.6 s
39. 6.50 m/s
41. 9.0 m/s to the west
43. (a) 2.41 m/s (b) 11.5 m
45. (a) 0 m; 5.0 m/s; -20 m/s^2
   (b)

(c) -5.0 m/s (d) 25 m/s
47. (a)

(b) 2.2 m/s (c) 2.8 m/s^2
49. 9.1 m/s
51. (a) 0.077 s (b) 0.46 μm
53. 3.8×10^4 m/s^2
55. (a) 21 m (b) greater than 6.0 m/s
   (c) 8.49 m/s
57. (a) 32 m/s^2 (b) less than 8.0 cm
   (c) 4.0 cm
59. (a) 5.7 s (b) -0.175 m/s^2 (c) 2.2 m/s
61. (a) 4.8 m/s^2 toward third base (b) 3.4 m
63. (a) 8.8 m (b) 4.4 m (c) 4.4 m/s
65. (a) $a_A = 1.71$ m/s^2, $a_B = 3.36$ m/s^2,
   $a_C = 4.91$ m/s^2 (b) $\theta_A = 10.0°$,
   $\theta_B = 20.0°$, $\theta_C = 30.0°$
67. 2.7 s
69. 11 m/s
71. (a) 8.4 m/s (b) -1.4 m/s
73. 0.10 s

75. (a) equal to (b) II
77. (a) 11 m (b) 15 m/s (c) 2.1 s
79. 13 m
81. (a) first (dropped) ball (b) ball 1: 24.5 m/s; ball 2: 15.7 m/s
83. (a) 0.49 s (b) 4.8 m/s
85. (a) 65 m (b) 36 m/s (c) 8.5 s
87. (a) 6.7 m/s (b) -5.9 m/s (c) 9.6 m/s
89. 1.01 m/s
91. 5.9 km
93. ball A v negative, a negative; ball B v positive, a negative
95. (a) greater than (b) less than
97. (a) 4.0 m/s^2 (b) 14.5 m (c) 64.5 m
99. (a) 9.81 m/s^2 downward (b) 13.9 m
   (c) 2.21 s (d) 16.5 m/s
101. 3.0 m
103. (a) 12.4 m/s^2 (b) -9.20 m/s^2
105. (a) $\sqrt{2gh}$ (b) $-gh/d$
107. (a) 0.60 s (b) 0.44 m
109. B. 6.78 ft/s
111. plot C
113. 4.3 m/s^2
115. (a) 2.3 s (b) 20 m/s

CHAPTER 3

1. (a) factor of 2 (b) factor of 1
3. $D < C < B < A$
5. 187 ft
7. 3°
9. (a) $(75 \text{ m})\hat{\mathbf{x}} + (47 \text{ m})\hat{\mathbf{y}}$
   (b) $(39 \text{ m})\hat{\mathbf{x}} + (79 \text{ m})\hat{\mathbf{y}}$
11. (a) 56 m (b) 130°
13. 42 ft
15. (a) $-34°$ (b) 17 m (c) direction remains same; magnitude doubles (d) $-34°$; 34 m
17. (a) $\vec{\mathbf{A}}$ (b) $\vec{\mathbf{A}}$
19. (a) 73.5 m (b) 202 m
21. (a) less than (b) equal to
23. (a)

(b) 320 km, 10° north of west
25. (a)

(b) about 120 m; about 130° (c) 121 m; 128°
27. (a) $-27°$; 11 units (b) 207°; 11 units
   (c) 27°; 11 units

29. $(40\,\text{m})\hat{\mathbf{x}} + (-36\,\text{m})\hat{\mathbf{y}}$
31. (a) 23 m (b) 23 m
33. (a) $-22°$; 5.4 m (b) 110°; 5.4 m
  (c) 45°; 4.2 m
35. (a) $(20\,\text{m})\hat{\mathbf{x}} + (-27\,\text{m})\hat{\mathbf{y}}$
  (b) $(-20\,\text{m})\hat{\mathbf{x}} + (27\,\text{m})\hat{\mathbf{y}}$
37. $(2.1\,\text{m})\hat{\mathbf{x}} + (0.74\,\text{m})\hat{\mathbf{y}}$
39. 40° south of west; 5.3 km
41. (a) 1.4 km
  (b) $(-490\,\text{m})\hat{\mathbf{x}} + (-950\,\text{m})\hat{\mathbf{y}}$
  (c) $(490\,\text{m})\hat{\mathbf{x}} + (950\,\text{m})\hat{\mathbf{y}}$
43. 59° north of east; 4.9 m/s
45. $(-9.8\,\text{m/s}^2)\hat{\mathbf{y}}$
47. 22.0 m/s
49. 15.3 m/s
51. (a) $(-4.5\,\text{m/s})\hat{\mathbf{x}}$ (b) $(4.5\,\text{m/s})\hat{\mathbf{x}}$
53. 18 s
55. (a) 11° west of north
  (b)

  (c) increased
57. 11 m/s
59. (a) Jet Ski A (b) $\Delta t_A / \Delta t_B = 0.82$
61. (a) equal to (b) II
63. $-53.1°$; 40.3 m
65. between 27° and 360°
67. (a) 326 km (b) 46.4° south of east
  (c) 40.3 min
69. (a) 10.4 m/s (b) 70.4° below horizontal
71. (a) $(16.6\,\text{m/s})\hat{\mathbf{x}} - (17.2\,\text{m/s})\hat{\mathbf{y}}$
  (b) 23.9 m/s; 46.0° below horizontal
73. (a) 53° north of east; 12 m/s (b) 53°
  south of west; 12 m/s
75. (a) 18 cm; 215° counterclockwise from
  $\vec{\mathbf{A}}$ (b) 0.0034 mm/s at 215°
77. (a) 3.7 (b) 67 m/s
79. (a) 3.5 cm (b) less than (c) 1.7 cm
81. $A = 28$ m; $B = 19$ m
83. C. 1.00 m/s
85. D. 52.1°
87. (a) 12° upstream (b) decrease

CHAPTER 4

1. (a) straight upward (b) III
3. (a) 12.0 s (b) 55.5 s
5. (a) $x = -55$ m; $y = 31$ m
  (b) $v_x = -22$ m/s; $v_y = 6.2$ m/s
  (c) increase with time
7. (a) 48 s (b) 0.021 m/s^2
  (c) $v_x = 2.5$ m/s; $v_y = 1.0$ m/s
9. (a) 27° (b) 0.87 m/s
11. (a) greater than (b) II
13. 46.2 m/s
15. 1.77 m/s^2
17. (a) 9.7 m (b) 1.4 m
19. (a) 3.1 m (b) 3.4 m beyond the far edge
21. 2.6 m/s
23. (a) 1.41 m (b) the same time

25. 1.20 m
27. C. 60°
29. (a) $A < B < C$ (b) $A = B = C$
31. (a) 14.3 m/s (b) 2.24 s
33. 1.3 m
35. $\theta_A = -90°$; $\theta_B = -47°$
37. 67.3 m
39. (a) 0.26 s (b) 2.2 m
41. 75.5°
43. 3.03 m
45. (a) 44.5° (b) 30.1 m/s
47. (a) 21.50 m/s (b) 0.18 m/s
49. (a) 849 m/s (b) less than
51. 63.1°
53. equal to
55. equal to
57. (a) less than (b) II
59. 0.786 m
61. 1.2 s
63. (a) $(15.6\,\text{m/s})\hat{\mathbf{x}} + (3.7\,\text{m/s})\hat{\mathbf{y}}$ (b) 17 m
65. (a) 1.52 m (b) $(4.03\,\text{m/s})\hat{\mathbf{y}}$
  (c) x (m)

67. 24.0 m
69. (a) $x = 0.950$ m; $y = 0.950$ m
  (b) $x = 0.950$ m; $y = 0.699$ m
71. 17 m
73. (a) 12.1 m/s; at the peak (b) 13.2 m/s
  at 23.5° above horizontal (c) 1.40 m
  (d) 1.40 m
75. 5.8 m
77. 5.8 m/s
79. (a) 6.53 m/s (b) 0.270 s
  (c) $(6.53\,\text{m/s})\hat{\mathbf{x}} + (-2.65\,\text{m/s})\hat{\mathbf{y}}$
81. (a) $v_1 \dfrac{X}{Y}$ (b) $\left(v_1 \dfrac{X}{Y}t - X\right)\hat{\mathbf{x}} + (Y - v_1 t)\hat{\mathbf{y}}$
83. 76.0°
87. (a) 5.60 m (b) 4.98 m (c) 5.20 m
89. D. 6.0 mm
91. C. 40 cm
93. (a) 18.2 m (b) 2.5 m/s
95. (a) 1.40 m (b) stay the same (c) 1.40 m

CHAPTER 5

1. $\frac{1}{2}T$
3. 2.72 m
5. 1.6 kN

7.

9. (a) $(827\,\text{N})\hat{\mathbf{y}}$ (b) greater than
11. 0.14 kg
13. (a) 54.3 m/s (b) 646 m
15. (a) $(-18.7\,\text{kN})\hat{\mathbf{x}}$ (b) Determine the
  acceleration from the speeds and
  displacement, and the force from the
  acceleration and mass.
17. (a) greater than (b) II
19. 9.2 km/s^2
21. (a) the same as (b) more than
  (c) 0.93 m/s^2
23. (a) 1.04 N (b) 3.59 N
25. (a) sometimes true (b) never true
  (c) always true (d) sometimes true
27. 7.8 kN
29. 0.10 kN
31. 1.5 m/s^2
33. 146°; 7.8 N
35. (a) 24.5° (b) 4.77×10^{20} N
  (c) 0.00649 m/s^2
37. (a) 0.023 N (b) 6.6 days
39. 2.4 m/s^2 upward
41. (a) upward (b) 1.9 m/s^2 (c) The elevator is either speeding up if it is traveling upward, or slowing down if it is
  traveling downward.
43. (a) $(0.0119\,\text{N})\hat{\mathbf{y}}$ (b) the same as
45. 0.11 kN
47. (a) 91 N (b) 0.13 kN
49. (a) $\frac{1}{2}mg$ (b) mg (c) $2mg$
51. (a) 23 N (b) increase
53. 60°
55. (a) higher than (b) I
57. (a) greater than (b) III
59. 69 m/s^2
61.

63. 1.57×10^5 N
65. $A = B = C$
67. $(5.29\,\text{N})\hat{\mathbf{x}} + (-6.00\,\text{N})\hat{\mathbf{y}}$
69. (a) 3.5 (b) 25 m/s^2 upward
  (c) 0.25 m/s
71. 2.70×10^4 kg
73. (a) $(-0.42\,\text{kN})\hat{\mathbf{x}}$ (b) 2

75. (a) 59 m/s^2 (b) 0.091 s
77. (a) 5.25 kg (b) 4.66 kg
79. (a) 4.2×10^5 N (b) Determine the acceleration during takeoff using the given data (x, t). Then calculate the force using $F = ma$.
81. (a) 2.06 kg (b) 5.00 kg
83. 74 kg
85. $F_1 = \frac{1}{2}m(a_1 + a_2)$; $F_2 = \frac{1}{2}m(a_1 - a_2)$
87. C. 1.05×10^4 N
89. D. 4.21×10^4 N
91. (a) decreased (b) 2.00 kg (c) 2.50 m/s^2
93. (a) greater than (b) 36° (c) 0.079 m/s^2

CHAPTER 6

1. (a) equal to (b) II
3. (a) 0.29 (b) equal to
5. 0.10 kN
7. box 3 < box 2 < box 1
9. (a) 0.42 (b) 23°
11. (a) 0.64 N (b) 0.25
13. $v^2/2\mu g$ (b) quadruples (c) the same
15. (a) greater than (b) greater than
17. 2.13 kN/m
19. 9.4 cm
21. (a) 4.8 cm (b) Yes; the spring compression would double.
23. (a) 0.33 kN (b) 0.66 kN (c) 0.66 kN
25. (a) 5.1 N (b) 15 N
27. 4.5 kg
29. (a) Graph B (b) II
31. $F_1 = F_2 = 0.85$ N
33. (a) 0.020 kN (b) remains the same
35. (a) stay the same (b) increase
37. (a) $T_1 < T_2 < T_3$ (b) $T_1 < T_2 < T_3$
39. (a) decrease (b) 1.4 m/s^2
41. (a) less than (b) 4.36 m/s^2 (c) 15.3 N
43. (a) 3.2 m/s^2 (b) greater than (c) 4.7 m/s^2
45. Path B
47. 6.1 kN
49. 101 m
51. (a) Your apparent weight is greater at the bottom of the Ferris wheel, where the normal force must both support your weight and provide the upward centripetal acceleration. (b) 0.52 kN at top; 0.56 kN at bottom
53. less than
55. inward
57. (a) 35 N (b) greater than (c) 40 N
59. 0.48 km/s^2
61. 0.285
63.

applied force	friction force magnitude	motion
0 N	0 N	at rest
5.0 N	5.0 N	at rest
11 N	11 N	at rest
15 N	8.8 N	accelerating
11 N	8.8 N	accelerating
8.0 N	8.8 N → 8.0 N	decelerating → at rest
5.0 N	5.0 N	at rest

65. (a) 58° (b) 1.4 N

67. (a) 32 N (b) increase
69. (a) static (b) 0.50
71. (a) 0.39 N (b) less than
73. 28°
75. 5.4 cm
77. $v = \sqrt{Mrg/m}$
79. (a) 0.163 s (b) 0.102 m
81. C. 2.7 N/m
83. A. 0.23 N
85. (a) 0.315 (b) 0.127
87. (a) 15 m/s (b) equal to

CHAPTER 7

1. (a) positive (b) zero
3. 0.34 kJ
5. (a) 38 J (b) zero
7. 1.2 J
9. 0.15 kJ
11. (a) 14 J (b) decrease
13. 35 m
15. (a) 0.55 J (b) 0.28 J (c) -1.4 J
17. (a) positive (b) negative
19. 12.7 MJ
21. (a) 1500 kg (b) 16 m/s
23. C < A = D < B
25. (a) -10 J (b) 0.63 N upward
27. (a) -587 J (b) 0.284
29. (a) -7.2 kJ (b) 28 m (c) 0.26 kN
31. (a) 44 J (b) 44 J
33. (a) 0.45 J (b) 0.24 J
35. Spring 1 is stiffer
37. (a) 0.80 m/s (b) 0.21 m
39. $F_4 < F_3 < F_1 < F_2$
41. 3.2 μW
43. 1.1 kN
45. (a) 0.22 W (b) 0.57 N
47. (a) 15.6 N (b) double the speed
49. (a) less than (b) 1.8 kW (c) 5.4 kW
51. (a) $3T$ (b) $v\sqrt{2}$
53. (a) $3\Delta x$ (b) III
55. 1.2×10^{-20} J
57. (a) 0.18 kW (b) 6.3 kJ
59. (a) $W_0/4$ (b) $3W_0/4$
61. $K/2$
63. (a) 13.1 J (b) 13.1 J (c) stay the same
65. 0.348
67. (a) 0.80 J (b) 1.4 J (c) -1.3 J
69. (a) 1.04 kN/m (b) 563 N
71. (a) 12 W (b) 1.1 kJ
73. (a) 15 h (b) 0.62 m/s (c) 1.6 s
75. (a) kg/m (b) $\sqrt[3]{p/b}$ (c) 1.26
77. $\frac{1}{2}(k_1 + k_2)x^2$
79. B. $7.7 - 15$ m/s
81. C. 490 J
83. (a) 0.93 kN/m (b) 1.7 cm
85. (a) 7.8 s (b) less than (c) 3.9 s

CHAPTER 8

1. (a) -3 J (b) -4 J
3. $W_1 = -0.10$ kJ; $W_2 = -47$ J; $W_3 = -66$ J
5. (a) $W_1 = 51$ J; $W_2 = 51$ J (b) increase
7. (a) equal to (b) II
9. 7.6 MJ
11. 538 kJ
13. (a) factor of 4 (b) 3.85 J
15. (a) 0.59 mm (b) 5.7 W
17. -0.70 J
19. (a) K (b) I

21. 6.73 m/s
23. 5.9 m/s
25.

y (m)	4.0	3.0	2.0	1.0	0
U (J)	8.2	6.2	4.1	2.1	0
K (J)	0	2.1	4.1	6.2	8.2
E (J)	8.2	8.2	8.2	8.2	8.2

27. (a) 15 m/s (b) 43 m
29. (a) 0; 113 J; 113 J (b) 113 J; 226 J; 113 J
31. (a) $v_f = \sqrt{2gh\left(\dfrac{m_2 - m_1}{m_1 + m_2}\right)}$ (b) 1.1 m/s
33. (a) less than (b) III
35. 2.7 m/s
37. -45 MJ
39. (a) decrease (b) stay the same (c) decrease
41.

	W_{nc}(J)	U (J)	K (J)	E (J)
(a)	0	34	0	34
(b)	-2.3	24	7.0	31
(c)	-4.6	15	14	29

43. (a) 0.76 kN (b) -38 kJ (c) 30 m/s
45. As the object travels from point A to point B, its speed increases. From point B to point C, its speed decreases, from point C to point D, the speed increases again, and from point D to point E, the speed decreases. At point E it comes to rest momentarily before retracing its path all the way back to point A, at which time the cycle begins again.
47. 0.6 m and 4.6 m
49. (a)

(b) ± 4.5 cm
51. (a) 500 N/m (b) 0.29 m
53. (a) disagree (b) agree (c) agree
55. 7.65 m/s
57. (a) 1.5 mm (b) 0.46 J
59. 1.04 m
61. -58 GJ
63. $\Delta y = h\left(\dfrac{m_2 - m_1}{m_1 + m_2}\right) = 6$ cm
65. 1.1 kN
67. 0.85 m
69. 6.3 cm
71. 48.2°
73. (a) $5r/2$ (b) Although a larger mass would have a larger gravitational force on it, it would also have a larger inertia, hence the effect of the extra mass cancels out.

75. 0.097 m
77. C. 290 N/m
79. A. 0.766 mJ
81. **(a)** 1.03 m/s **(b)** 1.19 m/s

CHAPTER 9

1. 1.2 kg
3. 53 kg·m/s
5. **(a)** 0.42 m/s **(b)** No **(c)** 0.19 J
7. **(a)** 1.3 kg·m/s **(b)** $-$ 0.14 kg·m/s
   (c) The quantity in part (a)
9. inelastic
11. **(a)** less than **(b)** I
13. D < C < B < A
15. 9.0 ms
17. **(a)** 0.33 kN **(b)** 73.6 kg
19. 2.7 kg·m/s
21. **(a)** 27° above horizontal; 5.6 kg·m/s
    (b) double in magnitude; no change in
    direction **(c)** no change
23. 31 kg
25. 2; The piece with the smaller kinetic
    energy has the larger mass.
27. $-$0.80 m/s
29. 12 m/s; 11 m/s
31. $-$0.067 J
33. **(a)** 5.5 m/s **(b)** 3.8 m/s **(c)** $-$0.38 kN
35. 1.32 m/s; 107°
37. 1.1 J
39. **(a)** 0.9978 **(b)** 1 × 10^{-6} **(c)** 0.9807
41. $v_{1,\mathrm{f}} = 0.40$ m/s; $v_{2,\mathrm{f}} = 0.16$ m/s;
    $v_{1,\mathrm{f}} = 0.64$ m/s
43. less than
45. 0.50 m
47. 1.05 m
49. $X_{\mathrm{cm}} = 0$; $Y_{\mathrm{cm}} = 0.036$ nm
51.

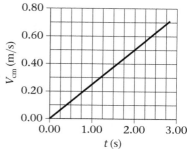

53. **(a)** zero **(b)** zero
55. 63 rocks/min
57. 0.23 kN
59. **(a)** 0.25 N **(b)** more than 2.5 N **(c)** 2.8 N
61. C < A < B
63. $-$0.10 kg·m/s
65. **(a)** green path **(b)** II
67. **(a)** 0.57 **(b)** $-$3.1 × 10^3 kg·m/s

(c) 1.4 × 10^3 kg·m/s
(d) 4.5 × 10^3 kg·m/s
69. 0.24 N
71. 0.48 kg
73. $R/2$
75. 3.80 cm/s
77. **(a)** No **(b)** 6.71 cm
79. **(a)** 65 N **(b)** 3.1 m/s² **(c)** 9.4 m/s
81. **(a)** $3v_0/7$ **(b)** 3/7
87. **(b)** The net force on the two masses
    must point downward to accelerate the
    center of mass downward, hence the
    upward scale force must be less than
    the total weight.
89. A 0.95 knot
91. B. 270 kN
93. **(a)** 0.160 km/s **(b)** 1.58 m/s
95. **(a)** 0.19 m/s **(b)** $-$0.022 J

CHAPTER 10

1. $\frac{\pi}{6}$rad; $\frac{\pi}{4}$rad; $\frac{\pi}{2}$rad; π rad
3. **(a)** 60 rev/h **(b)** 360 deg/min
   (c) $\pi/30$ rad/s
5. 9.4 rad
7. 0.018 s
9. **(a)** 31 rad/s **(b)** 8.3 rad/s **(c)** $-$4.4 rad/s
   (d) $-$43 rad/s
11. **(a)** 1.3 × 10² rad/s **(b)** 2.1 × 10² rad/s
    (c) 3.0 × 10² rad/s **(d)** positive **(e)** 85
    rad/s²; 85 rad/s²
13. $\omega/2$
15. **(a)** $-$0.226 rad/s² **(b)** 28.1 s
17. **(a)** 1.4 rev **(b)** 2.9 s
19. **(a)** 6.29 rev/s² **(b)** 327 rev
21. **(a)** $-$334 rad/s² **(b)** 33.1 m **(c)** 10.0 in
23. **(a)** equal to **(b)** III
25. 26 rad/s
27. **(a)** 1.3 × 10² rad/s **(b)** 690 m/s²
29. **(a)** 0.503 m/s² **(b)** 0.503 m/s²
31. **(a)** 0.29 m/s **(b)** The linear speed would
    also double.
33. 172° below the direction of motion;
    2.1 m/s²
35. **(a)** 4.12 × 10^{16} rad/s
    (b) 6.56 × 10^{15} orbits
    (c) 8.98 × 10^{22} m/s²
37. 0.5 rev
39. 1.12 m/s
41. **(a)** 0.069 s **(b)** 91 rad
43. 69 page/min
45. **(a)** greater than **(b)** II
47. 0.36 m
49. 1.4 mJ
51. **(a)** 12 mJ **(b)** doubled **(c)** 68 rad/s
53. **(a)** 1.2 MJ **(b)** 1.8 min
55. **(a)** 1.1 kJ **(b)** 0.17 km
57. 0.17 kg·m²
59. **(a)** Equate the initial and final mechan-
    ical energies, then solve the resulting
    expression for the mass of the pulley.
    (b) 2.2 kg
61. **(a)** 0.53 m/s **(b)** stay the same
63. 5.3 m/s
65. **(a)** 18 J **(b)** 5.3 J **(c)** 13 J
67. greater than
69. 874 m
71. **(a)** 2.0 kJ **(b)** 3.5% **(c)** 86%

73. $I_3 < I_2 < I_1$
75. **(a)** 2.07 rev **(b)** 0.501 rev
77. **(a)** decrease **(b)** 50.0 rad/s
    (c) 20.8 rad/s **(d)** $-$7.31 × 10^{-3} rad/s²
79. **(a)** 8.6 kJ **(b)** 3.5 kJ
81. 0.20 m/s
83. **(a)** 6.9 rad/s² **(b)** It doubles.
85. **(a)** 15 rad/s² **(b)** 4.0 × 10^3 rad
87. 750 rev
89. **(a)** $\sqrt{3g/L}$ **(b)** $\sqrt{3gL}$
91. **(a)** 0.50 m **(b)** 2.4 rotations
93. **(a)** 1.5 m **(b)** 2.7 rev **(c)** increase
95. C. 30.6 rpm
97. C. 3
99. **(a)** Object 2 first, object 1 second,
    object 3 third **(b)** $K_1 = K_2 = K_3$

CHAPTER 11

1. 60 N
3. 25 N·m
5. 68 N·m
7. **(a)** counterclockwise **(b)** counterclockwise **(c)** clockwise
9. C < B < A
11. 1.09 kg·m²
13. 4.05 N·m
15. **(a)** $-$0.25 N·m **(b)** decrease
17. C < B < A
19. **(a)** Greatest about the x axis; least
    about the z axis. **(b)** 7.4 rad/s²
    (c) 5.1 rad/s² **(d)** 3.0 rad/s²
21. 0.290 N·m
23. **(a)** more than **(b)** 90.3 N
25. 1.1 m
27. 9.14 cm
29. **(a)** $(-21.3$ N$)\,\hat{\mathbf{y}}$ **(b)** $(32.1$ N$)\,\hat{\mathbf{y}}$
31. **(a)** $F/\sqrt{2}$ **(b)** $-F/2$ **(c)** $F/2$
33. $F_{\mathrm{H}} = 330$ N; $F_{\mathrm{J}} = 534$ N
35. 49.7 cm from the milk
37. **(a)** less than **(b)** 0.080 kg
39. 1.05 kg
41. 208 kg
43. 18 kN·m
45. 2.5 kg·m²/s
47. **(a)** point B **(b)** the same as
    (c) $L_{\mathrm{A}} = 0$ kg·m²/s;
    $L_{\mathrm{B}} = 2.55 × 10^3$ kg·m²/s;
    $L_{\mathrm{O}} = 2.55 × 10^3$ kg·m²/s
49. 0.20 kN·m
51. **(a)** stay the same **(b)** III
53. **(a)** stay the same **(b)** decrease
55. 2.84 rad/s
57. **(a)** faster **(b)** 38 rev/min
59. **(a)** $\dfrac{I v}{I + mR^2}$ **(b)** As $I \to 0$, $v_{\mathrm{c,g}} \to 0$
    (c) As $I \to \infty$, $v_{\mathrm{c,g}} \to v$
61. 0.089 N·m
63. 24.8 J
65. **(a)** 33 J **(b)** 1.4 W
67. **(a)** 0.85 W **(b)** 1.2 W **(c)** 2.1 W
69. **(a)** counterclockwise **(b)** move to the right
71. **(a)** move toward the left **(b)** II
73. 5.5 mJ
75. **(a)** increase **(b)** $2v$
77. 4.4 kN/m
79. **(a)** $f_1 = 0.89$ kN; $f_2 = 0.15$ kN;
    $f_3 = 0.15$ kN **(b)** $f_1 = 0.89$ kN;
    $f_2 = 0.22$ kN; $f_3 = 0.22$ kN

81. (a) 2.49 m (b) 1.49 kN
83. 0.16 N
85. 962 N
87. $\sqrt{15}\,Mg$
89. 60°
91. $2L/3$
93. A. 0.0023 N·m
95. A. $F_{1y} = -1.7$ N, $F_{2y} = 3.5$ N
97. (a) increase (b) 1.0 N
99. 121°

CHAPTER 12

1. B < A < C < D
3. (a) 5.1 kN (b) 0.17 kN
5. (a) 3.4×10^{-12} N (b) 3.4×10^{-12} N
7. (a) 3.56×10^{22} N toward the Sun
   (b) 2.40×10^{20} N toward the Sun
   (c) 3.58×10^{22} N toward the Earth-Moon system
9. 3.58×10^{22} N at 0.147° toward Earth from the Sun-Moon line
11. (a) 4.7×10^{-8} N at 16° to the left of downward (b) All magnitudes reduced by a factor of 4, directions unchanged.
13. (a) 3.70 m/s² (b) 8.87 m/s²
15. 5.6×10^{-8} N
17. 13.5 m/s²
19. $M_E/96$
21. (a) Use $\frac{1}{2}m\,v_i^2 = m\,g\,h_f$ to find g, and use $g = GM/R^2$ to find M.
    (b) 8.94×10^{22} kg
23. (a) increase (b) stay the same
25. 1.98 h
27. 1.4×10^{11} m
29. (a) less than (b) 1.5×10^{17} kg
31. (a) satellite 2 (b) 5.59 km/s (c) 4.56 km/s
33. (a) farther from (b) 2.36×10^7 m
35. 1.91×10^{30} kg
37. 36.3 GJ
39. (a) -1.5×10^{-8} J (b) quadruple
    (c) double
41. -48.8 nJ
43. (a) 9.0×10^{10} J (b) 2.0×10^{12} J
45. less than
47. 5.03 km/s
49. 1.05 km/s
51. (a) less than (b) 783 m/s
53. 5.73×10^7 m
55. 2.7×10^5 m
57. (a) 2.07×10^{-7} m/s (b) increase
59. (a) 8.2 N (b) 1.7×10^8 m
63. increase
65. B < C < A
67. -6.34×10^{-10} J
69. (a) 1.98×10^{20} N (b) 4.36×10^{20} N
    (c) Orbiting the Sun, with a small effect due to the Earth.
71. 7.40×10^{-5} m/s
73. (a) more than (b) 39.2 m/s²
75. (a) 0.18 N (b) 1.7 m/s
77. $T = \sqrt{\dfrac{72\pi^2 r_1^3}{Gm_1}}$
79. 2.3 m/s
81. (a) greater than (b) westward (c) 25.4 h
83. $C = 1.00$
85. (a) No (b) 7.66 km/s (c) 1.55 h
87. (a) $\sqrt{GM_E/r}$ (c) Yes

89. $T = 2\pi\sqrt{\dfrac{\sqrt{3}\,R^3}{GM}}$
91. C. 2.7×10^{12} J
93. more than
95. 8.3×10^{10} m
97. (a) increase (b) 15.8 km/s
    (c) stay the same

CHAPTER 13

1. 1.75 s; 0.57 Hz
3. 6.3 s
5. (a) 0.029 s (b) 34 Hz
7. (a) 0.018 s; 55 Hz (b) 1800 rpm
9. (a) $T/2$ (b) $3T/4$
11. (a) 1.5 Hz (b) 0.34 s
13. (a) v_{max} (b) zero (c) zero (d) $-a_{max}$
15. (a) $x = (3.50$ nm$) \cos\left[(1.26 \times 10^{15}\text{ rad/s})\,t\right]$ (b) sine function
17. (a) 0.88 s (b) -2.7 cm
19. 10.7 s
21. (a) 20 m (b) 30 s
23. (a) $0.0495g$ (b) 0.128 m/s (c) minimum
25. $12g$
27. (a) 1.1 km/s² (b) 6.2 m/s
29. (a) 0.017 J (b) 0.34 N
31. $2T/3$
33. (a) greater than (b) III
35. (a) equal to (b) II
37. 2.28 kg
39. Block 1: 0.708 s; block 2: 0.708 s
41. (a) 13 rad/s (b) 0.45 m/s (c) 0.48 s
43. (a) 1.06×10^3 kg (b) 9.3×10^2 kg
45. 6.31 cm
47. (a) greater than (b) $\sqrt{2}\,T$
49. 1.3 J
51. (a) 0.26 m/s (b) 2.8 cm
53. (a) 13.1 N/m (b) 0.309 s (c) decrease
55. downward
57. 8.95 m
59. 2.4 s
61. $T = 2\sqrt{2}\,\pi\sqrt{\dfrac{R}{g}}$
63. (a) greater than (b) 0.667 m
65. (a) 1.2 s (b) 0.91 m/s
67. $2D$
69. (a) remain the same (b) remain the same
71. 0.52 m/s
73. 2.9×10^{-9} Hz
75. (a) zero (b) zero (c) positive
    (d) negative
77. (a) 0.384 Hz (b) 0.125 m/s (c) 5.20 cm
79. (a) less than (b) 0.240 km/s (c) 0.433 s
81. 0.34 s
83. (a) 0.107 m (b) 0.125 s
85. (a) greater than 0.50 m/s (b) 0.79 m/s
    (c) 1.2 J
87. (a) 7.9 N/m (b) 4.6 kg
89. (a) less than (b) $T = \pi\left(\sqrt{\dfrac{\ell}{g}} + \sqrt{\dfrac{L}{g}}\right)$
    (c) 1.5 s
91. $T = \pi\sqrt{\dfrac{2L}{g}\left(\dfrac{m_1 + m_2}{m_1 - m_2}\right)}$
93. B. 10
95. B. 2.2 Hz
97. 11 s
99. (a) stay the same (b) 0.199 s

CHAPTER 14

1. (a) 44 cm (b) 8.0 cm
3. 1.2 m
5. (a) 0.43 Hz (b) 3.7 m/s
7. (a) 2.7 m/s (b) 0.59 Hz
9. (a) 1.00 (b) 0.50
11. (a) less than (b) II
13. (a) B < C < A (b) A < C < B
    (c) A = B = C
15. (a) less time (b) 0.10 s (c) 0.083 s
17. (a) zero (b) zero (c) 18 N (d) 6.6 m/s
    (e) 1.4 s
19. $v \propto \sqrt{\dfrac{F}{R^2\rho}}$
21. $y = (0.11$ m$)\cos\left(\dfrac{\pi}{1.3\text{ m}}x - \dfrac{\pi}{0.091\text{ s}}t\right)$
23. 1.94×10^8 m/s
25. (a)

(b)

(c)

(d) 8.0 s
27. 266 m
29. lowest note 27.50 Hz, 12.5 m; highest note 4.187 kHz, 8.19 cm
31. (a) 0.847 m (b) decrease (c) 0.707 m
33. 1.01 m/s
35. 6.3×10^{-4} W/m²
37. 107 W/m²
39. (a) 104 dB (b) 99.6 dB (c) 2.0×10^6 m
41. (a) 69.5 dB (b) less than
43. (a) greater than (b) III
45. 19.5 m/s
47. (a) 35.3 kHz (b) higher (c) 35.4 kHz
49. 4.2 min
51. (a) 0.33 kHz (b) (i)
53. 15.7 m/s
55. (a) 2.0 cm (b) -2.0 cm
57. 2.1 kHz
59. 0.322 m

61. 147 Hz
63. 49 Hz
65. 2.6 kHz
67. (a) 50 Hz (b) 150 Hz
69. (a) third (b) 44 cm
71. (a) 2.60 Hz (b) 5.21 Hz (c) increase
73. (a) 92.5 Hz (b) 1.85 m
75. string B
77. 3 Hz, 4 Hz, 7 Hz
79. 469 Hz
81. 1.20 m
83. A, III; B, I; C, I; D, III; E, II; F, II
85. 6.72×10^6 m
87. 4 antinodes
89. 1.0×10^{-10} W/m^2
91. 43 m/s
93. (a) No (b) 875 Hz
95. 36.1 m/s
97. 6.73 m/s approaching, 7.00 m/s receding
99. (a) 2.00 (b) 1/3 (c) 2.00
101. (a) 0.0032 J (b) 0.0081 J (c) No
103. 7.06 Hz
105. (a) increase
107. C. 64 Hz
109. B. 2.3 m
111. (a) greater than (b) 30.1 m/s
113. (a) increase (b) 464 Hz

CHAPTER 15

1. 10^3 N
3. No
5. 5.8×10^8 N
7. 1.9×10^6 Pa
9. 601 N
11. (a) 145 cm^2 (b) decrease (c) 43.7 lb/in^2
13. A = B = C = D
15. 2.5 kN
17. 1.1 m
19. (a) 3.2 m (b) 129 kPa
21. (a) 1.45×10^5 Pa (b) 1.54×10^5 Pa
   (c) Because pressure increases with depth.
23. 1.03×10^5 Pa
25. (a) The pressure difference between the liquid surface and your mouth creates an upward force on the liquid. (b) 10.3 m
27. 0.74 m
29. (a) equal to (b) No
31. 8.4 kN
33. (a) remain the same (b) zero
35. 1.08 kg/m^3
37. (a) decrease (b) III
39. less than
41. at the same level
43. (a) 3.1×10^{-4} m^3 (b) 7.1×10^3 kg/m^3
45. (a) 1.04×10^3 kg/m^3 (b) 0.0744 m^3
   (c) 26 N
47. (a) 0.008 m^3 (b) 31 N
49. 2.1 cm
51. 62 m/s
53. 16 min
55. (a) 1.9 cm/s (b) 0.039 cm/s
57. 7.9 m/s
59. (a) 32 cm/s (b) 43 Pa
61. (a) 6.1 m/s (b) 1.4 cm
63. (a) The tube with the greater diameter
   (b) The tube with the smaller diameter
   (c) 0.90 m/s
65. 7.5 m/s

67. (a) 19 kPa (b) 2.0 kN
69. (a) 980 kN (b) upward
71. (a) less than (b) $P_{\text{narrow}} - P_{\text{wide}} = -\frac{15}{2}\rho\, v^2$
73. (a) 20 cm^3 (b) 2.2
75. (a) 2.0 cm^3/s (b) 1.4 L/s
77. move up
79. (a) higher (b) stay the same
81. greater than
83. 95.3 kPa
85. (a) 8.97×10^3 kg/m^3 (b) 820 kg/m^3
87. (a) 0.00141 m^3 (b) decrease
   (c) 0.00140 m^3
89. (a) 10.6 N
91. (a) 4.1 kPa (b) increase (c) 4.3 kPa
93. (a) 3.1×10^{-4} m^3 (b) drop (c) 1.5 cm
95. (a) 5.9 m/s (b) 119 kPa
97. (a) 9.58×10^{-3} m^3 (b) 10.1 kg
99. (a) 94 Pa (b) 0.0025 m^3/s
101. (a) 17 μm (b) yes (c) 21 kg
103. 0.28
107. B. 360 kg/m^3
109. C. 0.28H
111. (a) decrease (b) 0.20 cm
113. (a) 0.715 m (b) 0.104 m

CHAPTER 16

1. (a) less than (b) II
3. 2530°C
5. (a) 5.7×10^3 °C (b) 1.0×10^4 °F
7. (a) 79.3 kPa (b) 151°C
9. 0.23 K/s
11. 320°F = 160°C
13. (a) strip B (b) lead
15. (a) decreased (b) I
17. 1.6 m
19. (a) 96.23 mm (b) −45°C
21. (a) 83°C (b) −82°C
23. 0.0014 in
25. (a) 19×10^{-6} K^{-1} (b) brass
27. (a) water will overflow (b) 24 cm^3
29. 98 W, 0.13 hp
31. 2.6×10^4
33. 4.3 h
35. (a) greater than (b) I
37. 15 kJ
39. 13 K
41. 22.9°C
43. (a) 118 kJ (b) 16.4 kW
45. 837 J/(kg·K); glass
47. 70.5°C
49. $\Delta T_B < \Delta T_A < \Delta T_C$
51. (a) less than (b) III
53. 63 kJ/min
55. 1.6 J
57. (a) 0.800 m (b) decrease by a factor of 2
59. (a) greater than (b) 98°C
61. 49°C
63. (a) $9P/4$ (b) $9P$
65. (a) No (b) Yes (c) No
67. (a) greater than (b) II
69. 1 K
71. 43 J/s
73. (a) 591 steps (b) 18°C
75. 77.9°F
77. (a) heated (b) 860°C
79. 0.17 C°
81. (a) 637 kW (b) 664 kW
   (c) $T_{f,1} = 33.9°C$, $T_{f,2} = 34.1°C$

83. 2.4 min
85. (a) 3.9×10^{26} W (b) 1.4 kW/m^2
87. 9:59:56.8 A.M.
89. 1.1 m
91. C. Using cooled blankets above and below
93. A. 95 W
95. 0.93 kg
97. (a) increase (b) 35.2 J

CHAPTER 17

1. (a) equal to (b) less than
3. 1/4
5. 0.55 L
7. 1.33×10^3 kg
9. C < A < B < D
11. (a) 3.9×10^{24} atoms (b) 1.26
13. 15.6 g/mol
15. 0.8328
17. 1.60×10^5 Pa
19. (a) less than (b) III
21. 19.1 K
23. (a) 4.0×10^6 Pa (b) 1.18×10^{-20} J
   (c) The pressure decreases by a factor of 2 and the average kinetic energy of O$_2$ remains the same.
25. (a) decrease (b) 920 m/s
27. (a) 2.01% (b) 2.01%
29. 2.6×10^{-19} N
31. 51 kg
33. 1.1×10^6 N/m^2
35. $Y_3 < Y_1 = Y_2 < Y_4$
37. (a) 1.8×10^{-3} (b) 0.60 cm
39. 233 N
41. 4.2 kPa
43. (a) 2.3 kPa (b) 15°C (59°F)
45. (a) 3.5 MPa (b) increase
47. (a) gas (b) solid (c) 5707 kPa
49. 6.6×10^5 J
51. 528 kJ
53. D < C < B < A
55. (a) No (b) 0°C; 0.4 kg
57. (a) 27.3 s (b) 61.5 s (c) 246 s
   (d) The water is boiling.
59. 2.9 MJ
61. 44.2 g
63. 2.3 kg
65. (a) 3.6°C (b) No ice is present.
67. 50.3 h
69. (a) less than (b) III
71. 2.0×10^{22} molecules
73. (a) 0.21 kg (b) 93 kJ
75. 133°C
77. (a) 2.07×10^7 Pa (b) 369°C
79. (a) 0.034 (b) 0.034 (c) 0.043
81. (a) 4.2×10^{-7} m (b) 2.7×10^{-6} m
83. 1.1×10^6 students
85. (a) 25 W (b) 19 h
87. 15 s
89. A. 65.2 mol
91. A. 47 cm
93. (a) 0°C (b) 1.8°C

CHAPTER 18

1. (a) zero (b) zero (c) −100 J
3. $\Delta U = -11.6 \times 10^5$ J, $W = 7.7 \times 10^5$ J, $Q = -3.9 \times 10^5$ J
5. (a) 119 J (b) 35 J (c) zero
7. (a) −822 kJ (b) 196 C

9. (a) -4.24 MJ/mi (b) decrease
11. (a) zero (b) -136 J (c) -205 J (d) 341 J
   (e) 494 J
13. The pressure decreases by a factor of 3.
15. 60 Pa
17. (a) on the system (b) -670 J
19. (a) 1200 kJ (b) No
21. (a) 408 K (b) 555 kJ
23. (a) added to the system (b) 55 kJ
25. (a) 150 kJ (b) zero (c) 150 kJ
27. (a) $T_A = 267$ K, $T_B = 357$ K, $T_C = 89.1$ K
   (b) A→B heat enters, B→C heat leaves,
   C→A heat enters (c) A→B 376 kJ,
   B→C -375 kJ, C→A 150 kJ
29. (a) increase (b) 310 K
31. (a) (i) (b) (iii) (c) (i) (d) (iii)
33. (a) 3.1 kJ (b) 1.9 kJ
35. (a) 23 kJ (b) 14 kJ
37. (a) 8000 K (b) 2.88×10^7 Pa
39. (a) 0.315 (b) 0.630 (c) 1.0×10^5 Pa; 222 K

41. (a) Process 1: 159 kJ; process 2: 1060 kJ
   (b) 424 kJ (c) 795 kJ
43. 0.30
45. (a) 8.5 kJ (b) 6.0 kJ
47. (a) 1.16 GW (b) 1.71 GW
49. (a) 407 K (b) decreased (c) 339 K
51. (a) 0.660 (b) 872 MW (c) 0.297
53. (a) 1/3 (b) 2/3
55. (a) 0.59 kJ (b) 0.19
57. (a) 1.1×10^8 J (b) $340
59. 0.41 kW
61. 1.5×10^5 J
63. (a) increase (b) I
65. 1.12×10^4 J/K
67. D < C < B < A
69. 0.11 kJ/K
71. 2.6 J/K
73. (a) negative (b) positive (c) negative
75. 0.250
77. (a) 1900 m^3 (b) 2.6×10^7 J

79. (a) Engine A (b) 0.32
81. (a) 751 kJ (b) 194 kJ (c) 945 kJ
83. (a) $nR\Delta T$ (b) zero (c) $\frac{3}{2}nR\Delta T$ (d) zero
   (e) $\frac{3}{2}nR\Delta T$ (f) $nRT\ln(V_f/V_i)$ (g) zero
85. (a) -503 J/K (b) 168 kJ (c) 569 J/K
   (d) 66 J/K
87. (a) decreases (b) -0.65 kJ/K (c) 0.66 kJ/K
   (d) 12 J/K
89.

	Q	W	ΔU
A → B	806 J	806 J	0
B → C	-938 J	-375 J	-563 J
C → A	563 J	0	563 J

93. A. 6.10%
95. C. 500 m
97. (a) greater than (b) 0.488 (c) 512 J

CHAPTER 1

1. No. The factor of 2 is dimensionless.
3. (a) Not possible, since units have dimensions. For example, seconds can only have the dimension of time. (b) Possible, since different units can be used to measure the same dimensions. For example, time can be measured in seconds, minutes, or hours.
5. (a) Both Albert and Isaac are running with a speed of 3.4 m/s; thus, Albert's speed is equal to Isaac's speed. (b) Albert and Isaac have different velocities, because they are running in different directions.
7. To the nearest power of ten: (a) 1 m; (b) 10^{-2} m; (c) 10 m; (d) 100 m; (e) 10^7 m.

CHAPTER 2

1. The displacement is the same for you and your dog; the distance covered by the dog is greater.
3. No. After one complete orbit the astronaut's displacement is zero. The distance traveled, however, is roughly 25,000 miles.
5. A speedometer measures speed, not velocity. For example, if you drive with constant speed in a circular path, your speedometer maintains the same reading, even though your velocity is constantly changing.
7. Constant-velocity motion; that is, straight-line motion with constant speed.
9. (a) Yes. The object might simply be at rest. (b) Yes. An example would be a ball thrown straight upward; at the top of its trajectory its velocity is zero, but it has a nonzero acceleration downward.
11. Yes. A ball thrown straight upward and caught when it returns to its release point has zero average velocity, but it has been accelerating the entire time.
13. When she returns to her original position, her speed is the same as it was initially; that is, 4.5 m/s.

CHAPTER 3

1. (a) scalar; (b) vector; (c) vector; (d) scalar.
3. (a) $\vec{\mathbf{A}}$ and $\vec{\mathbf{B}}$ have the same magnitude. (b) $\vec{\mathbf{A}}$ and $\vec{\mathbf{B}}$ have opposite directions.
5. Yes, if they have the same magnitude and point in opposite directions.
7. Note that the magnitudes A, B, and C satisfy the Pythagorean theorem. It follows that $\vec{\mathbf{A}}$, $\vec{\mathbf{B}}$, and $\vec{\mathbf{C}}$ form a right triangle, with $\vec{\mathbf{C}}$ as the hypotenuse. Thus, $\vec{\mathbf{A}}$ and $\vec{\mathbf{B}}$ are perpendicular to one another.
9. $\vec{\mathbf{A}}$ and $\vec{\mathbf{B}}$ must be collinear and point in opposite directions. In addition,

the magnitude of $\vec{\mathbf{A}}$ must be greater than the magnitude of $\vec{\mathbf{B}}$, that is, $|A| > |B|$.
11. The vector $\vec{\mathbf{A}}$ points in one of two directions: (1) 135° counterclockwise from the x axis; (2) 45° clockwise from the x axis. Case (1) corresponds to $A_x < 0$ and case (2) corresponds to $A_x > 0$.
13. Tilt the umbrella forward so that it points in the opposite direction of the rain's velocity relative to you.

CHAPTER 4

1. Ignoring air resistance, the acceleration of a projectile is vertically downward at all times.
3. The projectile was launched at an angle of 30° above the positive x axis; when it landed its direction of motion was 30° below the positive x axis. Hence, its change in direction was 60° clockwise.
5. At its highest point, the projectile is moving horizontally. This means that gravity has reduced its y component of velocity to zero. The x component of velocity is unchanged by gravity, however. Therefore, the projectile has a velocity equal to $\vec{\mathbf{v}} = (4 \text{ m/s})\hat{\mathbf{x}}$ at the highest point in its trajectory.
7. Less than 45°. See Figure 4-14.
9. (a) From the child's point of view the scoop of ice cream falls straight downward. (b) From the point of view of the parents the scoop of ice cream follows a parabolic trajectory.
11. (a) At its highest point the projectile moves horizontally. Therefore, its launch angle was 50°. (b) In this case, the launch angle was 150°, or, equivalently, 30° above the negative x axis.

CHAPTER 5

1. The force exerted on the car by the brakes causes it to slow down, but your body continues to move forward with the same velocity (due to inertia) until the seat belt exerts a force on it to decrease its speed.
3. No. You are at rest relative to your immediate surroundings, but you are in motion relative to other objects in the universe.
5. When the magnitude of the force exerted on the girl by the rope equals the magnitude of her weight, the net force acting on her is zero. As a result, she moves with constant velocity.
7. The astronaut should push the jet pack away from him, in the opposite direction from the spaceship. As a result, the reaction force exerted on him by the pack will accelerate him toward the ship.

9. Mr. Ed's reasoning is incorrect because he is adding two action-reaction forces that act on *different* objects. Wilbur should point out that the net force exerted on the cart is simply the force exerted on it by Mr. Ed. Thus the cart will accelerate. The equal and opposite reaction force acts on Mr. Ed, and does not cancel the force acting on the cart.
11. The whole brick has twice the force acting on it, but it also has twice the mass. Since the acceleration of an object is proportional to the force exerted on it and inversely proportional to its mass, the whole brick has the same acceleration as the half brick.
13. Yes. An object with zero net force acting on it has a constant velocity. This velocity may or may not have zero magnitude.
15. The acceleration of an object is inversely proportional to its mass. The diver and the Earth experience the same force, but the Earth—with its much larger mass—has a much smaller acceleration.
17. An astronaut can tell which of two objects is more massive by pushing on both with the same force. Since acceleration is inversely proportional to mass, the more massive object can be recognized by its smaller acceleration.
19. On solid ground you come to rest in a much smaller distance than when you plunge into water. The smaller the distance over which you come to rest the greater the acceleration, and the greater the acceleration the greater the force exerted on you. Thus, when you land on solid ground the large force it exerts on you may be enough to cause injury.
21. Yes, it would still hurt. The reason is that even though the bucket is "weightless," it still has nonzero mass. Therefore, it has inertia, and when it is kicked it will resist having its state of motion changed. The way it resists a change in its motion is by exerting an equal-but-opposite reaction force on the object trying to make the change; in this case, your foot. This is the force that can make kicking the bucket painful.
23. In everyday units, 1 N is approximately a quarter pound. Objects with a weight of roughly a quarter pound include a quarter-pound hamburger, an apple, a glass of water, and half a dozen Fig Newtons.

CHAPTER 6

1. The clothesline has a finite mass, and so the tension in the line must have an upward component to oppose the

downward force of gravity. Thus, the line sags much the same as if a weight were hanging from it.

3. The braking distance of a skidding car depends on its initial speed and the coefficient of kinetic friction. Thus, if the coefficient of friction is known reasonably well, the initial speed can be determined from the length of the skid marks.

5. The force that ultimately is responsible for stopping a train is the frictional force between its metal wheels and the metal track. These are fairly smooth surfaces. In contrast, the frictional force that stops a car is between the rubber tires and the concrete roadway. These are rougher surfaces with a greater coefficient of friction.

7. As you brake harder your car has a greater acceleration. The greater the acceleration of the car, the greater the force required to give the flat of strawberries the same acceleration. When the required force exceeds the maximum force of static friction the strawberries begin to slide.

9. For a drop of water to stay on a rotating wheel an inward force is required to give the drop the necessary centripetal acceleration. Since the force between a drop of water and the wheel is small, the drop will separate from the wheel rather than follow its circular path.

11. A centripetal force is required to make the motorcycle follow a circular path, and this force increases rapidly with the speed of the cycle. If the necessary centripetal force exceeds the weight of the motorcycle, because its speed is high enough, then the track must exert a downward force on the cycle at the top of the circle. This keeps the cycle in firm contact with the track.

13. Since the passengers are moving in a circular path a centripetal force must be exerted on them. This force, which is radially inward, is supplied by the wall of the cylinder.

15. This helps because the students sitting on the trunk increase the normal force between your tires and the road. Since the force of friction is proportional to the normal force, this increases the frictional force enough (one hopes) to allow your car to move.

17. The normal force exerted on a gecko by a vertical wall is zero. If the gecko is to stay in place, however, the force of static friction must exert an upward force equal to the gecko's weight. For this to happen when the normal force is zero would require an infinite coefficient of static friction. Thus, we conclude that the physics of gecko feet is more complex than our simple models of friction.

19. Yes. The steering wheel can accelerate a car—even if its speed remains the same—by changing its direction of motion.

21. When a bicycle rider leans inward on a turn, the force applied to the wheels of the bicycle by the ground is both upward and inward. It is this inward force that produces the centripetal acceleration of the rider.

CHAPTER 7

1. No. Work requires that a force acts through a distance.

3. True. To do work on an object a force must have a nonzero component along its direction of motion.

5. The frictional force between your shoes and the ground does positive work on you whenever you begin to walk.

7. Gravity exerts an equal and opposite force on the package, and hence the net work done on it is zero. The result is no change in kinetic energy.

9. No, we must also know how much time it takes for engine 1 to do the work. For example, if engine 1 takes twice as much time to do twice the work of engine 2, it has the same power output as engine 2. If it takes more than twice as much time, then engine 1 actually produces less power than engine 2.

CHAPTER 8

1. The kinetic energy cannot be negative, since m and v^2 are always positive or zero. The gravitational potential energy can be negative since any level can be chosen to be zero.

3. The initial mechanical energy of the system is the gravitational potential energy of the mass-Earth system. As the mass moves downward, the gravitational potential energy of the system decreases. At the same time, the potential energy of the spring increases as it is compressed. Initially, the decrease in gravitational potential energy is greater than the increase in spring potential energy, which means that the mass gains kinetic energy. Eventually, the increase in spring energy equals the decrease in gravitational energy and the mass comes to rest.

5. The object's kinetic energy is a maximum when it is released, and a minimum when it reaches its greatest height. The gravitational potential of the system is a minimum when the object is released, and a maximum when the object reaches its greatest height.

7. The jumper's initial kinetic energy is largely converted to a compressional, spring-like potential energy as the pole bends. The pole straightens out, converting its potential energy into

gravitational potential energy. As the jumper falls, the gravitational potential energy is converted into kinetic energy, and finally, the kinetic energy is converted to compressional potential energy as the cushioning pad on the ground is compressed.

9. Zero force implies that the rate of change in the potential energy with distance is zero—that is, the potential energy is constant—but the value of the potential energy can be anything at all. Similarly, if the potential energy is zero, it does not mean that the force is zero. Again, what matters is the rate of change of the potential energy with distance.

11. The dive begins with the diver climbing the ladder to the diving board, which converts chemical energy in the muscles into an increased gravitational potential energy. Next, by jumping on the board the diver causes the board to flex and to store potential energy. As the board rebounds, the diver springs into the air, using the kinetic energy derived from the leg muscles and the potential energy released by the board. The diver's kinetic energy is then converted into an increased gravitational potential energy until the highest point of the dive is reached. After that, gravitational potential energy is converted back to kinetic energy as the diver moves downward. Finally, the kinetic energy of the diver is converted into heat, sound, and flowing water as splashdown occurs.

CHAPTER 9

1. The momentum of the keys increases as they fall because a net force acts on them. The momentum of the universe is unchanged because an equal and opposite force acts on the Earth.

3. If the kinetic energy is zero the speeds must be zero as well. This means that the momentum is zero.

5. Yes, in much the same way that a propeller in water can power a speedboat.

7. (a) The force due to braking—which ultimately comes from friction with the road—reduces the momentum of the car. The momentum lost by the car does not simply disappear, however. Instead, it shows up as an increase in the momentum of the Earth. (b) As with braking, the ultimate source of the force accelerating the car is the force of static friction between the tires and the road.

9. (a) Yes. Suppose two objects have momenta of equal magnitude. If these objects collide in a head-on, completely inelastic collision, they will be at rest after the collision. In this case, all of the initial kinetic energy is converted to other forms of energy. (b) No. In

order for its momentum to change, an external force must act on the system. We are given, however, that the system is isolated. Therefore, the only forces acting on it are internal forces.

11. The kinetic energy of the bullet is much greater than that of the gun. Thus, less energy is dissipated in stopping the gun.

13. The plane weighs the same whether the fly lands on the dashboard or flies about the cockpit. The reason is that the fly must exert a downward force on the air in the cockpit equal in magnitude to its own weight in order to stay aloft. This force ultimately acts downward on the plane, just as if the fly had landed.

15. Your center of mass is somewhere directly above the area of contact between your foot and the ground.

CHAPTER 10

1. All points on the rigid object have the same angular speed. Not all points have the same linear speed, however. The farther a given point is from the axis of rotation the greater its linear speed.

3. No. As long as you are driving in a circular path you will have a nonzero centripetal acceleration.

5. (a) No, because your direction of motion is constantly changing. (b) Yes, your linear speed is simply the radius of the wheel times your angular speed. (c) Yes. It is equal to your linear speed squared divided by the radius of the wheel. (d) No. Your centripetal acceleration is always toward the center of the wheel, but this is in a different direction as you rotate to different locations.

7. (a) A basketball thrown with no spin. (b) A spinning airplane propeller on a plane that is at rest. (c) A bicycle wheel on a moving bicycle.

9. When the chunky stew is rolled down the aisle, all of the contents of the can roll together with approximately the same angular speed. This is because the chunky stew is thick and almost solid. The beef broth, however, is little more than water. Therefore, when the broth is rolled down the aisle, almost all that is actually rolling is the metal can itself. It follows that the stew has the greater initial kinetic energy, and hence it rolls a greater distance.

CHAPTER 11

1. No. Torque depends both on the magnitude of the force and on the distance from the axis of rotation, or moment arm, at which it is applied. A small force can produce the same torque as a large force if it is applied farther from the axis of rotation.

3. The long pole has a large moment of inertia, which means that for a given applied torque the walker and pole have a small angular acceleration. This allows more time for the walker to "correct" his balance.

5. A force applied radially to a wheel produces zero torque, though the net force is nonzero.

7. No. In most cars the massive engine is located in the front, thus the car's center of mass is not in the middle of the car, but is closer to the front end. This means that the force exerted on the front tires is greater than the force exerted on the rear tires. (This situation is analogous to Active Example 11-1.)

9. You are in static equilibrium as you sit in your chair; so is the building where you have your physics class.

11. The angular speed of the dust cloud increases, just like a skater pulling in her arms, due to conservation of angular momentum.

13. Yes. Imagine turning on a ceiling fan. This increases the fan's angular momentum, without changing its linear momentum.

CHAPTER 12

1. No. The force of Earth's gravity is practically as strong in orbit as it is on the surface of the Earth. The astronauts experience weightlessness because they are in constant free fall.

3. As the tips of the fingers approach one another, we can think of them as two small spheres (or we can replace the finger tips with two small marbles if we like). As we know, the net gravitational attraction outside a sphere of mass is the same as that of an equivalent point mass at its center. Therefore, the two fingers simply experience the finite force of two point masses separated by a finite distance.

5. No. A satellite must be moving relative to the center of the Earth to maintain its orbit, but the North Pole is at rest relative to the center of the Earth. Therefore, a satellite cannot remain fixed above the North Pole.

7. Yes. The rotational motion of the Earth is to the east, and therefore if you launch in that direction you are adding the speed of the Earth's rotation to the speed of your rocket.

9. As the astronauts approach a mascon, its increased gravitational attraction would increase the speed of the spacecraft. Similarly, as they pass the mascon, its gravitational attraction would now be in the backward direction, which would decrease their speed.

11. It makes more sense to think of the Moon as orbiting the Sun, with the Earth providing a smaller force that

makes the Moon "wobble" back and forth in its solar orbit.

CHAPTER 13

1. The motion is periodic. It is not simple harmonic, however, because the position and velocity of the ball do not vary sinusoidally with time.

3. The motion of the air cart is periodic; it repeats after a fixed length of time. It is not simple harmonic motion, however, because the position and velocity of the cart do not vary sinusoidally with time.

5. The period remains the same because, even though the distance traveled by the object is doubled, its speed at any given time is also doubled.

7. The period of a pendulum is independent of the mass of its bob. Therefore, the period should be unaffected.

CHAPTER 14

1. To generate a longitudinal wave, hit the nail on the head in a direction parallel to its length. To generate a transverse wave, hit the nail in a direction that is perpendicular to its length.

3. This wave is longitudinal, because each cat moves in the same direction as the wave.

5. The Doppler effect applies to radar waves as well as to sound waves. In particular, the ball sees a Doppler-shifted radar frequency for the waves coming from the radar gun. Then, the ball acts as a moving source for waves of this frequency, giving a second Doppler shift to the echoes that are picked up by the radar gun. This provides a one-to-one correspondence between the final observed frequency and the speed of the ball.

7. The sliding part of a trombone varies the length of the vibrating air column that produces the trombone's sound. By adjusting this length, the player controls the resonant frequencies of the instrument. This, in turn, varies the frequency of sound produced by the trombone.

9. You hear no beats because the difference in frequency between these notes is too great to produce detectable beats.

CHAPTER 15

1. To draw a liquid up a straw you expand your lungs, which reduces the air pressure inside your mouth to less than atmospheric pressure. The resulting difference in pressure produces a net upward force on the liquid in the straw.

3. The pressure in a tank of water increases with depth, hence the pressure is greatest near the bottom. To provide sufficient support there, the metal

bands must be spaced more closely together.

5. This experiment shows that a certain pressure is needed at the bottom of the water column and not just a certain weight of water. To blow the top off the barrel it is necessary to increase the pressure in the barrel enough so that the increase in pressure times the surface area of the top exceeds 400 N. Thus, the required height of water is the height that gives the necessary increase in pressure. But the increase in pressure, $\rho g h$, depends only on the height of the water in the tube, not on its weight.

7. Two quantities are unknown; the object's density and its volume. The two weight measurements provide two independent conditions that can be solved for the two unknowns.

9. The Great Salt Lake has water with a higher salinity, and hence a higher density, than ocean water. In fact, the density of its water is somewhat greater than the density of a typical human body. This means that a person can float in the Great Salt Lake much like a block of wood floats in fresh water.

11. The problem is that as you go deeper into the water, the pressure pushing against your chest and lungs increases rapidly. Even if you had a long tube on your snorkel, you would find it difficult to expand your lungs to take a breath. The air coming through the snorkel is at atmospheric pressure, but the water pushing against your chest might have twice that pressure or more. Thus, scuba gear not only holds air for you to breathe in a tank, it also feeds this air to you under pressure.

13. As the water falls it speeds up. Still, the amount of water that passes a given point in a given time is the same at any height. If the thickness of the water stayed the same, and its speed increased, the amount of water per time would increase. Hence, the thickness of the water must decrease to offset the increase in speed.

15. If you takeoff into the wind the air speed over the wings is greater than if you takeoff with the wind. This means that more lift is produced when taking off into the wind, which is clearly the preferable situation.

17. The ball should spin so that the side facing the batter is moving upward and the side facing the pitcher is moving downward.

CHAPTER 16

1. The coffee is not in equilibrium because its temperature is different from that of its surroundings. Over time the temperature of the coffee will decrease, until finally it is the same as room temperature. At this point it will be in equilibrium—as long as the room stays at the same temperature.

3. No. Heat is not a quantity that one object has more of than another. Heat is the energy that is transferred between objects of different temperatures.

5. The mercury level drops at the beginning because the glass is the first to increase its temperature when it comes into contact with the hot liquid. Therefore, the glass expands before the mercury, leading to a drop in level. As the mercury attains the same temperature few moments later, its level will increase.

7. If the objects have different masses, the less massive object will have a greater temperature change, since it has the smaller heat capacity. On the other hand, the objects may have the same mass but differ in the material from which they are made. In this case, the object with the smaller specific heat will have the greater temperature change.

9. Both the metal and the wood are at a lower temperature than your skin. Therefore, heat will flow from your skin to both the metal and the wood. The metal feels cooler, however, because it has a greater thermal conductivity. This allows the heat from your skin to flow to a larger effective volume than is the case with the wood.

11. Two important factors work in favor of the water-filled balloon. First, the water has a large heat capacity, hence it can take on a large amount of heat with little change in temperature. Second, water is a better conductor of heat than air, hence the heat from the flame is conducted into a large volume of water—which gives it a larger effective heat capacity.

13. The hollow fibers of hair are effective insulators because the gas within the fibers has a low thermal conductivity. This is analogous to double-pane windows, which trap a layer of gas between the panes for a greatly enhanced insulating effect.

CHAPTER 17

1. If the temperature of the air in a house is increased, and the amount of air in the house remains constant, it follows from the ideal gas law that the pressure will increase as well.

3. Yes. If the pressure and volume are changed in such a way that their product remains the same, it follows from the ideal gas law that the temperature of the gas will remain the same. If the temperature of the gas is the same, the average kinetic energy of its molecules will not change.

5. No. The temperature at which water boils on a mountaintop is less than its boiling temperature at sea level due to the low atmospheric pressure on the mountain. Therefore, if the stove is barely able to boil water on the mountain, it will not be able to boil it at sea level, where the required temperature is greater.

7. If we look at the phase diagram in Figure 17-16, we can see that in order to move upward in the graph from the sublimation curve to the fusion curve, the pressure acting on the system must be increased.

9. As high-speed molecules leave the drop, it draws heat from its surroundings to keep its temperature constant. As long as the temperature of the drop is constant, it will continue to have the same fraction of molecules moving quickly enough to escape.

CHAPTER 18

1. (a) Yes. Heat can flow into the system if at the same time the system expands, as in an isothermal expansion of a gas. (b) Yes. Heat can flow out of the system if at the same time the system is compressed, as in an isothermal compression of a gas.

3. Yes. In an isothermal expansion, all the heat added to the system to keep its temperature constant appears as work done by the system.

5. (a) No; (b) yes; (c) no; (d) no; (e) yes; (f) no.

7. Yes, this is possible. The problem is that you would need low-temperature reservoirs of ever lower temperature to keep the process going.

9. The law of thermodynamics most pertinent to this situation is the second law, which states that physical processes move in the direction of increasing disorder. To decrease the disorder in one region of space requires work to be done, and a larger increase in disorder in another region of space.

Photo Credits

Chapter 1—p. 1 Andrew Cowie/AFP/GettyImage p. 3 **(center right)** John F. Kennedy Space Center/NASA p. 3 **(top right)** Oliver Meckes/Science Source p. 4 **(bottom)** U.S. Department of Commerce p. 4 **(top)** Bureau International des Poids et Mesures p. 6 Jeff Greenberg/PhotoEdit, Inc p. 8 Antonin Thuillier/AFP/GettyImages p. 9 **(top)** David R. Frazier/Science Source p. 10 CNRI/Science Photo Library/Science Source p. 11 US Department of Energy p. 15 **(center right)** Jeff Greenberg/PhotoEdit, Inc p. 15 **(top)** Andrew Cowie/AFP/GettyImage p. 17 **(bottom)** USDA Forest Service p. 19 **(left)** Vadim Petrakov/Shutterstock, Visions of America, LLC/Alamy, Eric Grave/Science Source

Chapter 2—p. 19 Dave Thompson-British Athletics/Contributor/Getty Images p. 26 James Bowyer/Getty Images p. 29 NASA p. 32 Chris Urso/AP Images p. 33 David Scharf/Science Source p. 43 **(bottom right)** Rainer Albiez/Shutterstock p. 43 **(center right)** Steve Kazlowski/Danita Delimont Photography/Newscom p. 43 **(top right)** Jim Sugar/Corbis p. 45 Julia Davila-Lampe/Flickr Open/Getty Images p. 54 Jill Morgan/Alamy p. 55 Stephen Dalton/Science Source, Pearson Education/PH College

Chapter 3—p. 60 Maksym Yemelyanov/Fotolia p. 61 Jeff Greenberg/Science Source p. 66 **(bottom left)** Delmas Lehman/Fotolia p. 72 George Whiteley/Science Source p. 76 Laurent Rebours/AP Images p. 84 US Department of Defense p. 86 **(center right)** Ezra Shaw/Getty Images p. 86 **(top left)** PIXOLOGICSTUDIO/Science Photo Library/Alamy

Chapter 4—p. 88 Richard Megna/Fundamental Photographs p. 94 **(top center)** NASA p. 94 **(top left)** Richard Megna/Fundamental Photographs p. 94 **(top right)** Richard Megna/Fundamental Photographs p. 97 **(top left)** Nick Turner/Alamy p. 97 **(top right)** Jon Bilous/Alamy p. 100 **(center)** John Terence Turner/Alamy p. 100 **(left)** NaturePL/Superstock p. 100 **(right)** Bernd Zoller/ImageBroker/Alamy p. 107 G. I. Bernard/Science Source, Gregory G. Dimijian/Science Source

Chapter 5—p. 112 **(top)** Tcly/Shutterstock p. 117 Alessandro Bianchi/Reuters p. 119 James S. Walker p. 122 **(bottom)** NASA p. 122 **(center)** Bruce Bennett/Getty Images p. 128 Federal Aviation Administration p. 136 Pete Saloutos/Corbis p. 140 Johnson Space Center/NASA p. 153 Frank Greenaway/Dorling kindersly, Ltd, United States Marines

Chapter 6—p. 155 Greg Epperson/iStock/Getty Image p. 156 **(left)** Ian Aitken/Dorling Kindersley, Ltd. p. 156 **(right)** Ernest Braun/Stone Allstock/Getty Images p. 160 James Marvin Phelps/Shutterstock p. 162 **(left)** Photodisc/Getty Images p. 162 **(right)** Andreas Bauer/iStock/Getty Images p. 164 **(bottom)** Richard Megna/Fundamental Photographs p. 164 **(top)** Richard Megna/Fundamental Photographs p. 168 **(bottom)** Science Museum/Science & Society/The Image Works p. 168 **(top)** Oote boe 3/Alamy p. 172 Oleg Zabielin/Shutterstock p. 173 Warner Bros pictures/Everett Collection p. 178 Brian McEntire/Shutterstock p. 181 **(bottom)** © Bombardier p. 181 **(center)** Mark Meyer/Navy Visual News Service p. 181 **(top)** Dave Martin/AP Images p. 182 Vasiliy Koval/Fotolia p. 190 Tsuneo Nakamura/Volvox Inc/Alamy p. 191 V. J. Caiozzo, University of California, Irvine, AF archive/Alamy

Chapter 7—p. 195 VisualCommunications/iStock/Getty Images p. 198 **(bottom)** AFP/Getty Images p. 198 **(center)** AFP/Getty Images p. 198 **(top)** holbox/Shutterstock p. 209 Dr T J McMaster, Wellcome Images p. 213 National Geographic/Getty Images

Chapter 8—p. 223 Alex Emanuel Koch/Shutterstock p. 225 **(left)** Herbert Esser/Fotolia p. 225 **(right)** Westend61/Getty Images p. 229 Wales News Service/Splash News/Newscom p. 231 Barry Bland/Alamy p. 233 Todd Winner/Stocktrek Images/Alamy p. 234 **(bottom)** Brisbane/Shutterstock p. 234 **(top)** hatchapong/Fotolia p. 239 I Love Images/Cultura/Getty Images p. 246 Federal Highway Administration, Colorado Division p. 249 **(center)** Stephen Derr/The Image Bank/Getty Images p. 249 **(right)** Dmitry Kalinovsky/Shutterstock

Chapter 9—Sasa Huzjak/Alamy, National Oceanic and Atmospheric Administration p. 257 stockshoppe/Shutterstock p. 263 **(left)** Kathy Willens/ AP Images p. 263 **(right)** Wally McNamee/Corbis p. 266 Oxford Scientific/Getty Images p. 268 NASA p. 270 NASA p. 272 **(bottom)** Franck Seguin/Corbis p. 272 **(bottom)** Jacqueline Larma/AP Images p. 272 **(top)** Bruce Bennett/Getty Images p. 280 **(bottom left)** Richard Megna/Fundamental Photographs p. 280 **(top left)** Richard Megna/Fundamental Photographs p. 280 **(top right)** Richard Megna/Fundamental Photographs p. 287 **(bottom)** Richard Megna/Fundamental Photographs p. 287 **(top)** Richard Megna/Fundamental Photographs p. 289 **(bottom)** Vittorio Bruno/Shutterstock p. 289 **(top)** US Air Force p. 290 Richard Megna/Fundamental Photographs p. 290 Richard Megna/Fundamental Photographs

p. 291 **(bottom)** David J. Phillip/AP Images p. 291 **(top)** Vittorio Bruno/Shutterstock

Chapter 10—p. 299 European Southern Observatory p. 300 **(bottom center)** Novastock/PhotoEdit Inc. p. 300 **(bottom left)** NASA p. 300 **(bottom right)** A. Barry Dowsett/Science Source p. 301 Daniel Dempster/Alamy p. 308 **(bottom)** Karl Weatherly/Photodisc/Getty Images p. 308 **(top)** Michael Putland/Getty Images p. 309 **(bottom)** Vasiliy Koval/Fotolia p. 309 **(top)** Pavel Losevsky/Fotolia p. 309 **(top)** Pavel Losevsky/Fotolia p. 312 Richard Megna/Fundamental Photographs p. 317 Mehau Kulyk/Science Photo Library/Science Source p. 318 Mark Scheuern/Alamy p. 326 Jeff Hester/Arizona State University/NASA p. 327 **(right)** Farmer/Fotolia p. 327 **(top left)** Paul Bricknell/Dorling Kindersley, Ltd p. 331 Yoshikazu Tsuno/AFP Photo/Getty Images p. 333 Ames Research Center/Ligo Project/NASA

Chapter 11—p. 334 Ursula Düren/picture-alliance/dpa/AP Images p. 335 Kevin M. Law/Alamy p. 336 Fred Winner/Science Source p. 337 US Coast Guard p. 340 Siegfried Haasch/Getty Images p. 342 Richard Megna/Fundamental Photographs p. 348 EggHeadPhoto/Shutterstock p. 353 **(top center)** Janelle Lugge/Shutterstock p. 353 **(top left)** altrendo images/Getty Images p. 353 **(top right)** Andreas Edelmann/Fotolia p. 360 Rogan Thomson/ActionPlus/Corbis p. 362 **(bottom)** NASA p. 362 **(top left)** NASA Earth Observatory p. 362 **(top right)** Travis Heying/The Wichita Eagle/AP Images p. 366 **(left)** Pearson Education/PH College p. 366 **(right)** Science Source p. 368 NASA p. 371 Polka Dot Images/Getty Images

Chapter 12—p. 381 Denis Sarrazin/NASA Earth Observatory p. 384 MShields-Photos/Alamy p. 387 **(bottom right)** European Space Agency p. 387 **(left)** Ralph White/NASA Earth Observing System p. 387 **(top right)** Philip A. Dwyer/MCT/Newscom p. 388 **(bottom)** NASA p. 388 **(top)** NASA Headquarters p. 394 **(left)** NASA Headquarters p. 394 **(right)** NASA Headquarters p. 396 **(left)** NASA Headquarters p. 396 **(right)** NASA/Johnson Space Center p. 402 **(bottom)** AP Images p. 402 **(center)** Mark Pilkington/Geological Survey of Canada/SPL/Science Source p. 402 **(top)** USGS p. 404 David Nunuk/Science Source p. 408 **(bottom)** David R. Frazier Photolibrary, Inc./Alamy p. 408 **(top)** David R. Frazier Photolibrary, Inc./Alamy p. 415 NASA/Jet Propulsion Laboratory, NASA/Johnson Space Center, NASA Headquarters

Chapter 13—USGS p. 418 Sergii Shcherbakov/Fotolia p. 419 Tewan Banditrukkanka/Shutterstock p. 420 FloridaStock/Shutterstock p. 431 NASA Headquarters p. 433 **(left)** Ted Kinsman/Science Source p. 433 **(right)** Terje Rakke/Getty Images p. 443 **(bottom)** Julianwphoto/Fotolia p. 443 **(top)** Marc Tielemans/Alamy p. 446 **(bottom)** AP Images p. 446 **(bottom)** michakloodwijk/Fotolia p. 446 **(top)** Balfour Studios/Alamy p. 450 **(left)** Pearson Education p. 450 **(right)** Pearson Education p. 451 Dmitry Vereshchagin/Fotolia p. 453 **(top left)** Dr. Harold G. Craighead

Chapter 14—p. 456 Jeremy Hoare/Alamy p. 457 **(bottom)** CoverSpot Photography/Alamy p. 457 **(top)** Laurent_Imagery/Flickr State/Getty Images p. 464 Leonard Lessin/Science Source p. 466 **(center)** Tsekhmister/Shutterstock p. 466 **(left)** Oxford Scientific/Getty Images p. 466 **(right)** Phanie/Science Source p. 467 **(bottom)** AP Images p. 467 **(center)** Patricio Robles Gil/AgeFotostock p. 467 **(top)** francescodemarco/Fotolia p. 477 **(bottom)** ESA&Hubble/NASA p. 477 **(center)** Chris Gallagher/Science Source p. 477 **(top)** NOAA p. 483 **(bottom)** Richard Megna/Fundamental Photographs p. 483 **(center)** Richard Megna/Fundamental Photographs p. 483 **(top)** Richard Megna/Fundamental Photographs p. 484 **(top center)** Richard Megna/Fundamental Photographs p. 484 **(top left)** Richard Megna/Fundamental Photographs p. 484 **(top right)** Richard Megna/Fundamental Photographs p. 485 **(left)** Netfalls/Fotolia p. 485 **(right)** kataijudit/Shutterstock p. 489 **(bottom)** epitavi/Fotolia p. 489 **(top)** Dave King/Dorling Kindersley/Getty Images, Daniel Ruta/iStock/Getty Images,

Chapter 15—p. 502 Violetstar/Fotolia p. 505 Michael Fritzen/Fotolia p. 506 **(left)** Richard Megna/Fundamental Photographs p. 506 **(right)** Michal Heron/Pearson Education/PH College p. 510 Richard Megna/Fundamental Photographs p. 516 Joe Traver/The LIFE Images Collection/Getty Images p. 517 Ruslan Dashinsky/iStockphoto/Getty Images p. 518 **(bottom)** UPPA/ZUMA Press/Newscom p. 518 **(top left)** Gary Blakeley/Shutterstock p. 518 **(top right)** Steve Bower/Shutterstock p. 519 Seth Resnick/RGB Ventures/SuperStock/Alamy p. 528 **(bottom)** Kim Schott/iStockphoto/Getty Images p. 528 **(top)** Judy Kennamer/Shutterstock p. 533 AlasdairJames/Vetta/Getty Images p. 540 Fabrication Enterprises Inc.

Chapter 16—p. 544 Fred Olivier/Nature Picture Library p. 552 **(top left)** wajan/Fotolia p. 553 **(center)** Joe Sohm/The Image Works p. 553 **(top)** Richard Choy/Photolibrary/Getty Images p. 556 **(bottom)** NHPA/Science Source p. 556

Chapter cover— **cover** Danny Hu/Getty Images **cover** Nishant Shah/Getty Images **cover** Joseph Giacomin/Getty Images **cover** Australian Land, City, People Scape Photographer/Getty Images **cover** US Navy **cover** NASA Goddard MODIS Rapid Response Team

Text and Illustration Credits

Chapter 3—p. 43 Based on Wald, G. The origins of life. Proc. Natl. Acad. Sci. USA 52 (1994): **595**.

Chapter 6—p. 188 Based on "Kwan, ASK, 2004, Mechanics and Structure of Spider Webs, in Proceedings of The Seventh International Conference on Computational Structures Technology, B.H.V. Topping and C.A. Mota Soares, (Editors), Civil-Comp Press, Stirling, United Kingdom, paper 216, ISBN-0-948749-95-4."

Chapter 10—p. 328 James Walker, Principles of Physics, 1/e, Benjamin Cummings, 371150855, Figure 08-46

Chapter 22—p. 784 Based on https://commons.wikimedia.org/wiki/File:Loudspeaker-bass.png.

Chapter 25—p. 895 NASA.

Chapter 28—p. 996 Based on http://commons.wikimedia.org/wiki/File:Comparison_CD_DVD_HDDVD_BD.svg.

constant, 121–122
contact, 131–133
deforming, 592–596
direction of, 118, 119, 121–123
electromotive, 732
electrostatic. *See* Electrostatic force
exerted by spheres, 669
free-body diagrams of external,
    124–125, 124f, 125f, 130
frictional, 119, 156–164
fundamental, of nature, 1147–1148,
    1148t, 1150–1151, 1151f
gravitational. *See* Gravity
with horizontal components,
    343–350, 348f
impulse and, 262–263, 262f
internal vs. external, effect on
    momentum of, 266–268
magnetic. *See* Magnetic force
magnitude of, 118, 142, 142f
measurement of, 122
net, 118, 119, 120, 122, 132f
nontangential, and torque, 336–337
normal, 124, 141–145, 141f, 142f
pairs, 129–130, 130f
path and, 224–225
per area, pressure and, 504
between point charges, 663, 664f
reduction of, 119
restoring, 421
spring, 166–169, 208–212
in strings, 164–166
strong nuclear, 1124
superposition of, 666
tangential, and torque, 335–336,
    335f, 336f
units of, 123t
variable, work done by, 207–212
as vectors, 118, 122, 134–137,
    134f
with vertical components, 343–350,
    348f
visualization of, 122f, 198f
work done by. *See* Work
zero net, 343–350, 343f
Force constant, 167, 168
Force constant, for mass on a spring,
    430–435
Force fields, 670–671. *See also* Electric
    field(s)
Force pairs
action-reaction, 129–131, 130f
internal vs. external forces and,
    266–268
Force per charge, electric, 671
Forensics, fluorescence use in, 1111
Fosbury flop, 285, 291f
Fountain jets, 97f
Frames of reference
accelerated, 1039, 1039f, 1040f
definition of, 119–120
inertial, 1018, 1018f
noninertial, 120
Franklin, Benjamin, 669, 693
Free fall
acceleration in, 43, 43f, 93, 93f
air resistance and, 43
characteristics of, 43–44
definition of, 43
examples of, 43f
gravity and, 44
of objects, 43–49
orbital motion and, 382, 382f
position and, 44, 44f
from rest, 45, 45f
separation, 45
speed and, 46–48
symmetry of, 43f, 47, 47–49
velocity and, 43f, 44
weight and, 137–138, 137f
weightlessness and, 140
Free space
permeability of, 789
permittivity of, 683
Free-body diagrams, 124–125, 124f,
    125f, 130
Frequency

angular, 425
beat, 489–492, 490f
cutoff, 1058, 1059f
definition of, 419
harmonic, 482–488
natural, 445
of periodic motion, 419–420
range, of human ear, 485
RLC circuits and, 866
of sound waves, 466–467
typical values for, 420t
units of, 419
wavelength and, 459
    for electromagnetic waves,
        889–890, 890f
    for sound waves, 459, 459f
Frets, guitar, 485, 485f
Friction, 156–164
calculation of, 156–157, 160–161
definition of, 155
functions of, 156
kinetic, 156–160
as nonconservative force, 224, 224f
origin of, 156, 156f
reduction of, 119
static, 160–164
surface and, 156, 156f
visualization of, 156f
work done by, 224, 225–227, 225f
Fringes. *See* Bright fringe; Dark fringe;
    Interference fringes
From the Earth to the Moon (Verne), 54,
    404
Frost heaving, 556, 556f
Frost wedging, 599
Fundamental mode, 482–484, 482f,
    483f, 486–489
Fuses, 849
Fusion, latent heat of, 602, 603t
Fusion curve, 598–599, 599f
Fusion reactions, 1034, 1034f
Fusion reactors, 1143–1144, 1144f

G

Galaxies
red shifted, 887, 887f
speed of, Doppler effect and, 472, 477
Galileo Galilei, 43, 407, 438–439, 885,
    972
Galileo thermometer, 951
Gallinule, 505, 505f
Galvanometers, 787, 787f
Gamma decay, 1130
Gamma rays, 890f, 891, 1126, 1130
Gas(es)
equations of state for, 582, 582f
ideal, 580–586
    definition of, 580
    equation of state for, 580, 580f, 582
    internal energy of, 591
    isotherms and, 583–584, 624, 624f
    molar specific heat for, 628–629,
        628f
    moles in, 582–583
    pressure of, 580–581, 580f, 583,
        583f, 585–586
    volume of, 585–586
isotherms for, 583–584, 583f
kinetic theory of, 586–592
molecular speed of, 587–590, 588f
moles in, 582
noble, 1103
pressure of, calculation of, 580–581
thermal behavior of, 548–549
volume of, temperature and, 585,
    585f
work done by expanding, 621–622,
    621f, 623, 623f, 624, 624f
Gauge pressure, 506
Gauss (G), 774
Gaussian surface, 684–685
Gauss's law, 683–684
Gay-Lussac, Joseph, 585
Geiger, Hans, 1083
Gell-Mann, Murray, 1148, 1149
General launch angle, 99–103, 99f, 100f
General theory of relativity, 1039–1043

gravity waves and, 1042–1043, 1042f
principle of equivalence in, 1039,
    1040f
Schwarzschild radius, 1042
Generators. *See* Electric generators
Geographic poles, 771–772, 772f
Geomagnetism, 771–772, 772f
Geometrical optics, 914, 954–975
Geosynchronous satellites, 396, 396f
Germer, L. H., 1067
Geysers, 598
GFCI (Ground fault circuit interrupter),
    850f, 851
GFP (Green fluorescent protein),
    1111–1112, 1112f
Glare, 902, 902f, 903f, 915
Glashow, Sheldon, 1151
Glass
electrical properties of, 658–659
light tunneling and, 1073, 1073f
slabs and refraction, 932f, 933
Global Positioning System (GPS), 395,
    395f
Golf
on Moon, 104
putt judgement in, 244
Gossamer Albatross, 213, 213f
GPS (Global Positioning System), 395,
    395f
GPS technology, 1039
Grams, 4
Grand unified theory (GUT), 1151
Graphical method, for vector addition,
    66–67, 66f, 67f
Graphs
acceleration, 29, 29f
average acceleration, 30f
average velocity, 24–25, 25f, 28, 28f, 35f
instantaneous acceleration, 30f
instantaneous velocity, 27f, 28f
one-dimensional motion, 25f
v-vs.-*t*, 29–30
x-vs.-*t*, 24
Grating spectroscope, 1007–1008, 1008f
Gratings. *See* Diffraction gratings;
    Reflection grating
Gravimeters, 387, 387f, 441
Gravitational field, of Sun, 1040, 1041f
Gravitational lensing, 407, 407f, 1041,
    1041f
Gravitational potential energy, 247f,
    398–401, 398f
altitude and, 229–231, 398–400
of fluids, 510, 510f
kinetic energy and, 402, 403f
of mass, 400–401, 400f
potential well of, 402, 403f
as scalar, 400
zero level of, 410
Gravitropism, 190, 384, 384f
Gravity, 381–409
altitude and, 386–387
atoms and, 664–665
as attractive force, 664
compared to electric force, 664
distance and, 393–394, 393f
on Earth, 381–385, 385f
effect of, on light, 1040
force of, 120–121
free fall and, 44–45
as fundamental force, 1148
inverse square dependence and, 383
Isaac Newton and, 382
law of universal gravitation and,
    381–385
magnitude of, 382
on Moon, 137, 387–388
orbital motion and, 391–394
radiation pressure and, 896, 896f
range and, 104
spherical bodies and, 385–391
superposition and, 383
tides and, 407–409, 408f
universal gravitational constant and,
    388–389, 389f
values of, based on latitude, 44f
vector addition and, 383

waves, 1042–1043, 1042f, 1043f
weight and, 137–139
work done by, 199–200, 203, 203f,
    205, 224, 224f–226f, 225–227
Gravity escape system, 199–200
*Gravity Field and Steady-State Ocean
    Circulation Explorer*, 387, 387f
Gravity maps, 441
Green fluorescent protein (GFP),
    1111–1112, 1112f
Ground fault circuit interrupter (GFCI),
    850f, 851, 851f
Ground state, of atoms, 1091
of multielectron atoms, 1100–1101
probability clouds for, 1098, 1098f
Grounded plugs, 850–851, 850f
Grounding, 681, 681f
Guitar frets, 485, 485f
Guitar pickups, 813, 814f
Gyroscopes, 367–368, 367f

H

Hadrons, 1148, 1149t, 1150t
Hafele, J. C., 1022
Hahn, Otto, 1139
Hale-Bopp comet, 404f
Half-life, of radioactive material, 1133f,
    1134
Halley's comet, 414
Halogens, 1103
Hang time, concept of, 100, 100f
Harmonic motion. *See* Simple harmonic
    motion
Harmonic waves, 457, 457f, 458f,
    463–464, 463f
Harmonics
of columns of air, 485–489, 486f,
    487f, 488f
on a string, 483–485, 483f
Headlights, resolution and, 1002
Headphones, active noise reduction, 481
Heads-up display, 918, 918f
Hearing
frequency range in, 466, 485, 487f
sound perception in, 468–469, 470,
    487
Heart
electrical activity in, 710–711, 710f,
    711f
pacemakers for, 757, 757f
Heartbeat recoil detectors, 269
Heat
calories in, 557
definition of, 545
energy and, 545–546
energy conservation and, 556–557
flow of, 562–566, 562f
latent, 602–606
measurement of, 558–560
mechanical equivalent of, 557, 557f
sign conventions for, 618
specific, 559–560
in thermal equilibrium, 559–560
thermal expansion and, 550–556
in thermodynamic systems, 618–620
types of, 558–562
units of, 557
work and, 556–557
Heat capacity, 558–559
Heat death, of Universe, 646
Heat engines, 632–638, 632f, 633f, 638f
Carnot's theorem and, 634–636
efficiency of, 633–636
entropy change and, 642
living systems as, 646–647
maximum efficiency of, 635
Heat exchange
by conduction, 562–566, 562f, 563f,
    566
by convection, 566–570, 566f, 567f
phase changes and, 606–608
by radiation, 567–568, 567f, 568f
Heat flow
in adiabatic processes, 624–626, 626f
entropy and, 645–646, 645f
spontaneous, 632, 637
in thermodynamic systems, 618–620

MULTIPLES AND PREFIXES FOR METRIC UNITS

Multiple	Prefix (Abbreviation)	Pronunciation
10^{24}	yotta- (Y)	yot'ta (*a* as in *a*bout)
10^{21}	zetta- (Z)	zet'ta (*a* as in *a*bout)
10^{18}	exa- (E)	ex'a (*a* as in *a*bout)
10^{15}	peta- (P)	pet'a (as in *peta*l)
10^{12}	tera- (T)	ter'a (as in *terra*ce)
10^{9}	giga- (G)	ji'ga (*ji* as in *jiggle*, *a* as in *a*bout)
10^{6}	mega- (M)	meg'a (as in *mega*phone)
10^{3}	kilo- (k)	kil'o (as in *kilo*watt)
10^{2}	hecto- (h)	hek'to (*heck-toe*)
10	deka- (da)	dek'a (*deck* plus *a* as in *a*bout)
10^{-1}	deci- (d)	des'i (as in *deci*mal)
10^{-2}	centi- (c)	sen'ti (as in *senti*mental)
10^{-3}	milli- (m)	mil'li (as in *mili*tary)
10^{-6}	micro- (μ)	mi'kro (as in *micro*phone)
10^{-9}	nano- (n)	nan'oh (*an* as in *ann*ual)
10^{-12}	pico- (p)	pe'ko (*peek-oh*)
10^{-15}	femto- (f)	fem'toe (*fem* as in *fem*inine)
10^{-18}	atto- (a)	at'toe (as in an*atomy*)
10^{-21}	zepto- (z)	zep'toe (as in *zep*pelin)
10^{-24}	yocto- (y)	yock'toe (as in *sock*)

THE GREEK ALPHABET

Alpha	A	α
Beta	B	β
Gamma	Γ	γ
Delta	Δ	δ
Epsilon	E	ε
Zeta	Z	ζ
Eta	H	η
Theta	Θ	θ
Iota	I	ι
Kappa	K	κ
Lambda	Λ	λ
Mu	M	μ
Nu	N	ν
Xi	Ξ	ξ
Omicron	O	o
Pi	Π	π
Rho	P	ρ
Sigma	Σ	σ
Tau	T	τ
Upsilon	Υ	υ
Phi	Φ	ϕ
Chi	X	χ
Psi	Ψ	ψ
Omega	Ω	ω

SI BASE UNITS

Physical Quantity	Name of Unit	Symbol
Length	meter	m
Mass	kilogram	kg
Time	second	s
Electric current	ampere	A
Temperature	kelvin	K
Amount of substance	mole	mol
Luminous intensity	candela	cd

SOME SI DERIVED UNITS

Physical Quantity	Name of Unit	Symbol	SI Unit
Frequency	hertz	Hz	s^{-1}
Energy	joule	J	$kg \cdot m^2/s^2$
Force	newton	N	$kg \cdot m/s^2$
Pressure	pascal	Pa	$kg/(m \cdot s^2)$
Power	watt	W	$kg \cdot m^2/s^3$
Electric charge	coulomb	C	$A \cdot s$
Electric potential	volt	V	$kg \cdot m^2/(A \cdot s^3)$
Electric resistance	ohm	Ω	$kg \cdot m^2/(A^2 \cdot s^3)$
Capacitance	farad	F	$A^2 \cdot s^4/(kg \cdot m^2)$
Inductance	henry	H	$kg \cdot m^2/(A^2 \cdot s^2)$
Magnetic field	tesla	T	$kg/(A \cdot s^2)$
Magnetic flux	weber	Wb	$kg \cdot m^2/(A \cdot s^2)$

SI UNITS OF SOME OTHER PHYSICAL QUANTITIES

Physical Quantity	SI Unit
Density (ρ)	kg/m^3
Speed (v)	m/s
Acceleration (a)	m/s^2
Momentum, impulse (p)	$kg \cdot m/s$
Angular speed (ω)	rad/s
Angular acceleration (α)	rad/s^2
Torque (τ)	$kg \cdot m^2/s^2 \ or \ N \cdot m$
Specific heat (c)	$J/(kg \cdot K)$
Thermal conductivity (k)	$W/(m \cdot K) \ or \ J/(s \cdot m \cdot K)$
Entropy (S)	$J/K \ or \ kg \cdot m^2/(K \cdot s^2) \ or \ N \cdot m/K$
Electric field (E)	$N/C \ or \ V/m$

APPENDICES IN THE TEXT

FUNDAMENTAL CONSTANTS

Quantity	Symbol	Approximate Value
Speed of light	c	3.00×10^8 m/s $= 3.00 \times 10^{10}$ cm/s $= 186,000$ mi/s
Universal gravitational constant	G	6.67×10^{-11} N·m²/kg²
Stefan-Boltzmann constant	σ	5.67×10^{-8} W/(m²·K⁴)
Boltzmann's constant	k	1.38×10^{-23} J/K
Avogadro's number	N_A	6.022×10^{23} mol⁻¹
Gas constant	$R = N_A k$	8.31 J/(mol·K) $= 1.99$ cal/(mol·K)
Coulomb's law constant	$k = 1/4\pi\epsilon_0$	8.99×10^9 N·m²/C²
Electron charge	e	1.60×10^{-19} C
Permittivity of free space	ϵ_o	8.85×10^{-12} C²/(N·m²)
Permeability of free space	μ_o	$4\pi \times 10^{-7}$ T·m/A $= 1.26 \times 10^{-6}$ T·m/A
Planck's constant	h	6.63×10^{-34} J·s
	$\hbar = h/2\pi$	1.05×10^{-34} J·s
Atomic mass unit	u	1.6605×10^{-27} kg $\leftrightarrow 931.5$ MeV
Electron mass	m_e	9.10939×10^{-31} kg $= 5.49 \times 10^{-4}$ u $\leftrightarrow 0.511$ MeV
Neutron mass	m_n	$1.675\ 00 \times 10^{-27}$ kg $= 1.008\ 665$ u $\leftrightarrow 939.57$ MeV
Proton mass	m_p	$1.672\ 65 \times 10^{-27}$ kg $= 1.007\ 267$ u $\leftrightarrow 938.28$ MeV

USEFUL PHYSICAL DATA

Acceleration due to gravity (surface of Earth)	9.81 m/s² $= 32.2$ ft/s²
Absolute zero	0 K $= -273.15$°C $= -459.67$°F
Standard temperature & pressure (STP)	0°C $= 273.15$ K
	1 atm $= 101.325$ kPa
Density of air (STP)	1.29 kg/m³
Speed of sound in air (20 °C)	343 m/s
Density of water (4 °C)	1.000×10^3 kg/m³
Latent heat of fusion of water	3.35×10^5 J/kg
Latent heat of vaporization of water	2.26×10^6 J/kg
Specific heat of water	4186 J/(kg·K)

SOLAR SYSTEM DATA*

Equatorial radius of Earth	6.37×10^3 km $= 3950$ mi
Mass of Earth	5.97×10^{24} kg
Radius of Moon	1740 km $= 1080$ mi
Mass of Moon	7.35×10^{22} kg $\approx \frac{1}{81}$ mass of Earth
Average distance of Moon from Earth	3.84×10^5 km $= 2.39 \times 10^5$ mi
Radius of Sun	6.95×10^5 km $= 432,000$ mi
Mass of Sun	2.00×10^{30} kg
Average distance of Earth from Sun	1.50×10^8 km $= 93.0 \times 10^6$ mi

*See Appendix C for additional planetary data.